T0205276

Lecture Notes in Artificial Intelligence 12672

Subseries of Lecture Notes in Computer Science

More information about this subseries at http://www.springer.com/series/1244

Ngoc Thanh Nguyen · Suphamit Chittayasothorn ·
Dusit Niyato · Bogdan Trawiński (Eds.)

Intelligent Information and Database Systems

13th Asian Conference, ACIIDS 2021
Phuket, Thailand, April 7–10, 2021
Proceedings

Editors
Ngoc Thanh Nguyen
Wrocław University of Science and Technology
Wrocław, Poland

Dusit Niyato
Nanyang Technological University
Singapore, Singapore

Suphamit Chittayasothorn
King Mongkut's Institute of Technology Ladkrabang
Bangkok, Thailand

Bogdan Trawiński
Wrocław University of Science and Technology
Wrocław, Poland

ISSN 0302-9743 ISSN 1611-3349 (electronic)
Lecture Notes in Artificial Intelligence
ISBN 978-3-030-73279-0 ISBN 978-3-030-73280-6 (eBook)
https://doi.org/10.1007/978-3-030-73280-6

LNCS Sublibrary: SL7 – Artificial Intelligence

This Springer imprint is published by the registered company Springer Nature Switzerland AG
The registered company address is: Gewerbestrasse 11, 6330 Cham, Switzerland

Preface

ACIIDS 2021 was the 13th edition of the Asian Conference on Intelligent Information and Database Systems. The aim of ACIIDS 2021 was to provide an international forum for research workers with scientific backgrounds on the technology of intelligent information and database systems and its various applications. The ACIIDS 2021 conference was co-organized by King Mongkut's Institute of Technology Ladkrabang (Thailand) and Wrocław University of Science and Technology (Poland) in cooperation with the IEEE SMC Technical Committee on Computational Collective Intelligence, the European Research Center for Information Systems (ERCIS), The University of Newcastle (Australia), Yeungnam University (South Korea), Leiden University (The Netherlands), Universiti Teknologi Malaysia (Malaysia), BINUS University (Indonesia), Quang Binh University (Vietnam), Nguyen Tat Thanh University (Vietnam), and the "Collective Intelligence" section of the Committee on Informatics of the Polish Academy of Sciences. ACIIDS 2021 was at first scheduled to be held in Phuket, Thailand during April 7–10, 2021. However, due to the COVID-19 pandemic, the conference was moved to the virtual space and conducted online using the ZOOM videoconferencing system.

The ACIIDS conference series is already well established. The first two events, ACIIDS 2009 and ACIIDS 2010, took place in Dong Hoi City and Hue City in Vietnam, respectively. The third event, ACIIDS 2011, took place in Daegu (South Korea), followed by the fourth event, ACIIDS 2012, in Kaohsiung (Taiwan). The fifth event, ACIIDS 2013, was held in Kuala Lumpur (Malaysia) while the sixth event, ACIIDS 2014, was held in Bangkok (Thailand). The seventh event, ACIIDS 2015, took place in Bali (Indonesia), followed by the eighth event, ACIIDS 2016, in Da Nang (Vietnam). The ninth event, ACIIDS 2017, was organized in Kanazawa (Japan). The 10th jubilee conference, ACIIDS 2018, was held in Dong Hoi City (Vietnam), followed by the 11th event, ACIIDS 2019, in Yogyakarta (Indonesia). The 12th and 13th events were initially planned to be held in Phuket (Thailand). However, the global pandemic relating to COVID-19 resulted in both editions of the conference being held online in virtual space.

For this edition of the conference, we received in total 291 papers whose authors came from 40 countries around the world. Each paper was peer reviewed by at least two members of the international Program Committee and one member of the international board of reviewers. Only 69 papers of the highest quality were selected for oral presentation and publication in this LNAI volume of the ACIIDS 2021 proceedings.

Papers included in these proceedings cover the following topics: data mining methods and applications, advanced data mining techniques and applications, decision support and control systems, natural language processing, cybersecurity intelligent methods, computer vision techniques, computational imaging and vision, advanced data mining techniques and applications, intelligent and contextual systems, commonsense knowledge, reasoning and programming in artificial intelligence, data modelling and processing for industry 4.0, and innovations in intelligent systems.

The accepted and presented papers focused on new trends and challenges facing the intelligent information and database systems community. The presenters showed how research work could stimulate novel and innovative applications. We hope that you found these results useful and inspiring for your future research work.

We would like to express our sincere thanks to the honorary chairs for their support: Prof. Suchatvee Suwansawat (President of King Mongkut's Institute of Technology Ladkrabang, Thailand), Prof. Arkadiusz Wójs (Rector of Wrocław University of Science and Technology, Poland), and Prof. Moonis Ali (President of the International Society of Applied Intelligence, USA). We would like to express our thanks to the keynote speakers for their world-class plenary speeches: Dr. Edwin Lughofer from Johannes Kepler University Linz (Austria), Prof. Manuel Núñez from Universidad Complutense de Madrid (Spain), and Prof. Agachai Sumalee from Chulalongkorn University (Thailand).

We cordially thank our main sponsors: King Mongkut's Institute of Technology Ladkrabang (Thailand) and Wrocław University of Science and Technology (Poland), as well as all aforementioned cooperating universities and organizations. Our special thanks are also due to Springer for publishing the proceedings and to all the other sponsors for their kind support.

Our special thanks go to the Program Chairs, Special Session Chairs, Organizing Chairs, Publicity Chairs, Liaison Chairs, and Local Organizing Committee for their work towards the conference. We sincerely thank all the members of the international Program Committee for their valuable efforts in the review process, which helped us to guarantee the highest quality of the selected papers for the conference. We cordially thank all the authors, for their valuable contributions, and the other participants of the conference. The conference would not have been possible without their support. Thanks are also due to the numerous experts who contributed to making the event a success.

April 2021

Ngoc Thanh Nguyen
Suphamit Chittayasothorn
Dusit Niyato
Bogdan Trawiński

Special Session Chairs

Krystian Wojtkiewicz — Wrocław University of Science and Technology, Poland
Rathachai Chawuthai — King Mongkut's Institute of Technology Ladkrabang, Thailand

Liaison Chairs

Ford Lumban Gaol — Bina Nusantara University, Indonesia
Quang-Thuy Ha — VNU-University of Engineering and Technology, Vietnam
Mong-Fong Horng — National Kaohsiung University of Applied Sciences, Taiwan
Dosam Hwang — Yeungnam University, South Korea
Le Minh Nguyen — Japan Advanced Institute of Science and Technology, Japan
Ali Selamat — Universiti Teknologi Malaysia, Malaysia
Paweł Sitek — Kielce University of Technology, Poland

Organizing Chairs

Wiboon Prompanich — King Mongkut's Institute of Technology Ladkrabang, Thailand
Amnach Khawne — King Mongkut's Institute of Technology Ladkrabang, Thailand
Adrianna Kozierkiewicz — Wrocław University of Science and Technology, Poland
Bogumiła Hnatkowska — Wrocław University of Science and Technology, Poland

Publicity Chairs

Natthapong Jungteerapani — King Mongkut's Institute of Technology Ladkrabang, Thailand
Marek Kopel — Wrocław University of Science and Technology, Poland
Marek Krótkiewicz — Wrocław University of Science and Technology, Poland

Webmaster

Marek Kopel — Wrocław University of Science and Technology, Poland

Conference Organization

Honorary Chairs

Suchatvee Suwansawat	President of King Mongkut's Institute of Technology Ladkrabang, Thailand
Arkadiusz Wójs	Rector of Wrocław University of Science and Technology, Poland
Moonis Ali	President of the International Society of Applied Intelligence, USA

General Chairs

Ngoc Thanh Nguyen	Wrocław University of Science and Technology, Poland
Suphamit Chittayasothorn	King Mongkut's Institute of Technology Ladkrabang, Thailand

Program Chairs

Dusit Niyato	Nanyang Technological University, Singapore
Tzung-Pei Hong	National University of Kaohsiung, Taiwan
Edward Szczerbicki	University of Newcastle, Australia
Bogdan Trawiński	Wrocław University of Science and Technology, Poland

Steering Committee

Ngoc Thanh Nguyen (Chair)	Wrocław University of Science and Technology, Poland
Longbing Cao	University of Science and Technology Sydney, Australia
Suphamit Chittayasothorn	King Mongkut's Institute of Technology Ladkrabang, Thailand
Ford Lumban Gaol	Bina Nusantara University, Indonesia
Tzung-Pei Hong	National University of Kaohsiung, Taiwan
Dosam Hwang	Yeungnam University, South Korea
Bela Stantic	Griffith University, Australia
Geun-Sik Jo	Inha University, South Korea
Le Thi Hoai An	University of Lorraine, France
Toyoaki Nishida	Kyoto University, Japan
Leszek Rutkowski	Częstochowa University of Technology, Poland
Ali Selamat	Universiti Teknologi Malaysia, Malaysia

Conference Organization

Honorary Chairs

Suchatvee Suwansawat	President of King Mongkut's Institute of Technology Ladkrabang, Thailand
Arkadiusz Wójs	Rector of Wrocław University of Science and Technology, Poland
Moonis Ali	President of the International Society of Applied Intelligence, USA

General Chairs

Ngoc Thanh Nguyen	Wrocław University of Science and Technology, Poland
Suphamit Chittayasothorn	King Mongkut's Institute of Technology Ladkrabang, Thailand

Program Chairs

Dusit Niyato	Nanyang Technological University, Singapore
Tzung-Pei Hong	National University of Kaohsiung, Taiwan
Edward Szczerbicki	University of Newcastle, Australia
Bogdan Trawiński	Wrocław University of Science and Technology, Poland

Steering Committee

Ngoc Thanh Nguyen (Chair)	Wrocław University of Science and Technology, Poland
Longbing Cao	University of Science and Technology Sydney, Australia
Suphamit Chittayasothorn	King Mongkut's Institute of Technology Ladkrabang, Thailand
Ford Lumban Gaol	Bina Nusantara University, Indonesia
Tzung-Pei Hong	National University of Kaohsiung, Taiwan
Dosam Hwang	Yeungnam University, South Korea
Bela Stantic	Griffith University, Australia
Geun-Sik Jo	Inha University, South Korea
Le Thi Hoai An	University of Lorraine, France
Toyoaki Nishida	Kyoto University, Japan
Leszek Rutkowski	Częstochowa University of Technology, Poland
Ali Selamat	Universiti Teknologi Malaysia, Malaysia

Special Session Chairs

Krystian Wojtkiewicz Wrocław University of Science and Technology, Poland
Rathachai Chawuthai King Mongkut's Institute of Technology Ladkrabang, Thailand

Liaison Chairs

Ford Lumban Gaol Bina Nusantara University, Indonesia
Quang-Thuy Ha VNU-University of Engineering and Technology, Vietnam
Mong-Fong Horng National Kaohsiung University of Applied Sciences, Taiwan
Dosam Hwang Yeungnam University, South Korea
Le Minh Nguyen Japan Advanced Institute of Science and Technology, Japan
Ali Selamat Universiti Teknologi Malaysia, Malaysia
Paweł Sitek Kielce University of Technology, Poland

Organizing Chairs

Wiboon Prompanich King Mongkut's Institute of Technology Ladkrabang, Thailand
Amnach Khawne King Mongkut's Institute of Technology Ladkrabang, Thailand
Adrianna Kozierkiewicz Wrocław University of Science and Technology, Poland
Bogumiła Hnatkowska Wrocław University of Science and Technology, Poland

Publicity Chairs

Natthapong Jungteerapani King Mongkut's Institute of Technology Ladkrabang, Thailand
Marek Kopel Wrocław University of Science and Technology, Poland
Marek Krótkiewicz Wrocław University of Science and Technology, Poland

Webmaster

Marek Kopel Wrocław University of Science and Technology, Poland

Local Organizing Committee

Pakorn Watanachaturaporn	King Mongkut's Institute of Technology Ladkrabang, Thailand
Sathaporn Promwong	King Mongkut's Institute of Technology Ladkrabang, Thailand
Putsadee Pornphol	Phuket Rajabhat University, Thailand
Maciej Huk	Wrocław University of Science and Technology, Poland
Marcin Jodłowiec	Wrocław University of Science and Technology, Poland
Marcin Pietranik	Wrocław University of Science and Technology, Poland

Keynote Speakers

Edwin Lughofer	Johannes Kepler University Linz, Austria
Manuel Núñez	Universidad Complutense de Madrid, Spain
Agachai Sumalee	Chulalongkorn University, Thailand

Special Sessions Organizers

1. ADMTA 2021: Special Session on Advanced Data Mining Techniques and Applications

Chun-Hao Chen	Tamkang University, Taiwan
Bay Vo	Ho Chi Minh City University of Technology, Vietnam
Tzung-Pei Hong	National University of Kaohsiung, Taiwan

2. CIV 2021: Special Session on Computational Imaging and Vision

Manish Khare	Dhirubhai Ambani Institute of Information and Communication Technology, India
Prashant Srivastava	NIIT University, India
Om Prakash	HNB Garwal University, India
Jeonghwan Gwak	Korea National University of Transportation, South Korea

3. CoSenseAI 2021: Special Session on Commonsense Knowledge, Reasoning and Programming in Artificial Intelligence

Pascal Bouvry	University of Luxembourg, Luxembourg
Matthias R. Brust	University of Luxembourg, Luxembourg
Grégoire Danoy	University of Luxembourg, Luxembourg
El-ghazil Talbi	University of Lille, France

4. DMPI-4.0 vol. 2 – 2021: Special Session on Data Modelling and Processing for Industry 4.0

Du Haizhou	Shanghai University of Electric Power, China
Wojciech Hunek	Opole University of Technology, Poland
Marek Krótkiewicz	Wrocław University of Science and Technology, Poland
Krystian Wojtkiewicz	Wrocław University of Science and Technology, Poland

5. ICxS 2021: Special Session on Intelligent and Contextual Systems

Maciej Huk	Wrocław University of Science and Technology, Poland
Keun Ho Ryu	Ton Duc Thang University, Vietnam
Goutam Chakraborty	Iwate Prefectural University, Japan
Qiangfu Zhao	University of Aizu, Japan
Chao-Chun Chen	National Cheng Kung University, Taiwan
Rashmi Dutta Baruah	Indian Institute of Technology Guwahati, India
Tetsuji Kuboyama	Gakushuin University, Japan

6. ISCEC 2021: Special Session on Intelligent Supply Chains and e-Commerce

Arkadiusz Kawa	Łukasiewicz Research Network – The Institute of Logistics and Warehousing, Poland
Justyna Światowiec-Szczepańska	Poznań University of Economics and Business, Poland
Bartłomiej Pierański	Poznań University of Economics and Business, Poland

7. MMAML 2021: Special Session on Multiple Model Approach to Machine Learning

Tomasz Kajdanowicz	Wrocław University of Science and Technology, Poland
Edwin Lughofer	Johannes Kepler University Linz, Austria
Bogdan Trawiński	Wrocław University of Science and Technology, Poland

Senior Program Committee

Ajith Abraham	Machine Intelligence Research Labs, USA
Jesus Alcala-Fdez	University of Granada, Spain
Lionel Amodeo	University of Technology of Troyes, France
Ahmad Taher Azar	Prince Sultan University, Saudi Arabia
Thomas Bäck	Leiden University, Netherlands
Costin Badica	University of Craiova, Romania
Ramazan Bayindir	Gazi University, Turkey
Abdelhamid Bouchachia	Bournemouth University, UK
David Camacho	Universidad Autonoma de Madrid, Spain

Leopoldo Eduardo Cardenas-Barron	Tecnologico de Monterrey, Mexico
Oscar Castillo	Tijuana Institute of Technology, Mexico
Nitesh Chawla	University of Notre Dame, USA
Rung-Ching Chen	Chaoyang University of Technology, Taiwan
Shyi-Ming Chen	National Taiwan University of Science and Technology, Taiwan
Simon Fong	University of Macau, Macau SAR
Hamido Fujita	Iwate Prefectural University, Japan
Mohamed Gaber	Birmingham City University, UK
Marina L. Gavrilova	University of Calgary, Canada
Daniela Godoy	ISISTAN Research Institute, Argentina
Fernando Gomide	University of Campinas, Brazil
Manuel Grana	University of the Basque Country, Spain
Claudio Gutierrez	Universidad de Chile, Chile
Francisco Herrera	University of Granada, Spain
Tzung-Pei Hong	National University of Kaohsiung, Taiwan
Dosam Hwang	Yeungnam University, South Korea
Mirjana Ivanovic	University of Novi Sad, Serbia
Janusz Jeżewski	Institute of Medical Technology and Equipment ITAM, Poland
Piotr Jedrzejowicz	Gdynia Maritime University, Poland
Kang-Hyun Jo	University of Ulsan, South Korea
Jason J. Jung	Chung-Ang University, South Korea
Janusz Kacprzyk	Systems Research Institute, Polish Academy of Sciences, Poland
Nikola Kasabov	Auckland University of Technology, New Zealand
Muhammad Khurram Khan	King Saud University, Saudi Arabia
Frank Klawonn	Ostfalia University of Applied Sciences, Germany
Joanna Kolodziej	Cracow University of Technology, Poland
Józef Korbicz	University of Zielona Gora, Poland
Ryszard Kowalczyk	Swinburne University of Technology, Australia
Bartosz Krawczyk	Virginia Commonwealth University, USA
Ondrej Krejcar	University of Hradec Králové, Czech Republic
Adam Krzyzak	Concordia University, Canada
Mark Last	Ben-Gurion University of the Negev, Israel
Le Thi Hoai An	University of Lorraine, France
Kun Chang Lee	Sungkyunkwan University, South Korea
Edwin Lughofer	Johannes Kepler University Linz, Austria
Nezam Mahdavi-Amiri	Sharif University of Technology, Iran
Yannis Manolopoulos	Open University of Cyprus, Cyprus
Klaus-Robert Müller	Technical University of Berlin, Germany
Saeid Nahavandi	Deakin University, Australia
Grzegorz J. Nalepa	AGH University of Science and Technology, Poland

Ngoc-Thanh Nguyen	Wrocław University of Science and Technology, Poland
Dusit Niyato	Nanyang Technological University, Singapore
Yusuke Nojima	Osaka Prefecture University, Japan
Manuel Núñez	Universidad Complutense de Madrid, Spain
Jeng-Shyang Pan	Fujian University of Technology, China
Marcin Paprzycki	Systems Research Institute, Polish Academy of Sciences, Poland
Bernhard Pfahringer	University of Waikato, New Zealand
Hoang Pham	Rutgers University, USA
Tao Pham Dinh	INSA Rouen, France
Radu-Emil Precup	Politehnica University of Timisoara, Romania
Leszek Rutkowski	Częstochowa University of Technology, Poland
Juergen Schmidhuber	Swiss AI Lab IDSIA, Switzerland
Björn Schuller	University of Passau, Germany
Ali Selamat	Universiti Teknologi Malaysia, Malaysia
Andrzej Skowron	Warsaw University, Poland
Jerzy Stefanowski	Poznań University of Technology, Poland
Agachai Sumalee	Chulalongkorn University, Thailand
Edward Szczerbicki	University of Newcastle, Australia
Ryszard Tadeusiewicz	AGH University of Science and Technology, Poland
Muhammad Atif Tahir	National University of Computing & Emerging Sciences, Pakistan
Bay Vo	Ho Chi Minh City University of Technology, Vietnam
Gottfried Vossen	University of Münster, Germany
Lipo Wang	Nanyang Technological University, Singapore
Junzo Watada	Waseda University, Japan
Michał Woźniak	Wrocław University of Science and Technology, Poland
Farouk Yalaoui	University of Technology of Troyes, France
Sławomir Zadrożny	Systems Research Institute, Polish Academy of Sciences, Poland
Zhi-Hua Zhou	Nanjing University, China

Program Committee

Muhammad Abulaish	South Asian University, India
Bashar Al-Shboul	University of Jordan, Jordan
Toni Anwar	Universiti Teknologi PETRONAS, Malaysia
Taha Arbaoui	University of Technology of Troyes, France
Mehmet Emin Aydin	University of the West of England, UK
Amelia Badica	University of Craiova, Romania
Kambiz Badie	ICT Research Institute, Iran
Hassan Badir	École Nationale des Sciences Appliquées de Tanger, Morocco
Dariusz Barbucha	Gdynia Maritime University, Poland

Paulo Batista	University of Évora, Portugal
Maumita Bhattacharya	Charles Sturt University, Australia
Bülent Bolat	Yildiz Technical University, Turkey
Mariusz Boryczka	University of Silesia, Poland
Urszula Boryczka	University of Silesia, Poland
Zouhaier Brahmia	University of Sfax, Tunisia
Stephane Bressan	National University of Singapore, Singapore
Peter Brida	University of Zilina, Slovakia
Andrej Brodnik	University of Ljubljana, Slovenia
Grażyna Brzykcy	Poznań University of Technology, Poland
Robert Burduk	Wrocław University of Science and Technology, Poland
Aleksander Byrski	AGH University of Science and Technology, Poland
Dariusz Ceglarek	WSB University in Poznań, Poland
Zenon Chaczko	University of Technology Sydney, Australia
Somchai Chatvichienchai	University of Nagasaki, Japan
Chun-Hao Chen	Tamkang University, Taiwan
Leszek J. Chmielewski	Warsaw University of Life Sciences, Poland
Kazimierz Choroś	Wrocław University of Science and Technology, Poland
Kun-Ta Chuang	National Cheng Kung University, Taiwan
Dorian Cojocaru	University of Craiova, Romania
Jose Alfredo Ferreira Costa	Federal University of Rio Grande do Norte (UFRN), Brazil
Ireneusz Czarnowski	Gdynia Maritime University, Poland
Piotr Czekalski	Silesian University of Technology, Poland
Theophile Dagba	University of Abomey-Calavi, Benin
Phuc Do	Vietnam National University, Ho Chi Minh City, Vietnam
Tien V. Do	Budapest University of Technology and Economics, Hungary
Rafal Doroz	University of Silesia, Poland
El-Sayed M. El-Alfy	King Fahd University of Petroleum and Minerals, Saudi Arabia
Keiichi Endo	Ehime University, Japan
Sebastian Ernst	AGH University of Science and Technology, Poland
Usef Faghihi	Université du Québec à Trois-Rivières, Canada
Rim Faiz	University of Carthage, Tunisia
Victor Felea	Alexandru Ioan Cuza University, Romania
Dariusz Frejlichowski	West Pomeranian University of Technology, Szczecin, Poland
Blanka Frydrychova Klimova	University of Hradec Kralove, Czech Republic
Janusz Getta	University of Wollongong, Australia
Gergő Gombos	Eötvös Loránd University, Hungary
Antonio Gonzalez-Pardo	Universidad Autónoma de Madrid, Spain

Quang-Thuy Ha	VNU-University of Engineering and Technology, Vietnam
Dawit Haile	Addis Ababa University, Ethiopia
Pei-Yi Hao	National Kaohsiung University of Applied Sciences, Taiwan
Marcin Hernes	Wrocław University of Economics and Business, Poland
Kouichi Hirata	Kyushu Institute of Technology, Japan
Bogumiła Hnatkowska	Wrocław University of Science and Technology, Poland
Huu Hanh Hoang	Posts and Telecommunications Institute of Technology, Vietnam
Quang Hoang	Hue University of Sciences, Vietnam
Van-Dung Hoang	Ho Chi Minh City University of Technology and Education, Vietnam
Jeongky Hong	Yeungnam University, South Korea
Maciej Huk	Wrocław University of Science and Technology, Poland
Zbigniew Huzar	Wrocław University of Science and Technology, Poland
Dmitry Ignatov	National Research University Higher School of Economics, Russia
Hazra Imran	University of British Columbia, Canada
Sanjay Jain	National University of Singapore, Singapore
Khalid Jebari	Abdelmalek Essaâdi University, Morocco
Joanna Jędrzejowicz	University of Gdansk, Poland
Przemysław Juszczuk	University of Economics in Katowice, Poland
Dariusz Kania	Silesian University of Technology, Poland
Mehmet Karaata	Kuwait University, Kuwait
Arkadiusz Kawa	Institute of Logistics and Warehousing, Poland
Zaheer Khan	University of the West of England, UK
Marek Kisiel-Dorohinicki	AGH University of Science and Technology, Poland
Attila Kiss	Eötvös Loránd University, Hungary
Jerzy Klamka	Silesian University of Technology, Poland
Shinya Kobayashi	Ehime University, Japan
Marek Kopel	Wrocław University of Science and Technology, Poland
Jan Kozak	University of Economics in Katowice, Poland
Adrianna Kozierkiewicz	Wrocław University of Science and Technology, Poland
Dalia Kriksciuniene	Vilnius University, Lithuania
Dariusz Król	Wrocław University of Science and Technology, Poland
Marek Krótkiewicz	Wrocław University of Science and Technology, Poland
Marzena Kryszkiewicz	Warsaw University of Technology, Poland

Jan Kubíček	VSB - Technical University of Ostrava, Czech Republic
Tetsuji Kuboyama	Gakushuin University, Japan
Elżbieta Kukla	Wrocław University of Science and Technology, Poland
Julita Kulbacka	Wrocław Medical University, Poland
Marek Kulbacki	Polish-Japanese Academy of Information Technology, Poland
Kazuhiro Kuwabara	Ritsumeikan University, Japan
Jong Wook Kwak	Yeungnam University, South Korea
Halina Kwaśnicka	Wrocław University of Science and Technology, Poland
Annabel Latham	Manchester Metropolitan University, UK
Yue-Shi Lee	Ming Chuan University, Taiwan
Florin Leon	Gheorghe Asachi Technical University of Iasi, Romania
Horst Lichter	RWTH Aachen University, Germany
Tony Lindgren	Stockholm University, Sweden
Sebastian Link	University of Auckland, New Zealand
Igor Litvinchev	Nuevo Leon State University, Mexico
Doina Logofătu	Frankfurt University of Applied Sciences, Germany
Lech Madeyski	Wrocław University of Science and Technology, Poland
Bernadetta Maleszka	Wrocław University of Science and Technology, Poland
Konstantinos Margaritis	University of Macedonia, Greece
Takashi Matsuhisa	Karelia Research Centre, Russian Academy of Science, Russia
Tamás Matuszka	Eötvös Loránd University, Hungary
Michael Mayo	University of Waikato, New Zealand
Vladimir Mazalov	Karelia Research Centre, Russian Academy of Science, Russia
Héctor Menéndez	University College London, UK
Mercedes Merayo	Universidad Complutense de Madrid, Spain
Jacek Mercik	WSB University in Wrocław, Poland
Radosław Michalski	Wrocław University of Science and Technology, Poland
Peter Mikulecky	University of Hradec Kralove, Czech Republic
Miroslava Mikusova	University of Zilina, Slovakia
Marek Milosz	Lublin University of Technology, Poland
Jolanta Mizera-Pietraszko	Wrocław University of Science and Technology, Poland
Dariusz Mrozek	Silesian University of Technology, Poland
Leo Mrsic	IN2data Ltd., Croatia
Agnieszka Mykowiecka	Institute of Computer Science, Polish Academy of Sciences, Poland

Pawel Myszkowski	Wrocław University of Science and Technology, Poland
Fulufhelo Nelwamondo	Council for Scientific and Industrial Research, South Africa
Huu-Tuan Nguyen	Vietnam Maritime University, Vietnam
Loan T. T. Nguyen	Ho Chi Minh City International University, Vietnam
Quang-Vu Nguyen	Vietnam-Korea University of Information and Communication Technology, Vietnam
Thai-Nghe Nguyen	Cantho University, Vietnam
Mariusz Nowostawski	Norwegian University of Science and Technology, Norway
Alberto Núñez	Universidad Complutense de Madrid, Spain
Tarkko Oksala	Aalto University, Finland
Mieczysław Owoc	Wrocław University of Economics and Business, Poland
Marcin Paprzycki	Systems Research Institute, Polish Academy of Sciences, Poland
Panos Patros	University of Waikato, New Zealand
Danilo Pelusi	University of Teramo, Italy
Marek Penhaker	VSB - Technical University of Ostrava, Czech Republic
Maciej Piasecki	Wrocław University of Science and Technology, Poland
Bartłomiej Pierański	Poznań University of Economics and Business, Poland
Dariusz Pierzchala	Military University of Technology, Poland
Marcin Pietranik	Wrocław University of Science and Technology, Poland
Elias Pimenidis	University of the West of England, UK
Jaroslav Pokorný	Charles University in Prague, Czech Republic
Nikolaos Polatidis	University of Brighton, UK
Elvira Popescu	University of Craiova, Romania
Petra Poulova	University of Hradec Kralove, Czech Republic
Om Prakash	University of Allahabad, India
Radu-Emil Precup	Politehnica University of Timisoara, Romania
Małgorzata Przybyła-Kasperek	University of Silesia, Poland
Paulo Quaresma	Universidade de Evora, Portugal
David Ramsey	Wrocław University of Science and Technology, Poland
Mohammad Rashedur Rahman	North South University, Bangladesh
Ewa Ratajczak-Ropel	Gdynia Maritime University, Poland
Alexander Ryjov	Lomonosov Moscow State University, Russia
Keun Ho Ryu	Chungbuk National University, South Korea
Virgilijus Sakalauskas	Vilnius University, Lithuania

Daniel Sanchez	University of Granada, Spain
Rafal Scherer	Częstochowa University of Technology, Poland
S. M. N. Arosha Senanayake	Universiti Brunei Darussalam, Brunei Darussalam
Natalya Shakhovska	Lviv Polytechnic National University, Ukraine
Andrzej Siemiński	Wrocław University of Science and Technology, Poland
Dragan Simic	University of Novi Sad, Serbia
Bharat Singh	Universiti Teknology PETRONAS, Malaysia
Paweł Sitek	Kielce University of Technology, Poland
Krzysztof Ślot	Łódź University of Technology, Poland
Adam Słowik	Koszalin University of Technology, Poland
Vladimir Sobeslav	University of Hradec Kralove, Czech Republic
Zenon A. Sosnowski	Bialystok University of Technology, Poland
Harco Leslie Hendric Spits Warnars	Binus University, Indonesia
Bela Stantic	Griffith University, Australia
Stanimir Stoyanov	Plovdiv University "Paisii Hilendarski", Bulgaria
Ja-Hwung Su	Cheng Shiu University, Taiwan
Libuse Svobodova	University of Hradec Kralove, Czech Republic
Jerzy Świątek	Wrocław University of Science and Technology, Poland
Julian Szymański	Gdansk University of Technology, Poland
Yasufumi Takama	Tokyo Metropolitan University, Japan
Maryam Tayefeh Mahmoudi	ICT Research Institute, Iran
Zbigniew Telec	Wrocław University of Science and Technology, Poland
Dilhan Thilakarathne	Vrije Universiteit Amsterdam, Netherlands
Satoshi Tojo	Japan Advanced Institute of Science and Technology, Japan
Bogdan Trawiński	Wrocław University of Science and Technology, Poland
Maria Trocan	Institut Superieur d'Electronique de Paris, France
Krzysztof Trojanowski	Cardinal Stefan Wyszyński University in Warsaw, Poland
Ualsher Tukeyev	Al-Farabi Kazakh National University, Kazakhstan
Olgierd Unold	Wrocław University of Science and Technology, Poland
Jørgen Villadsen	Technical University of Denmark, Denmark
Eva Volna	University of Ostrava, Czech Republic
Wahyono Wahyono	Universitas Gadjah Mada, Indonesia
Junzo Watada	Waseda University, Japan
Pawel Weichbroth	Gdansk University of Technology, Poland
Izabela Wierzbowska	Gdynia Maritime University, Poland

Krystian Wojtkiewicz	Wrocław University of Science and Technology, Poland
Krzysztof Wróbel	University of Silesia, Poland
Marian Wysocki	Rzeszow University of Technology, Poland
Xin-She Yang	Middlesex University, UK
Tulay Yildirim	Yildiz Technical University, Turkey
Jonghee Youn	Yeungnam University, South Korea
Drago Zagar	University of Osijek, Croatia
Danuta Zakrzewska	Łódź University of Technology, Poland
Constantin-Bala Zamfirescu	Lucian Blaga University of Sibiu, Romania
Katerina Zdravkova	Ss. Cyril and Methodius University in Skopje, Macedonia
Vesna Zeljkovic	Lincoln University, USA
Aleksander Zgrzywa	Wrocław University of Science and Technology, Poland
Jianlei Zhang	Nankai University, China
Maciej Zięba	Wrocław University of Science and Technology, Poland
Adam Ziębiński	Silesian University of Technology, Poland

Program Committees of Special Sessions

ADMTA 2021: Special Session on Advanced Data Mining Techniques and Applications

Tzung-Pei Hong	National University of Kaohsiung, Taiwan
Tran Minh Quang	Ho Chi Minh City University of Technology, Vietnam
Bac Le	VNUHCM-University of Science, Vietnam
Bay Vo	Ho Chi Minh City University of Technology, Vietnam
Chun-Hao Chen	National Taipei University of Technology, Taipei
Mu-En Wu	National Taipei University of Technology, Taipei
Wen-Yang Lin	National University of Kaohsiung, Taiwan
Yeong-Chyi Lee	Cheng Shiu University, Taiwan
Le Hoang Son	VNU-University of Science, Vietnam
Vo Thi Ngoc Chau	Ho Chi Minh City University of Technology, Vietnam
Van Vo	Ho Chi Minh University of Industry, Vietnam
Ja-Hwung Su	National University of Kaohsiung, Taiwan
Ming-Tai Wu	Shandong University of Science and Technology, China
Kawuu W. Lin	National Kaohsiung University of Science and Technology, Taiwan
Ju-Chin Chen	National Kaohsiung University of Science and Technology, Taiwan
Tho Le	Ho Chi Minh City University of Technology, Vietnam
Dang Nguyen	Deakin University, Australia

Hau Le	Thuyloi University, Vietnam
Thien-Hoang Van	Ho Chi Minh City University of Technology, Vietnam
Tho Quan	Ho Chi Minh City University of Technology, Vietnam
Ham Nguyen	University of People's Security, Vietnam
Thiet Pham	Ho Chi Minh University of Industry, Vietnam
Nguyen Thi Thuy Loan	Nguyen Tat Thanh University, Vietnam
C. C. Chen	National Cheng Kung University, Taiwan
Jerry Chun-Wei Lin	Western Norway University of Applied Sciences, Norway
Ko-Wei Huang	National Kaohsiung University of Science and Technology, Taiwan
Ding-Chau Wang	Southern Taiwan University of Science and Technology, Taiwan

CIV 2021: Special Session on Computational Imaging and Vision

Ishwar Sethi	Oakland University, USA
Moongu Jeon	Gwangju Institute of Science and Technology, South Korea
Benlian Xu	Changshu Institute of Technology, China
Weifeng Liu	Hangzhou Danzi University, China
Ashish Khare	University of Allahabad, India
Moonsoo Kang	Chosun University, South Korea
Sang Woong Lee	Gachon University, South Korea
Ekkarat Boonchieng	Chiang Mai University, Thailand
Jeong-Seon Park	Chonnam National University, South Korea
Unsang Park	Sogang University, South Korea
R. Z. Khan	Aligarh Muslim University, India
Suman Mitra	DA-IICT, India
Bakul Gohel	DA-IICT, India
Sathya Narayanan	NTU, Singapore
Jaeyong Kang	Korea National University of Transportation, South Korea
Zahid Ullah	Korea National University of Transportation, South Korea

CoSenseAI 2021: Special Session on Commonsense Knowledge, Reasoning and Programming in Artificial Intelligence

Roland Bouffanais	Singapore University of Technology and Design, Singapore
Ronaldo Menezes	University of Exeter, UK
Apivadee Piyatumrong	NECTEC, Thailand
M. Ilhan Akbas	Embry-Riddle Aeronautical University, USA
Christoph Benzmüller	Freie Universität Berlin, Germany

Bernabe Dorronsoro	University of Cadiz, Spain
Rastko Selmic	Concordia University, Canada
Daniel Stolfi	University of Luxembourg, Luxembourg
Juan Luis Jiménez Laredo	Normandy University, France
Kittichai Lavangnananda	King Mongkut's University of Technology Thonburi, Thailand
Boonyarit Changaival	University of Luxembourg, Luxembourg
Marco Rocchetto	ALES, United Technologies Research Center, Italy
Jundong Chen	Dickinson State University, USA
Emmanuel Kieffer	University of Luxembourg, Luxembourg
Fang-Jing Wu	Technical University Dortmund, Germany
Umer Wasim	University of Luxembourg, Luxembourg

DMPI-4.0 vol. 2 – 2021: Special Session on Data Modelling and Processing for Industry 4.0

Jörg Becker	University of Münster, Germany
Rafał Cupek	Silesian University of Technology, Poland
Helena Dudycz	Wrocław University of Economics and Business, Poland
Marcin Fojcik	Western Norway University of Applied Sciences, Norway
Du Haizhou	Shanghai University of Electric Power, China
Marcin Hernes	Wrocław University of Economics and Business, Poland
Wojciech Hunek	Opole University of Technology, Poland
Marcin Jodłowiec	Wrocław University of Science and Technology, Poland
Marek Krótkiewicz	Wrocław University of Science and Technology, Poland
Florin Leon	Gheorghe Asachi Technical University of Iasi, Romania
Rafał Palak	Wrocław University of Science and Technology, Poland
Jacek Piskorowski	West Pomeranian University of Technology, Szczecin, Poland
Khouloud Salameh	American University of Ras Al Khaimah, United Arab Emirates
Predrag Stanimirović	University of Niš, Serbia
Krystian Wojtkiewicz	Wrocław University of Science and Technology, Poland
Feifei Xu	Southeast University, China

ICxS 2021: Special Session on Intelligent and Contextual Systems

Adriana Albu	Polytechnic University of Timisoara, Romania
Basabi Chakraborty	Iwate Prefectural University, Japan
Chao-Chun Chen	National Cheng Kung University, Taiwan
Dariusz Frejlichowski	West Pomeranian University of Technology, Szczecin, Poland
Erdenebileg Batbaatar	Chungbuk National University, South Korea
Goutam Chakraborty	Iwate Prefectural University, Japan
Ha Manh Tran	Ho Chi Minh City International University, Vietnam
Hong Vu Nguyen	Ton Duc Thang University, Vietnam
Hideyuki Takahashi	Tohoku Gakuin University, Japan
Jerzy Świątek	Wrocław University of Science and Technology, Poland
Józef Korbicz	University of Zielona Gora, Poland
Keun Ho Ryu	Chungbuk National University, South Korea
Khanindra Pathak	Indian Institute of Technology Kharagpur, India
Kilho Shin	Gakashuin University, Japan
Lkhagvadorj Munkhdalai	Chungbuk National University, South Korea
Maciej Huk	Wrocław University of Science and Technology, Poland
Marcin Fojcik	Western Norway University of Applied Sciences, Norway
Masafumi Matsuhara	Iwate Prefectural University, Japan
Meijing Li	Shanghai Maritime University, China
Min-Hsiung Hung	Chinese Culture University, Taiwan
Miroslava Mikusova	University of Žilina, Slovakia
Musa Ibrahim	Chungbuk National University, South Korea
Nguyen Khang Pham	Can Tho University, Vietnam
Nipon Theera-Umpon	Chiang Mai University, Thailand
Plamen Angelov	Lancaster University, UK
Qiangfu Zhao	University of Aizu, Japan
Quan Thanh Tho	Ho Chi Minh City University of Technology, Vietnam
Rafal Palak	Wrocław University of Science and Technology, Poland
Rashmi Dutta Baruah	Indian Institute of Technology Guwahati, India
Sansanee Auephanwiriyakul	Chiang Mai University, Thailand
Sonali Chouhan	Indian Institute of Technology Guwahati, India
Takako Hashimoto	Chiba University of Commerce, Japan
Tetsuji Kuboyama	Gakushuin University, Japan
Thai-Nghe Nguyen	Can Tho University, Vietnam

ISCEC 2021: Special Session on Intelligent Supply Chains and e-Commerce

Arkadiusz Kawa	Łukasiewicz Research Network – The Institute of Logistics and Warehousing, Poland
Bartłomiej Pierański	Poznań University of Economics and Business, Poland
Carlos Andres Romano	Polytechnic University of Valencia, Spain
Costin Badica	University of Craiova, Romania
Davor Dujak	University of Osijek, Croatia
Miklós Krész	InnoRenew, Slovenia
Paweł Pawlewski	Poznań University of Technology, Poland
Adam Koliński	Łukasiewicz Research Network – The Institute of Logistics and Warehousing, Poland
Marcin Anholcer	Poznań University of Economics and Business, Poland

MMAML 2021: Special Session on Multiple Model Approach to Machine Learning

Emili Balaguer-Ballester	Bournemouth University, UK
Urszula Boryczka	University of Silesia, Poland
Abdelhamid Bouchachia	Bournemouth University, UK
Robert Burduk	Wrocław University of Science and Technology, Poland
Oscar Castillo	Tijuana Institute of Technology, Mexico
Rung-Ching Chen	Chaoyang University of Technology, Taiwan
Suphamit Chittayasothorn	King Mongkut's Institute of Technology Ladkrabang, Thailand
José Alfredo F. Costa	Federal University of Rio Grande do Norte (UFRN), Brazil
Ireneusz Czarnowski	Gdynia Maritime University, Poland
Fernando Gomide	State University of Campinas, Brazil
Francisco Herrera	University of Granada, Spain
Tzung-Pei Hong	National University of Kaohsiung, Taiwan
Piotr Jędrzejowicz	Gdynia Maritime University, Poland
Tomasz Kajdanowicz	Wrocław University of Science and Technology, Poland
Yong Seog Kim	Utah State University, USA
Bartosz Krawczyk	Virginia Commonwealth University, USA
Kun Chang Lee	Sungkyunkwan University, South Korea
Edwin Lughofer	Johannes Kepler University Linz, Austria
Hector Quintian	University of Salamanca, Spain
Andrzej Siemiński	Wrocław University of Science and Technology, Poland
Dragan Simic	University of Novi Sad, Serbia

Adam Słowik	Koszalin University of Technology, Poland
Zbigniew Telec	Wrocław University of Science and Technology, Poland
Bogdan Trawiński	Wrocław University of Science and Technology, Poland
Olgierd Unold	Wrocław University of Science and Technology, Poland
Michał Woźniak	Wrocław University of Science and Technology, Poland
Zhongwei Zhang	University of Southern Queensland, Australia
Zhi-Hua Zhou	Nanjing University, China

Contents

Decision Support and Control Systems

Advanced Data Mining Techniques and Applications

Intelligent and Contextual Systems

Data Mining Methods and Applications

Mining Partially-Ordered Episode Rules in an Event Sequence

Philippe Fournier-Viger[1(✉)], Yangming Chen[1], Farid Nouioua[2], and Jerry Chun-Wei Lin[3]

[1] Harbin Institute of Technology (Shenzhen), Shenzhen, China
[2] University of Bordj Bou Arreridj, El Anceur, Algeria
[3] Department of Computing, Mathematics and Physics, Western Norway University of Applied Sciences (HVL), Bergen, Norway
jerrylin@ieee.org

Abstract. Episode rule mining is a popular data mining task for analyzing a sequence of events or symbols. It consists of identifying subsequences of events that frequently appear in a sequence and then to combine them to obtain episode rules that reveal strong relationships between events. But a key problem is that each rule requires a strict ordering of events. As a result, similar rules are treated differently, though they in practice often describe a same situation. To find a smaller set of rules that are more general and can replace numerous episode rules, this paper introduces a novel type of rules called partially-ordered episode rules, where events in a rule are partially ordered. To efficiently find all these rules in a sequence, an efficient algorithm named POERM (Partially-Ordered Episode Rule Miner) is presented. An experimental evaluation on several benchmark dataset shows that POERM has excellent performance.

Keywords: Pattern mining · Sequence · Partially ordered episode rule

1 Introduction

Pattern mining is a sub-field of data mining, which aims at identifying interesting patterns in data that can help to understand the data and/or support decision-making. In recent years, numerous algorithms have been designed to find patterns in discrete sequences (a sequence of events or symbols) as this data type is found in many domains. For instance, text documents can be represented as sequence of words, customers purchases as sequences of transactions, and drone trajectories as sequence of locations. Whereas some pattern mining algorithms find similarities between sequences [6,7,15] or across sequences [16], others identify patterns in a single very long sequence. One of the most popular task of this type is *Frequent Episode Mining* (FEM) [9,11,12,14]. It consists of finding all frequent episodes in a sequence of events, that is all subsequences that have a *support* (occurrence frequency) that is no less than a user-defined *minsup* threshold. Two types of sequences are considered in FEM: *simple sequences* where

N. T. Nguyen et al. (Eds.): ACIIDS 2021, LNAI 12672, pp. 3–15, 2021.
https://doi.org/10.1007/978-3-030-73280-6_1

events have timestamps and are totally ordered, and *complex sequences* where simultaneous events are allowed. Many algorithms were designed for discovering frequent episodes such as MINEPI and WINEPI [12], EMMA and MINEPI+ [9], and TKE [8]. While some algorithms find *serial episodes* (ordered lists of events), others find *parallel episodes* (sets of simultaneous events) or *composite episode* (a combination of serial/parallel episodes). Though finding frequent episodes is useful, episodes are only discovered on the basis of their support (occurrence frequencies) [1]. Thus, some events may only appear together in an episode by chance. Moreover, frequent episodes do not provide information about how likely it is that some events will occur following some other events. To address these issues, a post-processing step can be applied after FEM, which is to combine pairs of frequent episodes to create *episode rules*. An episode rule is a pattern having the form $E_1 \rightarrow E_2$, which indicates that if some episode E_1 appears, it will be followed by another episode E_2 with a given confidence or probability [4,12].

Episode rule mining is useful as it can reveal strong temporal relationships between events in data from many domains [2–4,12]. For example, a rule $R_1 : \langle\{a\},\{b\},\{c\}\rangle \rightarrow \langle\{d\}\rangle$ could be found in moviegoers data, indicating that if a person watches some movies a, b and c in that order, s/he will then watch movie d. Based on such rules, marketing decisions could be taken or recommendation could be done. However, a major drawback of traditional episode rule mining algorithms is that events in each rule must be strictly ordered. As a result, similar rules are treated differently. For example, the rule R_1 is considered as different from rules $R_2 : \langle\{b\},\{a\},\{c\}\rangle \rightarrow \langle\{d\}\rangle$, $R_3 : \langle\{b\},\{c\},\{a\}\rangle \rightarrow \langle\{d\}\rangle$, $R_4 : \langle\{c\},\{a\},\{b\}\rangle \rightarrow \langle\{d\}\rangle$, $R_5 : \langle\{c\},\{b\},\{a\}\rangle \rightarrow \langle\{d\}\rangle$ and $R_6 : \langle\{a\},\{c\},\{b\}\rangle \rightarrow \langle\{d\}\rangle$. But all these rules contain the same events. This is a problem because all these rules are very similar and may in practice represents the same situation that someone who has watched three movies (e.g. Frozen, Sleeping Beauty, Lion King) will then watch another (e.g. Harry Potter). Because these rules are viewed as distinct, their support (occurrence frequencies) and confidence are calculated separately and may be very different from each other. Moreover, analyzing numerous rules representing the same situation with slight ordering variations is not convenient for the user. Thus, it is desirable to extract a more general and flexible type of rules where ordering variations between events are tolerated.

This paper addresses this issue by introducing a novel type of rules called *Partially-Ordered Episode Rules* (POER), where events in a rule antecedent and in a rule consequent are unordered. A POER has the form $I_1 \rightarrow I_2$, where I_1 and I_2 are sets of events. A rule is interpreted as if all event(s) in I_1 appear in any order, they will be followed by all event(s) from I_2 in any order. For instance, a POER $R_7 : \{a, b, c\} \rightarrow \{d\}$ indicates that if someone watches movies a, b and c in any order, s/he will watch d. The advantage of finding POERs is that a single rule can replace multiple episode rules. For example R_7 can replace $R_1, R_2, \ldots, R_6$. However, discovering POER is challenging as they are not derived from episodes. Thus, a novel algorithm must be designed to efficiently find POER in a sequence.

The contributions of this paper are the following. The problem of discovering POERs is defined and its properties are studied. Then, an efficient algorithm named POERM (Partially-Ordered Episode Rule Miner) is presented. Lastly, an experimental evaluation was performed on several benchmark datasets to evaluate POERM. Results have shown that it has excellent performance.

The rest of this paper is organized as follows. Section 2 defines the proposed problem of POER mining. Section 3 describes the POERM algorithm. Then, Sect. 4 presents the experimental evaluation. Finally, Sect. 5 draws a conclusion and discusses future work.

2 Problem Definition

The type of data considered in episode rule mining is a sequence of events with timestamps [9,12]. Let there be a finite set of **events** $E = \{i_1, i_2, \ldots, i_m\}$, also called items or symbols. In addition, let there be a set of **timestamps** $T = \{t_1, t_2, \ldots t_n\}$ where for any integers $1 \leq i < j \leq n$, the relationship $t_i < t_j$ holds. A time-interval $[t_i, t_j]$ is said to have a duration of $t_j - t_i$ time. Moreover, two time intervals $[t_{i1}, t_{j1}]$ and $[t_{i2}, t_{j2}]$ are said to be **non-overlapping** if either $t_{j1} < t_{i2}$ or $t_{j2} < t_{i1}$. A subset $X \subseteq E$ is called an **event set**. Furthermore, X is said to be a k-event set if it contains k events. A **complex event sequence** is an ordered list of event sets with timestamps $S = \langle (SE_{t_1}, t_1), (SE_{t_2}, t_2), \ldots, (SE_{t_n}, t_n) \rangle$ where $SE_{t_i} \subseteq E$ for $1 \leq i \leq n$. A **simultaneous event set** in a complex event sequence is an event set where all events occurred at the same time. If a complex event sequence contains no more than one event per timestamp, it is a **simple event sequence**. Data of various types can be represented as an event sequence such as cyber-attacks, trajectories, telecommunication data, and alarm sequences [8].

For example, a complex event sequence is presented in Fig. 1. This sequence contains eleven timestamps (T $= \{t_1, t_2, \ldots, t_{11}\}$) and events are represented by letters ($E = \{a, b, c, d\}$). That sequence indicates that event c appeared at time t_1, followed by events $\{a, b\}$ simultaneously at time t_2, followed by event d at time t_3, followed by a at t_5, followed by c at t_6, and so on.

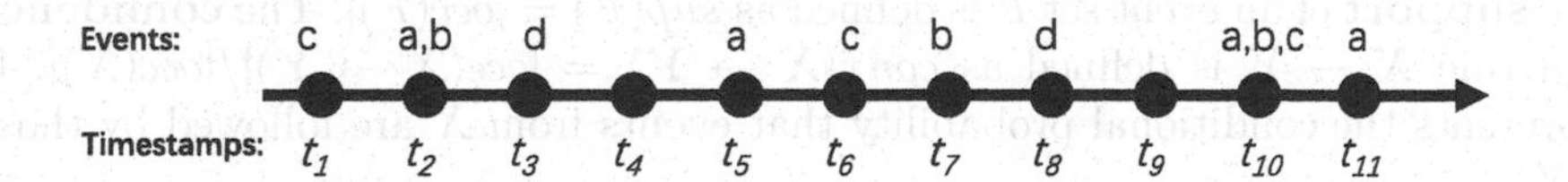

Fig. 1. An complex event sequence with 11 timestamps

This paper proposes a novel type of rules called **partially-ordered episode rule**. A POER has the form $X \rightarrow Y$ where $X \subset E$ and $Y \subset E$ are non empty event sets. The meaning of such rule is that if all events from X appear in any order in the sequence, they will be followed by all events from Y. To avoid finding rules containing events that are too far appart three constraints are specified: (1) events from X must appear within some maximum amount of time $XSpan \in \mathbb{Z}^+$, (2) events from Y must appear within some maximum amount of time $YSpan \in \mathbb{Z}^+$, and (3) the time between X and Y must be no less than a constant $XYSpan \in \mathbb{Z}^+$. The three constraints $XSpan$, $YSpan$ and $XYSpan$ must be specified by the user, and are illustrated in Fig. 2 for a rule $X \rightarrow Y$.

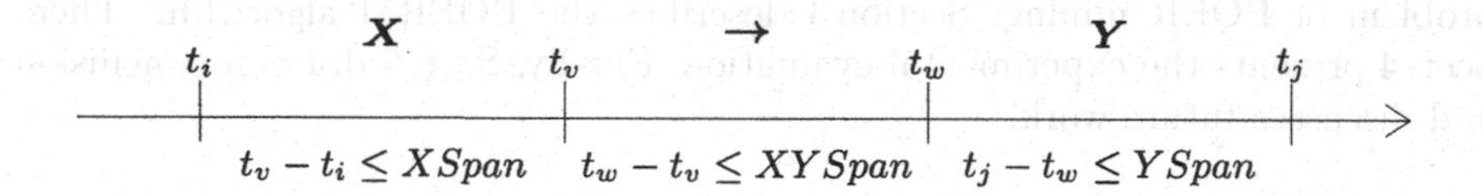

Fig. 2. The three time constraints on a POER

Furthermore, to select interesting rules, two measures are used called the support and confidence, which are inspired by previous work on rule mining. They are defined based on the concept of occurrence.

An **occurrence of an event set** $F \subset E$ in a complex event sequence S is a time interval $[t_i, t_j]$ where all events from F appear, that is $F \subseteq SE_i \cup SE_{i+1} \ldots \cup SE_j$. An **occurrence of a rule** $X \rightarrow Y$ in a complex event sequence S is a time interval $[t_i, t_j]$ such that there exist some timestamps t_v, t_w where $t_i \leq t_v < t_w \leq t_j$, X has an occurrence in $[t_i, t_v]$, Y has an occurrence in $[t_w, t_j]$, $t_v - t_i < XSpan$, $t_w - t_v < XYSpan$, and $t_j - t_w < YSpan$.

Analyzing occurrences of event sets or rules one can reveal interesting relationships between events. However, a problem is that some occurrences may overlap, and thus an event may be counted as part of multiple occurrences. To address this problem, this paper proposes to only consider a subset of all occurrences defined as follows. An occurrence $[t_{i1}, t_{j1}]$ is said to be **redundant** in a set of occurrences if there does exist an overlapping occurrence $[t_{i2}, t_{j2}]$ such that $t_{i1} \leq t_{i2} \leq t_{j1}$ or $t_{i2} \leq t_{i1} \leq t_{l2}$. Let $occ(F)$ denotes the **set of all non redundant occurrences of an event set** F in a sequence S. Moreover, let $occ(X \rightarrow Y)$ denotes the **set of non redundant occurrences of a rule** $X \rightarrow Y$ in a sequence S.

The **support** of a rule $X \rightarrow Y$ is defined as $sup(X \rightarrow Y) = |occ(X \rightarrow Y)|$. The **support** of an event set F is defined as $sup(F) = |occ(F)|$. The **confidence** of a rule $X \rightarrow Y$ is defined as $conf(X \rightarrow Y) = |occ(X \rightarrow Y)|/|occ(X)|$. It represents the conditional probability that events from X are followed by those of Y.

Definition 1 (Problem Definition). *XY Let there be a complex event sequence S and five user-defined parameters: $XSpan$, $YSpan$, $XYSpan$, minsup and minconf. The problem of mining POERs is to find all the valid POERs. A POER r is said to be frequent if $sup(r) \geq minsup$, and it is said to be valid if it is frequent and $conf(r) \geq minconf$.*

For instance, consider the sequence of Fig. 1, $minsup = 3$, $minconf = 0.6$, $XSpan = 3$, $XYSpan = 1$ and $YSpan = 1$. The occurrences of $\{a, b, c\}$

are $occ(\{a,b,c\}) = \{[t_1,t_2],[t_5,t_7],[t_{10},t_{10}]\}$ The occurrences of the rule $R : \{a,b,c\} \rightarrow \{d\}$ are $occ(R) = \{[t_1,t_3],[t_5,t_8]\}$. Hence, $supp(R) = 3$, $conf(R) = 2/3$, and R is a valid rule.

3 The POERM Algorithm

The problem of POER mining is difficult as if a database contains m distinct events, up to $(2^m - 1) \times (2^m - 1)$ rules may be generated. Moreover, each rule may have numerous occurrences in a sequence.

To efficiently find all valid POERs, this section describes the proposed POERM algorithm. It first finds event sets that may be antecedents of valid rules and their occurrences in the input sequence. Then, the algorithm searches for consequents that could be combined with these antecedents to build POERs. The valid POERs are kept and returned to the user. To avoid considering all possible rules, the POERM algorithm utilizes the following search space pruning property (proof is omitted due to space limitation).

Property 1 (Rule Event Set Pruning Property). An event set X cannot be the antecedent of a valid rule if $sup(X) < minsup$. An event set Y cannot be the consequent of a valid rule if $sup(Y) < minsup \times minconf$.

The POERM algorithm (Algorithm 1) takes as input a complex event sequence S, and the user-defined $XSpan$, $YSpan$, $XYSpan$, $minsup$ and $minconf$ parameters. The POERM algorithm first reads the input sequence and creates a copy $XFres$ of that sequence containing only events having a support no less than $minsup$. Other events are removed because they cannot appear in a valid rule (based on Property 1). Then, POERM searches for antecedents by calling the *MiningXEventSet* procedure with $XFres$, $XSpan$ and $minsup$. This procedure outputs a list $xSet$ of event sets that may be antecedents of valid POERs, that is each event set having at least $minsup$ non overlapping occurrences of a duration not greater than $XSpan$.

The *MiningXEventSet* procedure is presented in Algorithm 2. It first scans the sequence $XFres$ to find the list of occurrences of each event. This information is stored in a map $fresMap$ where each pair indicates an event as key and its occurrence list as value. Then, scan all the pairs of $fresMap$ and put the pairs that have at least $minsup$ non-overlapping occurrences are added to $xSet$. At this moment, $xSet$ contains all 1-event sets that could be valid rule antecedents based on Property 1. Then, the procedure scans the sequence again to extend these 1-event sets into 2-event sets, and then extends 2-event sets into 3-event sets and so on until no more event sets can be generated. During that iterative process, each generated event set having more than $minsup$ non-overlapping occurrences is added to $xSet$ and considered for extensions to generate larger event sets. The *MiningXEventSet* procedure returns $xSet$, which contains all potential antecedents of valid POERs (those having at least $minsup$ non-overlapping occurrences).

A challenge for implementing the *MiningXEventSet* procedure efficiently is that event sets are by definition unordered. Hence, different ordering of a same set represent the same event set (e.g. $\{a, b, c\}$,$\{b, a, c\}$ and $\{a, c, b\}$ are the same set). To avoid generating the same event set multiple times, event sets are in practice sorted by the lexicographic order (e.g. as $\{a, b, c\}$), and the procedure only extends an event set F with an event e if e is greater than the last event of F.

A second key consideration for implementing $MiningXEventSet$ is how to extend an l-event set F into $(l + 1)$-event sets and calculate their occurrences in the $XFres$ sequence. To avoid scanning the whole sequence, this is done by searching around each occurrence *pos* of F in its *OccurrenceList*. Let *pos.start* and *post.end* respectively denote the start and end timestamps of *pos*. The algorithm searches for events in the time intervals [*pos.end*− *XSpan* +1, *pos.start*), [*pos.end*+1, *pos.start*+*XSpan*), and [*pos.start*, *pos.end*]. For each event e that is greater than the last item of F, the occurrences of $F \cup \{e\}$ of the form [i, *pos.end*], [*pos.start*, i] or [*pos.start*, *pos.end*] are added in $fresMap$. Then, $fresMap$ is scanned to count the non-overlaping occurrences of each $(l + 1)$-node event set, and all sets having more than *minsup* non-overlapping occurrences are added to $xSet$.

Counting the maximum non-overlapping occurrences of an l-event set F is done using its occurrence list (not shown in the pseudocode). The procedure first applies the quick sort algorithm to sort the *OccurrenceList* by ascending ending timestamps (*pos.end*). Then, a set $CoverSet$ is created to store the maximum non-overlapping occurrences of F. The algorithm loops over the *OccurrenceList* of F to check each occurrence from the first one to the last one. If the current occurrence does not overlap with the last added occurrence in $CoverSet$, it is added to $CoverSet$. Otherwise, it is ignored. When the loop finishes, $CoverSet$ contains the maximum non-overlapping occurrences of F.

After applying the *MiningXEventSet* procedure to find antecedents, the POERM algorithm searches for event sets (consequents) that could be combined with these antecedents to create valid POERs. The algorithm first eliminates all events having less than $minsup \times minconf$ occurrences from the sequence $XFres$ to obtain a sequence $YFres$ (based on pruning Property 1). Then, a loop is done over each antecedent x in $xSet$ to find its consequents. In that loop, the time intervals where such consequents could appear are first identified and stored in a variable *xOccurrenceList*. Then, a map $conseMap$ is created to store each event e and its occurrence lists for these time intervals in $YFres$. The map is then scanned to create a queue *candidateRuleQueue* containing each rule of the form $x \rightarrow e$ and its occurrence list. If a rule $x \rightarrow e$ is such that $|occ(x \rightarrow e)| \geq minconf \times |occ(x)|$, then it is added to the set *POERs* of valid POERs. At this time, *candidateRuleQueue* contains all the candidate rule with a 1-event consequent.

The algorithm then performs a loop that pops each rule $X \rightarrow Y$ from the queue to try to extend it by adding an event to its consequent. This is done by scanning $conseMap$. The obtained rule extensions of the form $X \rightarrow Y \cup \{e\}$ and

Algorithm 1: POERM

Input: an event sequence S, the $XSpan$, $YSpan$, $XYSpan$, $minsup$ and $minconf$ parameters;
Output: the set $POERs$ of valid partially-ordered episode rules

```
1  XFres = loadFrequentSequence(S, minsup);
2  xSet = MiningXEventSet(XFres, XSpan, minsup);
3  YFres = loadFrequentSequence(XFres, minsup × minconf);
4  POERs ← ∅;
5  foreach event set x in xSet do
6      for i = 1 to (XYSpan + YSpan) do
7          xOccurrenceList ← {t|t = occur.end + i, occur ∈ x.OccurrenceList};
8      end
9      Scan each timestamp of YFres in xOccurrenceList to obtain a map
       conseMap that records each event e and its occurrence list;
10     Scan conseMap and put the pair (x → e, OccurrenceList) in a queue
       candidateRuleQueue (note: infrequent rules are kept because event e may
       be extended to obtain some frequent rules);
11     Add each rule x → e such that |occ(x → e)| ≥ minconf × |occur(x)| into the
       set POERs;
12     while candidateRuleQueue ≠ ∅ do
13         Pop a rule X → Y from candidateRuleQueue;
14         For each occurrence occur of X → Y, let;
15         start ← max(occur.X.end + 1, occur.Y.end − YSpan + 1);
16         end ← min(occur.X.end + XYSpan + YSpan, occur.X.start + YSpan);
17         Scan each timestamp in [start,end), add each candidate rule in
           candidateRuleQueue, and add each valid POER in POERs;
18     end
19 end
20 return POERs;
```

their occurrence lists are added to *candidateRuleQueue* to be further extended. Moreover, each rule such that $|occ(x \rightarrow e)| \geq minconf \times\ |occ(x)|$ is added to the set *POERs* of valid POERs. This loop continues until the queue is empty. Then, the algorithm returns all valid POERs.

It is worth noting that an occurrence of a rule $X \rightarrow Y$ having an l-event set consequent may not be counted for its support while it may be counted in the support of a $(l+1)$-event consequent rule that extends the former rule. For instance, consider the sequence of Fig. 1, that $XSpan = YSpan = 2$, and $XYSpan = 1$. The occurrence of $\{a\} \rightarrow \{b\}$ in $[t_5, t_7]$ is not counted in the support of $\{a\} \rightarrow \{b\}$ because the time between a and b is greater than $XYSpan$. But $\{a\} \rightarrow \{b\}$ can be extended to $\{a\} \rightarrow \{b, c\}$ and the occurrence $[t_5, t_7]$ is counted in its support. To find the correct occurrences of rules of the form $X \rightarrow Y \cup \{e\}$ extending a rule $X \rightarrow Y$, the following approach is used. For each occurrence *occur* of $X \rightarrow Y$, a variable *start* is set to max(occur.X.end+1,occur.Y.end-$YSpan$+1), a variable *end* is set to min(occur.X.end+$XYSpan$+$YSpan$, occur.Y.start+$YSpan$). Thereafter, three

Algorithm 2: MiningXEventSet

Input: $XFres$: the sequence with only events having $support \geq minsup$; $XSpan$: maximum window size; $minsup$ threshold;
Output: a list of event sets that may be antecedents of valid POERs

```
1  Scan the sequence XFres to record the occurrence list of each event in a map
   fresMap (key = event set, value = occurrence list);
2  xSet ← all the pairs of fresMap such that |value| ≥ minsup;
3  start ← 0;
4  while start < |xSet| do
5      F ← xSet[start].getKey;
6      OccurrenceList ← xSet[start].getValue;
7      Clear fresMap;
8      start = start +1;
9      foreach occurrence pos in OccurrenceList do
10         pStart ← pos.end− XSpan +1; pEnd ← pos.start+ XSpan;
11         Search the time intervals [pStart, pos.start), [pos.end + 1, pEnd),
           [pos.start, pos.end] to add each event set F ∪ {e} such that
           e > F.lastItem and its occurrences of the forms [i, pos.end],
           [pos.start, i] or [pos.start, pos.end], in the map fresMap;
12     end
13     Add each pair of fresMap such that |value| ≥ minsup into xSet;
14 end
15 return xSet;
```

intervals are scanned, which are $[start, occur.Y.end)$, $(occur.Y.end, end)$ and also $[occur.Y.start,\ occur.Y.end]$ to add each rule $X \rightarrow Y \cup \{e\}$ such that $e > Y.lastItem$ and its occurrences of the forms $[i, occur.Y.end]$, $[occur.Y.start, i]$ or $[occur.Y.start, occur.Y.end]$ in the $conseMap$. These three intervals are illustrated in Fig. 3 and allows to find the correct occurrences of $(l+1)$-node consequent rules. In that figure, t_Y denotes $occur.Y$ and t_X denotes $occur.X$.

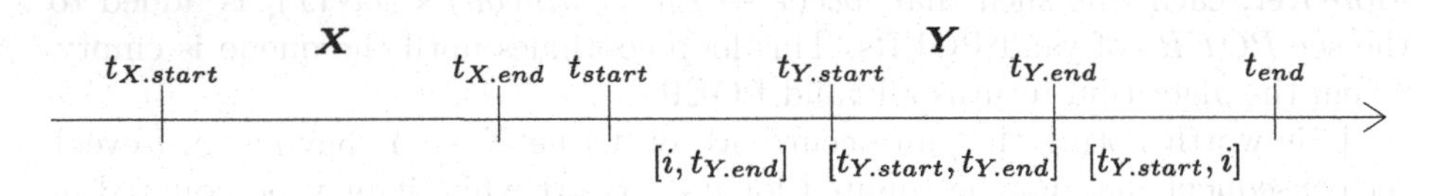

Fig. 3. The timestamps that are searched for a candidate rule $X \rightarrow Y$

The proposed POERM algorithm can find all valid POERs since it only prunes events that cannot be part of valid rules by Property 1. The following section presents an experimental evaluation of POERM, where its performance is compared with that of a baseline version of POERM, called POERM-ALL. The difference between POERM and POERM-ALL is that the latter finds all possible antecedents and consequents separately before combining them to generate rules, rather than using antecedents to search for consequents.

4 Experimental Evaluation

The proposed POERM algorithm's efficiency was evaluated on three benchmark sequence datasets obtained from the SPMF library [5], named *OnlineRetail, Fruithut* and *Retail. OnlineRetail* is a sequence of 541,909 transactions from a UK-based online retail store with 2,603 distinct event types indicating the purchase of items. *Fruithut* is a sequence of 181,970 transactions with 1,265 distinct event types, while *Retail* is a sequence of 88,162 customer transactions from an anonymous Belgian retail store having 16,470 distinct event types. Because there is no prior work on POER mining, the performance of POERM was compared with the POERM-ALL baseline, described in the previous section. Both algorithms were implemented in Java and source code and datasets are made available at http://philippe-fournier-viger.com/spmf/.

In each experiment, a parameter is varied while the other parameters are fixed. Because algorithms have five parameters and the space does not allow evaluating each parameter separately, the three time constraint parameters $XSpan$, $YSpan$ and $XYSpan$ were set to a same value called *Span*. The default values for ($minsup$, $minconf$, *Span*) on the *OnlineRetail*, *Fruithut* and *Retail* datasets are (5000, 0.5, 5), (5000, 0.5, 5) and (4000, 0.5, 5), respectively. These values were found empirically to be values were algorithms have long enough runtimes to highlight their differences.

Influence of *minsup*. Figure 4(a) shows the runtimes of the two algorithms when *minsup* is varied. As *minsup* is decreased, runtimes of both algorithms increase. However, that of POERM-ALL increases much more quickly on the *OnlineRetail* and *FruitHut* datasets, while *minsup* has a smaller impact on POERM-ALL for *Retail.* The reason for the poor performance of POERM-ALL is that as *minsup* decreases, POERM-ALL considers more antecedent event sets ($sup(X) \geq minsup$) and more consequent event sets ($sup(Y) \geq minsup * minconf$). Moreover, POERM-ALL combines all antecedents with all consequents to then filter out non valid POERs based on the confidence. For instance, on *OnlineRetail* and $minsup = 6000$, there are 1,173 antecedents and 5,513 consequents, and hence 6,466,749 candidate rules. But for $minsup = 4000$, there are 2,813 antecedents, 11,880 consequents and 33,193,400 candidate rules, that is a five time increase. The POERM algorithm is generally more efficient as it does not combine all consequents will all antecedents. Hence, its performance is not directly influenced by the number of consequent event sets. This is why *minsup* has a relatively small impact on POERM's runtimes on the *OnlineRetail* and *FruitHut* dataset. However, *Retail* is a very sparse dataset. As a result for $minsup = 3000$, POERM-ALL just needs to combine 190 antecedents with 747 consequents, and thus spend little time in that combination stage. Because there are many consequents for *Retail*, POERM scans the input sequence multiple times for each antecedent to find all possible consequents. This is why POERM performs less well than POERM-ALL on *Retail.*

Influence of $minconf$. Figure 4(b) shows the runtimes of the two algorithms for different $minconf$ values. It is observed that as $minconf$ is decreased, runtimes increase. POERM-ALL is more affected by variations of $minconf$ than POERM on *OnlineRetail* and *FruitHut* but the impact is smaller on *Retail*. The reason is that as $minconf$ decreases, POERM-ALL needs to consider more consequent event sets meeting the condition $sup(Y) \geq minsup \cdot minconf$, and thus the number of candidates rules obtained by combining antecedents and consequents increases. For instance, on *OnlineRetail* and $minconf = 0.6$, POERM-ALL finds 1,724 antecedents and 5,315 consequents. But for $minconf = 0.2$, the number of antecedents is the same, while POERM-ALL finds 46,236 consequents, increasing the number of candidate rules by nine times. On the other hand, $minconf$ does not have a big influence on POERM as it does not use a combination approach and hence is not directly influenced by the consequent count. On *Retail* the situation is different as it is a very sparse dataset. For each antecedent, POERM needs to scan the input sequence multiple time to find all the corresponding consequents. Thus, POERM performs less well on *Retail*.

Influence of *Span*. Figure 4(c) compares the runtimes of both algorithms when *Span* is varied. As *Span* is increased, runtimes increase, and POERM-ALL is more affected by an increase of *Span* than POERM for *OnlineRetail* and *FruitHut*. But the impact on *Retail* is very small. This is because, as *Span* is increased, the time constraints are more loose and POERM-ALL needs to find more antecedents and consequent event sets, which results in generating more candidate rules by the combination process. For instance, on *OnlineRetail* and *Span* = 3, POERM-ALL finds 44 antecedents and 2,278 consequents to generate 1,239,232 candidate rules. But for *Span* = 7, it finds 5,108 antecedents,

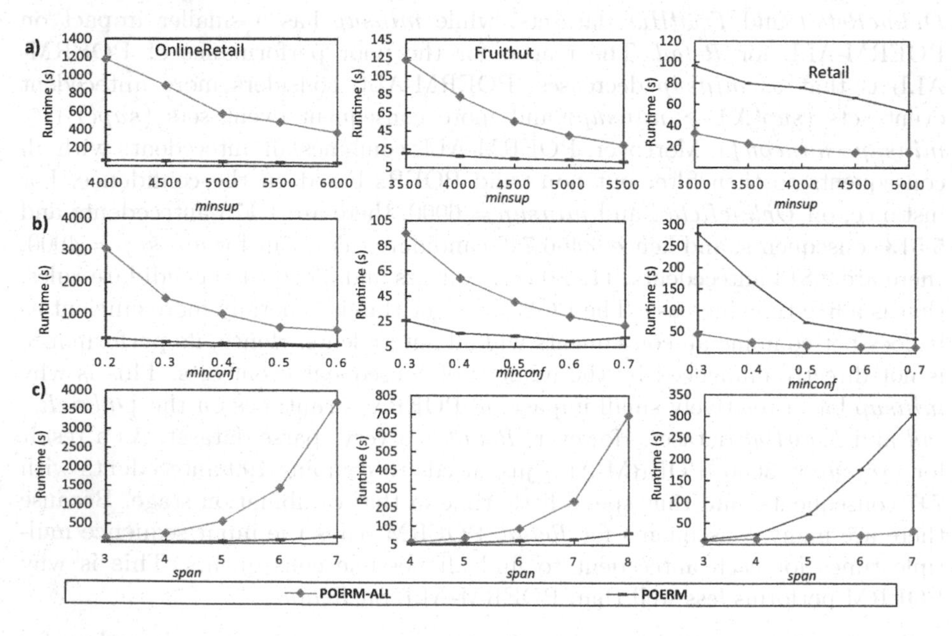

Fig. 4. Influence of (a) *minsup*, (b) $minconf$ and (c) *span* on runtime

26,009 consequents and 132,853,972 rules (100 times increase). As POERM does not use this combination approach, its runtime is not directly influenced by the consequent count, and its runtime mostly increases with the antecedent count. Thus, changes in *Span* have a relatively small impact on POERM for *OnlineRetail* and *FruitHut*. But *Retail* is a sparse dataset. For each antecedent episode, POERM needs to scan the input sequence multiple times to make sure it finds all the corresponding consequents. Thus, POERM performs less well.

Memory Consumption. Figure 5 shows the memory usage of both algorithms when *Span* is varied. Similarly to runtime, as *Span* is increased, memory usage increases because the search space grows. On *OnlineRetail* and *FruitHut*, POERM consumes from 25% to 200% less memory then POERM-ALL. But on sparse datasets like *Retail*, POERM needs to scan the input sequence multiple times and store antecedent episodes in memory to make sure it finds all the corresponding consequents. Thus, POERM consumes more memory then POERM-ALL on *Retail*.

Discovered Patterns. Using the POERM algorithm, several rules were discovered in the data. For instance, some example rules found in the FruitHut dataset are shown in Table 1. Some of these rules have a high confidence. For example, the rule $CucumberLebanese, FieldTomatoes \rightarrow BananaCavendish$ has a confidence of 4910/6152 = 79.8%. Note that only a subset of rules in the table due

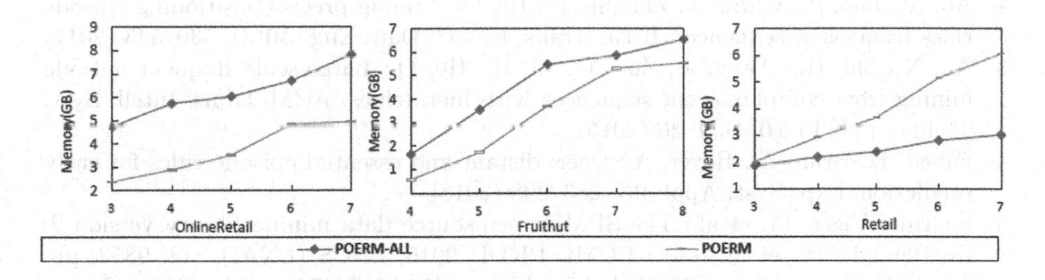

Fig. 5. Influence of *span* on Memory usage

Table 1. Example rules from *FruitHut*

Rule	occ(X → Y)	occ(X)
Cucumber Lebanese, Field Tomatoes→Banana Cavendish	4910	6152
Capsicum red, Field Tomatoes→Banana Cavendish	5033	6352
Broccoli, Capsicum red→Field Tomatoes	2343	4043
Nectarine White→Watermelon seedless	2498	5687
Garlic loose, Field Tomatoes→Capsicum red	1752	4409
Cucumber Lebanese, Capsicum red →Eggplant	1236	4098

to space limitation. The presented subset of rules is selected to give an overview of rules containing various items.

5 Conclusion

To find more general episode rules, this paper has proposed a novel type of rules called partially-ordered episode rules, where events in a rule are partially ordered. To efficiently find all these rules in a sequence, an efficient algorithm named POERM (Partially-Ordered Episode Rule Miner) was presented. An experimental evaluation on several benchmark dataset shows that POERM has excellent performance.

There are several possibilities for future work such as (1) extending POERM to process streaming data or run on a big data or multi-thread environment to benefit from parallelism, (2) considering more complex data such as events that are organized according to a taxonomy [3] or a stream [17], and (3) developing a sequence prediction model based on POERs. Other pattern selection functions will also be considered such as the utility [13,18] and rarity [10].

References

1. Ao, X., Luo, P., Li, C., Zhuang, F., He, Q.: Online frequent episode mining. In: 2015 IEEE 31st International Conference on Data Engineering, pp. 891–902. IEEE (2015)
2. Ao, X., Luo, P., Wang, J., Zhuang, F., He, Q.: Mining precise-positioning episode rules from event sequences. IEEE Trans. Knowl. Data Eng. **30**(3), 530–543 (2017)
3. Ao, X., Shi, H., Wang, J., Zuo, L., Li, H., He, Q.: Large-scale frequent episode mining from complex event sequences with hierarchies. ACM Trans. Intell. Syst. Technol (TIST) **10**(4), 1–26 (2019)
4. Fahed, L., Brun, A., Boyer, A.: Deer: distant and essential episode rules for early prediction. Exp. Syst. Appl. **93**, 283–298 (2018)
5. Fournier-Viger, P., et al.: The SPMF open-source data mining library version 2. In: Berendt, B., et al. (eds.) ECML PKDD 2016. LNCS (LNAI), vol. 9853, pp. 36–40. Springer, Cham (2016). https://doi.org/10.1007/978-3-319-46131-1_8
6. Fournier-Viger, P., Lin, J.C.W., Kiran, U.R., Koh, Y.S.: A survey of sequential pattern mining. Data Sci. Pattern Recogn. **1**(1), 54–77 (2017)
7. Fournier-Viger, P., Wu, C.W., Tseng, V.S., Cao, L., Nkambou, R.: Mining partially-ordered sequential rules common to multiple sequences. IEEE Trans. Knowl. Data Eng. **27**(8), 2203–2216 (2015)
8. Fournier-Viger, P., Yang, Y., Yang, P., Lin, J.C.-W., Yun, U.: TKE: mining top-k frequent episodes. In: Fujita, H., Fournier-Viger, P., Ali, M., Sasaki, J. (eds.) IEA/AIE 2020. LNCS (LNAI), vol. 12144, pp. 832–845. Springer, Cham (2020). https://doi.org/10.1007/978-3-030-55789-8_71
9. Huang, K., Chang, C.: Efficient mining of frequent episodes from complex sequences. Inf. Syst. **33**(1), 96–114 (2008)
10. Koh, Y.S., Ravana, S.D.: Unsupervised rare pattern mining: a survey. ACM Trans. Knowl. Disc. Data (TKDD) **10**(4), 1–29 (2016)

11. Lin, Y.F., Huang, C.F., Tseng, V.S.: A novel methodology for stock investment using high utility episode mining and genetic algorithm. Appl. Soft Comput. **59**, 303–315 (2017)
12. Mannila, H., Toivonen, H., Verkamo, A.I.: Discovering frequent episodes in sequences. In: Proceedings of 1st International Conference on Knowledge Discovery and Data Mining (1995)
13. Song, W., Huang, C.: Mining high average-utility itemsets based on particle swarm optimization. Data Sci. Pattern Recogn. **4**(2), 19–32 (2020)
14. Su, M.Y.: Applying episode mining and pruning to identify malicious online attacks. Comput. Electr. Eng. **59**, 180–188 (2017)
15. Truong, T., Duong, H., Le, B., Fournier-Viger, P.: Fmaxclohusm: an efficient algorithm for mining frequent closed and maximal high utility sequences. Eng. Appl. Artif. Intell **85**, 1–20 (2019)
16. Wenzhe, L., Qian, W., Luqun, Y., Jiadong, R., Davis, D.N., Changzhen, H.: Mining frequent intra-sequence and inter-sequence patterns using bitmap with a maximal span. In: Proceedings of 14th Web Information Systems and Applications Conference, pp. 56–61. IEEE (2017)
17. You, T., Li, Y., Sun, B., Du, C.: Multi-source data stream online frequent episode mining. IEEE Access **8**, 107465–107478 (2020)
18. Yun, U., Nam, H., Lee, G., Yoon, E.: Efficient approach for incremental high utility pattern mining with indexed list structure. Fut. Gener. Comput. Syst **95**, 221–239 (2019)

Investigating Crossover Operators in Genetic Algorithms for High-Utility Itemset Mining

M. Saqib Nawaz[1], Philippe Fournier-Viger[1(✉)], Wei Song[2], Jerry Chun-Wei Lin[3], and Bernd Noack[4]

[1] School of Humanities and Social Sciences, Harbin Institute of Technology (Shenzhen), Shenzhen, China
msaqibnawaz@hit.edu.cn

[2] School of Information Science and Technology, North China University of Technology, Beijing, China
songwei@ncut.edu.cn

[3] Department of Computing, Mathematics and Physics, Western Norway University of Applied Sciences (HVL), Bergen, Norway
jerrylin@ieee.org

[4] Center for Turbulence Control, Harbin Institute of Technology (Shenzhen), Shenzhen, China
bernd.noack@hit.edu.cn

Abstract. Genetic Algorithms (GAs) are an excellent approach for mining high-utility itemsets (HUIs) as they can discover most of the HUIs in a fraction of the time spent by exact algorithms. A key feature of GAs is crossover operators, which allow individuals in a population to communicate and exchange information with each other. However, the usefulness of crossover operator in the overall progress of GAs for high-utility itemset mining (HUIM) has not been investigated. In this paper, the headless chicken test is used to analyze four GAs for HUIM. In that test, crossover operators in the original GAs for HUIM are first replaced with randomized crossover operators. Then, the performance of original GAs with normal crossover are compared with GAs with random crossover. This allows evaluating the overall usefulness of crossover operators in the progress that GAs make during the search and evolution process. Through this test, we found that one GA for HUIM performed poorly, which indicates the absence of well-defined building blocks and that crossover in that GA was indeed working as a macromutation.

Keywords: High-utility itemsets · Genetic algorithms · Crossover

1 Introduction

High-utility itemset mining (HUIM) [1,11,17,20] is a popular data mining problem, which aims at discovering all important patterns in a quantitative database,

N. T. Nguyen et al. (Eds.): ACIIDS 2021, LNAI 12672, pp. 16–28, 2021.
https://doi.org/10.1007/978-3-030-73280-6_2

where pattern importance is measured using a numerical utility function. One of the main applications of HUIM is to enumerate all the sets of items (itemsets) purchased together that yield a high profit in customer transactions. An itemset is called a high-utility itemset (HUI) if its utility (profit) value is no less than a user-specified minimum utility threshold. Several exact HUIM algorithms have been designed to efficiently find all HUIs. However, these algorithms can still have very long runtimes because the search space size is exponential with the number of distinct items [1]. Long runtimes are inconvenient for users who often have to wait hours to obtain results even on small databases.

To address this issue, an emerging research direction is to design Nature-inspired Algorithm (NAs) for HUIM as they can solve hard optimization and computational problems. NAs have been used for problems ranging from bioinformatics and scheduling applications to artificial intelligence and control engineering. Some of the most popular NAs are Genetic Algorithms (GAs) [4], Simulated Annealing (SA) [8] and Particle Swarm Optimization (PSO) [7]. NAs were proposed to find HUIs in large databases based on GAs [6,9,15,21], PSO [9,10,15,16], Artificial Bee Colony (ABC) [14], the Bat algorithm [15] and Ant Colony System (ACS) [18]. In this paper, we are interested by GAs as they have excellent performance, are easy to implement and can discover most of the HUIs in a fraction of the time spent by exact algorithms.

A key feature that differentiates GAs from other NAs is crossover operators. The crossover process is simple: two chromosomes (solutions) are selected as parents and parts of them are combined to generate a new solution. The main *idea* of crossover is that such combination may yield better child solutions. This intuition was formalized by Holland [4] with the concept of building blocks used in schema theory. The *mechanics* of crossover provides a way to implement this *idea*. Thus, all types of crossover share the same *idea* but the *mechanics* to implement the *idea* can vary considerably. For example, single-point crossover (SPC) uses a single crossing point while two-point crossover (TPC) uses two.

Despite that GAs provide excellent performance for HUIM, the influence of crossover for that problem has not been investigated. Assessing the usefulness of crossover operators in GAs is an important research topic. Jones [5] argued that crossover *mechanics* alone can be used effectively for search and evolution even in the absence of the crossover *idea*. For this, a testing method (called headless chicken test) was proposed to examine the usefulness of crossover for a particular problem instance. In that test, a normal GA is compared with the same GA that uses a random version of crossover. Using this test, one can distinguish the gains that the GA makes through the idea of crossover from those made simply through the mechanics. If GA is not making any additional progress due to the idea of crossover, one might do as well simply by using macromutations. A poor performance of the original GA with normal crossover compared to the GA with random crossover indicates the absence of well-defined building blocks.

In this paper, we perform the headless chicken test to assess the usefulness of crossover in GAs for HUIM to better understand their performance. We applied the test on four GAs for HUIM [6,9,15,21]. In the test, original GAs for

HUIM (called normal GAs) were compared with the same GAs with randomized crossover operators (called randomized GAs). We found that three normal GAs for HUIM [6,9,21] that use normal crossover such as SPC and uniform crossover performed almost the same as randomized GAs. The GA for HUIM [15] that defined a new crossover performed worse than the randomized GA.

The remainder of this paper is organized as follows. Section 2 briefly discusses HUIM and GAs respectively. Section 3 provides the details for the headless chicken test performed on four GAs for HUIM. Section 4 presents the experiments performed in the test and discusses the obtained results. Finally, the paper is concluded with some remarks in Sect. 5.

2 Preliminaries

This section introduces preliminaries about high-utility itemset mining and genetic algorithms.

High-Utility Itemset Mining. Let $I = \{i_1, i_2, ..., i_m\}$ be a finite set of m distinct items and $TD = \{T_1, T_2, T_3, ..., T_n\}$ be a transaction database. Each transaction T_c in TD is a subset of I and has a unique integer identifier c ($1 \leq c \leq n$) called its TID. A set $X \subseteq I$ is called an *itemset* and an itemset that contains k items is called a k-itemset. Every item i_j in a transaction T_c has a positive number $q(i_j, T_c)$, called its *internal utility*. This value represents the purchase quantity (occurrence) of i_j in T_c. The *external utility* $p(i_j)$ is the unit profit value of the item i_j. A profit table $ptable = \{p_1, p_2, ..., p_m\}$ indicates the profit value p_j of each item i_j in I.

The overall utility of an item i_j in a transaction T_c is defined as $u(i_j, T_c) = p(i_j) \times q(i_j, T_c)$. The *utility of an itemset* X *in a transaction* T_c is denoted as $u(X, T_c)$ and defined as $u(X, T_c) = \sum_{i_j \subseteq X \wedge X \subseteq T_c} u(i_j, T_c)$. The *overall utility of an itemset* X in a database TD is defined as $u(X) = \sum_{X \subseteq T_c \wedge T_c \in TD} u(X, T_c)$ and represents the profit generated by X.

The *transaction utility* (TU) of a transaction T_c is defined as $TU(T_c) = u(T_c, T_c)$. The *minimum utility threshold* δ, specified by the user, is defined as a percentage of the sum of all TU values for the input database, whereas the *minimum utility value* is defined as $min\_util = \delta \times \sum_{T_c \in TD} TU(T_c)$. An itemset X is called an HUI if $u(X) \geq min\_util$.

The problem of HUIM is defined as follows [19]. Given a transaction database (TD), its profit table (*ptable*) and the minimum utility threshold, the goal is to enumerate all itemsets that have utilities equal to or greater than min_util.

To reduce the search space in HUIM, an upper bound on the utility of an itemset and its supersets called the *transaction-weighted utilization* (TWU) is often used [11]. The TWU of an itemset X is the sum of the transaction utilities of all the transactions containing X, which is defined as $TWU(X) = \sum_{X \subseteq T_c \wedge T_c \in TD} TU(T_c)$. An itemset X is called a *high transaction weighted-utilization itemset* (HTWUI) if $TWU(X) \geq min\_util$; otherwise, X is a low transaction weighted-utilization itemset (LTWUI). An HTWUI/LTWUI with k items is called a k-HTWUI/k-LTWUI.

Genetic Algorithms. GAs [4] are based on Darwin's theory (survival of the fittest) and biological evolution principles. GAs have the ability to explore a huge search space (population) to find nearly optimal solutions to difficult problems that one may not otherwise find in a lifetime. The foremost steps of a GA include: (1) population generation, (2) selection of candidate solutions from a population, (3) crossover and (4) mutation. Candidate solutions in a population are known as chromosomes or individuals, which are typically finite sequences or strings ($x = x_1, x_2 ..., x_n$). Each x_i (genes) refers to a particular characteristics of the chromosome. For a specific problem, GA starts by randomly generating a set of chromosomes to form a population and evaluates these chromosomes using a fitness function f. The function takes as parameter a chromosome and returns a score indicating how good the solution is. The general framework of GAs to mine HUIs is shown in Fig. 1. For HUIM, the utility of an itemset is used as the fitness function and the stopping criterion is the user-specified maximum number of generations.

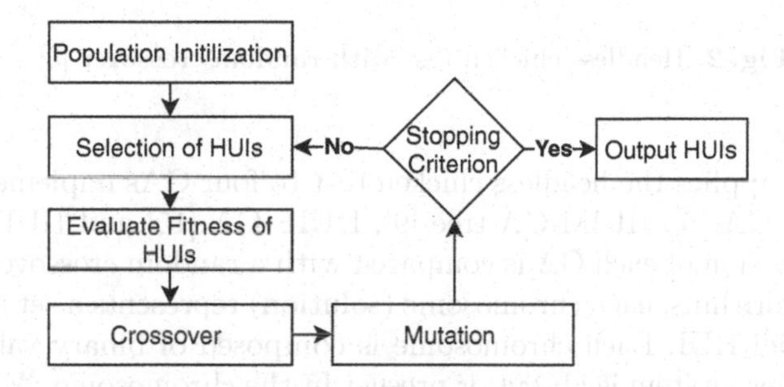

Fig. 1. General framework of GAs for HUIM

The crossover operator is used to guide the search toward the best solutions. If an appropriate crossing point is chosen, then the combination of sub-chromosomes from parent chromosomes may produce better child chromosomes. The mutation operator applies some random changes to one or more genes. This may transform a chromosome into a better chromosome.

3 The Headless Chicken Test for HUIM Using GAs

In this study, we assess the usefulness of crossover operators in GAs for the HUIM problem by applying the headless chicken test. In that test, a GA with normal crossover is compared with the identical GA with random crossover. The random crossover operator is illustrated in Fig. 2. While a normal crossover combines two parents to generate two new individuals (using either SPC or TPC), a random crossover generates two random individuals and uses them to do crossover with the parents. In the random crossover, there is no direct communication between parents. As individuals involved in random crossover are randomly selected, the operation is purely mechanical and does not carry the spirit of crossover. Despite the identical mechanical rearrangement, it is argued that random crossover is not a crossover [5]. A reason is that this operation does not requires two parents. For example, for TPC, one can simply select the crossing points and set the loci between the points to randomly chosen alleles. Hence, random crossover is a macromutation.

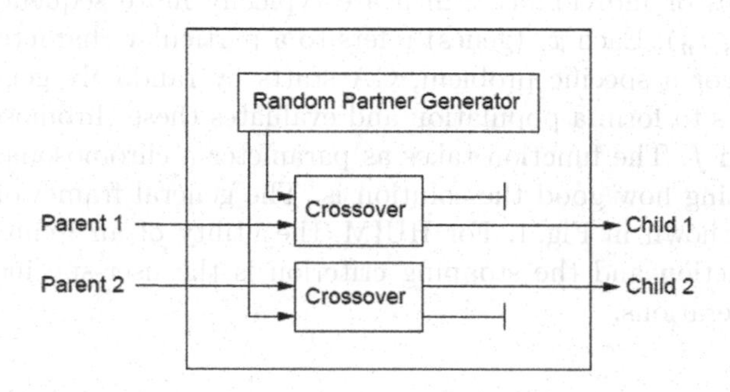

Fig. 2. Headless chicken test with random crossover [5]

This study applies the headless chicken test to four GAs implementations for HUIM: HUIM-GA [6], HUIM-GA-tree [9], HUIF-GA [15] and HUIM-IGA [21]. The original version of each GA is compared with a random crossover version. In all the four algorithms, each chromosome (solution) represents a set of items that form a potential HUI. Each chromosome is composed of binary values (0 or 1) that tell whether an item is absent or present in the chromosome. A value of 1 at the i-th position of a chromosome means that the corresponding item is present in the potential HUI, while a value of 0 indicates that it is absent. The total number of 1-HTWUIs in the database represent the size of the chromosome.

In HUIM-GA [6] two GAs (called HUPE$_{UMU}$-GARM and HUPE$_{WUMU}$-GARM) were proposed for HUIM. Applying HUPE$_{UMU}$-GARM requires setting a minimum utility threshold while HUPE$_{WUMU}$-GARM does not require a minimum utility threshold. Both GAs used the common operators (selection, single-point crossover, and mutation) iteratively to find HUIs.

Let there be two parent chromosomes $x = x_1, x_2, ..., x_n$ and $y = y_1, y_2, ..., y_n$ of length n [12]. Let position i $(1 \leq i \leq n)$ be a randomly selected crossing point in both parent chromosomes. The two new child chromosomes generated by the single-point crossover operator with i are:

$$x' = x_1, ..., x_i, y_{i+1}, ..., y_n$$

$$y' = y_1, ..., y_i, x_{i+1}, ..., x_n$$

Differently from single-point crossover, two-point crossover selects two crossing points i, j such that $1 \leq i \leq j \leq n$. The result is two new child chromosomes:

$$x' = x_1, ..., x_i, y_{i+1}, ..., y_j, x_{j+1}, ...x_n$$

$$y' = y_1, ..., y_j, x_{i+1}, ..., x_k, y_{j+1}, ...y_n$$

Examples of SPC and TPC, and their randomized versions are presented in Fig. 3(a) and Fig. 3(b), respectively. There, P_1, P_2 represent two parents, and C_1, C_2 represent the generated childs and R_1, R_2 are the random solutions.

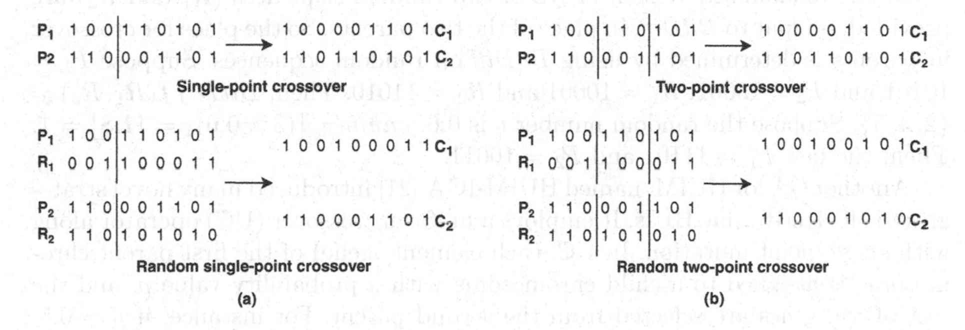

Fig. 3. SPC and TPC with randomized versions

HUIM-GA [6] cannot easily find the 1-HTWUIs initially to use them as chromosomes and thus perform a very large search for selecting appropriate chromosomes for mining valid HUIs. Additionally, setting the appropriate values for some specific parameters is a nontrivial task. The performance of HUIM-GA [6] was improved in HUIM-GA-tree [9] by using the OR/NOR-tree structure for pruning. SPC was also used in HUIM-GA-tree.

HUIF-GA [15] makes use of efficient strategies for database representation and a pruning process to accelerate HUI discovery. This GA does not use any normal crossover version such as SPC or TPC. For crossover, HUIF-GA uses a technique called *BittDiff* to first find the locations in two parents (bit vectors) where their values do no match. *BittDiff* is defined as follows:

Definition 1. *Let P_1 and P_2 be the two bit vectors with n bits. The bit difference set is defined as* [15]*:*

$$BitDiff(P_1, P_2) = \{i | 1 \leq i \leq n, b_i(P_1) \oplus b_i(P_2) = 1\} \tag{1}$$

where $b_n(P_1)$ is the n-th bit of P_1 and $\oplus$ denotes the exclusive disjunction.

The total number of points that will be used for crossover in the two selected chromosomes is:

$$cnum = \lfloor |BittDiff(P_1, P_2)| \times r \rfloor \tag{2}$$

where r is a random number in the range (0, 1), $|BittDiff(P_1, P_2)|$ is the number of elements in $BitDiff(P_1, P_2)$, and $\lfloor|BittDiff(P_1, P_2)| \times r\rfloor$ denotes the largest integer that is less than or equal to $|BitDiff(P_1, P_2)| \times r$. We call this crossover *BittDiff* crossover (BDC). It is explained with a simple example. Suppose that P_1 = 10010 and P_2 = 11000. As 10010 $\oplus$ 11000 = 01010, $BitDiff(P_1, P_2) = \{2, 4\}$, that is, P_1 and P_2 differ in their second and fourth positions. Suppose that the random number r is 0.5. Then, $cnum = \lfloor(2 \times 0.5)\rfloor = \lfloor 1 \rfloor = 1$. For P_1, one position (either the second or fourth) will be selected for crossover. Suppose the second position is selected to be changed. Then, the new P_1 is 11010 and new P_2 is 10000.

In the randomized version of BDC, two random sequences (R_1 and R_2) are provided as input to *BitDiff*, in place of the two parents. So the place for crossover in parents is determined by using *BitDiff* on random sequences. Suppose P_1 = 10101 and P_2 = 10001, R_1 = 10001 and R_2 = 11010. Then, $BitDiff(R_1, R_2) = \{2, 4, 5\}$. Suppose the random number r is 0.6, $cnum = \lfloor(3 \times 0.6)\rfloor = \lfloor 1.8 \rfloor = 1$. Then, the new P_1 = 11101 and P_2 = 10011.

Another GA for HUIM, named HUIM-IGA [21] introduced many novel strategies to efficiently mine HUIs. It employs a uniform crossover (UC) operator along with single-point mutation. In UC, each element (gene) of the first parent chromosome is assigned to a child chromosome with a probability value p, and the rest of the genes are selected from the second parent. For instance, if $p = 0.5$, the child has approximately half of the genes from the first parent and the other half from the second parent [13]. The randomized version of UC is depicted in Fig. 4. For UC, generated child chromosomes can be different for each run as it depends on the selection probability.

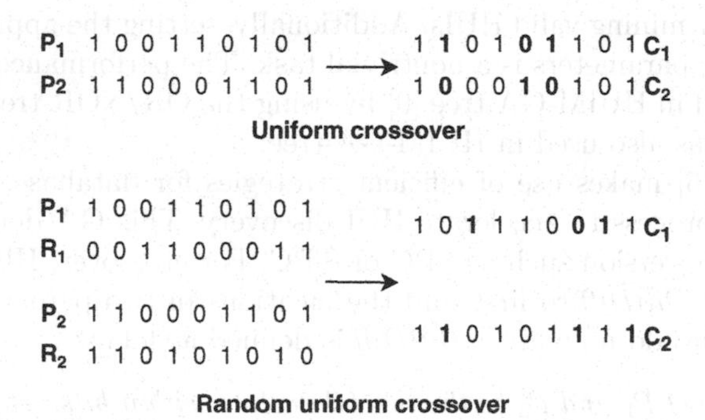

Fig. 4. Uniform crossover and its randomized version

Note that in HUPE$_{UMU}$-GARM and HUIM-GA-tree, we replaced SPC by random SPC and random TPC. The reason is to compare these two GAs with two randomized versions to check whether the change in the crossover operator has an effect on the overall performance of these two GAs.

4 Experiments and Results

This section presents the experimental evaluation of the headless chicken test on the four GAs. The experiments were performed on a computer with an 8-Core 3.6 GHz CPU and 64 GB memory running 64-bit Windows 10. The programs for randomized GAs were developed in Java. Five real standard benchmark datasets were used to evaluate the performance of the algorithms. The Foodmart dataset has real utility values while the remaining four datasets have synthetic utility values. The characteristics of the datasets are presented in Table 1.

Table 1. Characteristics of the datasets

Dataset	Avg. trans. len.	#Items	#Trans	Type
Foodmart	4.42	1,559	4,141	Sparse
Chess	37	75	3,196	Dense
Mushroom	23	119	8,124	Dense
Accidents_10%	34	468	34,018	Dense
Connect	43	129	67,557	Dense

All the datasets were downloaded from the SPMF data mining library [2]. The Foodmart dataset contains customer transactions from a retail store. The Chess dataset originates from game steps. The Mushroom dataset describes various species of mushrooms and their characteristics, such as shape, odor, and habitat. The Accident dataset is composed of (anonymized) traffic accident data. Similar to previous studies [9,14,15,21], only 10% of this dataset was used in experiments. The Connect dataset is also derived from game steps. For all experiments, the termination criterion was set to 10,000 iterations and the initial population size was set to 30.

In the experiments, the normal GA (that uses SPC) for HUIM [6] is called HUPE$_{UMU}$-GARM and the HUPE$_{UMU}$-GARM with random SPC and random TPC are named HUPE$_{UMU}$-GARM+ and HUPE$_{UMU}$-GARM++ respectively. The normal GA (that uses SPC) for HUIM in [9] is named HUIM-GAT and HUIM-GAT with random SPC and random TPC are named HUIM-GAT+ and HUIM-GAT++, respectively. Similarly, the original GA that uses BDC for HUIM in [15] is named HUIF-GA and the HUIF-GA with random BDC is called HUIF-GA+. The original GA for HUIM in [21] is named HUIM-IGA and the HUIM-IGA with random uniform crossover is called HUIM-IGA+.

4.1 Runtime

Experiments were first carried out to evaluate the efficiency of the algorithms in terms of runtime. The runtime was measured while varying the minimum utility value for each dataset. Figure 5 shows the execution time of algorithms for the five datasets.

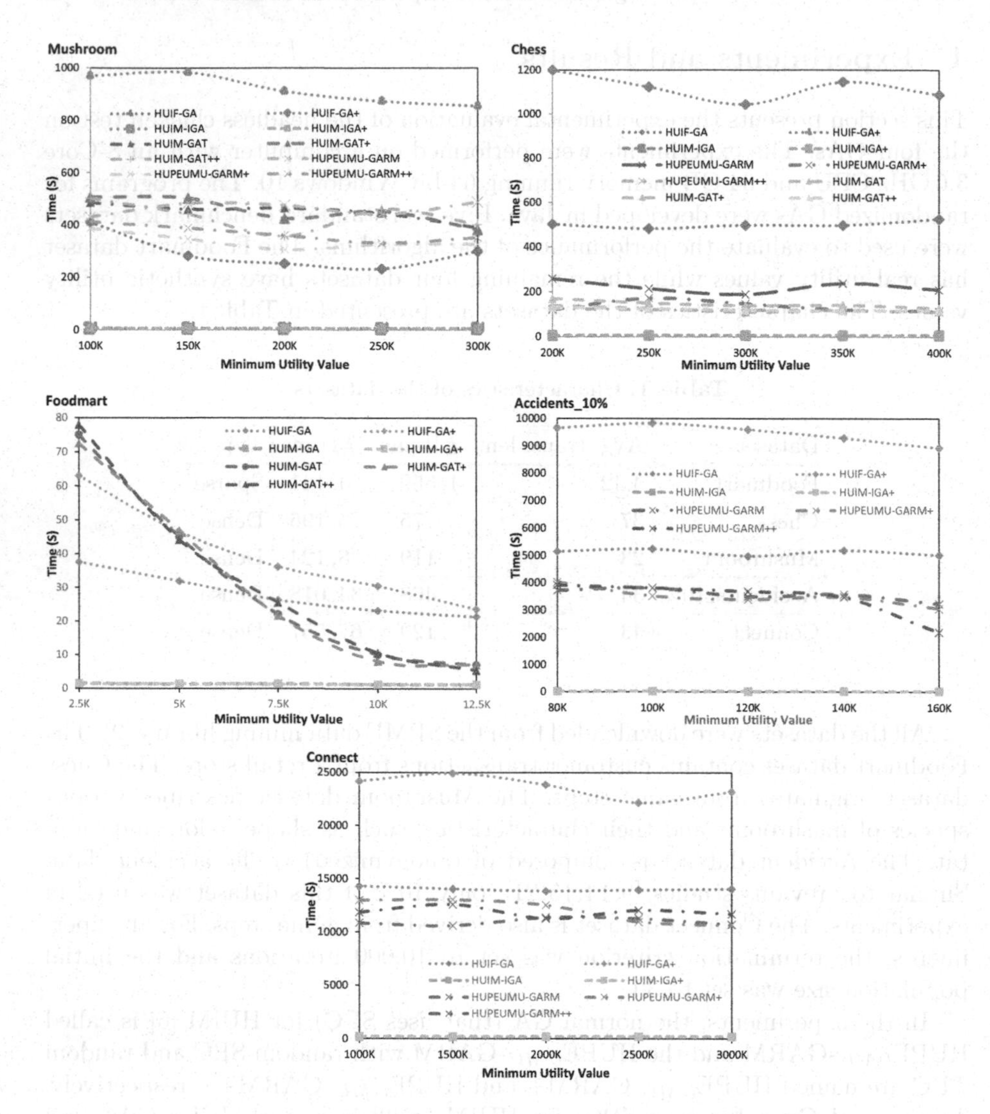

Fig. 5. Execution times of compared algorithms on five datasets

It is observed that the randomized HUIF-GA+ algorithm was slower than the normal HUIF-GA for all datasets. Moreover, it was slower than all other algorithms except for the Foodmart database, where HUIF-GA+ was faster than HUIM-GAT and its randomized versions (HUIM-GAT+ and HUIM-GAT++) at the start. However, HUIF-GA+ tends to become slower than all other algorithms as the minimum utility value is increased. On the other hand, all other algorithms and their randomized versions have almost the same execution time with negligible difference. Almost the same execution times for HUPE$_{UMU}$-GARM and HUIM-GAT and their randomized versions suggest that the use of SPC or TPC has no noticeable effect on the runtime of these algorithms. On the basis of runtime, HUIM-IGA and its randomized version (HUIM-IGA+) performed better than other algorithms on five datasets.

On the Foodmart dataset, HUPE$_{UMU}$-GARM and its randomized versions did not terminate after more than five hours. That is why their results are not shown in the chart. For the same reason, results of HUIM-GAT and its randomized versions are not shown for the Accidents and Connect datasets. They were unable to terminate in less than ten hours.

4.2 Discovered HUIs

Next, the numbers of HUIs discovered by the algorithms for the five datasets and parameter values are compared. The results are shown in Table 2.

HUIF-GA+ outperformed HUIF-GA on all datasets. This seems to be the reason why HUIF-GA+ was slower than the other algorithms. The performance of other normal GAs and their randomized versions were almost similar with negligible difference. The same performance of HUPE$_{UMU}$-GARM, HUIM-GAT, HUIM-IGA and their randomized versions indicate the presence of building blocks for the crossover operators (SPC and UC). On the other hand, the BDC in HUIG-GA was indeed working as a macromutation and failed to incorporate

Table 2. Discovered HUIs

D	MUV	HUIF-GA/GA+	HUIM-IGA/IGA+	GARM/GARM+/GARM++	HUIM-GAT/GAT+/GAT++
M	100K	15349/24846	5948/6221	66/75/96	68/78/94
	150K	10714/18945	4490/5450	57/53/85	61/62/75
	200K	8796/ 12285	3630/4568	45/31/65	47/43/57
	250K	6637/11731	3082/3791	33/28/51	34/21/24
	300K	5031/9498	2821/3534	22/16/47	23/31/15
Ch	200K	18269/33030	6773/6872	205/198/250	172/185/173
	250K	17501/29078	5912/6142	184/157/205	137/142/138
	300K	14151/25596	5120/5619	153/133/163	105/116/110
	350K	13437/21148	4814/4954	128/94/130	85/95/94
	400K	9828/17654	4211/4313	96/71/ 89	59/70/85
F	2.5K	2076/2436	1493/1515	x/x/x	72/77/75
	5K	921/1153	1085/1084	x/ x/x	41/46/40
	75.5K	508/748	707/705	x/x/x	18/19/16
	10K	312/465	424/423	x/x/x	5/4/0
	12.5K	196/236	207/209	x/x/x	0/0/0
A	80K	84856/98256	4836/4829	433/445/437	x/x/x
	100K	80214/90365	5021/4942	384/385/378	x/x/x
	120K	74785/84561	4694/4620	314/307/311	x/x/x
	140K	68646/76431	4482/4392	257/272/267	x/x/x
	160K	61915/69752	4577/4588	207/216/214	x/x/x
Co	1000K	58925/66154	5910/5879	137/142/147	x/x/x
	1500K	52589/59745	5864/5814	102/107/105	x/x/x
	2000K	48254/53856	5801/5694	88/92/98	x/x/x
	2500K	45812/52380	5635/5604	57/55/61	x/x/x
	3000K	40987/46982	5546/5484	32/34/37	x/x/x

D = Dataset, MUV = Minimum utility value, GARM = HUPE$_{UMU}$-GARM, Mushroom, Ch = Chess, F = Foodmart, A = Accidents, Co = Connect, x = run out of time the basic idea of crossover. The reason for this is the non-availability of building blocks that makes HUIF-GA to perform more poorly than HUIF-GA+. BDC failed to implement the crossover idea of exchanging the building blocks between individuals as it was changing the values of bits in individuals (HUIs) at specific locations.

Note that for the Foodmart dataset, the results for $HUPE_{UMU}$-GARM and its randomized versions ($HUPE_{UMU}$-GARM+ and $HUPE_{UMU}$-GARM++) were not included because they were unable to terminate after five hours of execution. For the same reason, no results is shown for HUIM-GAT on Accidents and Connect. HUIM-GAT and its randomized version were unable to terminate after

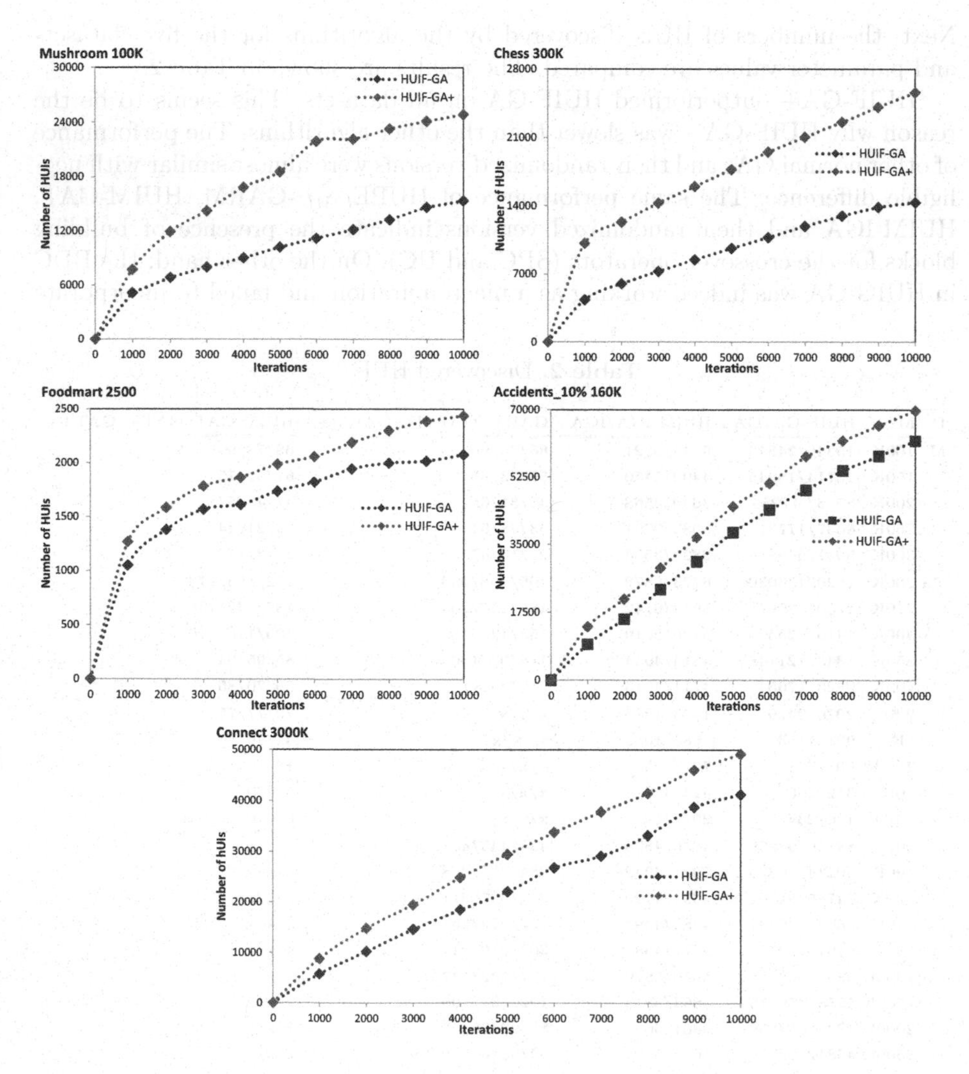

Fig. 6. Convergence performance of HUIF-GA and HUIF-GA+

ten hours of execution. Lastly, HUPE$_{UMU}$-GARM and HUIM-GAT and their randomized versions discovered almost the same number of HUIs. This suggests that the use of SPC or TPC has the same effect on the performance of these two algorithms in terms of discovered HUIs.

4.3 Convergence

As HUIF-GA+ outperformed HUIF-GA, we evaluate their convergence speed for all datasets. Obtained results are shown in Fig. 6.

HUIF-GA+ converged faster than HUIF-GA from the start on all datasets. The convergence speed of HUIF-GA and HUIF-GA+ for the Accidents and Connect datasets that contain a large number of transactions were linear. Whereas their convergence speed on other datasets (Mushroom, Chess and Foodmart) that have much less transactions (compared to Accidents and connect) were linear at the start (in the first 1000 iterations). However, as the number of iterations was increased, the convergence speed of both algorithms tend to decrease.

5 Conclusion

This paper investigated the performance of four GAs for HUIM by performing the headless chicken test. This test investigates the usefulness and worth of crossover operators in GA. Obtained results showed that three GAs for HUIM that employed normal crossover (such as single-point crossover and uniform crossover) were helping GAs to make progress. However one GA that was not employing any normal crossover was actually working as a macromutation that indicated the absence of well-defined building blocks. Thus, the efficacy of new crossover operators, particularly the specialized one for HUIM, can be investigated with the headless chicken test.

In the future, we intend to implement the PSO algorithms with headless chicken macromutation [3] for HUIM and compare the results with already implemented PSO algorithms to mine HUIs [9,10,15].

References

1. Fournier-Viger, P., Chun-Wei Lin, J., Truong-Chi, T., Nkambou, R.: A survey of high utility itemset mining. In: Fournier-Viger, P., Lin, J.C.-W., Nkambou, R., Vo, B., Tseng, V.S. (eds.) High-Utility Pattern Mining. SBD, vol. 51, pp. 1–45. Springer, Cham (2019). https://doi.org/10.1007/978-3-030-04921-8_1
2. Fournier-Viger, P., et al.: The SPMF open-source data mining library version 2. In: Berendt, B., et al. (eds.) ECML PKDD 2016. LNCS (LNAI), vol. 9853, pp. 36–40. Springer, Cham (2016). https://doi.org/10.1007/978-3-319-46131-1_8
3. Grobler, J., Engelbrecht, A.P.: Headless chicken particle swarm optimization algorithms. In: Tan, Y., Shi, Y., Niu, B. (eds.) Advances in Swarm Intelligence. ICSI 2016, vol. 9712, pp. 350–357. Springer, Cham (2016). https://doi.org/10.1007/978-3-319-41000-5_35

4. Holland, J.H.: Adaptation in Natural and Artificial Systems. University of Michigan Press, Ann Arbor (1975)
5. Jones, T.: Crossover, macromutation, and population-based search. In: 6th International Conference on Genetic Algorithms, pp. 73–80. Morgan Kaufmann (1995)
6. Kannimuthu, S., Premalatha, K.: Discovery of high utility itemsets using genetic algorithm with ranked mutation. Appl. Artif. Intell **28**(4), 337–359 (2014)
7. Kennedy, J., Eberhart, R.: Particle swarm optimization. In: International Conference on Neural Networks, pp. 1942–1948. IEEE (1995)
8. Kirkpatrick, S., Gelatt, C.D., Vecchi, M.P.: Optimization by simulated annealing. Science **220**(4598), 671–680 (1983)
9. Lin, J.C., Yang, L., Fournier-Viger, P., Hong, T., Voznák, M.: A binary PSO approach to mine high-utility itemsets. Soft Comput. **21**(17), 5103–5121 (2017)
10. Lin, J.C., et al.: Mining high-utility itemsets based on particle swarm optimization. Eng. Appl. Artif. Intell. **55**, 320–330 (2016)
11. Liu, Y., Liao, W., Choudhary, A.: A two-phase algorithm for fast discovery of high utility itemsets. In: Ho, T.B., Cheung, D., Liu, H. (eds.) PAKDD 2005. LNCS (LNAI), vol. 3518, pp. 689–695. Springer, Heidelberg (2005). https://doi.org/10.1007/11430919_79
12. Nawaz, M.S., Lali, M.I., Pasha, M.A.: Formal verification of crossover operator in genetic algorithms using prototype verification system (PVS). In: 9th International Conference on Emerging Technologies, pp. 1–6. IEEE (2013)
13. Nawaz, M.S., Sun, M.: A formal design model for genetic algorithms operators and its encoding in PVS. In: 2nd International Conference on Big Data and Internet of Things, pp. 186–190. ACM (2018)
14. Song, W., Huang, C.: Discovering high utility itemsets based on the artificial bee colony algorithm. In: Phung, D., Tseng, V.S., Webb, G.I., Ho, B., Ganji, M., Rashidi, L. (eds.) PAKDD 2018. LNCS (LNAI), vol. 10939, pp. 3–14. Springer, Cham (2018). https://doi.org/10.1007/978-3-319-93040-4_1
15. Song, W., Huang, C.: Mining high utility itemsets using bio-inspired algorithms: a diverse optimal value framework. IEEE Access **6**, 19568–19582 (2018)
16. Song, W., Huang, C.: Mining high average-utility itemsets based on particle swarm optimization. Data Sci. Pattern Recogn **4**(2), 19–32 (2020)
17. Truong, T., Tran, A., Duong, H., Le, B., Fournier-Viger, P.: EHUSM: mining high utility sequences with a pessimistic utility model. Data Sci. Pattern Recogn. **4**(2), 65–83 (2020)
18. Wu, J.M.T., Zhan, J., Lin, J.C.W.: An ACO-based approach to mine high-utility itemsets. Knowl. Based Syst. **116**, 102–113 (2017)
19. Yao, H., Hamilton, H.J.: Mining itemset utilities from transaction databases. Data Knowl. Eng. **59**(3), 603–626 (2006)
20. Yun, U., Kim, D., Yoon, E., Fujita, H.: Damped window based high average utility pattern mining over data streams. Knowl. Based Syst. **144**, 188–205 (2018)
21. Zhang, Q., Fang, W., Sun, J., Wang, Q.: Improved genetic algorithm for high-utility itemset mining. IEEE Access **7**, 176799–176813 (2019)

Complexes of Low Dimensional Linear Classifiers with L_1 Margins

Leon Bobrowski[1,2(✉)]

[1] Faculty of Computer Science, Bialystok University of Technology, Wiejska 45A, Bialystok, Poland
l.bobrowski@pb.edu.pl

[2] Institute of Biocybernetics and Biomedical Engineering, PAS, Warsaw, Poland

Abstract. Large data sets consisting of a relatively small number of multidimensional feature vectors are often encountered. Such a structure of data sets appears, inter alia, in in the case of genetic data. Various types of classifiers are designed on the basis of such data sets. Small number of multivariate feature vectors are almost always linearly separable. For this reason, the linear classifiers play a fundamental role in the case of small samples of multivariate vectors.

The maximizing margins is one of fundamental principles in classification. The Euclidean (L_2) margins is a basic concept in the support vector machines (*SVM*) method of classifiers designing. An alternative approach to designing classifiers is linked to the local maximization of the margins of the L_1 norm. This approach allows also designing complexes of linear classifiers on the basis of small samples of multivariate vectors.

Keywords: Data mining · Complexes of linear classifiers · Feature clustering · Convex and piecewise linear functions

1 Introduction

Data mining proceedurs are used to discover important new patterns in large data sets [1]. The term patterns means various types of regularity in the explored data set, such as decision rules, clusters, trends or interaction models [2]. The extracted patterns are used in many practical issues, e.g. related to scientific discoveries, supporting medical diagnosis or econometric forecasting [3].

Discovering crucial regularities and relationships on the basis of small samples of multidimensional vectors is a challenging task that requires the development of new inference models and computational tools. The source of additional difficulties is, among others, a large number of potentially important solutions [4].

The article proposes a new concept of complexes of low-dimensional linear classifiers based on the L_1 norm margins. Designing complexes of low-dimensional linear classifiers can be related to the ergodic theory [5]. The basic idea is that under certain conditions, averaging a few number of feature vectors (objects) can be replaced by averaging over a large number of subsets of features.

N. T. Nguyen et al. (Eds.): ACIIDS 2021, LNAI 12672, pp. 29–40, 2021.
https://doi.org/10.1007/978-3-030-73280-6_3

2 Linear Separability of Feature Vectors

We assume that the explored data set C consists of m feature vectors $\mathbf{x}_j = [x_{j,1}, \ldots, x_{j,n}]^T$ belonging to the n-dimensional feature space $F[n](\mathbf{x}_j \in F[n])$:

$$C = \{\mathbf{x}_j : j = 1, \ldots, m\} \tag{1}$$

The component $x_{j,i}$ of the vector $\mathbf{x}_j$ is the numerical value of the i-th feature X_i ($i = 1, \ldots, n$) of the j-th object O_j. We consider a situation in which the feature vectors x_j can be mixed - of the qualitative-quantitative type. Components $x_{j,i}$ of feature vectors $\mathbf{x}_j$ may be binary $(x_{j,i} \in \{0, 1\})$ or real numbers $(x_{j,i} \in R^1)$.

Two disjoint learning sets are considered: *positive* G^+ and *negative* G^- containing respectively m^+ and $m^- (m = m^+ + m^-)$ feature vectors $\mathbf{x}_j$ [3]:

$$G^+ = \{\mathbf{x}_j : j \in J^+\}, \quad and \quad G^- = \{\mathbf{x}_j : j \in J^-\} \tag{2}$$

where J^+ and J^- are non-empty sets of m^+ and m^- indices j respectively $(J^+ \cap J^- = \varnothing)$.

A possibility of separation of the learning sets G^+ and G^- (2) by the hyperplane $H(\mathbf{w}, \theta)$ in the feature space $F[n]$ is examined:

$$H(\mathbf{w}, \theta) = \{\mathbf{x} : \mathbf{w}^T\mathbf{x} = \theta\} \tag{3}$$

where $\mathbf{w}$ is the *weight vector* ($\mathbf{w} \in R^n$), θ is the *threshold* ($\theta \in R^1$), and $\mathbf{w}^T\mathbf{x} = \sum_i w_i x_i$ is the inner product.

***Definition* 1:** The learning sets G^+ and G^- (2) are *linearly separable* in the feature space $F[n]$, if and only if there exists such a *weight vector* $\mathbf{w} = [W_1, \ldots, W_n] \in R^n$, and a *threshold* $\theta \in R^1$ that these sets can be separated by the hyperplane $H(\mathbf{w}, \theta)$ (3):

$$\begin{aligned} (\exists \mathbf{w}, \theta) \quad (\forall \mathbf{x}_j \in G^+) \quad \mathbf{w}^T\mathbf{x}_j > \theta, \; and \\ (\forall \mathbf{x}_j \in G^-) \quad \mathbf{w}^T\mathbf{x}_j < \theta \end{aligned} \tag{4}$$

If the above inequalities are fulfilled, then all the feature vectors $\mathbf{x}_j$ from the set G^+ are located on the positive side of the hyperplane $H(\mathbf{w}, \theta)$ (3) and all the vectors $\mathbf{x}_j$ from the set G^- are located on the negative side of this hyperplane.

Augmented feature vectors $\mathbf{y}_j (\mathbf{y}_j \in F[n+1])$ are used to testing linear separability (4) [Duda]:

$$(\forall \mathbf{x}_j \in G^+) \mathbf{y}_j = [\mathbf{x}_j^T, 1]^T \quad and \quad (\forall \mathbf{x}_j \in G) \quad \mathbf{y}_j = -[\mathbf{x}_j^T, 1]^T \tag{5}$$

The *augmented* weight vector $\mathbf{v}(\mathbf{v} \in R^{n+1})$ is defined below [1]:

$$\mathbf{v} = [\mathbf{w}^T, -\theta]^T = [W_1, \ldots, w_n, -\theta]^T \tag{6}$$

The linear separability inequalities (4) now take the below form:

$$(\exists \mathbf{v}) \quad (\forall j \in \{1, \ldots, m\}) \quad \mathbf{v}^T\mathbf{y}_j > 0 \tag{7}$$

The linear separability inequalities (3) can also be represented with a *margin* as follows [4]:

$$(\exists \mathbf{v})\ (\forall j \in \{1, \ldots, m\})\ \mathbf{v}^T \mathbf{y}_j \geq 1 \tag{8}$$

The inequalities (8) were directly used to define the *perceptron criterion function* and other types of *convex and piecewise linear functions* (*CPL*) [3].

3 Dual Hyperplanes and Vertices in the Parameter Space

Each of the augmented feature vectors $\mathbf{y}_j$ (4) allows to define a dual hyperplane h_j^1 in the parameter space $\boldsymbol{R}^{n+1}$ [4]:

$$(\forall j \in \{1, \ldots, m\})\ h_j^1 = \left\{\mathbf{v} : \mathbf{y}_j^T \mathbf{v} = 1\right\} \tag{9}$$

Similarly, each of n unit vectors $\mathbf{e}_i$ defines a dual hyperplane h_i^0 in the parameter space $\boldsymbol{R}^{n+1}$ [4]:

$$(\forall i \in \{1, \ldots, n\})\ h_i^0 = \left\{\mathbf{v} : \mathbf{e}_i^T \mathbf{v} = 0\right\} \tag{10}$$

Each vertex $\mathbf{v}_k$ in the parameter space R^{n+1} $\left(\mathbf{v}_k \in R^{n+1}\right)$ is located at the intersection point of at least $(n + 1)$ hyperplanes h_j^1 (9) or h_j^0 (10).

The vertex $\mathbf{v}_k$ of the *rank* r_k is defined in the parameter space $\boldsymbol{R}^{n+1}$ by the following system of linear equations:

$$(\forall j \in J_k)\ \ \mathbf{v}_k^T \mathbf{y}_j = 1 \tag{11}$$

and

$$(\forall i \in I_k)\ \ \mathbf{v}_k^T \mathbf{e}_i = 0 \tag{12}$$

where J_k is the k-th subset of indices j of r_k linearly independent feature vectors $\mathbf{x}_j$ ($j \in J_k$), and I_k is the k-th subset of indices i of $n + 1 - r_k$ unit vectors $\mathbf{e}_i$ ($i \in I_k$) in the $(n + 1)$ - dimensional feature space $\boldsymbol{F}[n + 1]$.

The linear Eqs. (11) and (12) can be represented in the below matrix form [4]:

$$\mathbf{B}_k[n+1]\mathbf{v}_k = \mathbf{1}'[n+1] = [1, \ldots, 1, 0, \ldots, 0]^T \tag{13}$$

where $\mathbf{B}_k[n + 1]$ is a non-singular matrix constituting the k-th *basis* linked to the vertex $\mathbf{v}_k$:

$$\mathbf{B}_{\underline{k}}[n+1] = \left[\mathbf{y}_1, \ldots, \mathbf{y}_{rk}, \mathbf{e}_{i(rk+1)}, \ldots, \mathbf{e}_{i(n+1)}\right]^T \tag{14}$$

and

$$\mathbf{v}_k = \mathbf{B}_k[n+1]^{-1}\mathbf{1}'[n+1] \tag{15}$$

The inverse matrix $\mathbf{B}_k[n+1]^{-1}$ is represented as follows:

$$\mathbf{B}_k[n+1]^{-1} = [\mathbf{r}_1, \ldots, \mathbf{r}_{rk}, \ldots, \mathbf{r}_{n+1}]^T \tag{16}$$

Feature vectors $\mathbf{x}_j$ from the data set C (1) are *linearly independent* in the feature space $F[n]$ $(\mathbf{x}_j \in F[n])$ if none of the vectors $x_{j'}$ can be represented as a linear combination (2) of other vectors $\mathbf{x}_j (j \neq j')$ in this set. The maximal number m of feature vectors $\mathbf{x}_j$ (4) that can by linearly independent in the n - dimensional feature space $F[n]$ is equal to n. If data set C (1) is composed of a small number m of n - dimensional feature vectors $\mathbf{x}_j (m << n)$ then these vectors are usually linearly independent [1].

***Theorem* 1:** If m $(m \leq n)$ feature vectors $\mathbf{x}_j (j = 1, \ldots, m)$ are linearly independent in the n - dimensional feature space $F[n]$, then the learning sets G^+ and G^- (2) are linearly separable (4) regardless of the assignment of individual vectors $\mathbf{x}_j$ (1) to these sets.

4 Perceptron Criterion Function

The linear separability (8) of the learning sets G^+ and G^- (2) can be explored by minimizing the perceptron criterion function $\Phi(\mathbf{v})$ [3]. The perceptron criterion function $\Phi(\mathbf{v})$ is defined as the weighted sum of m penalty functions $\varphi_j(\mathbf{v}) \cdot (j = 1, \ldots, \cdot m)$:

$$\Phi(\mathbf{v}) = \sum_j \alpha_j \varphi_j(\mathbf{v}) \tag{17}$$

where

$$(\forall j \in \{1, \ldots, m\}) \quad \varphi_j(\mathbf{v}) = \begin{cases} 1 - \mathbf{y}_j^T\mathbf{v} & if\ \mathbf{y}_j^T\mathbf{v} < 1 \\ 0 & if\ \mathbf{y}_j\mathbf{y}^T\mathbf{v} \geq 1 \end{cases} \tag{18}$$

The *convex and piecewise-linear* (*CPL*) penalty functions $\varphi_j(\mathbf{v})$ (18) are intended to strengthen the linear separability inequalities (8).

The standard (default) values of the parameters α_j (17) are specified below [4]:

$$\begin{aligned} &if\ \mathbf{x}_j \in G^+, \quad then \quad \alpha_j = 1/(2m^+), \quad and \\ &if\ \mathbf{x}_j \in G^-, \quad then \quad \alpha_j = 1/(2m^-) \end{aligned} \tag{19}$$

The optimal vector $\mathbf{v}^*$ determines the minimum value Φ^* of the *CPL* criterion function $\Phi(\mathbf{v})$ (17) defined on m elements $\mathbf{x}_j$ from the learning sets G^+ and G^- (2):

$$(\exists \mathbf{v}^*) \quad (\forall \mathbf{v} \in R^{n+1}) \Phi(\mathbf{v}) \geq \Phi(\mathbf{v}^*) = \Phi^* \geq 0 \tag{20}$$

It has been proved that the minimum value $\Phi^* = \Phi(\mathbf{v}^*)$ (20) of the perceptron criterion function $\Phi(\mathbf{v})$ (17) is equal to zero $(\Phi^* = 0)$ if and only if the learning sets G^+ and G^- (2) are linearly separable (4) [4]. It can also be proved that the minimal value Φ^* (20) of the criterion function $\Phi(\mathbf{v})$ (17) with the parameters α_j specified by (19) is near to one ($\Phi^* \approx 1.0$) if the learning sets G^+ and G^- (2) are almost completely overlapped ($G^+ \approx G^-$). On this basis, the following standardization was introduced [3]:

$$0 \leq \Phi(\mathbf{v}^*) \leq 1.0 \tag{21}$$

The regularized criterion function $\Psi_\lambda(\mathbf{v})$ is the sum of the perceptron function $\Phi(\mathbf{v})$ (17) and additional *CPL* penalty functions in the form of absolute values $|w_i|$ of components w_i of the weight vector $\mathbf{w} = [W_1, \ldots, W_n]$ multiply by the *costs* γ_i ($\gamma_i > 0$) of particular features X_i [3]:

$$\Psi_\lambda(\mathbf{v}) = \Phi(\mathbf{v}) + \lambda \sum_{i=1,\ldots,n} \gamma_i |w_i| \tag{22}$$

where $\lambda \geq 0$ is the *cost level*. The values of the cost parameters γ_i may be equal to one:

$$(\forall i = 1, \ldots, n) \quad \gamma_i = 1.0 \tag{23}$$

The regularized criterion function $\Psi_\lambda(\mathbf{v})$ (22) is used in the *Relaxed Linear Separability* (*RLS*) method of feature subset selection [6]. The regularization component λ Σ γ_i $|w_i|$ used in the function $\Psi_\lambda(\mathbf{v})$ (22) is similar to that used in the *Lasso* method developed in the framework of the regression analysis for the model selection [7]. The main difference between the *Lasso* and the *RLS* methods is in the types of the basic criterion functions. The basic criterion function typically used in the *Lasso* method is the *residual sum of squares*, whereas the perceptron criterion function $\Phi(\mathbf{v})$ (17) is the basic function used in the *RLS* method. This difference effects among others the computational techniques used to minimize the criterion function. The criterion function $\Psi_\lambda(\mathbf{v})$ (22), similarly to the function $\Phi(\mathbf{v})$ (17), is convex and piecewise-linear (*CPL*). The basis exchange algorithm, similar to linear programming, allows for efficient and precise finding of the minimum of the criterion function $\Psi_\lambda(\mathbf{v})$ (22), even in the case of large data sets $\boldsymbol{C}$ (1) [8]:

$$(\exists \mathbf{v}_\lambda^*) \quad (\forall \mathbf{v}) \quad \Psi_\lambda(\mathbf{v}) \geq \Psi_\lambda(\mathbf{v}_\lambda^*) \tag{24}$$

where (6):

$$\mathbf{v}_\lambda^* = \left[(\mathbf{w}_\lambda^*)^T, -\theta_\lambda^*\right]^T = \left[w_{\lambda 1}^*, \ldots, w_{\lambda n}^*, -\theta_\lambda^*\right]^T \tag{25}$$

The optimal parameters $w_{\lambda 1}^*$ are used in the below feature *selection rule* [6]:

$$(w_{\lambda i}^* = 0) \Rightarrow (\text{ the } i\text{-th feature } X_i \text{ is reduced}) \tag{26}$$

The reduction of such feature X_i which is related to the weight $W_{\lambda i}^*$ equal to zero $\left(w_{\lambda,i}^* = 0\right)$ does not change the value of the inner product $(\mathbf{w}_\lambda^*)^T \mathbf{x}_j$. The positive location $\left((\mathbf{w}_\lambda^*)^T \mathbf{x}_j > \theta_\lambda^*\right)$ or the negative location $\left((\mathbf{w}_\lambda^*)^T \mathbf{x}_j < \theta_\lambda^*\right)$ of all feature vectors $\mathbf{x}_j$ in respect to the optimal separating hyperplane $\boldsymbol{H}(\mathbf{w}, \theta) = \left\{\mathbf{x} : (\mathbf{w}_\lambda^*)^T \mathbf{x} = \theta_\lambda^*\right\}$ (3) remains unchanged.

Based on the fundamental theorem of linear programming, it can be shown that the minimum value $\Phi^* = \Phi(\mathbf{v}_k^*)$ (20) of the perceptron criterion function $\Phi(\mathbf{v})$ (17) can be found in one of the vertices $\mathbf{v}_k$ (15) [9]. Similar property has the regularized criterion function $\Psi_{k,\lambda}(\mathbf{w})$ (22).

5 Basis Exchange Algorithms

The description of the basis exchange algorithm is presented here in the context of a small number m of high - dimensional feature vectors $\mathbf{x}_j$ (1) ($m << n$). With this assumption, we can expect that each of the m dual hyperplanes h_j^1 (9) defined by the augmented feature vector $\mathbf{y}_j(\mathbf{y}_j \in \boldsymbol{F}[n+1])$ (5) passes (11) through the vertex $\mathbf{v}_k$ [3]:

$$(\forall j \in \{1, \ldots, m\}) \quad \mathbf{v}_k^T \mathbf{y}_j = 1 \tag{27}$$

In this case, the k-th basis has the rank $r_k = m$ and can be represented as (14):

$$\mathbf{B}'_{\underline{k}}[n+1] = \left[\mathbf{y}_1, \ldots, \mathbf{y}_m, \mathbf{e}_{i(m+1)}, \ldots, \mathbf{e}_{i(n+1)}\right]^T \tag{28}$$

where $\mathbf{y}_j$ is the j - th augmented vector (5), and $i(l)$ is the index of such unit vector $\mathbf{e}_{i(l)}(i(l) \in I_k(12))$, which constitutes the l - th row of the k - th basis $\mathbf{B}'_{\underline{k}}[n+1]$.

The inverse matrix $\mathbf{B}'_{\underline{k}}[n+1]^{-1}$ has the following form:

$$\mathbf{B}'_{\underline{k}}[n+1]^{-1} = \left[\mathbf{r}_1, \ldots, \mathbf{r}_m, \ldots, \mathbf{r}_{n+1}\right]^T \tag{29}$$

The first m columns $\mathbf{r}_j(k)$ of the inverse matrix $\mathbf{B}'_{\underline{k}}[n+1]^{-1}$ (29) are linked to the augmented vectors $\mathbf{y}_j$ (5) in the basis $\mathbf{B}'_{\underline{k}}[n+1]^{-1}$ (28). The inverse matrix $\mathbf{B}'_{\underline{k}}[n+1]^{-1}$ (29) can be efficiently computed in a stepwise procedure starting from the unit matrix $\mathbf{I}[n+1] = \left[\mathbf{e}_1, \ldots, \mathbf{e}_m, \ldots, \mathbf{e}_{n+1}\right]$ [10]. In the successive steps l of this procedure we replace the unit vectors $\mathbf{e}_l$ with m augmented vectors $\mathbf{y}_l (l = 1, \ldots, m)$ (5). The replacement of the unit vectors $\mathbf{e}_l$ with the augmented vectors $\mathbf{y}_l$ can be described as follows by using the Gauss - Jordan transformation [11]:

$$\mathbf{r}_l(k+1) = \left(1/\mathbf{r}_l(k)^T \mathbf{y}_l\right) \mathbf{r}_l(k) \tag{30}$$

and

$$(\forall i \in \{1, \ldots, n; i \neq l\})$$

$$\mathbf{r}_i(k+1) = \mathbf{r}_i(k) - \left(\mathbf{r}_i(k)^T \mathbf{y}_l\right) \mathbf{r}_l(k+1) = \mathbf{r}_i(k) - \left(\mathbf{r}_i(k)^T \mathbf{y}_l / \mathbf{r}_l(k)^T \mathbf{y}_l\right) \mathbf{r}_l(k) \tag{31}$$

It can be seen that the columns $\mathbf{r}_i = \left[r_{i,1}, \ldots, r_{i,n+1}\right]^T$ of the inverse matrix $\mathbf{B}'_{\underline{k}}[n+1]^{-1}$ (29) have the last $n + 1$ - m components $r_{i,i'}$ equal to zero:

$$(\forall i \in \{1, \ldots, n+1\}) \ \mathbf{r}_i = \left[r_{i,1}, \ldots, r_{i,m}, 0, \ldots, 0\right]^T \tag{32}$$

The vertex $\mathbf{v}'_k[n+1]$ of the rank $r_k = m$ linked to the basis $\mathbf{B}'_{\underline{k}}[n+1]$ (28) has the below structure [ ]:

$$\mathbf{v}'_k[n+1] = \left[\left(\mathbf{w}'_k\right)^T, -\theta'_k\right]^T = \mathbf{B}'_{\underline{k}}[n+1]^{-1} \mathbf{1}'[n+1] \tag{33}$$

where the vector $\mathbf{1}'[n+1] = [1, \ldots, 1, 0, \ldots, 0]^T$ has the first m components equal to one and the last $n + 1 - m$ components are equal to zero. As a result:

$$\mathbf{v}'_k[n+1] = \mathbf{r}_1 + \ldots + \mathbf{r}_m \tag{34}$$

The vertex $\mathbf{v}'_k[n+1]$ (33) linked to the basis $\mathbf{B}'_{\underline{k}}[n+1]$ (36) has the below structure [3]:

$$\mathbf{v}'_k[n+1] = \left[\mathrm{v}_{k,1}, \ldots, \mathrm{v}_{k,n+1}\right]^T = \left[\mathrm{v}_{k,1}, \ldots, \mathrm{v}_{k,m}, 0, \ldots, 0\right]^T \tag{35}$$

The basis exchange algorithm is based on the vectoral Gauss - Jordan transformation resulting from the conversion of one vector in the basis $\mathbf{B}'_{\underline{k}}[n+1]$ (28) to another [11]. The assumption of a small number of multidimensional feature vectors $\mathbf{x}_j$ (1) indicates that all augmented vectors $\mathbf{y}_j$ (5) should remain in the basis $\mathbf{B}'_{\underline{k}}[n+1]$ (28). Only certain unit vectors $\mathbf{e}_i$ in the basis $\mathbf{B}'_{\underline{k}}[n+1]$ can be exchanged.

Assume that the basis $\mathbf{B}'_{\underline{k+1}}[n+1]$ is obtained as a result of the replacing the unit vectors $\mathbf{e}_{i(l)}$ in the basis $\mathbf{B}'_{\underline{k+1}}[n+1]$ (28) by the vector $\mathbf{e}_{i(k)}$. The unit vectors $\mathbf{e}_{i(l)}$ is removed from the basis $\mathbf{B}'_{\underline{k}}[n+1]$ (28) and replaced with the vector $\mathbf{e}_{i(k)}$. The Gauss - Jordan transformation resulting from such basis exchange can be given as follows [11]:

$$\mathbf{r}_{i(l)}(k+1) = \left(1/\mathbf{r}_{i(l)}(k)^T\mathbf{e}_{i(k)}\right)\mathbf{r}_{i(l)}(k) = \left(1/\mathrm{r}_{i(l),i(k)}(k)\right)\mathbf{r}_{i(l)}(k) \tag{36}$$

and

$$(\forall i \in \{1, \ldots, n; i \neq i(l)\})$$
$$\mathbf{r}_i(k+1) = \mathbf{r}_i(k) - \left(\mathbf{r}_i(k)^T\mathbf{e}_{i(k)}\right)\mathbf{r}_{i(l)}(k+1) = \mathbf{r}_i(k) - \left(\mathrm{r}_{i,i(k)}(k)/\mathrm{r}_{i(l),i(k)}(k)\right)\mathbf{r}_{i(l)}(k) \tag{37}$$

The symbols used in formulas (36) and (37) are explained as follows (29):

$$\mathbf{B}'_{\underline{k}}[n+1]^{-1} = \left[\mathbf{r}_1(k), \ldots, \mathbf{r}_m(k), \ldots, \mathbf{r}_{n+1}(k)\right] \tag{38}$$

where $(\forall i \in \{1, \ldots, n\})\mathbf{r}_i(k) = \left[\mathrm{r}_{i,1}(k), \ldots, \mathrm{r}_{i,m}(k), \ldots, \mathrm{r}_{i,m}(k)\right]^T$ is the i - th column of the inverse matrix $\mathbf{B}'_{\underline{k}}[n+1]^{-1}$ (38).

Properties of algorithms based on the Gauss - Jordan transformation depends on which vector $\mathbf{e}_{i(l)}$ leaves the basis and the vector $\mathbf{e}_{i(k)}$ which enters this basis. The *exit criterion* determines the unit vectors $\mathbf{e}_{i(l)}$ which leaves the basis $\mathbf{B}'_{\underline{k}}[n+1]$ (28). The *entry criterion* determines the unit vectors $\mathbf{e}_{i(k)}$ which enters the basis $\mathbf{B}'_{\underline{k}}[n+1]$.

6 Reduced Criterion Functions

Let us explain the construction of the exit criterion and the entry criterion of the basis exchange algorithm on the example of the regularized criterion function $\Psi_\lambda(\mathbf{v})$ (22).

***Lemma* 1:** If the basis $\mathbf{B}'_{\underline{k}}[n+1]$ (28) contains all m augmented vectors $\mathbf{y}_l$ ($l = 1, \ldots, m$) (5), then the perceptron criterion function $\Phi(\mathbf{v})$ (17) is equal to zero ($\Phi^* = 0$) [3].

So, under the assumption (27), the regularized criterion function $\Psi_\lambda(\mathbf{v})$ (22) can be *reduced* to $\Psi'(\mathbf{w})$, where $\mathbf{w} = [\mathrm{w}_1, \ldots, \mathrm{w}_n]^T$:

$$\Psi'(\mathbf{w}) = \sum_{i=1,\ldots,n} \gamma_i |\mathrm{W}_i| \tag{39}$$

or to $\Psi(\mathbf{w})$ (23):

$$\Psi(\mathbf{w}) = \sum_{i=1,\ldots,n} |\mathrm{W}_i| \tag{40}$$

The basis exchange algorithm allows for finding of the minimum value of the reduced criterion function $\Psi(\mathbf{w})$ (40) constrained by (27), even in the case of large data sets $\boldsymbol{C}$ (1) [8]:

$$(\exists \mathbf{w}_k^*)\ (\forall \mathbf{w})\ \Psi(\mathbf{w}) \geq \Psi(\mathbf{w}_k^*) \tag{41}$$

where $\mathbf{w}_k^* = \left[\mathrm{w}_{k,1}^*, \ldots, \mathrm{w}_{k,n}^*\right]^T$ (6).
The constraining condition (27) can be represented as follows (4) []:

$$(\forall \mathbf{x}_j \in \boldsymbol{G}^+(2))\mathbf{w}^T\mathbf{x}_j = \theta + 1 \quad and \quad (\forall \mathbf{x}_j \in \boldsymbol{G}^-(2))\mathbf{w}^T\mathbf{x}_j = \theta - 1 \tag{42}$$

The minimization of the reduced criterion function $\Psi(\mathbf{w})$ (40) with constraints (42), allows to characterize the optimal vertex $\mathbf{W}_k^*$ (41) by the following property [3]:

$$min\left\{\sum |\mathrm{w}_i| : (42)\right\} = \sum_{i=1,\ldots,n} \left|\mathrm{w}_{k,i}^*\right| \tag{43}$$

The property (43) means that the minimization of the criterion function $\Psi(\mathbf{w})$ (40) with constraints (42) leads to the optimal vertex $\mathbf{W}_k^*$, which is characterized by the minimal L_1 length $\|\mathbf{w}_k^*\|_{L1}$.

***Lemma* 2:** Minimization (41) of the reduced criterion function $\Psi(\mathbf{w})$ (40) constrained by (42) leads to such an optimal vertex $\mathbf{W}_k^*$ (41) which is characterized by the lowest L_1 norm $\|\mathbf{w}_k^*\|_{L1} = \sum_i \left|\mathrm{W}_{k,i}^*\right|$ among all such vertices $\mathbf{w}_l$ that satisfy the condition (42).

The condition (27) at the vertex $\mathbf{w}_l$ means that each of m dual hyperplanes h_j^1 (9) passes through the augmented vertex $\mathbf{v}_l$). The proof of this lemma can be based on the fact that the criterion function $\Psi(\mathbf{w})$ (40) is *convex and piecewise-linear* (*CPL*) [].

The linear decision rule based on the optimal vertex $\mathbf{W}_k^*[n]$ (41) is characterized by the largest L_1 margin $\delta_{L1(\mathbf{w})}$ among all vertices $\mathbf{w}_k[n]$ satisfying the condition (42):

$$\delta_{L1}(\mathbf{w}) = 1/\|\mathbf{w}\| = (|\mathrm{w}_1| + \ldots + |\mathrm{w}_n|)^{-1} \tag{44}$$

High quality of the *support vector machines* (*SVM*) linear classifiers is obtained by increasing the margin $\delta_{L2}(\mathbf{w}) = 1/\|\mathbf{w}\|_{L^2}$ based on the Euclidean (L_2) norm $\|\mathbf{w}\|_{L^2}$ of the vector $\mathbf{w} = [\mathrm{w}_1, \ldots, \mathrm{w}_n]^T$ [12]:

$$\delta_{L2}(\mathbf{w}) = 1/\|\mathbf{w}\|_{L2} = \left(\mathrm{w}_1^2 + \ldots + \mathrm{w}_n^2\right)^{-1/2} \tag{45}$$

Comparing linear classifiers with the L_1 and L_2 margins, we pay attention to the fundamental difference between these two types of classifiers.

In the case of the L_1 margins $\delta_{L1(\mathbf{w})}$ (44), the optimal linear classifier can be defined in one of the vertices $\mathbf{w}_k[n]$ of the solution region (42). In the case of the L_2 margins $\delta_{L2(\mathbf{w})}$ (45), the optimal linear classifier is defined either in one of the vertices $\mathbf{w}_k[n]$ or on some wall of the solution area (42). Such a property of L_2 margins classifiers may reduce the precision and efficiency of calculations, especially in the case of large data sets $\boldsymbol{C}$ (1).

7 Complexes of Linear Classifiers

Minimization (20) of the perceptron criterion function $\Phi(\mathbf{v})$ (17) defined on the learning sets $\boldsymbol{G}^+$ and $\boldsymbol{G}^-$ (2) allows to find optimal parameters $\mathbf{v}^* = \left[(\mathbf{w}^*)^T, -\theta^*\right]^T$ (6) of a linear classifier in the augmented feature space $\boldsymbol{F}[n+1]$ (5). The *working threshold* θ_w is defined on the basis of the optimal value θ^* (6) as follows :

$$\textit{if } \theta^* > 0, \textit{ then } \theta_w = 1, \text{ and } \textit{if } \theta^* < 0, \textit{ then } \theta_w = -1 \tag{46}$$

The minimization (41) of the reduced criterion function $\Psi(\mathbf{w})$ (40) with the constraints (42) allows to find the optimal vertex $\mathbf{W}_k^*[n]$ and the linear decision rule:

$$\begin{aligned} &(\forall \mathbf{x} \in \boldsymbol{F}[n]) \\ &\textit{if } \mathbf{w}_k^*[n]^T \mathbf{x} \geq \theta_w(46), \textit{ then } \mathbf{x} \text{ is in the category } \omega^+ \left(\mathbf{x} \in \omega^+\right) \\ &\textit{if } \mathbf{w}_k^*[n]^T \mathbf{x} < \theta_w(46), \textit{ then } \mathbf{x} \text{ is in the category } \omega^- \left(\mathbf{x} \in \omega^-\right) \end{aligned} \tag{47}$$

The optimal vertex $\mathbf{w}_k^*[n] = \left[\mathrm{w}_{k,1}^*, \ldots, \mathrm{w}_{k,n}^*\right]^T$ (41) is linked to the basis containing $n - m$ unit vectors $\mathbf{e}_i$ ($i \in I_k$ (12)). Each component $\mathrm{w}_{k,i}^*$ of the optimal vertex $\mathbf{W}_k^*[n]$ corresponding to the unit vector $\mathbf{e}_i$ in the basis $\mathbf{B}_{\underline{k}}'[n+1]$ (28) is equal to zero:

$$(\forall i \in I_k(12)) \quad \mathrm{w}_{k,i}^* = 0 \tag{48}$$

Such $n - m$ features X_i which are related to weights $\mathrm{w}_{k,i}^*$ equal to zero $\left(\mathrm{w}_{k,i}^* = 0\right)$ can be omitted (*feature selection rule*) because it does not change the decision rule (47):

$$\left(\mathrm{w}_{k,i}^* = 0\right) \Rightarrow (\text{feature } X_i \text{ is ignored in the feature subspace } \boldsymbol{F}_k[m]) \tag{49}$$

Positive location $\left(\mathbf{w}_k^*[n]^T \mathbf{x} \geq \theta_w\right)$ (47) or negative location $\left(\mathbf{w}_k^*[n]^T \mathbf{x} < \theta_w\right)$ of all m feature vectors $\mathbf{x}_j$ with respect to the optimal separating hyperplane $\boldsymbol{H}\left(\mathbf{w}_k^*[n], \theta_w\right) = \{\mathbf{x} : (\mathbf{w}_k^*[n]^T \mathbf{x} = \theta_w(46)\}$ (3) remains unchanged.

Vertexical feature subspace $\boldsymbol{F}_k[m]$ ($\boldsymbol{F}_k[m] \subset \boldsymbol{F}[n]$) is defined on the basis of the optimal vertex $\mathbf{w}_k^*[m] = \left[\mathrm{W}_{k,1}^*, \ldots, \mathrm{W}_{k,m}^*\right]^T$ using the feature selection rule (49) [6]. The subspace $\boldsymbol{F}_k[m]$ is composed of m features X_i selected in accordance with the rule

(47). The linear decision rules in the vertexical feature subspaces $\boldsymbol{F}_k[m]$ are given as follows:

$$(\forall k \in \{1, \ldots, K\})\ (\forall \mathbf{z}_k(\mathbf{x}) \in \boldsymbol{F}_k[m])$$
$$\textit{if } \mathbf{w}_k^*[\mathrm{m}]^T z_k(\mathbf{x}) \geq \theta_w(68),\ \textit{then } \mathbf{x} \text{ belongs to the category } \omega^+$$
$$\textit{if } \mathbf{w}_k^*[m]^T \mathbf{z}_k(\mathbf{x}) < \theta_w(68),\ \textit{then } \mathbf{x} \text{ belongs to the category } \omega^- \tag{50}$$

where $\mathbf{z}_k(\mathbf{x})$ ($\mathbf{z}_k(\mathbf{x}) \in \boldsymbol{F}_k[m]$) is the reduced vector obtained from the feature vector $\mathbf{x} = [x_1, \ldots, x_n]^T$ after neglecting $n - m$ components x_i linked to the weights $w_{k,1}^*$ equal to zero ($w_{k,i}^* = 0$) in the optimal vertex $\mathbf{w}_k^*[n] = \left[W_{k,1}^*, \ldots, W_{k,n}^*\right]^T$ (41).

The reduced decision rule (50) correctly allocates all feature vectors $\mathbf{x}_j$ from the learning sets $\boldsymbol{G}^+$ and $\boldsymbol{G}^-$ (2):

$$(\forall \mathbf{x}_j \in \boldsymbol{G}^+)\mathbf{w}_k^*[m]^T \mathbf{z}_k(\mathbf{x}_j) > \theta_w,\ \textit{and}\ (\forall \mathbf{x}_j \in \boldsymbol{G})\ \mathbf{w}_k^*[m]^T \mathbf{z}_k(\mathbf{x}_j) < \theta_w \tag{51}$$

where $\mathbf{z}_k(\mathbf{x}_j)$ ($j = 1, \ldots, m$) is the m-dimensional reduced vector obtained from the feature vector $\mathbf{x}_j = \left[x_{j,1}, \ldots, x_{j,n}\right]^T$ by neglecting the $n - m$ components $x_{j,i}$, which are linked to the weights $w_{k,i}^*$ (41) equal to zero.

The reduced feature subspaces $\boldsymbol{F}_k[m]$ ($k = 0, 1, \ldots, K$) of the dimension m can be characterized by the subsets $F_k(m)$ of m features X_i used in these subspaces. Let the symbol $F_0(n)$ denote the set of all n features X_i:

$$F_0(n) = \{X_1, \ldots, X_n\} \tag{52}$$

The *complex of linear classifiers* can be viewed as a layer of K linear classifiers with reduced decision rules (50). Such a layer is designed in multi-step procedure. One linear classifiers is added to the layer during each step of this procedure.

In the first step ($k = 1$) of the design procedure, the optimal vertex $\mathbf{w}_1^*[n]$ (41) and the subset R_1 of m features X_i are determined:

$$R_1(m) = \left\{X_{i(1)}, \ldots, X_{i(m)}\right\} \tag{53}$$

where the symbol $X_{i(l)}$ means a feature X_i that has not been reduced by the rule (49) because $w_{k,l}^* \neq 0$.

The set $F_0(n)$ (52) is reduced to $F_1(n_1) = F_1(n - m)$ by omitting the subset $R_1(m)$ (53):

$$F_1(n_1) = F_0(n) - R_1(m) \tag{54}$$

In the second step ($k = 2$), the optimal vertex $\mathbf{w}_2^*[n_1]$ (41) and the subset $R_2(m)$ (53) of m features X_i are computed in the feature subspace $\boldsymbol{F}_1[n_1]$ (54) of the dimension $n_1 = n - m$.

Successive steps k of design ($k = 3, \ldots, K$) contain the same elements: computation of the subset $R_k(m)$ (54) of m features X_i, reduction of the subset $F_k(n_k)$ (54) of n_k features X_i to the subset $F_{k+1}(n_k - m)$ and calculation of the optimal vertex $\mathbf{W}_{k+1}^*[n_k - m]$ (41) in the feature subspace $\boldsymbol{F}_{k+1}[n_k - m]$.

***Definition* 2:** A complex of linear classifiers is a layer of K linear classifiers with reduced decision rules (50) defined by optimal vertices $\mathbf{w}_k^*[n_k]$ (41) ($k = 1,\ldots, K$). The k-th optimal vertex $\mathbf{w}_k^*[n_k]$ is calculated by minimizing (41) with the constrain (41) of the reduced criterion function $\Psi_k(\mathbf{w})$ (40) defined in the k-th reduced feature subspace $\boldsymbol{F}_k[n_k]$ of the dimension $n_k = n - (k - 1)\, m$, where $k = 1,\ldots, K$.

***Example* 1:** The complex of four ($K = 4$) linear classifiers (50) designed on the base of $m = 3$ feature vectors $\mathbf{x}_j$ (1) with a large dimension $n = 1000$ ($n >> m$).

In this case, each of the four linear classifiers (50) of the complex is based on three selected features $X_{i(k,1)}, X_{i(k,2)}, X_{i(k,3)}$. The features $X_{i(k,\,l)}$ of the l-th classifier (50) are selected by minimizing the reduced criterion function $\Psi_k(\mathbf{w})$ (40) ($k = 1, 2, 3, 4$) with the constraints given by (42). Features $X_{i(1,l)}$ of the first classifier ($k = 1$) of this complex are selected from $n = 1000$ features X_i. The second classifier of this complex has the features $X_{i(2,l)}$ selected from $n - m = 997$ features X_i. Features $X_{i(3,l)}$ of the third classifier are selected from $n - 2\,m = 994$ features X_i. The features $X_{i(4,l)}$ of the fourth classifier are selected from $n - 3\,m = 991$ features X_i. The L_1 margines $\delta_{L1}(\mathbf{w}_k^*)$ (44) of these four classifiers (50) are arranged as follows: $\delta_{L1}(\mathbf{w}_1^*) \geq \delta_{L1}(\mathbf{w}_2^*) \geq \delta_{L1}(\mathbf{w}_3^*) \geq \delta_{L1}(\mathbf{w}_4^*)$, where the optimal vertex $\mathbf{w}_k^*$ constitutes the minimum value $\Psi_k(\mathbf{w}_k^*)$ (41).

The decision rules of the reduced linear classifier in the m - dimensional feature subspaces $\boldsymbol{F}_k[m]$ ($\mathbf{z}_k(\mathbf{x}) \in \boldsymbol{F}_k[m]$) are given by the formula (50). The decision rule of a complex of K linear classifiers (50) can be chosen according to the majority principle:

if most of the K linear classifiers (50) put the feature vector x in the category $\omega^+(\omega^-)$,
then the vector x belongs to the category $\omega^+(\omega^-)$. (55)

It can be expected that a quality of a complex rule (55) should be higher than the quality of a single classifiers (50) [13].

8 Concluding Remarks

The article presents the theoretical foundations of complex models of low-dimensional linear classifiers with L_1 margins. Complex models of linear classifiers are designed on the basis of a small number m of feature vectors $\mathbf{x}_j$ (1) with a large dimension n ($m << n$).Data sets with this property can be thought of as *dimensionally unbalanced* [14].

According to the proposed approach, the vertexical subspaces $\boldsymbol{F}_k[m]$ (54) with the dimensions m are extracted from the data set $\boldsymbol{C}$ (1) in subsequent steps k. The subspaces $\boldsymbol{F}_k[m]$ (54) are based on the optimal vertices $\mathbf{w}_k^*[m]$ (41) which are computed by minimizing (41) the convex and piecewise-linear (*CPL*) criterion function $\Psi(\mathbf{w})$ (40) with the constraints (42). Thus, the sequence of low-dimensional linear classifiers (50) with decreasing L_1 margins $\delta_{L1}(\mathbf{w}_k^*)$ (44) can be obtained.

Designing complexes of low-dimensional linear classifiers can be related to the ergodic theory [5]. According to the approach proposed here, averaging over a small number of feature vectors (objects) $\mathbf{x}_j$ ($j = 1, \ldots, m$) is replaced by averaging over a large number of subsets of features (genes) X_i ($i = 1, \ldots, n$).

Acknowledgments. The presented study was supported by the grant WZ/WI-IIT/3/2020 from Bialystok University of Technology and funded from the resources for research by Polish Ministry of Science and Higher Education.

References

1. Duda, O.R., Hart, P.E., Stork, D.G.: Pattern Classification. Wiley, New York (2001)
2. Bishop, C.M.: Pattern Recognition and Machine Learning. Springer, New York (2006)
3. Bobrowski, L.: Data Exploration and Linear Separability, pp. 1–172. Lambert Academic Publishing (2019)
4. Bobrowski, L.: Data mining based on convex and piecewise linear criterion functions. Technical University Białystok (2005). (in Polish)
5. Petersen, K.: Ergodic Theory. Cambridge Studies in Advanced Mathematics. Cambridge University Press, Cambridge (1990)
6. Bobrowski, L., Łukaszuk, T.: Relaxed linear separability (*RLS*) approach to feature (Gene) subset selection. In: Xia, X. (ed.) Selected Works in Bioinformatics, pp. 103–118. INTECH (2011)
7. Tibshirani, R.: Regression shrinkage and selection via the lasso. J. Roy. Stat. Soc. B **58**(1), 267–288 (1996)
8. Bobrowski, L.: Design of piecewise linear classifiers from formal neurons by some basis exchange technique. Pattern Recogn. **24**(9), 863–870 (1991)
9. Simonnard, M.: Linear Programming. Prentice Hall, New York (1966)
10. Bobrowski, L.: Large matrices inversion using the basis exchange algorithm. Br. J. Math. Comput. Sci. **21**(1), 1–11 (2017). http://www.sciencedomain.org/abstract/18203)
11. Bobrowski, L.: Symmetrical discrimination in pattern recognition - theory, algorithms, and applications in computer aided medical diagnosis, pp. 1–171. Ossolineum, Wroclaw (1987). (in Polish)
12. Vapnik, V.N.: Statistical Learning Theory. Wiley, New York (1998)
13. Blachnik, M.: Ensembles of instance selection methods: a comparative study. Int. J. Appl. Math. Comput. Sci. **29**(1), 151–168 (2019)
14. Janicka, M., Lango, M., Stefanowski, J.: Using information on class interrelations to improve classification of multiclass imbalanced data: a new resampling algorithm. Int. J. Appl. Math. Comput. Sci. **29**(4), 769–781 (2019)

Automatic Identification of Bird Species from Audio

Silvestre Carvalho[1] and Elsa Ferreira Gomes[1,2](✉)

[1] Instituto Superior de Engenharia do Porto, Rua Dr. Bernardino de, Almeida, 431, 4200-072 Porto, Portugal
efg@isep.ipp.pt
[2] INESC TEC, Campus da FEUP, Rua Dr. Roberto Frias, 4200-465 Porto, Portugal

Abstract. Bird species identification is a relevant and time-consuming task for ornithologists and ecologists. With growing amounts of audio annotated data, automatic bird classification using machine learning techniques is an important trend in the scientific community. Analyzing bird behavior and population trends helps detect other organisms in the environment and is an important problem in ecology. Bird populations react quickly to environmental changes, which makes their real time counting and tracking challenging and very useful. A reliable methodology that automatically identifies bird species from audio would therefore be a valuable tool for the experts in different scientific and applicational domains.

The goal of this work is to propose a methodology able to identify bird species by its chirp. In this paper we explore deep learning techniques that are being used in this domain, such as Convolutional Neural Networks and Recurrent Neural Networks to classify the data. In deep learning, audio problems are commonly approached by converting them into images using audio feature extraction techniques such as Mel Spectrograms and Mel Frequency Cepstral Coefficients. We propose and test multiple deep learning and feature extraction combinations in order to find the most suitable approach to this problem.

Keywords: Bird species classification · Deep learning · Audio feature extraction

1 Introduction

With climate change, the analysis of interactions between organisms and their environment is an important problem in ecology [1]. Bird behavior and population trends enable the detection of other organisms in the environment because birds react quickly to environmental changes. Bird identification can be done manually by experts in the area, either through audio or images [2], but the growing amounts of data makes this process tedious and time-consuming. It requires a lot of costly human effort [1]. The objective of this paper is to propose a methodology able to identify a bird species by its chirp. We intend to explore and improve the results of deep learning techniques that are being applied in this domain.

N. T. Nguyen et al. (Eds.): ACIIDS 2021, LNAI 12672, pp. 41–52, 2021.
https://doi.org/10.1007/978-3-030-73280-6_4

2 State of Art

Deep learning is commonly used for sound classification in many different domains. A typical approach is to convert the audio file into an image, such a spectrogram, and use a deep neural network to process that image [3].

2.1 Preprocessing and Feature Extraction

Regarding to the audio processing, it is common to normalize the sound signal and use filters for noise removal. In particular, the syllable segmentation algorithm has been used in frog sounds [4, 5] and in bird sounds [6].

The appropriate signal features extraction is crucial for the process of bird species recognition, as the acoustical environment of bird audio recordings contain noise in the signal, and there is a big diversity among bird species songs [7]. A Mel Spectrogram is a spectrogram where the frequencies are converted to the mel scale. This is a perceptual scale of pitches such that equal distances in pitch sound equally distant to the listener [8]. The Mel-Frequency Cepstral Coefficients (MFCC) are widely used in audio classification problems, due to its computational efficiency and noise robustness [9, 10].

2.2 Deep Learning for Sound Classification

Deep Learning is a sub-field of machine learning that exploits multiple flexible and stackable artificial neural networks solutions that learn successive layer of concepts [11].

A Convolutional Neural Network (CNN) is a deep learning architecture which is able to learn image classification from raw image data, using representation learning. Its architecture is analogous to the connectivity pattern of the neurons in the human brain, being inspired by the organization of the visual cortex [12]. A Recurrent Neural Network (RNN) is designed to recognize patterns in sequential data. RNN have feedback loops in order to keep the context of previously seen data and take those in consideration for classification or prediction. The algorithms used in RNN take time and sequence into account, meaning they have a temporal dimension. RNN are applicable to images since these can be decomposed into a series of patches and treated as a sequence [13]. Long Short-Term Memory (LSTM) networks are a modified version of recurrent neural networks. These networks were created with the intent of fixing computational issues of the original RNN [14]. Regular RNN have a vanishing gradient problem that may quickly lead the learning process to a halt. LSTM were designed with the intent of overcoming these error back-flow problems [15]. LSTM are capable of learning long-term dependencies. Retaining long-term temporal features in audio classification may help to improve the accuracy of the bird classification task given that context is kept. LSTM can remember information for long periods of time as their default behavior as they are explicitly designed to avoid the long-term dependency problem [16]. Gated Recurrent Unit Networks (GRU) are another improvement of standard RNN. GRU can be considered as a variant of LSTM since both are designed similarly and, in some cases, produce similar results [17]. Convolutional Recurrent Neural Network (CRNN) is a hybrid between CNN and RNN. A CRNN is a modified CNN with the last convolutional layers replaced with an RNN. In CRNN, CNN and RNN play the roles of feature extractor and temporal summarizer, respectively. Adopting an RNN for feature aggregation enables the networks to take the global structure into account while local features are extracted by the remaining convolutional layers [18].

2.3 Other Approaches

Lasseck [19] achieved top scores in the 2019 challenge BirdCLEF. The goal of the challenge was to localize and identify all audible birds within the provided soundscape test set. Each soundscape was divided into segments of 5 s [20]. The used evaluation metric was the Classification Mean Average Precision (cmAP). For each class 'c', all predictions classified as that class are obtained and ordered by decreasing probability to compute the average precision for that class. The mean for all classes is the main evaluation metric.

Hiatt [21], compiled the dataset used in this work and proposed an approach. This dataset contains a subset of records, labeled by species from California and Nevada. A pre-processing methodology was used where a data generator combines Mel Spectrograms with MFCC into a single 2-dimensional array. The deep learning model implemented is a Convolutional Neural Network with 3 stacks of 2-d Convolutions using ReLU activation and MaxPooling layers followed by dropout layers with a rate of 0.2. The model architecture is topped by a global average pooling layer, which is followed by a fully connected dense layer using the softmax activation. The accuracy metric is used to evaluate the performance of the models on a 33% holdout methodology. The documented results correspond to the model from the epoch with the minimum validation loss value. It achieved a validation accuracy of 19.27%, and a training accuracy of 20.44% [21].

3 Our Approach

The dataset used in this work contains a set of recordings, labeled by species, from California and Nevada, USA [21]. It contains 91 species, with 30 sound sample files per species. In total there are 2730 samples (MP3 file format), ranging from less than 1 s to 195 s. The sum of the duration of all samples from all the species is 20 h, 25 min, and 8 s (73508 s). The proposed methodology has three main steps:

- Pre-processing;
- Feature extraction;
- Deep learning modeling.

The methodology is described in the next sections.

3.1 Pre-processing

The audio files were converted from MP3 to WAV file format. Since the original files had sample rates between 22.5 kHz and 44.1 kHz, we normalized all the files to the minimum sample rate of 22.5 kHz. The file size difference is a downside, because while the dataset using the MP3 file format takes 1.4 GB, the WAV file format version takes 9 GB. The Python library Pydub was used to normalize the audio and export to the WAV file format.

Before the feature extraction, the audio data can be further improved using multi-threading support to speed up the pre-processing. This process starts with converting the audio channels from Stereo to Mono. Then, an envelope filter can be applied to the audio data. This envelope basically removes parts of the audio that are below a certain threshold and can be considered background noise. In Fig. 1 is presented the padding on a signal (of the species Black-tailed Gnatcatcher) to be used on a split with a window length of 3 s. Half of the window length is added to the beginning and the end of the signal (1.5 s).

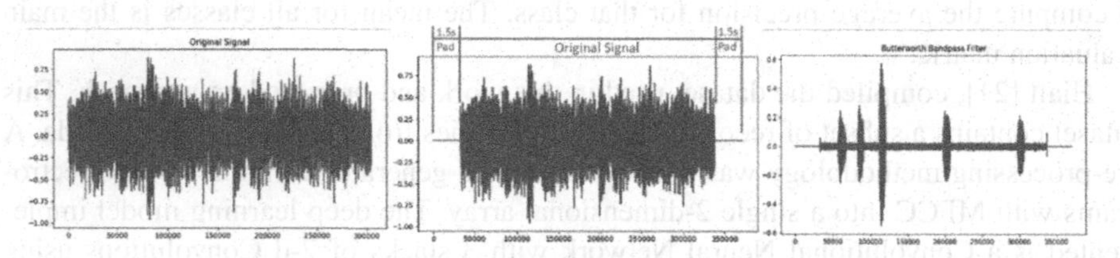

Fig. 1. Signal and Butterworth Bandpass filter applied to the signal

The Butterworth Bandpass filter [22] is used to remove frequencies outside of the bird vocalization range. The value used for the lowest frequency cut is 1500 Hz, and 8000 Hz for the highest. While testing the pre-processing of the audio data, the Butterworth bandpass filter attenuates most of the noise, such as rain or wind, which in some cases makes a big difference both visually and audibly regarding the clarity of the bird chirps.

In Fig. 1 we can also see the result of the Butterworth Bandpass filter applied to the signal. The final step is splitting the audio data into multiple samples. This process begins by finding the peaks in the sound. These peaks are obtained using a function from the Python library SciPy. This function has user definable parameters such as peak threshold and minimum distance between peaks. After getting a list of peaks, the audio data is then split into multiple samples, centered at the peak, and all with the same length of 3 s.

3.2 Feature Extraction

We used MFCC to extract features from the signal, using Python library python_speech_features [23] (Fig. 2). The parameters of this function are: sample_rate; numcep, the number of cepstrum (default is 13); nfilt, the number of filters in the filter-bank (default is 26); nfft, is FFT size (default is 512); and the other two corresponds to lowest band edge of mel filters (in Hz, default is 0) and highest band edge of mel filters (in Hz, default is samplerate/2).

Mel Spectrograms were extracted using the Python library LibROSA [24] (Fig. 3). The parameters of this function are: sample_rate; n_fft, length of the FFT window

```
1 from python_speech_features import mfcc
2
3 def get_mfccs(time_series, sample_rate):
4     nfft = (round(sample_rate / 40))
5     return mfcc(time_series, sample_rate, numcep=13, nfilt=26, nfft=nfft
    , lowfreq=1500, highfreq=8000).T
```

Fig. 2. MFCC extraction code

(default is 2048); hop_length, number of samples between successive frames (default is 512); n_mels, number of Mel bands to generate (default is 128) and the lowest and highest frequency (in Hz).

```
1 from librosa.feature import melspectrogram
2
3 def get_melspectrogram(time_series, sample_rate):
4     return melspectrogram(y=time_series, sr=sample_rate, n_fft=1024,
   hop_length=1024, n_mels=128, htk=True, fmin=1500, fmax=8000)
```

Fig. 3. Mel Spectrogram extraction code

In Fig. 4 we can see a visualization of the extracted features from the audio segment shown in Fig. 4. The obtained images are then saved into JPEG files. The Python library Matplotlib [25] was used to obtain the images.

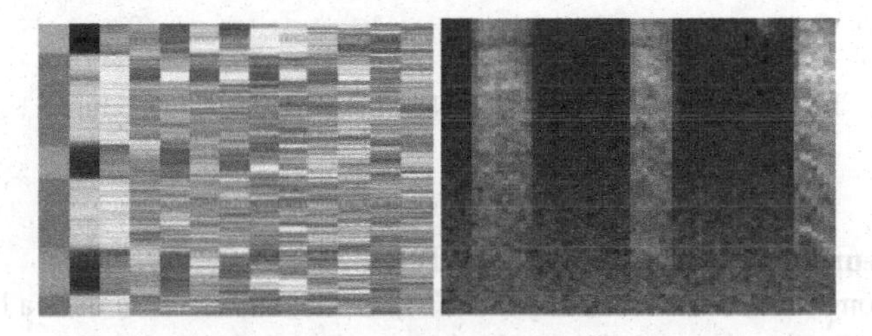

Fig. 4. Extracted MFCC (left) and Mel Spectrogram (right)

3.3 Deep Learning Model Architectures

In this section, we describe the deep learning models we have used. These models are implemented using the TensorFlow framework [26] with the Keras back-end [27] for Python. These models are the result of analyzing and testing multiple existing deep learning model architectures available online, such as the implementations by Adams [28]. The models were adapted and optimized to our problem.

Convolutional Neural Network

The selected Convolutional Neural Network (CNN) architecture is composed by 6 convolutional layers (Conv2D), each followed by a pooling layer (MaxPooling2D). As we can see in Fig. 5 between each convolutional and pooling layer pair, a batch normalization layer is added to reduce the amount of shift on the values of the hidden layers and increase the learning speed and reducing overfitting [29]. After a flatten layer, a dropout layer with a rate of 50% is added to reduce overfitting. The last layer is the fully connected layer, which is a dense layer with the softmax activation. In total, this model has 933211 trainable parameters.

	Type	Kernel	Kernel Size	Notes	Input Shape
1	*Conv2D*	8	7x7	Activation = *relu*	224 x 224 x 3
2	*MaxPooling2D*		2x2		224 x 224 x 8
3	*BatchNormalization*				112 x 112 x 8
4	*Conv2D*	16	5x5	Activation = *relu*	112 x 112 x 8
5	*MaxPooling2D*		2x2		112 x 112 x 16
6	*BatchNormalization*				56 x 56 x 16
7	*Conv2D*	32	3x3	Activation = *relu*	56 x 56 x 16
8	*MaxPooling2D*		2x2		56 x 56 x 32
9	*BatchNormalization*				28 x 28 x 32
10	*Conv2D*	64	3x3	Activation = *relu*	28 x 28 x 32
11	*MaxPooling2D*		2x2		28 x 28 x 64
12	*BatchNormalization*				14 x 14 x 64
13	*Conv2D*	128	3x3	Activation = *relu*	14 x 14 x 64
14	*MaxPooling2D*		2x2		14 x 14 x 128
15	*BatchNormalization*				7 x 7 x 128
16	*Conv2D*	256	3x3	Activation = *relu*	7 x 7 x 128
17	*Dropout*			Rate = 0.5	7 x 7 x 256
18	*MaxPooling2D*		2x2		7 x 7 x 256
19	*Flatten*				4 x 4 x 256
20	*Dropout*			Rate = 0.5	4096
21	*Dense*	128		Activation = *relu*	4096
22	*Dense*	91		Activation = *softmax*	128

Fig. 5. Convolutional Neural Network - Model Architecture

Recurrent Neural Network - Long Short-Term Memory

The chosen Long Short-Term Memory (LSTM) model architecture uses a bidirectional LSTM layer. The first layer reshapes the image data to have the correct dimensions for the LSTM layer. The implemented Long Short-Term Memory model architecture is presented in Fig. 6. In total, this model has 376955 trainable parameters.

Recurrent Neural Network - Gated Recurrent Unit

The Gated Recurrent Unit model architecture is similar to the presented Long Short-Term Memory model architecture but with the LSTM layer switched to the GRU layer and different units on the Dense layers after the MaxPooling1D layer. Figure 7 presents the implemented Gated Recurrent Unit model architecture. In total, this model has 465627 trainable parameters.

Convolutional Recurrent Neural Network - Long Short-Term Memory

The Convolutional Recurrent Neural Network (CRNN) using Long Short-Term Memory model architecture is based on merging CNN and LSTM model architectures. Layers have been added, removed, and modified in order to get better results. A dropout layer

	Type	Kernel	Notes	Input Shape
1	*Reshape*		*TimeDistributed*; *Target* = -1	224 x 224 x 3
2	*Dense*	64	*TimeDistributed*; Activation = *tanh*	224 x 672
3	*LSTM*	128	*BiDirectional*	224 x 64
4	*Dense*	64	Activation = *relu*	224 x 256
5	*MaxPooling1D*			224 x 64
6	*Dense*	32	Activation = *relu*	112 x 64
7	*Flatten*			112 x 32
8	*Dropout*		Rate = 0.5	3584
9	*Dense*	32	Activation = *relu*	3584
10	*Dense*	91	Activation = *softmax*	32

Fig. 6. Long Short-Term Memory - Model Architecture

	Type	Kernel	Notes	Input Shape
1	*Reshape*		*TimeDistributed*; *Target* = -1	224 x 224 x 3
2	*Dense*	64	*TimeDistributed*; Activation = *tanh*	224 x 672
3	*GRU*	128	*BiDirectional*	224 x 64
4	*Dense*	128	Activation = *relu*	224 x 256
5	*MaxPooling1D*			224 x 128
6	*Dense*	64	Activation = *relu*	112 x 128
7	*Flatten*			112 x 64
8	*Dropout*		Rate = 0.5	7168
9	*Dense*	32	Activation = *relu*	7168
10	*Dense*	91	Activation = *softmax*	32

Fig. 7. Gated Recurrent Unit - Model Architecture

is added between the Convolutional layers and the Long Short-Term Memory layers in order to reduce overfitting. In Fig. 8 we present the implemented CRNN–LSTM model architecture. In total, this model has 8981307 trainable parameters.

	Type	Kernel	Notes	Input Shape
1	*Conv2D*	32	Activation = *relu*	224 x 224 x 3
2	*MaxPooling2D*			224 x 224 x 32
3	*Conv2D*	64	Activation = *relu*	112 x 112 x 32
4	*MaxPooling2D*			112 x 112 x 64
5	*Conv2D*	128	Activation = *relu*	56 x 56 x 64
6	*MaxPooling2D*			56 x 56 x 128
7	*Conv2D*	256	Activation = *relu*	28 x 28 x 128
8	*MaxPooling2D*			28 x 28 x 256
9	*Dropout*		Rate = 0.5	14 x 14 x 256
10	*Reshape*		*TimeDistributed*; *Target* = -1	14 x 14 x 256
11	*LSTM*	256	*BiDirectional*	14 x 3584
12	*MaxPooling1D*			7 x 512
13	*Dropout*		Rate = 0.5	7 x 512
14	*LSTM*	128	*BiDirectional*	7 x 512
15	*MaxPooling1D*			7 x 256
16	*Flatten*			3 x 256
17	*Dropout*		Rate = 0.5	768
18	*Dense*	32	Activation = *relu*	768
19	*Dense*	91	Activation = *softmax*	64

Fig. 8. CRNN-LSTM - Model Architecture

Convolutional Recurrent Neural Network - Gated Recurrent Unit

The Convolutional Recurrent Neural Network - Gated Recurrent Unit (CRNN-GRU) model architecture is based on the presented CRNN-LSTM but with the LSTM layer switched to the GRU layer, with double the amount of kernel units. The second to last Dense layer also has double the amount of kernel units. In Fig. 9 we present the implemented CRNN-GRU model architecture. In total, this model has 15065915 trainable parameters.

	Type	Kernel	Notes	Input Shape
1	*Conv2D*	32	Activation = *relu*	224 x 224 x 3
2	*MaxPooling2D*			224 x 224 x 32
3	*Conv2D*	64	Activation = *relu*	112 x 112 x 32
4	*MaxPooling2D*			112 x 112 x 64
5	*Conv2D*	128	Activation = *relu*	56 x 56 x 64
6	*MaxPooling2D*			56 x 56 x 128
7	*Conv2D*	256	Activation = *relu*	28 x 28 x 128
8	*MaxPooling2D*			28 x 28 x 256
9	*Dropout*		Rate = 0.5	14 x 14 x 256
10	*Reshape*		*TimeDistributed*; *Target* = -1	14 x 14 x 256
11	*GRU*	512	*BiDirectional*	14 x 3584
12	*MaxPooling1D*			14 x 1024
13	*Dropout*	.	Rate = 0.5	7 x 1024
14	*GRU*	256	*BiDirectional*	7 x 1024
15	*MaxPooling1D*			7 x 512
16	*Flatten*			3 x 512
17	*Dropout*		Rate = 0.5	1536
18	*Dense*	64	Activation = *relu*	1536
19	*Dense*	91	Activation = *softmax*	64

Fig. 9. CRNN-GRU- Model Architecture

4 Experiments and Evaluation

Multiple combinations of feature extraction methods and deep learning models were tested. In total, there are 10 combinations to be tested, 5 deep learning models and 2 feature extraction methods: CNN with MFCC and Mel Spectrogram; RNN LSTM with MFCC and Mel Spectrogram; RNN GRU with MFCC and Mel Spectrogram; CRNN LSTM with MFCC and Mel Spectrogram and CRNN GRU with MFCC and Mel Spectrogram.

4.1 Evaluation Metrics

The methodology evaluation in the experiments was holdout. The data was randomly split in two folds, the training set (80% of the data) and the validation set (20% of the data). The main metric used to compare the deep learning models was the average recall per bird species and accuracy.

4.2 Experiments

We evaluated 10 combinations: 2 feature extraction methods and 5 deep learning models. These combinations are:

- CNN with: MFCC and Mel Spectrogram
- RNN LSTM with: MFCC and Mel Spectrogram
- RNN GRU with: MFCC and Mel Spectrogram
- CRNN LSTM with: MFCC and Mel Spectrogram
- CRNN GRU with: MFCC and Mel Spectrogram

The experiments were conducted in the following environment:

- OS: Ubuntu 20.04 LTS (Focal Fossa)
- Python: version 3.6
- TensorFlow: version 2.2
- Processor: AMD Ryzen 9 3900X
- RAM: 32 GB
- Graphics Card: AMD RX 580X 8 GB (RadeonOpenCompute 3.5.0)

4.3 Best Results

In Fig. 10 we can see the summary of the results for the proposed models.

From the experiments, the feature extraction Mel Spectrogram with 3 s sample lengths obtained better results than MFCCs. The best model architecture is the CRNN using Gated Recurrent Unit layers, as it achieved the best results when compared to the other implementations in terms of recall and accuracy.

Our best methodology obtained in this work was also compared with a previous approach by Hiatt [21]. We have used the same holdout split (33% for test set). However, we do not have specific information about which examples were used for training and testing by Hiatt. The results obtained are presented in Fig. 11. As we can see, our methodology performed better, achieving an accuracy of 44.26% on the validation set, while Hiatt's obtained 19.27%.

Comb.		At Maximum Recall (Validation Set)							
		Validation Set			Epoch	Time	Training Set		
F.	Models	Recall	Acc.	Loss			Recall	Acc.	Loss
MFCCs 3s	CNN	40.50%	45.41%	3.040	47	57m	98.82%	99.13%	0.0787
	LSTM	32.44%	38.01%	3.778	47	91m	77.86%	84.65%	0.6491
	GRU	35.50%	39.53%	3.979	46	107m	85.69%	89.77%	0.4622
	CRNN LSTM	**44.76%**	46.10%	4.233	42	74m	96.17%	97.01%	0.1556
	CRNN GRU	44.40%	47.81%	3.470	41	88m	90.80%	92.79%	0.3498
MFCCs 1.5s	CNN	38.82%	44.44%	3.134	45	52m	98.19%	98.76%	0.1116
	LSTM	27.27%	33.65%	3.909	45	89m	71.99%	80.79%	0.7931
	GRU	31.36%	36.23%	4.097	46	111m	79.56%	85.26%	0.6354
	CRNN LSTM	42.36%	43.95%	4.404	43	76m	91.97%	93.81%	0.2796
	CRNN GRU	**43.35%**	46.47%	3.539	40	82m	90.82%	92.64%	0.3476
Mel Spectrogram 3s	CNN	47.51%	43.05%	2.574	40	68m	99.05%	99.37%	0.0761
	LSTM	26.27%	33.32%	3.832	44	95m	65.85%	75.80%	0.9608
	GRU	28.23%	33.48%	4.271	43	110m	75.96%	83.36%	0.7034
	CRNN LSTM	47.95%	50.30%	3.475	48	93m	94.40%	95.60%	0.2140
	CRNN GRU	**50.17%**	53.13%	3.003	45	98m	90.64%	92.70%	0.3659
Mel Spectrogram 1.5s	CNN	44.21%	49.28%	2.957	41	68m	98.76%	99.14%	0.0813
	LSTM	24.62%	32.51%	3.790	47	101m	62.68%	73.29%	1.051
	GRU	27.56%	32.51%	4.071	46	115m	79.56%	85.46%	0.6354
	CRNN LSTM	46.46%	48.48%	3.763	40	78m	91.30%	93.39%	0.3109
	CRNN GRU	**47.92%**	50.71%	3.286	46	102m	89.74%	91.95%	0.3935

Fig. 10. Extracted features and deep learning model architecture combination comparison

Implementation	At Minimum Loss (Validation Set)			
	Validation Set		Epoch	Training Set
	Accuracy	Loss		Accuracy
This Work	44.26%	2.657	6	70.74%
Hiatt 2019	19.27%	3.605	85	20.44%

Fig. 11. Results comparison

5 Conclusions and Future Work

The goal of this paper is to contribute to the development of a reliable methodology that classifies bird species by their chirp, using deep learning techniques. We have presented an audio pre-processing methodology that normalized the audio, removed noise outside of the bird vocalization range and split an audio file into multiple equal length samples by detecting syllables or peaks. The audio samples then have their features extracted in one of two methods. These are the Mel Spectrograms and the Mel Frequency Cepstrum Coefficients. The classification methodology for this solution uses one of the two types of extracted features to train a Deep Learning model for predicting the species of a bird, by first pre-processing the audio file using the same audio pre-processing methodology that was used on the dataset used to train the model. An optimization methodology was created with the intent of testing and refining both the pre-processing methodology and the deep learning models. During the experiments conducted using this methodology, multiple alternative deep learning model architectures were tested with alternatives of Convolutional Neural Networks, Recurrent Neural Networks and Convolutional Recurrent Neural Networks. The best results were obtained with the Convolutional Recurrent Neural Network using Gated Recurrent Unit layers, the Convolutional Recurrent Neural Network using Long Short-Term Memory layers and the Convolutional Neural Network. The final experiment is comparing the best methodology obtained in this work with another implementation [21], while using the same experiment environment. Our methodology performed better, achieving an accuracy of 44.26% on the validation set, while Hiatt's obtained 19.27%.

In the future we intend to improve the audio pre-processing methodology with other audio pre-processing and feature extraction techniques such as data preparation and data augmentation techniques and other feature extraction methods.We also intend to use further optimized deep learning models with different architectures and hyper-parameters. Additionally, we will work on the discrepancy between validation and training results to reduce the possibility of overfitting. Transfer learning could also be used [30]. Furthermore, we intend to conduct more experiments using different types of bird sound datasets.

Acknowledgements. This work is financed by National Funds through the Portuguese funding agency, FCT - Fundação para a Ciência e a Tecnologia, within project UIDB/50014/2020.

References

1. Martinsson, J.: Bird Species Identification using Convolutional Neural Networks. Ph.D. thesis (2017). https://odr.chalmers.se/handle/20.500.12380/249467
2. Gavali, P., et al.: Bird species identification using deep learning. Int. J. Eng. Res. Technol. **8**(4) (2019). ISSN 2278-0181. https://www.ijert.org/bird-species-identification-using-deep-learning
3. Boddapati, V., et al.: Classifying environmental sounds using image recognition networks. Procedia Comput. Sci. **112**, 2048–2056 (2017). https://doi.org/10.1016/j.procs.2017.08.250. ISSN 1877-0509
4. Huang, C.-J., et al.: Frog classification using machine learning techniques. Expert Syst. Appl. **36**(2), 3737–3743 (2009). https://doi.org/10.1016/j.eswa.2008.02.059. ISSN 0957-4174
5. Colonna, J., et al.: Automatic classification of anuran sounds using convolutional neural networks. In: ResearchGate, pp. 73–78 (2016). https://doi.org/10.1145/2948992.2949016
6. Fagerlund, Seppo: Bird species recognition using support vector machines. EURASIP J. Adv. Signal Process. **2007**(1), 1–8 (2007). https://doi.org/10.1155/2007/38637. ISSN 1687-6180
7. Wielgat, R., et al.: HFCC based recognition of bird species. In: Signal Processing Algorithms, Architectures, Arrangements, and Applications SPA 2007, pp. 129–134 (2007). ISSN 2326-0319. https://doi.org/10.1109/spa.2007.5903313
8. Roberts, L.: Understanding the mel spectrogram - analytics vidhya - medium. In: Medium (2020). https://medium.com/analytics-vidhya/understanding-the-melspectrogram-fca2afa2ce53

9. Davis, S., Mermelstein, P.: Comparison of parametric representations for monosyllabic word recognition in continuously spoken sentences. IEEE Trans. Acoust. Speech Signal Process. **28**, 357–366 (1980). https://doi.org/10.1109/tassp.1980.1163420. ISSN 0096-3518
10. Kortas, M.: Sound-based bird classification. In: Medium (2020). https://towardsdatascience.com/sound-based-bird-classification-965d0ecacb2b
11. https://www.deeplearningbook.org/contents/intro.htmls. Accessed 14 Nov 2020
12. Saha, S.: A comprehensive guide to convolutional neural networks the ELI5 way. In: Medium (2018). ISSN 3211-6453. https://towardsdatascience.com/a-comprehensive-guide-to-convolutional-neural-networks-the-eli5-way-3bd2b1164a53. Accessed 23 Dec 2019
13. Nicholson, C.: A beginner's guide to LSTMs and recurrent neural networks (2019). https://pathmind.com/wiki/lstm. Accessed 27 Dec 2019
14. Nguyen, M.: Illustrated guide to LSTM's and GRU's: a step by step explanation. In: Medium (2019). https://towardsdatascience.com/illustrated-guideto-lstms-and-gru-s-a-step-by-step-explanation-44e9eb85bf21. Accessed 28 Dec 2019
15. Hochreiter, S., Schmidhuber, J.: Long short-term memory. Neural Comput. **9**(8), 1735–1780 (1997). https://doi.org/10.1162/neco.1997.9.8.1735. ISSN 0899-7667
16. Olah, C.: Understanding LSTM networks (2019). https://colah.github
17. Kostadinov, S.: Understanding GRU networks. In: Medium. https://towardsdatascience.com/understanding-gru-networks-2ef37df6c9be. Accessed 28 Dec 2019
18. Choi, K., et al.: Convolutional recurrent neural networks for music classification (2016). arXiv:1609.04243[cs.NE]
19. Lasseck, M.: Bird species identification in soundscapes. In: CLEF (2019)
20. Kahl, S., et al.: Overview of BirdCLEF 2019: large-scale bird recognition in soundscapes. In: CLEF (2019)
21. Hiatt, S.: Avian vocalizations - report. In: Kaggle (2019).https://www.kaggle.com/samhiatt/avian-vocalizations-report
22. Butterworth, S., et al.: On the theory of filter amplifiers. Wirel. Eng. **7**(6), 536–541 (1930)
23. Lyons, J., et al.: "jameslyons/python_speech_features: release v0.6.1". In: Zenodo. https://doi.org/10.5281/zenodo.3607820. Ph.D. Thesis. https://odr.chalmers.se/handle/20.500.12380/249467
24. McFee, B., et al.: "librosa/librosa: 0.7.2" (2020). https://doi.org/10.5281/zenodo.3606573
25. Hunter, J.D.: Matplotlib: A 2D graphics environment. Comput. Sci. Eng. **9**(3), 90–95 (2007). https://doi.org/10.1109/mcse.2007.55
26. Abadi, M., et al.: TensorFlow: large-scale machine learning on heterogeneous systems (2015). Software available from tensorflow.org. https://www.tensorflow.org/
27. Chollet, F., et al.: Keras (2015). https://keras.io
28. Adams, S.: Audio-classification (2020). https://github.com/seth814/AudioClassiftion/tree/2f0032d81dcfa3d662cab1c1c4e7e30520f7edd6. Accessed 7 Jun 2020
29. Doukkali, F.: Batch normalization in neural networks - towards data science. In: Medium (2019). https://towardsdatascience.com/batch-normalization-in-neural-networks-1ac91516821c
30. Xie, J., Ding, C., Li, W., Cai, C.: Audio-only bird species automated identification method with limited training data based on multi-channel deep convolutional neural networks (2018). arXiv:abs/1803.01107

Session Based Recommendations Using Recurrent Neural Networks - Long Short-Term Memory

Michal Dobrovolny[1], Ali Selamat[1,2,3], and Ondrej Krejcar[1](✉)

[1] Faculty of Informatics and Management, Center for Basic and Applied Research, University of Hradec Kralove, Hradec Kralove, Czech Republic
{michal.dobrovolny,ondrej.krejcar}@uhk.cz

[2] Malaysia Japan International Institute of Technology (MJIIT), Universiti Teknologi Malaysia Kuala Lumpur, Jalan Sultan Yahya Petra, 54100 Kuala Lumpur, Malaysia

[3] School of Computing, Faculty of Engineering, Universiti Teknologi Malaysia (UTM), 81310 Skudai, Malaysia
aselamat@utm.my

Abstract. This paper describes the use of long short-term memory (LSTM) for session-based recommendations. This paper aims to test and propose the best solution using word-level LSTM as a real-time recommendation service. Our method is for general use. Our model is composed of embedding, two LSTM layers and dense layer. We employ the mean of squared errors to assess the prediction results. Also, we tested our prediction of recall and precision metrics. The best performing network has been a trainer for the last year of likes on an image-based social platform and contained about 2000 classes. Our best model has resulted in recall value 0.0213 and precision value 0.0052 on twenty items.

Keywords: Neural networks · Collaborative filtering · Deep learning · Recommender systems · Long short-term memory

1 Introduction

In many modern applications, recommendation systems now expose the consumer to a large selection of products. Usually, such systems provide the consumer with a list of suggested items that they might prefer or estimate how much each item might prefer. These systems help users decide on suitable items and ease selecting the collection's desired items [9].

One of the essential activities that humans are trying to accomplish is forecasting. Using a modern power system, we try to understand the patterns of repetition and forecast the next cycle. Predicting a long-term or mid-term bid can be very difficult. The approaches are very rapidly going forward. Algorithms of neural networks (NN) can search and represent both structured and not structured data – for instance, natural language processing, time series or image data [1,19]. In image data processing can be found examples about fixing an image

N. T. Nguyen et al. (Eds.): ACIIDS 2021, LNAI 12672, pp. 53–65, 2021.
https://doi.org/10.1007/978-3-030-73280-6_5

[23,24], compression [21], super-resolution [7,15], image classification [5,18] or forecasting [8]. Algorithms of Neural network (NN) more often occur in the first position in many competitions. Image processing and time series are a very developed field. Some examples are image understanding, fixing old image damage, styling image and fake image generating. Researchers try to improve their models with many variations.

In this work, we aimed to test LSTM as a deep learning algorithm to automatically, and real-time recommend content. We developed a deep learning algorithm to suggest user content based on collaborative filtering and evaluated its performance with precision and recall metrics.

The paper's organisation is structured as follows: In Sect. 2, we discuss the methods of content recommendation. Section 3 describes the evaluation metrics we used. Section 4 outlines our approach to employ LSTM for session-based recommendations and our dataset. Section 5 elaborates on the results of our experiments. Finally, in Sect. 6, we conclude the paper and provide an outlook on our future work plans.

2 Methods of Content Recommendation

At the heart of any modern recommendation system, which has seen tremendous success in businesses such as Amazon, Netflix and Spotify, is collaborative filtering. It works by gathering human judgments (known as ratings) for products in a given domain and matching individuals with the same needs for knowledge or the same tastes. Collaborative filtering system users share their analytical judgments and views on each item they consume so that other system users can better assess which items to consume. In return, for new products, the collaborative filtering framework offers useful personalised recommendations.

There are several methods, more and less advanced. The methods commonly used for a recommendation are matrix factorisation, multi-layer perceptrons, autoencoders, and Boltzmann machines.

2.1 Multi-Layer Perceptron

This subsection describes the simplest method of content recommendation. As this method, we consider the multi-layer perceptron.

Wide and Deep Learning. For recommendation systems, memorisation, and generalisation are also important. A paradigm to combine the strengths of large linear models and deep neural networks to solve both issues is proposed in the paper "Wide and Deep Learning for Recommender Systems" [4]. This framework has been developed and evaluated on the Google Play recommender scheme, a massive-scale retail app store.

Deep Factorization Machines. As an extension of the Wide and Deep Learning approach, "DeepFM: A Factorisation Machine-Based Neural Network for CTR Prediction" by Huifeng Guo et al. is an end-to-end model that seamlessly integrates Factorisation Machine (wide components) and Multi-Layer Perceptrons (deep component). Compared to the Wide and Deep Model, DeepFM does not require tedious feature engineering [10].

To capture the linear and pairwise interactions between characteristics, the Factorization Machine uses addition and internal product operations. To model high-order interactions, the Multi-Layer Perceptron leverages nonlinear activation and deep structure.

Extreme Deep Factorization. As a Deep Factorization System extension, Extreme Deep Factorization Machine (xDeepFM) can model explicit and implicit function interactions together. Via a Compact Interaction Network, direct high-order feature interactions are learned, while implicit high-order feature violations are learned via a Multi-Layer Perceptron. This model also involves no manual engineering of functionality and releases repetitive feature search work from data scientists [17].

Neural Collaborative Filtering. Neural Factorization Machines for Sparse Predictive Analytics [12] is another parallel work that combines Factorisation Machines and Multi-Layer Perceptron seamlessly. This model brings together linear factorisation machines' efficacy for sparse predictive analytics with the excellent representation potential of nonlinear neural networks.

An activity called bilinear interaction pooling, which enables a neural network model to learn more informative feature interactions at the lower level, is the key to its architecture. The authors could deepen the shallow linear Factorisation Machine by stacking nonlinear layers above the bilinear interaction layer, effectively modelling higher-order and nonlinear feature interactions to enhance the Factorization Machine's expressiveness. This use of bilinear interaction pooling encodes more insightful feature interactions than conventional deep learning techniques that merely concatenate or average low-level embedding vectors, significantly facilitating the following deep layers to learn meaningful information [12].

2.2 Autoencoders

Autoencoders, which are a form of neural network that Kramer [14] has popularised as a more efficient method for the representation and reduction of data dimensionality than primary component analysis. Autoencoders are trained in an unsupervised way in which the network tries to recreate the input bypassing data through a bottleneck architecture to the output layer. There are some architectures we did not describe below. We do not target to describe them all.

Autoencoder. With just one hidden layer (i.e., the latent layer) between the input and output is the simplest type of autoencoder. The vector is taken from the latent layer, the Session and maps it to a latent representation [3].

Denoising Autoencoder. Extending autoencoders by corrupting the input may demonstrate unexpected benefits. The principle of autoencoder denoising is to learn representations that are resilient to minor, irrelevant input changes. Until we measure the final output, corrupting the input can be achieved with either one or more layers [16].

2.3 Boltzmann Machines

Boltzmann machines are stochastic and generative neural networks capable of learning internal representations and representing difficult combinatorial problems and solving them (given ample time) [2,20].

Restricted Boltzmann Machines. Non-deterministic (or stochastic) generative Deep Learning models (Boltzmann machines) have only two types of nodes - hidden and visible nodes. No output nodes are present! This fact may seem unusual, but this is what gives them this non-deterministic trait. Using Stochastic Gradient Descent, they do not have the usual 1 or 0 form output from which patterns are learned and optimised. Without that ability, they learn patterns, and this is what makes them so special! [20].

Explainable Restricted Boltzmann Machines. There can be several advantages to the reasons for recommendations, including efficacy (helping users make the right decisions), performance (assisting users to make faster decisions), and accountability (revealing the justification behind the recommendations). It is hard to interpret these learned characteristics in the case of Restricted Boltzmann Machine (RBM), which assigns a low-dimensional set of attributes to objects in a latent space. Therefore, choosing an interpretable approach with moderate prediction accuracy for RBM is a huge challenge.

The key idea is that if the suggested item has been rated by several neighbours, this might provide a basis for explaining the recommendations using neighbourhood-style clarification frameworks. The Explainability Ranking for user-based neighbour-style explanations [2].

3 Evaluation of Recommendation Systems

As evaluation metrics, we decided to measure different metrics. We are calculation the mean absolute error (MAE) and the mean squared error (MSE), which is also calculated during training on the training set. We also decided to measure decision support metrics like precision and recall.

3.1 MAE

Mean absolute error is the average difference between the value predicted by the recommender and the user's actual value. First, we compute the error by subtracting the predicted ratings and actual rating for each user, and then we take the mean of all errors to calculate MAE.

MAE shows how much the predicted score is far from the actual score. We take absolute (as the name suggests) to cancel the negative sign. As we are not interested in positive or negative scores, we only want to know the difference between real and predicted values.

Zero MAE means there was no difference between predicted and actual ratings and that the model predicted accurately. Therefore, smaller the MAE the better. In our case, MAE is 1.5, which is close to zero, indicating that our model will predict the ratings for movies for any given user accurately.

3.2 MSE

MSE is similar to MAE and the only difference is that instead of taking the absolute error to cancel the negative sign, we square it.

MAE helps penalise the results, so even a small change will result in a big difference. This also suggests that if MSE is close to zero, the recommender system did well because otherwise, the MSE will not be this small.

3.3 Precision

Precision is the number of selected items that are relevant. Suppose our recommender system selects three items to recommend to users out of which two are relevant then the precision will be 66%. As shown in Fig. 1.

Precision is about retrieving the best items to the user, assuming that there are more useful items available then you want [9].

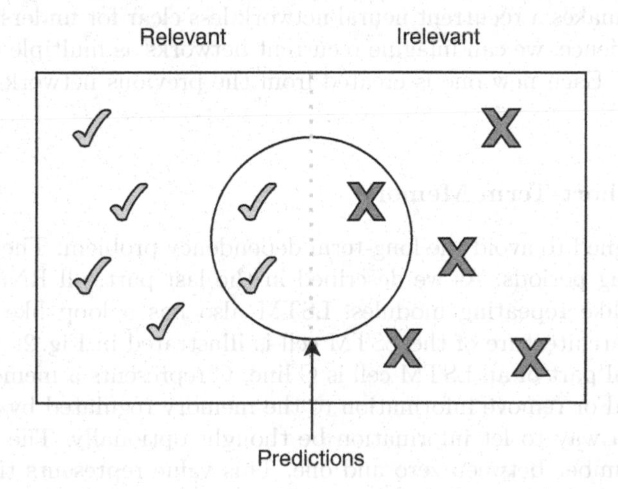

Fig. 1. Precision is the number of correctly selected items from space. On the left side are relevant items and on the right side are irrelevant items.

3.4 Recall

The recall is the number of relevant items that are selected. Suppose there are six relevant items out of which the recommender selects two relevant items, then the recall will be 33%. As shown in Fig. 1. The recall is about not missing useful items [9].

4 Using the LSTM as a Recommender System

For instance, we have a text: "I love the university, the university is life". Separate them by "," into two sentences and separate each one by space into words. We will have the following result of two sequences with length equals three. LSTM results will be like "I love the university is life". The result is silly and pure from the grammar point of view. However, the logic is still here.

4.1 Recurrent Neural Networks

Humans do not start their thinking from scratch every second. As we do anything, we understand we build our experiences on previous knowledge. We do not throw everything away and start thinking from scratch again. During the reading, we understand every word in the context of other words.

Traditional neural networks do not have memory. For instance, imagine the network is trying to solve a classification task. To be correct, classify every frame of a movie. Every point of the timeline has to be described. It is impossible to understand a movie in context if the network does not have any information about previous events in the movie.

RNN is a network architecture which addresses this issue. RNN contains a loop to allow the network to persist information. A loop allows information to be passed from one step to the next one [13].

This loop makes a recurrent neural network less clear for understanding. For a better experience, we can imagine recurrent networks as multiple copies of the same network. Each new one is created from the previous network. Consider it as a loop.

4.2 Long Short-Term Memory

LSTM is designed to avoid the long-term dependency problem. They keep information for long periods. As we described in the last part, all RNN have some form of loop-like repeating modules. LSTM also has a loop like a repeating module. The architecture of the LSTM cell is illustrated in Fig. 2.

The critical part of an LSTM cell is C line. C represents a memory pipeline. LSTM can add or remove information to the memory regulated by gates. Gates x and + are a way to let information be thought optionally. The σ layer outputs a real number between zero and one. This value represents the impact of information. Zero is for no impact, and one is for very important.

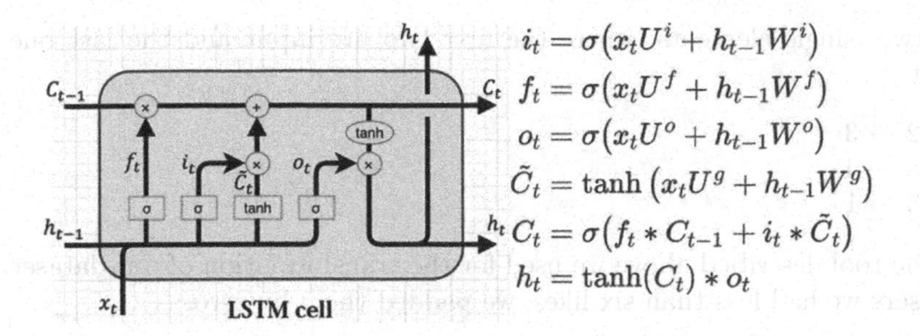

Fig. 2. An illustration of LSTM cell with memory pipe [22].

4.3 Data Preparation

As were described the LSTM is ideal for time series. Therefore, the trick is to transform our data into a sequence of steps.

For instance, if we have a list of pictures the user liked, we have to transform them from a list into a sequence. We can look at a user's likes on an image as a sentence of IDs if we have three users. Each one has a history of likes for a certain period. We want to predict what will be the next one?

Let us make a text story about a shopping list. There are thee customers who bought for items - milk, bread, cheese, eggs.

Table 1. Shopping list example.

Name	Items			Next item
Joey	Milk	Bred	Cheese	?
Ross	Bred	Cheese	Eggs	?
Rachel	Eggs	Cheese	Milk	?

In Table 1 is described the idea of a sentence which continues below.

- Joey: "I want milk. I want bread and cheese."
- Ross: "I want bread. I want cheese and eggs."
- Rachel: "I want eggs. I want cheese and milk."

If we transform our sentences into simple informations, we get rows from Table 1. Now we have to transform words into ids. The result of this transformation will be:

- "1,2,3"
- "2,3,4"
- "4,3,1"

Supervised observation is a way of data preparation when we have some inputs and corresponding outputs (targets or labels). Let us split each sequence into two, single elements, where the first two are input and the last one is a target.

- $1, 2 \rightarrow 3$
- $2, 3 \rightarrow 4$
- $4, 3 \rightarrow 1$

The tool described above we used for the transformation of our dataset. For the users we had less than six likes, we padded them by zero.

Dataset. We are using a dataset of 1890 stories from the social platform. Where we have ids of the stories, a user liked. Our dataset contains about 101 430 connections. We transformed the list of likes into sentences of five likes pointing to the sixth one.

4.4 Model Definition

Our starting architecture contains the embedding stateless LSTM layer and a dense layer.

Embedding. Word embedding is a feature learning technique in natural language processing natural language processing (NLP) where words are mapped to vectors of real numbers [6]. We defined the embedding dimensions to 448 regarding the test result of Table 2.

LSTM. Long short-term memory is described in Sect. 4.2. We decided to start with a single layer; then we changed to two layers as the best result of testing regarding Table 2.

Dense Layer. The dense layer is defined by the size of the number of classes. Dense layer purpose is to predict which class is the next one to recommend.

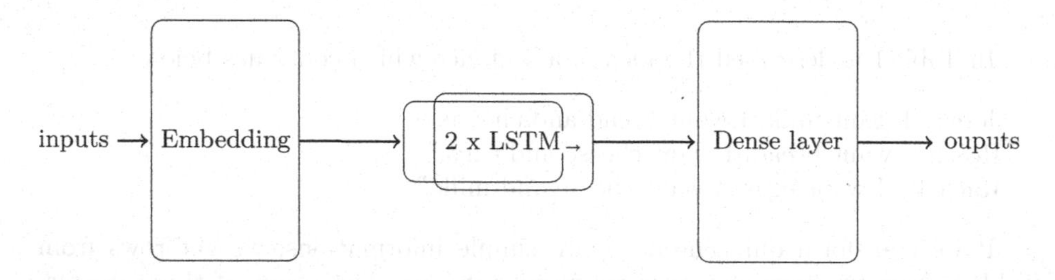

Fig. 3. Our model architecture contains embedding, two LSTM layers and dense layer.

For the architecture described in Fig. 3 we used the optimisation framework optuna. We run ten trials to find the best architecture. In Table 2 are described the results in numbers.

Table 2. Optuna hyperparameter optimisation result table with tested parameters were embedding, number of hidden LSTM layers, number of LSTM units and learning rate.

Trial number	Embedding dim	Hidden dim	Number of layers	Learnin rate	loss
1	128	320	3	0.03	0.0217
2	256	320	3	0.03	0.0231
3	448	384	2	0.03	NaN
4	64	256	1	0.03	NaN
5	320	384	2	0.03	0.0246
6	64	512	3	0.01	0.0208
7	448	448	2	0.02	0.0314
8	256	256	1	0.03	0.0232
9	**448**	**256**	**2**	**0.03**	**0.0199**
10	448	448	1	0.03	0.0474

In Table 2 are described trials of the models we tested. The best performing one we selected for future testing (number nine). All models were trained for six epochs and had a running time between five and eight minutes.

5 Results and Discussions

Our experiment can is replicable on any dataset according to the following description bellow. This section also describes our results.

5.1 Experiment

The first step is to create a dataset. The dataset can be formed from any positive user feedback. In our case, they were lychees. An alternative could be, for example, purchasing products. We group this list of products according to users as a sorted list of identifiers. We have chosen the postcreation date as the sorting method, as our aim to recommend new content.

We will transform this dataset according to the procedure described in chapter ref s: data. So let us quantify the classes. This number gives us a size of input and output neurons.

From the obtained number of classes, we define our model. That is if the number of classes is 1000. Our model will have 1000 input and output neurons. The hidden layers will consist of two LSTM layers.

Subsequently, we will perform supervised training, which will teach our network the links.

5.2 Recall and Precision Results

Table 3 shows result of our model. We evaluated on our dataset with metrics as precision described in Sect. 3.3 and recall described in Sect. 3.4.

During the training process, we tried this model fit on movielens database [11]. There was a problem with the model size. Since the input and output size contains the number of classes, we got a hardware memory error.

```
Tried to allocate 47.81 GiB (
 GPU 0;
 14.73 GiB total capacity;
 11.64 GiB already allocated;
 2.30 GiB free;
 11.66 GiB reserved in total by PyTorch
)
```

This error leads us to the limitation of a dataset. We decided to use a smaller dataset with only almost 2000 classes. With this number of classes, the model has around 2 GB of GPU memory.

LSTM, RNN and NN generally are very performance demanding. Our computation computer has two dedicated cards with a total of 7 934 CUDA cores. These cards are one of the top-performing gaming cards of current NVIDIA cards. Because of the framework support, we decided to use NVIDIA cards only. One of our cards is a 1080TI with 11 176 MB graphic memory and 1607 MHz max clock rate. Another one is a 2080TI with 11 019 MB of graphics memory and 1545 max clock rate. The used processor is i7-8700 with 3.20 GHz clock.

We used as a programming language Python in version 3.7.3. For programming in Python, we used a web-based interactive computational environment, Jupyter Notebook. PyTorch is our main framework. For fast prototyping, we used PyTorch Lightning. PyTorch Lightning is a top-level framework with higher abstraction on top of PyTorch. For the finetuning of our model, we use the framework Optuna.

Table 3. Table of recall and precision results on 20 predicted classes.

Recall@20	Precision@20
0.0213	0.0052

6 Conclusions and Future Work

Collaborative filtering is a significant domain of machine learning and has a big number of possible architectures. All of them are very advanced. All of them have some advantages and disadvantages. The LSTM can be a way to introduce a solution eliminating some of the disadvantages.

LSTM models might be used in recommendation systems. However, they also have their limitations – for instance, the mentioned number of classes.

To use LSTM as a recommender, we have to, first of all, you need to extend your knowledge about its structure and know all processes inside them.

Regarding our data preparation, we did a text trick to make a list into a sentence.

For predictions made on LSTM network, we must have enough data, or more importantly, connections between classes. If we have a small dataset, it will fail by overfitting or getting a week model.

If we train our model on users' correct data previously marked, we will get the model to make predictions in a real-time mode.

Future work will focus on an experiment on char-level LSTM. We think that if we transform the data to a more relative way for char-level LSTM, we can get even better predictions. Our future work will also focus on the more relative comparison, including movielens [11] and defined metrics on a related dataset.

Acknowledgment. The work and the contribution were supported by the SPEV project "Smart Solutions in Ubiquitous Computing Environments", University of Hradec Kralove, Faculty of Informatics and Management, Czech Republic.

References

1. Abdel-Nasser, M., Mahmoud, K.: Accurate photovoltaic power forecasting models using deep LSTM-RNN. Neural Comput. Appl **31**(7), 2727–2740 (2019). https://doi.org/10.1007/s00521-017-3225-z, wOS:000478687000053
2. Abdollahi, B., Nasraoui, O.: Explainable Restricted Boltzmann Machines for Collaborative Filtering (2016). arXiv:1606.07129 [cs, stat], http://arxiv.org/abs/1606.07129
3. Bengio, Y., Lamblin, P., Popovici, D., Larochelle, H., Montreal, U.: Greedy layer-wise training of deep networks. Adv. Neural Inf. Process. **19**, 153 (2007)
4. Cheng, H.T., et al.: Wide & Deep Learning for Recommender Systems(2016). arXiv:1606.07792 [cs, stat], http://arxiv.org/abs/1606.07792
5. Ciresan, D., Meier, U., Schmidhuber, J.: Multi-column deep neural networks for image classification. In: 2012 IEEE Conference on Computer Vision and Pattern Recognition (CVPR), pp. 3642–3649. IEEE, New York (2012) wOS:000309166203102
6. Cui, P., Wang, X., Pei, J., Zhu, W.: A survey on network embedding. IEEE Trans. Knowl. Data Eng. **31**(5), 833–852 (2019). https://doi.org/10.1109/TKDE.2018.2849727, https://ieeexplore.ieee.org/document/8392745/
7. Dobrovolny, M., Mls, K., Krejcar, O., Mambou, S., Selamat, A.: Medical image data upscaling with generative adversarial networks. In: Rojas, I., Valenzuela, O., Rojas, F., Herrera, L.J., Ortuño, F. (eds.) IWBBIO 2020. LNCS, vol. 12108, pp. 739–749. Springer, Cham (2020). https://doi.org/10.1007/978-3-030-45385-5_66
8. Dobrovolny, M., Soukal, I., Lim, K.C., Selamat, A., Krejcar, O.: Forecasting of FOREX price trend using recurrent neural network - long short-term memory, pp. 95–103 (2020). https://doi.org/10.36689/uhk/hed/2020-01-011, http://hdl.handle.net/20.500.12603/212
9. Gunawardana, A., Shani, G.: Evaluating recommender systems. In: Ricci, F., Rokach, L., Shapira, B. (eds.) Recommender Systems Handbook, pp. 265–308. Springer, Boston, MA (2015). https://doi.org/10.1007/978-1-4899-7637-6_8
10. Guo, H., Tang, R., Ye, Y., Li, Z., He, X.: DeepFM: A Factorization-Machine based Neural Network for CTR Prediction (2017). arXiv:1703.04247 [cs], http://arxiv.org/abs/1703.04247

11. Harper, F.M., Konstan, J.A.: The MovieLens datasets: history and context. ACM Trans. Interact. Intell. Syst. **5**(4), 1–19 (2016). https://doi.org/10.1145/2827872, https://dl.acm.org/doi/10.1145/2827872
12. He, X., Liao, L., Zhang, H., Nie, L., Hu, X., Chua, T.S.: Neural Collaborative Filtering (2017). arXiv:1708.05031 [cs], http://arxiv.org/abs/1708.05031
13. Hochreiter, S., Schmidhuber, J.: Long short-term memory. Neural Comput. **9**(8), 1735–1780 (1997). https://doi.org/10.1162/neco.1997.9.8.1735, WOS:A1997YA04500007
14. Kramer, M.A.: Nonlinear principal component analysis using autoassociative neural networks. AIChE J. **37**(2), 233–243 (1991). https://doi.org/10.1002/aic.690370209, https://aiche.onlinelibrary.wiley.com/doi/abs/10.1002/aic.690370209, _eprint: https://doi.org/10.1002/aic.690370209
15. Ledig, C., et al.: Photo-realistic single image super-resolution using a generative adversarial network. In: 30th IEEE Conference on Computer Vision and Pattern Recognition (CVPR 2017), pp. 105–114. IEEE, New York (2017). wOS:000418371400012
16. Li, S., Kawale, J., Fu, Y.: Deep collaborative filtering via marginalized denoising auto-encoder. In: Proceedings of the 24th ACM International on Conference on Information and Knowledge Management, CIKM 2015, pp. 811–820. Association for Computing Machinery, New York (2015). https://doi.org/10.1145/2806416.2806527
17. Lian, J., Zhou, X., Zhang, F., Chen, Z., Xie, X., Sun, G.: xDeepFM: combining explicit and implicit feature interactions for recommender systems. In: Proceedings of the 24th ACM SIGKDD International Conference on Knowledge Discovery & Data Mining, pp. 1754–1763 (2018). https://doi.org/10.1145/3219819.3220023, http://arxiv.org/abs/1803.05170, arXiv: 1803.05170
18. Mambou, S., Krejcar, O., Selamat, A., Dobrovolny, M., Maresova, P., Kuca, K.: Novel thermal image classification based on techniques derived from mathematical morphology: case of breast cancer. In: Rojas, I., Valenzuela, O., Rojas, F., Herrera, L.J., Ortuño, F. (eds.) IWBBIO 2020. LNCS, vol. 12108, pp. 683–694. Springer, Cham (2020). https://doi.org/10.1007/978-3-030-45385-5_61
19. Pena-Barragan, J.M., Ngugi, M.K., Plant, R.E., Six, J.: Object-based crop identification using multiple vegetation indices, textural features and crop phenology. Remote Sens. Environ. **115**(6), 1301–1316 (2011). https://doi.org/10.1016/j.rse.2011.01.009, wOS:000290011200001
20. Salakhutdinov, R., Mnih, A., Hinton, G.: Restricted Boltzmann machines for collaborative filtering. In: Proceedings of the 24th international conference on Machine learning - ICML 2007, pp. 791–798. ACM Press, Corvalis (2007). https://doi.org/10.1145/1273496.1273596, http://portal.acm.org/citation.cfm?doid=1273496.1273596
21. Sun, Y., Chen, J., Liu, Q., Liu, G.: Learning image compressed sensing with sub-pixel convolutional generative adversarial network. Pattern Recognition **98**, (2020). https://doi.org/10.1016/j.patcog.2019.107051, http://www.sciencedirect.com/science/article/pii/S003132031930353X
22. Varsamopoulos, S., Bertels, K., Almudever, C.G.: Designing neural network based decoders for surface codes, p. 13 (2018)

23. Wolterink, J.M., Leiner, T., Viergever, M.A., Isgum, I.: Generative adversarial networks for noise reduction in low-dose CT. IEEE Trans. Med. Imaging **36**(12), 2536–2545 (2017). https://doi.org/10.1109/TMI.2017.2708987, wOS:000417913600013
24. Yang, Q., et al.: Low-dose CT image denoising using a generative adversarial network with wasserstein distance and perceptual loss. IEEE Trans. Med. Imaging **37**(6), 1348–1357 (2018). https://doi.org/10.1109/TMI.2018.2827462, wOS:000434302700006

UVDS: A New Dataset for Traffic Forecasting with Spatial-Temporal Correlation

Khac-Hoai Nam Bui, Hongsuk Yi(✉), and Jiho Cho

Korea Institue of Science and Technology Information, Daejeon 34141, Korea
{hsyi,jhcho}@kisti.re.kr

Abstract. This paper introduces UVDS, a traffic flow dataset from the vehicle detection system (VDS) in an urban area of South Korea. Specifically, with the rapid growth of computer vision for intelligent transportation systems, using detection systems for estimating traffic flow become an emergent issue. In this study, we first discuss the main differences between UVDS and existing datasets in terms of spatial-temporal dependencies for accurate traffic prediction. Then, preliminary work for construct a graph structure of the VDS data based on the geometric information is presented. The objective is to provide a benchmark dataset for exploring the capabilities of graph neural networks for traffic forecasting. Consequently, we present baseline results by adopting state-of-the-art models in this research field and discuss some future work for exploring the UVDS dataset.

Keywords: Traffic dataset · Graph neural network · Traffic forecasting · Spatial-temporal correlation

1 Introduction

Traffic forecasting is an essential role in the Intelligent Transportation System (ITS) [27]. For instance, estimating accurate multi-step traffic flow is able to enable the dynamic traffic light control system [1]. Recently, with the rapid development of Deep Learning (DL) models, a significant amount of research efforts have been introduced for improving the prediction abilities. Specifically, DL models are able to achieve better performance, which exploit much more features and complex architectures of traffic data than the classical method using statistical and Machine Learning (ML) models such as ARIMA [17], Feature-based methods [10] and Gaussian process models [6]. Particularly, several well-known DL models have been proposed for improving traffic forecasting such as Deep Belief Networks (DBNs) [8], Stacked Autoencoder (SAE) [12], Convolution Neural Networks (CNN) [13], and Recurrent Neural Network - Long Short Term Memory (RNN-LSTM) [22,23].

N. T. Nguyen et al. (Eds.): ACIIDS 2021, LNAI 12672, pp. 66–77, 2021.
https://doi.org/10.1007/978-3-030-73280-6_6

Recent state-of-the-art methods in this research field mainly focus on applying Graph Neural Network (GNN) to model non-euclidean spatial structure data, which is able to get more accurate with the structure of traffic road network [24]. Specifically, GNN-based methods are able to model the dependencies between nodes in a graph which enables the breakthrough in the research area related to graph analysis [18]. Subsequently, several variants of GNN have been proposed (e.g., Spatio-Temporal Graph Convolutional Networks (STGCN) [25], Traffic Graph Convolutional Long Short-Term Memory Neural Network (TGC-LSTM) [4], and Graph Multi-Attention Network (GMAN) [26]) which provide the promising results compared with previous works by adopting spatial-temporal correlation for traffic flow prediction. Figure 1 depicts the general framework for traffic forecasting using GNN models. Particularly, due to the time-varying traffic patterns and the complicated spatial dependencies on road networks, representing traffic data as graph-structured data is able to significantly improve the performance of traffic forecasting. In this regard, several well-known real-world traffic datasets with spatial-temporal information have been introduced for exploiting the capabilities of GNN-based methods such as PeMS, METR-La, LOOP, NYC Taxi, and so on. In this study, we introduce a new traffic dataset, which is named as UVDS, for further research on traffic forecasting using graph representation learning approaches. Specifically, traffic data have collected from VDS in urban areas which enforce the high complexity and diversity of the road network. Furthermore, the UVDS dataset obtains various traffic characteristics such as speed, volume, vehicle type, and occupancy for further research on multivariate time series forecasting using graph learning, which is one of the emergent issues in this research field [2,19]. In particular, the main contribution of this paper is threefold as follows:

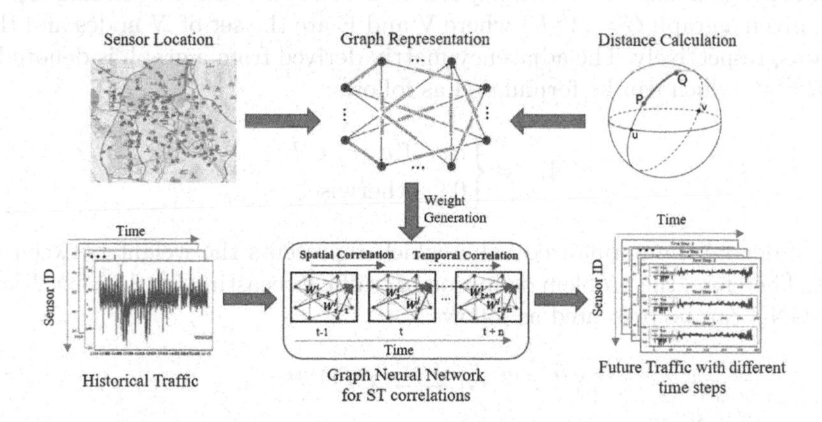

Fig. 1. General framework for traffic forecasting using spatial-temporal correlation.

- We provide the first version of the UVDS dataset which includes 104 sensors that cover whole the main roads in Daejeon City, an urban area in

South Korea. Specifically, the raw data are collected and preprocessed during 3 months of time duration which are aggregated every 5 min from the publicly available web of transportation in South Korea.
- We present a preliminary work for representing the traffic road network of UVDS in graph structure for learning based on the geometric information. Moreover, further directions for exploring this issue are discussed in the experiment section.
- We provide the benchmark evaluation and baseline results using UVDS dataset by implementing state-of-the-art GNN models and further discussions on how to exploit the dataset in this research field.

The rest of this paper is organized as follows: We take a brief review of traffic forecasting using GNN in Sect. 2. Moreover, previous public traffic datasets are summarized in this section. In Sect. 3, we introduce our dataset in terms of data collection and processing for using graph representation learning. Section 4 provides baseline results and discussions for open research issues using UVDS dataset. The future work of this study is concluded in Sect. 5.

2 Literature Review

2.1 Traffic Forecasting Using Graph Neural Network

In recent years, GNN has emerged as a promising DL method for handling spatial dependencies among entities in a network which assume that the state of a node depends on the states of its neighbors in the network. Specifically, by propagating information through graph structures, GNN with spatial-temporal dependence modeling becomes a key research issue for traffic forecasting. Specifically, given a graph $G = (V, E)$ where V and E are the set of N nodes and their relations, respectively. The adjacency matrix derived from a graph is denoted by $A \in R^{N \times N}$, which can be formulated as follows:

$$A_{i,j} = \begin{cases} c, & \text{if } v_i, v_j \in E \\ 0, & \text{otherwise.} \end{cases} \tag{1}$$

where c denote the normalized value which represents the weight between two nodes. Therefore, the problem definition of traffic forecasting for the next T steps using GNN can be presented as follows:

$$[X^{(t-S):t}, G] \xrightarrow{f(x)} X^{(t+1):(t+T)} \tag{2}$$

where $X^t \in R^{N \times F}$ represent the dynamic feature matrix at each time step t and S denotes the historical steps. Technically, current state-of-the-art models highly focus on Graph Convolutional Networks (GCNs) and Graph Attention Networks (GATs) for exploiting the spatial-temporal correlation.

2.1.1 Graph Convolutional Network CNN is limited to model Euclidean data. Therefore, GCN is introduced for modeling non-Euclidean spatial structure data which is suitable for traffic road networks. Sequentially, there are two types of GCN which are spectral-based and spatial-based methods. Specifically, in spectral networks, the convolution operation performs graph data from spectral-domain by computing the eigendecomposition of the graph Laplacian matrix. According to the study in [9], the transformed signal matrix Z can be generalized as follows:

$$\begin{aligned} Z &= \tilde{D}^{-1/2}\tilde{A}\tilde{D}^{-1/2}X\Theta \\ s.t.: \tilde{A} &= A + I_N \\ \tilde{D}_{ii} &= \sum \tilde{A}_{ij} \end{aligned} \quad (3)$$

where $D \in R^{N\times N}$, I_N, and Θ denote diagonal degree matrix, identity matrix, and filter parameters, respectively. Sequentially, authors in [25] present STGCN model by incorporating ChebNet graph convolution for extract spatial dependencies and 1D convolution for temporal correlations to enable faster training. Sequentially, ASTGCN [7], a new model based on the concept of STGCN with adding two attention layers, has introduced for capturing the dynamic spatial-temporal correlation.

On the other hand, spatial-based methods define convolutions directly on the graph by aggregating its neighboring information. Specifically, comparing with spectral-based methods, this method is able to support the multi-dimensional inputs. Li et al. [11], utilize the adjacency matrix to propose DCRNN, which was a cutting edge model for prediction task, using diffusion GCN and RNN for incorporating spatial and temporal dependencies, respectively. Sequentially, Graph WaveNet [20], that combines diffusion graph convolution with dilated casual convolution, have introduced for capturing the dynamic feature of the graph structure. Particularly, the diffusion graph convolution layers of DCRNN and Graph WaveNet are defined in Eq. 4 and Eq. 5 (for undirected graphs), respectively.

$$Z = \sum_{k=0}^{K} P_k X W_k \quad (4)$$

$$\begin{aligned} Z &= \sum_{k=0}^{K} P_k X W_{k1} + \hat{A}_k X W_{k2} \\ s.t.: \hat{A} &= SoftMax(ReLU(E_1 E_2^T)) \end{aligned} \quad (5)$$

where P_k, $W \in R^{F\times C}$, and E denote the power series of the transition matrix of k diffusion step, model parameter matrix, and node embedding dictionaries, respectively.

2.1.2 Graph Attention Network Attention mechanisms have been widely applied for natural language processing. The core idea of the attention method is to focus on the relevant features from input data which is properly suitable with the complex spatial-temporal correlation of traffic road networks. Sequentially, recent studies try to adopt attention mechanisms, which based on the concept of GAT [16], for traffic forecasting with promising results, especially in the case of the long-term prediction. For instance, Pan et al. [15] presents a deep meta learning-based model, ST-MetaNet, which combines Meta-GAT and Meta-RNN (i.e., GRU) to capture the diversities of spatial and temporal correlations, respectively. Recently, Zheng et al. [26] exploit the dynamic of traffic condition with the spatial-temporal correlation by proposing GMAN, a graph multi-attention network including a transform attention layer between the encoder and the decoder for improving the performance of long-term prediction.

2.2 Public Traffic Flow Datasets

High-quality datasets are essential for traffic flow prediction task. In this section, we take a brief review of several real-word public datasets that are well-known and widely used in this research field. Specifically, Table 1 shows the open datasets that have frequently used in recent works of traffic forecasting with spatial-temporal correlation [24].

Table 1. Public datasets for traffic prediction task

Datasets	Characteristic	Time interval	Source (https://github.com)
PeMS-BAY	Speed	5 min	/liyaguang/DCRNN (2018)
METR-LA	Speed	5 min	/liyaguang/DCRNN (2018)
LOOP	Speed	5 min	/zhiyongc/Seattle-Loop-Data (2018)
PeMSD4,8	Volume	5 min	/Davidham3/ASTGCN (2018)
Q-Traffic	Speed	15 min	/JingqingZ/BaiduTraffic (2018)
NYC Taxi	Demand	30 min	/toddwschneider/nyc-taxi-data (2019)
PeMSD3,7	Volume	5 min	/Davidham3/STSGCN (2020)

Accordingly, these datasets are sequentially described as follows:

- **PEMS**: the data have collected from California Transportation Agency Performance Measurement System in which the locations of sensors cover all major metropolitan areas of the state of California. As shown in Table 1, depending on the scope of each study, there are several different sub-datasets from PeMS in terms of traffic characteristic (e.g., traffic speed, volume, and demand), time interval and duration, and location such as PeMS-BAY (1–6/2017, 325 sensors), PeMSD3 (9–11/2018, 358 sensors), PeMSD4 (1–2/2018, 3848 sensors), PeMSD7 (7–8/2016, 883 sensors), PeMSD8 (7–8/2016, 1979 sensors).
- **METRA-LA**: the data have collected from the highways system of Los Angeles County and pre-processed by authors in [11] which includes 207 sensors during 6 months of time duration (1–6/2012).

- **LOOP**: the data have collected from loop detectors of Greater Seattle Area which includes 323 sensors during 1 year of the time duration (2015).
- **Q-Traffic**: the data have collected in Beijing city for two months of the time duration (4–5/2017).
- **NYC Taxi**: the data have collected based on trajectory data of taxi GPS in New York city with around billions of taxi and for-hire vehicles since 2009.

3 UVDS: Urban Vehicle Detection System Dataset

3.1 Data Description and Research Challenges

In this study, the new traffic data (i.e., UVDS) have collected based on VDS locating in an urban area, which measures the traffic passing over specific points. Specifically, traffic information is collected by forming one or more detecting areas at specific points on the road. In this study, the first version of the UVDS dataset contains traffic information from 104 VDS sensors that cover whole the main roads of an urban area. Consequentially, we collected the raw VDS data from the publicly available web of transportation in South Korea and pre-process the data into multiple traffic features such as traffic flow, types of vehicles (i.e., small, middle and large size of vehicles), traffic speed, and occupancy. More details of the considered area and the format of data are shown in Fig. 2 [21].

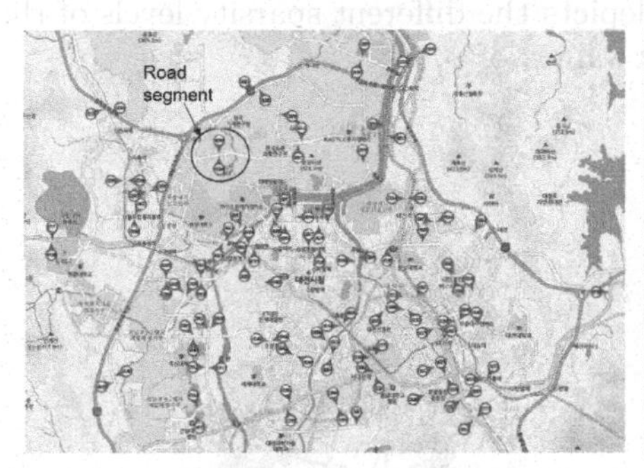

a) Map location of 104 sensors

Variable	Type	Notation
Sensor ID	String(ID)	VDSID
Time	dd/mm/yyyyhh:mm:ss	REG-HM
Traffic Flow	Integer	TR-VOL
Type-Small	Integer	SM-TR-VOL
Type-Middle	Integer	MD-TR-VOL
Type-Large	Integer	LG-TR-VOL
Speed	float	TRVL-SPD
Occupancy	float	OCCUPY-RATE

b) Data format

Fig. 2. VDS map and data format.

Specifically, comparing with previous public traffic datasets, there are several research issues using UVDS dataset for traffic forecasting as follows:

- **Complexity of the Spatial Information**: as shown in Fig. 2(a), 104 sensors span across all the main road of the whole city, which is very difficult to determine and calculate the real distance among VDS sensors. Furthermore, most road segments are asymmetric and the vehicles from different directions are detected by different sensors. Therefore, generating the weight adjacency matrix based on the spatial information and learning the dynamics of spatial dependency are challenging issues.

- **Multivariate Time Series of the Temporal Information**: most previous public traffic datasets are processed with univariate time series for the prediction task. However, traffic flow is often influenced by a number of complex factors. UVDS provides multivariate time series traffic data for further learning on the impact between features. Specifically, multivariate time series analysis using GNN is an emergent issue for traffic forecasting.

3.2 Graph Construction

In this section, preliminary work for representing the graph structure of the UVDS dataset based on the geometric information of VDS sensors is presented. Particularly, given an undirected graph $G(V, E, W)$ for UVDS datasets, where each sensor represents a node (N = 104), a set of edge E and their weight W are computed based on the distances between sensors. In this study, following previous works [11], we define the adjacency matrix by using the thresholded Gaussian Kernel which is formulated as follows:

$$W_{i,j} = \begin{cases} exp(-\frac{d^2_{v_i,v_j}}{\sigma^2}), & \text{if } exp(-\frac{d^2_{v_i,v_j}}{\sigma^2}) \geq \beta \\ 0, & \text{otherwise.} \end{cases} \quad (6)$$

where $\beta(\beta \in [0, 1])$ denotes the threshold for controlling the sparsity of the adjacency matrix. For instance, Fig. 3 depicts the different sparsity levels of the constructed graph by adopting different values of β.

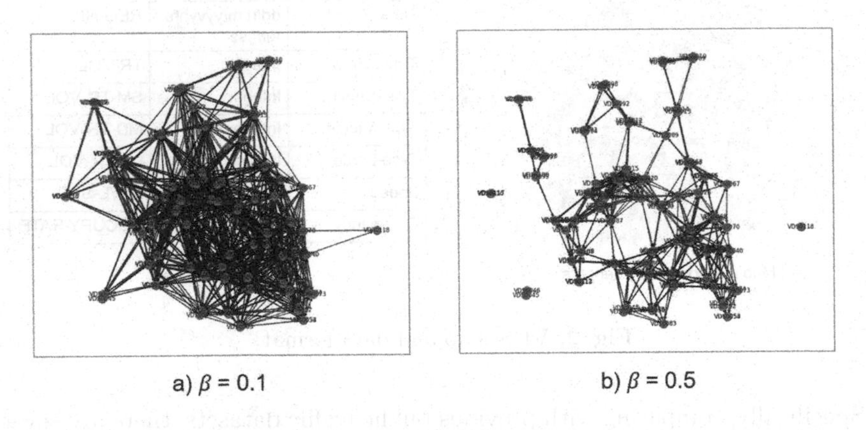

Fig. 3. Graph construction with different threshold values.

Furthermore, d_{v_i,v_j} denotes the road distance between two nodes v_i and v_j. As we mentioned, it is very difficult to calculate the real distance d between two nodes. Therefore, in this study, we calculate the distances using Haversine formula [3], given the longitudes and latitudes of the sensors' locations. Specifically, Haversine formula is an accurate method for computing distances between two points/nodes on the surface of a sphere, especially in terms of small angles and distances. Particularly, the distance between two points is calculated as follows:

$$d_{v_i,v_j} = 2r sin^{-1}\left(\sqrt{sin^2(\frac{\phi_j - \phi_i}{2}) + cos(\phi_i)cos(\phi_j)sin^2(\frac{\varphi_j - \varphi_i}{2})}\right) \quad (7)$$

where ϕ and φ denote the latitude and longitude of two nodes, and r is the radius of the Earth.

4 Baseline Results and Discussion

4.1 Baseline Models

This section describes the baseline model that we use for evaluating the performance on UVDS dataset. Specifically, we adopt four state-of-the-art models that represent different approaches in GNN and use the multi-step traffic speed prediction problem as the benchmark task. In particular, the baseline models are sequentially described as follows:

- **STGCN** [25]: is a spectral model which incorporates ChebNet graph convolution to extract spatial dependencies and 1D convolution for the temporal correlation.
- **Graph Wavenet (GraphWN)** [20]: is a spatial model that integrates diffusion graph convolution and 1D dilated casual convolution.
- **GMAN** [26]: is a graph multi-attention network that uses spatial and temporal attention mechanisms to model the complex Spatio-temporal correlations.
- **MTGNN** [19]: is a recent GNN model to deal with undefined graph structures, where the dependencies are not known in advance, by adding a graph learning layer to capture the hidden relationships among time series data.

4.2 Experimental Results

Regarding the evaluation of the performance, three widely used metrics are used in this study, which are Mean Absolute Error (MAE), Root Mean Squared Error (RMSE), and Mean Absolute Percentage Error (MAPE). Specifically, these metrics are sequentially formulated as follows:

$$\begin{aligned} MAE &= \frac{1}{n}\sum_{i=1}^{n} \mid f_i - \widehat{f}_i \mid \\ MAPE &= \frac{1}{n}\sum_{i=1}^{n} \frac{\mid f_i - \widehat{f}_i \mid}{f_i} \\ RMSE &= \sqrt{\frac{1}{n}\sum_{i=1}^{n} (\mid f_i - \widehat{f}_i \mid)^2} \end{aligned} \quad (8)$$

where f and $\widehat{f}$ are observation and prediction values, respectively. Consequently, the experimental results with multi-step (i.e., 15, 30, 45, and 60 min) of the prediction are shown in Table 2. Accordingly, based on the observation of the experimental results, there are several main points that we can conclude as follows:

Table 2. Evaluated results of baseline models using UVDS dataset

Step	Metric	STGCN	GraphWN	GMAN	MTGNN
15 min	MAE	3.755	**3.642**	3.695	3.648
	RMSE	5.765	5.583	5.692	**5.577**
	MAPE	**7.716%**	7.947%	8.130%	7.992%
30 min	MAE	3.992	3.782	**3.770**	3.774
	RMSE	6.148	5.822	5.860	**5.807**
	MAPE	**8.254%**	8.337%	8.343%	8.342%
45 min	MAE	4.182	3.854	**3.821**	3.845
	RMSE	6.430	5.936	5.955	**5.924**
	MAPE	8.655%	8.542%	**8.477%**	8.548%
60 min	MAE	4.332	3.923	**3.869**	3.927
	RMSE	6.649	6.034	**6.027**	6.039
	MAPE	8.970%	8.707%	**8.613%**	8.804%

- There is no model that can achieve the best results for all cases of the multi-step prediction. For instance, the performance of GMAN model is significantly better than others in terms of long-term prediction by adopting a transform attention mechanism between the encoder and the decoder. However, it takes more time to train the model comparing with other models as shown in Table 3.
- The performance of the STGCN, which is a spectral-based method, is generally lower than spatial-based and attention-based methods. Moreover, the attention-based methods are able to improve the performance of the long-term prediction.

Table 3. The computation time on UVDS dataset

Model	Time consumption	
	Training (s/epoch)	Inference(s)
STGCN	5.234	3.884
GraphWN	26.563	0.789
GMAN	208.750	8.778
MTGNN	25.529	0.784

- The slight difference results between MTGNN and other models indicate that using a graph learning layer as the input for GNN is able to overcome the lacking of the graph structure. In the other perspective, more studies to exploit the graph structure of the UVDS dataset are required to improve the performance of the prediction.

4.3 Open Research Issues Using UVDS Dataset

The experimental results for multi-step traffic prediction using state-of-the-art GNN models have shown promising results. However, there are still open issues for further exploitation in this research field, especially, using the UVDS dataset as follows:

- GNN-based traffic prediction with multivariate time series data: exploiting latent dependencies among variables integrating GNN-based methods is able to improve the performance of the prediction. The current version of the UVDS dataset provides six variables which are traffic volume, speed, types of vehicles (i.e., small, middle, large vehicles), and occupancy for multivariate time series forecasting.
- Graph theory-based road network graph construction: Most state-of-the-art GNN require well-defined graph structures for information propagation. Therefore, building an appropriate adjacency matrix of the road network graph plays an important role. Consequently, using graph theory for representing traffic road networks, especially the UVDS dataset which includes very complex spatial information among VDS sensors, is able to improve significantly the performance of the prediction.

5 Conclusion

Recently, GNN-based methods with the spatial-temporal correlation have made great results for traffic forecasting. However, there are still many open challenges for further investigation in this research field. One of the most important issues is traffic datasets. In this study, we present a new traffic dataset, UVDS, and explain the advantage of the datasets that can support further study in this research field. Specifically, the complexity of spatial information and multivariate time series data of UVDS can bring many research issues for the further exploitation of traffic forecasting using graph representation learning.

For the scalability problem, the size of the dataset should be updated and increased in terms of both spatial (i.e., extending considered areas) and temporal (i.e., increasing the time duration for collecting data) information. Furthermore, replication data is an importance issue in multivariate analyses. Therefore, considering the consensus-based methods for analyzing multivariate time-series dataset are able to enable the further learning on the impact between features [5,14]. There are interesting issues that we take into account regarding the future work of this study.

Acknowledgment. This work was partly supported by Institute for Information & communications Technology Promotion(IITP) grant funded by the Korean Ministry of Science and ICT (MSIT) (No.2018-0-00494, Development of deep learning-based urban traffic congestion prediction and signal control solution system) and Korea Institute of Science and Technology Information(KISTI) grant funded by the Korean Ministry of Science and ICT (MSIT) K-20-L02-C09-S01).

References

1. Bui, K.-H.N., Lee, O.-J., Jung, J.J., Camacho, D.: Dynamic traffic light control system based on process synchronization among connected vehicles. Ambient Intelligence- Software and Applications – 7th International Symposium on Ambient Intelligence (ISAmI 2016). AISC, vol. 476, pp. 77–85. Springer, Cham (2016). https://doi.org/10.1007/978-3-319-40114-0_9
2. Cao, D., et al.: Spectral temporal graph neural network formultivariate time-series forecasting. In: Proceedings of the 33rd Annual Conference on Neural Information Processing Systems (NeurIPS) (2020)
3. Chopde, N.R., Nichat, M.K.: Landmark based shortest path detection by using a* and haversine formula. Int. J. Innov. Res. Comput. Commun. Eng **1**(2), 298–302 (2013)
4. Cui, Z., Henrickson, K., Ke, R., Wang, Y.: Traffic graph convolutional recurrent neural network: a deep learning framework for network-scale traffic learning and forecasting. IEEE Trans. Intell. Transp. Syst (2020). https://doi.org/10.1109/TITS.2019.2950416, (Early Access)
5. Danilowicz, C., Nguyen, N.T.: Consensus-based methods for restoring consistency of replicated data. Intell. Inf. Syst **4**, 325–335 (2000)
6. Diao, Z., et al.: A hybrid model for short-term traffic volume prediction in massive transportation systems. IEEE Trans. Intell. Transp. Syst. **20**(3), 935–946 (2019)
7. Guo, S., Lin, Y., Feng, N., Song, C., Wan, H.: Attention based spatial-temporal graph convolutional networks for traffic flow forecasting. In: Proceedings of the 33rd AAAI Conference on Artificial Intelligence (AAAI), pp. 922–929. AAAI Press (2019)
8. Huang, W., Song, G., Hong, H., Xie, K.: Deep architecture for traffic flow prediction: deep belief networks with multitask learning. IEEE Trans. Intell. Transp. Syst. **15**(5), 2191–2201 (2014)
9. Kipf, T.N., Welling, M.: Semi-supervised classification with graph convolutional networks. In: Proceedings of the 5th International Conference on Learning Representation (ICLR). OpenReview.net (2017)
10. Li, W., et al.: A general framework for unmet demand prediction in on-demand transport services. IEEE Trans. Intell. Transp. Syst. **20**(8), 2820–2830 (2019)
11. Li, Y., Yu, R., Shahabi, C., Liu, Y.: Diffusion convolutional recurrent neural network: Data-driven traffic forecasting. In: Proceedings of the 6th International Conference on Learning Representation (ICLR). OpenReview.net (2018)
12. Lv, Y., Duan, Y., Kang, W., Li, Z., Wang, F.: Traffic flow prediction with big data: a deep learning approach. IEEE Trans. Intell. Transp. Syst. **16**(2), 865–873 (2015)
13. Ma, X., Dai, Z., He, Z., Ma, J., Wang, Y., Wang, Y.: Learning traffic as images: a deep convolutional neural network for large-scale transportation network speed prediction. Sensors **18**(7), 818 (2017)

14. Nguyen, N.T., Sobecki, J.: Using consensus methods to construct adaptive interfaces in multimodal web-based systems. Univers. Access Inf. Soc. **2**(4), 342–358 (2003)
15. Pan, Z., Liang, Y., Wang, W., Yu, Y., Zheng, Y., Zhang, J.: Urban traffic prediction from spatio-temporal data using deep meta learning. In: Proceedings of the 25th International Conference on Knowledge Discovery & Data Mining (KDD), pp. 1720–1730. ACM (2019)
16. Velickovic, P., Cucurull, G., Casanova, A., Romero, A., Liò, P., Bengio, Y.: Graph attention networks. In: Proceedings of the 6th International Conference on Learning Representation (ICLR). OpenReview.net (2018)
17. Williams, B.M., Hoel, L.A.: Modeling and forecasting vehicular traffic flow as a seasonal arima process: theoretical basis and empirical results. J. Transp. Eng. **129**(6), 664–672 (2003)
18. Wu, Z., Pan, S., Chen, F., Long, G., Zhang, C., Yu, P.S.: A comprehensive survey on graph neural networks. IEEE Trans. Neural Netw. Learn. Syst (2020). https://doi.org/10.1109/TNNLS.2020.2978386, (Early Access)
19. Wu, Z., Pan, S., Long, G., Jiang, J., Chang, X., Zhang, C.: Connecting the dots: multivariate time series forecasting with graph neural networks. In: Proceedings of the 26th International Conference on Knowledge Discovery & Data Mining (KDD), pp. 753–763. ACM (2020)
20. Wu, Z., Pan, S., Long, G., Jiang, J., Zhang, C.: Graph wavenet for deep spatial-temporal graph modeling. In: Proceedings of the 28th International Joint Conference on Artificial Intelligence (IJCAI), pp. 1907–1913. ijcai.org (2019)
21. Yi, H., Bui, K.-H.N.: VDS data-based deep learning approach for traffic forecasting using LSTM network. In: Moura Oliveira, P., Novais, P., Reis, L.P. (eds.) EPIA 2019. LNCS (LNAI), vol. 11804, pp. 547–558. Springer, Cham (2019). https://doi.org/10.1007/978-3-030-30241-2_46
22. Yi, H., Bui, K.H.N.: An automated hyperparameter search-based deep learning model for highway traffic prediction. IEEE Trans. Intell. Transp. Syst. (2020). https://doi.org/10.1109/TITS.2020.2987614, (Early Access)
23. Yi, H., Bui, K.N., Jung, H.: Implementing a deep learning framework for short term traffic flow prediction. In: Proceedings of the 9th International Conference on Web Intelligence, Mining and Semantics (WIMS), pp. 7:1–7:8. ACM (2019)
24. Yin, X., Wu, G., Wei, J., Shen, Y., Qi, H., Yin, B.: A comprehensive survey on traffic prediction. CoRR abs/2004.08555 (2020). https://arxiv.org/abs/2004.08555
25. Yu, B., Yin, H., Zhu, Z.: Spatio-temporal graph convolutional networks: a deep learning framework for traffic forecasting. In: Proceedings of the 27th International Joint Conference on Artificial Intelligence (IJCAI), pp. 3634–3640. ijcai.org (2018)
26. Zheng, C., Fan, X., Wang, C., Qi, J.: GMAN: a graph multi-attention network for traffic prediction. In: Proceedings of the 34th AAAI Conference on Artificial Intelligence (AAAI), pp. 1234–1241. AAAI Press (2020)
27. Zhu, L., Yu, F.R., Wang, Y., Ning, B., Tang, T.: Big data analytics in intelligent transportation systems: a survey. IEEE Trans. Intell. Transp. Syst. **20**(1), 383–398 (2019)

A Parallelized Frequent Temporal Pattern Mining Algorithm on a Time Series Database

Nguyen Thanh Vu[1](✉) and Chau Vo[2](✉)

[1] HCMC University of Foreign Language and Information Technology, Ho Chi Minh City, Vietnam
thanhvu1710@huflit.edu.vn
[2] Ho Chi Minh City University of Technology, Vietnam National University–HCMC, Ho Chi Minh City, Vietnam
chauvtn@hcmut.edu.vn

Abstract. For many years, time series data have been considered as one of the most popular significant data forms in our daily life. Time series exist in many application domains such as finance, medicine, geology, meteorology, and telecommunication. Among the time series mining tasks, frequent temporal pattern discovery is an interesting task because this task brings us a deep insight view of relationships between many objects and events through time. However, it is challenging when a combinatorial explosion occurs with many longer time series. It is more challenging if more informative patterns are required from a time series database. In this paper, we propose a parallel algorithm, called PTP, to cope with the frequent temporal pattern mining task. Our PTP is developed with multithreading on a frequent temporal pattern tree where each branch is processed in parallel. In addition, our PTP maintains the details of each frequent temporal pattern not only from its frequent occurrences in an individual time series but also from its frequent inter-time series associations. As a result, frequent temporal patterns discovered by PTP are more informative with explicit and exact temporal information, showing the relationships among the objects/events corresponding to the time series. Through the experimental results on real time series, PTP outperforms the brute force algorithm and the existing non-parallel algorithm in terms of both time and space. The found frequent temporal patterns can be further analyzed for other tasks such as prediction, classification, and clustering.

Keywords: Frequent temporal inter-object pattern · Temporal pattern mining · Time series mining · Parallel temporal pattern tree · Multithreading

1 Introduction

Due to high availability and popularity of time series in many application domains, time series analysis plays an important role in machine learning and statistics when bringing us a deep insight view of objects and their behaviors, which are reflected via the knowledge hidden in their corresponding time series. Many mining tasks have been investigated for knowledge discovery in time series. Among these tasks, frequent pattern mining is a significant one because discovered frequent patterns provide us with descriptive

N. T. Nguyen et al. (Eds.): ACIIDS 2021, LNAI 12672, pp. 78–91, 2021.
https://doi.org/10.1007/978-3-030-73280-6_7

information about underlying time series and their corresponding objects [4, 15, 19]. They also help further examine associations between the time series [20–22, 32] and support other tasks on time series such as prediction via fuzzy association rules [24], classification [25], and trend analysis [26]. Therefore, in this paper, our work is dedicated to frequent temporal pattern mining on time series.

Although frequent itemset mining on transactional databases was very well studied with Apriori [1], FP-Growth [8], and many other variants and similarly, sequential pattern mining on sequence databases was very much investigated, transformation their solutions to those on time series is non-trivial. This is because of the combinatorial explosion problem as discussed in [19]. Even with frequent itemset mining and sequential pattern mining, more efficiency is still a need for improvement. A promising approach to efficiency enhancement is parallel processing. For example, [29] introduced a parallel apriori algorithm for frequent itemset mining, while [13] proposed a parallel FP-growth algorithm. As for the other existing works in [5, 9, 10, 18, 27], different approaches, algorithms, and frameworks were considered for frequent itemset mining. In the meantime, many works were proposed for parallel sequential pattern mining as reviewed in [7]. Also many algorithms, approaches, and frameworks have been defined on sequence databases like those in [11, 12, 28, 31]. As previously mentioned, differences in data characteristics between transactional databases, sequence databases, and time series databases make the existing solutions hardly adapted to resolving the task on the others.

On time series, we beware of just a few related works exploiting parallel processing such as [22, 23] for parallel rule mining in time series. In these works, a parallel algorithm based on a lattice theoretic approach was proposed for association rule mining on multi-stream time series data from human motions on a distributed shared memory multiprocessors system. The experimental results showed that parallel processing could speed up the mining process.

According to the existing works for pattern and rule mining on different database categories, the frequent pattern mining task is a significant one on time series databases which still needs more investigation from two following various perspectives.

First, patterns and rules discovered from time series databases vary work from work. Therefore, its complexity is also different from case study to case study. Sometimes the combinatorial explosion problem needs to be solved accordingly. This becomes a more important issue when not only intra-relationships in each individual time series but also their inter-relationships need to be discovered.

Second, parallel processing for the frequent pattern mining task on time series databases is still in its initial stage. It needs to be taken into account more so that the combinatorial explosion problem can be handled well for finding more informative complex patterns hidden in larger time series databases.

Based on the aforementioned motivations, we propose a parallel tree-based structure and a corresponding frequent pattern mining algorithm for mining frequent temporal patterns in a time series database. Compared to the existing works such as [22, 23], our work is novel in the following points. First, frequent temporal patterns discovered from our work are more informative with detailed relationships between different time series. They are called frequent temporal inter-object patterns. These patterns are expected to be more helpful for decision making support when temporal aspects and more explicit exact information are extracted and maintained along with the found patterns. Second, BranchTree is a novel tree structure that helps manage frequent temporal inter-object patterns in an effective and efficient parallel processing manner. This tree is also nicely defined to save both processing time and memory storage when multithreading is deployed branch by branch and the memory of each branch is freed as soon as its process completes. This enables our resulting parallelized frequent temporal inter-object pattern mining algorithm, called PTP, to discover the required patterns with no full exploitation of BranchTree. Therefore, when many different longer time series are processed and a larger number of frequent temporal patterns are discovered, BranchTree grows very much but these factors have little impact on the processing of our algorithm in terms of both time and memory space.

As a result, our algorithm outperforms the brute force algorithm and another non-parallel one in an empirical study on real time series. When the number of time series increases and longer time series are considered, our algorithm can return all the frequent temporal inter-object patterns in less execution time with less consumed memory as compared to the other algorithms. These experimental results confirm that our parallel algorithm is defined efficiently for the frequent temporal pattern mining task on time series.

The rest of this paper is structured as follows. Section 2 defines our frequent temporal pattern mining task on time series in detail. Section 3 presents our BranchTree and proposes PTP algorithm. In Sect. 4, an empirical evaluation is given for our algorithm. Finally, we conclude our paper and state some future works in Sect. 5.

2 A Frequent Temporal Inter-object Pattern Mining Task

In this section, a temporal inter-object pattern mining task is defined on a time series database. The input of this task is a time series database which is composed of different time series. The output of this task is a complete set of all the frequent temporal inter-object patterns which express the relationships between multiple objects represented by the given different time series. The task is to discover all the frequent temporal inter-object patterns hidden in the given time series database.

Formally, let *DB* be an input time series database and *TS* be one of the time series in *DB*. *TS* is defined as $TS = (v_1, v_2, \ldots, v_n)$. In our work, *TS* is a univariate time series in an *n*-dimension space. $v_1, v_2, \ldots$, and v_n are time-ordered real numbers, corresponding to *n* dimensions. Regarding semantics, time series is understood as the recording of a quantitative characteristic of an object or phenomenon of interest observed regularly over the time. Thus, each time series in the task corresponds to an object of interest which can be some phenomenon or physical object in our real life.

As for the output, we refer to a notion of temporal inter-object pattern as a temporal relationship among various objects being considered. Our work aims to capture more temporal aspects of their relationships so that discovered patterns can be more informative and applicable to decision making support. We formally define a temporal inter-object pattern at level *k* for $k > 1$ in the following form:

"$\mathrm{m_1 - m_1.ID}$ <operator $\mathrm{type_1}$: delta $\mathrm{time_1}$> $\mathrm{m_2 - m_2.ID} \ldots \mathrm{m_{k-1} - m_{k-1}.ID}$ <operator $\mathrm{type_{k-1}}$: delta $\mathrm{time_{k-1}}$> $\mathrm{m_k - m_k.ID}$".

In this pattern form, m_k for $k \geq 1$ are repeating trend-based subsequences whose frequencies are greater than or equal to a pre-specified *minimum support count*. These subsequences can be considered motifs defined in [14]. Each m_k.ID for $k \geq 1$ is an identifier of the time series which m_k belongs to. Each *operator type* is an Allen's temporal operator [2] showing an interval-based relationship among the trend-based subsequences in a single time series or in many different time series. Furthermore, each *delta time* is the time information of each relationship expressed via its corresponding *operator type*. A *temporal inter-object pattern* is said to be *frequent* if and only if its frequency is greater than or equal to the given *minimum support count*.

As described in [16, 17], there are three main phases processed in the task. Phase 1 transforms the given raw time series into trend-based time series. Phase 2 discovers a set of repeating trend-based subsequences, which are regarded as motifs from each time series. Finally, phase 3 discovers all the frequent temporal inter-object patterns.

In our work, we define trend-based time series using short-term and long-term moving averages [4, 30]. In particular, a trend based time series is a symbolic sequence with six trend indicators: A (a weak increasing trend), B (a strong increasing trend), C (starting an increasing trend), D (starting a decreasing trend), E (a strong decreasing trend), and F (a weak decreasing trend). As a result, a time series is represented as a trend-based time series: *TS*' = $(s_1, s_2, \ldots, s_n)$ where $s_i \in \{A, B, C, D, E, F\}$ for $i = 1 \ldots n$.

For example, let NY and SH be two time series corresponding to daily stock prices of two corresponding organizations. A frequent temporal inter-object pattern discovered from these two time series is: AAC – NY <e : 3> AAC – SH {0, 10, 25}, where *AAC* represents weak increasing trends followed by starting a strong increasing trend, *e* represents "equal" to imply the same duration of time, and {0, 10, 25} shows particular points in time in these two time series. Via this pattern, we can state about the relationship between NY and SH as follows. NY has weak increasing trends before starting a strong increasing trend and so does SH in the same duration of 3 time points. Such changes occur frequently three times in their lifetime at points 0, 10, and 25. Based on the discovered patterns, more investigation can be made to confirm the relationships between time series. Further analysis can be done to exploit the patterns.

In [16], a temporal pattern tree was constructed and exploited to gradually derive frequent temporal inter-object patterns in a level-wise manner. Using the temporal pattern tree, a tree-based algorithm was proposed for the task. However, this algorithm faces a great challenge when a combinatorial explosion emerges with many longer time series. The main reason to make this algorithm suffer from the combinatorial explosion problem is that the entire temporal pattern tree maintains in the main memory. The increasing growth of the tree along with a larger set of found frequent temporal inter-object patterns leads to the time consuming and "out of memory" case.

Moreover, none of the existing works on time series mining handles this task. Some works similar to ours are [3, 15, 19–23, 26, 32]. However, their algorithms cannot produce frequent temporal inter-object patterns like those in our work. As for the existing works on sequential pattern mining, different kinds of sequential patterns were supported and discovered. Some examples are given in [7, 11, 12, 31]. Discussed in [19], adapting a solution with sequential pattern mining on sequences to frequent temporal pattern mining on time series is non-trivial due to the combinatorial explosion problem.

Therefore, to tackle this frequent temporal inter-object pattern mining task on a time series database, a parallel approach is appropriate as it has been examined efficiently for frequent itemsets in [5, 9, 10, 18, 27, 29]. In particular, we define a novel parallelized frequent temporal pattern mining algorithm on time series as shown next.

3 The Parallelized Frequent Temporal Inter-object Pattern Mining Algorithm on a Time Series Database

In [16], a tree-based structure was defined for frequent temporal pattern mining on time series. Although efficiently processing with generation of fewer candidates compared to the brute force process, the resulting tree-based algorithm does not use the power of parallel processing, leading to its limited response to many longer time series.

In this paper, we present an updated tree, called BranchTree, and a parallel algorithm, called PTP (Parallelized frequent Temporal inter-object Pattern mining algorithm). To the best of our knowledge, PTP is the first parallel algorithm for the task defined above. As developed in an initial stage, our parallel mechanism is defined with multithreading programming on a multicore computer. This parallel mechanism is embedded in the tree-based structure so that no real entire frequent temporal pattern tree needs to be generated and stored in the main memory. At the same time, parallel processing is designed to perform on each branch of the tree. As a result, both processing time and memory storage can be saved while all the frequent temporal inter-object patterns can be generated with no loss. Its details are presented as follows.

3.1 BranchTree

For a self-contained work, we first introduce Temporal Pattern Tree, called TP-tree, in [16]. TP-tree is a tree with n nodes. Each node corresponds to a component of a frequent temporal inter-object pattern. The node structure is defined below.

- **ParentNode** – a pointer that points to the parent node,
- **OperatorType** – an operator about the temporal relationship between the current node and its parent node,
- **DeltaTime** – an exact time interval associated with the temporal relationship in OperatorType field,
- **ParentLength** – the length of the pattern counting up to the current node,
- **Info** – information about the pattern that the current node represents,
- **ID** – an object identifier of the object which the current node stems from,
- ***k*** – the level of the current node,
- **List of Instances** – a list of all instances from all positions of the pattern that the current node represents,
- **List of ChildNodes** – a hash table that contains pointers pointing to all children nodes of the current node at level $(k + 1)$. Key information of an element in the hash table is: [*OperatorType* associated with a child node + *DeltaTime* + *Info* of a child node + *ID* of a child node].

In addition, we create a BaseHeader structure that stores references to all nodes at each level. This structure helps create all possible combinations of nodes at level (k − 1) to prepare nodes at level k. Nodes at level k associated with support counts greater than or equal to the *minimum support count* are then inserted into TP-tree.

Based on TP-tree, BranchTree is built and operated in parallel processing to discover frequent temporal patterns on each branch.

In particular, BranchTree is built in a combined depth-first-search and parallel approach from level 0 (*root* level) up to level k corresponding to the way we discover frequent patterns at level (k − 1) first and then use them to discover frequent patterns at level k. It is realized that a pattern at level k is only generated from all nodes at level (k − 1) which belong to the same parent node. This feature helps us avoid traversing the entire tree so far built to discover and create frequent patterns at higher levels and expand the rest of the tree. Indeed, from level 3, parallel processing can be conducted branch by branch as shown in Fig. 1. This allows us to start a single thread on each branch to discover frequent patterns from level 3 simultaneously. By doing that, processing time is now saved. On the other hand, due to independence of processing on each branch, as soon as completely processed, a branch can be freed. Therefore, memory space can be saved and our algorithm never explores BranchTree in its entirety. As a result, BranchTree can overcome the combinatorial explosion problem.

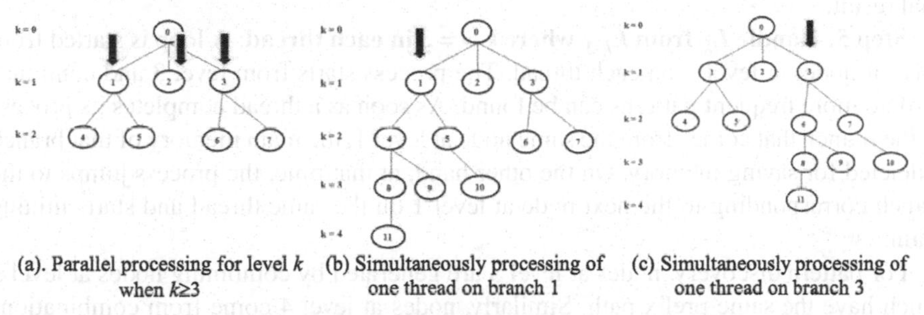
(a). Parallel processing for level k when k≥3 (b) Simultaneously processing of one thread on branch 1 (c) Simultaneously processing of one thread on branch 3

Fig. 1. Parallel processing on BranchTree

3.2 From BranchTree to PTP

Using BranchTree and its parallel processing mechanism, PTP is detailed step by step as follows to build BranchTree and discover frequent temporal inter-object patterns at the same time. The input includes repeating trend-based subsequences (aka motifs).

Step 1 – Initialize BranchTree: Create the root of BranchTree labeled 0 at level 0.

Step 2 – Handle L_1: From the input L_1 which contains m motifs from different trend-based time series with a support count satisfying the *minimum support count*, create m nodes and insert them into BranchTree at level 1. *Delta time*s of these nodes are 0 and no value is given to their Allen's *OperatorType* fields. Furthermore, we create a BaseHeader structure to store references to all nodes at level 1.

Step 3 – Handle L_2 from L_1: Generate all possible combinations between the nodes at level 1 as all nodes at level 1 belong to the same parent node which is the root. This step is performed with seven Allen's temporal operators as follows:

Let m and n be two instances in L_1. With no loss of generality, these two instances are considered for a valid combination if m.StartPosition $\leq$ n.StartPosition where m.StartPosition and n.StartPosition are starting points in time of m and n. A combination process to generate a candidate in C_2 is conducted below. If any combination has a satisfied support count, it is a frequent pattern at level 2 and added into L_2.

- If m and n come from the same object, m must precede n. A combination is in the form of: m-m.ID<*precedes* : *delta*>n-n.ID where m.ID = n.ID.
- If m and n come from two different objects, i.e. m-m.ID $\neq$ n-n.ID, it is considered to generate a combination for the additional six Allen's operators: *meets* (m), *overlaps* (o), *finished by* (f), *contains* (d), *starts* (s), and *equal* (e).

Step 4 – Divide and assign nodes at level 1 to multiple threads: At this step, we begin to use multiple threads for parallel processing to handle L_k for $k \geq 3$. First we divide and assign all nodes at level 1 from BaseHeader to each thread equally for load balancing. At the same time (in parallel), all the threads work on branches in the same manner. After the mining process in each thread finishes, we join all threads and get the final result.

Step 5: Handle L_k from $L_{k\text{-}1}$ where $k >= 3$ in each thread: A loop is started from a list of nodes at level 1 on each thread. The process starts from level 3 and continues until no more frequent patterns can be found. As soon as a thread completes its process on the branch that comes from the same node at level 1, the main memory of that branch is deleted for saving memory. On the other hand, at that time, the process jumps to the branch corresponding to the next node at level 1 on the same thread and starts mining again.

For pattern discovery, nodes at level 3 are generated by combining nodes at level 2 which have the same prefix path. Similarly, nodes at level 4 come from combinations of nodes at level 3 and so on. This can be done efficiently because using information available in BranchTree, we do not need to generate all possible combinations between patterns at level 2 as candidates for patterns at level 3. Instead, we simply traverse BranchTree to generate all combinations from branches sharing the same prefix path one level right before the level being considered. In short, the number of combinations can be greatly reduced and thus, processing time can be decreased. For more efficiency, some other tree-based parallel approaches like FPO Tree and DP3 [10] in which GPU and Cluster are combined or partitioning input data in [18, 29] can be exploited for mining frequent patterns.

4 Empirical Evaluation

In this section, we conduct an empirical study to evaluate our proposed algorithm with the following aspects:

– Question 1: Does our algorithm outperform the existing ones on various datasets?

– Question 2: How stable is our algorithm with different inter-object associations?
– Question 3: How stable is our algorithm with varying *minimum support counts*?

These questions are examined by carrying out the experiments with C# programming language on a 3.0 GHz Intel Core i5 PC with 8.00 GB RAM, using 4 threads.

In these experiments, two groups of real-life stock datasets of the daily closing stock prices from 01/01/2020 [6] are used. The first group, called Technology, contains time series of 5 stocks: S&P 500, FB from Facebook, MSFT from Microsoft, TSLA from Tesla, and TWTR from Twitter. The second one, called Healthcare, includes time series of 5 stocks: S&P 500, ABT from Abbott Laboratories, DXCM from DexCom, HCA from HCA Healthcare, Inc., and JNJ from Johnson & Johnson.

For Question 1, we compare our algorithm (PTP) with the brute-force (BF) and TP-tree-based (TP) ones [16] when finding frequent patterns from all the five datasets with *minimum support count* = 6 and *time series length* varying from 60 to 210. It is noted that *time series length* corresponds to the number of real-world transaction days. Their differences in execution time (second) and main memory used (kilobytes) are recorded as average values from 30 runs in Table 2 and Table 3 for each group. We also report the number of motifs (M#) and the number of found frequent temporal patterns (FTP#) in each group in Table 1.

Table 1. Motif and frequent temporal pattern results from our algorithm on 5 inter-object associations with varying *time series lengths* and *minimum support count* = 6

Length	Healthcare		Technology	
	M#	FTP#	M#	FTP#
60	28	250	37	397
90	93	2,051	99	2,481
120	166	10,193	163	16,889
150	224	38,612	244	112,525
180	282	121,107	284	672,646
210	328	767,744	329	2,499,277

In our experiments, we ran the mining process for each algorithm independently and saved the final results. After that, we checked the number of frequent patterns and contents to make sure that all the results from three algorithms have been the same. In this paper, we do not consider the scalability by increasing the number of cores or CPUs since we do not have multiple computers with the same RAMs and Core version. We will prepare for this kind of experiments in the near future.

From the results in Table 1, changes in the number of motifs and the number of frequent temporal patterns are very large when the *time series length* changes a little. If a brute-force manner is conducted, too many candidates are generated. That would lead to the combinatorial explosion problem mentioned in [19]. A remarkable point is with the change from length = 180 to length = 210. In this case, the increase in M# is just about

15% for both groups. However, unfortunately, the number of frequent temporal patterns increases about 6 times for Healthcare group and about 3.5 times for Technology one. Along with them is the resource that each algorithm needs to handle the task efficiently. This also shows how challenging the task is.

In Tables 2 and 3, there are several cases where N/A is used to mark the unavailability of an algorithm as it cannot successfully run. For our algorithm, there is no N/A in any case because our algorithm can sustain its process with multithreading whenever the TP-tree grows with the increasing number of frequent temporal patterns. By contrast, BF (the brute-force algorithm) and TP (the non-parallel TP-tree-based algorithm) are unable to run with a huge number of frequent temporal patterns. Nevertheless, when the *time series length* is short, e.g. 60, TP performs better ours and BF in terms of both time and space. This is understandable because our algorithm needs to initialize parallel settings before its multithreading process starts. If just a few branches in TP-tree exist, multithreading is exploited a little as compared to the whole process.

Table 2. Time results from 30 executions of three algorithms on 5 inter-object associations with varying *time series lengths* and *minimum support count* = 6

Length	Healthcare					Technology				
	BF	TP	PTP	BF/PTP	TP/PTP	BF	TP	PTP	BF/PTP	TP/PTP
60	0.36	0.17	0.14	2.53	1.24	0.39	0.13	0.14	2.96	0.96
90	3.92	1.24	0.70	5.58	1.76	4.46	1.06	0.87	4.22	1.22
120	38.05	6.59	3.33	11.42	1.98	70.19	6.66	3.97	10.53	1.68
150	N/A	33.33	14.92	N/A	2.23	N/A	42.01	23.53	N/A	1.79
180	N/A	178.33	71.33	N/A	2.50	N/A	476.98	196.71	N/A	2.42
210	N/A	N/A	414.49	N/A	N/A	N/A	N/A	1,548.04	N/A	N/A

Table 3. Memory results from 30 executions of three algorithms on 5 inter-object associations with varying *time series lengths* and *minimum support count* = 6

Length	Healthcare					Technology				
	BF	TP	PTP	BF/PTP	TP/PTP	BF	TP	PTP	BF/PTP	TP/PTP
60	4,381	4,658	4,631	0.95	1.01	4,564	4,583	4,648	0.98	0.99
90	5,547	5,375	5,138	1.08	1.05	6,104	5,219	5,171	1.18	1.01
120	15,589	12,649	6,704	2.33	1.89	25,154	20,401	7,017	3.58	2.91
150	N/A	34,491	11,372	N/A	3.03	N/A	149,016	13,542	N/A	11.00
180	N/A	111,258	21,835	N/A	5.10	N/A	1,194,533	31,401	N/A	38.04
210	N/A	N/A	51,458	N/A	N/A	N/A	N/A	74,392	N/A	N/A

As *time series length* gets increasing, our algorithm outperforms BF and TP. Indeed, its differences in performance from BF and TP become clearer and clearer, showing that it has better efficiency on a consistent basis. Such better results reflect its theoretical design by parallel processing on each branch when it discovers frequent temporal patterns simultaneously and frees the underlying memory once completed, and thus saving time and memory space very much.

In Table 2, it is realized that although 4 threads are used, the speed-up is just about twice better (considered TP/PTP). The reason is that the processing time is not the same for all threads. Some threads take more time because the number of frequent patterns can be found from the branches handled by these threads are greater than the others. Furthermore, in C# language we must wait until all threads finish and join them together before going to the next instruction.

In Table 3, the parallel processing of PTP takes less main memory than the sequential version of TP because in the sequential version, we create the tree using a breadth-wise approach and store the entire tree structure and its nodes in the main memory. By contrast, in PTP, we distribute the mining process into multiple branches, free the main memory used for each branch after the mining process on each branch completes, and then jump into the next one. Therefore, in PTP, we can save more memory space for the remaining mining processes.

For the second question, we examine the results from one execution of our algorithm on varying inter-object associations of each group from one to five time series with *time series length* = 180 and *minimum support count* = 6. The results are presented in Table 4 where time is recorded in second and memory in kilobytes.

Table 4. Results from one execution of our algorithm on varying inter-object associations with *time series length* = 180 and *minimum support count* = 6

TS#	Healthcare					Technology				
	Association	M#	FTP#	Time	Memory	Association	M#	FTP#	Time	Memory
1	S&P500	53	467	0.68	5,737.20	S&P500	53	467	0.68	5,737.20
2	S&P500, ABT	131	6,194	4.60	6,309.68	S&P500, FB	103	4,661	4.04	7,132.76
3	S&P500, ABT, DXCM	185	19,766	18.10	9,113.61	S&P500, FB, MSFT	150	33,379	23.36	10,943.53
4	S&P500, ABT, DXCM, HCA	231	63,650	46.65	15,110.57	S&P500, FB, MSFT, TSLA	209	185,865	76.36	19,614.16
5	S&P500, ABT, DXCM, HCA, JNJ	282	121,107	80.64	21,715.16	S&P500, FB, MSFT, TSLA, TWTR	284	672,646	162.96	31,923.34

The experimental results in Table 4 show that our algorithm can efficiently deal with the complexity of many inter-object associations that increase the number of motifs and the number of frequent temporal patterns very much. Despite such increases on both groups, our algorithm consumes time and space at a slow pace. For example, although for Healthcare group, the number of frequent temporal patterns almost doubles when 4 inter-object associations change to 5 ones and even for Technology group, that number almost triples, the execution time of our algorithm nearly doubles and the used memory of our algorithm does not double. Instead, it increases just about 45% to 65%. This is because our algorithm performs the parallel processing on each branch of the TP-tree and frees the memory as soon as done. Such a process helps the TP-tree cover the discovered patterns while more associations are required with more time series added. Therefore, for Question 2, via illustrations on these two groups, it is said that our algorithm is stable with different inter-object associations.

For Question 3, we keep 5 inter-object associations with *time series length* = 180 while varying *minimum support count* from 6 to 12. The results from one execution of our algorithm are recorded with time (second) and memory (kilobytes) in Table 5.

Table 5. Results from one execution of our algorithm on 5 inter-object associations with *time series length* = 180 and varying *minimum support count* values

Minimum support count	Healthcare				Technology			
	M#	FTP#	Time	Memory	M#	FTP#	Time	Memory
6	282	121,107	80.64	21,715.16	284	672,646	162.96	31,923.34
7	232	24,110	28.97	12,009.95	242	57,207	45.15	15,533.50
8	205	8,312	13.81	8,721.01	210	14,912	19.74	10,123.14
9	169	3,701	9.26	7,321.11	175	6,034	11.40	8,089.31
10	156	1,980	6.50	6,654.44	157	2,984	7.65	7,108.53
11	127	1,140	4.28	6,280.54	130	1,653	5.03	6,546.84
12	105	669	2.93	6,041.54	113	965	3.60	6,212.06

From the results in Table 5, it is worth noting that the larger a *minimum support count*, the fewer number of combinations that need to be checked for frequent temporal patterns. This leads to the less processing time required by our algorithm. Indeed, once a *minimum support count* is set high, a pattern is required to be more frequent, implying that a pattern needs to occur more along the time series. In our real world, the event/action corresponding to the pattern happens more frequently in the lifespan of the corresponding object. Such a requirement makes fewer patterns returned. A consistent result is obtained for both Technology and Healthcare groups. Therefore, Question 3 is answered "Yes" for the stable efficiency of our algorithm.

In summary, our PTP algorithm outperforms the existing ones when discovering frequent temporal inter-object patterns in a time series database in most of the cases, especially where longer time series are considered. With more inter-object associations, our algorithm still runs in a stable manner although the number of temporal patterns gets larger and larger. The same conclusion is drawn with more *minimum support count* values specified for our algorithm. These confirm the theoretical aspects of our PTP algorithm. As a result, our PTP algorithm is an efficient parallel solution to frequent temporal inter-object pattern mining in a time series database.

5 Conclusion

In this paper, we have proposed a parallel temporal pattern tree-based algorithm, called PTP, for frequent temporal pattern mining on a time series database. PTP is novel in the following aspects. Firstly, it constructs a temporal tree of frequent temporal patterns in a level-wise manner, however discovers frequent temporal patterns at the same time in a parallel processing manner. Multithreading is utilized to simultaneously handle each branch of the tree. As a result, it can speed up the pattern mining process and save main memory for more processing when the tree grows. Secondly, it supports frequent temporal inter-object patterns in a time series database with a point-based representation, using Allen's temporal operators. Such frequent temporal patterns are more informative than frequent patterns considered in the existing works in transactional, temporal, sequential, and time series databases. Moreover, these frequent temporal patterns can be further processed to derive temporal association rules which can be revised for prediction rules as time is included. Above all, our PTP algorithm is the first parallel solution to frequent temporal pattern mining on time series.

In the future, we will define the next step for mining temporal association rules from our frequent temporal inter-object patterns and apply them to time series prediction for more interpretability and accuracy. Application development from the discovered patterns and rules is also taken into account in some application domains like healthcare, finance, and e-commerce.

References

1. Agrawal, R., Srikant, R.: Fast algorithms for mining association rules. In: Proceedings of VLDB, pp. 487–499. Springer (1994)
2. Allen, J.F.: Maintaining knowledge about temporal intervals. Commun. ACM **26**, 832–843 (1983)
3. Batal, I., Fradkin, D., Harrison, J., Mörchen, F., Hauskrecht, M.: Mining recent temporal patterns for event detection in multivariate time series data. In: Proceedings of KDD, pp. 280–288. ACM (2012)
4. Batyrshin, I., Sheremetov, L., Herrera-Avelar, R.: Perception based patterns in time series data mining. In: Batyrshin, I., Kacprzyk, J., Sheremetov, L., Zadeh, L.A. (eds.) Perception-based Data Mining and Decision Making in Economics and Finance. SCI, vol. 36, pp. 85–118. Springer, Heidelberg (2007). https://doi.org/10.1007/978-3-540-36247-0_3
5. Djenouri, Y., Djenouri, D., Belhadi, A., Cano, A.: Exploiting GPU and cluster parallelism in single scan frequent itemset mining. Inf. Sci. **496**, 363–377 (2019)
6. Financial time series, http://finance.yahoo.com/, Historical Prices tab, 01/01/2020
7. Gan, W., Lin, J.C.-W., Fournier-Viger, P., Chao, H.-C., Yu, P.S.: A survey of parallel sequential pattern mining. ACM Trans. Knowl. Discov. Data. **13**(3), 25:1–25:34 (2019)
8. Han, J., Pei, J., Yin, Y.: Mining frequent patterns without candidate generation. In: Proceedings of SIGMOD, pp. 1–12. ACM (2000)
9. Huynh, B., Trinh, C., Dang, V., Vo, B.: A parallel method for mining frequent patterns with multiple minimum support thresholds. Int. J. Innov. Comput. Inf. Control **15**(2), 479–488 (2019)
10. Huynh, V.Q.P., Küng, J.: FPO tree and DP3 algorithm for distributed parallel frequent itemsets mining. Expert Syst. Appl. **140**, 1–13 (2020)

11. Kim, B., Yi, G.: Location-based parallel sequential pattern mining algorithm. IEEE Access **7**, 128651–128658 (2019)
12. Le, B., Huynh, U., Dinh, D.-T.: A pure array structure and parallel strategy for high-utility sequential pattern mining. Expert Syst. Appl. **104**, 107–120 (2018)
13. Li, H., Wang, Y., Zhang, D., Zhang, M., Chang, E.: PFP: parallel FP-Growth for query recommendation. In: Proceedings of the 2008 ACM Conference on Recommender Systems, pp. 107–114. ACM (2008)
14. Lin, J., Keogh, E., Lonardi, S., Patel, P.: Mining motifs in massive time series databases. In: Proceedings of ICDM, pp. 370–377. IEEE (2002)
15. Mörchen, F., Ultsch, A.: Efficient mining of understandable patterns from multivariate interval time series. Data Min. Knowl. Disc. **15**, 181–215 (2007). https://doi.org/10.1007/s10618-007-0070-1
16. Vu, N.T., Chau, V.T.N.: Frequent temporal inter-object pattern mining in time series. In: Huynh, V., Denoeux, T., Tran, D., Le, A., Pham, S. (eds.) Knowledge and Systems Engineering. AISC, vol. 244, pp. 161–174. Springer, Cham (2014). https://doi.org/10.1007/978-3-319-02741-8_15
17. Nguyen, T.V., Vo, T.N.C.: An efficient tree-based frequent temporal inter-object pattern mining approach in time series databases. VNU J. Sci. Comput. Sci. Comput. Eng. **31**(1), 1–21 (2015)
18. Özkural, E., Uçar, B., Aykanat, C.: Parallel frequent item set mining with selective item replication. IEEE Trans. Parallel Distrib. Syst. **22**(10), 1632–1640 (2011)
19. Tran, P.T.B., Chau, V.T.N., Anh, D.T.: Towards efficiently mining frequent interval-based sequential patterns in time series databases. In: Bikakis, A., Zheng, X. (eds.) MIWAI 2015. LNCS (LNAI), vol. 9426, pp. 125–136. Springer, Cham (2015). https://doi.org/10.1007/978-3-319-26181-2_12
20. Pradhan, G.N., Prabhakaran, B.: Association rule mining in multiple, multidimensional time series medical data. J. Healthc. Inform. Res. **1**(1), 92–118 (2017). https://doi.org/10.1007/s41666-017-0001-x
21. Qin, L.-X., Shi, Z.-Z.: Efficiently mining association rules from time series. Int. J. Inf. Technol. **2**(4), 30–38 (2006)
22. Sarker, B.K., Mori, T., Hirata, T., Uehara, K.: Parallel algorithms for mining association rules in time series data. In: Guo, M., Yang, L.T. (eds.) ISPA 2003. LNCS, vol. 2745, pp. 273–284. Springer, Heidelberg (2003). https://doi.org/10.1007/3-540-37619-4_28
23. Sarker, B.K., Uehara, K., Yang, L.T.: Exploiting efficient parallelism for mining rules in time series data. In: Yang, L.T., Rana, O.F., Di Martino, B., Dongarra, J. (eds.) HPCC 2005. LNCS, vol. 3726, pp. 845–855. Springer, Heidelberg (2005). https://doi.org/10.1007/11557654_95
24. Srivastava, D.K., Roychoudhury, B., Samalia, H.V.: Fuzzy association rule mining for economic development indicators. Int. J. Intell. Enterp. **6**(1), 3–18 (2019)
25. Ting, J., Fu, T., Chung, F.: Mining of stock data: intra- and inter-stock pattern associative classification. In: Proceedings of ICDM, pp. 30–36. IEEE (2006)
26. Wan, Y., Lau, R.Y.K., Si, Y.-W.: Mining subsequent trend patterns from financial time series. Int. J. Wavelets Multiresolut. Inf. Process. **18**(3), 1–38 (2020)
27. Xun, Y., Zhang, J., Qin, X.: FiDoop: parallel mining of frequent itemsets using MapReduce. IEEE Trans. Syst. Man Cybern. Syst. **46**(3), 313–325 (2016)
28. Yan, D., Qu, W., Guo, G., Wang, X.: PrefixFPM: a parallel framework for general-purpose frequent pattern mining. In: Proceedings of ICDE, pp. 1938–1941. IEEE (2020)
29. Ye, Y., Chiang, C.-C.: A parallel apriori algorithm for frequent itemsets mining. In: Proceedings of the 4th International Conference on Software Engineering Research, Management and Applications, pp. 87–93. IEEE (2006)

30. Yoon, J.P., Luo, Y., Nam, J.: A bitmap approach to trend clustering for prediction in time-series databases. In: Data Mining and Knowledge Discovery: Theory, Tools, and Technology II (2001)
31. Yu, X., Li, Q., Liu, J.: Scalable and parallel sequential pattern mining using spark. World Wide Web **22**(1), 295–324 (2018). https://doi.org/10.1007/s11280-018-0566-1
32. Zhuang, D.E.H., Li, G.C.L., Wong, A.K.C.: Discovery of temporal associations in multivariate time series. IEEE Trans. Knowl. Data Eng. **26**(12), 2969–2982 (2014)

An Efficient Approach for Mining High-Utility Itemsets from Multiple Abstraction Levels

Trinh D. D. Nguyen[1], Loan T. T. Nguyen[2,3](✉), Adrianna Kozierkiewicz[4], Thiet Pham[1], and Bay Vo[5]

[1] Faculty of Information Technology, Industrial University of Ho Chi Minh City, Ho Chi Minh City, Vietnam
20126291.trinh@student.iuh.edu.vn
[2] School of Computer Science and Engineering, International University, Ho Chi Minh City, Vietnam
nttloan@hcmiu.edu.vn
[3] Vietnam National University, Ho Chi Minh City, Vietnam
[4] Faculty of Computer Science and Management, Wroclaw University of Science and Technology, Wrocław, Poland
Adrianna.kozierkiewicz@pwr.edu.pl
[5] Faculty of Information Technology, Ho Chi Minh City University of Technology (HUTECH), Ho Chi Minh City, Vietnam
vd.bay@hutech.edu.vn

Abstract. The goal of the high-utility itemset mining task is to discover combinations of items which that yield high profits from transactional databases. HUIM is a useful tool for retail stores to analyze customer behaviors. However, it ignores the categorization of items. To solve this issue, the ML-HUI Miner algorithm was presented. It combines item taxonomy with the HUIM task and is able to discover insightful itemsets, which are not found in traditional HUIM approaches. Although ML-HUI Miner is efficient in discovering itemsets from multiple abstraction levels, it is a sequential algorithm. Thus, it cannot utilize the powerful multi-core processors, which are currently available widely. This paper addresses this issue by extending the algorithm into a multi-core version, called the MCML-Miner algorithm (Multi-Core Multi-Level high-utility itemset Miner), to help reduce significantly the mining time. Each level in the taxonomy will be assigned a separate processor core to explore concurrently. Experiments on real-world datasets show that the MCML-Miner up to several folds faster than the original algorithm.

Keywords: Hierarchical database · Multi-level database · High-utility itemset · Data mining · Multi-core algorithm

1 Introduction

In the field of knowledge management [1], especially data mining, one of the major topics is pattern mining, which aims to discover patterns that match a specific type of interest. This in turn serves as the input for other knowledge discovery tasks. The

N. T. Nguyen et al. (Eds.): ACIIDS 2021, LNAI 12672, pp. 92–103, 2021.
https://doi.org/10.1007/978-3-030-73280-6_8

problem of Frequent Itemset Mining (FIM), was first proposed by Agrawal et al. in 1993 [2]. FIM is used to analyzing customer behaviors from transactional databases in retail stores. It discovers items that are often purchased together, call Frequent Itemsets (FIs). Several efficient approaches were developed to mine FIs that rely on a powerful property to reduce the search space, called the downward closure property on the support measure. However, FIM has a major drawback, it only returns itemsets that have high occurrence frequency and ignores other important criteria, such as the profit generated by the itemsets. As a result, FIM often returns many itemsets that are frequent but have low profit while leaving out several infrequent itemsets having high profit.

Table 1. A sample transaction database $\mathcal{D}$

TID	Transaction
T_1	(*Pepsi* : 1), (*Pizza* : 2)
T_2	(*Coke* : 2), (*Pizza* : 2)
T_3	(*Coke* : 3), (*Donut* : 2), (*Burger* : 1)
T_4	(*Pepsi* : 2), (*Pizza* : 2), (*Burger* : 1)

Table 2. Unit profit of items in $\mathcal{D}$

Items	*Pepsi*	*Coke*	*Pizza*	*Donut*	*Burger*
Profit	$1	$1	$10	$2	$5

Table 3. Discovered MLHUIs from $\mathcal{D}$ with $\mu = 50$

MLHUI	Utility
{*Snack*, *Drink*}	$82
{*Snack*}	$74
{*Pizza*}	$60

The task of mining high-utility itemsets (HUIM) is then proposed to address this drawback [3]. While FIM considers an item to appear almost once in a transaction, HUIM overcomes this limitation by allowing each item to have a quantity associated with the transaction it appears. Besides, an item also has another associated value, called its unit profit. By taking the product of these two values produces the profit or *utility* generated by an item in a transaction. Table 1 presents a sample transaction database containing 4 transactions. Each of them has a unique identifier called TID and provides information about the quantity of the items purchased. For instance, T_2 denotes that a customer has purchased of two *Cokes* and two *Pizzas*. The unit price of these items is provided in Table 2, whereas *Coke* has the price of $1 per can and $10 for a *Pizza*, etc. Since HUIM is a more generalized problem of FIM, it is also a more challenging task.

It is because the utility measure does not anti-monotonic. And thus, efficient pruning techniques used in FIM cannot be applied in HUIM.

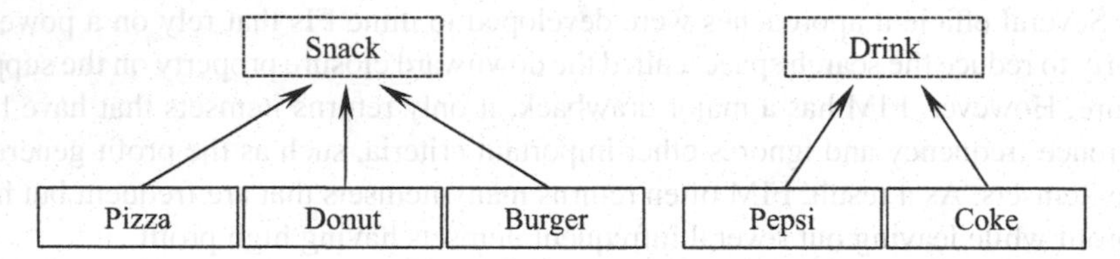

Fig. 1. An example taxonomy of items from a customer transaction database

Furthermore, traditional HUIM algorithms often ignore the categorization of items into a taxonomy. For example, "*Pepsi*" and "*Coke*" can be generalized as "*Soft drinks*", "*Burger*", "*Pizza*", "*Donut*" is the specialized items of the category "*Snack*". Then, "Snack" and "Soft drinks" can be categorized into a higher level of abstraction as "*Fast food*". Although some specialized items may not be a part of HUIs, their generalized items could be a part of some HUIs.

Several approaches have been introduced to solve the FIM problem when considering multiple levels of abstractions of items [4–6] and they all rely on the support measure, which is downward closure satisfied. Unfortunately, this cannot be applied to the utility measure. So far, there are only two known algorithms developed to solve this problem in HUIM. The first one is introduced by Cagliero et al. [7] with the proposing of the ML-HUI Miner algorithm. The algorithm is capable of mining HUIs where the items can be appeared in different abstraction levels, independently. The latter is the CLH-Miner, presented by Fournier-Viger et al. [8], which extends the original problem to mine HUIs where items can be cross-level. However, they are sequential algorithms, which are considered not efficient in the era of multi-core processors. They only use a single core in a multiple cores CPU, leaving all the remaining cores unoccupied, and thus they have a long execution time. To address this issue, this work aims to harness the full power of those widely available processors to reduce the mining time. This can be achieved by applying multi-core parallelism to concurrently exploring the search space to reduce the mining time. Evaluations on real-world datasets show that the speed-up factor of the parallel algorithm is up to several folds better than the sequential algorithm in terms of runtime.

The rest of the paper is organized as follows. Related works are reviewed in Sect. 2. Section 3 outlines some preliminaries and defines the problem of mining HUIs using multiple abstraction levels. Section 4 presents the MCML-Miner algorithm, which adapts multi-core parallelism to perform the multi-level HUIs mining task. Section 5 presents the experiments used to evaluate the performance of the proposed algorithm. Finally, conclusions are drawn in Sect. 6.

2 Related Work

Since the introduction of FIM, many algorithms were presented to efficiently discover the frequent itemsets from customer transactional databases [2, 9, 10]. The downward closure property of the *support* measure was heavily used to reduce the search space. However, a major drawback of FIM is that it ignores the profit yield by the items in transactions. And thus, the output of FIM often contains itemsets that have high appearance frequency but have low profit. The infrequent itemsets with high profit are often ignored by the FIM mining approaches. To address this major drawback, it has been generalized into HUIM to discover itemsets having the high utility [3]. With HUIM, each item is now having two associated values, called its internal utility and external utility. Several approaches have been introduced to efficiently mine HUIs from transactional databases. Some notable state-of-the-art works are Two-Phase [11], UP-Growth [12], HUI-Miner [13], FHM [14], EFIM [15], iMEFIM [16]. Two-Phase, as its name implies, the algorithm requires two phases to complete. The authors introduced the TWU (Transaction Weighted Utilization) as an anti-monotonic upper-bound to reduce the search space of the problem. However, it requires several database scans and the TWU upper-bound is not tight enough to prune the candidates efficiently. UP-Growth, which adopted the pattern growth approach, extended the FP-Tree structure into the UP-Tree structure and used it to mine HUIs. UP-Growth still relies on the TWU to prune the candidates and still a two-phase algorithm. HUI-Miner is then proposed to solve the issues by introducing a tighter upper-bound called the remaining utility, which helps eliminates a large number of unpromising candidates. Furthermore, HUI-Miner also introduced the utility-list structure to reduce the database scans and allow fast utility calculations. HUI-Miner is the first one-phase algorithm to mine HUIs. Extended from the HUI-Miner algorithm, the FHM algorithm introduced the effective pruning strategy named EUCP and the EUCS structured used by the strategy. Later on, the EFIM algorithm introduced two new and much tighter upper-bounds, named local utility and sub-tree utility, to prune more candidates and improve the efficiency. Besides, the authors also proposed two new strategies to reduce the database scan costs and memory consumption, named high-utility transaction merging and high-utility database projection. The iMEFIM algorithm enhances further EFIM by proposing a new framework for calculating dynamic utility values of items, a new P-Set structure to reduce further the cost of database scans.

However, none of the mentioned HUIM approaches consider the taxonomy of item within the transactional databases. The concept of items taxonomy was brought up in the early days of FIM. In 1997, Skirant and Agrawal proposed the Cumulate algorithm [4] to mine association rules from databases using item taxonomy. The algorithm Prutax [5], proposed by Hipp et al., combines item taxonomy with the use of the vertical database format and two new pruning strategies to eliminate infrequent itemsets. It returns the frequent itemsets containing cross-level items. The algorithm MMS_GIT-tree [6], proposed by Vo and Le in 2009, requires only one database scan to discover generalized association rules. When it comes to HUIM, there are only two known algorithms that use the item taxonomy to solve this task, they are ML-HUI Miner [7] and CLH-Miner [8]. ML-HUI Miner introduced the concept of generalized HUIs based on taxonomy and extends the FHM algorithm to perform the MLHUI mining task. While ML-HUI Miner only considers mining for HUIs containing items from the same level, the CLH-Miner algorithm considers cross-level itemsets. CLH-Miner also extends the FHM algorithm. The authors introduced a new upper-bound for generalized HUI called GWU (Generalized-Weighted Utilization). However, CLH-Miner still requires a long execution time to perform the task. A variation of CLH-Miner was also introduced to mine top-k cross-level HUIs [17].

All the mentioned algorithms are designed to work as sequential, and thus they are time-consuming. One of the approaches to boost the performance of the mining process, parallel computing is considered. With the increase in popularity and the availability of the multi-core processors, which allow tasks to be executed simultaneously, several works have presented parallel approaches for FIM, such as pSPADE [18], Par-CSP [19], Par-ClosP [20]. On the HUIM side, approaches are pEFIM [21], MCH-Miner [22], and pHAUI-Miner [23]. On the distributed systems, Nguyen has presented several approaches to solve the conflicts in distributed systems [24, 25]. Chen and An has introduced a distributed version of the HUI-Miner algorithm, named PHUI-Miner [26], Sethi et al. has presented the P-FHM+ [27], which is an extended version of the FHM+ using distributed computing.

This work aims to extend the ML-HUI Miner into a multi-core version to utilize as many as possible available processor cores to reduce the execution time of the MLHUIs mining process.

3 Preliminaries

This section presents several key definitions and formulates the task of mining multi-level HUIs from transactional databases enriched with taxonomy data.

Let there be a finite set of m distinct items, denoted $\mathcal{I}$. A transaction database, denoted $\mathcal{D} = \{T_1, T_2, \ldots, T_n\}$, is a multiset of transaction T_q. In which, each transaction T_q has a unique identifier called TID and $T_q \subseteq \mathcal{I}$. Each item $i \in T_q$ has an associated positive value called its internal utility in transaction T_q, denoted $ui(i, T_q)$. Also, for each item $i \in \mathcal{I}$, it has an associated positive value called its external utility, denoted $ue(i)$. The utility value of an item i in transaction T_q, denoted $u(i, T_q)$, is defined as the product of its external utility and internal utility in transaction T_q, $ue(i) \times ui(i, T_q)$. The utility of an itemset X (a set of items of arbitrary size, $X \subseteq \mathcal{I}$) in transaction T_q, denoted as $u(X, T_q)$, is defined as $u(X, T_q) = \sum_{i \in X} u(i, T_q)$. The utility value of an itemset X within the database $\mathcal{D}$, denoted $u(X)$, is defined as $u(X) = \sum_{T_i \in \mathcal{T}(X)} u(X, T_i)$, whereas $\mathcal{T}(X)$ is the set of all transactions containing itemset X.

Let τ be a taxonomy, is a tree defined on the database $\mathcal{D}$. In which, every leaf node in τ is an item $i \in \mathcal{I}$. Each inner node in τ aggregates all its descendant nodes or descendant categories into an abstract category at a higher level of abstraction. The set of all generalized items in τ is denoted as $\mathcal{G}$. It is assumed that each item $i \in \mathcal{I}$ in the taxonomy τ is generalized into one and only one specific generalized item $g \in \mathcal{G}$. Let g. be a generalized item in the taxonomy τ, $g \in \mathcal{G}$, the set of all descendant nodes of g is a subset of $\mathcal{I}$, denoted as $Desc(g, \tau) \subseteq \mathcal{I}$. Figure 1 shows an example of taxonomy constructed from the sample database given in Table 1. In this example, *Pizza*, *Donut* and *Burger* are categorized into a generalized item named *Snack*. Whereas, *Snack* has the level of 1 and the level of *Pizza*, *Donut*, *Burger* has the level of 0.

Let g be a generalized item in database $\mathcal{D}$ using taxonomy τ. The utility of g in transaction T_q is defined as $u(g, T_q) = \sum_{i \in Desc(g,\tau)} ue(i) \times ui(i, T_q)$.The utility of a generalized itemset GX in T_q is defined as $u(GX, T_q) = \sum_{i \in GX} u(i, T_q)$. The utility of a generalized itemset GX in database $\mathcal{D}$ is defined as $u(GX) = \sum_{T_q \in G(GX)} u\left(GX, T_q\right)$ whereas $G(GX) = \left\{T_q \in \mathcal{D} | \exists X \subset T_q \wedge X \textit{ is a descendant of } GX\right\}$. Let $\mathcal{D}$ be a transaction database using taxonomy τ and a user-specified minimum utility threshold μ. A (generalized) itemset X is called a multi-level high-utility itemset (MLHUI) if and only if $u(X) \geq \mu$. The task of mining multi-level high-utility itemsets is to return the complete set of all HUIs and MLHUIs. The utility value of a generalized itemset aggregates those of all its descendant itemsets. Thus, the higher the abstraction level an itemset are, the more likely it satisfies the μ threshold than lower-level ones. To avoid this issue, every level of abstraction of a given itemset has its own utility threshold. Itemsets including items at a higher level of abstraction will have a higher utility threshold and all itemsets in the same level have the same utility threshold. The utility threshold of level l of the generalized items, denoted as $\gamma(l)$, is μ if $l = 0$, otherwise $\gamma(l) = l \times \mu$. $\gamma(l)$ is called per-function threshold level. Consider the example transactional database $\mathcal{D}$ given in Table 1, Table 2, taxonomy τ in Fig. 1, when the μ threshold was set to 50, the set of all MLHUIs satisfy μ are shown in Table 3.

4 Proposed Algorithm

This section presents an approach to improve the MLHUI mining performance by extending the ML-HUI Miner using multi-core parallelism architecture. The proposed algorithm is named MCML-Miner (Multi-Core Multi-Level high-utility itemset Miner). The MCML-Miner algorithm inherits all the characteristics of the original algorithm ML-HUI Miner. Furthermore, MCML-Miner applies multi-core parallel processing to harness the power of the multicore processors, which are widely available, to efficiently mining all MLHUIs simultaneously. The features of the MCML-Miner can be summarized as follows:

- *Taxonomy-based algorithm:* Mining multi-level high-utility itemsets from a transaction database using enhanced with taxonomy data.
- *Single-phase algorithm*: The algorithm extracts all MLHUIs in just one phase.
- *Avoiding the generation of uninteresting itemsets*: The algorithm focuses on mining the itemsets containing only items at the same abstraction level.
- *Multi-core algorithm*: The algorithm performs the search space exploration in parallel to reduce the overall mining time.

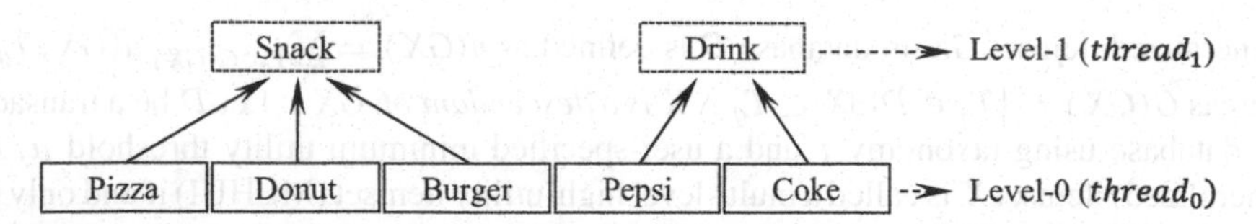

Fig. 2. The parallelized search space of the MCML-Miner algorithm.

Algorithm 1. The MCML-Miner algorithm

Input: $\mathcal{D}$ - transaction database; τ - taxonomy; μ threshold;
$\gamma(l)$ - per-level threshold function
Output: $\mathcal{R}$ - the set of all discovered MLHUIs
1: $\mathcal{R} \leftarrow \emptyset$
2: Constructs the set $\mathcal{I}$ and $\mathcal{G}$ from $\mathcal{D}$ and taxonomy τ.
3: Scans $\mathcal{D}$ using τ to calculate TWU values for $i \in \mathcal{I}$.
4: Calculates TWU values for generalized item $g \in \mathcal{G}$.
5: $\mathcal{I}^+ \leftarrow \{i \in \mathcal{I} \mid TWU(i) \geq \mu\}$
6: $\mathcal{G}^+ \leftarrow \{g \in \mathcal{G} \mid TWU(g) \geq \gamma(l) \times \mu\}$
7: Constructs the utility-list and EUCS for items in $\mathcal{I}^+ \cup \mathcal{G}^+$.
8: **FOREACH** level l in τ **DO PARALLEL**
9: $\mathcal{GI}_l^+ \leftarrow$ Items in $\mathcal{I}^+ \cup \mathcal{G}^+$ on level l.
10: $\mathcal{R}_l \leftarrow$ FHM-Search($\emptyset, \mathcal{GI}_l^+, \gamma(l) \times \mu$, *EUCS*)
11: **END FOR**
12: **RETURN** $\mathcal{R} \leftarrow \bigcup_{l \in \tau} \mathcal{R}_l$

To perform the mining task in parallel, we apply the multi-threading strategy to the algorithm. As mentioned, the search space of the problem is explored using level wise and no cross-level itemsets are be considered, thus the whole search space can be safely divided into separated sub-spaces and each sub-space is then explored in parallel by a working thread assigned, as illustrated in Fig. 2. The more levels the taxonomy τ has, the more threads will be allocated to perform the task and thus offer much better speed up to the algorithm. The only drawback of this approach is that if the given transactional database has no given taxonomy, the problem becomes the traditional HUIM problem with only the most specialized items. Then the algorithm only allocated only one thread to perform the mining task and thus completely disable the parallelism.

During the MLHUI mining process, TWU is used to prune the unpromising candidates, utility-list and EUCS are used to speed up the utility calculations. The MCML-Miner algorithm explores each sub-space in a mechanism similar to the FHM.

The pseudo-code of the MCML-Miner is presented in Algorithm 1. The algorithm first initializes the set $\mathcal{R}$ as an empty set at line #1. $\mathcal{R}$ contains all the discovered MLHUIs. Then at line #2, using the given taxonomy τ and $\mathcal{D}$, the algorithm constructs the set of all specialized items $\mathcal{I}$ and the set of all generalized items $\mathcal{G}$. Line #3 calculates the TWU values for all items in $\mathcal{I}$ by scanning through $\mathcal{D}$. Based on the computed TWU values, line #4 calculates the TWU values for all items in $\mathcal{G}$. These TWU values are then used to prune all items from $\mathcal{I}$ that do not satisfy the μ threshold at level 0 to form the set $\mathcal{I}^+$, which is done in line #5. The same is happened on line #6 to construct the set $\mathcal{G}^+$ based on the per-level function $\gamma(l)$ and the μ threshold. Line #7 constructs the utility-lists and the EUCS structure for the items found in $\mathcal{I}^+ \cup \mathcal{G}^+$ to speed up the candidates checking process.

The major difference between the proposed MCML-Miner algorithm and the original ML-HUI Miner algorithm comes from lines #8 to #12. Line #8 starts by allocating and assigning a working thread for each level l in the taxonomy τ. Line #9 extracts from $\mathcal{I}^+ \cup \mathcal{G}^+$ the items that exist at level l. Then at line #10, each allocated thread performs its job by exploring the assigned sub-space recursively using the search function that comes from the FHM algorithm, *FHM-Search* [14] (Algorithm 2). It achieves the goal by combining items i from the set $\mathcal{I}^+$ with the generalized items g from the set $\mathcal{G}^+$ in the same level and satisfy the per-level threshold specified via the function $\gamma(l)$ and the μ threshold. The discovered MLHUIs at level l are stored in the subset $\mathcal{R}_l$. Finally, line #12 aggregates all the subset $\mathcal{R}_l$ into the final output set $\mathcal{R}$ and returns it to the user.

Algorithm 2. The FHM-Search algorithm

```
Input: P - an itemset; ExtP - set of extensions of P;
       μ threshold; EUCS - per-level threshold function
Output: O - the set of MLHUIs
 1: FOR EACH X ∈ ExtP DO
 2:    IF SUM(X.UL.iutils) ≥ μ THEN O ← O ∪ X.
 3:    IF SUM(X.UL.ituils) + SUM(X.UL.rutils) ≥ μ THEN
 4:       ExtX ← ∅
 5:       FOR EACH Y ∈ ExtP AND Y ≻ X DO
 6:          IF EXIST (x,y,c) ∈ EUCS AND c ≥ μ THEN
 7:          XY ← X ∪ Y
 8:          Construct utility-list of XY using utility-list
             of X and utility-list of Y.
 9:          ExtX ← ExtX ∪ XY.
10:       END IF
11:       END FOR
11:    R_l ← FHM-Search(∅, GI_l^+, γ(l) × μ, EUCS)
12:    END IF
13: END FOR
14: RETURN O
```

5 Evaluation Studies

To have a clearer view of how the proposed algorithm MCML-Miner performs, this section evaluates the performance of the MCML-Miner in terms of run time and peak memory usage. The experiments were conducted on a workstation equipped with an Intel® Xeon® E5-2678v3 (12 cores) clocked at 2.5 GHz; using 32 GB RAM and running Windows 10 Pro Workstation. The performance of MCML-Miner is tested against the original ML-HUI Miner algorithm to mine multi-level HUIs. All the tested algorithms were implemented using the Java programming language. The measured running time and peak memory usage were obtained by using standard Java API. Two real-world datasets were used to evaluate the performance of the algorithms, namely *Fruithut* and *Foodmart*. Their characteristics are presented in Table 4. In which, $|\mathcal{D}|$ denotes the total transactions contained in each dataset, $|\mathcal{I}|$ denotes the number of distinct items, $|GI|$ denotes the number of generalized items, *Depth* represents the number of levels in each dataset's taxonomy, $Trans_{MAX}$ and $Trans_{AVG}$ denote the maximum transaction length and average transaction length, respectively.

The number of threads used by the MCML-Miner was automatically determined by the maximum number of levels from each dataset's taxonomy, which were given in the *Depth* column of Table 4. E.g., for the *Fruithut* dataset, 4 threads were used and for the *Foodmart* dataset, the algorithm allocated 5 threads to perform the mining task.

Table 4. Dataset characteristics

Dataset	$\|\mathcal{D}\|$	$\|\mathcal{I}\|$	$\|GI\|$	*Depth*	$Trans_{MAX}$	$Trans_{AVG}$
Fruithut	181,970	1,265	43	4	36	3.58
Foodmart	53,537	1,560	102	5	28	4.6

To evaluate the performance of the MCML-Miner, the μ threshold was varied from $1K$ to $5K$ on both datasets. The results are shown in Fig. 3. It can be seen from the results, MCML-Miner has a much better execution time than the original algorithm ML-HUI Miner. In the *Foodmart* dataset (Fig. 3a), as the μ threshold decreased, the runtime of the two algorithms increased. By using 5 threads, the average speed-up factor for the MCML-Miner is up to 5.6 times compared to the ML-HUI Miner throughout all tested μ thresholds. The overall speed-up factor of MCML-Miner against ML-HUI Miner is shown in Fig. 4. For instance, at $\mu = 1K$, MCML-Miner was only required 16.05 s to complete while the ML-HUI Miner took 89.99 s to return the complete set of MLHUIs, which is 5 times faster. And at $\mu = 5K$, MCML-Miner finished in 1.73 s and ML-HUI Miner finished in 3.05 s, which is also offered the speed-up factor of 1.8. Despite of being a small dataset, *Foodmart* contains more generalized items and has the average transaction length longer than *Fruithut*. As a result, when the μ threshold decreased, the number of discovered MLHUIs in this dataset is rapidly increased, thus increase the mining time. For instance, at $\mu = 5K$, the number of MLHUIs found is only over $33K$, but at $\mu = 1K$, it is roughly $91.4M$. The MCML-Miner deals with this effectively by performing the mining task parallely at each taxonomy level. For the *Fruithut* dataset (Fig. 3b), which is a large dataset with over $180K$ transactions with 4-level taxonomy, 4 threads were allocated for the MCML-Miner algorithm. The average speed-up factor for the MCML-Miner is 2 times only. Throughout the μ thresholds, the MCML-Miner took average 5.61 s to complete, while the ML-HUI Miner required average 10.40 s to complete. This can be explained by the short average transaction length, low number of generalized item and low taxonomy level. It is also showed that MCML-Miner has higher effectiveness when applied to datasets containing a large number of generalized items, high number of levels in taxonomy and long transactions, such as those like *Foodmart*.

The peak memory usage of the two test algorithms is presented in Fig. 5. It can be seen that the MCML-Miner has slightly higher memory consumption than the original

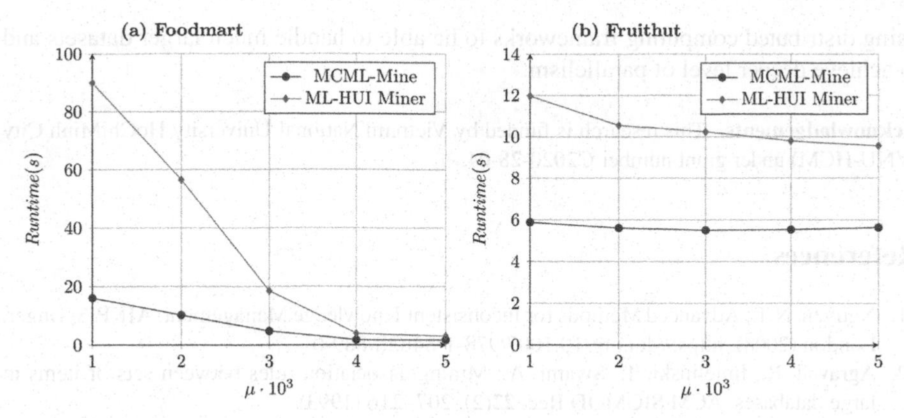

Fig. 3. Runtime comparison

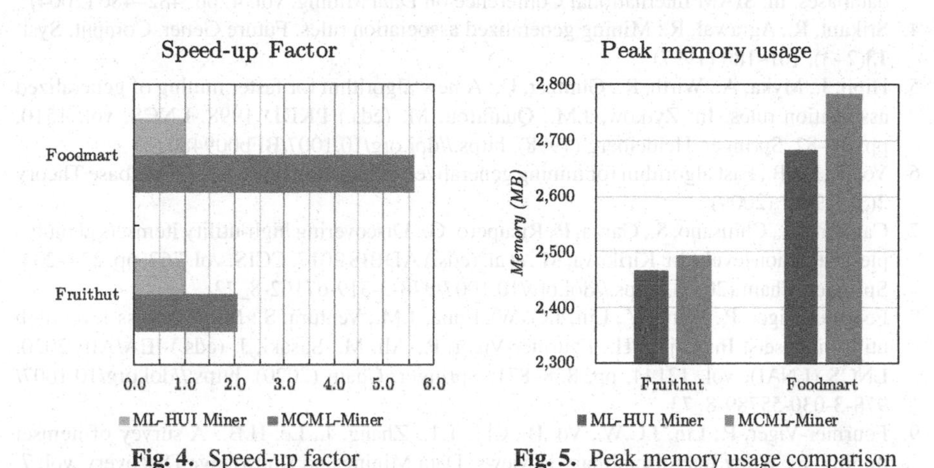

Fig. 4. Speed-up factor

Fig. 5. Peak memory usage comparison

ML-HUI Miner algorithm, 1.011 times and 1.038 times for the *Fruithut* and *Foodmart* dataset, respectively. MCML-Miner has the peak memory usage higher than ML-HUI Miner since it is a parallelized algorithm and thus, it requires more resources to maintain concurrently executed threads. However, this is considered acceptable since most of the current computers have the minimum memory amount installed at 4 GB.

6 Conclusion and Future Work

This work presented an efficient approach to mine multi-level high-utility itemsets. It applied parallel processing to extend the state-of-the-art algorithm ML-HUI Miner for mining MLHUIs. The proposed method in this work is named the MCML-Miner algorithm. It makes use of the multicore processors, which are available widely, to boost the performance of the MLHUI mining process. Evaluation studies have shown that MCML-Miner outperformed the original ML-HUI Miner up to 5.6 times and consumed slightly higher memory than its original version. In the future, we will expand the task using distributed computing frameworks to be able to handle much larger datasets and to achieve deeper level of parallelism.

Acknowledgements. This research is funded by Vietnam National University HoChiMinh City (VNU-HCM) under grant number C2020-28-04.

References

1. Nguyen, N.T.: Advanced Methods for Inconsistent Knowledge Management. AIKP. Springer, London (2008). https://doi.org/10.1007/978-1-84628-889-0
2. Agrawal, R., Imieliński, T., Swami, A.: Mining association rules between sets of items in large databases. ACM SIGMOD Rec. **22**(2), 207–216 (1993)
3. Yao, H., Hamilton, H.J., Butz, G.J.: A foundational approach to mining itemset utilities from databases. In: SIAM International Conference on Data Mining, vol. 4, pp. 482–486 (2004)
4. Srikant, R., Agrawal, R.: Mining generalized association rules. Future Gener. Comput. Syst. **13**(2–3), 161–180 (1997)
5. Hipp, J., Myka, A., Wirth, R., Güntzer, U.: A new algorithm for faster mining of generalized association rules. In: Żytkow, J.M., Quafafou, M. (eds.) PKDD 1998. LNCS, vol. 1510, pp. 74–82. Springer, Heidelberg (1998). https://doi.org/10.1007/BFb0094807
6. Vo, B., Le, B.: Fast algorithm for mining generalized association rules. Int. J. Database Theory **2**(3), 19–21 (2009)
7. Cagliero, L., Chiusano, S., Garza, P., Ricupero, G.: Discovering high-utility itemsets at multiple abstraction levels. In: Kirikova, M., et al. (eds.) ADBIS 2017. CCIS, vol. 767, pp. 224–234. Springer, Cham (2017). https://doi.org/10.1007/978-3-319-67162-8_22
8. Fournier-Viger, P., Wang, Y., Lin, J.C.-W., Luna, J.M., Ventura, S.: Mining cross-level high utility itemsets. In: Fujita, H., Fournier-Viger, P., Ali, M., Sasaki, J. (eds.) IEA/AIE 2020. LNCS (LNAI), vol. 12144, pp. 858–871. Springer, Cham (2020). https://doi.org/10.1007/978-3-030-55789-8_73
9. Fournier-Viger, P., Lin, J.C.W., Vo, B., Chi, T.T., Zhang, J., Le, H.B.: A survey of itemset mining. In: Wiley Interdisciplinary Reviews: Data Mining and Knowledge Discovery, vol. 7, no. 4 (2017)
10. Han, J., Pei, J., Yin, Y., Mao, R.: Mining frequent patterns without candidate generation: a frequent-pattern tree approach. Data Min. Knowl. Disc. **8**(1), 53–87 (2004). https://doi.org/10.1023/B:DAMI.0000005258.31418.83
11. Liu, Y., Liao, W.-k., Choudhary, A.: A two-phase algorithm for fast discovery of high utility itemsets. In: Ho, T.B., Cheung, D., Liu, H. (eds.) PAKDD 2005. LNCS (LNAI), vol. 3518, pp. 689–695. Springer, Heidelberg (2005). https://doi.org/10.1007/11430919_79
12. Tseng, V.S., Wu, C.W., Shie, B.E., Yu, P.S.: UP-Growth: an efficient algorithm for high utility itemset mining. In: ACM SIGKDD International Conference on Knowledge Discovery and Data Mining, pp. 253–262 (2010)
13. Liu, M., Qu, J.: Mining high utility itemsets without candidate generation. In: 21st ACM International Conference on Information and Knowledge Management, pp. 55–64 (2012)
14. Fournier-Viger, P., Wu, C.-W., Zida, S., Tseng, V.S.: FHM: faster high-utility itemset mining using estimated utility co-occurrence pruning. In: Andreasen, T., Christiansen, H., Cubero, J.-C., Raś, Z.W. (eds.) ISMIS 2014. LNCS (LNAI), vol. 8502, pp. 83–92. Springer, Cham (2014). https://doi.org/10.1007/978-3-319-08326-1_9
15. Zida, S., Fournier-Viger, P., Lin, J.C.-W., Wu, C.-W., Tseng, V.S.: EFIM: a fast and memory efficient algorithm for high-utility itemset mining. Knowl. Inf. Syst. **51**(2), 595–625 (2016). https://doi.org/10.1007/s10115-016-0986-0

16. Nguyen, L.T.T., Nguyen, P., Nguyen, T.D.D., Vo, B., Fournier-Viger, P., Tseng, V.S.: Mining high-utility itemsets in dynamic profit databases. Knowl.-Based Syst. **175**, 130–144 (2019)
17. Nouioua, M., Wang, Y., Fournier-Viger, P., Lin, J.C.-W., Wu, J.M.-T.: TKC: mining top-k cross-level high utility itemsets. In: 3rd International Workshop on Utility-Driven Mining (UDML 2020) (2020)
18. Alias, S., Norwawi, N.M.: pSPADE: mining sequential pattern using personalized support threshold value. In: Proceedings - International Symposium on Information Technology 2008, ITSim, vol. 2, pp. 1–8 (2008)
19. Cong, S., Han, J., Padua, D.: Parallel mining of closed sequential patterns. In: Proceedings of the ACM SIGKDD International Conference on Knowledge Discovery and Data Mining, pp. 562–567 (2005)
20. Zhu, T., Bai, S.: A parallel mining algorithm for closed sequential patterns. In: Proceedings - 21st International Conference on Advanced Information Networking and Applications Workshops/Symposia, AINAW 2007, vol. 2, pp. 392–395 (2007)
21. Nguyen, T.D.D., Nguyen, L.T.T., Vo, B.: A parallel algorithm for mining high utility itemsets. In: Świątek, J., Borzemski, L., Wilimowska, Z. (eds.) ISAT 2018. AISC, vol. 853, pp. 286–295. Springer, Cham (2019). https://doi.org/10.1007/978-3-319-99996-8_26
22. Vo, B., Nguyen, L.T.T., Nguyen, T.D.D., Fournier-Viger, P., Yun, U.: A multi-core approach to efficiently mining high-utility itemsets in dynamic profit databases. IEEE Access **8**, 85890–85899 (2020)
23. Nguyen, L.T.T., et al.: Efficient method for mining high-utility itemsets using high-average utility measure. In: Nguyen, N.T., Hoang, B.H., Huynh, C.P., Hwang, D., Trawiński, B., Vossen, G. (eds.) ICCCI 2020. LNCS (LNAI), vol. 12496, pp. 305–315. Springer, Cham (2020). https://doi.org/10.1007/978-3-030-63007-2_24
24. Nguyen, N.T.: Using consensus methods for solving conflicts of data in distributed systems. In: Hlaváč, V., Jeffery, K.G., Wiedermann, J. (eds.) SOFSEM 2000. LNCS, vol. 1963, pp. 411–419. Springer, Heidelberg (2000). https://doi.org/10.1007/3-540-44411-4_30
25. Nguyen, N.T.: Consensus system for solving conflicts in distributed systems. Inf. Sci. **147**(1), 91–122 (2002)
26. Chen, Y., An, A.: Approximate parallel high utility itemset mining. Big Data Res. **6**, 26–42 (2016)
27. Sethi, K.K., Ramesh, D., Edla, D.R.: P-FHM+: parallel high utility itemset mining algorithm for big data processing. Procedia Comput. Sci. **132**, 918–927 (2018)

Mining Class Association Rules on Dataset with Missing Data

Hoang-Lam Nguyen[1,2], Loan T. T. Nguyen[1,2](✉), and Adrianna Kozierkiewicz[3]

[1] School of Computer Science and Engineering, International University, Ho Chi Minh City, Vietnam
ITITIU16038@student.hcmiu.edu.vn, nttloan@hcmiu.edu.vn

[2] Vietnam National University, Ho Chi Minh City, Vietnam

[3] Faculty of Computer Science and Management, Wroclaw University of Science and Technology, Wrocław, Poland
Adrianna.kozierkiewicz@pwr.edu.pl

Abstract. Many real-world datasets contain missing values, affecting the efficiency of many classification algorithms. However, this is an unavoidable error due to many reasons such as network problems, physical devices, etc. Some classification algorithms cannot work properly with incomplete dataset. Therefore, it is crucial to handle missing values. Imputation methods have been proven to be effective in handling missing data, thus, significantly improve classification accuracy. There are two types of imputation methods. Both have their pros and cons. Single imputation can lead to low accuracy while multiple imputation is time-consuming. One high-accuracy algorithm proposed in this paper is called Classification based on Association Rules (CARs). Classification based on CARs has been proven to yield higher accuracy compared to others. However, there is no investigation on how to mine CARs with incomplete datasets. The goal of this work is to develop an effective imputation method for mining CARs on incomplete datasets. To show the impact of each imputation method, two cases of imputation will be applied and compared in experiments.

Keywords: Missing value · Class association rules · Incomplete instance · Imputation method

1 Introduction

In the field of knowledge management, the task of machine learning and data mining often the same techniques to achieve their goal and thus, they share many common aspects [1]. Machine learning focuses on prediction using known properties obtained from training data, data mining focuses on the discovery of unknown properties in data (or often referred as knowledge discovery from databases). Data mining uses several machine learning algorithms and also, machine learning uses many data mining techniques, mostly in its preprocessing steps to improve the learner accuracy. Classification is one of the most important tasks in the field of machine learning and data mining. Two main processes of classification are training and application. In the training process, a

N. T. Nguyen et al. (Eds.): ACIIDS 2021, LNAI 12672, pp. 104–116, 2021.
https://doi.org/10.1007/978-3-030-73280-6_9

classifier is built and will be used later in the application (or test) process. In reality, there are several applications for classification, such as face [1] or fingerprint recognition [2], movie rating [3], healthcare [4], etc. Among classification algorithms, CAR results in higher accuracy compared with others. However, the main issue with CAR is that CAR can only be used effectively on complete datasets.

Unfortunately, numerous real-world datasets contain missing value. The majority of datasets in the UCI machine learning repository are incomplete. Industrial datasets might contain missing values as a result of a machine malfunction during the data collection process. In social surveys, data collection is often insufficient because respondents might refuse to answer personal questions. In the field of medical, the data can be missing since not all patients did all the given tests. Researchers cannot always collect data due to undesired conditions (for example, unsatisfactory weather conditions).

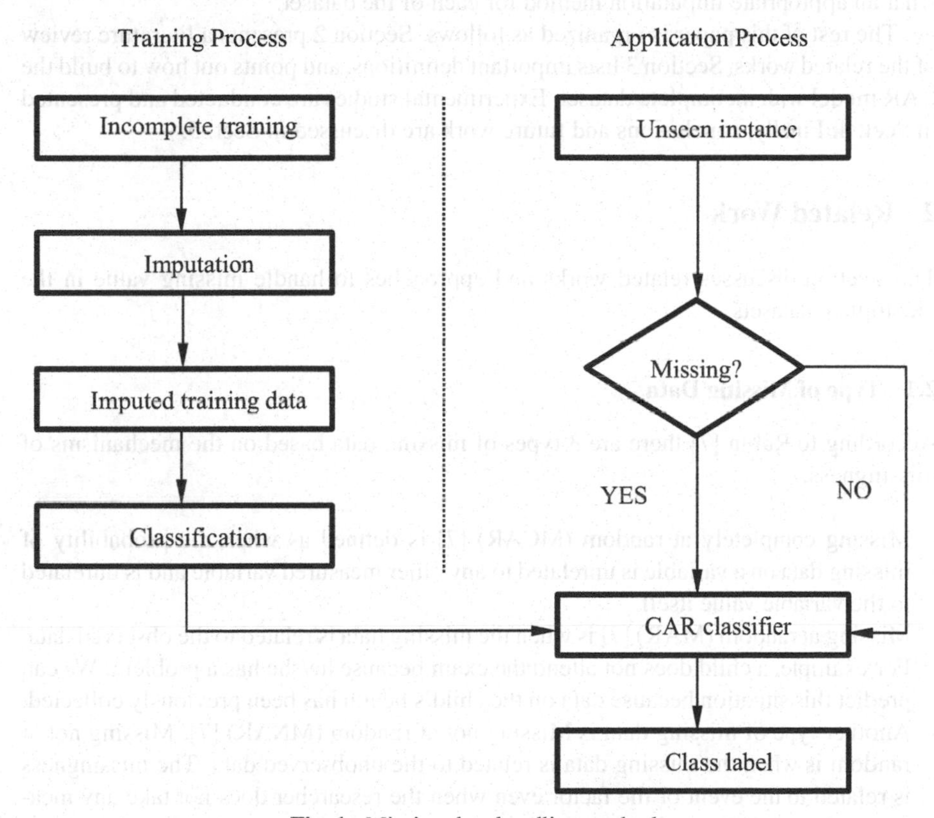

Fig. 1. Missing data handling method

One of the most common approaches to handle incomplete dataset is using an imputation method to fill the missing values. For example, mean imputation has been used commonly to process incomplete dataset [5]. The idea of mean imputation is that it replaces each missing value with the average of existing values in the features. For that reason, single imputation can be applied to almost any dataset. It (such as mean imputation) is very efficient but the accuracy is inadequate and it produced biased results [5]. In contrast, multiple imputation has been proven to be more accurate than single imputation but the major drawback is its high computational cost [6]. There are no certain ways to determine a suitable imputation method for CARs to enhance its accuracy. One of the possible methods to examine it is discussed in this article. In general, all imputation methods help generate a complete dataset which can be used to build a CAR model.

This work aims to examine and develop an effective method for classification based on association rules with incomplete dataset. Imputation methods are used to fill missing values on training datasets. Each imputation method is compared with the others for CAR classification. The main contributions of this paper are as follow:

Effectively and efficiently apply imputations with incomplete data for CARs.

Providing a comparison of imputation methods in terms of accuracy to determine what an appropriate imputation method for each of the dataset.

The rest of this paper is organized as follows. Section 2 presents a literature review of the related works. Section 3 lists important definitions, and points out how to build the CAR model with incomplete dataset. Experimental studies are conducted and presented in Sect. 4. Finally, conclusions and future work are discussed in Sect. 5.

2 Related Work

This section discusses related works and approaches to handle missing value in the incomplete datasets.

2.1 Type of Missing Data

According to Rubin [7] there are 3 types of missing data based on the mechanisms of missingness.

- Missing completely at random (MCAR) [7] is defined as when the probability of missing data on a variable is unrelated to any other measured variable and is unrelated to the variable value itself.
- Missing at random (MAR) [7] is when the missing data is related to the observed data. For example, a child does not attend the exam because he/she has a problem. We can predict this situation because data on the child's health has been previously collected.
- Another type of missing data is Missing not at random (MNAR) [7]. Missing not at random is when the missing data is related to the unobserved data. The missingness is related to the event or the factor even when the researcher does not take any measurements on it. For example, a person cannot attend a drug test because the person took the drug test the night before. MNAR is called "non-ignorable", it is necessary to use a suitable imputation method to find out what likely the value is and why it is missing.

The reason we need to thoroughly examine the mechanism is that some imputation method can only be applied to a suitable dataset. One example is that Multiple Imputation assumes the data are at least MAR. One record in the dataset can have missing data in many features and they may not have the same mechanism. Therefore, it is fundamental to investigate the mechanism of each feature before selecting a suitable approach.

Table 1. The *example of mean imputation*

Name	Age
A	17
B	22
C	23
D	?
E	25
F	27
G	?
H	32
I	32
J	35
K	?

Although the simplest way is to delete the missing value, this approach is not rational as it can result in an enormous loss of missing value and the consequence might be a decrease in classification accuracy. For that reason, the imputation methods are the most common way to handle missing values. Imputation method transforms original data with missing values into complete data before training a model. Then when it detects a new incomplete instance, it can be classified directly. The advantage of this method is that it is applicable to any classification algorithm. It can also deal with a large number of missing values.

Figure 1 shows the way to classify incomplete datasets, imputation methods are used to preprocess data. This step generates suitable values for each missing value. After that, Imputed training data can be used to build a model. After this process, new instances can be classified directly whether they are missing or not by using the CAR model built beforehand.

2.2 Imputation Method

There are two traditional imputation methods, single imputation and multiple imputation.

Single imputation means each missing value is filled with one value. Mean imputation is the most popular way to process incomplete data [5]. It fills all the missing values with the mean of the columns. Sometimes if the data are categorical, mode and median imputation should be considered. For example, if Male is 1 and Female is 0, mean imputation cannot be used as it will yield a meaningless number like 0.5. Hence, in some instance's mode and median imputation can be considered as a better option.

Mean/mode/median imputation should not be used in MNAR. These methods can handle good MCAR and MAR. Consider the dataset presented in Table 1.

For mean imputation, the mean value will be generated by using the following calculation:

$$Value = \frac{17 + 22 + 23 + 25 + 27 + 32 + 32 + 35}{11} = 19$$

The obtained dataset after performing imputation is given in Table 2 with the imputation values highlighted in bold text.

Since 32 appears twice and others appear only once. Mode imputation will replace all missing values with 32. With median imputation, it picks the element in the middle position. For example, in this dataset, the data series is numeric. The length of the series is equal to 8, the median index would be 4, and thus the median value equals 27. Median imputation will then replace all missing values with 27.

One disadvantage of mean/mode/median imputation is that it can lead to distortions in the histogram and underestimated variance because the method generates the same value for all the missing variables [5].

Table 2. After filling imputation method

Name	Age
A	17
B	22
C	23
D	**19**
E	25
F	27
G	**19**
H	32
I	32
J	35
K	**19**

K-nearest neighbors (KNN) imputation has been proved to be one of the most powerful single imputation [8]. The idea of KNN is that it searches for K-nearest neighbors to fill missing values. After that it will look for one value for the missing value by computing the average. The problem with KNN is its time-consuming nature compared to other single imputation methods (such as mean, mode, median imputation). This is because identifying *K* neighbors is required before the calculation process. The Euclidean metric is often used by KNN to determine the neighborhood.

Depending on the dataset, each single imputation method will affect the classification algorithm differently (in this paper, it is CAR). Thus, selecting a suitable imputation method is heavily based on the given dataset. Single imputation has an advantage in terms of running time over Multiple imputation. Besides KNN has been shown that it outperforms others single imputation methods.

Multiple imputations were introduced by Rubin [9]. Multiple imputation generates a set of values for one missing value as opposed to a single imputation, which only calculates one value for each missing value. Although it requires more time to calculate one value, multiple imputation produces more accurate results than single imputation [10–13].

The advantages of multiple imputation include:

- Reducing bias. Bias refers to the error that affects the analysis.
- Increasing precision, meaning how close two or more measurements to each other.
- Resistance to outliers.

Multiple imputations use Chain Equation (MICE). MICE has been used in many classification algorithms. The main idea of MICE is that it uses a regression method in order to estimate missing value. First, each missing value will be replaced by a random value in the same feature. Next, each incomplete feature is regressed on the other features to compute a better estimate for the feature. This process is repeated several times until the whole incomplete feature is imputed. Then the whole procedure is again repeated several times to provide imputed datasets. Finally, the result is calculated by the average of the imputed datasets previously.

Many studies show that MICE outperforms single imputation. MICE is a powerful imputation method. However, in reality, MICE requires long execution time in the process of estimating the missing values [14]. Therefore, further investigations are required for an effective and efficient use of this method.

2.3 Mining Class Association Rules

In 1998, Liu and partners proposed the CBA method [15] (Classification based on association) for mining class association rule. CBA includes 2 main stage:

- The stage to generate the rule – CBA-RG algorithm.
- The stage to build a classifier.

In 2001, Liu et al. proposed the CMAR algorithm CMAR (classification based on multiple association rules) [16]. This method is based on the FP–tree structure to compress the data and use projection on the tree to find association rules. In 2004, Thabtah et al. proposed the MMAC (multi-class, multi-label associative classification) [17]. In 2008, Vo and Le proposed ECR–CARM (equivalent class rule – class association rule mining) [18]. The authors have proposed the ECR tree structure, based on this tree, they presented the ECR-CARM algorithm to mine CARs in only one dataset scan. The object identifiers were used to quickly determine the support of the itemset. However, the biggest disadvantage of ECR-CARM is time-consuming for generating-and-test candidates because all itemsets are grouped into one node in the tree. When the two nodes I_i and I_j are joined to generate a new node, each element of I_i will be checked with each element of I_j to determine if their prefix is the same or not. In 2012, Nguyen et al. proposed a new method for pruning redundant rules based on lattice [19].

3 Model for Mining Class Association Rules with Incomplete Datasets

This section presents in detail on how to apply the CAR model to incomplete datasets. Furthermore, the process of applying imputation methods to improve the performance of CAR is also discussed. It describes the details of the training process and the application process.

3.1 Definition

Let $D = \{(X_i, c_i)|(i = 1, \ldots m)\}$ be a dataset. X_i represents an input instance with its associated class label c_i, and m is the number of instances in the dataset. The subset of features is denoted by $F = \{F_1, ...F_n\}$. An instance X_i is represented by a vector of n values $(x_{i1}, x_{i2}, ..., x_{in})$ where an x_{ij} is a value of j feature or "?". It means the value is unknown (or is called the missing value).

An instance X_i is called an incomplete instance if and only if it contains at least one missing value. A dataset is called an incomplete dataset if it has at least one incomplete instance. A feature (F_j) is defined as an incomplete feature if it contains one incomplete instance, X_i with a missing value x_{ij}, The dataset shown in Table 3 contains 5 incomplete instances. It has 4 incomplete features.

3.2 Method

The main idea is to use imputation during the training progress, not in the application progress. The goal of applying imputation is to generate a complete dataset to improve the accuracy of the classifiers. Good imputation methods such as multiple imputation are computationally expensive. However, in the training process, there is no time limit in any application. Therefore, the use of multiple imputation in this case is acceptable.

3.3 Training Process

The training process has 2 main purposes. The first purpose is to build complete datasets. It first splits a dataset into m folds (depend on the user). It takes $m-1$ fold for the training process and the remaining fold is a test set. The process starts by using an imputation method to estimate missing values on a training dataset. It first begins with a single imputation method (KNN and mean/mode/median). The imputation will be used to generate an imputed dataset. After having complete datasets, CAR will be applied on complete dataset in order to build a classifier. A test set is used to evaluate the competency of the classifier without having any imputation on that. The whole process is repeated with a new imputation method in order to find the best methods which can lead the construction of a good classifier.

Table 3. Sample dataset

	F_1	F_2	F_3	F_4	F_5
X_1	1	?	12	18	23
X_2	2	7	13	?	?
X_3	3	8	?	19	25
X_4	4	9	?	20	26
X_5	5	?	16	21	27
X_6	6	11	17	22	28

Implementation Steps

- **Step 1:** Divide the dataset into *m* folds (usually *m* is 10). Take *m* – 1 for a training set and the last one is used as the test set.
- **Step 2:** Use imputation method on training set only. Imputation methods include single imputation (mean and *KNN* imputation)
- **Step 3:** Use a test set to evaluate the model.
- **Step 4:** Repeat all steps with different imputation methods in order to find the models with highest accuracy.

3.4 Application Process

An application process is to classify new instances using the learnt classifier without having any imputation on this. An input in an application process is an instance with some missing attributes. The algorithm will output the most suitable class label for that instance.

4 Experimental Studies

The algorithms used in the experiments were coded on a personal computer with Weka 3.8.4 using Windows 10, Intel® Core® i5 9600K (6 CPUs @ 3.7 GHz) and 16 GB RAM. The experiments were conducted on the datasets collected from UCI Machine Learning Repository. The characteristics of the experimental datasets are shown in the Table 4. The first column presents a name of the dataset. The second columns show the attribute. The third columns show the class. The fourth columns show the number of instances of each datasets. And the final column contain % missing value that the dataset has.

There are different features in the experimental datasets. Some datasets have many attributes with several instances while others have average and large one (*mushroom* dataset). Missing values can have on multiple attributes. However, *mushroom* dataset only has one attribute with missing value. The datasets also have varying types of features including real, integer and nominal. The choice of datasets is intended to reflect incomplete problems of varying difficulty, size, dimensionality, and feature types.

The experimental used two imputation method. KNN based imputation and Mean/mode/median imputation. The implement of KNN imputation choose the number of neighbors k with yield the best accuracy for the model to compare with mean/mode/median imputation. Both of the imputation methods were performed by Weka 3.8.4. Ten-fold cross validation was used to separate each dataset into different training and test sets.

We performed experiments to evaluate the effectiveness of the imputation methods. In Table 5, the first column shows the datasets, and the second column shows classification algorithm with KNN imputation. The third column shows classification algorithm with mean imputation. From Figs. 2, 3, 4, 5, 6, 7 and 8 show the details of each fold of each dataset. The blue line shows the accuracy of CAR algorithm with mean imputation. And the red line shows the accuracy of CAR algorithm with KNN imputation.

In further investigation, House vote and *mammographic masses* dataset contains some folds that mean imputation yield better result than KNN. In *house vote*, the difference seems insignificant. However, the difference between mean and KNN imputation for each fold in *mammographic_masses* can be up to 10%. Overall, In Fig. 2, the KNN imputation can increase the accuracy up to 2.8%.

Table 4. The characteristic of the experimental datasets.

Dataset	Attribute	Classes	#instances	%missing value
House vote	16	2	435	66.2
CRX	15	2	690	9.7
mammographic_masses	5	2	961	16.9
chronic_kidney_disease	24	2	400	74
Hepatitis	19	2	155	72.3
Mushroom	22	3	8124	30.5

Overall, the results, as presented from Figs. 2, 3, 4, 5, 6, 7 and 8, Table 5, have shown that KNN seems to produce higher accuracy than mean imputation. Even though some results indicate that the accuracy of both methods are the same. However, KNN performs better in most tested situations.

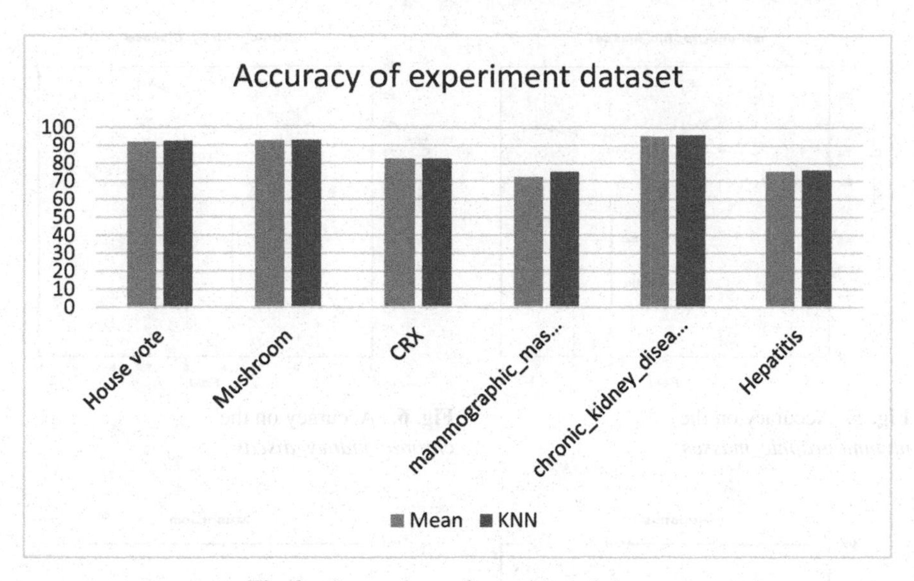

Fig. 2. Comparison of experimental dataset

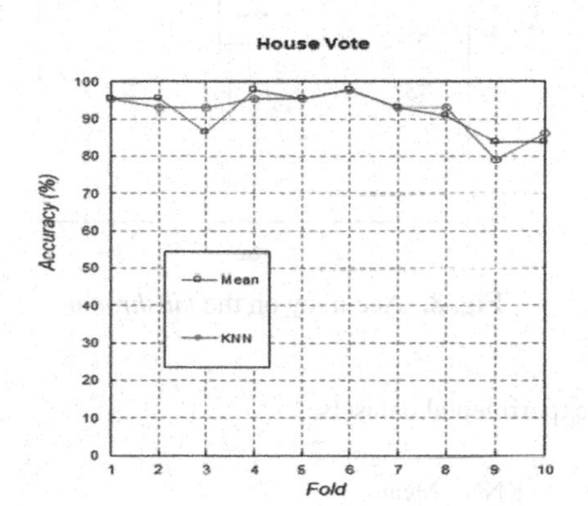

Fig. 3. Accuracy on the *House Vote*

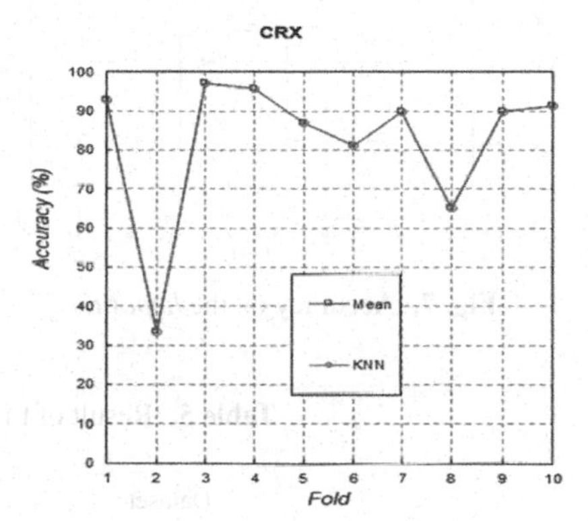

Fig. 4. Accuracy on the *CRX*

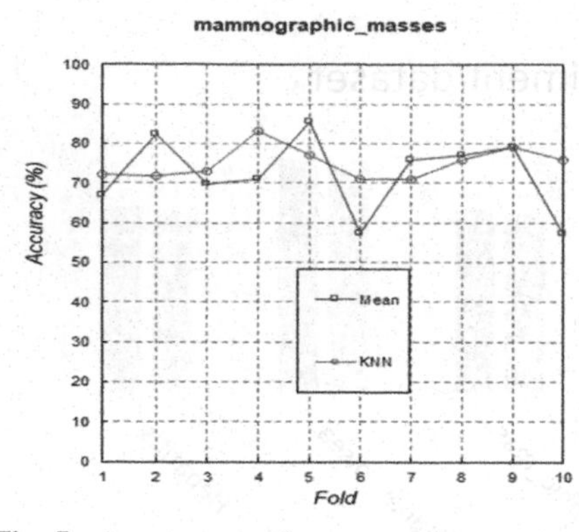

Fig. 5. Accuracy on the *mammographic_masses*

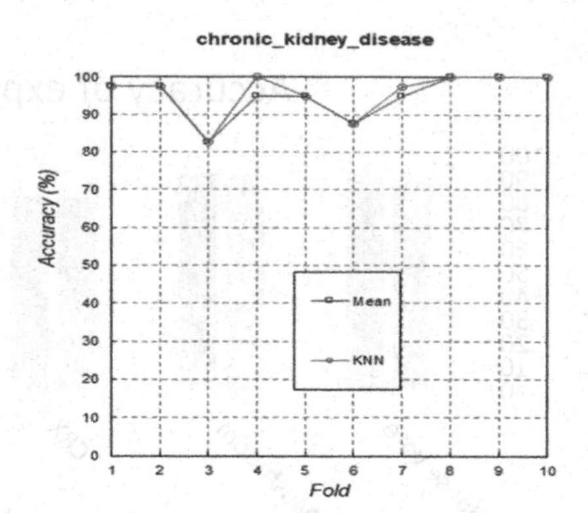

Fig. 6. Accuracy on the *chronic_kidney_diseas*

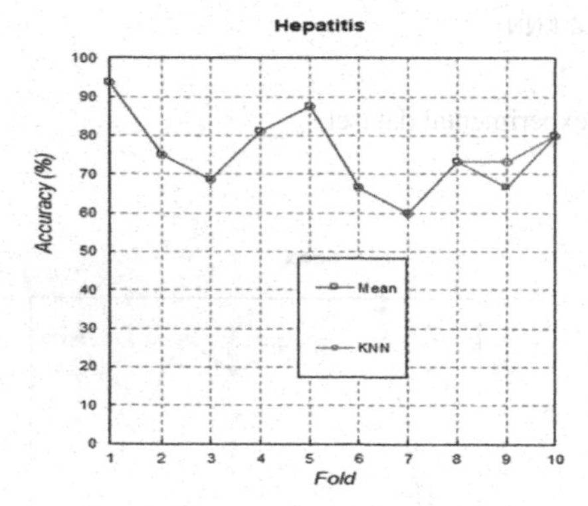

Fig. 7. Accuracy on the *hepatitis*

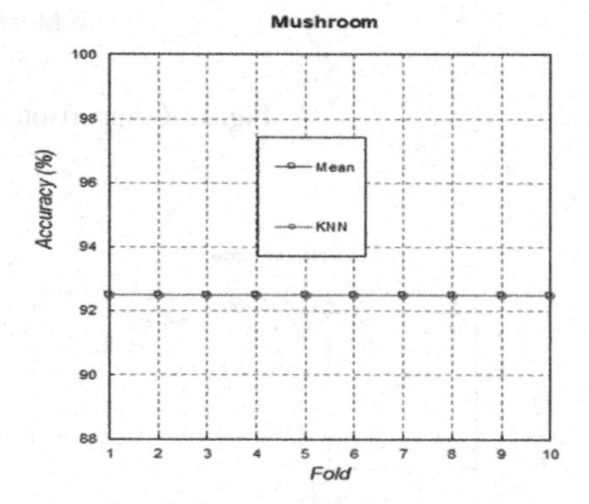

Fig. 8. Accuracy on the *mushroom*

Table 5. Result of the experimental datasets.

Dataset	KNN	Mean
House vote	92.16	91.93
CRX	82.32	82.32
mammographic_masses	75.03	72.22
chronic_kidney_disease	95.75	95
Hepatitis	75.96	75.29
Mushroom	92.49	92.49

5 Conclusion and Future Work

This paper proposed a method for mining incomplete datasets with CAR by using single imputation. In the training process, the imputation method is used to generate a complete training dataset. The experiments show the comparison between KNN and

Mean/Mode/Median imputation. The use of the KNN imputation gives a better result than Mean/Mode/Median imputation. Even though the percentage of missing value is different on each dataset, the CAR with the use KNN imputation yield higher accuracy. In addition, in some datasets, the computational time between KNN and Mean/Mode/Median imputation has no difference.

Missing values are a common issue in many datasets. In addition, mining dataset with CAR has been developed in recent years. However, there has not been much work on handling missing data in other CAR algorithms. In the future, further research with different CAR methods will be conducted on incomplete dataset.

Acknowledgements. This research is funded by International University, VNU-HCM under grant number SV2019-IT-03.

References

1. Nguyen, N.T.: Advanced Methods for Inconsistent Knowledge Management. Springer, London (2008). https://doi.org/10.1007/978-1-84628-889-0
2. Zong, W., Huang, G.B.: Face recognition based on extreme learning machine. Neurocomputing **74**(16), 2541–2551 (2011)
3. Adiraju, R.V., Masanipalli, K.K, Reddy, T.D., Pedapalli, R., Chundru, S., Panigrahy, A.K.: An extensive survey on finger and palm vein recognition system. Mater. Today Proc. (2020). https://doi.org/10.1016/j.matpr.2020.08.742
4. Wei, S., Zheng, X., Chen, D., Chen, C.: A hybrid approach for movie recommendation via tags and ratings. Electron. Commer. Res. Appl. **18**, 83–94 (2016)
5. Ahmad, M.A., Teredesai, A., Eckert, C.: Interpretable machine learning in healthcare. In: Proceedings of 2018 IEEE International Conference on Healthcare Informatics, ICHI 2018, p. 447 (2018)
6. Donders, A.R.T., van der Heijden, G.J.M.G., Stijnen, T., Moons, K.G.M.: Review: a gentle introduction to imputation of missing values. J. Clin. Epidemiol. **59**(10), 1087–1091 (2006)
7. Darmawan, I.G.N.: NORM software review: handling missing values with multiple imputation methods. Eval. J. Australas. **2**(1), 51–57 (2002)
8. Little, R.J.A., Rubin, D.B.: Statistical Analysis with Missing Data, 3rd edn. Wiley (2019)
9. Jadhav, A., Pramod, D., Ramanathan, K.: Comparison of performance of data imputation methods for numeric dataset. Appl. Artif. Intell. **33**(10), 913–933 (2019)
10. Rubin, D.B.: An overview of multiple imputation. In: Proceedings of the Survey Research Methods Section, pp. 79–84. American Statistical Association (1988)
11. Gómez-Carracedo, M.P., Andrade, J.M., López-Mahía, P., Muniategui, S., Prada, D.: A practical comparison of single and multiple imputation methods to handle complex missing data in air quality datasets. Chemom. Intell. Lab. Syst. **134**, 23–33 (2014)
12. Nguyen, N.T.: Consensus systems for conflict solving in distributed systems. Inf. Sci. **147**(1–4), 91–122 (2002)
13. Nguyen, N.T.: Using consensus methods for solving conflicts of data in distributed systems. In: Hlaváč, V., Jeffery, K.G., Wiedermann, J. (eds.) SOFSEM 2000. LNCS, vol. 1963, pp. 411–419. Springer, Heidelberg (2000). https://doi.org/10.1007/3-540-44411-4_30
14. Musil, C.M., Warner, C.B., Yobas, P.K., Jones, S.L.: A comparison of imputation techniques for handling missing data. West. J. Nurs. Res. **24**(7), 815–829 (2002)
15. Liu, B., Hsu, W., Ma, Y., Ma, B.: Integrating classification and association rule mining. In: Knowledge Discovery and Data Mining, pp. 80–86 (1998)

16. Li, W., Han, J., Pei, J.: CMAR: accurate and efficient classification based on multiple class-association rules. In: Proceedings of IEEE International Conference on Data Mining, ICDM, pp. 369–376 (2001)
17. Thabtah, F.A., Cowling, P., Peng, Y.: MMAC: a new multi-class, multi-label associative classification approach. In: Proceedings of Fourth IEEE International Conference on Data Mining, ICDM 2004, pp. 217–224 (2004)
18. Vo, B., Le, B.: A novel classification algorithm based on association rules mining. In: Richards, D., Kang, B.-H. (eds.) PKAW 2008. LNCS (LNAI), vol. 5465, pp. 61–75. Springer, Heidelberg (2009). https://doi.org/10.1007/978-3-642-01715-5_6
19. Nguyen, L.T.T., Vo, B., Hong, T.P., Thanh, H.C.: Classification based on association rules: a lattice-based approach. Expert Syst. Appl. **39**(13), 11357–11366 (2012)

Machine Learning Methods

Clustering Count Data with Stochastic Expectation Propagation

Xavier Sumba[1(✉)], Nuha Zamzami[2,3], and Nizar Bouguila[3]

[1] Department of Electrical and Computer Engineering, Concordia University, Montreal, Canada
xavier.sumba93@ucuenca.edu.ec

[2] Department of Computer Science and Artificial Intelligent, University of Jeddah, Jeddah, Saudi Arabia

[3] Concordia Institute for Information Systemes Engineering, Concordia University, Montreal, Canada

Abstract. Clustering count vectors is a challenging task given their sparsity and high-dimensionality. An efficient generative model called EMSD has been recently proposed, as an exponential-family approximation to the Multinomial Scaled Dirichlet distribution, and has shown to offer excellent modeling capabilities in the case of sparse count data and to overcome some limitations of the frameworks based on the Dirichlet distribution. In this work, we develop an approximate Bayesian learning framework for the parameters of a finite mixture of EMSD using the Stochastic Expectation Propagation approach. In this approach, we maintain a global posterior approximation that is being updated in a local way, which reduces the memory consumption, important when making inference in large datasets. Experiments on both synthetic and real count data have been conducted to validate the effectiveness of the proposed algorithm in comparison to other traditional learning approaches. Results show that SEP produces comparable estimates with traditional approaches.

Keywords: Mixture model · Emsd distribution · Stochastic expectation propagation

1 Introduction

Statistical methods are excellent at modeling semantic content of text documents [9]. More specifically, document clustering is widely used in a variety of applications such as text retrieval or topic modeling, (see e.g. [3]). Words in text documents usually exhibit appearance dependencies, *i.e.*, if word w appears once, it is more probable that the same word w will appear again. This phenomenon is called burstiness, which has shown to be addressed by introducing the prior information into the construction of the statistical model to obtain several computational advantages [15]. Given that the Dirichlet distribution is

N. T. Nguyen et al. (Eds.): ACIIDS 2021, LNAI 12672, pp. 119–129, 2021.
https://doi.org/10.1007/978-3-030-73280-6_10

generally taken as a conjugate prior to the multinomial, the most popular hierarchical approach is the Dirichlet Compound Multinomial (DCM) distribution [14]. While the Multinomial distribution fails to model the words burstiness given its dependency assumption, the DCM distribution not only captures this behavior but also models text data better [14]. However, The Dirichlet distribution has its own limitations due to is negative covariance structure and equal confidence [11,24]. Hence, a generalization of it called the Scaled Dirichlet (SD) distribution has shown to be a good alternative as a prior to the multinomial resulting in the Multinomial scaled Dirichlet (MSD) distribution recently proposed in [25]. Indeed, MSD has shown to have high flexibility in count data modeling with superior performance in several real-life challenging application [25–28]. Despite its flexibility, MSD distribution shares similar limitations to the one with DCM since its parameter estimation is slow, especially in high-dimensional spaces. Thus, [28] proposed a close exponential-family approximation called EMSD to combine the flexibility and efficiency of MSD with the desirable statistical and computational properties of the exponential family of distributions, including sufficiency. EMSD has shown to reduce the complexity and computational efforts, considering the sparsity and high-dimensionality nature of count data.

In this work, we study the application of the Bayesian framework for learning the exponential-family approximation to the Multinomial Scaled Dirichlet (EMSD) mixture model which has been shown to be an appropriate distribution to model the burstiness in high-dimensional feature space. In particular, we propose a learning approach for an EMSD mixture model using Stochastic Expectation Propagation (SEP) [10] for parameter estimation. Indeed, SEP combines both Assumed Density Filtering (ADF) and Expectation Propagation (EP) in order to scale to large datasets while mantaining accurate estimations. Only EP is usually more accurate than methods such as variational inference and MCMC [1,18], and SEP solves some of the problems encountered when using EP given that the number of parameters increase according to number of datapoints. Thus, SEP is a deterministic approximate inference method that prevents memory overheads when increasing the number of data points. EP has shown to be an appropriate generalization in the case of Gaussian mixture model [20], hierarchical models such as LDA [18] or even infinite mixture models [6]. Furthermore, SEP has been used with Deep Gaussian process [4], showing the benefits of scalable Bayesian inference and outperforming traditional Gaussian process. The contributions of this paper are summarized as follows: 1) we show that SEP can provide effective parameter estimates when dealing with large datasets; 2) we derive foundations to learn an EMSD mixture model using SEP; 3) we exhaustively evaluate the proposed approach on synthetic and real count data and compare the performance with other models and learning approaches.

2 The Exponential-Family Approximation to MSD Distribution

In the clustering setting, We are given a dataset $\mathcal{X}$ with D samples $\mathcal{X} = \{\mathbf{x_i}\}_{i=1}^{D}$, each $\mathbf{x}_i$ is a vector of count data (*e.g.* a text document or an image, represented as a frequencies vector of words or visual words, respectively). We assume that each data set has a vocabulary of size V.

The the Multinomial Scaled Dirichlet (MSD) is the marginal distribution defined by integrating out the probability parameter of scaled Dirichlet over all possible multinomials, and it is given by [25]:

$$\mathcal{MSD}(\mathbf{x} \mid \boldsymbol{\rho}, \boldsymbol{\nu}) = \frac{n!}{\prod_{w=1}^{V} x_w!} \frac{\Gamma(s)}{\Gamma(s+n) \prod_{w=1}^{V} \nu_w^{x_w}} \prod_{w=1}^{V} \frac{\Gamma(x_w + \rho_w)}{\Gamma(\rho_w)} \tag{1}$$

Note that the authors in [25] use the approximation $\left(\sum_{w=1}^{V} \nu_w p_w\right)^{\sum_{w=1}^{V} x_w} \approx \prod_{w=1}^{V} \nu_w^{x_w}$. It is worth mentioning that DCM is a special case of MSD, such that when $\boldsymbol{\nu} = 1$ in Eq. (1), we obtain the Dirichlet Compound Multinomial (DCM) distribution [14]. Similar to DCM, the considered model MSD, has an intuitive interpretation representing the Scaled Dirichlet as a general topic and the Multinomial as a document-specific subtopic, making some words more likely in a document $\mathbf{x}$ based on word counts.

The representation of text documents is very sparse as many words in the vocabulary do not appear in most of the documents. Thus, in [28], the authors note that using only the non-zero values of $\mathbf{x}$ is computationally efficient since $x_w! = 1$, $\nu_w^{x_w} = 1$ and $\Gamma(x_w + \rho_w)/\Gamma(\rho_w) = 1$ when $x_w = 0$. Moreover, since in high dimensional data the parameters are very small, [5], the following fact for small values of ρ when $x \geq 1$ was used in [28]:

$$\lim_{\rho \to 0} \frac{\Gamma(x+\rho)}{\Gamma(\rho)} - \Gamma(x)\rho = 0 \tag{2}$$

Thus, being able to approximate $\Gamma(x_w + \rho_w)/\Gamma(\rho_w) = \Gamma(x_w)\rho_w$ and using the fact that $\Gamma(x_w) = (x_w - 1)!$ leads to an approximation of the MSD distribution known as the Exponential-family approximation to the MSD distribution (EMSD), given by:

$$\mathcal{EMSD}(\mathbf{x} \mid \boldsymbol{\alpha}, \boldsymbol{\beta}) = \frac{n!}{\prod_{w:x_w \geq 1}^{V} x_w} \frac{\Gamma(s)}{\Gamma(s+n)} \prod_{w:x_w \geq 1}^{V} \frac{\alpha_w}{\beta_w^{x_w}} \tag{3}$$

The parameters of the EMSD distribution are denoted by $\boldsymbol{\alpha}$ and $\boldsymbol{\beta}$ to distinguish them from the MSD parameters for clarity.

3 Stochastic Expectation Propagation

Efficient inference and learning for probabilistic models that scale to large datasets are essential in the Bayesian setting. Thus, a variety of methods have been proposed from sampling approximations [17] to distributional approximations such as stochastic variational inference [8].

Another deterministic approach is Expectation Propagation (EP) that commonly provides more accurate approximations compared to sampling methods [21] and variational inference [19,20]. Yet, the number of parameters grows with the number of data points, causing memory overheads and making it difficult to scale to large datasets. Besides, Assumed Density Filtering (ADF) [22], which has been introduced before EP, maintains a global approximating posterior; however, it results in poor estimates. Therefore, [10] proposed an alternative to push EP to large datasets denominated Stochastic Expectation Propagation (SEP). SEP takes the best of these two methods by maintaining a global approximation that is updated locally. It does this by introducing a global site that captures the average effect of the likelihood sites and, as a result avoiding memory overheads.

Given a probabilistic model $p(\mathcal{X} \mid \boldsymbol{\theta})$ with parameters $\boldsymbol{\theta}$ drawn from a prior $p_0(\boldsymbol{\theta})$, SEP approximates a target distribution $p(\boldsymbol{\theta} \mid \mathcal{X})$, which is commonly the posterior, with a global approximation $q(\boldsymbol{\theta})$ that belongs to the exponential family. The target distribution must be factorizable such that the posterior can be split into D sites $p(\boldsymbol{\theta} \mid \mathcal{X}) \propto p_0(\boldsymbol{\theta}) \prod_{i=1}^{D} p_i(\boldsymbol{\theta})$; the initial site p_0 is commonly interpreted as the prior distribution and the remaining p_i sites represent the contribution of each ith item to the likelihood. The approximating distribution must admit a similar factorization, $q(\boldsymbol{\theta}) \propto p_0(\boldsymbol{\theta})\tilde{p}(\boldsymbol{\theta})^D$.

Unlike EP, the SEP maintains a global approximating site, $\tilde{p}(\boldsymbol{\theta})^D$, to capture the average effect of a likelihood on the posterior. Thus, we only have to maintain the parameters of the approximate posterior and approximate global site that commonly belongs to the exponential family. Consequently, each site is refined to create a cavity distribution by dividing the global approximation over one of the copies of the approximate site, $q^{\setminus 1}(\boldsymbol{\theta}) \propto q(\boldsymbol{\theta})/\tilde{p}(\boldsymbol{\theta})$.

Additionally, in order to approximate each site, a new tilted distribution is introduced using the cavity distribution and the current site $\hat{p}_i(\boldsymbol{\theta}) \propto p_i(\boldsymbol{\theta})q^{\setminus 1}(\boldsymbol{\theta})$.

Subsequently, a new posterior is found by minimizing the Kullback Leibler divergence $D_{KL}(\hat{p}_i(\boldsymbol{\theta}) \mid\mid q^{new}(\boldsymbol{\theta}))$ such that $\tilde{p}_i(\boldsymbol{\theta}) \approx p_i(\boldsymbol{\theta})$. This minimization is equivalent to match the moments of those distributions [1,20]. Finally, the revised approximate site is updated by removing the remaining terms from the current approximation by employing damping [7,18] in order to make a partial update since $\tilde{p}_i$ captures the effect of a single likelihood function:

$$\tilde{p}(\boldsymbol{\theta}) = \tilde{p}(\boldsymbol{\theta})^{1-\eta} \left(\frac{q^{new}(\boldsymbol{\theta})}{q^{\setminus w}(\boldsymbol{\theta})} \right)^{\eta} = \tilde{p}(\boldsymbol{\theta})^{1-\eta} \tilde{p}_i(\boldsymbol{\theta})^{\eta} \tag{4}$$

Notice that η is the step size, and when $\eta = 1$, no damping is applied. A natural choice is $\eta = 1/D$.

4 EMSD Mixture Model

4.1 Clustering Model

We assume that we are given D documents drawn from a finite number of EMSD distributions, and each $\mathbf{x}_i$ document is composed of V words. $K \geq 1$ represents the number of mixture components. Thus, a document is drawn from its respective component j as follows: $\mathbf{x_i} \sim \mathcal{EMSD}(\boldsymbol{\alpha}_j, \boldsymbol{\beta}_j)$.

In a mixture model, a latent variable $\mathcal{Z} = \{\mathbf{z}_i\}_{i=1}^{D}$ is introduced for each $\mathbf{x}_i$ document in order to represent the component assignment. We posit a Multinomial distribution for the component assignment such that $\mathbf{z}_i \sim Mult(1, \boldsymbol{\pi})$ where $\boldsymbol{\pi} = \{\pi_j\}_{j=1}^{K}$ represents the mixing weights, and they are subject to the constraints $0 < \pi_j < 1$ and $\sum_j \pi_j = 1$. In other words, $\mathbf{z}_i$ is a K-dimensional indicator vector containing a value of one when document $\mathbf{x}_i$ belongs to the component j, and zero otherwise. Note that in this setting the value of $z_{ij} = 1$ acts as the selector of the component that generates $\mathbf{x}_i$ document with parameters $\boldsymbol{\alpha}_j$ and $\boldsymbol{\beta}_j$; hence, $p(\mathbf{z}_i \mid \boldsymbol{\pi}) = \pi_j$. Thus, the full posterior is in Eq. 5.

$$p(\boldsymbol{\pi}, \boldsymbol{\alpha}, \boldsymbol{\beta} \mid \mathcal{X}) \propto p(\boldsymbol{\pi}) p(\boldsymbol{\alpha}) p(\boldsymbol{\beta}) \prod_{i}^{D} \sum_{j}^{K} \pi_j p(\mathbf{x}_i \mid \boldsymbol{\alpha}_j, \boldsymbol{\beta}_j) \tag{5}$$

4.2 Parameter Learning

We use SEP in order to learn the parameters of the mixture model. We start by partitioning the likelihood in D sites and define a global approximating site for each of the latent variables ($\boldsymbol{\pi}$, $\boldsymbol{\alpha}$, and $\boldsymbol{\beta}$). Theoretically, any distribution belonging to the exponential family can be used for the sites. We use a Gaussian distribution for the parameters of the EMSD distribution in order to facilitate calculations [12]. For the mixture weights, we use a Dirichlet distribution since it belongs to the $K-1$ simplex and fits the constraints imposed by the mixing weights. Equations 6 illustrate the choices for the approximate sites.

$$\tilde{p}(\boldsymbol{\pi}) \propto \prod_{j}^{K} \pi_j^{a_j} \quad \tilde{p}(\boldsymbol{\alpha}) = \prod_{j}^{K} \mathcal{N}(\boldsymbol{\alpha}_j \mid \boldsymbol{m}_j, p_j^{-1}) \quad \tilde{p}(\boldsymbol{\beta}) = \prod_{j}^{K} \mathcal{N}(\boldsymbol{\beta}_j \mid \boldsymbol{n}_j, q_j^{-1}) \tag{6}$$

Once the global approximate site has been defined, we compute the approximate posterior $q(\boldsymbol{\pi}, \boldsymbol{\alpha}, \boldsymbol{\beta})$ by introducing the priors and the average effect of the global site:

$$\begin{aligned} q(\boldsymbol{\pi}, \boldsymbol{\alpha}, \boldsymbol{\beta}) \propto & p(\boldsymbol{\pi}, \boldsymbol{a}^0) \tilde{p}(\boldsymbol{\pi} \mid \boldsymbol{a})^D \prod_{j}^{K} p\left(\boldsymbol{\alpha}_j \mid \boldsymbol{m}_j^0, (p_j^0)^{-1}\right) \tilde{p}\left(\boldsymbol{\alpha}_j \mid \boldsymbol{m}_j, (p_j)^{-1}\right)^D \\ & p\left(\boldsymbol{\beta}_j \mid \boldsymbol{n}_j^0, (q_j^0)^{-1}\right) \tilde{p}\left(\boldsymbol{\beta}_j \mid \boldsymbol{n}_j, q_j^{-1}\right)^D \end{aligned}$$

The approximate posterior distribution has the following parameters:

$$\begin{aligned} \boldsymbol{a}' = 1 + \boldsymbol{a}^0 + D\boldsymbol{a} \quad & (p_j^{'})^{-1} = (p_j^0 + Dp_j)^{-1} \quad (q_j^{'})^{-1} = (q_j^0 + Dq_j)^{-1} \\ \boldsymbol{m}_j^{'} = (p_j^{'})^{-1}(p_j^0 \boldsymbol{m}_j^0 + Dp_j \boldsymbol{m}_j) \quad & \boldsymbol{n}_j^{'} = (q_j^{'})^{-1}(q_j^0 \boldsymbol{n}_j^0 + Dq_j \boldsymbol{n}_j) \end{aligned} \tag{7}$$

Consequently, we introduce a cavity distribution by removing the contribution of one of the copies of the global site. The cavity distribution has parameters

$\boldsymbol{a}^{\setminus 1}$, $\left(p_j^{\setminus 1}\right)^{-1}$, $\boldsymbol{m}_j^{\setminus 1}$, $\left(q_j^{\setminus 1}\right)^{-1}$, and $\boldsymbol{n}_j^{\setminus 1}$ illustrated in Eq. 8 that are calculated as follows: $q(\boldsymbol{\pi}, \boldsymbol{\alpha}, \boldsymbol{\beta})/\tilde{p}_i(\boldsymbol{\pi}, \boldsymbol{\alpha}, \boldsymbol{\beta})$

$$\boldsymbol{a}^{\setminus 1} = \boldsymbol{a}' - \boldsymbol{a} \quad \left(p_j^{\setminus 1}\right)^{-1} = \left(p_j' - p_j\right)^{-1} \quad \left(q_j^{\setminus 1}\right)^{-1} = \left(q_j' - q_j\right)^{-1}$$
$$\boldsymbol{m}_j^{\setminus 1} = \left(p_j^{\setminus 1}\right)^{-1}\left(p_j'\boldsymbol{m}_j' - p_j\boldsymbol{m}_j\right) \quad \boldsymbol{n}_j^{\setminus 1} = \left(q_j^{\setminus 1}\right)^{-1}\left(q_j'\boldsymbol{n}_j' - q_j\boldsymbol{n}_j\right) \tag{8}$$

We use the cavity distribution and incorporate the *ith* site, resulting in the tilted distribution $\hat{p} = \frac{1}{Z_i} p_i q^{\setminus 1}$. We use this distribution to compute the KL divergence with the approximate distribution, which is equivalent to matching the moments. However, in this case, matching the moments leads to another analytically intractable integral (i.e. $Z_i = \sum_j^K \frac{a_j^{\setminus 1}}{\sum_k^K a_k^{\setminus 1}} \mathbb{E}_{p(\alpha_j, \beta_j)}\left[p(x_i \mid \alpha_j, \beta_j)\right]$). Thus, we compute this integral via Monte Carlo sampling. After matching the moments, we obtain the parameters for an updated approximate posterior.

$$\Psi(a_j') - \Psi(\sum_j^K a_j') = \Psi(a_j^{\setminus 1}) - \Psi(\sum_j^K a_j^{\setminus 1}) + \nabla_{a_j^{\setminus 1}} \log Z_i$$
$$p_j' = \left(p_j^{\setminus 1}\right)^{-1}\left(2\nabla_{\left(p_j^{\setminus 1}\right)^{-1}} \log Z_i + p_j^{\setminus 1}\right)\left(p_j^{\setminus 1}\right)^{-1} - \left(\boldsymbol{m}_j' - \boldsymbol{m}_j^{\setminus 1}\right)\left(\boldsymbol{m}_j' - \boldsymbol{m}_j^{\setminus 1}\right)^{\intercal}$$
$$q_j' = \left(q_j^{\setminus 1}\right)^{-1}\left(2\nabla_{\left(q_j^{\setminus 1}\right)^{-1}} \log Z_i + q_j^{\setminus 1}\right)\left(q_j^{\setminus 1}\right)^{-1} - \left(\boldsymbol{n}_j' - \boldsymbol{n}_j^{\setminus 1}\right)\left(\boldsymbol{n}_j' - \boldsymbol{n}_j^{\setminus 1}\right)^{\intercal}$$
$$\boldsymbol{m}_j' = \boldsymbol{m}_j^{\setminus 1} + \left(p_j^{\setminus 1}\right)^{-1} \nabla_{\boldsymbol{m}_j^{\setminus 1}} \log Z_i \quad \boldsymbol{n}_j' = \boldsymbol{n}_j^{\setminus 1} + \left(q_j^{\setminus 1}\right)^{-1} \nabla_{\boldsymbol{n}_j^{\setminus 1}} \log Z_i \tag{9}$$

The values of $\boldsymbol{a}'$ are calculated using fixed point iteration as described in [16]. Using this updated approximate posterior, we remove the cavity distribution in order to obtain an approximation to the *ith* site.

$$\boldsymbol{a} = \boldsymbol{a}' - \boldsymbol{a}^{\setminus 1} \quad (p_j)^{-1} = (p_j' - p_j^{\setminus 1})^{-1} \quad \boldsymbol{m}_j = (p_j)^{-1}\left(p_j'\boldsymbol{m}_j' - p_j^{\setminus 1}\boldsymbol{m}_j^{\setminus 1}\right)$$
$$(q_j)^{-1} = (q_j' - q_j^{\setminus 1})^{-1} \quad \boldsymbol{n}_j = (q_j)^{-1}\left(q_j'\boldsymbol{n}_j' - q_j^{\setminus 1}\boldsymbol{n}_j^{\setminus 1}\right) \tag{10}$$

Finally, we use damping to partially update the global approximate site. First, we update the parameters of the global site as follows $\Theta^{new} = (1-\eta)\Theta^{old} + \eta\Theta_i$ where Θ^{old} are the current parameters of the global site, and Θ_i are the parameters for the approximation of a single likelihood. Then, we introduce the global approximate site in the approximate distribution. The learning approach is described in the Algorithm 1.

5 Experimental Results

In this section, we describe the experiments carried out to test the validity of the proposed method on both synthetic and real count data.

Algorithm 1: Stochastic Expectation Propagation (SEP) algorithm for learning a EMSD Mixture model

Input : K: number of clusters; $\mathcal{X} = \{\mathbf{x}_1, \ldots, \mathbf{x}_D\}$: corpus; $p_0(\boldsymbol{\pi}, \boldsymbol{\alpha}, \boldsymbol{\beta})$: prior knowledge

1 Initialize the approximate site $\tilde{p}(\boldsymbol{\pi}, \boldsymbol{\alpha}, \boldsymbol{\beta})$.
2 If priors are not provided, initialize them to 1 (i.e. $p_0(\boldsymbol{\pi}, \boldsymbol{\alpha}, \boldsymbol{\beta})$=1)
3 Compute the approximate distribution $q(\boldsymbol{\pi}, \boldsymbol{\alpha}, \boldsymbol{\beta})$ by calculating the average effect $\tilde{p}(\boldsymbol{\pi}, \boldsymbol{\alpha}, \boldsymbol{\beta})^D$ of the likelihood and introducing the priors p_0
4 **while** *not convergence* **do**
5 **for** x_i *in* $\mathcal{X}$ **do**
6 Compute the cavity distribution $q^{\setminus 1}(\boldsymbol{\pi}, \boldsymbol{\alpha}, \boldsymbol{\beta})$ by removing the contribution of one of the copies of the approximate site.
7 Match moments of the tilted distribution $\hat{p}(\boldsymbol{\pi}, \boldsymbol{\alpha}, \boldsymbol{\beta})$ and approximate posterior $q^{new}(\boldsymbol{\pi}, \boldsymbol{\alpha}, \boldsymbol{\beta})$ by minimizing $D_{KL}(\hat{p} \parallel q^{new})$.
8 Compute the parameters of a revised approximate site after matching the moments.
9 Make a partial update to the approximate site and include the approximate site in the approximate distribution.
10 **end**
11 **end**
12 Estimate mixing weights π_j

5.1 Synthetic Dataset

We create a synthetic dataset $\mathcal{X} = \{\mathbf{x}_i\}_{i=1}^{D}$ by using the probabilistic mixture model with $D = 210$ data points. We use $K = 3$ components, each of which is an EMSD distribution where the mixing weights are uniformly sampled. For simplicity, we set a fixed value of 1 for the scale parameter of the Scaled Dirichlet. Since the shape parameter is commonly $\alpha_w \ll 1$ [5], we sample from a Beta distribution.

We initialize the priors of the model with covariance matrix $5\boldsymbol{I}$ and $3\boldsymbol{I}$ for the scale and shape parameter. Random values are used for the prior means and mixing weights parameter. We set a step size of $\eta = 0.1$ and approximate the posterior using SEP. Table 1 show the obtained estimates. The mixing weights can be estimated using the expected value of π_j; for instance, $\mathbb{E}\left[\pi_j\right] = a_j^{'} / \sum_{j=1}^{K} a_j^{'}$.

The used parameters as well as the estimated values are shown in Table 1. We notice that estimates are very close to the target values. Since we need to store only the local and global parameters, we emphasize the fact that SEP reduces memory consumption allowing us to scale EP.

5.2 Sentiment Analysis

We analyze the problem of sentiment analysis in the setting when online users employ online platforms to express opinions or experiences regarding a product or service through reviews. We exploit these data to investigate the validity

Table 1. Original parameters and estimated parameters for the mixture of EMSD using the proposed approach.

j	π	α	β
Real			
1	0.333	[0.610, 0.318, 0.646]	1
2	0.333	[0.556, 0.188, 0.848]	1
3	0.334	[0.129, 0.891, 0.507]	1
Estimation			
1	0.335	[0.663, 0.305, 0.676]	[1.082, 1.055, 1.062]
2	0.332	[0.573, 0.098, 0.720]	[0.963, 1.027, 0.996]
3	0.333	[0.193, 0.858, 0.527]	[1.087, 0.976, 1.002]

of our framework where we know the right number of components (i.e. positive/negative, $K = 2$). We use three benchmark datasets [13,29]: 1) Amazon Review Polarity; 2) Yelp review Polarity; 3) IMDB Movie Reviews. This section presents the details of our experimentation and its results.

Before describing the experimental results, we first outline the key properties of the datasets and the performed setup. We pre-process the dataset as follows: 1) lowercase all text; 2) remove non-alphabetical characters; 3) lemmatize text. All datasets are reviews and contain two labels indicating whether the post has a positive or negative sentiment.

Amazon Review Polarity contains $180k$ customer reviews that span a period of 18 years, for products on the *Amazon.com* website. The dataset has an average of 75 words per review with a vocabulary size of over $55k$ unique words.

Yelp Review Polarity contains $560k$ user reviews from *Yelp* with an average of 133 words with $> 85k$ unique words. The Yelp dataset contains a polarity label by considering stars 1 and 2 negative, and 3 and 4 positive reviews about local businesses.

IMDB movie reviews this dataset consists of $50K$ movie reviews with an average 231 words per review and a vocabulary size of over $76k$ unique words. Ratings on IMDB are given as star values $\in [1, 10]$ which were linearly mapped to $[0, 1]$ to use as document labels; negative and positive, respectively.

We compare the clustering performance of EMSD mixture model using the proposed SEP to different models with the same approach and different learning techniques such as Expectation Propagation (EP), and maximum-likelihood (ML) for parameter estimation. More precisely, we compared the performance of EMSD models to the following models that use maximum-likelihood for estimating its parameters. Firstly, we have a mixture of Multinomials (MM) [2]. Even though the MM is appropriate for modeling common words, not words burstiness problem, we add it to the comparison to evaluate its predictive power. Next, we make a comparison with different models that capture the words bustiness problem such as Dirichlet Compound Multinomial (DCM) [14], the Exponential approximation to the Dirichlet Compound Multinomial (EDCM) [5], the Multinomial Scaled Dirichlet (MSD) [25], and the Exponential approximation to the Multinomial Scaled Dirichlet (EMSD) [28]. Furthermore, we compare to the performance of EDCM mixture model in case of considering EP for parameter estimation as we have recently proposed in [23]. We evaluate the performance of the considered models according to precision and recall as illustrated in Table 2.

Table 2. Results on the three text datasets. Comparison using precision and recall. ML: maximum-likelihood; EP: expectation propagation; SEP: sthocastic expectation propagation.

		Dataset		
Metrics		Amazon	Yelp	IMDB
Precision	ML-MM	50.83	89.12	64.18
	ML-DCM	55.65	91.01	71.14
	ML-EDCM	80.65	89.25	78.54
	EP-EDCM	86.91	80.50	86.36
	ML-MSD	82.21	86.96	84.00
	ML-EMSD	83.31	87.23	85.00
	SEP-EMSD (ours)	86.35	82.83	86.83
Recall	ML-MM	51.99	89.20	64.40
	ML-DCM	63.94	91.01	89.45
	ML-EDCM	80.88	89.28	89.33
	EP-EDCM	84.82	93.83	85.94
	ML-MSD	82.21	87.09	84.00
	ML-EMSD	83.57	87.28	86.00
	SEP-EMSD (ours)	83.91	90.02	87.64

In general, most models are superior to the Multinomial mixture model (except for Yelp dataset). We notice that SEP gives comparable results to the EDCM model in terms of precision and recall. Additionally, we evaluate an EDCM mixture that uses EP for parameter learning where we can assume that SEP is computing similar approximations to EP with the advantage that there is no need to store the parameters for each of the approximate sites. One of the main advantages is that we only store the local and global parameters, reducing memory usage. More specifically, for the Amazon dataset, EP and SEP are superior in terms of precision and recall compared with most models that use maximum-likelihood estimation. Our intuition is that the length of documents plays a critical role in parameter estimation. That is, in the Amazon dataset, for example, we obtain better precision and recall using a Bayesian approach given that the document length is relatively shorter than in the other two datasets.

6 Conclusions

In this paper, we propose a Stochastic Expectation Propagation (SEP) algorithm to learn a finite EMSD mixture model. We derive the mathematical framework using SEP, and since performing moment matching leads to an intractable integral, we use sampling in order to compute its moments. Then, we evaluate the proposed approach on both synthetic and real data and notice that SEP-EMSD provides comparable results to traditional approaches and in some cases is superior. Although we evaluated the proposed learning method with text data, we can use any type of count data such as a clustering of visual words for images or videos. It is noticeable that SEP does not need a site per data point and similar to variational inference maintains a global posterior approximation that is updated locally and reduces memory consumption.

References

1. Bishop, C.M.: Pattern Recognition and Machine Learning. Springer, Heidelberg (2006)
2. Bouguila, N., Ziou, D.: Unsupervised learning of a finite discrete mixture model based on the multinomial dirichlet distribution: Application to texture modeling. In: Fred, A.L.N. (ed.) Pattern Recognition in Information Systems, Proceedings of the 4th International Workshop on Pattern Recognition in Information Systems, PRIS 2004, in conjunction with ICEIS 2004, Porto, Portugal, April 2004, pp. 118–127. INSTICC Press (2004)
3. Boyd-Graber, J., Hu, Y., Mimno, D., et al.: Applications of topic models. Found. Trends® Inf. Retrieval **11**(2–3), 143–296 (2017)
4. Bui, T.D., Hernández-Lobato, J.M., Li, Y., Hernández-Lobato, D., Turner, R.E.: Training deep gaussian processes using stochastic expectation propagation and probabilistic backpropagation. arXiv preprint arXiv:1511.03405 (2015)
5. Elkan, C.: Clustering documents with an exponential-family approximation of the dirichlet compound multinomial distribution. In: Proceedings of the 23rd International Conference on Machine Learning, pp. 289–296. ACM (2006)
6. Fan, W., Bouguila, N.: Expectation propagation learning of a dirichlet process mixture of beta-liouville distributions for proportional data clustering. Eng. Appl. Artif. Intell. **43**, 1–14 (2015)
7. Gelman, A., et al.: Expectation propagation as a way of life: a framework for bayesian inference on partitioned data. arXiv preprint arXiv:1412.4869 (2017)
8. Hoffman, M.D., Blei, D.M., Wang, C., Paisley, J.: Stochastic variational inference. J. Mach. Learn. Res. **14**(1), 1303–1347 (2013)
9. Lafferty, J., McCallum, A., Pereira, F.C.: Conditional random fields: Probabilistic models for segmenting and labeling sequence data (2001)
10. Li, Y., Hernández-Lobato, J.M., Turner, R.E.: Stochastic expectation propagation. In: Advances in Neural Information Processing Systems, pp. 2323–2331 (2015)
11. Lochner, R.H.: A generalized dirichlet distribution in bayesian life testing. J. Roy. Stat. Soc. Ser. B (Methodol.) **37**(1), 103–113 (1975)
12. Ma, Z., Leijon, A.: Expectation propagation for estimating the parameters of the beta distribution. In: 2010 IEEE International Conference on Acoustics, Speech and Signal Processing, pp. 2082–2085. IEEE (2010)

13. Maas, A.L., Daly, R.E., Pham, P.T., Huang, D., Ng, A.Y., Potts, C.: Learning word vectors for sentiment analysis. In: Proceedings of the 49th Annual Meeting of the Association for Computational Linguistics: Human Language Technologies, vol. 1, pp. 142–150. Association for Computational Linguistics (2011)
14. Madsen, R.E., Kauchak, D., Elkan, C.: Modeling word burstiness using the dirichlet distribution. In: Proceedings of the 22nd International Conference on Machine Learning, pp. 545–552. ACM (2005)
15. Margaritis, D., Thrun, S.: A bayesian multiresolution independence test for continuous variables. arXiv preprint arXiv:1301.2292 (2013)
16. Minka, T.: Estimating a dirichlet distribution (2000)
17. Minka, T.: Power ep. Technical report, Microsoft Research, Cambridge (2004)
18. Minka, T., Lafferty, J.: Expectation-propagation for the generative aspect model. In: Proceedings of the Eighteenth Conference on Uncertainty in Artificial Intelligence, pp. 352–359. Morgan Kaufmann Publishers Inc. (2002)
19. Minka, T.P.: Expectation propagation for approximate bayesian inference. In: Proceedings of the Seventeenth Conference on Uncertainty in Artificial Intelligence, pp. 362–369. Morgan Kaufmann Publishers Inc. (2001)
20. Minka, T.P.: A family of algorithms for approximate Bayesian inference. Ph.D. thesis, Massachusetts Institute of Technology (2001)
21. Neal, R.M.: Probabilistic inference using markov chain monte carlo methods (1993)
22. Opper, M., Winther, O.: A bayesian approach to on-line learning. In: On-line Learning in Neural Networks, pp. 363–378 (1998)
23. Sumba, X., Zamzami, N., Bouguila, B.: Improving the edcm mixture model with expectation propagation. In: 2020 Association for the Advancement of Artificial Intelligence AAAI. FLAIRS 33 (2020)
24. Wong, T.T.: Alternative prior assumptions for improving the performance of naïve bayesian classifiers. Data Min. Knowl. Disc. **18**(2), 183–213 (2009)
25. Zamzami, N., Bouguila, N.: Text modeling using multinomial scaled dirichlet distributions. In: Mouhoub, M., Sadaoui, S., Ait Mohamed, O., Ali, M. (eds.) IEA/AIE 2018. LNCS (LNAI), vol. 10868, pp. 69–80. Springer, Cham (2018). https://doi.org/10.1007/978-3-319-92058-0_7
26. Zamzami, N., Bouguila, N.: An accurate evaluation of msd log-likelihood and its application in human action recognition. In: 2019 IEEE Global Conference on Signal and Information Processing (GlobalSIP), pp. 1–5. IEEE (2019)
27. Zamzami, N., Bouguila, N.: Hybrid generative discriminative approaches based on multinomial scaled dirichlet mixture models. Appl. Intell **49**(11), 3783–3800 (2019)
28. Zamzami, N., Bouguila, N.: A novel scaled dirichlet-based statistical framework for count data modeling: unsupervised learning and exponential approximation. Pattern Recogn. **95**, 36–47 (2019)
29. Zhang, X., Zhao, J., LeCun, Y.: Character-level convolutional networks for text classification. In: Advances in Neural Information Processing Systems, pp. 649–657 (2015)

Entropy-Based Variational Learning of Finite Generalized Inverted Dirichlet Mixture Model

Mohammad Sadegh Ahmadzadeh[1], Narges Manouchehri[2(✉)], Hafsa Ennajari[2], Nizar Bouguila[2], and Wentao Fan[3]

[1] Department of Electrical Engineering, Concordia University, Montreal, Canada
m_hmadza@encs.concordia.ca
[2] Concordia Institute for Information Systems Engineering, Concordia University, Montreal, Canada
narges.manouchehri@mail.concordia.ca, h_ennaja@encs.concordia.ca, nizar.bouguila@concordia.ca
[3] Department of Computer Science and Technology, Huaqiao University, Xiamen, China
fwt@hqu.edu.cn

Abstract. Mixture models are considered as a powerful approach for modeling complex data in an unsupervised manner. In this paper, we introduce a finite generalized inverted Dirichlet mixture model for semi-bounded data clustering, where we also developed a variational entropy-based method in order to flexibly estimate the parameters and select the number of components. Experiments on real-world applications including breast cancer detection and image categorization demonstrate the superior performance of our proposed model.

Keywords: Unsupervised learning · Clustering · Generalized inverted dirichlet distribution · Entropy-based variational learning

1 Introduction

Nowadays, large amounts of complex data in various formats (e.g., image, text, speech) are generated increasingly at a bottleneck speed. This increase motivated data scientists to develop tactical models in order to automatically analyze and infer useful knowledge from these data [1]. In this context, statistical modeling plays a significant role in helping machines interpret data with statistics. An essential approach in statistical modeling is finite mixture models that are effectively used for clustering purposes, separating heterogeneous data into homogeneous groups [2]. The usefulness of mixture models has been widely demonstrated in many application areas including pattern recognition, text and image analysis [3]. However, there exist several challenges to address when working with mixture models: (1) Standard finite mixture models assume that the

N. T. Nguyen et al. (Eds.): ACIIDS 2021, LNAI 12672, pp. 130–143, 2021.
https://doi.org/10.1007/978-3-030-73280-6_11

observed data are normally distributed [4]. This is not always the case, in several applications. Lately, multiple studies have shown that other non-Gaussian statistical models (e.g., Dirichlet, inverted Dirichlet and Gamma) are effective in modeling data [5–15]. Thus, choosing a suitable probability distribution that better describes the nature and the properties of the observed data is crucial to the assessment of the validity of the model. For instance, the inverted Dirichlet mixture, has good flexibility in accepting different symmetric and asymmetric forms that results in better generalization capabilities. But, the model usually supposes that the features of the vectors are positively correlated, and that is not always applicable for real-life applications. (2) In most cases, the mixture model fitting is not straightforward and analytically intractable. Methods like Expectation-Maximization (EM) and Maximum likelihood [1] are widely used in this context, but they remain impractical as they are sensitive to initialization and usually lead to over-fitting [16]. An alternative approach to solve these problems is Bayesian learning, particularly, variational inference has made the parameter estimation process more computationally efficient. (3) The selection of the number of components is an important issue to consider in the design of mixture models, because a high number of components may lead to learning the data too much, whereas inference under a model with a small number of components can be biased. To this end, multiple effective methods have been proposed, like minimum message length criterion [17,18]. To overcome the aforementioned challenges, we introduce a novel finite variational Generalized Inverted Dirichlet Mixture Model for data clustering, which learns the latent parameters based on the variational inference algorithm. Our work is motivated by the success of the Generalized Inverted Dirichlet (GID) distribution [19]. The GID has great efficiency in comparison to Gaussian distribution when dealing with positive vectors and has been shown to be more practical due to its higher general covariance structure. Also the GID samples can be represented in a transformed space where features are independent and follow the inverted Beta distribution [20–22]. Moreover, the use of the variational inference algorithm allows us to minimize the Kullback–Leibler divergence between the true posterior and the approximated variational distribution, leading to accurate and computationally efficient parameter estimation of our proposed mixture model [22]. The main challenge here, is to design a good mixture model that better fits the observed semi-bounded data with the right number of components. We propose to apply an entropy-based variational inference combined with our GID Mixture Model. We started with one component and proceed incrementally to find the best number of components and we will explain the model complexity and approximate the perfect number of components by a compression between the estimated and theoretical entropy [23] similar to researches that have been successful on distributions like the Dirichlet mixture model [24]. To demonstrate the effectiveness of the proposed approach, we evaluate the Entropy-Based Variational Learning of Finite Generalized Inverted Dirichlet Mixture Model (EV-GIDMM) on real-world applications including breast cancer detection and image categorization.

The remainder of this paper is organized as follows. We provide an overview of the statistical background of our GID mixture model in Sect. 2. Section 3 is assigned to the variational inference process of our model. We explain the entropy-based variational inference of EV-GIDMM in Sect. 4. The results of our experiments on real data are provided in Sect. 5. Finally, we conclude the paper in Sect. 6.

2 Model Specification

2.1 Finite Generalized Inverted Dirichlet Mixture

Lets us assume $\boldsymbol{Y} = (\boldsymbol{Y}_1, \ldots, \boldsymbol{Y}_N)$ is a set of N independent identically distributed vectors, where every single $\boldsymbol{Y}_i$ is defined as $\boldsymbol{Y}_i = (Y_{i1}, \ldots, Y_{iD})$, where D is the dimensionality of the vector. We are assuming that each $\boldsymbol{Y}_i$ follows a mixture of GIDs, where the probability density function of the GID is given by [19]:

$$p(\boldsymbol{Y_i} \mid \boldsymbol{\alpha}_j, \boldsymbol{\beta}_j) = \prod_{d=1}^{D} \frac{\Gamma(\alpha_{jd} + \beta_{jd})}{\Gamma(\alpha_{jd})\Gamma(\beta_{jd})} \frac{Y_{id}^{\alpha_{jd}-1}}{(1 + \sum_{l=1}^{d} Y_{il})^{\gamma_{jd}}} \tag{1}$$

where $\boldsymbol{\alpha}_j$ and $\boldsymbol{\beta}_j$ are the parameters of the GID, and they are defined as $\boldsymbol{\alpha}_j = (\alpha_{j1}, \ldots, \alpha_{jd})$ and $\boldsymbol{\beta}_j = (\beta_{j1}, \ldots, \beta_{jd})$ with constraints $\alpha_{jd} > 0$ and $\beta_{jd} > 0$. We can find γ_{id} according to $\gamma_{id} = \beta_{jd} + \alpha_{jd} - \beta_{j(d+1)}$. Supposing that the model consists of M different components [1], we are able to define the GID mixture model as follows:

$$p(\boldsymbol{Y}_i \mid \boldsymbol{\pi}, \boldsymbol{\alpha}, \boldsymbol{\beta}) = \sum_{j=1}^{M} \pi_j p(\boldsymbol{Y}_i \mid \boldsymbol{\alpha}_j, \boldsymbol{\beta}_j) \tag{2}$$

where $\boldsymbol{\pi}$ represents its mixing coefficients correlated with the components, where, $\boldsymbol{\pi} = (\pi_1, \ldots, \pi_M)$ with constrains $\pi_j \geq 0$ and $\sum_{j=1}^{M} \pi_j = 1$, and the shape parameters of the distribution are denoted as $\boldsymbol{\alpha} = (\boldsymbol{\alpha}_1, \ldots, \boldsymbol{\alpha}_M)$, $\boldsymbol{\beta} = (\boldsymbol{\beta}_1, \ldots, \boldsymbol{\beta}_M)$ and $j = 1, \ldots, M$. According to [21], we can replace the GID distribution with a product of D Inverted Beta distributions, considering that it does not change the model, therefore, Eq. (2) can be rewritten as:

$$p(\mathcal{X} \mid \pi, \alpha, \beta) = \prod_{i=1}^{N} \left(\sum_{j=1}^{M} \pi_j \prod_{l=1}^{D} P_{iBeta}(X_{il} \mid \alpha_{jl}, \beta_{jl}) \right) \tag{3}$$

By considering that $\mathcal{X} = (\boldsymbol{X}_1, \ldots, \boldsymbol{X}_N)$ where $\boldsymbol{X}_i = (X_{i1}, \ldots, X_{iD})$, we have $X_{il} = Y_{il}$ and $X_{il} = \frac{Y_{il}}{1+\sum_{k=1}^{l-1} Y_{ik}}$ for $l > 1$. The inverted Beta distribution is defined by $P_{iBeta}(X_{il}|\alpha_{jl}, \beta_{jl})$ with the parameters α_{jl} and β_{jl} and given by:

$$P_{iBeta}(X_{il} \mid \alpha_{jl}, \beta_{jl}) = \frac{\Gamma(\alpha_{jl} + \beta_{jl})}{\Gamma(\alpha_{jl})\Gamma(\beta_{jl})} \frac{{X_{il}}^{\alpha_{jl}-1}}{(1 + X_{il})^{\alpha_{jl}+\beta_{jl}}} \tag{4}$$

In proportion to this design, we are able to estimate the parameters from Eq. (3) instead of the Eq. (2). We define the latent variables as $\mathcal{Z} = (\boldsymbol{Z}_1, \dots, \boldsymbol{Z}_N)$ where $\boldsymbol{Z}_i = (Z_{i1}, \dots, Z_{iM})$ with the conditions $Z_{ij} \in \{0, 1\}$ that Z_{ij} is equal to 1 if $\boldsymbol{X}_i$ is assigned to cluster j and zero otherwise, and $\sum_{j=1}^{M} Z_{ij} = 1$. The conditional probability for the latent variables $\mathcal{Z}$ given $\boldsymbol{\pi}$ can be written as:

$$p(\mathcal{Z} \mid \boldsymbol{\pi}) = \prod_{i=1}^{N} \prod_{j=1}^{M} {\pi_j}^{Z_{ij}} \tag{5}$$

We write the probability of the observed data vectors $\mathcal{X}$ given the latent variable and component parameters as:

$$p(\mathcal{X} \mid \mathcal{Z}, \boldsymbol{\alpha}, \boldsymbol{\beta}) = \prod_{i=1}^{N} \prod_{j=1}^{M} \Big(\prod_{l=1}^{D} p_{iBeta}(X_{il} \mid \alpha_{jl}, \beta_{jl}) \Big)^{Z_{ij}} \tag{6}$$

By assuming that the parameters are independent and positive, we can suppose that the priors of these parameters are Gamma distributions $\mathcal{G}(.)$. According to [25], we can describe them as:

$$p(\alpha_{jl}) = \mathcal{G}(\alpha_{jl} \mid u_{jl}, \nu_{jl}) = \frac{\nu_{jl}^{u_{jl}}}{\Gamma(u_{jl})} {\alpha_{jl}}^{u_{jl}-1} e^{-\nu_{jl}\alpha_{jl}} \tag{7}$$

$$p(\beta_{jl}) = \mathcal{G}(\beta_{jl} \mid g_{jl}, h_{jl}) = \frac{h_{jl}^{g_{jl}}}{\Gamma(g_{jl})} {\beta_{jl}}^{g_{jl}-1} e^{-h_{jl}\beta_{jl}} \tag{8}$$

We define the joint distribution including all random variables , as follows:

$$p(\mathcal{X}, \mathcal{Z}, \boldsymbol{\alpha}, \boldsymbol{\beta} \mid \boldsymbol{\pi}) = p(\mathcal{X} \mid \mathcal{Z}, \boldsymbol{\alpha}, \boldsymbol{\beta}) p(\mathcal{Z} \mid \boldsymbol{\pi}) p(\boldsymbol{\alpha}) p(\boldsymbol{\beta}) \tag{9}$$

$$\begin{aligned} p(\mathcal{X}, \mathcal{Z}, \boldsymbol{\alpha}, \boldsymbol{\beta} \mid \boldsymbol{\pi}) = & \prod_{i=1}^{N} \prod_{j=1}^{M} \Big(\prod_{l=1}^{D} \frac{\Gamma(\alpha_{jl} + \beta_{jl})}{\Gamma(\alpha_{jl})\Gamma(\beta_{jl})} \frac{X_{il}^{\alpha_{jl}-1}}{(1 + X_{il})^{\alpha_{jl}+\beta_{jl}}} \Big)^{Z_{ij}} \Big(\prod_{i=1}^{N} \prod_{j=1}^{M} \pi_j^{Z_{ij}} \Big) \\ & \prod_{j=1}^{M} \prod_{l=1}^{D} \Big(\frac{\nu_{jl}^{u_{jl}}}{\Gamma(u_{jl})} {\alpha_{jl}}^{u_{jl}-1} e^{-\nu_{jl}\alpha_{jl}} \times \frac{h_{jl}^{g_{jl}}}{\Gamma(g_{jl})} {\beta_{jl}}^{g_{jl}-1} e^{-h_{jl}\beta_{jl}} \Big) \end{aligned} \tag{10}$$

3 Model Learning with Variational Inference

The GID mixture model contains hidden variables that can not be estimated directly. In order to estimate them, we apply the variational inference method, in which we aim to find an approximation of the posterior probability distribution of $p(\Theta|\mathcal{X}, \boldsymbol{\pi})$ by having $\Theta = \{\mathcal{Z}, \boldsymbol{\alpha}, \boldsymbol{\beta}\}$. Inspired by [24], we introduce $Q(\Theta)$ as an approximation of the true posterior distribution $p(\Theta|\mathcal{X}, \boldsymbol{\pi})$. We make use of the Kullback-Leibler (KL) divergence in order to minimize the difference between the true posterior distribution and the approximated one, which can be expressed as follows:

$$KL(Q \,||\, P) = -\int Q(\Theta) \ln \left(\frac{p(\Theta \mid \mathcal{X}, \boldsymbol{\pi})}{Q(\Theta)} \right) d\Theta = \ln\, p(\mathcal{X} \mid \boldsymbol{\pi}) - \mathcal{L}(Q) \quad (11)$$

where $\mathcal{L}(Q)$ is defined as:

$$\mathcal{L}(Q) = \int Q(\Theta) \ln \left(\frac{p(\mathcal{X}, \Theta \mid \boldsymbol{\pi})}{Q(\Theta)} \right) d\Theta \quad (12)$$

Starting from the fact that $\mathcal{L}(Q) \leq \ln p(\mathcal{X}|\boldsymbol{\pi})$, we can see that $\mathcal{L}(Q)$ is the lower bound of the log likelihood. Thus, we have to maximize $\mathcal{L}(Q)$ in order to minimize the KL divergence. We assume a factorization assumption around $Q(\Theta)$ to apply it in variational inference. This assumption is called the Mean Field Approximation. We can factorize the posterior distribution $Q(\Theta)$ as $Q(\Theta) = Q(\mathcal{Z})Q(\boldsymbol{\alpha})Q(\boldsymbol{\beta})Q(\boldsymbol{\pi})$ [26,27]. In order to obtain a variational solution for the lower bound with respect to all the model parameters, we consider an optimal solution for a fix parameter s that is defined as $\ln Q_s^*(\Theta_s) = \langle \ln p(\mathcal{X}, \Theta) \rangle_{i \neq s}$ where $\langle \cdot \rangle_{i \neq s}$ refers to the expectation with respect to all the parameters apart from Θ_s, if an exponential is taken from both sides, the normalized equation is as follows.

$$Q_s(\Theta_s) = \frac{exp\langle \ln\, p(\mathcal{X}, \Theta) \rangle_{i \neq s}}{\int exp\langle \ln\, p(\mathcal{X}, \Theta) \rangle_{i \neq s} d\Theta} \quad (13)$$

We obtain the optimal variational posteriors solution that are formulated as:

$$Q(\mathcal{Z}) = \prod_{i=1}^{N} \prod_{j=1}^{M} r_{ij}^{Z_{ij}} \quad (14)$$

$$Q(\boldsymbol{\alpha}) = \prod_{j=1}^{M} \prod_{l=1}^{D} \mathcal{G}(\alpha_{jl} \mid u_{jl}^*, \nu_{jl}^*)\ ,\ Q(\boldsymbol{\beta}) = \prod_{j=1}^{M} \prod_{l=1}^{D} \mathcal{G}(\beta_{jl} \mid g_{jl}^*, h_{jl}^*) \quad (15)$$

$$r_{ij} = \frac{\tilde{r}_{ij}}{\sum_{j=1}^{M} \tilde{r}_{ij}} \quad (16)$$

$$\ln \tilde{r}_{ij} = \ln \pi_j + \sum_{l=1}^{D} \tilde{R}_{jl} + (\bar{\alpha}_{jl} - 1) \ln X_{il} - (\bar{\alpha}_{jl} + \bar{\beta}_{jl}) \ln(1 + X_{il}) \quad (17)$$

$$\begin{aligned} \tilde{R} = \ln \frac{\Gamma(\bar{\alpha} + \bar{\beta})}{\Gamma(\bar{\alpha})\Gamma(\bar{\beta})} &+ \bar{\alpha}[\psi(\bar{\alpha} + \bar{\beta}) - \psi(\bar{\alpha})](\langle \ln \beta \rangle - \ln \bar{\beta}) + 0.5\alpha^2[\psi'(\bar{\alpha} + \bar{\beta}) \\ &- \psi'(\bar{\alpha}))]\langle(\ln \alpha - \ln \bar{\alpha})^2\rangle + 0.5\beta^2[\psi'(\bar{\alpha} + \bar{\beta}) - \psi'(\bar{\beta}))]\langle(\ln \beta - \ln \bar{\beta})^2\rangle \\ &+ \bar{\alpha}\bar{\beta}\psi'(\bar{\alpha} + \bar{\beta})(\langle \ln \alpha \rangle - \ln \bar{\alpha})(\langle \ln \beta \rangle - \ln \bar{\beta}) \end{aligned} \quad (18)$$

$$u_{jl}^* = u_{jl} + \sum_{i=1}^{N} \langle Z_{ij} \rangle \bar{\alpha}_{jl} \left[\psi(\bar{\alpha}_{jl} + \bar{\beta}_{jl}) - \psi(\bar{\alpha}_{jl}) + \bar{\beta}_{jl} \psi'(\bar{\alpha}_{jl} + \bar{\beta}_{jl})(\langle \ln \beta_{jl} \rangle - \ln \bar{\beta}_{jl}) \right] \quad (19)$$

$$\nu_{jl}^* = \nu_{jl} - \sum_{i=1}^{N} \langle Z_{ij} \rangle \ln \frac{X_{il}}{1 + X_{il}} \quad (20)$$

$$g_{jl}^* = g_{jl} + \sum_{i=1}^{N} \langle Z_{ij} \rangle \bar{\beta}_{jl} \left[\psi(\bar{\alpha}_{jl} + \bar{\beta}_{jl}) - \psi(\bar{\beta}_{jl}) + \bar{\alpha}_{jl} \psi'(\bar{\alpha}_{jl} + \bar{\beta}_{jl})(\langle \ln \alpha_{jl} \rangle - \ln \bar{\alpha}_{jl}) \right] \quad (21)$$

$$h_{jl}^* = h_{jl} - \sum_{i=1}^{N} \langle Z_{ij} \rangle \ln \frac{1}{1 + X_{il}} \quad (22)$$

Furthermore $\psi(\cdot)$ and $\psi^{'}(\cdot)$ are representing the Digamma and Trigamma functions, respectively. As $R = \langle \ln \frac{\Gamma(\bar{\alpha}+\bar{\beta})}{\Gamma(\bar{\alpha})\Gamma(\bar{\beta})} \rangle$ is intractable, we have used the second order Taylor expansion for its approximation. The expected values of the above equations are as follows:

$$\langle Z_{ij} \rangle = r_{ij} \quad (23)$$

$$\bar{\alpha}_{jl} = \langle \alpha_{jl} \rangle = \frac{u_{jl}^*}{\nu_{jl}^*} \ , \ \langle \ln \alpha_{jl} \rangle = \psi(u_{jl}^*) - \ln \nu_{jl}^* \quad (24)$$

$$\bar{\beta}_{jl} = \langle \beta_{jl} \rangle = \frac{g_{jl}^*}{h_{jl}^*} \ , \ \langle \ln \beta_{jl} \rangle = \psi(g_{jl}^*) - \ln h_{jl}^* \quad (25)$$

$$\pi_j = \frac{1}{N} \sum_{i=1}^{N} r_{ij} \quad (26)$$

4 Entropy-Based Variational Model Learning

In this section, we develop an entropy-based variational inference to learn the generalized inverted Dirichlet mixture model, that is mainly motivated by [23]. The core idea is to evaluate the quality of fitting of a component of our mixture model. Hence, we do a comparison between the theoretical maximum entropy and the MeanNN entropy [28]. In case of a significant difference, we proceed with a splitting process to fit the component, which consists in splitting the component into two new clusters.

4.1 Differential Entropy Estimation

The probability density function of an observation $\boldsymbol{X}_i = (\boldsymbol{X}_1, \ldots, \boldsymbol{X}_D)$ is defined as $p(\boldsymbol{X}_i)$, with a set of N samples $\{\boldsymbol{X}_i, \ldots, \boldsymbol{X}_N\}$, the differential entropy can be defined as:

$$H(\boldsymbol{X}_i) = - \int p(\boldsymbol{X}_i) log_2 P(\boldsymbol{X}_i) d\boldsymbol{X}_i \quad (27)$$

We introduce the maximum differential entropy of the GID as follows:

$$H_{GID}(\boldsymbol{X}_i \mid \alpha_j, \beta_j) = \sum_{l=1}^{D} \Big[-\ln\Gamma(\alpha_{jl}+\beta_{jl}) + \ln\Gamma(\alpha_{jl}) + \ln\Gamma(\beta_{jl}) \tag{28}$$
$$-(\alpha_{jl}-1)\big[-\psi(\alpha_{jl}+\beta_{jl}) + \psi(\alpha_{jl})\big] + (\alpha_{jl}+\beta_{jl})[-\psi(\alpha_{jl}+\beta_{jl})]\Big]$$

4.2 MeanNN Entropy Estimator

In order to make sure that the specified component is indeed distributed according to a generalized inverted Dirichlet distribution, we choose the MeanNN entropy estimator [23], to estimate $H(\boldsymbol{X}_i)$ for random variable $\boldsymbol{X}_i$ with D dimensions, that has an unknown density function $P(\boldsymbol{X}_i)$ [29]. By considering the fact that the Shannon entropy estimator in (27) can be considered equal to the average of $-\log P(\boldsymbol{X}_i)$, we can exploit an unbiased estimator by estimating $\log P(\boldsymbol{X}_i)$ [28,29]. We assume that $\boldsymbol{X}_i$ is the center of a ball with diameter ϵ, and that there is a point within the distance $[\epsilon, \epsilon + d_\epsilon]$ from $\boldsymbol{X}_i$. We have $\hat{k}-1$ points in a smaller distance, and the other $N-\hat{k}-1$ points are within a large distance from $\boldsymbol{X}_i$. Consequently, we can define the probability of the distances and the k-th nearest neighbor as follows:

$$p_{i\hat{k}}(\epsilon) = \frac{(N-1)!}{\left(\hat{k}-1\right)!\left(N-\hat{k}-1\right)!}\frac{dp_i(\epsilon)}{d\epsilon}p_i^{\hat{k}-1}(1-p_i)^{N-\hat{k}-1} \tag{29}$$

where $p_i(\epsilon)$ denotes the mass of the ϵ-ball centered on $\boldsymbol{X}_i$:

$$p_i(\epsilon) = \int_{||\boldsymbol{X}-\boldsymbol{X}_i||<\epsilon} p(\boldsymbol{X}_i)d\boldsymbol{X}_i \tag{30}$$

We can easily define the expectation of $\log p_i(\epsilon)$ with respect to $p_i(\epsilon)$ as mentioned in Eq. (31):

$$\mathbb{E}\big(\log p_i(\epsilon)\big) = \int_0^{\infty} p_{i\hat{k}} \log p_i(\epsilon)d\epsilon = \psi(\hat{k}) - \psi(N) \tag{31}$$

Imagine $P(\boldsymbol{X}_i)$ is unchanging in the center of the ϵ-ball, we have $p_i(\epsilon) \simeq V_d\epsilon^d p(\boldsymbol{X}_i)$, where d corresponds to the dimension of $\boldsymbol{X}_i$, and V_d is the unit ball volume calculated by $V_d = \pi^{\frac{d}{2}}\Gamma(1+d/2)$. Now, we are able to approximate $-\log P(\boldsymbol{X}_i)$ by substituting (30) into (31) we can get the Eq. (32). Hence, we get the unbiased K-NN estimator of the differential entropy, expressed in (33):

$$-\log p(\boldsymbol{X}_i) \simeq \psi(N) - \psi(\hat{k}) + dE(\log\epsilon) + \log V_d \tag{32}$$

$$H_{\hat{k}}(\boldsymbol{X}) = \psi(N) - \psi\left(\hat{k}\right) + \frac{d}{N}\sum_{i=1}^{N} log\epsilon_i + \log V_d \tag{33}$$

To reduce the high computational expenses of the K-NN estimator, we use an extension of the K-NN estimator called MeanNN, proposed in [30]. The main idea behind the MeanNN entropy estimator is to average the $\hat{k}$ nearest neighbor statics for all feasible values of order k in the range of $[1, N-1]$. The MeanNN estimator for the differential entropy is calculated according to (34).

$$H_M(X) = \frac{1}{N-1}\sum_{\hat{k}=1}^{N-1} H_{\hat{k}}(X) = \log V_d + \psi(N) + \frac{1}{N-1}\sum_{\hat{k}=1}^{N-1}\left[\frac{d}{N}\sum_{i=1}^{N}\log \epsilon_{i,\hat{k}} - \psi\left(\hat{k}\right)\right] \quad (34)$$

where $\epsilon_{i,\hat{k}}$ determines the $\hat{k}$-th nearest neighbor of $\boldsymbol{X}_i$. To find the maximum differential entropy of each individual cluster, we use:

$$H_{GID} = \sum_{j=1}^{M} \pi_j H_{GID}(j) \quad (35)$$

At this point, we are able to give an accurate evaluation of the model fitting, by evaluating and comparing the MeanNN and the theoretical maximum differential entropy [30]. Afterwards, we define Ω_{GID}, which is the normalized weighted sum of the difference between the theoretical and the estimated entropy of every component correlated with the generalized inverted Dirichlet mixture model, as expressed bellow:

$$\Omega_{GID} = \sum_{j=1}^{M} \pi_j \left[\frac{H_{GID}(j) - H_M(j)}{H_{GID}(j)}\right] = \sum_{j=1}^{M} \pi_j \left[1 - \frac{H_M(j)}{H_{GID}(j)}\right] \quad (36)$$

The normalized weight Ω_{GID} operates in the range of $[0, 1]$ and it is equal to zero, only if the data was genuinely distributed. The splitting process is performed by choosing the cluster j^* with the highest Ω_{GID} according to Eq. (37), and split the chosen component j^* into two new components.

$$j^* = \arg\max_j \left[\Omega_{GID}(j)\right] = \arg\max_j \left[\pi_j \frac{H_{GID}(j) - H_M(j)}{H_{GID}(j)}\right] \quad (37)$$

The overall entropy-based variational learning algorithm of the GID mixture model is illustrated in Algorithm 1.

Algorithm 1. Entropy-based variational learning of GID mixture models

1. Initialization: Set $M = 1$, $j^* = M$, $\pi_1 = 1.$ and initialize hyperparameters u_{jl},ν_{jl},g_{jl},h_{jl}.
2. The splitting process.
   - Split j^* into two new components j_1 and j_2 with equal proportion $\pi^*/2$.
   - Set $M = M + 1$.
   - Initialize the parameters of j_1 and j_2 using the same parameters of j^*.
3. Apply standard variational Bayes until convergence.
4. Determine the number of components through the evaluation of the mixing coefficients π_j according to 26.
5. If $\pi_j \approx 0$. where $j \in 1, \ldots, M$ then set $M = M - 1$ and terminate the program.
6. Else evaluate Ω_{MD}, choose j^* according to 37 and go back to the splitting process in step 2.

5 Experimental Results

In order to demonstrate the effectiveness of the proposed model, Entropy-Based Variational Learning of Finite Generalized Inverted Dirichlet Mixture Model (EV-GIDMM), we conduct several experiments on two real-world challenging applications, including breast cancer detection and image categorization. In the first one, we used the standard breast cancer (Wisconsin Prognostic) dataset with numerical features, whereas in the second one, we run our experiments on two other popular datasets, namely, Caltech101 and Describable Texture Dataset (DTD). To validate the performance of our model, we compared our proposed EV-GIDMM against three unsupervised state-of-the-art mixture models, including the Entropy-based variational inference on Multivariate Beta Mixture Model (EV-MBMM) [23], variational Dirichlet Mixture Model (varDMM) [25] and Entropy-based variational on Dirichlet Mixture Model (E-DMM) [24].

5.1 Breast Cancer

The first application that we considered to evaluate the performance of our proposed model is breast cancer detection. According to the WHO (World Health Organization), breast cancer has been declared as the most frequent cancer among women that affects about 2.1 million women every year. Machine learning techniques can be of great help in this context, in early detection of women breast cancer, thus, they can have a great impact on the breast cancer treatment. To this end, we applied our proposed model on the breast cancer Wisconsin dataset that is publicly available[1]. This dataset includes 569 data samples of patients seen by Dr. Wolberg, that have been diagnosed with either malignant or benign cancer. The number of patients having a benign tumor is 357, whereas 212 cases with malignant tumor cancer. This dataset was obtained by applying the Fine Needle Aspiration (FNA) method [31,32], and it contains cases showing invasive

[1] https://archive.ics.uci.edu/ml/datasets/Breast+Cancer+Wisconsin+(Diagnostic).

breast cancer and no sign of distant metastases. The first 30 features describe the characteristics of each nuclei cell in the images of the tissue. Table 1 shows the experimental results of our model as well as the baseline methods for the breast cancer detection task. We can see that our proposed EV-GIDMM successfully achieved the best accuracy on this task.

Table 1. Accuracy performance of our model and the baselines on the breast cancer dataset

Method	Accuracy(%)
EV-GIDMM	**92.6**
EV-MBMM	90.8
E-DMM	89.7
varDMM	63.5

5.2 Image Analysis

We are now ready to evaluate the performance of the proposed approach on the image categorization task, which is a significant research topic and aims at classifying images into their corresponding category. To do so, we used two popular image datasets, namely, Caltech101 and Describable Texture Dataset (DTD). In this experiment, we first considered the Caltech101 image dataset[2] [33], which originally contains a set of images depicting objects belonging to 101 classes, from which we selected three main object categories: Airplane, Sea Horse and Brain. Some sample images from this dataset are illustrated in Fig. 1.

Fig. 1. Sample images of each cluster from the Caltech101 dataset.

In order to use our model for the selected dataset, we need to form a bag of visual words model (BoVW) [34]. Before applying the BoVW, we first need to apply some descriptor extraction method, that, we choose SIFT [35]. Therefore we extract the features with the help of SIFT and then apply K-means clustering on the descriptors extracted with SIFT from the image. As a result a BOVW feature vector is formed for each image. Our experiments revealed that the SIFT

[2] http://www.vision.caltech.edu/Image_Datasets/Caltech101.html.

method is more suitable for our selected dataset, resulting in more discriminative descriptors. After applying SIFT to all images, we obtain a matrix that serves as an input for our model. We report the results of this experiment in Table 2, which shows that our proposed model outperformed all the baseline methods in image clustering, with a considerable accuracy margin of almost 6.6%.

Table 2. Accuracy comparison of our proposed model and the baseline methods on the Caltech101 dataset.

Method	Accuracy(%)
EV-GIDMM	**90.9**
EV-MBMM	84.3
E-DMM	74.9
varDMM	40.3

In the second part of our experiments, we focus on texture differentiation. This dataset will be a good challenge for our model as images are very similar. In order to show how machines are becoming more capable of detecting and recognizing fine-grained images, in this experiment, we chose to use the Describable Texture Dataset[3] that includes 120 images per class where each class consists of different types of textures. We have chosen Dotted, Frilly and Meshed image categories to evaluate our model as illustrated in Fig. 2.

Fig. 2. Sample images of each cluster from the DTD dataset.

Similarly, we performed the BoVW and used SIFT, to generate a discriminative input for our EV-GIDMM. The results of clustering evaluation on DTD are listed in Table 3. From this table it can be confirmed that our proposed mixture model achieves the best accuracy performance among all the other mixture models.

[3] https://www.robots.ox.ac.uk/~vgg/data/dtd/.

Table 3. Accuracy comparison of our EV-GIDMM approach and the baseline methods on the DTD dataset.

Method	Accuracy(%)
EV-GIDMM	**85.5**
EV-MBMM	65.3
E-DMM	65.8
varDMM	71.9

6 Conclusion

In this paper, we introduced an unsupervised entropy-based variational framework that effectively learns the finite generalized inverted Dirichlet mixture model. In our method, we used a splitting technique called Entropy, where we started by comparing the theoretical maximum entropy and the resulting entropy from MeanNN. Thereafter, we proceeded to split the component that has the highest difference into two smaller components, since it was concluded that the mixture model is not describing the component properly. Our experimental results have demonstrated that EV-GIDMM works very well and has outperformed other models on two real-world applications, namely, breast cancer detection and image categorization, across three different benchmark data sets. The results indicate that our proposed mixture model is able to produce high quality data clusters.

Acknowledgment. The completion of this research was made possible thanks to the Natural Sciences and Engineering Research Council of Canada (NSERC) and the National Natural Science Foundation of China (61876068).

References

1. Bishop., C.M : Pattern recognition and machine learning. Information science and statistics. Springer, New York, NY, (2006). Softcover published in 2016
2. McLachlan, G.J., Peel, D.: Finite Mixture Models. John Wiley & Sons, Hoboken (2004)
3. Ho, T.K., Baird, H.S.: Large-scale simulation studies in image pattern recognition. IEEE Trans. Pattern Analy. Mach. Intell. **19**(10), 1067–1079 (1997)
4. Fan, W., Bouguila, N.: Non-Gaussian data clustering via expectation propagation learning of finite Dirichlet mixture models and applications. Neural Process. Lett. **39**(2), 115–135 (2014)
5. Bdiri, T., Bouguila, N.: Positive vectors clustering using inverted Dirichlet finite mixture models. Expert Syst. Appl. **39**(2), 1869–1882 (2012)
6. Bouguila, N., Ziou, D.: A countably infinite mixture model for clustering and feature selection. Knowl. Inf. Syst. **33**(2), 351–370 (2012)
7. Bouguila, N., Amayri, O.: A discrete mixture-based kernel for SVMs: application to spam and image categorization. Inf. Process. Manag. **45**(6), 631–642 (2009)

8. Sefidpour, A., Bouguila, N.: Spatial color image segmentation based on finite non-Gaussian mixture models. Expert Syst. Appl. **39**(10), 8993–9001 (2012)
9. Bouguila, N.: A model-based approach for discrete data clustering and feature weighting using MAP and stochastic complexity. IEEE Trans. Knowl. Data Eng. **21**(12), 1649–1664 (2009)
10. Bdiri, T., Bouguila, N.: Bayesian learning of inverted Dirichlet mixtures for SVM kernels generation. Neural Comput. Appl. **23**(5), 1443–1458 (2013)
11. Fan, W., Bouguila, N.: Online learning of a Dirichlet process mixture of beta-liouville distributions via variational inference. IEEE Trans. Neural Netw. Learn. Syst. **24**(11), 1850–1862 (2013)
12. Mashrgy, M.A.I., Bdiri, T., Bouguila, N.: Robust simultaneous positive data clustering and unsupervised feature selection using generalized inverted Dirichlet mixture models. Knowl. Based Syst. **59**, 182–195 (2014)
13. Bdiri, T., Bouguila, N.: Learning inverted Dirichlet mixtures for positive data clustering. In: Kuznetsov, S.O., Ślęzak, D., Hepting, D.H., Mirkin, B.G. (eds.) RSFDGrC 2011. LNCS (LNAI), vol. 6743, pp. 265–272. Springer, Heidelberg (2011). https://doi.org/10.1007/978-3-642-21881-1_42
14. Bdiri, T., Bouguila, N.: An infinite mixture of inverted Dirichlet distributions. In: Lu, B.-L., Zhang, L., Kwok, J. (eds.) ICONIP 2011. LNCS, vol. 7063, pp. 71–78. Springer, Heidelberg (2011). https://doi.org/10.1007/978-3-642-24958-7_9
15. Tirdad, P., Bouguila, N., Ziou, D.: Variational learning of finite inverted Dirichlet mixture models and applications. In: Laalaoui, Y., Bouguila, N. (eds.) Artificial Intelligence Applications in Information and Communication Technologies. SCI, vol. 607, pp. 119–145. Springer, Cham (2015). https://doi.org/10.1007/978-3-319-19833-0_6
16. Fukumizu, K., Amari, S.: Local minima and plateaus in hierarchical structures of multilayer perceptrons. Neural Netw. **13**(3), 317–327 (2000)
17. Bouguila, N., Ziou, D.: High-dimensional unsupervised selection and estimation of a finite generalized Drichlet mixture model based on minimum message length. IEEE Trans. Pattern Anal. Mach. Intell. **29**(10), 1716–1731 (2007)
18. Bouguila, N., Ziou, D.: MML-based approach for finite Dirichlet mixture estimation and selection. In: Perner, P., Imiya, A. (eds.) MLDM 2005. LNCS (LNAI), vol. 3587, pp. 42–51. Springer, Heidelberg (2005). https://doi.org/10.1007/11510888_5
19. Maanicshah, K., Bouguila, N., Fan, W.: Variational learning for finite generalized inverted Dirichlet mixture models with a component splitting approach. In: 2019 IEEE 28th International Symposium on Industrial Electronics (ISIE), pp. 1453–1458. IEEE (2019)
20. Bourouis, S., Mashrgy, M.A.L., Bouguila, N.: Bayesian learning of finite generalized inverted Dirichlet mixtures: application to object classification and forgery detection. Expert Syst. Appl. **41**(5), 2329–2336 (2014)
21. Bdiri, T., Bouguila, N., Ziou, D.: Variational Bayesian inference for infinite generalized inverted Dirichlet mixtures with feature selection and its application to clustering. Appl. Intell. **44**(3), 507–525 (2016)
22. Mashrgy, M.A.L., Bdiri, T., Bouguila, N.: Robust simultaneous positive data clustering and unsupervised feature selection using generalized inverted Dirichlet mixture models. Knowl.-Based Syst. **59**, 182–195 (2014)
23. Manouchehri, N., Rahmanpour, M., Bouguila, N., Fan, W.: Learning of multivariate beta mixture models via entropy-based component splitting. In: 2019 IEEE Symposium Series on Computational Intelligence (SSCI), pp. 2825–2832. IEEE (2019)

24. Fan, W., Al-Osaimi, F.R., Bouguila, N., Du, J.: Proportional data modeling via entropy-based variational Bayes learning of mixture models. Appl. Intell. **47**(2), 473–487 (2017). https://doi.org/10.1007/s10489-017-0909-0
25. Fan, W., Bouguila, N., Ziou, D.: Variational learning for finite Dirichlet mixture models and applications. IEEE Trans. Neural Netw. Learn. Syst. **23**(5), 762–774 (2012)
26. Chandler, D.: Introduction to Modern Statistical. Mechanics. Oxford University Press, Oxford, UK (1987)
27. Celeux, G., Forbes, F., Peyrard, N.: Em procedures using mean field-like approximations for Markov model-based image segmentation. Pattern Recogn. **36**(1), 131–144 (2003)
28. Faivishevsky, L., Goldberger, J.: ICA based on a smooth estimation of the differential entropy. In: Advances in Neural Information Processing Systems, pp. 433–440 (2009)
29. Leonenko, N., Pronzato, L., Savani, V., et al.: A class of rényi information estimators for multidimensional densities. Ann. Stat. **36**(5), 2153–2182 (2008)
30. Penalver, A., Escolano, F.: Entropy-based incremental variational Bayes learning of Gaussian mixtures. IEEE Trans. Neural Netw. Learn. Syst. **23**(3), 534–540 (2012)
31. Dua, D., Graff, C.: UCI machine learning repository (2017)
32. Wolberg, W.H., Street, W.N., Mangasarian, O.L.: Machine learning techniques to diagnose breast cancer from image-processed nuclear features of fine needle aspirates. Cancer Lett. **77**(2–3), 163–171 (1994)
33. Fei-Fei, L., Fergus, R., Perona, P.: Learning generative visual models from few training examples: an incremental Bayesian approach tested on 101 object categories. In: 2004 Conference on Computer Vision and Pattern Recognition Workshop, pp. 178–178. IEEE (2004)
34. Li, T., Mei, T., Kweon, I.-S., Hua, X.-S.: Contextual bag-of-words for visual categorization. IEEE Trans. Circ. Syst. Video Technol. **21**(4), 381–392 (2010)
35. Lowe, D.G.: Distinctive image features from scale-invariant keypoints. Int. J. Comput. Vis. **60**(2), 91–110 (2004)

Mixture-Based Unsupervised Learning for Positively Correlated Count Data

Ornela Bregu[1(✉)], Nuha Zamzami[2], and Nizar Bouguila[1]

[1] Concordia Institute for Information Systems Engineering (CIISE), Concordia University, Montreal, QC, Canada
{ornela.bregu,nizar.bouguila}@concordia.ca, nezamzami@uj.edu.sa

[2] Department of Computer Science and Artificial Intelligence, University of Jeddah, Jeddah, Saudi Arabia

Abstract. The Multinomial distribution has been widely used to model count data. However, its Naive Bayes assumption usually degrades clustering performance especially when correlation between features is imminent, i.e., text documents. In this paper, we use the Negative Multinomial distribution to perform clustering based on finite mixture models, where the mixture parameters are to be estimated using a novel minorization-maximization algorithm, thriving in high-dimensionality optimization settings. Furthermore, we integrate a model-based feature selection approach to determine the optimal number of components in the mixture. To evaluate the clustering performance of the proposed model, three real-world applications are considered, namely, COVID-19 analysis, Web page clustering and facial expression recognition.

Keywords: Negative multinomial distribution · Positive correlation · Minorization-maximization · Minimum message length · Overdispersion

1 Introduction

Data mining has gained significant momentum during the past few years, triggered by the immediate need to extract useful information from large datasets [8]. It is estimated that 80% of the enormous volume of generated data is unstructured, in the form of document or image collections, audio, video, log files, sensor, or social media posts [40]. It is common for unstructured datasets not to have associated labels, which brings numerous challenges during the knowledge discovery process. Since labeled data is both tedious and costly to obtain, cluster analysis serves as an important tool to perform unlabeled data exploration, based on the similarity measures between objects. More specifically, clustering, as an unsupervised learning approach, assigns data into several clusters in such a way that the inter-class variability is maximized while the intra-class variability is minimized. Therefore, cluster analysis has been widely applied in the past in many applications, from document/image organization, summarization and retrieval [1,19,32] to scene classification in computer vision [4,5,49] and speech recognition in human-machine interaction [3].

N. T. Nguyen et al. (Eds.): ACIIDS 2021, LNAI 12672, pp. 144–154, 2021.
https://doi.org/10.1007/978-3-030-73280-6_12

Most existing approaches use the Bag-of-Words (BOW) representation, where each document or image can be represented by a vector corresponding to the appearance frequencies of words [44] or visual words [16,22], respectively. Also, in speech recognition systems, human speech is converted into textual data, where user queries are clustered in order to provide an appropriate response [7]. A major characteristic of this representation is the generation of sparse, high-dimensional count data [15]. Traditional clustering methods, such as K-means, spectral clustering, etc., perform poorly in computing distance measures in a high dimensional setting [12,20]. Matrix factorization methods such as Latent Semantic Indexing [26], Non-Negative Matrix Factorization [31,42], Concept Factorization [36], etc., have been used to reduce the dimensionality, reduce the noise of similarity and magnify the semantic effects of data. Despite the numerous benefits, it is impossible to perform dimensionality reduction without potential useful information loss [10,46]. Therefore, finite mixture models have been widely used to provide a formal framework for clustering, where data is considered to arise from two or more underlying groups with common distributional form but different parameters [9,11,34].

To model discrete data sets, Multinomial Distribution (MN) [14,28] has been widely in the past, under the assumption that features are independent from each other, which is not the case when dealing with high-dimensional vectors generated from the BOW representation [30]. Indeed, the generated feature vectors have several major characteristics to be considered before choosing a distribution to model the data. For example, words in text documents appear in bursts [33], i.e., if a word appears once, it is more likely to appear again. Also, it is common for the data to be overdispersed, owing to the large size of the vocabulary and the sparsity level (more than 90%). As a solution, the Dirichlet Compound Multinomial Distribution [24,35] was build by introducing the Dirichlet distribution as a prior to the multinomial parameters. However, the Dirichlet prior has several limitations, including that it has a strictly negative covariance matrix. Words in text documents are expected to be positively correlated due to linguistic properties. Intuitively, the probability of a particular word to appear as a singular noun is likely tied to the probability of the same word appearing in plural. A natural candidate for a distribution that models the covariance, as well as the overdispersion, is the negative multinomial distribution (NMD) [21,39].

In this paper, we use the Minorization-Maximization (MM) [27] framework to estimate the parameters of the NDM mixture model. MM is a generalization of the well-known Expectation-Maximization (EM) algorithm and among many other benefits, has a better performance in high dimensional optimization algorithms [48]. Furthermore, finding the optimal number of clusters, which offers the best representation of the data, remains a challenging task in unsupervised learning [6,41]. Therefore, we have adopted the component-wise MM approach, which allows initializing the algorithm with a large number of components and iteratively eliminates the weak components, until reaching a termination condition based on MML criterion [17,29,45].

The remaining parts of this paper are organized as follows. In Sect. 2, we introduce the Negative Multinomial mixture model and adapt the minorization-maximization framework to estimate the model parameters. In Sect. 3, we use component-wise MM, based on MML criterion, to select the optimal number of components. Section 4 demonstrates the experimental results. Finally, we summarize the findings in Sect. 5.

2 Negative Multinomial Mixture Model

The Negative Multinomial Distribution is the multivariate generalization of Negative Binomial Distribution [21], which is predicated upon the fact that a random variable X denotes the number of failures that occur before β successes are obtained in a sequence of *Bernoulli*(p) trials, also known as the inverse sampling. To construct the multivariate analogue of the NBD, we run D independent Poisson processes with intensities $p_1 \dots p_D$ until a certain length of time, which follows a gamma distribution with shape parameter $\beta > 0$ and intensity parameter p_{D+1}. At the expiration of this waiting time, the frequency for each random events X_d, among the first D categories, constitutes the input vector for our model. Therefore, the vector of word frequencies, $\mathbf{X} = (X_1 \dots X_D)$, has the Negative Multinomial density function, given by:

$$\mathcal{NMD}(\mathbf{X}|\beta, p) = \frac{\Gamma(\beta+m)}{\prod_{d=1}^{D} x_d!} \prod_{d=1}^{D} p_d^{X_d} p_{D+1}^{\beta} \tag{1}$$

where $\Gamma(\beta+m)$ denotes the Gamma function; $m = \sum_{d=1}^{D} X_d$ represents the length of the document; p_d is the probability of emitting a word X_d which is subject to the constraints $p_d > 0$ and $\sum_{d=1}^{D+1} p_d = 1$.

The Poisson process perspective yields the moments:

$$\mathbf{E}(X_i) = \beta \frac{p_i}{p_{D+1}}, \quad i = 1, \cdots, D \tag{2}$$

$$\mathbf{Var}(X_i) = \beta \frac{p_i}{p_{D+1}} \left(1 + \frac{p_i}{p_{D+1}}\right), \quad i = 1, \cdots, D \tag{3}$$

$$\mathbf{Cov}(X_i, X_j) = \beta \frac{p_i}{p_{D+1}} \frac{p_j}{p_{D+1}}, \quad i \neq j; \quad i, j = 1, \cdots, D \tag{4}$$

By comparing Eqs. 2 and 3, it is clear that the variance is always greater than the mean, resulting in overdispersed component X_i. Also, the covariance between counts in Eq. 4 is always positive, appealing to the positively correlated nature of counts in natural language processing, image summarization and other applications.

Let $\mathcal{X} = \{\mathbf{X}_1, \dots, \mathbf{X}_N\}$ be a set of N independently and identically distributed documents or images, where each can be represented as a sparse D dimensional vector of cell counts $\mathbf{X}_i = (X_{i1}, \dots, X_{iD})$, assumed to follow a

Negative Multinomial distribution. Then, a finite mixture model of K negative multinomial distributions is denoted as follows:

$$P(\mathbf{X}_i|\Theta) = \sum_{k=1}^{K} \pi_k \mathcal{NMD}(\mathbf{X}_i|\beta_k, p_k) \tag{5}$$

where π_k are the mixing weights, which must satisfy the condition $\sum_{k=1}^{K} \pi_k = 1$, $K > 1$ is number of components in the mixture; β_k and p_k are the parameters defining the kth component; and $\Theta = \{\beta_1, \ldots \beta_K, p_1, \ldots, p_K, \pi_1, \ldots, \pi_K\}$ is the set of all latent variables.

2.1 Maximum Likelihood Parameter Estimation

The Expectation-Maximization algorithm is widely used to estimate the mixture's parameters in presence of unobserved latent variables. For ease of calculation, we maximize the log of the likelihood function instead of the likelihood such that:

$$\mathcal{L}(\mathcal{X}|\Theta) = \prod_{i=1}^{N} \sum_{k=1}^{K} \log \Big(\pi_k \mathcal{NMD}(\mathbf{X}_i|\beta, p_k) \Big) \tag{6}$$

A membership vector $Z_i = (z_{i1}, \ldots, z_{iK})$ is assigned to each observation $\mathbf{X}_i$ such that $z_{ik} = 1$ if the object i belongs in the cluster k and all other elements equal to 0. Thus, we can rewrite the complete data log likelihood in the given form:

$$\mathcal{L}(\mathcal{X}, \mathcal{Z}|\Theta) = \prod_{i=1}^{N} \sum_{k=1}^{K} z_{ik} \Big(\log \pi_k + \log NMD(\mathbf{X}_i|\beta_k, p_k) \Big) \tag{7}$$

In EM, the learning of the parameters of a mixture model is done by a two-step iteration. In the E-step, the posterior probabilities of the missing variables $P(\mathcal{Z}|\mathcal{X}, \Theta^{(t)})$ are evaluated using the current values of the parameters, as:

$$\hat{z}_{ik} = P(\mathcal{Z}|\mathcal{X}, \Theta^{(t)}) = \frac{\pi_k^{(t)} NMD(\mathbf{X}_i|\beta_k^{(t)}, p_k^{(t)})}{\sum_{j=1}^{K} \pi_j^{(t)} NMD(\mathbf{X}_i|\beta_j^{(t)}, p_j^{(t)})} \tag{8}$$

Then, in the M-step, the expectation of the complete-data log likelihood with respect to the missing variables is maximized. By setting the derivative of the complete log-likelihood function equal to zero, we obtain:

$$\hat{\pi}_k = \frac{1}{N} \sum_{i=1}^{N} \hat{z}_{ik} \tag{9}$$

However, finding a solution for the β_k and p_k parameters is slightly more complicated. Therefore, to calculate maximum likelihood estimates of the model parameters we use the MM framework, which relies on constructing an appropriate surrogate function minorizing the log-likelihood function, in such a way that the surface of the surrogate function lies below the surface of the objective function and they are tangent at the point $\Theta = \Theta^{(t)}$, where $\Theta^{(t)}$ represents the current iterate [47]. MM relies on recognizing and manipulating inequalities after close examination of the log-likelihood, given by:

$$\begin{aligned}
\mathcal{L}(\mathcal{X}, \mathcal{Z}|\Theta) &= \sum_{i=1}^{N}\sum_{k-1}^{K} \hat{z}_{ik} \left(\log \pi_k + \log NMD(\mathbf{X}_i|\beta_k, p_k)\right) \\
&= \sum_{i=1}^{N}\sum_{k=1}^{K} \hat{z}_{ik} \Big(log \pi_k + \sum_{l=0}^{m_i-1} \ln(\beta_k + l) + \sum_{d=1}^{D} X_{id} \ln p_{kd} \\
&\quad + \beta_k \ln p_{k,D+1} - \sum_{d=1}^{D} \ln X_{id}! \Big)
\end{aligned} \tag{10}$$

In this work, we strategically minorize parts of the overall objective function while leaving the other parts untouched. Thus, to construct an MM algorithm, we need to minorize terms such as $\ln(\beta_k + l)$ for $l = (0, \ldots, m_i)$. Note how the term $\ln(\beta_k + l)$ occurs in the data log likelihood only when the condition $m_i \geqslant l + 1$ is met. Therefore, we can rewrite the term:

$$\begin{aligned}
\sum_{i=1}^{N}\sum_{k-1}^{K} \hat{z}_{ik} \sum_{l=0}^{m_i-1} \ln(\beta_k + l) &= \sum_{k-1}^{K} \sum_{l=0}^{max_i m_i-1} \ln(\beta_k + l) \sum_{i=1}^{N} \hat{z}_{ik(m_i-1 \geqslant l)} \\
&= \sum_{k-1}^{K} \sum_{l=0}^{max_i m_i-1} r_{kl} \ln(\beta_k + l)
\end{aligned} \tag{11}$$

where $r_{kl} = \sum_{i=1}^{N} \hat{z}_{ik}(m_i - 1 \geqslant l)$ represents the sum of responsibilities the of data points X_{id} where the batch size m_i is bigger than the variable l. Making use of Jensen inequality, the surrogate function is built:

$$\begin{aligned}
\mathcal{G}\left(\theta|\theta^{(t)}\right) &= \sum_{k-1}^{K} \sum_{l=0}^{max_i m_i-1} r_{kl} \frac{\beta^{(t)}}{\beta^{(t)} + l} + \sum_{i=1}^{N}\sum_{k=1}^{K} \hat{z}_{ik} \sum_{d=1}^{D} X_{id} \ln p_{kd} \\
&\quad + \sum_{i=1}^{N}\sum_{k=1}^{K} \hat{z}_{ik} \beta_k \ln p_{k,D+1}
\end{aligned} \tag{12}$$

up to an irrelevant constant. Maximizing the surrogate function under the parameters β_k and p_{kd} generates the following updates [23]:

$$\hat{\beta}_k = -\Big(\sum_{l=0}^{max_i m_i-1} r_{kl} \frac{\beta_k}{\beta_k + l} \Big) / \Big(\sum_{i=1}^{N} \hat{z}_{ik} \beta_k \ln p_{k,D+1} \Big) \tag{13}$$

$$\hat{p}_{k,D+1} = -\frac{\sum_{i=1}^{N} \hat{z}_{ik} \beta_k}{\sum_{j=1}^{D} \sum_{i=1}^{N} \hat{z}_{ik} X_{ij} + \sum_{i=1}^{N} \hat{z}_{ik} \beta_k} \tag{14}$$

$$\hat{p}_{kd} = -\frac{\sum_{i=1}^{N} \hat{z}_{ik} X_{id}}{\sum_{j=1}^{D} \sum_{i=1}^{N} \hat{z}_{ik} X_{ij} + \sum_{i=1}^{N} \hat{z}_{ik} \beta_k} \quad (15)$$

where $p_{D+1} = 1 - \sum_{d=1}^{D} p_d$.

It is obvious that MM updates, respect the parameter conditions, offering great numerical stability for the algorithms.

3 Model Selection

Among the most widely used model selection criteria are the deterministic methods based on information theory, such as Minimum Message Length (MML) [25], Akaike's Information Criterion (AIC) [2], Minimum Description Length (MDL) [37], etc. The MML implementation as a model selection criterion has proven to be efficient, statistically consistent and easily integrated with mixtures models, in such a way that the parameter estimations and models selection can happen simultaneously [6,18].

According to information theory, the optimal number of clusters K is the candidate value which minimizes the message length [6], given by:

$$\hat{\Theta} = \arg\min_{\Theta} \left\{ -\log(h(\Theta)) - \log(P(\mathcal{X}|\Theta)) + \frac{1}{2}\log|F(\Theta)| + \frac{D}{2}\left(1 + \log\frac{1}{12}\right) \right\} \quad (16)$$

where $h(\Theta)$ is the prior probability, $P(\mathcal{X}|\Theta)$ is the likelihood for the complete data set, and $|F(\Theta)|$ is the determinant of the expected Fisher information matrix.

However, initializing and running the algorithm several times with random initial number of components is required, making the approach time consuming and unfeasible when number of clusters might vary from a few to thousands, such as the case of topics in document collections. Therefore, we use a component-wise approach which allows the model to be initialized with a large value of K and remove empty components iteratively. After several transformations [29], we gain an easier form representing the message length:

$$\hat{\Theta} = \arg\min_{\Theta} \left\{ \frac{D}{2} \sum_{k=1}^{K} \log\left(\frac{N\pi_k}{12}\right) + \frac{K}{2}\log\frac{N}{12} + \frac{K(D+1)}{2} - \log p(\mathcal{X} \mid \Theta) \right\} \quad (17)$$

Starting with a large value of K may lead to several empty components and there will be no need to estimate, and transmit, their parameters. Thus, we need to update the component's weight in the M-step as:

$$\hat{\pi}_k = \frac{\max\left(\sum_i \hat{z}_{ik} - \frac{D}{2}, 0\right)}{\sum_j \max\left(\sum_i \hat{z}_{ij} - \frac{D}{2}, 0\right)} \quad (18)$$

The model built this way can be initialized with a large value of K, thus surpassing the limitation of initialization dependency. The advantage of the new update formula is its pruning behavior, that when some of the π_k go to zero they will be removed.

4 Experimental Results

To test the clustering performance of our algorithms we have chosen three different applications. First, we perform document clustering in a recently available dataset, COVID-19 Open Research Dataset (CORD-19)[1]. The aim of the data set collection is to facilitate information retrieval, extraction, knowledge-based discovery, and text mining efforts focused on COVID-19 management policies and effective treatment [38]. **The CORD-19 database** contains 52481 articles in English language, divided as follows: 26434 with general information, 10639 related to business, 2601 articles in technology, 506 in the science domain and 12266 finance-related. The amount of the articles increases daily, therefore, we apply the proposed online algorithm to the dataset in order to test its effectiveness. Here, each article is represented as a vector of word counts, achieved after performing the Bag-Of-Words approach.

Secondly, motivated by the enormous amount of articles published on the World Wide Web everyday, we use the WebKB[2] data set, containing WWW-pages collected from computer science departments of various universities. **The WebKB database** considered here is a subset which contains 4,199 Web pages limited to the four most common categories: Course, Faculty, Project, and Student. The first step in our preprocessing is removing all stop and rare words from the vocabularies. Then, each web page is represented as a vector containing the frequency of occurrence of each word from the term vector.

Finally, one of the most popular application in computer vision, is image clustering [13]. As already stated, we can represent the image as a vector of visual word counts [43]. To demonstrate the clustering performance of images, we have chosen the MMI dataset, built for the Facial Expression Recognition purpose. Understanding emotional states of users is one of the most attractive and growing topics nowadays in various research areas, but nevertheless it remains a challenging task. **The MMI[3] database** contains 1,140 images with a size of 720×576 pixels, where each of them belongs in one of 6 basic categories of facial recognition: Anger, Disgust, Fear, Happiness, Sadness, and Surprise). Sample images from the MMI database with different facial expressions are shown in Fig. 1. Participants, where 66% are male, range in age from 19 to 62, having either a European, Asian, or South American ethnic background.

The performance of our model, evaluated using the average classification accuracy, is shown in Table. 1. The Negative Multinomial Mixture model achieves significantly higher levels of accuracy as compared to its baseline model, the Multinomial Mixture model. The improvement in accuracy is almost 10% for the CORD-19 and MMI datasets, and around 5% for the WebKB dataset. Therefore, experiments support our claim that NMD can achieve a better modeling of sparse datasets with correlated features and variance greater than the mean. Moreover, to estimate the number of components for NMD model, we run the algorithm 10

[1] https://www.kaggle.com/allen-institute-for-ai/CORD-19-research-challenge.

[2] http://www.cs.cmu.edu/afs/cs.cmu.edu/project/theo-20/www/data/webkb-data.

[3] https://mmifacedb.eu/.

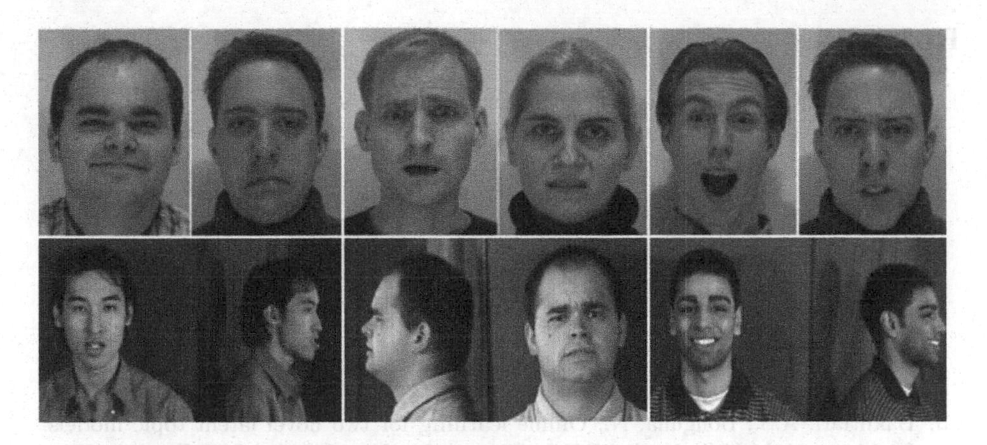

Fig. 1. Different sample frames on facial expressions in MMI database.

Table 1. Average accuracy (in %) of NMD vs MN algorithm, and estimated number of components $\hat{K}$ of NMD algorithm over different runs; K represents the true number of components

Data set	MN	NMD	K	$\hat{K}$
CORD-19	65.62	76.14	5	5
WebKB	81.16	86.72	4	4
MMI	67.74	77.41	6	5

times and round the mean value. The estimated number of components found was close to the true number, showing premise for a good performance when clustering higher number of topics or images.

5 Conclusion

In this paper, we have built a mixture model based on Negative Multinomial Distribution for clustering high-dimensional discrete datasets. To tackle the optimization challenges of large and high dimensional datasets, the Minorization-Maximization framework is used. The optimal number of clusters is simultaneously determined along with the estimation of the posterior probabilities and the mixture's parameters, by virtue of a component-wise MM, based on the MML model selection criteria. Through experiments on three data sets, we demonstrate the capability of our model in clustering of documents, web pages and images, prone to overdispersion and high positive correlation between features. The Negative Multinomial mixture model achieves better clustering results than the baseline model, the Multinomial mixture model.

References

1. Aggarwal, C.C., Zhai, C.: A survey of text clustering algorithms. In: Aggarwal, C., Zhai, C. (eds.) Mining Text Data, pp. 77–128. Springer, London (2012). https://doi.org/10.1007/978-1-4614-3223-4_4
2. Akaike, H.: A new look at the statistical model identification. IEEE Trans. Autom. Control **19**(6), 716–723 (1974)
3. Azam, M., Bouguila, N.: Bounded generalized Gaussian mixture model with ICA. Neural Process. Lett. **49**(3), 1299–1320 (2019)
4. Bakhtiari, A.S., Bouguila, N.: An expandable hierarchical statistical framework for count data modeling and its application to object classification. In: IEEE 23rd International Conference on Tools with Artificial Intelligence, ICTAI 2011, Boca Raton, FL, USA, 7–9, November 2011, pp. 817–824. IEEE Computer Society (2011)
5. Bakhtiari, A.S., Bouguila, N.: Online learning for two novel latent topic models. In: Linawati, M.M.S., Neuhold, E.J., Tjoa, A.M., You, I. (eds.) ICT-EurAsia 2014. LNCS, vol. 8407, pp. 286–295. Springer, Heidelberg (2014)
6. Baxter, R.A., Oliver, J.J.: Finding overlapping components with mml. Stat. Comput. **10**(1), 5–16 (2000)
7. Bijl, D., Hyde-Thomson, H.: Speech to text conversion, Jan 9 2001, uS Patent 6,173,259
8. Bouguila, N.: A data-driven mixture kernel for count data classification using support vector machines. In: 2008 IEEE Workshop on Machine Learning for Signal Processing. pp. 26–31 (2008). https://doi.org/10.1109/MLSP.2008.4685450
9. Bouguila, N.: Clustering of count data using generalized Dirichlet multinomial distributions. IEEE Trans. Knowl. Data Eng. **20**(4), 462–474 (2008)
10. Bouguila, N.: A model-based approach for discrete data clustering and feature weighting using MAP and stochastic complexity. IEEE Trans. Knowl. Data Eng. **21**(12), 1649–1664 (2009)
11. Bouguila, N.: Count data modeling and classification using finite mixtures of distributions. IEEE Trans. Neural Networks **22**(2), 186–198 (2011)
12. Bouguila, N., Amayri, O.: A discrete mixture-based kernel for SVMs: application to spam and image categorization. Inf. Process. Manag. **45**(6), 631–642 (2009)
13. Bouguila, N., ElGuebaly, W.: A generative model for spatial color image databases categorization. In: Proceedings of the IEEE International Conference on Acoustics, Speech, and Signal Processing, ICASSP 2008, March 30–April 4, 2008, Caesars Palace, Las Vegas, Nevada, USA, pp. 821–824. IEEE (2008)
14. Bouguila, N., ElGuebaly, W.: On discrete data clustering. In: Washio, T., Suzuki, E., Ting, K.M., Inokuchi, A. (eds.) PAKDD 2008. LNCS (LNAI), vol. 5012, pp. 503–510. Springer, Heidelberg (2008). https://doi.org/10.1007/978-3-540-68125-0_44
15. Bouguila, N., ElGuebaly, W.: Discrete data clustering using finite mixture models. Pattern Recognit. **42**(1), 33–42 (2009)
16. Bouguila, N., Ghimire, M.N.: Discrete visual features modeling via leave-one-out likelihood estimation and applications. J. Vis. Commun. Image Represent. **21**(7), 613–626 (2010)
17. Bouguila, N., Ziou, D.: MML-based approach for finite Dirichlet mixture estimation and selection. In: Perner, P., Imiya, A. (eds.) MLDM 2005. LNCS (LNAI), vol. 3587, pp. 42–51. Springer, Heidelberg (2005). https://doi.org/10.1007/11510888_5
18. Bouguila, N., Ziou, D.: Unsupervised selection of a finite Dirichlet mixture model: an mml-based approach. IEEE Trans. Knowl. Data Eng. **18**(8), 993–1009 (2006)

19. Bouguila, N., Ziou, D.: Unsupervised learning of a finite discrete mixture: Applications to texture modeling and image databases summarization. J. Vis. Commun. Image Represent. **18**(4), 295–309 (2007)
20. Chakraborty, S., Paul, D., Das, S., Xu, J.: Entropy weighted power k-means clustering. In: International Conference on Artificial Intelligence and Statistics, pp. 691–701. PMLR (2020)
21. Chiarappa, J.A.: Application of the negative multinomial distribution to comparative Poisson clinical trials of multiple experimental treatments versus a single control. Ph.D. thesis, Rutgers University-School of Graduate Studies (2019)
22. Csurka, G., Dance, C., Fan, L., Willamowski, J., Bray, C.: Visual categorization with bags of keypoints. In: Workshop on statistical learning in computer vision, ECCV, vol. 1, pp. 1–2. Prague (2004)
23. De Leeuw, J.: Block-relaxation algorithms in statistics. In: Bock, HH., Lenski, W., Richter, M.M. (eds) Information Systems and Data Analysis. Studies in Classification, Data Analysis, and Knowledge Organization, pp. 308–324. Springer, Heidelberg (1994). https://doi.org/10.1007/978-3-642-46808-7_28
24. Elkan, C.: Clustering documents with an exponential-family approximation of the Dirichlet compound multinomial distribution. In: Proceedings of the 23rd International Conference on Machine Learning, pp. 289–296 (2006)
25. Figueiredo, M.A.T., Jain, A.K.: Unsupervised learning of finite mixture models. IEEE Trans. Pattern Anal. Mach. Intell. **24**(3), 381–396 (2002)
26. Hofmann, T.: Probabilistic latent semantic indexing. In: Proceedings of the 22nd annual international ACM SIGIR Conference on Research and Development in Information Retrieval, pp. 50–57 (1999)
27. Hunter, D.R., Lange, K.: A tutorial on MM algorithms. Am. Stat. **58**(1), 30–37 (2004)
28. Kesten, H., Morse, N.: A property of the multinomial distribution. Ann. Math. Stat. **30**(1), 120–127 (1959)
29. Law, M.H., Figueiredo, M.A., Jain, A.K.: Simultaneous feature selection and clustering using mixture models. IEEE Trans. Pattern Anal. Mach. Intell. **26**(9), 1154–1166 (2004)
30. Li, T., Mei, T., Kweon, I.S., Hua, X.S.: Contextual bag-of-words for visual categorization. IEEE Trans. Circuits Syst. Video Technol. **21**(4), 381–392 (2010)
31. Li, Z., Tang, J., He, X.: Robust structured nonnegative matrix factorization for image representation. IEEE Trans. Neural Networks Learn. Syst. **29**(5), 1947–1960 (2017)
32. Lu, Y., Mei, Q., Zhai, C.: Investigating task performance of probabilistic topic models: an empirical study of PLSA and LDA. Inf. Retrieval **14**(2), 178–203 (2011)
33. Madsen, R.E., Kauchak, D., Elkan, C.: Modeling word burstiness using the Dirichlet distribution. In: Proceedings of the 22nd International Conference on Machine Learning, pp. 545–552 (2005)
34. McLachlan, G.J., Peel, D.: Finite Mixture Models. Wiley, New York (2004)
35. Minka, T.: Estimating a Dirichlet distribution (2000)
36. Pei, X., Chen, C., Gong, W.: Concept factorization with adaptive neighbors for document clustering. IEEE Trans. Neural Networks Learn. Syst. **29**(2), 343–352 (2016)
37. Rissanen, J.: Modeling by shortest data description. Automatica **14**(5), 465–471 (1978)
38. Shuja, J., Alanazi, E., Alasmary, W., Alashaikh, A.: Covid-19 open source data sets: a comprehensive survey. Applied Intelligence, pp. 1–30 (2020)

39. Sibuya, M., Yoshimura, I., Shimizu, R.: Negative multinomial distribution. Ann. Inst. Stat. Math. **16**(1), 409–426 (1964). https://doi.org/10.1007/BF02868583
40. Taleb, I., Serhani, M.A., Dssouli, R.: Big data quality assessment model for unstructured data. In: 2018 International Conference on Innovations in Information Technology (IIT), pp. 69–74. IEEE (2018)
41. Wallace, C.S., Dowe, D.L.: MMl clustering of multi-state, poisson, von mises circular and gaussian distributions. Stat. Comput. **10**(1), 73–83 (2000)
42. Xu, W., Liu, X., Gong, Y.: Document clustering based on non-negative matrix factorization. In: Proceedings of the 26th Annual International ACM SIGIR Conference on Research and Development in Informaion Retrieval, pp. 267–273 (2003)
43. Yang, J., Jiang, Y.G., Hauptmann, A.G., Ngo, C.W.: Evaluating bag-of-visual-words representations in scene classification. In: Proceedings of the International Workshop on Workshop on Multimedia Information Retrieval, pp. 197–206 (2007)
44. Zamzami, N., Bouguila, N.: Text modeling using multinomial scaled Dirichlet distributions. In: Mouhoub, M., Sadaoui, S., Ait Mohamed, O., Ali, M. (eds.) IEA/AIE 2018. LNCS (LNAI), vol. 10868, pp. 69–80. Springer, Cham (2018). https://doi.org/10.1007/978-3-319-92058-0_7
45. Zamzami, N., Bouguila, N.: Model selection and application to high-dimensional count data clustering - via finite EDCM mixture models. Appl. Intell. **49**(4), 1467–1488 (2019)
46. Zenil, H., Kiani, N.A., Tegnér, J.: Quantifying loss of information in network-based dimensionality reduction techniques. J. Complex Networks **4**(3), 342–362 (2016)
47. Zhou, H., Lange, K.: MM algorithms for some discrete multivariate distributions. J. Comput. Graph. Stat. **19**(3), 645–665 (2010)
48. Zhou, H., Zhang, Y.: EM VS MM: a case study. Comput. Stat. Data Analy. **56**(12), 3909–3920 (2012)
49. Zhu, J., Li, L.J., Fei-Fei, L., Xing, E.P.: Large margin learning of upstream scene understanding models. In: Advances in Neural Information Processing Systems, pp. 2586–2594 (2010)

Phase Prediction of Multi-principal Element Alloys Using Support Vector Machine and Bayesian Optimization

Nguyen Hai Chau[1(✉)], Masatoshi Kubo[2], Le Viet Hai[1], and Tomoyuki Yamamoto[2]

[1] Faculty of Information Technology, VNU University of Engineering and Technology, 144 Xuan Thuy, Cau Giay, Hanoi, Vietnam
{chaunh,16020936}@vnu.edu.vn

[2] Graduate School of Fundamental Science and Engineering, Waseda University, Tokyo 169-8050, Japan
masa104k@ruri.waseda.jp, tymmt@waseda.jp

Abstract. Designing new materials with desired properties is a complex and time-consuming process. One of the challenging factors of the design process is the huge search space of possible materials. Machine learning methods such as *k*-nearest neighbours, support vector machine (SVM) and artificial neural network (ANN) can contribute to this process by predicting materials properties accurately. Properties of multi-principal element alloys (MPEAs) highly depend on alloys' phase. Thus, accurate prediction of the alloy's phase is important to narrow down the search space. In this paper, we propose a solution of employing support vector machine method with hyperparameters tuning and the use of weight values for prediction of the alloy's phase. Using the dataset consisting of the experimental results of 118 MPEAs, our solution achieves the cross-validation accuracy of 90.2%. We confirm the superiority of this score over the performance of ANN statistically. On the other dataset containing 401 MPEAs, our SVM model is comparable to ANN and exhibits 70.6% cross-validation accuracy.

Keywords: Multi-principal element alloys · High-entropy alloys · Phase prediction · Support vector machine · Bayesian optimization.

1 Introduction

Alloys are compounds that are made by mixing two or more metallic or non-metallic elements. Conventional alloys are based on one principal element and have been developed with the aim of obtaining the desirable properties by adding other elements for thousands of years [2,26,33,36]. However, these traditional alloys are located on the corners of the composition space [19,35] and the majority of alloys in the interior region were neglected. In recent decades, the subject of researchers' interest have been shifted to multi-principal element alloys

N. T. Nguyen et al. (Eds.): ACIIDS 2021, LNAI 12672, pp. 155–167, 2021.
https://doi.org/10.1007/978-3-030-73280-6_13

(MPEAs) [35,37], which occupy the central area of composition space. MPEAs are defiened as the alloys composed of five or more principal elements, each of which has a concentration ranging from 5% to 35% (additional minor elements are also allowed) [19,32,36]. MPEAs are expected to exhibit superior properties such as high strength/hardness, beneficial corrosion and fatigue resistance, good wear resistance, and high thermal stability in comparison to conventional alloys [1,12,32,33]. Mixing the constituent elements in the equimolar or near-equimolar compositions leads to increase of configurational entropy (also referred to as mixing entropy), which indicates the randomness of the atomic arrangement. Therefore, alloys with five or more principal elements are also called as high entropy alloys (HEAs) [16,19], and the terms HEAs and MPEAs are used interchangeably in this field [10].

The mechanical and physical properties of HEAs highly depend on their phase and microstructure [3,15]. The phases of HEAs are roughly classified into solid solution (SS), intermetallic compound (IM), amorphous (AM), and the mixture of SS and IM (SS + IM) [10,15], and these indicate the order of atomic arrangement and the types of alloys' crystal structure. While they can be identified by experiment, it is quite impractical to synthesize alloys with various compositions and analyze their phases individually because of the time-consuming and costly process. Therefore, prediction of the possible phase of HEAs is crucial for designing efficiently new applicable HEAs to various purposes with desired properties [3,10,15]. There are three approaches to predict phases of alloys and resulting properties. The first approach is parametric phase prediction [6,8,10,15,31,32] and the second one employs density functional theory (DFT) calculation [9,12,13,17,22,29,30,34], while the third strategy applies machine learning (ML) methods [5,8,10,15,18,23–25,33].

In the parametric approach, the possible or stable phase of HEAs is judged by their physical and structural properties. Alloys are plotted in the two dimensional scatter diagram, in which the horizontal and vertical axes correspond to the quantities mentioned above such as mixing enthalpy (ΔH_{mix}), atomic size difference (δ), and mixing entropy (ΔS_{mix}). Then, the empirical rules for phase formation are extracted from the plots. For instance, SS phase is observed when the following conditions are satisfied simultaneously: $-22 \leq \Delta H_{\text{mix}} \leq 7$ kJ/mol, $\delta \leq 8.5$, and $11 \leq \Delta S_{\text{mix}} 14$ J/(K·mol) [6,32]. However, it is problematic to establish the empirical rules including the multiple variables because of the difficulty in visualization of the trend of data points in high-dimensional space [10].

DFT is a kind of the methods of the first principles calculation, which is the quantum-mechanical technique to simulate the electronic structure and the resultant properties of materials. The DFT itself cannot suggest or search for the most probable phase of HEAs, because it only perform the computation for the given crystal structure required as the input condition. However, it is the effective tool to estimate the properties of the materials. For example, Huhn et al. [9] discuss the occurrence of the order/disorder phase transition in HEA MoNbTaW, while it is considered which structure is the most stable through estimation of the energy of each system in Ref. [17,29].

ML is the methodology to construct the predictive model based on the available data and improve its accuracy by autonomous "learning" of the computer. In recent years, ML has been increasingly used in the field of materials science [8,10,15,23,33] and contributes to the discovery of hidden patterns behind huge amount of data and efficient prediction of materials' properties. The general design process of new materials with the aid of ML can be expressed in the form of Pseudocode 1 [33].

Pseudocode 1: A general design process of a new material.

Input: A desired property P of the new material, a training data set D with known properties.
Output: A candidate material with P property.

```
1  while material with property P not found AND search space ≠ ∅ do
2  |  Train a model m on D using method M;
3  |  Apply model m on a materials space S to predict the properties P;
4  |  Base on the prediction results and apply a utility function, select a
   |  subset S_P of S;
5  |  Perform synthesis and experiments for materials in S_P;
6  |  Record experimental values P of materials in S_P and add them to D;
7  |  if a material with property P is found then
8  |  |  Indicate that the desired material is found;
9  |  end
10 end
```

This novel technique is useful for reducing the time required in three steps of the design strategy, i.e. lines 2–4. Accurate prediction of the property P in line 3 enables us to narrow down the search space in line 4 and thus reduce the total number of iterations of the **while** loop. Line 3 is actually the direct consequence of the model m trained in line 2. Hence, appropriate selection of the method M is crucial to successful optimization of the ML-assisted material design process. In general, the method M should possess as many merits as possible like the following features: (1) ability to predict accurately, (2) little training time and (3) high interpretability.

There are some attempts to find suitable methods for prediction of MPEAs' phases. Islam et al. [10] trained the artificial neural network (ANN) model on the dataset containing 118 MPEAs (hereafter MPEA-118) and achieved the average accuracy of 84.3% for cross-validation. Huang et al. [8] compared average cross-validation accuracy of phase prediction on the dataset composed of 401 HEAs (hereafter HEA-401). They applied three methods: k-nearest neighbours (KNN), support vector machine (SVM) and ANN. Consequently, the scores (average cross-validation accuracy) of the three methods were 68.6%, 64.3% and 74.3%, respectively. Though ANN were better than SVM with respect to accuracy, SVM is more beneficial in terms of less risk of overfitting [21] and high interpretability [14].

In this paper, therefore, we employ SVM method to predict the phase selection of MPEAs. Bayesian optimization [20] is used to tune hyperparameters of

SVM, related to its predictive performance, and achieve the highest possible prediction accuracy. The benchmark tests are performed using the existing datasets MPEA-118 and HEA-401 mentioned above.

2 Description of Datasets

Physics and materials science literatures show that the selection of MPEA phase is correlated with elemental features including valence electron concentration (VEC) and electronegavity difference ($\Delta\chi$). In addition, the phase is determined by parameters such as the atomic size difference (δ), mixing entropy (ΔS_{mix}) and mixing enthalpy (ΔH_{mix}). These five features are usually used in the previously mentioned approaches of phase prediction and are defined as follows:

$$\text{VEC} = \sum_{i=1}^{n} c_i \text{VEC}_i, \tag{1}$$

$$\Delta\chi = \sqrt{\sum_{i=1}^{n} c_i(\chi_i - \bar{\chi})^2} \tag{2}$$

$$\delta = 100 \times \sqrt{\sum_{i=1}^{n} c_i \left(1 - \frac{r_i}{\bar{r}}\right)^2} \tag{3}$$

$$\Delta H_{\text{mix}} = \sum_{i=1, i<j}^{n} 4H_{ij} c_i c_j, \tag{4}$$

$$\Delta S_{\text{mix}} = -R \sum_{i=1}^{n} c_i \ln c_i, \tag{5}$$

where c_i ($0 < c_i < 1$), VEC_i, and r_i the atomic concentration, VEC and atomic radius of each element in an MPEA, respectively. H_{ij} is the enthalpy of atomic pairs calculated by Miedema's model [28] and R is the gas constant. $\bar{\chi}$ and $\bar{r}$ are weighted Pauling electronegavity and atomic radius written as

$$\bar{\chi} = \sum_{i=1}^{n} c_i \chi_x \tag{6}$$

and

$$\bar{r} = \sum_{i=1}^{n} c_i r_i, \tag{7}$$

respectively.

The MPEA-118 and HEA-401 datasets' structures are exactly the same. Each dataset has seven variables where the first one is alloy's chemical formula (or composition), the next five are features given in Eqs. (1–5) and the last is the alloy's phase. The MPEA-118 dataset is extracted from Table 2 of Ref. [6]. Obtaining the HEA-401 dataset is less direct. At first, data of 648 complex, concentrated alloys studied in 2015 from Table S1, available as the supplemental document of Ref. [19], is extracted; of which 247 compositions are removed because of duplication or unknown phase. We then write a Python program to create HEA-401 by applying Eqs. (1–5) and Miedema's model [28] for the remaining 401 compositions.

Number of alloys of MPEA-118 and HEA-401 are 118 and 401, respectively. Tables 1 and 2 show samples of MPEA-118 and HEA-401.

Table 1. Samples of MPEA-118 dataset.

Composition	Vec	Deltachi	Delta	Deltahmix	Deltasmix	Phase
FeNi2CrCuAl0.6	8.36	0.12	4.49	−3.27	12.72	SS
AlCoCrCu0.5FeNi1.5	7.75	0.12	5.35	−8.28	14.53	SS
CoCrFeNiTi	7.4	0.14	6.68	−16.32	13.38	IM
Al0.5CoCrCuFeNiV1.0	7.77	0.12	4.04	−5.25	16.01	IM
Pd40Cu30Ni10P20	9.3	0.14	9.08	−24.88	10.64	AM
Cu46Zr42Al7Y5	7.1	0.28	11.84	−24.88	8.79	AM

Table 2. Samples of HEA-401 dataset.

Composition	Vec	Deltachi	Delta	Deltahmix	Deltasmix	Phase
Al0.25CoCrCuFeMnNiTiV	7.36	0.148	5.68	−9.52	17.89	SS
AlCo3.5CrCu0.5FeNi	8	0.108	4.75	−7.03	13.09	SS
Al2.8CoCrCuFeNi	6.72	0.131	6.57	−10.28	14	IM
AlCo1.5Cr2Fe1.5Mn2NiV	6.95	0.133	4.84	−10.15	15.83	IM
AlCoCrFeNb0.75Ni	6.91	0.127	6.5	−18.02	14.85	SS+IM
CoFeNiSi0.5	8.29	0.032	2.66	−19.43	11.24	SS+IM

3 Building Prediction Models with SVM

There are three consecutive steps to build a prediction model with SVM. At the first step, predictors and an outcome of the model must be defined. Next, search operations are performed to tune hyperparameters of SVM. Finally, cross-validation is applied to assess accuracy of prediction models that employ SVM with the tuned hyperparameters. The steps are presented in Sects. 3.1, 3.2 and 3.3, respectively.

3.1 Defining Predictors and an Outcome

In both MPEA-118 and HEA-401 datasets, five numerical variables including VEC, $\Delta\chi$, δ, ΔH_{mix} and ΔS_{mix} are used as predictors. The **phase** variable is the outcome. Since it is a categorical variable, the alloy's phase prediction is a classification task.

3.2 Tuning Hyperparameters Using Bayesian Optimization

There are two hyperparameters of SVM: C - the cost of misclassification and γ - a parameter for the RBF (Radial Basis Function) kernel [14]. Pseudocodes 2 and 3 are used to tune hyperparameters as follows:

Pseudocode 2: Objective function O.

Input: Dataset D, hyperparameters of SVM: C, γ and optional weighted values vector w.

Output: Mean of prediction accuracy.

1 Randomly split D into 4 equal folds;
2 **for** *i from 1 to 4* **do**
3   Use i-th fold as a test dataset; other folds form a train dataset;
4   Build a model m on the train dataset using SVM method. SVM parameters are: kernel = RBF, scaling data = Yes, C, γ and w, where w is a vector of weighted values of the phases. Components of w are set inversely proportional to the number of phases in the train dataset if w is used;
5   Predict alloy's phase on the test dataset;
6   Calculate Accuracy measure and save it into an array A;
7 return mean(A); // This is the result of the objective function

---

In Pseudocode 2, an objective function O is defined. In this function, 4-fold cross-validation is used for fair comparison with prediction results in Ref. [10] and [8]. Return value of O is prediction accuracy of the model. Pseudocode 3 employs Bayesian optimization [20] to search for values of C and γ that maximize O. Tuned hyperparameters are listed in Table 3. The **Weighted** column indicates if weighted parameters are used. The last column contains maximum values of the objective function O.

Pseudocode 3: Tuning hyperparameters C, γ.

Input: Dataset D, objective function O.
Output: Tuned hyperparameters C, γ and O's corresponding value.

1 $C_{min} = 0$; $C_{max} = 30$; // Initial range for C;
2 $\gamma_{min} = 0$; $\gamma_{max} = 30$; // Initial range for γ;
3 $S = \emptyset$;
4 **for** *j from 1 to 10* **do**
5 Run Bayesian optimization [20] with the following parameters: objective function O, number of random initial points = 3 and number of iterations = 10, $C \in [C_{min}, C_{max}]$ and $\gamma \in [\gamma_{min}, \gamma_{max}]$;
6 Bayesian optimization returns a set $S_j = \{(C_i, \gamma_i, O_i), i = \overline{1, 13}\}$;
7 Calculate 95% high density intervals [27] (HDI95) $[C_l, C_h]$ of $\{C_i, i = \overline{1, 13}\}$ and $[\gamma_l, \gamma_h]$ of $\{\gamma_i, i = \overline{1, 13}\}$;
8 $C_{min} = C_l$; $C_{max} = C_h$; // Narrow down C's range for the next search;
9 $\gamma_{min} = \gamma_l$; $\gamma_{max} = \gamma_h$; // Narrow down γ's range for the next search;
10 $S = S \cup S_j$;
11 return $(C, \gamma, O_{\max})$ where $O_{\max} = \max\{O_i, \forall (C_i, \gamma_i, O_i) \in S\}$

Table 3. SVM tuned hyperparameters found by Bayesian optimization.

Dataset	Weighted	C	γ	O_{max}
MPEA-118	Yes	12.464	0.606	92.36%
MPEA-118	No	15.745	1.100	92.47%
HEA-401	Yes	26.295	3.462	74.10%
HEA-401	No	11.614	4.284	72.84%

3.3 Cross-Validation of Prediction Models Employing Tuned Hyperparameters

To perform cross-validation on the MPEA-118 and HEA-401 for assessment of phase prediction accuracy, Pseudocode 4 is applied. Cross-validation results are presented in Fig. 1 and Table 4. On the MPEA-118 dataset, the mean and median of accuracy are 90% regardless of using weighted values. On the HEA-401 dataset, mean and median of accuracy are 70.2% and 70.3% if no weighted values used. If weighted values are used, the mean and median are 70.6% and 70.7%, respectively.

To compare accuracy of phase prediction employing Bayesian optimization with results of Islam et al. [10] and Huang et al. [8], statistical tests are performed. The criterion of comparison in the tests is prediction accuracy. Test results are presented in the next section.

Pseudocode 4: Cross-validation.

Input: Dataset D, tuned hyperparameters C_o, γ_o.
Output: Array $A, Prec, Recall, F1$ containing prediction accuracy, precision, recall and F1 measures [7].

```
1 for l from 1 to 1000 do
2     Randomly split D into 4 equal folds;
3     for i from 1 to 4 do
4         Use i-th fold as a test dataset; other folds form a train dataset;
5         Build a model m on the train dataset using SVM method with the
          tuned hyperparameters C_o, γ_o and optional w;
6         Predict alloy's phase on the test dataset;
7         Calculate Accuracy measure and save them into an array A;
8         Calculate Precision, Recall and F1 measures and save them into arrays
          Prec, Recall, F1, respectively;
9 return A, Prec, Recall, F1;
```

Table 4. Summary of SVM's cross-validation accuracy.

Dataset	Weighted	Min	Q1	Median	Mean	Q3	Max
MPEA-118	No	66.7%	86.7%	90.0%	90.2%	93.3%	100.0%
MPEA-118	Yes	66.7%	86.7%	90.0%	90.0%	93.3%	100.0%
HEA-401	No	55.4%	67.3%	70.3%	70.2%	73.0%	85.3%
HEA-401	Yes	56.6%	68.0%	70.7%	70.6%	73.3%	85.3%

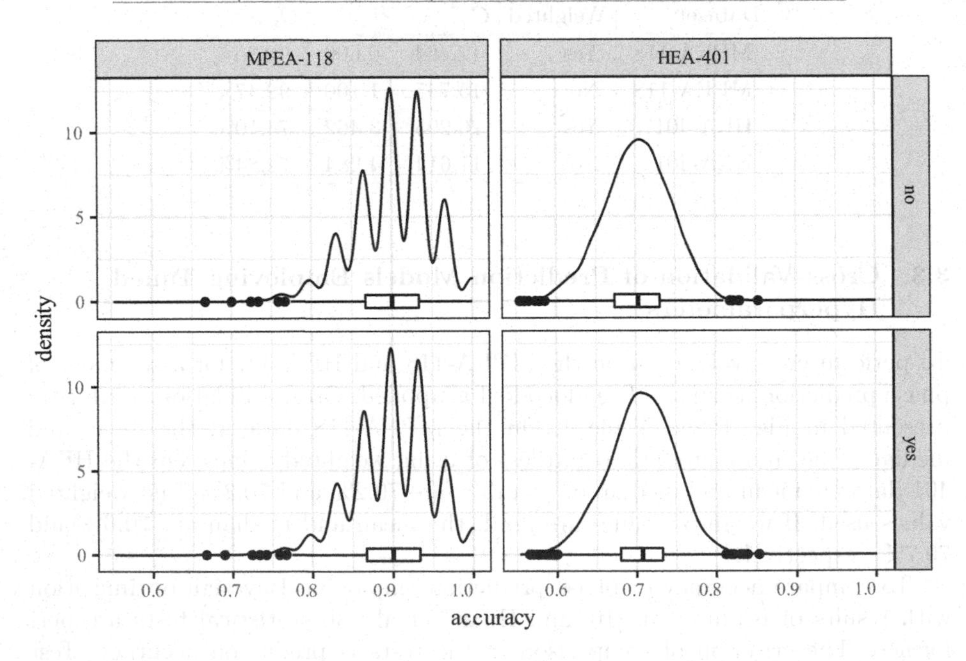

Fig. 1. SVM's cross-validation accuracy for phase prediction shown in density and box plots. Columns represent datasets, rows indicate the use of weighted values.

4 Statistical Analysis of Experimental Results

In this section, statistical tests to compare cross-validation accuracy of SVM and ANN methods on the two datasets are conducted. Then other tests are performed to observe effects of weighted values in SVM to accuracy and F1 measures. In statistics terms, the tests are called group comparison and a group in this context is a sample from a population.

Firstly, cross-validation accuracy of using SVM with hyperparameters proposed in this paper is compared with accuracy of Islam et al. [10] and Huang et al. [8]. Since the cross-validation accuracy is not normally distributed, all of the tests will be non-parametric [4], i.e. they are median comparison of two groups. The Wilcoxon statistical test [11] is used in the tests. Group 1 is cross-validation accuracy of SVM approaches and group 2 is cross-validation accuracy of ANN approaches of either Islam et al. [10] or Huang et al. [8]. The null hypothesis H_0 of all of the tests is *median of group 1 is equal to median of group 2*. Statistical test results in Table 5 show that prediction accuracy of SVM with hyperparameters regardless of weighted values is 6.7% higher than that of Islam et al. [10]. This difference is statistical significant because p-value < 0.05 and the corresponding CI95 (95% confidence interval) does not contain 0. The prediction accuracy of SVM with hyperparameters regardless of weighted values is comparable to that of Huang et al. [8] because of the insignificance of the median difference (p-value > 0.05 and the corresponding CI95 contains 0).

Table 5. Comparison of cross-validation accuracy. The **Diff.** column contains the difference in median of accuracy.

Group 1	Group 2	Diff.	p-value	CI95	Decision
SVM/MPEA-118	ANN/MPEA-118 Islam et al.	6.7%	0.01	[3.0%,13.2%]	Reject H_0
SVM weighted/MPEA-118	ANN/MPEA-118 Islam et al.	6.7%	0.01	[2.9%,13.2%]	Reject H_0
SVM/HEA-401	ANN/HEA-401 Huang et al.	−4.0%	0.23	[−18.9%,11.0%]	Fail to reject H_0
SVM weighted/HEA-401	ANN/HEA-401 Huang et al.	−3.6%	0.26	[−17.7%,11.0%]	Fail to reject H_0

Secondly, other groups' comparison tests are performed to observe effect of weighted values to cross-validation accuracy and F1 of the most minority phase - the IM. To perform the tests, cross-validation accuracy or F1 of SVM without weight values is compared with those with weighted values. Test results in Table 6 show that the use of weighted values increases prediction accuracy on HEA-401 by a small amount and has no effect on MPEA-118. In spite of that, results in Table 7 show that the use of weighted values also increase F1 measure on the minority class IM by a small amount.

Table 6. Difference of accuracy with and without SVM weighted values. The **Diff.** column contains the difference in median of accuracy.

Group 1	Group 2	Diff.	p-value	CI95	Decision
SVM weighted/MPEA-118	SVM/MPEA-118	−0.01%	1e−01	[−0.003%,0.004%]	Fail to reject H_0
SVM weighted/HEA-401	SVM/HEA-401	0.31%	1e−04	[0.223%,0.600%]	Reject H_0

Table 7. Difference of F1 with and without SVM weighted values. The **Diff.** column contains the difference in median of F1 (for IM phase).

Group 1	Group 2	Diff.	p-value	CI95	Decision
SVM weighted/MPEA-118	SVM/MPEA-118	1.29%	3e−16	[1.015%,1.769%]	Reject H_0
SVM weighted/HEA-401	SVM/HEA-401	0.44%	9e−08	[0.261%,0.668%]	Reject H_0

5 Conclusions

In this paper, we constructed the SVM models to predict the possible phase of MPEAs. Tuning of the hyperparameters were performed to maximize the prediction accuracy efficiently by Bayesian optimization, and each of the classes (i.e. phases) was weighted to avoid the biased prediction in the imbalanced dataset. Then, cross-validation was performed on the datasets MPEA-118 and HEA-401 to assess the values of prediction accuracy. We obtained the mean values of cross-validation accuracy of 90.2% (not weighted) and 70.6% (weighted) for MPEA-118 and HEA-401, respectively. The results of the Wilcoxon test indicates that our solution outperforms or is comparable to the other approach that employs ANN method. Furthermore, it is confirmed that the use of weighted values improves F1 measures of phase prediction. Accurate prediction of alloys' structure (SS, IM, and AM) contributes to the efficient search for the new materials by offering the feasible candidates for the various uses, which require the materials with a specific phase. Therefore, it is worth utilizing the SVM method for predicting the phase of MPEAs in conjunction with other ML algorithms.

Acknowledgement. This work was partly carried out at the Joint Research Center for Environmentally Conscious Technologies in Materials Science (Project No. 02007, Grant No. JPMXP0618217637) at ZAIKEN, Waseda University.

References

1. Abu-Odeh, A., et al.: Efficient exploration of the high entropy alloy composition-phase space. Acta Mater. **152**, 41–57 (2018). https://doi.org/10.1016/j.actamat.2018.04.012
2. Cantor, B., Chang, I.T.H., Knight, P., Vincent, A.J.B.: Microstructural development in equiatomic multicomponent alloys. Mater. Sci. Eng. A **375–377**(1–2), 213–218 (2004). https://doi.org/10.1016/j.msea.2003.10.257
3. Chattopadhyay, C., Prasad, A., Murty, B.S.: Phase prediction in high entropy alloys - a kinetic approach. Acta Mater. **153**, 214–225 (2018). https://doi.org/10.1016/j.actamat.2018.05.002

4. Cohen, P.R.: Empirical Methods for Artificial Intelligence. The MIT Press, Cambridge (1995)
5. Dai, D., et al.: Using machine learning and feature engineering to characterize limited material datasets of high-entropy alloys. Comput. Mater. Sci **175**, 109618 (2020). https://doi.org/10.1016/j.commatsci.2020.109618
6. Guo, S., Liu, C.T.: Phase stability in high entropy alloys: formation of solid-solution phase or amorphous phase. Progress Natural Sci. Materials Int. **21**(6), 433–446 (2011). https://doi.org/10.1016/S1002-0071(12)60080-X
7. Han, J., Kamber, M., Pei, J.: Data Mining. Elsevier (2012). https://doi.org/10.1016/C2009-0-61819-5. https://linkinghub.elsevier.com/retrieve/pii/C20090618195
8. Huang, W., Martin, P., Zhuang, H.L.: Machine-learning phase prediction of high-entropy alloys. Acta Mater. **169**, 225–236 (2019). https://doi.org/10.1016/j.actamat.2019.03.012
9. Huhn, W.P., Widom, M.: Prediction of A2 to B2 phase transition in the high-entropy alloy Mo-Nb-Ta-W. JOM **65**(12), 1772–1779 (2013). https://doi.org/10.1007/s11837-013-0772-3
10. Islam, N., Huang, W., Zhuang, H.: Machine learning for phase selection in multi-principal element alloys. Comput. Mater. Sci. **150**, 230–235 (2018). https://doi.org/10.1016/j.commatsci.2018.04.003
11. Kanji, G.: 100 Statistical Tests. SAGE Publications Ltd, 1 Oliver's Yard, 55 City Road, London EC1Y 1SP United Kingdom (2006). https://doi.org/10.4135/9781849208499, http://methods.sagepub.com/book/100-statistical-tests
12. Kim, G., Diao, H., Lee, C., Samaei, A., Phan, T., de Jong, M., An, K., Ma, D., Liaw, P.K., Chen, W.: First-principles and machine learning predictions of elasticity in severely lattice-distorted high-entropy alloys with experimental validation. Acta Mater. **181**, 124–138 (2019). https://doi.org/10.1016/j.actamat.2019.09.026
13. Koval, N.E., Juaristi, J.I., Díez Muiño, R., Alducin, M.: Elastic properties of the TiZrNbTaMo multi-principal element alloy studied from first principles. Intermetallics **106**, 130–140 (2019). https://doi.org/10.1016/j.intermet.2018.12.014
14. Kuhn, M., Johnson, K.: Applied Predictive Modeling. Springer, New York (2013). https://doi.org/10.1007/978-1-4614-6849-3
15. Lee, S.Y., Byeon, S., Kim, H.S., Jin, H., Lee, S.: Deep learning-based phase prediction of high-entropy alloys: optimization, generation, and explanation. Mater. Des. **197**, 109260 (2021). https://doi.org/10.1016/j.matdes.2020.109260
16. Lyu, Z., Lee, C., Wang, S.-Y., Fan, X., Yeh, J.-W., Liaw, P.K.: Effects of constituent elements and fabrication methods on mechanical behavior of high-entropy alloys: a review. Metall. Mater. Trans. A. **50**(1), 1–28 (2018). https://doi.org/10.1007/s11661-018-4970-z
17. Ma, D., Grabowski, B., Körmann, F., Neugebauer, J., Raabe, D.: Ab initio thermodynamics of the cocrfemnni high entropy alloy: importance of entropy contributions beyond the configurational one. Acta Mater. **100**, 90–97 (2015). https://doi.org/10.1016/j.actamat.2015.08.050
18. Manzoor, A., Aidhy, D.S.: Predicting vibrational entropy of fcc solids uniquely from bond chemistry using machine learning. Materialia **12**, 100804 (2020). https://doi.org/10.1016/j.mtla.2020.100804
19. Miracle, D.B., Senkov, O.N.: A critical review of high entropy alloys and related concepts. Acta Mater. **122**, 448–511 (2017). https://doi.org/10.1016/j.actamat.2016.08.081
20. Mockus, J.: Bayesian Approach to Global Optimization: Theory and Applications. Kluwer Academic Publishers, New York (1989)

21. Niu, X., Yang, C., Wang, H., Wang, Y.: Investigation of ANN and SVM based on limited samples for performance and emissions prediction of a CRDI-assisted marine diesel engine. Appl. Therm. Eng. **111**, 1353–1364 (2017). https://doi.org/10.1016/j.applthermaleng.2016.10.042
22. Nong, Z., Wang, H., Zhu, J.: First-principles calculations of structural, elastic and electronic properties of $(TaNb)_{0.67}(HfZrTi)_{0.33}$ high-entropy alloy under high pressure. Int. J. Miner. Metall. Mater. **27**(10), 1405–1414 (2020). https://doi.org/10.1007/s12613-020-2095-z
23. Nong, Z.S., Zhu, J.C., Cao, Y., Yang, X.W., Lai, Z.H., Liu, Y.: Stability and structure prediction of cubic phase in as cast high entropy alloys. Mater. Sci. Technol. (United Kingdom) **30**(3), 363–369 (2014). https://doi.org/10.1179/1743284713Y.0000000368
24. Pei, Z., Yin, J., Hawk, J., Alman, D.E., Gao, M.C.: Machine-learning informed prediction of high-entropy solid solution formation: Beyond the Hume-Rothery rules. NPJ Comput. Mater. **6**(1), 50 (2020). https://doi.org/10.1038/s41524-020-0308-7
25. Roy, A., Babuska, T., Krick, B., Balasubramanian, G.: Machine learned feature identification for predicting phase and Young's modulus of low-, medium- and high-entropy alloys. Scripta Mater. **185**, 152–158 (2020). https://doi.org/10.1016/j.scriptamat.2020.04.016
26. Senkov, O.N., Miller, J.D., Miracle, D.B., Woodward, C.: Accelerated exploration of multi-principal element alloys with solid solution phases. Nat. Commun. **6**, 6529 (2015). https://doi.org/10.1038/ncomms7529
27. Stanton, J.M.: Reasoning with Data: An Introduction to Traditional and Bayesian Statistics Using R. The Guilford Press, New York (2017)
28. Takeuchi, A., Inoue, A.: Classification of bulk metallic glasses by atomic size difference, heat of mixing and period of constituent elements and its application to characterization of the main alloying element. Mater. Trans. **46**(12), 2817–2829 (2005). https://doi.org/10.2320/matertrans.46.2817, https://www.jstage.jst.go.jp/article/matertrans/46/12/46_12_2817/_article
29. Tian, F., Delczeg, L., Chen, N., Varga, L.K., Shen, J., Vitos, L.: Structural stability of $NiCoFeCrAl_x$ high-entropy alloy from AB initio theory. Phys. Rev. B **88**(8), 085128 (2013). https://doi.org/10.1103/PhysRevB.88.085128
30. Tian, F., Varga, L.K., Chen, N., Shen, J., Vitos, L.: Ab initio design of elastically isotropic $TiZrNbMoV_x$ high-entropy alloys. J. Alloy. Compd. **599**, 19–25 (2014). https://doi.org/10.1016/j.jallcom.2014.01.237
31. Troparevsky, M.C., Morris, J.R., Daene, M., Wang, Y., Lupini, A.R., Stocks, G.M.: Beyond atomic sizes and Hume-Rothery rules: understanding and predicting high-entropy alloys. JOM **67**(10), 2350–2363 (2015). https://doi.org/10.1007/s11837-015-1594-2
32. Tsai, M.H., Yeh, J.W.: High-entropy alloys: a critical review. Mater. Res. Lett. **2**(3), 107–123 (2014). https://doi.org/10.1080/21663831.2014.912690
33. Wen, C., et al.: Machine learning assisted design of high entropy alloys with desired property. Acta Mater. **170**, 109–117 (2019). https://doi.org/10.1016/j.actamat.2019.03.010
34. Yao, H.W., Qiao, J.W., Hawk, J.A., Zhou, H.F., Chen, M.W., Gao, M.C.: Mechanical properties of refractory high-entropy alloys: experiments and modeling. J. Alloy. Compd. **696**, 1139–1150 (2017). https://doi.org/10.1016/j.jallcom.2016.11.188

35. Ye, Y.F., Wang, Q., Lu, J., Liu, C.T., Yang, Y.: High-entropy alloy: challenges and prospects. Mater. Today **19**(6), 349–362 (2016). https://doi.org/10.1016/j.mattod.2015.11.026
36. Yeh, J.W., et al.: Nanostructured high-entropy alloys with multiple principal elements: Novel alloy design concepts and outcomes. Adv. Eng. Mater. **6**(5), 299–303 (2004). https://doi.org/10.1002/adem.200300567
37. Zhou, Z., Zhou, Y., He, Q., Ding, Z., Li, F., Yang, Y.: Machine learning guided appraisal and exploration of phase design for high entropy alloys. NPJ Comput. Mater. **5**(1), 128 (2019). https://doi.org/10.1038/s41524-019-0265-1

VEGAS: A Variable Length-Based Genetic Algorithm for Ensemble Selection in Deep Ensemble Learning

Kate Han[1], Tien Pham[2], Trung Hieu Vu[3], Truong Dang[1], John McCall[1], and Tien Thanh Nguyen[1](✉)

[1] School of Computing, Robert Gordon University, Aberdeen, UK
t.nguyen11@rgu.ac.uk

[2] College of Science and Engineering, Flinders University, Adelaide, Australia

[3] School of Electronics and Telecommunications, Hanoi University of Science and Technology, Hanoi, Vietnam

Abstract. In this study, we introduce an ensemble selection method for deep ensemble systems called VEGAS. The deep ensemble models include multiple layers of the ensemble of classifiers (EoC). At each layer, we train the EoC and generates training data for the next layer by concatenating the predictions for training observations and the original training data. The predictions of the classifiers in the last layer are combined by a combining method to obtain the final collaborated prediction. We further improve the prediction accuracy of a deep ensemble model by searching for its optimal configuration, i.e., the optimal set of classifiers in each layer. The optimal configuration is obtained using the Variable-Length Genetic Algorithm (VLGA) to maximize the prediction accuracy of the deep ensemble model on the validation set. We developed three operators of VLGA: roulette wheel selection for breeding, a chunk-based crossover based on the number of classifiers to generate new offsprings, and multiple random points-based mutation on each offspring. The experiments on 20 datasets show that VEGAS outperforms selected benchmark algorithms, including two well-known ensemble methods (Random Forest and XgBoost) and three deep learning methods (Multiple Layer Perceptron, gcForest, and MULES).

Keywords: Deep learning · Ensemble learning · Ensemble selection · Classifier selection · Ensemble of classifiers · Genetic algorithm.

1 Introduction

In recent years, deep learning has emerged as a hot research topic because of its breakthrough performance in diverse learning tasks. For instance, in computer vision, Convolutional Neural Network (CNN), a deep neural network (DNN), significantly outperforms traditional machine learning algorithms on the image classification task on some large scale datasets. Despite its many successes, there are

N. T. Nguyen et al. (Eds.): ACIIDS 2021, LNAI 12672, pp. 168–180, 2021.
https://doi.org/10.1007/978-3-030-73280-6_14

some limitations of DNNs. Firstly, these deep models are usually very complex with many parameters and can only be trained on specially-designed hardware. Secondly, DNNs require a considerable amount of labeled data for the training process. When the cost of labeled data is too prohibitive, deep models might not bring about the expected gains in performance. It is well-recognized that there are many learning tasks where DNNs are poorer than traditional machine learning methods, especially state-of-the-art methods like Random Forest and XgBoost [12].

Meanwhile, ensemble learning is developed to obtain a better result than using single classifiers. By using an ensemble of classifiers (EoC), the poor predictions of several classifiers are likely to be compensated by those of others, which boosts the performance of the whole ensemble. Ensemble systems have been applied in many areas such as computer vision, software engineering, and bioinformatics. Traditionally, ensemble systems have been constructed with one layer of EoC. A combining algorithm combines the predictions of EoC for the final collaborated prediction [10].

The term *"deep learning"* makes the crowd only think of DNNs, which include multiple layers of parameterized differentiable nonlinear modules. In 2014, Zhou and Feng introduced a deep ensemble system called gcForest, including several layers of Random Forest-based classifiers [22]. The introduction of gcForest has shown that DNN is only a subset of deep models or deep learning models can be constructed with multiple layers of non-differentiable learning modules. Experiments on some popular datasets have shown that deep ensemble models outperform not only DNNs but also several state-of-the-art ensemble methods [15,22].

The *predictive performance* and *computational/storage efficiency* of ensemble systems can be further improved by obtaining a subset of classifiers that performs competitively to the whole ensemble. This research topic is known as ensemble selection (aka ensemble pruning or selective ensemble), an ensemble design stage to enhance ensemble performance based on searching for an optimal EoC from the whole ensemble. In this study, we introduce an ensemble selection method to improve the performance and efficiency of deep ensemble models. A configuration of the deep model is encoded in the form of binary encoding, showing which classifiers are selected or not. It is noted that the length of the proposed encoding depends on the number of layers and the number of classifiers in each layer that we use to construct the deep model. The configuration of the deep model thus is given in variable-length encoding. To find the optimal set of classifiers, we consider an optimization problem by maximizing the classification accuracy of the deep ensemble system on the validation data. In this work, we develop VEGAS: a Variable-length Genetic Algorithm (VLGA) to solve this optimization problem of the ensemble selection. The main contributions of our work are as follows:

- We propose a classifier selection approach for deep ensemble system
- We propose to encode classifiers in all layers of the deep ensemble system in a variable-length encoding

- We develop a VLGA to search for the optimal number of layers and the optimal classifiers.
- We experimentally show that VEGAS is better than some well-known benchmark algorithms on several datasets.

2 Background and Related Work

2.1 Ensemble Learning and Ensemble Selection

Ensemble learning refers to a popular research topic in machine learning in which multiple classifiers are combined to obtain a better result than using single classifiers. Two main stages in building an ensemble system are somehow to generate diverse classifiers and then combine them to make a collaborated decision. For the first stage, we train a learning algorithm on multiple training sets generated from the original training data [11] or train different learning algorithms on the original training data [10] to generate EoC. For the second stage, a combining method works on the predictions of the generated classifiers for the final decision. Based on experiments on diverse datasets, Random Forest [1] and XgBoost [3] were reported as the top-performance ensemble methods.

Inspired by the success of DNNs in many areas, several ensemble systems have been constructed with a number of layers of EoCs. Each layer receives outputs of the subsequent layer as its input training data and then outputs the training data for the next layer. The first deep ensemble system called gcForest was proposed in 2014, including many layers of two Random Forests and two Completely Random Tree Forests working in each layer. After that, several deep ensemble systems were introduced such as deep ensemble models of incremental classifiers [7], an ensemble of SVM classifiers with AdaBoost in finding model parameters [17], and deep ensemble models for multi-label learning [21]. Nguyen et al. [15] proposed MULES, a deep ensemble system with classifier and feature selection in each layer. The optimization problem was considered under bi-objectives: maximizing classification accuracy and diversity of EoC in each layer.

Meanwhile, ensemble selection is an additional intermediate stage of the ensemble design process that aims to select a subset of classifiers from the ensemble to achieve higher *predictive performance* and *computational/storage efficiency* than using the whole ensemble. Ensemble selection can be formulated as an optimization problem that can be solved by either Evolutionary Computation methods or greedy search approach. Nguyen et al. [13] used Ant Colony Optimisation (ACO) to search for the optimal combining algorithm and the optimal set from the predictions for the selected classifier in the ensemble system. Hill climbing search-based ensemble pruning, on the other hand, greedily selects the next configuration of selected classifiers around the neighborhood of the current configuration. Two crucial factors of the hill-climbing search-based ensemble pruning strategy are the searching direction and the measure to evaluate the different branches of the search process [16]. For an EoC, the measures determine the best single classifier to be added to or removed from the ensemble to optimize either performance or diversity of the ensemble. Examples of evaluation measures are *Accuracy, Mean Cross-Entropy, F-score, ROC area* [16], and *Concurrency* [2]. The direction of the search process can be conducted in the forward selection which starts from the empty EoC and adds a base classifier in sequence, or backward selection which prunes a base classifier from the whole set of classifiers until reaching the optimal subset of classifiers. Dai et al. [4] introduced the Modified Backtracking Ensemble Pruning algorithm to enhance the search processing in the backtracking method. The redundant solutions in the search space are reduced by using a binary encoding for the classifiers.

2.2 Variable Learning Encoding in Evolutionary Computation

There are some variable length encoding-based algorithms introduced recently. In [19], a VLGA is proposed to search for the best CNN structure. Each chromosome consists of three types of units corresponding to convolutional layers, pooling layers, and full connection layers. Later in [20], a non-binary VLGA was also proposed to search for the best CNN structure. This variable-length encoding strategy used different representations for different layer types. A skipping layer consists of two convolutional layers and one skipper connection; its encoding is the number of feature maps of the two convolutional layers within this skip layer. The encoding of the pooling layers is the pooling operation type, i.e. mean pooling or maximum pooling.

For applications, in [18], the GA with a variable-length chromosome was used to solve the path optimization problems. The path optimization problem is modeled as an abstract graph. Each chromosome is a set of nodes consisting of a feasible solution and therefore has a length equal to node amount. In [6], a GA with variable length chromosomes was also used to solve path planning problems for the autonomous mobile robot. Each chromosome is a position set that represents a valid path solution. The length of the chromosome is the number of the intermediate nodes. In [9], the proposed VLGA was also implemented to solve the multi-objective multi-robot path planning problems.

3 Proposed Method

3.1 Ensemble Selection for Deep Ensemble Systems

Let $\mathcal{D}$ be the training data of N observations $\{(\mathrm{x}_n, \hat{y_n})\}$, where x_n is the D-feature vector of the training instance and y_n be its corresponding label. True label $\hat{y_n}$ belongs to label set $\mathcal{Y}, |\mathcal{Y}| = M$. We aim to learn a hypothesis $\mathbf{h}$ (i.e., classifier) to approximate unknown relationship between the feature vector and its corresponding label $g : \mathrm{x}_n \rightarrow \hat{y_n}$ and then use this hypothesis to assign a label for each unlabeled instance. We also denote $\mathcal{K} = \mathcal{K}_k$ as the set of K learning algorithms. In deep ensemble learning consisting of s layers, we train an EoC $\{\mathbf{h}_k^{(i)}\}$ on i^{th} layer ($i = 1, \ldots, s$ and $k = 1, \ldots, K$) and then use a combining algorithm C on the hypothesis of the sth layer $\tilde{\mathbf{h}} = C\left\{\mathbf{h}_k^{(s)}, k = 1, \ldots, K\right\}$ for final decision making.

A deep ensemble system has multiple layers where each of them consists of EoCs. In the first layer, we obtain EoC $\left\{\mathbf{h}_k^{(1)}, k = 1, \ldots, K\right\}$ by training K learning algorithms on the original training data $\mathcal{D}$. The first layer also generates input data for the second layer by using the Stacking algorithm with the set of learning algorithms $\mathcal{K}$. Specifically, $\mathcal{D}$ is divided into T_1 disjoint parts in which the cardinality of each part is nearly similar. For each part, we train classifiers on its complementary and use these classifiers to predict for observations of this part. Thus, each observation in $\mathcal{D}$ will be tested one time. For observation x_n, we denote $p_{k,m}^{(1)}(\mathrm{x}_n)$ is the prediction of the kth classifier in the first layer that observation belongs to the class label y_m. The predictions in terms of M class labels are given in the form of probability: $\sum_{m=1}^{M} p_{k,m}^{(1)}(\mathrm{x}_n = 1; k = 1, \ldots, K$ and $n = 1, \ldots, N$. We denote $P^{(1)}(\mathrm{x}_n) = \left[p_{1,1}^{(1)}(\mathrm{x}_n), p_{1,2}^{(1)}(\mathrm{x}_n), \ldots, p_{K,M}^{(1)}(\mathrm{x}_n)\right]$ is a (MK) prediction vector of the EoC in the first layer for x_n. The prediction vectors for all observations in $\mathcal{D}$ is given in the form of a $N \times (MK)$ matrix.

$$\mathcal{P}_1 = \left[P^{(1)}(\mathrm{x}_1) P^{(1)}(\mathrm{x}_2) \ldots P^{(1)}(\mathrm{x}_N)\right]^T \tag{1}$$

We denote $\mathcal{L}_1$ denotes the new data generated by the 1st layer as the input for the 2nd layer. Normally, $\mathcal{L}_1$ is created by concatenating the original training data and the predictions classifiers as below:

$$\mathcal{L}_1 = \mathcal{D} \bigoplus \mathcal{P}_1 \tag{2}$$

in which $\bigoplus$ denotes the concatenation operator between two matrices $\mathcal{D}$ of size $N \times D + 1$ and $\mathcal{P}_1$ of size $N \times (MK)$. Thus $\mathcal{L}_1$ is obtained in the form of a $N \times (D + MK + 1)$ matrix including D features of original data, MK features of predictions, and ground truth of observations. A similar process conducts on the next layers until reaching the last layer in which at the ith layer, we train the EoC of K classifiers $\mathbf{h}_k^{(i)}, k = 1, \ldots, K$ on the input data $\mathcal{L}_{i-1}$ generated by $(i-1)^{th}$ layer and generate input data $\mathcal{L}_i$ for the $(i+1)^{th}$ layer

$$\mathcal{L}_i = \mathcal{D} \bigoplus \mathcal{P}_i \tag{3}$$

The predictions of EoC of the last layer i.e. s^{th} layer are combined for the collaborated decision. In this study, we use the Sum rule for combining [14]. For an instance x, the Sum rule summarizes the predictions of EoC of the last layer concerning each class label. The label associated with the maximum value is assigned to this instance as follows:

$$\tilde{\mathbf{h}} : \mathrm{x} \in y_t \text{ if } t = argmax_{m=1,\ldots,M}\left\{\sum_{k=1}^{K} p_{k,m}^{(s)}(\mathrm{x})\right\} \tag{4}$$

In the classification process, each unseen instance is fed forward through the layers until reaching the last layer. The predictions of K classifier at the last layer i.e. $P^{(s)}(.) = \left[p_{1,1}^{(s)}(.), p_{1,2}^{(s)}(.), \ldots p_{K,M}^{(s)}(.)\right]$ are combined by Sum Rule in (4) to obtain the predicted label.

It is recognized that there is existing a subset of EoC that performs better than using the whole ensemble. Moreover, storing a subset of the ensemble will save the computational cost and storage cost. In this study, we propose an ensemble selection approach for deep ensemble systems. We propose to encode classifiers in the deep ensemble system using binary encoding in (5), showing which classifiers are presented or absent. For a deep ensemble system of s layers, since there are K classifiers in each layer, the encoding associated with the model of s layers has $s \times K$ binary elements. It is noted that the length of the proposed encoding is not fixed and depends on the number of layers that we use to construct the deep model. If the number of layers is chosen by $1 \leq s \leq S$, we have S groups of encoding with the lengths of $\{K, 2 \times K, \ldots, S \times K\}$. By using these groups of encoding, we aim to search for the optimal number of layers and the optimal set of classifiers in each layer for the deep ensemble system.

$$\mathbf{h}_k^{(i)} = \begin{cases} 1, & k^{th} \text{ classifier at } i^{th} \text{ layer is selected} \\ 0, & \text{otherwise} \end{cases} \tag{5}$$

3.2 Optimization Problems and Algorithm

We consider optimization for the model selection problem. The objective is maximizing the accuracy of the classification task on a validation set $\mathcal{V}$:

$$max_E \left\{ \frac{1}{|\mathcal{V}|} \sum\nolimits_{n=1}^{|\mathcal{V}|} \left\| \tilde{\mathbf{h}}_E(\mathrm{x}_n) = \hat{y}_n \right\| \right\} \tag{6}$$

where $\tilde{\mathbf{h}}_E$ is the combining model using the Sum Rule in (4) associated with encoding E, $|.|$ denotes the cardinality of a set, and$\|.\|$ is equal 1 if the condition is true, otherwise equal 0. In this study, we develop a VLGA to solve this optimization problem. Genetic Algorithm (GA) is a search heuristic inspired by Charles Darwin's theory of natural evolution. It is widely recognized that GA commonly generates high-quality solutions for search problems [8]. Three operators of GA are considered in this study:

Selection: We apply the roulette wheel selection approach to select a pair of individuals for breeding. The probability of choosing an individual from a population is proportional to its fitness as an individual has a higher chance of being chosen if its fitness is higher than those of others. Probability of choosing individual i^{th} is equal to:

$$\mathrm{p}_i = \frac{f_i}{\sum_{j=1}^{popSize} f_j} \tag{7}$$

where f_i is the accuracy of the deep ensemble model with the corresponding configuration of the ith individual and $popSize$ is the size of the current generation.

Crossover: We define the probability P_c for the crossover process in which crossover occurs if the generated crossover probability is smaller than P_c. Here

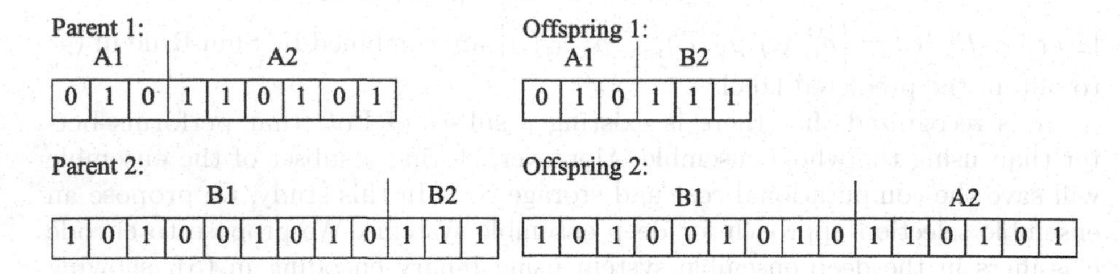

Fig. 1. The illustration of chunk-based crossover operator

we develop a chunk-based crossover operator to generate new offsprings. As mentioned before, since there are at most K classifiers in each layer of a deep ensemble system with s layers, each chromosome is given in the form of s-chunk in which the chunk size is K. On two selected parents with s_1 and s_2 layers, we generate two random numbers r_1 and r_2 which are the multiple of s_1 and s_2 i.e. $r_1 \in \{K, 2 \times K, \ldots, s_1 \times K\}$ and $r_2 \in \{K, 2 \times K, \ldots, s_2 \times K\}$. r_1 and r_2 will divide each parent into two parts. Each parent exchanges its tail with the other while retains its head. After crossover is performed, we have two new offspring chromosomes. We illustrate in Fig 1 how chunk-based crossover works on a deep ensemble model with 3 classifiers in each layer. Parent 1 encodes a 3-layer deep ensemble model, while parent 2 encodes a 4-layer deep ensemble model. On parent 1 and 2, two random numbers are generated as $r_1 = 3$ and $r_1 = 9$. By retaining heads and exchanging tails on these parents, we obtain two new offsprings, the first one encodes a 2-layer deep ensemble, and the second encodes a 5-layer deep ensemble. By using this crossover operator, we can generate the offsprings with different sizes compared to those of their parents, thus improving exploration of the searching process.

Mutation: Mutation operators introduce genetic diversity from one generation in a population to the next generation. It also prevents the algorithm from falling into local minima or maxima by making the population of chromosomes different from each other. We define the probability P_m for the mutation process in which mutation occurs if the generated mutation probability is smaller than P_m. In this study, we propose to apply a multiple point-based mutation operator on an offspring. First, we generate several random numbers which show the position of mutated genes in a chromosome. The values of these mutated genes will be flipped, i.e., from 0 to 1 or 1 to 0. By doing this way, we obtain a new offspring, which may change entirely from the previous one; consequently, GA can escape from local minima or maxima and reach a better solution.

The pseudo-code of VLGA is present in Algorithm 1. The algorithm gets the inputs including the training data $\mathcal{D}$, the validation data $\mathcal{V}$ and some parameters for the evolutionary process (the population size *popSize*, the number of generations *nGen*, crossover probability P_c and mutation probability P_m) We first randomly generate a population with *popSize* individuals and then calculate the fitness of each individual on $\mathcal{V}$ by using Algorithm 2 (Step 1 and 2).

The probabilities for individual selection are computed by using (7) in Step 3. Two selected parents will bread a pair of offsprings if they satisfy the crossover check (Step 7–8). These offsprings will pass through mutation in which some random positions of them are changed if mutation occurs (Step 15–21). We also calculate the fitness of each offspring on $\mathcal{V}$ by using Algorithm 2 before adding them to the population. The step 6–23 are repeated until we generate a new *popSize* offsprings. From the population of 2 × *popSize* individuals, we keep *popSize* best individuals for the next generation. The algorithm runs until it reaches the number of generations. We select the candidate from the last population, which is associated with the best fitness value as the solution of the problem.

Algorithm 2 aims to calculate the fitness and deep model generation associated with an encoding. The algorithm inputs training data $\mathcal{D}$, validation data $\mathcal{V}$, an encoding E, and the number of T-folds. From the configuration of E, we can obtain the number of layers and which classifiers are selected in each layer. On the ith layer, we do two steps (i) train selected classifiers at the on whole $\mathcal{L}_{i-1}$denoted by $\{h_k^{(i)}\}$ (Step 4) and (ii) generate training data for the $(i+1)^{th}$ layer by using T-fold Cross-Validation and concatenation operator between prediction data and original training data (Step 7–14). The classifier $\{h_k^{(i)}\}$ predicts on $\mathcal{V}_{i-1}$ which is the prediction matrix for observations of $\mathcal{V}$ at the $(i-1)^{th}$ layer, to obtain the prediction $\mathcal{P}_i(\mathcal{V})$ (Step 6). $\mathcal{P}_i(\mathcal{V})$ is also concatenated with $\mathcal{V}$ to obtain the validation data for the $(i+1)^{th}$ layer. After running through the last layer i.e. the sth layer, we apply the Sum Rule on the prediction $\mathcal{P}_s(V)$ to obtain the fitness value of E. We also obtain the classifiers $\{h_k^{(i)}\}$ associated with E.

In the classification process, we assign the class label to an unlabeled test sample. In each layer, the input test data will be predicted by classifiers and then be concatenated with the original test sample to generate new test data for the next layer. The combining function in (4) is applied to the outputs of classifiers of the last layer to give the final prediction.

4 Experimental Studies

4.1 Experimental Settings

We conducted the experiments on the 20 datasets collected from different sources such as the UCI Machine Learning Repository and OpenML. We used 5 classifiers in each layer of VEGAS in which these classifiers were generated by using learning algorithms: Naïve Bayes classifiers with Gaussian distribution, XgBoost with 200 estimators, Random Forest with 200 estimators, and Logistic Regression. We used the 5-fold Cross-Validation in each layer to generate the new training data for the next layer. 20% of the training data is used for validation purposes [22]. For VLGA, the maximum number of generations was set to 50, the population size was set to 100, and the crossover and mutation probability was set to 0.9 and 0.1, respectively.

Algorithm 1. Variable length Genetic Algorithm

Require: Training data $\mathcal{D}$, Validation data $\mathcal{V}$ population size: *popSize*, number of generations *nGen*, crossover probability: P_c, mutation probability: P_m
Ensure: Optimal configuration
1: Randomly generate population
2: Calculate fitness on $\mathcal{V}$ of each individual using Algorithm 2
3: Calculate selection probabilities by (7)
4: **for** $i \leftarrow 1, nGen$ **do**
5:   **while** *currentpopulationsize* $< 2 \times popSize$ **do**
6:     Select a pair of individuals based on selection probabilities
7:     Generate a random number $r_c \in [0, 1]$
8:     **if** $r_c \leq P_c$ **then**
9:       Generate two random number r_1, r_2 which are multiple of K
10:       Divide parents to head and tail based on r_1 and r_2
11:       Swap tails of two parents to create two new offsprings
12:     **else**
13:       Continue
14:     **end if**
15:     Generate a random number $r_m \in [0, 1]$
16:     **if** $r_m \leq P_m$ **then**
17:       **for** each offspring **do**
18:         Generate random number r of mutation points
19:         Flip the binary value associated with mutation points
20:       **end for**
21:     **end if**
22:     Calculate the fitnesses of two new offsprings using Algorithm 2
23:     Add two new offsprings to the population
24:   **end while**
25:   Keep *popSize* best individuals for the next generation
26:   Calculate selection probabilities by (7)
27: **end for**
28: Return individual (encoding and associated deep model) with the best fitness from the last generation

VEGAS was compared to some algorithms, including the ensemble methods and deep learning models. Three well-known ensemble methods were used as the benchmark algorithms: Random Forest, XgBoost, and Rotation Forest. All these methods were constructed by using 200 learners. Three deep learning models were compared with VEGAS: gcForest (4 forests with 200 trees in each forest) [22], MULES [15], and Multiple Layer Perceptron (MLP). For MULES, we used parameter settings like in the original paper [15]. It is noted that the performance of MLP significantly depends on the network structure. To ensure a fair comparison, we experimented with MLP on a number of different network configurations: input-30-20-output, input-50-30-output, and input-70-50-output by referencing the experiments [22]. We then reported the best performance of MLP among all configurations and used this result to compare with VEGAS.

We used Friedman test to compare performance of experimental methods on experimental datasets. If the P-Value of this test is smaller than a significant threshold, e.g. 0.05, we reject the null hypothesis and conduct the Nemenyi post-hoc test to compare each pair of methods [5].

4.2 Comparison to the Benchmark Algorithms

Table 1 shows the prediction accuracy of VEGAS and the benchmark algorithms. Based on the Friedman test, we reject the null hypothesis that all methods per-

Algorithm 2. Fitness calculation and deep model generation on an encoding

Require: Training Data $\mathcal{D}$, validation data $\mathcal{V}$, and encoding E, number of T-fold
Ensure: the fitness value of E

1: Get number of layer s and selected classifiers in each layer from E
2: $\mathcal{L}_0 = \mathcal{D}, \mathcal{V}_0 = \mathcal{V}$
3: **for** $i \leftarrow 1, s$ **do**
4: Train selected classifiers at the ith layer on whole $\mathcal{L}_{i-1}$: $\{h_k^{(i)}\}$
5: $\mathcal{P}_i = \emptyset$
6: Use $\{h_k^{(i)}\}$ to predict for $\mathcal{V}_{i-1}$ to obtain $\mathcal{P}_i(\mathcal{V})$
7: **for** $t \leftarrow 1, T$ %generate the running data for the next layer **do**
8: $\mathcal{L}_{i-1} = \bigcup_{j=1}^{T} \mathcal{L}_{i-1}^{(j)}, \mathcal{L}_{i-1}^{(j_1)} \bigcap \mathcal{L}_{i-1}^{(j_2)} = \emptyset, \left|\mathcal{L}_{i-1}^{(j_1)}\right| \approx \left|\mathcal{L}_{i-1}^{(j_2)}\right|,$
9: $1 \leq j_1, j_2 \leq T, j_1 \neq j_2$
10: **for all** $\mathcal{L}_{i-1}^{(j)}$ **do**
11: Train Selected Classifiers on $\mathcal{L}_{i-1} - \mathcal{L}_{i-1}^{(j)}$
12: Use these classifiers to predict on $\mathcal{L}_{i-1}^{(j)}$ to obtain $\mathcal{P}_i^{(j)}$
13: $\mathcal{P}_i = \mathcal{P}_i \bigcup \mathcal{P}_i^{(j)}$
14: **end for**
15: $\mathcal{L}_i = \mathcal{L}_0 \bigoplus \mathcal{P}_i$
16: **end for**
17: $\mathcal{V}_i = \mathcal{V}_0 \bigoplus \mathcal{P}_i(\mathcal{V})$
18: **end for**
19: Using combing method (4) on $\mathcal{P}_i(\mathcal{V})$
20: Calculate fitness f of E by using (6)
21: Return f and $\{h_k^{(i)}\} i = 1, \ldots, s$

form equally. The Nemenyi test in Fig 2 shows that VEGAS is better than all benchmark algorithms. In detail, VEGAS performs the best among all methods on 15 datasets. VEGAS ranks second on 5 datasets, and the prediction accuracy of VEGAS and the first rank method are not significant differences (0.9610 vs 0.9756 of gcForest on the Breast-cancer dataset, for example). The outstanding performance of VEGAS over the benchmark algorithms comes from (i) the use of multi-layer architecture over one-layer ensembles (ii) the use of ensemble selection to search for optimal configuration for VEGAS on each dataset.

Surprisingly, Random Forest ranks higher than the other benchmark algorithms in our experiment. Random Forest ranks the first on two datasets Hayes-Roth and Wine white (about 2% better than VEGAS for prediction accuracy). In contrast, VEGAS is significantly better than Random Forest on some datasets such as Hill-valley (about 30% better), Sonar (about 6% better), Vehicle (about 6% better), Tic-Tac-Toe (about 8% better).

MULES and XgBoost rank the middle in our experiment. VEGAS outperforms MULES on all datasets. MULES looks for optimal EoC of each layer by considering two objectives: maximising accuracy and diversity. Meanwhile, VEGAS learns the optimal configuration for all layers of the deep ensemble. It demonstrates the efficiency of the optimisation method, i.e. VLGA of VEGAS. For MLP, although we ran the experiments on its 3 different configurations and reported its best result for the comparison, this method is worse than VEGAS on up to 18 datasets. gcForest is worst among all methods on the experimental datasets. On some datasets such as Conn-bench-vowel, Hill-valley, Sonar, Texture, Tic-Tac-Toe, Vehicle, and Wine-white, gcForest performs poorly and by far worse than VEGAS.

Table 1. The classification accuracy and ranking of all methods

	gcForest	MLP	Random forest	XgBoost	MULES	VEGAS
Artificial	0.7905 (2)	0.6476 (6)	0.7619 (3)	0.7571 (4)	0.7238 (5)	**0.8000 (1)**
Breast-cancer	**0.9756 (1)**	0.9512 (5.5)	0.9561 (3)	0.9512 (5.5)	0.9512 (4)	0.9610 (2)
Cleveland	0.6000 (2.5)	0.5000 (6)	0.6000 (2.5)	0.5556 (5)	0.5889 (4)	**0.6111 (1)**
Conn-bench-vowel	0.6289 (6)	0.8365 (4)	0.9119 (2)	0.8428 (3)	0.8050 (5)	**0.9245 (1)**
Hayes-roth	0.8333 (3)	0.7708 (6)	**0.8750 (1)**	0.8125 (5)	0.8333 (4)	0.8542 (2)
Hill-valley	0.5852 (6)	0.8942 (3)	0.6593 (5)	0.6786 (4)	0.9766 (2)	**0.9849 (1)**
Iris	0.9556 (4)	**1.0000 (1)**	0.9333 (5.5)	0.9333 (5.5)	0.9556 (3)	0.9778 (2)
Led7digit	0.7067 (3)	0.7200 (2)	0.6933 (4)	0.6600 (5)	0.4400 (6)	**0.7333 (1)**
Musk1	0.8462 (3)	0.8252 (6)	0.8462 (4)	0.8322 (5)	0.8671 (2)	**0.8951 (1)**
Musk2	0.9515 (6)	0.9798 (3)	0.9773 (4)	0.9929 (2)	0.9727 (5)	**0.9939 (1)**
Newthyroid	0.9692 (4)	0.9846 (2.5)	0.9846 (2.5)	0.9538 (5)	0.9385 (6)	**1.0000 (1)**
Page-blocks	0.9464 (6)	0.9629 (4)	0.9695 (2)	0.9683 (3)	0.9568 (5)	**0.9720 (1)**
Pima	0.7229 (5)	0.7316 (4)	0.7403 (2)	0.7359 (3)	0.6840 (6)	**0.7576 (1)**
Sonar	0.8095 (6)	0.8889 (2)	0.8413 (3.5)	0.8413 (3.5)	0.8254 (5)	**0.9048 (1)**
Spambase	0.9392 (5)	0.9356 (6)	0.9551 (2)	0.9522 (3)	0.9508 (4)	**0.9594 (1)**
Texture	0.8436 (6)	**0.9933 (1)**	0.9752 (5)	0.9848 (4)	0.9855 (3)	0.9915 (2)
Tic-Tac-Toe	0.8368 (6)	0.9271 (4)	0.9236 (5)	**1.0000 (1.5)**	0.9792 (3)	**1.0000 (1.5)**
Vehicle	0.7283 (5)	0.6890 (6)	0.7638 (4)	0.7717 (3)	0.8189 (2)	**0.8307 (1)**

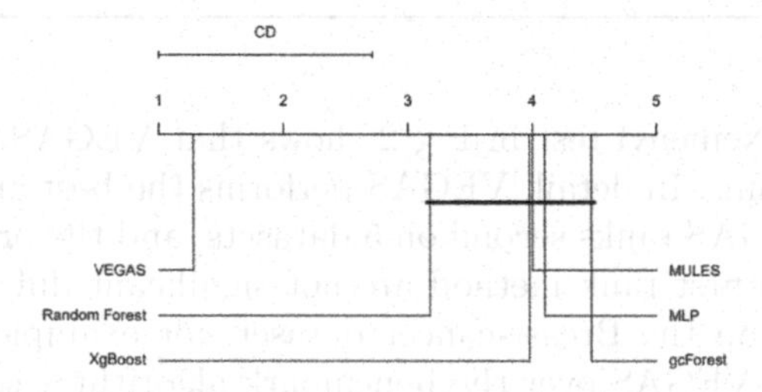

Fig. 2. The Nemenyi test result

4.3 Discussions

VEGAS takes higher training time compared to two deep ensemble models i.e. gcForest and MULES. On the Tic-Tac-Toe dataset, for example, gcForest used only 311.78 s for the training process compared to 15192.08 of VEGAS (100 individuals in each generation). Meanwhile, MULES (50 individuals in each generation) used 3154.86 s for its training process. It is noted that the training time of VEGAS can reduce based on either parallel implementation or early stopping of VLGA. Figure 3 shows the average and global best of the fitness function in the generations of VEGAS on 4 selected datasets. It is observed that the global bests converge quickly after several generations. On Balance and Artificial dataset, for example, their global bests converge after 8 iterations. Thus, the training time on some datasets can reduce by an early stopping based on the convergence of the global best.

On the other hand, although VEGAS creates more layers than gcForest and MULES (6.4 vs 3.8 and 2.75 on average), the classification time of VEGAS is

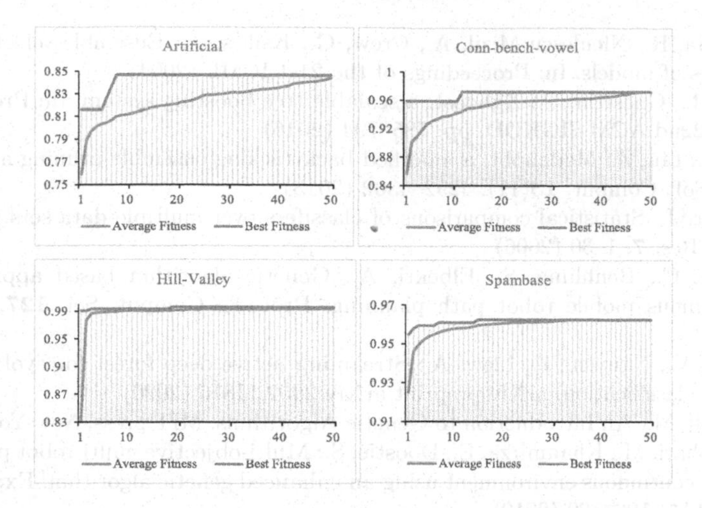

Fig. 3. The average and global best of the fitness function in the generations of VLGA on 4 datasets

competitive to those of gcForest and MULES. That is because on some datasets, VEGAS selects a small number of classifiers in each layer. On the Tic-Tac-Toe dataset, VEGAS takes only 0.016 s for classification with 4 layers and 6 classifiers in total. Meanwhile, gcForest (6 layers and 4800 classifiers in total) used 0.62 s to classify all test instances, and MULES (4 layers with 11 classifiers in total) used 0.26 s with its selected configuration [15].

5 Conclusions

The deep ensemble models have further improved the predictive accuracy of one-layer ensemble models. However, the appearance of unsuitable classifiers in each layer reduces predictive performance and the *computational/storage efficiency* of the deep models. In this study, we have introduced an ensemble selection method for the deep ensemble systems called VEGAS. We design the deep ensemble system involving multiple layers of the EoC. The training data is populated through layers by concatenating the predictions of classifiers in the subsequent layer and the original training data. The predictions of the classifiers in the last layer are combined by a combining method to obtain the final collaborated prediction. We proposed the VLGA to search for the optimal configuration, which maximizes the prediction accuracy of the deep ensemble model on each dataset. Three operators of VLGA were considered in this study, namely selection, crossover, and mutation. The experiments on 20 datasets show that VEGAS is better than both well-known ensemble methods and other deep ensemble methods.

References

1. Breiman, L.: Random forests. Mach. Learn. **45**(1), 5–32 (2001)

2. Caruana, R., Niculescu-Mizil, A., Crew, G., Ksikes, A.: Ensemble selection from libraries of models. In: Proceedings of the 21st ICML (2004)
3. Chen, T., Guestrin, C.: Xgboost: a scalable tree boosting system. In: Proceedings of the 22nd ACM SIGKDD, pp. 785–794 (2016)
4. Dai, Q., Liu, Z.: Modenpbt: a modified backtracking ensemble pruning algorithm. Appl. Soft Comput. **13**(11), 4292–4302 (2013)
5. Demsar, J.: Statistical comparisons of classifers over multiple data sets. J. Mach. Learn. Res. **7**, 1–30 (2006)
6. Lamini, C., Benhlima, S., Elbekri, A.: Genetic algorithm based approach for autonomous mobile robot path planning. Procedia Comput. Sci. **127**, 180–189 (2018)
7. Luong, V., Nguyen, T., Liew, A.: Streaming active deep forest for evolving data stream classification. arXiv preprint arXiv:2002.11816 (2020)
8. Mitchell, M.: An Introduction to Genetic Algorithms. MIT press, New York (1998)
9. Nazarahari, M., Khanmirza, E., Doostie, S.: Multi-objective multi-robot path planning in continuous environment using an enhanced genetic algorithm. Expert Syst. Appl. **115**, 106–120 (2019)
10. Nguyen, T., Dang, T., Baghel, V., Luong, V., McCall, J., Liew, A.: Evolving interval-based representation for multiple classifier fusion. Knowledge-based systems, p. 106034 (2020)
11. Nguyen, T., Dang, T., Liew, A., Bezdek, J.C.: A weighted multiple classifier framework based on random projection. Inf. Sci. **490**, 36–58 (2019)
12. Nguyen, T., et al.: Deep heterogeneous ensemble. Australian J. Intell. Inf. Process. Syst. **16**(1), (2019)
13. Nguyen, T., Luong, V., Liew, A., McCall, J., Van Nguyen, M., Ha, S.: Simultaneous meta-data and meta-classifier selection in multiple classifier system. In: Proceedings of the GECCO, pp. 39–46 (2019)
14. Nguyen, T., Pham, C., Liew, A., Pedrycz, W.: Aggregation of classifiers: a justifiable information granularity approach. IEEE Trans. Cybern. **49**(6), 2168–2177 (2018)
15. Nguyen, T., Van, N., Dang, T., Luong, V., McCall, J., Liew, A.: Multi-layer heterogeneous ensemble with classifier and feature selection. In: Proceedings of the GECCO, pp. 725–733 (2020)
16. Partalas, I., Tsoumakas, G., Vlahavas, I.: An ensemble uncertainty aware measure for directed hill climbing ensemble pruning. Mach. Learn. **81**(3), 257–282 (2010)
17. Qi, Z., Wang, B., Tian, Y., Zhang, P.: When ensemble learning meets deep learning: a new deep support vector machine for classification. Knowl.-Based Syst. **107**, 54–60 (2016)
18. Qiongbing, Z., Lixin, D.: A new crossover mechanism for genetic algorithms with variable-length chromosomes for path optimization problems. Expert Syst. Appl. **60**, 183–189 (2016)
19. Sun, Y., Xue, B., Zhang, M., Yen, G.G.: Evolving deep convolutional neural networks for image classification. IEEE Trans. Evol. Comput, **24**(2), 394–407 (2019)
20. Sun, Y., Xue, B., Zhang, M., Yen, G.G., Lv, J.: Automatically designing CNN architectures using the genetic algorithm for image classification. IEEE Trans. Cybern. (2020)
21. Yang, L., Wu, X.Z., Jiang, Y., Zhou, Z.H.: Multi-label learning with deep forest. arXiv preprint arXiv:1911.06557 (2019)
22. Zhou, Z.H., Feng, J.: Deep forest. arXiv preprint arXiv:1702.08835 (2017)

Demand Forecasting for Textile Products Using Statistical Analysis and Machine Learning Algorithms

Leandro L. Lorente-Leyva[1,2](✉), M. M. E. Alemany[1], Diego H. Peluffo-Ordóñez[2,3], and Roberth A. Araujo[4]

[1] Universitat Politècnica de València, Camino de Vera S/N 46022, València, Spain
lealo@doctor.upv.es
[2] SDAS Researh Group, Ibarra, Ecuador
[3] Modeling, Simulation and Data Analysis (MSDA) Research Program, Mohammed VI Polytechnic University, Ben Guerir, Morocco
[4] North Technical University, Ibarra, Ecuador

Abstract. The generation of an accurate forecast model to estimate the future demand for textile products that favor decision-making around an organization's key processes is very important. The minimization of the model's uncertainty allows the generation of reliable results, which prevent the textile industry's economic commitment and improve the strategies adopted around production planning and decision making. That is why this work is focused on the demand forecasting for textile products through the application of artificial neural networks, from a statistical analysis of the time series and disaggregation in different time horizons through temporal hierarchies, to develop a more accurate forecast. With the results achieved, a comparison is made with statistical methods and machine learning algorithms, providing an environment where there is an adequate development of demand forecasting, improving accuracy and performance. Where all the variables that affect the productive environment of this sector under study are considered. Finally, as a result of the analysis, multilayer perceptron achieved better performance compared to conventional and machine learning algorithms. Featuring the best behavior and accuracy in demand forecasting of the analyzed textile products.

Keywords: Demand forecasting · Textile products · Statistical analysis · Machine learning algorithms · Artificial neural networks · Temporal hierarchies

1 Introducción

Ecuador's textile and apparel industry, today, represents the fifth most important and representative manufacturing industry in the country. The sales generated by this sector within the Ecuadorian domestic market for the year 2019, according to the Association of Textile Industries of Ecuador, bordered the 1500 million dollars and, at an external level, the 115 million dollars, which makes this productive sector a determining factor in the economic development of Ecuador.

N. T. Nguyen et al. (Eds.): ACIIDS 2021, LNAI 12672, pp. 181–194, 2021.
https://doi.org/10.1007/978-3-030-73280-6_15

Currently, the uncertainty lies in the various sources of demand, the variability in the availability of raw materials, limited delivery times and the uncertain determination of the quantities in inventory that have also caused uncertainty throughout the production process of the textile sector. Economically compromising the companies that currently compete aggressively within demanding markets, becoming a rather uncertain, variable, and complex scenario. The instability of such behavior has forced manufacturing companies in this sector to optimally forecast demand for a given horizon, in order to favor the development of effective production plans. When the efficient determination of the various production variables (production levels, inventories, labor, among others) involved in this process, to respond to such instability, are not economically compromised [1].

Several researches have directed their efforts to the development of methods, based on Statistical and Machine learning (ML) methods, to optimize the estimation of demand [2–5]. All this, to increase the chances of survival of the companies, by responding efficiently to the fluctuations in demand within the totally changing, demanding and aggressive global markets in which they operate. The general objective of these methods is to reduce, in an intelligent manner, the level of uncertainty to which the demand forecasting is subject, so that, with the obtaining of forecasts adjusted to reality, a greater degree of flexibility is granted to the productive system to comply with the levels of production of textile products demanded from period to period.

Therefore, this work focuses on developing the demand forecasting of textile products through the application of Artificial Neural Networks (ANN), considering all the variables and scenarios present in this manufacturing sector. With this, a statistical analysis of the time series is carried out to analyze the correlation, study the variability and describe the dataset obtained, with a disaggregation in different time horizons by means of time hierarchy and achieve greater accuracy in the developed forecast. From this, a set of forecasting, statistical and ML methods are applied where the performance is compared with different indicators that evaluate the forecast error.

The rest of the document is structured as follows: Sect. 2 presents the related works of this research. Section 3 describes the case study, data and statistical analysis. The results of demand forecasting and discussion are presented in Sect. 4. Finally, the conclusions and future works are shown in Sect. 5.

2 Related Works

Several techniques have been used for the development of accurate forecasts, which have provided a clear vision of the future behavior of demand under uncertain and diffuse environments, where the data are often noisy [6–8]. This is where machine learning methods have achieved accurate results, with minimal margins of error, and greater accuracy compared to other alternative and conventional forecasting methods [9–11].

Within the textile and clothing industry, there have been several investigations on the demand forecasting. In [12] the authors make a comparison between several classical and ML forecasting methods. Where the artificial neural networks (ANN) present the best results, with the best performance in the forecast of each analyzed textile product, favoring the efficient development of the productive operations of the Ecuadorian textile industry.

Other authors [13] address the problem of inventory planning and forecasting future demand for fashion products in retail operations through a case study. DuBreuil and Lu [14] predict traditional fashion trends versus Big data. They suggest the viability of using Big data techniques to help these companies in the creation of new products. In other research [15] they propose a new measure of error prediction to consider the cost that causes the error, allowing to calculate the real effect of the forecast error.

Different techniques have been used to forecast demand, such as in the metal-mechanical industry [16], in forecasting the price of horticultural products [17], in forecasting urban water demand [18] and short-term water demand by applying LASSO (Least Absolute Shrinkage and Selection Operator) [19], and forecasting customer demand for production planning using k-Nearest Neighbor (kNN) [20]. Also, in the combination of specific forecasts [21], to generate an increase in the expected profit in relation to other methods. In [22] the authors develop a set of safe hybrid rules for efficient optimization of LASSO-type problems. This model has been used continuously for model selection in ML, data mining and high dimensional statistical analysis [23, 24].

In 2018, Loureiro et al. [25] propose a deep artificial neural network model for sales forecasting in textile retailing. Comparing the proposed method with a set of techniques such as Decision tree, Random forest, ANN and Linear Regression. The results of the study show that the proposed model has a good performance with respect to other alternatives. In [26], they develop a multi-objective artificial neural network model based on optimization, to address the problem of short-term replacement forecast in the fashion industry.

Other research [27] proposes hybrid models to forecast demand in the supply chain, showing that demand volatility significantly impacts on forecast accuracy. Similarly, in [28] they evaluate various macroeconomic indicator models to improve tactical sales forecasts with a statistical approach that automatically selects the indicator and the order of potential customers.

From the analysis of the related literature in the textile industry, mainly in Ecuador, the use of machine learning in demand forecasting is still limited, even though the results derived from their application are satisfactory. The use of classic methodologies, in the best scenario, or the use of the experience of planners to carry out the forecast calculation, is quite common. Sometimes, it compromises not only the accuracy of the results, but also the prevention of additional excessive costs and inefficiencies in the decision-making process in the short, medium, and long-term production planning. For this reason, the application of ANN is proposed, for demand forecasting of textile products, since its behavior assures results that contribute effectively to the decision making in changing and uncertain scenarios, by not eluding the restrictions and factors to which the sector under study is subject.

3 Materials and Methods

3.1 Case Study Description

The company under study is a medium enterprise in the province of Imbabura, Ecuador, whose raison is the production and commercialization of textile products. The development of this research will be carried out based on the application of artificial neural networks, to demand forecasting for textile products, and data from historical series of short-sleeved shirts, heaters, long-sleeved shirts, short-sleeved blouses, and long-sleeved blouses, defined as SKU (Stock Keeping Unit), in a consecutive way. The objective of the above is to obtain accurate and precise forecasts, where the error rate generated is minimal and the decisions, which are generated from these results, increase the chances of company success under study and strengthen the production planning process at different levels.

For this purpose, we have data from various time periods provided by the company under study, corresponding to three years of sales, from January 2017 to December 2019. It is important to note that, for this case study, the sales that have been invoiced, for each of the 5 products analyzed, can be considered representative of the company's demand, given that invoicing is done in the month in which an order is given.

Figure 1 shows in detail the behavior of the historical data series for each of the products analyzed. In this study, a set of causal variables will be considered that influence the predicted quantities, such as sales prices and manufacturing costs, the consumer price index, manufacturing GDP and the unemployment index. This will ensure the quality and accuracy of the results obtained.

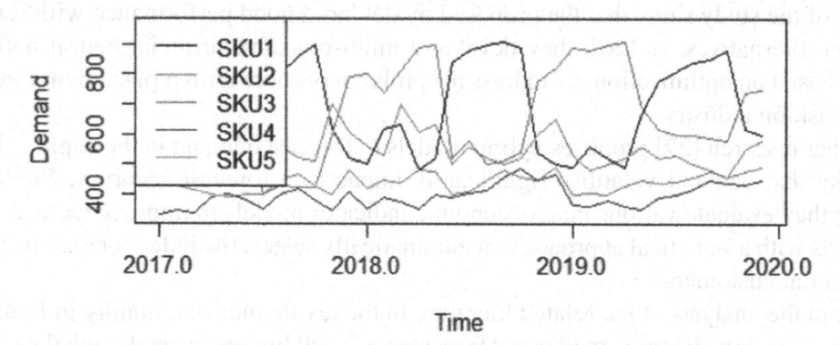

Fig. 1. Historical time series of the dataset.

The study of the demand seeks to minimize the degree of uncertainty of the input information, in order to predict its behavior in the near future, so that the decision-making process is objective and based on robust analytical calculations. Each of the stages involved in the construction of the proposed demand forecasting model, for the company under study, are developed through the programming language R.

From this, a statistical analysis of the historical series used will be developed, to determine the holding and behavior of the data, in order to detect some type of predominant pattern, and determine with an analysis of the correlation coefficients for each product, greater accuracy of the data patterns.

3.2 Statistical Analysis

Data pattern analysis is performed for each of the products analyzed. Figure 1 shows the complete time series of monthly sales. While it is useful to graph the behavior of the series to detect predominant pattern, this is not always feasible. In the case of SKU 1 and 3 no significant behavior can be observed that leads to specific decisions. However, the opposite is true for SKU 2, 4 and 5, where there is a slight negative trend for SKU 2 and a positive one for SKU 4 and 5.

To determine more accurately the data patterns, the analysis of the autocorrelation coefficients for each type of SKU is carried out. Table 1 contains the values of these coefficients for 15-time delays.

Table 1. Correlation coefficients for the dataset.

Delays	Correlation coefficients				
	SKU 1	SKU 2	SKU 3	SKU 4	SKU 5
1	0,642	0,328	0,681	0,566	0,365
2	0,207	0,052	0,294	0,362	0,095
3	−0,006	−0,035	−0,012	0,175	0,023
4	−0,278	0,101	−0,283	−0,059	0,010
5	−0,613	−0,097	−0,584	−0,188	−0,094
6	−0,666	0,017	−0,667	−0,290	−0,178
7	−0,462	0,049	−0,463	−0,334	0,099
8	−0,234	0,112	−0,206	−0,225	0,106
9	−0,084	0,050	−0,026	0,018	−0,126
10	0,166	−0,108	0,212	0,143	−0,282
11	0,497	−0,057	0,498	0,105	−0,135
12	0,632	0,216	0,521	0,355	0,208
13	0,377	0,274	0,312	0,269	−0,008
14	0,163	0,058	0,053	0,217	0,048
15	0,027	−0,060	−0,093	0,146	0,080

If we analyze the values obtained for the SKU 1 and SKU 3 the autocorrelation coefficients decrease rapidly, to be exact, from the second time delay, in both cases. Evidence that the series is stationary, however, and to reinforce this conclusion, the Dickey-Fuller test is applied, which gives results, see Table 2, that support this last decision.

In addition, the presence of a quite remarkable repetitive behavior in the months of April and October for the SKU 1, and in April for the SKU 3, as shown in Fig. 2.

Table 2. SKU 1 and SKU 3 dataset stationary test.

Hypothesis test	$H_0 = p - value > 0.05$ Time series is not stationary $H_1 = p - value < 0.05$ Time series is stationary		
SKU to evaluate	Dickey Fuller: *p-value*	Level of significance	Condition of H_0
SKU 1	0,0212	0,05	Rejected
SKU 3	0,0206	0,05	Rejected

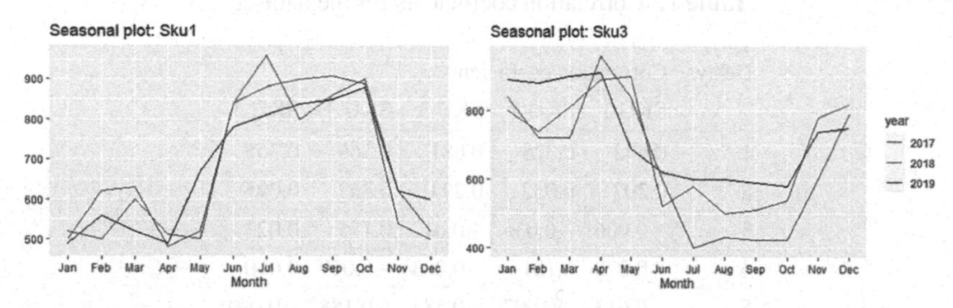

Fig. 2. Analysis of repetitive behaviors in the SKU 1 and SKU 3 dataset.

On the other hand, the autocorrelation coefficients presented in Table 1, belonging to SKU 2, SKU 4 and SKU 5, reveal the presence of a slight negative trend for SKU 2 and positive for SKU 4 and SKU 5. To reinforce such conclusion, the Dickey-Fuller test is applied that demonstrates the nonexistence of some stationary pattern, see Table 3.

Table 3. SKU 2, SKU 4 and SKU 5 dataset stationary test.

Hypothesis test	$H_0 = p - value > 0.05$ Time series is not stationary $H_1 = p - value < 0.05$ Time series is stationary		
SKU to evaluate	Dickey Fuller: *p-value*	Level of significance	Condition of H_0
SKU 2	0,2352	0,05	Rejected
SKU 4	0,2488	0,05	Rejected
SKU 5	0,3488	0,05	Rejected

As for the existence of repetitive behaviors, in the case of the SKU 2 there is no doubt that point in time that presents such conditions can be ruled out. However, this is not the case if we analyze the SKU 4 and SKU 5, where sales in the months of June and September show a high degree of similarity, as shown in Fig. 3.

With this developed statistical analysis, essential to generate an accurate demand forecasting, where the past behavior of the data is taken into consideration. From this analysis we proceed to develop a demand forecasting for textile products.

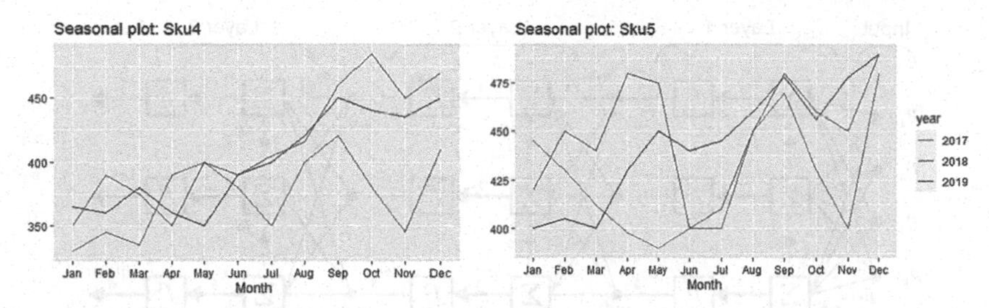

Fig. 3. Analysis of repetitive behaviors in the SKU 4 and SKU 5 dataset.

3.3 Demand Forecasting

The development of a model that adapts to the existing patterns, detailed in the previous point, in an efficient manner, will derive in results adjusted to reality, even if multiple challenges arise around the specification of the network architecture in accordance with the time series structure [29]. From this, the collection, manipulation, and cleaning of the data will be carried out, then the construction and evaluation of the model, developing the implementation of this one, with the obtaining of the real forecast of textile products and the evaluation of the forecast by means of the comparison with several metrics that express the accuracy as a percentage of the error and indicate the best adjustment.

This study focuses on the application of ANN, specifically Multilayer Perceptron (MLP) for the demand forecasting of textile products from historical data series. From this, an analysis of the time series by temporal hierarchies will be performed, to disaggregate in several time horizons and improve the forecast performance. This will be compared with other statistical and machine learning algorithms such as Linear regression, ARIMA, Holt-Winters, Exponential smoothing, LASSO, Random forest and kNN. And in this way analyze the quality of the developed forecast. Below is a brief description of how MLP works.

Multilayer Perceptron for Time Series Forecasting

Artificial neural networks are intelligent, non-parametric mathematical models [30] composed of a set of simple processing elements operating in parallel; these elements are inspired by biological nervous systems and have demonstrated advanced behavior in demand forecasting. Figure 4 shows the MLP structure proposed in this research for textile demand forecasting.

The notation that describes the structure detailed in Fig. 4 is presented in Eq. (1), where R is the number of input elements, S the number of neurons per layer, p are the input values, w the weight of each neuron, b represents the value of the bias of each neuron, n is the net input of the transfer function denoted as f and a the output value.

$$a^3 = \mathrm{f}^3\left(LW_{3.2}\mathrm{f}^2\left(LW_{2.1}\mathrm{f}^1(IW_{1.1}p + b_1) + b_2\right) + b_3\right) = y \quad (1)$$

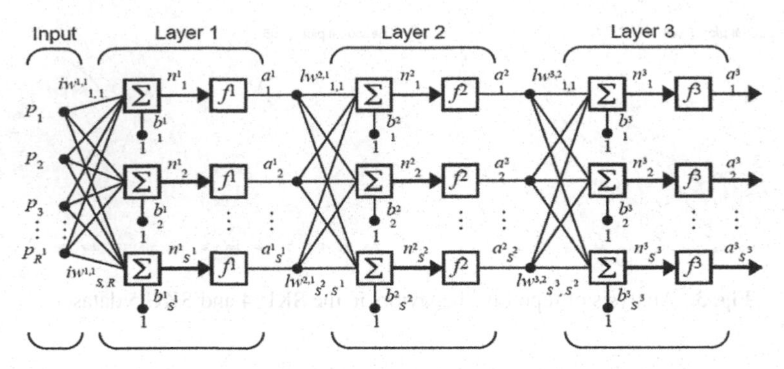

Fig. 4. MLP neural network structure [31].

4 Results and Discussion

In this section, the results obtained with the MLP and the methods applied will be presented, developing a comparison based on evaluation metrics that analyze the forecasting error, in order to determine the result with better behavior and accuracy. The company handles a detailed record of monthly sales for each SKU analyzed. The sample with which the forecast was made corresponds to 3 years, that is, in monthly terms the equivalent of 36 past sales data.

By applying the MLP neural network for the demand forecasting of the textile products under study, the forecast is obtained for the year 2020. The summary of the MLP configuration and the results are presented in Table 4. Where it was possible to determine the optimal network configuration using the pyramid rule and the optimal forecast result.

Table 4. MLP configuration.

Hidden layers	(22,16,10)
Network structure	Input layer: 23 neurons Hidden layer 1: 22 neurons Hidden layer 2: 16 neurons Hidden layer 3: 10 neurons

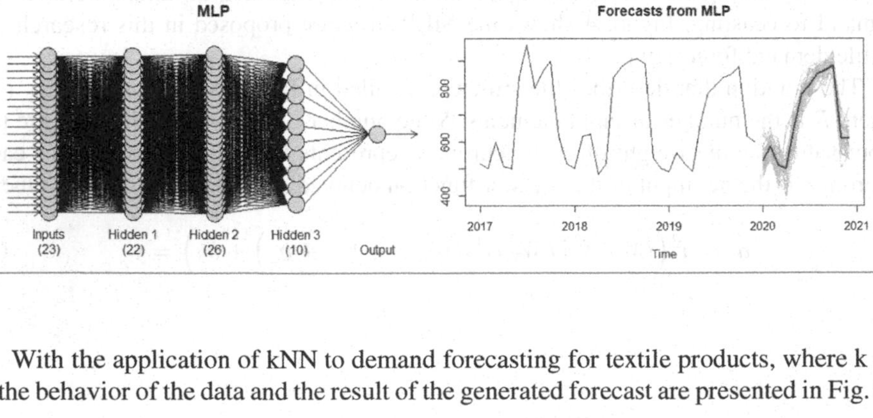

With the application of kNN to demand forecasting for textile products, where k = 2, the behavior of the data and the result of the generated forecast are presented in Fig. 5.

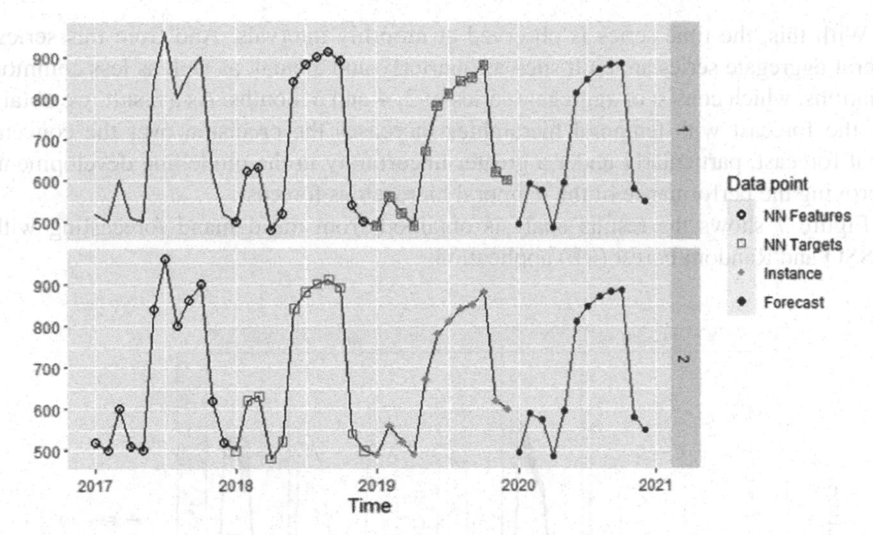

Fig. 5. Demand forecasting with kNN.

Figure 6 shows the developed forecast of the temporal hierarchies. The predictions are plotted in the grey filled in area. The continuous line is used for baseline predictions and the line in red, in the gray area for the temporal hierarchy predictions. With this analysis and by merging it with other forecasting methods, more accurate results are obtained at different time horizons [32, 33].

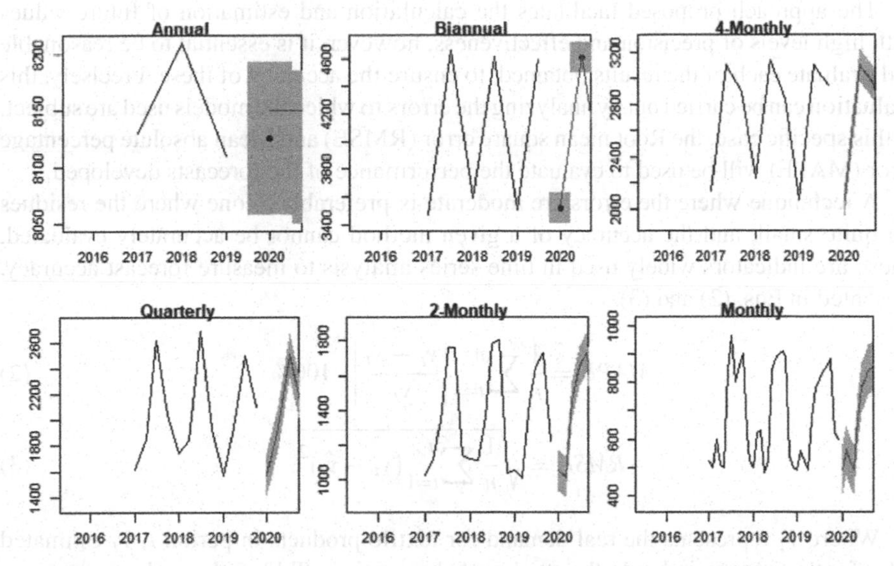

Fig. 6. Forecasts for the time series of textile products with temporal hierarchies.

With this, the time series is observed at monthly intervals. And from this series, several aggregate series are built such as quarterly and annual, as well as less common variations, which consist of aggregate series of 2, 4 and 6 months. As a result, we obtain that the forecast with temporal hierarchies increases the precision over the conventional forecast, particularly under a greater uncertainty in the modelling development, improving the performance of the temporal hierarchies forecast.

Figure 7 shows the results analysis obtained from the demand forecasting with LASSO and Random Forest (RF) application.

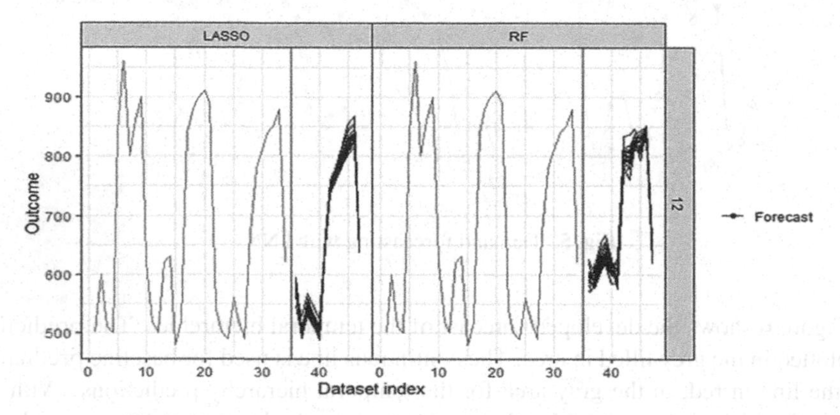

Fig. 7. Demand forecasting with LASSO and Random Forest.

The approach proposed facilitates the calculation and estimation of future values with high levels of precision and effectiveness, however, it is essential to be reasonable and evaluate each of the results obtained, to ensure the accuracy of these. Precisely, this evaluation can be carried out by analyzing the errors to which the models used are subject. In this specific case, the Root mean square error (RMSE) and Mean absolute percentage error (MAPE) will be used to evaluate the performance of the forecasts developed.

A technique where the errors are moderate is preferable to one where the residues are quite small, and the accuracy of a given method cannot be accurately evaluated. These are indicators widely used in time series analysis to measure forecast accuracy, presented in Eqs. (2) and (3).

$$MAPE = \frac{1}{n}\sum\nolimits_{t=1}^{n}\left|\frac{y_t - \widehat{y_t}}{y_t}\right| \cdot 100\% \tag{2}$$

$$RMSE = \sqrt{\frac{1}{n}\sum\nolimits_{t=1}^{n}\left(y_t - \widehat{y_t}\right)^2} \tag{3}$$

Where Y_t represents the real demand for textile products in period t, $\widehat{Y_t}$ estimated value for the same period and n the number of observations. Table 5 shows the comparison of the forecasting results and performance in relative % for each forecasting technique.

With this analysis it can be shown that MLP have excellent accuracy compared to alternative methods, which demonstrates the quality of the predicted values for the next

Table 5. Forecasting error results and performance of each forecasting technique.

Evaluation metrics	Methods	Products				
		SKU 1	SKU 2	SKU 3	SKU 4	SKU 5
MAPE	Linear regression	0.161	0.051	0.134	0.130	0.209
	ARIMA (2,1,2) (1,1,1)	0.171	0.0569	0.166	0.125	0.303
	Holt-Winters	0.153	0.0716	0.154	0.208	0.184
	Exponential smoothing	0.112	0.0586	0.152	0.231	0.143
	MLP	0.105	0.050	0.131	0.103	0.143
	LASSO	0.493	0.490	0.579	0.476	0.399
	Random Forest	1.153	0.621	0.735	0.667	0.379
	kNN	0.481	0.837	0.892	0.418	0.776
RMSE	Linear regression	106.0	39.0	75.2	53.6	50.1
	ARIMA (2,1,2) (1,1,1)	113.0	36.3	74.0	52.1	75.5
	Holt-Winters	117.0	58.3	67.8	81.1	54.7
	Exponential smoothing	88.9	44.6	73.0	88.0	57.4
	MLP	83.9	36.9	58.5	44.8	57.0
	LASSO	43.8	33.6	51.6	26.9	18.0
	Random Forest	75.6	47.3	87.1	35.4	23.5
	kNN	37.4	47.5	78.6	22.0	42.6

period. The results achieved show that the MLP learning capacity is so effective that MAPE is closer to the ideal value than the results obtained by the other methods used for forecasting.

In summary, it has been shown how multilayer neural networks are effectively applicable to the problem of predicting time series of textile products. Favoring the objective, solid and sustained decision-making, with a view to improving the processes immersed in the company under study, such as production planning, inventory management, among others, from the minimization of the uncertainty to which they are exposed.

5 Conclusions and Future Work

The competitiveness of a company is often conditioned by its ability to respond in a timely manner to demand, based on accurate and successful forecasts. Where sales prices and manufacturing costs, consumer price index, manufacturing GDP and unemployment index, as main factors, determine the uncertain, variable, and complex behavior of the demand for textile products in Ecuador. This has unleashed several problems in companies dedicated to the production of this type of goods, such as excess inventories, increased costs, and reduced service levels, among others. All the above justifies the need to use robust forecasting methods that provide feasible and accurate solutions.

This research shows that MLP can effectively respond to the influence of the different patterns involved in the time series of each of the products analyzed. Its adaptation and generalization capacity, over the other techniques used, such as Linear regression, ARIMA, Holt-Winters, Exponential smoothing, LASSO, Random Forest and kNN, allowed obtaining a high performance, where the results obtained reached minimum errors.

As a result of the analyses carried out in the study, it can be concluded that the application of multilayer artificial neural networks can reach quite considerable levels of precision, when there is a limited amount of data and they present several behavioral patterns. With this method, the best performance is obtained, where the forecast errors reach the most ideal values and the error percentage, on average, is very small, with a high forecast accuracy. This prevents any company from being economically compromised and even more makes it vulnerable in an environment where the competition is increasingly adopting rigid strategies based on the application of methodologies that undermine the possibility of failure.

As future work, it is proposed to develop a more in-depth analysis with the application of temporal hierarchies, for any time series independently of the forecast model used for the demand forecasting for textile products. Where uncertainty is mitigated and hybridized with machine learning algorithms, to improve the forecast results. Considering all the variables that influence an accurate forecast development, as well as the case we are currently facing, with the COVID-19 pandemic, which directly affects the behavior of all industry sectors and mainly, the textile industry.

Acknowledgment. The authors are greatly grateful by the support given by the SDAS Research Group (www.sdas-group.com).

References

1. Lorente-Leyva, L.L., et al.: Optimization of the master production scheduling in a textile industry using genetic algorithm. In: Pérez García, H., Sánchez González, L., Castejón Limas, M., Quintián Pardo, H., Corchado Rodríguez, E. (eds.) HAIS 2019. LNCS (LNAI), vol. 11734, pp. 674–685. Springer, Cham (2019). https://doi.org/10.1007/978-3-030-29859-3_57
2. Ren, S., Chan, H.-L., Ram, P.: A comparative study on fashion demand forecasting models with multiple sources of uncertainty. Ann. Oper. Res. **257**(1–2), 335–355 (2016). https://doi.org/10.1007/s10479-016-2204-6
3. Bruzda, J.: Quantile smoothing in supply chain and logistics forecasting. Int. J. Prod. Econ. **208**, 122–139 (2019). https://doi.org/10.1016/j.ijpe.2018.11.015
4. Silva, P.C.L., Sadaei, H.J., Ballini, R., Guimaraes, F.G.: Probabilistic forecasting with fuzzy time series. IEEE Trans. Fuzzy Syst. (2019). https://doi.org/10.1109/TFUZZ.2019.2922152
5. Trull, O., García-Díaz, J.C., Troncoso, A.: Initialization methods for multiple seasonal Holt-Winters forecasting models. Mathematics **8**(2), 1–6 (2020). https://doi.org/10.3390/math8020268
6. Murray, P.W., Agard, B., Barajas, M.A.: Forecast of individual customer's demand from a large and noisy dataset. Comput. Ind. Eng. **118**, 33–43 (2018). https://doi.org/10.1016/j.cie.2018.02.007
7. Prak, D., Teunter, R.: A general method for addressing forecasting uncertainty in inventory models. Int. J. Forecast. **35**(1), 224–238 (2019). https://doi.org/10.1016/j.ijforecast.2017.11.004
8. Fabianova, J., Kacmary, P., Janekova, J.: Operative production planning utilising quantitative forecasting and Monte Carlo simulations. Open Engineering **9**(1), 613–622 (2020). https://doi.org/10.1515/eng-2019-0071
9. Bajari, P., Nekipelov, D., Ryan, S.P., Yang, M.: Machine learning methods for demand estimation. Am. Econ. Rev. **105**(5), 481–485 (2015). https://doi.org/10.1257/aer.p20151021

10. Villegas, M.A., Pedregal, D.J., Trapero, J.R.: A support vector machine for model selection in demand forecasting applications. Comput. Ind. Eng. **121**, 1–7 (2018). https://doi.org/10.1016/j.cie.2018.04.042
11. Han, S., Ko, Y., Kim, J., Hong, T.: Housing market trend forecasts through statistical comparisons based on big data analytic methods. J. Manage. Eng. **34**(2) (2018). https://doi.org/10.1061/(ASCE)ME.1943-5479.0000583
12. Lorente-Leyva, L.L., Alemany, M.M.E., Peluffo-Ordóñez, D.H., Herrera-Granda, I.D.: A comparison of machine learning and classical demand forecasting methods: a case study of Ecuadorian textile industry. In: Nicosia, G., et al. (eds.) LOD 2020. LNCS, vol. 12566, pp. 131–142. Springer, Cham (2020). https://doi.org/10.1007/978-3-030-64580-9_11
13. Ren, S., Chan, H.-L., Siqin, T.: Demand forecasting in retail operations for fashionable products: methods, practices, and real case study. Ann. Oper. Res. **291**(1–2), 761–777 (2019). https://doi.org/10.1007/s10479-019-03148-8
14. DuBreuil, M., Lu, S.: Traditional vs. big-data fashion trend forecasting: an examination using WGSN and EDITED. Int. J. Fashion Des. Technol. Educ. **13**(1), 68–77 (2020). https://doi.org/10.1080/17543266.2020.1732482
15. Ha, C., Seok, H., Ok, C.: Evaluation of forecasting methods in aggregate production planning: a Cumulative Absolute Forecast Error (CAFE). Comput. Ind. Eng. **118**, 329–339 (2018). https://doi.org/10.1016/j.cie.2018.03.003
16. Lorente-Leyva, L.L., et al.: Artificial neural networks in the demand forecasting of a metal-mechanical industry. J. Eng. Appl. Sci. **15**, 81–87 (2020). https://doi.org/10.36478/jeasci.2020.81.87
17. Weng, Y., et al.: Forecasting horticultural products price using ARIMA model and neural network based on a large-scale data set collected by Web Crawler. IEEE Trans. Comput. Soc. Syst. **6**(3), 547–553 (2019). https://doi.org/10.1109/TCSS.2019.2914499
18. Lorente-Leyva, L.L., et al.: Artificial neural networks for urban water demand forecasting: a case study. J. Phys. Conf. Ser. **1284**(1), 012004 (2019). https://doi.org/10.1088/1742-6596/1284/1/012004
19. Kley-Holsteg, J., Ziel, F.: Probabilistic multi-step-ahead short-term water demand forecasting with Lasso. J. Water Resour. Plan. Manage. **146**(10), 04020077 (2020). https://doi.org/10.1061/(ASCE)WR.1943-5452.0001268
20. Kück, M., Freitag, M.: Forecasting of customer demands for production planning by local k-nearest neighbor models. Int. J. Prod. Econ. **231**, 107837 (2021). https://doi.org/10.1016/j.ijpe.2020.107837
21. Gaba, A., Popescu, D.G., Chen, Z.: Assessing uncertainty from point forecasts. Manage. Sci. **65**(1), 90–106 (2019). https://doi.org/10.1287/mnsc.2017.2936
22. Zeng, Y., Yang, T., Breheny, P.: Hybrid safe–strong rules for efficient optimization in lasso-type problems. Comput. Stat. Data Anal. **153**, 107063 (2021). https://doi.org/10.1016/j.csda.2020.107063
23. Coad, A., Srhoj, S.: Catching Gazelles with a Lasso: big data techniques for the prediction of high-growth firms. Small Bus. Econ. **55**(3), 541–565 (2019). https://doi.org/10.1007/s11187-019-00203-3
24. Li, M., Guo, Q., Zhai, W.J., Chen, B.Z.: The linearized alternating direction method of multipliers for low-rank and fused LASSO matrix regression model. J. Appl. Stat. **47**(13–15), 2623–2640 (2020). https://doi.org/10.1080/02664763.2020.1742296
25. Loureiro, A.L., Miguéis, V.L., da Silva, L.F.: Exploring the use of deep neural networks for sales forecasting in fashion retail. Decis. Support Syst. **114**, 81–93 (2018). https://doi.org/10.1016/j.dss.2018.08.010
26. Du, W., Leung, S.Y.S., Kwong, C.K.: A multiobjective optimization-based neural network model for short-term replenishment forecasting in fashion industry. Neurocomputing **151**, 342–353 (2015). https://doi.org/10.1016/j.neucom.2014.09.030

27. Abolghasemi, M., Beh, E., Tarr, G., Gerlach, R.: Demand forecasting in supply chain: the impact of demand volatility in the presence of promotion. Comput. Ind. Eng. **142**, 106380 (2020). https://doi.org/10.1016/j.cie.2020.106380
28. Sagaert, Y.R., Aghezzaf, E.-H., Kourentzes, N., Desmet, B.: Tactical sales forecasting using a very large set of macroeconomic indicators. Eur. J. Oper. Res. **264**(2), 558–569 (2018). https://doi.org/10.1016/j.ejor.2017.06.054
29. Crone, S.F., Kourentzes, N.: Feature selection for time series prediction–a combined filter and wrapper approach for neural networks. Neurocomputing **73**(10–12), 1923–1936 (2010). https://doi.org/10.1016/j.neucom.2010.01.017
30. Ewees, A., et al.: Improving multilayer perceptron neural network using chaotic grasshopper optimization algorithm to forecast iron ore price volatility. Resour. Policy **65**, 101555 (2020). https://doi.org/10.1016/j.resourpol.2019.101555
31. Beale, M.H., Hagan, M.T., Demuth, H. B.: Neural network toolbox™ user's guide. In: R2012a, The MathWorks, Inc., 3 Apple Hill Drive Natick, MA, 01760-2098 (2012). www.mathworks.com.
32. Athanasopoulos, G., Hyndman, R.J., Kourentzes, N., Petropoulos, F.: Forecasting with temporal hierarchies. Eur. J. Oper. Res. **262**(1), 60–74 (2017). https://doi.org/10.1016/j.ejor.2017.02.046
33. Spiliotis, E., Petropoulos, F., Assimakopoulos, V.: Improving the forecasting performance of temporal hierarchies. PLoS ONE **14**(10), e0223422 (2019). https://doi.org/10.1371/journal.pone.0223422

Parallelization of Reinforcement Learning Algorithms for Video Games

Marek Kopel(✉) and Witold Szczurek

Faculty of Computer Science
and Management, Wroclaw University of Science and Technology, wybrzeze
Wyspiańskiego 27, 50-370 Wroclaw, Poland
marek.kopel@pwr.edu.pl

Abstract. This paper explores a new way of parallelization of Reinforcement Learning algorithms - simulation of environments on the GPU. We use the recently proposed framework called CUDA Learning Environment as a basis for our work. To prove the approach's viability, we performed experimentation with two main class of Reinforcement Learning algorithms - value based (Deep-Q-Network) and policy based (Proximal Policy Optimization). Our results validate the approach of using GPU for environment emulation in Reinforcement Learning algorithms and give insight into convergence properties and performance of those algorithms.

Keywords: Reinforcement learning · Video game · Parallel computing · Artificial intelligence

1 Introduction

The main disadvantage of Reinforcement Learning (RL) algorithms are their long training times and much compute required for those algorithms to provide good results. To increase innovation in the field, the researchers need to have faster turnaround times when developing new algorithms. Because most of today's improvements in processing speeds come from parallelization, it is the first consideration when improving algorithm convergence times.

Here we research the influence of parallelization of training environments on two well-known RL algorithms. The first one, from the family of value-based, the Deep Q Network (DQN) algorithm, and the second from policy gradient methods, the Proximal Policy Optimization (PPO) algorithm. We use the CUDA Learning Environment (CuLE) [5] framework that emulates ATARI environments on GPU, which provides a way to simulate many more environments compared to classical CPU emulation engines.

Thanks to that property, we can evaluate classical algorithms' performance in a highly parallel setting, similar to those used in cluster computing. Insight into the performance of those algorithms on such a highly parallel scale, which is at the same time available on most computing machines, should be a valuable tool in future RL research.

N. T. Nguyen et al. (Eds.): ACIIDS 2021, LNAI 12672, pp. 195–207, 2021.
https://doi.org/10.1007/978-3-030-73280-6_16

The rest of the paper is divided as follows:

- Background and related work - provides a background on Reinforcement Learning itself and methods of parallelization already employed in previous research.
- Experiments - Explains the aim of the experiments and describes hyperparameters for algorithms and measures used.
- Results - Describes the result of experiments and provides with the commentary regarding those.
- Conclusions - Concludes the work done and provides future research directions.

2 Background and Related Work

2.1 Reinforcement Learning

RL is a subtype of Machine Learning (ML). It is concerned with the sequential decision making of an actor in an environment. The decision making on behalf of the actor is made by a piece of software called "software agent" or just "agent" for short. It is often viewed as an optimization task and formalized as a Markov Decision Process (MDP) [16]. The agent is trying to maximize a reward signal that comes from the environment. It is often better for an agent to take actions that result in immediate rewards over actions that result in rewards that are further in time. This notion of an agent caring more about the present than the future is represented by a total discounted return (1). It is the sum of all rewards from time step t onward, discounted by a discount factor γ:

$$G_t = \sum_{k=0}^{\infty} \gamma^k * R_{t+k} \tag{1}$$

Total return G_t is used to estimate how good is some state in MDP. Function that calculates the value of some state is called value function (2). It is the base for a family of reinforcement learning algorithms called value-based methods.

$$v(s) = E[G_t|S_t = s] \tag{2}$$

Variation of the value function is an action-value function (3). It is an estimate not only for a given state in MDP but also for all actions chosen from this state. The action-value function is often more useful than the solely value function because it directly ties action to states. Q value optimization is basis for DQN algorithm [8] used in experiments in this paper.

$$q(a, s) = E[G_t|S_t = s, A_t = a] \tag{3}$$

Some algorithms instead of using value or action-value function, use so-called policy (4). It is essentially a mapping from states to actions without a notion of a given state's value. Algorithms optimizing policy functions are called policy gradient methods.

$$\pi(a|s) = Pr(A_t = a|S_t = s) \tag{4}$$

2.2 Methods for Solving RL Problems

There are two main classes of methods for solving RL problems: value-based and policy gradients. Value-based strategies optimize value function and implicitly build policy for acting in the environment based on optimal value function. Policy gradients methods optimize policy function directly. The hybrid approaches that utilize both of those functions are called Actor-Critic (AC) methods.

The first notable advancements in value-based RL were achieved with DQN algorithm. Since then, many variations of this algorithm have been developed to improve its sample efficiency. Double DQN [8], Dueling networks [17], Prioritized Replay [13], n-step learning [9], and Noisy Nets [7].

Double DQN overcomes the existing limitations of classical DQN, particularly the overestimation of the action-value function. Prioritized experience replay, instead of sampling data from replay buffer uniformly uses, prioritization to replay only essential transitions. Dueling Network is a type of neural network architecture proposed for the estimation of action-value function. This approach uses classical value function and advantage function to achieve better performance.

In n-step learning approach from classical RL for deep backups is combined with function approximation techniques used in DQN. In Noisy Nets, random parametric noise is added to neural network weights. It provides significantly better exploration than classical epsilon greedy techniques. State-of-the-art algorithmic improvements have been achieved by combining all of those techniques into one meta-algorithm called RAINBOW [10].

Algorithmic improvements to policy gradient methods included Trust Region Policy Optimization (TRPO) [14] and PPO [15]. Those methods use advantage estimates and, therefore, are considered AC methods. Another AC method, without special gradients as in TRPO or PPO, is Advantage Actor-Critic (A2C) [11], and it is asynchronous variation Asynchronous Advantage Actor-Critic (A3C). Recently published is a policy gradient method called Agent57 [2], which achieved super-human performance on all 57 ATARI games, which have not been done before.

2.3 Parallelization of RL Methods

Since RL methods' main disadvantage is long training times, it is essential to use available computing efficiently. Most of today's processing platforms uses parallel processing for scaling of computationally demanding workloads. Massively parallel architectures have been used to speed-up value-based algorithms. Gorilla [12] architecture uses multiple CPUs to simulate similar agents that collect data from episodes and then update a single server with neural network parameters.

In the domain of policy gradient algorithms: GA3C [1] is a GPU accelerated version of A3C. It uses multiple agents that are run on the CPU to collect data. GPU is used to train neural networks based on data gathered by agents. Data gathered by agents are stored in queues to help keep GPU busy. Including queues makes this algorithm offline and may lead to destabilization in learning, as GPU optimizes neural networks in the direction of "old" data gathered by agents. Synchronized learning without queues had been implemented in PAAC [4]. It stabilizes learning at the cost of GPU underutilization.

As in DQN, massively parallel methods had been employed to achieve the speed-up of those algorithms. One such system is called IMPALA [6]. It uses many CPUs to accelerate the A2C algorithm. As part of system-level efforts, the work regarding the use of specialized chips called FPGA (Field Programming Gate Arrays) promises improvements in hardware efficiency and power consumption. FA3C [3] is state of the art A3C implemented on FPGA. It provides better performance than powerful GPU platforms such as NVIDIA Tesla P100.

As a framework for experimentation, we use CuLE [5], which enables the emulation of ATARI environments on GPU. Such emulation pattern is different from typical emulation engines, which uses CPU, and allows simulation of much more environments, which was crucial for experimentation done in this research. Such a workload assignment enables us to simulate much more environments and reduce communication between CPU and GPU. Because both simulation and training is done on GPU, this provides better GPU utilization.

3 Experiment

The experiments aimed to validate the new approach for parallelization and compare convergence properties of classical RL algorithms. To do this, we measured training times and corresponding frames per second for different numbers of parallel environments, which will be denoted n_e. Additionally, we compared GPU processors and memory utilization for CPU-GPU and GPU only training to compare both approaches' benefits.

3.1 Environment and Algorithms Parameters

Hyperparameters necessary for training are provided to the program using an initialization file. For the DQN, Adam optimizer is used instead of RMSProp as in [10], because it can help reduce algorithm sensitivity to the learning rate. The learning rate for DQN was chosen to be 0.0000625 and for the Adam optimizer 0,0015. Batch size across all runs of the algorithm was set to be 32 transitions. Due to memory constraints on the GPU device, memory capacity was set to be 50000 transitions. Experiments involving the PPO algorithm also use Adam optimizer, but the learning rate is set to 0.00001, while the PPO learning rate is 0.00025. Both algorithms limit the maximum length of episodes, which is set to 18000 transitions. Hyperparameters explained above and additional ones, along with system specifications, are included in appendices.

3.2 Measures

To evaluate tested algorithms' performance, we use the games' score, which for ATARI pong is a scalar value between −21 and 21. The comparison between algorithms' running times was made by finding the convergence curve's first step when the algorithm scored 19 points or higher. Such comparison promotes algorithm runs that converge faster. As a throughput measure during algorithms' training, we use frames per second (FPS) generated by the CuLE engine. To measure GPU's utilization and memory consumption, we use a program called "nvidia-smi," which was run as a daemon during training.

4 Results

4.1 Influence of Parallelization on Training Times

Figure 1 shows convergence curves for PPO. There are mean rewards gathered from the evaluation stage on the vertical axis, as described in the previous section. The horizontal axis shows wall time. For clarity's sake, time is represented on a logarithmic scale. Optimal performance is achieved with the n_e equal to 256. When a smaller number is used, the training times are significantly longer. When a bigger number is used, the training times are slightly higher than with 256 parallel environments.

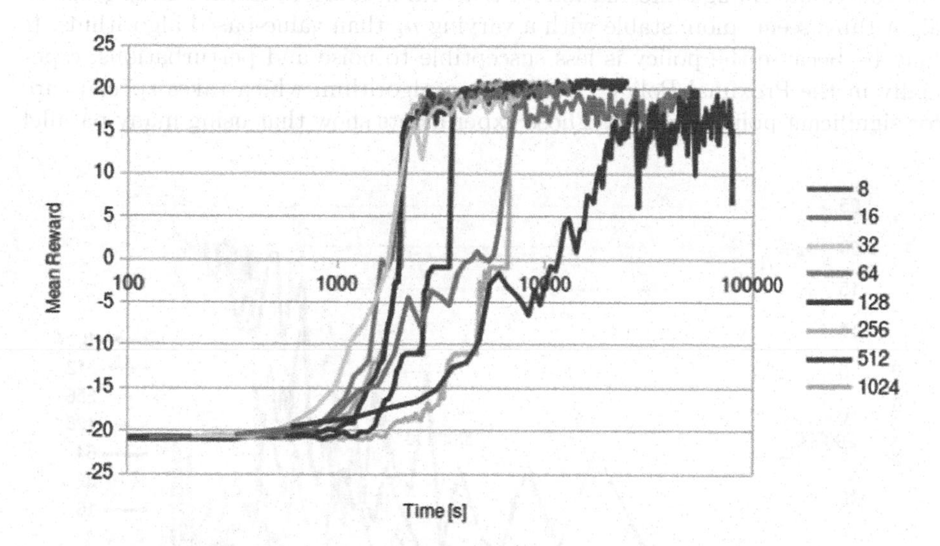

Fig. 1. PPO convergence curves for different n_e.

The same experiments were performed for the DQN algorithm. Convergence curves are shown in Fig. 2. Surprisingly training times increase with a higher n_e. The algorithm learning stops completely with the n_e equal to 512 and 1024.

This effect can be associated with the quality of data that is being gathered into a replay buffer. Each agent can write only a small number of frames that do not capture enough information for an algorithm to learn. This property shows that DQN is more susceptible to the n_e used in training and may not be as suitable for parallelization as policy gradient techniques. It may still be useful to run the DQN algorithm with a moderately high n_e because it can speed up the training time. Nevertheless, a large n_e impacts training negatively. For the DQN algorithm, it may be useful to have a variable n_e to provide a run-time controlled trade-off between convergence speed and an agent's performance.

In Fig. 3, aggregated results are shown for both PPO and DQN. The horizontal axis represents the time it took for an agent to score 19 or higher. The DQN algorithm takes much more time to converge to optimal policy than PPO. Additionally, DQN training times increase with the n_e, and the learning stops entirely with the n_e higher than 256 (those measurements are not shown in the figure). The PPO's best performance can be achieved with the number of agents equal to 256, as was stated earlier. For the number of parallel agents higher than 512, no significant speed-up can be achieved, similarly, for the smaller n_e.

Bad performance of DQN can be associated with throughput oriented nature of CuLE, as stated in [5]. The time-domain of each parallel environment is being explored less efficiently, which leads to no further improvements in convergence. Such behavior is that when more agents are playing in environments, they can gather only short episodes of play to a replay buffer. Those short episodes do not contribute enough information for a neural network to learn. Policy gradient algorithms seems more stable with a varying n_e than value-based algorithms. It may be because the policy is less susceptible to noise and perturbations, especially in the Proximal Policy Optimization algorithm, which takes special care for significant policy updates. Those experiments show that using many parallel

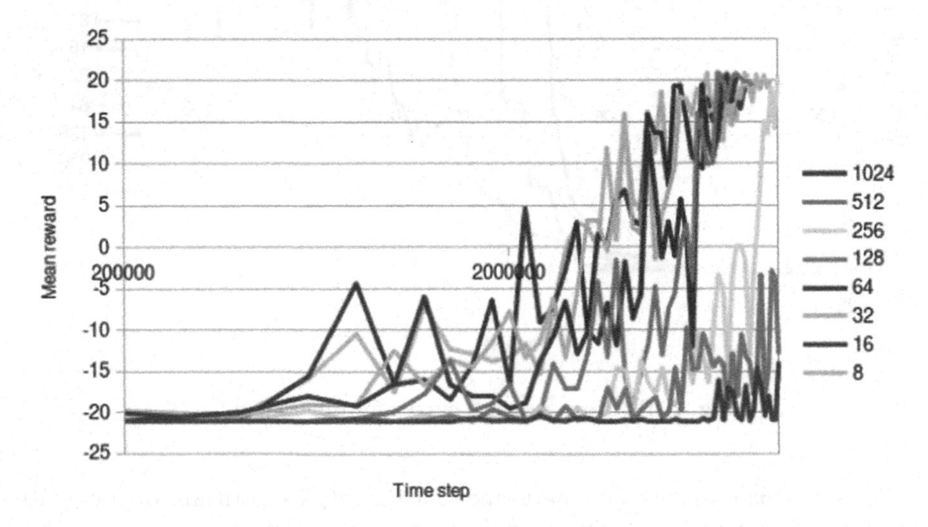

Fig. 2. DQN convergence curves for different n_e.

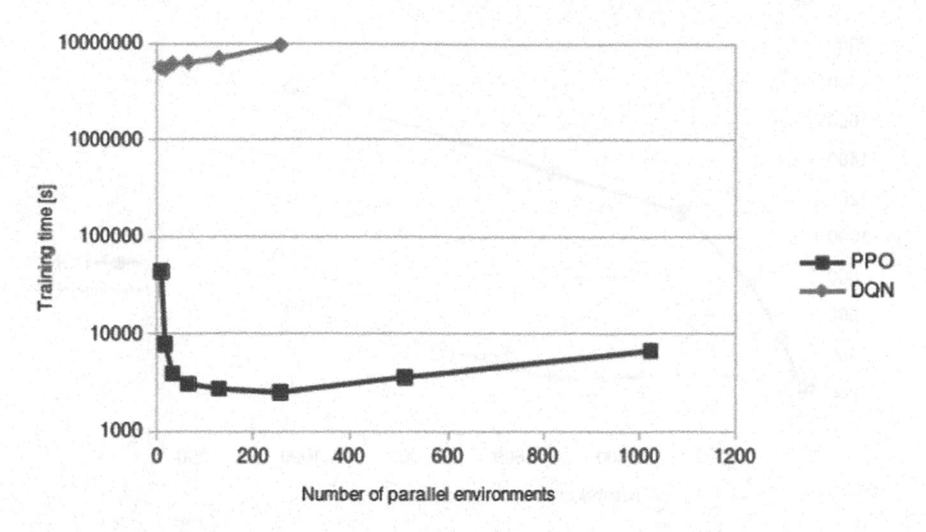

Fig. 3. PPO and DQN training times as a function of n_e (lower is better). Note that values of DQN that are not shown resulted in no convergence.

environments can significantly speed up the training times of RL algorithms. Performing both simulation and training on GPU enables simulation of much higher parallel games at once. Thanks to this, researchers can benefit from faster workflows when developing new algorithms for RL. Additionally, experiments show that the PPO algorithm can benefit more from parallelization than the DQN, which may be right for those algorithms' general family (value-based and policy gradients) though some further experimentation is needed to prove that.

4.2 Throughput Measurements

Here we compare throughput generated by the CuLE framework for DQN and PPO as a function of the number of training environments. The generated FPS differs between algorithms because, in DQN, parallel environments correspond to only the data acquisition stage, while in PPO, parallel agents are also responsible for advantage estimate computation. Though CuLE is capable of much higher throughput, it does not directly correlate with better convergence of algorithms. For example, the DQN algorithm performed worse with the increasing n_e, even though it can achieve higher throughput than PPO. Simultaneously, increasing the number of parallel agents in PPO corresponds to faster convergence times, though this speed up is not linear, and the throughput is worse than in the DQN case (Fig. 4).

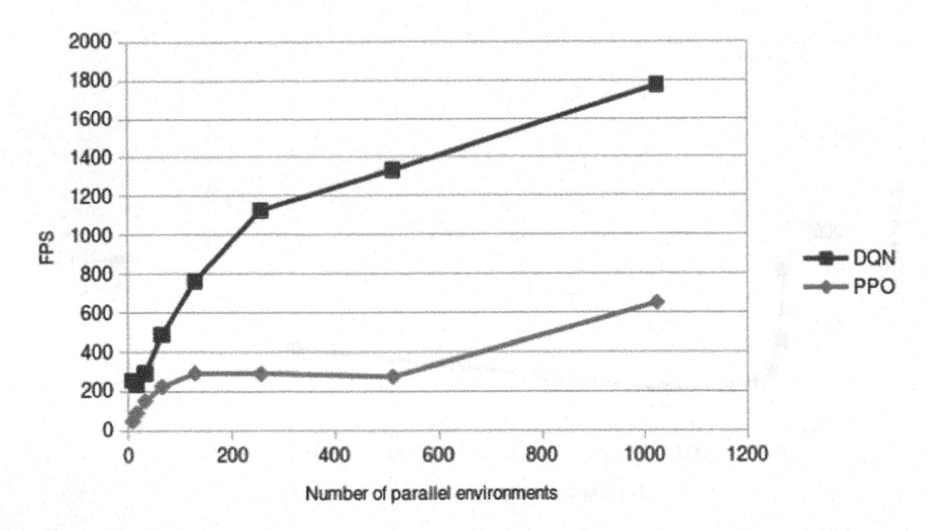

Fig. 4. PPO and DQN FPS for varying n_e (higher means better throughout).

4.3 System Performance Measurements

In this section, two parallelization approaches were tested (CPU-GPU vs. GPU only) to check which of them has better resource utilization. In the CPU-GPU approach, environments are simulated on processors, while learning is done on GPU. The second approach performs both steps on the GPU. This experiment does not require a comparison between DQN and PPO algorithms because CPU-GPU approaches and GPU-only approach should be independent of the algorithm being used. As such, for this experiment, the PPO algorithm was chosen,

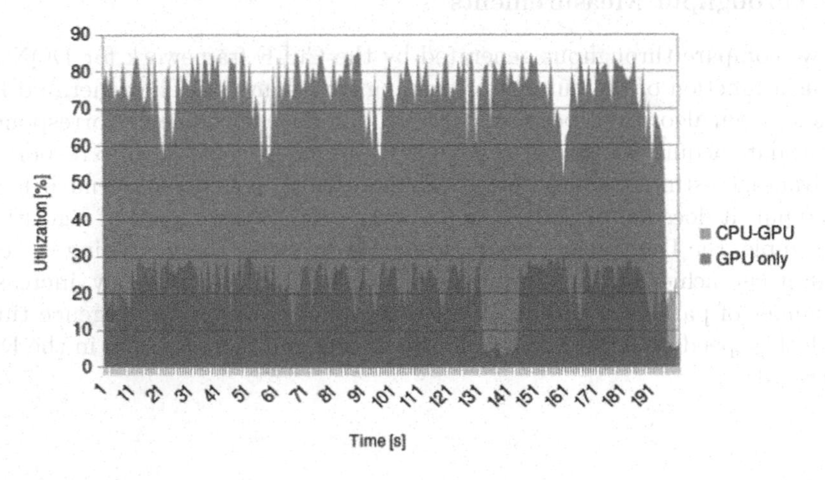

Fig. 5. Comparison of PPO GPU processors utilization in CPU-GPU and GPU only version.

with a constant n_e, namely 32. Only several iterations of running algorithms were measured because the system resources footprint should remain similar throughout the training process.

As a metric of resources, utilization GPU memory and processor utilization were measured. Figure 5 compares processors utilization for CPU-GPU and GPU only. As can be seen, GPU only approach outperforms CPU-GPU. Better compute utilization is achieved because this approach can keep GPU busy by switching between environment simulation and learning steps. A higher peaks of utilization seen in Fig. 5 corresponds to subsequent training results. Lower utilization intervals mark the end of each iteration of training. In the GPU-only approach, the system manages to perform several iterations during the measurement period. The end of iterations can be seen around 21, 61, 111, 141, 161 s of measurement. At the same time, the CPU-GPU approach manages to perform only the first iteration and starts second. The end of the first iteration is marked with a sudden drop in GPU utilization around 141 s of measurement. During this time, CPU emulated environments are being reset, and GPU remains almost idle. This experiment shows clearly that the GPU-only approach provides better system utilization and, thus, higher throughput. The GPU-only approach can perform around four iterations of training. At the same time, the CPU-GPU approach can only do one iteration of training.

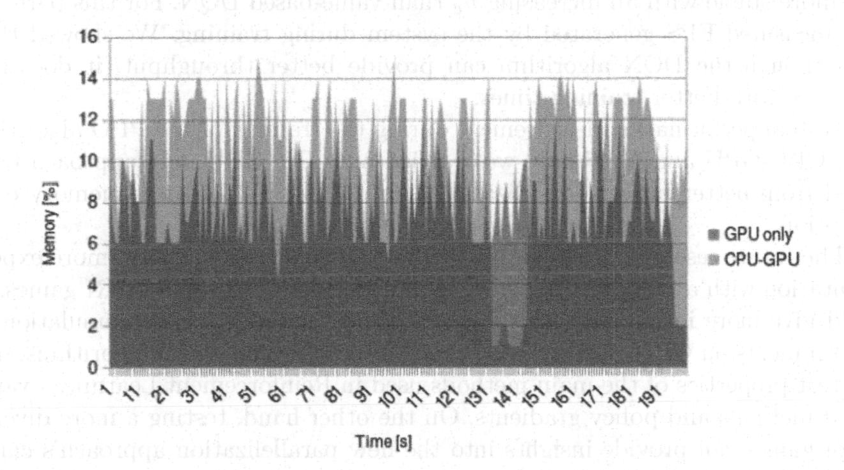

Fig. 6. Comparison of PPO GPU memory utilization in CPU-GPU and GPU only version.

In Fig. 6, the memory usage of both approaches is being compared. Surprisingly, the CPU-GPU approach uses more memory than the GPU only approach. Such a result may be because of memory copies that CPUs need to perform to transfer data to GPU. In GPU only approach, there is no need to make those copies, and thus less memory is used. Interestingly, in the CPU-GPU approach, an end of iteration can be seen on memory measurements. It can be marked at around the same location as in utilization measurements, namely around 141 s of the experiment. The sudden drop in memory utilization comes from the fact that the algorithm resets environments emulated on the CPU, and during this time, GPU remains almost idle. Such precise iterations cannot be seen in GPU-only measurements. Presumably, even when iteration ends and environments are reset, GPU memory is still being utilized in the same percentage, but for a different purpose.

5 Conclusions

In this work, we researched the influence of parallelization on RL algorithms on the ATARI domain. The agents were tasked with learning to play ATARI games from raw visual input. As a base for our experimentation, we have chosen a recently proposed CuLE framework, enabling the emulation of ATARI environments on GPU. We tested two algorithms from the branch of model-free reinforcement learning methods, namely Deep Q-Networks and Proximal Policy Optimization. Those algorithms represent two categories of methods for solving RL problems: value-based and policy gradients.

Using GPU for simulation and training proved to be a viable strategy for reducing RL algorithms' training times on the ATARI domain. PPO algorithm was more stable with an increasing n_e than value-based DQN. For this part, we also measured FPS generated by the system during training. We showed that even though the DQN algorithm can provide better throughput, it does not correlate with better training times.

System performance measurement during the training of the PPO algorithm with CPU-GPU and GPU only was carried out. The GPU only approach benefited from better processor utilization than CPU-GPU and less memory consumption.

The future research direction for this work would be to perform more experimentation with a more diverse set of algorithms and different ATARI games. It could give more insight into this new method of parallelization - simulation of environments on GPU. Additionally, using a more diverse set of algorithms, one can test properties of the main methods used in Reinforcement Learning - value based methods and policy gradients. On the other hand, testing a more diverse set of games can provide insights into the new parallelization approach's capabilities. It may be so that more extended and more complex games suffer more from such an approach because of GPU thread divergence, which was suggested in the original work regarding the CuLE framework.

Acknowledgements. We want to thank the authors of CuLE for making their work freely available, which helped us do our research.

Appendicies

See Tables 1, 2 and 3.

A Hyperparameters

Table 1. PPO hyperparameters

Hyperparameter	Value
alpha	0.99
Clip ϵ	0.1
Adam ϵ	0.00001
Discount factor	0.99
Learning rate	0.00025
Max episode length	18000

Table 2. DQN hyperparameters

Hyperparameter	Value
Adam ϵ	0,0015
Batch size	32
Discount factor	0,99
Hidden layer size	512
History length	4
Learning rate	0,0000625
Max episode length	18000
Memory capacity	5000

B System specification

Table 3. Hardware and software specification

OS	ubuntu 18.04
Libraries	cuda 10.0, cudnn-7, python 3.6.9, pytorch 1.2.0
Graphics card	NVIDIA GeForce GTX 960M
CPU	Intel Skylake i5-6300HQ, 2,3 GHz
RAM	16 GB DDR3, 1600 MHz

References

1. Babaeizadeh, M., Frosio, I., Tyree, S., Clemons, J., Kautz, J.: GA3C: GPU-based A3C for Deep Reinforcement Learning. CoRR abs/1611.06256 (Nov 2016)
2. Badia, A.P., et al.: Agent57: outperforming the Atari Human Benchmark. arXiv:2003.13350 [cs, stat], March 2020, http://arxiv.org/abs/2003.13350, arXiv: 2003.13350
3. Cho, H., Oh, P., Park, J., Jung, W., Lee, J.: FA3C: FPGA-accelerated deep reinforcement learning. In: Proceedings of the Twenty-Fourth International Conference on Architectural Support for Programming Languages and Operating Systems, ASPLOS 2019, pp. 499–513. Association for Computing Machinery, Providence, RI, USA, April 2019. https://doi.org/10.1145/3297858.3304058, https://doi.org/10.1145/3297858.3304058
4. Clemente, A.V., Castejón, H.N., Chandra, A.: Efficient parallel methods for deep reinforcement learning. arXiv:1705.04862 [cs], May 2017, http://arxiv.org/abs/1705.04862, arXiv: 1705.04862
5. Dalton, S., Frosio, I., Garland, M.: GPU-Accelerated Atari Emulation for Reinforcement Learning. arXiv:1907.08467 [cs, stat] (Jul 2019), http://arxiv.org/abs/1907.08467, arXiv: 1907.08467
6. Espeholt, L., Soyer, H., Munos, R., Simonyan, K., Mnih, V., Ward, T., Doron, Y., Firoiu, V., Harley, T., Dunning, I., Legg, S., Kavukcuoglu, K.: IMPALA: Scalable Distributed Deep-RL with Importance Weighted Actor-Learner Architectures. arXiv:1802.01561 [cs] (Jun 2018), http://arxiv.org/abs/1802.01561, arXiv: 1802.01561
7. Fortunato, M., Azar, M.G., Piot, B., Menick, J., Osband, I., Graves, A., Mnih, V., Munos, R., Hassabis, D., Pietquin, O., et al.: Noisy networks for exploration. arXiv preprint arXiv:1706.10295 (2017)
8. van Hasselt, H., Guez, A., Silver, D.: Deep Reinforcement Learning with Double Q-learning. arXiv:1509.06461 [cs], December 2015, http://arxiv.org/abs/1509.06461, arXiv: 1509.06461
9. Hernandez-Garcia, J.F., Sutton, R.S.: Understanding multi-step deep reinforcement learning: a systematic study of the dqn target. arXiv preprint arXiv:1901.07510 (2019)
10. Hessel, M., et al.: Rainbow: combining improvements in deep reinforcement learning. arXiv:1710.02298 [cs], October 2017, http://arxiv.org/abs/1710.02298, arXiv: 1710.02298
11. Mnih, V., et al.: Asynchronous methods for deep reinforcement learning. arXiv:1602.01783 [cs], June 2016, http://arxiv.org/abs/1602.01783, arXiv: 1602.01783
12. Nair, A., et al.: Massively parallel methods for deep reinforcement learning. arXiv:1507.04296 [cs] (Jul 2015), http://arxiv.org/abs/1507.04296, arXiv: 1507.04296
13. Schaul, T., Quan, J., Antonoglou, I., Silver, D.: Prioritized Experience Replay. arXiv:1511.05952 [cs], February 2016, http://arxiv.org/abs/1511.05952, arXiv: 1511.05952
14. Schulman, J., Levine, S., Moritz, P., Jordan, M.I., Abbeel, P.: Trust Region Policy Optimization. arXiv:1502.05477 [cs], April 2017, http://arxiv.org/abs/1502.05477, arXiv: 1502.05477
15. Schulman, J., Wolski, F., Dhariwal, P., Radford, A., Klimov, O.: Proximal policy optimization algorithms. arXiv:1707.06347 [cs], August 2017, http://arxiv.org/abs/1707.06347, arXiv: 1707.06347

16. Sutton, R.S., Barto, A.G.: Reinforcement Learning: An Introduction, 2nd edn. Adaptive Computation and Machine Learning Series. The MIT Press Cambridge, Massachusetts (2018)
17. Wang, Z., Schaul, T., Hessel, M., van Hasselt, H., Lanctot, M., de Freitas, N.: Dueling network architectures for deep reinforcement Learning. arXiv:1511.06581 [cs], April 2016, http://arxiv.org/abs/1511.06581, arXiv: 1511.06581

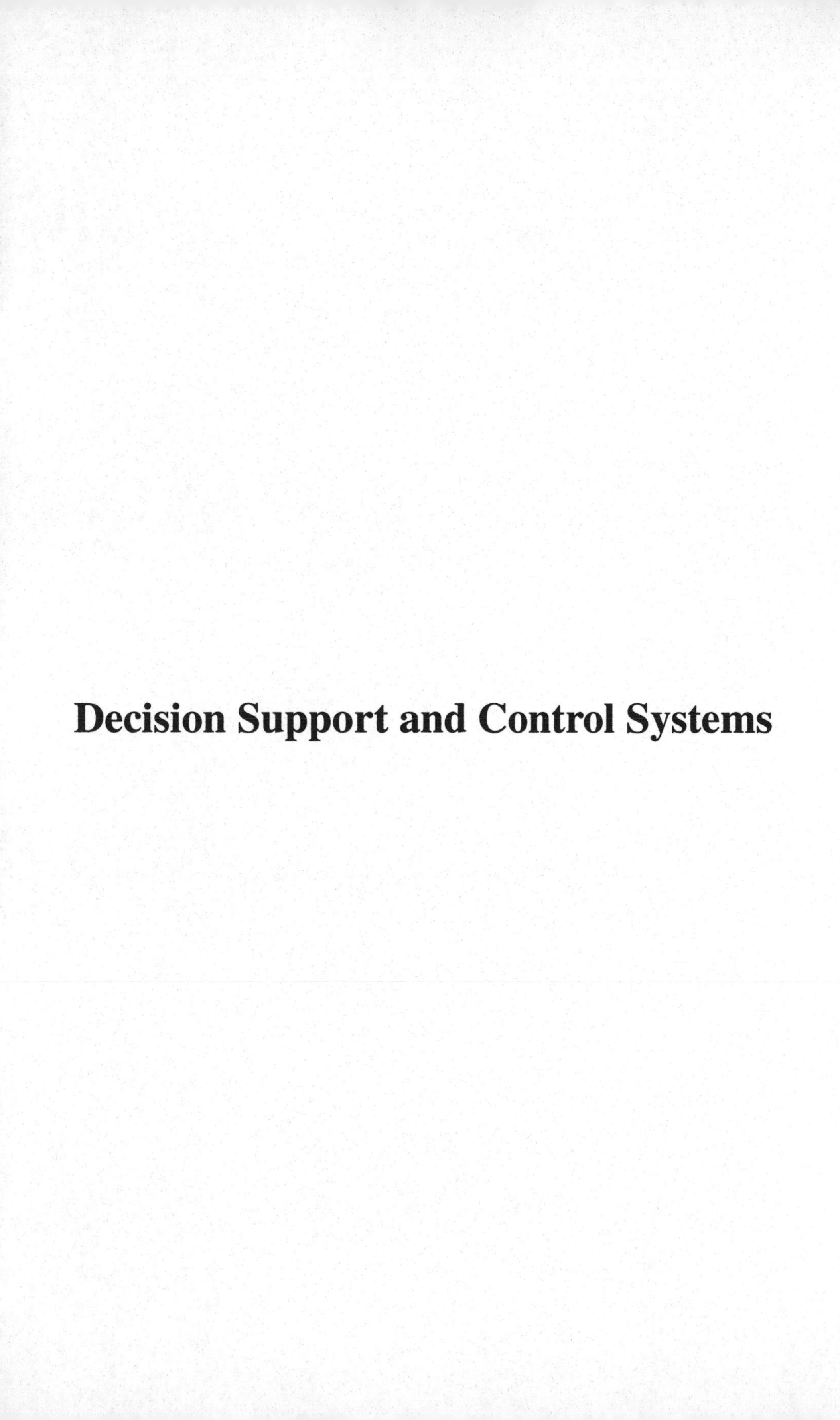

Decision Support and Control Systems

A Gap–Based Memetic Differential Evolution (GaMeDE) Applied to Multi–modal Optimisation – Using Multi–objective Optimization Concepts

Maciej Laszczyk(✉) and Paweł B. Myszkowski(✉)

Faculty of Computer Science and Management, Wrocław University of Science and Technology, Wrocław, Poland
{maciej.laszczyk,pawel.myszkowski}@pwr.edu.pl

Abstract. This paper presents a method that took second place in the GECCO 2020 Competition on Niching Methods for Multimodal Optimization. The method draws concepts from combinatorial multi–objective optimization, but also adds new mechanisms specific for continuous spaces and multi–modal aspects of the problem. GAP Selection operator is used to keep a high diversity of the population. A clustering mechanism identifies promising areas of the space, that are later optimized with a local search algorithm. The comparison between the top methods of the competition is presented. The document is concluded by the discussion on various insightson the problem instances and the methods, gained during the research.

Keywords: Multi–modal optimization · Memetic algorithm · Gap selection · Multi–objective optimization

1 Introduction

In many practical applications, optimization problems have multiple optimal solutions. The goal of the optimization is to find all those solutions and present the user with a choice. This concept manifests in multi–objective optimization, as the objectives are not assigned weights and a set of solutions is considered Pareto–Optimal. On the other hand, in multi–modal optimization there exists multiple solutions with identical values of the fitness function.

Participation in the GECCO 2020 Competition on Niching Methods for Multimodal Optimization allowed for comparison of the results with the state-of-the-art methods and provided plethora of information from the previous editions of the competition. It's been ran each consecutive year starting in 2016, while the earliest edition can be traced back to year 2013.

In 2019 the competition has been dominated by a Hill-Valley Evolutionary Algorithm [1]. It is an iterative approach, which divides the population into

N. T. Nguyen et al. (Eds.): ACIIDS 2021, LNAI 12672, pp. 211–223, 2021.
https://doi.org/10.1007/978-3-030-73280-6_17

niches, which are then optimized using an adapted maximum-likelihood Gaussian model iterated density-estimation evolutionary algorithm (AMaLGaM) [2]. Other competing approaches included methods based on Differential Evolution and on K-nearest-neighbour clustering.

During GECCO 2020 both the first and the third places in the competition were taken by methods based on a Covariance matrix self-adaptation Evolution Strategy [5]. It is an evolutionary approach, which uses a covariance matrix with an adaptive step to learn and utilize the shape of the solution space. The former method included a repelling subpopulations, while the latter used a dynamic initial population selection.

There is a recent trend in literate to use Differential Evolution (DE) metaheuristic to solve multi–objective optimization and multi modal problems. For example, [8] presents an application of Binary DE to solve multi–objective optimization problem of feature selection (objectives: classification accuracy and the number of selected features). In [9] the double–layered-clustering DE application to multi–modal optimization is described – the approach is successfully compared with 17 state-of-art niching algorithms on 29 multi–modal problems. In [10] multi–modal optimization is solved by DE extended by niching methods and clustering of individuals to create various niches by division and applying different guiding strategies. Moreover, [11] presents a DE approach enhanced by Crowding Distance clustering algorithm applied to solve multi–objective multi–objective optimization problems.

Section 2 presents why the multi–objective concepts have been applied to multi–modal optimization and how they work, while Sect. 3 provides a thorough description of the novel method. Section 4 describes the setup and the results of the competition along with a discussion of the results. The paper is concluded in Sect. 5.

2 Contribution and Motivation

The proposed GAP-based Memetic Differential Evolution (GaMeDE) is the main contribution of this paper. It started as an attempt to adapt a combinatorial multi–objective NTGA2 (Non-Dominated Tournament Genetic Algorithm) [3] with GAP selection operator to multi–modal optimization in continuous spaces. Newly created method needed some modifications to be effective in multi–modal optimization and continuous spaces. Table 1 shows the similarities between NTGA2 and GaMeDE, and enhancements introduced in GaMeDE.

NTGA2 is an evolutionary metaheuristic, which keeps an archive of non-dominated solutions. The method attempts to keep the points in the archive well distributed by utilizing a selection operator based on the distances between the points. A Pareto–dominance approach was not suitable for continuous space, as it was unlikely for two points to have the same fitness value. There was always a non-zero difference in their fitness values, hence an approach based on distance, inspired by an existing ϵ–dominance approach [4] was created. The archive in GaMeDE stores the points on different optima (global or local ones) and uses the same selection scheme as NTGA2.

Table 1. Comparison of NTGA2 [3] and GaMeDE

	NTGA2 [3]	GaMeDE
Archive	+	+
Two phases	+	+
Gap selection	+	+
Tournament	+	+
Dominance	Pareto	ϵ-dom. inspired
Landscape	Combinat. (GA, TSP)	Continuous (DE)
Archive management	Simple	Advanced
Local search	–	+
Clustering	–	+

An improved archive management schema has been used in the GaMeDE. NTGA2 stores only non-dominated individuals, whereas GaMeDE often adds and removes the individuals, trying to keep only the global optima, and the points that are most likely going to lead to finding other global optima in the next generations.

Additionally, to adapt GaMeDE to continuous spaces the underlying method has been changed from Genetic Algorithm (TSP–like representation and genetic operators) to Differential Evolution. Both NTGA2 and GaMeDE methods have two alternating phases, where they focus on either spreading out the solutions in the archive or on their convergence.

GaMeDE has been additionally enhanced by a clustering scheme, which selects a subspace of the solution space that shows promise. Each cluster aims to find a single global optimum. Moreover, continuous optimization can benefit from local search methods significantly. Hence, two local optimization mechanisms have been added to GaMeDE to improve its efficiency.

The main strength of GaMeDE lies in the ability to quickly cover the solution space and identify the local optima. At the same time, method eliminates parts of the solution space, where there are no good solutions. Whereas it is common for other methods to either increase the granularity of the search in those areas or to revisit them, which leads to wasted computation.

3 Approach

GaMeDE uses the Differential Evolution [6] at its core, but the main search mechanism is based on multi–objective optimisation method – NTGA2 [3]. First, a population of random individuals is generated, where each individual represents a solution to the given problem. Then, in an iterative evolutionary process new populations are generated, where single iteration is called a generation. Due to the local search, the number of fitness evaluations per generation may differ, the evolution stops when the limit of fitness evaluations is reached. During each generation, a selection scheme is used to select two parental individuals, which then undergo a mutation and crossover. Two child individuals created in this manner are added to the next population. The generation concludes when the population is filled. The general GAMEDE pseudocode is presented in pseudocode 1.

Algorithm 1: Pseudocode of GaMeDE

```
   Data: Archive
   Result: Global Optima
 1 init(Pop_i);
 2 evaluate(Pop_i);
 3 while not stopCondition() do
 4     Pop_{i+1} := ∅ ;
 5     while Pop_{i+1}.size() ¡ Pop_i.size() do
 6         archive.getDistances();
 7         if Phase == WIDE then
 8             parents := randomSelect(Pop_i);
 9         else
10             if gapFlag(i) then
11                 parents := gap(archive);
12             else
13                 parents := tournament(archive);
14             end
15         end
16         mutants := mutate(parents);
17         children := crossover(mutants);
18         Pop_{i+1}.add(children);
19     end
20     archive.update(Pop_{i+1});
21     i + +;
22     optima := archive.getOptima();
23     archive.resize(min, optima);
24     Phase := archive.switchPhase();
25     if Phase == WIDE then
26         clusters = clustering(archive);
27         localOptimization(clusters, archive, Pop_{i+1});
28     else
29         hillClimbing(archive);
30     end
31 end
```

Figure 1 presents the GaMeDE scheme, while pseudocode 1 shows its pseudocode. First, the population is randomly initialized and evaluated. Then the loop starts, which terminates when the available number of fitness evaluations have been carried out. Then, an empty population is created and filled with the loop in line 5. The distances between the individuals are calculated in line 6.

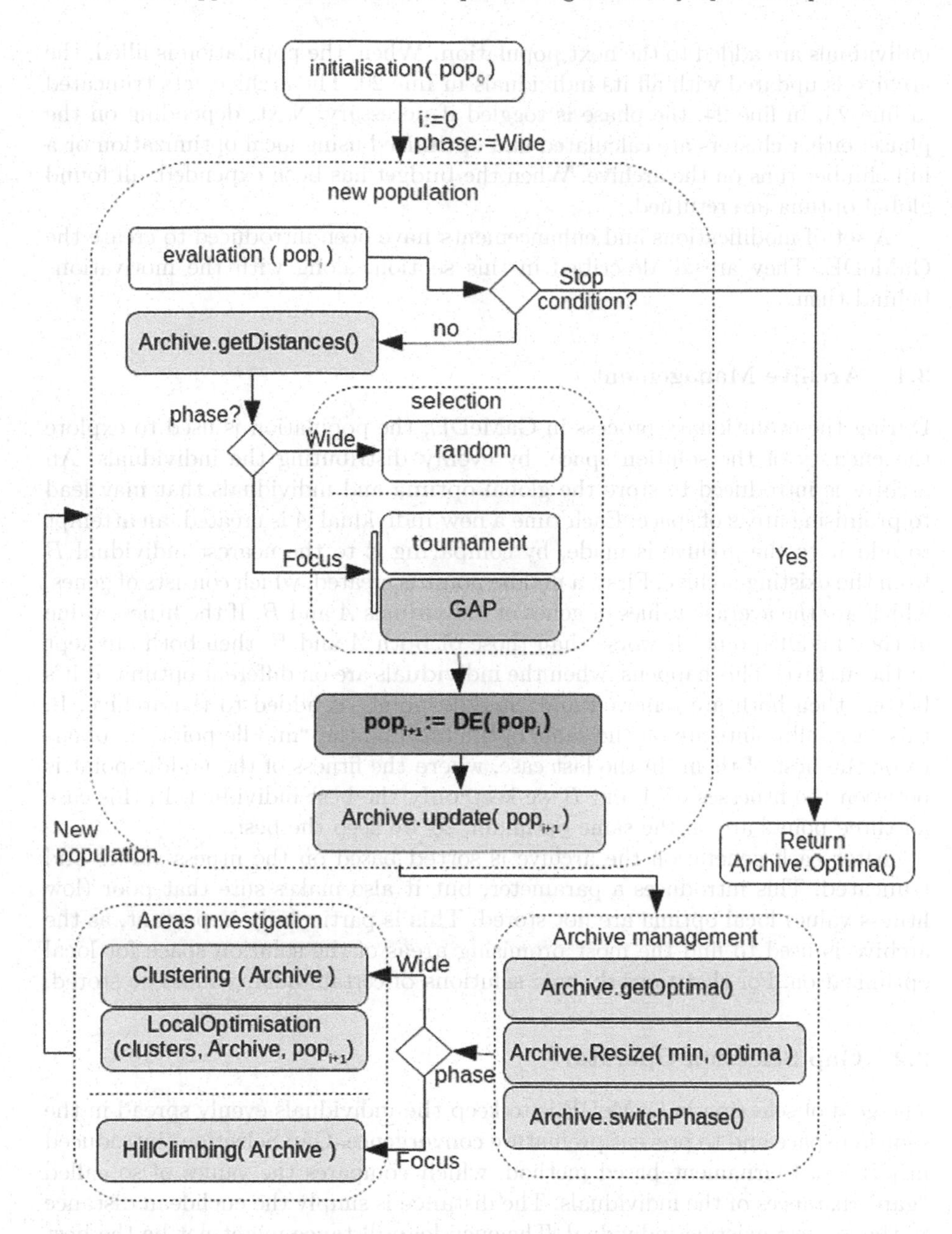

Fig. 1. A general GaMeDE schema

Then, depending on the phase, parents are selected either randomly, or using gap or tournament selection – it depends on the *gapFlag* value. The latter two are used in an alternating manner. The parents are mutated to create the mutant individuals, which in turn undergo a crossover. Finally, the resulting children individuals are added to the next population. When, the population is filled, the archive is updated with all its individuals in line 20. The archive gets truncated in line 23. In line 24, the phase is toggled if necessary. Next, depending on the phase, either clusters are calculated and optimized using local optimization or a hill climber runs on the archive. When the budget has been expended, all found global optima are returned.

A set of modifications and enhancements have been introduced to create the GaMeDE. They are all described in this section, along with the motivations behind then.

3.1 Archive Management

During the evolutionary process in GaMeDE, the population is used to explore the entirety of the solution space, by evenly distributing the individuals. An archive is introduced to store the global optima and individuals that may lead to promising areas of space. Each time a new individual A is created, an attempt to add it to the archive is made, by comparing it to the nearest individual B from the existing archive. First, a middle point is created, which consists of genes, which are the average values of genes of individuals A and B. If the fitness value of the "middle point" is worse than those of both A and B, then both are kept in the archive. This happens, when the individuals are on different optima. If it's better, then both are removed and "middle point" is added to the archive. In this case, all points are on the same optimum, and the "middle point" happens to be the best of them. In the last case, where the fitness of the middle point is between the fitnesses of A and B we keep only the best individual. In this case all three points are on the same optimum, so we keep the best.

After each generation the archive is sorted based on the fitness values and truncated. This introduces a parameter, but it also makes sure that poor (low fitness value) local optima are not stored. This is particularly important, as the archive is used to find the most promising areas of the solution space for local optimization. For that to work, only solutions of certain quality must be stored.

3.2 Gap Selection Operator

The goal of selection in GaMeDE is to keep the individuals evenly spread in the solution space and to prevent premature convergence. Gap Selection (introduced in [3]) is a tournament–based method, which compares the values of so called "gap" distances of the individuals. The distance is simply the euclidean distance to the nearest existing individual. The euclidean distance might not be the best fit for high dimensions, but at this stage no experiments have been done to verify that. GAP Selection uses archive content (and the current population) and returns the individual with the highest "gap" distance. Its goal is to explore the unexplored parts of the space and increase the diversity of the individuals.

Introduction of the GAP Selection changes the purpose of the population. With this approach, the archive is the main driving force of the improvement during the evolution. The population exists solely to provide a diverse set of genes, which can be used to probe the solution space.

3.3 Two–Phased GaMeDE

During the experiments it became evident that many of the global optima are rather easy to find, regardless of the used method. The most prominent example can be found in problem no.9 from the Table 2. There are 216 global optima in total, where each of them is trivial to be found. The main difficulty lies in finding all of them.

Two distinct phases in GaMeDE (see Figure 1 or pseudocode 1) are utilized. The first phase $FOCUS$ attempts to find all "easy" optima. Also, it alternates between a Gap Selection and a classical Tournament Selection. The former keeps the diversity, while the latter improves the existing solutions. In this phase, both selections work directly on the archive, as many solutions in the archive are sub–optimal and can be easily improved in this phase. If the first phase no longer finds new optima, the second phase $WIDE$ "kicks" in. The goal in the second phase is to look for the promising areas of space, where the remaining "difficult optima may lie. The random selection is used, which draws the individuals from the population, and the responsibility for finding the optima is shifted to the local optimizers.

3.4 Clustering

A clustering mechanism is used in the $WIDE$ phase of GaMeDE. Its primary goal is to identify the promising areas for local optimization. First, each individual from the archive is assigned to its closest found global optimum. At the beginning of evolution, the best individual is considered to be a global optimum, even though it might not actually be true. The result is a set of clusters, of a size equal to the number of found global optima. From each of these clusters we choose a point, that is the furthest away from its global optimum. These are the promising points to be optimized later on. That way, local optimization will cover different parts of space each time. Additionally, since the points are only taken from the archive, the parts of space, which have no good solutions will be omitted altogether.

This approach does not account for the global optima that are close together. This has been evident in the composite functions with 8 global optima (problems no. 12, 15, 17, 19 and 20 from the Table 2). Hence, an additional promising points must be chosen. If one of the clusters contains a solution much better in terms of fitness, than all other solutions, it becomes a promising point for optimization.

At this point, previously created clusters are no longer necessary and so they are discarded. Each of the promising points creates its own, new cluster, which will be optimized. The optimization method will require some information about the landscape, so n (as parameter) points closest to the promising point is drawn from the population to fill the cluster.

3.5 Local Optimization

GaMeDE uses two, different local optimizer. In the first phase, it is a simple hill climber, with a dynamic, decreasing step. In the second phase an AMaLGaM-Univariate [2] adaptation is used. Each of the created clusters undergoes the process, and the resulting, single point is attempted to be added to the archive. Local optimization is the last step of the evolutionary process. Assuming there are still available fitness function evaluations (*stopCondition*), the GaMeDE progresses to the next generation.

4 Experiments

The performance of the proposed GaMeDE is evaluated on the test instances from the CEC2013 niching benchmark [7]. Table 2 describes all 20 of utilized problems. The table describes each instance with a name, number of dimensions, number of global and local optima and a budget, which is the maximum number of fitness function evaluations allowed for a given problem. The number of optima is known, however it is not used in the optimization process. The information is only used for performance metrics.

Table 2. CEC2013/GECCO2020 Benchmark suite on multi–modal optimization

#	Function name	d	#gopt	#lopt	Budget
1	Five-Uneven-Peak Trap	1	2	3	50K
2	Equal maxima	1	5	0	50K
3	Uneven decreasing maxima	1	1	4	50K
4	Himmelblau	2	4	0	50K
5	Six-hump camel back	2	2	5	50K
6	Shubert	2	18	Many	200K
7	Vincent	2	36	0	200K
8	Shubert	3	81	Many	400K
9	Vincent	3	216	0	400K
10	Modifier rastrigin	2	12	0	200K
11	Composite function 1	2	6	Many	200K
12	Composite function 2	2	8	Many	200K
13	Composite function 3	2	6	Many	200K
14	Composite function 3	3	6	Many	400K
15	Composite function 4	3	8	Many	400K
16	Composite function 3	5	6	Many	400K
17	Composite function 4	5	8	Many	400K
18	Composite function 3	10	6	Many	400K
19	Composite function 4	10	8	Many	400K
20	Composite function 4	20	8	Many	400K

4.1 Setup

The experiments are carried out on all 20 problems. The GaMeDE is ran 50 times for each instance. Optimization is terminated when given number of fitness function evaluations is reached. At this point, currently found global optima are returned, along with the information how many fitness evaluations were required to find each optimum. This information is used to calculate the values of 3 performance metrics and the results are averaged over 50 runs. The results of the top 3 methods of the GECCO 2020 Competition on Niching Methods for Multimodal Optimization are compared. The results used in this paper are the same as in the competition itself. They are available under [12].

GaMeDE used a population size equal to 1000. The number of births was constrained by the problem instance and is available in Table 2. Since the number of births can differ between the generations, the number of generations could also be different for each run. A random initial population generation was used along with random mutation (often denoted DE/rand/1/bin) and a uniform crossover. The probabilities were set to 0.7 and 0.3 for mutation and crossover respectively, while the value of F, used in mutation was set to 0.01. During $WIDE$ phase a random selection is used, while the $FOCUS$ phase uses the Gap selection in addition to the Tournament selection. The latter utilizes a tournament size of 10.

The parameter configurations have not been disclosed for the other methods in the competition.

4.2 Metrics

Three performance metrics – proposed in GECCO2020 competition [12] – were used in the experiments. The first metric ($S1$) is simply the number of detected peaks (which is known as Recall in Information Retrieval). The second metric ($S2$) is a static F_1 measure defined as $F_1 = \frac{2*Recall*Precision}{Recall+Precision}$. Precision is calculated as the number of relevant detected optima, divided by all found optima.

The first two metrics measure effectiveness, whereas the third ($S3$) measures efficiency. It is a dynamic F_1 measure, which is calculated by computing the F_1 for each point in time, when new solution is found. It results in a curve, where each point on the curve provides the value of F_1 measure up to given point in time. The area under that curve is the third metric.

4.3 Results

Table 3 contains the values of all 3 metrics for all problem instances. The results of the top 3 methods of the GECCO2020 competition are presented: 1nd place – RS-CMSA-ESII, 2nd place – proposed GaMeDE, and 3-rd place CMSA-ES-DIPS.

On average, GaMeDE achieves slightly worse values of $S1$ and $S2$ metrics than the winner of the competition. Interestingly, it consistently finds all global optima for problems 8 and 9. Both those have a high number of optima, which are relatively easy to find. All composite function problems (11–20) contain 2

Table 3. Comparison of the GECCO2020 competition top 3 method' results.

	S1			S2			S3		
p	RS-CMSA-ESII	GaMeDE	CMSA-ES-DIPS	RS-CMSA-ESII	GaMeDE	CMSA-ES-DIPS	RS-CMSA-ESII	GaMeDE	CMSA-ES-DIPS
1	1.000	1.000	1.000	1.000	1.000	1.000	0.990	0.969	0.987
2	1.000	1.000	1.000	1.000	1.000	1.000	0.990	0.966	0.617
3	1.000	1.000	1.000	1.000	1.000	1.000	0.990	0.565	0.672
4	1.000	1.000	1.000	1.000	1.000	1.000	0.982	0.852	0.965
5	1.000	1.000	0.998	1.000	1.000	0.998	0.990	0.888	0.940
6	1.000	1.000	1.000	1.000	1.000	1.000	0.963	0.894	0.917
7	1.000	1.000	0.991	1.000	1.000	0.995	0.960	0.884	0.900
8	0.996	**1.000**	0.594	0.998	1.000	0.744	0.827	0.791	0.440
9	0.985	**1.000**	0.686	0.992	1.000	0.812	0.760	0.628	0.640
10	1.000	1.000	1.000	1.000	1.000	1.000	0.989	0.950	0.858
11	1.000	1.000	0.987	1.000	1.000	0.993	0.990	0.941	0.966
12	1.000	1.000	0.995	1.000	1.000	0.997	0.981	0.859	0.922
13	0.977	**1.000**	0.920	0.987	1.000	0.953	0.951	0.919	0.895
14	0.847	0.847	0.829	0.916	0.916	0.903	0.907	0.841	0.842
15	0.750	0.750	0.675	0.857	0.857	0.803	0.836	0.804	0.747
16	**0.833**	0.667	0.671	0.909	0.800	0.802	0.871	0.766	0.768
17	0.748	**0.750**	0.651	0.855	0.857	0.783	0.805	0.799	0.705
18	0.667	0.667	0.667	0.800	0.800	0.800	0.781	0.748	0.644
19	**0.708**	0.665	0.538	0.827	0.797	0.697	0.745	0.655	0.646
20	**0.618**	0.500	0.468	0.763	0.667	0.635	0.643	0.577	0.544
Avg	**0.906**	**0.892**	0.833	0.945	0.935	0.896	0.898	0.815	0.781

very steep global optima, which posed difficulties for all methods. In case of 2-dimensional problems (11–13) the solution space was small enough, that all optima could be found. Problem 14 was the first, where GaMeDE struggled. It consistently could find 4 easy optima, but of the remaining two only 1 was found on average. For problems 15–18 GaMeDE was not able to find any of the 2 difficult optima, all the easy ones were found in all cases. Problems with 10 and more dimensions (19 and 20) caused problems even with the easy optima, probably due to the size of the solution space. RS-CMSA-ESII was able to find more global optima in those cases and that's where it's found its edge over GaMeDE.

4.4 Discussion

GaMeDE attempts to identify promising areas of space and then focus only on those areas. It spends relatively high amount of time on looking for areas worth optimizing. It might be the cause of lower value of $S3$ metric.

None of the 3-top GECCO2020 approaches can consistently find 2 very steep optima, which are present in all composite functions. RS-CMSA-ESII performed better in this context, which is shown by the results on problem 16th. Nevertheless, it found only 1 of 2 optima, which shows that finding such optima require more research and maybe an additional optimization mechanism.

GaMeDE utilizes a 2-phase scheme, which allows it to solve problems with a very high number of global optima (no.8 and 9), while remaining competitive for other problems. First phase of the algorithm performs a wide sweep of the solution space, which creates a high diversity and often lands on many of the yet unknown global optima.

Problem no.19 is the first, where GaMeDE could not consistently find all easy optima. Moreover, for problem no.20, it was only able to find 4 out of 6 easy optima and it never found any of the 2 difficult ones. Due to high dimensionality of those problems and their large solution space, it might indicate that, the number of evaluations was simply too small - it was equal to the number of evaluations allowed for 3 dimensional problems. It might be interesting to experiments with those problems and a larger budget.

Interestingly, both RS-CMSA-ESII and CMSA-ES-DIPS often found points that were duplicated or were simply too far away from the global optimum to be considered a valid solution. This was never the case for GaMeDE, which might indicate that the clustering mechanism was always able to identify the solutions close to the same global optimum. Additionally, used optimizer had good enough convergence to avoid suboptimal solutions.

During the research various parameter configurations have been tested. Interestingly, they had very little impact on the results. Be it type of mutation, or a tournament size, no change significantly impacted the results. The only significant parameters are the ones connected to the archive. Too small archive, would result in ignoring parts of the space, that would eventually lead to finding an optimum solution. Too large archive, and GaMeDE would identify the incorrect areas for optimization, which would waste a lot of the budget. For larger problems, the population size was also important. Too low values would create a very sparse coverage of the solution space and clusters would contain the points, that often belonged to different global optima. In consequence, the same global optima were being found multiple times.

5 Conclusions

The main novelty lies in the proposed evolutionary GaMeDE method. It incorporates concepts from multi–objective optimization, namely NTGA2 method, to solve a multi–modal problem. A multi–objective GAP Selection operator enforces high diversity of the population, which allows for finding multiple optima. Moreover, a clustering and a local search algorithms are used to subsequently identify an area of space that may contain global optimum and then find said optimum.

GaMeDE is a competitive, multi–modal optimization method, which achieved good results on the CEC2013 Benchmark suite on multi–modal optimization. It attempts to create a good coverage of the solution space, identify the promising areas and then focus the local optimization on those areas.

The method was able to find all optima apart from those, that were very steep. When the point landed nearby, its fitness value was so low, that it was being truncated from the archive, and so was not considered for the optimization. It worked well, was when an individual happened to be so close to the global optimum, that its fitness value was relatively high. In case of steep "hills", this approach lost its consistency in 3 dimensional problems, and failed altogether in 5 and more dimensional problems. In those cases, the fitness value is not a good indicator of whether the area of the point is worth exploring. Hence, the method had to rely on random traversal. To make it work in higher dimensions, the population size had to be significantly increased, but then large part of the budget was spent on processing the population and there was not enough for local optimization.

GaMeDE method relies on its first phase to create a good spread of the individuals. Its random initialization method could be changed to one, that would provide the method with a good diversity from the very beginning. Moreover, the archive proved to be the most important part of the method. Its management scheme could be improved. Each generation the archive was sorted by the fitness value of its individuals and then truncated. This resulted in many individuals being close together, when the hill was relatively wide. A better approach could be to also enforce some diversity, so worse values are only preserved, provided they are far away from existing individuals.

Another possibility, would be to create an adaptation scheme for various parameters of the method. The most promising one, being the tournament size, as it is the most significant one. The size could change based on the features of the data instance, or even during the run. Early on, large number of individuals would allow for a wider search, while later on smaller size would ensure that only individuals close to the global optima are stored.

A smarter truncating mechanism might also be beneficial. One that considers, not only fitness of the individuals, but for example their closeness to other archived points. Individuals that are close to already found optima might also not be necessary as they will most probably lead to discovering the same optimum again. Additional heuristics could be considered.

All composite functions used in the competition are quite similar. One avenue of future work might be to propose a set of additional instances, that have more variety. Instances, that have particular features would allow for more thorough evaluations of the methods and deeper insight into their inner workings.

References

1. Maree, S. C., Alderliesten, T., Bosman, P.A.N.: Benchmarking HillVallEA for the GECCO 2019 competition on multimodal optimization. preprint arXiv:1907.10988 (2019)
2. Bosman, P.A.N., Grahl, J.: Matching inductive search bias and problem structure in continuous estimation-of-distribution algorithms. Eur. J. Oper. Res. **185**(3), 1246–1264 (2008)
3. Myszkowski, P.B., Laszczyk, M.: Diversity based selection for many-objective evolutionary optimisation problems with constraints. Inf. Sci. (in Press)
4. Köppen, M., Yoshida, K.: Substitute distance assignments in NSGA-II for handling many-objective optimization problems. In: Obayashi, S., Deb, K., Poloni, C., Hiroyasu, T., Murata, T. (eds.) EMO 2007. LNCS, vol. 4403, pp. 727–741. Springer, Heidelberg (2007). https://doi.org/10.1007/978-3-540-70928-2_55
5. Spettel, P., Beyer, H.-G., Hellwig, M.: A Covariance matrix self-adaptation evolution strategy for optimization under linear constraints. IEEE Trans. Evol. Comput. **23**(3), 514–524 (2018)
6. Price, K.V.: Differential evolution. In: Handbook of Optimization, pp. 187–214. Springer, Heidelberg (2013). https://doi.org/10.1007/978-3-642-30504-7_8

7. Li, X., Engelbrecht, A., Epitropakis, M.G.: Benchmark functions for CEC 2013 special session and competition on niching methods for multimodal function optimization. RMIT University, Evolutionary Computation and Machine Learning Group, Australia, Technical report (2013)
8. Zhang, Y., Gong, D., Gao, X., Tian, T., Sun, X.: Binary differential evolution with self-learning for multi-objective feature selection. Inf. Sci. **507**, 67–85 (2020)
9. Liu, Q., Du, S., van Wyk, B.J., Sun, Y.: Double-layer-clustering differential evolution multimodal optimization by speciation and self-adaptive strategies. Inf. Sci. **545**, 465–486 (2021)
10. Hong, Z., Chen, Z.-G., Liu, D., Zhan, Z.-H., Zhang, J.: A multi-angle hierarchical differential evolution approach for multimodal optimization problems. IEEE Access **8**, 178322–178335 (2020)
11. Liang, J., et al.: A clustering-based differential evolution algorithm for solving multimodal multi-objective optimization problems, warm and evolutionary computation, vol. 60, p. 100788 (2021)
12. https://github.com/mikeagn/CEC2013/

Simulating Emergency Departments Using Generalized Petri Nets

Ibtissem Chouba[1,2](✉), Lionel Amodeo[1](✉), Farouk Yalaoui[1](✉), Taha Arbaoui[1](✉), and David Laplanche[2](✉)

[1] Institut Charles Delaunay, LOSI, Université de Technologie de Troyes, 12 Rue Marie Curie, CS 42060, 10004 Troyes Cedex, France
{ibtissem.chouba,lionel.amodeo,farouk.yalaoui, taha.arbaoui}@utt.fr

[2] Pôle d'Information Médicale et Évaluation des Performances, Centre Hospitalier de Troyes, 101 Avenue Anatole France, 10000 Troyes, France
david.laplanche@ch-troyes.fr

Abstract. Emergency departments (ED) face significant challenges to deliver high quality and timely patient care while facing rising costs. Prediction models in healthcare are becoming a trend allowing healthcare institutions to better manage the flow of patients through the services. In this work, a Generalized Stochastic Petri net (GSPN) simulation engine is used as a simulator (and performance evaluation) and a waiting time prediction model for the ED of Troyes hospital in France. Compared to existing literature, this work puts more emphasis on using features on GSPN (time, hierarchy, concurrency, parallelism) to capture the complex nature of the system. A new concept of real-time estimation of waiting time is presented. We validated the simulation model with a gap less than 10% between real and simulated performances. To the best of our knowledges, this work is the first to use a GSPN model as a management and waiting time's forecasting tool for emergency departments.

Keywords: Emergency department · Simulation · Generalized Stochastic Petri Net · Waiting time prediction

1 Introduction

With an increasing population worldwide, Emergency Departments (ED) are overflowing. This problem requires complex management and forecasting tools to improve the performance of ED. Thus, inadequate resources allocation, staff shift organization and increase in demand for ED service can cause resource contention and delay the accomplishment of service goal [2, 4, 6]. Moreover, queues quickly overflow due to a mismatch between demand and the system's capacity. Often-times, potential delays can be averted or minimized by using resource analysis to identify strategies in which tasks can be executed in parallel and in efficient ways. Several modeling techniques have been used in different ways to predict and improve the ED system. Some hospitals have resorted to manage bed capacity [7] or resource allocation [6] to deal with growing demand, or using

N. T. Nguyen et al. (Eds.): ACIIDS 2021, LNAI 12672, pp. 224–234, 2021.
https://doi.org/10.1007/978-3-030-73280-6_18

queueing theory to improve patient flow [12]. For a good management strategy, other approaches used to forecast daily patients flow in the long and short term [3]. Recently, the medical industry has started to publish data on waiting times related to services provided to their patients. Using different methods [1, 5], many studies reported that real-time prediction of waiting time improved patient satisfaction and service quality.

Implementing these strategies does not always guarantee a better performance because of the random variations that can still result in periods of overcrowding. Moreover, experimenting new strategies on the existing system is considered as too risky, as it might disrupt the overall operations, increase the costs and causing congestion if errors occur. Therefore, the management should first study in detail the ED operation system before implementing any improvements [10].

Simulation is well-suited to tackle this kind of complex and dynamic problems. It provides a useful tool for capacity planning and efficiency improvement. The ED may be effectively described as a Discrete Event System (DES) whose dynamics depend on the interaction of discrete events with a high degree of concurrency and parallelism [9]. Among the DES models, Petri nets (PNs) have been a popular formalism for modeling systems exhibiting behaviors of competition and concurrency, particularly in the modeling and analysis of systems sharing resources [11].

Despite the diversity and richness of the literature around the use of the PN's to model and analyze several health systems, few works are interested in the simulation of emergency services with PNs taking into account all of their aspects. However, in [2], the authors apply timed PNs to model and evaluate performance measurement in cardiological ED. Furthermore, they predict the impact of variations of medical staff and service facilities on performance selected. In [13], with a timed PN framework, authors focused on the pulmonology department workflow and the drug distribution system. Then, they used the simulation as a decision system. Recently, Zhou et al. [8] presented a Stochastic Timed Petri nets (STPNs) simulation engine. It provides all critical features necessary for the analysis of ED quality and resource provisioning. This STPN model supports non-Markovian transition firing times, race and preselection transition firing policies, and both single-server and multi-server transitions.

The objective of this study is to develop a simulation framework to model the structure and the dynamics of the ED of the hospital center of Troyes (CHT), France. For this purpose, a GSPN-based healthcare workflow and resource modelling technique is presented. This simulation framework's objective is threefold. First, it describes in a concise and detailed way the structure of the ED of CHT. The model describes the complete workflow and management of patients from their arrival to the discharge. The variations of processing times depending on patient state [3], radiology and labs tests are considered in our simulation model. Second, it evaluates the ED performances such as the average waiting time (AWT) and the average inpatient stay (AS) and foresees the impact of variations in the medical and para-medical staff then the treatment room on the selected performances. Third, it estimates the average patient waiting time in real time. The main contribution of this work is the introduction of a GSPN-based model to be used as a short and mid-term optimization tool to deal with both the scheduling and the allocation of resources and to estimate in real-time the patient's waiting time. To the best of our knowledge, this work is the first to consider these aspects simultaneously.

The remainder of this paper is organized as follows. IN Section two, the proposed modelling approach is exposed. Section 3 details the proposed GSPN. In Sect. 4, we present the experimental results of the proposed model. The conclusion is to be found in Sect. 5.

2 Modelling Approach

In this paper, a GSPN-based model is used to create ED system. Our choice of using GSPN stems from the fact that it captures structural properties of the system which we can study and use, further it provides the basic primitives for modeling concurrency and synchronization, conditions that are common in our model. Based on the initial model (Fig. 1), the generic conceptual model was developed (Figs. 2, 3 and 4). Both models are based on a huge amount of data that has been treated in a preprocessing procedure to be usable in the simulation.

2.1 Data Collection and Processing

Troyes Hospital Center (CHT) is the second largest health care center in the Champagne-Ardenne territory of France serving a population of 130194 [3]. The ED of CHT provides treatment services 24 h a day and receives an average of 61497 patients per year. The data on which we based our analysis and the construction of the simulation models are extracted from hospital information system (ResUrgence). This latter records every action of the patient from their entrance until they quit the ED. The inputs data required by our system are: date and time of patient arrivals, patient conditions, the frequency of each pathway and the processing times of services according to patient conditions. We collected the data from the first January 2019 to the 30st November 2019 which counted 56097 entry, sampling in ward, and open interview with employees. Then, we model its probability distribution. The ED consists of two main circuits: the short circuit (includes pediatric, gynecological and psychiatric patients who do not need major paramedical care) and the long circuit. The resources available within the ED are: a triage nurse accompanied by a caregiver that are available 24 h a day. A second triage nurse is available from 11 AM to 8 PM. In the short circuit, 2 emergency physicians, 1 resident physician, 1 nurse and 1 caregiver are available 24 h a day. In the long circuit, 2 emergency physicians, 2 resident physicians, 2 nurses and 2 caregivers available 24 h a day. A third nurse is available from 1:30 PM to 7 PM.

2.2 Simulation Model

When a patient arrives at the ED, they are first registered at reception. Then the triage nurse makes an initial evaluation. According to the patient's state noted EP (ETAT Patient), the nurse decides whether the patient is in a serious health conditions, in order to specify the flow that they will go through (short or long circuit). The patient's state classification assembles multiple crossed CCMU and GEMSA categories, then, EP is categorized from the least urgent to the most urgent by triage nurses. This system categorizes ED patients both by acuity and resource needs [3]. In addition, not all patients are treated equally. The patient's flow is completely based on the EP level. Nevertheless, a patient with a stable state may wait for longer because of the permanent presence of patients with a more severe state. Therefore, the priority of passage in the ED is a compromise between the severity of the patients and their arrival time. All these variables were taken into account in our simulation model. Following this, the patient has recourse to the next stage: the examination that starts when an appropriate box gets free and the right emergency physician becomes available. A nurse or a caregiver accompanies the patient to the box. At that moment, the patient's Waiting Time will be recorded (WT = entrance time in the box – time arrival). The emergency physician makes a first assessment and decides if the patient needs further exams to confirm their diagnosis. If not needed, the patient is discharged to return home. In the case where additional exams are requested, the patient can wait for the test results in their box or can be transferred, if possible, to an internal waiting room in order to make the box available for other patients. Figure 1 shows an overall patient flow which was transformed to a GSPN-based model (Figs. 2,3 and 4).

3 Simulation Modeling

3.1 The Proposed GSPN-Based Model

GSPN are considered among PN models in which non-deterministic firing delays, associated with transitions, coexist with immediate transitions. The widely used GSPNs model enable both types of transitions immediate and timed. Once enabled, immediate transitions fire in zero time. Timed transitions fire after a random exponentially distributed firing delay as in the case of STPNs. The firing of immediate transitions always has priority over that of timed transitions.

Formally, GSPN is a 6-uplet: GSPN = (P, T, Pre, Post, μ, M_0) where:

- $P = [P_1, P_2, \ldots P_m]$ is a finite set of places;
- $T = [T_1, T_2, \ldots, T_m]$ is a finite set of transitions;
- Pre: P × T is the place application;
- Post: P × T is the following place application
- $M_0 = \{0, 1, 2, 3, \ldots, i^{th}\}$ is the initial marking of the PN and represents the number of tokens in the ith place.
- $\mu = \{\mu_1, \mu_2, \ldots, \mu_n\}$ where μ_i is the random distribution of the firing time of the transition T_i.

A transition $T_j \in T$ is enabled at a marking M if and only if $M_{(Pi)} \geq Pre_{(Pi,Tj)}$, $\forall P_i \in {}^{\circ}T_j$ where ${}^{\circ}T_j = \{P_i \in P: Pre_{(Pi,Tj)} > 0\}$. From the marking M, firing a transition Tj leads to the new marking M' defined as follows: $M'_{(Pi)} = M_{(Pi)} - Pre_{(Pi,Tj)} + Post_{(Pi,Tj)}$, $\forall$ Pi $\in$ P. Timed transition sequence of a GSPN system can be defined as a transition sequence augmented with a set of nondecreasing real values describing the epochs of firing of each transition (τ_i). Such a timed transition sequence is denoted as follows: $[(\tau_{(1)}, T_{(1)}), \cdots, (\tau_{(j)}, T_{(j)}); \cdots]$.

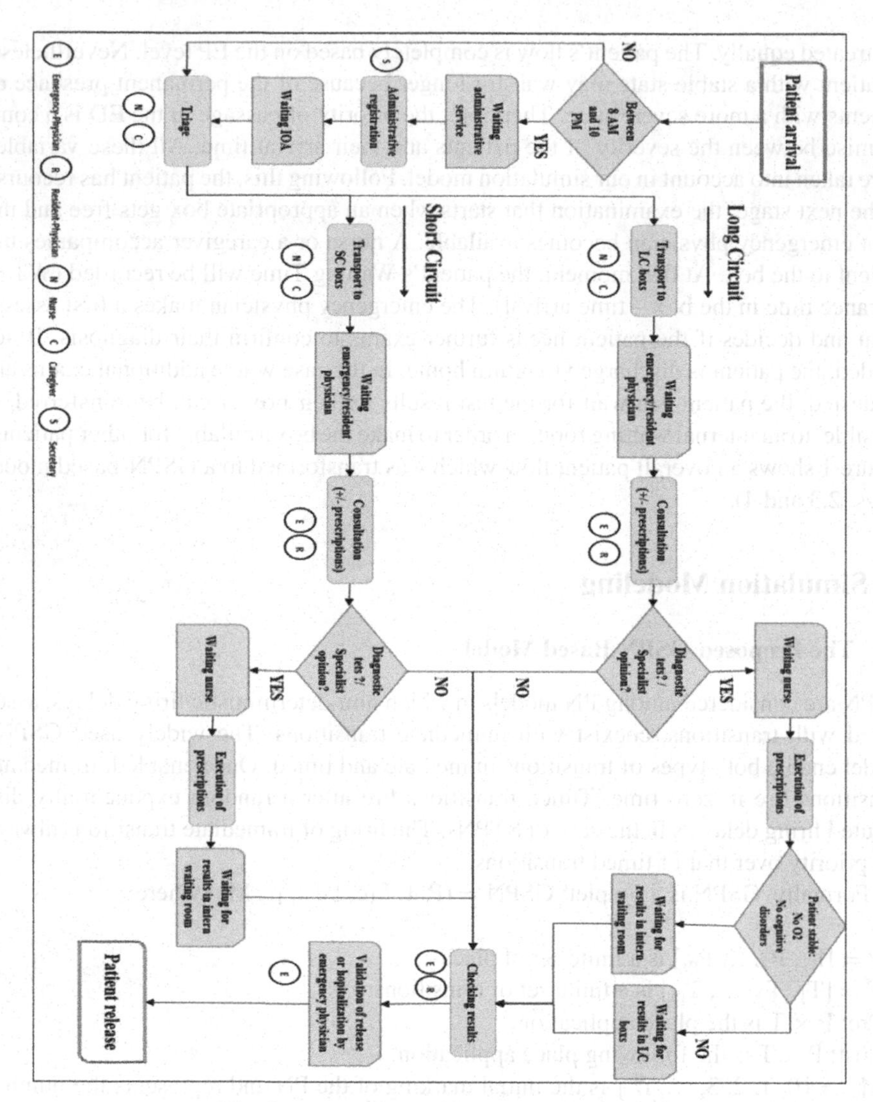

Fig. 1. Diagram of patient flow in ED service

The time intervals $[\tau_{(i)}, \tau_{(i+1)})$ between consecutive epochs represent the periods in which the PN sojourns in marking $m_{(i)}$ and it is denoted as $\sigma_{(i)} = \tau_{(i+1)} - \tau_{(i)}$. This sojourn time corresponds to a time elapsed in which the execution of one or more activities is in progress and the state of the system does not change [7]. Based on markings and sojourn times, the timed transition sequence yields also a timed marking sequence $[(\sigma_{(1)}, m_{(1)}); \cdots; (\sigma_{(j)}, m_{(j)}); \cdots]$. Thus, $\sigma_{(1)}$ is the time spent in marking $m_{(1)}$ that was reached after the firing of transition $T_{(1)}$ and was left upon firing of transition $T_{(2)}$.

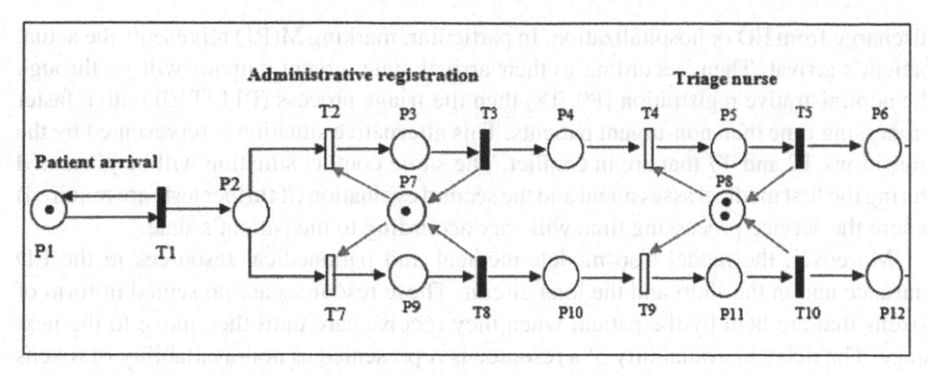

Fig. 2. Patient process from arrival to triage nurse

3.2 The Hierarchical GSPN Model of the ED

We used GSPN to model the patient flow of the ED of CHT, where places with capacities model the medical and paramedical staff as well as the available boxes. In addition, transitions describe the flow of patients and the actions of the emergency staff. Figure 2 describes the patient's process from their arrival to the triage process. Figure 3 describes the patient's process from the arrival to initial consultation in both the short and the long circuit. Figure 4 describes the patient's process from the diagnosis tests to their

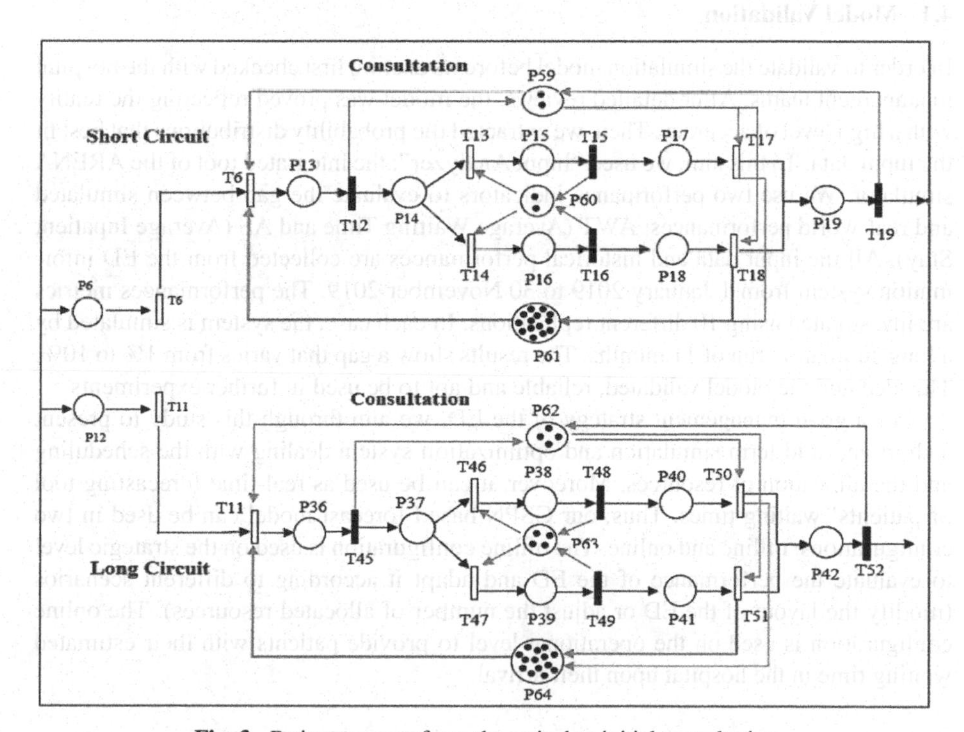

Fig. 3. Patient process from the arrival to initial consultation

discharge from ED or hospitalization. In particular, marking M(P1) represents the actual patient's arrival. Then, according to their arrival state, urgent patients will go through the administrative registration (P9, T8) then the triage process (P11, T10) with a faster processing time than non-urgent patients. This alternative situation is represented by the transitions T2 and T7 that are in conflict. The same conflict situation will be presented during the first medical assessment and the second evaluation (if further tests are required) where the service processing time will vary according to the patient's state.

Moreover, the model also models medical and paramedical resources in the ED entrance and in the short and the long circuit. These resources are presented in form of tokens that are held by the patient when they receive care until they move to the next stage. The delay in availability of a resource is represented as non-availability of tokens to advance the patient tokens through the net. This delay increases the number of patient tokens in the system waiting for a resource [3].

An important aspect represented by the GSPN model is the simulation of the behavior details of the patient's flow. For example, the model captures the time that patients spend waiting for the doctor in the box for a consultation or evaluation. It also models the behavior of the human resources in each service whose absence due to long treatment may cause a slowdown in the flow of patients.

4 Simulation Results and Discussion

4.1 Model Validation

In order to validate the simulation model before its use, we first checked with the hospital management teams. After detailed reviews, the model was proved reflecting the reality with a high level of accuracy. Then, we extracted the probability distributions that best fit the input data. To this aim, we used "Input Analyzer": the integrated tool of the ARENA simulator. We use two performance indicators to evaluate the gap between simulated and real-world performances: AWT (Average Waiting Time and AS (Average Inpatient Stay). All the input data and historical performances are collected from the ED information system from 1 January 2019 to 30 November 2019. The performances metrics are investigated using 10 different replications. In each case, the system is simulated by a long simulation run of 11 months. The results show a gap that varies from 1% to 10%. This deemed the model validated, reliable and apt to be used in further experiments.

For a good management strategy of the ED, we aim through this study to present a short and mid-term simulation and optimization system dealing with the scheduling and the allocation of resources. Moreover, it can be used as real-time forecasting tool of patients' waiting times. Thus, our GSPN-based forecast model can be used in two configurations: offline and online. The offline configuration is used on the strategic level to evaluate the performance of the ED and adapt it according to different scenarios (modify the layout of the ED or adjust the number of allocated resources). The online configuration is used on the operational level to provide patients with their estimated waiting time in the hospital upon their arrival.

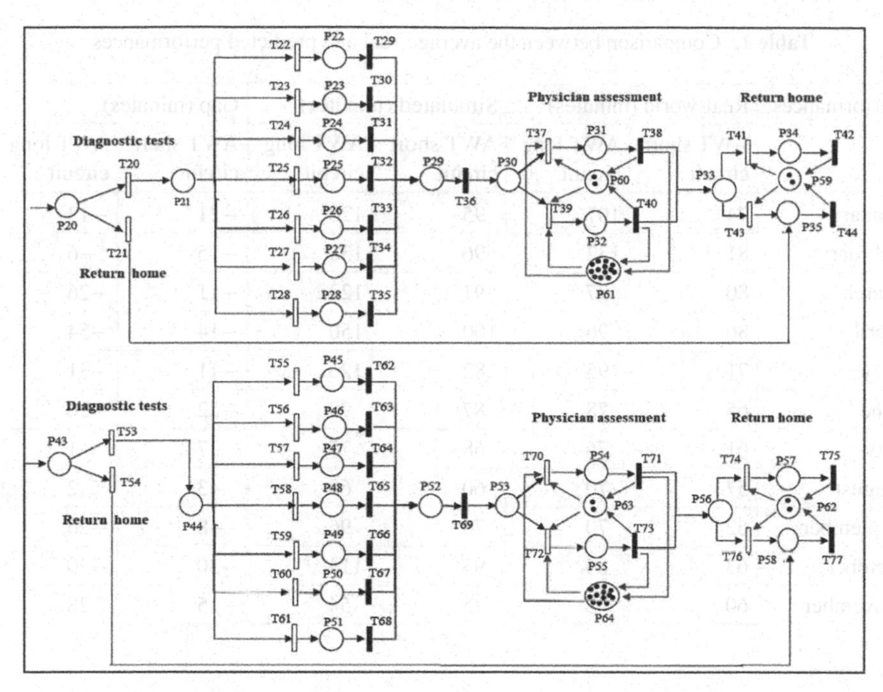

Fig. 4. Patient process from the diagnosis test to his exit from ED

4.2 Forecasting Patients' Waiting Times

The model proceeds by taking into account different real-time data that are obtained from the hospital's information system for N patients. The simulation is run with the current system's state, e.g. the number of patients in each circuit, their positions in each service and the available human and material resources. This data allows us to have a system with a steady state. Hence, our simulation does not consider the warm-up period.

When a new patient (N + 1) arrives to the ED, a simulation is run to obtain their expected waiting time. The simulation considers the N patients in the system and the newly arrived patient and attempts to forecast their waiting time (see Fig. 3). Table 1 shows the difference between the average real and estimated wait time performance of the N patients in the short and long circuit. In addition, our ongoing work aim to refine the model and minimize as possible the gap for the performance of each patient (Fig. 5).

Table 1. Comparison between the average real and predicted performances

Performances	Real-world (minutes)		Simulated (minutes)		Gap (minutes)	
	AWT short circuit	AWT long circuit	AWT short circuit	AWT long circuit	AWT short circuit	AWT long circuit
January	74	107	95	120	−21	−13
February	81	132	96	138	−15	−6
March	80	97	91	123	−11	−26
April	86	96	100	150	−14	−54
May	71	95	82	126	−11	−31
June	65	78	87	98	−22	−20
July	61	76	68	77	−7	−1
August	57	70	60	68	−3	2
September	62	70	70	96	−8	−26
October	63	84	93	114	−30	−30
November	60	82	75	54	−15	28

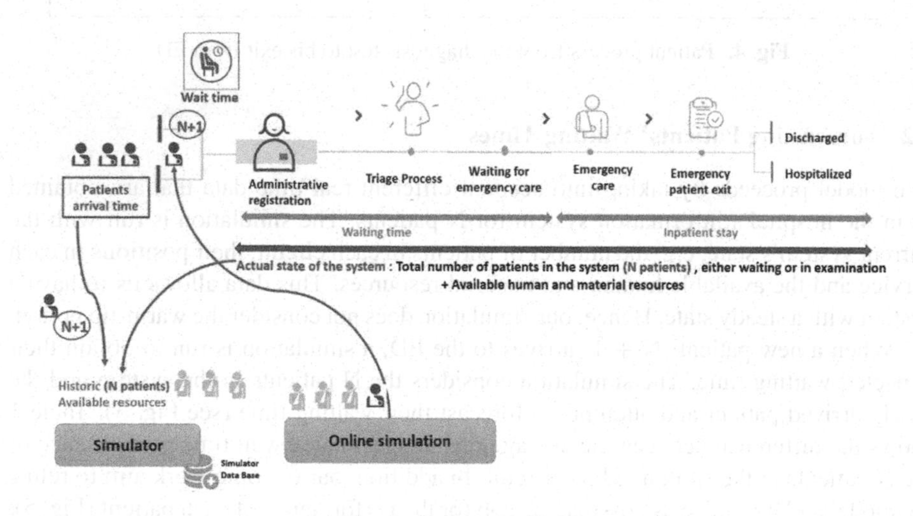

Fig. 5. Online simulation

5 Conclusion

This paper tackles the simulation and the performance evaluation of the ED of the hospital center of Troyes, France. In this study, a GSPN model was proposed as a short and a mid-term simulation system dealing with both the scheduling and the allocation of resources and the real-time forecasting of the patient's waiting time. The model was validated with a gap less than 10% between simulated and observed performances. This model objectives are twofold. First, it allows to describe in a concise and detailed way the complex structure and dynamics of the complete workflow and management of patients from their arrival to the discharge. Moreover, the variations of processing time, radiology and labs tests are taken into account. Second, it evaluates the ED performance indicators such as AWT and AS and foresees the impact of variations in the medical and paramedical staff on the selected performances. Finally, the proposed model provides a good forecast for the average patient waiting time in the long and the short circuit of the ED.

References

1. Ang, E., Kwasnick, S., Bayati, M., Plambeck, E.L., Aratow, M.: Accurate emergency department wait time prediction. Manuf. Serv. Oper. Manage. **18**(1), 141–156 (2016). https://doi.org/10.1287/msom.2015.0560
2. Amodio, G., Fanti, M.P., Martino, L., Mangini, A., Ukovich, W.: A Petri net model for performance evaluation and management of an emergency cardiology departments. In: Proceedings of the XXXV ORHAS Conference, p. 97 (2009)
3. Afilal, M., Yalaoui, F., Dugardin, F., Amodeo, L., Laplanche, D., Blua: Emergency department flow: a new practical patients classification and forecasting daily attendance. IFAC-PapersOnLine **49**(12), 721–726 (2016). https://doi.org/10.1016/j.ifacol.2016.07.859
4. Kang, C.W., Imran, M., Omair, M., Ahmed, W., Ullah, M., Sarkar, B.: Stochastic-petri net modeling and optimization for outdoor patients in building sustainable healthcare system considering staff absenteeism. Mathematics **7**(6), 499 (2019). https://doi.org/10.3390/math7060499
5. Sun, Y., Teow, K.L., Heng, B.H., Ooi, C.K., Tay, S.Y.: Real-time prediction of waiting time in the emergency department, using quantile regression. Ann. Emerg. Med. **60**(3), 299–308 (2012). https://doi.org/10.1016/j.anne-mergmed.2012.03.011
6. Zeinali, F., Mahootchi, M., Sepehri, M.M.: Resource planning in the emergency departments: a simulation-based metamodeling approach. Simul. Model. Pract. Theory **53**, 123–138 (2015). https://doi.org/10.1016/j.sim-pat.2015.02.002
7. Balbo, G.: Introduction to Generalized Stochastic Petri Nets. In: Bernardo, M., Hillston, J. (eds.) SFM 2007. LNCS, vol. 4486, pp. 83–131. Springer, Heidelberg (2007). https://doi.org/10.1007/978-3-540-72522-0_3
8. Zhou, J., Wang, J., Wang, J.: A simulation engine for stochastic timed Petri nets and application to emergency healthcare systems. IEEE/CAA J. Automatica Sin. **6**(4), 969–980 (2019). https://doi.org/10.1109/jas.2019.1911576
9. Gunal, M.M., Pidd, M.: Interconnected DES models of emergency, outpatient, and inpatient departments of a hospital. In: 2007 Winter Simulation Conference, pp. 1461–1466. IEEE (2007). https://doi.org/10.1109/wsc.2007.4419757
10. Ibrahim, I.M., Liong, C.Y., Bakar, S.A., Ahmad, N., Najmuddin, A.F.: Minimizing patient waiting time in emergency department of public hospital using simulation optimization approach. In: AIP Conference Proceedings, p. 060005. AIP Publishing LLC (2017). https://doi.org/10.1063/1.4980949
11. David, R., Alla, H.L.: Du Grafcet aux réseaux de Petri (1992)

12. Vass, H., Szabo, Z.K.: Application of queuing model to patient flow in emergency department. Case Study Procedia Econ. Finance **32**, 479–487 (2015). https://doi.org/10.1016/s2212-5671(15)01421-5
13. Wang, J., Tian, J., Sun, R.: Emergency healthcare resource requirement analysis: a stochastic timed Petri net approach. In: 2018 IEEE 15th International Conference on Networking, Sensing and Control (ICNSC), pp. 1–6. IEEE (2018). https://doi.org/10.1109/icnsc.2018.8361301

What Do You Know About Your Network: An Empirical Study of Value Network Awareness in E-commerce

Bartłomiej Pierański(✉) and Arkadiusz Kawa

The Poznan University of Economics and Business, Al. Niepodległości 10, 61-875 Poznań, Poland

bartlomiej.pieranski@ue.poznan.pl

Abstract. Creating value for customers in e-commerce is a relatively new research area. The majority of the studies conducted to date focus only on the relationship between the customer and the online seller, ignoring the role and participation of other entities involved in the process of customer value creation. Therefore, this paper identifies a broad spectrum of entities involved in creating customer value, based on the model developed by Nalebuff and Branderburger. The empirical material was obtained through qualitative research using the Focus Group Interview method. The research was conducted in three groups of entities: suppliers and complementors; online sellers; and end users. Its aim was to diagnose network awareness; relating to the perception of the study participants regarding how many and which entities are involved in creating value for buyers.

Keywords: Value network · Network perception · E-commerce · FGI

1 Introduction

It has been typically assumed that the burden of creating value rests exclusively on entities offering goods directly to end customers because it is this group of entities that is responsible for the preparation and appropriate presentation of the product range, and later for the processing of customer orders. It seems, however, that this view is not entirely justified because value is also created and delivered by entities that often do not conduct commercial activities. These include, among others, financial institutions, logistics companies, warehousing service providers and companies dealing with website positioning; so although these entities play a significant role, their importance in creating value for customers in e-commerce is often overlooked in the literature [1]. Thus, there is a shortage of scholarly research that would examine the issue of creating customer value by a range of different entities. In order to fill this gap, this paper will present the results of qualitative empirical studies diagnosing network awareness. The perception of the network of entities creating value for customers was diagnosed from three perspectives. The first two are suppliers and complementors (the first group of entities) and online sellers (the second group of entities). Additionally, the perspective of the entity that ultimately evaluates the actions of many enterprises striving to create value, i.e. end

N. T. Nguyen et al. (Eds.): ACIIDS 2021, LNAI 12672, pp. 235–245, 2021.
https://doi.org/10.1007/978-3-030-73280-6_19

customers, was also analysed. The theoretical framework for the considerations is the model developed by [2], which shows the relationships between entities cooperating with each other and creating a kind of network for creating customer value. According to this model, value is created, delivered and appropriated by various entities that together form a network [3]. These entities include not only enterprises, but also individual customers [4].

The remainder of the paper is organized as follows: first, the theoretical background is presented, which focuses on determining the essence and components of a value network as well as defining the concept of network awareness, very important from the point of view of the conducted research. Next, the process and results of qualitative studies conducted using the FGI method are presented, followed by a discussion and the conclusions.

2 Theoretical Background

2.1 Value Network

The concept of "value for the customer" was first introduced by Drucker in The Practice of Management in 1954 [5]. Over time, the concept of value for the customer has become the subject of many publications in the field of management sciences. It is assumed that value for the customer (also referred to as "perceived value" [6] is the difference between the subjectively perceived benefits associated with the purchase and use of a product, and the subjective costs incurred in obtaining it (e.g. in: [7, 8]).

Value for the customer can be created by a company through the company's internal resources as well as through the relational resources that stem from the partnership between the supplier and the customer [9]. However, creating value through the company's internal resources is extremely difficult. Obtaining such resources is often costly and time-consuming, sometimes even impossible. Therefore, in order to create value for the customer, close cooperation between the company and external entities is essential. The cooperation of a large number of various entities, connected with one another by means of various relationships, may take the form of an inter-organizational network. According to the Industrial Marketing and Purchasing Group "(…) a business network is a collection of long-term formal and informal as well as direct and indirect relationships that exist between two or more entities" [10]. A model of cooperation within a business network, presenting the relationships between the entities cooperating with each other in order to create customer value, was proposed by Nalebuff and Branderburger [2]. This model distinguishes four groups of entities that can engage in cooperation with the aim of creating value for the customer (the customer is the fifth element of this model). Apart from a specific company, these entities are suppliers, competitors and complementors (see Fig. 1). The participation of suppliers in creating customer value seems indisputable. However, the presence of complementors and, above all, competitors in this model requires some explanation. Complementors are companies which offer products that complement the offer of the main company. These products can be sold to customers together with this company's product, or offered to customers as independent complementary products. Complementors' activities undoubtedly provide value for buyers. The Nalebuff and Branderburger model also assumes that value can be delivered to customers as a result of cooperation between competitors. Such cooperation is referred to as coopetition and concerns the relationship between the company and its competitors. Competitors are considered to be other entities from which customers can buy the same or similar products.

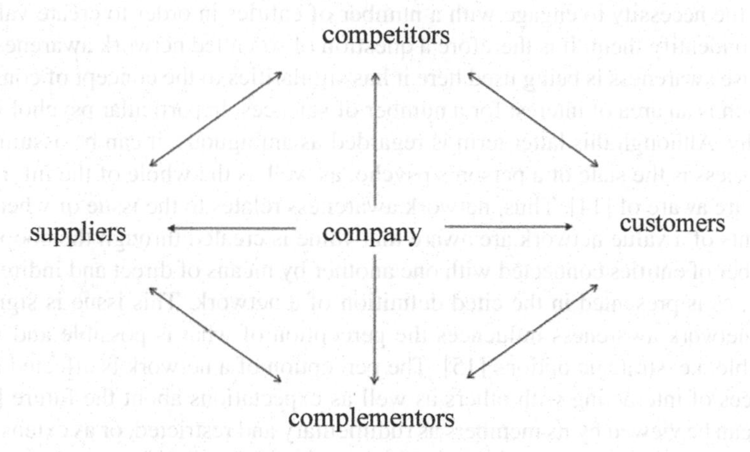

Fig. 1. Value network model. Source: [2]

Although the model was developed before e-commerce became one of the most dynamically developing forms of sales, its versatility means that it can be adapted to the needs of research on creating value networks in online commerce. The model has served as a basis for creating assumptions relating to the identification of entities involved in creating customer value in e-commerce. In the present study, the company was represented by online sellers, understood as both online stores; and entities without an online store which offer products, for example auction sites. Customers are defined as individual buyers, and entities representing suppliers in e-commerce are enterprises that provide goods to sellers (e.g. manufacturers, distributors, wholesalers). Finally, complementors are companies offering products that complement the basic product of an online seller (e.g. logistics services, online payment providers, IT services, marketing services, etc.).

2.2 Network Awareness

The Nalebuff and Brandenburg model belongs to the category of normative models. Such models present economic phenomena, in this case multi-entity value creation, in an idealized, theoretical way, indicating how market entities should behave if the conditions described in such models were met. However, normative models offer only limited possibilities for explaining the behaviour of market players in real situations. The actual behaviour of market entities tends to differ markedly from theoretical predictions. This, however, does not undermine the sense of certain theories [11]; it only indicates the necessity to search for the sources of the observed anomalies [12]. Hence, describing entities through the prism of how they should behave (based on a specific model) should be replaced with identifying the determinants of the actual behaviour of entities in specific market situations.

As mentioned earlier, the Nalebuff and Brendenburg model assumes the cooperation of a number of entities (suppliers, complementors, competitors and customers) in order to create a certain value. However, the question arises whether these entities actually cooperate with each other or, looking at the issue in a different light, whether they are aware of the necessity to engage with a number of entities in order to create value and are able to identify them. It is therefore a question of so-called network awareness [13]. In the sense awareness is being used here it has similarities to the concept of consciousness, which is an area of interest for a number of sciences, in particular psychology and philosophy. Although this latter term is regarded as ambiguous, it can be assumed that consciousness is the state of a person's psyche, as well as the whole of the information that they are aware of [14]. Thus, network awareness relates to the issue of whether the participants of a value network are aware that value is created through the cooperation of a number of entities connected with one another by means of direct and indirect relationships, as is presented in the cited definition of a network. This issue is significant because network awareness influences the perception of what is possible and what is not possible, i.e. strategic options [15]. The perception of a network is affected by past experiences of interacting with others as well as expectations about the future [16]. A network can be viewed by its members as rudimentary and restricted, or as extensive and complex. Network awareness is essentially subjective, although it is possible (and desirable) for network perceptions to overlap as this indicates a certain cognitive community and mutual awareness of the network participants [17]. Moreover, the perception of a network is multi-dimensional [18]. One of the dimensions is the extent of the network, relating to the diversity and number of entities cooperating within the network. It can be assumed that entities with a low level of network awareness (in terms of its extent) will cooperate with a narrow circle of entities (e.g. only with those with whom they have direct relationships). Such limited cooperation will result in creating less value. On the other hand, cooperating with a wide range of entities may lead to creating unique value that will be difficult to reproduce by competitors.

In view of the above, the awareness of a value network among its members seems to be an important research topic, all the more so because this area has not yet been adequately explored despite being very important for understanding the behaviour of market entities.

3 Overview of the Studies

The aim of the research was to examine the condition of so-called network awareness among entities which, according to the Nalebuff and Brandenburg model, constitute a value network.

Empirical research made use of the Focus Group Interview (FGI) method. Three separate studies were conducted, based on the Nalebuff and Branderburger model [2]. The first study involved suppliers and complementors; the second study involved online sellers; and the third study focused on end customers. In order to achieve the aim of the study, the participants had to perform the following tasks:

- in the first study, suppliers and complementors were to identify entities that in their opinion co-create value within the value network
- in the second study, online sellers were to identify entities that in their opinion co-create value within the value network
- in the third study, end customers were to identify entities that in their opinion co-create value within the value network.

All three FGIs were conducted according to a specifically designed interview scenario. The scenario was developed by the project team and given to a research and advisory company. The scenario together with detailed directions for the implementation of the studies was agreed in consultation with the moderator. The scenario assumed that some of the tasks will be performed individually in order to activate all the participants of the study and avoid certain negative intra-group processes, for example conformism and imitation. On the other hand, some tasks were performed together by all the participants of each study group in the hope of achieving a synergistic effect [19]. Only open-ended questions were used, giving the participants the freedom to express their views.

Each of the three interviews began with greeting the respondents. The moderator presented the purpose of the study, the title of the project and the name of the institution for which the study was being done. The moderator informed the participants that they were taking part in a focus group interview and communicated the basic assumptions of this type of study. The participants were asked to express their opinion freely. Next, the moderator informed the participants that the study was being recorded but that its results with their statements would be completely anonymous. The participants were asked not to speak at the same time so that all statements could be recorded. Throughout the study, only the participants' first names were used to ensure their anonymity. The study was conducted in the form of a discussion, during which the participants expressed their opinions with regard to the specific questions. If a group had any doubts about any of the questions, the moderator helped by providing a context or clarifying certain terms.

A transcript of the audio-visual material recorded during each study was made. The transcript was prepared by the person recording the sessions and then checked by the moderator (cross-check). Appropriate adjustments were made where necessary. Then the transcripts were submitted to the research team, who analysed them using the inductive approach generally accepted in qualitative research [20]. The analysis consisted in reading the transcript multiple times for the purpose of coding them (marking them with short words or phrases to describe specific features of the statements). After analysing the transcripts, a set of codes corresponding to the content of the statements was established. The codes were grouped according to certain features, such as the types of entities indicated in the participants' statements. Then the statements were marked with the codes. The coding made it possible to examine the respondents' statements both in terms of their answers to individual questions and in terms of the context of the individual issues throughout each interview.

The analysis of the transcripts showed that the organisers of the studies were able to avoid some possible unfavourable phenomena, such as unauthorised conversations between the participants (both related and unrelated to the subject of the study); excessive activity or complete inactivity of individual participants; or shouting speakers down [19].

3.1 Sampling of Study 1 – Suppliers and Complementors

The study made use of purposive sampling. The participants were selected in such a way that the study group included representatives of various types of entities who performed managerial functions within their organizations. The main selection criterion was operating a business for at least 12 months. Taking into account recommendations regarding the number of participants in an in-depth group interview [19], the sample size was set at 7 people (N = 7). Recruitment was done by the research company and the characteristics of participants of Study 1 are presented in Table 1.

Table 1. Characteristics of participants – Study 1

No.	Area of activity	Position/function	Age
1	Shopping platform	Product manager	26
2	IT services	n/a	55
3	Courier services	Marketing manager	32
4	Online payments	Implementation department manager	n/a
5	Marketing and positioning services	n/a	n/a
6	Marketing and positioning services	n/a	30
7	Courier services	Owner	n/a

Source: Own compilation

3.2 Sampling of Study 2 – Online Sellers

The study made use of purposive sampling. Each of the participants had to fulfil the condition of trading on the Internet for at least 12 months. The sampling also ensured the greatest possible variety in terms of the products offered. Taking into account recommendations regarding the number of participants in an in-depth group interview [19], the sample size was set at 5 people (N = 5). Recruitment was done by the research company and the characteristics of participants of Study 2 are presented in Table 2.

3.3 Sampling of Study 3 – End Customers

The study made use of purposive sampling. The condition for taking part in the interview was making at least one online purchase per month during the 12 months preceding the survey. In addition, it was ensured that the study participants purchased products from the broadest possible spectrum of assortment groups. Efforts were also made to make the group as diverse as possible in terms of gender and age. Taking into account recommendations regarding the number of participants in an in-depth group interview [19], the sample size was set at 7 people (N = 7). Recruitment was done by the research company and the characteristics of participants of Study 3 are presented in Table 3.

Table 2. Characteristics of participants – Study 2

No.	Products offered	Position/ffunction	Age	Monthly orders
1	Brand clothing, mainly women's, sometimes also children's	Owner	47	800
2	A wide range of different products	Owner	63	4
3	Children's clothes	Owner	28	10
4	Pet products	Sales specialist	32	3000
5	Second-hand goods: clothing, furniture, ceramics	Owner	–	n/a

Source: Own compilation

Table 3. Characteristics of participants – Study 3

No.	Gender	Age	Most frequently purchased products (assortment groups)	Average number of purchases per month
1	Woman	58	Footwear, over-the-counter medicines, plants	4–5
2	Woman	44	Clothing, optical goods (glasses, lenses)	1–2
3	Woman	28	The respondent did not indicate the most frequently purchased products	2
4	Woman	25	Clothing, cosmetics, books, over-the-counter medicines	More than 3
5	Man	45	Computer products, sports products	2–3
6	Man	29	Clothing	2–4
7	Man	19	Cosmetics, perfume	4

Source: Own compilation

4 Results

Suppliers and complementors named 26 different entities that are involved in creating value for customers in e-commerce. The vast majority of them mentioned entities belonging to the group of complementors. The largest number (7) of these complementors were various types of IT service providers, including web hosting services, programmers and website developers. The second largest group of these entities were companies providing marketing services. The participants indicated 4 such entities: marketing agencies, search engine optimisation (SEO) services, copywriters and photographers. Other entities mentioned by the respondents which belonged to the group of complementors were companies providing administrative and accounting services as well as online payment service providers. Respondents also mentioned suppliers and end customers. However, the lists of entities indicated by the study participants differed significantly from one another. All the participants (7 out of 7) indicated numerous complementors, often forgetting about the other entities. End customers were indicated by 3 out of 7 respondents; suppliers (understood as suppliers of goods) were mentioned by 3 respondents; and online sellers were indicated by only 2 respondents (Table 4).

Table 4. Number of participants in studies 1–3 indicating the entities included in the Nalebuff and Brandenburger model

Groups of entities	Number of study participants (FGI) who at least once indicated an entity belonging to the following groups (according to the Nalebuff and Brandenburger model)			
	Suppliers	Complementors	Online sellers	End customers
Suppliers/Complementors (N = 7)	3/7	7/7	2/7	3/7
Online sellers (N = 5)	5/5	5/5	5/5	5/5
End customers (N = 7)	0/7	7/7	7/7	0/7

Source: Own compilation

The second group of respondents, online sellers, mentioned 14 different entities that take part in creating value for customers in e-commerce. This group of study participants indicated all groups of entities included in the Nalebuff and Brandenburger model, namely suppliers (including producers' suppliers, producers, wholesalers); complementors (e.g. banks, payment service providers, couriers, programmers, accounting services); online sellers; and end customers. The lists of entities indicated by the survey participants did not differ significantly between the respondents. All the participants indicated suppliers (5 out of 5 participants), complementors, online sellers and end customers (Table 4).

Finally, end customers indicated the fewest, only 10, entities participating in creating value for the buyer. All end customers indicated online sellers and complementors. However, none of them indicated suppliers (understood as the suppliers of goods to online sellers). Interestingly, the customers did not indicate themselves as participants in the value network (Table 4). All the participants indicated online stores and two entities included in the group of complementors: banks and companies responsible for delivering orders. Moreover, four participants indicated payment operators (e.g. PayU). The entities from the following list were indicated once: lending companies other than banks, companies providing hotline services for the seller, opinion-forming portals, advertising agencies, and Internet service providers.

5 Discussion

The aim of the research was to investigate the network awareness of the entities which, according to the Nalebuff and Brandenburg model, form the value network. The presented research takes into account the perspective of suppliers/complementors, online sellers and end customers. On the one hand, it can be said that the study participants were generally aware that value is created and delivered not only by online sellers but also by many other entities; such as product manufacturers, financial institutions, logistics companies, warehousing service providers and customers themselves. On the other hand, this awareness was found to differ significantly between the surveyed groups. These differences related to both the type of entities mentioned and to the horizon of network perception.

The largest number of entities was indicated by the suppliers and complementors participating in Study 1. They had no problems with identifying the entities involved in creating value for the customer. The most frequently indicated entities were those that could be defined as complementors; such as IT service providers, companies offering marketing services and logistics service providers. At the same time, respondents had considerable difficulty in discerning more distant entities, those not directly related to the area or industry of the respondents' business operations. In particular, it is surprising that only 3 out of 7 respondents indicated end customers, with whom they mainly have indirect relationships (e.g. web developers). Thus, it can be concluded that the network awareness of Study 1 participants is restricted mainly to entities of the same type as those represented by the study participants (suppliers and complementors), which shows that the network awareness of these entities is relatively limited. This phenomenon is referred to in the literature as network myopia [21] and understood as deficiencies in the perception of important elements of the environment and their mutual relationships [17]. It should also be noted that the images of incomplete networks presented by the surveyed suppliers/complementors overlapped to a very small extent. This indicates very high subjectivity and a limited cognitive community among this group of network participants. Based on the research conducted, it can be concluded that the limited network awareness of the suppliers and complementors participating in the study makes it difficult for them to perceive themselves as a group of entities that have a significant impact on creating value in e-commerce. This may be the reason, for example, for very negative assessments of courier companies belonging to the group of complementors: negative ratings in this case being given by both online sellers and end buyers.

Online sellers have considerably greater network awareness. This is probably due to their "central" position in the value network, being a link between production and purchase (individual buyers). All the participants of the study indicated all the entities forming the value network in the Nalebuff and Brandenburger model. Thus, this group cannot be said to be affected by network myopia. It should be said, however, that even though online sellers acknowledge various network participants, this recognition is rather superficial. Online sellers indicated 14 entities belonging to all the groups of network participants, in relation to the 26 indicated by suppliers and complementors belonging mainly to these two groups of companies. Very probably, this superficiality of perception is the source of a very similar image of the network presented by the online sellers surveyed. This shows, in contrast to suppliers/complementaries, a high level of cognitive community among this group of entities. In consequence, this community leads to a more objectified image of the value network.

The smallest number of entities, only 10, was selected by the participants of Study 3; i.e. end customers. Additionally, the vast majority of these respondents focused on only three of them. They restricted the number and type of entities mainly to those with whom they have direct contact during each transaction. Thus, it can be assumed that these respondents are not aware of the necessity of cooperation between a number of various entities connected by direct and indirect relationships in order to create value. In the case of this group of entities, however, it is difficult to speak of network myopia as this notion can be assigned to enterprises and not to individual buyers. Nevertheless, the limited awareness of the value network among end customers, restricted only to the entities closest to them, is an extremely important factor in the functioning of online sellers as it means that the consequences of any actions of complementors or producers that reduce the value provided are likely to be attributed to online sellers. A slow website, difficulty in making payments or a product of unsatisfactory quality will be blamed on the sellers instead of on the producers or complementors. This proves that there is high risk involved in online retail operations, the main goal of which is to provide the highest possible value for end customers.

6 Limitations and Future Research

Due to the characteristics of the qualitative research conducted by means of FGI, the presented results may suffer from certain limitations. First of all, the results should not be generalized; they can only be considered in relation to the studied entities. This is connected with the specificity of qualitative methods, in which attention is focused on discovering phenomena rather than on testing theoretical assumptions. This study can therefore be used as a source of information for the development of appropriate tools for future research that will be conducted using the quantitative method. Based on such research, it will be possible to draw conclusions relating to the studied populations.

Acknowledgment. The work presented in this paper has been supported by the National Science Centre, Poland [Narodowe Centrum Nauki], grant No. DEC-2015/19/B/HS4/02287.

References

1. Kawa, A.: Sieć wartości w handlu elektronicznym. Manage. Forum **5**(3), 7–12 (2017)
2. Nalebuff, B.J., Brandenburger, B.M.: Coopetition – kooperativ konkurrieren. Frankfurt am Main und New York (1996)
3. Lusch, R., Vargo, S., Tanniru, M.: Service, value networks and learning. J. Acad. Mark. Sci. **38**(1), 19–31 (2010)
4. McColl-Kennedy, J.R., Vargo, S.L., Dagger, T.S., Sweeney, J.C., van Kasteren, Y.: Health care customer value co-creation practice styles. J. Serv. Res. **15**(4), 370–389 (2012)
5. Doligalski, T.: Internet w zarządzaniu wartością klienta. Oficyna Wydawnicza-Szkoła Główna Handlowa, Warszawa (2013)

6. Agarwal, S., Teas, R.K.: Perceived value: mediating role of perceived risk. J. Mark. Theory Pract. **9**(4), 1–14 (2001)
7. Monroe, K.B.: Pricing – Making Profitable Decisions. McGraw-Hill, New York (1991)
8. Szymura-Tyc, M.: Marketing we współczesnych procesach tworzenia wartości dla klienta i przedsiębiorstwa. Wydawnictwo Akademii Ekonomicznej, Katowice (2005)
9. Światowiec-Szczepańska, J.: Tworzenie i zawłaszczanie wartości na rynku B2B. Handel wewnętrzny **4**(363), 313–324 (2016)
10. Ratajczak-Mrozek, M.: Sieci biznesowe a przewaga konkurencyjna przedsiębiorstw zaawansowanych technologii na rynkach zagranicznych. Wydawnictwo Uniwersytetu Ekonomicznego w Poznaniu, Poznań (2010)
11. Bruni, L., Sugden, R.: The road not taken: how psychology was removed from economics, and how it might be brought back. Econ. J. **117**, 146–173 (2007)
12. Thaler, R.: The psychology of choice and the assumptions of economics. In: Thaler, R. (ed.) Quasi Rational Economics, pp. 48–73. Russell Sage Foundation, New York (1991)
13. Kawa, A., Pierański, B.: Świadomość sieciowa we współpracy gospodarczej przedsiębiorstw - wyniki badań. Przegląd Organizacji **12**, 21–27 (2015)
14. Gerrig, R.J., Zimbardo, P.G.: Psychologia i życie. Wydawnictwo Naukowe PWN, Warszawa (2009)
15. Czakon, W.: Obrazy sieci w zarządzaniu strategicznym. ZN WSB w Poznaniu 19 (2007)
16. Abrahamsen, M.H., Henneberg, S.C., Huemer, L., Naude, P.: Network picturing: an action research study of strategizing in business network. Ind. Mark. Manage. **59**, 107–119 (2016)
17. Leek, S., Mason, K.: Network pictures: Building an holistic representation of a dyadic business-to-business relationship. Ind. Mark. Manage. **38**(6), 599–607 (2009)
18. Corsaro, D., Ramos, C., Henneberg, S.C., Naude, P.: Actor network pictures and networking activities in business networks: an experimental study. Ind. Mark. Manage. **40**(6), 919–932 (2011)
19. Olejnik, I., Kaczmarek, M., Springer, A.: Badania jakościowe. Metody i zastosowania. Wydawnictwo CeDeWu, Warszawa (2018)
20. Thomas, D.R.: A general inductive approach for analyzing qualitative evaluation data. Am. J. Eval. **27**(2), 237–246 (2016). https://doi.org/10.1177/109821400528374
21. Czakon, W., Kawa, A.: Network myopia: an empirical study of network perception. Ind. Market. Manage. **73**, 116–124 (2018). https://doi.org/10.1016/j.indmarman.2018.02.005

A Robust Approach to Employee Competences in Project Management

Jarosław Wikarek and Paweł Sitek(✉)

Kielce University of Technology, Kielce, Poland
{j.wikarek,sitek}@tu.kielce.pl

Abstract. In modern knowledge-based economy most of new undertakings are accomplished in the form of projects. The critical elements in project management used to be resources and time. Due to continuous technological development of automation, robotization and IT, the access to the resources ceases to be the key issue. Nevertheless, the employees with specific competences engaged in realization of a given project have become a critical resource. It is so mainly due to the fact that not all employee competences can be replaced by a machine or software. Additionally, employees can be temporarily unavailable, change the terms of employment, etc. Prior to starting any new project, the manager has to answer a few key questions concerning this resource, e.g.: Does the team accomplishing the project possess sufficient set of competences? If not, which and how many competences are missing? The answer to these questions can determine the success or failure of a project even prior to its start. The paper presents an employee competences configuration model which makes it possible to answer the above-mentioned questions and quickly verifies the feasibility of a given project in terms of the set of the team competences. Another possible way the proposed model might be used is to evaluate the robustness of the project realization taking into account the team competences.

Keywords: Project management · Employee competences · Decision support · Mathematical programming · Constraint logic programming

1 Introduction

A project approach is characteristic of the innovative modern knowledge-based economy. Introduction of new technologies into production and distribution, deployment of innovations, product launching, introduction of new services, etc., usually requires a project approach. Numerous methodologies of project management have been developed (PMBOK, Prince2, Scrum, CPM, PMI, Agile, etc.) [1, 2]. These methodologies involve, to a different extent and in different forms, planning and scheduling, allocation of resources, control and risk management, and more. Nevertheless, even if all the approaches take into account resources, they do not consider their traits, e.g. employee competences, despite the fact, that the key element of the success of any project is proper selection of the team of employees possessing competences suitable for its completion.

N. T. Nguyen et al. (Eds.): ACIIDS 2021, LNAI 12672, pp. 246–258, 2021.
https://doi.org/10.1007/978-3-030-73280-6_20

The paper presents a model of employee competences configuration in a project team which allows to answer the key questions associated with managing a team accomplishing a given project. The questions include, among others: (a) *Does the project team possess competences suitable to accomplish a project in accordance with a given schedule?* (b) *How many employees and with what competences are missing to accomplish the project?* (c) *What is the cost of supplementing the competences required to accomplish the project?* and many more.

In the process of the project development an assumption was made that the plan/schedule of project completion was defined and all sufficient resources necessary for its completion were available, with the exception of employees. The suggested model can also be used to estimate the robustness of the proposed way of project accomplishment to the absenteeism of employees with particular competences.

2 Illustrative Example

In order to bring closer the problem of configuring employee competences in the con-text of a robust project management an illustrative example *Sample_1* is presented which concerns implementation of the information system of MRP II (Manufacturing Resource Planning) class in a production company. The implementation is being accomplished in the form of a project consisting of stages/tasks (Table 1) with a defined order of accomplishing presented in Fig. 1. Figure 2 presents the schedule of accomplishing tasks in compliance with a graph shown in Fig. 1. The graph was obtained with the use of CPM (Critical Path Method) [3, 4]. In practice, the presented schedule (Fig. 2) and the way of project accomplishment (Fig. 1) are imposed or agreed between the parties of the project even prior to its start.

Table 1. Characteristics of particular tasks of the project for illustrative example *Sample_1*

Task	Name	Time	Max number of employees
T01	Approval of the budget and schedule of the project	4	1
T02	Analytical works on network infrastructure	5	1
T03	Analytical works on processes in company	4	1
T04	Analytical works on specification of hardware and auxiliary software	2	1
T05	Preparation of an implementation plan according to the assumed schedule and results of analytical works	3	1
T06	Installation and configuration of the auxiliary software, the databases and the network system.	3	2

(continued)

Table 1. (*continued*)

Task	Name	Time	Max number of employees
T07	Installation and configuration of the modules of the MRP II system	8	1
T08	Training of employees	6	1
T09	Data migration and start of a test system	4	1
T10	System testing, bug fixes, configuration corrections	3	1
T11	Configuration of reports and analyses	2	2
T12	System commissioning and production start	2	1

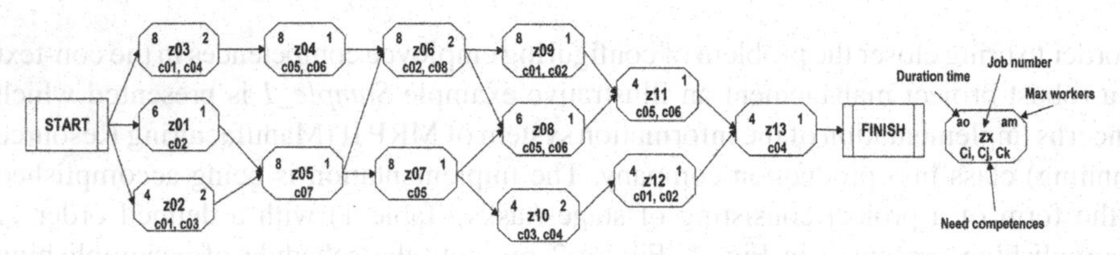

Fig. 1. Graph representing a network of activities for the illustrative example (ERP class system implementation project)

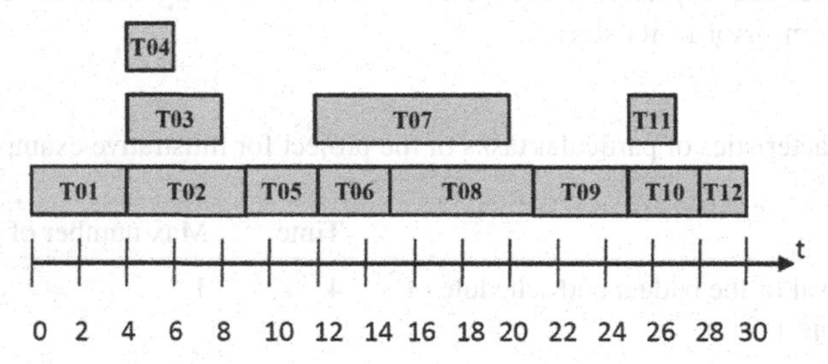

Fig. 2. Example of a schedule (Gantt chart) – for the illustrative example *Sample_1* without taking into account resources (competences)

Each task within the project *Sample_1* is characterized by a specific time of completion and requires a specific number of employees with specific competences (Table 2 - Competences). The competences of employees (Table 5) are presented in Table 6 - Competences of employees.

In order to estimate and emphasize the importance of the problem, it is assumed that other resources, except for employees with specific competences, are not critical, i.e. they are available in sufficient number and do not influence feasibility of the schedule in any way, contrary to the employees with specific competences, who can determine successful accomplishment of the project in accordance with a given schedule. In such

Table 2. Employee competences

C	Name
c01	Project manager
c02	Analyst
c03	Administrator of databases
c04	Computer networks and hardware specialist
c05	Implementer
c06	Instructor

context, it is important to define robustness of the project completion against absence of some employees or insufficient competences of the team accomplishing the project. For this reason the project manager prior to start of the project should know the answer to a number of questions relating to the team assigned to accomplish the project. The most important ones include the following:

- Q_1: Is the possessed team of employees with specific competences sufficient to accomplish a given project in compliance with a given schedule?
- Q_2: How many employees and with what competences are missing to accomplish a given project in compliance with a given schedule?
- Q_3A: What is the minimum cost of accomplishing the project in compliance with a given schedule by a given set of employees with specific competences?
- Q_3B: What is the minimum cost of accomplishing the project in compliance with a given schedule when supplementation of the set of employees and/or their competences is required?
- Q_4: Is accomplishment of the project in compliance with a given schedule possible if employee *e05* should not work with employee *e06*?
- Q_5: Is accomplishment of the project in compliance with a given schedule possible in absence of employee *e01/e02/…/e06*?
- Q_6: What configuration of the employee competences guarantees completion of the project in compliance with a given schedule in the absence of any employee?

To get the answers to the above questions automatically, a mathematical model of an employee competence configuration for projects was proposed (Sect. 3). The model is formulated as a set of constraints and a set of questions Q1…Q6. On the basis of this model, a method for verification of the set of employee competences was developed, which was implemented with the use of the GUROBI [5] and Eclipse [6] environments (Sect. 4).

3 Model of Employee Competence Configuration Problem

During development of the model of employee competence configuration problem the following assumptions were made:

(a) Each employee of the project team possesses specific competences;
(b) The schedule of the project completion is known prior to start of project accomplishment;
(c) In order to accomplish a task, an employee/employees with specific competences are required;
(d) An employee is assigned to a task throughout the whole time of task accomplishment;
(e) For particular tasks one or more employees with specific competences can be assigned;
(f) Employees are able to acquire some new competences, but not all of them;
(g) The cost of acquiring a competence by a given employee and the cost of work of a given employee are known;
(h) Working time of employees is limited.

The model was formulated as a (1)…(13) set of constraints and a Q1…Q6 set of questions. The proposed model can be classified as CSP (Constraint Satisfaction Problem) [7] or BIP (Binary Integer Programming) [8] depending on the question under consideration. For questions Q3A and Q3B, a discrete optimization model is obtained. For the

Table 3. Constraints of the employee competence configuration problem

Constraint	Description
(1)	The constraint guarantees that the employees with the appropriate competences are assigned to each task
(2)	The constraint guarantees that the employees concerned can only obtain the competences that are permitted for them
(3)	The constraint states that if the selected employee is assigned to a specific task, they must have the required competences
(4)	The constraint determines the cost of obtaining of certain competences by the employees
(5)	The constraint guarantees that a given employee works only during the permitted working hours
(6,7)	The constraints specify binding of the variables $Y_{e,z}$ and $X_{e,z,c}$
(8)	The constraint guarantees that overlapping/simultaneous tasks cannot be performed by the same employee
(9)	Two employees who should not work together are working on different tasks if possible
(10)	The constraint ensures that only the allowed number of employees perform the task
(11)	The constraint determines the cost of the employees' labor
(12)	The constraint determines the number of employees who have to work together, although they should not
(13)	Binarity constraint of the decision variables

Table 4. Indices, parameters and decision variables

Symbol	Description
Indices	
E	Set of employees
Z	Set of tasks
C	Set of competences
e	Employee index $(e \in E)$
z	Task index $(z \in Z)$
c	Competence index $(c \in C)$
Parameters	
$ro_{e,c}$	If the employee *e* has the competence *e* then $ro_{e,c}= 1$ otherwise $ro_{e,c}= 0$ $(e \in E, c \in C)$
$rw_{e,c}$	If the employee *e* can acquire the competence *c* then $rw_{e,c}= 1$ otherwise $rw_{e,c}= 0$ $(e \in E, c \in C)$
$rc_{e,c}$	Cost of acquisition by the employee *e* of the competence *c* $(e \in E, c \in C)$
$so_{z,c}$	If competence *c* is necessary for completion of task *z* then $so_{z,c}= 1$, otherwise $so_{z,c}= 0$
ao_z	Task *z* completion time $(z \in Z)$
am_z	Maximum number of employees that can be assigned to the task *z* $(z \in Z)$
bo_e	Maximum working time of the employee *e* $(e \in E)$
bc_e	Work cost of the employee *e* per unit of time
$fo_{z1,z2}$	If the schedule assumes the implementation of task *z1*, which coincides with the implementation of task *z2* then $fo_{z1,z2}= 1$, otherwise $fo_{z1,z2}= 0$ $(z1, z2 \in Z)$
$go_{e1,e2}$	If employee *e1* cannot work together with *e2* then $go_{e1,e2} = 0$, otherwise $go_{e1,e2} = 1$
st	Arbitrarily large constant
Decision variables	
$X_{e,z,c}$	If the employee *e* performs task *z* using the competence *c* then $X_{e,z,c} = 1$, otherwise $X_{e,z,c} = 0$ $(e \in E, z \in Z, c \in C)$
$U_{e,c}$	If the realization sets of tasks require that the employee *e* acquire the competence *c* then $U_{e,c} = 1$, otherwise $U_{e,c} = 0$ $(e \in E, c \in C)$
$Y_{e,z}$	If the employee *e* executes task *z* then $Y_{e,z} = 1$, otherwise $Y_{e,z} = 0$ $(e \in E, z \in Z)$
$W_{e1,e2}$	If employee *e1* has to work together with e2 then $W_{e1,e2} = 1$, otherwise $W_{e1,e2} = 0$
Cost1	The cost of changing the qualifications of employees
Cost2	Work cost of employees
Count1	Number of employees who have to work together, although they should not

remaining questions, a constraint satisfaction model is obtained, which enables the generation of feasible solutions. Table 4 shows the basic indices, parameters and decision variables of the model. The description of the constraints (1)...(13) of the problem under consideration is presented in Table 3.

$$\sum_{e \in E} X_{e,z,c} = so_{z,c} \forall z \in Z, c \in C \tag{1}$$

$$U_{e,c} \leq rw_{e,c} \forall e \in E, c \in C \tag{2}$$

$$X_{e,z,c} \leq U_{e,c} + ro_{e,c} \forall e \in E, z \in Z, c \in C \tag{3}$$

$$Cost1 = \sum_{e \in E} \sum_{c \in C} (U_{e,c} \cdot rc_{e,c}) \tag{4}$$

$$\sum_{z \in Z} Y_{e,z} \cdot ao_z \leq bo_e \forall e \in E \tag{5}$$

$$\sum_{c \in C} X_{e,z,c} \leq st \cdot Y_{e,z} \forall e \in E, z \in Z \tag{6}$$

$$\sum_{c \in C} X_{e,z,z} \geq Y_{e,z} \forall e \in E, z \in Z \tag{7}$$

$$Y_{e,z1} + Y_{e,z2} \leq 1 \forall z1,z2 \in Z \wedge fo_{z1,z2} = 1, e \in E \tag{8}$$

$$Y_{e1,z} + Y_{e2,z} \leq st \cdot (go_{e21,e2} + W_{e1,e2}) \forall e1,e2 \in E, z \in Z \tag{9}$$

$$\sum_{e \in E} Y_{e,z} \leq am_z \; z \in Z \tag{10}$$

$$Cost2 = \sum_{e \in E} \sum_{z \in Z} ao_z \cdot bc_e \cdot Y_{e,z} \tag{11}$$

$$Cost3 = \sum_{e1 \in E} \sum_{e2 \in E} W_{e1,e2} \tag{12}$$

$$X_{e,z,c} \in \{0,1\} \forall e \in E, z \in Z, c \in C$$
$$U_{e,c} \in \{0,1\} \forall e \in E, c \in C$$
$$Y_{e,z} \in \{0,1\} \forall e \in E, z \in Z$$
$$W_{e1,e2} \in \{0,1\} \forall e1.e2 \in E \tag{13}$$

Table 5. Employee parameters

e	Name	bo_e	bc_e
e01	Employer e01	15	90
e02	Employer e02	15	90
e03	Employer e03	15	70
e04	Employer e04	15	70
e05	Employer e05	15	50
e06	Employer e06	15	50
e07	Employer e07	15	60

Table 6. Structure of employee competences

	c01			c02			c03			c04			c05			c06		
	ro	rw	rc	ro	rw	rc	ro	rw	rc	ro	rw	rc	ro	rw	rc	ro	rw	rc
e01	1	0	0	0	1	800	0	1	900	0	1	900	0	1	700	0	1	700
e02	0	1	2000	1	0	0	0	1	900	0	1	900	1	0	0	0	1	700
e03	0	0	X	1	0	0	0	1	900	0	1	900	1	0	0	0	1	700
e04	0	0	X	0	1	800	0	1	900	0	1	900	1	0	0	0	1	400
e05	0	0	X	0	1	800	1	0	0	1	0	0	1	0	0	0	1	700
e06	0	0	X	0	X	800	0	1	900	1	0	900	1	0	0	0	1	700
e07	0	0	X	0	1	800	0	1	900	0	1	900	1	0	0	1	0	200

X- acquisition of competences impossible – undefined/unavailable cost

4 Method for Verifying the Set of Employee Competences

The model of the employee competence configuration presented in Sect. 3 became the basis for the development of the method for verification of such competences. The outline

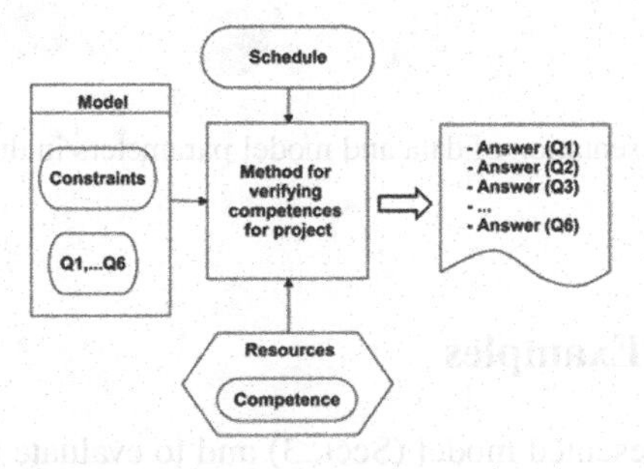

Fig. 3. Concept of the method for verifying the set of employee competences

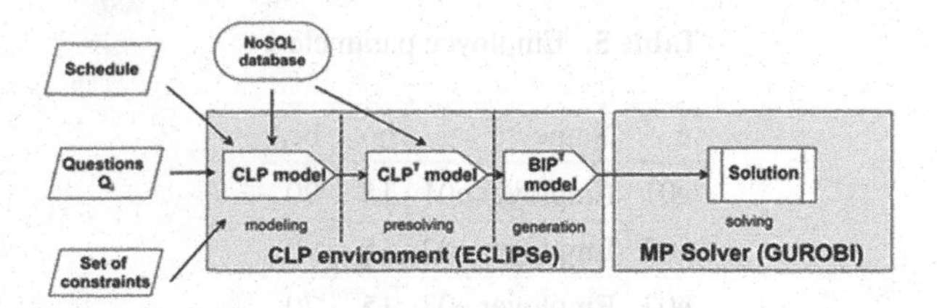

Fig. 4. Diagram of hybrid approach

of this method is shown in Fig. 3. Based on the constraints and questions of the model and the set of competences according to a specified schedule, the method allows for answers to set of questions.

The implementation of the proposed method was based on the author's hybrid approach [9, 10], the diagram of which is shown in Fig. 4.

Two environments were used to implement the method using the hybrid approach. To model the problem, its transformation (presolving) and the generation of the model transformed in the MP (Mathematical Programming) solver format, the ECL^{i}PS^e [6] environment was used. ECL^{i}PS^e is one of the CLP (Constraint Logic Programming) tools. CLP works very well for problems with a large number of constraints and discrete variables. The GUROBI environment was used to solve the transformed model. A detailed description of the transformation of the model that constitutes its presolving is presented in [10]. On the other hand, the manner of saving the model data, parameters and coefficients in the NoSQL database (Neo4j) [11] is shown in Fig. 5. The use of the Neo4j graph database for the implementation of the method facilitates its use with project management systems and implementation in the cloud.

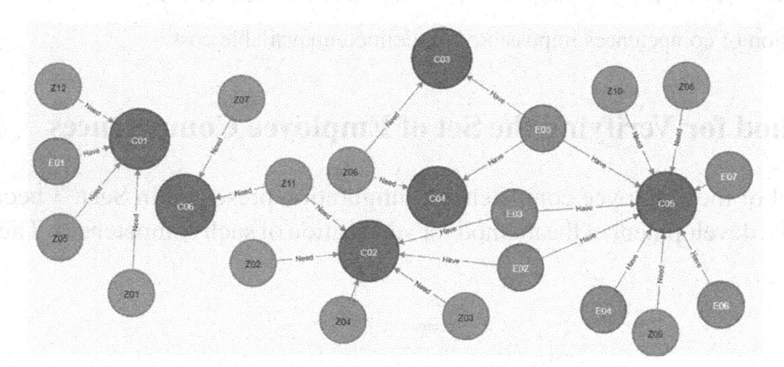

Fig. 5. Graphical representation of data and model parameters in the Neo4j database [11]

5 Computational Examples

In order to verify the presented model (Sect. 3) and to evaluate the hybrid approach to implementation of the method for verifying the set of employee competences (Fig. 3), a number of computational experiments were carried out. The computational experiments were carried out in two stages. The first stage involved finding answers to questions Q_1…Q_6 for an illustrative example Sample_1 (Sect. 2).

Instances of data shown in Tables 1, 2, 3 and 4 were used for these experiments. The results are presented in Table 7 as An(Q_1)…An (Q_6). Additionally, the obtained answers to the questions Q_2, Q_3B and the corresponding values of the decision variables have been illustrated in the form of Gantt charts (Fig. 6 and Fig. 7).

Table 7. Results for the illustrative example

Question	Answer - An(Q_i)
Q_1	No
Q_2	e04 needs c02; e06 needs c06 cost1 = 1500 cost2 = 3180
Q_3a	No
Q_3b	e05 needs c02; e06 needs c06 cost1 = 1500; cost2 = 2950
Q_4	Yes e03 needs c06; e07 needs c02; e07 needs c03; e07 needs c04 cost1 = 3300 cost2 = 3260
For proactive questions (Q_5, Q_6) it was assumed that e05 has c02 and e06 has c06 which ensures the minimum initial cost of the implementation of the schedule and its admissibility	
Q_5	e01 – e02 needs c01; cost1 = 2000 cost2 = 2950 e02 – e07 needs c02; cost1 = 800 cost2 = 2860 e03 – e07 needs c02; cost1 = 800 cost2 = 2900 e04 – cost2 = 2950 e05 – e07 needs c02; e07 need 3; e07 needs 4; cost1 = 2600 cost2 = 3130 e06 – e03 needs c06; cost1 = 700 cost2 = 3050 e07 – cost2 = 2096
Q_6	e02 needs c0; e05 needs c06; e07 needs c02; e07 need 3; e07 needs 4

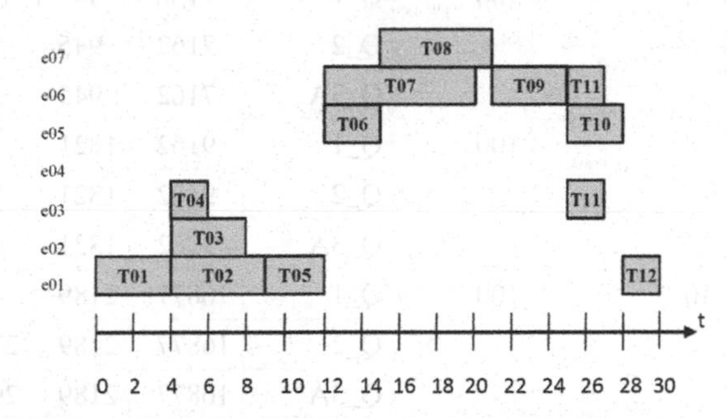

Fig. 6. Assignment of employees to tasks in accordance with the answer to question Q_2

The second stage included a series of computational experiments aimed at assessing the effectiveness of the hybrid approach to the implementation of the presented method (Fig. 3). The implementation was performed for many questions (Q_1, Q_2, Q_3B) and various data instances. Individual data instances (D1…D6) differed in the number of tasks, employees and competences. In order to evaluate the effectiveness of the proposed

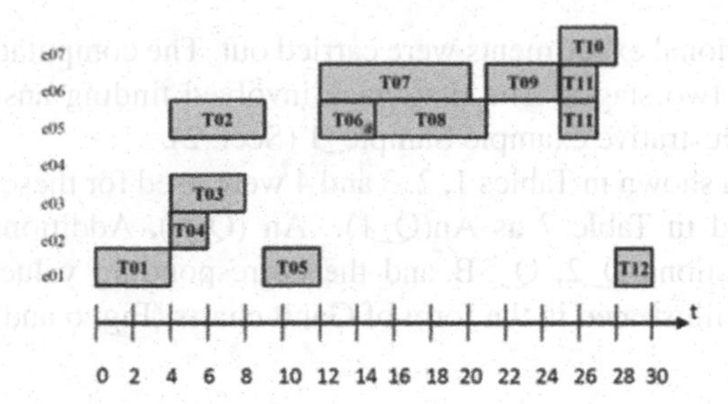

Fig. 7. Assignment of employees to tasks in accordance with the answer to question Q_3B

Table 8. Results for hybrid approach and mathematical programming

Dn	Number of employee	Number of competences	Number of Tasks	Question	Number of variables		Computation time	
					MP	Hyb	MP	Hyb
D1	7	6	12	Q_1	681	250	1	1
				Q_2	681	250	2	1
				Q_3A	681	250	2	1
D2	8	6	30	Q_1	1794	342	7	1
				Q_2	1794	342	21	3
				Q_3A	1794	342	25	4
D3	10	6	40	Q_1	2962	456	18	2
				Q_2	2962	456	54	11
				Q_3A	2962	456	51	10
D4	10	6	100	Q_1	7162	945	29	6
				Q_2	7162	945	345	35
				Q_3A	7162	945	389	48
D5	10	8	100	Q_1	9162	1321	87	12
				Q_2	9162	1321	594	51
				Q_3A	9162	1321	610	66
D6	15	10	100	Q_1	16877	2189	476	15
				Q_2	16877	2189	2789	137
				Q_3A	16877	2189	2934	167

hybrid approach in the implementation of the presented method, it was simultaneously implemented using also mathematical programming for the same data instances. The obtained results have been shown in Table 8. Analysis of the results shows that the hybrid approach enables a significant reduction in the number of decision variables of the model (up to 3 times), which results in a reduction of the computation time (up to 20 times).

6 Conclusions

The presented employee competence configuration model is a base in a method of verifying the competences of a team accomplishing a project. The method, apart from the model, also uses a given schedule of project completion. With this method the project manager even prior to start of the project is able to estimate whether the team possesses sufficient competences or whether it is possible to complete to project with a given team in compliance with a given schedule, etc.

In case of absence of employees with specific competences or lack of competences in case of sufficient number of employees, the suggested method will indicate what competences which employees should supplement and/or what other employees should be engaged in the project. Additionally, it is possible to con-figure employees competences in a project team in such a way that the project accomplishment is robust against the absence of any employee (question Q_6). It applies to frequent situations of temporary or permanent absence of an employee with specific competences which would not compromise completion of the project in accordance with the assumed schedule. That is a significant characteristics of the suggested method of competences verification (Fig. 3) as it allows for pro-active project management. The proposed implementation (Fig. 4) proved to be effective compared to classical model of mathematical programming. It is due to the fact that the hybrid approach allows for reduction of the number of constraints and variables of the modelled problem [9, 12]. In case of the implemented model the use of the hybrid approach made it possible to decrease the number of decision variables by three times compared to implementation of the model with the use of technics of mathematical programming. That decrease, in turn, results in the reduction of the calculation time by twenty times for the analyzed data instances (Table 8). Proposing a graph database NoSQL for recording the model data instances, parameters and coefficients allows for extremely flexible integration of the proposed method with existing project management systems.

Further research will focus on the development of the method through introduction of new questions, extension of the proactive approach including new situations, e.g. absence of any two employees, temporary absence of employees, etc. [13]. It is planned to apply the proposed method to other problems concerning production [14, 15], logistics [16, 17], transport [18], etc. For larger projects, a multi-agent approach is planned to be implemented [19, 20].

References

1. Hughes, B.(ed.), Ireland, R., West, B., Smith, N., Shepherd, D.I.: Project management for it-related projects. BCS, The Chartered Institute for IT, 2nd edn. (2012)
2. Kuster, J., et al.: Project Management Handbook, 1st edn. Springer, Berlin, Heidelberg (2015). https://doi.org/10.1007/978-3-662-45373-5
3. Punmia, B.C., Khandelwal, K.K.: Project Planning and Control with PERT & CPM. Firewall Media (2002)
4. Averous, J., Linares, T.: Advanced Scheduling Handbook for Project Managers. Fourth Revolution (2015)
5. Gurobi. http://www.gurobi.com/. Accessed 10 Oct 2019

6. Eclipse - The Eclipse Foundation open source community website. www.eclipse.org. Accessed on 19 Oct 2019
7. Rossi, F., Van Beek, P., Walsh, T.: Handbook of Constraint Programming, in Foundations of Artificial Intelligence. Elsevier Science Inc., New York, NY, USA (2006)
8. Schrijver, A.: Theory of Linear and Integer Programming. Wiley (1998). ISBN 0-471-98232-6
9. Sitek, P., Wikarek, J.: A multi-level approach to ubiquitous modeling and solving constraints in combinatorial optimization problems in production and distribution. Appl. Intell. **48**, 1344 (2018). https://doi.org/10.1007/s10489-017-1107-9
10. Sitek, P., Wikarek, J.: A hybrid programming framework for modeling and solving constraint satisfaction and optimization problems. Sci. Program. **2016**, 13 p. (2016). https://doi.org/10.1155/2016/5102616. Article ID 5102616
11. Neo4j Graph Platform. https://neo4j.com/. Accessed on 19 Oct 2020
12. Sitek, P., Wikarek, J.: Capacitated vehicle routing problem with pick-up and alternative delivery (CVRPPAD): model and implementation using hybrid approach. Ann. Oper. Res. **273**, 257 (2019). https://doi.org/10.1007/s10479-017-2722-x
13. Szwarc, E., Bocewicz, G., Banaszak, Z., Wikarek, J.: Competence allocation planning robust to unexpected staff absenteeism. Eksploatacja i Niezawodnosc - Maintenance Reliab. **21**, 440–450 (2019). https://doi.org/10.17531/ein.2019.3.10
14. Gola, A.: Reliability analysis of reconfigurable manufacturing system structures using computer simulation methods. Eksploatacja I Niezawodnosc – Maintenance Reliab. **21**(1), 90–102 (2019). https://doi.org/10.17531/ein.2019.1.11
15. Janardhanan, M.N., Li, Z., Bocewicz, G., Banaszak, Z., Nielsen, P.: Metaheuristic algorithms for balancing robotic assembly lines with sequence-dependent robot setup times. Appl. Math. Model. **65**, 256–270 (2019)
16. Dang, Q.V., Nielsen, I., Yun, W.Y.: Replenishment policies for empty containers in an inland multi-depot system. J. Maritime Econ. Logist. Macmillan Publishers Ltd. **15**(1), 1479–2931, 120–149 (2013)
17. Grzybowska, K., Kovács, G.: Developing agile supply chains - system model, algorithms, applications. In: Jezic, G., Kusek, M., Nguyen, N.T., Howlett, R.J., Jain, L.C. (eds.) Agent and Multi-Agent Systems. Technologies and Applications. KES-AMSTA 2012. Lecture Notes in Computer Science: Lecture Notes in Artificial Intelligence and Lecture Notes in Bioinformatics (LNAI), vol. 7327, pp. 576–585. Springer, Berlin, Heidelberg (2012). https://doi.org/10.1007/978-3-642-30947-2_62
18. Wirasinghe, S.C.: Modeling and optimization of transportation systems. J. Adv. Transp. **45**(4), 231–347 (2011)
19. Lasota, T., Telec, Z., Trawiński, B., Trawiński, K.: A multi-agent system to assist with real estate appraisals using bagging ensembles. In: Nguyen, N.T., Kowalczyk, R., Chen, S.M. (eds.) Computational Collective Intelligence. Semantic Web, Social Networks and Multiagent Systems. Lecture Notes in Computer Science (LNAI), vol. 5796, pp. 813–824. Springer, Berlin, Heidelberg (2009). https://doi.org/10.1007/978-3-642-04441-0_71
20. Rossi, F., Bandyopadhyay, S., Wolf, M., Pavone, M.: Review of multi-agent algorithms for collective behavior: a structural taxonomy. IFAC-PapersOnLine **51**(12), 112–117 (2018). https://doi.org/10.1016/j.ifacol.2018.07.097. ISSN 2405-8963

Key Aspects of Customer Intelligence in the Era of Massive Data

Nguyen Anh Khoa Dam[1,2](✉), Thang Le Dinh[1], and William Menvielle[1]

[1] Université du Québec à Trois-Rivières, Trois-Rivières, QC G8Z 4M3, Canada
{Nguyen.Anh.Khoa.Dam,Thang.Ledinh,William.Menvielle}@uqtr.ca
[2] The University of Danang, University of Science and Technology, Da Nang, Vietnam

Abstract. The era of massive data has changed the manner that customer intelligence is examined and applied to intelligent information systems. Customer intelligence is the ability to acquire knowledge and skills from massive data through customer analytics then apply them to the process of creating, communicating, delivering, and co-creating to offer value. Considering the vast nature of this research stream, the paper sheds light on investigating key aspects of customer intelligence with relevance to marketing solutions. The objective of the study is to conduct a critical literature review and develop a theoretical framework on customer intelligence in the context of massive data to support marketing decisions. The results of the paper indicate various applications of customer intelligence through the lens of the marketing mix. Accordingly, customer intelligence is applied to the aspects of the extended marketing mix (7P's), including Product/Service, Price, Promotion, Place, People, Process, and Physical evidence. The paper makes major contributions to the application of customer intelligence for intelligent information systems in supporting marketing decisions.

Keywords: Customer intelligence · Massive data · Literature review

1 Introduction

The era of massive data has created numerous myths around customer intelligence, particularly opportunities that customer intelligence can offer for enterprises [1, 2]. Accordingly, customer intelligence is believed to support intelligent information systems in developing products/services, understanding customer behaviors, and improving marketing strategies [3–5]. Customer intelligence is applied to different aspects associated with the decision-making process, including predicting customers' needs, segmentation, service innovation, promotional strategies, pricing strategies, and customer lifetime value estimation [6, 7]. With the support of customer intelligence, intelligent information systems can provide tailored and smart products/services (for example: recommender systems) and optimize marketing decisions [5, 8]. Enterprises, therefore, can improve their marketing and financial performance whereas customers can increase their satisfaction and experience with products/services [9, 10].

Regardless of the attention on this research stream, most enterprises are not clear on the notion of customer intelligence in the context of massive data as it lies at the

N. T. Nguyen et al. (Eds.): ACIIDS 2021, LNAI 12672, pp. 259–271, 2021.
https://doi.org/10.1007/978-3-030-73280-6_21

junction of massive data and customer insights [10, 11]. In fact, the majority of definitions of customer intelligence has been outdated due to the massive data revolution [6, 7]. It is noted that enterprises are obsessed to take advantage of customer intelligence; nevertheless, they lose track due to this vast research domain [2, 12]. Bearing in mind the diverse sources of massive data, the transformation of customer data from various sources to customer intelligence is not a trivial task [11]. This prompts the motivation to study key aspects of customer intelligence in the age of massive data and develop a framework to manage and leverage its value.

Building on these reflections, the primary research objective of this paper is to conduct a critical literature review on customer intelligence with relevance to massive data and to identify key aspects for supporting marketing decisions. The secondary objective is to develop a theoretical framework for customer intelligence management. To respond to the research objective, the following research questions (RQ) are explored:

RQ 1: *How does literature conceptually approach customer intelligence in the era of massive data?*
RQ 2: *What are the key aspects of customer intelligence concerning value creation for marketing solutions?*

To carry out this research, the paper adopted the exploratory research approach, which is defined as the process of investigating a problem that has not yet been studied or thoroughly investigated [13]. The paper uses the literature as existing resources on the subject under study. Consequently, the remaining structure of the paper continues with the literature review which clarifies the research process and the revolution of customer intelligence. Then the resource-based theory is examined to set the foundation for the theoretical framework. Through the framework, significant aspects of customer intelligence are discussed. The last section of the paper indicates an in-depth discussion of theoretical and practical contributions.

2 Literature Review

2.1 Research Process

To ensure a comprehensive review, the paper follows the systematic approach by Webster, Watson [14]. Searches are conducted in databases, including Scopus, Science Direct, Emerald Insight, SpringerLink, and ProQuest [7, 15]. Different keywords such as "customer intelligence", "customer knowledge", "customer information", "customer insight", and "customer intimacy" are used to look for articles from these databases. As a synonym for "customer", the noun "consumer" is also used to combine with other words such as intelligence, knowledge, and insights to identify relevant articles.

Subsequently, inclusion and exclusion criteria are applied to filter out relevant articles. Assuming the majority of research on customer intelligence is written in English, only publications in this language are chosen. The time span from 2000 to 2020 is applied to filter out articles as it is broad enough to view changes in this research stream. Articles that are not peer-reviewed are excluded. Finally, the selected literature is also chosen with relevance to massive data.

Through the literature, the overview of customer intelligence over the past 20 years is clarified. With the purpose to synthesize and extend existing studies, the paper continues by proposing a theoretical framework for customer intelligence. A theoretical framework is based on theories to develop factors, constructs, variables, and their relationships [14]. In this paper, the theoretical framework is derived from the literature review and is based on the resource-based view [16]. The following part of the paper presents the revolution of customer intelligence.

2.2 The Revolution of Customer Intelligence

This section illustrates the revolution of customer intelligence over the past 20 years. Table 1 summarizes relevant studies of customer intelligence based on three dimensions: management, organization, and technology [10, 15, 17]. The management dimension focuses on the applications of customer intelligence. The organizational dimension touches upon changes in organizational structure, culture, and business process to support customer intelligence. The technological dimension spotlights information technology and analytics to transform data into customer intelligence.

Table 1. A comparison of definitions of customer intelligence.

Article	Organization	Management	Technology
Rygielski et al. [18]		Customer targeting Product sales	CRM
Chen, Popovich [17]	Business process	Understand customer behaviors	CRM, ERP
Keskin [19]	Organizational learning	Innovation	
Lewrick et al. [20]		Product innovation and development	
Guarda et al. [21]		Customer service	Data system
Singh, Verma [22]	Business Process	Customer experience Customer behaviors	
Samuel [23]			Social media data
Gobble [24]		Customer relationships Decision-making	
Rakthin et al. [25]		Customer needs Buying decision model	
Stone et al. [26]	Human capital Business process		Social media data Massive data
López-Robles et al. [27]		Decision making	Massive data
Yan et al. [5]		Recommendations	Internet of things Massive data

Literature hardly recognizes an official definition of customer intelligence, especially a definition that can comprehensively cover the three dimensions of management, organization, and technology [5, 24, 25]. The majority of research on customer intelligence concentrates the management dimension with various streams, including customer target [18], innovation [19, 20], customer service [21], customer experience [22], customer behaviors [22, 25], customer relationships [24], decision-making [27], and recommendations [5]. Table 1 reveals that the majority of definitions of customer intelligence focus on the organizational dimension. The studies by Keskin [19] and Singh and Verma [22] made a difference by highlighting the dimensions of organizational learning and the business process of customer intelligence. On the other side, the technological dimension has witnessed the revolution from traditional Enterprise Resource Planning (ERP), Customer Relationship Management (CRM) to massive data platforms [23, 26]. Customer intelligence is moving to a higher-level due to the support of data mining techniques for the collaborative decision-making process [24, 27]. Therefore, customer intelligence is capable of enhancing personalized customer experience through analytics and the excavation of massive data [5].

As the majority of definitions on customer intelligence is outdated in the context of the massive data revolution [8, 18, 24], there is a need for an updated definition to adapt to these changes. In fact, the era of massive data has reshaped the definition of customer intelligence [9, 10]. This study relies on the definition of intelligence from the Oxford dictionary as the "ability to acquire and apply knowledge and skills" [28]. To amplify the value of customer intelligence for marketing solutions, the authors also examine the role of marketing in "creating, communicating, delivering, and exchanging offerings" [29]. According to the American Association of Marketing (AMA), customers are "actual or prospective purchasers of products or services" [30]. In the era of massive data, the proliferation of customer-generated data on digital platforms along with analytic techniques has changed the way that marketing to acquire and apply knowledge and skills from customers [9, 10]. In accordance with this view, contemporary marketing relies on analytic techniques, including descriptive, predictive, prescriptive to transform data into intelligence [12, 15]. Building on these reflections, this paper reshapes the definition of customer intelligence in the age of massive data as follows:

Customer intelligence is the ability to acquire knowledge and skills from massive data through customer analytics then apply them to the process of creating, communicating, delivering, and co-creating to offer value.

The next part of the paper continues with the examination of the resource-based theory [16] to set the foundation for the development of the framework for customer intelligence management.

3 Theoretical Framework

3.1 Resource-Based Theory

This section of the paper addresses the resource-based view [16] which serves as an anchor for the development of the theoretical framework. The study relies on the resource-based theory [16] as this theory is well adopted by researchers in explaining the key to success for enterprises [9, 25, 31]. The resource-based theory supports customer intelligence due to its emphasis on resources, capabilities, and particularly knowledge as the key indicator to create competitive advantages for enterprises [5, 31].

The wide variety of resources is divided into four types: physical, human, organizational, and network resources. In the context of massive data, physical resources include equipment, software, and technologies for the collection, storage, and analysis of customer data [15, 32]. Human resources reflect the capability of data scientists or managers in interpreting and applying customer intelligence for value creation [5, 33]. Organizational resources deal with the organizational structure/hierarchy, culture, and business process to facilitate customer intelligence [10]. Network resources emphasize the role of customers in co-creating values with enterprises.

3.2 A Resource-Based Framework of Customer Intelligence Management

Set the foundation upon the resource-based theory [16], this section presents the framework of customer intelligence management, which is adapted from different studies in the literature [31, 34]. According to the resource-based theory [16], enterprises can take advantage of physical, human, organizational, and network resources to facilitate customer intelligence. In terms of physical resources (a), enterprises can provide digital engagement platforms to interact with customers for massive data [35, 36]. Concerning organizational resources (b), enterprises can adjust organizational culture, structure, communication, and policy to stimulate the transformation for customer intelligence [3, 37]. Regarding human resources (c), managers can leverage values of customer intelligence through various applications, especially marketing solutions [10, 38]. The subsection presents other components of the framework as illustrated in Fig. 1.

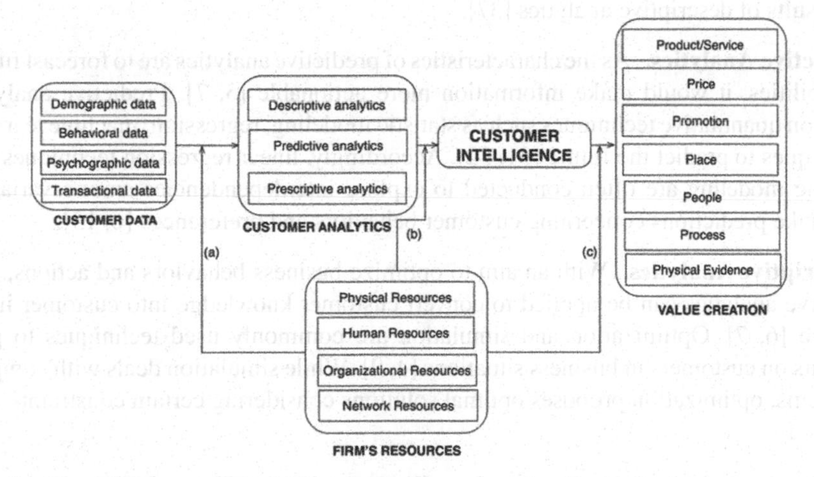

Fig. 1. A resource-based framework of customer intelligence management.

Customer Data. As illustrated in Fig. 1, customers create a significant amount of data, which is also known as massive data, through interactions with enterprises and other customers on digital platforms [39]. Massive data are extremely large data sets in volume, velocity, variety, and veracity, and are considered a great source for customer intelligence [34, 37]. Through the lens of customer intelligence, customer data can be divided into demographic, behavioral, and psychographic data. Demographic data reveal *Who* customers are [7]. Demographic data contain data on age, gender, profession, location, income, and marital status [7, 31]. Primary sources of demographic data are the U.S Census Bureau, CRM systems, and social media. On the other hand, behavioral data show *How* customers interact with enterprises and products [8, 40]. Customer interactions on websites and mobile devices - particularly clickstream data, add-to-favorites, add-to-cart data can disclose customer behaviors [15, 34]. Behavioral data can be exploited via various sources such as websites, social media, and mobile devices. Finally, transactional data unveil *What* customers buy [4, 9]. Various types of transaction data are historical purchases, invoices, payments [31, 34]. Transactional data can be found in transaction records, sales reports, billing records, CRM systems [6, 15].

Customer Analytics. Customer analytics focuses on the techniques which transform customer data into customer intelligence. This part clarifies customer analytics including descriptive, predictive, and prescriptive [7, 9].

Descriptive Analytics. Descriptive analytics is ideal to explore historical data and transform them into information. Common techniques of descriptive analytics are business reporting, descriptive statistics, and visualization [6, 41]. Business reporting involves the process of generating standard reports, ad hoc reports, query/drill down, and alerts [6, 15]. Different descriptive statistical methods such as association, clustering, regression, decision trees, etc. are applied to analyze trends of customers from historical data [9, 31]. Visual presentation of customer intelligence is frequently used to better communicate the results of descriptive analytics [37].

Predictive Analytics. As the characteristics of predictive analytics are to forecast future possibilities, it would make information more actionable [5, 7]. Predictive analytics relies on quantitative techniques such as statistic modeling, regression, machine learning techniques to predict the future [31, 32]. Accordingly, linear regression techniques and statistic modeling are often conducted to explore interdependencies among variables and make predictions concerning customer behaviors and preferences [6, 15].

Prescriptive Analytics. With an aim to optimize business behaviors and actions, prescriptive analytics can be applied to convert customer knowledge into customer intelligence [6, 7]. Optimization and simulation are commonly used techniques to gain insights on customers in business situations [4, 9]. While simulation deals with complex problems, optimization proposes optimal solutions considering certain constraints.

4 Key Aspects of Customer Intelligence in the Era of Massive Data

To leverage the value creation of customer intelligence, the concept of the marketing mix is integrated to stipulate key aspects relevant to customer intelligence. Approaching customer intelligence from the marketing mix approach helps enterprises identify significant aspects that support marketing decisions [12, 34]. The part of this paper presents how customer intelligence is applied through the lens of the marketing mix. The marketing mix consists of the 7 Ps: product, price, promotion, place, people, process, and physical evidence [34, 42]. The functionality of each P is specified as follows.

i) ***Product/Service*** includes features and characteristics that a product or service can satisfy the need of customers. The Product aspect responds to *what to offer customers.*
ii) ***Price*** indicates the amount that customers are willing to pay for a product/service [42]. Gaining insights on *What value customers offer* helps enterprises make better decisions and adjust their marketing strategies.
iii) ***Promotion*** consists of advertising, public relations, direct marketing, and sales promotion. Facilitating *How to interact with customers interact* reinforces the significance of customer intelligence for marketing solutions.
iv) ***Place*** refers to physical or virtual locations/channels that customers can access products/services. The aspect of Place in the marketing mix responds to the question *"Where are customers?"* by clarifying distribution channels.
v) ***People*** are human actors, including employees, leaders, managers of enterprises. The People aspect emphasizes the role of staffs in service delivery. Consequently, it involves *How staffs support customers.*
vi) ***Process*** focuses on *How customers experience* products and services. This paper gives prominence to customer intelligence to improve their experience.
vii) ***Physical evidence*** is defined as the appearance of the environment in which products and services are delivered. Physical evidence can be an ambiance, a spatial layout, or surroundings. This P aspect replies to the question *What is the physical environment for customers*?

4.1 Product/Service Aspect

From the Product/Service aspect, it is significant to apply customer intelligence to determine *What to offer* to customers to satisfy their needs. Products and services are considered solutions to deliver value for customers [34]. Therefore, enterprises are under pressure of development and innovation of products and services [35, 43].

Customer Product/Service Development. The application of customer intelligence in service and product development would yield profitable outcomes [11, 31]. It keeps track of changes in markets and customer preferences as it is extracted from customer interactions on digital platforms [25, 36]. Nowadays, customers increasingly express their voices on social media, which creates a significant amount of real-time user-generated data for product development and customer services [22, 34].

Customer Product/Service Innovation. Customers can involve as co-ideators for product/service innovation [35, 43]. In fact, customers with product/service interests and passions often participate in the process of idea generation for the product/service conceptualization and improvement [44, 45]. The interaction between an enterprise and customers facilitates improvement for existing products/services and gradually stimulates ideas for product/service innovation [8].

4.2 Price Aspect

In the context of customer intelligence, the Price aspect indicates the value of customers. Gaining insights on *What customers offer* helps enterprises make better decisions and adjust their marketing strategies. Correspondingly, two significant concepts of the Price aspect are customer values and customer lifetime value.

Customer Values. Customers can offer various values for enterprises. Customer values can be categorized into three types: i) Economic value – the measure of profits), ii) Social value – how customers influence other customers, and iii) Cognitive value – value gained from customers' knowledge and experience [36, 44]. In the era of massive data, knowledge and skills from customers are considered more significant as they are a great source for customer intelligence.

Customer Lifetime Value. Customer lifetime value is a prediction of the total monetary value that customers are expected to spend for an enterprise during their lifetime [46]. With the support of customer intelligence, service providers will have sufficient data and information to estimate customer lifetime value [5, 46]. Therefore, they can fine-tune their marketing strategies, particularly strategies related to segmentation, targeting, and positioning, for optimal outcomes [4, 5, 18].

4.3 Promotion Aspect

Through the lens of customer intelligence, the Promotion aspect highlights the importance of interactions with customers. Facilitating *How customers interact* reinforces the significance of customer intelligence for insights on three relevant concepts: customer behaviors, customer recommender systems, and customer co-creation.

Customer Behaviors. Integrated from diverse sources of data, customer intelligence can reveal attributes of customer behaviors along with changes in their preferences [4]. Understanding customer behaviors is significant in responding to customer needs with relevant promotional strategies and tactics [3, 4]. Customer intelligence enables the analysis of recency, frequency, and monetary of purchases, which can be applied to comprehend customer behaviors and improve direct marketing strategies [8].

Customer Recommender Systems. The application of customer intelligence in recommender systems for recommendation service promotes the optimization of customer experience from finding to engaging with products/services [5, 47]. Recommender systems take advantage of data related to products and customers to predict and recommend the most relevant services or products. It is noted that customer intelligence is applied as an important part of the development of recommender systems [8].

Customer Co-creation. Customer co-creation is described as the joint creation of value by the service providers and customers through the mutual application of operant resources [44, 45]. From the standpoint of service providers, customer value co-creation provides a significant source of customer data for product and service innovation [8]. From the perspective of customers, customer value co-creation improves customer experiences and knowledge [39].

4.4 Place Aspect

The Place aspect responds to the question *"Where are customers?"*. Customer intelligence supports enterprises in two key concepts, including customer identification and segmentation, customer profiling.

Customer Identification and Segmentation. Customer identification focuses on how to identify the most profitable customers [34, 41]. In fact, it is challenging to identify the most lucrative customer segments due to the huge volume and variety of massive data. Customer intelligence is applied to identify customer segments with similar interests and profitability [7, 46]. Various demographic, psychographic, behavioral, or geographic criteria are used for segmentation [7, 18]. Customer segmentation divides customers into homogenous segments and builds customer profiles [34, 41].

Customer Profiling. Customer profiles contain information on demography, buying behaviors, purchasing attributes, product category, and estimated customer lifetime values [7, 9]. Due to customer intelligence, enterprises can implement different data mining techniques related to target customer analysis to choose the most profitable segment [34, 46]. Enterprises rely on customer profiling to develop marketing campaigns and maintain a long-term relationship with customers [11, 48].

4.5 People Aspect

The People aspect of the marketing mix deals with *how staffs support customers*. This subsection puts forward to leadership and culture with customer orientation.

Customer-Oriented Culture. A customer-oriented culture clarifies the shared purpose and values among employees within an enterprise [38, 49]. Specifically, directors, managers, and subordinates are aligned with a customer-oriented mindset to be aware of the visions and missions of enterprises [3, 38]. A customer-oriented culture can also clarify communication across departments for data-driven decisions [10, 38].

Customer-Oriented Leadership. To instill a customer-oriented culture, enterprises establish a clear vision and leadership among their employees [10, 38]. Customer-oriented leaders are in charge of connecting and inspiring employees across an enterprise to engage with the decision-making process based on customer intelligence [11, 50]. Leadership with customer orientation involves transforming people so that they have the capacity to shape customer intelligence projects [10, 31].

4.6 Process Aspect

The Process aspect reflects *how customers experience*. Customer intelligence clarifies customer journeys from pre-purchase to post-purchase. Furthermore, customer intelligence elucidates customer experience in each journey to facilitate engagement.

Customer Journeys. Customer intelligence can be applied to design and optimize customer journeys, which consist of *recognizing* customer needs, *requesting* a product or service that might meet their needs, and *responding* to the delivery of service and products [50]. In other words, the three stages represent the pre-purchase, purchase, and post-purchase of a customer journey [40]. Accordingly, service providers attempt to map customer journeys with key touchpoints in each stage [51, 52].

Customer Experience. Customer experience catches the attention of researchers and practitioners as it reflects the extent of how customers engage with products [9, 43]. Customer intelligence is capable of enhancing personalized customer experience through analytics and excavation of massive data [5]. In particular, analyzing customer-generated contextual data is significant to manage customer experience and to reinvent customer journeys from pre-purchases to post-purchases [40, 48]. Furthermore, integrating and interpreting different sources of customer data help enterprises identify and prioritize key customer journeys to optimize customer experience [10].

4.7 Physical Evidence Aspect

Finally, the Physical Evidence aspect acts upon *what the physical environment for customers is*. In light of customer intelligence, the Physical Evidence aspect involves the design and layout of an enterprise's website or online store.

Customer-Friendly Interface. The design and layout of websites, media pages, or online stores need to be simple and straightforward so that customers can easily navigate and find relevant information [26, 37]. This would encourage customers to interact with enterprises on digital platforms [39, 45]. In an online setting, the physical evidence is characterized by customer's subjective evaluations and the level of technology acceptance [36]. Enterprises should add more rigors on customer-friendly interfaces by considering the level of emotional engagement of machines, and the level of technology acceptance by users [8, 15].

5 Conclusion

The proliferation of massive data has changed the definition of customer intelligence [5, 9]. This study aims at carrying out an exploratory research to address those changes and the need to redefine customer intelligence in the age of massive data. Set the solid foundation on the resource-based theory, the paper provides a critical literature review on significant aspects of customer intelligence in the context of massive data. The authors thoroughly examined those aspects through the lens of the marketing mix. Each aspect of Ps reveals interesting findings associated with customer intelligence.

In terms of theoretical contributions, a comprehensive review of customer intelligence over the past 20 years was conducted to reveal the revolution in this research stream. Accordingly, the paper proposed a revised definition of customer intelligence with modifications from massive data. The paper has also enriched the literature by exploring different aspects of customer intelligence in light of the marketing mix in developing intelligent information systems. Each aspect can be served as a research direction for future scholars. Finally, the resource-based framework of customer intelligence management has made a significant theoretical contribution by bridging the gap between information systems (particularly, customer intelligence) and marketing. The framework is promised to be a source of reference for both practitioners and researchers to further their studies.

Regarding practical contributions, the resource-based framework of customer intelligence management holds important implications for enterprises. Enterprises can rely on the framework to identify relevant physical, human, and organizational resources in order to take advantage of customer intelligence in the era of massive data. Furthermore, key aspects of customer intelligence through the lens of the marketing mix can support intelligent information systems with marketing decisions related to product/service, price, promotion, place, people, process, and physical evidence [12, 34]. These aspects will enable intelligent information systems to clarify what to offer customers, what values customers can offer, how customers interact, where customers are, how staffs support customers, how customers experience, and what the physical environment for customers is. Considering enormous applications of customer intelligence for intelligent information systems, these key aspects would assist enterprises to stay on track in reaching marketing-oriented goals.

References

1. Gandomi, A., Haider, M.: Beyond the hype: big data concepts, methods, and analytics. Int. J. Inf. Manage. **35**(2), 137–144 (2015)
2. Jagadish, H.V.: Big data and science: Myths and reality. Big Data Res. **2**(2), 49–52 (2015)
3. Zerbino, P., Aloini, D., Dulmin, R., Mininno, V.: Big data-enabled customer relationship management: a holistic approach. Inf. Process. Manage. **54**(5), 818–846 (2018)
4. Anshari, M., Almunawar, M.N., Lim, S.A., Al-Mudimigh, A.: Customer relationship management and big data enabled: personalization & customization of services. Appl. Comput. Inf. **15**(2), 94–101 (2019)
5. Yan, Y., Huang, C., Wang, Q., Hu, B.: Data mining of customer choice behavior in internet of things within relationship network. Int. J. Inf. Manage. **50**, 566–574 (2020)
6. Sivarajah, U., Kamal, M.M., Irani, Z., Weerakkody, V.: Critical analysis of big data challenges and analytical methods. J. Bus. Res. **70**, 263–286 (2017). https://doi.org/10.1016/j.jbusres.2016.08.001
7. France, S.L., Ghose, S.: Marketing analytics: methods, practice, implementation, and links to other fields. Exp. Syst. Appl. **119**, 456–475 (2018). https://doi.org/10.1016/j.eswa.2018.11.002
8. Dam, N.A.K., Le Dinh, T., Menvielle, W.: A service-based model for customer intelligence in the age of big data (2020)
9. Holmlund, M., et al.: Customer experience management in the age of big data analytics: a strategic framework. J. Bus. Res. (2020)
10. Tabrizi, B., Lam, E., Girard, K., Irvin, V.: Digital transformation is not about technology. Harvard Bus. Rev. (2019)
11. Davenport, T.H., Spanyi, A.: Digital transformation should start with customers. MIT Sloan Manage. Rev. (2019)

12. Lau, R.Y.K., Zhao, J.L., Chen, G., Guo, X.: Big data commerce. Inf. Manage. **53**(8), 929–933 (2016). https://doi.org/10.1016/j.im.2016.07.008
13. Stebbins, R.A.: Exploratory Research in the Social Sciences, vol. 48. Sage (2001)
14. Webster, J., Watson, R.T.: Analyzing the past to prepare for the future: writing a literature review. MIS Q. **26**(2), xiii–xxiii (2002)
15. Chen, H., Chiang, R.H.L., Storey, V.C.: Business intelligence and analytics: from big data to big impact. MIS Q. **36**, 1165–1188 (2012)
16. Barney, J.: Firm resources and sustained competitive advantage. J. Manage. **17**(1), 99–120 (1991)
17. Chen, I.J., Popovich, K.: Understanding customer relationship management (CRM) People, process and technology. Bus Process Manage. J. **9**(5), 672–688 (2003)
18. Rygielski, C., Wang, J.-C., Yen, D.C.: Data mining techniques for customer relationship management. Technol. Soc. **24**(4), 483–502 (2002)
19. Keskin, H.: Market orientation, learning orientation, and innovation capabilities in SMEs. Eur. J. Innov. Manage. **9**(4), 396–417 (2006)
20. Lewrick, M., Omar, M., Williams Jr., R.L.: Market orientation and innovators' success: an exploration of the influence of customer and competitor orientation. J. Technol. Manage. Innov. **6**(3), 48–62 (2011)
21. Guarda, T., Santos, M.F., Pinto, F., Silva, C., Lourenço, J.: A conceptual framework for marketing intelligence. Int. J. e-Educ. e-Bus. e-Manage. e-Learn. **2**(6), 455 (2012)
22. Singh, P., Verma, G.: Mystery shopping: measurement tool for customer intelligence management. IOSR J. Bus. Manage. Sci. **16**(6), 101–104 (2014)
23. Samuel, A.: Data is the next big thing in content marketing. Harvard Bus. Rev. (2015)
24. Gobble, M.M.: From customer intelligence to customer intimacy. Res. Technol. Manage. **58**(6), 56–61 (2015)
25. Rakthin, S., Calantone, R.J., Wang, J.F.: Managing market intelligence: the comparative role of absorptive capacity and market orientation. J. Bus. Res. **69**(12), 5569–5577 (2016)
26. Stone, M., et al.: How platforms are transforming customer information management. Bottom Line **30**(3) (2017)
27. López-Robles, J.-R., Otegi-Olaso, J.-R., Gómez, I.P., Cobo, M.-J.: 30 years of intelligence models in management and business: a bibliometric review. Int. J. Inf. Manage. **48**, 22–38 (2019)
28. Oxford Dictionary: Intelligence. In: Oxford Dictionary (2014)
29. Rownd, M., Heath, C.: The American marketing association releases new definition for marketing. AMA, Chicago, Il (2008)
30. Association, A.M.: Customer definition. In: AMA Dictionary. (2015)
31. Erevelles, S., Fukawa, N., Swayne, L.: Big data consumer analytics and the transformation of marketing. J. Bus. Res. **69**(2), 897–904 (2016)
32. Baesens, B., Bapna, R., Marsden, J.R., Vanthienen, J., Zhao, J.L.: Transformational issues of big data and analytics in networked business. MIS Q. **40**(4) (2016)
33. LaValle, S., Lesser, E., Shockley, R., Hopkins, M.S., Kruschwitz, N.: Big data, analytics and the path from insights to value. MIT Sloan Manage. Rev. **52**(2), 21 (2011)
34. Fan, S., Lau, R.Y.K., Zhao, J.L.: Demystifying big data analytics for business intelligence through the lens of marketing mix. Big Data Res. **2**(1), 28–32 (2015). https://doi.org/10.1016/j.bdr.2015.02.006
35. Frow, P., Nenonen, S., Payne, A., Storbacka, K.: Managing co-creation design: a strategic approach to innovation. Br. J. Manage. **26**(3), 463–483 (2015)
36. Verleye, K.: The co-creation experience from the customer perspective: its measurement and determinants. J. Serv. Manage. **26**(2), 321–342 (2015)
37. Rao, T.R., Mitra, P., Bhatt, R., Goswami, A.: The big data system, components, tools, and technologies: a survey. Knowl. Inf. Syst. Manage. 1–81 (2018)

38. Yohn, D.L.: 6 ways to build a customer-centric culture. Havard Bus. Rev. (2018)
39. Ramaswamy, V., Ozcan, K.: Digitalized interactive platforms: turning goods and services into retail co-creation experiences. NIM Mark. Intell. Rev. **11**(1), 18–23 (2019)
40. Rawson, A., Duncan, E., Jones, C.: The truth about customer experience. Harvard Bus. Rev. (2013)
41. Amado, A., Cortez, P., Rita, P., Moro, S.: Research trends on big data in marketing: a text mining and topic modeling based literature analysis. Eur. Res. Manage. Bus. Econ. **24**(1), 1–7 (2018). https://doi.org/10.1016/j.iedeen.2017.06.002
42. McCarthy, E.J., Shapiro, S.J., Perreault, W.D.: Basic Marketing. Irwin-Dorsey (1979)
43. Burrell, L.: Co-creating the employee experience. Harvard Bus. Rev. (2018)
44. Oertzen, A.-S., Odekerken-Schröder, G., Brax, S.A., Mager, B.: Co-creating services—conceptual clarification, forms and outcomes. J. Serv. Manage. **29**(4), 641–679 (2018)
45. Rashid, Y., Waseem, A., Akbar, A.A., Azam, F.: Value co-creation and social media. Eur. Bus. Rev. (2019)
46. Ngai, E.W.T., Xiu, L., Chau, D.C.K.: Application of data mining techniques in customer relationship management: a literature review and classification. Expert Syst. Appl. **36**(2), 2592–2602 (2009). https://doi.org/10.1016/j.eswa.2008.02.021
47. Dam, N.A.K., Le Dinh, T.: A literature review of recommender systems for the cultural sector. In: Paper presented at the Proceedings of the 22nd International Conference on Enterprise Information, Czech Republic (2020)
48. Lemon, K.N., Verhoef, P.C.: Understanding customer experience throughout the customer journey. J. Mark. **80**(6), 69–96 (2016)
49. Dam, N.A.K., Le Dinh, T., Menvielle, W.: A systematic literature review of big data adoption in internationalization. J. Mark. Anal. **7**(3), 182–195 (2019). https://doi.org/10.1057/s41270-019-00054-7
50. Siggelkow, N., Terwiesch, C.: The age of continuous connection. Harvard Bus. Rev. (2019)
51. Maechler, N., Neher, K., Park, R.: From touchpoints to journeys: Seeing the world as customers do. Retrieved **18** Mar 2017 (2016)
52. Halvorsrud, R., Kvale, K., Følstad, A.: Improving service quality through customer journey analysis. J. Serv. Theor. Pract. (2016)

How Spatial Data Analysis Can Make Smart Lighting Smarter

Sebastian Ernst(✉) and Jakub Starczewski

Department of Applied Computer Science, AGH University of Science and Technology, Al. Mickiewicza 30, Kraków, Poland
ernst@agh.edu.pl

Abstract. The paper proposes the use of spatial data analysis procedures to automatically determine lighting situations to allow for easy estimation of the predicted power efficiency of lighting retrofit projects. The presented method has been applied to real-world data for a large city, covering over 50,000 lamps. Results show that the accuracy of the algorithm is promising but, more importantly, demonstrate that the traditional estimation methods, based on the existing infrastructure parameters, can lead to highly biased results compared to the proposed method.

Keywords: Spatial data analysis · Road lighting optimisation · Geographic data processing

1 Introduction

Nowadays, many cities are undergoing changes aimed at improving the comfort and safety of their citizens, while lowering the use of resources such as power. These modernisations are often connected to introduction of various ICT (information and communication technology) solutions, such as IoT (Internet of Things) devices. The concept of *smart cities* is often focused around the collection of diverse sensor data and using it to support various ongoing operations [5].

However, while sensor data can be useful for short-term decisions, it can rarely be used directly for strategic planning. Let us focus on the example of street lighting, which constitutes a significant part of energy consumption in cities [2].

Replacing old light fixtures with new ones based on LEDs (light emitting diodes) can yield significant savings, often reducing power consumption by more than half [9]. However, if the existing infrastructure and its energy consumption are used as the basis for the new one, it is very easy to carry over mistakes made years earlier when these old installations were being designed. This can affect both the decisions (e.g. where and in what order to perform the modernisations) and execution (e.g. what fixtures should be installed in given locations) of such projects.

Supported by AGH University of Science and Technology grant number 16.16.120.773.

N. T. Nguyen et al. (Eds.): ACIIDS 2021, LNAI 12672, pp. 272–285, 2021.
https://doi.org/10.1007/978-3-030-73280-6_22

In order to accurately determine the lighting (and power consumption) needs of each street in the city, one needs to assess the lighting requirements, analyse the spatial properties of these situations and choose the optimal fixture models for each one. However, this process is usually costly and work-intensive, and cannot be justified during the planning phase; often, even in the execution phase, it is performed with significant simplifications.

This paper proposes the use of georeferenced (spatial) data processing and analysis to estimate the light and energy needs of streets in urban areas. The proposed process is automatic, and therefore only requires (costly and time-consuming) human activity when the model is tuned, but can later be scaled to arbitrarily large areas, or even other cities. It is demonstrated in the area of Washington, D.C., using open data provided by the District of Columbia[1] and that that available within the OpenStreetMap project.[2]

The methods described in this paper were researched and developed as part of a MsC thesis project [11].

2 Problem Statement

The problem discussed by this paper concerns the common case when old, legacy and power-consuming lighting devices (fixtures) are replaced with more efficient LED counterparts. The types and models of the new fixtures are chosen from some group of already-existing, available products. It does not concern the process of designing the lamps themselves, as this scenario only occurs rarely, e.g. when it is necessary to provide devices which visually match historic surroundings.

In practice, this selection process is usually performed using one of the following approaches:

1. A LED replacement is selected based on the power and type (road or park lamp) of the existing fixture; this approach is referred to as using a *conversion chart*, as the replacements are deterministic and based only on the parameters of the existing devices.
2. A designer analyses the characteristics of roads in the area concerned and groups similar ones, defining so-called *road profiles*. Then, these profiles are entered into photometric calculation software such as DIALux[3] which allows the designer to verify the performance of different lamp types in the given situation.

Both of these approaches have some problems. In the former case (approach 1), the workload is relatively low, which results in this approach being often used when performing preliminary assessment of the power savings, e.g. to prioritise the streets with the highest potential. However, this approach also carries over any possible problems of the existing installation.

[1] https://opendata.dc.gov.
[2] https://www.openstreetmap.org.
[3] https://www.dial.de/en/dialux/.

The required lighting parameters for a given road depend on its characteristics and are specified by standards appropriate for a given region, such as CEN/TR 13201:2014 in Europe [3] or ANSI/IES RP-8-14 in the U.S. [4]. Problems of existing installations may include underlighting (when the lighting parameters are insufficient, e.g. due to excessive spacings between poles) or overlighting (e.g. due to old lamps being available in only a few power versions). In the *conversion chart*-based approach, these problems are carried over to the new installation. Even for the purpose of initial estimation and prioritisation, this can bias the results and decisions.

In the latter case (approach 2), there are two main problems. The first one is related to the complexity of the task. Accurate selection of the new fixtures depend on a number of parameters, such as:

- width of the road,
- pole setback (i.e. the distance of the pole from the side of the road),
- pole spacing (i.e. the distance between subsequent poles),
- pole height,
- arm length.

These parameters vary for each street in a city, and their accurate assessment is a difficult task to be performed "by hand." The designer needs to perform measurements on maps or in the field, and even for moderately-sized areas and rough aggregation, the process may yield several dozens of different profiles [1]. Processing that many profiles in DIALux is a time-consuming task as well; however, this can be remedied by using batch processing photometric calculation tools [10]. Nevertheless, the workload related to the data preparation phase itself may result in this approach being inappropriate for investment planning or in introduction of simplifications which affect the accuracy of the process.

In this paper, we propose utilisation of various GIS datasets to automate the data preparation phase for modernisation planning of road lighting in urban areas. The datasets used include:

1. Streetlight data [7]; this dataset contains points representing streetlights, along with attributes such as lamp type, power, number of fixtures, but without a precise indication of the area being lit by each lamp;
2. Road shape data [7]; this dataset contains polygons representing the precise shapes of roads and other areas, such as median islands, but without additional attributes, obtained as a fusion of LiDAR scans and other survey methods;

Table 1. Number of items in datasets used for experiments

Dataset	Original	Filtered
Streetlights	70,956	55,020
Road shapes	52,339	21,398
Road lines	32,841	24,662

3. Road lines and parameters [8]; this dataset contains lines representing roads, along with attributes such as road type, name, width, etc.

The objects present in the datasets have been filtered to only include the relevant ones (e.g., lamps illuminating roadsigns have been excluded from the streetlights dataset). The number of objects used for experiments is shown in Table 1.

3 Methodology

The proposed method involved the development of a general algorithm, which was iteratively run against the examined sets of data. After each iteration, the outcomes were examined in order to identify specific issues, which were not addressed by the standard procedure. The algorithm was then either tuned by adjusting already-existing parameters or extended with additional heuristics to manage situations in which it previously failed. The iterations were performed until the satisfactory results were obtained. The following methodology was developed as the general algorithm:

- identify the relationships between datasets (OSM Ways–Roads, Roads–Lamps),
- determine the data sources for parameters (road type – from OSM, lamp power – from Streetlights),
- identify spatial relationships between objects in different datasets using spatial joins (candidate assignments).

This process, however, requires a toolkit capable of both manipulating the data and performing spatial operations on the geo-referenced data. Thus, the Python programming language and its ecosystem had been selected as the primary technology used in the approach, due to the variety of open-source libraries designated to execute the aforementioned tasks. Three main Python packages were used during the development of the proposed approach:

- pandas[4], to clean and manipulate the data,
- Shapely[5], to perform geospatial operations on a various geometry types,
- GeoPandas[6] (built on top of the two preceding packages), to perform either bulk operations on geometric attributes or spatial joins.

4 Implementation

This section describes the implementation details of individual algorithms. Each one is presented according to the aforementioned schema, i.e. the description of the general algorithm is followed by heuristics developed for special cases.

4 https://pandas.pydata.org/.
5 https://shapely.readthedocs.io/en/stable/manual.html.
6 https://geopandas.org/.

4.1 Mapping of Linestrings to Road Shapes

General Algorithm. Integration of OSM *ways* with *roads* was performed based on the spatial intersection relationship between the datasets, i.e. a way linestring crossed multiple polygons (Fig. 1). The OSM dataset contains information about the street name and the road type; therefore, merging this data into a set of road shapes will allow the introduction of additional heuristics for further calculations. The integration process was initialised by performing a spatial join operation on aforementioned datasets, which determined all *roads* intersecting with a particular *way*. These polygons were combined into a single geometry; however, it was crucial to split the shapes in accordance with the peripheral points of the linestring they crossed. This operation ensured the homogeneity of attributes in the created shapes.

To do so, a geometric buffer of a certain size was applied to the OSM dataset and was confronted against the previously-created union of *road* shapes. Finally, only the areas which intersected with the *way's* buffer (further referred to as road segments) were considered viable for further examination (see Fig. 2).

Specific Cases. The procedure resulted in the creation of a total of 18,998 road segments. A vast majority of the constructed objects were valid (see Fig. 2). However, one notable problem was observed: a substantial number of multipolygons (sets of polygon shapes) was formed during this process (12.5%); this

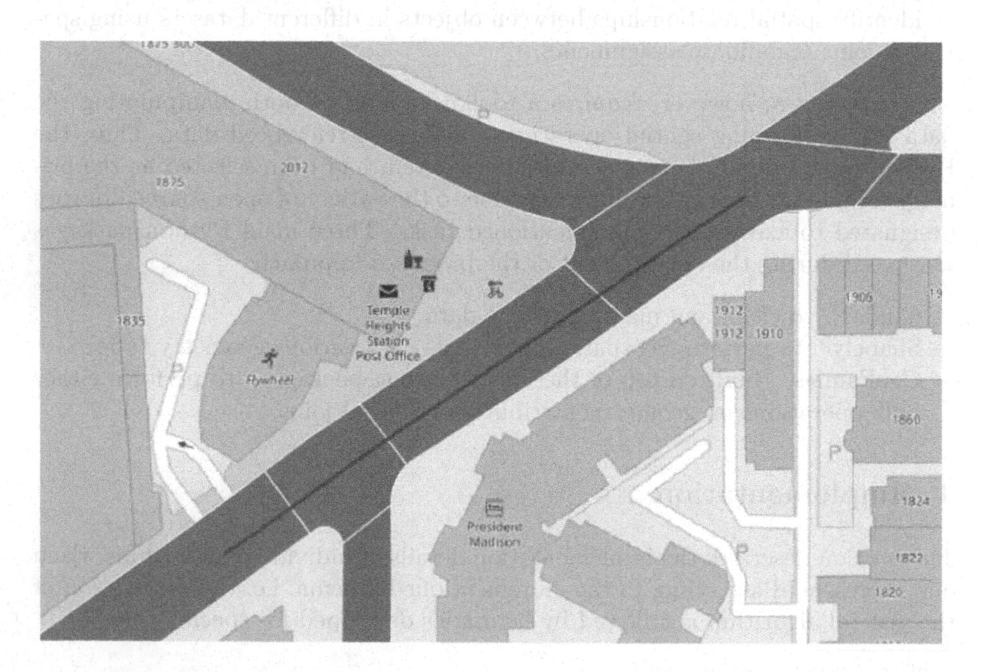

Fig. 1. The visual representation of a group of road polygons (grey) intersecting with a specific OSM way (blue) (source: [11]) (Color figure online)

is highly undesirable for further calculations. An in-depth analysis determined the following reasons for creation of problematic shapes:

- An OSM way may intersect with roads only on its ends – Fig. 3 displays a problem from which most of the unwanted structures originated. There, an OSM way crosses the carriageway which does not take part in the analysis, but it intersects with other roads on its ends. To address this issue, the percentage of the intersection between the OSM way and the road segment in relation to the total shape area was calculated. Then, the minimum acceptable percentage was empirically derived (by manually reviewing the cases), and the shapes with values below the determined level were dropped. Finally, as it comes to multi-polygon objects above the aforementioned value, only the polygon with the longest intersection with the linestring was kept.
- Occurrence of gaps between road polygons – Another portion of problematic unions have been created in a manner depicted in Fig. 4. To overcome this issue, a buffer was implemented to polygons in a road segment, and the geometries were unified afterwards. Finally, a buffer with an opposite distance parameter was applied to retain the initial form of the shape.
- Inaccurate road shapes – Last but not least, the invalid structure of road polygons was also the reason for multi-polygon creation (see Fig. 5). Instead of being represented as a separate carriageways, these shapes consist of multiple roads. Therefore, if a buffer of a linestring crosses these areas, an unwanted object is formed. To address this issue, a heuristic, which removed all non-intersecting polygons (no common points with OSM way) was applied. This algorithm extensions produced single polygons only.

In the second iteration, the solutions to the aforementioned issues were applied, which eliminated all issues with multi-polygon geometries. As a result, a total of 18,188 viable areas were determined.

4.2 Assignment of Lights to Lit Areas

General Algorithm. This section describes the integration process of road segments (created as described in Sect. 4.1) and street lights. The main goal here was to accurately determine the area lit by each fixture in the dataset.

An area, together with the assigned light poles, constitute a lighting situation. Their creation allows to define the mutual relationships between the lamps within each segment, and consequently calculate their spacing values, which are derived from the distances between the consecutive lamps. This parameter, other than being crucial for final computations, also serves as a road segment quality indicator – inconsistencies imply possible problems.

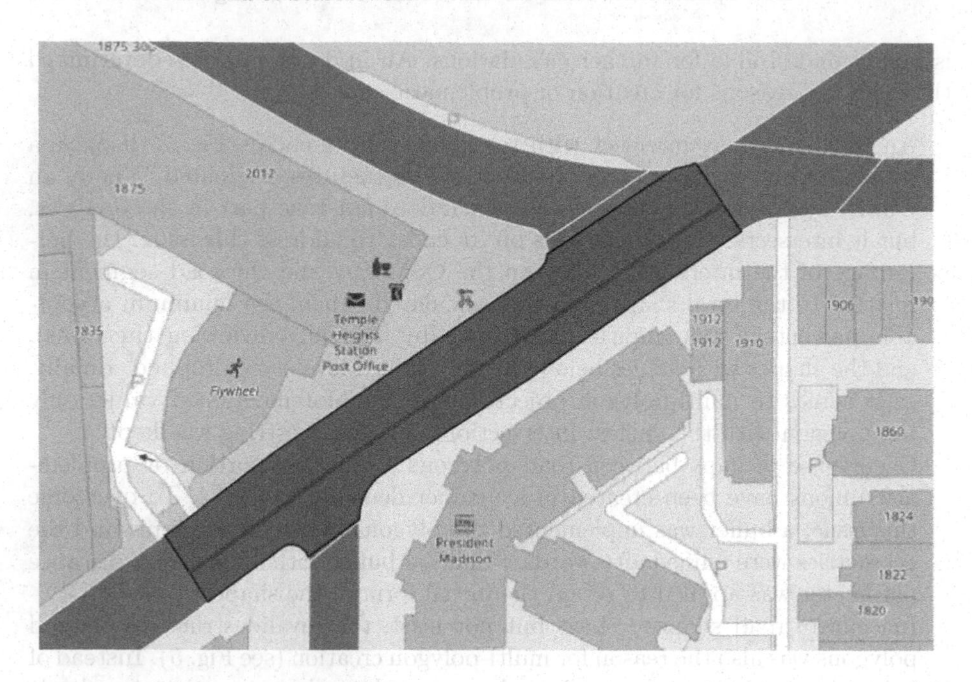

Fig. 2. The union of polygons (violet) intersecting with the buffer of an OSM way buffer (blue) (source: [11]) (Color figure online)

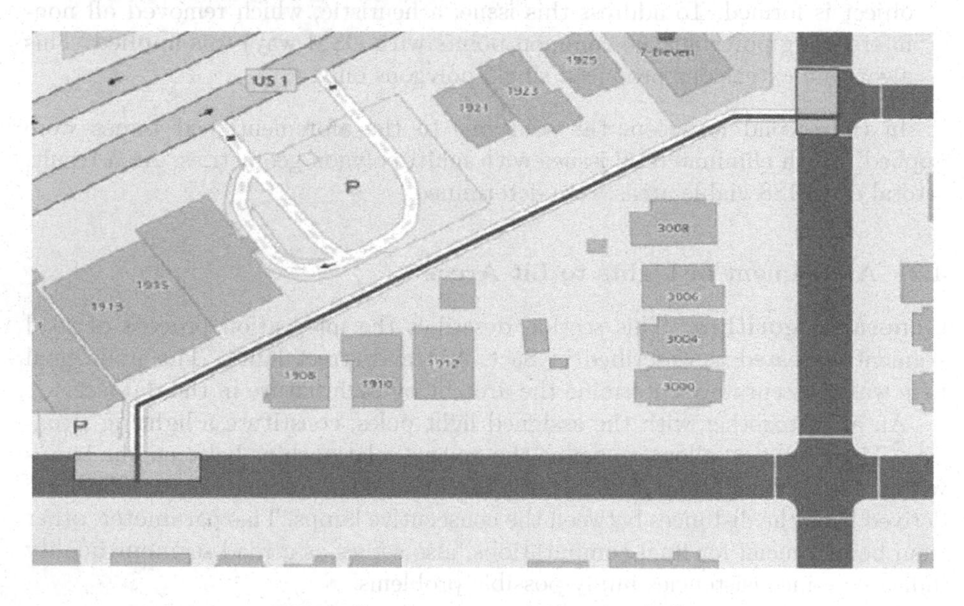

Fig. 3. Issue 1 – example (source: [11])

The basis for integration was also the intersection relation between the aforementioned datasets; however, before spatially joining them, the points which

Fig. 4. Issue 2 – example (source: [11])

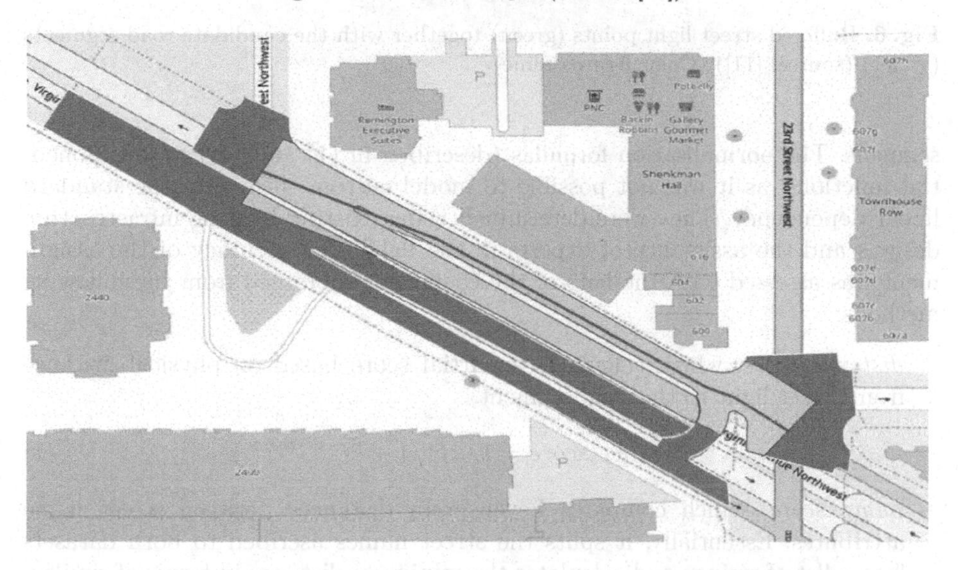

Fig. 5. Issue 3 – example (source: [11])

refer to the street lights were buffered by a certain distance. This operation enabled identification of potential assignment candidates for each lamp (Fig. 6).

Specific Cases. After performing a spatial join, the created table was grouped by the identification number of the street light and, for each road segment in a group, a several factors were calculated, including the setback (physical distance from the pole to the egde of the road segment) and the Levenshtein distance [6] between street name attributes of both dataset.[7] Based on these values, a normalised score within the range of $(0; 1)$ was obtained for each candidate road

[7] Levenshtein distance is defined as a minimum number of character operations (e.g. substitute, delete or insert) to transform an expression into another.

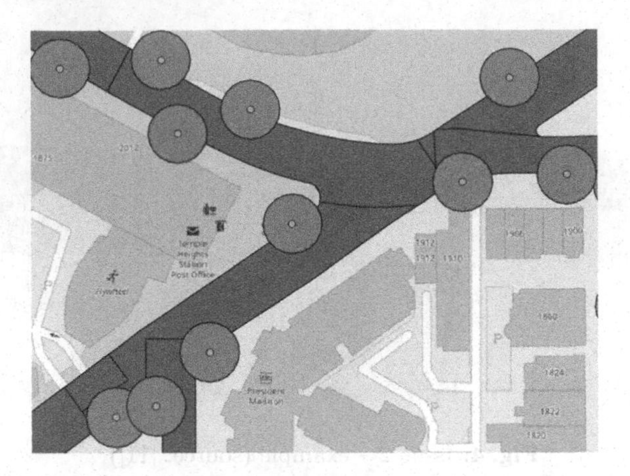

Fig. 6. Buffered street light points (green) together with the candidate road segments (violet) (source: [11]) (Color figure online)

segment. The normalisation formulas (described in (1) and (2)) were exponential functions, as it was not possible to model custom rules with the standard linear dependency. They were determined using existing lighting infrastructure designs and the assistance of experts in the field. The accuracy of the assignment was assessed with the help of the coefficients obtained from the following methods:

- *distance score*, which calculates a partial score based on physical distance from street light to the road segment.

$$distance\_score = 1 - (1/1 + e^{5-distance}) \tag{1}$$

- *name score*, which compares Levenshtein distances between street name attributes. Essentially, it splits the street names ascribed to both datasets into a list of strings and calculates the minimum distance in terms of similarity between any of the examined elements in the collection.

$$name\_score = 1 - (1/1 + e^{5-(2.5*min\_lv\_distance})) \tag{2}$$

- *type score*, which evaluates the type of road the light is being assigned to. It is based on the assumption that the lamps will most likely illuminate the streets of dominant categories (primary, residential etc.), while the others will receive a slightly lower rating (e.g. service road).

The weights for each of the partial scores were determined empirically, the overall score was computed and the road segment with the highest overall score was assigned to the street light. Additionally, a score threshold, below which an assignment was considered wrong, was implemented. This helped to further improve the accuracy of the results i.e. in case of areas with only one assignment candidate. The outcome of the automatic process was evaluated by reviewing the problematic cases, which was followed by iterative adjustment of both weights and the minimum score in order to achieve possibly best assignment without any manual corrections. Finally, 8,259 of initial 18,188 areas were assigned to one or more lighting points (omitted road segments did not intersect with any lamp within a specified distance). Moreover, nearly all street lights (99%) have been assigned to the most relevant road segment. The unassigned 1% either did not acquire the minimum score or were to far from the road.

5 Results

5.1 Quality of Modelled Situations

After the adjustment of the assignments, several factors resulting from mutual relationships between the datasets were calculated per each road segment. These included the the distance from the street light to the edge of road segment and the spacing between each consecutive pair of lamps. The latter property, or rather its variance, is being used as a measure of how well-modelled a lighting situation is. Lower variance values are considered superior, as the spacing median of particular lighting situation is the computation basis during the optimisation process, rather than its individual values. The results were analysed and consequently categorised into the following groups:

- N/A value (31.7%) – which are connected with the lighting situations with less than 3 street lights assigned. The variance is not calculated as it is impossible to derive variance from less than 2 values, and thus, quality evaluation of these situations was impossible.
- Value lower than 100 (48.3%) – identified for almost half of examined situations, indicates their exceptional quality (Fig. 7).

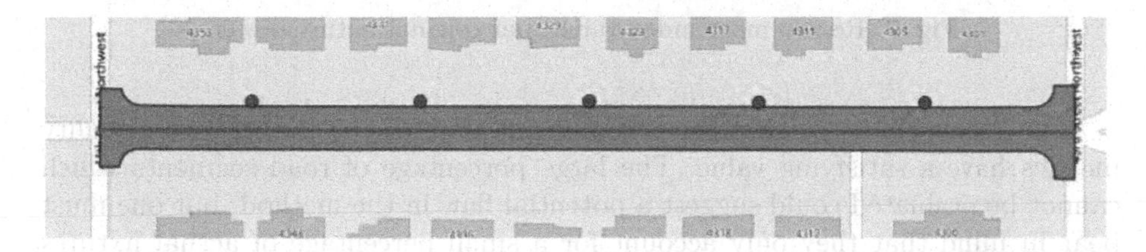

Fig. 7. Example of a road segment with spacing variance below 100 (source: [11])

- Value between 100 and 200 (10.5%) – situations with negligible differences in spacing values (Fig. 8)
- Value above 200 (9.5%) – which suggests spacing issues within a lighting situation. One of the identified causes of the problem, was a bilateral arrangement of poles. As spacing is calculated between the consecutive street lights, around 50% of these distances are close to 0, which can lead to an extremely small value of spacing median. Around 1/5 of these segments were further resolved by applying this metod.

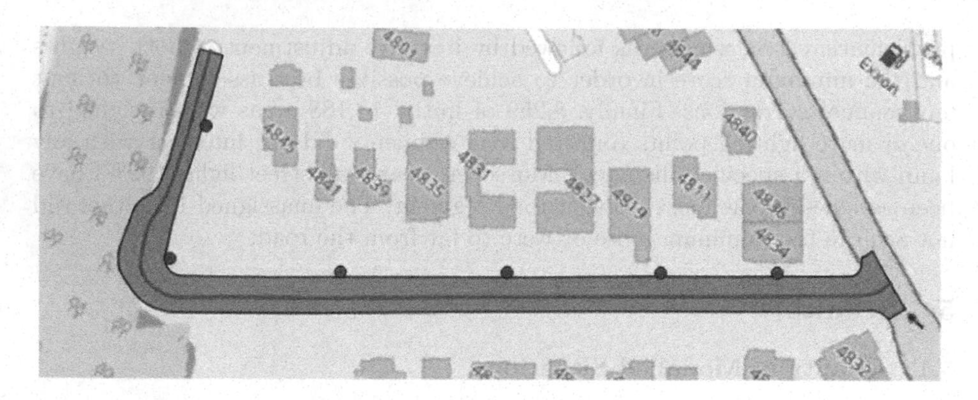

Fig. 8. Example of a road segment with spacing variance between 100 and 200 (source: [11])

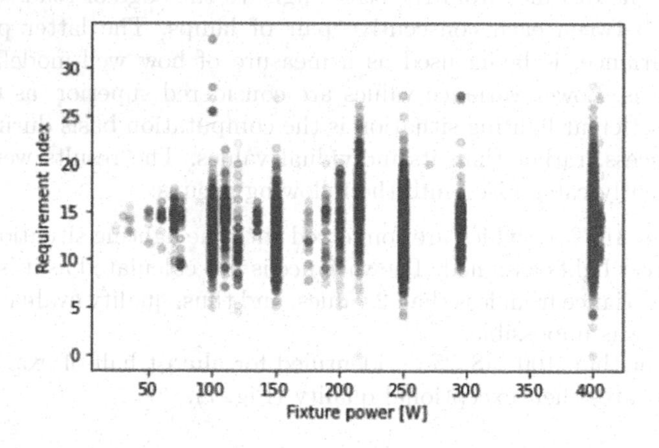

Fig. 9. Requirement index values per original fixture power

The results indicate that for a significant part of the fixtures, the quality metrics have a satisfying value. The large percentage of road segments which cannot be evaluated could suggest a potential flaw in the method, but one must bear in mind that they only account for a small percentage of actual fixtures and their impact is insignificant.

5.2 Estimation Efficiency

Please note that the power consumption of the resulting installation is not a measure of the efficiency or accuracy of the proposed method. The existing installation may be grossly insufficient, which may be due to existing poles being available at very large spacings (e.g. 100 m). Such cases are very common especially in rural areas, where power network poles are used to install fixtures instead of deploying dedicated infrastructure for lighting. This would result in severe underestimation of the resulting power.

Conversely, areas may be overlit, e.g. due to the fact that legacy devices were available with much more coarse granulation of available power. In this case, the resulting power would probably be overestimated.

Since the goal is to provide a more accurate estimation of the resulting installation power, the efficiency has to be measured with regard to that parameter. However, it is difficult to define a reasonable metric for this real-world accuracy without actually performing the photometric optimisation, which is beyond the scope of this paper.

Instead, let us propose a coefficient, requirement index r_i, which is an estimation of the difficulty of illuminating a given road segment, taking into account the factors with the biggest impact on the required fixture power. It is defined as:

$$r_i = \sqrt[4]{w_i \cdot h_i \cdot s_i \cdot c_i} \tag{3}$$

where:

- w_i is the width of the road lit by lamp i, in meters,
- h_i is the height of the pole of lamp i, in meters,
- s_i is the median of the spacing of lamp poles in street illuminated by lamp i, in meters,
- c_i is the lighting requirement imposed by the lighting class of the street according to [4], L_{avg} in cd/m^2.

The obtained results are presented in Fig. 9, a plot of r_i values in relation to the power of the original fixture. Had the original, legacy infrastructure been designed optimally, only using inefficient fixtures, one would expect a monotonic dependence between the power and the requirement index values. However, the results show that each fixture power had been used of a wide variety of lighting situations, and while lower-powered fixtures were clearly unable to handle "difficult" ones, some high-power fixtures were used for situations which should not have required them.

This signifies the necessity for real-life evaluation of lighting situations, as one cannot rely on the accuracy of existing, legacy infrastructure.

6 Conclusions

In this paper, we proposed a new approach to estimating the efficiency of road lighting modernisation projects in urban areas. Instead of relying on the existing information parameters (e.g., the power of the existing, legacy fixtures), we analyse the actual parameters of the roads to allow for more accurate estimation of these parameters.

The method has been applied to spatial data for Washington, D.C., covering 55,020 lamps. Spatial analysis procedures have been applied to datasets for streetlights, road shapes and road lines, allowing for estimation of parameters crucial for designing road lighting installations, such as road width, pole setback and pole spacing.

As shown in Sect. 5.1, the results are reasonably reliable, as the quality indicators for a vast majority of roads (and fixtures) have promising values.

To evaluate the real-world accuracy of the method and to estimate the inadequacies of existing infrastructure, a new coefficient called *requirement index* was calculated. Results presented in Sect. 5.2 indicate that the power of the existing lamps was not proportional to the estimated difficulty of illuminating their corresponding areas.

These preliminary results show that estimation of the efficiency (power consumption reduction) based just on the parameters of the existing infrastructure can be misleading. Furthermore, if these estimates were used to determine the types and wattages of new luminaires, this could result in significant under- or overlighting of streets and pavements, since the inadequacies of the old designs would be carried over to the new setup.

Some issues need to be addressed in future work. Firstly, the iterative procedure based on identifying issues and applying heuristics to resolve them results in complicated sets of routines, which could be difficult to maintain or debug. Research is being done regarding spatially-triggered graph transformations (STGT) and graph structures which could alleviate this problem by allowing systematic verification of hypotheses in the process.

Secondly, in order to obtain the actual power of the new fixtures, one needs to use the results of the presented work as input for a photometric optimisation engine. For large datasets, containing dozens of thousands of records, the optimisation process can take days. To improve the applicability of the method, means of quick estimation of the required lamp power could be useful.

References

1. Ernst, S., Łabuz, M., Środa, K., Kotulski, L.: Graph-based spatial data processing and analysis for more efficient road lighting design. Sustainability **10**(11), 3850 (2018). https://doi.org/10.3390/su10113850
2. European Commission: Lighting the Future Accelerating the deployment of innovative lighting technologies. Green Paper COM/2011/0889, Publications Office of the European Union, December 2011
3. European Committee for Standarization: EN 13201–2: Road lighting - Part 2: Performance requirements. Technical report, December 2014
4. Illuminating Engineering Society of North America: Roadway Lighting (ANSI/IES RP-8-14), October 2014
5. Lai, C.S., et al.: A review of technical standards for smart cities. Clean Technol. **2**(3), 290–310 (2020). https://doi.org/10.3390/cleantechnol2030019
6. Levenshtein, V.I.: Binary codes capable of correcting deletions, insertions and reversals. Sov. Phys. Dokl. **10**(8), 707–710, February 1966
7. Office of the Chief Technology Officer: Spatial data retrieved from https://opendata.dc.gov (2020)
8. OpenStreetMap contributors: Planet dump retrieved from https://planet.osm.org (2020)
9. Sędziwy, A., Basiura, A.: Energy reduction in roadway lighting achieved with novel design approach and LEDs. LEUKOS **14**(1), 45–51 (2018). https://doi.org/10.1080/15502724.2017.1330155

10. Sędziwy, A., Kotulski, L., Basiura, A.: Multi-agent support for street lighting modernization planning. In: Nguyen, N.T., Gaol, F.L., Hong, T.-P., Trawiński, B. (eds.) ACIIDS 2019. LNCS (LNAI), vol. 11431, pp. 442–452. Springer, Cham (2019). https://doi.org/10.1007/978-3-030-14799-0_38
11. Starczewski, J.: Road lighting efficiency improvement through data processing and automated computation. Master's thesis, AGH University of Science and Technology; IST Instituto Superior Técnico (2020)

Natural Language Processing

Convolutional Neural Networks for Web Documents Classification

Codruț-Georgian Artene[1,2](✉), Marius Nicolae Tibeică[2], Dumitru Daniel Vecliuc[1,2], and Florin Leon[1]

[1] Department of Computer Science and Engineering, "Gheorghe Asachi" Technical University of Iași, Iași, Romania
{codrut-georgian.artene,dumitru-daniel.vecliuc, florin.leon}@academic.tuiasi.ro

[2] Bitdefender, Iași, Romania
{cartene,mtibeica,dvecliuc}@bitdefender.com

Abstract. Web page classification is an important task in the fields of information retrieving and information filtering. Text classification is a well researched topic and recent state-of-the-art results involve deep learning algorithms. However, few works use text classification deep learning algorithms for web page classification. This work describes an experimental study on classifying web pages using convolutional neural networks (CNNs). An in-house multi-label multi-language dataset is used. We apply text classification CNNs on the textual information extracted from the body, the title and the meta description tag of web pages. Overall, our CNN text classification model achieves good results on web documents classification and one can conclude that it is suited to be part of an automatic web documents classification system.

Keywords: Web documents classification · Text classification · Deep learning · Convolutional neural networks

1 Introduction

Web page classification is the process of assigning one or multiple predefined category labels to a web page. Classification is a supervised machine learning task where a model or classifier is constructed to predict class labels. Based on the number of labels that the model assigns to a given sample, the classification can be *multi-class classification*, when the categories are mutually exclusive and the output is a single category label, or *multi-label classification*, when there are no constraints regarding the number of category labels the model assigns to a specific input sample. When the total number of category labels is two, the multi-class classification problem becomes a *binary classification* one. We can look at the multi-label classification problem as a collection of binary classification problems, one for each category label, the output of each model being whether or not it assigns the corresponding label to the given sample.

Supported by Bitdefender.

N. T. Nguyen et al. (Eds.): ACIIDS 2021, LNAI 12672, pp. 289–302, 2021.
https://doi.org/10.1007/978-3-030-73280-6_23

There are millions of web pages available on the Internet and the set of topics is very large and diverse, making the classification a challenging task. Web page classification can be divided into two more specific problems: *subject classification*, when the classification is made based on the subject or the topic of web pages and *functional classification*, when the classification is made based on the role that web pages play (i.e., register, contact, payment, etc.) [16].

The web page classification problem is a variant of the text classification problem, considering that the input is the textual information from the web page. Text classification is a long-standing research field within the area of Natural Language Processing (NLP) and researchers achieved state-of-the-art results in recent years, especially using deep learning algorithms. However, there are only a few works that explored the effectiveness of deep learning text classification methods on textual information from web documents.

Our work consists of an experimental study on web page classification using Convolutional Neural Networks (CNNs). The goal is to find out if it is possible to use a CNN text classification approach to construct an automatic web documents classification system. We experiment with an in house multi-label multi-language web documents dataset and we train our classifier with text input from the body, the title and the meta description tag of web pages. Using pretrained multilingual aligned word embedding vectors we are able to expand the vocabulary, enabling this way the model to classify web pages containing words other than the ones in the training dataset.

The rest of the paper is organized as follows. In Sect. 2 a literature study relevant to our problem is discussed. In Sect. 3 we address the text classification problem. In Sect. 4 we present the methodology of this work, including a dataset analysis and an overview of the classification model and the evaluation process. The experiments conducted with the classifier and the results are discussed in detail in Sect. 5. We summarize our findings and conclude this work in Sect. 6.

2 Related Work

Web page classification is different when compared to the text classification, mainly due to the structure of the input. In [16], a survey on features and algorithms used for web page classification is presented and the authors note the importance of web-specific features. Web pages are written in HTML and, besides the textual information, may contain elements that bring additional input for the classification process, such as HTML tags, the URL, visual elements (i.e., images, logos, etc.) or even the relation with other web resources represented by hyperlinks.

In [7], the authors present a review of web page classification methods, including text-based classification, image-based classification and combined methods. Text-based classification of web pages involves the semantic information and the structural information of web pages and includes methods that use term frequency representation of the textual information for subject classification, phishing detection and content filtering.

The most related to our work are approaches that apply text classification on text extracted from web pages. A Bayesian approach for phishing web page detection based on textual and visual contents analysis is used in [20]. For the text classification task the authors apply the Bayes classifier on the term frequencies representation of the text extracted from web pages. In [10], a simplified swarm optimization for binary classification of web pages is used. The authors consider the HTML tags and terms extracted from web pages as classification features. The content, the URL and the structural information of web pages are used in [1] to identify the news pages from a web corpus. The authors of [11] apply the Naïve Bayes classifier for subject classification of massive web pages based on N-gram feature vectors extracted from the title and the main text.

There is a large corpus of work in web page classification and researchers achieved good results using ML classifiers. However, while the vast majority of new text classification methods use deep learning algorithms, few works address the issue of using them to classify web pages.

3 Text Classification

Text classification is a technique used to classify textual information into a predefined set of categories. Text classification has been widely studied by researchers and it is used in many applications, such as information retrieval, information filtering, sentiment analysis, recommender systems, knowledge management or document summarization [9]. Most text classification systems follow the pipeline illustrated in Fig. 1, including the following processing steps: text preprocessing, text representation, classification and evaluation.

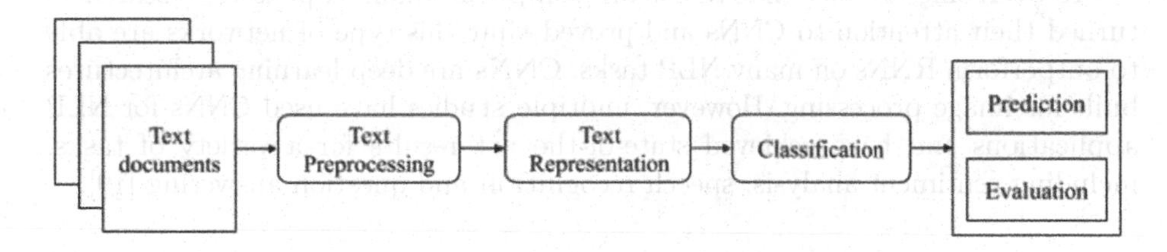

Fig. 1. Text classification pipeline: text preprocessing, text representation, classification and evaluation.

Text Preprocessing. To use text as input for a classifier, first, the text has to be preprocessed and redundant textual features have to be removed. The preprocessing phase usually consists of the following steps: tokenization, noise removal, stop words removal, words filtering, capitalization and stemming.

Text Representation. After the preprocessing, the text is transformed into a feature vector that can be passed to the classifier. The bag-of-terms and the term frequency - inverse document frequency (TFIDF) representations are commonly used in methods of document classification.

Bag-of-terms models do not capture the order and semantics of words, limiting this way the performance of the text classifier. Word embedding addresses these problems through a model that allows the representation of semantic and syntactic similarity between words. *Word2Vec* [13,14], *GloVe* [15] and *FastText* [2] are three of the most common word embedding methods used for NLP tasks. Most recent works involve state-of-the-art attention based networks for text representation, such as Transformers [18] and BERT [5].

Classification. The results of the classification may vary depending on the used classifier. Therefore, choosing the right classifier is an important task of the classification pipeline and depends on both, the structure of the input data and the type of classification. A text classification algorithms survey that covers different existing algorithms and their applications is presented in [9]. Naïve Bayes, K-Nearest Neighbors (KNN), Support Vector Machine (SVM), Decision Trees and Random Forest are ML algorithms widely used for text classification.

In recent years, deep learning approaches achieved state-of-the-art results on a wide variety of NLP tasks, including text classification, mainly due to the introduction of Recurrent Neural Networks (RNNs). Long Short-Term Memory (LSTM) and Gated Recurrent Unit (GRU) networks are the types of RNN mostly used for text classification. These networks are able to model the semantics of the text very well, producing good classification results. However, their performance comes with a high computational cost.

To overcome the fact that RNNs are computationally expensive, researchers turned their attention to CNNs and proved that this type of networks are able to outperform RNNs on many NLP tasks. CNNs are deep learning architectures built for image processing. However, multiple studies have used CNNs for NLP applications and have achieved state-of-the art results for a variety of tasks, including sentiment analysis, speech recognition and question answering [19].

Evaluation. Evaluation is the final step of the text classification pipeline. Since model interpretability in case of deep learning models is a limiting factor [9] and we cannot determine how a model performs analytically, the performance of the classifiers is evaluated experimentally.

To evaluate the performance of a classifier, we first split the dataset into the train set, the validation set and the test set. We use the train set to train our classifier and then we measure its performance by calculating some metrics using the test set. The validation set is used to evaluate these metrics during the training process. Usually, the test set and the validation set are the same, especially when only small datasets are available.

Depending on the type of classification problem, binary, multi-class or multi-label, there are multiple metrics that enable us to measure the efficiency of a classifier. For the specific case of binary classification problem, a *confusion matrix* is structured as shown in Table 1. We denote by $n_{samples}$ the total number of samples in the dataset, by $n_{classes}$ the total number of classes, by $\boldsymbol{C} = \{c_1, c_2, ..., c_{n_{classes}}\}$ the set of classes and by $\boldsymbol{D} = \boldsymbol{d_i},\ i = \overline{1, n_{samples}}$ the dataset.

Table 1. Confusion matrix structure

Confusion Matrix		Predicted	
		Positive	*Negative*
Actual	*Positive*	True Positives (TP)	False Negatives (FN)
	Negative	False Positives (FP)	True Negatives (TN)

Based on the *True Positives* (TP), *True Negatives* (TN), *False Positives* (FP), *False Negatives* (FN) rates from the confusion matrix, the following metrics can be defined:

$$accuracy = \frac{TP + TN}{n_{samples}}, \tag{1}$$

$$precision = \frac{TP}{TP + FP}, \tag{2}$$

$$recall = \frac{TP}{TP + FN}, \tag{3}$$

$$F1 = \frac{2 \times precision \times recall}{precision + recall}. \tag{4}$$

There are some methods to average binary metric calculations across the set of classes, in order to extend them to multi-label or multi-class problems: *micro-averaging*, *macro-averaging* and *example based averaging* [17].

Micro-averaging calculates $TP(c_i)$, $FP(c_i)$, $TN(c_i)$, $FN(c_i)$ for each class across the dataset and sums the dividends and divisors that make up the per-class metric to calculate the overall value. In this way, each sample-class pair has an equal contribution to the overall metric, making it more suitable for the case of an imbalanced dataset, where the classes are not equally represented. The micro-averaged precision and recall are defined as follows:

$$Precision = \frac{\sum_{c_i \in \boldsymbol{C}} TP(c_i)}{\sum_{c_i \in \boldsymbol{C}} (TP(c_i) + FP(c_i))}, \tag{5}$$

$$Recall = \frac{\sum_{c_i \in \boldsymbol{C}} TP(c_i)}{\sum_{c_i \in \boldsymbol{C}} (TP(c_i) + FN(c_i))}. \tag{6}$$

The micro-averaged $F1$ score is defined as the harmonic mean (4) of the micro-averaged precision (5) and micro-averaged recall (6).

In multi-class classification problems, the subset accuracy is used, meaning that for a dataset sample, if the set of the predicted labels is the exact match of the true set of labels, the subset accuracy is 1, otherwise it is 0. If we denote with $\boldsymbol{y_i}$ the true set of labels for the i-th example from the dataset and with $\hat{\boldsymbol{y}}_{\boldsymbol{i}}$ the corresponding predicted set of labels, the multi-class accuracy is defined as:

$$Accuracy = \frac{1}{n_{samples}} \sum_{\boldsymbol{d_i} \in \boldsymbol{D}} 1_{\boldsymbol{y_i} = \hat{\boldsymbol{y}}_{\boldsymbol{i}}}. \quad (7)$$

4 Methodology

Inspired by the work done so far in text classification and web classification, we propose a series of experiments with CNNs for web page classification, aiming to explore the effectiveness of deep learning text classification on textual information from web documents.

4.1 Dataset

In our experiments, we use a manually labeled dataset of web page documents. The dataset contains web documents with textual information written in four languages: English (*en*), French (*fr*), Arabic (*ar*) and Portuguese (*pt*). We consider that the text in each web document is written only in one language and we mark the words from other languages as being unknown. The corresponding number of web documents for each language available in our dataset is presented in Table 2, along with the vocabulary size including all words from the text in all web documents after capitalization, tokenization, stopwords removal and punctuation removal.

Table 2. Summary statistics of the multi-language dataset

Language	The number of web documents	The size of the vocabulary
en	4291	303461
fr	2017	189679
ar	1155	173298
pt	1335	128818
Total	8798	795256

Besides the textual information extracted from the body of web documents, we include the title and the text from the <*meta*> tag from the *HTML*. We refer to these attributes as *body*, *title* and *meta*. In Table 3, we show some statistics about the length of the text for the above mentioned attributes: l_{max} - the maximum length, in words. l_{avg} - the averaged length, in words.

Table 3. Statistics about text from body, title and meta tag

Language	Body		Meta		Title	
	l_{max}	l_{avg}	l_{max}	l_{avg}	l_{max}	l_{avg}
en	12970	859	2089	8	120	15
fr	10650	872	159	7	596	16
ar	8805	956	585	8	497	22
pt	12419	822	2114	8	85	14
Overall	12970	869	2114	8	596	16

Since our model requires fixed size inputs, the analysis of the text length distribution in the labeled dataset for the three attributes helped us in choosing the maximum length for the text corresponding to each attribute. Therefore, we choose 5000 words for the body, 100 words for the title and 100 words for the meta, since the majority of web documents fall within these limits.

We prepared our dataset for two different classification problems: multi-class classification and multi-label classification. For each document in the dataset we assigned the main category. This category is meant to be used for multi-class classification. The number of documents for each category is not the same, meaning that we are dealing with an imbalanced dataset and this could affect our classifier.

The dataset contains web documents labeled with 69 different categories[1] and there is no differentiation between the topic based categories (i.e., *astrology*, *medical*, *auto*, etc.) and the functionality based categories (i.e., *filesharing*, *onlinepay*, *onlineshop*, etc.). Furthermore, the categories are not mutually exclusive (i.e., *games_kids* and *entertainment_kids*).

Besides the main category, we assigned other secondary labels to the web documents, preparing the dataset for the multi-label classification problem. More than half of the documents have one label, a small majority have two labels and only a few web documents have three and four labels. This may be a positive aspect since the classifier will learn features from samples that only have one label without having to distinguish what features correspond to each label.

4.2 Model

Our CNN model architecture, shown in Fig. 2, is an adapted variant of the text classification CNN introduced by Kim [8].

Input Layer. The input is represented by the lists of tokens extracted from the text in the body, title and meta tag.

[1] The distribution of web documents per category label in the multi-class setup and in the multi-label setup is available at http://tibeica.com/webcateg/categories.html.

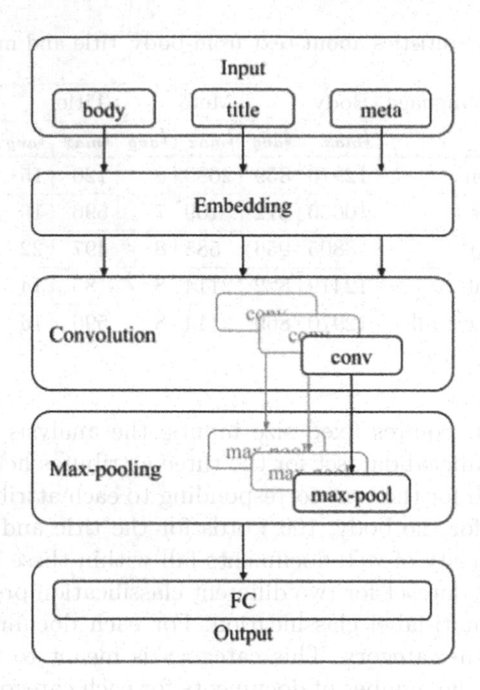

Fig. 2. CNN model architecture.

Embedding Layer. This is a shared layer across the three inputs which maps each word into an embedding vector. We experiment here with pretrained word vectors from [6] and [4]. This layer remains fixed during the training process and the values of the embedding vectors are not modified.

An input of n words is represented by an embedding matrix $\boldsymbol{X} \in \mathbb{R}^{n \times k}$, where k is the dimension of the word embedding and $\boldsymbol{x_i} \in \mathbb{R}^k$ represents the word embedding for the i-th word in the input. Since CNNs are restricted to fixed size input and the length of sentences varies, longer sentences have to be truncated and shorter sentences padded.

Convolution Layer. Shared convolution layers across the inputs with *rectified linear unit* (ReLU) activation and variable number and heights (512 of height 1, 256 of height 2, 128 of height 3). Because each row represents the embedding vector of one word, the width of the filters is equal to the length of the embedding vector (300 in our case).

Convolution is performed on the input embedding layer, using a filter, also called kernel, $\boldsymbol{w} \in \mathbb{R}^{h \times k}$, applied to a window of h words to produce a new *feature*, c_i:

$$c_i = f(\boldsymbol{w} \cdot \boldsymbol{x_{i:i+h-1}} + b), \tag{8}$$

where $\boldsymbol{x}_{i:i+h-1}$ refers to the window of words $x_i, x_{i+1}, \ldots, x_{i+h-1}$, h is the height of the filter, $b \in \mathbb{R}$ is a bias term and f is a non-linear activation function, such as the *hyperbolic tangent*. The filter $\boldsymbol{w}$ is applied to all possible windows of h words in the sentence to produce a *feature map*, $\boldsymbol{c} \in \mathbb{R}^{n-h+1}$:

$$\boldsymbol{c} = [c_1, c_2, \ldots, c_{n-h+1}]. \tag{9}$$

Max-Pooling Layer. The size of the feature map resulted from each convolution filter varies in function of the input length and the filter height. We use here max over time pooling [3] to extract the maximum value $c = max(\boldsymbol{c})$ from each feature map. This operation is carried on, first to reduce the size of the output produced by the convolution layer by keeping only the feature with the maximum value, and then to map this output to a fixed length input required for classification [19]. We concatenate the resulting outputs into a feature vector which is then fed to the *Output* layer.

Output Layer. Final *fully connected* (FC) layer with dropout and *sigmoid* (for multi-label classification) or *softmax* (for multi-class classification) activation. For regularization, we employ dropout with a rate of 0.5 and a constraint on l_2-norms of the weight vectors. The output is the probability distribution over labels.

4.3 Evaluation

For the multi-label setup, the classifiers are evaluated using the micro-averaged precision (5), recall (6) and F1 score (4) metrics. For the multi-class setup, we add the multi-class accuracy metric (7).

All experiments are carried out using stratified 5-fold cross-validation. We chose 5-fold cross-validation instead of the more common 10-fold cross-validation because the model is computationally expensive and the dataset is imbalanced and includes categories with a small number of documents. The stratification of the folds is made based on the main category for both setups, multi-class and multi-label. For each experiment, we use the same indices of the samples for each fold.

We do not perform dataset specific tuning, but we include early stopping on validation sets based on the F1 score, meaning that if the F1 score does not improve after ten consecutive training epochs the training process on that fold is stopped and the weights for the best F1 score are restored before performing the evaluation on the validation dataset. For training we use back propagation over shuffled batches of size 10 with the *Adam* optimizer. The loss function for the multi-label setup is binary cross-entropy and for the multi-class setup we use categorical cross-entropy.

5 Experimental Results

5.1 Baseline Setup

First, we evaluate and analyze the performances of our baseline CNN model on the multi-label classification task. For this experiment the weights of the *Embedding* layer are initialized using pretrained word vectors from [6]. The results are listed in Table 4.

Table 4. Micro-averaged F1, recall and precision of our baseline CNN model on multi-label classification

Model	F1	Recall	Precision
$CNN_{multi-label}$	0.79	0.71	0.90

A higher value of the micro-averaged precision means that the classifier does not label as positive a pair (web document, category) which is negative. In other words, the false positive rate is small. The micro-averaged recall describes the ability of the classifier to find all the positive pairs (web document, category) and we consider that the classifier achieved satisfactory results. Simply put, our classifier is more predisposed to false negatives than to false positives.

5.2 Multi-class Problem

Despite being counter-intuitive, the authors of [12] claim that categorical cross-entropy loss worked better than binary cross-entropy loss in their multi-label classification problem. In this experiment we changed the activation of the output layer from sigmoid to softmax and the loss function from binary cross-entropy to categorical cross-entropy. We also reduced the labeling of our dataset to the multi-class setup using only the main category. Besides the F1, recall and precision metrics, we use here the multi-class accuracy.

The results are presented in Table 5 and they show an increase in the values of the micro-averaged F1, recall and precision, in comparison with the results from Table 4. Furthermore, the classifier achieves a good multi-class accuracy. However, this setup is not suitable for our problem, since the content from a web document usually belongs to multiple categories.

Table 5. Micro-averaged F1, recall and precision and the multi-class accuracy of the CNN model on multi-class classification

Model	F1	Recall	Precision	Accuracy
$CNN_{multi-class}$	0.90	0.87	0.93	0.90

5.3 Input Variation

In our baseline architecture we use as input the textual information from the body, the title and the meta tag from web documents. We intend to find out if the title and the meta are redundant information and in this experiment we use the three inputs as separate inputs for three individual models. The models are initialized with the same values. The results are listed in Table 6 and, comparing with the results from Table 4, we conclude that using the text from body, meta, and title as individual inputs at the same time is the best approach.

Table 6. Evaluation results of the CNN model using different setups for the input

Input	F1	Recall	Precision
Body	0.72	0.62	0.87
Title	0.70	0.58	0.86
Meta	0.67	0.56	0.84

5.4 Embedding Vectors Variation

In our baseline architecture we use word vectors from [6] without adjusting their values during the training process. In this experiment we evaluate the results for four variations of the Embedding Layer. We include embedding vectors from [4] and we test the following variants: non-static random initialized word vectors, static and non-static vectors from [6] and static and non-static vectors from [4]. To ensure the consistency of results we keep the same indices of the samples for the 5-fold cross-validation.

The results of this experiment are listed in Table 7. We find that the usage of pretrained word vectors improves the results. Also, using pretrained multilingual aligned word embedding vectors we are able to expand the vocabulary, enabling this way the model to classify web pages containing words other than the ones in the training dataset. However, there are no further improvements when these vectors are fine-tuned during the training process and we believe that this may not be the case if the training dataset will be larger.

Table 7. Evaluation results of the CNN model using different setups for the word embedding vectors

Vectors source	Model	F1	Recall	**Precision**
Randomly generated	Non-static	0.74	0.65	0.86
[6]	Static	0.79	0.71	0.90
	Non-static	0.79	0.73	0.87
[4]	Static	0.79	0.71	0.90
	Non-static	0.79	0.72	0.88

5.5 Language Transfer Learning

In multilingual aligned word embedding vectors, translation words from different languages have similar representations. In this experiment we aim to find out if it is possible to extend our classifier to other languages without having to create a dataset and to train the classifier. Thus, after we initialize the embedding layer with multilingual pretrained aligned word vectors, we train the model on a subset of languages and evaluate it using the remaining languages.

The results listed in Table 8 show that we cannot extend our models to unseen languages only by using multilingual aligned word embedding vectors and we believe that there are multiple factors that influenced the results: our dataset is too small and imbalanced, multilingual aligned word vectors have similar representations but not exactly the same and the word order typology differs from one language to another. However, the higher values of the precision indicate a low false positive rate, meaning that we can use this method in a hybrid system including other classification techniques.

Table 8. Language transfer learning results of the CNN model

Train languages	Test language	F1	Recall	**Precision**
English	French	0.25	0.16	0.50
	Arabic	0.22	0.14	0.55
	Portuguese	0.22	0.13	0.64
French Arabic Portuguese	English	0.33	0.22	0.65
English Arabic Portuguese	French	0.49	0.34	0.86
English French Portuguese	Arabic	0.33	0.20	0.88
English French Arabic	Portuguese	0.55	0.40	0.90

6 Conclusions

In this work, we explored the idea of using CNNs for web page classification. Through a series of experiments we showed that text classification CNNs produce good results when they are applied to the textual information from web pages. Including the text from the title and from the meta description tag alongside the text from the body as inputs to our model made consistent improvements to the results. The quality of the results was also enhanced when pretrained word embedding vectors were employed. Overall, although there appears to be room for improvement, we think that our web page classification CNN model is suited to be part of an automatic web documents classification system. As future work, we plan to explore the use of state-of-the-art attention based networks The next step is to perform comparative experiments between Transformers networks [18] and CNNs for web page classification.

References

1. Arya, C., Dwivedi, S.K.: News web page classification using URL content and structure attributes. In: 2016 2nd International Conference on Next Generation Computing Technologies (NGCT), pp. 317–322. IEEE (2016)
2. Bojanowski, P., Grave, E., Joulin, A., Mikolov, T.: Enriching word vectors with subword information. Trans. Assoc. Comput. Linguist. **5**, 135–146 (2017)
3. Collobert, R., Weston, J., Bottou, L., Karlen, M., Kavukcuoglu, K., Kuksa, P.: Natural language processing (almost) from scratch. J. Mach. Learn. Res. **12**, 2493–2537 (2011)
4. Conneau, A., Lample, G., Ranzato, M., Denoyer, L., Jégou, H.: Word translation without parallel data. arXiv preprint arXiv:1710.04087 (2017)
5. Devlin, J., Chang, M.W., Lee, K., Toutanova, K.: BERT: pre-training of deep bidirectional transformers for language understanding. arXiv preprint arXiv:1810.04805 (2018)
6. Grave, E., Bojanowski, P., Gupta, P., Joulin, A., Mikolov, T.: Learning word vectors for 157 languages. In: Proceedings of the International Conference on Language Resources and Evaluation (LREC 2018) (2018)
7. Hashemi, M.: Web page classification: a survey of perspectives, gaps, and future directions. Multimed. Tools Appl. **79**, 1–25 (2020). https://doi.org/10.1007/s11042-019-08373-8
8. Kim, Y.: Convolutional neural networks for sentence classification. arXiv preprint arXiv:1408.5882 (2014)
9. Kowsari, K., Jafari Meimandi, K., Heidarysafa, M., Mendu, S., Barnes, L., Brown, D.: Text classification algorithms: a survey. Information **10**(4), 150 (2019)
10. Lee, J.H., Yeh, W.C., Chuang, M.C.: Web page classification based on a simplified swarm optimization. Appl. Math. Comput. **270**, 13–24 (2015)
11. Li, H., Xu, Z., Li, T., Sun, G., Choo, K.K.R.: An optimized approach for massive web page classification using entity similarity based on semantic network. Futur. Gener. Comput. Syst. **76**, 510–518 (2017)
12. Mahajan, D., et al.: Exploring the limits of weakly supervised pretraining. In: Proceedings of the European Conference on Computer Vision (ECCV), pp. 181–196 (2018)
13. Mikolov, T., Chen, K., Corrado, G., Dean, J.: Efficient estimation of word representations in vector space. arXiv preprint arXiv:1301.3781 (2013)
14. Mikolov, T., Sutskever, I., Chen, K., Corrado, G.S., Dean, J.: Distributed representations of words and phrases and their compositionality. In: Advances in Neural Information Processing Systems, pp. 3111–3119 (2013)

15. Pennington, J., Socher, R., Manning, C.: GloVe: global vectors for word representation. In: Proceedings of the 2014 Conference on Empirical Methods in Natural Language Processing (EMNLP), pp. 1532–1543 (2014)
16. Qi, X., Davison, B.D.: Web page classification: features and algorithms. ACM Comput. Surv. (CSUR) **41**(2), 1–31 (2009)
17. Tsoumakas, G., Katakis, I., Vlahavas, I.: Mining multi-label data. In: Data Mining and Knowledge Discovery Handbook, pp. 667–685. Springer (2009). https://doi.org/10.1007/978-0-387-09823-4_34
18. Vaswani, A., et al.: Attention is all you need. In: Advances in Neural Information Processing Systems, pp. 5998–6008 (2017)
19. Young, T., Hazarika, D., Poria, S., Cambria, E.: Recent trends in deep learning based natural language processing. IEEE Comput. Intell. Mag. **13**(3), 55–75 (2018)
20. Zhang, H., Liu, G., Chow, T.W., Liu, W.: Textual and visual content-based anti-phishing: a Bayesian approach. IEEE Trans. Neural Netw. **22**(10), 1532–1546 (2011)

Sequential Model-Based Optimization for Natural Language Processing Data Pipeline Selection and Optimization

Piyadanai Arntong and Worapol Alex Pongpech(✉)

Faculty of Applied Statistics, National Institute of Development Administration, Bangkok, Thailand
piyadanai.arnt@stu.nida.ac.th, worapol@as.nida.ac.th

Abstract. Natural language processing (NLP) aims to analyze a large amount of natural language data. The NLP computes textual data via a set of data processing elements which is sequentially connected to a path data pipeline. Several data pipelines exist for a given set of textual data with various degrees of model accuracy. Instead of trying all the possible paths, such as random search or grid search to find an optimal path, we utilized the Bayesian optimization to search along with the space of hyper-parameters learning. In this study, we proposed a data pipeline selection for NLP using Sequential Model-based Optimization (SMBO). We implemented the SMBO for the NLP data pipeline using Hyperparameter Optimization (Hyperopt) library with Tree of Parzen Estimators (TPE) model and Adaptive Tree of Parzen Estimators (A-TPE) model for a surface model with expected improvement (EI) acquired function.

Keywords: Natural language processing · Sequential Model-based Optimization · Tree of Parzen Estimators

1 Introduction

In the past few years, text data generated from social networks has been steadily and rapidly increasing. Given the current situation of the novel coronavirus (Covid-19), text data potentially increase at an exponential rate. Moreover, these text data extracted from social networks is not only a valuable source of social network analysis but is also utilized to understand people's behaviours, and provides a clearer picture of how people express their opinions.

The method of extracting information from given sets of textual data is a crucial part of NLP, which is a branch of artificial intelligence dealing with the interaction between computers and humans using the natural language. It is composed of three major processes:

N. T. Nguyen et al. (Eds.): ACIIDS 2021, LNAI 12672, pp. 303–313, 2021.
https://doi.org/10.1007/978-3-030-73280-6_24

1. The preprocessing step involving cleaning data before extracting features.
2. The information retrieval process involving transforming text into a modelable format.
3. The machine learning modeling process to understand the language.

The preprocessing component of NLP involves several complex sequential tasks, such as removing unwanted characters, converting all text into lowercase, splitting text from a given sentence into words (tokenization), removing the most generally common words, spelling check of words, and converting words to a derivation word (stemming and lemmatization).

Uysal [1] demonstrated in the study of Emails and News in English and Turkish Language Classification that the preprocessing step in the text classification is as essential as feature extraction, feature selection, and classification steps. The study of spam detection on hotels reviews by Etaiwi [2] found that not only preprocessing steps is crucial but also impacts each preprocessing step on each classification method.

Alam [3] performed a sentiment analysis on Twitter API and suggested that text preprocessing impacts model accuracy, and each preprocessing step also impacts the accuracy of a machine learning algorithm.

In the information retrieval, in which the text is converted into a modelable format, the complexity is in the parameter tuning, such as the number of words (n) in the TF term and the limitation of the top-k frequency term per text document to reduce the size of the TF-IDF in order to solve the overfitting problem.

Tripathy [4] carried out a sentiment analysis of IMDB. They experimented the tuning of n-gram of TF-IDF with Machine Learning Algorithms Naive Bayes, Maximum Entropy, Stochastic Gradient Descent, and SVM. The study showed that the n-gram, which produced a high accuracy, is varied among different Machine Learning Algorithms. For example, the combinations of uni-gram, bi-gram, tri-gram gave better results on Naive Bayes and SVM, while uni-gram worked better with the Maximum Entropy n-gram classifier and Stochastic Gradient Descent.

Gaydhani [5] who studied the Detection of Hate Speech and Offensive Language on Twitter determined that the combination of uni-gram, bi-gram, tri-gram provided a high accuracy with TF-IDF L2 Norm on machine learning algorithms, Naive Bayes, Logistic Regression and Support Vector Machines(SVM).

Most models of the machine learning model process usually need one or more parameters depending on the model's complexity. Complex models require more parameters to be tuned. This tuning of the parameters in each step is required for selecting the most accurate result. Finding the optimal parameters in each step for each textual dataset and model is however highly complicated and time-consuming.

Steps and their chosen sequences are essential and directly impact model accuracy. Finding the optimal steps and sequences is a problem of data pipeline optimization. This problem is a part of the Hyperparameter Tuning problem where more than one parameter must be tuned simultaneously. Our literature review summarized most of the works in data pipeline optimization on structure data pipeline optimization [6], but do not apply to unstructured data pipeline for the NLP.

In this paper, we focused mainly on the standard NLP processing requirements which remove unwanted characters, convert all text to lower case, split text from sentence to words, and remove the most generally common words (stop word). The hyper-parameters that we will be considered are given as follows:

- Method for converting words to a derivation word (Stemming and Lemmatization) choosing one on Porter Stemming, Snowball Stemming, Lancaster Stemming, or not do this step.
- Parameter of the TF-IDF language model by setting N-Gram to Unigram (1-gram), Bigram (2-gram), Combination Unigram+Bigram (1,2-gram), and Combination of Unigram+Bigram+Trigram (1,2,3-gram).
- Min_df parameter ,cut-off threshold of vocabulary ignore terms, choosing 1 of 14 on this set [1, 2, 3, 4, 5, 6, 7, 8, 9, 10 ,20, 30, 40, 50].

We used Language Model TF-IDF, with model Multinomial Naive Bay, Logistic Regression, Support Vector Classifier, and Random Forest to classify a class of text, then evaluated with the weight F1-score.

The objective of data pipeline tuning is to maximize a weighted F1-score. We tested 224 possible hyper-parameters and compared with a model baseline of the data pipeline by choosing parameters via random process.

This paper used Distributed Hyperparameter Optimization (Hyperopt) library to experimental SMBO with NLP data pipeline optimization, and using Random search, Tree of Parzen Estimators (TPE) and Adaptive-Tree of Parzen Estimators (A-TPE) model for an acquisition model with expected improvement (EI) as acquired function for performing SMBO optimization with 20 iterations limited resources or 9% of possible hyper-parameters. The contributions are summarized as follows:

1. Impact analysis of different preprocessing steps and methods.
2. Using Sequential Model-Based Optimization (SMBO) performs NLP data pipeline hyperparameter tuning, compared with the random search on limited time resources.

The second section of the paper presents the background of SMBO and related works. The third section discusses our finding on the data pipeline optimization for NLP using SMBO. The last section presents our conclusion and future work.

2 Related Works

Models with few parameters benefit most from Hyperparameter tuning in which they can utilize grid search to test every possible parameter. In the case of large searching space, the tuning process uses Random Search instead of picking only a certain set of parameters to test depending on the limited time [7].

We found many studies on the Tuning data pipeline for Structure Data and Un-Structure Data. Our literature review does not show research on Tuning NLP for the whole data pipeline. Specifically, the study of PolyMage Automatic Optimization for Image Processing Pipelines Optimization [8], the study of TPOT (A Tree-Based Pipeline Optimization Tool), and the AutoML for structure data [9], are prevalent in the research field.

There are many studies on Tuning Pipeline of Structure Data, for example, Quemy [6] used Sequential Model-Based Optimization (SMBO) to perform Data Pipeline Optimization on the preparation of structure data. Subsequently, the result shows that the SMBO cold identify optimized parameters faster than random search within the same number of limited resources.

SMBO is a technique for tuning of data pipeline hyperparameter based on Bayesian Optimization. This technique is designed to select the best hyperparameter from many sets of parameters at each step to construct a data pipeline to gives a high value of accuracy or objective function, according to the number of rounds specified [10].

SMBO is a state-of-the-art hyperparameter optimization method as given in Fig. 1. SMBO does not test all hyper-parameters because each single trial uses high resources to complete and return due to a functional loss. SMBO can choose the best hyper-parameters minimizing a loss function in the limited resources. The SMBO is based on the continual statistics research on experimental study for black-box function optimization [10]. SMBO is also used as the basis of Automatic machine learning (AutoML) optimization [11].

Input: *Hyperparameter space* Λ, *observation history* $\mathcal{H}$, *transfer function* $\mathcal{T}$, *acquisition function* a, *surrogate model* Ψ, *trade off parameter* α, *hyperparameter response function* f, *to be minimized, total number of HOM iterations* T
Output: *Best hyperparameter configuration found.*

1. $\Lambda_0 \leftarrow \varnothing, f_{best} \leftarrow \infty$
2. **for all** $t = 1...T$ **do**
3. $\quad$ *Fit* Ψ *to* $\mathcal{H}$
4. $\quad$ $\lambda \leftarrow arg\ min_{\lambda' \in \Lambda}(1-\alpha_t)\mathcal{T}(\lambda', \Lambda_{t-1}) - \alpha_t a(\lambda', \Psi)$
5. $\quad$ $\Lambda_t \leftarrow \Lambda_{t-1} \cup \{\lambda\}$
6. $\quad$ **Evaluate** $f(\lambda)$
7. $\quad$ $\mathcal{H} \leftarrow \mathcal{H} \cup \{(\lambda, f(\lambda))\}$
8. $\quad$ **if** $f(\lambda) < f_{best}$ **then**
9. $\quad\quad$ $\lambda_{best}, f_{best} \leftarrow \lambda, f(\lambda)$
10. **return** λ_{best}

Fig. 1. Hyperparameter optimization machines

The SMBO focuses on the construction of a regression model, usually called a response surface model, that captures the dependence of the loss function of the various hyper-parameters and then uses this model for optimization.

The SMBO initialized state takes some data gathered from trialing randomized hyper-parameters of the real model for fitting a response surface model. The response surface model has many options; for instance, the Gaussian process (GP) and Tree of Parzen Estimators (TPE) [12].

The SMBO uses this predictive performance model as an acquisition function to select a promising parameter configuration for the next run. An acquisition function combined with two objectives: firstly the one with high in regions of low predictive mean, and secondly the one with high predictive variance, or a so-called expected improvement (EI) criterion.

The SMBO iterates between fitting a model and gathering additional data based on this model until the time budget has been exhausted.

This paper discusses Distributed Hyperparameter Optimization (Hyperopt) library [13] to experimental SMBO with NLP pipeline optimization, and using Tree of Parzen Estimators (TPE) model and Adaptive Tree of Parzen Estimators (A-TPE) model, for a surface model with expected improvement (EI) acquired function.

3 Experiments

3.1 Impact Analysis of Applied Different Preprocessing Step and Method

We used the K-fold technique to split the training and testing data to minimize the bias. We used K = 5 throughout this study. We chose the weighted F1-Score average on each K-Fold to evaluate a model accuracy because this score can be utilized to evaluate both binary and multiple classification models. We calculated the weighted F1-Score using an average of F1-Score on each class multiplied by the number of samples.

We tested our SMBO optimization using the three commonly utilized machine learning models commonly in the NLP: the Multinomial Naive Bayes model, the Logistic Regression model, and the Random Forest model. We tested our optimization using the SMS Spam Detection dataset to represent a small size dataset, the IMDB Review Sentiment Analysis dataset to represent a medium-sized data set, and the BBC News Category Classification as a large dataset.

We also tested Stemming, N-Gram, and min_df parameter NLP preprocessing and feature extraction, whether they could affect the model accuracy (weighted F1-Score). We then determined whether the parameter changing in each step would affect the F1-score of each model and dataset.

Figure 2(a) displays the result of changing a parameter on the small dataset SMS. We found that the stemming does not obviously impact the models. Figure 2(b) illustrates that N-Gram does have impacts in all of the models. Figure 2(c), min_df does present a clear impact on all models. We here show that the highest weighted F1-score model is the Random Forest.

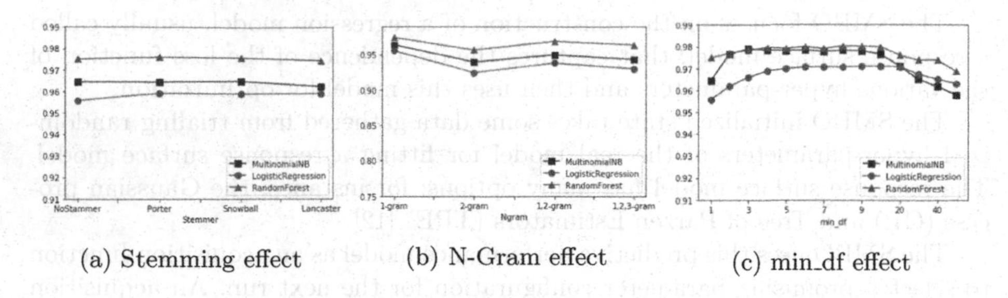

(a) Stemming effect (b) N-Gram effect (c) min_df effect

Fig. 2. The preprocessing effect on SMS dataset

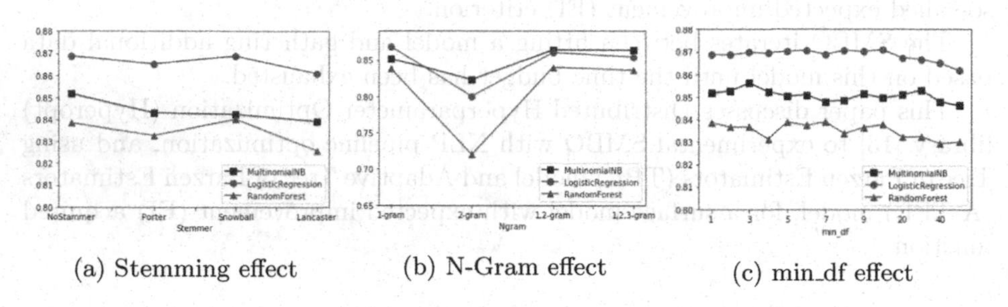

(a) Stemming effect (b) N-Gram effect (c) min_df effect

Fig. 3. The preprocessing effect on IMDB dataset

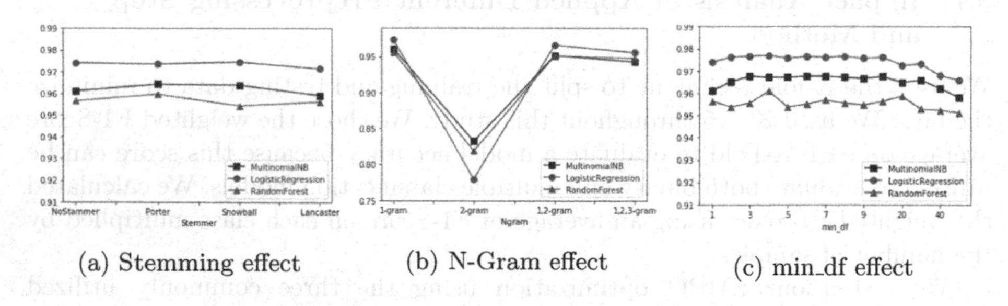

(a) Stemming effect (b) N-Gram effect (c) min_df effect

Fig. 4. The preprocessing effect on BBC dataset

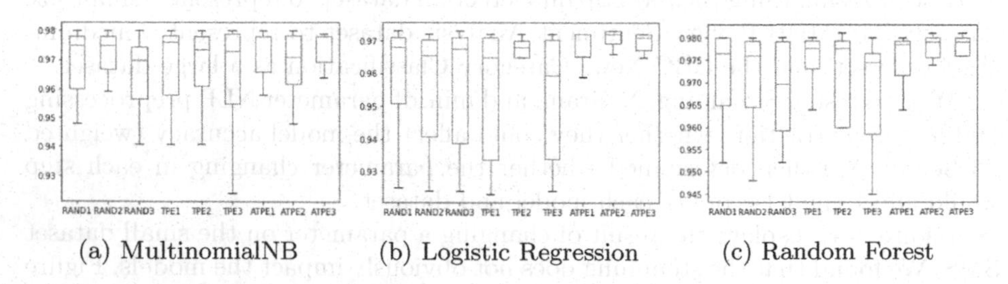

(a) MultinomialNB (b) Logistic Regression (c) Random Forest

Fig. 5. Data-pipeline optimization result on each machine learning model using SMS dataset

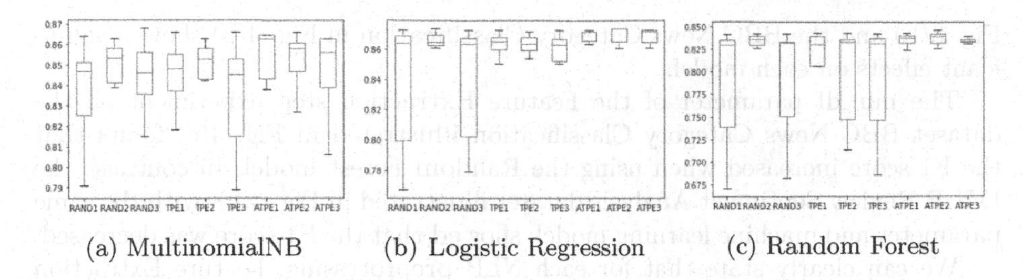

(a) MultinomialNB (b) Logistic Regression (c) Random Forest

Fig. 6. Data-pipeline optimization result on each machine learning model using IMDB dataset

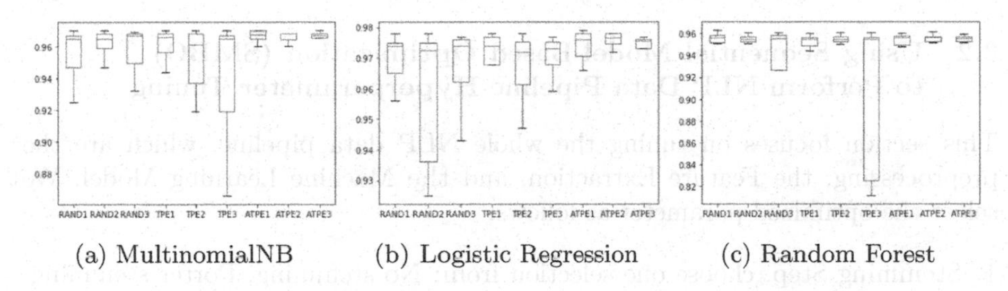

(a) MultinomialNB (b) Logistic Regression (c) Random Forest

Fig. 7. Data-pipeline optimization result on each machine learning model using BBC dataset

Figure 3 shows the changes in parameter on the dataset IMDB Review Sentiment Analysis. Figure 3(a) shows that the stemming does have a concrete impact only on Random Forest Model and Logistic Regression. Figure 3(b) shows that the N-Gram does impact every tested model. Finally, Fig. 3(c) shows that the min_df also does have some impacts on every tested model. We found that the Logistic Regression model provides the highest weighted F1-score.

Figure 4 illustrates the changes in parameter on dataset BBC News Category Classification. Figure 4(a) shows that doing stemming for each method did not significantly impact any models. Figure 4(b) shows that the N-Gram does have a clear impact on all of the models. Figure 4(c) shows that the min_df has some impacts on each model. We also show that the Logistic Regression model shows highest weighted F1-score model.

The stemming methods experiment showed that each dataset and each model have different effects on each stemming method. For instance, dataset SMS Spam Detection in Fig. 2(a) rarely has any effect on Model Multinomial Naive Bayes and Random Forest. On the other hand, the IMDB Review Sentiment Analysis dataset in Fig. 3(a) significantly affects each model.

The N-Gram step experiment showed that each dataset and each model have a different effect on each N-Gram, similar to the stemming result. For instance, on a dataset SMS Spam Detection in Fig. 2(b) has a slight effect on all machine learning models. Conversely, the IMDB Review Sentiment Analysis dataset in

Fig. 3(b) and the BBC News Category Classification in Fig. 4(b) showed significant effects on each model.

The min_df parameter of the Feature Extraction step experiment on the dataset BBC News Category Classification illustrated in Fig. 4(c) found that the F1-score increased when using the Random Forest model. In contrast, the IMDB Review Sentiment Analysis dataset illustrated in Fig. 3(c), with the same parameter and machine learning model, showed that the F1-score was decreased.

We can clearly state that for each NLP preprocessing, Feature Extraction step, the stemming, N-gram, and min_df parameter, with a different parameter, have different effects on each dataset and machine learning models.

3.2 Using Sequential Model-Based Optimization (SMBO) to Perform NLP Data Pipeline Hyperparameter Tuning

This section focuses on tuning the whole NLP data pipeline, which are the preprocessing, the Feature Extraction, and the Machine Learning Model. We select the optimized parameter as follows.

1. Stemming Step, choose one selection from: No stemming, Porter stemming, Snowball stemming, and Lancaster stemming.
2. N-gram on Feature Extraction step, choose one selection from: Unigram (1-gram), Bigram (2-gram), Combination Unigram+Bigram (1,2-gram), and Combination of Unigram+Bigram+Trigram (1,2,3-gram).
3. Min_df parameter on Feature Extraction step, choose one selection from: 1, 2, 3, 4, 5, 6, 7, 8, 9, 10 ,20, 30, 40, 50

We tuned the whole NLP data pipeline using the Sequential Model-Based Optimization (SMBO) Technique from Distributed Hyperparameter Optimization (Hyperopt) library.

We tested three methods: Random Search, SMBO using Tree of Parzen Estimators (TPE) as response surface model, and SMBO using Adaptive-Tree of Parzen Estimators (A-TPE) as response surface model, with the following machine learning models: Multinomial Naive Bayes, Logistic Regression and Random Forest.

We used twenty iterations per round, and we executed each method and each model three rounds to reduce bias and increasing clarity for a better comparison using the same three datasets as the previous testing.

Figure 5 shows the experiment of Data Pipeline Optimization on Dataset SMS Spam Detection. We found that on the Model Multinomial Naive Bayes illustrated in Fig. 5(a), the used technique of Random and SMBO with TPE shows the not-so-different value of the median and the distribution. Whereas the SMBO with A-TPE has a better median, and the distribution is not different from the aforementioned value. We found that the chosen Parameter by A-TPE in each iterator gives the value of F1-Score to be in an upward trend.

Model Logistic Regression illustrated in Fig. 5(b) shows the use of the Random Search technique and SMBO with TPE has a slight difference in median and distribution value. However, the SMBO with A-TPE shows a better value of median and distribution.

Model Random Forest illustrated in Fig. 5(c) illustrates the use of the SMBO technique with TPE gives a better median than Random Search but rarely shows any different distribution value. Nevertheless, the SMBO with A-TPE provides the best Median and Distribution when compared to the previous models.

Figure 6 shows our experiment on Data Pipeline Optimization on the IMDB Review Sentiment Analysis dataset. The finding is that the Model Multinomial Naive Bayes illustrated in Fig. 6(a) on technique Random and SMBO with TPE yield not much different value of Median and Distribution. However, SMBO with A-TPE has a better Median, while Distribution shows the same value.

Model Logistic Regression illustrated in Fig. 6(b) found that the use of technique SMBO with TPE provides a better Distribution than the Random Search. However, the value of their Median is indifferent. However, the SMBO with A-TPE has a better Median and Distribution Value.

Model Random Forest illustrated in Fig. 6(c) demonstrates that the Random Search and the SMBO with TPE has shown a slightly different value of Median and Distribution. On the other hand, the SMBO with A-TPE has a better Distribution, but the Median is not that much different.

Figure 7 depicts the Data Pipeline Optimization on BBC News Category Classification dataset. We found that on Model Multinomial Naive Bayes illustrated in Fig. 7(a), the use of the Random Search technique and the SMBO with TPE gives no difference in the value of median and distribution. Conversely, the SMBO with A-TPE has a better Distribution but not much varied in Median.

Model Logistic Regression illustrated in Fig. 7(b) shows the use of the SMBO with TPE that gives a better Distribution than the Random Search but shows less likely differences of the Median value. Overall, the SMBO with A-TPE has a better Median and Distribution.

Model Random Forest illustrated in Fig. 7(c), the use of the techniques: Random, SMBO with TPE, and SMBO with A-TPE yields little or no differences in result of both Median and Distribution.

We have shown that SMBO can be utilized to facilitate NLP data pipeline selection and optimization. Each section of our experiment demonstrated the benefits of using SMBO.

The data pipeline optimization using sequential Model-based Optimization (SMBO) experiment found that, with the Multinomial Naive Bayes model, the distribution and F1-score median of Random Search and Tree of Parzen Estimators (TPE) are not significantly different on every dataset illustrated in Fig. 5–7(a). However, the distribution of Adaptive-Tree of Parzen Estimators (A-TPE) is better than the other methods on the BBC News Category Classification dataset illustrated in Fig. 7(a).

We found that, in the Model Logistic Regression, the distribution of Tree of Parzen Estimators (TPE) on the dataset IMDB Review Sentiment Analysis and BBC News Category Classification performed better than the Random Search as shown in Fig. 6(b), 7(b). Furthermore, the distribution and F1-score median of Adaptive-Tree of Parzen Estimators (A-TPE) are higher than other methods in all datasets as shown in Fig. 5–7(b).

We also found that, on the Model Random Forest, the distribution of Random Search and Tree of Parzen Estimators (TPE) on the Dataset BBC News Category Classification and IMDB Review Sentiment Analysis are not significantly different as shown in Fig. 6–7(c). However, the Dataset SMS Spam Detection where the median of F1-Score of Tree of Parzen Estimators (TPE) is better than Random Search shown in Fig. 5 (c). The Adaptive-Tree of Parzen Estimators (A-TPE) provided the best result compared to the other methods with the same dataset shown in Fig. 5–7(c).

Consequently, every method could identify a parameter of a data pipeline that generates the best F1-Score when it has enough iteration, in this paper is 20 iteration from 224 possible parameter or 9% of a total possible parameter.

4 Conclusion

This study successfully used sequential Model-based Optimization (SMBO) to perform NLP Data Pipeline Selection and Optimization. We showed that SMBO made most of the datasets and machine learning models, had better mean and distribution scores than Random Searches. The SMBO allowed a better understanding of selecting the next set of parameters on the next iteration likely to make a better score. This demonstrated that the SMBO performs better than the Random Search.

We anticipate that the sequential Model-based Optimization (SMBO) has a high potential for improvement. This paper describes the building of a surface model from Random, Tree of Parzen Estimators (TPE), and Adaptive-Tree of Parzen Estimators (A-TPE). However, the other surface models, such as Gaussian Process or Random Forest could be investigated further for the NLP data pipeline optimization. Moreover, experiments on a model for initializing the first step of SMBO, such as Meta-Learning will benefit the study for the data pipeline optimization.

References

1. Uysal, A.K., Gunal, S.: The impact of preprocessing on text classification. Inf. Process. Manage. **50**(1), 104–112 (2014)
2. Naymat, G., Etaiwi, W.: The impact of applying different preprocessing steps on review spam detection. In: The 8th International Conference on Emerging Ubiquitous Systems and Pervasive Networks (2017)
3. Alam, S., Yao, N.: The impact of preprocessing steps on the accuracy of machine learning algorithms in sentiment analysis. Comput. Math. Organ. Theory **25**(3), 319–335 (2019)
4. Tripathy, A., Agrawal, A., Rath, S.K.: Classification of sentiment reviews using N-gram machine learning approach. Expert Syst. Appl. **57**, 117–126 (2016)

5. Gaydhani, A., Doma, V., Kendre, S., Bhagwat, L.: Detecting hate speech and offensive language on twitter using machine learning: an N-gram and TFIDF based approach. arXiv preprint arXiv:1809.08651 (2018)
6. Quemy, A.: Data pipeline selection and optimization. Design, Optimization, Languages and Analytical Processing of Big Data (2019)
7. Bergstra, J., Bengio, Y.: Random search for hyper-parameter optimization. J. Mach. Learn. Res. **13**(1), 281–305 (2012)
8. Mullapudi, R.T., Vasista, V., Bondhugula, U.: PolyMage: automatic optimization for image processing pipelines. ACM SIGARCH Comput. Archit. News **43**(1), 429–443 (2015)
9. Moore, J.H., Olson, R.S.: TPOT: a tree-based pipeline optimization tool for automating machine learning, pp. 151–160 (2019)
10. Hutter, F., Hoos, H.H., Leyton-Brown, K.: Sequential model-based optimization for general algorithm configuration. In: Coello, C.A.C. (ed.) LION 2011. LNCS, vol. 6683, pp. 507–523. Springer, Heidelberg (2011). https://doi.org/10.1007/978-3-642-25566-3_40
11. Chauhan, K., et al.: Automated machine learning: the new wave of machine learning, pp. 205–212 (2020)
12. Bergstra, J.S., Bardenet, R., Bengio, Y., Kégl, B.: Algorithms for hyper-parameter optimization. In: Advances in Neural Information Processing Systems, pp. 2546–2554 (2011)
13. Bergstra, J., Yamins, D., Cox, D.D.: Hyperopt: A python library for optimizing the hyperparameters of machine learning algorithms. In: Proceedings of the 12th Python in Science Conference, vol. 13, p. 20. Citeseer (2013)

Automatic Cyberbullying Detection on Twitter Using Bullying Expression Dictionary

Jianwei Zhang[1](✉), Taiga Otomo[1], Lin Li[2], and Shinsuke Nakajima[3]

[1] Iwate University, Morioka, Japan
zhang@iwate-u.ac.jp
[2] Wuhan University of Technology, Wuhan, China
cathylilin@whut.edu.cn
[3] Kyoto Sangyo University, Kyoto, Japan
nakajima@cc.kyoto-su.ac.jp

Abstract. Cyberbullying has become a serious problem with the spread of personal computers, smartphones and SNS. In this paper, for automatic cyberbullying detection on Twitter, we construct a bullying expression dictionary, which registers bullying words and their degrees related to bullying. The words registered in the dictionary are those that appear in the collected bullying-related tweets, and the bullying degrees attached to the words are calculated using SO-PMI. We also construct models to automatically classify bullying and non-bullying tweets by extracting multiple features including the bullying expression dictionary and combining them with multiple machine learning algorithms. We evaluate the classification performance of bullying and non-bullying tweets using the constructed models. The experimental results show that the bullying expression dictionary can contribute to cyberbullying detection in most of the machine learning algorithms and that the best model can obtain an evaluation of over 90%.

Keywords: Cyberbullying · Bullying expression dictionary · Twitter · Machine learning

1 Introduction

In recent years, cyberbullying has become a serious problem with the spread of personal computers, smartphones and SNS. American national statistics show that about 15% of high school students were bullied online or by text in 2017 [1]. In Japan, according to a survey conducted by the MEXT, the number of cases of cyberbullying recognized in 2017 was 12,632, which is increasing year by year [2]. At present, much research on cyberbullying is being studied in the education field. In the information technology field, most of the research related to automatic cyberbullying detection focus on English data, and the analysis of Japanese data is still limited. Moreover, the analysis of features for automatic cyberbullying detection is insufficient. Therefore, there is a need for a technology

N. T. Nguyen et al. (Eds.): ACIIDS 2021, LNAI 12672, pp. 314–326, 2021.
https://doi.org/10.1007/978-3-030-73280-6_25

that can extract predictive features and detect cyberbullying automatically and accurately for Japanese data.

In this paper, we utilize machine learning methods to realize automatic cyberbullying detection. Two important aspects of classifying cyberbullying based on machine learning are to extract what features and to select what machine learning algorithms. Targeting the texts on Twitter (hereinafter referred to as "tweet"), we intend to find the features that mostly contribute to cyberbullying detection. In our previous work [3], we have analyzed a range of textual features. Although we consider that emotion feature is useful for cyberbullying detection, the existing resources of Japanese emotion dictionary are constructed on newspaper articles [4] or national language dictionaries [5]. Some words specific to the internet and new slang phrases are not registered in these dictionaries. Therefore, in this paper, we construct a bullying expression dictionary that includes the words specific to the internet and new slang phrases and apply it to cyberbullying detection. The bullying expression dictionary is constructed based on Twitter texts and registers bullying words and their degrees related to bullying. The words registered in the dictionary are those that appear in the tweets collected based on some specific bullying words, and the bullying degrees attached to the words are calculated using SO-PMI. Other features such as n-gram, Word2vec, Doc2vec except for the bullying expression dictionary are also tested for cyberbullying detection. Moreover, we combine the features with multiple machine learning algorithms to find an optimal model. The results of evaluating the automatic classification of bullying and non-bullying tweets show that the bullying expression dictionary can contribute to cyberbullying detection in most of the machine learning algorithms and the best model can obtain an evaluation of over 90%.

2 Related Work

Most research on automatic cyberbullying detection focused on English resources and intended to improve the accuracy of cyberbullying detection classifiers. Burnap et al. used hashtags to collect tweets related to a murder incident and used n-grams and grammatical dependencies between words to automatically detect hate speech focusing on race, ethnicity, or religion [6]. Rafiq et al. collected Vine video sessions and proposed to automatically detect instances of cyberbullying over Vine by using the features such as the information of the video, the user who posted it, the emotion of the comment, n-grams [7]. Hosseinmardi et al. collected Instagram data consisting of images and their associated comments and proposed to automatically detect incidents of cyberbullying over Instagram by using the features such as the information of the image, the user who posted it, the interval between comments, n-grams [8]. Nobata et al. proposed to automatically detect abusive language from comments on Yahoo! Finance and Yahoo! News by using n-grams, linguistic, syntactic and distributional semantics features [9]. Chatzakou et al. analyzed bullying and aggressive behaviours on Twitter and distinguished bullies and aggressors from regular users by using text, user, and network-based attributes [10].

In contrast with the work of solving the accuracy issue, Rafiq et al. aimed to address the issues of scalability and timeliness of cyberbullying detection systems by proposing a multi-stage method comprised of a dynamic priority scheduler and an incremental classification mechanism [11]. Li proposed graph neural networks to improve the performance of identifying cyberbullying behaviours and provided model explainability to understand the language use of anti-social behaviours [12]. Cheng et al. introduced an unsupervised cyberbullying detection model consisting of a representation learning network and a multi-task learning network, which can jointly optimize the parameters of both components [13].

Both a survey on cyberbullying detection [14] and the work of Notata et al. [9] pointed out that it remains to be investigated whether the approaches examined on English resources are equally effective in other languages. Some work of cyberbullying detection on the languages other than English targeted Dutch [15] by Van Hee et al. and German [16] by Ross et al. Little work has been done with cyberbullying detection on Japanese. Ptaszynski et al. used a Brute-Force search algorithm to extract patterns and classify Japanese harmful expressions [17]. Using deep learning models, they further improved the results [18].

However, to the best of our knowledge, the investigated aspects for cyberbullying detection on Japanese resources are limited to only some basic features such as tokens, lemmas and POS. Our work aimed at cyberbullying detection from Japanese text on Twitter by studying not only surface features such as n-grams, but also deeper textual features including Word2vec [19], Doc2vec [20], and bullying expressions extracted from Twitter, and investigating the classification performance of multiple machine learning algorithms with these features.

3 Method Overview

Figure 1 shows the overview of the proposed method.

1. Data collection: First, tweets are collected from Twitter for the bullying expression dictionary construction and bullying/non-bullying classification.
2. Bullying expression dictionary construction: The words that should be registered in the bullying expression dictionary are extracted from the tweets, and their bullying degrees are calculated using SO-PMI.
3. Data annotation: The tweets used for classification are specified and labeled as bullying or non-bullying according to their contents by crowdsourcing.
4. Feature extraction: Features are extracted from the collected tweets, including bullying expressions, n-grams, Word2vec, and Doc2vec.
5. Model generation: The labeled tweets are divided into training data, verification data and test data. Models are generated using those features and each machine learning algorithm based on the training and verification data. In order to generate better models and fairly evaluate the models, we also perform cross-validation and grid search.
6. Model evaluation: Using F-measure, we evaluate how well the generated models can classify bullying and non-bullying tweets based on the test data.

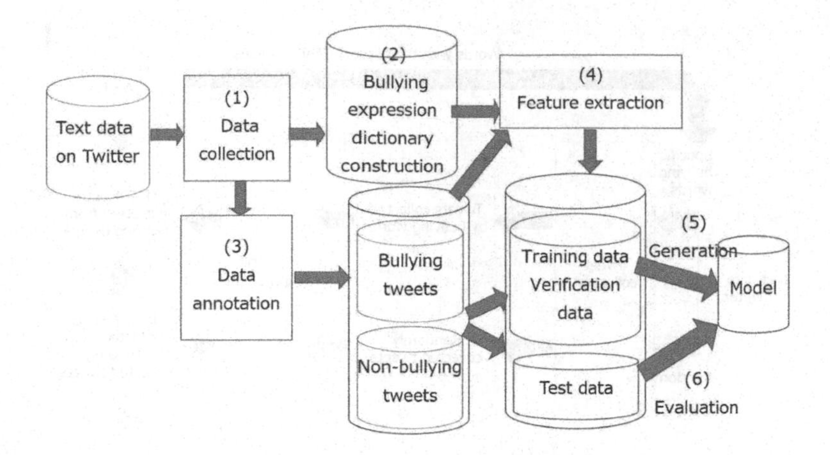

Fig. 1. Overview of the proposed method

4 Bullying Expression Dictionary Construction

There are two issues for constructing a bullying expression dictionary: how to automatically extract words registered in the bullying expression dictionary and how to calculate the bullying degrees of the registered words. We describe the solution to these two requirements.

4.1 Automatic Extraction of Bullying Words

We consider that two kinds of words should be registered in the bullying expression dictionary: one is the words that are abusive, and the other is the words that are not abusive alone but become abusive expressions when co-occurring with other words. An example for the latter is that although the word "monkey" itself is not abusive, the expression "the owner of the brain below the monkey" is abusive. To improve the comprehensiveness of the bullying expression dictionary, we extract both kinds of words. Figure 2 shows the flow of bullying expression extraction.

Tweet Collection. We use a Python library called GetOldTweets3 to collect tweets. The collection period is one week from June 16, 2019, to June 23, 2019. We firstly use 37 specific bullying words introduced in previous research [21,22] to collect tweets as bullying-related tweets, and randomly select tweets as non-bullying tweets. Some characters such as "RT", "FAVORITES", URL, etc. that are unique to Twitter and redundant for bullying classification are excluded from the collected tweets.

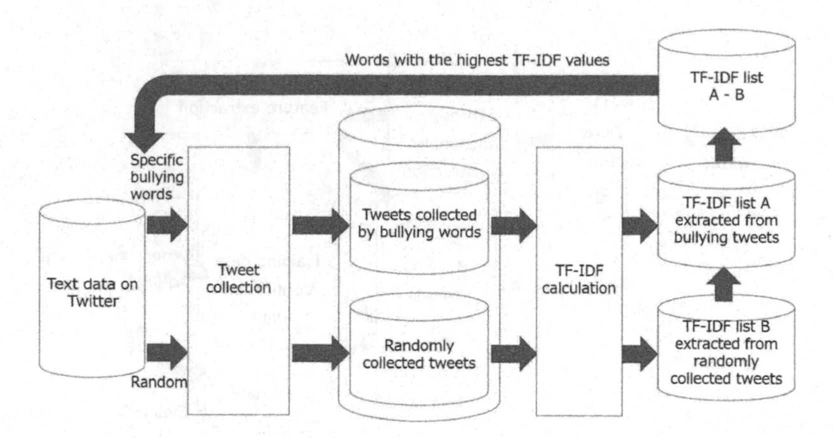

Fig. 2. The flow of bullying expression extraction

TF-IDF Calculation. We use JUMAN++ [23] to perform morphological analysis on the collected tweets. JUMAN++ is a high-performance morphological analysis system that uses the Recurrent Neural Network Language Model (RNNLM) and considers the word sequence. Based on the JUMAN++ analysis results, the morphemes (or words) that have a part of speech of "noun", "verb", "adjective", "undefined word" are extracted. Undefined words are also targeted because new words or slang phrases are likely to be identified as "undefined word" by JUMAN++. We calculate the TF-IDF values for the words respectively from the bullying-related tweets and non-bullying tweets. TF-IDF is one of the methods to evaluate the importance of words contained in documents, and it is mainly used in the fields of information retrieval and topic analysis.

Tweet Recollection. The words with high TF-IDF values in TF-IDF list A extracted from bullying-related tweets also contain some daily words. For automatically excluding these daily words, we extract the words with high TF-IDF values from the randomly selected non-bullying tweets, which constitute TF-IDF list B. We delete the words in B from A and further manually check whether the remaining words are related to bullying. Those words related to bullying are used to recollect new tweets. Since the initial number of bullying words is 37 words, we recollect tweets using the top 30 words with the highest TF-IDF values in the word list (A - B). This recollection process is repeated three times so that a large amount and various kinds of bullying-related tweets can be obtained. The total number of specific bullying words is 97, and the total number of collected tweets is 2,827,284. We use 10% of the collected tweets and the morphemes (or words) extracted from them by JUMAN++ are registered to the bullying expression dictionary.

4.2 Calculation of Bullying Degrees of Registered Words

SO-PMI (Semantic Orientation Using Pointwise Mutual Information) [24] is used to calculate the bullying degrees of the registered bullying expressions. For calculating SO-PMI, two basic words are prepared in advance, and a target word's co-occurrence with the two basic words is measured. The formula for SO-PMI calculation is as follows.

$$C(w) = \log \frac{hit(w, w_p) * hit(w_n)}{hit(w, w_n) * hit(w_p)}$$

$$f(\alpha) = \alpha * \log \frac{hit(w_p)}{hit(w_n)}$$

$$SO-PMI(w) = C(w) + f(\alpha)$$

w: target word
$C(w)$: a value reflecting the co-occurrence of w with w_p and w_n
w_p: bullying basic words
w_n: non-bullying basic words
$hit(w)$: number of tweets containing w
$hit(w, w_p)$: number of tweets containing both w and w_p
$f(\alpha)$: a function that alleviates the impact of the difference in the numbers of $hit(w_p)$ and $hit(w_n)$
α: a constant for setting the weight of the function

In order to select an appropriate value of α for setting the weight of the function $f(\alpha)$, we test four values: 0.9, 0.6, 0.3, and 0.0 in the preliminary experiment. We finally select $\alpha = 0.6$ since in this case, the calculated bullying degrees are the closest to human senses.

The selection of basic words w_p and w_n is also important. The original SO-PMI definition [24] only used one basic bullying word and one basic non-bullying word respectively for w_p or w_n. However, co-occurrence may be insufficient for tweets that are usually short texts if only two basic words are given initially. Therefore, we set multiple basic words for w_p and w_n. For the bullying basic words w_p, we select five words that frequently occur in bullying texts: "Shine (Die)", "Kasu (Scum)", "Kimoi (Disgusting)", "Kuzu (Waste)", "Busu (Ugly woman)". For the non-bullying basic words w_n, we select five words that are commonly used for compliments: "Kawai (Cute)", "Suteki (Nice)", "Ikemen (Handsome man)", "Subarashi (Great)", "Utsukushi (Beautiful)".

4.3 Result

As the result, 86,906 words are registered to the bullying expression dictionary. A word with a larger bullying degree indicates that it is more relevant to cyberbullying. The bullying expression dictionary is used as features for bullying and non-bullying classification in the experiments.

5 Data Annotation

For generating the learning data, Yahoo! Japan crowdsourcing [25] is used to label tweets as bullying or non-bullying. Since the number of the tweets collected in Sect. 4 is too large, we need to consider the crowdsourcing cost and thus select the tweets that will be labeled by crowd workers by ranking the collected tweets. For each of 97 bullying words that are used for collecting tweets in Sect. 4, the tweets containing this word are given a score based on how many specific bullying words are contained in it, i.e., the tweets containing multiple bullying words have higher scores than those containing few bullying words. After that, for each bullying word, from the ranked tweets containing it, 50 tweets with the highest scores and 25 tweets with the lowest scores are extracted. In this way, we intend to obtain tweets with various combinations of bullying words.

After deleting the duplicated tweets obtained respectively by 97 bullying words, 5,399 tweets are collected and sent to Yahoo! Japan crowdsourcing to get bullying or non-bullying labels from crowd workers. Each crowd worker is asked to label ten tweets and each tweet is evaluated by three crowd workers. As a result from the crowds, 660 tweets are judged as bullying by all the three workers, 1,106 tweets are judged as bullying by two of the three workers, 1,524 tweets are judged as bullying by only one worker, and 2,109 tweets are judged as non-bullying by all the three workers. In our experiments, 660 bullying tweets labeled by all the three workers and 1,106 bullying tweets labeled by two of the three workers are used as bullying tweets. For generating features and negative learning data, we also randomly collect 1,766 tweets from Twitter as non-bullying tweets since the probability that randomly selected tweets from large Twitter are bullying is small.

6 Feature Extraction

For applying machine learning to cyberbullying detection, it is important to select features that are effective for the classification of bullying and non-bullying texts. Except for bullying expressions, we also extract n-grams, Word2vec, and Doc2vec from the collected tweets, and investigate which features are predictive.

6.1 Bullying Expression

In this paper, we construct the bullying expression dictionary maintaining the knowledge of the bullying words and their bullying degrees and consider it may be a kind of features that can contribute to cyberbullying detection. We generate a matrix in which each row represents a tweet and each column represents a word registered in the bullying expression dictionary. The value of each element in the matrix is calculated with multiplying the bullying degree of the word (the column) by the occurrence frequency of the word in the tweet (the row).

However, if all the words in the bullying expression dictionary are used as feature vectors, the number of features becomes too large. To solve this problem, we use principal component analysis to reduce the dimension of the feature vector. Principal component analysis uses an orthogonal transformation to convert possibly correlated features into linearly independent features. Generally, after the transformation, only important features that can represent the data characteristic are left for further analysis. The algorithm firstly finds the "first principle component" that is the direction with the highest variance. In this direction the data have the most information. Next, the algorithm finds the "second principle component" that is the direction having the highest variance among the directions orthogonal to the first principle component. The directions found in this way are called "principal components". The number of all the principal components is same as that of original features and the principal components are sorted in the descending order of importance for explaining the data. Since the number of principal components used for further analysis is decided by a parameter, the user can reduce the dimensions of feature space by only leaving the top components. In the experiments, we reduce the feature vector of bullying expressions to 100 dimensions.

6.2 N-gram

Another kind of features is n-gram. N-gram divides a string or document into n continuous characters or words and counts their occurrence times in the string or document. It is considered that the occurrence probability of a character or word depends on the previous character or word. We consider the characters or words often used in bullying text may be predictive features for classification. We use both character division (hereinafter character n-grams, $n = 2{\sim}5$) and word division (hereinafter word n-grams, $n = 1{\sim}5$). A vector for a tweet is generated by using each character n-gram or word n-gram as the element and taking their occurrence frequency in the tweet as the element value.

However, if all the n-grams are used as features, the number of the features will be too large and obviously many elements are not important at all. Especially, most character n-grams are some meaningless combinations of characters and tend to create noise. Therefore, we also use principal component analysis to reduce each kind of n-grams to 100 dimensions.

6.3 Word2vec

We also apply Word2vec to contribute to cyberbullying classification. Word2vec vectorizes the meaning of a word based on the analysis on a large corpus of text data. It can be used to estimate semantic similarity between two words by calculating the inner product of the two word vectors or to understand the relationship between two words by performing addition/subtraction of the two word vectors. In our experiment, Word2vec is trained based on 2,827,284 tweets collected in Sect. 4. The word vector is set to 100 dimensions. For each tweet, its Word2vec feature is generated by averaging the word vectors of the words that appear in the tweet.

6.4 Doc2vec

In contrast with Word2vec that vectorizes the meaning of a word, Doc2vec vectorizes the meaning of a document. Doc2vec also considers the order of words appearing in a sentence. The relationship between two documents can be calculated with Doc2vec. In our experiment, Doc2vec is also trained based on 2,827,284 tweets, same with those for Word2vec. The dimension number of the document (tweet) vector is also set to 100. Since Doc2vec analyzes text in a document unit, each tweet can directly be represented by a Doc2vec vector.

7 Machine Learning Algorithm Selection

For better cyberbullying detection, it is also important to select what machine learning algorithms. We compare six machine learning algorithms including two linear models: linear support vector machine (SVM), logistic regression (LR), three tree-based models: decision tree (DT), random forest (RF), gradient boosting decision tree (GBDT), and one deep learning model: multilayer perceptron (MLP).

8 Experimental Evaluation

8.1 Experiment Setup

Using the labeled tweets, experiments are conducted to confirm how well cyberbullying texts can be classified with each type of features and each type of machine learning algorithms, and which features and which machine learning algorithms are especially effective for classification. The tweets used for classification are 3,532 tweets consisting of 1,766 bullying tweets and 1,766 non-bullying tweets described in Sect. 5.

First, for testing the predictive effects of features, we individually use each of the five kinds of features: bullying expression, character n-gram (n = 2∼5), word n-gram (n = 1∼5), Word2vec, and Doc2vec. Next, for investigating whether the bullying expression dictionary can help improve bullying classification or not, we combine the bullying expression feature with the individual best feature. Finally, we compare the classification results between the case where all the features are used and the case where all but bullying expression features are used.

As the evaluation criterion, F-measure is used. For all types of machine learning algorithms, we perform nested stratified 5-fold cross-verification including the grid search of model parameters. The values of F-measure of five times are averaged so that each model can obtain a more robust evaluation.

8.2 Experimental Results

The results of the classification evaluation are shown in Table 1, Table 2, and Table 3.

Table 1. Classification evaluation (individual feature and all features)

	Bullying expression	Character n-gram	Word n-gram	Word2vec	Doc2vec	All features
SVM	0.822	0.870	0.812	**0.902**	0.881	0.819
LR	0.838	0.870	0.813	0.882	**0.883**	**0.921**
DT	**0.805**	0.759	0.669	0.774	0.706	**0.827**
RF	0.801	0.802	0.695	**0.839**	0.767	**0.840**
GBDT	0.828	0.830	0.760	**0.868**	0.810	**0.892**
MLP	0.845	0.869	0.816	**0.910**	0.882	**0.914**

Table 2. Classification evaluation ("individual best feature" vs "individual best feature + bullying expression")

	Word2vec	Word2vec+ Bullying expression	Doc2vec	Doc2vec+ Bullying expression
SVM	**0.902**	0.894	**0.881**	0.872
LR	0.882	**0.899**	0.883	**0.899**
DT	0.774	**0.812**	0.706	**0.804**
RF	0.839	**0.853**	0.767	**0.813**
GBDT	0.868	**0.890**	0.810	**0.844**
MLP	**0.910**	0.907	0.882	**0.892**

Table 1 shows the values of F-measure that are obtained by applying each machine learning algorithm with each individual feature or all the features. For each machine learning algorithm, the cell of the individual feature leading to the best score is shown in bold type. If using all the features leads to a higher score than using any individual feature, the last column for using all the features is also shown in bold type. The results show that when the individual feature is used, the combination of MLP and Word2vec can obtain the best F-measure 0.910. Checking each individual feature, Word2vec obtains the best scores for SVM, RF, GBDT, and MLP, Doc2vec obtains the best score for LR, and bullying expression obtains the best score for DT. Therefore, the feature of Word2vec greatly contributes to cyberbullying detection in most of the machine learning algorithms, and the feature of bullying expression is especially predictive when the machine learning algorithm is a decision tree.

Table 2 shows the effect of bullying expression by combining it with Word2vec and Doc2vec that are two mostly contributing features. For each machine learning algorithm, we compare the results between individual Word2vec or Doc2vec feature and their feature combination with bullying expression. By combining Word2vec with bullying expression, the F-measure of LR, DT, RF and GBDT are larger than individual Word2vec. By combining Doc2vec with bullying expression, five of the six machine learning algorithms except SVM perform better

Table 3. Classification evaluation ("all features" vs "all features - bullying expression")

	All features	All features - Bullying expression
SVM	0.819	**0.833**
LR	**0.921**	0.920
DT	**0.827**	0.783
RF	**0.840**	0.824
GBDT	**0.892**	0.881
MLP	**0.914**	**0.914**

than individual Doc2vec. Even for the machine learning algorithms that do not improve F-measure with adding bullying expression, the before-after difference is less than 1%. Therefore, it can be said that bullying expression can contribute to cyberbullying detection.

Table 3 also shows the effect of bullying expression by comparing the results between using all the features and using all but the bullying expression features. When the bullying expression feature is removed from all the features, the F-measure values for four of the six machine learning algorithms decrease. This indicates that bullying expression combined with other multiple features can help improve cyberbullying detection. Overall, the best F-measure is 0.921 obtained by using LR and all the features.

8.3 Discussion

We discuss the improvement of the bullying expression dictionary. For the individual feature, Word2vec shows the best scores for most of the machine learning algorithms, while bullying expression obtains the best score only when the machine learning algorithm is a decision tree. Word2vec is generated based on all the collected 2,827,284 tweets while the current bullying expression dictionary is constructed based on 10% of all the collected tweets. Bullying expression may better contribute to cyberbullying detection if more tweets are analyzed and more words are registered to the bullying expression dictionary. Moreover, for calculating the co-occurrence of target words, we adopt five basic bullying words and five basic non-bullying words by referring to previous research. Deliberation on these basic words may further improve the bullying degrees of words registered in the bullying expression dictionary. Besides, since new words and slang phrases appear continuously, it is necessary to periodically reconstruct the bullying expression dictionary.

9 Conclusions

In this paper, we aimed at achieving automatic cyberbullying detection on Twitter. We constructed a bullying expression dictionary consisting of 86,906 words and their bullying degrees from texts on Twitter. Combining six machine learning algorithms and five kinds of features including the bullying expression, we constructed the models to automatically classify bullying and non-bullying tweets. The experimental results showed that classification evaluation can be improved by adding the bullying expression to features and the best model can obtain an evaluation of over 90%.

In the future, we plan to improve the bullying expression dictionary as described in Sect. 8.3 and test the classification performance with the strengthened dictionary.

Acknowledgments. This research was supported by JSPS 19K12230.

References

1. stopbullying.gov. Facts About Bullying (2017). https://www.stopbullying.gov/media/facts/index.html Accessed 27 Dec 2020
2. Bullying investigation results in Japan (2018). https://www.mext.go.jp/content/1410392.pdf Accessed 27 Dec 2020
3. Zhang, J., Otomo, T., Li, L., Nakajima, S.: Cyberbullying detection on twitter using multiple textual features. In: iCAST 2019, pp. 1–6 (2019)
4. Zhang, J., Minami, K., Kawai, Y., Shiraishi, Y., Kumamoto, T.: Personalized web search using emotional features. CD-ARES **2013**, 69–83 (2013)
5. Takamura, H., Inui, T., Okumura, M.: Extracting semantic orientations of words using spin model. ACL **2005**, 133–140 (2005)
6. Burnap, P., Williams, M.L.: Cyber hate speech on twitter: an application of machine classification and statistical modeling for policy and decision making. Policy Int. **7**(2), 223–242 (2015)
7. Rafiq, R.I., Hosseinmardi, H., Han, R., Lv, Q., Mishra, S., Mattson, S.A.: Careful what you share in six seconds: detecting cyberbullying instances in vine. ASONAM **2015**, 617–622 (2015)
8. Hosseinmardi, H., Mattson, S.A., Ibn Rafiq, R., Han, R., Lv, Q., Mishra, S.: Analyzing labeled cyberbullying incidents on the Instagram social network. SocInfo 2015. LNCS, vol. 9471, pp. 49–66. Springer, Cham (2015). https://doi.org/10.1007/978-3-319-27433-1_4
9. Nobata, C., Tetreault, J., Thomas, A., Mehdad, Y., Chang, Y.: Abusive language detection in online user content. WWW **2016**, 145–153 (2016)
10. Chatzakou, D., Kourtellis, N., Blackburn, J., Cristofaro, E.D., Stringhini, G., Vakali, A.: Mean birds: detecting aggression and bullying on Twitter. WebSci **2017**, 13–22 (2017)
11. Rafiq, R.I., Hosseinmardi, H., Han, R., Lv, Q., Mishra, S.: Scalable and timely detection of cyberbullying in online social networks. SAC **2018**, 1738–1747 (2018)
12. Li, C.: Explainable detection of fake news and cyberbullying on social media. WWW **2020**, 398 (2020)
13. Cheng, L., Shu, K., Wu, S., Silva, Y.N., Hall, D.L., Liu, H.: Unsupervised cyberbullying detection via time-informed gaussian mixture model. CIKM **2020**, 185–194 (2020)
14. Schmidt, A., Wiegand, M.: A survey on hate speech detection using natural language processing. In: SocialNLP@EACL 2017, pp. 1–10 (2017)

15. Van Hee, C., et al.: Detection and fine-grained classification of cyberbullying events. RANLP **2015**, 672–680 (2015)
16. Ross, B., et al.: Measuring the reliability of hate speech annotations: the case of the European refugee crisis. In: NLP4CMC 2017, pp. 6–9 (2017)
17. Ptaszynski, M., Masui, F., Kimura, Y., Rzepka, R., Araki, K.: Automatic extraction of harmful sentence patterns with application in cyberbullying detection. LTC **2015**, 349–362 (2015)
18. Ptaszynski, M., Eronen, J.K.K., Masui, F.: Learning deep on cyberbullying is always better than brute force. LaCATODA **2017**, 3–10 (2017)
19. Mikolov, T., Sutskever, I., Chen, K., Corrado, G., Dean, J.: Distributed representations of words and phrases and their compositionality. NIPS **2013**, 3111–3119 (2013)
20. Le, Q., Mikolov, T.: Distributed representations of sentences and documents. ICML **2014**, 1188–1196 (2014)
21. Nitta, T., Masui, F., Ptaszynski, M., Kimura, Y., Rzepka, R., Araki, K.: Detecting cyberbullying entries on informal school websites based on category relevance maximization. IJCNLP **2013**, 579–586 (2013)
22. Hatakeyama, S., Masui, F., Ptaszynski, M., Yamamoto, K.: Statistical analysis of automatic seed word acquisition to improve harmful expression extraction in cyberbullying detection. IJETI **6**(2), 165–172 (2016)
23. Morita, H., Kawahara, D., Kurohashi, S.: Morphological analysis for unsegmented languages using recurrent neural network language model. EMNLP **2015**, 2292–2297 (2015)
24. Wang, G., Araki, K.: Modifying SO-PMI for Japanese weblog opinion mining by using a balancing factor and detecting neutral expressions. ACL **2007**, 189–192 (2007)
25. Yahoo! Japan crowdsourcing. https://crowdsourcing.yahoo.co.jp/ Accessed 27 Dec 2020 from

Development of Morphological Segmentation for the Kyrgyz Language on Complete Set of Endings

Aigerim Toleush[1], Nella Israilova[2], and Ualsher Tukeyev[1](✉)

[1] Al-Farabi Kazakh National University, Almaty, Kazakhstan
[2] Kyrgyz State Technical University named after I.Razzakov, Bishkek, Kyrgyzstan

Abstract. The problem of word segmentation of source texts in the training of neural network models is one of the actual problems of natural language processing. A new model of the morphology of the Kyrgyz language based on a complete set of endings (CSE) is developed. Based on the developed CSE-model of the morphology of the Kyrgyz language, a computational data model, algorithm and a program for morphological segmentation are developed. Experiments on morphological segmentation of Kyrgyz language texts showed 82% accuracy of text segmentation by the proposed method.

Keywords: Morphological segmentation · Kyrgyz language · Morphology · Model

1 Introduction

The problem of the word segmentation in neural network processing of natural language texts is a very urgent problem in connection with the problems of unknown words and rare words, problems of overflowing the memory of dictionaries of the neural machine translation (NMT) system. Especially the problem of overflowing the memory of dictionaries is relevant for agglutinative languages, since when processing texts of agglutinative languages, each word form of the language is written into the system's dictionary, which leads to overflow of the system's memory. In addition, since for training systems of neural network text processing of natural languages, the larger the amount of initial data for training, the better the quality of training the system. The Kyrgyz language belongs to agglutinative languages and is a low-resource language since there are low language resources for natural language processing (NLP).

This paper proposes a solution to the problem of morphological segmentation for the Kyrgyz language based on a new morphology CSE (Complete Set of Endings) – model [1]. The advantage of computational data model of word segmentation based on morphology's CSE-model is its tabular form, which makes it possible to obtain a solution in one-step. Moreover, computational solutions for CSE-model are the universal. Necessary for computational processing of segmentation of a particular language is the construction of a tabular data model of word segmentation of this language.

N. T. Nguyen et al. (Eds.): ACIIDS 2021, LNAI 12672, pp. 327–339, 2021.
https://doi.org/10.1007/978-3-030-73280-6_26

2 Related Works

In recent years, the NMT systems output quality is significantly improved. The main problem of morphologically rich language translation is a fixed-size vocabulary of NMT consisting high frequency words to train the neural network. The using a fixed number of words in language pairs leads to out-of-vocabulary problems. Therefore, neural systems are often mistaken in the translation of rare and unknown words [2].

In addition, the large-size vocabularies are difficult to process because they affect the computational complexity of the model, taking up a large amount of memory. These lead to difficulties when translating the morphological-rich agglutinative languages that need a large vocabulary size.

The agglutinative languages in Turkic languages group have a complex structure of words and many ordering the affixes, which can produce a large vocabulary. An example of such languages with complex morphology and low-resourced is the Kyrgyz language. Because of the inflectional and derivational morphology, it has a large volume of word list. For the Kyrgyz language, the above presented issues are significant because of the inability of the NMT to determine the complex structure of words in the training corpus.

The morphologically rich languages need a deep analysis at the word level. In order to reduce the vocabulary size and improve the translation of morphologically rich languages, morphological word segmentation is used. The morphological segmentation is a process of dividing words in sub-units. As the result of morphological segmentation, the words in the agglutinative languages splits to the word stem and affixes that reduces the vocabulary size for model training. Previous studies had shown that a morpheme-based modeling of the language can provide a better performance [3].

Two methods for unsupervised segmentation of words into small units in Finnish and English languages were introduced in [4]. They used Minimum Description Length (MDL) principle with recursive segmentation in the first method. The linear segmentation and Maximum Likelihood (ML) optimization are used in the second method. For the morphologically rich Finnish language, the recursive segmentation with the MDL principle had shown better results than sequential segmentation.

The BPE (byte pair encoding) method was proposed as rare word segmentation and translation of them by NMT [2]. In this paper was showed that the NMT was able to translate rare words by their subunits. To increase the quality of word segmentation they had used list of stop words, words that cannot be divided into segments. At the result, they demonstrated that NMT models with open vocabulary could better translate rare words than models with large-size vocabularies.

A group of scientists represented word segmentation method that reduces the vocabulary of agglutinative languages and maintains the linguistically word properties [5]. They called this method Linguistically Motivated Vocabulary Reduction (LMVR). To develop the LMVR they used unsupervised segmentation framework Morfessor Flatcat and can be used for any language pairs where a source language is morphologically rich. The LMVR outputs better results than the byte pair encoding technique [6].

It should be noted that the task of word segmentation differs from the task of morphological analysis. Therefore, we do not provide an analysis of works on morphological analysis. Although, of course, the problem of word segmentation can be solved using adapted methods and algorithms for morphological analysis [7, 8]. However, the authors believe that this way of solving the word segmentation problem is computationally expensive.

In this paper, the word segmentation of the Kyrgyz language texts through creation of morphology's CSE-model is realized.

3 General Characteristics of the Kyrgyz Language

The Kyrgyz language is a one of the Turkic languages and the national language of the indigenous population of the Kyrgyz republic. Native speakers of Kyrgyz language are ethnic Kyrgyz living in the bordering regions of Kazakhstan, in the Namangan, Andijan and Ferghana regions of Uzbekistan, in some mountain regions of Tajikistan and the Xinjiang Uygur Autonomous Region of the China and some other countries [9].

According to the morphological classification, the Kyrgyz language refers to agglutinative languages. They are characterized by system of word-formation and inflectional affixation, a single type of declension and conjunction, grammatical uniqueness of affixes and the absence of significant alternations [8, 9].

The part of speech that indicates an action, state of the person and object is called a verb. The verb has conjugated and non-conjugated forms. The non-conjugated forms include the participles, the action names. Conjugate forms in a sentence act as predicate, non-conjugate forms as any member of a sentence. The verb has grammatical categories of time, person, number, mood, voice.

The affixes may be used in combination as: plural-case (ата-лар-да (fathers have), stem-plural affix-locative case affix); plural-possessive (ата-лар-ы (their fathers), stem-plural affix-possessive 3rd person affix); plural-person (ата-лар-быз (we are fathers), stem-plural affix-1 person plural affix); plural-possessive-case (ата-лар-ыбыз-дын (of our fathers), stem-plural affix-plural affix-possessive affix 1st person plural form); plural-possessive-person (ата-лар-ы-быз (we are their fathers), stem-plural affix-possessive affix 3rd person-1st person plural type affix); person-case; possessive-case (бала-м-ды (my child), stem-possessive 1st person type affix-accusative affix), possessive-person (бала-ӊ-мын (I am your child)).

In the morphological segmentation, we do not consider derivational affixes, because they change the meaning of words and the grammatical word class by producing new word.

4 Development of Kyrgyz Morphology's CSE-Model

For developing the complete system of endings of the Kyrgyz language, the approach proposed by Tukeyev U. in [1] is used.

Consider the Kyrgyz language ending system of two classes: nominal parts of speech affixes (nouns, adjectives, numerals) and endings of verb base words (verbs, participles, moods, voices). In this section, we will consider the complete set of endings in Kyrgyz language.

The Kyrgyz language has the four types of endings: plural endings (denoted by K); possessive endings (denoted by T); case endings (denoted by C); personal endings (denoted by J); a stem denoted by S.

There are four variants of placing the types of endings: from one type, from two types, from three types and from four types. A number of placements defines by following formula:

$$A_n^k = n!/(n - k)! \tag{1}$$

Then, the number of placements will be determined as follows:

$$\begin{aligned} A_4^1 &= 4!/(4-1)! = 4, \\ A_4^2 &= 4!/(4-2)! = 12, \\ A_4^3 &= 4!/(4-3)! = 24, \\ A_4^4 &= 4!/(4-4)! = 24. \end{aligned} \tag{2}$$

Total possible placements are 64.

Consider which of them are semantically valid.

Placements of one type of ending (K, T, C, J) are all semantically valid by definition.

Placements of two types of endings can be as follows: **KT, TC, CJ**, JK **KC, TJ**, CT, JT **KJ**, TK, CK, and JC.

An analysis of the placement semantics by two types of endings shows that the bold placements are valid (**KT, TC, CJ, KC, TJ, KJ**) and other placements are invalid. For instance, JK – there is no use of plural endings after personal endings. Therefore, the number of permissible placements of two types of endings will be 6.

Acceptable placements of three types of endings will be 4 from 24 possible variants: **KTC, KTJ, TCJ, and KCJ**.

Placement of four types of endings will be 1: **KTCJ.**

In total, there are 4 allowed placements from one type, 6 from two types, 4 from three types and 1 from four types of endings. The total number of all acceptable placements is 15.

The set of the endings of verbs in Kyrgyz language contains following types:

- set of verbs endings,
- set of participle endings (атоочтук),
- set of gerund endings (чакчылдар),
- set of mood endings,
- set of voice endings.

The set of endings to verb stems includes following types: tenses (9 types); person (3 types); negative type. Then the number of possible types of verb endings is 28.

The set of endings to verb stems of participle includes endings of participle (denoted by R), plural endings (denoted by K), possessive endings (denoted by T), case endings (denoted by C), personal endings (denoted by J).

Then, the possible variants of ending types will be:

1) with one type of endings: **RK, RT, RC, RJ,**

2) with two types of endings: **RKT, RTC,** RCJ, RJK, **RKC**, RTJ, RCT, RJT, **RKJ**, RTK, RCK, RJC,

3) with three types of endings: **RKTC**, RTCJ, RCJK, RJKT, RKTJ, RTCK, RCJT, RJKC, RKCJ, RTJK, RCTK, RJTK, RKCT, RTJC, RCTJ, RJTC, RKJT, RTKC, RCKT, RJCK, RKJC, RTKJ, RCKJ, RJCT,

4) with four types of endings: RKTJC, RTKJC, RCKTJ, RJKTC, RKTCJ, RTKCJ, RCKJT, RJKCT, RKJTC, RTJKC, RCTKJ, RJTKC, RKJCT, RTJCK, RCTJK, RJTCK, RKCTJ, RTCJK, RCJKT, RJCKT, RKCJT, RTCKJ, RCJTK, RJCTK.

In total, the number of acceptable ending types is 9.

Consider the ending types of gerund. They contain endings followed by personal endings: PJ, where P is a base of gerund, J is the personal ending. Therefore, there is 1 type of gerund ending.

There are five forms of voice in Kyrgyz language, however we will consider four of them: reflexive, passive, joint, compulsory. The basic voice has no special endings, and it is the initial form for the formation of other voice forms using affixes. Respectively, there are 4 types of voice endings.

Table 1. Number of endings type K and T.

Suffixes type K	Suffixes type T	
	Singular	Plural
-лар- -лер- -лор- -лөр- -дар- -дер- -дор- -дөр- -тар- -тер- -тор- -төр-	м, -ым (-им, -ум, -үм),	-быз (-биз, -буз, -бүз), ыбыз (-ибиз, -убуз, -үбүз),
	-ң, -ың (-иң, -уң, -үң),	-ңыз (-ңиз, -ңуз, -ңүз), -ыңыз (-иңиз, -уңуз, -үңүз),
	-ыңыз, -иңиз, -уңуз, -үңүз	-ңар (-ңер, -ңор, -ңөр), -ыңар (-иңер, -ңар, -үңөр), -ңыздар (-ңиздер, -ңуздар, -ңүздөр), -ыңыздар (-иңиздер, -уңуздар, -үңүздөр)
	-сы (-си, -су, -сү), -ы (-и, -у, -ү) -ныкы (-ники, -нуку, -нүкү); -дыкы (-дики, -дуку, -дүкү); -тыкы (-тики, -туку, -түкү).	-сы (-си, -су, -сү), -ы (-и, -у, -ү), -ныкы (-ники, -нуку, -нүкү); -дыкы (-дики, -дуку, -дүкү); -тыкы (-тики, -туку, -түкү).
12	62 (different)	

Moods in Kyrgyz language have five forms: imperative, conditional, desirable, subjunctive and indicative. The formal indicators of indicative mood are the affixes of the verb tense. Therefore, indicative mood affixes will not be considered. Types of mood endings are also determined according to the previous scheme: the basic endings of moods followed by personal endings. In total, there are 6 types of mood endings.

As a result, the total number of types of endings of verb stems will be 48. According to these calculations, the total number of types of endings with noun stems and with verb stems will be 63.

The next task is to determine the forms of endings and their number from the received types of endings. In this direction, finite sets of endings for all the main parts of the Kyrgyz language were constructed. In first, it needs to define the number of endings of single types K, T, C, J (Tables 1, 2 and 3).

Table 2. Number of endings type C.

Suffixes types C	
1. Атооч (nominative)	-
2. Илик (genitive)	- нын, нин, нун, нүн, дын, дин, дун, дүн, тын, тин, тун, түн;
3. Барыш (dative)	- га, ге, го, гө, ка, ке, ко, кө, а, е, о, ө, на, не, но, нө; - ны, ни,ну, нү, ды, ди, ду, дү, ты, ти, ту, тү, н;-да, де, до, дө, та, те, то, тө
4. Табыш (accusative)	-да, де, до, дө, та, те, то, тө, нда, нде, ндо, ндө; -дан, ден, дон, дөн, тан, тен, тон, төн, нан, нен, нон, нөн.
5. Жатыш (locative)	
6. Чыгыш (ablative)	
	65 (different)

Table 3. Number of endings type J.

Suffixes type J	
Singular	Plural
J1)-мын, мин, мун, мүн; J2)-сың, сиң, суң, сүң; J3)-сыз, сиз, суз, сүз; J4) -	-быз, (-пыз), -биз, (-пиз), -буз, (-пуз), -бүз, (-пүз), -сыңар, сиңер, суңар, сүңер; - сыздар, сиздер, суздар, сүздер; -
28	

The sum of all four types of endings, specifically K, T, C, J is 167 (different). Combinations in placement KT: K * T = (12 suffixes K) * (5 suffixes different T) = 60 endings KT. For choosing of suffix in T for each suffix of K is working harmony rules of the Kyrgyz language. In addition, a similar suffix of T not take to account, therefore are 5 suffixes T. The Table 4 shows the number of possible endings for a pair of plural and possessive endings denoted as KT.

In the Table 4 the example has five endings: ата-лар-ым (my fathers), ата-лар-ың (your fathers), ата-лар-ыңыз (your fathers polite), ата-лар-ы (their fathers), ата-лар-ыбыз (our fathers).

Here the law of vowel harmony: hard vowels with hard vowels, and soft vowels with soft vowels. The vowel in a root defines a tone of affix. This means that vowel of affixes should be in harmony with the last syllable of the word.

Calculating the remaining types of endings, we will get that type placement KC has 60 endings and type placement KJ has 36 endings.

Table 4. Number of endings of KT

<table>
<tr><th rowspan="2">Example</th><th rowspan="2">Suffixes type K</th><th colspan="2">Suffixes type T</th><th rowspan="2">Number of endings</th></tr>
<tr><th>Singular</th><th>Plural</th></tr>
<tr><td rowspan="4"></td><td rowspan="4">-лар-
-лер-
-лор-
-лөр-
-дар-
-дер-
-дор-
-дөр-
-тар-
-тер-
-тор-
-төр-</td><td>-ым, -им, -ум, -үм</td><td>-ыбыз, -ибиз, -убуз, -үбүз</td><td rowspan="4">12 * 5 = 60</td></tr>
<tr><td>-ың, -иң, -уң, -үң</td><td>-ыңыз, -иңиз, -уңуз, -үңүз</td></tr>
<tr><td>-ыңыз, -иңиз, -уңуз, -үңүз</td><td>-ыңыз, -иңиз, -уңуз, -үңүз</td></tr>
<tr><td>-ы, -и, -у, -ү</td><td>-ы, -и, -у, -ү</td></tr>
<tr><td>ата (grandfather)</td><td>-лар-</td><td>-ым, -ың, -ыңыз, -ы</td><td>-ыбыз</td><td>5</td></tr>
</table>

In pair of CJs: the genitive, dative and accusative cases are not used, thus, we consider only locative and ablative cases. We cannot consider the nominative case because it does not have any suffixes and represents the initial form of the word. The locative and ablative cases have 8 endings for each of them. So, there are 96 endings of CJ combination.

The possessive T endings have singular and plural types: -м, -ң, -ңыз, -сы, -быз, -ңар, -ңыздар. Therefore, for singular and plural types the number of endings is considered separately. There are 14 types of singular T endings and 16 types of plural T endings. As the 5 types case suffixes can be used with possessive endings the total number of TC possible endings will be 150 (14 * 5 + 16 * 5).

In the TJ placement the T1 cannot be used with the first side J ending: *"ата-м-мын"* is wrong, because we cannot say "me is my grandfather". Also, T2 and T3 connects only with J1 because there is no use of T2 with J2 as: *"ата-ң-сың"* – *"you are your grandfather"*. The possible combination of T and J endings and the total number 62 of TJ is shown in Table 5 below.

Considering the endings of type placement KTC we will get next: there are 12 types of plural, 4 types of possessive and 5 types of case endings. As a result, the possible number of endings of KTC is 240.

In the Kyrgyz language, there is possible only one combination of endings from four types: KTCJ. There can be a combination of endings from 12 affixes type K, 4 affixes type T, 2 affixes type C and 6 affixes type J. At the result, we will get 576 endings.

The verbs can be used in three tenses as future, present and past. There are four types of past tense definite, accidental, general, habit. The accidental past tense has 172 possible affixes in total with its negative affixes. The habit past tense is formed by attaching affix -чу, -чү to the word stem. So, there are 20 possible combination of habit past tense and possessive, however, it does not have a negative type.

Table 5. Number of endings of TJ

Suffixes type T				Suffixes type J		Number of endings
Vowels		Consonants		Singular	Plural	
Hard	Soft	Hard	Soft			
T1-м T2-ң T3-ңыз T4-сы	-м, -ң, -ңи, -си	-ым -ы -ыңыз -ы	-им -иң -иңиз -и	J1)-мын, мин, мун, мүн; J2)-сың, сиң, суң, сүң J3)-сыз, сиз,суз, сүз; J4) -	-быз, биз, буз, бүз, пуз, пүз; -сыңар, сиңер, суңар, сүңер; -сыздар, сиздер, суздар, сүздер; -	T1-J2:8; T1-J3:8; T2-J1:10; T3-J1:10; T4-J1:10; T4-J2:8; T4-J3:8; Total: 62

The all types of past tense on 3rd person plural form are made by attaching affix -ыш (-иш, - уш, -үш, -ш) to the word stem following tense affixes. For example: *бар-ыш-ты (went), иште-ш-ти (worked).*

The Table 6 below shows the definite past tense suffixes with personal endings.

Table 6. Definite past tense affixes

Number	Person	Suffixes				Number of endings
Singular	1 person	-ды, ты-м	-ди, ти-м	-ду, ту-м	-дү, тү-м	8
	2 person	-ды, ты-ң	-ди, ти-ң	-ду, ту-ң	-дү, тү-ң	8
	2 person polite	-ды, ты-ңыз	-ди, ти-ңиз	-ду, ту-ңуз	-дү, тү-ңүз	8
	3 person	-ды, ты	-ди, ти	-ду, ту	-дү, тү	8
Plural	1 person	-ды, ты-к	-ди, ти-к	-ду, ту-к	-дү, тү-к	8
	2 person	-ды, ты-ңар	-ди, ти-ңер	-ду, ту-ңар	-дү,тү-нөр	8
	2 person polite	-ды, ты-ңыздар	-ди, ти-ңиздер	-ду, ту-ңуздар	-дү, тү-ңүздөр	8
	3 person	ыш, ш- ты	-иш, ш-ти	-уш, ш-ту	-үш, ш-тү	8
Total						64

The negative form of definite past tense is formed by adding after the stem of the verb the affix -ба (-бе, -бо, -бө), if the stem ends with a vowel or voiced consonant. If the stem ends to a deaf consonant then the negative form of definite past tense is made by attaching the affix -па (-пе, -по, -пө). For example, *мен бар-ба-ды-м (I did not go), сиздер кел-бе-ди-ңиздер.* So, the number of endings for negative form of definite past tense is 64 too.

An accidental past tense can be formed through attaching affix -ып (-ип, -уп, -үп, -п) or affix -ыптыр (-иптир, -уптур, -үптүр, -птыр, -птир, -птур, -птүр) and possessive endings after the verb stem. Number of endings of accidental past tense is 112. Moreover, other types of past tense have 64 affixes each.

Present simple and complex tenses have 12 affixes in total. The present simple forms by -ууда (-оодо, -өөдө, -үүдө) affix and affixes of definite future tense. While the complex present tense is formed by participle 2 affix -а (-е, -й) and -ып (-ип, -уп, -үп, -п) attached to the base of the verb and auxiliary verbs *"жат" – "lie", "тур" – "stand)", "отур" – "sit", "жүр" – "go"* in the present simple tense. For example: ачыл-ууда; жаз-а-м; кел-е жат-а-мын; күт-үп жүр-бүз.

In the Kyrgyz language, verbs have three types of future tense. The definite future tense *(Айкын келер чак)* is formed by joining the participial affix -а with phonetic variants and personal affixes to the base (root). So, there are 69 positive endings of definite future tense. In the 1st person, the full (а-мын) and incomplete (а-м) forms are possible: *бар-а-мын, бар-а-м -I will go.* If the last syllable of the stem ends with a vowel, then the participle affix -й added: *оку-й-мун – I will read.* If the last syllable of the stem ends with a consonant, then the phonetic variants of the participle affix -а (-е, -о, -ө) are added.

The negative type of verb in future tense is formed using affix -ба (-бе, -бо, -бө, -па, -пе, -по, -пө) after the word stem. There are 72 forms of endings of definite future tense negative type.

Indefinite future tense is formed by joining an affix -ар with phonetic variants followed by personal affixes to the word stem. Also, its negative type is formed using an affix -бас with phonetic variants. As the result of calculation, there are 61 possible affixes of indefinite future tense and 64 affixes of its negative type. The complex future tense has 8 types of affixes.

Thus, the total number of affixes of the verb in Kyrgyz language is 721.

The participles have nine acceptable placements of affixes. The participle 1 has 19 affixes (-ган, -ген, -гон, -гөн, -кан, -кен, -кон, -көн,-оочу, -уучу, -өөчү, -үүчү, -чу, -чү-, ар, -ер, -ор, -өр, -р), but only first 14 of them are used with plural affixes. The example of the word *"бар" – "go"* in the form of RKC is *"бар-ган-дар-га" – "to people who is went".*

The affixation of word *"айт"(say)* with RJ endings will be as following: *айт-кан-мын (I said); айт-кан-сың (You said); айт-кан-сыз (You said (polite)); айт-кан-быз (We said); айт-кан-сыңар (You said (plural)); айт-кан-сыздар (You said (plural polite)).*

In RKJ endings, the participle has 14 forms of suffixes because of the suffixes (-ар, -ер, -ор, -өр, -р) do not use with plural suffixes. The Table 7 shows the analysis of RKJ endings.

In the Kyrgyz language, the following moods of the verb are distinguished: Imperative mood; Conditional mood; Desirable mood; Subjunctive mood.

The imperative mood is made from verbal stem through attaching the affixes -гын, -кын, -ңыз, -ыңыз, -сын, -гыла, -кыла, -ңыздар, -ыңыздар and their phonetic variants, also, negative type is formed by attaching the -ба, (-бе, -бо, -бө, -па, -пе, -по, -пө) affix after the stem. As a result, there are 78 forms of imperative mood suffixes. There

Table 7. Number endings of RKJ

Example	Suffixes type R	Suffixes type K	Suffixes type J	Number of endings
	-ган, -ген, -гон, -гөн, -кан, -кен, -кон, -көн- -оочу, -уучу, -өөчү, -үүчү, -чу, -чү-	-лар- -лер- -лор- -лөр- -дар- -дер- -дор- -дөр- -тар- -тер- -тор- -төр-	-мын, -мин, -мун, -мүн, -быз, -биз, -буз, -бүз, -пыз, -пиз, -пуз, -пүз	14 * 3 = 42
			-сың, -сиң, -суң, -сүң, -сыңар, -сиңер, -суңар, -сүңер	
			-сыз, -сиз, -суз, -сүз, -сыздар, -сиздер, -суздар, -сүздер	
Бар-	-ган-	-дар-	-быз, -сыңар, -сыздар	

are 84 of conditional mood suffixes that are formed by using affix -са and its phonetic variants with possessive affixes. For example: *жаан жааса – if it rains, барсам – if I go, барбаса – if he/she do not go, жазсаңыз – if you write.*

The desirable mood has 31 suffixes and subjunctive mood affixes consists of 8 forms of affixes. The positive type suffixes of conditional mood have 69 forms of suffixes.

At the result of the constructed complete sets of endings, for nominal stems is 2 096, and the number of endings for verb stems is: verbs – 721, participles 1–1 325, gerund – 25, moods – 201, voices – 189, action name – 8, derivative adverbs – 36, and number of base suffixes – 167 (K-12, T-62, C-65, J-28). Total, there are 4 768 endings in Kyrgyz.

The developed CSE model of the Kyrgyz language is a computational morphology model based on the use of the complete set of language endings [10]. This computational model belongs to the "Item and Arrangement" (IA) morphology model [11, 12]. IA-model is convenient for agglutinative languages. The CSE computational model is an alternative to the TWOL computational model of morphology [13], which is more focused on describing the morphology of inflected languages. TWOL computational model is useful for description of the dynamic nature of allomorphs, uses two levels of word forms representation and refers to the morphology model "Item & Process" (IP-model).

5 Developing a Computational Data Models and an Algorithm of Morphological Segmentation

The principle of the Kyrgyz language word segmentation is next. The morphological segmentation using the CSE model is that a tabular data structure is built, which is a decision table. This decision table has two columns: the first column consists the word's endings; the second column is the segmented endings. The morphological segmentation algorithm of the current word will consist of:

1) finding the ending of the word,
2) finding the segmentation of the found word's ending.

To finding the ending of word, stemming algorithms are used without a stem dictionary and with a stem dictionary [14, 15]. To find the segmentation of the endings of words, a decision table of segmentation of word endings is used: the word's ending found during stemming of a word is searched for in the first column of the decision table, then the selected value of the second column in the found row of the table is the segmentation of this word's ending. For experiments are used the universal program[1] of stemming and segmentation based on CSE model.

6 Results of the Experiments

This works main target was to develop morphological segmentation for the Kyrgyz language through a vocabulary of endings, created by technology of CSE-model, and to use a list of stop words for improving of result of segmentation. The segmentation program was developed taking into account the above-mentioned algorithms, schemes and all the rules. Pronouns, conjunctions, postpositions, interjections, particles, modal words and auxiliary verbs were taken as stop words. These are the words that have no endings. Also, stop word list contains names of people and place names. The total number of stop words is 600.

In order to verify the operation of the algorithm and the program, the following testing was done. For segmentation, a texts were taken from the official web pages of the government, ministries, news agencies and the grammar books. This was done to mix different genres of publications. Common volume of text is 6450 words. The part sentences of them with segmentation is provided below:

- *"Атам келгенге чейин күтө турамын"(Wait till my father comes)*, segmented as *"Ата-м кел-ген@@ге чейин күт-ө тур-а@@мын."*
- *"Алар жолго камынып жатышат"(They are getting ready for the road)*, segmented as *"Ал-ар жол-го камын-ып жат-ыш@@а@@т."*
- *"Биздин үйгө коноктор келишти"(Guests came to our house)*, segmented as *"Биз-дин үй-гө конок-тор кел-иш@@ти."*
- *"Апамдан китептер кайда турганын сурадык"(We asked my mother where books)*, segmented as *"Апам-дан китеп-тер кайда тур-ган@@ы@@н сура-ды@@к."*
- *"Мындай дарылар дарыканаларда рецептсиз сатылбайт" (These drugs are not sold without prescription in pharmacies)*, segmented as *"Мындай дары-лар дарыкана-лар@@да рецепт-сиз сатыл-ба@@й@@т."*

[1] https://github.com/NLP-KazNU.

- *"Мен компьютерде отурганды анча жактырбайм" (I do not like sitting at the computer), segmented as "Мен компьютер-де отур-ган@@ды анча жактыр-ба@@й@@м."*
- *"Биздин көчөнүн балдарынын көбүрөөгү мектепте окушат" (Most of the children on our street go to school), segmented as "Биз-дин көчө-нүн бал-дар@@ы@@нын көбүрөө-гү мектеп-те оку@@ш@@а@@т."*

The results show that 5 264 words out of 6 450 are segmented correctly. This means that the word segmentation accuracy is 82%. The analysis of incorrect segmentations show that causes of incorrect segmentation are: 1) for single letter endings - accepting the last letter of stem as an ending of participle, gerund and possessive ending; 2) for endings from 2-d and 3-d letters - accepting the last letters of stem as a possessive, past and future tense ending.

7 Conclusion and Future Works

In this paper, we propose a solution to the word segmentation of the Kyrgyz language by using the CSE - morphology model. A computational model on base of CSE - model of the morphology of the Kyrgyz language has been built, a computational data model for morphological segmentation as decision table has been developed, and experiments have been carried out with texts of the Kyrgyz language. The results obtained show 82% accuracy of morphological text segmentation. To improve the quality of morphological segmentation of the text, it is planned to conduct research on the improving of the dictionary's volume of Kyrgyz language stems in the proposed scheme of morphological segmentation. In future works is planned to use received results for preprocessing stage for an investigation of neural machine translation for the Kyrgyz language pair with different languages.

References

1. Tukeyev, U.: Automation models of the morphological analysis and the completeness of the endings of the Kazakh language. In: Proceedings of the International Conference "Turkic Languages Processing: TurkLang-2015". Publishing house Academy of Sciences of the Republic of Tatarstan, Kazan, pp. 91–100 (2015). (in Russian)
2. Sennrich, R., Haddow, B., Birch, A.: Neural machine translation of rare words with subword units. In: Proceedings of the 54th Annual Meeting of the Association for Computational Linguistics, vol. 1, pp. 1715–1725 (2016)
3. Pan, Y., Li, X., Yang, Y., Dong, R.: Morphological Word Segmentation on Agglutinative Languages for Neural Machine Translation. arXiv:abs/2001.01589 (2020)
4. Creutz, M., Lagus, K.: Unsupervised discovery of morphemes. In: Proceedings of the ACL-02 Workshop on Morphological and Phonological Learning, vol. 6, pp. 21–30 (2002)
5. Ataman, D., Negri, M., Turchi, M., Federico, M.: Linguistically motivated vocabulary reduction for neural machine translation from Turkish to English. Prague Bull. Math. Linguist. **108**, 331–342 (2017)
6. Ataman, D., Federico, M.: An evaluation of two vocabulary reduction methods for neural machine translation. In: Proceedings of the 13th Conference of the Association for Machine Translation in the Americas, vol. 1, pp. 97–110 (2018)

7. Washington, J.N., Ipasov, M., Tyers, F.M.: A finite-state morphological analyser for Kyrgyz. In: Proceedings of the 8th Conference on Language Resources and Evaluation, LREC2012-Istanbul, Turkey, pp. 934–940 (2012)
8. Israilova, N.A., Bakasova, P.S.: Morphological analyzer of the Kyrgyz language. In: Proceedings of the V International Conference on Computer Processing of Turkic Languages Turklang 2017 - Conference Proceedings, vol. 2, Publishing House Academy of Sciences of the Republic of Tatarstan, Kazan , pp. 100–116 (2017). (in Russian)
9. Biyaliev K.A.: Guide to the Grammar of the Kyrgyz Language, 128 p. Kyrgyz-Russian Slavic University. Bishkek. (in Russian)
10. Tukeyev, U.: Computational models of Turkic languages morphology on complete sets of endings. QS Subject Focus Summit 'Modern Languages and Linguistics', section 'Linguistics and Artificial Intelligence', report (2020). https://qssubjectfocus.com/moscow-2020/
11. Spencer, A.: Morphological Theory: An Introduction to Word Structure in Generative Grammar, 512 p. Blackwell Publishers (1991)
12. Plungyan, V.A.: Common Morphology: Introduction to Problematics: Educational Manual, 2nd edn., edited.-M.: Editorial UPCC, 384 p. (2003). (in Russian)
13. Koskenniemi, K.: Two-level morphology: a general computational model of word-form recognition and production. Technical report publication No. 11. Department of General Linguistics. University of Helsinki, 160 p. (1983)
14. Tukeyev, U.A., Turganbayeva, A.: Lexicon - free stemming for the Kazakh language. In: Proceedings of the international scientific conference "Computer science and Applied Mathematics" dedicated to the 25th anniversary of Independence Of the Republic of Kazakhstan and the 25th anniversary of the Institute of Information and Computing Technologies. - Almaty, pp. 84–88 (2016). (in Russian)
15. Tukeyev, U., Karibayeva, A., Zhumanov, Z.: Morphological segmentation method for turkic language neural machine translation. Cogent Eng. 7(1), (2020). https://doi.org/10.1080/23311916.2020.1856500

Empirical Study of Tweets Topic Classification Using Transformer-Based Language Models

Ranju Mandal[1], Jinyan Chen[2], Susanne Becken[2], and Bela Stantic[1](✉)

[1] School of Information and Communication Technology, Griffith University, Brisbane, Australia
{r.mandal,b.stantic}@griffith.edu.au

[2] Griffith Institute for Tourism, Griffith University, Brisbane, Australia
{jinyan.chen,s.becken}@griffith.edu.au

Abstract. Social media opens up a great opportunity for policymakers to analyze and understand a large volume of online content for decision-making purposes. People's opinions and experiences on social media platforms such as Twitter are extremely significant because of its volume, variety, and veracity. However, processing and retrieving useful information from natural language content is very challenging because of its ambiguity and complexity. Recent advances in Natural Language Understanding (NLU)-based techniques more specifically Transformer-based architecture solve sequence-to-sequence modeling tasks while handling long-range dependencies efficiently, and models based on transformers setting new benchmarks in performance across a wide variety of NLU-based tasks. In this paper, we applied transformer-based sequence modeling on short texts' topic classification from tourist/user-posted tweets. Multiple BERT-like state-of-the-art sequence modeling approaches on topic/target classification tasks are investigated on the Great Barrier Reef tweet dataset and obtained findings can be valuable for researchers working on classification with large data sets and a large number of target classes.

Keywords: Transformer · Natural language processing · Topic classification · Target classification · Deep learning

1 Introduction

During the past few years, there has been a colossal increase in social media websites, and personalized websites, which in turn has popularized the platforms/forums for the expression of public opinions. It has expanded the scope of commercial activities by enabling the users to discuss, share, analyze, criticize, compare, appreciate and research about products, brands, services through various social media platforms. Social media platforms like Twitter, Facebook, Pinterest, and LinkedIn, Amazon review, IMDB, and Yelp have become popular sources for retrieving public opinions and sentiment (Tang et al. [21]).

N. T. Nguyen et al. (Eds.): ACIIDS 2021, LNAI 12672, pp. 340–350, 2021.
https://doi.org/10.1007/978-3-030-73280-6_27

The influence of social media, mobile, analytics and cloud has offered the new technology paradigm and has transformed the operative environment and user engagement on the web. Similarly, this pool of information can be explored for the mutual benefit of both the user and the organization. Analyzing sentiments of this extremely large corpus of opinions can help an organization in realizing the public opinion and user experiences of the products or services (Kumar & Jaiswal [11]). There is also increasing interest in using social media data for the monitoring of nature experience and environmental changes. Sometimes this is coupled with citizen science where common citizens are specifically encouraged to contribute data (Lodi and Tardin [15]). Further evidence is needed to assess how well citizen science, collective sensing and social media data integrate with professional monitoring systems (Becken et al. [3]). Social media data are often only indirectly relevant to a particular research question, for example, the way people perceive a natural phenomenon, where they go or what they do. However, with appropriate filtering rules, it is possible to convert these unstructured data into a more useful set of data that provides insights into people's opinions and activities (Becken et al. [4]). Using Twitter as a source of data, Daume and Galaz [7] concluded that Twitter conversations represent "embryonic citizen science communities".

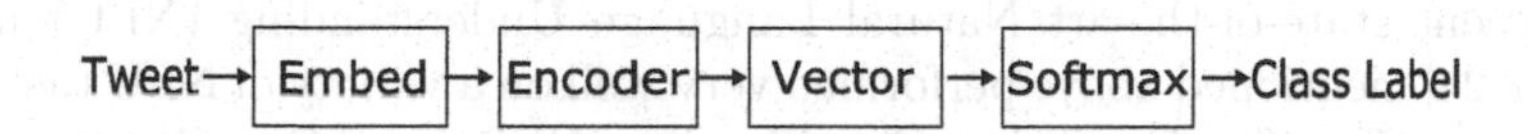

Fig. 1. Flow of target classification pipeline. The system takes a tweet text as input and returns a label from 11 predefined target categories.

Similarly, we capitalize on social media data and state-of-the-art Natural Language Processing (NLP) approach to assess how social media data sources, namely Twitter, can be used to better understand human experiences related to the Great Barrier Reef (GBR), Australia. An automated tool based on a machine learning system could be useful to capture the user experiences and potentially changing environmental conditions on the GBR. Sentiment analysis has been studied to assess the polarities of user opinion in a particular context, and many publications can be found in the literature [10,19]. However, social media platform such as Twitter, information filtering is a necessary step to select the relevant tweets as the users may become overwhelmed by the raw data [20]. One solution to this problem is the automatic classification of tweet text. Automatic detection of target or topic of tweets enables applications featuring opinion target/aspect extraction, polarity assessment as well as target/aspect-oriented sentiment analysis. Target-specific polarity detection is a key challenge in the field of sentiment analysis, as the sentiment polarity of words and phrases may depend on the aspect [1]. Other reasons for tweet classification include identifying trending topics (Lee et al. [13], Allan [2,25]). Real-time detection of breaking news or relevant target and tracking them is essential due to the real-time nature of Twitter data. However, we are interested in classifying a predefined set of topics instead of a dynamic set (topic classes change over time) of topics.

Text classification is the process of assigning labels or categories to text according to its content. It's one of the fundamental tasks in Natural Language Processing (NLP) with broad applications such as sentiment analysis, topic labeling, spam detection, and intent detection. The proposed work for developing target classification method is based on deep learning techniques and sophisticated language modeling. To address this problem, transformer-based BERT [8] encoder model, and three more improved BERT models [12,14,24] are used for feature extraction and classifications. We also have adopted a transfer learning technique where instead of training a model from scratch, models used in our experiments are pre-trained on a large dataset and then the GBR tweet dataset is used to fine-tune all the models for our specific text classification task. During the investigation, a comparative study is pursued based on the model's performance on prediction accuracy and computational complexity. Figure 1 shows the diagram of our classification pipeline which takes a tweet text as input and produces a class label as output. The proposed approach effectively classifies the text to a predefined set of generic classes such as politics, travel, safety, culture, climate, terrestrial, coastal, etc.

The motivation behind our proposed research work has many folds. All of these recent state-of-the-art Natural Language Understanding (NLU) models [8,12,14,24] mentioned above performed very well on a variety of NLU tasks such as Sequence Classification, Token Classification (NER), and Question Answering on a large dataset of English text corpus contains a large number of tokens in each sequence. First, we investigate the recent NLU models to find out the best suitable and effective solution on our short tweet text classification problem. Our dataset has a limited number of sequences with a small number of tokens (i.e. most samples have less than 32 tokens and few samples contain more than 32 tokens but less than 64 tokens) in each sequence in contrast to the popular NLU dataset such as GLUE, RACE. A detailed description of data collection, preliminary data analysis, annotation procedure and the steps involved in the experimentation on social media data (Twitter) to producing results are presented in respective sections.

The rest of this paper is organized as follows: Section 2 provides a review of the relevant works of literature on text classification. A brief description of the transformer-based architectures used in our experiment is presented in Sect. 3. We have demonstrated the experimental results and analyzed the performance in Sect. 4. Finally, conclusion and future work are presented in Sect. 5.

2 Related Work

Research articles on the problem of text classification can be found decades back [16,20], and until recently it transformed [8] by a giant stride in terms of efficiently handling challenging not only the classification problem instead of a wide variety of NLU tasks. Nigam et al. [16] proposed a text classification algorithm based on the combination of Expectation-Maximization (EM) and a naive Bayes classifier for learning from labeled and unlabeled documents. Sriram et al. [20] proposed method classifies incoming tweets into multiple categories such as news, events, opinions, deals, and private messages based on the author's information and 7 other binary features. Chen et al. [5] produced discriminatively features by leveraging topics at multiple granularities, which can model the short text more precisely. Vo et al. [23] proposed topic models-based feature enhancing method which makes the short text seem less sparse and more topic-oriented for classification. The topic model analysis was based on Latent Dirichlet Allocation, and finally, features were enhanced by combining external texts from topic models that make documents more effective for classification, and a large-scale data collection consisting of Wikipedia, LNCS, and DBLP were explored for discovering latent topics. In another method, a cluster-based representation enrichment strategy was adopted to obtain a low dimensional representation of each text for short text classification [6].

Recently proposed methods such as BERT [8], ULMFiT [9], GPT-2 [18], and ELMo [17] attain a significant milestone on the NLP domain, outperforming the state-of-the-art on several NLP tasks including text classification. Universal Language Model Fine-tuning (ULMFiT [9]) is an effective transfer learning method that can be applied to any task in NLP. GPT-2 [18] is a large transformer-based language model with 1.5 billion parameters, trained on a large dataset of 8 million web pages. GPT-2 is trained with a simple objective: predict the next word, given all of the previous words within some text. The introduction of a new dataset which emphasizes the diversity of content, and has been curated/filtered manually is also a significant contribution. ELMo [17] is a deep contextualized word-embeddings technique where representation for each word depends on the entire context of the text unlike using a fixed embedding for each word. A bi-directional LSTM model is trained to be able to create those embeddings to accomplish a specific task. These methods also have revolutionized the field of transfer learning in NLP by using language modeling during pre-training. Further enhancement of Google BERT [8] and many improved versions of BERT inspired transformer-based sequence modeling techniques such as XLNet [24], RoBERTa [14], ALBERT [12] has significantly improved on the for a variety of tasks in natural language understanding. Recent works on BERT performance enhancement are achieved by increasing training data, computation power, or training procedure. Further advancement that can enhance performance while using fewer data and compute resources will be a step forward.

3 Methodology

In the transfer learning technique, the pre-trained weights of an already trained model on a large dataset can be used to fine-tune the model for a specific Natural Language Processing (NLP) tasks. We adopted a transfer learning strategy to leverage from pre-trained models that use language models pre-trained on exceptionally large curated datasets and the models have demonstrated state-of-the-art performance in text classification tasks. The pre-trained model weights already have enormous encoded information of English language, and it takes much less time to fine-tuned the model with the new dataset to obtain the features required for classification. Semantic understanding of language has been significantly improved by the language modeling approaches. A brief description of the state-of-the-art language modeling approaches that we have adapted to our problem is presented in this section, and all of these language models are pre-trained with very large curated datasets.

3.1 BERT [8]

It stands for Bidirectional Encoder Representations from Transformers (Vaswani et al. [22]) pre-trained over a large volume of unlabeled textual data to learn a language representation that can be used to fine-tune for specific machine learning tasks. The general transformer represents an encoder-decoder network architecture, but BERT is a pre-training model, that uses the encoder to learn a latent representation of the input text. It builds upon recent work in pre-training contextual representations such as GPT [18], Elmo [17], and ULMFit [9] (these three methods were significant milestones before the BERT method had been introduced). Pre-trained representations can either be context-free or contextual, and contextual representations can further be unidirectional or bidirectional. It is the first deeply bidirectional (learn sequence from both ends), unsupervised language representation, pre-trained using only a plain text corpus (Wikipedia). It is designed to pre-train deep bidirectional representations from an unlabelled text by jointly conditioning on both the left and right contexts. It is considered a key technical innovation is applying the bidirectional training of Transformer, a popular attention model, to language modeling. This is in contrast to previous efforts that looked at a text sequence either from left to right or combined left-to-right and right-to-left training. The experimental outcomes of BERT show that a language model that is bidirectionally trained can have a deeper sense of language context and flow than single-direction language models. As a result, the pre-trained BERT model can be fine-tuned with an extra additional output layer to create state-of-the-art models for a wide range of NLP tasks. This bidirectional Transformer model redefines the state-of-the-art for a range of natural language processing tasks (e.g. Text classification), even surpassing human performance in the challenging area.

3.2 XLNet [24]

XLNet is a large bidirectional transformer that uses improved training methodology, larger data and more computational power to achieve better than BERT prediction metrics on 20 language tasks. To improve the training, XLNet introduces permutation language modeling, where all tokens are predicted but in random order. This is in contrast to BERT's masked language model where only the masked (15%) tokens are predicted. This is also in contrast to the traditional language models, where all tokens were predicted in sequential order instead of random order. This helps the model to better handle dependencies and relations between words. Transformer XL was used as the base architecture, which showed good performance even in the absence of permutation-based training. XLNet was trained with over 130 GB of textual data and 512 TPU chips running for 2.5 days.

3.3 RoBERTa [14]

RoBERTa modifies the BERT [8] pretraining procedure that improves end-task performance and these improvements were aggregated and evaluated to obtain combined impact. The authors coined this modified configuration as RoBERTa for Robustly optimized BERT approach. Few research works have been published as an improvement of BERT method performance on either its prediction accuracy or computational speed. RoBERTa improves the BERT performance on prediction accuracy. It is a replication study of BERT pretraining with a robust and improved training methodology training as BERT was significantly undertrained according to the authors. The modifications include hyperparameter tuning such as bigger batch sizes (8k) with 10 times more data compare to BERT training, longer training sequences, dynamic masking strategy during training step, etc. RoBERTa's experimental dataset consists of 160 GB of text for pre-training, including 16 GB of books corpus and English Wikipedia used in BERT. The additional data included CommonCrawl News dataset (63 million articles, 76 GB), Web text corpus (38 GB) and Stories from Common Crawl (31 GB). RoBERTa outperforms both BERT [8] and XLNet [24] on the GLUE benchmark dataset.

3.4 ALBERT [12]

ALBERT stands for A Lite BERT for self-supervised learning of language representations. This method outperformed BERT on both metrics, its prediction accuracy, and computational complexity. The performance of the ALBERT language model superior to BERT on memory optimization and training time by parameter-reduction techniques and it also achieved state-of-the-art results on the three most popular NLP datasets (i.e. GLUE, RACE, and SQuAD 2.0). Recent state-of-the-art language models have hundreds of millions of parameters which make these model less convenient on low or limited hardware (i.e. low GPUs or TPUs computing) resource environments. Besides, researchers also have investigated that stacking more layers in the BERT-large model can lead to a negative impact on overall performance. These obstacles motivated Google to thorough analysis into parameter reduction techniques that could reduce the size of models while not affecting their performance. ALBERT is an attempt to scale down the BERT model by reducing parameters. For example, the ALBERT-large model has about 18 Million parameters which are 18x fewer compared to BERT-large (334 Million).

3.5 Implementation Details

All experiments have been conducted on an Intel(R) Xeon(R) CPU E5-2609 v3 @ 1.90 GHz Linux cluster node with a GeForce GTX 1080 GPU installed.

The PyTorch-based Transformers library[1] was used to develop code for all the experiments. The models are trained with the AdamW optimizer, learning rate of $4e-5$, and token size of 32 or 64. We train each model for 2 to 5 epochs with a batch size of 32. The results for each of the models are reported in Table 2 and Table 3.

4 Results and Discussions

Dataset: Twitter API with restrictions to capture geo-tagged tweets posted from the GBR geographic area was used to download user tweets. A rectangular bounding box was defined (Southwest coordinates: 141.459961, −25.582085 and Northeast coordinates: 153.544922, −10.69867) that broadly represents the GBR region. For the investigation, we extracted 13,000 relevant tweets to our research work for human annotation. All downloaded tweets go through a multi-step pre-processing stage before used as training samples. In Table 1, a detail description of 11 target classes and indicator words for tweet annotation are presented. Table 1 shows the number of tweets and percentages for each of the 11 target classes present in our dataset. It can be observed that 70.46% of tweets are labeled as "other" category. We created a smaller dataset with 4,342 tweet samples where the "other" category samples are reduced to 500 to have balance in numbers, and we refer the dataset as the "balanced" dataset. The experimental results on two different datasets are reported in this section in Table 2 and Table 3.

Table 1. Target classes with the number of annotated tweets for each class.

Target class	Tweets	Percentage (%)
Accom	292	2.25
Climate	204	1.57
Coastal	511	3.93
Culture	127	0.98
GBRact	185	1.42
Landact	907	6.97
Politics	903	6.94
Safety	181	1.39
Terrestrial	196	1.51
Travel	336	2.58
Other	9164	70.46

[1] https://huggingface.co/.

Table 2. Results obtained from the experiment on the Great Barrier Reef (GBR) full dataset. This dataset contains 13000 tweet samples (15% test and 85% train) and 11 classes.

Architecture	Model	Parameters	Sequence	Batch	Epoch	MCC	Accuracy
BERT	Bert-base	110M	32	32	2	0.555	0.786
BERT	Bert-base	110M	64	32	2	0.545	0.780
BERT	Bert-large	340M	32	32	3	0.561	0.781
XLNet	Xlnet-base	110M	32	32	3	0.555	0.777
RoBERTa	Roberta-base	125M	32	32	2	0.548	0.774
ALBERT	Albert-base-v1	11M	32	32	2	0.549	0.788
ALBERT	Albert-base-v2	11M	32	32	3	0.530	0.768

Table 3. Results obtained from the experiments on balanced Great Barrier Reef (GBR) dataset. This dataset contains 4342 tweet samples (3690 & 652 for train and test respectively) and 11 classes.

Architecture	Model	Parameters	Sequence	Batch	Epoch	MCC	Accuracy
BERT	Bert-base	110M	32	32	3	0.634	0.682
BERT	Bert-base	110M	64	32	3	0.633	0.680
BERT	Bert-large	340M	32	32	4	0.660	0.705
XLNet	Xlnet-base	110M	32	32	5	0.628	0.676
RoBERTa	Roberta-base	125M	32	32	5	0.646	0.693
ALBERT	Albert-base-v1	11M	32	32	4	0.587	0.642
ALBERT	Albert-base-v2	11M	32	32	5	0.630	0.679

Preprocessing: Twitter data are often messy and contain a lot of redundant information. Also, several other steps need to be put in place to make subsequent analysis easier. In other words, to eliminate text/data which is not contributing to the assessment is important to pre-process the tweet. Initial data cleaning involved the following:

a) Removing Twitter Handles (@user): The Twitter handles do not contain any useful information about the nature of the tweet, so they can be removed.
b) Removing Punctuations, Numbers, and Special Characters: The punctuations, numbers and even special characters are removed since they typically do not contribute to differentiating tweets.
c) Tokenization: We split every tweet into individual words or tokens which is an essential step in any NLP task. The following example shows a tokenization result,

Input: [sunrise was amazing]

After tokenization: [sunrise, was, amazing]

The respective models' tokenizer (e.g. BertTokenizer, XLNetTokenizer, RobertaTokenizer, AlbertTokenizer) are used in our data preprocessing step.

d) Stemming: It is a rule-based process of stripping the suffixes ("ing", "ly", "es", "s" etc.) from a word. For example: "play", "player", "played", "plays" and "playing" are the different variations of the word - "play". The objective of this process is to reduce the total number of unique words in our data without losing a significant amount of information.

The classification results on GBR full and balanced datasets from a range of experiments along with the hyperparameters are detailed in Table 2 and Table 3. We present our test's accuracy using F1 score or F-measure and the Matthews correlation coefficient (MCC) score. F1 score is based on precision (p) and the recall (r) values and MCC is calculated from the confusion matrix as follows:

$$F1 = \frac{2 * Precision * Recall}{Precision + Recall}$$

$$MCC = \frac{TP \times TN - FP \times FN}{\sqrt{(TP + FP)(TP + FN)(TN + FP)(TN + FN)}}$$

In our experiments, due to relatively small annotated sample size, based on empirical evaluation, we have found that of split data into 85% for training and 15% for testing is performing the best. We have considered 7 different models with a parameter range of 11 million (i.e. "albert-base-v1") to 340 million (i.e. "bert-large"). In Table 2, the best MCC of 0.549 and F1 score of 0.788 are obtained from the "albert-base-v1" model after 2 epochs. In Table 3 which represents the result from a smaller balanced dataset, the "bert-large" model obtained the best MCC score of 0.660 and F1 score of 0.705 after 4 epochs. However, model "bert-large" has 31 times more parameter compared to "albert-base-v1". We adapted the sliding window-based approach for long sequences during the evaluation step. It is an experimental feature moves a sliding window over each sequence and generates sub-sequences with length equal to the "max-seq-length" parameter. The model output for each sub-sequence is averaged into a single vector before the classification step. We obtained a 0.01 accuracy boost using this method. Also, we observed from our experiments that setting Max Sequence length to 64 tokens does not improve performance as we use a sliding window-based evaluation method. The full dataset achieved overall the higher accuracy, which is due to the bigger training set and higher accuracy of 'other' target class. However, balanced dataset was introduced to balance number of targets and achieve better accuracy for targets with smaller sample sizes.

5 Conclusion

In this work we developed a multi-class tweet (i.e. short text) classification system to identify the pre-defined topic of a tweet and as case study we used tweets sent from the Great Barrier Reef region. The importance of language modeling on text classification systems are studied in detail from the literature, and recently develop state-of-the-art transformer-based sequence modeling architectures are adapted to for the modeling of our GBR tweet text classification task. A wide range of experiments and comprehensive analysis of performance is presented on short text sequence classification. A number of parameters have been varied and the findings can be valuable for researchers working on classification with large data sets and a large number of target classes.

References

1. Alaei, A.R., Becken, S., Stantic, B.: Sentiment analysis in tourism: capitalizing on big data. J. Travel Res. **58**(2), 175–191 (2019)
2. Allan, J.: Introduction to topic detection and tracking. The Information Retrieval Series, vol. 12 (2012)
3. Becken, S., Connolly, R.M., Chen, J., Stantic, B.: A hybrid is born: integrating collective sensing, citizen science and professional monitoring of the environment. Ecol. Inform. **52**, 35–45 (2019)
4. Becken, S., Stantic, B., Chen, J., Alaei, A., Connolly, R.M.: Monitoring the environment and human sentiment on the great barrier reef: assessing the potential of collective sensing. J. Environ. Manag. **203**, 87–97 (2017)
5. Chen, M., Jin, X., Shen, D.: Short text classification improved by learning multi-granularity topics. In: IJCAI (2011)
6. Dai, Z., et al.: Crest: cluster-based representation enrichment for short text classification. In: Advances in Knowledge Discovery and Data Mining, pp. 256–267 (2013)
7. Daume, S., Galaz, V.: "Anyone know what species this is?" - twitter conversations as embryonic citizen science communities. Plos One **11**, 1–25 (2016)
8. Devlin, J., Chang, M., Lee, K., Toutanova, K.: BERT: pre-training of deep bidirectional transformers for language understanding. CoRR abs/1810.04805 (2018). http://arxiv.org/abs/1810.04805
9. Howard, J., Ruder, S.: Universal language model fine-tuning for text classification. CoRR abs/1801.06146 (2018). http://arxiv.org/abs/1801.06146
10. Hutto, C.J., Gilbert, E.: Vader: a parsimonious rule-based model for sentiment analysis of social media text. In: Proceedings of the 8th International AAAI Conference on Weblogs and Social Media, pp. 216–225 (2014)
11. Kumar, A., Jaiswal, A.: Systematic literature review of sentiment analysis on twitter using soft computing techniques. Concurrency and Computation: Practice and Experience, vol. 32, no. 1 (2019)
12. Lan, Z., Chen, M., Goodman, S., Gimpel, K., Sharma, P., Soricut, R.: Albert: a lite bert for self-supervised learning of language representations (2019)
13. Lee, K., Palsetia, D., Narayanan, R., Patwary, M.M.A., Agrawal, A., Choudhary, A.: Twitter trending topic classification. In: International Conference on Data Mining Workshops, pp. 251–258 (2011)
14. Liu, Y., et al.: Roberta: a robustly optimized BERT pretraining approach. CoRR abs/1907.11692 (2019). http://arxiv.org/abs/1907.11692
15. Lodia, L., Tardin, R.: Citizen science contributes to the understanding of the occurrence and distribution of cetaceans in south-eastern brazil - a case study. Ocean Coast. Manag. **158**, 45–55 (2018)
16. Nigam, K., Mccallum, A.K., Thrun, S.: Text classification from labeled and unlabeled documents using EM. Mach. Learn. **39**(2), 103–134 (2000)

17. Peters, M.E., et al.: Deep contextualized word representations. In: Proceedings of NAACL (2018)
18. Radford, A., Wu, J., Child, R., Luan, D., Amodei, D., Sutskever, I.: Language models are unsupervised multitask learners (2019)
19. Ribeiro, F.N., Araújo, M., Gonçalves, P., Benevenuto, F., Gonçalves, M.A.: A benchmark comparison of state-of-the-practice sentiment analysis methods. CoRR abs/1512.01818 (2015)
20. Sriram, B., Fuhry, D., Demir, E., Ferhatosmanoglu, H., Demirbas, M.: Short text classification in twitter to improve information filtering. In: Proceedings of the 33rd International ACM SIGIR Conference on Research and Development in Information Retrieval, pp. 841–842 (2010)
21. Tang, D., Qin, B., Liu, T.: Deep learning for sentiment analysis: successful approaches and future challenges. WIREs Data Min. Knowl. Disc. **5**(6), 292–303 (2015)
22. Vaswani, A., et al.: Attention is all you need. CoRR abs/1706.03762 (2017). http://arxiv.org/abs/1706.03762
23. Vo, D.T., Ock, C.Y.: Learning to classify short text from scientific documents using topic models with various types of knowledge. Expert Syst. Appl. **42**, 1684–1698 (2015). https://doi.org/10.1016/j.eswa.2014.09.031
24. Yang, Z., Dai, Z., Yang, Y., Carbonell, J.G., Salakhutdinov, R., Le, Q.V.: Xlnet: generalized autoregressive pretraining for language understanding. CoRR abs/1906.08237 (2019). http://arxiv.org/abs/1906.08237
25. Yüksel, A.E., Türkmen, Y.A., Özgür, A., Altınel, B.: Turkish tweet classification with transformer encoder. In: Proceedings of the International Conference on Recent Advances in Natural Language Processing (RANLP 2019), pp. 1380–1387. INCOMA Ltd. (2019). https://doi.org/10.26615/978-954-452-056-4_158

A New Approach for Measuring the Influence of Users on Twitter

Dinh Tuyen Hoang[1], Botambu Collins[1], Ngoc Thanh Nguyen[2], and Dosam Hwang[1(✉)]

[1] Department of Computer Engineering, Yeungnam University, Gyeongsan, South Korea
[2] Faculty of Computer Science and Management, Wroclaw University of Science and Technology, Wroclaw, Poland
Ngoc-Thanh.Nguyen@pwr.edu.pl

Abstract. Social networks are increasingly proving to be the core of today's web. Identifying the influence on social networks is an area of research that presents many open issues. The challenge is finding ways that can effectively calculate and classify users according to criteria that suit them closer to reality. In this paper, we proposed a new method for measuring user influence on social networks. The influence of a user measures by taking into account the activity and the popularity of the user. We use Twitter as a case study for our method. Experiments show that our method achieves promising results in comparison to other methods.

Keywords: Twitter influence · Influential user · Twitter analytics

1 Introduction

With the development of technology and Internet, Social Networks have become a challenging case of studies where big data content and relationships among users in the network are interesting topic researchers. The analysis of social networks to identify the most relevant agents of a social network is getting more and more attention. Identifying the influence of users on the network can be useful for many applications, such as the searching engine, viral marketing, recommendations, managing social customer relations, and information sharing [14,16,18].

Twitter is one of the leading online social networking services with many different types of users. Usually, with other social networking services, users are often granted friends links for other users to make friends with them. However, Twitter uses a social network model called "following", where each user is authorized to determine the person she/he wants to follow without asking permission. On the contrary, she/he can also be followed by other users without earlier permission [8].

In Twitter, user A is called a "follower" of user B, if user A is following user B and also we call user B is a "following" of user A. Twitter has become widespread in recent years as shown in Fig. 1. Twitter claims 330 million monthly active

N. T. Nguyen et al. (Eds.): ACIIDS 2021, LNAI 12672, pp. 351–361, 2021.
https://doi.org/10.1007/978-3-030-73280-6_28

users, and more than 145 million use the service daily[1]. Therefore, measuring user influence on Twitter is becoming increasingly important. There are four kinds of relationships in Twitter such as user-to-tweet, user-to-user, tweet-to-user, tweet-to-tweet. The relationship between the user-to-tweet and tweet-to-user is symmetrical. The are several actions that we can consider them belongs to four kinds of relationships such as post, retweet, like, reply, mention, follower, following. A Twitter network can be considered as a graph $G = (V, E)$ where V is a set of nodes can represent users, and E is a set of edges represent the relationship between users. The centrality of a node is the most common measure of the relative importance of that node in the network [13].

There are many studies on centrality measures to identify the most important social media actors, where each method is based on different eligibility criteria [11,12,19]. The standard measures are *degree*, *betweenness*, *closeness*, and *eigenvector. Degree centrality* indicates an essential score based merely on the number of links contained by a specific node. *Degree centrality* is the most straightforward measure for node connectivity. It is helpful to view at *in-degree* and *out-degree* as separate measures. *Betweenness centrality* is to compute the number of times that a node is on the shortest path between nodes in a network. Betweenness helps to analyze dynamic communication. *Closeness centrality* computes the shortest paths among all nodes and assigns each node a score using its sum of shortest paths. It is useful for obtaining the nodes that are best placed to impact on the entire network. *Eigen Centrality* measures the influence of a node using the number of links that connect from the node to other nodes in a network. *Eigen Centrality* takes a further step by considering how well connected a node is and also how many links their connections have over the whole network. Other methods such as PageRank [10], TwitterRank [17] were applied to find Twitter communities. However, the exploitation of user behavior and relationships to find appropriate methods to measure influence is still little known. In this study, we proposed a new method for measuring user influence on social networks by taking into consideration relationships between users as well as users' behaviors. The activity, popularity, and influence measurements are computed based on the user's activities. The ranking score is calculated to find the communities. The rest of this paper is organized as follows. In the next Section, related work on Twitter measurement is reviewed. In Sect. 3, we detail the proposed method. Experimental results are described in Sect. 4. Lastly, the conclusions and future work are shown in Sect. 5.

2 Related Works

2.1 PageRank

The PageRank algorithm is a model algorithm that was pioneered by Larry Page which was unique in the ranking of pages on the web [10]. The more connected a page or link is on the web determine the influence of such a link, this idea is

[1] https://www.oberlo.com/blog/twitter-statistics.

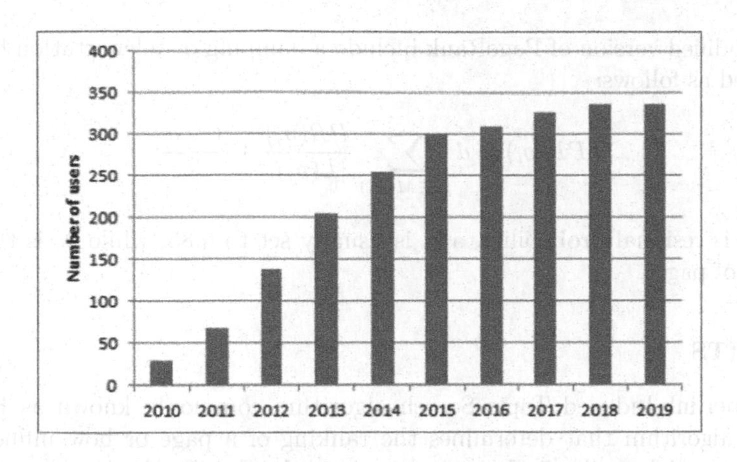

Fig. 1. Number of monthly active Twitter users

also useful in measuring the influence of users on social media especially with regard to Twitter users. The PageRank method is useful in determining how connected a user is on the twitter network, a more connected user will mean such a user is influential base on number followers or number of a retweet. A page having multiple links pointing to it makes that an important page in the Google PageRank. Once an important page points to another page in a directed graph, it automatically makes that page useful meaning that so many pages pointing to a page do not simply mean such a page is an important page. This premise makes Google PageRank unique when compared with other ranking algorithms such as the eigenvector algorithm which measures the degree of nodes pointing to another node in a network. Although the PageRank is a variant of the Eigenvector centrality measures in determining users influence, the significant difference lies in the fact that Google PageRank does not depend on the number of incoming links but only if the link is coming from an important page, also if a page sends out multiple links(outbound links) it also degrade such a link, in other words, an influential user does not contact many users rather many users try to contact such a user and once such as user in return contact a user, such a user become an important user thereby commanding a great deal of influence. In terms of influence on twitter, celebrities, politicians, business tycoons will have many incoming links from multiple users but they will send out very few links, their ranking automatically makes them higher or influential more than other users even if having the same incoming link. Contrary to the eigenvector degree of ranking, it measures the users ranking base on the number of incoming links not taking into cognizance the outgoing links as obtained in the PageRank.

$$PR(p_i) = \sum_{p_j \in M(p_i)} \frac{PR(p_j)}{|L(p_j)|} \tag{1}$$

where $PR(p_i)$ is the PageRank value of webpage p_i, and $M(p_j)$ is the set of webpages linking to p_i; $L(p_j)$ is the set of pages linked from p_i.

A modified version of PageRank include a damping or teleportation factor is computed as follows:

$$PR(p_i) = d \sum_{p_j \in M(p_i)} \frac{PR(p_j)}{|L(p_j)|} + \frac{1-d}{N} \tag{2}$$

where d is residual probability, and is usually set to 0.85, while N is the total number of pages.

2.2 HITS

The Hyperlink-Induced Topic Search algorithm commonly known as HITS is another algorithm that determines the ranking of a page or how influential a user is on social media [5]. It is analogous to the PageRank algorithm but can only work on a directed graph while PageRank can work on both directed and undirected graphs. Conversely, while the PageRank work on the entire web, the HITS works within a particular query on a subgraph within the web linked to a particular keyword. The method is based on the supposition that there exist two unique types of pages on the web namely; authorities and hubs. Authorities correlate with a page with many inbound links that maintain the fact that a page with many links to it commands high authority which depicts its influence on the web ecosystem. While Hubs are made up of links that have access to authorities on the webs. Authority and hub scores can be computed as follows:

$$autho(p) = \sum_{q \in P_{to}} hub(q) \tag{3}$$

where P_{to} is all pages which link to page p.

$$hub(p) = \sum_{q \in P_{from}} autho(q) \tag{4}$$

where P_{from} is all pages which page p link to.

2.3 TunkRank

The TunkRank [15] name after Daniel Tunkelang who championed the idea of ranking user on twitter based on the probability that a given user will read a tweet of another user which is not limited to likes, retweet and may follow such an individual base on his tweets. The main tenet of TunkRank is that although user i may follow user j, the influence of j to i is mutually exclusive since user j may exert influence over i not necessarily meaning i will do the same thereby implying different ranking scores.

2.4 TwitterRank

This method ranks user base on certain trending topics in which a given user introduce on the twitter network in which many users will retweet or comment on such a topic, thus, if a given user peddles in a certain sensitive topic which is making headlines within the twitter network, the user who initiated such a topic is given a higher score within the twitter rank method. This method is often characterized by a lot of abuses as some malicious users introduce unrealistic or false discussions in order to score points within the twitter ranking system [17].

2.5 Other Methods

Meeyoung Cha and colleagues [4] peddled in a method to measure user influence on Twitter by focusing on indegree, retweets, and mentions. The amount of time a user is mentioned on a given or a topic introduced by such a user is retweeted signifies the influence such a user. This method is different from our work in that it ranks every individual equally from the beginning and compute their interaction on twitter network and the opinions of other users vis-a-vis their tweets. In this regard, a post by an individual regarding a trending topic with few followers may induce or command significant attention within the Twitter ecosystem thereby enhancing the influence of such a user, while a celebrity may become a topic of discussion who is also a member of the same network, we assume a celebrity, politicians or users with huge numbers of followers command a great influence and hence should not be given equal point when computing the user influence which makes our work different.

The work of Huberman et al. [6], on influence and passivity in social media focuses on ranking users base on their ability to influence the opinions of other users. They coined the term IP algorithm to indicate an algorithm that examines user activities on twitter by computing influence score and a passivity score. This could be in the form of like or retweet. The passivity of a user is believed to be one who is not prone to being influence by others in the network.

Bakshy et al. [2] He took a different approach to calculate user influence on Twitter: they tracked the spread of URLs across user communities along with the relationship between the length of the user. Each diffuser floor as well as some features of the user that started it. Features include the number of followers and following, number of tweets, date of account creation, and a user's past influences based on the length of tiers initiated by that user in the past. By training on historical data, they tried to predict the length that a given user-initiated floor would reach. They found that past influence and a large follower base were needed to initiate the large falls.

3 Proposed Methodology

This section presents the method for measuring user influence on social networks by taking into consideration the activity and popularity. The ranking score is calculated to find the communities. The detail of the proposed system is shown in Fig. 2.

3.1 Activity Measures

Users are considered active when their participation in social networks is continuous and regular over a specified period. User participation on social networks is viewed through actions such as tweets, retweets, replies, mentions, etc. Of course, some users read Twitter very actively, but they leave no trace online, then, we can not collect any data. The general active of user u can be computed as follows:

$$GeneralActive(u) = tweet(u) + reply(u) + retweet(u) + like(u) \tag{5}$$

Definition 1. *The active of user u in given topic i, is defined as follows:*

$$Active(u|Topic(i)) = \frac{1}{1 + e^{-GeneralActive(u)}} \tag{6}$$

We can use the Latent Dirichlet Allocation (LDA) [3] algorithm for topical classification. However, the classical LDA algorithms are applied for more astronomical texts than tweets. Thus, the Divergence From Randomness retrieval model [1] is used to determine the topical relevance of each tweet.

3.2 Popularity Measures

A user is considered as a popular user when he/ she is followed by many other users on the network. A popular user is usually a celebrity, a politician, an expert in their field, etc.

The most common method of popularity measure is FollowerRank, proposed by Nagmoti et al. [9], as follows:

$$FollowerRank(u) = \frac{|Follower(u)|}{|Follower(u) + Following(u)|} \tag{7}$$

The main drawback of these measures is the big difference between the number of followers of Twitter users that is too high compared to the rest. We minimize the differences by converting these values between 0 and 1.

Definition 2. *The popularity measure of user u is defined as follows:*

$$Popular(u) = \frac{1}{1 + e^{-FollowerRank(u)}} \tag{8}$$

The value of $Popular(u)$ ranges in $(0, 1)$, which is the same domain as other factors.

3.3 Influence Measure

Definition 3. *The influence of user u can be computed as follows:*

$$Influence(u) = \alpha.Active(u) + (1 - \alpha).Popular(u) \tag{9}$$

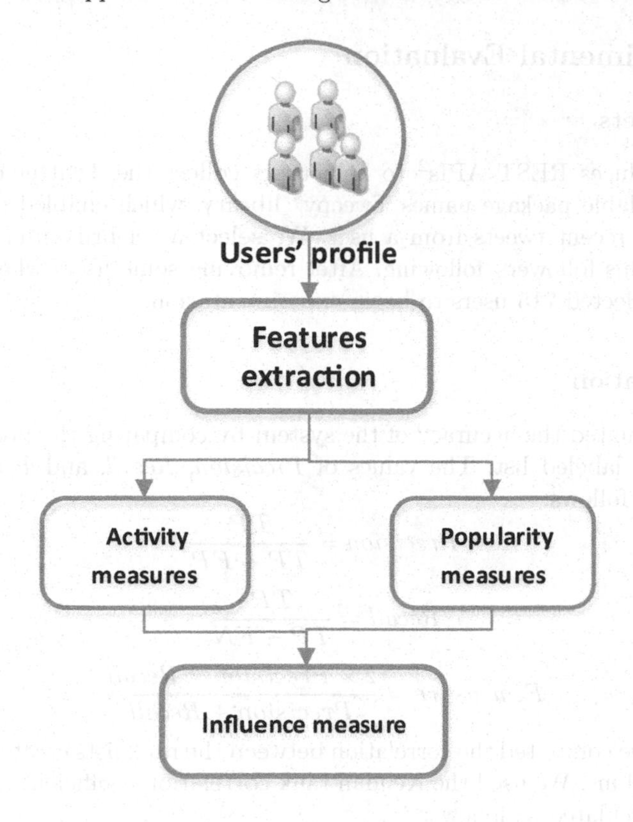

Fig. 2. The proposed method

where $\alpha \in [0, 1]$, which controls the weight between $Active(u)$ and $Popular(u)$.

Definition 4. *The transition probability from user u_j to user u_i is computed as follows:*

$$P(u_j, u_i) = \frac{|T_i|}{\sum_{u_j \in Followers(u_i)}} sim(u_j, u_i) \quad (10)$$

where $|T_i|$ represents for the number of tweets posted by user u_i. $sim(u_j, u_i)$ is the similarity value between users u_i and u_j. We used Word2Vec model [7] to convert a Tweet to a vector and using Cosine similarity to compute the value of $sim(u_j, u_i)$.

Definition 5. *The influence score of user u_i in the next work can be computed as follows:*

$$Score(u_i) = (1 - \gamma) \sum_{u_j \in Followers(u_i)} P(u_j, u_i).Influence(u_j) + \gamma.\frac{|T_i|}{|T|} \quad (11)$$

where the value of γ belong to $[0, 1]$, which can adjust by using experiments. $|T|$ represents for the number of tweets posted by all the users.

4 Experimental Evaluation

4.1 Datasets

Twitter produces REST APIs[2] to help users collect the Twitter dataset. We used an available package names Tweepy[3] library, which enabled downloading around 3200 recent tweets from a user. We select a celebrity user and collect tweets from his followers/following. After removing some users who have a few tweets, we selected 715 users to begin experimentation.

4.2 Evaluation

First we evaluated the accuracy of the system by comparing the predicted rank list with the labeled list. The values of *Precision*, *Recall*, and *F_measure* are computed as follows:

$$Precision = \frac{TP}{TP + FP} \tag{12}$$

$$Recall = \frac{TP}{TP + FN} \tag{13}$$

$$F\_measure = \frac{2 \times Precision \times Recall}{Precision + Recall} \tag{14}$$

Second, we computed the correlation between the rank lists created by the different algorithms. We used the Kendall rank correlation coefficient. The Kendall value τ is calculated as follows:

$$\tau = \frac{2}{n(n-1)} \sum_{i<j} sgn(x_i - x_j) sgn(y_i - y_j) \tag{15}$$

The value of τ is in the range of [-1, 1]. If the predicted lists are the same, then $\tau = 1$; otherwise if one predicted list is the contrary of the other, $\tau = -1$. The greater value of τ indicates a larger agreement between the two predicted lists. As shown in Fig. 4 the accuracy of the system got the best result at γ =[0.4, 0.5, 0.6]. The results in Fig. 3 shows that our method is more agreement with PageRank than In-degree method.

[2] https://dev.twitter.com/rest/public.
[3] http://tweepy.readthedocs.io/en/v3.5.0/.

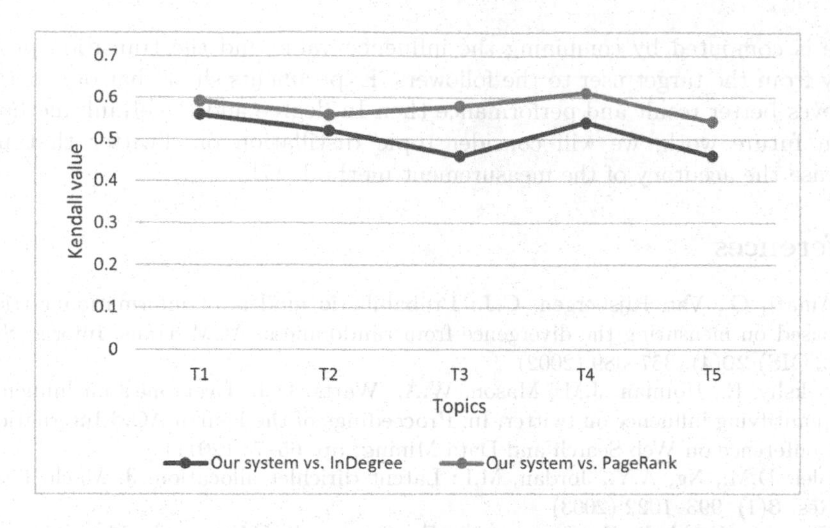

Fig. 3. The results of the proposed system use Kendall

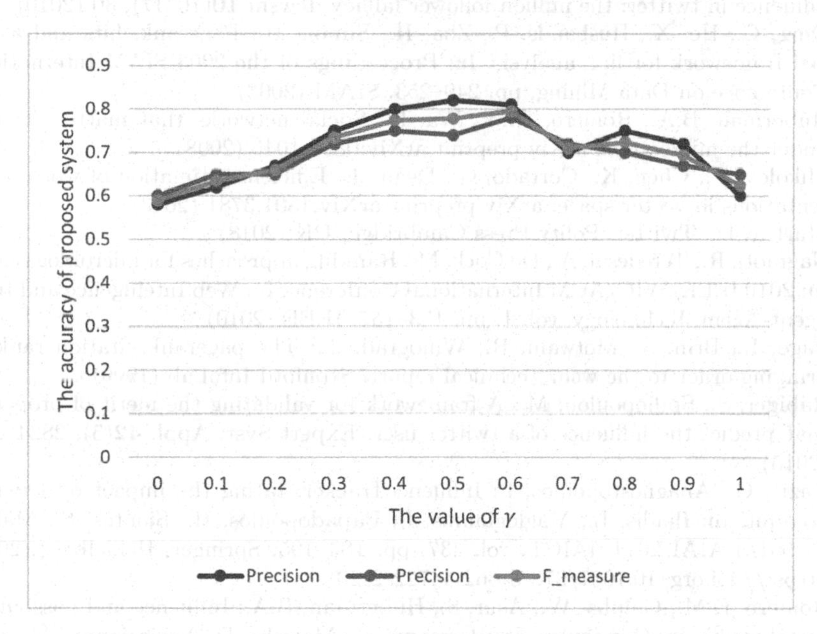

Fig. 4. The results of the proposed system use Precision

5 Conclusion and Future Work

This work proposed a new method for measuring the influence of users on Twitter. Firstly, the activity of the user is computed by considering the number of *Tweet*, *reply*, *retweet*, *like*. Secondly, the popularity of the user is measured by taking into account the number of *follower* and *following*. Lastly, the influence score is computed by combining the influence value and the transition probability from the target user to the followers. Experiments show that our method achieves better result and performance then In-degree and PageRank methods.

In future work, we will consider topic distillation on Twitter that may increase the accuracy of the measurement method [17].

References

1. Amati, G., Van Rijsbergen, C.J.: Probabilistic models of information retrieval based on measuring the divergence from randomness. ACM Trans. Inform. Syst. (TOIS) **20**(4), 357–389 (2002)
2. Bakshy, E., Hofman, J.M., Mason, W.A., Watts, D.J.: Everyone's an influencer: quantifying influence on twitter. In: Proceedings of the Fourth ACM International Conference on Web Search and Data Mining, pp. 65–74 (2011)
3. Blei, D.M., Ng, A.Y., Jordan, M.I.: Latent dirichlet allocation. J. Mach. Learn. Res. **3**(1), 993–1022 (2003)
4. Cha, M., Haddadi, H., Benevenuto, F., Gummadi, P.K., et al.: Measuring user influence in twitter: the million follower fallacy. Icwsm **10**(10–17), 30 (2010)
5. Ding, C., He, X., Husbands, P., Zha, H., Simon, H.: Pagerank, hits and a unified framework for link analysis. In: Proceedings of the 2003 SIAM International Conference on Data Mining, pp. 249–253. SIAM (2003)
6. Huberman, B.A., Romero, D.M., Wu, F.: Social networks that matter: Twitter under the microscope. arXiv preprint arXiv:0812.1045 (2008)
7. Mikolov, T., Chen, K., Corrado, G., Dean, J.: Efficient estimation of word representations in vector space. arXiv preprint arXiv:1301.3781 (2013)
8. Murthy, D.: Twitter. Polity Press Cambridge, UK (2018)
9. Nagmoti, R., Teredesai, A., De Cock, M.: Ranking approaches for microblog search. In: 2010 IEEE/WIC/ACM International Conference on Web Intelligence and Intelligent Agent Technology, vol. 1, pp. 153–157. IEEE (2010)
10. Page, L., Brin, S., Motwani, R., Winograd, T.: The pagerank citation ranking: bringing order to the web. Technical report, Stanford InfoLab (1999)
11. Räbiger, S., Spiliopoulou, M.: A framework for validating the merit of properties that predict the influence of a twitter user. Expert Syst. Appl. **42**(5), 2824–2834 (2015)
12. Razis, G., Anagnostopoulos, I.: InfluenceTracker: rating the impact of a twitter account. In: Iliadis, L., Maglogiannis, I., Papadopoulos, H., Sioutas, S., Makris, C. (eds.) AIAI 2014. IAICT, vol. 437, pp. 184–195. Springer, Heidelberg (2014). https://doi.org/10.1007/978-3-662-44722-2_20
13. Romero, D.M., Galuba, W., Asur, S., Huberman, B.A.: Influence and passivity in social media. In: Gunopulos, D., Hofmann, T., Malerba, D., Vazirgiannis, M. (eds.) ECML PKDD 2011. LNCS (LNAI), vol. 6913, pp. 18–33. Springer, Heidelberg (2011). https://doi.org/10.1007/978-3-642-23808-6_2
14. Sung, J., Moon, S., Lee, J.-G.: The influence in twitter: are they really influenced? In: Cao, L., et al. (eds.) BSI/BSIC -2013. LNCS (LNAI), vol. 8178, pp. 95–105. Springer, Cham (2013). https://doi.org/10.1007/978-3-319-04048-6_9
15. Tunkelang, D.: Tunkrank: A twitter analog to pagerank. ttp. thenoisychannel. com/2009/01/13/atwitter-analog-to-pagerank (2009)
16. Waugh, B., Abdipanah, M., Hashemi, O., Rahman, S.A., Cook, D.M.: The influence and deception of twitter: the authenticity of the narrative and slacktivism in the australian electoral process (2013)

17. Weng, J., Lim, E.P., Jiang, J., He, Q.: Twitterrank: finding topic-sensitive influential twitterers. In: Proceedings of the Third ACM International Conference on Web Search and Data Mining, pp. 261–270 (2010)
18. Xiao, C., Zhang, Y., Zeng, X., Wu, Y.: Predicting user influence in social media. J. Netw. 8(11), 2649–2655 (2013)
19. Ye, S., Wu, S.F.: Measuring message propagation and social influence on Twitter.com. In: Bolc, L., Makowski, M., Wierzbicki, A. (eds.) SocInfo 2010. LNCS, vol. 6430, pp. 216–231. Springer, Heidelberg (2010). https://doi.org/10.1007/978-3-642-16567-2_16

Building a Domain-Specific Knowledge Graph for Business Networking Analysis

Didier Gohourou[1(✉)] and Kazuhiro Kuwabara[2]

[1] Graduate School of Information Science and Engineering, Ritsumeikan University, Kyoto, Japan
gr0259sh@ed.ritsumei.ac.jp

[2] College of Information Science and Engineering, Ritsumeikan University, Kusatsu, Shiga 525 -8577, Japan

Abstract. The emergence of graph learning supports the study of phenomenon that are best represented by knowledge graphs, using deep learning techniques. However, deep learning is mostly built around supervised learning which requires datasets for training and testing. This paper presents BusinessNet, a domain-specific knowledge graph to examine the business networking phenomenon and offers a novel dataset for the graph learning community. BusinessNet is built from unstructured news wire text data from which entities such as organizations, people, and places, as well as the relationships among them are extracted to build the graph. The resulting graph is a heterogeneous network of organizations, people, and places as nodes and that will include different types relations. The analysis of the resulting dataset shows interesting characteristics for graph learning models evaluations.

Keywords: Dataset · Knowledge graph · Graph learning

1 Introduction

Deep learning brought record-breaking performances for tasks that were previously difficult to automatize. This is in part thanks to the datasets on which models are trained and tested. Widely used datasets such as WordNet [20] and ImageNet [6] offers unprecedented support for research in their respective domain: Natural Language Processing and Computer Vision. Deep learning achievements also gave rise to work that revisited existing research fields with graph theory being one such field recently.

Graphs are useful data structures to represent interconnected entities such as social networks, molecules, traffic maps, among others. The emergence of graph learning [2,4,10,25,28] which is the study of deep learning algorithms for graphs, provides new insights for the above-mentioned domains. Another interesting domain for graph learning applications is business networking. Business networking can be defined as a socio-economic process through which businesses establish relationships among them. Many social networks bring together

N. T. Nguyen et al. (Eds.): ACIIDS 2021, LNAI 12672, pp. 362–372, 2021.
https://doi.org/10.1007/978-3-030-73280-6_29

business professionals and enable them to build relationships through interpersonal connections, In many cases, the resulting interpersonal relationships from those social networks give substance for social network analysis. However, to the best of our knowledge, no learning model studies the business to business networking phenomenon itself. While interpersonal relationships involve friends, followers, subordinates, superiors, colleagues. business-to-business relationships involve customers, suppliers, owners and investors. Compared to the former type of social networks and relationships, the deep learning literature has barely covered how they are created, how they evolve, and what kind of predictions one can make regarding their dynamics.

A previous study showed that deep learning models from the natural language processing field can provide hidden insights for business to business networking decision making [9]. The conception of a learning model specific to business networking will require a domain-specific dataset. Leveraging the intuitive graph structure of the business networking phenomenon, a knowledge graph upon which graph learning models will be trained and tested will lead to more in-depth decision making insights for business networking. However it is difficult to find such datasets that will help study and understand businesses, their interconnections, and patterns that arise from their behavior.

This paper presents BusinessNet a domain-specific knowledge graph for business analysis. The primary goal of BusinessNet is to serve as a knowledge ground for a business market analysis and recommendation platform. BusinessNet is also intended to help study behaviors exhibited by businesses during their life cycle, including the relationships they built at different stages of their evolution, and how those relationships impact their performances and other aspects.

The contributions of this study are as follows:

- A domain-specific dataset for business market analysis: This study describes the primary dataset for a business market analysis and recommendation tool.
- A dataset for graph learning research: The proposed knowledge graph is developed on the basis of the proposed graph neural network benchmarking mechanisms from the literature [7].

The remainder of the paper is structured as follows: Section 2 describes related works. Section 3 exposes the construction and the current state of the dataset. Section 4 presents a simple usage application of the obtained knowledge graph. Section 5 discusses the current state of the dataset and its viability. Section 6 concludes the paper and lays down future study directions.

2 Related Work

This section presents influential datasets for the deep learning community, discusses domain-specific datasets frequently used with graph learning models, and finally introduces how this study fits into the body of work on datasets.

The Modified National Institute of Standards and Technology (MNIST) dataset is probably one of the most influential datasets in the deep learning community. It is a large database of handwritten digits, created using samples from the NIST's original datasets [11] used for training image processing models. It has become the defacto dataset used to teach introductory notions of deep learning. MNIST consists of 60,000 training images and 10,000 testing images.

WordNet is a lexical database of semantic relation between words [20]. It was initiated in 1985 by the Cognitive Science Laboratory at Princeton University, for the English language. WordNet links words into semantic relations including synonyms, hyponyms and meronyms. The synonyms are grouped into synsets. The database consists of 155,327 words organized in 175,979 synsets for a total of 207,016 word-sense pairs. WordNet is also available in more than 200 other languages.

ImageNet, a large scale image database built upon the hierarchical structure provided by WordNet [6], was created and is mainly maintained by the Stanford Vision Lab at Stanford University. ImageNet was initiated to expand and improve the data available to train AI algorithms and was introduced to the research community in 2009. ImageNet aims to provide the most comprehensive and diverse coverage of the image world. In doing that, it succeeded to datasets such as ESP, Tiny Image, Label me, and Lotus Hills dataset among other to become the defacto dataset for computer vision related tasks. ImageNet contains more than 20 millions images hand-annotated by crowdworkers through Amazon Mechanical Turk[1]. The images are organized into more than 20,000 categories. ImageNet is at the center of the annual ImageNet Large Scale Visual Recognition Challenge (ILSVRC).

To study phenomenon such as social interactions and user preferences, other datasets have been built by observing those domain specific phenomenon.

The **Zachary karate club** [27] is a social network that represents the members of a karate club. The dataset was initially created for a behavioral study to understand fissions in a small bounded group. Data were gathered by observing the karate club members for three years. The obtained social network has been used as a dataset represented as a graph to test graph learning algorithms such as Deepwalk [22] and Graph Convolutional Network [16].

MovieLens is a movie recommendation service created in 1997. The Movielens dataset was first released in 1998 [13]. It contains users' preferences for movies. Those preferences are represented with tuples that take the form of `<user, item, timestamp>`. These tuples express the 0–5 star rating of a movie by a user. The MovieLens 25M Dataset[2] format recommended for research and released in December 2019, contains 25 million ratings and 1 million tag applications applied to 62,000 movies by 162,000 users. MovieLens datasets are widely used in many recommendation studies, including some that use Graph Learning [26].

CiteSeer is an autonomous citation indexing system for academic literature [8]. CiteSeer is built by crawling academic documents over the web, extracting and parsing the citations section. This enables interconnection between

[1] https://www.mturk.com/.
[2] https://grouplens.org/datasets/movielens/25m/.

research documents, creating a network of academic literature. CiteSeer has been used for various applications such as classification [18,23] and recommendation [5]. Because CiteSeer is built autonomously, it may contain duplicates or some ambiguous information. Studies address those shortcomings to improve the dataset [3].

The primary goal of the BusinessNet dataset developed in this work is for to train and test of algorithms that study the business-to-business networking phenomenon in a way similar to how the Zachary karate club represents to social connections, MovieLens represents to movie recommendations and CiteSeer represents to the interconnections of research articles. BusinessNet is represented using a knowledge graph, obtained by observing business networking phenomenon through business news wire related text. A secondary goal of BusinessNet includes completing the dataset collection available for graph learning research. Our aim is for BusinessNet to be a benchmark graph dataset like thoses included in the Open Graph Benchmark [15]. It will do so by following practices and methodologies from influential datasets such as ImageNet and WordNet among others.

3 BusinessNet Knowledge Graph

This section presents our knowledge graph building methodology and the resulting dataset when applied to the Reuters corpus. The Reuters corpus is a collection of over 800,000 manually categorized news wire stories made available by Reuters Ltd. for research purpose [17]. The version used in this study is accessible as a dataset within the NLTK Natural Language Processing python library [1], or can be downloaded as a text dataset on the web[3].

3.1 Knowledge Graphs

A knowledge graph can be defined as a graph of data intended to accumulate and convey knowledge of the real world whose nodes represent entities of interest and whose edges represent relations between these entities [14].

Knowledge graphs can be obtained by manual data gathering and linking, an automated process, or an hybrid one that combines manual labor and automated data processing. Knowledge graphs are used in a various of applications, including question answering, natural language generation, and commonsense reasoning.

Specifically, knowledge graphs are represented using the **graphs** data structure. Below are a subset of definitions of a graph and some of its properties used for analysis in this work.

Definition 1. *A Graph G given a pair:* $G = (V, E)$*, comprises a set of vertices (or nodes)* $V = \{v_1, ..., v_{|V|}\}$ *connected by edges* $E = \{e_1, ..., e_{|E|}\}$*, where each edge* e_k *is a pair* (v_i, v_j) *with* $v_i, v_j \in V$.

[3] https://www.kaggle.com/nltkdata/reuters.

In our study we use undirected graph (with undirected edges).

Definition 2. *A path P is a sequence of edges* $(u_{i1}, u_{i2}), (u_{i2}, u_{i3}), ...(u_{ik}, u_{ik+1})$ *of length k.*

Definition 3. *Given two nodes* (u, v) *in a graph G the distance from u to v, denoted* $d_G(u, v)$*, to be the length of the shortest path from u to v, or* ∞ *is there exist no path from u to v.*

Definition 4. *The degree,* $deg(v_i)$*, of a vertex* v_i *is the number of edges incident to it.*

Definition 5. *A* ***component****, of an undirected graph is an induced subgraph in which any two vertices are connected to each other by paths, and which is connected to no additional subgraph.*

3.2 Methodology

Similar to a methodology proposed [26] to build a knowledge graph for tasks such as questions-answering, we built our knowledge graph from unstructured text document from the Reuters corpus.

Preprocessing. From the documents that compose the Reuters corpus, we extracted the text contained in the <BODY> tag. The text was later concatenated and prepared as a one sentence per line format.

Identifying Entities. For a business to business networking study, we decided to focus on three types of nodes: organisations, people and places (locations). The named entity recognition feature of the spaCy[4] NLP library was used to identify the entities of interest within the preprocessed text corpus. A unique identifier was also generated for each one of the identified named entity. Table 1 shows the resulting number of entities extracted.

Table 1. Numbers of entities extracted by entity types, for the full graph and the graph component with the highest number of nodes (giant component).

Entity type	Full graph	Giant component
Organizations	17,679	3,367
People	6,732	1,669
Places	2,205	604
Total	26,616	5,640

[4] https://spacy.io/.

Finding Relationships. OpenIE [19] was used to extract triplets relationships from the corpus. The triplets were filtered to keep relationships among named entities of our interest: organizations, people and places. In the presented knowledge graph, we consider an edge to be a relationship triplet between two entities of interest.

Building the Graph. NetworkX [12], a Python library for creating and manipulating graph networks, was used for the resulted knowledge graph dataset representation and manipulation. Figure 1 is an overview of the component with the highest number of nodes in the obtained graph. Figure 2 shows the frequency of node degrees within the knowledge graph.

Selecting Content. The resulting graph contains 1133 components of which more than 95% are composed of less than 5 connected nodes. Those can be hardly used for a connected graph analysis, while they can be useful to understand how business communities are formed. Thus, according to the type of analysis to be conducted, the graph can be pruned. Figure 3 shows the number of different component size that can be found in the obtained graph dataset.

In this paper, we use the giant graph component (see Fig. 1) for a sample visualization on the obtained graph and graph learning experiment.

4 Sample Usage

4.1 Node Classification Experiment

We use the giant component from the graph generated by our methodology to perform a node classification task using a graph learning model. The learning model we utilize in our experiment is the graph convolutional network (GCN) [16]. The GCN is a semi-supervised learning algorithm that proceed in a message-passing fashion.

We sampled 10% of the nodes representing organisations, 10% of the nodes representing people and 10% of the nodes representing places as labelled nodes for the semi-supervised classification.

We used the Deep Graph Library (DGL) [24] implementation of the GCN, running Pytorch [21] as a backend. We built a simple network with two graph convolutional layers of size 50 and an output layer of size 3 for the number of classes. After 100 epochs of training using an Adam loss function, the GCN model is able to correctly classify 90% of the nodes in the giant component. We can argue that the model is able to learn some structure out of the graph we propose.

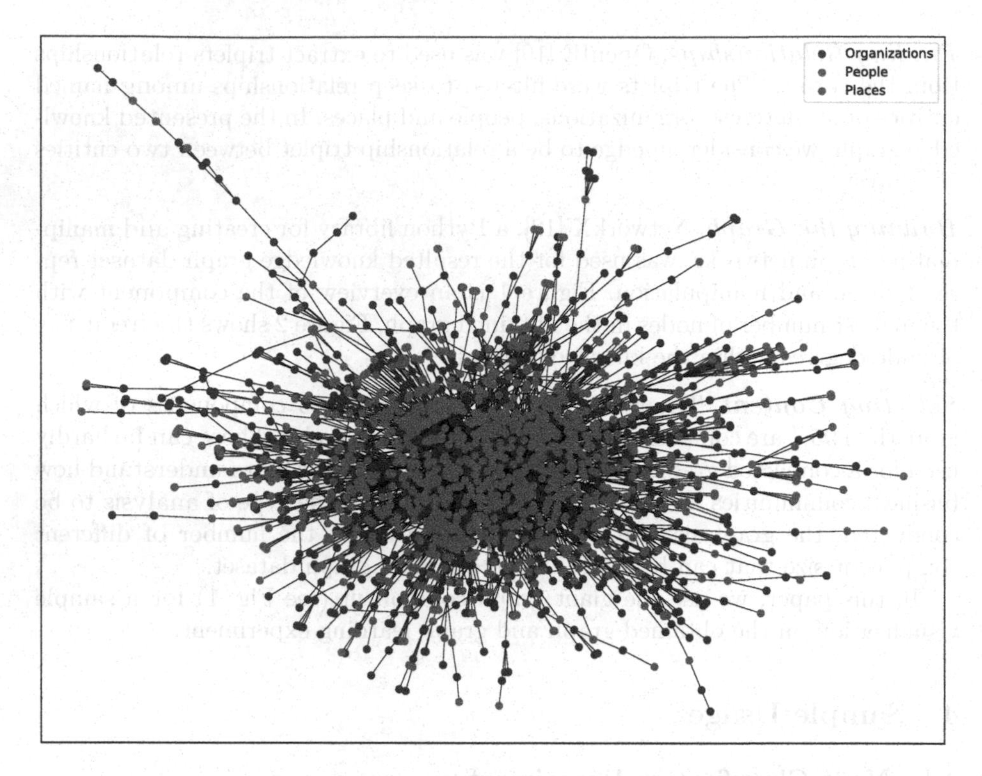

Fig. 1. The resulted graph component with the highest number of nodes after applying our proposed building mechanism on the Reuters corpus.

4.2 Further Experimentation with BusinessNet

Further experiments and analysis can be conducted with the tools used to construct BusinessNet. The source code of the tool-chain used to build BusinessNet will be made publicly available[5].

The released version will contain the Python implementation of the tools with running instructions that will include the required libraries (cited in this paper) such as NetworkX, spaCy, Numpy, Pytorch and DGL.

[5] https://github.com/semlab/businessnet.

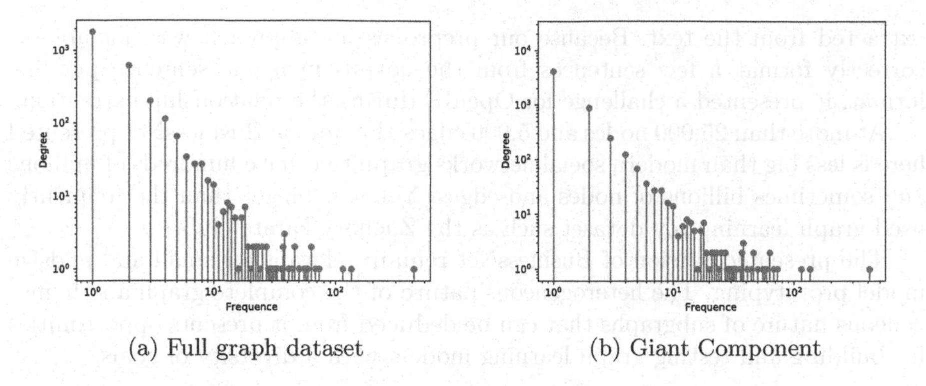

(a) Full graph dataset (b) Giant Component

Fig. 2. Degree histogram for the resulting graph dataset. Figure 2a represent the one for the entire graph, while Fig. 2b is the one for the component with the highest number of nodes.

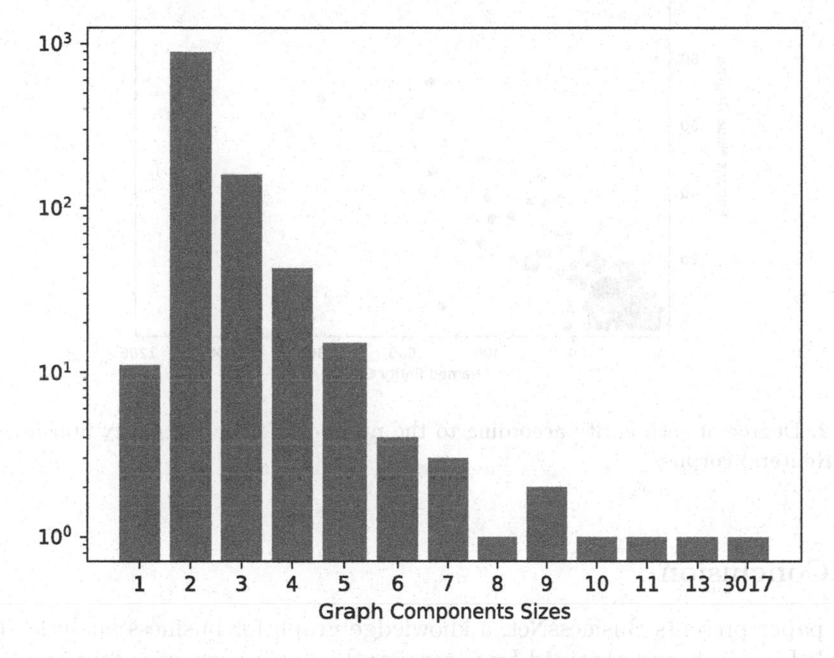

Fig. 3. Histogram of the number of different component size within the resulting graph.

5 Discussion

The current entity detection and extraction mechanism is able to identify most entities (organisations, people and places) within the corpus. We can naturally note that the more an entity is mentioned in the corpus, the more likely we will discover its related entities. Figure 4 shows the degree of node entities according to the number of times that entity is mentioned in the Reuters corpus. For the edges, we relied upon OpenIE. Most relationship triplets were successfully extracted from the text. Because our preprocessing approach was not able to correctly format a few sentences from the corpus in a one sentence per line format, it presented a challenge for OpenIE during the relationship extraction.

At more than 25,000 nodes and 5,000 edges the sample BusinessNet presented here is less big than modern social networks graph that have hundreds of millions and sometimes billions of nodes and edges. Yet it is bigger than the frequently used graph learning toy dataset such as the Zachary karate club.

The presented version of BusinessNet remain relevant for small and midsize model prototyping. The heterogeneous nature of the complete graph and homogeneous nature of subgraphs that can be deduced from it presents opportunities for building and testing graph learning models with a diversity of goals.

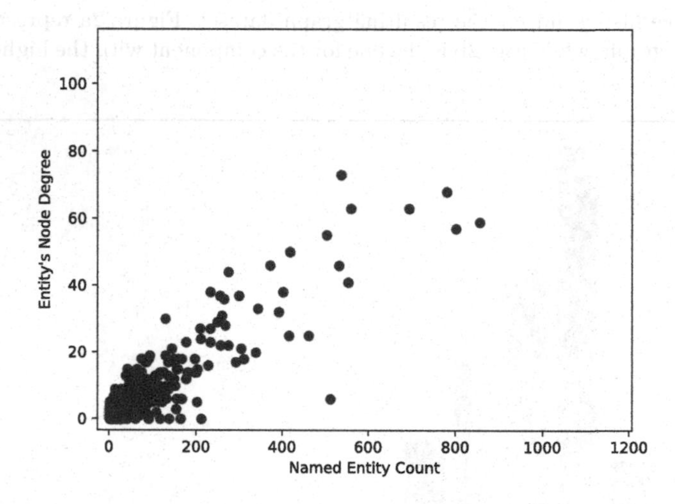

Fig. 4. Degree of each entity according to the number of time the entity appeared in the (Reuters) corpus.

6 Conclusion

This paper presents BusinessNet, a knowledge graph for business analytic. The knowledge graph was obtained by processing business news wire text from the Reuters corpus to extract entities (organizations, people and places), and relationships triplets among them. Experiments were conducted to observe how a proven graph learning algorithm behaved when trained and ran using the resulting knowledge graph as dataset. Observations supports that the knowledge graph built in this study is a viable training dataset for graph learning algorithms.

Future works aim at completing the dataset, expanding it by running our methodology over other text corpora such as Wikipedia dumps[6], and news wire datasets such as the CNN and DailyMail news wire corpora[7]. More experiments

[6] https://dumps.wikimedia.org/backup-index.html.
[7] https://cs.nyu.edu/~kcho/DMQA/.

of graph learning applications, including link prediction, community detection algorithms and graph classification, will be run on the dataset to further evaluate how suitable it is for those tasks. The knowledge graph will also be enriched with hierarchical node linking (e.g. industry/sector of the organisation nodes), and a temporal dimension for the study of the dynamic nature of business networking over time.

References

1. Bird, S., Klein, E., Loper, E.: Natural Language Processing with Python. O'Reilly Media, Sebastopol (2009)
2. Cai, H., Zheng, V., Chang, K.: A comprehensive survey of graph embedding: problems, techniques, and applications. IEEE Trans. Knowl. Data Eng. **30**, 1616–1637 (2018)
3. Caragea, C., et al.: CiteSeerx: a scholarly big dataset. In: de Rijke, M., et al. (eds.) ECIR 2014. LNCS, vol. 8416, pp. 311–322. Springer, Cham (2014). https://doi.org/10.1007/978-3-319-06028-6_26
4. Chami, I., Abu-El-Haija, S., Perozzi, B., Ré, C., Murphy, K.: Machine learning on graphs: a model and comprehensive taxonomy. ArXiv abs/2005.03675 (2020)
5. Chen, H.H., Gou, L., Zhang, X., Giles, C.: Collabseer: a search engine for collaboration discovery, pp. 231–240 (2011). https://doi.org/10.1145/1998076.1998121
6. Deng, J., Dong, W., Socher, R., Li, L.J., Li, K., Fei-Fei, L.: Imagenet: a large-scale hierarchical image database. In: IEEE Conference on Computer Vision and Pattern Recognition, 2009. CVPR 2009, pp. 248–255. IEEE (2009)
7. Dwivedi, V.P., Joshi, C.K., Laurent, T., Bengio, Y., Bresson, X.: Benchmarking graph neural networks. ArXiv abs/2003.00982 (2020)
8. Giles, C., Bollacker, K., Lawrence, S.: Citeseer: an automatic citation indexing system. In: Proceedings of the ACM International Conference on Digital Libraries, Proceedings of the 1998 3rd ACM Conference on Digital Libraries; Conference date: 23-06-1998 Through 26-06-1998, pp. 89–98. ACM (1998)
9. Gohourou, D., Kurita, D., Kuwabara, K., Huang, H.-H.: International business matching using word embedding. In: Nguyen, N.T., Tojo, S., Nguyen, L.M., Trawiński, B. (eds.) ACIIDS 2017. LNCS (LNAI), vol. 10191, pp. 181–190. Springer, Cham (2017). https://doi.org/10.1007/978-3-319-54472-4_18
10. Goyal, P., Ferrara, E.: Graph embedding techniques, applications, and performance: a survey. Knowl.-Based Syst. **151**, 78–94 (2018). https://doi.org/10.1016/j.knosys.2018.03.022
11. Grother, P.: NIST special database 19 handprinted forms and characters database (1995). https://doi.org/10.18434/T4H01C
12. Hagberg, A.A., Schult, D.A., Swart, P.J.: Exploring network structure, dynamics, and function using networkx. In: Varoquaux, G., Vaught, T., Millman, J. (eds.) Proceedings of the 7th Python in Science Conference, pp. 11–15. Pasadena, CA USA (2008)
13. Harper, F., Konstan, J.: The movielens datasets. ACM Trans. Interact. Intell. Syst. **5**, 1–19 (2015). https://doi.org/10.1145/2827872
14. Hogan, A., et al.: Knowledge graphs. ArXiv abs/2003.02320 (2020)
15. Hu, W., et al.: Open graph benchmark: datasets for machine learning on graphs. ArXiv abs/2005.00687 (2020)

16. Kipf, T., Welling, M.: Semi-supervised classification with graph convolutional networks. In: 5th International Conference on Learning Representations, ICLR 2017 - Conference Track Proceedings, pp. 1–14 (2017)
17. Lewis, D., Yang, Y., Russell-Rose, T., Li, F.: Rcv1: a new benchmark collection for text categorization research. J. Mach. Learn. Res. **5**, 361–397 (2004)
18. Lu, Q., Getoor, L.: Link-based classification. In: Proceedings of the 20th International Conference on Machine Learning, pp. 496–503 (2003)
19. Schmitz, M., Bart, R., Soderland, S., Etzioni, O.: Open language learning for information extraction. In: Proceedings of the 2012 Joint Conference on Empirical Methods in Natural Language Processing and Computational Natural Language Learning, pp. 523–534. EMNLP-CoNLL '12, Association for Computational Linguistics, USA (2012)
20. Miller, G.A.: Wordnet: a lexical database for english. Commun. ACM **38**, 39–41 (1995)
21. Paszke, A., et al.: Pytorch: an imperative style, high-performance deep learning library. In: Advances in Neural Information Processing Systems, vol. 32, pp. 8024–8035. Curran Associates, Inc. (2019)
22. Perozzi, B., Al-Rfou, R., Skiena, S.: Deepwalk: online learning of social representations. In: Proceedings of the 20th ACM SIGKDD International Conference on Knowledge Discovery and Data Mining, pp. 701–710. KDD '14, Association for Computing Machinery, New York, NY, USA (2014). https://doi.org/10.1145/2623330.2623732
23. Sen, P., Namata, G., Bilgic, M., Getoor, L., Gallagher, B., Eliassi-Rad, T.: Collective classification in network data articles. AI Mag. **29**, 93–106 (2008)
24. Wang, M., et al.: Deep graph library: a graph-centric, highly-performant package for graph neural networks. arXiv preprint arXiv:1909.01315 (2019)
25. Wu, Z., Pan, S., Chen, F., Long, G., Zhang, C., Yu, P.S.: A comprehensive survey on graph neural networks. IEEE Trans. Neural Netw. Learn. Syst. **32**(1), 4–24 (2021). https://doi.org/10.1109/tnnls.2020.2978386
26. Yu, S., He, T., Glass, J.: Constructing a knowledge graph from unstructured documents without external alignment. arXiv 2008.08995 (2020)
27. Zachary, W.: An information flow model for conflict and fission in small groups. J. Anthropol. Res. **33**, 452–473 (1976)
28. Zhang, Z., Cui, P., Zhu, W.: Deep learning on graphs: a survey. In: IEEE Transactions on Knowledge and Data Engineering, pp. 1–1 (2020). https://doi.org/10.1109/tkde.2020.2981333

Cybersecurity Intelligent Methods

N-Tier Machine Learning-Based Architecture for DDoS Attack Detection

Thi-Hong Vuong(✉), Cam-Van Nguyen Thi, and Quang-Thuy Ha

VNU-University of Engineering and Technology (UET), Vietnam National University, Hanoi (VNU), No. 144, Xuan Thuy, Cau Giay, Hanoi, Vietnam
{hongvt57,vanntc,thuyhq}@vnu.edu.vn

Abstract. Distributed Denial of Service (DDoS) attack is a menace to network security that aims at exhausting the target networks with malicious traffic. With simple but powerful attack mechanisms, it introduces an immense threat to the current Internet community. In this paper, we propose a novel multi-tier architecture intrusion detection model based on a machine learning method that possibly detects DDoS attacks. We evaluate our model using the newly released dataset CICDDoS2019, which contains a comprehensive variety of DDoS attacks and address the gaps of the existing current datasets. Experimental results indicated that the proposed method is more efficient than other existing ones. The experiments demonstrated that the proposed model accurately recognize DDoS attacks outperforming the state-of-the-art by F1-score.

Keywords: DDoS attacks · CICDDoS2019 · Machine learning methods · Intrusion detection

1 Introduction

Cyber security is a significant role in secure communication to prevent services from being paralyzed by network intrusions. Intruders often exploit popular rogue software to carry out multiple attacks against networked computer systems. The damage done in a cyber attack can range from a slight disruption in service to huge financial losses. Furthermore, today's escalation of Internet of Things (IoT) devices and services has dramatically changed our daily lives. A large number of advanced IoT technology-based applications have been successfully built and deployed, such as smart cities, smart health care, smart home and vehicle networks [2]. These systems more opportunities for attackers to easily break into the network.

Recently, intrusion detection systems (IDSs) detect and eliminate malicious activities in cyber security [21]. As the number of malicious attacks constantly increases exponentially, IDSs have a duty to deal with eliminating such attacks before they cause massive destruction on a large area of cyberspace. In the passive of time, these systems are designed and combined with machine learning

N. T. Nguyen et al. (Eds.): ACIIDS 2021, LNAI 12672, pp. 375–385, 2021.
https://doi.org/10.1007/978-3-030-73280-6_30

techniques to handle unauthorized use and access to network resources. According to [15], there are different techniques to propose detection and defense mechanisms such as machine learning, knowledge-based, and statistical. The statistical methods are not allowed to determine with certainty the normal network packet distribution. The machine learning approach is better than as they do not have any prior known data distribution, but defining the best feature-set is one of the main concerns for them [17].

There are two challenges for machine learning-based IDSs. Firstly, datasets were shortcomings and problems [16,18,23]. Machine learning models face a major challenge of having to develop a model from a limited set of data called training data sets. The ideal case is to look for the best models to classify intrusive and non-intrusive network activities that have the features such as high detection accuracy, low false positive rate (FPR), activation which is related to the IDS alert enhancement and IDS must be adaptable in relation to constantly changing network environments. Scale ability in relation to the challenges ahead is tracking increasingly complex and heterogeneous. IDSs cannot deal with it. In [12], it is challenging to develop such models with the above criteria in real life. On the other hand, Shiravi et al. [16], Singh and De [18] and Yu et al.[23] tried to develop DDoS dataset. But Iman et al. [15] showed that there are many shortcomings and problems, such as incomplete traffic, anonymized data, and out-dated attack scenarios, still researchers struggle to find comprehensive and valid dataset to test and evaluate their proposed detection and defense models. Therefore, having a suitable dataset is a significant challenge. Secondly, the model needs to identify new attacks. A different important issue of characteristics of the DDoS attacks can be manually set, controlled by the attacker, or it can be automated. Therefore, there is a need to identify new attacks and come up with new taxonomies in [15]. Detection of the attack is an important step to DDoS mitigation. Detection is very simple as the performance of the service or system degrades dramatically when an attack occurs. However, it is always challenging to differentiate malicious flows from legitimate flows.

In this paper, we propose a novel multi-tier architecture intrusion detection model base on the machine learning approach to resolve the above challenges. The proposed method use different level classifies to detect attacks. The novel proposed method present on the newly released dataset CICDDoS2019[1] [15], which remedies the shortcomings and limitations of previous datasets. We also compare the baseline methods such as Naive Bayes (NB), Support Vector Machine (SVM), Decision Tree (DT) in [7,15] to indicate the efficacy of the proposed method on CICDDoS2019.

The rest of the paper is presented as follows. Section 2 describes the related work and reviews two approach of IDSs along with two detection mechanisms: signature-based attack detection and anomaly-based attack detection. Section 3 introduces our proposed framework, a brief description of the dataset, preprocess data, extract features and presents n-tier machine learning-based architecture for DDoS attack detection. Compared with the well-known methods and the

[1] https://www.unb.ca/cic/datasets/ddos-2019.html.

state-of-the-art classification models, the experiment results of the proposed method are show in Sect. 4. Finally, the paper provides several conclusions and further work in Sect. 5.

2 Related Work

Attacks can be determined as examined by two approaches: signature-based detection, anomaly-based detection [3,13]. While anomaly-based detection systems determine the traffic which has harmful content as result of the analysis of network traffic, signature-based detection systems examine based the previous data, which were recorded on the system and confirms the attack. Signature-based detection mechanisms rely on known DDoS attacks to identity the attack signatures [5,6,9]. It is successful in detecting known DDoS attacks. However, any variations in already existing attacks remain unnoticed by these detection mechanisms. Indeed, as the techniques unaware of the new signatures, they do not work for any new attack types that have not been seen previously. Anomaly-based detection mechanism can handle attacks with new signatures as well as newly appear attacks [1,10,19]. However, the selection of threshold value to differentiate between attack traffic and normal traffic is an open challenge for these techniques. There are lot of the other methods try to differentiate attack traffic and legitimate traffic based on the analysis and detection of anomalies in traffic flows [4,8,20].

There are the most recent and popular mechanisms that have been used for the detection of DDoS attacks in software-defined network environments [11,14,22]. Ye et al. in [11] proposed Support Vector Machine (SVM) classification algorithm to detect DDoS attack including UDP, TCP SYN and ICMP flood traffic. Rahman et al. in [14] used four machine learning techniques to detect DDoS under context of software-defined network. The experiments showed that the J48 has better accuracy in the baseline methods with two DDoS attack including TCP and ICMP floods. The three different machine learning algorithms: SVM, Navie Bayes (NB) and Neural Network to detect flow-table overflow attack were used in [22]. One of the challenges to detect application layer DDoS attacks is high similarity of attacks and benign behaviors. Therefore, there is a lack of available features to define as attacks. Indeed, many detection systems are not suitable to identify it. In addition, the previous proposed techniques that used public datasets to train anomaly detection systems suffer of several issues such as incomplete traffic, anonymized data and out of date attack scenarios. According to [7,15], a comprehensive and valid dataset has a great impact on the evaluation of detection algorithms and techniques systems. Thus, to evaluate our proposed model, we use the newest public available dataset CICDDoS2019 [15]. The CICDDoS2019 dataset contain a comprehensive variety of DDoS attacks and address the gaps of the existing current datasets. With CICDDoS 2019, we approach an anomaly-based detection. In this paper, we propose n-tier machine learning-based architecture for DDoS attack detection.

3 Proposed Framework

According to [13], the basic structure of a DDoS attack has four different components such as attacker, multiple control masters or handlers, multiple slaves, agents or zombies and a victim or target machine. Different phases and characteristics of the DDoS attacks can be manually set, controlled by the attacker, or it can be automated. Therefore, there is a need to identify new attacks and come up with new taxonomies in [15]. We propose a novel n-tier machine learning-based architecture intrusion for DDoS attack detection as Fig. 1. N-tier architecture intrusion detection model is also called as multiple classifier system that uses a set of classifiers as base machine learning techniques to build training data to classify unknown data. The proposed framework includes two phases. In the first phase, we preprocess CICDDoS2019 and select features. Then, feeding processing data into the second phase to build an attack model base on machine learning technique.

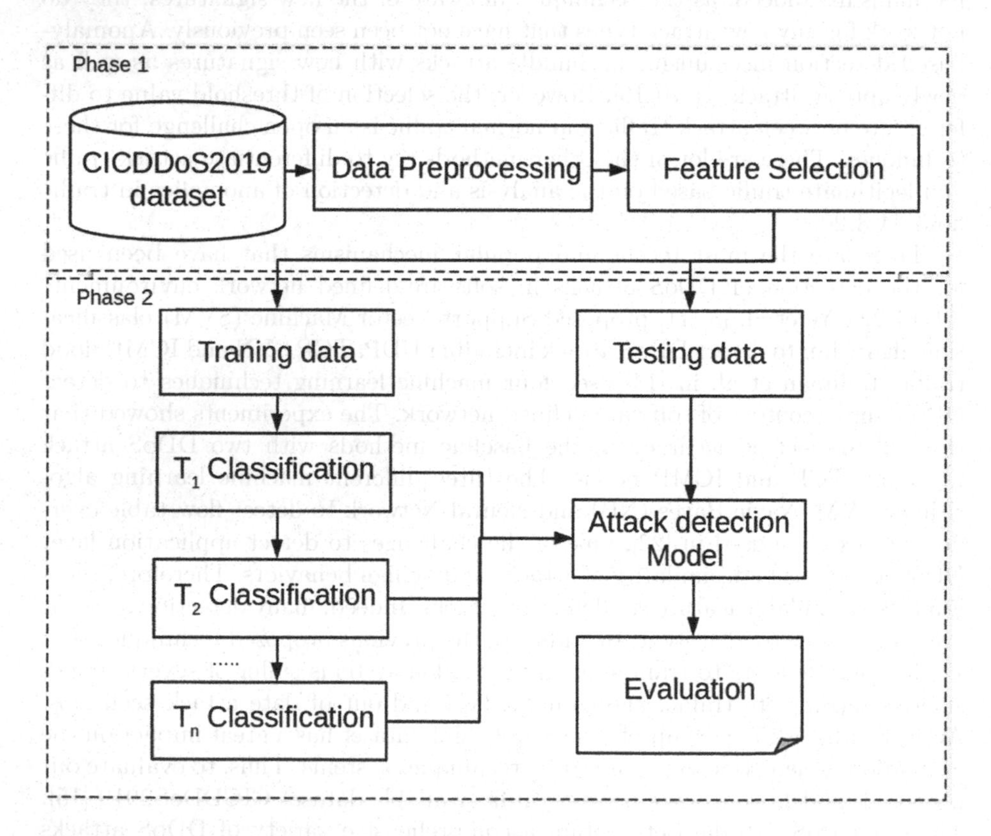

Fig. 1. The general flowchart of the proposed method. Given CICDDoS2019 dataset, preprocess data and extract features in the first phase. In the second phase, use n-tier classification architecture and evaluate for DDoS attack detection model.

3.1 Dataset Preprocessing

The first phase before training the IDS models is to preprocess the dataset to make it more suitable for the training phase and avoid the overfitting problem. The steps are taken for preprocessing as follows:

- The CICDDoS2019 dataset contains the socket information such as source IP, destination IP, flowID, etc. The original dataset includes 88 features when removed to all socket features. One-hot encoding scheme is used convert the labeled string to numerical values.
- In the CICDDoS2019, the following class labels are employed: UDP, BENIGN, UDP-Lag, SYN, MSSQL, NetBIOS, LDAP. According to [15], the capturing period for the training day on January 12th started at 10:30 and ended at 17:15, and for the testing day on March 11th started at 09:40 and ended at 17:35. The statistics of CICDDoS2019 with 7 class labels details in Table 1.
- The class labels in CICDDoS2019 were classified on terms of reflection-based and exploitation-based attacks [15]. Because of imbalanced class labels, we group the class labels into 4 based on the two above terms of attacks: 1 (UDP, UDP-Lag, SYN), 2 (NetBIOS, LDAP), 3 BENIGN, and 4 MSSQL in Table 2.

Table 1. Detail statistics CICDDoS2019 with seven class labels

Label	Training day Jan 12th	Percentage	Testing day Mar 11th	Percentage
UDP	3134645	19,67%	3867155	19,17%
BENIGN	56863	0,36%	56965	0,28%
UDPLag	366461	2,30%	1873	0,01%
SYN	1582289	9,93%	4891500	24,24%
MSSQL	4522492	28,38%	5787453	28,68%
NetBIOS	4093279	25,68%	3657497	18,13%
LDAP	2179930	13,68%	1915122	9,49%
Total	15935959	100%	20177565	100%

Table 2. Detail statistics CICDDoS2019 with group class labels

Label	Training day Jan 12th	Percentage	Testing day Mar 11th	Percentage
(1) UDP, UDP-Lag, SYN	5083395	31,90%	8760528	43,42%
(2) BENIGN	56863	0,36%	56965	0,28%
(3) NetBIOS, LDAP	6273209	39,36%	5572619	27,62%
(4) MSSQL	4522492	28,38%	5787453	28,68%
Total	15935959	100%	20177565	100%

3.2 Feature Selection

In the feature selection, we focus on selecting the best features to predict DDoS attacks instead of all features in the original data. The Random Forest Regressor was used to calculate the importance of each feature among 88 features in the dataset [15]. We select a subset of 24 features from the original CICDDoS2019 dataset to train our learning model as in Table 3.

Table 3. Feature set used in the intrusion detection system

1	Fwd packet length max	13	Subflow fwd bytes
2	Fwd packet length min	14	Destination port
3	Min packet length	15	Protocol
4	Max packet length	16	Packet length std
5	Average packet size	17	Flow duration
6	Fwd packets/s	18	Fwd IAT total
7	Fwd header length	19	ACK flag count
8	Fwd header length 1	20	Init_Win_Bytes_Forward
9	Min_Seg_Size_Forward	21	Flow IAT mean
10	Total length of fwd packet	22	Flow IAT max
11	Fwd packet length std	23	Fwd IAT mean
12	Flow IAT min	24	Fwd IAT max

3.3 Multi-tier Architecture Intrusion Detection Model

We propose a n-tier classification architecture for DDoS attacks detection. The pseudo-code of the proposed method for attack detection shown in Algorithm 1. In the one step, we used binary classifiers T_1, T_2,..., T_n to classify the input traffic into normal and malicious types by Random Forest (RF). Then, these learner outputs are integrated for the second step to detect DDoS attacks as multi-class classification framework.

Algorithm 1. N-tier architecture intrusion detection model

```
1: Input: CICDDoS2019 atfer preprocessing and extracting
       D= (x1,y1),(x2,y2),...(xn,yn) , n learning datasets
2: Step 1: Learn binary classifier
3: for Each class label do
4:     Implement RF for attacked or normal classification
5:     Ti Classification, i: attack class label
6:     Save Ti classification model
7: end for
8: Step 2: Learn N-tier classifier
9: Integrate Ti classification model into n-tier classifier
10: Classify xj as attacked or normal
11: Evaluate model
```

4 Experiments

4.1 Evaluation Metrics

In this experiment, we have evaluated the effectiveness of our IDS with the use of the Confusion Matrix as shown in Table 4 below:

Table 4. Confusion matrix

	Predicted attack	Predicted normal
Actual attack	True positive (TP)	False negative (FN)
Actual normal	False positive (FP)	True negative (TN)

- True positive (TP): the number of rightly recognized malicious code.
- True negative (TN): the number of rightly recognized benign code.
- False positive (FP): the number of incorrectly identified benign code, when a detector recognizes a benign code as a Malware.
- False negative (FN): the number of incorrectly recognized malicious code, when a detector recognizes a Malware as a benign code

We use precision (P), recall (R) and F1 score (F1) to evaluate the proposed framework with the baseline methods. P, R, F1 are define in Eq. 1, 2, 3.

$$P = \frac{TP}{TP + FP} \tag{1}$$

$$R = \frac{TP}{TP + TN} \tag{2}$$

$$F1 = 2 * \frac{P * R}{P + R} \tag{3}$$

4.2 Experiment Cases

We present the proposed method with two cases and compare it with the baseline methods such as NB, SVM, DT in [7,15]. In the first experiment case, the proposed method is implemented on the CICDDoS2019 with seven class labels as the following in Table 1. In the second experiment case, we implement the proposed method above dataset with group labels in Table 2. We use the training day for training data, and 20% of the testing day for testing data for the proposed method and baseline methods.

1. Comparison with the baseline methods on CICDDoS2019 in Table 1
   The experiments indicated that the proposed method is more efficient than the baseline methods in Table 5. The performance of the proposed framework overall measure such as P, R, and F1-score is higher than the comparison

methods. F1-score of the proposed method is extremely higher than NB from 30% to 90% on labels. The results know that F1-score is higher than SVM and DT from 5% to 30% on labels in Fig. 2.
The experiments show that the proposed method is more efficient on labels such as SYN, BENIGN, UDP, MSSQL and NetBIOS from 60% to 99%. However, the proposed method also limited detection labels as UDPLag and LDAP. Because the UPDLag percentage is extremely smaller only 2,30% on training data. It is not enough to learn for DDoS attack detection. LDAP percentage also is smaller than others only 13,68%. To improve the efficiency of the proposed framework, we group the class labels into four labels as Table 2. After grouping, the percentage of class labels is approximate each other except BENIGN.

Table 5. Comparison the baseline methods of seven class labels

Measure	NB			SVM			DT			Proposed method		
	P	R	F1	P	R	F1	P	R	F1	P	R	F1
UDP	0,951	0,027	0,053	0,974	0,477	0,641	0,986	0,497	0,661	0,982	0,526	**0,685**
BENIGN	0,078	0,100	0,087	0,939	0,464	0,621	0,989	0,912	0,949	0,995	0,992	**0,993**
UDPLag	0,000	0,000	0,000	0,000	0,000	0,000	0,002	0,316	0,004	0,843	0,178	**0,294**
SYN	0,983	0,099	0,179	0,991	0,548	0,706	1,000	0,953	0,976	0,860	0,998	**0,924**
MSSQL	0,000	0,000	0,000	0,444	0,869	0,588	0,512	0,626	0,563	0,506	0,698	**0,587**
NetBIOS	0,002	0,000	0,000	0,205	0,225	0,215	0,375	0,555	0,447	0,511	0,630	**0,564**
LDAP	0,108	1,000	0,195	0,000	0,000	0,000	0,507	0,198	0,284	0,622	0,286	**0,392**

2. Comparison with the baseline methods on CICDDoS2019 in Table 2
The proposed framework is implemented on CICDDoS2019 after grouping labels with the same term. The result of experiments indicated that the proposed method on grouping labels is the best on both P, R and F1-score than the baseline methods in Table 6. F1-score of (1 UDP, UDP-Lag, SYN), (2 BENIGN), (3 NetBIOS, LDAP) and (4 MSSQL) are about 87,6%, 99,3%, 67,8% and 53%, correspondingly. The results demonstrated that the proposed framework accurately recognize DDoS attacks outperforming the baseline methods in Fig. 3. Indeed, by the grouping labels, the efficiency of the proposed method is better.

Table 6. Comparison with the baseline methods of four class labels

Measure	NB			SVM			DT			Proposed method		
	P	R	F1	P	R	F1	P	R	F1	P	R	F1
1	0,790	0,004	0,008	0,988	0,459	0,627	0,997	0,704	0,825	0,995	0,782	**0,876**
2	0,030	0,104	0,047	0,979	0,495	0,658	0,991	0,918	0,953	0,995	0,991	**0,993**
3	0,016	0,000	0,000	0,472	0,956	0,632	0,457	0,680	0,547	0,587	0,803	**0,678**
4	0,288	0,999	0,447	0,535	0,099	0,167	0,589	0,482	0,530	0,589	0,482	**0,530**

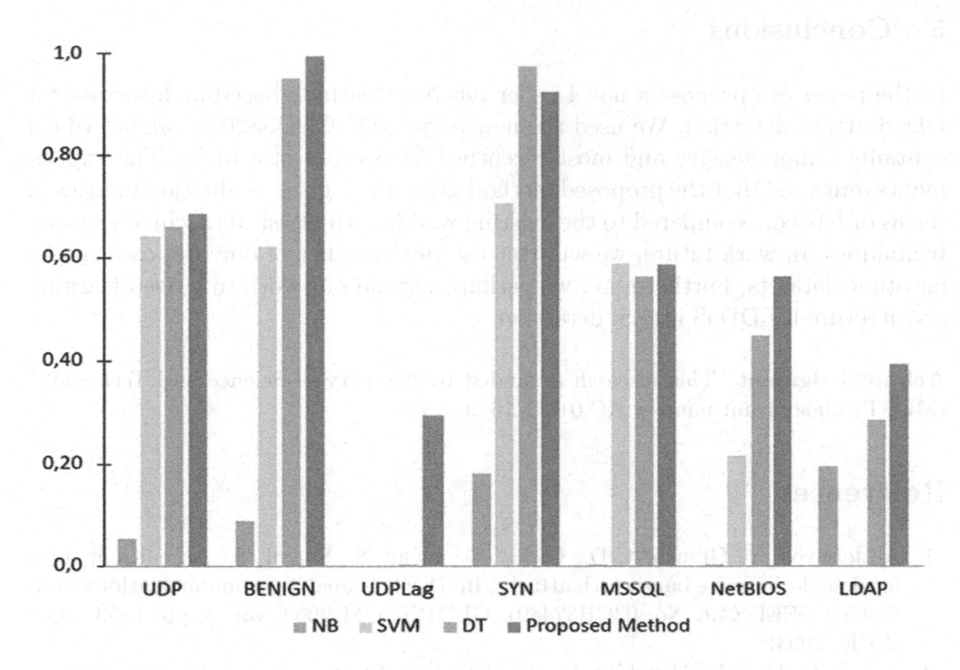

Fig. 2. Comparison the baseline methods on F1-score with seven class labels

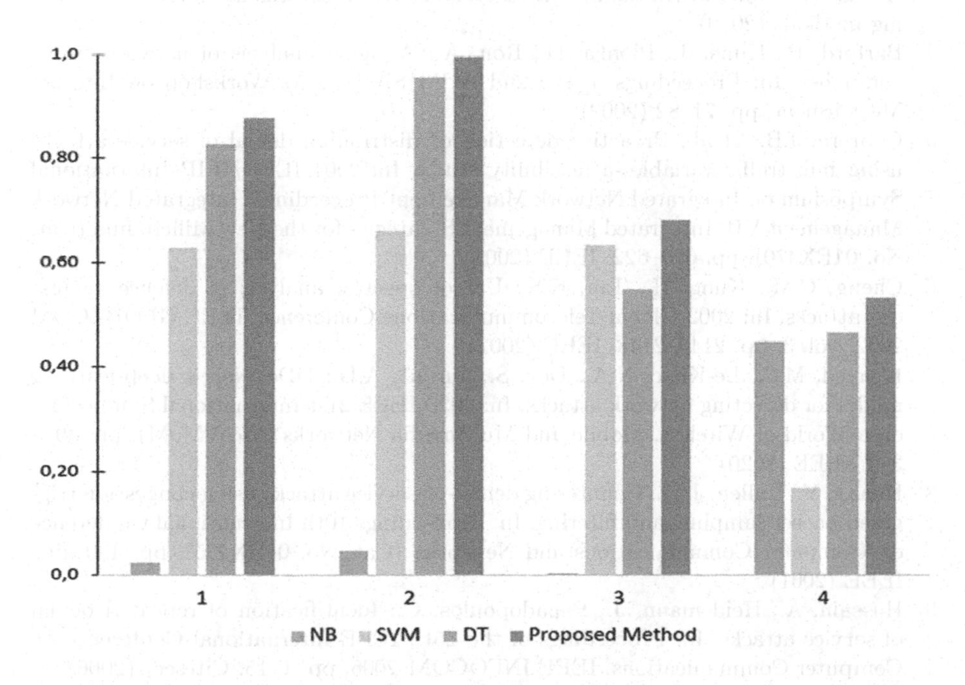

Fig. 3. Comparison the baseline methods on F1-score with four class labels

5 Conclusions

In the paper, we propose a novel n-tier machine learning-based architecture for DDoS attack detection. We used the new released CICDDoS2019 dataset which contains comprehensive and most recently DDoS types of attacks. The experiments indicated that the proposed method gives the highest evaluation metrics in terms of F1-score compared to the existing well known classical machine learning techniques. In work future, we will test the performance of our proposed model on other datasets. Furthermore, we perhaps extend our work to a deep learning architecture for DDoS attack detection.

Acknowledgment. This research is funded by Ministry of Science and Technology (MOST) under grant number KC.01.28/16-20.

References

1. Abdelsayed, S., Glimsholt, D., Leckie, C., Ryan, S., Shami, S.: An efficient filter for denial-of-service bandwidth attacks. In: IEEE Global Telecommunications Conference (IEEE Cat. No. 03CH37489), GLOBECOM 2003. vol. 3, pp. 1353–1357. IEEE (2003)
2. Atzori, L., Iera, A., Morabito, G.: The internet of things: a survey. Comput. Netw. **54**(15), 2787–2805 (2010)
3. Aytaç, T., Aydın, M.A., Zaim, A.H.: Detection DDoS attacks using machine learning methods (2020)
4. Barford, P., Kline, J., Plonka, D., Ron, A.: A signal analysis of network traffic anomalies. In: Proceedings of the 2nd ACM SIGCOMM Workshop on Internet Measurment, pp. 71–82 (2002)
5. Cabrera, J.B., et al.: Proactive detection of distributed denial of service attacks using mib traffic variables-a feasibility study. In: 2001 IEEE/IFIP International Symposium on Integrated Network Management Proceedings. Integrated Network Management VII. Integrated Management Strategies for the New Millennium (Cat. No. 01EX470), pp. 609–622. IEEE (2001)
6. Cheng, C.M., Kung, H., Tan, K.S.: Use of spectral analysis in defense against dos attacks. In: 2002 Global Telecommunications Conference, IEEE GLOBECOM 2002, vol. 3, pp. 2143–2148. IEEE (2002)
7. Elsayed, M.S., Le-Khac, N.A., Dev, S., Jurcut, A.D.: DDoSNet: a deep-learning model for detecting network attacks. In: 2020 IEEE 21st International Symposium on a World of Wireless, Mobile and Multimedia Networks (WoWMoM), pp. 391–396. IEEE (2020)
8. Huang, Y., Pullen, J.M.: Countering denial-of-service attacks using congestion triggered packet sampling and filtering. In: Proceedings 10th International Conference on Computer Communications and Networks (Cat. No. 01EX495), pp. 490–494. IEEE (2001)
9. Hussain, A., Heidemann, J., Papadopoulos, C.: Identification of repeated denial of service attacks. In: Proceedings of the 25th IEEE International Conference on Computer Communications, IEEE INFOCOM 2006, pp. 1–15. Citeseer (2006)
10. Jow, J., Xiao, Y., Han, W.: A survey of intrusion detection systems in smart grid. Int. J. Sens. Netw. **23**(3), 170–186 (2017)

11. Karan, B., Narayan, D., Hiremath, P.: Detection of DDoS attacks in software defined networks. In: 2018 3rd International Conference on Computational Systems and Information Technology for Sustainable Solutions (CSITSS), pp. 265–270. IEEE (2018)
12. Kumar, G., Thakur, K., Ayyagari, M.R.: MLEsIDSs: machine learning-based ensembles for intrusion detection systems—a review. J. Supercomput. **76**(11), 8938–8971 (2020). https://doi.org/10.1007/s11227-020-03196-z
13. Mahjabin, T., Xiao, Y., Sun, G., Jiang, W.: A survey of distributed denial-of-service attack, prevention, and mitigation techniques. Int. J. Distrib. Sens. Netw. **13**(12), 1550147717741463 (2017)
14. Rahman, O., Quraishi, M.A.G., Lung, C.H.: DDoS attacks detection and mitigation in SDN using machine learning. In: 2019 IEEE World Congress on Services (SERVICES), vol. 2642, pp. 184–189. IEEE (2019)
15. Sharafaldin, I., Lashkari, A.H., Hakak, S., Ghorbani, A.A.: Developing realistic distributed denial of service (DDoS) attack dataset and taxonomy. In: 2019 International Carnahan Conference on Security Technology (ICCST), pp. 1–8 (2019)
16. Shiravi, A., Shiravi, H., Tavallaee, M., Ghorbani, A.A.: Toward developing a systematic approach to generate benchmark datasets for intrusion detection. Comput. Secur. **31**(3), 357–374 (2012)
17. Jin, S Yeung, D.S.: A covariance analysis model for DDoS attack detection. In: 2004 IEEE International Conference on Communications (IEEE Cat. No.04CH37577), vol. 4, pp. 1882–1886 (2004)
18. Singh, K.J., De, T.: An approach of DDoS attack detection using classifiers. In: Shetty, N.R., Prasad, N.H., Nalini, N. (eds.) Emerging Research in Computing, Information, Communication and Applications, pp. 429–437. Springer, New Delhi (2015). https://doi.org/10.1007/978-81-322-2550-8_41
19. Sun, B., Xiao, Y., Wang, R.: Detection of fraudulent usage in wireless networks. IEEE Trans. Veh. Technol. **56**(6), 3912–3923 (2007)
20. Talpade, R., Kim, G., Khurana, S.: NOMAD: traffic-based network monitoring framework for anomaly detection. In: Proceedings of the IEEE International Symposium on Computers and Communications (Cat. No. PR00250), pp. 442–451. IEEE (1999)
21. Tama, B.A., Rhee, K.H.: An extensive empirical evaluation of classifier ensembles for intrusion detection task. Comput. Syst. Sci. Eng. **32**(2), 149–158 (2017)
22. Ye, J., Cheng, X., Zhu, J., Feng, L., Song, L.: A ddos attack detection method based on svm in software defined network. Secur. Commun. Netw. **2018**, 8 (2018)
23. Yu, S., Zhou, W., Jia, W., Guo, S., Xiang, Y., Tang, F.: Discriminating DDoS attacks from flash crowds using flow correlation coefficient. IEEE Trans. Parallel Distrib. Syst. **23**(6), 1073–1080 (2011)

Detection of Distributed Denial of Service Attacks Using Automatic Feature Selection with Enhancement for Imbalance Dataset

Duy-Cat Can(✉), Hoang-Quynh Le, and Quang-Thuy Ha

Faculty of Information Technology, University of Engineering and Technology, Vietnam National University Hanoi, Hanoi, Vietnam
{catcd,lhquynh,thuyhq}@vnu.edu.vn

Abstract. With the development of technology, the highly accessible internet service is the biggest demand for most people. Online network, however, has been suffering from malicious attempts to disrupt essential web technologies, resulting in service failures. In this work, we introduced a model to detect and classify Distributed Denial of Service attacks based on neural networks that take advantage of a proposed automatic feature selection component. The experimental results on CIC-DDoS 2019 dataset have demonstrated that our proposed model outperformed other machine learning-based model by large margin. We also investigated the effectiveness of weighted loss and hinge loss on handling the class imbalance problem.

Keywords: DDoS attack · Multi-layer perceptron · Automatic feature selection · Class imbalance · Multi-hinge loss · Weighted loss

1 Introduction

A distributed denial-of-service (DDoS) attack is a malicious attempt to disrupt regular traffic of a targeted server, service, or network by overwhelming the target or its surrounding infrastructure with a flood of traffic from illegitimate users [15]. It aims at depleting network bandwidth or exhausting target's resources with malicious traffic. This attack causes the denial of service to authenticated users and causes great harm to the security of network environment.

A DDOS attack is relatively simple but often brings a disturbing effect to Internet resources. Together with the popularity and low-cost of the Internet, DDoS attacks have become a severe Internet security threat that challenging the accessibility of resources to authorized clients. According to the forecast of the Cisco Visual Networking Index, the number of DDoS attacks grew 172% in 2016, and expects that this will increase 2.5-fold to 3.1 million by 2021 globally[1].

[1] https://www.infosecurity-magazine.com/news/cisco-vni-ddos-attacks-increase/, archived on 11 November, 2020.

N. T. Nguyen et al. (Eds.): ACIIDS 2021, LNAI 12672, pp. 386–398, 2021.
https://doi.org/10.1007/978-3-030-73280-6_31

Basically, DDoS attacks are based on the same techniques as another regular denial of service (DoS) attacks. The differences are, *(i)* it uses a single network connection to flood a target with malicious traffic, and (ii) it uses botnets to perform the attack on a much larger scale than regular DOS attacks [4]. A botnet is a combination of numerous remotely managed compromised hosts (i.e., bots, zombies, or other types of slave agents), often distributed globally. They are under the control of one or more intruders. This work focuses on attack detection, that identifying the attacks immediately after they actually happen to attack a particular victim with different types of packets. The experts define several kinds of DDoS attacks; examples include UDP Flood, ICMP (Ping) Flood, SYN Flood, Ping of Death, Slowloris, HTTP Flood, and NTP Amplification [13].

DDoS defense system consists of four phases: Attack prevention, Attack detection and characterization, Traceback and Attack reaction [4]. This work focuses on attack detection, that identifying the attacks immediately after they actually happen. In the case of a system is under DDoS attack, there are unusual fluctuations in the network traffic. The attack detector must automatically monitor and analyze these abrupt changes in the network to notice unexpected problems [8]. In this work, We consider this problem as a classification problem, i.e., classifies DDoS attacks packets and legitimate packets.

Although many statistical methods have been designed for DDoS attack detection, designing an effective detector with the ability to adapt to change of DDoS attacks automatically is still challenging so far. This paper proposes a deep learning-based model for DDoS detection that selects the features automatically. The experiments were conducted on CIC-DDoS 2019 dataset [20]. Evaluation results show that our proposed model achieved notable performance improvement in terms of $F1$ compared to the baseline model and several existing DDoS attack detection methods.

The main contributions of our work can be concluded as:

i. We represent a deep neural network model to detect DDoS attacks.
ii. We propose and demonstrate the effectiveness of automatic feature selection method.
iii. We investigate the contributions of multi-hinge loss and weighted loss to handling class imbalance problem.

2 Related Work

Kaur et al. [13] classifies DDOS detection methods into two main groups: signature-based detection and anomaly-based detection. Detection with signature-based makes utilization of 'signs' about different attacks. This approach is only operative in case of known attack; it works by matching the arriving traffic with the previously-stored pattern. Anomaly-based detection methods are more commonly used since it is fit for recognizing unknown attack. The main strategy is comparing standard network performance with arriving information to detect anomalies, i.e., when a system is under DDoS attack, unexpected fluctuations in the network traffic need to be noticed. Since DDoS attacks are still growing year by year, knowledge-based methods are inflexible to adapt to their growth and change. The research community has been paying attention to DDoS detection for years, provided several different methods for recognizing DDoS attacks based on statistical and Machine Learning techniques.

Statistical methods are basically done by measuring statistical properties (i.e., means and variances) of various normal traffic parameters. Three of the most widely used techniques in these approaches are ARIMA [23], SSM [18] and CAT-DCP [2]. The limitation of these methods is they are not able to determine with certainty the normal network packet distribution.

Machine learning-based techniques are useful as they do not have any prior known data distribution. The machine learning methods used for DDoS detection are very diverse: Support Vector Machine [12,20], Naive Bayes [6], Decision Tree [22]. Tradition machine learning methods require selecting the best feature-set to bring good performance i.e., they often require the contribution of human experts high level to define patterns. This process is labor-intensive, comparatively expensive but often provide much error-prone [21].

Neural networks are introduced as an alternative to traditional machine learning methods that can handle complex data by automatically select useful features without human intervention. Many deep neural network-based methods have been successfully applied in many works for generating intrusion detection classifiers in general and DDOS detection in particular. Examples include Artificial Neural Network [19], Convolutional Neural Network, Recurrent Neural Network (RNN) and its improvements Long Short-Term Memory Unit (LSTM) and Gated Recurrent Units (GRUs) [1], Hopfield Networks and Radial Basis Function based Neural Networks [11], Replicator Neural Networks [18], Convolutional Neural Network [16], etc.

Although the deep learning-based model has recently achieved great success due to its high learning capacity, it still cannot escape from imbalanced data [9, 10]. To overcome this problem, two methods that have been successfully applied in other domains are using hinge loss [9] and applying class-weight to give priority to the minor classes [10].

3 Proposed Model

Figure 1 depicts the overall architecture of our proposed model DDoSNet. DDoSNet mainly consists of two components: a feature selection layer, and a classification layer using fully-connect multi-layer perceptron (MLP). Given a set of traffic features as input, we build an automatic feature selection model to calculate a weight for each feature. An MLP model is applied to capture higher abstract features, and a softmax layer is followed to perform a (K+1)-class distribution. The details of each layer are described below.

3.1 Data Preprocessing

In the first step of implementation, the preprocessing on our datasets is exerted. We follow four preprocessing operations to prepare the training data.

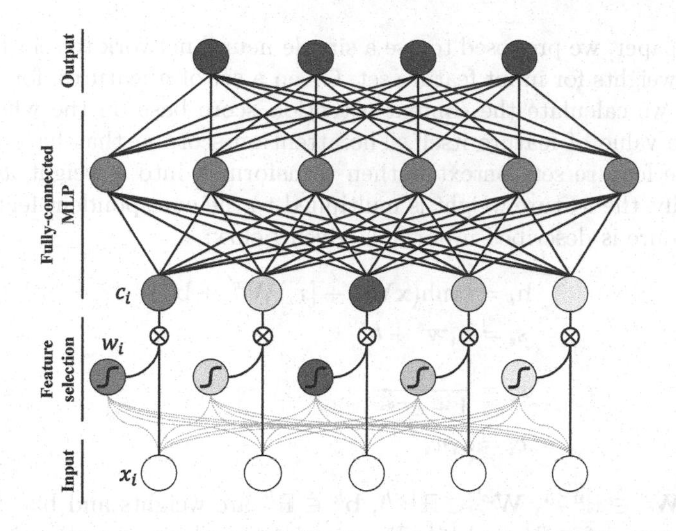

Fig. 1. An overview of proposed model.

- **Removal of irrelevant features**: we remove all of the attributes which are non informative such as `Unnamed: 0`, `Flow ID`, `Timestamp`, and all of the socket features like `Source IP`, `Destination IP`.
- **Cleaning the data**: we have convert invalid values such as `NaN` and `inf` or `SimillarHTTP` to corresponding value for efficient running of algorithms.
- **Label Encoding**: One-hot encoding is used to convert categorical label into numerals. In addition, we use another binary label 1 and 0 to denote if an example is an DDoS attack or not.
- **Normalization**: the features data have different numerical range value that make training process biasing on large values. For the random features, we normalize these features to normal distribution, as follow:

$$x_i = \frac{f_i - \mu_i}{\sigma_i} \tag{1}$$

where μ and σ are feature mean and standard deviation respectively. For the fixed value features, we normalize the data using min-max scaling as follow:

$$x_i = \frac{f_i - f_{min}}{f_{max} - f_{min}} \tag{2}$$

3.2 Feature Selection

Feature selection is one of the key problems for machine learning and data mining that selecting a feature subset that performs the best under a certain evaluation criterion. Sharafaldin et al. [20] have used Random Forest Regressor to examine the performance of the selected features and have selected 24 best features with corresponding weight for each DDoS attack.

In this paper, we proposed to use a simple neural network to select and learn important weights for input feature set. Given a set of n features, for each input feature x_i, we calculate the context attention score base on the whole feature set and the value of feature itself. The attention score s_i that have taken into account the feature set context is then transformed into a weight w_i in range $[0, 1]$. Finally, the feature weight is multiplied to the corespondent feature value. This procedure is described in formula given below:

$$\mathbf{h}_i = \tanh(\mathbf{x}\mathbf{W}^x + [x_i]\mathbf{W}^{x'} + \mathbf{b}^h) \tag{3}$$

$$s_i = \mathbf{h}_i \mathbf{w}^s + b^s \tag{4}$$

$$w_i = \frac{1}{1 + e^{-s_i}} \tag{5}$$

$$c_i = x_i w_i \tag{6}$$

where $\mathbf{W}^x \in \mathbb{R}^{n \times h}$, $\mathbf{W}^{x'} \in \mathbb{R}^{1 \times h}$, $\mathbf{b}^h \in \mathbb{R}^h$ are weights and bias for hidden attention layer; $\mathbf{w}^e \in \mathbb{R}^h$ and $b^e \in \mathbb{R}$ are weights and bias for attention score.

3.3 Classification

The features from the penultimate layer are then fed into a fully connected multi-layer perceptron network (MLP). We choose hyperbolic tangent as the non-linear activation function in the hidden layer. i.e.

$$\mathbf{h} = \tanh(\mathbf{c}\mathbf{W}_h + \mathbf{b}_h) \tag{7}$$

where $\mathbf{W}_h$ and $\mathbf{b}_h$ are weight and bias terms. We apply multi hidden layer to produce higher abstraction-level features The output h of the last hidden layer is the highly abstract representation of input features, which is then fed to a softmax classifier to predict a (K+1)-class distribution over labels $\hat{\mathbf{y}}$:

$$\hat{\mathbf{y}} = \text{softmax}(\mathbf{h}\mathbf{W}_y + \mathbf{b}_y) \tag{8}$$

3.4 Objective Function and Learning Method

We compute the penalized cross-entropy, and further define the training objective for a data sample as:

$$L(\hat{\mathbf{y}}) = -\sum_{i=0}^{K} \mathbf{y}_i \log \hat{\mathbf{y}}_i \tag{9}$$

where $\mathbf{y} \in \{0, 1\}^{(K+1)}$ indicating the one-hot vector represented the target label. In addition to categorical cross entropy losses, we use hinge loss to generate a decision boundary between classes. We use the formula defined for a linear classifier by Crammer and Singer [3] as follow:

$$\ell(y) = \max(0, 1 + \max_{y \neq t} \mathbf{x}\mathbf{w}_y - \mathbf{x}\mathbf{w}_t) \tag{10}$$

where t is the target label, $\mathbf{w}_t$ and $\mathbf{w}_y$ are the model parameters. We further add L^1-norm and L^2-norm of model's weights and L^2-norm of model's biases to model objective function to keep parameter in track and accelerate model training speed.

$$L(\theta) = \alpha \|\mathbf{W}\|_2 + \beta \|\mathbf{W}\|_1 + \lambda \|\mathbf{b}\|_2 \tag{11}$$

where α, β and λ are regularization factors.

The model parameters $\mathbf{W}$ and $\mathbf{b}$ are initialized using Xavier normal initializer [7] that draws samples from a truncated normal distribution centered on 0. To compute these model parameters, we minimize $L(\theta)$ by applying Stochastic Gradient Descent (SGD) with Adam optimizer [14] in our experiments.

To handle the class imbalance problem, we drive our model to have the classifier heavily weight the few examples that are available by using weighted loss. We calculate the weight for each class as follow:

$$w_i = \frac{1}{n_i}\frac{1}{2}\sum_j n_j \tag{12}$$

where n_i is number of class i examples. Scaling by total/2 helps keep the loss to a similar magnitude. The sum of the weights of all examples stays the same.

4 Experiment and Discussion

4.1 CIC-DDoS 2019 Dataset

Many data sets are using in studies that are made using different algorithms in Intrusion Detection System designs. In this paper, we evaluate our proposed classifier using the new released CIC-DDoS 2019 dataset [20] which was shared by Canadian Institute for Cybersecurity. The dataset contains a large amount of different DDoS attacks that can be carried out through application layer protocols using TCP/UDP.

Table 1 summarizes the distribution of the different attacks in the CIC-DDoS 2019 dataset. The dataset was collected in two separated days for training and testing evaluation. The training set was captured on January 12th, 2019, and contains 12 different kinds of DDoS attacks, each attack type in a separated PCAP file. The attack types in the training day includes DNS, LDAP, MSSQL, NetBIOS, NTP, SNMP, SSDP, Syn, TFTP, UDP, UDPLag, and WebDDoS DDoS based attack. The testing data was created on March 11th, 2019, and contains 7 DDoS attacks LDAP, MSSQL, NetBIOS, Portmap, Syn, UDP, and UDPLag.

The training and testing datasets vary in distribution of data. For example, two minor classes MSSQL and NetBIOS in training dataset are major class in testing dataset with percentage of 28.42% and 17.96% respectively. The class imbalance is also a challenge of this dataset in which minor classes account for less than 1%. Another notable remark is more than 68% of training dataset belong to the classes are totally absent from testing dataset. The Portmap attack in the testing set does not present in the training data for intrinsic evaluation of detection system.

Table 1. Statistic of training and testing dataset.

Label	Training dataset		Testing dataset	
	Num.	Per	Num.	Per
BENIGN	56863	0.11%	56965	0.28%
DNS	5071011	10.13%	–	–
LDAP	2179930	4.35%	1915122	9.40%
MSSQL	4522492	9.03%	5787453	28.42%
NetBIOS	4093279	8.18%	3657497	17.96%
NTP	1202642	2.40%	–	–
Portmap	–	–	186960	0.92%
SNMP	5159870	10.31%	–	–
SSDP	2610611	5.22%	–	–
Syn	1582289	3.16%	4891500	24.02%
TFTP	20082580	40.12%	–	–
UDP	3134645	6.26%	3867155	18.99%
UDPLag	366461	0.73%	1873	0.01%
WebDDoS	439	0.00%	–	–
Total	**50063112**	**100%**	**20364525**	**100%**

Experimental Configuration: In the experiments, we train our model on 90% of training dataset and report the results on the testing dataset, which is kept secret with the model. We leave 10% of training dataset for validation dataset to fine-tune model's hyper-parameters. We conduct the training and testing process 10 times and calculate the averaged results. For evaluation, the predicted labels were compared to the golden annotated data with common machine learning evaluation metrics: precision (P), recall (R), and F1 score.

4.2 System's Performance

We compared our model with various common machine learning algorithms namely decision tree, random forest, Naïve Bayes and logistic regression that reported by Sharafaldin et al. [20]. These performance examination results are in terms of the weighted average of the evaluation metrics with five-fold cross validation. For a fair comparison, we re-implemented these models and evaluate on separated training and testing datasets. Table 2 represents the classification metrics of our six model variants with different comparative models.

According to benchmark results, decision tree (ID3) performed the best with the fastest training time. Random forest is follow with the result of 69% on more than 15 h of training. The Naïve Bayes classifier performed poorly, primarily because the NB assumed that all attributes are independent of each other. Finally, logistic regression, with more than 2 d of training process, did not meet the expectation with 5% F1 score.

Table 2. System's performance on CIC-DDoS 2019 dataset.

Model		Average			Binary		
		P	R	F1	P	R	F1
Benchmark [20]†	Decision tree	78.00	65.00	69.00	–	–	–
	Random forest	77.00	56.00	62.00	–	–	–
	Naïve Bayes	41.00	11.00	5.00	–	–	–
	Logistic regression	25.00	2.00	4.00	–	–	–
Elsayed et al. [5]	Random forest	–	–	–	100.00	74.00	85.00
	SVM	–	–	–	99.00	88.00	93.00
	Logistic regression	–	–	–	93.00	99.00	96.00
	RNN-Autoencoder	–	–	–	99.00	99.00	99.00
Baseline‡	Naïve Bayes	30.30	17.51	7.35	–	–	–
	SVM	62.44	57.97	55.50	–	–	–
	Decision Tree	61.15	58.32	55.15	–	–	–
	Random Forest	50.76	36.91	39.57	–	–	–
Our model‡	24 features	91.12	72.91	74.00	99.99	99.93	99.96
	24 features + FS	85.19	76.51	75.44	99.99	99.89	99.94
	82 features	88.97	70.61	71.09	99.99	99.94	99.96
	82 features + FS	**91.16**	**79.41**	**79.39**	**99.98**	**99.89**	**99.93**
	- hinge loss	82.06	73.60	74.29	99.99	99.93	99.96
	- weighted loss	60.51	67.34	63.60	99.97	99.94	99.95

† *5-fold cross validation, weighted average*
‡ *train-test split, macro average*

Our reproduced baseline results on separated training and testing data have similarities with the benchmark results. Decision tree gives high performance at 55.15% F1, followed by random forest and Naïve Bayes with 39.57% and 7.35% respectively. In addition, we also try applying support vector machine (SVM) and have slightly better results than other methods.

The obtained results show that our model outperforms the other machine learning algorithms by large margin. Firstly, we apply our deep learning model directly on the input examples without automatic feature selection component. We observe that our proposed model produces better result on 24 selected features introduced in study of Sharafaldin et al. [20] with 2.91% gap with the model applied on all 82 features. However, when applied the automatic feature selection based on feature context vector, we notice the complete opposite results. With the improvement of 8.3%, 82-feature model (DDoSNet) yield the best F1 result. Meanwhile, the 24-feature model showed only a small improvement of about 1.44%. One possible reason is the feature weights are calculated based on the feature context vector, 82 features, therefor, give more information.

We also considered the binary result and compared our model with another neural network-based model (RNN-Autoencoder) that proposed by Elsayed et al. [5]. In this experiment, we have witnessed the dominance of deep learning models. The logistic regression model that gave poor results with imbalanced multi-class data has been re-vital that gave high results with binary data. RNN-autoencoder model as well as the our proposed deep learning models performed excelent on

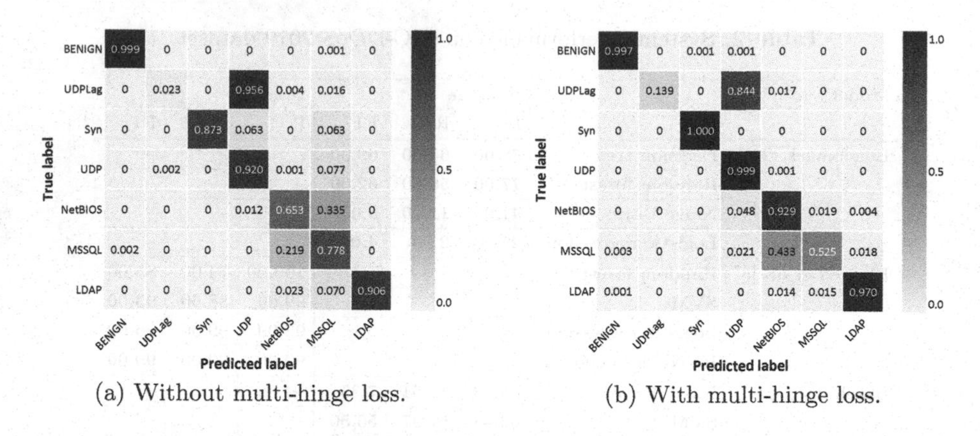

(a) Without multi-hinge loss.

(b) With multi-hinge loss.

Fig. 2. Model's prediction confusion matrix.

this binary data with over 99% of F1. Deep learning models rarely misclassified which example is DDoS attack.

4.3 Result Analysis

Class Imbalance Problem: CIC-DDoS 2019 is an imbalanced dataset, in which 2 major classes account for over 50%, the ratio between the largest and smallest class in the test set is more than 3000 times. We have done some further investigations into the experimental results. Figure 2b presents the confusion matrix of DDoSNet model's prediction on validation dataset. As we observe on the confusion matrix, examples of the BENIGN class - not a DDoS attack - are rarely confused with attack classes and vice versa. Among the attack classes, the syn and LDAP classes also performed well without being misclassified with the other classes. In contrast, 84.8% of inputs from UDPLag class were mistakenly classified as a UDP attack, causing the recall metric of UDPLag to drop to 11.28%. This can be explained by two reasons: (i) UDPLag is a minor label - the percentages in training and testing set are only 0.73% and 0.01% respectively - so classifiers are difficult to recognize the data belongs to this class; (ii) on the DDoS attack taxonomy tree, UDPLag is a child-node of UDP so UDPLag examples collapsed into UDP class is reasonable. Another class did not perform well is MSSQL with 43.3% of the input being mistaken to NetBIOS.

Another analysis on the results of each classes with different variations of the proposed model is summarized in Fig. 3. According to the statistics of model variants' results, class weight plays an important role in training the model to predict minor classes. When removing the weighted loss, the results of two classes UDPLag and LDAP dropped to 0.0%. The automatic feature selection component also plays a certain role in solving the class imbalance problem, the most obvious demonstration shown in the LADP class result. Another interesting observation is that although Syn was a minor class in the training set, the test results of this label exceeded our expectation. One possible reason is Syn label

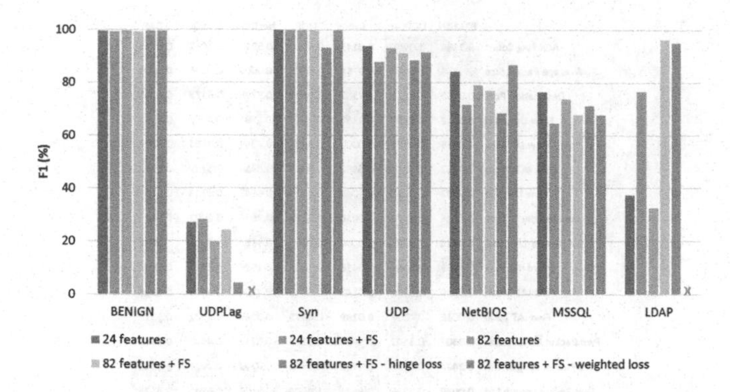

Fig. 3. Comparison of each label's F1 score of 6 model variants. X columns denote 0.0% of F1 score.

is on a separate branch on the taxonomy tree, so the features of this class are obvious making machine learning models easy to detect.

Experiment of Automatic Feature Selection: In this experiment, to analyze the efficiency of automatic feature selection module, we re-executed our model on 100,000 random validation examples and extracted the weight for each feature. The arithmetic mean of the weights of each feature by classes is presented in Fig. 4. Observing the weighted heat-map of input features, we have seen that the important levels that our model learned for BENIGN label is often in opposition to attack labels. The `ACK Flag Count`, `Destination Port`, `Init_Win_bytes_forward`, `min_seg_size_forward` and `protocol` features have been highlighted as the most important features for distinguishing types of DDoS attacks. When compared with the weights that have been meticulously selected through experiments in the study of Saharafaldin et al. [20], our automatically selected features have a lot of similarities. However, some of our weights are in stark contrast to the above study. For example, `Flow IAT Min` and `Fwd Packet Length Std` for BENIGN class, `ACK Flag Count` and `Fwd IAT Total` for Syn class, and `Average Packet Size` and Fwd `Packet Length Max` for LADP class.

Experiment of Multi-Hinge Loss: In this experiment, to analyze the effect of hinge loss in model training, we visualized 10,000 random inputs in validation set represented by two models. Input examples are fed-forward into the deep learning model, extracted the final hidden layer representation, transformed into lower-dimensional space via t-SNE [17] and plotted in Fig. 5. In Fig. 5a, data is represented by the without-hinge-loss model, all data points distributed into a sphere. The data has a certain cluster resolution, but there is large interference between clusters. In Fig. 5b, data is represented by the with-hinge-loss model, the data representation space has doubled from $[-2, 2]$ to $[-4, 4]$. We have observed that data belonging to the same class has been clustered closer together and these clusters also tend to be further apart. This is consistent with the idea of

	BENIGN	UDPLag	Syn	UDP	NetBIOS	MSSQL	LDAP
ACK Flag Count	**0.9194**	**1.0000**	**0.0145**	0.4956	0.5243	0.5658	0.5741
Average Packet Size	0.9938	0.0000	0.0764	0.0000	0.0005	0.0036	**0.0009**
Destination Port	0.0062	0.6214	0.9382	**1.0000**	0.3798	0.6173	0.5148
Flow Duration	0.0062	0.5235	**0.9382**	0.7849	0.5794	0.4563	0.5089
Flow IAT Max	0.0016	0.4878	0.0000	0.5848	0.4339	0.4963	0.2997
Flow IAT Mean	0.0062	0.4714	0.5462	0.4112	0.3913	0.5299	0.4268
Flow IAT Min	**0.0042**	0.5291	0.5048	0.4419	0.6048	0.6311	0.6151
Fwd Header Length	0.0016	0.4171	0.0050	0.5338	**0.9995**	0.4002	0.5097
Fwd Header Length.1	0.0016	0.4793	0.0000	0.4069	**0.9995**	0.4347	0.4856
Fwd IAT Max	0.7699	**0.8352**	0.3418	0.1108	0.1929	0.2244	0.1947
Fwd IAT Mean	0.0016	**0.9731**	0.0145	0.5085	0.4264	0.3819	0.3716
Fwd IAT Total	0.0016	0.4807	**0.0189**	0.4975	0.3946	0.5392	0.5523
Fwd Packet Length Max	0.0307	0.3542	0.5488	0.5197	0.5232	0.4901	**0.0983**
Fwd Packet Length Min	0.2444	0.7643	0.4765	0.7650	0.7779	0.7917	**0.8063**
Fwd Packet Length Std	**0.0060**	0.3566	0.8841	**1.0000**	0.6157	0.6241	0.4528
Fwd Packets/s	0.5628	0.4382	0.3884	0.4369	**0.4783**	**0.4914**	0.5018
Init_Win_bytes_forward	**0.0062**	**1.0000**	**0.9361**	0.5547	0.3279	0.5133	0.5226
Max Packet Length	0.4945	0.4986	0.7448	0.4997	0.5094	0.5020	**0.4957**
Min Packet Length	0.0068	0.4994	0.4899	0.4698	0.5684	0.4081	**0.9991**
min_seg_size_forward	0.0040	**1.0000**	**0.4880**	**1.0000**	**0.9995**	0.6211	0.6160
Packet Length Std	**0.4066**	0.6011	0.1491	**0.6021**	0.6234	0.6462	0.6668
Protocol	0.0062	0.4678	0.5121	**1.0000**	**0.9995**	**0.9964**	0.5558
Subflow Fwd Bytes	0.9938	0.0000	0.0619	0.0000	0.0005	0.0036	0.0009
Total Length of Fwd Packets	0.0061	0.4225	0.4380	0.4015	0.7212	0.5502	0.3591

0.0 — 1.0

Fig. 4. Weight of 24 features corresponding to each label. Feature weights are calculated by average of 100,000 random validation example. The weights in bold blue are for the best selected features according to Sharafaldin et al. [20].

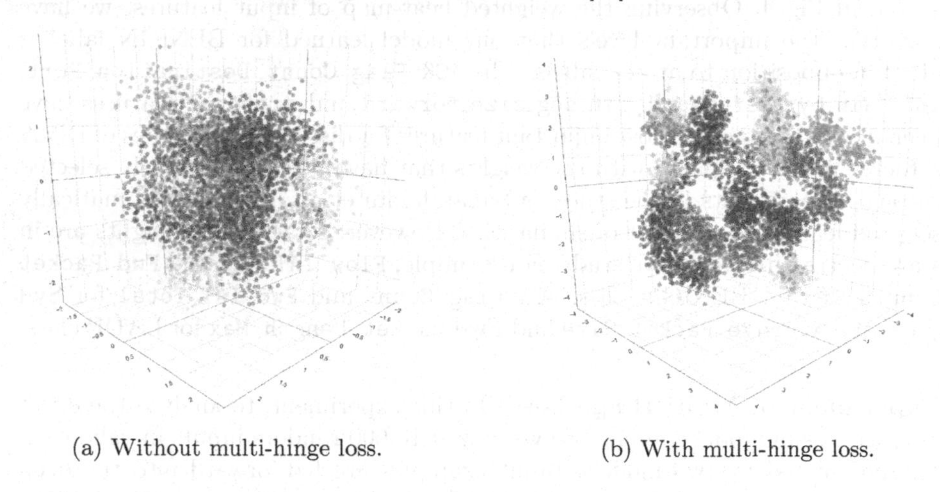

(a) Without multi-hinge loss.

(b) With multi-hinge loss.

Fig. 5. Visualization of data before softmax layer.

hinge loss, which is to maximize the margin between the data classes. As shown in Fig. 2, the hinge loss-trained model is less likely to misclassify classes when compared to the model that trained on cross entropy loss only.

5 Conclusion

We have proposed a DDoS attack detection and classification model that takes advantage of advanced deep learning techniques. It starts with an automatic feature-selection component based on the context of the input feature set. The weighted features are classified with a fully-connected MLP with softmax activation. Our models has been trained with objective function is combination of cross entropy and hinge loss.

The experiments on CIC-DDoS 2019 datasets has demonstrated the effectiveness of our proposed model when compared with other comparative machine learning-based and neural neural network-based models. We also investigated and verified the rationality and contributions of automatic feature selection models. Results have also shown the effectiveness of weighted loss and hinge loss in dealing with class imbalance problems.

Our limitation rare and hierarchical labels is highlighted since it resulted in low performance on UDPLag class. Furthermore, the discrepancy between the data distribution in training and testing data also led to poor results for some labels. We aim to address these problems in our future works.

Acknowledgments. This research is funded by Ministry of Science and Technology (MOST) under grant number KC.01.28/16-20

References

1. Aldweesh, A., Derhab, A., Emam, A.Z.: Deep learning approaches for anomaly-based intrusion detection systems: a survey, taxonomy, and open issues. Knowl.-Based Syst. **189**, 105124 (2020)
2. Chen, Y., Hwang, K., Ku, W.S.: Collaborative detection of DDoS attacks over multiple network domains. IEEE Trans. Parallel Distrib. Syst. **18**(12), 1649–1662 (2007)
3. Crammer, K., Singer, Y.: On the algorithmic implementation of multiclass kernel-based vector machines. J. Mach. Learn. Res. **2**, 265–292 (2001)
4. Douligeris, C., Mitrokotsa, A.: DDoS attacks and defense mechanisms: classification and state-of-the-art. Comput. Netw. **44**(5), 643–666 (2004)
5. Elsayed, M.S., Le-Khac, N.A., Dev, S., Jurcut, A.D.: Ddosnet: a deep-learning model for detecting network attacks. In: WoWMoM, pp. 391–396. IEEE (2020)
6. Fouladi, R.F., Kayatas, C.E., Anarim, E.: Frequency based DDoS attack detection approach using naive bayes classification. In: 2016 39th International Conference on Telecommunications and Signal Processing (TSP), pp. 104–107. IEEE (2016)
7. Glorot, X., Bengio, Y.: Understanding the difficulty of training deep feedforward neural networks. In: Proceedings of the Thirteenth International Conference on Artificial Intelligence and Statistics, pp. 249–256 (2010)
8. Gyanchandani, M., Rana, J., Yadav, R.: Taxonomy of anomaly based intrusion detection system: a review. Int. J. Sci. Res. Publ. **2**(12), 1–13 (2012)
9. Huang, C., Li, Y., Loy, C.C., Tang, X.: Learning deep representation for imbalanced classification. In: Proceedings of the IEEE Conference on Computer Vision and Pattern Recognition, pp. 5375–5384 (2016)

10. Johnson, J.M., Khoshgoftaar, T.M.: Survey on deep learning with class imbalance. J. Big Data **6**(1), 1–54 (2019). https://doi.org/10.1186/s40537-019-0192-5
11. Karimazad, R., Faraahi, A.: An anomaly-based method for DDoS attacks detection using RBF neural networks. In: Proceedings of the International Conference on Network and Electronics Engineering, vol. 11 (2011)
12. Kato, K., Klyuev, V.: An intelligent DDoS attack detection system using packet analysis and support vector machine. IJICR **14**(5), 3 (2014)
13. Kaur, P., Kumar, M., Bhandari, A.: A review of detection approaches for distributed denial of service attacks. Syst. Sci. Control Eng. **5**(1), 301–320 (2017)
14. Kingma, D.P., Ba, J.: Adam: A method for stochastic optimization. arXiv preprint arXiv:1412.6980 (2014)
15. Li, Q., Meng, L., Zhang, Y., Yan, J.: DDoS attacks detection using machine learning algorithms. In: Zhai, G., Zhou, J., An, P., Yang, X. (eds.) IFTC 2018. CCIS, vol. 1009, pp. 205–216. Springer, Singapore (2019). https://doi.org/10.1007/978-981-13-8138-6_17
16. Liu, Y.: DDoS attack detection via multi-scale convolutional neural network. Comput. Mater. Continua **62**(3), 1317–1333 (2020)
17. Maaten, L.V.D., Hinton, G.: Visualizing data using t-SNE. J. Mach. Learn. Res. **9**, 2579–2605 (2008)
18. Prasad, K.M., Reddy, A.R.M., Rao, K.V.: Dos and DDoS attacks: defense, detection and traceback mechanisms-a survey. Global J. Comput. Sci. Technol. (2014)
19. Saied, A., Overill, R.E., Radzik, T.: Detection of known and unknown DDoS attacks using artificial neural networks. Neurocomputing **172**, 385–393 (2016)
20. Sharafaldin, I., Lashkari, A.H., Hakak, S., Ghorbani, A.A.: Developing realistic distributed denial of service (DDoS) attack dataset and taxonomy. In: 2019 ICCST, pp. 1–8. IEEE (2019)
21. Shone, N., Ngoc, T.N., Phai, V.D., Shi, Q.: A deep learning approach to network intrusion detection. IEEE Trans. Emerg. Top. Comput. Intell. **2**(1), 41–50 (2018)
22. Wu, Y.C., Tseng, H.R., Yang, W., Jan, R.H.: DDoS detection and traceback with decision tree and grey relational analysis. Int. J. Ad Hoc Ubiquitous Comput. **7**(2), 121–136 (2011)
23. Zhang, G., Jiang, S., Wei, G., Guan, Q.: A prediction-based detection algorithm against distributed denial-of-service attacks. In: Proceedings of the 2009 International Conference on Wireless Communications and Mobile Computing: Connecting the World Wirelessly, pp. 106–110 (2009)

AntiPhiMBS: A New Anti-phishing Model to Mitigate Phishing Attacks in Mobile Banking System at Application Level

Tej Narayan Thakur and Noriaki Yoshiura(✉)

Department of Information and Computer Sciences, Saitama University, Saitama 338-8570, Japan
yoshiura@fmx.ics.saitama-u.ac.jp

Abstract. A main challenge in the mobile banking system is to mitigate security risks such as phishing attacks, man in the middle attacks, replay attacks, etc. Verizon's 2019 Data Breach Investigations Report (DBIR) reveals that nearly one-third of all data breaches involved phishing attacks in many kinds of ways. The phishing attack is a type of social engineering attack to steal secret information from users. This paper proposes a new **anti-phi**shing model for **M**obile **B**anking **S**ystem (**AntiPhiMBS**) that prevents mobile users from phishing attacks in the mobile banking system at the application level. The model prevents mobile users from phishing app installation by using application id, token number, and a unique id for application received from the bank to operate the mobile banking system. The phisher does not know the application id, token number, and unique id and the relationship among them and is unable to install phishing apps on the user's mobile. This paper develops the new anti-phishing model **AntiPhiMBS** with system properties specified using Process meta language (PROMELA) that are successfully verified using Simple PROMELA Interpreter (SPIN). The SPIN verification results show that the proposed anti-phishing model is error-free, and the financial institutions can implement the verified model for mitigating phishing attacks in the mobile banking system at the application installation level.

Keywords: Mobile banking system · Phishing · Verification

1 Introduction

According to the forecast number of mobile users worldwide 2019-2023 (published by Statista O'Dea, Feb 28, 2020), the number of mobile users worldwide is forecast to grow to 7.26 billion for 2020. According to the Juniper research's digital transformation readiness index 2020 [29], it is found that the total number of digital banking users will exceed 3.6 billion by 2024, up from 2.4 billion in 2020 that is a 54% increase and mobile banking users are exceeded 1.75 billion by 2019, representing 32% of the global adult population. People have increased the use of the mobile banking system in the daily activities for financial transactions and cyber-attacks are increasing globally. Cybercriminals are targeting mobile banking users with different attacks for money and

N. T. Nguyen et al. (Eds.): ACIIDS 2021, LNAI 12672, pp. 399–412, 2021.
https://doi.org/10.1007/978-3-030-73280-6_32

are transferring millions of dollars from user accounts to their accounts. It has become necessary for the financial institution to use enhanced secure mobile banking system to perform financial transactions securely. The challenges in the mobile banking system are security risks. Users suffer from different kinds of attacks such as phishing attacks, man in the middle attack, replay attack, the man in the browser attack, denial of service (DoS) attack, distributed DoS (DDoS) attack, etc. Among them, a phishing attack is one of the main challenges for security among mobile banking users. Phishing is an attack that uses social engineering and technology to steal financial account credentials from users. According to the anti-phishing working group's (APWG et al. 2020) report [28], the number of phishing sites detected in the first quarter of 2020 was 165,772, up from the 162,155 observed in the fourth quarter of 2019. According to Verizon's 2019 Data Breach Investigations Report (DBIR) [30], nearly one-third of all data breaches involved are the type of phishing in many kinds of ways.

Phishers can use a phishing app for mobile banking users to install on their mobile and can accumulate vital data from the users. The phishers can use a phishing login interface to collect mobile banking account credentials. The phishers can try for transactions using login credentials in the mobile banking systems. Mobile users are not willing to use mobile banking systems for financial transactions because of fear of attacks. Phishing attacks have become one of the main problems for implementations of the mobile banking system globally. The first known phishing attack was reported in September 2003 after which researchers have proposed various anti-phishing models to mitigate phishing attacks, but variations in procedures of phishing attacks have not been stopped and an anti-phishing model built may not be effective with time.

Zahid Hasan, Sattar, Mahmud, and Talukder [1] proposed a multifactor authentication model to mitigate the phishing attack of e-service systems and the use of multifactor (user id, security image and one-time password) authentication will help the users for the prevention of the phishing attacks in e-service systems. An authentication protocol dealing with an Authentication Server (AS) is proposed in [3], which sends a nonce message to the mobile customer device to be signed to avoid phishing attacks. The use of authentication server in [3] is better suited for browsing safe Webpages in mobile device and prevents users from visiting the phishing Webpages. Megha, Ramesh babu, and Sherly [5] developed an intelligent system for phishing attack detection and prevention with the help of different agents such as monitoring, message passer, and decision-maker agents. The advantage of [5] is that the model can work well for the known phishing attacks using machine learning classifier, but it is difficult to detect and prevent the newly designed and variated phishing attacks. A phishing detection model with a multi-filter approach proposed in [6] can detect phishing attack with the help of filtering in different layers but disadvantage of this model is that phishing attack will be detected till the whitelist layer is updated frequently. Researchers discussed the social engineering attacks [15] utilizing bidirectional communication, unidirectional communication, or indirect communication but it is not easy in reality to detect the phishing attack because of the complex psychological behavior of the people participating in the attack. Cheves [15] and Aburrous, Hossain, Dahal, Thabtah [19] proposed models that can suit for the prevention of phishing attacks for Internet banking systems only. Moreover, Yeop, Kim and Lee [12] also presented anti-phishing model that seems to be effective for only Internet banking systems only by using server authentication schemes.

Banking users must install genuine banking application (app) on their mobile to use mobile banking system. Phishers have already started using phishing application to install on the users' mobile to perform phishing attacks. Such attacks can be mitigated only if some anti-phishing models are developed for application level in mobile banking system. Unfortunately, anti-phishing model has not been developed yet especially for mitigating phishing attacks at the application level in the mobile banking system. This paper presents a new anti-phishing model for Mobile Banking System (AntiPhiMBS) to overcome this gap, and the objective of this research is to build a new anti-phishing model to mitigate phishing attacks in the mobile banking system at application level.

This model can be implemented by the developers to mitigate the phishing attacks in the mobile banking system. Mobile users will be able to download only genuine bank apps on their mobile after the implementation of this model. The phisher will not be able to steal user credentials using a phishing app. The phisher will not be able to succeed in the transactions using the mobile banking system, either. This model can play a significant role in the enhancement of mobile commerce (M-commerce) globally.

The paper is further structured as follows: Sect. 2 describes the related studies, Sect. 3 presents the new anti-phishing model for the mobile banking system, Sect. 4 presents the results and discussion, and Sect. 5 describes conclusions and future work.

2 Background

Current researchers have mentioned different phishing attack techniques and have proposed various anti-phishing models to mitigate the phishing attacks in different environments. Zahid Hasan, Sattar, Mahmud, and Talukder [1] proposed a multifactor authentication model to mitigate the phishing attack of e-service systems from Bangladesh's perspective. The model uses user Id, a secured image with a caption, and a one-time password for authentication in e-service systems. Shankar, Shetty, and Badrinath [2] provided insight into phishing, the mechanism of the attack, the types of attacks and the possible solutions to overcome them. Bojjagani, Brabin, and Rao [3] proposed a novel authentication protocol that deals with an Authentication Server (AS), which sends a nonce message to the mobile customer device to be signed to avoid phishing attacks. Aribake and Aji [4] developed a conceptual model based on technology threat avoidance theory (TTAT) and modified TTAT to evaluate the phishing attack among Internet banking users in Nigeria and to enhance avoidance behavior. Megha, Ramesh babu, and Sherly [5] developed an intelligent system for phishing attack detection and prevention with agents in which the first agent is responsible for extracting URLs. Authors of [6, 7] proposed models for detection and prevention of phishing attacks. Khalid, Jalil, Khalid, Maryam, Shafique, and Rasheed [6] presented a detailed discussion on several anti-mobile phishing models based on various methods for preventing users to evade phishing attacks. Glavan, Racuciu, Moinescu, and Eftimie [7] proposed an anti-phishing model framework to detect phishing attacks by analyzing various anti-phishing methods. Doke, Khismatrao, Jambhale, and Marathe [8] proposed a system with an extension to a web browser that made use of a machine-learning algorithm to extract various features to help the users to distinguish between legitimate website and phishing website. Various phishing attacks and their mitigation techniques have been explained in [9–12]. Yeop Na, Kim and Lee [12] focused on prevention schemes against phishing attacks on Internet banking systems. Lacey, Salmon, and Glancy [13] applied work domain analysis (WDA) to understand the functional structure of phishing attacks and the online transactional environment which they target as a sociotechnical system. Bann, Singh, and Samsudin [14] addressed the advanced persistent threat (APT) issue via spear-phishing attacks within the bring your own device (BYOD) environment through the mediation provided by security policies.

Authors of [17, 21] proposed anti-phishing models to mitigate phishing attacks. Shashidhar, and Chen [17] proposed a model which makes a list of phishing sites and the model checks the messages accordingly for detection of phishing attacks. Cheves [15] proposed a research model that can be used to evaluate the significance of cybercrime in deterring the use of e-banking in the financial sector. Mouton, Leenen, and Venter [16] proposed a social engineering attack detection model: SEADMv2 to cater for social engineering attacks that use bidirectional communication, unidirectional communication, or indirect communication. Aburrous, Hossain, Dahal, and Thabtah [19] mentioned about the investigation of phishing techniques and attack strategies for E-banking. Oh, and Obi [20] discussed the identification of phishing threats in government web services. Authors of [23–25] discussed the securities of the online banking system. Similarly, Akinyede and Esese [26] pointed out the development of a secure mobile e-banking system.

Our paper proposes a new anti-phishing model for the mobile banking system that prevents phishing attacks in the mobile banking systems at the application installation level. The model will be useful for banks and financial institutions globally for security in Electronic banking (E-Banking).

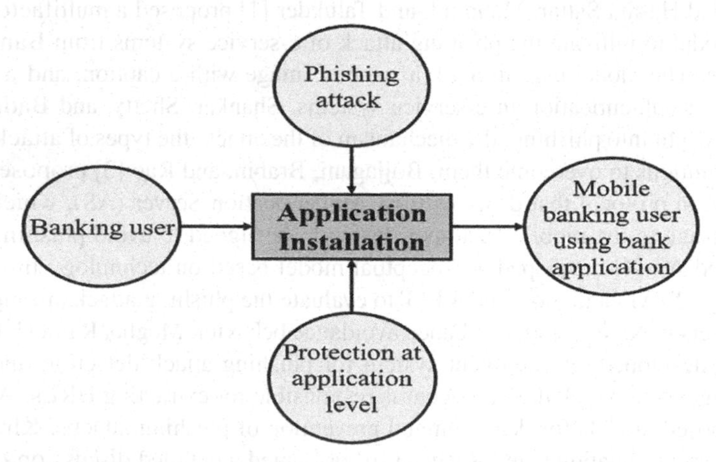

Fig. 1. Phishing attack protection at application level

3 Proposed Anti-phishing Model For Mobile Banking System

The proposed model is an **anti-phi**shing model for **M**obile Banking **S**ystem (**AntiPhiMBS**). AntiPhiMBS aims to mitigate phishing attacks in the mobile banking system at the application level. Banking users install mobile applications and may install phishing apps instead of genuine bank mobile applications by phishing attacks. AntiPhiMBS protects the banking users from installing phishing apps in their mobile phones and helps to use genuine bank applications in their mobile as shown in Fig. 1.

This paper proposes AntiPhiMBS and verifies that AntiPhiMBS satisfies the security properties.

3.1 Architecture of Anti-phishing Model AntiPhiMBS

The architecture of the anti-phishing model AntiPhiMBS consists of the model for defending against phishing attacks at the mobile application level. Participating entities in the model are mobile user, bank, bank application website, mobile banking system, and phishing application website. A mobile customer user opens an account in a bank and receives application installation parameters for the operation of the mobile banking system. The bank maintains a bank application website which offers users services of downloading bank applications for the mobile banking system. The bank shares application installation parameters with the mobile banking system. The phisher might make a phishing application website similar to that of the bank application website and might develop a phishing app for phishing attacks in the mobile banking system. The model uses various parameters for the operation of this model. Table 1 shows the notations for entities and parameters used in AntiPhiMBS.

Table 1. Notations used in AntiPhiMBS

Notation	Description	Notation	Description
U	Mobile User	app	Application
B	Bank	appId	Application Id
MBS	Mobile Banking System	tktNo	Ticket Number
BAW	Bank App Website	unqIdApp	Unique Id Application
PAW	Phishing App Website	mobNo	Mobile Number

Entities and Initial Conditions for Working of AntiPhiMBS

- A mobile user (U) opens an account in the Bank (B).
- Bank provides appId, tktNo and unqIdApp to the user for the operation of Mobile Banking System (MBS).
- Bank shares appId, tktNo, unqIdApp and mobNo of each user to MBS.

- Bank maintains an application website for users to download genuine mobile banking application. Each of the applications is identified by appId. Bank, bank app website and MBS only know the relationship of tktNo, appId, and unqIdApp.
- Users do not share information provided by the bank with any other entities.

Model for Defending Against Phishing Attacks in Mobile Banking System At Application Level

Generally, a bank should conduct phishing awareness training for mobile users when they open an account in the bank. Phishing training should be a part of the bank's cyber security business plan. The bank should give training to mobile users about the procedure of the installation of the mobile app and the use of the system. The application level of prevention against a phishing attack is for all the users who download the mobile app to use the mobile banking system. This method proposes a novel application id (appId), ticket number (tktNo), and a unique id for application (unqIdApp) system to prevent phishing app installation in the user's mobile. Correct steps are necessary for the successful operation of AntiPhiMBS and the bank, user, bank application website, and the mobile banking system should follow all the steps to mitigate the phishing attacks at application level. The scenario of bank app installation is shown in Fig. 2.

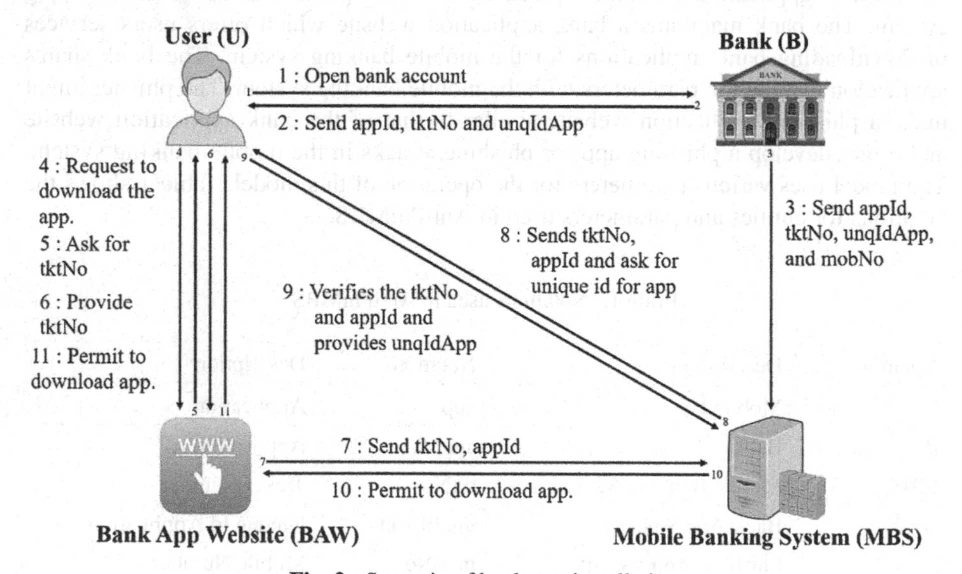

Fig. 2. Scenario of bank app installation

Scenario of Bank Mobile App Installation

The following steps are necessary for a user to install bank mobile app successfully.

- Step 1. A mobile user (U) opens a bank account in the Bank (B).

- Step 2. Bank sends application id (appId), ticket number (tktNo), and unique id for application installation (unqIdApp) to the user for the operation of Mobile Banking System (MBS).
- Step 3. Bank sends appId, tktNo, unqIdApp, and mobile number (mobNo) to MBS.
- Step 4. User visits bank's app website (BAW) and requests to download the app.
- Step 5. Bank's app website asks for ticket number to the user.
- Step 6. The user provides ticketNo to BAW.
- Step 7. BAW knows the relationship between tktNo and appId. BAW searches the appId based on ticketNo and sends both tktNo and appId to MBS.
- Step 8. MBS verifies the tktNo and appId, sends both to the user and asks for a unique Id for app.
- Step 9. User verifies tktNo and appId, and provides a unqIdApp to MBS.
- Step 10. If the unqIdApp from the user is valid, MBS informs the BAW that the user can download the app.
- Step 11. BAW permits the user to download the app.

Bank manages ticket number, application id, and a unique id for each user for the operation of the mobile banking system. Bank also manages the bank application website (BAW) for the management of downloading of a genuine app to the mobile banking system users. Bank communicates with the mobile banking system for the proper installation of the mobile application. BAW maintains the banking apps which are identified by the application id. Bank, BAW and MBS know the relationship between the ticket number, the application id, and the unique id for application allocated for each user in the bank. A mobile user opens an account in the bank and receives an application id, ticket number, and unique id from the bank. The user visits the bank application website and requests to download the app. BAW asks the user to input the correct ticket number and searches the application id based on the ticket number received from the user. BAW sends the ticket number and application id of that user together to the mobile banking system for verification of the application installation. The mobile banking system verifies the ticket number and application id with the help of the database already received from the bank. MBS shows the ticket number and application id to the user and asks to enter the valid unique id for the application. Users can verify the MBS supplied ticket number and application id and have the option of downloading the application or not. The user enters the unique id for the application if both of the ticket number and application id are correct. MBS verifies the unique id and sends permission to BAW to allow users to download the app if the unique id supplied by the user is valid. Finally, BAW allows users to download the app after getting a positive response from the mobile banking system.

Scenario of Phishing App Installation

Even though users know the correct procedure for downloading the app from the bank application website, they may receive phishing emails or SMS from the phishers. Phishers might develop a phisher application website (PAW) similar to that of the bank application website (BAW) and phisher mobile banking system. Users may visit PAW and may click the link unknowingly to download the mobile application on their mobile.

Thus, users may start phishing app installation unknowingly. Scenario of phishing app installation is shown in Fig. 3.

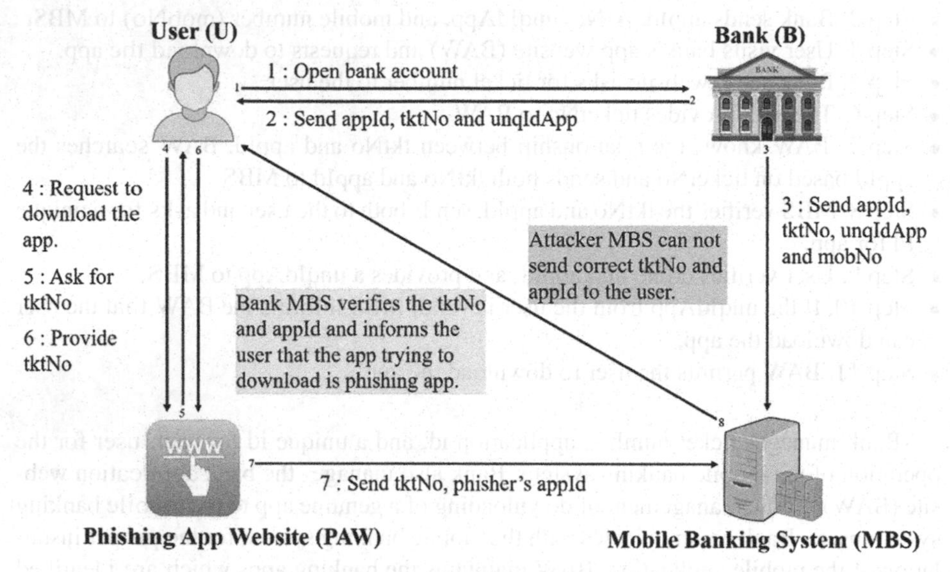

Fig. 3. Scenario of phishing app installation

The following steps may be executed during phishing app installation.

- Step 1. A mobile user (U) opens a bank account in the Bank (B).
- Step 2. Bank sends application id (appId), ticket number (tktNo), and unique id for application installation (unqIdApp) to the user for the operation of Mobile Banking System (MBS).
- Step 3. Bank sends appId, tktNo, unqIdApp, and mobile number (mobNo) to MBS.
- Step 4. User may receive phishing email or phishing SMS with links to download the app. User may click the provided links to download the app.
- Step 5. Phisher's website asks for ticket number to user.
- Step 6. User may provide tktNo to the Phisher's website.
- Step 7. Phisher's website sends tktNo and phisher's appId to MBS. MBS verifies the tktNo and appId and informs the user that the app trying to download is not bank's app. If the phisher act as MBS too, fake MBS cannot send the correct tktNo and appId to the user.

User may click phishing app download links unknowingly. PAW asks for the ticket number to the user and the user may input the ticket number to PAW. PAW does not know the true application id for that ticket number and sends a fake application id along with the ticket number to MBS. MBS verifies the ticket number and phisher application id and detects that the application download request is from the phisher application website. After that, MBS informs the user about the phishing app and prevents the user from downloading the phishing app. If the phisher uses fake MBS, fake MBS does not know the application id for the ticket number and cannot send it to the user. Fake MBS may send ticket number and fake application id to the user, but user verifies the ticket number and fake application id and knows about the phishing attack. Users do not provide a unique id and do not download the app until they receive the correct ticket number and the application id from MBS. Users have the option to continue downloading the app or cancel the process and are prevented from phishing attacks. Even though phishers collect ticket numbers using the phisher application website and try to download the bank app using the genuine ticket number, they fail in downloading the application because they do not know the unique id for the application. Hence, AntiPhiMBS prevents the users from installing phishing apps on their mobile and forbids the phishers from downloading the bank app on their mobile. Thus, the model prevents phishing attacks in the mobile banking system at the application level.

3.2 Verification of Proposed Anti-phishing Model AntiPhiMBS

We chose SPIN to verify the proposed model AntiPhiMBS because of its graphical user interface and counterexample generation during the verification and developed the verification model of AntiPhiMBS using PROMELA. This paper does not show the PROMELA codes because of space limitations. The PROMELA code in this paper has 201 lines. The PROMELA verification model of AntiPhiMBS consists of the processes, message channels, and data types. This paper explains the overview of the codes. The verification model of AntiPhiMBS consists of the following processes.

- mobileUser: The process represents the end user of the mobile banking system.
- bank: The process represents the bank where the user opens the bank account, and the bank manages the mobile banking system for the user.
- mobileBankingSystem: The process represents the mobile banking system which offers the services of banking operations in the mobile.
- bankAppWebsite: The process represents the bank's authorized application website which manages the downloading of mobile application for the user.
- phisherAppWebsite: The process represents the phisher's phishing website which imitates the bank application website and tries to fool the users to install a phishing app on their mobile.

Above mentioned processes communicate using message channels that are specified in the PROMELA code of AntiPhiMBS.

We also specified the following security properties using linear temporal logic (LTL) in the verification model of AntiPhiMBS.

[](((usrTktNo ==bankTktNo)&&(websiteAppId ==bankAppId)&&(usrUnqIdApp ==bankUnqIdApp))- > <>(appDownloadSuccess ==true))

Download of mobile banking application is successful only if (i) the ticket number provided by the user and received from bank is equal (ii) the application id provided by the bank application website and received from bank is equal (iii) the unique id for application provided by the user and received from the bank is equal.

4 Results and Discussion

This paper verifies the safety properties and LTL properties of the proposed model AntiPhiMBS. Safety properties include the checking of deadlocks and assertion violations in AntiPhiMBS and ensure that nothing bad ever happens in the proper functioning of the model. We accomplished experiments using SPIN Version 6.4.9 running on a computer with the following specifications: Intel® Core (TM) i5-6500 CPU@3.20 GHz, RAM 16 GB and windows10 64bit. We applied bitstate to save memory during the verification of the model. We set the advanced parameters in the SPIN for the verification of AntiPhiMBS and made physical memory available as 4096 (in Mbytes), maximum search depths (steps) as 1000000, and estimated state space size as 1000 for the experiments. Besides, we set extra compile-time directives to DVECTORSZ as 9216 to avoid the memory error during the experiments. After that, we ran SPIN to verify the safety properties of AntiPhiMBS for various users. SPIN checked the state space for invalid end states, assertion violations, and acceptance cycles. The SPIN verification results for safety properties are in Table 2.

Table 2. Verification results for safety properties

No. of users	Time (in seconds)	Memory (in Mbytes)	Transitions	States stored	Depth	Safety property verification status
1	1.43	39.026	2634164	196203	65223	Verified
5	4.26	39.026	4177108	283820	112921	Verified
10	24.1	156.799	6537677	431452	197962	Verified
20	49.6	320.276	8058943	511312	273755	Verified
50	126	788.733	9324748	574242	322689	Verified
80	201	1235.217	9578183	582609	324933	Verified
100	255	1549.866	9723016	588837	330954	Verified

The results in Table 2 shows the elapsed time, total memory usage, states transitioned, states stored, depth reached, and verification status for different numbers of users. The SPIN verification results show that the verification time has been increased consistently with the rise in the number of users during the verification of AntiPhiMBS. However, the memory required for the verification remained approximately equivalent for up to 5 users followed by an increment with the increase in the number of users. Moreover, the states stored, depth reached and transitions for various users increased significantly for up to 10 users and afterwards slightly for the rest of the users. The safety property is satisfied in each run of the SPIN and neither any deadlock nor any errors are detected during the verification of AntiPhiMBS.

After verification of the safety properties, we ran SPIN in the same computing environment to verify the LTL properties for various users. SPIN checked the statespace for never claim and assertion violations in run of LTL property. The SPIN verification result for LTL properties is in Table 3. Table 3 shows the results obtained from SPIN showing the elapsed time, total memory usage, states transitioned, states stored and verification status for various users. The LTL property of AntiPhiMBS is verified by the SPIN. The memory required for LTL property for all the users are same. The verification time has been increased proportionally with the rise in the number of users during the verification of AntiPhiMBS. Furthermore, the states stored, and elapsed time increased with the increase in the number of users.

Table 3. Verification results for LTL properties

No. of users	Time (in seconds)	Memory (in Mbytes)	Transitions	States stored	Depth	LTL property verification status
1	0.522	39.026	954795	78200	6525	Verified
5	1.96	39.026	1913065	139316	13532	Verified
10	5.83	39.026	3411072	245312	22076	Verified
20	16.1	39.026	5629439	402333	29722	Verified
50	45	39.026	7054808	525783	29214	Verified
80	68	39.026	6919770	541326	45642	Verified
100	83.1	39.026	6819757	551746	54444	Verified

SPIN displays execution paths as counterexamples if some situations do not meet the properties specified in the design of the model. In our case, SPIN did not generate any counterexample during the verification of AntiPhiMBS in any of the experiments. Hence, the verified AntiPhiMBS seems to be applicable for the development and implementation of the anti-phishing system within the banks and financial institutions globally to countermeasure the continued phishing attacks in Electronic Banking (E-Banking).

5 Conclusion and Future Work

Since phishing attack creates a negative impact on the E-banking, establishes relationships with other attacks, and creates financial risks for emerging digital era at the national and international level, this paper chose to develop a new anti-phishing model for Mobile Banking System (AntiPhiMBS) to mitigate phishing attacks in the mobile banking system at the application level. Our experimental results showed that the proposed AntiPhiMBS is free from errors and deadlocks. Furthermore, AntiPhiMBS is verified using SPIN for different numbers of users to assure the practical implementation of the model in the financial institutions. Moreover, the study presented the mitigation of phishing attacks at the application level in AntiPhiMBS. Phisher may send phishing emails or SMS (Short Messaging Service) to the users and may redirect them to download the phishing app. Phishers might use the phishing app to install it on the user's mobile to steal login credentials from the user. However, AntiPhiMBS uses the ticket number, application id, and a unique id for the application installation and prevents the user from installing the phishing app. Hence, financial institutions can implement this verified AntiPhiMBS model to mitigate the ongoing phishing attacks in the world of E-Banking and can save millions of dollars annually globally. Our research will also be beneficial to mobile app developers, end users, researchers, and bankers.

AntiPhiMBS model mitigates the phishing attacks in the mobile banking system at the application level only. In our future research, we will propose new anti-phishing models to mitigate phishing attacks in the mobile banking system at the authentication level and transaction level. Furthermore, this research could be further extended through the verification of new models to mitigate man in the middle (MITM) attack, man in the browser (MITB) attack, replay attack, SQL injection attack, and other probable attacks in E-Banking.

References

1. Zahid Hasan, M., Sattar, A., Mahmud, A., Talukder, K.H.: A multifactor authentication model to mitigate the phishing attack of e-service systems from Bangladesh perspective. In: Shetty, N.R., Patnaik, L.M., Nagaraj, H.C., Hamsavath, P.N., Nalini, N. (eds.) Emerging Research in Computing, Information, Communication and Applications. AISC, vol. 882, pp. 75–86. Springer, Singapore (2019). https://doi.org/10.1007/978-981-13-5953-8_7
2. Shankar, A., Shetty, R., Badrinath, K.: A review on phishing attacks. Int. J. Appl. Eng. Res. **14**(9), 2171–2175 (2019)
3. Bojjagani, S., Brabin, D., Rao, V.: PhishPreventer: a secure authentication protocol for prevention of phishing attacks in mobile environment with formal verification. In: Thampi, S., Madria, S., Fernando, X., Doss, R., Mehta, S., Ciuonzo, D. (eds.) Third International Conference on Computing and Network Communications 2019, vol. 171, pp. 1110–1119. Elsevier B.V, Heidelberg (2020)
4. Aribake, F.O., Aji, Z.M.: Modelling the phishing avoidance behavior among internet banking users in Nigeria: the initial investigation. J. Comput. Eng. Technol. **4**(1), 1–17 (2020)
5. Megha, N., Ramesh babu, K.R., Sherly, E.: An intelligent system for phishing attack detection and prevention. In: Proceedings of the Fourth International Conference on Communication and Electronics Systems, pp. 1577–1582. IEEE Xplore, Coimbatore, India (2019). https://doi.org/10.1109/icces45898.2019.9002204
6. Khalid, J., Jalil, R., Khalid, M., Maryam, M., Shafique, M.A., Rasheed, W.: Anti-phishing models for mobile application development: a review paper. In: Bajwa, I.S., Kamareddine, F., Costa, A. (eds.) INTAP 2018. CCIS, vol. 932, pp. 168–181. Springer, Singapore (2019). https://doi.org/10.1007/978-981-13-6052-7_15
7. Glavan, D., Racuciu, C., Moinescu, R., Eftimie, S.: Detection of phishing attacks using the anti-phishing framework. Sci. Bull. Naval Acad. **13**(1), 208–212 (2020)
8. Doke, T., Khismatrao, P., Jambhale, V., Marathe, N.: Phishing-inspector: detection & prevention of phishing websites. In: Patil, D.Y. (ed.) International Conference on Automation, Computing and Communication 2019, vol. 32, pp. 1–6. EDP Sciences, India (2020). https://doi.org/10.1051/itmconf/20203203004
9. Meena, K., Kanti, T.: A review of exposure and avoidance techniques for phishing attack. Int. J. Comput. Appl. **107**(5), 27–31 (2014)
10. Naidu, G.: A survey on various phishing detection and prevention techniques. Int. J. Eng. Comput. Sci. **5**(9), 17823–17826 (2016)

11. Akinyede, R.O., Adelakun, O.R., Olatunde, K.V.: Detection and prevention of phishing attack using linkguard algorithm. J. Inf. **4**(1), 10–23 (2018). https://doi.org/10.18488/journal.104.2018.41.10.23
12. Yeop Na, S., Kim, H., Lee, D.H.: Prevention schemes against phishing attacks on internet banking systems. Int. J. Adv. Soft Comput. Appl. **6**(1), 1–15 (2014)
13. Lacey, D., Salmon, P., Glancy, P.: Taking the bait: a systems analysis of phishing attacks. In: 6th International Conference on Applied Human Factors and Ergonomics and the Affiliated Conferences, vol. 3, pp. 1109–1116. Elsevier, Australia (2015). https://doi.org/10.1016/j.promfg.2015.07.185
14. Bann, L.L., Singh, M.M., Samsudin, A.: Trusted security policies for tackling advanced persistent threat via spear phishing in BYOD environment. In: The Third Information Systems International Conference, vol. 72, pp. 129–136. Elsevier ScienceDirect, Penang Malaysia (2015). https://doi.org/10.1016/j.procs.2015.12.113
15. Cheves, D.A.: The impact of cybercrime on e-banking: a proposed model. In: International Conference on Information Resources Management, pp. 1–10. Association for Information Systems AIS Electronic Library, West Indies (2019)
16. Mouton, F., Leenen, L., Venter, H.S.: Social engineering attack detection model: SEADMv2. In: International Conference on Cyberworlds, pp. 216–223. IEEE, Pretoria, South Africa (2015). https://doi.org/10.1109/cw.2015.52
17. Shashidhar, N., Chen, L.: A phishing model and its applications to evaluating phishing attacks. In: Proceedings of the 2nd International Cyber Resilience Conference, pp. 63–69. Edith Cowan University, Perth Western Australia (2011)
18. Rajalingam, M., Alomari, S.A., Sumari, P.: Prevention of phishing attacks based on discriminative key point features of webpages phishing attacks. Int. J. Comput. Sci. Secur. **6**(1), 1–18 (2012)
19. Aburrous, M., Hossain, M.A., Dahal, K., Thabtah, F.: Experimental case studies for investigating e-banking phishing techniques and attack strategies. Cogn. Comput. **2**, 242–253 (2010). https://doi.org/10.1007/s12559-010-9042-7
20. Oh, Y., Obi, T.: Identifying phishing threats in government web services. Int. J. Inf. Netw. Secur. **2**(1), 32–42 (2013)
21. Jakobsson, M.: Modeling and Preventing Phishing Attacks. In: Patrick, Andrew S., Yung, M. (eds.) FC 2005. LNCS, vol. 3570, p. 89. Springer, Heidelberg (2005). https://doi.org/10.1007/11507840_9
22. Jagadeesh Kumar, P.S., Nedumaan, J., Tisa, J., Lepika, J., Wenli, H., Xianpei, L.: New malicious attacks on mobile banking applications. Mob. Netw. Appl. **21**(3), 561–572 (2016). Special Issue on Mobile Reliability, Security and Robustness, Springer
23. Yildirim, N., Varol, A.: A research on security vulnerabilities in online and mobile banking systems. In: 7th International Symposium on Digital Forensics and Security, pp. 1–5. IEEE, Barcelos, Potugal (2019)
24. Hammood, W.A., Abdulla, R., Hammood, O.A., Asmara, S.M., Al-Sharafi, M.A., Hasan, A.M.: A review of user authentication model for online banking system based on mobile imei number. In: The 6th International Conference on Software Engineering & Computer Systems, IOP Conf. Series: Materials Science and Engineering 769, Kuantan, Pahang, Malaysia (2020). https://doi.org/10.1088/1757-899x/769/1/012061
25. Dhoot, A., Nazarov, A.N., Koupaei, A.N.A.: A security risk model for online banking system. In: 2020 Systems of Signals Generating and Processing in the Field of on Board Communications, pp. 1–4. IEEE, Moscow, Russia (2020)
26. Akinyede, R.O., Esese, O.A.: Development of a secure mobile e-banking system. Int. J. Comput. **26**(1), 23–42 (2017)

27. Ahmed, A.A., Adullah, A.N.: Real time detection of phishing websites. In: 7th Annual Information Technology, Electronics and Mobile Communication Conference, pp. 1–6. IEEE, Vancouver, Canada(2016)
28. APWG Homepage. https://docs.apwg.org/reports/apwg_trends_report_q1_2020.pdf. Accessed 11 Nov 2020
29. JUNIPER Research. https://www.juniperresearch.com/press/press-releases/digital-banking-users-to-exceed-3-6-billion. Accessed 12 Nov 2020
30. VERIZON data breach investigation report. https://docs.apwg.org/reports/apwg_trends_report_q1_2020.pdf. Accessed 11 Nov 2020

Missing and Incomplete Data Handling in Cybersecurity Applications

Marek Pawlicki[1,2(✉)], Michał Choraś[1,2], Rafał Kozik[1,2], and Witold Hołubowicz[2]

[1] ITTI Sp. Z O.O., Poznań, Poland
mpawlicki@itti.com.pl
[2] UTP University of Science and Technology, Bydgoszcz, Poland

Abstract. In intelligent information systems data plays a critical role. Preparing data for the use of artificial intelligence is therefore a substantial step in the processing pipeline. Sometimes, modest improvements in data quality can translate into a vastly superior model. The issue of missing data is one of the commonplace problems occurring in data collected in the real world. The problem stems directly from the very nature of data collection. In this paper, the notion of handling missing values in a real-world application of computational intelligence is considered. Six different approaches to missing values are evaluated, and their influence on the results of the Random Forest algorithm trained using the CICIDS2017 intrusion detection benchmark dataset is assessed. In result of the experiments it transpired that the chosen algorithm for data imputation has a severe impact on the results of the classifier used for network intrusion detection. It also comes to light that one of the most popular approaches to handling missing data - complete case analysis - should never be used in cybersecurity.

Keywords: Data science · Missing data · Incomplete data · Cybersecurity · Data imputation

1 Introduction

In machine learning and its applications, data preparation is crucial and often can take up a significant portion of effort in the data engineering endeavour. A slight data quality adjustment can potentially contribute to a compelling boost in the performance of ML [16].

One of the most prevailing flaws of the data coming from the real-world is the issue of missing data. The notion of properly handling missing data remains a relevant issue for most of Machine Learning (ML) applications [5]. The problem stems directly from the very nature of data collection and assumes different forms. Each form - or pattern - of missingness, results with some ramifications for the ML algorithms used down the road [3]. Properly handling missing data is of utmost importance and has to be put under particular scrutiny to make sure

N. T. Nguyen et al. (Eds.): ACIIDS 2021, LNAI 12672, pp. 413–426, 2021.
https://doi.org/10.1007/978-3-030-73280-6_33

ML methods are going to construct accurate reasoning [4]. Data that is missing may hold valuable information on the dataset [19]. Moreover, many ML methods do not support data with values missing [7].

There are different approaches to handling the problem; some of them are prevalent in contemporary applications, despite their flaws. The research community also offers more robust methods, such as multiple imputations, applicable to many situations [3]. This process starts with an attentive inspection of the data [5]. The premise of handling missing data is that the missing features can be inferred from the remainder of the dataset. In fact, in the example described in [12], some imputations are fairly straightforward. In the context of 'nonresponse' to a survey, where the missing response is part of several waves of measurements, they can be inferred from the measurements before and after the missing values with exceptional accuracy through interpolation [12]. Similarly, if the respondent stopped providing measurements after a certain measurement, a situation referred to as 'attrition', the missing data can be approximated via extrapolation [12]. When measurements are missing in a dataset, the processes necessary for pattern classification increase in complexity [27]. In intrusion detection, there are sometimes values missing in the NetFlow reports, which in turn disable the ablility to calculate certain features. Imputation is one of the ways to handle this predicament.

In this work, the problem of handling missing values in network intrusion detection has been addressed. The text is structured as follows: Sect. 2 opens with an introduction to the problem of data quality and missing data, the missingness mechanisms are introduced and discussed. This is followed by introducing visual analytics for handling missing data, discussing the related work in this subject, found in Sect. 3. Then, the used visual analytics for the selected case are shown in the same section. The work continues with an inventory of approaches to dealing with missing data in Sect. 4, along with the most applicable algorithms in the context of network intrusion detection. The pros and cons of the methods were disclosed. Then, the details of the experiments are given and the results obtained are communicated in Sect. 5. Finally, conclusions are drawn.

2 What Missing Data Is, and the Notion of Data Quality

In the broad sense, the quality of data can be characterised as '*good*' if, whenever used, the data is capable of producing a satisfactory result [9]. This might not be an exact definition, but opens up the way to articulating data quality using metrics suitable for the use case. In [8], the data quality metrics of accuracy, coherence, uniqueness, compliance and completeness are discussed. The last criterion, completeness, pertains to the fact that the data is not null and provides sufficient information to formulate an efficient ML model. This goes along with the definition of missingness found in [12], where missing data is quite simply the values that are not present in the dataset. The author proposes a convenient way to think about missing values as a binary variable, which is either missing or not [8,12].

2.1 Patterns of Missingness

Having established that a certain value is missing from the dataset, a question unfolds on the reason as to why the value is actually missing. This is referred to as the mechanism of missingness, or the pattern of missingness. The most commonly used classification of the patterns of missingness in contemporary research [10–13,15–17,21,26–28,30,32,34–36,39] follows the outline from [24], that is:

Missing Completely at Random (MCAR): In this situation, the null values are completely isolated and independent. To be more specific, this means that the values in question do not depend on other values that are missing, or on other values that are present in the dataset. MCAR data can inflate the standard errors, but do not contribute to systematic errors (bias - overestimation of benefits). MCAR, though incomplete, can be used to represent the entirety of the dataset. According to [13], this scenario is a rare case in the real world.

Missing At Random (MAR): In this less restrictive pattern, the values in question are not related to the missing data, but they are dependent on the data that is observed. In [27], this mechanism is illustrated by an example of a questionnaire inquiry of how many cigarettes per day the respondent smokes when the questionnaire is answered by a representative of a younger audience. Or the presence of the null response to a question about age being dependent on the sex variable [4,5]. This mechanism opens up the way to inferring values of missing data from the complete cases.

Missing Not At Random (MNAR): When the values in question depend on the missing values themselves, the pattern is known as MNAR. By definition, the missing values are unknown, therefore recognising this pattern cannot be accomplished from the observed data. A respondent dropping out of a drug survey because of the effect the drug has on them is an example proposed by [3]. This mechanism is sometimes referred to as 'informative missingness' [4].

The patterns of missingness represent completely different mechanisms of missing data. This form of understanding of the patterns of missingness bears serious ramifications [3]. First of all, for MNAR data the fact of missingness itself holds information. Additionally, MNAR data cannot be inferred from the data in the dataset, hence there is no straightforward way to distinguish between MNAR and MAR patterns. The MAR data can be, up to a point, recovered from the rest of the dataset [4,13]. MAR data is not like MCAR, as it has a high chance to be the mechanism actually existing in real-life scenarios and research [3]. Every pattern has its own potential solution [3]. The potential for systematic error is influenced by both the mechanism of missingness and the volume of missing data. Since the specific pattern of missingness is hard to ascertain, so is the risk of bias. It is of utmost importance to take the steps to limit the damage missing data can have on the performance of the data-related inference systems [4]. A formal description and understanding of the patterns of missingness enables the anticipation of potential consequences the null variables have [3] on the performance of ML.

With regard to the amount of missing data [7] provides a brief guideline, where less than 1% of missing values are considered trivial, the 1–5% range is referred to as 'flexible'. Missing between 5–15% of data calls for advanced methods. Anything above 15% bears immense impact on the results. In [4] the author recommends utmost care when dealing with large amounts of missing data, which is disclosed to be between 5–10% of the volume of the dataset.

The substantial volume, velocity, veracity and variety of data are the characteristics often described as the four V's of Big Data. The mentioned veracity hints at the noisy, incomplete, inaccurate or redundant samples of data. Variety suggests heterogeneity, which causes challenges in data integration and analytics. The multitude of sources and formats of data, along with the co-occurrence of the information pertaining different aspects of a certain object in multiple modalities can allow to compensate for incomplete data coming from a single source [38].

2.2 Dealing with Missing Values

A short primer of the general approach to researching the effects of handling of missing values can be found in [27]. Four main steps are outlined.

1. The researcher is advised to collect more than one complete set of data on the investigated domain, with varied feature types and different lengths of the feature vector.
2. The null variables can be generated synthetically. In a single (univariate) variable, or in more than one variable (multivariate), with a range of missing rates. The synthetic generation can follow the missingness mechanisms discussed earlier.
3. Dealing with the imputation approaches, either with statistical methods or with more intricate, ML-based algorithms. This step will be discussed at length.
4. Assessment of the results the chosen imputation methods had in the process of improving the results of the original task, in this case – intrusion detection, using available performance metrics and comparing the results to the ground truth [27].

3 Exploring Solutions

Complete Case Analysis (CCA). It only makes sense to start with one of the most popular approaches, as observed during the literature review. The reason for its popularity might be due to the fact that it is suggested as the default option by many statistical software pieces [4].

This approach, however, is one of the most problematic. Complete Case Analysis only uses the fully described datapoints as the basis for further analysis. This basically means that if a datapoint has a null record in any of its features, it is deleted. Hence, the method is also known under the name of 'Listwise Deletion' [12,21]. While it is the easy way out of the problem of missing values and is simple to implement, it comes with a significant price one has to be aware of. CCA forces any analysis performed on the newly formed dataset, including the ML algorithms, to effectively work with a reduced sample size, causing a higher propensity to worse performance metrics [4].

It goes without saying that deleting the incomplete cases discards any information that might have been contained in those records. The loss of statistical power can, in some cases, be devastating to the analysis. The example showcased in [12] illustrates that in an unfortunate distribution of missing values across the dataset, the null records in 16% if cases can result in the deletion of 80% of the dataset. This does not seem like a viable option. The size of the dataset and the relevance of attributes need to be assessed before attempting deletion [28].

One more issue is whether the complete cases are a representative sample of the problem at hand. This could be true if the data are MCAR, but as mentioned earlier, in a real-world situation it is rarely the case [12].

Another deletion form is **pairwise deletion**. This method is rooted in the variance-covariance matrix. As such, it is a more complete option than listwise deletion, as it relies on all the available data. However, as noted by [12], since the employed statistics use the records that exist, but omit the ones that are missing, variances and covariances are derived from different samples and thus the approximated parameters might be biased [12,23].

Single Imputation. Single imputation refers to a process where the missing variables are estimated using the observed values. By definition the single imputation approaches are geared towards handling data complying with the MAR mechanism [12,13]. There is a range of possible choices in this category. The premise of the 'last observation carried forward' approach is that, as the name suggests, the last observation of a respondent is used to fill the missing value. This strategy is reserved to longitudinal studies. Similarly, the 'worst observation carried forward' method pertains to using the worst observation of a subject [13].

Central Tendency Measures Substitution. Based on the reviewed literature, the three measures of central tendency - mean, median and mode - are used for imputation almost as frequently as listwise deletion [3,34]. Mean substitution puts in the place of the null record the mean computed taking into account the distribution of all the existing inputs across one feature [12,21,28], which is considered a plausible value [5]. According to [12], this is the worst of all the possible strategies, and is on par with pretending that no data is missing. The reason for this assessment is the way this procedure disturbes the variance of the variable along with covariances and correlations in the dataset [12,32]. In [13], it is noted that using the single mean imputation dilutes the dataset as every missing value bears the same weight as the known one. Using the single mean imputation also presupposes that the unknown values are similar to the observed values. Using mean imputation on data not fulfilling that assumption might lead to bias. For missing categorical data, mode can be employed, while in a situation where the distribution is skewed, median is a better option [28]. The two main advantages of this imputation method are that it returns a complete dataset, allowing researchers to use all the available data, and it is very easy to implement [34,37].

Hot-Deck Imputation. The term hot-deck is a reference back to the times of punch cards, where the currently processed deck would literally be 'hot'. The name suggests using data from the current dataset, as opposed to the 'cold-deck' imputation, where suitable candidates would be selected from external sources - data collected elsewhere, like a previous survey, or a different dataset. Hot-deck imputation methods rely on finding samples similar to the one with the null record and using the values copied from those samples [1,36]. The donor samples are found through the auxiliary values via algorithms like the K-Nearest Neighbours.

Regression Imputation. Mean imputation disregards the remaining features of a datapoint, thus not taking its nature into account. Mean imputation treats all classes as if they were the same. To handle this shortcoming regression, imputation can be used [4]. Regression imputation relies on considering the null value a dependent variable and using the remaining variables as features to form a regression model [13]. This definition presupposes there are relationships between the observed variables and the missing values [36], making it an approach suitable for MAR data. The drawback of regression imputation is the loss of variability, which can, however, be restored by using stochastic regression imputation. This approach adds a stochastic value to the regression result, to make the imputed value more life-like [36].

Multiple Imputation. Multiple Imputation is an attempt to answer the drawbacks of single imputation. The approach relies on filling the missing values a number of times, which creates a number of datasets, and the results of this process are combined for processing [34]. The imputation step is done with a chosen imputation algorithm, drawing the results from a distribution of possible values. Multiple Imputation has been validated to handle null records in randomised clinical trials [13]. One drawback of using multiple imputation is its computational overhead [2,13].

Multiple Imputation: Matrix Completion by Iterative Soft Thresholding of Singular Value Decompositions (Soft Impute). Soft Impute iteratively replaces the null values in a dataset with values from a soft-thresholded Singular Value Decomposition (SVD). Using Soft Impute allows for smoothly calculating an entire regularization path of solutions on a grid of values of the regularization parameter, with only the SVD of a dense matrix being a computationally-heavy task, making it a scalable method of multiple imputation [18].

Multiple Imputation: MICE. MICE, an acronym of Multiple Imputation with Chained Equations, is a multivariate technique for multiply imputing missing values using a sequence of regression models is a technique presented in [22,32]. The algorithm uses a sequence of regression models in order to model each missing value with regard to its distribution and achieve imputation [32].

Multiple Imputation: Matrix Completion by Iterative Low-rank SVD Decomposition (IterativeSVD/SVDimpute). The SVDimpute algorithm was developed by [33] in order to help facilitate gene expression microarray experiments, where fairly frequently the datasets come with multiple missing values. The authors, after multiple experiments, assess the robustness of their proposed methods as good as long as the amount of missing data does not exceed 20%. The method leverages singular value decomposition to acquire values equal to principal components of the gene expression matrix. The most significant of those values express most of the data. The procedure of IterativeSVD is redone up to the point where the total change is below a certain predetermined value [33].

K-Nearest Neighbour Imputation. This approach utilizes the KNN algorithm, calculating a choice distance measure (euclidean, jaccard, minkowski etc.) on variables observed in both donor and recipient samples [21,28,34,37]. The algorithm uses k most similar donor samples with non-null records to the recipient sample of the variable in question. The imputed value is either the most common value in the neighbours, or an average of those values [39]. The main drawback of this approach is poor scalability [32]. KNN Imputation is a very resource-heavy process and thus it is not usable in situations where the dataset is sizable. For this reason, the KNN imputation method is omitted from this study.

MissForest. MissForest is an adaptation of the Random Forest algorithm to an imputation method capable of dealing with mixed-type data. The Random Forest algorithm is known for good performance in difficult conditions, like high dimensions, complex interactions and non-linear data structures. The authors of [31] perform a series of experiments and conclude that missForest outperforms both KNN-Imputation and MICE [31].

4 Experimental Setup and Results

4.1 Dataset Preparation

To conduct the experiments, the CICIDS2017 dataset was used [29]. Dataset Description - Intrusion Detection Evaluation Dataset - CICIDS2017 is the effect of the endeavour to construct a reliable and current dataset in the domain of network intrusion detection. Datasets in the cybersecurity domain are particularly cumbersome to acquire. The ones that are accessible display some of discouraging perturbations, like insufficiency of traffic diversity, attack variety, superfluous features and so on. The authors of CICIDS2017 attempt to answer this need for realistic cyber datasets. CICIDS is an annotated capture of benign and malicious traffic. The attack captures contain malware, DoS attacks, web attacks and others. CICIDS2017 is currently one of the rare state-of-the-art datasets available to researchers.

The dataset has an unbalanced distribution of classes, with the BENIGN class being the majority class. The lack of class balance can become a problem for ML classifiers [14]. To counter this problem and prevent the impact an imbalanced dataset can have on the classifier, the majority class was randomly subsampled.

To prepare the dataset for imputation, a number of values was removed based on conditions pertaining different values (for example, in the feature '*Destination Port*' values where deleted if the value of '*Bwd Packet Length Max*' exceeded 443, which constitutes about 25% of the '*Destination Port*' variables). This is to imitate MAR missingness mechanism. Collectively, about 15% of the values in the testing part of the dataset were deleted.

4.2 Results

The effect of the imputation methods on the variable distribution of two of the affected features is illustrated by the density plots in Fig. 1 and 2. Even a very brief inspection of the distributions makes it clear to what extent the mean imputation distorts the nature of the dataset. The remaining imputation methods at least to some extent follow the distribution pattern of the original, full dataset. A clear winner comes in the performance of MissForest, which was able to retrieve the original distribution of values with surprising accuracy. This fact is also reflected in the results of the classifier.

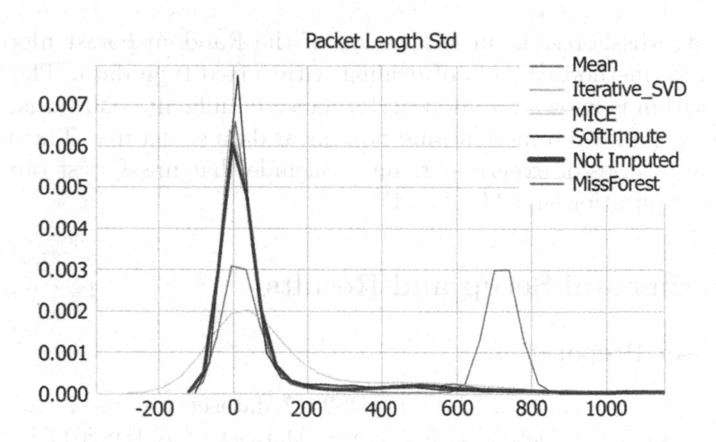

Fig. 1. The density plot of the Packet Length Std feature, with different imputation methods influencing the distrubution of feature values in different ways

The features considered in the experiment were selected using the feature_importances_ property from sklearn.ensemble ExtraTreesClassifier. Top ten most important features in CICIDS2017 were picked. These are: 'PSH Flag Count', 'Destination Port', 'Packet Length Std', 'min_seg_size_forward', 'Bwd Packet Length Std', 'Avg Bwd Segment Size', 'Bwd Packet Length Mean', 'Bwd Packet Length Min', 'Average Packet Size', 'Bwd Packet Length Max'. The class labels were accessible to the imputation algorithms, only to the classifier at test time.

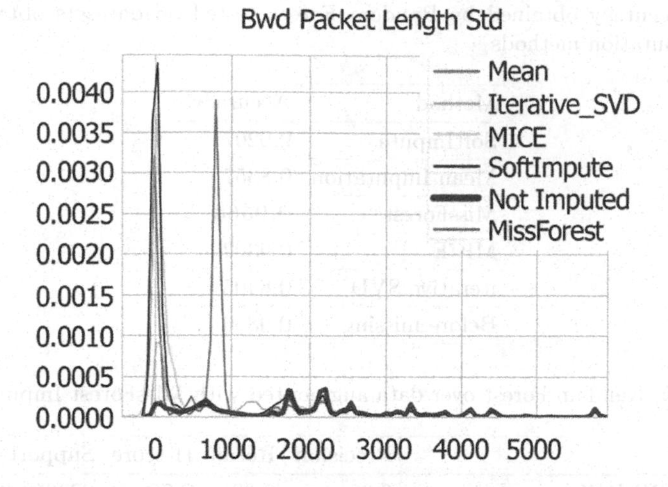

Fig. 2. The density plot of the 'Bwd Packet length Std' feature, with different imputation methods influencing the distribution of feature values in different ways

In the experiment, the chosen classifier - RandomForest - was trained using the complete cases in the training part of the dataset. The training set was randomly sampled from the whole dataset; 60% of the dataset was used.

The iterative_SVD, MICE, SoftImpute, MissForest Mean Imputation and Listwise Deletion approaches were tested. The influence over the IDS classifier the selected approaches had can be found in Tables 1, 2 and 3. The imputation methods were implemented using missingpy, fancyimpute [25] and sklearn.impute [20] packages.

The impact of imputation is measured by the classical metrics: accuracy, precision, recall and f-1 score. The imputation methods are the only variable in the experiment - the classifier and the data used are exactly the same, with the exception of listwise deletion, which, by its very nature, deletes the incomplete cases and thus makes the obtained results hard to compare. These can be investigated in Table 3. It is critical to note that listwise deletion eliminated 10 out of 14 classes of the attacks available in the CICIDS2017 dataset. The accuracy obtained by the classifier on the testing dataset before missing values were created are given in Table 1, along with the accuracy of the other methods.

4.3 Summary of Results

As evident by inspecting the accuracy of the RF classifier over the dataset obtained by different imputation methods, the choice of the method has serious impact over the results achieved by the classifier. From the evaluated methods, the **MissForest** imputation method provided the data that allowed the classifier to reach the highest accuracy - **96%** (Table 2). The next best result came with the data obtained by using MICE - 93%, closely followed by 92% obtained by SoftImpute. Of the two most often employed missing data handling strategies,

Table 1. Accuracy obtained by Random Forest tested on datasets obtained using different imputation methods

Method	Accuracy
SoftImpute	0.9207
Mean Imputation	0.8357
MissForest	**0.9565**
MICE	0.9302
iterative_SVD	0.8405
Before_missing	0.9856

Table 2. Random Forest over data augmented with MissForest Imputation

	Precision	Recall	f1-score	Support
BENIGN	0.98	0.99	0.98	223059
Bot	0.71	0.46	0.56	786
DDoS	1.00	0.82	0.90	51211
DoS GoldenEye	1.00	0.49	0.65	4117
DoS Hulk	0.88	0.98	0.93	92429
DoS Slowhttptest	0.75	0.92	0.82	2200
DoS slowloris	0.99	0.68	0.81	2319
FTP-Patator	1.00	1.00	1.00	3175
Heartbleed	0.00	0.00	0.00	4
Infiltration	1.00	0.07	0.13	14
PortScan	1.00	1.00	1.00	63572
SSH-Patator	0.83	0.49	0.62	2359
Web Attack Brute Force	1.00	0.04	0.07	603
Web Attack Sql Injection	0.00	0.00	0.00	8
Web Attack XSS	0.00	0.00	0.00	261
Accuracy			**0.96**	446117
Macro avg	0.74	0.53	0.57	446117
Weighted avg	0.96	0.96	0.95	446117

one gave sub-par results (Mean imputation), the other completely decimated the dataset (Listwise deletion/Complete case analysis - Table 3).

Table 3. Random Forest over data obtained by complete case analysis

	Precision	Recall	f1-score	Support
BENIGN	1.00	1.00	1.00	52707
DDoS	1.00	1.00	1.00	18840
DoS Hulk	0.98	0.99	0.98	4073
DoS slowloris	0.00	0.00	0.00	2
PortScan	1.00	1.00	1.00	2344
Accuracy			1.00	77966
Macro avg	0.79	0.80	0.80	77966
Weighted avg	1.00	1.00	1.00	77966

5 Conclusion

In this work, the problem of handling missing values in intrusion detection has been addressed. The work opens with an introduction to the problem of data quality and missing data, the missingness mechanisms are introduced and discussed. The work continues with an inventory of approaches to dealing with missing data, along with the most applicable algorithms in the context of network intrusion detection. The pros and cons of the methods were disclosed. Then, the details of the experiments are given and the obtained results are communicated. As presented in Table 1, the choice of the imputation method can have a critical impact on the results of the classification method employed in intrusion detection. Especially in IDS, deleting samples just because they contain some missing values can have serious consequences. On top of that, simply substituting the missing values with a central tendency measure not only distorts the nature of the data, but produces sub-par results when compared to more robust methods. The difference between the worst and the best obtained accuracies exceedes 12% points. Using complete case analysis in the evaluated case eliminated exactly ten attack classes with only 15% missing values in the dataset.

In future work the influence of imputation mechanisms over other classifiers - like optimised neural networks [6] - will be tested.

Acknowledgement. This work is funded under the PREVISION project, which has received funding from the European Union's Horizon 2020 research and innovation programme under grant agreement No. 833115.

References

1. Andridge, R.R., Little, R.J.A.: A review of hot deck imputation for survey non-response. Int. Stat. Rev. **78**(1), 40–64 (2010)
2. Azur, M.J., Stuart, E.A., Frangakis, C., Leaf, P.J.: Multiple imputation by chained equations: what is it and how does it work? Int. J. Meth. Psychiatr. Res. **20**(1), 40–49 (2011)
3. Baguley, T., Andrews, M.: Handling missing data. In: Modern Statistical Methods for HCI, pp. 57–82 (2016)
4. Baio, Gianluca, Leurent, Baptiste: An introduction to handling missing data in health economic evaluations. In: Round, Jeff (ed.) Care at the End of Life, pp. 73–85. Springer, Cham (2016). https://doi.org/10.1007/978-3-319-28267-1_6
5. Benferhat, S, Tabia, K., Ali, M.: Advances in artificial intelligence: from theory to practice. In: 30th International Conference on Industrial Engineering and Other Applications of Applied Intelligent Systems, IEA/AIE, pages Proceedings, Part I (2017)
6. Choraś, M., Pawlicki, M.: Intrusion detection approach based on optimised artificial neural network. Neurocomputing (2020)
7. Doreswamy, I.G., Manjunatha, B.R.: Performance evaluation of predictive models for missing data imputation in weather data. In: 2017 International Conference on Advances in Computing, Communications and Informatics (ICACCI), pp. 1327–1334 (2017)
8. Ezzine, I., Benhlima, L.: A study of handling missing data methods for big data. In: 2018 IEEE 5th International Congress on Information Science and Technology (CiSt), pp. 498–501 (2018)
9. Fan, W., Geerts, F.: Foundations of data quality management. Synth. Lect. Data Manage. **4**(5), 1–217 (2012)
10. Chang, G., Ge, T.: Comparison of missing data imputation methods for traffic flow. In: Proceedings 2011 International Conference on Transportation, Mechanical, and Electrical Engineering (TMEE), pp. 639–642 (2011)
11. Gleason, T.C., Staelin, R.: A proposal for handling missing data. Psychometrika **40**(2), 229–252 (1975)
12. Graham, J.W.: Missing Data. Springer, New York, New York, NY (2012)
13. Jakobsen, J.C., Gluud, C., Wetterslev, J., Winkel, P.: When and how should multiple imputation be used for handling missing data in randomised clinical trials - a practical guide with flowcharts. BMC Med. Res. Meth. **17**(1), 162 (2017)
14. Ksieniewicz, P., Woźniak, M.: Imbalanced data classification based on feature selection techniques. In: International Conference on Intelligent Data Engineering and Automated Learning, pp. 296–303. Springer (2018)
15. Li, Q., Tan, H., Wu, Y., Ye, L., Ding, F.: Traffic flow prediction with missing data imputed by tensor completion methods. IEEE Access **8**, 63188–63201 (2020)
16. Liu, S., Dai, H.: Examination of reliability of missing value recovery in data mining. In: 2014 IEEE International Conference on Data Mining Workshop, pp. 306–313 (2014)
17. Lu, X., Si, J., Pan, L., Zhao, Y.: Imputation of missing data using ensemble algorithms. In: 2011 Eighth International Conference on Fuzzy Systems and Knowledge Discovery (FSKD), vol. 2, pp. 1312–1315 (2011)
18. Mazumder, R., Hastie, T., Tibshirani, R.: Spectral regularization algorithms for learning large incomplete matrices. J. Mach. Learn. Res. **11**(80), 2287–2322 (2010)

19. Nogueira, B.M., Santos, T.R.A., Zarate, L.E.: Comparison of classifiers efficiency on missing values recovering: application in a marketing database with massive missing data. In: 2007 IEEE Symposium on Computational Intelligence and Data Mining, pp. 66–72 (2007)
20. Pedregosa, F., et al.: Scikit-learn: machine learning in Python. J. Mach. Learn. Res. **12**, 2825–2830 (2011)
21. Prince, J., Andreotti, F., De Vos, M.: Evaluation of source-wise missing data techniques for the prediction of parkinson's disease using smartphones. In: ICASSP 2019–2019 IEEE International Conference on Acoustics, Speech and Signal Processing (ICASSP), pp. 3927–3930 (2019)
22. Raghunathan, T.E., Lepkowski, J.M., Hoewyk, J.V., Solenberger, P.: A multivariate technique for multiply imputing missing values using a sequence of regression models. Surv. Methodol. **27**(1), 85–96 (2001)
23. Rana, S., John, A.H., Midi, H.: Robust regression imputation for analyzing missing data. In: 2012 International Conference on Statistics in Science, Business and Engineering (ICSSBE), pp. 1–4 (2012)
24. Rubin, D.B.: Inference and missing data. Biometrika **63**(3), 581–592 (1976)
25. Rubinsteyn, A., Feldman, S., O'Donnell, T., Beaulieu-Jones, B.: Hammerlab/fancyimpute: Version 0.2. 0. Zenodo. doi, **10** (2017)
26. Sakurai, D., et al.: Estimation of missing data of showcase using artificial neural networks. In: 2017 IEEE 10th International Workshop on Computational Intelligence and Applications (IWCIA), pp. 15–18 (2017)
27. Santos, M.S., Pereira, R.C., Costa, A.F., Soares, J.P., Santos, J., Abreu, P.H.: Generating synthetic missing data: a review by missing mechanism. IEEE Access **7**, 11651–11667 (2019)
28. Sessa, J., Syed, D.: Techniques to deal with missing data. In: 2016 5th International Conference on Electronic Devices, Systems and Applications (ICEDSA), pp. 1–4 (2016)
29. Sharafaldin, I., Lashkari, A.H., Ghorbani, A.A.: Toward generating a new intrusion detection dataset and intrusion traffic characterization. In: ICISSP, pp. 108–116 (2018)
30. Shi, W., et al.: Effective prediction of missing data on apache spark over multivariable time series. IEEE Trans. Big Data **4**(4), 473–486 (2018)
31. Stekhoven, D.J., Buhlmann, P.: MissForest-non-parametric missing value imputation for mixed-type data. Bioinformatics **28**(1), 112–118 (2012)
32. Tripathi, A.K., Rathee, G., Saini, H.: Taxonomy of missing data along with their handling methods. In: 2019 Fifth International Conference on Image Information Processing (ICIIP), pp. 463–468 (2019)
33. Troyanskaya, O., et al.: Missing value estimation methods for DNA microarrays. Bioinformatics **17**(6), 520–525 (2001)
34. Umathe, V.H., Chaudhary, G.: Imputation methods for incomplete data. In: 2015 International Conference on Innovations in Information, Embedded and Communication Systems (ICIIECS), pp. 1–4 (2015)
35. Wang, H., Shouhong.: A knowledge acquisition method for missing data. In: 2008 International Symposium on Knowledge Acquisition and Modeling, pp. 152–156 (2008)
36. Yeon, H., Son, H., Jang, Y.: Visual imputation analytics for missing time-series data in Bayesian network. In: 2020 IEEE International Conference on Big Data and Smart Computing (BigComp), pp. 303–310 (2020)

37. Zeng, D., Xie, D., Liu, R., Li, X.: Missing value imputation methods for TCM medical data and its effect in the classifier accuracy. In: 2017 IEEE 19th International Conference on e-Health Networking, Applications and Services (Healthcom), pp. 1–4 (2017)
38. Zhang, L., Xie, Y., Xi-dao, L., Zhang, X.: Multi-source heterogeneous data fusion. In: 2018 International Conference on Artificial Intelligence and Big Data (ICAIBD), pp. 47–51 (2018)
39. Zhang, Y., Kambhampati, C., Davis, D.N., Goode, K., Cleland, J.G F.: A comparative study of missing value imputation with multiclass classification for clinical heart failure data. In: 2012 9th International Conference on Fuzzy Systems and Knowledge Discovery, pp. 2840–2844 (2012)

An Experimental Study of Machine Learning for Phishing Detection

Dumitru-Daniel Vecliuc[1,2](✉), Codruţ-Georgian Artene[1,2], Marius-Nicolae Tibeică[2], and Florin Leon[1]

[1] Department of Computer Science and Engineering, "Gheorghe Asachi" Technical University of Iaşi, Iaşi, Romania
{dumitru-daniel.vecliuc,codrut-georgian.artene, florin.leon}@academic.tuiasi.ro

[2] Bitdefender, Iaşi, Romania
{dvecliuc,cartene,mtibeica}@bitdefender.com

Abstract. We approach the phishing detection problem as a data science problem with three different sub-models, each designed to handle a specific sub-task: URL classification, webpage classification based on HTML content and logo detection and recognition from the screenshot of a given webpage. The combined results from the sub-models are used as input for an ensemble model designed to do the classification. Based on the analysis performed on the results, one may conclude that ML techniques are suitable to be part of a system designed for automatic phishing detection.

Keywords: Phishing detection · Machine learning · Deep learning · URL classification · Webpage classification · Logo detection

1 Introduction

Phishing represents one of the most common types of cyber attacks. Various methods to detect this type of malicious actions exist: heuristic based detection, blacklist/whitelist based detection, ML, etc. The ML approach overtakes the trade-offs from the previous existing methodologies.

Phishing detection is transposed to a binary classification problem. Based on a given input (for example a URL) a decision whether the input represents a phishing or a *benign* (term used for legit, non-phishing sample) type is made. In this current approach, given the fact that the focus is divided into three parts (URL, visual representation-screenshot and the HTML content of the page), the initial classification problem represents a combination between two classification problems applied to the URL and to the HTML content of the webpage and a regression used for detection and localization of the logos and copyrights from a given screenshot. The final decision, whether a sample is phishing or not, is

Supported by Bitdefender.

N. T. Nguyen et al. (Eds.): ACIIDS 2021, LNAI 12672, pp. 427–439, 2021.
https://doi.org/10.1007/978-3-030-73280-6_34

made by a model which gathers the results from the specialized submodels. Text classification and object detection are well researched topics over the literature and there are multiple papers about these topics applied in different contexts.

Our work represents an experimental study of ML techniques applied on phishing detection problems and the goal is to determine if these techniques can be successfully applied to create an automatic phishing detection system.

We have summarized the main contributions in this work as follows:

- For the specific subtask of logo and copyright detection, to overcome the small dataset dimension we decided to tackle the specific brand detection task with a hybrid solution based on both deep learning and image processing techniques;
- We proposed an ensemble model for phishing webpage detection to benefit from the advantages offered by the special attention on each area of interest in phishing detection problems;
- We implemented a system that is able to perform real-time phishing detection on webpages.

The rest of the paper presents the following sections. Section 2 presents an overview over the literature relevant to our problem. In Sect. 3.1 and Sect. 3.2 the two main ML tasks are addressed. Next, Sect. 3.3 offers a brief introduction over the image matching concept. Section 3 presents the methodology of this work: dataset analysis, models overview and the evaluation process. In Sect. 4 we present the conducted experiments and the achieved results and in the last Sect. 5, we conclude with the main findings of this work.

2 Related Work

Phishing webpage detection benefits from an increased research attention in the past few years in both academia and in the tech industry. Based on the methodological approach, two categories of methods are distinguished: rule-based methods and ML-based methods.

2.1 Rule-Based Methods

One of the most common methods to use when it comes to phishing websites detection is the approach based on a set of rules created by the security experts. Based on these heuristics a webpage can be classified into *phishing* or *benign*.

A system based on a set of multiple heuristics working at the URL level is presented in [1], together with a component that uses an approximate matching of the URLs with a set of entries from an existing blacklist.

In [2] it is proposed a rule-based system that analyzes a set of predefined features, such as: the length of the URL, the presence of specific characters in the URL, the protocol used, domain based features and the HTML and Javascript behavior and based on the proposed heuristics the final decision was taken.

Although easy to use, these types of approaches come with a potential drawback because they are entangled in communication with an existing database or an external service.

2.2 ML-Based Methods

For the ML-based methods, there are many approaches that tackles the website phishing detection problem from different perspectives.

An efficient solution for malicious URL detection is presented in [3] and it uses a mix between a character-level CNN and a word-level CNN. The word embedding layer represents a combination of the individual embedding of the words and the character-level embedding of that word.

The approach presented in [4] uses the HTML content of a webpage for the classification. The proposed method is based on a CNN trained either on a character-level embedding or on a word-level embedding.

A solution with an approach similar to ours is the one proposed in [5]. The proposed solution performs classification based on the extracted features from the URL and HTML. The extracted features are combined and forwarded to three different stacking models whose predictions represent new features. The newly obtained features serve as the dataset used by the second layer of the stacking model whose final prediction represents the final decision. Authors also explored the impact of adding the visual features, represented by the image screenshot and they observed that produced an improvement compared to initial results.

Some notable mentions from the more general areas of text classification and object detection, which are related to our work are going to be briefly mentioned. For the text classification problem, BERT [6] represents a state-of-the-art solution based on the Transformers [7]. In the area of object detection, FCOS [8] comes with a new convolutional approach that eliminates the need of anchor boxes and it works fully based on the convolutions.

There are a lot of works in text classification and object detection areas with good experimental results. However, the focus of the new works from these sections are not concentrated in direction of phishing detection task so there is room for further research and improvement.

3 Methodology

Inspired by the work done so far in the areas of text classification, object detection and phishing detection, we propose a set of experiments to test the capability of the ML methods to solve the phishing detection problem. As we split the main problem into multiple subtasks, our plan is to evaluate the capabilities of the ML methods on each subtask and afterwards test the effectiveness of such an ML based system to see if it is feasible to use this type of solution for the problem that we tackle.

3.1 Text Classification

Text classification is one of the most important tasks in ML and it can be described as a process of automatically assigning a tag or category to a text based on its content. Usually, text classification systems respect the following steps:

- *Text preprocessing.* In order to use text as input for a model it has to be preprocessed and the redundant information has to be removed;
- *Text representation.* Some of the most common methods to represent textual data are Bag-of-Words (BoW) and Term Frequency-Inverse Document Frequency (TF-IDF). Other notable mentions for our work are word embedding techniques (e.g. *GloVe*, *FastText* and *Word2Vec*) and character embedding techniques (*Chars2vec* and *one-hot encoding*).
- *Classification Algorithms.* For text classification problem there are many ML algorithms that are widely used: logistic regression, support vector machine (SVM), decision trees, random forest and also deep learning approaches [9]; From the field of deep learning, notable mentions are: recurrent neural networks(RNNs) and convolutional neural networks (CNNs), with the mention that CNNs are suitable at extracting features that are invariant to location while RNNs perform better in a context where the actual step is related to previous steps;
- *Evaluation.* Once the training process is done, to determine the performance of the model it is necessary to evaluate it experimentally and decide based on a set of specific metrics. For the particular case of binary classification, a *confusion matrix* is used. Based on *true positives* (TP), *true negatives* (TN), *false positives* (FP) and *false negatives* (FN) rates, the following metrics are computed: *accuracy*, *precision*, *recall* and *F1-score*.

3.2 Object Detection

Object detection is a common Computer Vision task that identifies and locates objects from an image or a video. In our particular scenario, object detection is used in detection of logos and copyrights from a given image which represents the screenshot of a webpage.

Most object detection systems contain the following major steps:

- *Data Gathering and Annotation.* Some datasets suitable for generic object detection problems are: *IMAGENET* and *COCO*. For our logo and copyright detection problem, data comes from an in-house dataset. The preferred annotation format used to store the data is *Pascal VOC*;
- *Image preprocessing.* To be able to use images as input for a model, some preprocessing operations are needed to bring them to a standard format. This preprocessing phase usually consists of the following steps: image scaling and uniformization, image normalization, dimensionality reduction and data augmentation;
- *Image representation.* After the preprocessing step, the data is ready to be fed to the model. The input consists of a list of pairs: (*image*, *annotation*);
- *Object detection model.* Notable object detection models are: SSD [10], RetinaNet [11] and YOLOv3 [12]. YOLOv3 benefits from the advantages offered by the multi-scale detection and can detect more small objects, but comes with the trade-off on detecting medium and large objects. In our context the focus is based on reduced size (logo) object detection;

– *Evaluation.* For the evaluation of the object detection model in our scenario we decided to evaluate the performance of the model using the following metrics: *number of the predicted boxes reported to the number of ground truth boxes*, *TP*, *FP* and a metric that we define, called $almost\_TP$. The $Almost\_TP$ metric refers to those predicted boxes that intersect a ground truth box with an Intersection Over Union (IOU) lower than a fixed threshold. IOU computes the grade of overlapping between two boxes.

3.3 Image Matching

Image matching represents the process of locating the characteristic features (*keypoints*) from images and then comparing their descriptors through feature matching algorithms. The most common algorithms for feature detection are: SIFT, SURF and ORB, respectively: Brute-Force Matcher and FLANN for feature matching.

3.4 Datasets

The datasets presented in the following sections are from in-house sources and they were manually labeled.

URL Dataset. The dataset used for the training and evaluation of the URL model represents a list of samples of the two classes of interest, *phishing* and *benign* and the associated label. The examples are distributed in a balanced with phishing and benign in a 1:1 relationship. The dataset contains aproximatively 120k samples, distributed as it follows: 80% train, 5% validation and 15% test.

URLs have different lengths and based on the analysis of the URLs we decided to set the maximum length of a URL to 256 (bigger URLs are truncated). This is because a longer URL did not improve the performances of the model and it also came with training time drawbacks.

HTML Dataset. The dataset contains pairs of raw HTML content of the webpages from the same two classes, *phishing* and *benign* associated with their corresponding labels.

The dataset contains around 12k samples, with an imbalance factor of 10. This means that benign samples are ten times the number of phishing samples. Although the dataset is imbalanced, the test dataset contains an equal amount of samples for each class. The dataset is split in 96% training data and the last 4% are used for evaluating. The most encountered languages are: English, French, Portuguese and Arabic.

Object Detection Dataset. The dataset used for object detection contains only 1031 screenshots, with some duplicate images which were not removed. These samples are labeled as screenshots of the pages which were reported as being *phishing*. The pairs (*image*, *annotation*) store the following information:

- *images* are represented as a tuple: (W, H, NC), where W and H represents the width and height of the image after the preprocessing step and NC represents the number of channels;
- *annotations* represent tuples with the following format: $(x_{min}, y_{min}, x_{max}, y_{max}, C)$, where C represents the class of the box determined by the given coordinates.

Ensemble Model Dataset. The dataset used to train and evaluate the ensemble model contains samples for the two classes of interest, *phishing* and *benign*. The dataset contains labeled data, representing the predictions of each submodel.

Predictions generated by each of the submodels are initially combined in a (label-score) format: $((url\_pred_{label}, url\_pred_{score}), (html\_pred_{label}, html\_pred_{score}), (yolo\_pred_{label\,i}, yolo\_pred_{score\,i})_{i=\overline{1,10}})$, and are afterwards transformed in the following format, by summing, in order to reduce the dimensionality of the data: $((url\_pred_{label} + url\_pred_{score}), (html\_pred_{label} + html\_pred_{score}), (yolo\_pred_{label\,i} + yolo\_pred_{score\,i})_{i=\overline{1,10}}))$, all the *scores* are from the interval [0, 1].

The number of samples from the two classes is distributed as it is presented in Table 1. A notable mention is that the dataset contains all the features present (URL, HTML content and screenshot).

Table 1. Statistics regarding the initial dataset used for the ensemble model

Type	Train	Test
Benign	7494	200
Phishing	847	200
Total	8341	400

For the experiments section, we adjusted the dataset to cover multiple possible scenarios. These scenarios refer to one missing feature: URL prediction, HTML prediction or image prediction. The final configuration of the used dataset contains a number of samples equal with the number of samples from the Table 1 multiplied by 4. The new dataset maintains the same data distribution as in the mentioned table.

3.5 Models

URL Model. The model used for URL classification is based on the architecture presented in [13] and it is a CNN based architecture with five components:

- *Input layer*: in this case, the input represents the list of URLs;
- *Embedding layer*: maps each character from the URL to its corresponding one-hot representation;

- *Convolution layer*: performs six convolution operations with variable filters (two of 256 with size 7 and another four of 256 with size 3) with *ReLU* activation;
- *Max-pooling layer*: performs a downsampling operation and extracts for each performed convolution a scalar representing the maximum value for each feature map;
- *Output layer*: represents the final layer. It is a fully connected layer with sigmoid activation function, a dropout rate of 0.5 used for regularization and binary crossentropy loss.

HTML Model. The model proposed for the classification based on the HTML content uses a classic SVM decision model. The architecture of the model consists of:

- *Input layer*: the input is represented by a list of HTML page contents;
- *Preprocessing layer*: performs the transformation of the HTML page from the raw format to a model comprehensive representation. Preprocessed data is transformed in TF-IDF format and it is feed to the model;
- *Training layer*: this step involves only the training of the SVM model.

Object Detection Model. For logo and copyright detection, the architecture proposed in [12] serves as a source of inspiration. The reason why logos and copyrights are chosen as representative features for a phishing website is because according to some statistics pointed out in [14] they are very common in phishing attacks.

The object detection model contains a submodule responsible for matching the predicted bounding boxes with a list of predefined images containing logos, in order to identify the name of the brand. The components of the model are listed and explained below:

- *Input layer*: the input is represented by a list of pairs which contains the actual image and the corresponding annotation;
- *Preprocessing layer*: at this step, over the received pair (image, annotation), the following operations: *image scaling and uniformization* and *image normalization*;
- *YOLO network*: contains a lighter version of the YOLO model. The version is called *YOLOv3-tiny* [15]. It has a lower number of convolution steps. It comes with a signifiant speed improvement and with a trade-off when it comes to accuracy. The tiny YOLOv3 version uses 7 convolutional layers followed by max-pooling layers. Convolutional layers use the following structure: Convolution2D, Batch Normalization and LeakyRelu as activation. The sigmoid function is used to predict the *objectness score*, the *class score* and the *bounding box center coordinates*. The parameters learned by the network are the center of a bounding box together with its height and width, the confidence of the prediction and the class of the predicted bounding box;

- *Prediction adjustment layer*: refers to the process of determining the brands of the predicted bounding boxes which contain a logo. The *brand detection* step is made using image matching methods based on keypoints detection and matching. This approach was preferred due to lack of existing data for each logo.

Ensemble Model. The final decision whether a page represents a *phishing* or not is made by a classic kNN model. Based on the inputs gathered from all the individual specialized models, an output class is predicted. This model is called ensemble model and the architecture we propose for it is presented in Fig. 1, including the following components:

- *Input layer*: contains a list of features extracted from the URL that is going to be analyzed. The features consist of the actual URL, the HTML content for the given URL and the corresponding screenshot;
- *Classification and ensembling layer*: in this layer, each feature is classified with the corresponding submodel and afterwards the results are merged;
- *KNN model*: the kNN model receives the data containing all the information of each branch of interest and performs the classification.

The classification module contains a module that checks if the URL that is scanned is found in a *whitelist* or not.

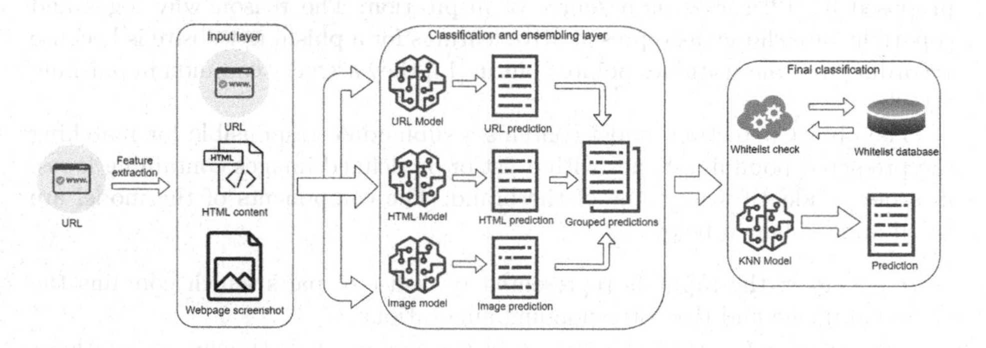

Fig. 1. Ensemble model architecture.

3.6 Evaluation

Each model is evaluated using its specific described metrics.

For the training and evaluation of the URL, HTML and YOLO models, 5-fold cross-validation is used. This method is used to reduce the bias but it also offers a broader view over the performance of the models in different situations.

The ensemble model is trained and evaluated with different configurations of the dataset, as it is going to be presented in Sect. 4.

The training process for the models with architectures based on CNNs use the *Adam* method as optimizer and binary crossentropy as loss function.

4 Experimental Results

4.1 Classic ML Model vs. Deep Learning Model for the URL Classification Problem

The CNN model that is used for this experiment is chosen based on multiple experiments with variation of the: alphabet, maximum URL size and the threshold used.

The results for the comparison between the SVM model and the character-level CNN are noted down in Table 2.

Table 2. Evaluation of ML and CNN models for URL based detection problem

Model type	Accuracy	Precision	Recall	F1
SVM	0.94	0.96	0.93	0.94
Char-level CNN	0.96	0.97	0.95	0.96

We believe that CNN performs better due to the way the data is represented. TF-IDF computation refers to other URLs when it is calculated. This may add a small bias in the SVM model. The URLs that cannot be easily classified have a similar representation. The reason why SVM performs worse is because the determined hyperplane maximizes the overall gain of the model but misclassifies the examples that are closer to the edge.

4.2 Classic ML Model vs. Deep Learning Model for the HTML Classification Problem

The CNN model used for this experiment is based on Kim's [16] network and it is chosen based on multiple experiments with variation of the: maximum input size, number of filters, size of the filters and the threshold that is used.

The results for the comparison between the CNN and the SVM model with input in TF-IDF format are noted in Table 3.

Table 3. Evaluation of SVM and CNN models for HTML based detection problem

Model type	Accuracy	Precision	Recall	F1
SVM	0.89	0.88	0.92	0.90
CNN	0.93	**0.89**	**0.98**	0.93

The metrics obtained by the CNN shows that the model performs good on our particular task. However, the gap between *recall* and *precision* represents a problem because this means that there is not a balance between the FP and

FN ratios. The model is predisposed to a higher FP ratio and we believe that this happens because the samples that represent the phising class contain some very similar examples. Also, the fact that CNN is dependant on some specific languages represents another important trade-off.

4.3 Object Detection Experiments

We evaluated the YOLO model using 5-fold cross-validation with input images at different scales as it is presented in Table 4.

Table 4. Number of ground truth boxes, number of predicted boxes, TP, FP for images evaluated with the mean of a 5-fold validation at different scales

Image size	Gt_count	Pred_count	TP	Almost_TP	FP
416	**408**	**549**	368	66	110
448	**408**	**569**	281	162	126
512	**408**	**655**	395	125	136

We chose to present the number of ground truth boxes and the number of the predicted boxes to highlight that the model is predisposed to predict more bounding boxes than they really are.

The inspection of the results shows that the overlapping predicted boxes are not properly eliminated by Non-maximum Suppression (NMS). Even with the high ratio of FPs and the misclassified boxes we are using this method because it offers a good output regarding the presence of a logo in the screenshot albeit it lacks at predicting an accurate location of the logo. We are keen to improve this method in order to increase the accuracy of the boxes localization. We strongly believe that a bigger dataset would increase the performance of the model.

The results of the YOLO model affects the performance of the brand identification. Because the predicted boxes are not properly identified and the brand identification task is based on the detected boxes, the evaluation and the results of the brand similarity method applied on our particular problem are not conclusive. However, we chose this similarity method which relies on keypoints detection and features matching based on the results from an experiment with 584 samples (non-unique logo images) from which 455 logos were properly identified.

4.4 Ensemble Model Configurations

The ensemble model takes as input the outputs from the previous submodels but there may be situations when one feature is missing. There are three experiments proposed for the ensemble model and they are presented in the following sections.

Train and Evaluate Using All Features. In this scenario, we train and evaluate the model with the dataset presented in Table 1. We consider that all the features are present both in train and evaluation datasets. The results of the experiment are presented in Table 5.

Table 5. Accuracy, precision, recall and F1 for models using only train and test samples that includes all features

Model type	Accuracy	Precision	Recall	F1	TP	FP	TN	FN
Naïve bayes	0.965	0.96	0.97	0.965	194	8	192	6
Logistic regression	0.965	0.96	0.97	0.965	194	8	192	6
SVM	0.965	0.96	0.97	0.965	194	8	192	6
Decision trees	0.965	0.955	0.975	0.965	195	9	191	5
Decision regression trees	0.965	0.955	0.975	0.965	195	9	191	5
Random forest	0.965	0.96	0.97	0.965	194	8	192	6
Random forest	0.965	0.96	0.97	0.965	194	8	192	6
KNN	0.965	0.96	0.97	0.965	194	8	192	6

The results are very similar because the test dataset is small and the variation of the FP and FN is almost unnoticeable in the metrics. The table shows that if all the features are available the ML models are able to perform *phishing detection* with impressive results.

Train with All the Features and Evaluate with only Two of Them Present. We are using this model in an automatic architecture therefore this experiment was specially designed to cover the situation when one of the features could not be obtained. For this example, the model is fed with all the features in the training process, but it is evaluated with one missing feature at a time (Table 6).

Table 6. Accuracy, precision, recall and F1 for models with all the features present in train and one feature missing in test

Model type	Accuracy	Precision	Recall	F1
Naïve bayes	0.82	0.74	0.98	0.85
Logistic regression	0.79	0.98	0.59	0.73
SVM	0.82	0.98	0.66	0.79
Decision trees	0.82	0.77	0.92	0.84
Decision regression trees	0.75	0.70	0.88	0.78
Random forest	0.91	0.98	0.83	0.90
KNN	0.88	0.99	0.77	0.86

This experiment shows the impact of a train dataset that does not properly cover all the situations to which the model will be exposed, in our scenario a missing feature.

Based on the metrics analysis, one can conclude that even in this scenario the model is capable to produce trustworthy results. Exploring the recall and precision metrics we can see that the model correctly identifies a high number of phishing samples with an even higher confidence.

Train and Evaluate with Possible Missing Features in Both Train and Test Datasets. This experiment exposes the models to situations when a feature may be missing, but unlike the previous experiment, models are trained with this type of behavior present in the training set (Table 7).

Table 7. Accuracy, precision, recall and F1 for models with possible missing features in both train and test datasets

Model type	Accuracy	Precision	Recall	F1
Naïve bayes	0.87	0.94	0.78	0.85
Logistic regression	0.85	0.95	0.73	0.83
SVM	0.84	0.95	0.73	0.82
Decision trees	**0.95**	**0.95**	**0.95**	**0.95**
Decision regression trees	**0.95**	**0.95**	**0.95**	**0.94**
Random forest	**0.96**	**0.96**	**0.95**	**0.96**
KNN	**0.96**	**0.96**	**0.96**	**0.96**

The results show that the models performance is increasing when it is trained with a proper dataset. We can observe an improvement for all the models compared to the previous situation when one feature was missing.

5 Conclusions

In this work we described the approaches that we tried in order to successfully create a model that is going to be used in the phishing detection problem. We proposed and evaluated multiple models and architectures for each subproblem in which the main phishing detection issue was divided. The attention was shifted towards the comparison between classic ML models and some deep learning architectures.

Based on the results of the proposed experiments we showed that generic subjects as text classification and object detection can be successfully applied to our particular scenarios of text classification based on the URL and the HTML content of a webpage respectively logo and copyright detection from the screenshot of a webpage.

Taking into account that e-mails usually represent the first vector of attack in phishing, our method can be easily extended and used to predict on e-mails. In order to solve this task it needs the features to be extracted from the e-mail.

Despite the fact that there is still room for improvement and there are more hypothesis to test, based on the results from the conducted experiments (especially on the performances of the ensemble model) we think that our phishing detection solution is suitable to be part of an automatic system.

References

1. Prakash, P., Kumar, M., Kompella, R., Gupta, M.: PhishNet: predictive blacklisting to detect phishing attacks. In: 2010 Proceedings IEEE INFOCOM, pp. 1–5 (2010)
2. Mohammad, R.M., Thabtah, F.A., McCluskey, T.: Intelligent rule-based phishing websites classification. IET Inf. Secur. **8**, 153–160 (2014)
3. Le, H., Pham, Q., Sahoo, D., Hoi, S. URLNet: Learning a URL Representation with Deep Learning for Malicious URL Detection. arXiv, abs/1802.03162 (2018)
4. Opara, C., Wei, B., Chen, Y.: HTMLPhish: enabling phishing web page detection by applying deep learning techniques on HTML analysis. In: 2020 International Joint Conference on Neural Networks (IJCNN), Glasgow, UK, 2020, pp. 1–8 (2020). https://doi.org/10.1109/IJCNN48605.2020.9207707
5. Li, Y., Yang, Z., Chen, X., Yuan, H., Liu, W.: A stacking model using URL and HTML features for phishing webpage detection. Future Gener. Comput. Syst. **94**, 27–39 (2019)
6. Devlin, J., Chang, M.-W., Lee, K., Toutanova, K.: BERT: Pre-training of Deep Bidirectional Transformers for Language Understanding, p. 13. arxiv:1810.04805Comment (2018)
7. Vaswani, A., et al.: Attention is all you need. In: Advances in Neural Information Processing Systems, pp. 5998–6008 (2017)
8. Tian, Z., Shen, C., Chen, H., He, T.: FCOS: Fully Convolutional One-Stage Object Detection. CoRR, abs/1904.01355 (2019)
9. Kowsari, K., Jafari Meimandi, K., Heidarysafa, M., Mendu, S., Barnes, L., Brown, D.: Text classification algorithms: a survey. Information **10**(4), 150 (2019)
10. Liu, W., et al.: SSD: Single Shot MultiBox Detector. arXiv, abs/1512.02325 (2016)
11. Lin, T.-Y., Goyal, P., Girshick, R., He, K., Dollár, P.: Focal Loss for Dense Object Detection. arXiv:1708.02002 (2017)
12. Redmon, J., Farhadi, A.: YOLOv3: An Incremental Improvement. arXiv, abs/1804.02767 (2018)
13. Zhang, X., Zhao, J., Lecun, Y.: Character-level convolutional networks for text classification. In: Advances in Neural Information Processing Systems, Jan 2015, pp. 649–657 (2015)
14. Geng, G., Lee, X., Zhang, Y.: Combating phishing attacks via brand identity and authorization features. Security Commun. Netw. **8**, 888–898 (2015). https://doi.org/10.1002/sec.1045
15. Darknet: Open Source Neural Networks in C. http://pjreddie.com/darknet/. Accessed 12 Oct 2020
16. Kim, Y.: Convolutional neural networks for sentence classification. In: Proceedings of the 2014 Conference on Empirical Methods in Natural Language Processing, EMNLP 2014, 25–29 Oct 2014, Doha, Qatar. A meeting of SIGDAT, a Special Interest Group of the ACL, pp. 1746–1751 (2014)

Computer Vision Techniques

Model for Application of Optical Passive SFM Method in Reconstruction of 3D Space and Objects

Ante Javor[1], Goran Dambic[2], and Leo Mrsic[2](✉)

[1] Bjelovar University of Applied Sciences, Trg E. Kvaternika 4, 43000 Bjelovar, Croatia
ajavor@vub.hr

[2] Algebra University College, Ilica 242, 10000 Zagreb, Croatia
{goran.dambic,leo.mrsic}@algebra.hr

Abstract. Reconstruction of objects and space-based on images is the topic of active research in a recent decade. The reason for that is because high-quality reconstruction is hard to achieve, but if achieved successfully there is a wide range of possible applications. Particularly in disruptive technologies such as virtual reality and augmented reality. Advancement in computer and optical hardware has enabled development of acceptable reconstruction that can be applied for some type of purposes. Depending on the goal of reconstruction there is a wide variety of possible approaches and methods. This thesis is based on structure from a motion approach that can be used for reconstruction of a single object, room, building, street, or city. The method uses a set of images that have a targeted object or space for achieving reconstruction. Structure for motion is based on feature extraction, camera registration and stereo vision. The method is not designed for real-time reconstruction.

Keywords: Structure from Motion · SFM · 3D reconstruction · Feature extraction · SIFT · KAZE · Stereo vision

1 Introduction

From the very beginning of the development of computer graphics, one of the important goals was to achieve the quality of computer-generated content so it looks as similar as possible to real world. By mixing realistic scenes, computer-generated 3D models, and digital effects in new movies, it is hard to separate computer-generated models from real physical objects that were on the scene during filming. Such advances in computer graphics have opened up new possibilities for the application of diverse digital content not only to the entertainment industry but also to many others. The problem with creating digital realistic 3D models is that they require a large amount of manual human work. Creating digital copies based on real objects requires additional work because it is necessary to pay attention to the details of the original model. This adds more work to an already demanding and complex process that is only partially automated. With the development of computer vision in the last two decades, high-quality, applicable and

N. T. Nguyen et al. (Eds.): ACIIDS 2021, LNAI 12672, pp. 443–453, 2021.
https://doi.org/10.1007/978-3-030-73280-6_35

robust algorithms have been developed that enable the creation of more or less realistic 3D models based on images or a sequence of images from video. Generating 3D models from images makes the process of generating the model cheaper and faster. With the mentioned automation and cheaper manufacturing, space is opened for the application of methods in various fields such as augmented reality, robotics, virtual reality, autonomous systems, tourism and the like. The aim of this paper is to propose a model for 3D reconstruction of space and/or objects based on a simple and accessible method called Structure from Motion (SFM) and to prototype an actual system based on the proposed model. The characteristic of this method is that it can make a 3D reconstruction entirely based on a collection of images, without the use of any active elements (laser, sonar, radar, etc.). The SFM method is based on camera registration, feature extraction and stereo vision.

2 Background

SFM is a method that requires a series of images of a space or object, which does not need to be sorted or pre-processed. It is crucial that there is an overlap between the images of the scene or part of the scene that is to be reconstructed. Each image contains a number of features that are specific to the specified scene to be reconstructed. The variety of shapes, textures, lines, and colors in an image will allow for a large number of specific image features to be extracted by the feature extraction process. Based on the overlap of the scene on the images, a match between the features is sought. By matching specific features between images and tracking them across multiple images, it is possible to obtain a 3D geometric relationship of features, i.e. the images themselves. This geometric relationship is further used for the reconstruction process, i.e. stereo vision that will allow the creation of the final 3D model.

2.1 Stereo Vision

The intrinsic matrix is made up of internal camera parameters and represents a mathematical model of a simple camera. Some of the common parameters are focal length, main point, and camera sensor distortion [2]. A matrix is used to describe the process of creating values in a single pixel of an image based on the rays of light falling on the sensor. However, to understand the scene and solve the problem of 3D space reconstruction, it is necessary to understand the position of the camera in relation to objects on the scene and other cameras. Parameters that describe the position of the camera in global space are called external or extrinsic parameters [8]. The extrinsic parameters of the camera are described with 6 degrees of freedom. Extrinsic parameters will allow the conversion of a point from one coordinate system to another using the parameters for rotation and translation. "The parameters for rotation and translation are described by a 3×3 rotation matrix R and a 3×1 translation vector t" [1]. The intrinsic and extrinsic parameters of the camera form a complete matrix of the camera, the projection matrix. These parameters can be obtained by automatic or manual calibration of the camera. For successful 3D reconstruction of space or objects, it is crucial to have precise information about the position of individual points in space in relation to the camera or the coordinate system of the scene. Using the position of the pixels in the image and the projection matrix, it is possible to obtain information about the direction in which a particular point on the scene can be located. The introduction of a second camera allows triangulation, but for the triangulation process to work, a composition of two or more cameras needs to be in a known geometric relationship. To ensure a geometrically ideal camera composition and to facilitate the search for the same pixels in both images, epipolar geometry is used. Stereo correspondence is the process of comparing the values of a pixel or group of pixels of the left and right cameras looking at the same scene or object [7]. A pixel comparison will be successful if the pixel in the left image is in the right image and vice versa. Once a match is found, a triangulation process is followed to obtain the distance from the stereo composition of the cameras to the point on the scene. In an ideal stereo composition, the pixel positions in which the projection axis of the left and right cameras passes through a point on the scene are known. However, in the real case, these points are not known and it is necessary to find two identical or as similar points as possible in the left and right images. The previous steps for camera calibration and stereo rectification should significantly facilitate this process. Epipolar geometry ensures that for a known left camera reference point, the search on the right camera is reduced to a one-dimensional series of pixels or an epipolar line [1]. In order to mitigate the negative effects, algorithms are used for correspondence that further ensure the robustness of the correspondence. "One of the standard ways of stereo correspondence is the block matching algorithm" [6]. This paper uses the Semi-Global Matching (SGM) method, which gives better results in practice [10]. Adjacent pixels on the epipolar line are not only compared horizontally but in multiple directions which ensures a more robust disparity calculation. Disparity defines the difference between the pixel position on the left camera and the pixel position in the right image. Disparity significantly affects triangulation and the calculation of the pixel distance from the camera.

2.2 Image Features

Image features represent all the specifics associated with the image. These specifics can represent the whole image or just a specific segment made up of a few pixels. These areas make the image unique and the amount of specificity, or features of the image, will depend on the content of the image itself. Images with more texture, color, and, in general, a variety of intensities will have many more potential features compared to images with a dominant intensity and little texture. In order for an image to be compared based on features, they need to be quantified on the image itself. "The process of quantifying image features is called feature extraction. The feature extraction process includes two sub-processes, namely feature detection and feature description" [11]. When feature extraction is applied, both feature detection and description processes are applied. Feature detection is an automated process of finding quality and robust features that are easy for a computer to recognize. Feature description is the process of describing a feature in an understandable way so that the same feature can be compared to a number of other features from images. Since features are used to recognize images and different segments of an image, it is crucial that the features detected be robust. This means they are independent of location change, rotation, size, lighting and the like. When viewed from a 3D reconstruction perspective, the most important feature of the model is the robustness to change in viewing angle between the feature and the camera [7]. In order to establish a mathematical relationship between the cameras and enable stereo vision, it is necessary to calculate the camera matrices that represent the positions and orientations during image acquisition. This is achieved by matching the features that are tracked on multiple consecutive images. Feature matching involves comparing an individual feature of the left image with the right image. The result of the match is the features visible in both images. It is crucial that the images of the scene overlap to find a sufficient number of the same features. By applying the RANSAC method to features that match on a pair or sequence of images, the necessary extrinsic parameters about the cameras are obtained to establish epipolar geometry, i.e. stereo vision.

3 Methodology

The proposed model and system for 3D space reconstruction is based on the open source libraries OpenMVG [4] and OpenMVS [5]. OpenMVG provides the ability to geometrically understand the relationship between a scene and a camera based on multiple images of the same scene. In this paper, the result of this library is the SFM method which ultimately provides a sparse reconstruction of key points and the necessary camera parameters. The above library and its pre-built modules are sufficient to accomplish the standard steps of SFM reconstruction. Using SFM camera output information, OpenMVS will construct a complete model of a 3D object or space in several different phases. The proposed model first uses the OpenMVG_ImageListing module. It receives a path to images from which the space is to be reconstructed, and it also receives a database with data on a number of popular cameras that could potentially be used to acquire said images. The database includes sensor width information along with EXIF data of a single image such as focal length in mm, width and height of the image and the like. With the help of the above information, the focal length in pixels is obtained, which makes the intrinsic matrix of the camera. In addition, the mentioned module will prepare a list of images for further processing, make an analysis of which image was taken by which camera and prepare intrinsic parameters for each camera [4]. In this paper, all images are acquired by the same camera, so the same intrinsic parameters are used for each image. Once the image list and partial intrinsic parameters are known, the proposed model uses the OpenMVG_ComputeFeatures module to extract image features. Detection and description of features is performed for each image delivered to the module using SIFT and AKAZE algorithms. Using the selected feature extraction procedure, the module

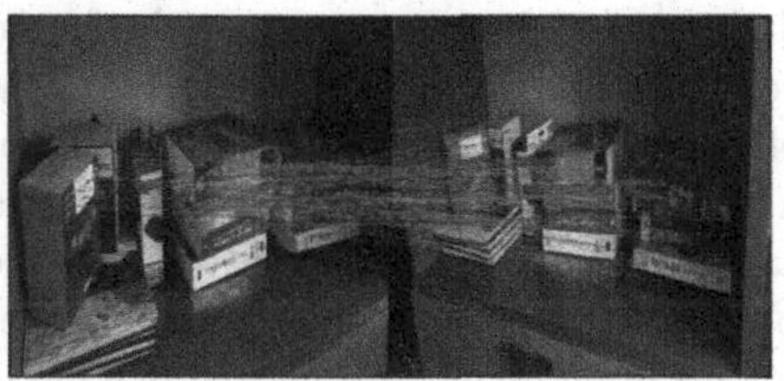

Fig. 1. Features found by the AKAZE algorithm and matching of features between the two images

will generate a description of the features used in the next steps of the proposed system to calculate feature matches in the images.

Figure 1 shows an example of feature extraction using the AKAZE algorithm. The algorithm recognized a total of 9,535 features highlighted in yellow in the image. Next, the proposed model uses methods to match features between individual images. In this step, image pairs are prepared for the SFM method, and this procedure will be done using the OpenMVG_ComputeMatches module. The result of this step are pairs of images that have overlap and interrelationship in space. The best pairs make up the images with the most matching features. The Fig. 1 shows the lines indicating the matching of features between the two images. Both images display the same shelf and are taken from different angles. The total number of features of the left image is 9,535, the right image 8,861 while the number of matches is 1,356. Among the matches we can find good and bad features (bad ones are lines that are not parallel). Once the feature matches between the images are known, the proposed model uses the OpenMVG_GlobalSfM module to initiate camera position acquisition and sparse reconstruction. The module applies the global SFM method [19] over matching image pairs. The input data to the module consists of an intrinsic camera matrix and pairs of images on which there are overlaps. The module returns a 3D model in the form of a sparse point cloud that represent key points, or features that match the images but also with depth in the scene. Based on the key points, the position of the camera at the time of shooting was obtained for each image, i.e. the extrinsic matrix of each camera. In this step, it is possible to configure the type of algorithm to estimate the rotations and translations of the cameras in the scene.

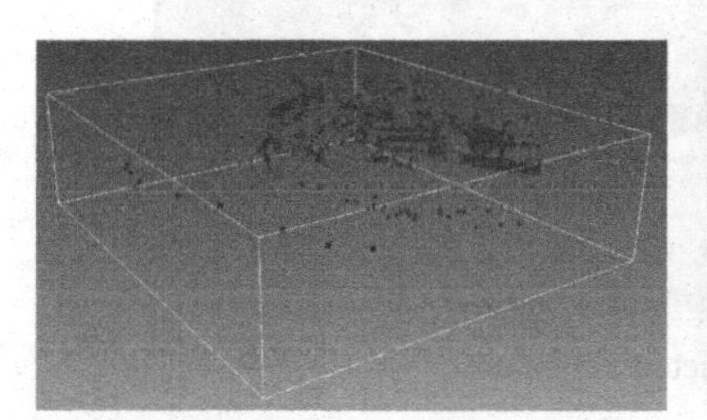

Fig. 2. Sparse point cloud with camera positions and results of dense reconstruction

This procedure can be seen in Fig. 2, which shows a sparse point cloud in gray and the position of the camera in green. The total number of points of the 3D model is 1,830. These matches were obtained based on 9 images and the global SFM method. The role of the OpenMVG_openMVG2openMVS module in the proposed model is the transformation of data for the needs of the OpenMVS so immediately after it the OpenMVS_DensifyPointCloud module is used. This module will create a dense reconstruction of point cloud, using previously obtained information from the SFM method. Since the camera parameters are known, the establishment of epipolar geometry and stereo correspondence is possible. The reconstruction process is performed using the stereo correspondence of the SGM method. First, the images that are most similar to each other according to the category of the number of features or the angle between the cameras themselves are taken. A disparity map is obtained for the specified pair of images. The disparity map is further improved by comparison with other folders of similar image pairs. Then, all these pairs of images are merged into a final 3D model with the rejection of redundant data that are duplicated [22]. This step is shown in Fig. 2 and represents a dense point cloud (the total number of points of this 3D model is 1,144,200). The OpenMVS_ReconstructMesh module will generate a 3D model without a texture. This procedure is based on using a cloud of dense reconstruction points and on using support for poorly supported and visible areas [23]. The reconstruction takes place using the Delaunay triangulation technique, which deals with the approximation of faces based on points of sparse or dense reconstruction. Method [23] under poorly supported areas implies parts that have a smaller amount of texture, reflective surface, and transparent surface or are obscured by an object so in these areas there will be significantly fewer points. Using Delaunay triangulation, potential objects that create occlusion and block segments of the model are calculated. The hidden segments are then reconstructed under special conditions to create a consistent model. An example of such reconstructed model is shown in Fig. 3.

Fig. 3. Reconstructed 3D model

Finally, for the purpose of visually enhancing the model, the proposed model uses the OpenMVS_RefineMesh and OpenMVS_TextureMesh modules that will apply textures on the resulting 3D model (Fig. 4).

Fig. 4. Up – sparse point cloud; middle – dense point cloud; down – final 3D model

4 Results

Four sets of pictures were used to test the proposed model for reconstruction: in set S1, a bookshelf was used; in set S2, a room in the house was used; in set S3, a car was used while in set S4, the house was used. Data on the reconstruction of each set by the SFM method is shown in Table 1. The tests were performed on a computer with an Intel i7-8750 h processor and 16 GB of RAM. Since only the 3D object model enhancement module has support for execution on the GPU so all tests were performed exclusively on the CPU and RAM, it is to be expected that the results would be better time-wise if executed on the GPU. Table 1 shows data on the total number of images used for the reconstruction, the total number of points representing the features and overlaps between the images, the points that make up the cloud of sparse reconstruction, and the reconstruction time in minutes.

Table 1. SFM image set reconstruction data

Set	Images	Total initial points	Sparse reconstruction points	Duration (min)
S1	9	8,447	1,830	1
S2	79	93,457	18,989	180
S3	77	144,726	39,522	48
S4	93	175,994	30,084	46

After applying the SFM method, the proposed model performs stereo correspondence, reconstruction and improvement of the obtained model. Data on this reconstruction is shown in Table 2. The data include the total number of pixels of images on which stereo correspondence is performed, the number of points of dense reconstruction, the number of vertices, the number of faces that make up the model and the time required to build the model.

Table 2. Results of stereo reconstruction

Set	Sum of all pixels in all images (millions)	Dense reconstruction points	Vertices	Faces	Duration (min)
S1	67.83	1,144,200	137,024	272,914	14
S2	595.39	2,760,764	473,650	944,062	124
S3	580.32	9,646,052	768,348	1,529,038	231
S4	700.91	9,759,339	559,785	1,116,603	285

The results obtained by the reconstruction of set S1 are shown in Fig. 5. The results show particular areas with features on the books themselves. Given that all the features are on the books, the reconstructed segment does not contain the background completely nor in good quality, but contains only its small parts. It is specific for this set that the reconstruction was done on the basis of only 9 images and yet certain segments were reconstructed with quality. Reconstruction of set S2 based on 79 images is shown in Fig. 5. Comparing the images from set S2 (part of the set is shown in Fig. 5) and the obtained model, it can be seen that areas without expressive textures were not successfully reconstructed. The reason is the lack of features and clear differences between pixels in areas of similar intensity. An example of this is a large area of a white wall or brown closet. Stereo correspondence is not able to distinguish the difference by epipolar lines between the pixels themselves, so these areas were not reconstructed in the final model.

Fig. 5. Part of the set S2 and final 3D model of the reconstructed room from set S2

This shortcoming can be visualized by merging the points of sparse reconstruction and the final 3D model. Sparse reconstruction points represent the available features in the image. Figure 6 shows the overlap between the image features located in the texture-rich areas and the actually reconstructed model. Green points are the positions of the cameras in space for each image while white points are features of sparse point cloud. There are almost no features in the wall areas, while a large number of features are visible on the bookshelf from the S1 set. The number of features themselves is often an indicator of the diversity of the scene itself. When comparing the results of the reconstruction of sets S2 and S3 (Fig. 5) which have 79 and 77 images in the set, a significant difference in the number of points and features can be noticed. The S3 car set contains two images less than the S2 room set, but according to Table 1, the total number of sparse reconstruction points is 20,563 points larger. The interior spaces have a significantly smaller number of different textures, edges and colors compared to outdoor images. The number of sparse points that mostly represent the reference points for the reconstruction significantly influences the further steps of the reconstruction and the result of the final 3D model. In addition to the number of features, the positions of the points are also important. Better distribution of points will result in better reconstruction. Figure 7 shows the result of the reconstruction of set S3. The image segment at the top left represents the original image, at the top right the reconstructed model, at the bottom left a dense reconstruction of points, and at the bottom right a cloud of sparse points. The mentioned reconstruction was achieved on the basis of 93 images. The model made up of sparse cloud clouds has a total of 39,522 different points, although most goes to the concrete around the car.

Fig. 6. Sparse point cloud and final room model from set S2

The model made up of dense reconstruction clouds has a total of 9,646,052 points. The final 3D model is made up of 768,348 vertices and 1,529,038 faces. This set shows the additional disadvantages of this method. Namely, methods based exclusively on images have a problem with the reconstruction of reflective surfaces, so it is evident that the roof of the car and the glass were not properly reconstructed. Since the reflection changes as the camera moves, it is difficult to find reference points due to the different reflections. This problem is also visible in the segment of the image showing the cloud of sparse points. It has a large number of features around the wheels that do not change depending on the viewing angle, while in the area of glass and metal surface there are significantly fewer points due to viewing angle of the camera.

Fig. 7. Set S3/car reconstruction and Set S4/reconstruction of the house

Figure 7 shows the result of the reconstruction of set S4. The image segment at the top left represents the original image from the set (although most of the images are from another angle), at the top right there is a reconstructed model, at the bottom left there is a dense reconstruction of points, and at the bottom right there is a cloud of sparse points. This reconstruction was done on the basis of 93 images. The model made of sparse cloud points has a total of 30,084 different points, and the model made of dense cloud points has a total of 9,759,339 points. The final 3D model is made up of 559,785 vertices and

1,116,603 faces. Although the set S4 has 16 images more than the set S3, the set S3 has a larger number of points of sparse and almost equal of the dense reconstruction. The reason is that the images from the set S4 set are mostly taken from the same angle, so a number of the features themselves are actually repeated.

5 Conclusion

The development of cameras and computer components has enabled the wide application and availability of computer vision algorithms. Reconstruction of objects or spaces 15 years ago was not possible on personal computers, which is not the case today. Increasing the availability of hardware opens up the potential for new research and development of computer vision applications. A number of passive and active methods for reconstruction are being developed. In addition, widely used fundamental computer vision techniques are being improved. An example of the development of new methods based on the improvement of existing techniques is the SFM method, or Structure from Motion. The development of feature extraction, stereo correspondence, and camera understanding has enabled the application of the SFM method to a large number of images. The robustness of the features in the images will significantly affect the numerical stability of the camera position calculation and orientation, so robust methods are used to extract features such as the SIFT and AKAZE algorithms. In addition, stereo correspondence algorithms should ensure triangulation of 3D points and finding the same pixels in different images which is a computationally demanding task. Hardware development has made it possible to use more advanced and robust stereo correspondence algorithms such as the SGM algorithm. All these improvements have enabled the usage of SFM method, which gives better results for the reconstruction of objects or spaces. However, this strongly depends on the content of the image, so passive methods of reconstruction do not give the desired results on a set of images with few features, uneven distribution of features, reflective, transparent surfaces and the like. This mainly includes interiors that contain a lot of white paint surfaces or glass partitions. In outdoor conditions, this method gives significantly better results due to the greater number of textures and the overall variety of the scene. To achieve a quality and robust reconstruction solution, procedures based on active methods still remain viable. Active methods give better results due to a more robust perception of depth and resistance to low texture areas, so alternative active methods with a laser-shaped light source are imposed as more precise and accurate methods for interior reconstruction than methods based on passive vision.

References

1. Moons, T., Gool, L.V., Vergauwen, M.: 3D reconstruction from multiple images. Found. Trends Comput. Graph. Vis. **4**(4), 287–404 (2010)
2. Szeliski, R.: Computer Vision: Algorithms and Applications. Texts in Computer Science. Springer, London (2010). https://doi.org/10.1007/978-1-84882-935-0
3. OPENCV, About OpenCV, May 2020. https://opencv.org/about/
4. OPENMVG, Multiple View Geometry documentation, May 2020. https://openmvg.readthedocs.io/en/latest/

5. OPENMVS, Open Multi-View Stereo reconstruction library, May 2020. http://cdcseacave.github.io/openMVS/
6. Bradski, G., Kaehler, A.: Learning OpenCV 3: Computer Vision in C++ with the OpenCV Library. O'Reilly Media, Sebastopol (2017)
7. Mangor, M.A., Grau, O., Sorkine-Hornung, O., Theobalt, C.: Digital Representations of the Real World: How to Capture, Model, and Render Visual Reality. CRC Press, Boca Raton (2015)
8. Hartley, R., Zisserman, A.: Multiple View Geometry in Computer Vision. Cambridge University Press, New York (2004)
9. The Middlebury Computer Vision, The Middlebury Stereo Vision Page, May. 2020. http://vision.middlebury.edu/stereo/
10. Hirschmuller, H.: Stereo processing by semiglobal matching and mutual information. Pattern Anal. Mach. Intell. PAMI **30**(2), 328–341 (2008)
11. Gonzalez, R.C., Woods, R.E.: Digital Image Processing. Pearson, London (2018)
12. Harris, C., Stephens, M.: A combined corner and edge detector. In: Proceedings of the 4th Alvey Vision Conference, pp. 147–151 (1988)
13. Lowe, D.G.: Distinctive image features from scale-invariant keypoints. Int. J. Comput. Vis. **60**(2), 91–110 (2004)
14. Alcantarilla, P.F., Bartoli, A., Davison, Andrew J.: KAZE features. In: Fitzgibbon, A., Lazebnik, S., Perona, P., Sato, Y., Schmid, C. (eds.) ECCV 2012. LNCS, vol. 7577, pp. 214–227. Springer, Heidelberg (2012). https://doi.org/10.1007/978-3-642-33783-3_16
15. Alcantarilla, P.F., Nuevo, J., Bartoli, A.: Fast explicit diffusion for accelerated features in nonlinear scale spaces. Trans. Pattern Anal. Mach. Intell. **34**, 1281–1298 (2011)
16. Tareen, S.A.K., Saleem, Z.: A comparative analysis of SIFT, SURF, KAZE, AKAZE, ORB, and BRISK. In: International Conference on Computing, Mathematics and Engineering Technologies (iCoMET), pp. 1–10 (2018)
17. Cheng, J., Leng, C., Wu, J., Cui, H., Lu, H.: Fast and accurate image matching with cascade hashing for 3D reconstruction. In: Computer Vision and Pattern Recognition, pp. 1–8 (2014)
18. Snavely, N., Seitz, S.M., Szeliski, R.: Photo tourism: exploring photo collections in 3D. ACM Trans. Graph. **25**, 835–846 (2006)
19. Moulon, P., Monasse, P., Marlet, R.: Global fusion of relative motions for robust, accurate and scalable structure from motion. IEEE Int. Conf. Comput. Vis. **2013**, 3248–3255 (2013)
20. Fischler, M.A., Bolles, R.C.: Random sample consensus: a paradigm for model fitting with applications to image analysis and automated cartography. Commun. ACM **24**, 381–395 (1981)
21. Agarwal, S., Snavely, N., Simon, I., Seitz, S.M., Szeliski R.: Building rome in a day. In: IEEE 12th International Conference on Computer Vision, Kyoto, pp. 72–79 (2009)
22. Shen, S.: Accurate multiple view 3D reconstruction using patch-based stereo for large-scale scenes. IEEE Trans. Image Process. **22**, 1901–1914 (2013)
23. Jancosek, M., Pajdla, T.: exploiting visibility information in surface reconstruction to preserve weakly supported surfaces. In: International Scholarly Research Notices, pp. 1–20 (2014)

Ground Plane Estimation for Obstacle Avoidance During Fixed-Wing UAV Landing

Damian Pęszor[1,3](✉), Konrad Wojciechowski[1], Marcin Szender[2], Marzena Wojciechowska[1], Marcin Paszkuta[1,3], and Jerzy Paweł Nowacki[1]

[1] Research and Development Center of Polish-Japanese Academy of Information Technology, Al. Legionów 2, 41-902 Bytom, Poland
{damian.peszor,konrad.wojciechowski,marcin.paszkuta}@polsl.pl, {dpeszor,mwojciechowska}@pjwstk.edu.pl

[2] MSP InnTech sp. z o.o., Poligonowa 1/81, 04-051 Warsaw, Poland
info@uav.com.pl

[3] Department of Computer Graphics, Vision and Digital Systems of Silesian University of Technology, Akademicka 16, 44-100 Gliwice, Poland

http://bytom.pja.edu.pl, http://uav.com.pl/en, http://kgwisc.aei.polsl.pl

Abstract. The automatic crash-landing of a fixed-wing UAV is challenging due to high velocity of approaching aircraft, limited manoeuvrability, lack of ability to hover and low quality of textural features in common landing sites. Available algorithms for ground plane estimation are designed for automatic cars, robots, and other land-bound platforms, where textural features can be easily compared due to low altitude, or for rotor-wing UAV that can hover. Their usefulness is limited when quick manoeuvre is needed to avoid collision with obstacles. Due to developments in parallelised, mobile computational architectures, an approach based on dense disparity estimation becomes available assuming proper constraints on ground plane transformation phase space. We propose an algorithm utilising such constraints for ground plane estimation on the often prohibitively time-consuming task of disparity calculation as well as plane estimation itself using a pyramid-based approach and random sample consensus in order to discard pixels belonging to obstacles. We use Inertial Navigation System, commonly available in fixed-wing UAVs, as a source of such constraints and improve our estimation in subsequent frames allowing for stable flight trajectory as well as detection of obstacles protruding from the ground plane.

Keywords: Ground plane estimation · Obstacle avoidance · Fixed wing · UAV · Unmanned aerial vehicle · Disparity

1 Introduction

Obstacle detection is crucial for navigation of unmanned aerial vehicles (UAVs) and ground plane estimation is an effective way to detect obstacles. The understanding of the environment allows for safe operations - for the drone itself, but

N. T. Nguyen et al. (Eds.): ACIIDS 2021, LNAI 12672, pp. 454–466, 2021.
https://doi.org/10.1007/978-3-030-73280-6_36

most importantly, for the surroundings, especially living creatures in general and people in specific. While the aircraft has a high altitude, the danger of collision is low, and potential landing sites can be found [1]. However, during the landing procedure, the probability of collision increases significantly, as not every obstacle would be found from above. Among the sensors that can provide information about ground plane transformation, imaging sensors prove to be attractive due to their low-cost, which is a substantial advantage for UAV, even if other solutions might provide better results in manned aircraft. Recent developments in massively parallelised processors both in terms of performance and weight, allow for processing of visual information onboard of an aircraft removing the main disadvantage of such systems. Presence of obstacles is both the most common reason for ground plane estimation as well as the main disruptor of it. This is especially important whenever the time that can be utilized for computations is highly restricted by the high velocity of a moving platform. Such is the case of a landing fixed-wing UAV. Unlike rotary-wing drones, for which the problem of ground plane detection is commonly described in the literature, fixed-wing platforms cannot hang in the air while waiting for calculations to improve the probability of correct plane estimation. Landing fixed-wing aircraft are restricted to estimating the transformation of a ground plane using the limited time during high-velocity descent, based on the current frame since the next one might be done after traversing a significant distance. This is true especially in case of a crash-landing, in which the best landing site might be too far and so proper decision about how to steer the flight needs to be made as fast as possible to avoid collisions while the control over the aircraft is limited. While waiting for the next frame is a useful strategy in terms of rotary-wing UAV, and can be used for optical flow-based solutions, it is not suitable for fixed-wing aircraft. Cameras mounted on the aircraft can be, however, synchronized to obtain multiple perspectives from a single moment in time. This allows for the use of stereovision to obtain a disparity image, which can be used to estimate ground plane transformation. While multivision systems are possible, the amount of calculations needed to utilize additional cameras is prohibiting, and the quality of estimation is not improving sufficiently due to the construction of aircraft, which is extended in just two axes, unusable one along the hull and the extent of wingspan (up to a point of wings being relatively fixed). Not much is gained by placing additional ones between extreme points on wings, hence the viability approach.

2 Related Work

The problem of ground detection has been addressed in the literature using various approaches. One of the usual contexts is the navigation of an autonomous robot, wherein the colour is often used to detect ground plane on an image, as in case of [2] and [3] where the playing field is detected based on its colour. Similarly, [4] uses the edges, RGB and HSV colour spaces to estimate the surface of a ground plane. During the emergency landing of a fixed-wing UAV, the colour composition of the landing surface is neither known nor easy to distinguish from the colour of obstacles, therefore such approaches prove to be unreliable. [5–7] use optical flow to obtain transformation of a ground plane. Such methods obtain better information than colour-based ones, due to multiple viewpoints. The scenario of a fixed-wing aircraft is connected to high velocities, so the dependency on consecutive frames being similar enough to properly establish correspondences between two images makes this approach not robust enough in face of surfaces with a high degree of similarity, such as grass field. While optical flow is computationally expensive, one might assume that faster methods will be able to handle the high velocity. However, such methods achieve higher performance due to usage of features such as detected corners [8–10], SIFT features [11] or otherwise established feature points [12], which are highly unreliable in the analysed, noisy scenario. Similarly to the proposed approach, some solutions use Inertial Navigation Systems (INS) to obtain constraints for a solution based on an acquired image. However, using high-quality laser-based altimeter as proposed in [13,14] in the context of solar system exploration is often cost-prohibitive for UAVs. Methods that do not rely on altimeter have been developed, such as [15], however, they often assume that the camera is looking down, in parallel to the gravity vector. This does not provide information about obstacles ahead on the trajectory and thus is not useful in this context. Similarly to the proposed approach, stereovision is used in the context of humanoid robots in [16], where the disparity is found and is converted into 3D information and then Hough transformation is used to extract ground plane. While this approach is sound, it is not robust enough to deal with both range of pitch angles and the noisy nature of natural landing plane (i.e. grass, vegetation, snow, mounds), which makes voting a highly unreliable procedure. A dynamic control loop used to update the hypothetical 3D reconstruction based on disparity values as presented in [17] is similar to our approach in terms of constant update. Contrary to the previous approach, it is designed for off-road context, so the nature of the ground plane is similar as in our case. However, the disparity values at an altitude of the land-based vehicle allow for a much higher level of accuracy in 3D reconstruction than in case of a landing aircraft which records the plane from a higher altitude and therefore the task of differentiating between a plane and an obstacle standing on the plane is much more difficult. Chumerin and Van Hulle [18] propose a similar approach, wherein the plane is fitted directly into the disparity data of 11 points using Iteratively Reweighted Least-Squares (IRLS) [19]. In the proposed scenario, however, the amount of noise is big enough for that approach to yield highly unstable results.

3 Proposed Approach

Let us consider the phase space of ground plane transformations with relation to the camera system. One can easily notice, that such a phase space is 4-dimensional, wherein 3 dimensions correspond to orientation in 3D space and the last one corresponds to distance from the camera. One can therefore notice that the orientational parameters are cyclic and that the distance can be limited in case of a real-life scenario, so the viable subspace is limited to a 4-ball of radius corresponding to the distance between ground plane and camera, which is synonymous with the altitude of the camera above the ground plane. While the yaw angle is a property of such transformation, it is irrelevant for ground plane detection, which allows the dimensionality reduction to a 3-ball defined by angles of pitch and roll and a radius of altitude. The assumption that the landing aircraft is flying properly, one can constrain the phase space to half of a 3-ball by reduction of range of possible pitch. The same holds for reduction of a possible roll, which together yields a quarter of a 3-ball, the spherical wedge of angle $\epsilon = \frac{\pi}{2}$ and a radius of maximum altitude at which the landing procedure begins. Let us, therefore, define the parameters required to describe the position in phase space corresponding to the transformation of the ground plane with relation to the camera system as $\theta_{est}, \phi_{est}, d_{est}$, as per Table 1.

Table 1. Parameters.

Parameter	*est*	*ins*	*err*
θ	Estimated pitch angle	INS-provided pitch angle	INS pitch error
ϕ	Estimated roll angle	INS-provided roll angle	INS roll error
d	Estimated altitude	INS-provided altitude	INS altitude error

Using INS, one can obtain information about current altitude d_{ins}. While this information does not take into account the ground plane for the current landing site but is commonly just an altitude above sea level, one can assume that $\Delta d_{ins} = \Delta d_{est}$ holds and therefore can be used to provide constraints and to initialize the true altitude d_{est} estimation. For the first analyzed frame, no such estimation exists, so we assume that the INS altitude error is zero $d_{err} = 0$. The maximum possible altitude d_{max}, and minimum possible altitude, below which no further calculations are necessary, d_{min}, serve as constraints in each frame, as follows in Eq. 1.

$$d_{est} = \min(d_{max}, \max(d_{min}, d_{ins} + d_{err})) \tag{1}$$

Since the main problem in ground plane estimation in case of landing plane is the noise in face of limited calculation time, one must limit the scope of analyzed possible values of disparity. The values themselves are dependant on the height of possible obstacles. The assumption of such a constraint does not mean that taller obstacles will not be detected, only that their parts above the assumed maximal height h_{max} will correspond to the incorrect disparity. While this introduces some error to the disparity values, it limits the number of necessary calculations and removes the outlying data that would affect the ground plane estimation. One can therefore assume minimal and maximal altitudes that need to be taken into account as seen on Eq. 2.

$$d_o = \max(d_{min}, d_{est} - h_{max}) \tag{2}$$

Similarly, INS provides information about current pitch θ_{ins} and roll ϕ_{ins}. While frame of reference is not necessarily the same as ground plane, due to possibility of slopes on the ground, one can assume that the pitch error θ_{err} and roll error ϕ_{err} are small enough to remain useful and therefore both $\Delta\theta_{est} = \Delta\theta_{ins}$ and $\Delta\phi_{est} = \Delta\phi_{phi}$ hold while $\theta_{est} = \theta_{ins} + \theta_{err}$ and $\phi_{est} = \phi_{ins} + \phi_{err}$. Since no information of such error is present at the initial frame, $\theta_{err} = 0$ and $\phi_{err} = 0$ are assumed. INS data can then be used to obtain estimated plane normal (A, B, C), based on plane equation, as in Eq. 3.

$$\begin{aligned} a &= \cos(-(\theta_{ins} + \theta_{err})\sin(\phi_{ins} + \phi_{err}) \\ b &= \cos(-(\theta_{ins} + \theta_{err})\cos(\phi_{ins} + \phi_{err}) \\ c &= \sin(-(\theta_{ins} + \theta_{err}) \end{aligned} \tag{3}$$

The fourth plane parameter, D, the distance between camera and ground plane is constrained between previously calculated altitude values, as in Eq. 4.

$$D \in [d_o, d_{est}] \tag{4}$$

Having bounds for expected plane transformation in three dimensions, one can find out the constraints for the plane in disparity space, similarly to [18], wherein this transformation from plane Π in three-dimensional space (Eq. 5) and plane Δ in disparity space (Eq. 6) is defined.

$$\Pi : aX + bY + cZ + D = 0 \tag{5}$$

$$\Delta : \delta = \alpha x + \beta y + \gamma \tag{6}$$

Such constraints can therefore be expressed as in Eq. 7, wherein B denotes the length of baseline - the distance between cameras and f denotes the focal length of both cameras.

$$\begin{aligned} \alpha &\in [\min(\frac{-aB}{-d_o}, \frac{-aB}{-d_{est}}), \max(\frac{-aB}{-d_o}, \frac{-aB}{-d_{est}})] \\ \beta &\in [\min(\frac{-bB}{-d_o}, \frac{-bB}{-d_{est}}), \max(\frac{-bB}{-d_o}, \frac{-bB}{-d_{est}})] \\ \gamma &\in [\min(\frac{-cBf}{-d_o}, \frac{-cBf}{-d_{est}}), \max(\frac{-cBf}{-d_o}, \frac{-cBf}{-d_{est}})] \end{aligned} \tag{7}$$

Since the range of possible disparity values can still be quite high, to minimize the size of the sliding window, a Gaussian pyramid [21] of L levels is built. Due to the fact, that γ depends on the focal length f, the range of possible values has to be adjusted by $2^{-(L-1)}$ to apply to the new resolution. This leads to the constraint of pixels above the horizon having zero disparity, as in Eq. 8.

$$
\begin{aligned}
v_{nx} &= \begin{cases} \gamma max + \alpha_{max} u_{max}, & \text{if } \alpha_{max} > 0 \\ \gamma max + \alpha_{max} u_{min}, & \text{if } \alpha_{max} \leq 0 \end{cases} \\
v_{nn} &= \begin{cases} \gamma min + \alpha_{min} u_{min}, & \text{if } \alpha_{min} > 0 \\ \gamma min + \alpha_{min} u_{max}, & \text{if } \alpha_{min} \leq 0 \end{cases} \\
v_{\delta 0} &= \begin{cases} \lfloor \frac{v_{nx}}{-\beta_{min}} \rfloor, & \text{if } \beta_{min} > 0 \\ \min(\frac{v_{nx}}{-\beta_{min}}, \frac{v_{nn}}{-\beta_{max}}), & \text{if } \beta_{max} > 0 \wedge \beta_{min} \leq 0 \\ \frac{v_{nn}}{-\beta_{max}}, & \text{otherwise} \end{cases} \\
&\forall u, \forall 0 \leq v \leq v_{\delta 0}, \delta(u, v) = 0
\end{aligned}
\tag{8}
$$

Having calculated minimal and maximal plane parameters, each (u, v) coordinate has disparity $\delta(u, v)$ restricted to the range of values based on Eq. 6 and Eq. 8, i.e. Eq. 9.

$$
\begin{aligned}
&\forall u, \forall 0 \leq v \leq v_{\delta 0}, \delta(u, v) = 0 \\
&\forall u, \forall v_{\delta 0} < v, \alpha_{min} u + \beta_{min} v + \gamma_{min} \leq \delta \leq \alpha_{max} u + \beta_{max} v + \gamma_{max}
\end{aligned}
\tag{9}
$$

Having a set of constraints, we use integral images [20] to calculate the square root of the sum of squares for each pixel of each colour channel within each image window as in Eq. 10.

$$
\begin{aligned}
I_{\Sigma^2}(u, v) = &\sum_{c \in C} \sum_{u' \leq u + r_u, v' \leq v + r_v} I_l(u', v', c)^2 \\
&- \sum_{c \in C} \sum_{u' \leq u - r_u, v' \leq v + r_v} I_l(u', v', c)^2 \\
&- \sum_{c \in C} \sum_{u' \leq u + r_u, v' \leq v - r_v} I_l(u', v', c)^2 \\
&+ \sum_{c \in C} \sum_{u' \leq u - r_u, v' \leq v - r_v} I_l(u', v', c)^2
\end{aligned}
\tag{10}
$$

To calculate disparity within a given window of size $2r_u + 1$ by $2r_v + 1$ one has to calculate some measure of similarity, of which normalized cross-correlation tends to give best results in the context of autonomous UAV landing in unknown, uniformly textured environment. Assuming that left image I_l is analyzed image, and the right image I_r is the source of the template, the normalized cross-correlation is calculated using a fragment of I_r, sliding over I_l. Then, one can define the disparity as in Eq. 11. Notice, that δ can (and should) be examined as a two-dimensional variable in case of not perfect rectification.

$$
\begin{aligned}
u' &\in [u - r_u, u + r_u] + \delta \\
v' &\in [v - r_v, v + r_v] + \delta \\
C(u, v, \delta) &= \frac{\sum\limits_{u',v'} (I_r(u', v') I_l(u + u', v + v'))}{\sqrt{\sum\limits_{u',v'} I_r(u', v')^2 \sum\limits_{u',v'} I_l(u + u', v + v')^2}} \\
\delta(u, v) &= \underset{\alpha_{min} u + \beta_{min} v + \gamma_{min} \leq \delta \leq \alpha_{max} u + \beta_{max} v + \gamma_{max}}{\arg\max} C(u, v, \delta)
\end{aligned}
\tag{11}
$$

However, one can easily notice that the order of sorted values of normalized cross-correlations is not affected by the division by $\sqrt{\sum_{u',v'} I_r(u', v')^2}$, so $\delta(u, v)$ may be calculated without it. Furthermore, $\sqrt{\sum_{u',v'} I_l(u + u', v + v')^2}$ is actually the previously calculated $I_{\Sigma^2}(u + u', v + v')$, so once I_{Σ^2} is calculated in vectorized form it can be reused for multiple templates. Fast Fourier Transform [23] and its inverse can be used to calculate cross-correlation, so the disparity can be calculated as in Eq. 12. In fact, one can omit the usual division by window area $(2r_v + 1)(2r_u + 1)$ in inverse Discrete Fourier Transform as it does not change the order of correlations.

$$
\begin{aligned}
F_r(u, v, m, n, c) &= \sum_{x=-r_u}^{r_u} \sum_{y=-r_v}^{r_v} I_{r,c}(u + x, v + y) e^{-2i\pi(\frac{ny}{2r_v+1} + \frac{mx}{2r_u+1})} \\
F_l(u, v, m, n, c) &= \sum_{x=-r_u}^{r_u} \sum_{y=-r_v}^{r_v} I_{l,c}(u + x, v + y) e^{-2i\pi(\frac{ny}{2r_v+1} + \frac{mx}{2r_u+1})} \\
F_I(u, v, \delta_u, \delta_v) &= \sum_{n=-r_u}^{r_u} \sum_{m=-r_v}^{r_v} (e^{2i\pi(\frac{\delta_v n}{2r_v+1} + \frac{\delta_u m}{2r_u+1})} \sum_{\forall c} (F_r(u, v, m, n, c) F_l(u, v, m, n, c))) \\
C(u, v, \delta_u, \delta_v) &= \frac{F_I(u, v, \delta_u, \delta_v)}{I_{\Sigma^2}(u + \delta_u, v + \delta_v)} \\
\begin{bmatrix} \delta_u(u, v) \\ \delta_v(u, v) \end{bmatrix} &= \underset{\substack{\alpha_{min} u + \beta_{min} v + \gamma_{min} \leq \delta_u \leq \alpha_{max} u + \beta_{max} v + \gamma_{max}; \\ \delta_{v,min} \leq \delta_v \leq \delta_{v,max}}}{\arg\max} C(u, v, \delta_u, \delta_v)
\end{aligned}
\tag{12}
$$

Once disparity δ is established for top pyramid level, it is upscaled and more accurate prediction can be calculated within a range of 1 pixel on lower pyramid level, as the resolution was reduced by previous downscaling. The final, obtained values are then used to fit the disparity plane using the least squares method. The main reason for ground plane estimation in case of fixed-wing unmanned aerial vehicle crash-landing is obstacle detection. However, obstacles are the source of disparities that do not correspond to the ground plane. This means, that in order to best estimate ground plane, one has to limit the fitting to the part of the image that corresponds to the plane itself. In order to do so, we use Random Sample Consensus, or RANSAC [22]. We choose it over Iteratively Reweighted Least Squares [19] because of two main factors. First, in case of analyzed scenario, the main part of the image is the ground plane itself, so there is a high probability of sampling the ground plane, and therefore there is a high probability that obstacle will not be sampled. The high noise in ground plane disparity together with the disparity of obstacles can easily lead to improper reweighting in case of IRLS. Second, since the disparity calculation is the most demanding part of the algorithm, greatly outweighing the fitting itself, one can limit the number of calculations by calculating only the parts of disparity image wherein the random sampling occurs. The vectorized nature of parallel computations leads to calculations in patches rather than single points, so the entirety of disparity patches can be used. Each subsequent iteration of RANSAC might reuse already calculated disparity patches or demand calculation of new ones. RANSAC converges in just a few iterations. If there are no sampled obstacles, as it is most often the case, one can assume convergence after the second iteration due to low change between found planes, affected only by disparity noise. Once the plane is fitted, the estimated ground plane parameters, θ_{est}, ϕ_{est}, d_{est} together with data from INS in the form of θ_{ins}, ϕ_{ins}, d_{ins} can be used to calculate the difference that corresponds to θ_{err}, ϕ_{err}, d_{err}. The next frame is then considered with updated error values, which allows for yet better estimation of ground plane transformation and thus better obstacle detection.

4 Results

To the author knowledge, at the time of publishing, no publicly available dataset consists of synchronized stereo images and INS data acquired from fixed-wing UAV during landing approach (which defines the altitude, orientation and velocity of an aircraft) with visible obstacles and with calibration recording necessary to obtain the calibration data which is useful yet not perfect. Although a 3D simulation could be used, the spatial resolution is insufficient to properly model the apparent noise of ground grass, which hinders the correspondences estimation. That is especially true when one considers the lighting conditions that affect the images as acquired by cameras. While creating such a public dataset is out of the scope of this paper, 119 recordings were taken during the project, including calibration recordings, test recordings and most importantly, recordings that compose of multiple landing approaches using fixed-wing UAV. Following results, as well as images, were obtained based on those recordings. Let us consider sample images from two cameras mounted on the wings of the unmanned aerial vehicle as part of experiments carried out during presented research, illustrated on Fig. 1 and Fig. 2.

Note, that orientation of both cameras is different. While precise alignment of cameras can be done using precise procedures, in practice, such close alignment is seldom achieved. The need to transport, assemble, disassemble and store the platform is a source of changes in relative transformation. However, in order to limit the scale of the problem, the calibration of both internal and external parameters is required. Such calibration is time-consuming, however, once initial calibration is obtained, one can do quick additional calibration right before the flight in order to improve the estimation of external parameters. Having calculated the stereovision systems parameters, rectified images in each frame can be calculated, with results as in Fig. 3 and in Fig. 4.

Fig. 1. Image captured from a left camera mounted on UAV.

Fig. 2. Image captured from a right camera mounted on UAV.

Fig. 3. Rectified image captured from a left camera mounted on UAV.

Fig. 4. Rectified image captured from a right camera mounted on UAV.

While images look rectified, closer inspection will prove, that in some areas, there is a vertical disparity of the scale of at most few pixels. This can be initially attributed to imperfect calibration, however, a closer analysis will prove, that this effect occurs even in case of very precise calibration occurring right before the flight. While the platform is in flight, there are forces pushing onto the wings, which affect the relative position and orientation of cameras slightly. This change in transformation results in imperfect rectification process. The epipolar constraint is commonly used to limit the number of disparity calculations, however in this situation, one needs to allow for the vertical disparity. Use of Gaussian pyramid allows for the lower level of the pyramid to accept single pixel of vertical disparity so that this effect is limited, however in case of very flexible wings or low precision of calibration, one has to allow for δ_v to have values different than 0. Nevertheless, the disparity is calculated on the basis of previous estimation and INS signal, as shown on Fig. 5.

Based on disparity and INS readings, the plane is fitted to the disparity values. Fitted plane corresponds to values of disparity as presented in Fig. 6.

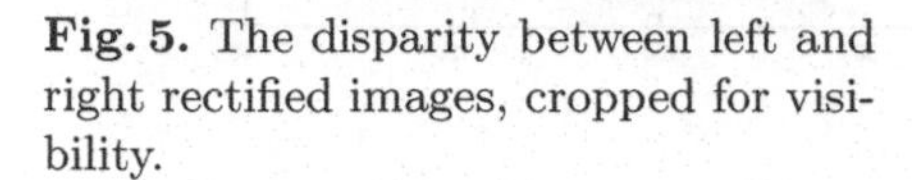

Fig. 5. The disparity between left and right rectified images, cropped for visibility.

Fig. 6. The disparity of plane fitted to the disparity image as presented in Fig. 5.

On the basis of the found plane, disparity corresponding to a ground plane can be removed from disparity image, so that the remaining values correspond to obstacles. In Fig. 7 we present segmented ground plane overlaid on the rectified image. One can easily see segmented out trees. While some mounds are visible, they are not tall enough to be dangerous for the UAV. A human is detected in the middle of the image, allowing for a change of flight path and thus avoiding potentially fatal damage.

Fig. 7. Figure 4 with the overlaid segmented ground plane, cropped for visibility.

With no ground truth present, we follow [18] and present in Fig. 8 the norm of the deviation of the two consecutive estimations of the ground plane normal vectors compared with our implementation of stabilized method as described in [18].

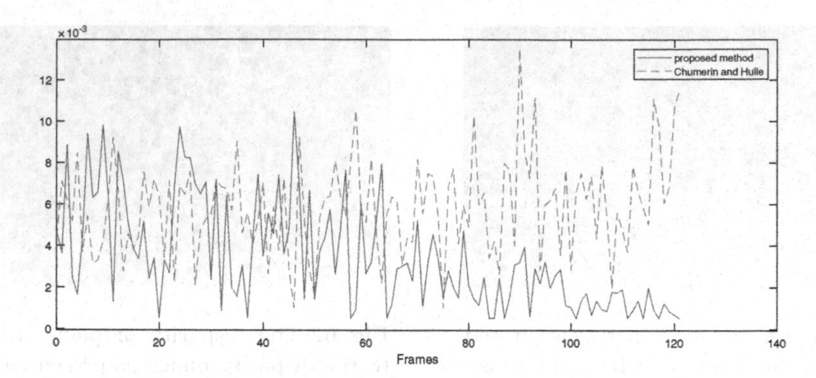

Fig. 8. Norm of the deviation of the two consecutive ground plane normal vectors

5 Conclusions and Future Work

In this work, we introduced a framework for estimating the ground plane accurately from high altitude platform moving at high velocity using stereovision supported by Inertial Navigation System in order to avoid obstacles during the landing procedure. The proposed approach takes into account the development in parallel mobile computational infrastructures. The algorithm is robust against highly noisy nature of disparity estimated on uniformly textured landing sites, such as grassland, and avoids the false altitude estimation based on obstacles on the ground plane. Unlike previous methods, which estimate ground plane from low altitude and the lower angle between camera and ground surface, the presented method can be used in unmanned aerial vehicles in order to land safely in unknown and unprepared conditions without collision with obstacles such as trees or humans. Further research is however needed in order to estimate the usefulness of the proposed approach in scenarios wherein multiple obstacles are visible from low altitudes, such as a robot or autonomous car navigation, where a single plane cannot be effectively fitted to the disparity data and the segmentation is needed.

Acknowledgements. This work has been supported by the National Centre for Research and Development, Poland in the frame of project POIR.01.02.00-00-0009/2015 "System of the autonomous landing of a UAV in unknown terrain conditions on the basis of visual data".

References

1. Rosner, J., et al.: A system for automatic detection of potential landing sites for horizontally landing unmanned aerial vehicles. AIP Conf. Proc. **1978**(1), 110006 (2018) https://doi.org/10.1063/1.5043764

2. Hoffmann, J., Jüngel, M., Lötzsch, M.: A vision based system for goal-directed obstacle avoidance. In: Nardi, D., Riedmiller, M., Sammut, C., Santos-Victor, J. (eds.) RoboCup 2004. LNCS (LNAI), vol. 3276, pp. 418–425. Springer, Heidelberg (2005). https://doi.org/10.1007/978-3-540-32256-6_35
3. Lenser, S., Veloso, M.: Visual sonar fast obstacle avoidance using monocular vision. In: IEEE/RSJ IROS 2003 Proceedings, pp. 886–891. IEEE, Las Vegas (2003). https://doi.org/10.1109/IROS.2003.1250741
4. Lorigon, L.M., Brooks, R.A., Grimson, W.E.L.: Visually-guided obstacle avoidance in unstructured environments. In: Proceedings of the 1997 IEEE/RSJ International Conference on Intelligent Robots and Systems. Innovative Robotics for Real-World Applications. IROS 1997, pp. 373–379. IEEE, Grenoble (1997). https://doi.org/10.1109/IROS.1997.649086
5. Kim, Y., Kim, H.: Layered ground floor detection for vision-based mobile robot navigation. In: IEEE International Conference on Robotics and Automation, 2004, Proceedings. ICRA 2004, vol. 1, pp. 13–18. IEEE, New Orleans (2004). https://doi.org/10.1109/ROBOT.2004.1307122
6. Pęszor, D., Wojciechowska, M., Wojciechowski, K., Szender, M.: Fast moving UAV collision avoidance using optical flow and stereovision. In: Nguyen, N.T., Tojo, S., Nguyen, L.M., Trawiński, B. (eds.) ACIIDS 2017. LNCS (LNAI), Part II, vol. 10192, pp. 572–581. Springer, Cham (2017). https://doi.org/10.1007/978-3-319-54430-4_55
7. Pęszor, D., Paszkuta, M., Wojciechowska, M., Wojciechowski, K.: Optical flow for collision avoidance in autonomous cars. In: Nguyen, N.T., Hoang, D.H., Hong, T.-P., Pham, H., Trawiński, B. (eds.) ACIIDS 2018, Part II. LNCS (LNAI), vol. 10752, pp. 482–491. Springer, Cham (2018). https://doi.org/10.1007/978-3-319-75420-8_46
8. Zhou, J., Li, B.: Robust ground plane detection with normalized homography in monocular sequences from a robot platform. In: 2006 International Conference on Image Processing, pp. 3017–3020. IEEE, Atlanta (2006). https://doi.org/10.1109/ICIP.2006.312972
9. Zhou, J., Li, B.: Homography-based Ground Detection for A Mobile Robot Platform Using a Single Camera. In: Proceedings 2006 IEEE International Conference on Robotics and Automation, 2006, ICRA 2006, pp. 4100–4105. IEEE, Orlando (2006) https://doi.org/10.1109/ROBOT.2006.1642332
10. Zhou, H., Wallace, A.M., Green, P.R.: A multistage filtering technique to detect hazards on the ground plane. Pattern Recogn. Lett. **24**(9–10), 1453–1461 (2003). https://doi.org/10.1016/S0167-8655(02)00385-9
11. Conrad, D., DeSouza, G.N.: Homography-based ground plane detection for mobile robot navigation using a modified EM algorithm. In: 2010 IEEE International Conference on Robotics and Automation, pp. 910–915. IEEE, Anchorage (2010). https://doi.org/10.1109/ROBOT.2010.5509457
12. Yamaguchi, K., Watanabe, A., Naito, T.: Road region estimation using a sequence of monocular images. In: 2008 19th International Conference on Pattern Recognition, pp. 1–4. IEEE, Tampa (2008). https://doi.org/10.1109/ICPR.2008.4761571
13. Roumeliotis, S.I., Johnson, A.E., Montgomery, J.F.: Augmenting inertial navigation with image-based motion estimation. In: Proceedings 2002 IEEE International Conference on Robotics and Automation, vol. 4, pp. 4326–4333. IEEE, Washington (2002). https://doi.org/10.1109/ROBOT.2002.1014441

14. Johnson, E., Mathies, H.: Precise image-based motion estimation for autonomous small body exploration. In: Artificial Intelligence, Robotics and Automation in Space, Proceedings of the Fifth International Symposium, ISAIRAS 1999, vol. 440, pp. 627–634 (2000)
15. Panahandeh, G., Jansson, M.: Vision-aided inertial navigation based on ground plane feature detection. IEEE/ASME Trans. Mech. **19**(4), 1206–1215 (2014). https://doi.org/10.1109/TMECH.2013.2276404
16. Sabe, K., Fukuchi, M., Gutmann, J.-S., Ohashi, T., Kawamoto, K., Yoshigahara, T.: Obstacle avoidance and path planning for humanoid robots using stereo vision. In: Proceedings on International Conference on Robotics and Automation, pp. 592–597. IEEE, New Orleans (2004). https://doi.org/10.1109/ROBOT.2004.1307213
17. Mandelbaum, R., McDowell, L., Bogoni, L., Beich, B., Hansen, M.: Real-time stereo processing, obstacle detection, and terrain estimation from vehicle-mounted stereo cameras. In: Proceedings on 4th IEEE Workshop on Applications of Computer Vision, pp. 288–289. IEEE, Princenton (1998). https://doi.org/10.1109/ACV.1998.732909
18. Chumerin, N., Van Hulle, M.M.: Ground plane estimation based on dense stereo disparity. In: Proceedings of the 5th International Conference on Neural Networks and Artificial Intelligence, pp. 209–213, Minsk, Belarus (2008)
19. Holland, P.W., Welsch, R.E.: Robust regression using iteratively reweighted least-squares. Commun. Stat. Theory Methods **6**(9), 813–827 (1977). https://doi.org/10.1080/03610927708827533
20. Crow, F.: Summed-area tables for texture mapping. In: SIGGRAPH 1984, pp. 207–212. Association for Computing Machinery, Minneapolis (1984). https://doi.org/10.1145/800031.808600
21. Burt, P.J., Adelson, E.H.: The Laplacian pyramid as a compact image code. IEEE Trans. Commun. **31**(4), 671–679 (1983). https://doi.org/10.1109/TCOM.1983.1095851
22. Fischler, M.A., Bolles, R.C.: Random sample consensus: a paradigm for model fitting with applications to image analysis and automated cartography. Commun. ACM **24**(6), 381–395 (1981). https://doi.org/10.1145/358669.358692
23. Frigo, M., Johnson, S.G.: The design and implementation of FFTW3. Proc. IEEE **93**(2), 216–231 (2005). https://doi.org/10.1109/JPROC.2004.840301

UAV On-Board Emergency Safe Landing Spot Detection System Combining Classical and Deep Learning-Based Segmentation Methods

Marcin Paszkuta[1,4](✉), Jakub Rosner[1], Damian Pęszor[1,4], Marcin Szender[3], Marzena Wojciechowska[1], Konrad Wojciechowski[1], and Jerzy Paweł Nowacki[2]

[1] Research and Development Center of Polish-Japanese Academy of Information Technology, al. Legionów 2, 41-902 Bytom, Poland
marcin.paszkuta@polsl.pl, mwojciechowska@pja.edu.pl

[2] Polish-Japanese Academy of Information Technology, Koszykowa 86, 02-008 Warszawa, Poland
nowacki@pjwstk.edu.pl

[3] MSP InnTech sp. z o.o., ul. Poligonowa 1/81, 04-051 Warsaw, Poland
info@uav.com.pl

[4] Department of Computer Graphics, Vision and Digital Systems of Silesian University of Technology, ul. Akademicka 16, 44-100 Gliwice, Poland

Abstract. This article proposes the system designed for automatic detection of emergency landing sites for horizontally landing unmanned aerial vehicles (UAV). The presented solution combines the classic computer vision algorithms and novel segmentation methods based on deep learning techniques using the U-Net inspired network architecture. The presented system uses a single nadir camera mounted on a UAV and the energy-efficient compute module capable of highly-parallelized calculations.

Keywords: Unmanned aerial vehicle · UAV · Landing sites detection · Convolutional neural network · Semantic segmentation · U-Net · 3D simulation · Unreal engine · AirSim

1 Introduction

Currently, UAV platforms are used in various industries and entertainment. Most of the platforms available on the market can fly without direct eye contact between the pilot and the drone (BVLOS) - the communication distance even in the simplest consumer models is counted in kilometers. Often the role of the pilot is limited only to flight planning, the rest is performed by the autopilot module. Immediate determination of a safe landing site is critical to minimize emergency landing losses. Our research focuses on fail-safe mode improvements

N. T. Nguyen et al. (Eds.): ACIIDS 2021, LNAI 12672, pp. 467–478, 2021.
https://doi.org/10.1007/978-3-030-73280-6_37

for horizontally landing unmanned aerial vehicles. In this article, we present the novel system that is a fusion between the traditional approach presented in [8] and the solution based on deep neural networks. The good result of landing site detection is offered the LiDAR scanner data analyses in combination with Convolutional Neural Networks like system described by Maturana and Scherer in [6]. The additional sensors like laser scanners, depth cameras, or radar significantly increase weight as well as raised cost. Unlike the solution presented by Hinzmann in [4] where the monocular camera is used for 3D depth estimation and the landing sites detection methods are similar like in LiDARs based system, we decide to analyze the 2D image. One of the most problematic areas to segmentation were water reservoirs that change the colors and texture during different weather conditions, sun position, and as well as camera angles. The sea and land segmentation method based on U-Net architecture were presented by Li et al. in [5]. Research results presented by Bhatnagar et al. in [1] show that U-Net architecture could be successfully used for drone image segmentation in terms of ecotopes classification. The U-Net in combination with ResNet-50 achieved one of the best results.

Past experience has shown that developing and testing vision algorithms for autonomous platforms in the real world is an expensive and time-consuming process. Moreover, in order to obtain satisfactory results of the deep learning-based segmentation we needed to collect a large amount of annotated training data in all popular seasons and weather conditions. We decided to use one of the most popular game engines that offer physically and visually realistic environments that could be used for computer vision algorithms testing and as well as training datasets creation. Nowadays the open-source simulation systems such as AirSim [9] or CARLA [2] that based on the 3D engines provides the state-of-the-art rendering quality could be successfully used for validation and testing vision algorithms. Moreover, they could be used as a high-quality deep learning data source due to multiple sensors simulation and various data types generation like segmentation or depth view. The example screenshot from the AirSim simulator was presented in Fig. 1. The main view is from the FPV camera, below was placed the additional insets with looking down RGB camera, object segmentation view, and depth camera view.

2 System Architecture

The assist system for emergency landing during an emergency situation is based on the energy-efficient compute module integrating the multi-core ARM architecture CPU and the powerful Graphics Processing Unit (GPU) optimized for parallel computing and deep learning. More and more often, the GPU-powered on-board image processing from the single RGB camera could take over expensive and complex scanning sensors like LiDAR. Parallel computing performed with the use of lightweight and efficient embedded modules that integrate multicore CPUs and GPU modules enables the use of a group of algorithms that until recently were beyond the scope of application in small flying platforms.

The solution consists of the customized industrial-grade autopilot module, highly accurate inertial navigation solution, compute module, and RGB camera.

Fig. 1. A snapshot from the simulation shows an FPV view from the aerial vehicle. Below the insets with looking down RGB camera, object segmentation and depth view generate in real-time.

The part described in the publication concerns the detection stage, the landing is performed by the autopilot module.

The presented algorithm version was optimized for the NVIDIA Jetson TX2 module that integrates hex-core ARMv8 64-bit CPU with 256-core NVIDIA Pascal GPU. One limitation was an 8GB build-in LPDDR4 operating memory that was shared between GPU and CPU. The used version of a module offered 32GB fast MMC memory, which was sufficient for normal operation but was extended by a fast SSD hard drive during data acquisition and result evaluation.

The system uses a single camera with a global shutter. The 2/3" CMOS sensor was of 8.5 mm by 7.1 mm within a pixel size of 3.45 µm. One of the problems was choosing the right lens, after some experiments, we decided for a fixed focal length optic 8 mm that offered low distortion and moderate vignetting effect. We would like to emphasize that the selection of optics parameters has been optimized for the limitations of the test platform used. The optimal flight altitude for our test platform was at the level of 300 m above the ground. We assumed 7 pixels per meter as a minimum information level resulting in 1 pixel corresponds approximately to 15 cm of terrain. The sensor combined with 8 mm lens offered the horizontal field of view at the level of 56°. Flying parallel to the ground at the assumed maximum altitude, the visibility area will be around 85 000 m^2 (318.8 m 266.3 m).

One of the most important issues for autonomous flight is the precise determination of the platform coordinates in the global frame and as well as its altitudes and orientations. While determining the position using a Global Navigation Satellite System (GNSS) receiver is not a problem, determining the altitudes is associated with certain approximations and limitations. We could estimate the height above means sea level but without the digital elevation model (DEM), we can't calculate the height above ground level. We used the Inertial Navigation Systems (INS) that integrates the GPS receiver, temperature calibrated industrial-grade Inertial Measurement Unit (IMU), and precise pressure sensor. The INS module shared data for the vision system and autopilot's Attitude and Heading Reference System (AHRS). GPS-Aided INS used the data fusion algorithm based on the extended Kalman filter. The solution combines a high update rate of local inertial measurements with global low update rate GNSS positioning. As a result, we get the high update rate UAVs orientation and position data in the global coordinate system.

We couldn't use the native synchronization mechanism, because we sharing a communication interface between autopilot. The update rate of the camera is 3 times lower than the INS data, we took the data with the nearest timestamp corresponding to the image acquisition time. The used camera system could capture images in native resolution at 20 frame/s maximum. The computation time was fickle and was partially dependent on detected content.

Fig. 2. The first stage of unsafe areas detection - potentially unsafe regions marked in white. The RGB input image b) and binarized gradient maps for gray-level a) and saturation c) analysis. (Color figure online)

3 Obtaining Map of Areas Suitable for Landing

The algorithm starts with performing bilateral filtering on a current source video frame to smooth out surfaces while trying to preserve edges. Such an image is then converted to a 3-channel HSV image for color analysis, however, only hue and saturation channels are used in further processing along with a gray-level image obtained from an equally-weighted source RGB image. We calculate gray-level and saturation domains based gradient maps based on horizontal and vertical Sharr gradients. In order to filter out relatively small frequencies the values are thresholded using $\Psi_{grey-level}$ and $\Psi_{saturation}$ parameters. A Low-frequency Gaussian filter was applied to filtering out richly-structured areas that could be potentially unsuitable for UAV landing. We applied parameter $\sigma = 2$ with different radius $r_{saturation} = 5$ and $r_{grey-level} = 7$. Figure 2 shows gradient maps in gray-level and saturation domains.

In the next step, we focus on hue and saturation components in HSV space. Due to high dependency on lighting conditions and high fluctuations the value of the HSV component is ignored on this analysis level. The input image is divided into relatively large overlapping blocks using radius $r_{block} = 10$ px. After that, we calculate histograms for each block with 72 and 50 histogram bins value for Hue and Saturation respectively. In each histogram, we looking for the most populated bin and analyze its value. If the value is greater than $\Theta_{hue} = 3\%$ of all pixels in Hue histogram and $\Theta_{saturation} = 11\%$ for Saturation, we could mark the corresponding pixel as area safe for landing. Note, that using a much smaller block size would result in very underpopulated histograms unsuitable for further analysis, while significantly bigger blocks would generalize the data too much.

The next steps depend on aircraft limitations, we need to filter out unwanted areas unsuitable for landing, e.g. water. At this level, the classic computer vision algorithms have been replaced by deep neural networks. In the previous implementation, [8], based on a rich set of data obtained under various weather conditions and during different seasons with the target aircraft's mounted video recording system we distinguished ranges of hue and saturation for the unsafe areas. Those areas include water since the plane can't land on it, concrete roads, parking, football fields, etc. as landing there might lead to life-threatening situations and other areas specified by professional pilots.

For our target platform, those areas include water, since the airplane's avionic could be damaged and access to the platform may be difficult. We also excluded areas such as concrete roads, parking, football fields, etc. as landing there might lead to life-threatening situations and other areas specified by professional pilots as potentially risky.

The typical use case of convolutional networks is on classification tasks, where the network response to an image is a single class label. However, for our needs, a class label should be supposed to each analyzed pixel. We decide to use the architecture similar to the U-net network presented by Ronneberger et al. [7]. Moreover, the proper datasets with thousands of training tagged images were beyond our reach. In the training and validation process, we decided to mix the real and simulated data obtained using the virtual 3D environments.

As a deep learning framework, we used the latest TensorFlow that was natively supported by compute module Jetson TX2 with GPU compute capability.

Figure 3 presents the resulting map incorporating all steps described above like, gradient analysis, HS subspace filtration, and color homogeneousness analysis. As results, we get the final map of safe areas as shown in Fig. 3b. Note, that this merging is performed using the OR operator i.e. any pixel is considered safe if and only if it is considered safe after each analysis step.

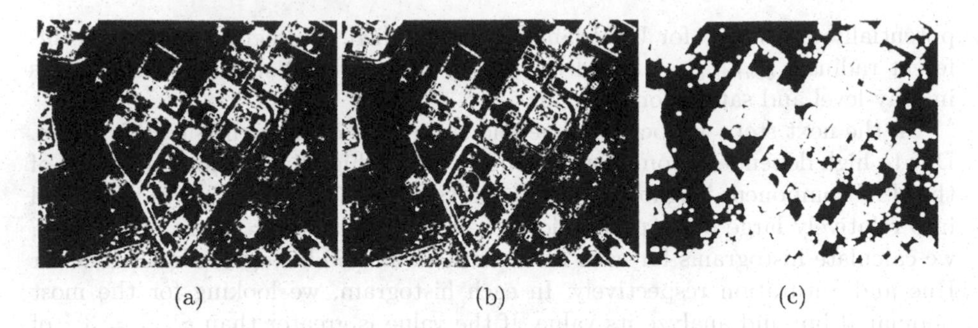

Fig. 3. a) Colour homogeneousness analysis and HS subspace filtration results. b) Map of areas suitable for landing. c) Safe landing area map dilated with an ellipse-shaped kernel with appropriate dimensions. White areas in are not suitable for landing.

4 Landing Site Detection

As a result of the final segmentation, we result in a map that contains two classes, the suitable and unsuitable for landing. For platforms capable of vertical landing such as multirotor we try to find the biggest safe area shaped to circle meeting the minimum radius requirements dependent on platform type and the adopted safety margin. The center of the circle could be reported as a potential landing site and added to the database at the coordinate conversion step. For fixed-wing platforms, landing is one of the most difficult stages of flight. The landing site should be a flat area with an unobstructed landing approach corridor. The minimal width, length, and surface type of airfield depend on the used platform type. Typically the minimum airstrip width is the plane width increased by a safe margin that depends on weather conditions (wind speed and direction). The total length of the analyzed landing site should include the obstacle-free corridor. We estimate the length using Eq. 1, where L_{land} is the maximum length needed by the plane to decelerate and stop after touching the ground, α_{desc} is the plane descent angle, and O_{height} is the assumed maximum obstacle height (see Fig. 3c). To minimize the weight and increase the applicability of the system we decided to use a camera mounted rigidly to the fuselage of the plane by a bracket with a vibration damper. The orientation and position of the camera are reported by an industrial-grade IMU sensor with integrated GNSS. The system uses only one inertial unit, the data from the sensor are shared between all subsystems. The photos whose orientation towards the ground differs from the adopted range are rejected. Based on intrinsic camera calibration parameters and extrinsic parameters estimated like height above ground and INS orientation we obtained the means pixel-per-meter ratios that we used to approximate landing site dimensions in pixels. Since the plane's altitude in relation to the ground could not be precisely defined using only the GNSS receiver we decide to add 5% safety margins to those dimensions. This approximation is sufficient for initial landing site parameter estimation.

$$L_{total} = L_{land} + \frac{O_{height}}{\tan \alpha_{desc}} \tag{1}$$

We decide to use morphology operation to enforce the safety margins. Dilation with an ellipse-shaped kernel with the a and b parameters equal to half of the airstrip widths in pixels was used. The ellipse eccentricity equals 0 ($a = b$) only if the pixel-per-meter values in both horizontal and vertical dimensions are identical. The example results of morphologic operation applied for an initial segmentation of areas suitable for landing could be seen in Fig. 3c). We can subtract the smaller dilate kernel dimension size from L_{total} and still be on the safe side. The classic segmentation approach finishes at this stage. Any found a straight line through the detected area should have appropriate safe width.

The next step is to find the longest straight lines that can be inscribed in the safe area. We trace the line in multiple directions using angles in the range of $\alpha = [0°, 180°)$ from each pixel marked as safe with unsafe neighbors. The angular search resolution is one of the algorithm parameters, interval $10°$ brings satisfying results. Higher resolution does not bring any noticeable improvement and could negatively affect performance. For the path tracing algorithm, we based on the angular interval to compute for each search angle the horizontal and vertical trace steps u_x and u_y, wherein $(u_x)^2 + (u_y)^2 = 1$. Start from the edge pixel we follow the line using precomputed steps and finding the nearest neighbors of our current position. If the pixel is marked as unsafe we have to check if the line is longer than L_{total}. If so, we put the line in a temporary set. We repeat the steps until we process the entire image. As result, we obtain the pixel coordinates of lines meeting the minimum length criterion. One of the landing site's measurable quality parameters its length. We sort descending the temporary set by length value. Going from longest to shortest record, we copy n initial lines to the final set. We skip landing sites that have the same angle and their beginning and end pixel coordinates are relatively close to any site that is already in the final set.

The algorithm parameters mentioned throughout this article were placed in the Table 1. The example algorithm results were shown in Fig. 4, the detected safe landing places marked as violet lines.

Table 1. The main parameters used to obtain the results described in this article.

Threshold	Value	Parameter	Value	Parameter	Value	Parameter	Value
Initial $\Psi_{gray-level}$	250	Θ_{hue}	3%	$\Theta_{saturation}$	11%	Plane descent angle	10%
Final $\Psi_{gray-level}$	180	r_{block}	10 px	Gauss kernel σ	2.0	Deceleration length	15 m
Initial $\Psi_{saturation}$	100	$r_{gray-level}$	7 px	Assumed height H	20 m	Distance enforced	7 m
Final $\Psi_{saturation}$	50	$r_{saturation}$	5 px	Landing site width	5 m	between landing sites	

(a) (b) (c)

Fig. 4. Example results obtained during the test flight. Safe airfields detected by the system were marked as violet lines.

5 From Image Pixel Coordinates to the Geographic Coordinate System

As a result of the image segmentation, we obtain pixel coordinates of potential safe landing sites defined as the parallel line that runs through the middle of it. The autopilot navigation system is based on the popular WGS 84 geographical coordinate system. Coordinate conversion is a multi-step process that could be realized based on intrinsic and extrinsic camera calibration parameters. We used the standard pinhole camera calibration method based on the checkerboard pattern detection described in OpenCV documentation. One of the main problems during coordinate conversion is the inability to determine the height above ground level based only on GNSS data.

In the first step, we remove distortion caused by imperfections in the lens using the distortion correction as in [3] and apply Eq. 2, wherein (x_d, y_d) is the point on the distorted image, (x_u, y_u) is the point on the undistorted image, (c_x, c_y) is the center of distortion and s_x is distortion coefficient.

$$r_d = \sqrt{(\frac{x_d - c_x}{s_x})^2 + (y_d - c_y)^2} \qquad y_u = \begin{cases} x_u = x_d + (x_d - c_x)k_1 r_d^2 \\ y_u = y_d + (y_d - c_y)k_1 r_d^2 \end{cases} \tag{2}$$

The goal is to find values of (c_x, c_y), k_1 and s_x such that the points indicated to be colinear were actually such after undistortion.

As we mention in the previous chapter the GNSS receiver does not determine the height above the ground but could return the altitude that is relative to the simple ellipsoid model of Earth defined by the WGS84 system. This altitude could be different relative to the real height above Mean Sea Level (MSL) even by 100 m. The EGM96 model can be used to define altitude relative to Mean Sea Level (MSL), which is an equipotential surface of the EGM96 gravitational model. However, this does not solve the problem of determining the height above the ground. We used the Digital Elevation Model (DEM) which allowed us to estimate the height of the terrain. The corrections of translation and rotation of the INS sensor relative to the camera mounting point were calculated during the calibration process. The transformation matrix was used to construct the extrinsic matrix of the camera based on the actual GNSS position and orientation. The intrinsic matrix (the focal length, image skew, and the center point) is fetched once because these parameters of the camera are fixed. The longitude, latitude, and altitude were computed by multiplying the point coordinates on the image by the multiplication of intrinsic and extrinsic matrix. From each analyzed image we selecting n best landing spots based on selected criteria. The designated position of the landing sites in the WGS-84 coordinates system is saved in the database. For each airstrip, we save information about the position, length, and estimated altitude. The results as shown in Fig. 5 could confirm the system's usefulness in the real environment.

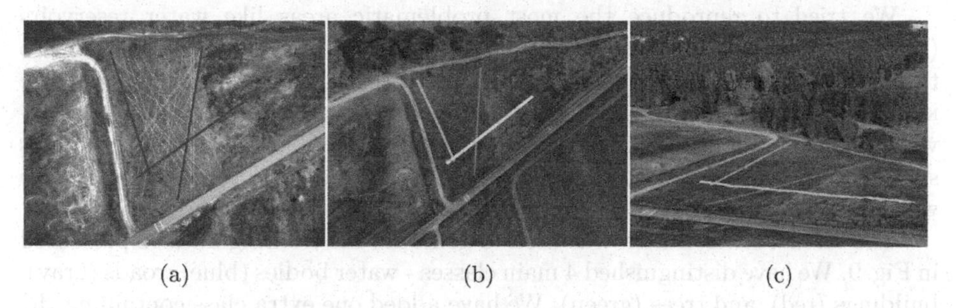

(a) (b) (c)

Fig. 5. The example results: a) view from NADIR camera with all detected landing places - yellow lines and top best places saved to database - red lines; b) google map view with detected landing spots - yellow and orange lines; c) google Earth 3D view with the results. (Color figuer online)

6 Testing Environments

In the first stage of project realization, we tried to collect a database that could reflect the target environment. The database consists of images with additional data from the INS system like GNSS position and camera orientation. Even though the data was collected over the year during various weather conditions and seasons then the database could never fully fill our needs. As a solution, we decided to use the popular and available for free Unreal Engine in combination with the Microsoft AirSim plugin [9]. This composition could be successfully used for AI research to experiment with deep learning, computer vision, and as well as reinforcement learning algorithms. We had to create our own scenes based on free elements available from UE Marketplace, because ones available as a binary release with the AirSim, did not fulfill our needs. The most difficult challenge was proper water simulation. Translucent materials like water or glass were invisible to the segmentation and depth camera view. The solution was to create a duplicate of the problematic object with a custom depth implementation and base material that was disabled from the main rendering process. The example materials represent various water types used for generating synthetic data was presented in Fig. 6.

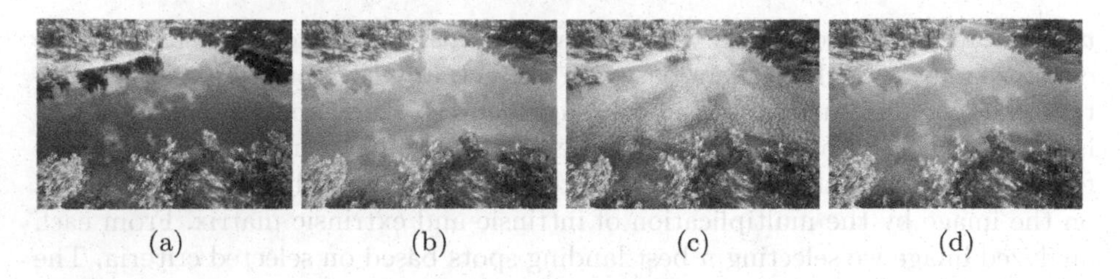

(a) (b) (c) (d)

Fig. 6. Examples of variate water types that was applied in the simulator for synthetic data generation.

We tried to reproduce the most problematic areas like water reservoirs (Fig. 8), which can be incorrectly classified as a proper landing place due to texture or color analysis. Our test environment covers an area of over 1.5 square kilometers and consists of various static elements such as vegetation, water reservoirs, roads, buildings, tennis courts, football fields, and as well dynamic objects such as vehicles and pedestrians. The example screens from the synthetic scene were presented in Fig. 7.

The virtual environment was used to generate the training data as presented in Fig. 9. We have distinguished 4 main classes - water bodies (blue), roads (gray), buildings (red), and trees (green). We have added one extra class containing the sky view because in some situations during dynamic maneuvers the excessive roll or pitch of the airplane could lead that sky will be visible at the level of the NADIR camera.

(a) (b)

Fig. 7. Simulation environment: a) terrain map view; b) an exemplary view from the FPV camera placed on the drone.

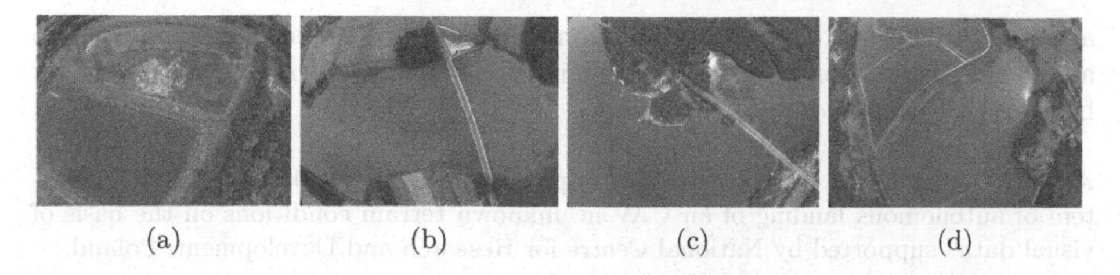
(a) (b) (c) (d)

Fig. 8. The example datasets contain water bodies that were used for training deep learning and result validation.

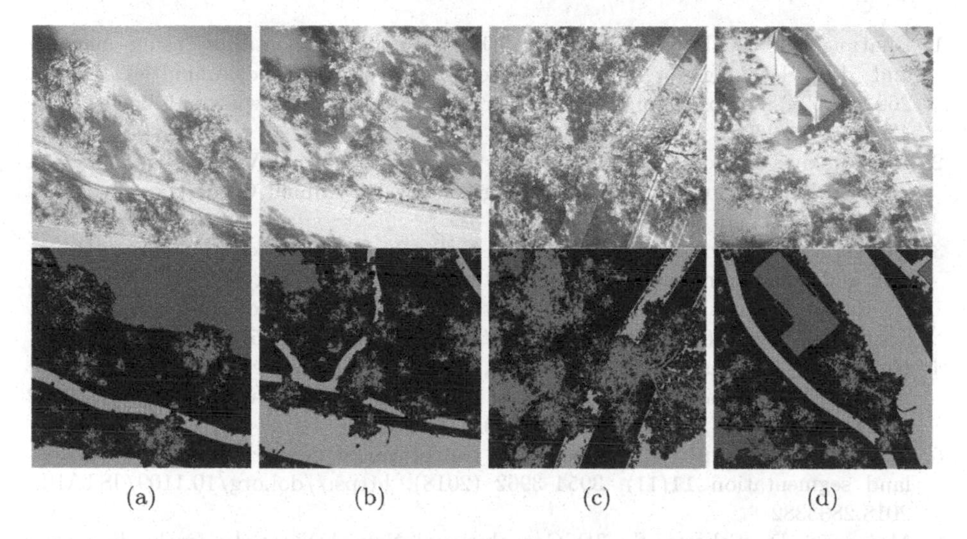
(a) (b) (c) (d)

Fig. 9. The example synthetic datasets with segmentation masks that were used for training deep learning and result validation. (Color figure online)

7 Conclusion and Future Work

In this paper, a system for finding safe landing sites for UAVs in an emergency situation has been presented. The necessary steps involved in distinguishing safe and unsafe locations and the stage of transition from the image to the real-world coordinates have been presented. The presented system was designed specifically for fixed-wing aircraft, but after selecting the appropriate landing site parameters, it can be used with any type of flying platforms. The main algorithm is consists of segmentation steps resulting in a map of areas suitable for landing and detection step during which the longest straight strips (landing sites) that meet certain criteria are returned. Obtained results show that combining traditional computer vision algorithms with deep learning-based semantic segmentation has reduced the risk of misclassification for problematic areas like water reservoirs. The solution was designed and tested with a custom industrial-grade autopilot and a highly accurate inertial navigation solution, but the authors would like to adopt the system for popular commercial open-source autopilots like Pixhawk based on PX4 or Ardupilot firmware.

Acknowledgments. This work is part of project POIR. 01.02.00-00-0009/2015 "System of autonomous landing of an UAV in unknown terrain conditions on the basis of visual data" supported by National Centre for Research and Development, Poland.

References

1. Bhatnagar, S., Gill, L., Ghosh, B.: Drone image segmentation using machine and deep learning for mapping raised bog vegetation communities **12**(16), 2602 (2020). https://doi.org/10.3390/rs12162602, https://www.mdpi.com/2072-4292/12/16/2602
2. Dosovitskiy, A., Ros, G., Codevilla, F., Lopez, A., Koltun, V.: CARLA: an open urban driving simulator. In: Proceedings of the 1st Annual Conference on Robot Learning, pp. 1–16 (2017)
3. Gudyś, A., Wereszczyński, K., Segen, J., Kulbacki, M., Drabik, A.: Camera calibration and navigation in networks of rotating cameras. In: Nguyen, N.T., Trawiński, B., Kosala, R. (eds.) ACIIDS 2015, Part II. LNCS (LNAI), vol. 9012, pp. 237–247. Springer, Cham (2015). https://doi.org/10.1007/978-3-319-15705-4_23
4. Hinzmann, T., Stastny, T., Cadena, C., Siegwart, R., Gilitschenski, I.: Free LSD: prior-free visual landing site detection for autonomous planes **3**(3), 2545–2552 (2018). https://doi.org/10.1109/LRA.2018.2809962
5. Li, R., et al.: DeepUNet: a deep fully convolutional network for pixel-level sea-land segmentation **11**(11), 3954–3962 (2018). https://doi.org/10.1109/JSTARS.2018.2833382
6. Maturana, D., Scherer, S.: 3D Convolutional Neural Networks for landing zone detection from LiDAR. In: 2015 IEEE International Conference on Robotics and Automation (ICRA), pp. 3471–3478 (2015). https://doi.org/10.1109/ICRA.2015.7139679
7. Ronneberger, O., Fischer, P., Brox, T.: U-Net: convolutional networks for biomedical image segmentation. In: Navab, N., Hornegger, J., Wells, W.M., Frangi, A.F. (eds.) MICCAI 2015, Part III. LNCS, vol. 9351, pp. 234–241. Springer, Cham (2015). https://doi.org/10.1007/978-3-319-24574-4_28
8. Rosner, J., et al.: A system for automatic detection of potential landing sites for horizontally landing unmanned aerial vehicles **1978**(1), 110006 (2018). https://doi.org/10.1063/1.5043764
9. Shah, S., Dey, D., Lovett, C., Kapoor, A.: AirSim: high-fidelity visual and physical simulation for autonomous vehicles. In: Field and Service Robotics (2017). https://arxiv.org/abs/1705.05065

Wasserstein Generative Adversarial Networks for Realistic Traffic Sign Image Generation

Christine Dewi[1,2], Rung-Ching Chen[1,2](✉), and Yan-Ting Liu[1]

[1] Department of Information Management, Chaoyang University of Technology, Taichung, Taiwan, R.O.C.
{s10714904,crching,s10414141}@cyut.edu.tw

[2] Faculty of Information Technology, Satya Wacana Christian University, Salatiga, Indonesia

Abstract. Recently, Convolutional neural networks (CNN) with properly annotated training data and results will obtain the best traffic sign detection (TSD) and traffic sign recognition (TSR). The efficiency of the whole system depends on the data collection, based on neural networks. The traffic sign datasets in most countries around the world are therefore difficult to identify because they are so different from one to others. To solve this issue, we need to generate a synthetic image to enlarge our dataset. We employ Wasserstein generative adversarial networks (Wasserstein GAN, WGAN) to synthesize realistic and various added training images to supply the data deficiency in the original image distribution. This research explores primarily how the WGAN images with different parameters are generated in terms of consistency. For training, we use a real image with a different number and scale. Moreover, the Image quality was measured using the Structural Similarity Index (SSIM) and the Mean Square Error (MSE). The SSIM values between images and their respective actual images were calculated in our work. The images generated exhibit high similarity to the original image when using more training images. Our experiment results find the most leading SSIM values reached when using 200 total images as input, images size 32 × 32, and epoch 2000.

Keywords: WGAN · GAN · Synthetic images · TSR · TSD

1 Introduction

The identification of road signals plays a significant role in driver assistance and autonomous driving technologies. However, this job is not easy for a computer since the visual appearance of traffic sign images differs widely. Hence, traffic sign detection (TSD) [1] and traffic sign recognition (TSR) technology have achieved great success in Convolutional Neural Networks (CNN).Training on the neural network requires a lot of data, so a good model can be trained easily and used for identification and recognition of signals. It is not difficult to design a classification system for CNN traffic signs if a sufficient number of labeled training data are available. Researchers have studied and explored traffic sign recognition systems widely in recent years. Many datasets for traffic

N. T. Nguyen et al. (Eds.): ACIIDS 2021, LNAI 12672, pp. 479–493, 2021.
https://doi.org/10.1007/978-3-030-73280-6_38

signs were also collected such as the German Traffic Sign Dataset (GTSRB) Collection [2], the Chinese Traffic Sign Database (TSRD), and the Tsinghua-Tencent 100 K (TT100 K) [3]. The researcher typically uses the accessible data collection to perform the experiment and to capture the traffic sign picture on his own. However, a huge amount of high-quality image data is not easy to access. Besides, several countries have different traffic sign types, ensuring that different data for different countries must be obtained and marked. For the purposes of this paper, we will focus on the Taiwan traffic sign image, as we can see in Fig. 1. Moreover, Fig. 1 shows an example of a unique Taiwan traffic sign image [4, 5]. Our motivation comes from the lack of such a database, image, and research system for Taiwan traffic sign recognition. Also, our work aims to create synthetic images where reference images are difficult to obtain.

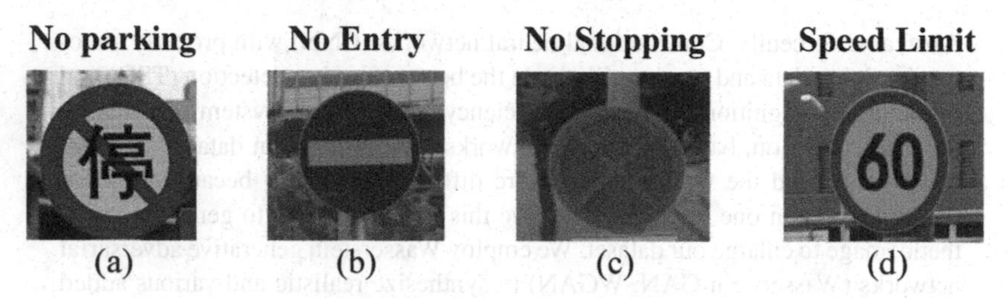

Fig. 1. Original Taiwan traffic sign image.

Generating images has recently obtained impressive results in the Generative Adversarial Networks (GAN) [6, 7]. The original GAN generate images from random vectors. On another hand, the conditional GAN [6], image-to-image GAN [8, 9] have been introduced to enlarge the original GANs to generate images under some specific circumstances. The synthetic image is an important issue in computer vision [10]. Traffic sign images synthesized from standard templates have been widely used for training machine learning-based classification algorithms [11–13] to acquire more training data with diversity and low cost. Wasserstein generative adversarial networks (Wasserstein GAN, WGAN) is used to generate complicated images in this work. The synthetic image is a resolution for holding a small amount of data. WGAN has accomplished outstanding results in image reproduction. Our experiment favors using synthetic images by WGAN to collect image data because it does not depend on a large number of datasets for training.

This works main contributions can be summarized as follows: First, in this research work, we implement WGAN to generate realistic traffic signs image. Second, WGAN analysis efficiency was performed in this work to generate a simulated image with various parameters like numbers and scale. The remainder of the paper is structured accordingly. The related works are introduced in Sect. 2. Section 3 discusses the proposed method. Section 4 includes descriptions of the tests, including experimental configuration and discussion of results. Section 5 outlines the conclusion and future work research.

2 Related Work

2.1 Generative Adversarial Networks (GAN) and (WGAN)

Generative Adversarial Network (GAN) was proposed in 2014 by Goodfellow [10]. GAN consists of two important elements, which are the discriminator D and the generator G and it is learned through competition between D and G. Furthermore, the role of D is to differentiate real images from generated images and the role of G is to create images that D cannot distinguish from the real images. The goal of GAN shown in Eq. (1) [14].

$$minmax(D, G) = E_{x-pdata(x)}\left[logD(x)\right] + E_{x-p(x)}\left[\log(1 - D(G(z)))\right] \tag{1}$$

In the past few years, the rapid development of convolutional neural networks (CNN) has achieved great success in object detection and classification. Researchers began to focus on solving image generation tasks [9, 15]. Under the experiments with GAN [12], which have been known to be unstable to train, the generators often produce meaningless outputs. The research trying to figure out and visualize what GAN learn as well as the intermediate representation of multilayer GAN is very limited. Wasserstein generative adversarial networks (Wasserstein GANs, WGAN) improve the performance of GAN significantly by imposing the Lipschitz constraints on the critic, which is implemented by weight clipping [16]. To make stable training process, WGAN [17, 18] is proposed. It has a new loss function derived from the Wasserstein distance which is a measure of the distance between two probability distributions. Wasserstein distance is expressed as Eq. (2).

$$W\left(P_r, P_g\right) = inf_{\gamma \in \Pi(P_r, P_g)} E_{(x,y)\sim\gamma}\left[||x - y||\right] \tag{2}$$

where $\Pi\left(P_r, P_g\right)$ denotes the set of all joint distributions $\gamma(x, y)$. Then, $\gamma(x, y)$ represents distance for transform the distribution P_r into the distribution P_g. The loss function of WGAN can be expressed by the Kantorovich Rubinstein duality in Eq. (3) [19].

$$\min \max E_{x\sim P_r}[D(x)] - E_{\tilde{x}\sim P_g}\left[D(G(\tilde{x}))\right] \tag{3}$$

where D is set of Lipschitz functions and discriminator D estimates distance between joint distribution samples. To enforce the Lipschitz constraint in WGAN, Gradient Penalty [20] is added and it can be expressed as follows.

$$\lambda E_{\tilde{x}\sim P_{\tilde{x}}}\left[(\|\nabla_{\tilde{x}} D(\tilde{x})\|_2 - 1)^2\right] \tag{4}$$

Furthermore, Gradient penalty borders gradient weights to the area [−c, c], where c is the threshold, for preventing gradient vanishing and exploding the problem. The architecture of WGAN is shown in Fig. 2, where z represents random noise, G represents generator, G(z) represents samples generated by generator, C represents discriminator, C* represents approximate expression of Wasserstein-1 distance.

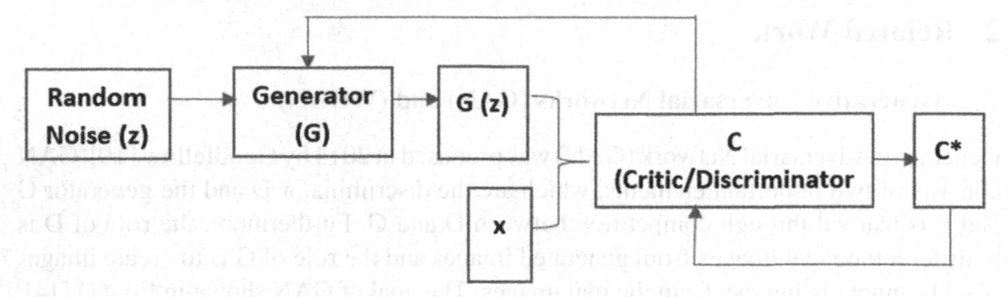

Fig. 2. Schematic of WGAN.

2.2 Performance Evaluation Method

The Structural Similarity (SSIM) index is a method for estimating the relationship between the two images. The SSIM index can be observed as a quality standard of one of the images being analyzed, presented the other image is observed as of excellent quality. It is an upgraded version of the universal image quality index suggested previously [21, 22]. The SSIM formula is based on three example measurements within the samples of x and y: luminance (l), contrast (c), and structure (s). The functions are shown in the formula (5–8).

$$l(x,y) = \frac{(2\mu_x\mu_y + C_1)}{\left(\mu_x^2 + \mu_y^2 + C_1\right)} \tag{5}$$

$$c(x,y) = \frac{(2\sigma_{xy} + C_2)}{(\sigma_x + \sigma_y + C_2)} \tag{6}$$

$$s(x,y) = \frac{(\sigma_{xy} + C_3)}{(\sigma_x\sigma_y + C_3)} \tag{7}$$

$$SSIM(x,y) = \frac{(2\mu_x\mu_y + C_1)(2\sigma_{xy} + C_2)}{\left(\mu_x^2 + \mu_y^2 + C_1\right)\left(\sigma_x^2 + \sigma_y^2 + C_2\right)} \tag{8}$$

Where μ_x is the average of *x*, μ_y is the average of *y*, σ_x^2 is the variance of x, σ_y^2 is the variance of y, and σ_{xy} is the covariance of *x* and *y*. The SSIM input is two images, one is an unpacked image and the other is a distorted image. The structural similarity between the two can be used as a measure of image quality. Contrary to Peak Signal-to-Noise Ratio (PSNR), SSIM is focused on visible structures in the image. Although PSNR is no longer an accurate image deterioration indicator, it is available in the SSIM module as an alternative calculation. The relationship between the SSIM index and more traditional quality metrics may be illustrated geometrically in a vector space of image components. These image components can be either pixel intensities or other extracted features such as transformed linear coefficients [23].

The Mean Squared Error (MSE) is a measure of the quality of an estimator. MSE value is always non-negative, and values closer to zero are better. The smaller values of MSE indicate a more satisfactory result [24, 25]. The equation of this as follows [26–28].

$$MSE = \frac{\sum_{i=1}^{n}(P_i - Q_i)^2}{n} \tag{9}$$

3 Methodology

3.1 WGAN Architecture

In this section, we will describe our proposed method to generate traffic sign images using WGAN. Figure 3 shows our system architecture to generate synthetic Taiwan traffic sign images using WGAN. Further, WGAN determines the real/fake of the generated images through Wasserstein distance. Two important contributions for WGAN are as follows: (1) WGAN has no sign of mode collapse in experiments. (2) The generator can still learn when the critic works well.

To estimate the Wasserstein distance, we need to find a 1-Lipschitz function. This experiment builds a deep network to learn the problem. Indeed, this network is very similar to the discriminator *D*, just without the sigmoid function and outputs a scalar score rather than a probability. This score can be explained as to how real the input images are. In WGAN the discriminator is changed to the critic to reflect its new role. The difference between GAN and general WGAN is to change discriminator become critic and the cost function. For both of them, the network design is almost the same except the critic does not have an output sigmoid function. The cost function in critic and generator for WGAN could be seen in the formula 10 and formula 11, respectively.

$$\nabla_w \frac{1}{m} \sum_{i=1}^{m} \left[f\left(x^{(i)}\right) - f\left(G\left(z^{(i)}\right)\right) \right] \tag{10}$$

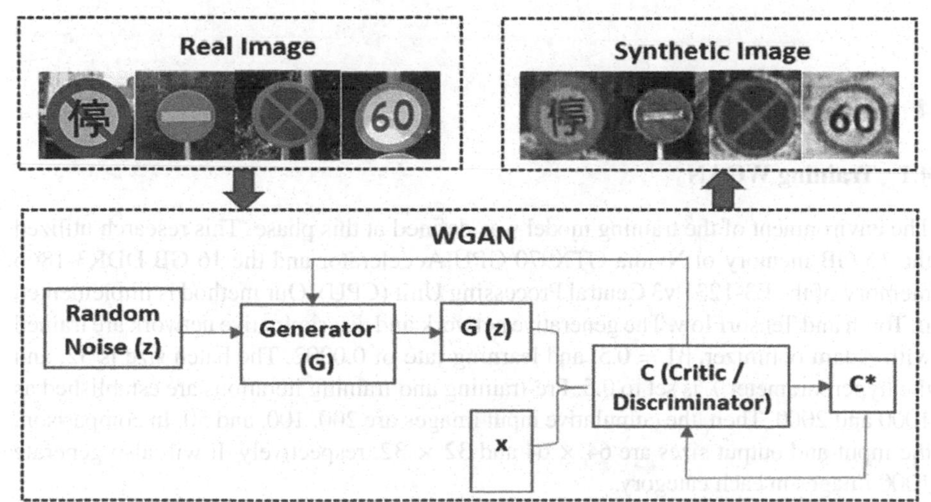

Fig. 3. WGAN architecture to generate synthetic images.

$$\nabla_\theta \frac{1}{m} \sum_{i=1}^{m} f\left(G\left(z^{(i)}\right)\right) \tag{11}$$

However, f has to be a 1-Lipschitz function. To enforce the constraint, WGAN applies a very simple clipping to restrict the maximum weight value in f. The weights of the discriminator must be within a certain range controlled by the hyperparameters c.

3.2 Experiment Setting for WGAN

In this work, several experiments with different settings were performed to produce a realistic WGAN synthesis picture. Besides, we concentrate only on Taiwanese prohibitive signs that do not consist of Class T1 (no entry images), Class T2 (no stopping images), Class T3 (no parking images), and Class T4 (speed limit images). Table 1 displays the specifics of each experiment setting.

Table 1. Experiment setting.

No	Total image	Image size (px)	Total generate image
1	200	64 × 64	1000
2	200	32 × 32	1000
3	100	64 × 64	1000
4	100	32 × 32	1000
5	50	64 × 64	1000
6	50	32 × 32	1000

4 Experiment and Result

4.1 Training WGAN

The environment of the training model was defined at this phase. This research utilized the 16 GB memory of Nvidia GTX970 GPU Accelerator and the 16 GB DDR3-1866 memory of the E3-1231 v3 Central Processing Unit (CPU). Our method is implemented in Torch and TensorFlow. The generative network and discriminative network are trained with Adam optimizer, $\beta 1 = 0.5$, and learning rate of 0.0002. The batch size is 32, and the hyperparameter λ is set to 0.5. Pre-training and training iterations are established as 1000 and 2000. Then, the cumulative input images are 200, 100, and 50. In comparison, the input and output sizes are 64 × 64 and 32 × 32, respectively. It will also generate 1000 images in each category.

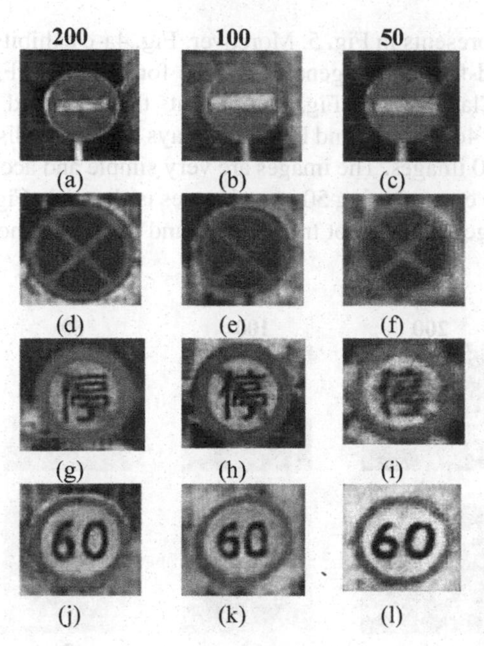

Fig. 4. Synthetic traffic sign images all class with size 64 × 64 and 1000 epoch generated by WGAN.

4.2 Discussions and Results

The result which is received during DCGAN training was explained in this section.

First, we analyze pictures generated by WGAN from Fig. 4, Fig. 5, Fig. 6, and Fig. 7. Figure 4 displays the synthetic traffic sign images all class with size 64 × 64 and 1000 epoch generated by WGAN. Next, Synthetic image generation using the same epoch

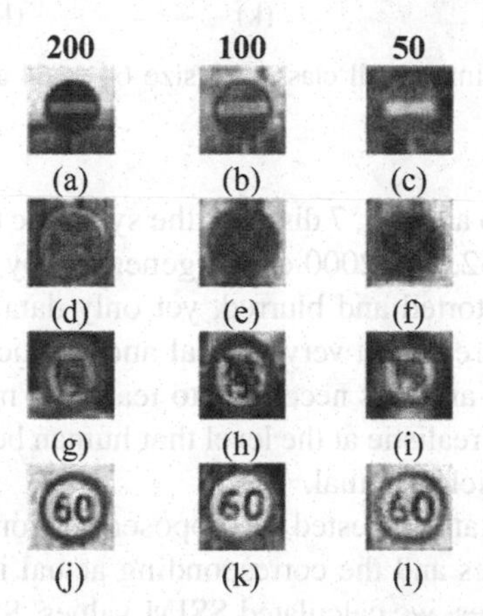

Fig. 5. Synthetic traffic sign images all class with size 32 × 32 and 1000 epoch generated by WGAN.

with size 32 × 32 represents in Fig. 5. Moreover, Fig. 4a-c exhibits the generate image for Class T1, Fig. 4d-f show the generate image for Class T2, Fig. 4g-i presents the generate image for Class T3, and Fig. 4j-l presents the generated image for Class T4. Besides, Fig. 4a, Fig. 4d, Fig. 4g, and Fig. 4j displays the best realistic synthetic version of our model with 200 images. The images are very simple and accurate. Therefore, the worst generate image created using 50 input images is shown in Fig. 4c, Fig. 4f, Fig. 4i, and Fig. 4l. The image is noisy, not transparent, and has lots of noises associated with others.

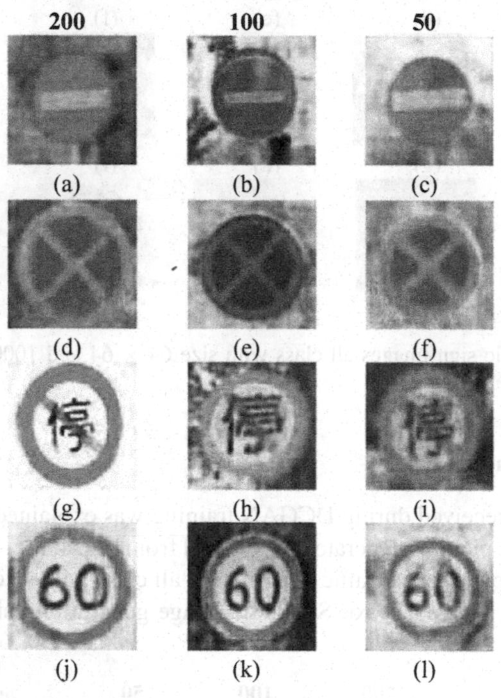

Fig. 6. Synthetic traffic sign images all class with size 64 × 64 and 2000 epoch generated by WGAN.

As a comparison, Fig. 6 and Fig. 7 displays the synthetic traffic sign images all class with size 64 × 64, 32 × 32, and 2000 epoch generated by WGAN. On the one side, certain pictures appear distorted and blurred, yet only data encrypted in the sign can be identified. Several pictures seem very normal and practical, on the other hand. The training took 50 iterations, and it is necessary to teach the model. However, the image performance is sufficiently realistic at the level that human beings cannot discern which picture is synthetic and which is actual.

Our studies have quantitatively tested the proposed network by analyzing correlations between synthesized images and the corresponding actual images. Between synthetic images and authentic images, we calculated SSIM values. SSIM integrates masking of luminance and contrast. The error estimation also involves significant interdependencies of closer pixels and the measurements are measured on narrow image frames. SSIM

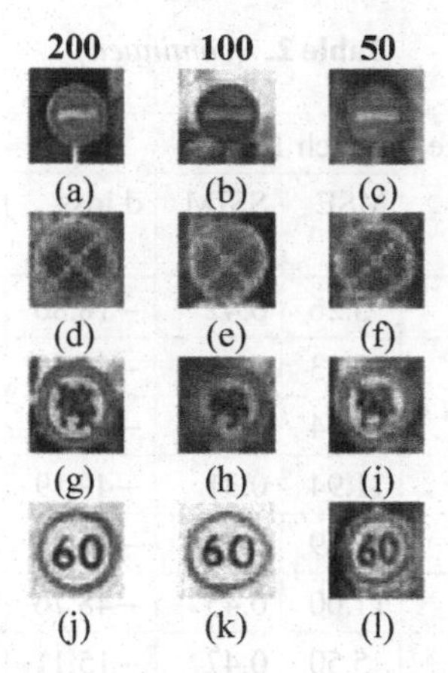

Fig. 7. Synthetic traffic sign images all class with size 32 × 32 and 2000 epoch generated by WGAN.

values are between −1 and 1 (better). Also, usually for real image the MSE is 0. Similarity effects calculated in Table 2, Table 3, and Fig. 8 have been recorded.

Table 2. Evaluation performance synthetic images all class generated by WGAN with 1000 epoch.

No	Class	Total image	Image size (px)	Epoch 1000					
				MSE	SSIM	d loss	Gradient penalty	g loss	Time
1	P1	200	64 × 64	8.12	0.48	−41.49	1.28	13.36	14 m 36 s
2	P1	200	32 × 32	4.18	0.51	−14.64	0.25	4.77	6 m 21 s
3	P1	100	64 × 64	8.76	0.46	−43.61	1.51	30.73	6 m 31 s
4	P1	100	32 × 32	4.65	0.49	−14.07	0.22	8.56	3 m 36 s
5	P1	50	64 × 64	10.44	0.35	−37.51	0.52	46.88	3 m 38 s
6	P1	50	32 × 32	5.38	0.41	−17.44	0.23	16.28	1 m 34 s
7	P2	200	64 × 64	9.51	0.44	−48.85	1.77	25.99	13 m 2 s
8	P2	200	32 × 32	4.91	0.40	−14.11	0.45	6.07	5 m 49 s
9	P2	100	64 × 64	9.55	0.40	−39.91	1.22	32.77	6 m 14 s
10	P2	100	32 × 32	4.89	0.35	−12.84	0.16	7.87	3 m 17 s
11	P2	50	64 × 64	10.38	0.32	−44.52	1.11	45.76	3 m 34 s
12	P2	50	32 × 32	5.19	0.27	−13.08	0.19	17.58	1 m 39 s
13	P3	200	64 × 64	10.32	0.41	−45.39	1.11	24.86	17 m 1 s

(continued)

Table 2. (*continued*)

No	Class	Total image	Image size (px)	Epoch 1000					
				MSE	SSIM	d loss	Gradient penalty	g loss	Time
14	P3	200	32 × 32	5.26	0.42	−18.80	0.50	5.77	6 m 46 s
15	P3	100	64 × 64	9.33	0.41	−52.52	1.56	41.68	8 m 11 s
16	P3	100	32 × 32	4.84	0.44	−16.76	0.26	7.59	4 m 18 s
17	P3	50	64 × 64	11.94	0.31	−49.19	0.90	42.90	4 m 6 s
18	P3	50	32 × 32	5.89	0.35	−15.21	0.28	14.02	2 m 7 s
19	P4	200	64 × 64	11.00	0.45	−48.76	1.31	65.49	12 m 44 s
20	P4	200	32 × 32	5.50	0.47	−15.11	0.23	8.49	5 m 9 s
21	P4	100	64 × 64	12.06	0.42	−63.20	2.20	42.65	6 m 10 s
22	P4	100	32 × 32	5.95	0.45	−16.05	0.17	9.35	3 m 11 s
23	P4	50	64 × 64	13.66	0.36	−52.89	1.10	51.98	3 m 44 s
24	P4	50	32 × 32	6.90	0.42	−14.96	0.16	14.67	1 m 34 s

Table 3. Evaluation performance synthetic images all class generated by WGAN with 2000 epoch.

No	Class	Total image	Image size (px)	Epoch 2000					
				MSE	SSIM	d loss	Gradient penalty	g loss	Time
1	P1	200	64 × 64	7.48	0.51	−37.20	0.78	11.54	20 m 25 s
2	P1	200	32 × 32	4.01	0.53	−15.35	0.44	1.88	9 m 25 s
3	P1	100	64 × 64	7.65	0.50	−44.83	1.28	13.92	12 m 49 s
4	P1	100	32 × 32	4.08	0.53	−15.09	0.40	3.19	7 m 9 s
5	P1	50	64 × 64	9.39	0.41	−55.02	1.41	47.84	7 m 27 s
6	P1	50	32 × 32	4.88	0.45	−16.78	0.34	6.47	3 m 32 s
7	P2	200	64 × 64	8.39	0.47	−50.86	1.45	9.06	19 m 58 s
8	P2	200	32 × 32	4.64	0.43	−12.73	0.30	3.20	8 m 46 s
9	P2	100	64 × 64	8.09	0.45	−52.11	2.03	20.58	12 m 24 s
10	P2	100	32 × 32	4.42	0.39	−15.13	0.34	3.22	6 m 12 s
11	P2	50	64 × 64	9.01	0.38	−48.88	1.19	48.10	7 m 22 s
12	P2	50	32 × 32	4.60	0.34	−15.69	0.24	11.26	3 m 5 s
13	P3	200	64 × 64	8.98	0.47	−48.33	1.29	11.63	24 m 28 s

(*continued*)

Table 3. (*continued*)

No	Class	Total image	Image size (px)	Epoch 2000					
				MSE	SSIM	d loss	Gradient penalty	g loss	Time
14	P3	200	32 × 32	4.87	0.46	−17.41	0.51	1.69	13 m 15 s
15	P3	100	64 × 64	8.32	0.48	−46.35	1.19	14.53	14.57 s
16	P3	100	32 × 32	4.58	0.47	−13.49	0.24	0.40	9 m 42 s
17	P3	50	64 × 64	9.94	0.38	−62.28	1.13	49.38	8 m 12 s
18	P3	50	32 × 32	5.50	0.39	−20.72	0.37	9.43	4 m 0 s
19	P4	200	64 × 64	9.65	0.48	−59.38	1.75	35.11	20 m 57 s
20	P4	200	32 × 32	4.89	0.50	−16.99	0.47	6.06	11 m 15 s
21	P4	100	64 × 64	10.83	0.46	−57.56	0.92	41.17	12 m 48 s
22	P4	100	32 × 32	5.53	0.47	−18.42	0.49	8.11	6 m 51 s
23	P4	50	64 × 64	13.03	0.40	−67.08	1.88	56.88	7 m 36 s
24	P4	50	32 × 32	6.10	0.45	−21.91	0.47	14.61	3 m 41 s

Based on the Table 2 and Table 3 result we can conclude as follows: (1) The maximum SSIM values reached while using a lot of images with small size images for input training. (2) The larger image size will produce a high MSE value and require longer training time. (3) The highest SSIM values reached when using 200 total images as input, images size 32 × 32, and epoch 2000. For instance, Table 3 display the optimum SSIM values is 0.53 for class P1, and SSIM value 0.50 for class P4.

Moreover, Fig. 8 represents the evaluation performance of SSI calculation for original image compare to synthetic image. All original image in Fig. 8a-d indicate the same MSE = 0 and SSI = 1. Hence, Fig. 8a reveal MSE = 3.49 and SSI = 0.62 for class P1. Next, Fig. 8b exhibit MSE = 7.25 and SSI = 0.61 for class P2. Class P3 reach MSE = 4.37 and SSI = 0.63 shown in Fig. 8c. Therefore, Fig. 8d represent MSE = 7.74 and SSI = 0.72 for class P4.

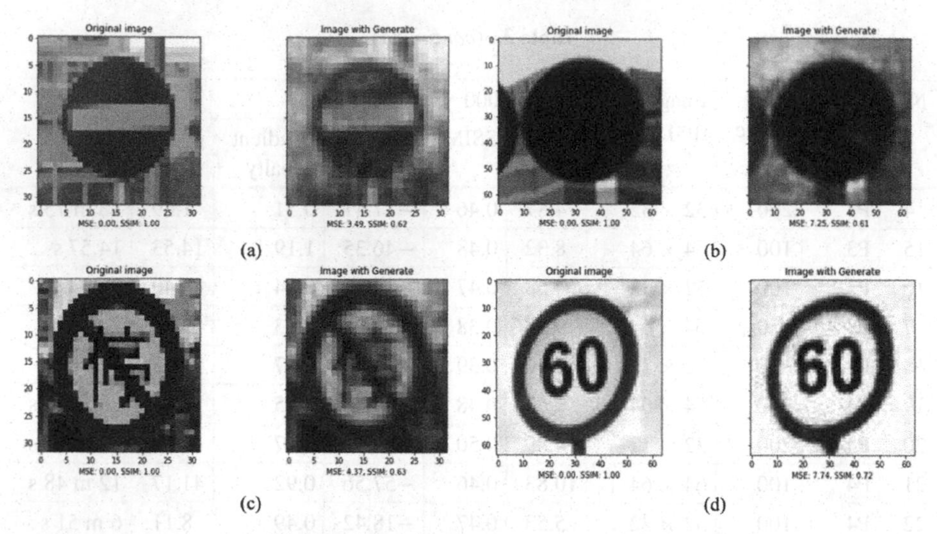

Fig. 8. Evaluation performance of SSIM.

Many loss functions have been developed and evaluated in an effort to improve the stability of training WGAN models. Further, Fig. 9 displays WGAN training loss value with d loss, gradient penalty and g loss. The loss function may be applied by measuring the average score over true and false images and by multiplying the average score by 1 and −1. This has the desired result of discriminating between true and false images. The advantage of Wasserstein's loss is that it creates a valuable gradient almost everywhere that allows the model to be constantly conditioned. It also means that a lower Wasserstein loss coincides with the higher picture quality of the generator, meaning that we specifically want to minimize the generator loss.

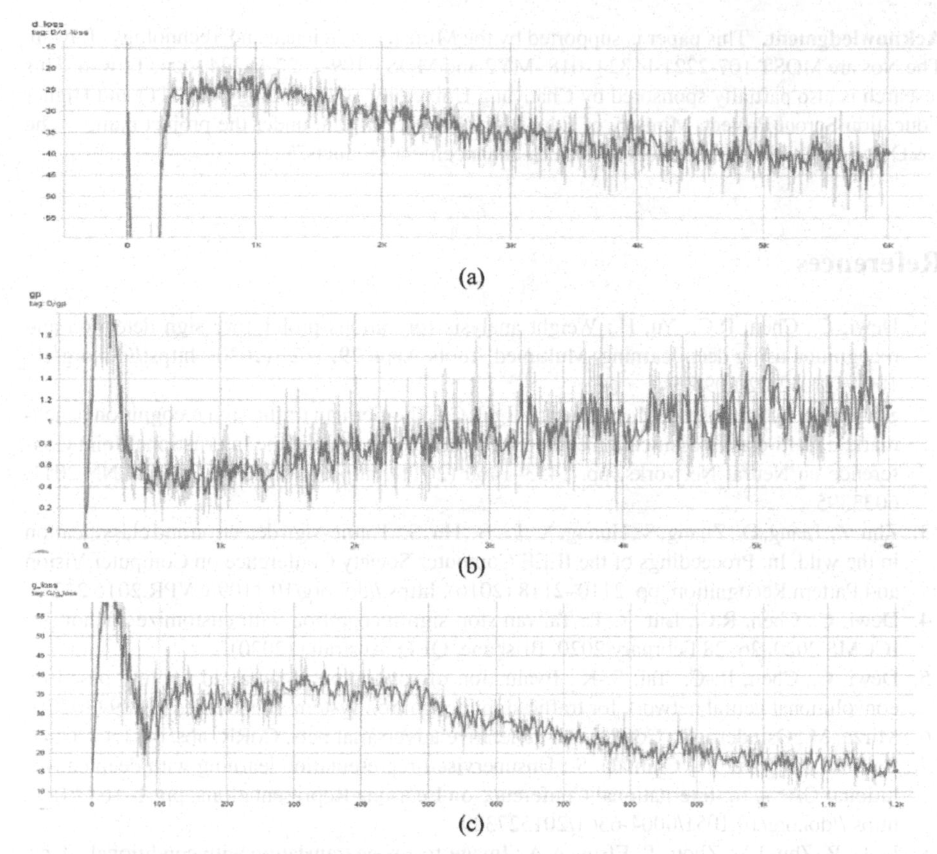

Fig. 9. WGAN training loss value. (a) d loss, (b) gradient penalty, (c) g loss.

5 Conclusions

A series of experiments producing synthetic traffic sign images with WGAN have been analyzed in this work. Based on the test findings in Table 2, Table 3, Fig. 4, Fig. 5, Fig. 6, Fig. 7, Fig. 8, and Fig. 9, it can be inferred that the trend of accuracy improves with the MSE value and training time. The highest SSIM values reached when using 200 total images as input, images size 32 × 32, and epoch 2000. Based on all the experimental results, it can be concluded that synthetic traffic sign images can still be produced with a small training image. In this experiment, we try to produce synthetic images using 50 real images and produce 1000 synthetic images. Although the quality is not as good as a class that uses lots of pictures for training, it can still be used to solve problems where the reference or real images are difficult to gain. We intend to build a complete Taiwan traffic sign dataset with more images in future work. We are also working on some modifications to the current Yolo V3, SPP, and Yolo V4. In future research, we will create synthetic pictures based on other GAN methods such as DCGAN and LSGAN. We reserve to practice a model with higher dimensional images to advance the efficiency of traffic detection and recognition.

Acknowledgment. This paper is supported by the Ministry of Science and Technology, Taiwan. The Nos are MOST-107-2221-E-324 -018 -MY2 and MOST-109-2622-E-324 -004, Taiwan. This research is also partially sponsored by Chaoyang University of Technology (CYUT) and Higher Education Sprout Project, Ministry of Education (MOE), Taiwan, under the project name: "The R&D and the cultivation of talent for health-enhancement products."

References

1. Dewi, C., Chen, R.C., Yu, H.: Weight analysis for various prohibitory sign detection and recognition using deep learning. Multimed. Tools Appl. **79,** 1–21 (2020). https://doi.org/10.1007/s11042-020-09509-x
2. Stallkamp, J., Schlipsing, M., Salmen, J., Igel, C.: The German traffic sign recognition benchmark: a multi-class classification competition. In: Proceedings of the International Joint Conference on Neural Networks, pp. 1453–1460 (2011). https://doi.org/10.1109/IJCNN.2011.6033395
3. Zhu, Z., Liang, D., Zhang, S., Huang, X., Li, B., Hu, S.: Traffic-sign detection and classification in the wild. In: Proceedings of the IEEE Computer Society Conference on Computer Vision and Pattern Recognition, pp. 2110–2118 (2016). https://doi.org/10.1109/CVPR.2016.232
4. Dewi, C., Chen, R.C., Liu, Y.-T.: Taiwan stop sign recognition with customize anchor. In: ICCMS 2020, 26–28 February 2020, Brisbane, QLD, Australia (2020)
5. Dewi, C., Chen, R.-C., Tai, S.-K.: Evaluation of robust spatial pyramid pooling based on convolutional neural network for traffic sign recognition system. Electronics **9**, 889 (2020)
6. Mirza, M., Osindero, S.: Conditional generative adversarial nets. CoRR. abs/1411.1 (2014)
7. Radford, A., Metz, L., Chintala, S.: Unsupervised representation learning with deep convolutional GANs. In: International Conference on Learning Representations, pp. 1–16 (2016). https://doi.org/10.1051/0004-6361/201527329
8. Isola, P., Zhu, J.Y., Zhou, T., Efros, A.A.: Image-to-image translation with conditional adversarial networks. In: Proceedings - 30th IEEE Conference on Computer Vision and Pattern Recognition, CVPR 2017, pp. 5967–5976 (2017). https://doi.org/10.1109/CVPR.2017.632
9. Shrivastava, A., Pfister, T., Tuzel, O., Susskind, J., Wang, W., Webb, R.: Learning from simulated and unsupervised images through adversarial training. In: Proceedings - 30th IEEE Conference on Computer Vision and Pattern Recognition, CVPR 2017, pp. 2242–2251. Institute of Electrical and Electronics Engineers Inc. (2017). https://doi.org/10.1109/CVPR.2017.241
10. Zhang, H., Goodfellow, I., Metaxas, D., Odena, A.: Self-attention generative adversarial networks. In: 36th International Conference on Machine Learning, ICML 2019, pp. 12744–12753 (2019)
11. Luo, H., Yang, Y., Tong, B., Wu, F., Fan, B.: Traffic sign recognition using a multi-task convolutional neural network. IEEE Trans. Intell. Transp. Syst. **19**, 1100–1111 (2018). https://doi.org/10.1109/TITS.2017.2714691
12. Dewi, C., Chen, R.-C., Hendry, H., Liu, Y.-T.: Similar music instrument detection via deep convolution YOLO-generative adversarial network. In: 2019 IEEE 10th International Conference on Awareness Science and Technology (iCAST), pp. 1–6 (2019)
13. Tai, S., Dewi, C., Chen, R., Liu, Y., Jiang, X.: Deep learning for traffic sign recognition based on spatial pyramid pooling with scale analysis. Appl. Sci. (Switzerland) **10**, 6997 (2020)
14. Cao, C., Cao, Z., Cui, Z.: LDGAN: a synthetic aperture radar image generation method for automatic target recognition. IEEE Trans. Geosci. Remote Sens. **58**, 3495–3508 (2020). https://doi.org/10.1109/TGRS.2019.2957453

15. Yang, J., Kannan, A., Batra, D., Parikh, D.: LR-GAN: layered recursive generative adversarial networks for image generation. In: 5th International Conference on Learning Representations, ICLR 2017 - Conference Track Proceedings, pp. 1–21 (2019)
16. Zhou, C., Zhang, J., Liu, J.: Lp-WGAN: using Lp-norm normalization to stabilize Wasserstein generative adversarial networks. Knowl. Based Syst. **161**, 415–424 (2018). https://doi.org/10.1016/j.knosys.2018.08.004
17. Arjovsky, M., Chintala, S., Bottou, L.: Wasserstein generative adversarial networks. In: 34th International Conference on Machine Learning, ICML 2017 (2017)
18. Lu, Y., Wang, S., Zhao, W., Zhao, Y.: WGAN-based robust occluded facial expression recognition. IEEE Access **7**, 1–17 (2019). https://doi.org/10.1109/ACCESS.2019.2928125
19. Lions, J.L.: Optimal control of systems governed by partial differential equations (1971). https://doi.org/10.1007/978-3-642-65024-6
20. Gulrajani, I., Ahmed, F., Arjovsky, M., Dumoulin, V., Courville, A.: Improved training of wasserstein GANs. In: Advances in Neural Information Processing Systems (2017)
21. Wang, Z., Bovik, A.C., Sheikh, H.R., Simoncelli, E.P.: Image quality assessment: from error visibility to structural similarity. IEEE Trans. Image Process. **13**, 600–612 (2004). https://doi.org/10.1109/TIP.2003.819861
22. Dewi, C., Chen, R.C.: Random forest and support vector machine on features selection for regression analysis. Int. J. Innov. Comput. Inf. Control **15**, 2027–2037 (2019). https://doi.org/10.24507/ijicic.15.06.2027
23. Wang, Z., Bovik, A.C.: A universal image quality index. IEEE Signal Process. Lett. (2002). https://doi.org/10.1109/97.995823
24. Dewi, C., Chen, R.-C.: Human activity recognition based on evolution of features selection and random forest. In: 2019 IEEE International Conference on Systems, Man and Cybernetics (SMC), pp. 2496–2501 (2019). https://doi.org/10.1109/SMC.2019.8913868
25. Dewi, C., Chen, R.C., Hendry, H., Hung, H.T.: Comparative analysis of restricted boltzmann machine models for image classification. In: Nguyen N., Jearanaitanakij K., Selamat A., Trawiński B., Chittayasothorn S. (eds.) Asian Conference on Intelligent Information and Database Systems ACIIDS 2020. LNCS, vol. 12034, pp. 285–296 (2020). https://doi.org/10.1007/978-3-030-42058-1_24
26. Shamshirband, S., Petkovic, D., Javidnia, H., Gani, A.: Sensor data fusion by support vector regression methodology - a comparative study. IEEE Sens. J. (2015). https://doi.org/10.1109/JSEN.2014.2356501
27. Chen, R.C., Dewi, C., Huang, S.W., Caraka, R.E.: Selecting critical features for data classification based on machine learning methods. J. Big Data **7**, 1–26 (2020). https://doi.org/10.1186/s40537-020-00327-4
28. Dewi, C., Chen, R.C.: Decision making based on IoT data collection for precision agriculture. Stud. Comput. Intell. (2020). https://doi.org/10.1007/978-3-030-14132-5_3

An Approach to Automatic Detection of Architectural Façades in Spherical Images

Marcin Kutrzyński, Bartosz Żak, Zbigniew Telec, and Bogdan Trawiński(✉)

Faculty of Computer Science and Management, Wrocław University of Science and Technology, Wrocław, Poland

{marcin.kutrzynski,zbigniew.telec,bogdan.trawinski}@pwr.edu.pl

Abstract. The fast and automatic façade extraction method is an important step to retrieve actual information about cities. It helps to maintain a public registry, may be used to examine the technical condition of buildings or in process of city planning. We propose an automatic method to detect and retrieve building façades from spherical images. Our method uses deep learning models trained on automatically labelled spherical images collected in the virtual city we have generated. Finally, we compare the proposed solution using three different deep learning models with a classic method. Our experiment revealed that the proposed approach has a similar or better performance than the current methods. Moreover, our solution works with unprepared data, while existing methods require data pre-processing.

Keywords: Deep learning · Façade segmentation · Object detection · Spherical images · Region-based convolutional neural networks

1 Introduction

Recording urban fabric is an important task addressed to municipal services in the field of monument protection, property valuation, habitats - taking into account the natural human needs, or the smart city trend: investment in human capital, sustainable development and improvement of the quality of life. One of such tasks is to register the façades of dense urban buildings and historic city centres.

Current methods take different approaches to solve this problem, such as using both aerial and street-level images, as well as multi-perspective stereo images [1]. Data for these experiments were collected using cameras or *LiDAR* devices [1–3]. These solutions used different algorithms for façade separation: repetitive patterns detection [4, 5], *RANSAC* (*Random Sample Consensus*), a watershed with gradient accumulation analysis [6].

Rapid advances in deep learning and new object detection model architectures that have recently been developed in the family of *Region-based Convolutional Neural Networks* (*R-CNNs*) are revealing new application areas. Our approach also uses a deep learning algorithm to detect and separate façades. The idea of using deep learning in this field is not new, but the presented method may focus on two new aspects: an inexpensive

N. T. Nguyen et al. (Eds.): ACIIDS 2021, LNAI 12672, pp. 494–504, 2021.
https://doi.org/10.1007/978-3-030-73280-6_39

learning process in fully artificial, city-like games while the evaluation is done on real city photos; and spherical photo processing, allowing real city photos to be used as such without pre-processing. The presented method simplifies the recording of information about façades, and what is more, it allows data to be captured by inexpensive universal 360° cameras mounted on cars, bikes or hand sticks.

Finally, the main deep learning models were compared by using *Mask R-CNN*, Mask *Scoring R-CNN* and *YOLACT*. The proposed deep learning methods surpass non-deep learning methods in terms of object identification and delineation of their boundaries, and perform well in object separation.

2 Architectural Façade Detection, Separation and Rectification

In recent years, several solutions for façade extraction from ground-level images have been proposed. Most of them use computer vision algorithms and focus on extracting only one façade from a given image. Sümer and Türker in [4] propose a method based on the repetitive Watershed segmentation, a technique described in [7, 8]. They claim that their solution allows the texture of the façade to be distinguished from the background, such as the sky, street or cars, using an approach based on automatic region growing.

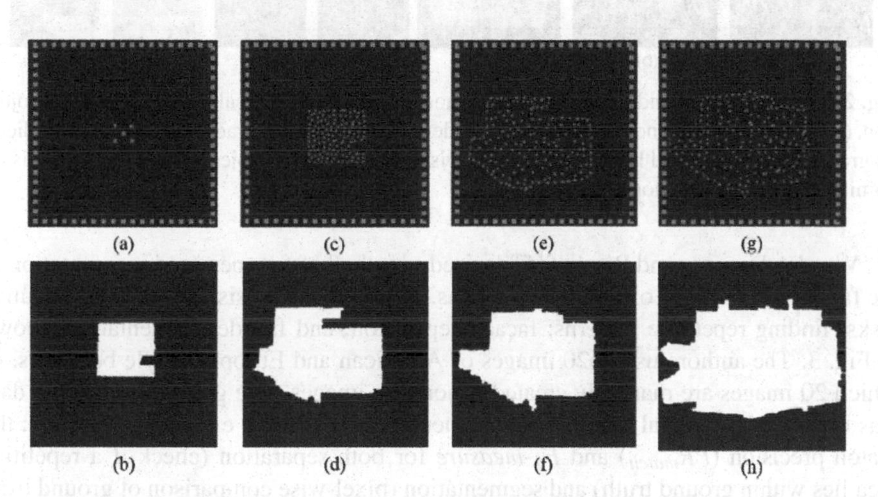

Fig. 1. The proposed repetitive watershed segmentation technique: (a) initial markers, (b) segmented region after the first iteration. (c), (e), (d) – markers in successive iterations, (d), (f), (h) – respective segmented regions for successive iterations. Source: [4] (Color figure online).

The authors propose the position of the initial markers, which are necessary to start segmentation, as in Fig. 1. The green points on the edges correspond to the background markers. The red points in the center correspond to the foreground object (façade). Segmentation begins with the initial markers and then the closest regions are iteratively combined based on gradient magnitude. For each iteration, new façade markers are arbitrarily placed inside the segmented regions. Segmentation is completed based on two-step criteria, which are the stability of the growth rate and the measure of the maximum repetition count.

Hernández and Marcotegui in [6] present an automatic method of elevation extraction based on morphological segmentation and gradient accumulation analysis. It is claimed that this technique is capable of extracting a single façade from a street-level image containing a block of buildings. For this purpose, a morphological opening in the horizontal or vertical direction is used, and a color gradient is generated using appropriate structuring elements and projected onto the profile (Fig. 2). Façade division is found with watershed segmentation of the projected profile. The algorithm finds two types of intra- and inter-façade divisions. The authors assume that intra-façade divisions are characterized by lower profile values due to elements included in the façade, such as floor separation or ornaments.

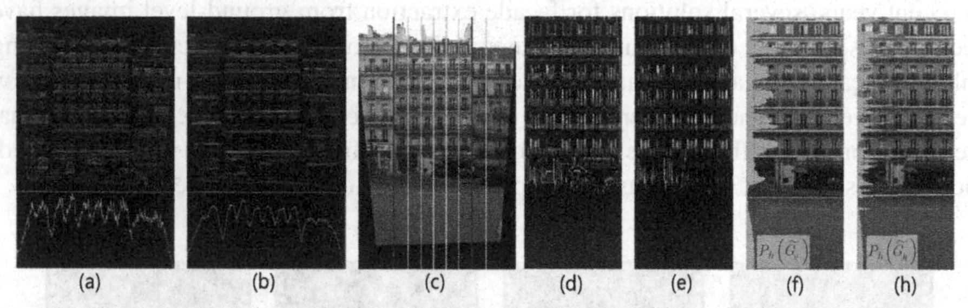

Fig. 2. Color gradients and projections: a–c – horizontal, d–h – vertical, (a) gradient and projection, (b) filtered gradient and projection, (c) façade division, (d) color gradient, (e) filtered gradient, (f) gradient projection and horizontal façade division, (h) gradient projection and façade division refining. Source: [6]. (Color figure online)

Wendel, Donoser and Bischof [5] devised a method of unsupervised segmentation of the façade with the use of repeating patterns. This method consists of three consecutive tasks: finding repetitive patterns; façade separation; and façade segmentation, shown in Fig. 3. The authors used 620 images of American and European style buildings, of which 20 images are manually created panoramic images. The ground truth (*GT*) data was obtained by manual labeling. Three metrics were used to evaluate the results: the match precision (PR_{match}) and F_1*-measure* for both separation (check if a repetitive area lies within ground truth) and segmentation (pixel-wise comparison of ground truth and obtained façade segments) individually. The achieved average result for test dataset is 72,8% PR_{match}, 94,0% $F_{1,sep}$, and 85,4% $F_{1,seg}$. This method can highlight multiple façades in an image with remarkable results. However, this technique is highly parameter dependent. The authors mention that the Harris corner detector must have different parameters for different architectural styles.

The *Region-Based Convolutional Neural Network* (*R-CNN*) method introduced in 2014 by Girshick [10] is one of the first techniques that successfully applied *CNNs* in the instance segmentation task. Girshick idea evolves and nowadays four methods perform high efficiency and accuracy: *Mask R-CNN* [11], *Mask Scoring R-CNN* [12], *YOLACT* [13], *TensorMask* [14]. Accuracy refers to the exact location and classification of objects in the images. Its improvement can be achieved by obtaining better distinctiveness (ability to discriminate between objects sharing common features) or better robustness (ability to detect and categorize objects despite the intra-class appearance variation). The efficiency depends on the computational cost of the segmentation technique [9].

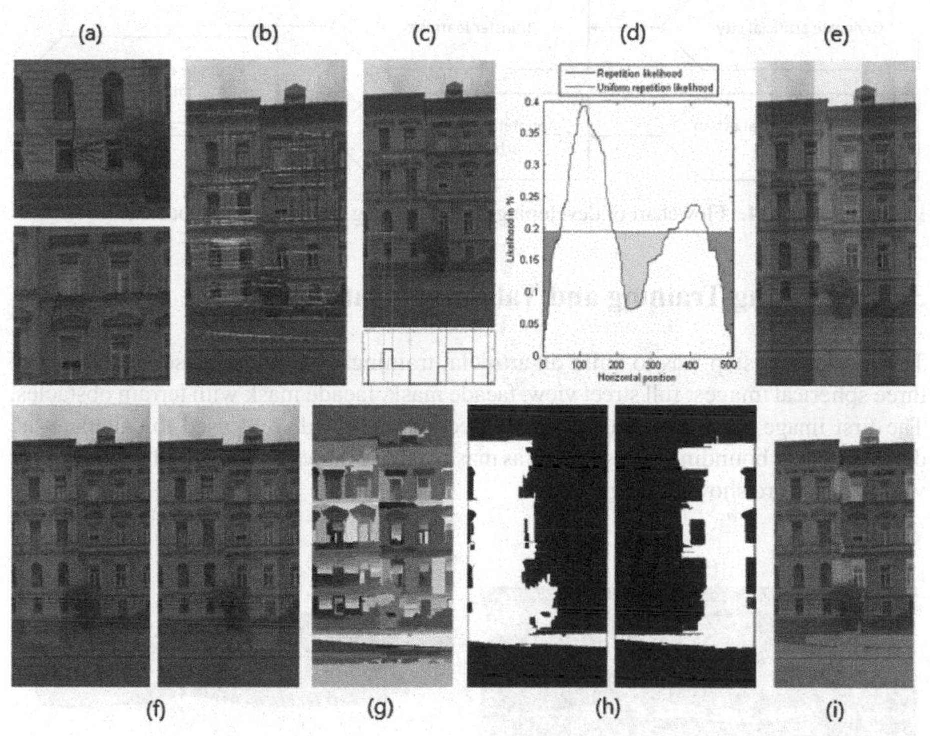

Fig. 3. Façade segmentation: (a) arbitrary areas matching, (b) detected repetitive patterns, (c) projection results in a match count along the horizontal axis, (d) thresholding the repetition likelihood with the uniform repetition likelihood, (e) resulting repetitive areas, green – separation, red – unknown, (f) convex hull of repetitive points as segmentation prior, (g) superpixel segmentation, (h) combination of prior and appearance, (i) final result. Source: [5]. (Color figure online)

Modern deep learning instance segmentation solutions are only promising when training on large datasets. Stratification on training elevations must be done in terms of architectural style, population density, visual obstacles, and weather. Even using online sources to obtain training photos, the task is difficult as it requires labeling, outlining and annotating all objects in the images. Our goal is a no-image-preparing approach to final segmentation that requires 360° spherical image processing. Current open-source online databases are not rich enough in such images.

The method proposed in this paper is consistent with the diagram shown in Fig. 4. The key was to build data sets for training and validation from an artificial city (a) and to obtain real data for model testing (b). Then, three deep learning models were developed and compared.

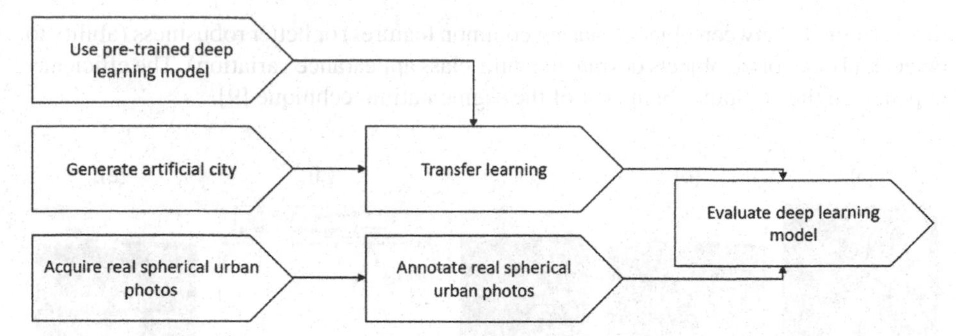

Fig. 4. Flowchart of developing and evaluating deep learning models.

3 Generating Training and Validation Datasets

The aim of this step was to build an artificial training/validation dataset consisting of three spherical images: full street view, façade mask, façade mask with terrain obstacles. The first image was the input image, the second and third were used to calculate the desired output: bounding boxes as well as masks. These three images with their combined visualization are shown in Fig. 5.

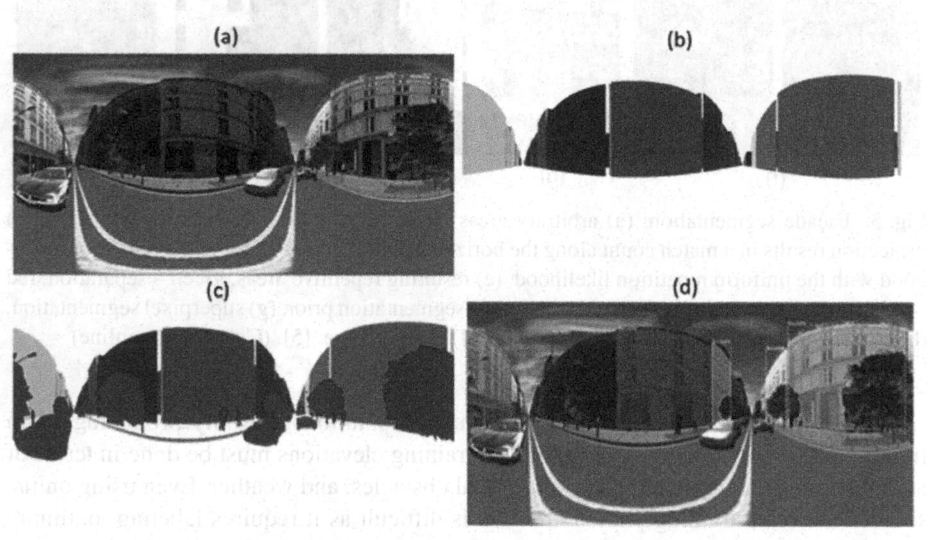

Fig. 5. Training data - instance mask labelling: (a) real-life-like view, (b) marked façade instances – each façade in different shade of green, (c) obstacle mask – blue, (d) real-life-like view with façade instance masks. (Color figure online)

To capture tens of thousands of street views, it was necessary to generate a street network to obtain a "new city" each time data was prepared. This was done in three steps as shown in Fig. 6: (a) creating three parallel main streets; (b) random nodes outside high streets were selected and joined; (c) the resulting areas were divided into parcels according to random rules. The proposed process allowed for creating not only a simple building passage view, but also parks, green areas and city squares. The spherical camera always moved only along the central, main road.

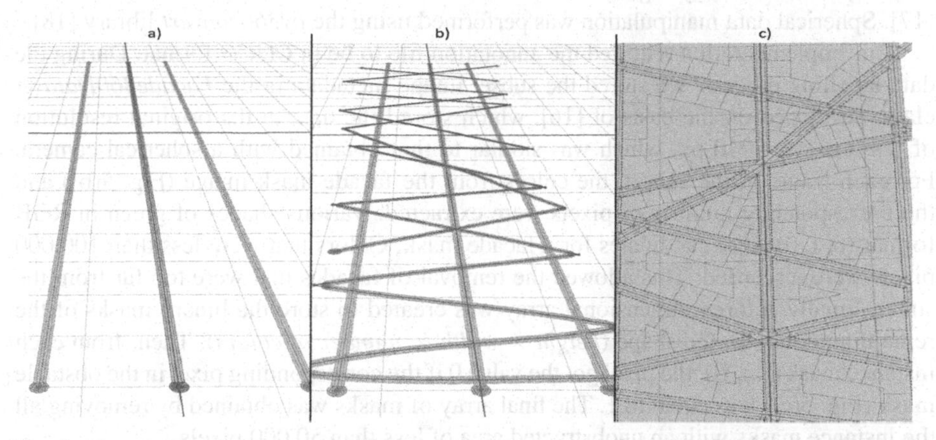

Fig. 6. The process of street network generation: (a) three parallel main streets, (b) streets created by connecting random nodes outside main streets, (c) generating parcels within obtained street network.

Building façades were created in accordance with the known rules of *ESRI CityEngine® Façade Wizard* [17] (Fig. 7). The buildings were of different heights and the façades were divided into random levels. Moreover, each floor was divided into sections. In the next stage, the texture was selected for each floor. The presented photos were based on the *ESRI Parisian Roof* collection, but we also used our own textures. The last step was to assign random UV filters to each section to distinguish between them. This method allowed the creation of an infinite number of artificial façades.

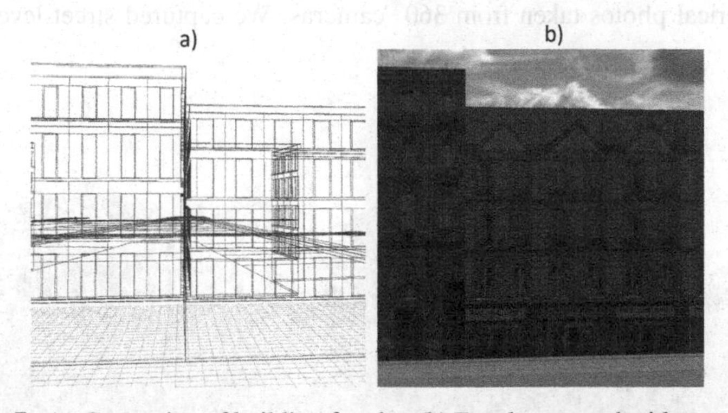

Fig. 7. (a) Generation of building façades, (b) Façades covered with textures.

The camera moved along the central main road and took six photos at each point using a 90° FoV lens to fill a standard cubemap. The cubemap was then converted to an equirectangular image required by deep learning models. In addition, at each point, photos were taken with a random rotation angle, tilt and vertical camera position to make it as distinct as possible. Artificial training and validation datasets were made at E*SRI CityEngine*®, *ESRI CityEngine Procedural Runtime*® and by the *PyPRT Phyton* library [17]. Spherical data manipulation was performed using the *py360convert* library [18].

Our implementation required the annotation file to be in *COCO* format. During the data labelling process, we stored the subset image metadata in the *Façade360Datase*t class, developed on the basis of [16], which stored the data in the original resolution of 6080 px vs. 3040 px, which was similar to that obtained with a spherical camera. For each frame in the subset, the colors from the façade-mask image (Fig. 5(b)) and the corresponding number of pixels were extracted. Various shades of green in RGB format (0, i, 0) were candidates for a façade mask. Colors that were less than 100,000 pixels were discarded. This allowed the removal of façades that were too far from the street. Finally, a three-dimensional array was created to store the binary masks of the remaining colors in the shape (*height* × *width* × *number of colors*). Then, from each instance mask ([:,:, i]), the pixel got the value 0 if the corresponding pixel in the obstacle mask (Fig. 5(c)) was equal to 1. The final array of masks was obtained by removing all the instance masks with an unobstructed area of less than 50,000 pixels.

We generated 1000 street view frames from our virtual city, which contained 5,532 annotated façade instances. We split the transformed data into a training and validation subset. The first one comprised 800 images with 4380 façade instances and was used to refine the pre-trained model. The second included 200 images with 1,152 façade instances and was employed to fine-tune the model. We did not create any test subset from artificial data because we decided to evaluate the models on real data, i.e. photos taken in the city.

4 Acquiring Real Spherical Photos

Unlike the training process in which artificial data was used, the evaluation was based on real spherical photos taken from 360° cameras. We captured street-level videos in

Fig. 8. Annotation of spherical photos.

Wroclaw city in Poland. The city is well known for its architectural style variation so it is a perfect place to evaluate algorithms for façade extraction.

We gathered the data with an omnidirectional camera – *Insta360 One X* attached to a folding stick. The camera was operated through a mobile application. The recordings were made by the operator holding a camera while walking around a city. The digital material was recorded in the native formats of the camera: *insp* for photos and *insv* for videos. However, it could be converted to popular *jpeg* and *mp4* formats with provided by manufacturer software *Insta360 STUDIO*. Each photo was annotated manually in *VIA VGG Image Annotator* [15], excluding obstacles such as lamps, cars, trees or pedestrians from the façade areas (Fig. 8). As a result, we labelled 50 images that contained 267 annotated façade instances, and they formed our test set.

5 Analysis of Experimental Results

Measures proposed in [4] to evaluate predicted mask quality were adopted. The *Façade Detection Percentage* (*FDP*) is a measure of object detection performance that describes how much of the *GT* façade has been detected. It can be calculated using Formula (1). The *Branching Factor* (*BF*), which can be calculated by Formula (2), measures over-segmentation. It is equal to 0 if all detected pixels refer to the *GT* mask area. The *Quality Percentage* (*QP*) is a measure of the absolute quality of the extracted façade. It can be calculated with Formula (3). If QP equals 100%, it means a perfect segmentation of the façade with respect to the corresponding *GT* data.

$$FDP = \frac{100 \times TP}{TP + FN} \tag{1}$$

$$BF = \frac{FP}{TP} \tag{2}$$

$$QP = \frac{100 \times TP}{TP + FP + FN} \tag{3}$$

The green areas shown in Fig. 9 are masks labeled *GT*, while the red areas are detected façades. *Mask R-CNN* (Fig. 9(a)) and *MS R-CNN* (Fig. 9(b)), which are based on a similar model architecture, exhibit similar behavior. However, *Mask R-CNN* shows better obstacle avoidance performance. This is clearly visible in the middle building in the left photo, for which the resulting mask effectively ignores the plants in front of the building. *MS R-CNN* and *YOLACT* (Fig. 9(c)) do not show similar "flexibility" in mask generation. In the case of the *MS R-CNN*, this results in the inclusion of obstacles in the elevation mask, which explains the higher *BF* than in the other models. In the case of *YOLACT*, this results in the cropping of part of the image with the problematic area, which explains the lower *BF* with a slightly lower *FDP* than *MS R-CNN*.

Table 1 summarizes the results compared to the non-deep learning models. The results of the *Region Growing* method significantly outperform the other methods, however, this method is limited to only one façade in the photo. The results of the deep learning method correspond to the visual comparison in Fig. 9. The high *R-CNN Mask* score on the *FDP* average is due to the visual obstacle avoidance feature, which allows

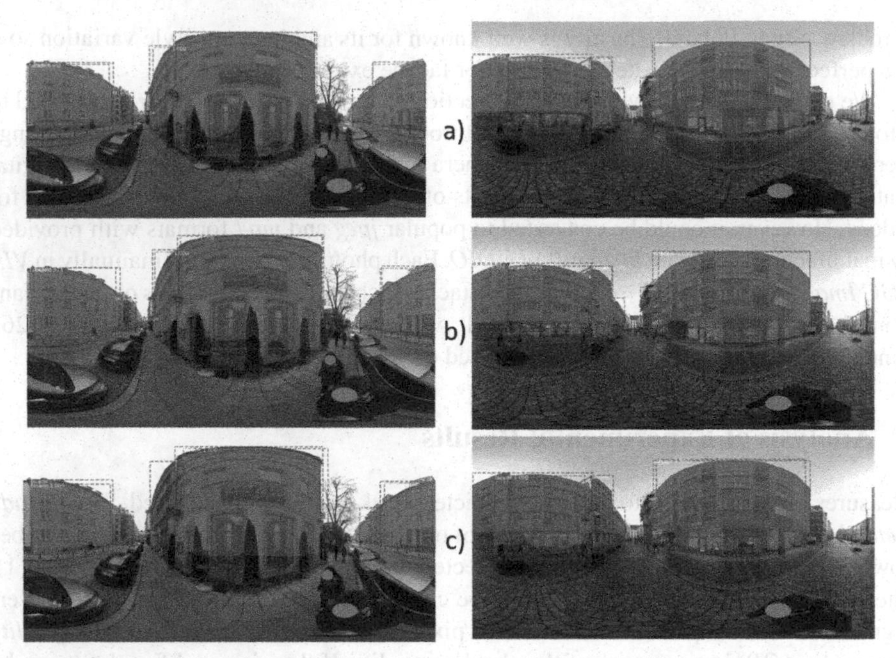

Fig. 9. Differences in detection: green – ground truth instances. Red – detected instances: (a) *Mask R-CNN* (b) *Mask Scoring R-CNN* (c) *YOLACT*. (Color figure online)

Table 1. Comparison of model performance

Method	Avg FDP	Avg BF	Avg QP
Mask R-CNN	0.846	0.079	0.797
Mask scoring R-CNN	0.821	0.091	0.766
YOLACT	0.806	0.077	0.761
Gradient accumulation	0.888	0.255	0.729
Region growing	0.876	0.068	0.821

the *GT* to conform. Additionally, a low *BF* average score makes *Quality Percentage* (*QP*) better than other methods can achieve.

Given that the images for the non-deep learning models were pre-processed for evaluation and not bare as in the proposed method, the results may not be fully comparable. Note, however, that the goal is to deny any preprocessing, so making any comparisons with post-processed data is not a thesis.

6 Conclusions and Future Work

Three modern instance segmentation solutions based on deep learning were used for automatic detection and separation of architectural façades. Automation of the segmentation process was achieved through the use of spherical images at the training/validation stage. This allows raw 360° photos to be used for evaluation without any preparation prior to testing. The creation of training data on a 3D model of an artificial city, represented by various architectural styles, urban density, city objects as visual obstacles, weather and daylight intensity, allows the preparation of a universal data set for the development of deep learning for segmentation of architectural façade instances.

Based on the results of the experiments, the following general conclusions can be drawn. The selected deep learning methods allow for façade segmentation and masking as well as the corresponding non-machine learning methods. However, the proposed methods work on unobstructed spherical images, in contrast to the previous methods, whose data is highly data limited and must be prepared or pre-processed before use.

Further research is planned to test the efficiency of deep learning methods in detecting instances. During the experiments, it was noted that YOLACT is not good enough at segmentation, but outperforms other methods at instance detection. Full automation of the façade register assumes instance detection, segmentation and rectification. One stage has already been completed, the rest are under development.

References

1. Meixner, P., Wendel, A., Bischof, H., Leberl, F.: Building façade separation in vertical aerial images. In: Annals of the International Society for Photogrammetry, Remote Sensing and Spatial Information Sciences (ISPRS), pp. 239–243 (2012). https://doi.org/10.5194/isprsannals-I-3-239-2012
2. Wang, R., Xia, S.: Façade separation in ground-based LiDAR point clouds based on edges and windows. IEEE J. Sel. Top. Appl. Earth Obs. Remote Sens. **12**(3), 1041–1052 (2019)
3. Recky, M., Wendel, A., Leberl, F.: Façade segmentation in a multi-view scenario. In: International Conference on 3D Imaging, Modeling, Processing, Visualization and Transmission (3DIMPVT), pp. 358–365 (2011). https://doi.org/10.1109/3DIMPVT.2011.52
4. Sümer, E., Türker, M.: An automatic region growing based approach to extract façade textures from single ground-level building images. J. Geodesy Geoinf. **2**(1), 9–17 (2013). https://doi.org/10.9733/jgg.061213.2
5. Wendel, A., Donoser, M., Bischof, H.: Unsupervised façade segmentation using repetitive patterns. In: Goesele, M., Roth, S., Kuijper, A., Schiele, B., Schindler, K. (eds.) DAGM 2010. LNCS, vol. 6376, pp. 51–60. Springer, Heidelberg (2010). https://doi.org/10.1007/978-3-642-15986-2_6
6. Hernández, J., Marcotegui, B.: Morphological segmentation of building façade images. In: 16th IEEE International Conference on Image Processing (ICIP), pp. 4029–4032 (2009). https://doi.org/10.1109/ICIP.2009.5413756
7. Eddins, S.: The watershed transform: strategies for image segmentation. https://www.mathworks.com/company/newsletters/articles/the-watershed-transform-strategies-for-image-segmentation.html. Accessed 15 Jan 2021
8. Szelinski, R.: Computer Vision: Algorithms and Applications. Springer, London (2011). https://doi.org/10.1007/978-1-84882-935-0

9. Hafiz, A.M., Bhat, G.M.: A survey on instance segmentation: state of the art. Int. J. Multimed. Inf. Retr. **9**(3), 171–189 (2020). https://doi.org/10.1007/s13735-020-00195-x
10. Girshick, R.: Fast R-CNN. In: IEEE International Conference on Computer Vision (ICCV), pp. 1440–1448 (2015). https://doi.org/10.1109/ICCV.2015.169
11. He, K., Gkioxari, G., Dollár, P., Girshick, R.: Mask R-CNN. In: IEEE International Conference on Computer Vision (ICCV), pp. 2980–2988 (2017). https://doi.org/10.1109/ICCV.2017.322
12. Huang, Z., Huang, L., Gong, Y., Huang, C., Wang, X.: Mask scoring R-CNN. In: IEEE/CVF Conference on Computer Vision and Pattern Recognition (CVPR), pp. 6402–6411 (2019). https://doi.org/10.1109/CVPR.2019.00657
13. Bolya, D., Zhou, C., Xiao, F., Lee, Y.J.: YOLACT: real-time instance segmentation. In: IEEE/CVF International Conference on Computer Vision (ICCV), pp. 9156–9165 (2019). https://doi.org/10.1109/ICCV.2019.00925
14. Chen, X., Girshick, R., He, K., Dollar, P.: TensorMask: a foundation for dense object segmentation. In: IEEE/CVF International Conference on Computer Vision (ICCV), pp. 2061–2069 (2019). https://doi.org/10.1109/ICCV.2019.00215
15. Dutta, A., Zisserman, A.: The VIA annotation software for images, audio and video. In: Proceedings of the 27th ACM International Conference on Multimedia (MM 2019), pp. 2276–2279 (2019). https://doi.org/10.1145/3343031.3350535
16. Lin, T.-Y., et al.: Microsoft COCO: common objects in context. In: Fleet, D., Pajdla, T., Schiele, B., Tuytelaars, T. (eds.) ECCV 2014. LNCS, vol. 8693, pp. 740–755. Springer, Cham (2014). https://doi.org/10.1007/978-3-319-10602-1_48
17. Zhao, M., Wu, J.: Automatic building modeling based on CityEngine. J. Geomatics **42**, 92–95 (2017). https://doi.org/10.14188/j.2095-6045.2015330
18. py360convert: https://github.com/sunset1995/py360convert. Accessed 15 Jan 2021

Computational Imaging and Vision

Real-Time Multi-view Face Mask Detector on Edge Device for Supporting Service Robots in the COVID-19 Pandemic

Muhamad Dwisnanto Putro(✉), Duy-Linh Nguyen(✉), and Kang-Hyun Jo(✉)

Department of Electrical, Electronics, and Computer Engineering, University of Ulsan, Ulsan, Korea
{dputro,ndlinh301}@mail.ulsan.ac.kr, acejo@ulsan.ac.kr

Abstract. The COVID-19 pandemic requires everyone to wear a face mask in public areas. This situation expands the ability of a service robot to have a masked face recognition system. The challenge is detecting multi-view faces. Previous works encountered this problem and tended to be slow when implemented in practical applications. This paper proposes a real-time multi-view face mask detector with two main modules: face detection and face mask classification. The proposed architecture emphasizes light and robust feature extraction. The two-stage network makes it easy to focus on discriminating features on the facial area. The detector filters non-faces at the face detection stage and then classifies the facial regions into two categories. Both models were trained and tested on the benchmark datasets. As a result, the proposed detector obtains high performance with competitive accuracy from competitors. It can run 20.60 frames per second when working in real-time on Jetson Nano.

1 Introduction

The technology of robots is developing rapidly in the industrial and medical fields. The Industrial Revolution 5.0 supports to encourage the implementation of robots in the public area. Service robots are one type used by humans to help with daily activities [11]. This robot has human-like abilities that can walk, see, talk, and understand the environment. Since the emergence of COVID-19 spread like a pandemic globally, prevention of this virus is the first step to reduce its impact by wearing face masks. It is useful for protecting the transmission of the virus through droplets in the mouth, and nose area [4]. So everyone to be required to wear a mask when in a public environment. This situation recommends service robots can detect and classify face masks in public areas. It is useful for warning people who don't use it.

Several previous studies have succeeded in classifying face masks. Ejaz et al. used the Principal Component Analysis (PCA) to recognize the face masks [3]. This study uses statistical differences of accuracy to measure the performance

N. T. Nguyen et al. (Eds.): ACIIDS 2021, LNAI 12672, pp. 507–517, 2021.
https://doi.org/10.1007/978-3-030-73280-6_40

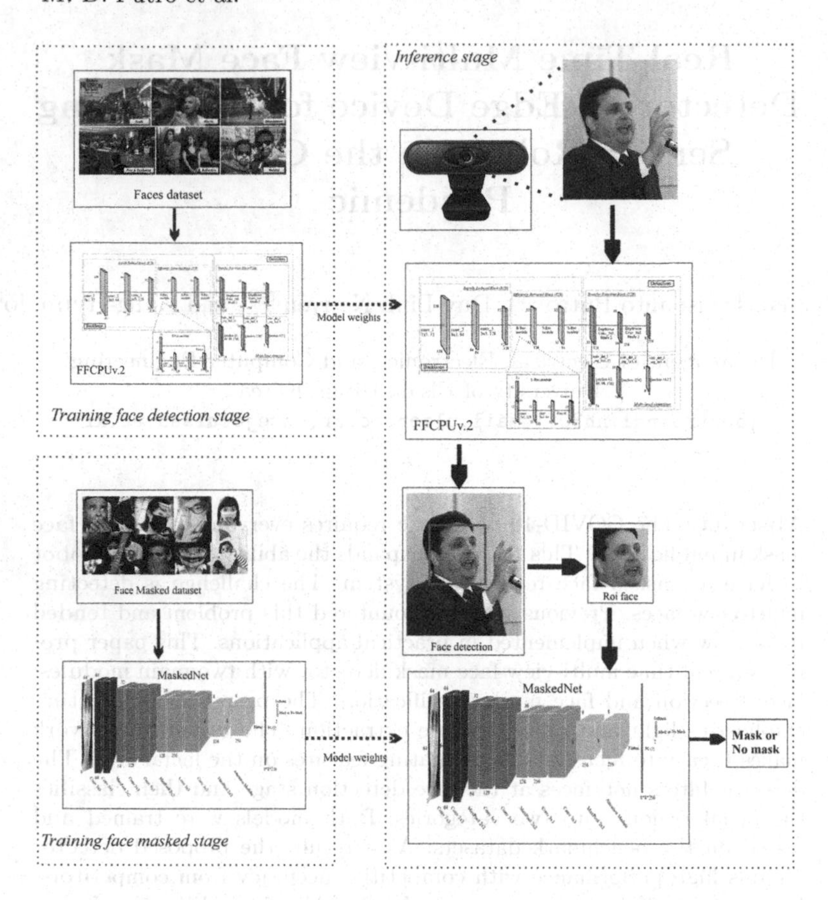

Fig. 1. Flowchart of overall face masks detector.

of the classification system. In addition, the face detection of Viola-Jones was employed for detecting the face region. Another work is applying the Gaussian Mixture Model (GMM) to build a faces model [1]. This system predicts face masks by analyzing and learning typical facial features in the eyes, nose, and mouth area. This model was tested on various face challenges using accessories such as respirators and sunglasses as an evaluation. However, both works are weak when classifying face masks on non-frontal faces. It is caused by the limitation of face detection and feature extraction.

Convolutional Neural Network (CNN) has been proven as a robust extractor feature [8]. In general, CNN architecture consists of feature extractor, and classifier [13]. This method can classify various images by updating the filter weights to produce an output that matched the ground truth [2]. However, this reliability is not supported by efficiency when run in real-time. Deep CNN tends to generate a large number of parameters and heavyweights [6]. Loey et al. used Resnet-50 as a backbone for feature extraction followed by decision trees, Support Vector Machine (SVM), and ensemble algorithm as the classifier [9]. The model produces high accuracy but is slow, while the practical application requires an algorithm to work in real-time on a portable device. Jetson Nano is a mini-computer supported by a 128-core graphic accelerator. This edge device is recommended for use as a robot processor because it supports sensor acquisition and actuator control functions.

This paper aims to build real-time masked face recognition on edge devices implemented in service robots. The contributions of this paper are as follows:

1. Fast face detector was developed to detect multi-view faces and occlusion challenges (FFDMASK). The process helps to get the RoI (Region of Interest) from the face.
2. Slim CNN architecture to fast and accurately classifies the face masks (masks or none). The attention module is applied to improve the quality of shallow feature maps.

As a result, it achieves competitive performance with state-of-the-art algorithms on several datasets and can work in real-time on the Jetson Nano without lack.

2 Proposed Architecture

The proposed detector consists of two-stage, including face detector and classification module. Figure 1 shows the overall diagram of the proposed system. A face detector works to detect faces in an image to produce bounding boxes as the face area [14]. The crop technique is used in each box to generate facial RoI. Then a classification module is employed to extract information and classify masked faces.

2.1 Face Detector

Backbone Module. FFDMASK develops the architecture of FFCPU [10]. This network consists of two main parts, including the backbone and detection block, as shown in Fig. 2. This model functions as a feature extractor to produce a clear feature map on the detection module. Rapidly Reduced Block (R2B) emphasizes reducing the dimension of the feature map with a convolutional layer and an Efficiently Extracted Block (E2B) to separate facial and non-facial distinctive features effectively. FFDMASK employs the S-Res (Split-residual) module to upgrade the E2B. This block divides the feature input into two parts based on the number of channels, then employs a bottleneck convolution on the first part and passes to the end of the module for the other chunk. Three modules are used to increase the detector's performance, which outperforms other competitors.

Detection Module and Anchor. The detection module is responsible for predicting the facial area, which includes three layers. This module employs a depthwise convolution between layers. This block saves computing power and increases speed. Besides, the decrease in accuracy does not have a significant impact. Furthermore, the assignment of anchors with scale variations at each detection layer can adjust the bounding box based on the size of the feature map, large anchors for small feature maps, and vice versa. In order to optimize training process, it still uses balanced loss which refers to the FFCPU.

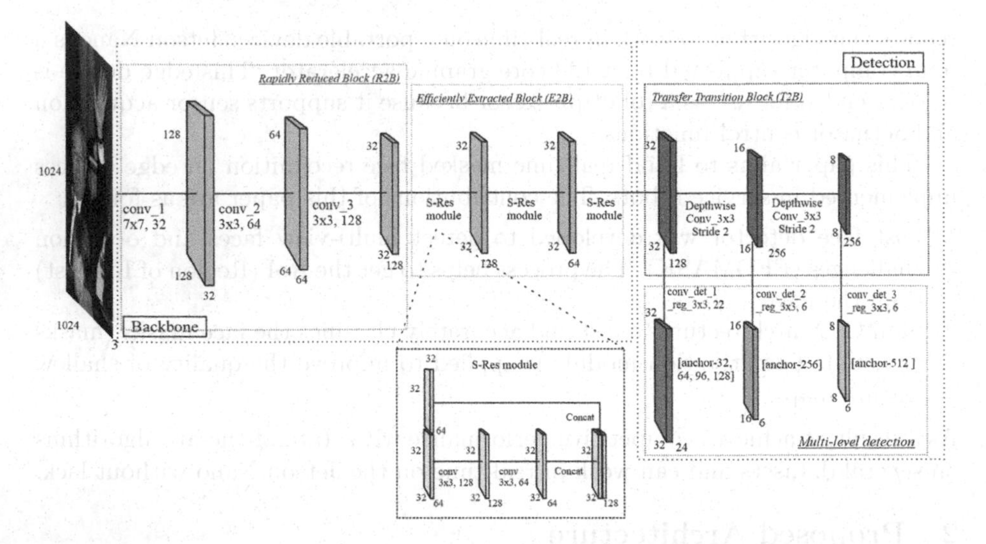

Fig. 2. The proposed architecture of face detector, including Rapidly Reduced Block (R2B), Efficiently Extracted Block (E2B), Transfer Transition Block (T2B), and Multi-level Detection. The training process requires 1024 × 1024 as the input image size.

2.2 Face Masks Classification Module

Baseline Module. The baseline module uses the convolutional neural network to extract distinctive features, and the max-pooling layers are used to shrink the feature map size. The ReLU and Batch Normalization are used to prevent saturation of the network. In order for the detector to work fast, this slim architecture emphasizes the shallow layers and narrow channels, as shown in Fig. 3. Each stage uses two 3 × 3 convolutions and a pool. This convolution has proven that it effectively separates distinctive features from the background (VGG) [12]. The input of this module is an RGB image with a size of 64 × 64. It produces 8 × 8 at the last of the feature map. Furthermore, this module employs an attention module to increase the discrimination power of facial features that are covered by masks and normal faces. Fully connected focuses on vectors with two categories and generates the final probability of predictions.

Attention Module. The shallow layer CNNs tend to produce low-level features. Even this architecture is underperforming when it discriminates against complex features [7]. This problem can be solved by employing the attention module to improve the quality of the feature map representing the global context of an input image. This technique can highlight the differences in facial features covered and without a mask. The position attention module is applied to capture context-based information and separate between interest and useless facial features [5].

The first step is to apply the 1 × 1 convolution as a simple buffer for the feature map output of the baseline module ($fm_{(b)}$), as illustrated by Fig. 4.

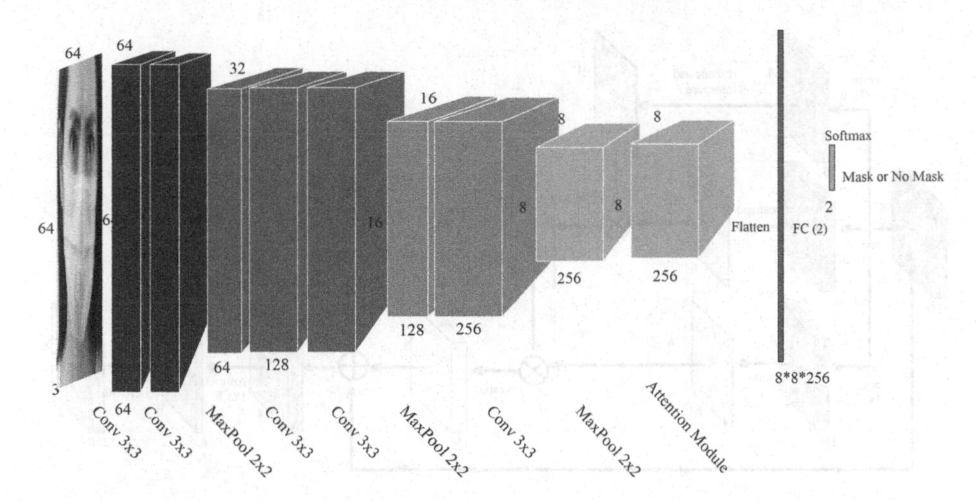

Fig. 3. The slim architecture of the face masks classification. This module consists of seven convolution layers, three max-pooling layers, an attention module, and a fully connected layer at the end of the network.

This operation generates new feature maps $fm_{(k)}$ and $fm_{(l)}$ with size H × W × C. Reshaping technique is required to obtain a single sized feature map (HW × HW). The probability of spatial weights is obtained to represent global information on a spatial scale, as shown in the following equation:

$$Att = fm_{(b)} + \frac{exp(fm_{(k)} \cdot fm_{(l)})}{\sum exp(fm_{(k)} \cdot fm_{(l)})} \cdot fm_{(m)}, \tag{1}$$

where Att measures each position pixel of the local features map with aggregate results from spatial attention and original maps. Furthermore, the module bottleneck is used as a simple feature extractor (1 × 1) without adding a significant amount of computation.

$$Att_{full} = W_{c2}ReLU(LN(W_{c1}Att)), \tag{2}$$

where it takes two convolutions (C_1 and C_2). The linear activation and normalization layers are only placed at the initial convolution. This module shrinks the channel size in the middle and then restores at the end of the module.

3 Implementation Setup

FFDMASK uses the WIDER FACE dataset as a knowledge of facial features to recognize the facial location in a set of images. Meanwhile, the Simulated Masked Face Dataset (SMFD) and the Labeled Faces in the Wild (LFW) are used for the training dataset of the face masks classification model. The detailed configuration of each training stage is shown in Table 1. The training was conducted on the Core I5-6600 CPU @ 3.30 Hz with GTX 1080Ti as an accelerator and Jetson Nano with 128 NVIDIA CUDA as edge devices for testing of the detector.

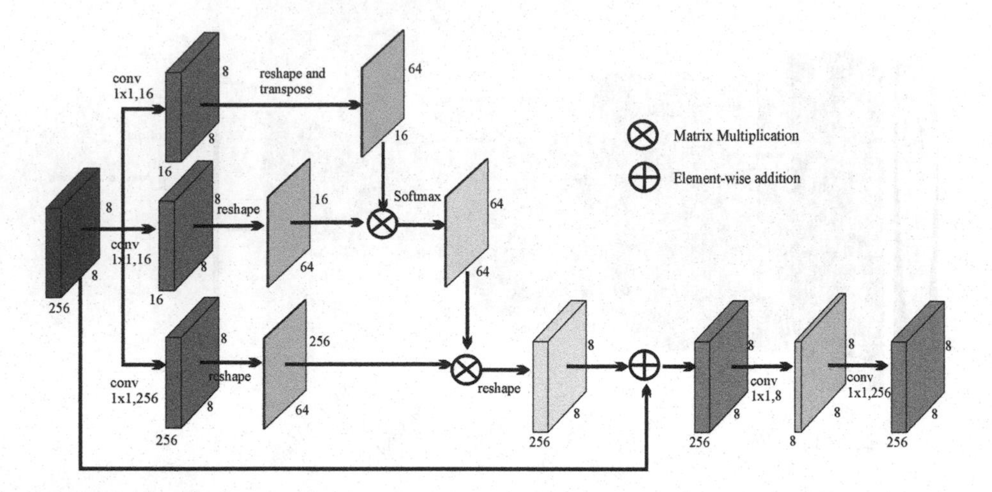

Fig. 4. Attention module.

Table 1. Implementation detail of face detection and face masks classification

Setting	Face detector	Face masks classification
Input image	1024 × 1024	64 × 64
Optimizer	Stochastic Gradient Descent (SGD)	Adam
Learning rate	10^{-5}–10^{-3}	10^{-7}–10^{-4}
Batch size	16	8
Total epoch	350	500
Loss function	L1 smooth loss	Categorical Cross-Entropy loss
Epsilon	–	10^{-7}
Weight decay	$5 \cdot 10^{-4}$	–
Momentum	0.9	–
IoU threshold	0.5	–
Framework	Pytorch	Keras

4 Experimental Results

In this section, the proposed architecture of the face detector and face masks classification is evaluated on several datasets. This evaluation shows the qualitative and quantitative results of each dataset. Additionally, another experiment has shown the runtime efficiency of a detector when tested on an edge device.

4.1 Face Detector Results

The FFDMASK detector's evaluation is carried out on the Face Detection Data Set and Benchmark (FDDB) dataset. It is a benchmark dataset consisting of 5,171 faces on 2,845 images. A variety of challenges are provided by this

Table 2. Accuracy of face masks classification on SMFD and LFW datasets

Model	Num of parameter	ACC (%) in SMFD	ACC (%) in LFW
Loey et al. [9]	23,591,816	99.49	100
Proposed	668,746	99.72	100

dataset, including scales, poses, lighting, and complex background. Discrete criteria were chosen as evaluations by comparing the IoU between prediction and ground truth. Figure 5(a) shows that the detector outperforms other competitors (FFCPU and Faceboxes). It is slightly superior to the leading competitors (FFCPU) by 0.2%. Besides, FFDMASK has faster data processing speed on different video input sizes, as shown in Fig. 5(b). Especially for the VGA input size, FFDMASK obtains 51.31 FPS while the FFCPU is 48.27 FPS on the Jetson Nano. These results compare the average speed of each detector when tested at 1000 frames. The quality of the detector performance is also shown in Fig. 5(c). It indicates that the proposed detector can overcome the challenges of occlusion, expressions, accessories, and complex backgrounds.

4.2 Face Masks Classification Results

Evaluation of SMFD Dataset. The dataset consists of 1,376 images, 690 for simulated masked faces, 686 for unmasked faces. It is used for the training and testing phases. This face dataset contains portrait images of male and female faces with a variety of poses and sizes. Face detection is applied to obtain a facial RoI measuring 64 × 64. This process helps the slim model to focus on learning facial features without being affected by background noise. In this dataset, the proposed architecture obtains an accuracy of 99.72%. This result is superior to Loey et al., which only achieves 99.49% on the same dataset. This success is supported by feature extraction suitable for preprocessed datasets. Figure 6(a) shows the qualitative results of the proposed detector. This result proves that the variations in facial poses in the dataset are not a barrier for the model to get high performance.

Evaluation of LFW Dataset. The benchmark dataset contains 13,000 masked faces for celebrities around the round. The training and testing process uses this dataset separately. The face dataset consists of facial images that were manipulated with an artificial mask that referred to work [9], as shown in Fig. 6(b). LFW masked has a size of 128 × 128, which instantly provides the RoI of a face that is avoided from the background. Our model uses 64 × 64 as the input size of RoI, which is reshaped from the original size. As a result, the proposed detector achieves perfect and competitive results with Loey et al., as shown in Table 2. The majority of this dataset is in the frontal pose. It is more comfortable than the SMFD dataset. The proposed model explicitly discriminates against nose

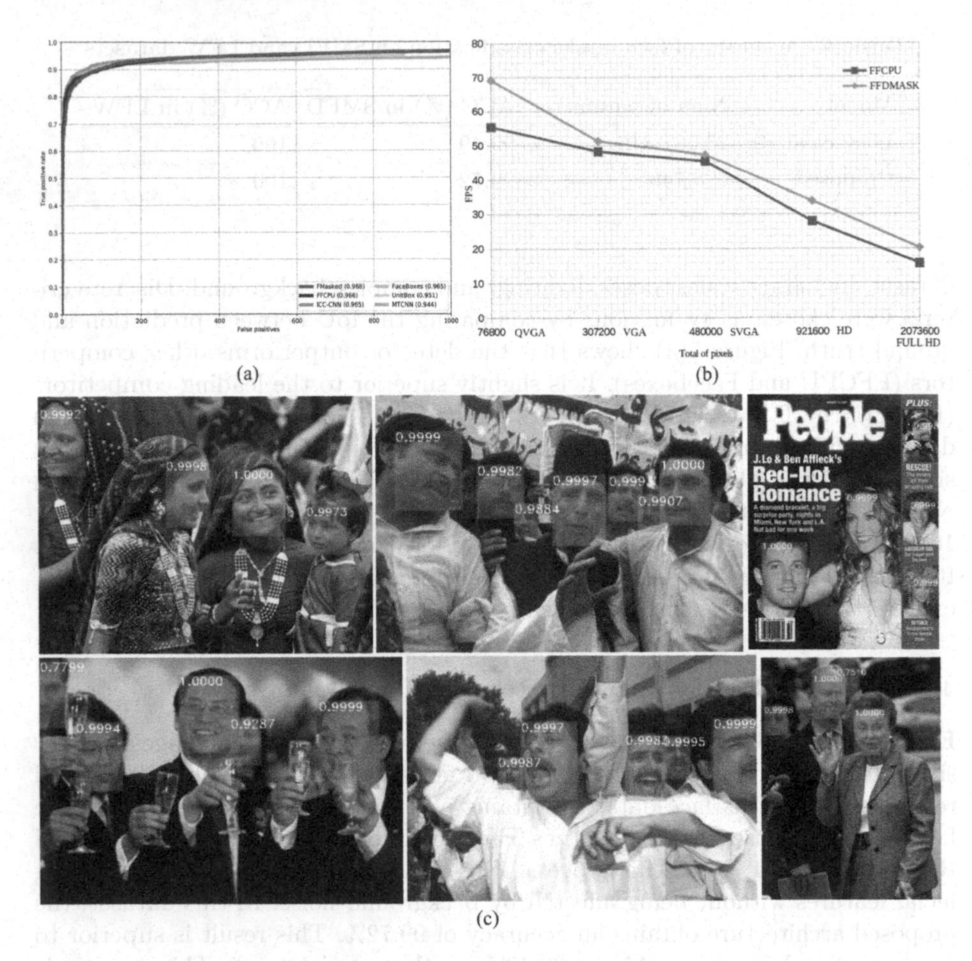

Fig. 5. ROC (Receiver Operating Characteristics) discrete evaluation curve on the FDDB dataset (a), comparison of mean detector speed at different video input sizes (b), qualitative results on FDDB datasets (c).

and mouth features from other features. These features tend to be closed and undetectable when the face is wearing a mask.

4.3 Runtime Efficiency

The practical application recommends a computer vision method to run real-time on portable devices. In general, service robots use mini-computers to process intelligent algorithms and computer vision. It requires a small accelerator to process the computation of the algorithm. Therefore, the proposed architecture is tested on a Jetson Nano to find out the efficiency of the model. Table 3 shows that face detector and face masks classification models have less computing power than competitors. FFDMASK produces a smaller number of parameters than

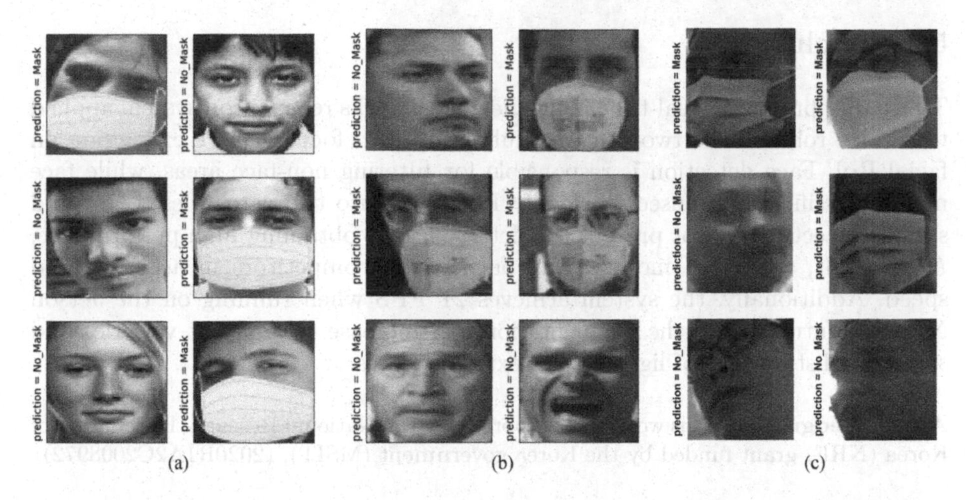

Fig. 6. Qualitative results on SMFD (a), LFW dataset (b), and running real-time (c).

Table 3. Comparison of detector speeds with competitors on the Jetson Nano.

Model	Input size	Num of parameter	Accuracy (%)	Speed(FPS)
Face detector				
FFCPU [10]	640 × 480	715,844	96.60	48.27
FFDMASK	640 × 480	602,310	96.80	51.31
Face masks classification				
Loey et al. [9]	64 × 64	23,591,816	99.49	5.80
Proposed	64 × 64	668,746	99.72	20.60

FFCPU. It also impacts the speed of the detector. The S-Res module emphasizes computational savings for the residual method without compromising the quality of feature extraction.

Furthermore, the proposed model of face masks classification produces a faster speed than Loey et al. This competitor uses the Resnet-50 backbone module, which generates many parameters and heavyweights. Meanwhile, the proposed module only requires a slim architecture to obtain superior results. Table 3 shows that this model produces 668,746 as of the number of parameters and achieves 20.60 FPS on the Jetson Nano. These results are an accumulation of face detection (VGA-resolution) and masks classification speed (RoI size of 64). Real-time detectors use training data on the SMFD dataset, which tends to have a variety of poses. As a result, the system achieves high performance when it recognizes multi-view masked faces, as shown in Fig. 6(c). Besides, this detector also obtained satisfactory results for the challenge of various colored masks.

5 Conclusion

This paper presents a real-time multi-view face masks recognition system applied to service robots. The two-stage module is used to focus feature extraction on facial RoI. Face detection is responsible for filtering non-face areas, while face masks classification is used to classify Roi faces into two categories. Light and slim architecture do not prevent the detector from obtaining high performance. As a result, two CNN modules can outperform competitors in accuracy and speed. Additionally, the system achieves 21 FPS when running on the Jetson Nano. In future work, the augmentation can increase the dataset varieties and solve the disturbance of lighting and extreme poses.

Acknowledgment. This work was supported by the National Research Foundation of Korea (NRF) grant funded by the Korea government (MSIT). (2020R1A2C2008972).

References

1. Chen, Q., Sang, L.: Face-mask recognition for fraud prevention using Gaussian mixture model. J. Vis. Commun. Image Represent. **55**, 795–801 (2018). http://www.sciencedirect.com/science/article/pii/S1047320318302050
2. Ejaz, M.S., Islam, M.R.: Masked face recognition using convolutional neural network. In: 2019 International Conference on Sustainable Technologies for Industry 4.0 (STI), pp. 1–6 (2019)
3. Ejaz, M.S., Islam, M.R., Sifatullah, M., Sarker, A.: Implementation of principal component analysis on masked and non-masked face recognition. In: 2019 1st International Conference on Advances in Science, Engineering and Robotics Technology (ICASERT), pp. 1–5 (2019)
4. Fadare, O.O., Okoffo, E.D.: Covid-19 face masks: a potential source of microplastic fibers in the environment. Sci. Total Environ. **737** (2020). http://www.sciencedirect.com/science/article/pii/S0048969720338006
5. Fu, J., Liu, J., Tian, H., Li, Y., Bao, Y., Fang, Z., Lu, H.: Dual attention network for scene segmentation. In: 2019 IEEE/CVF Conference on Computer Vision and Pattern Recognition (CVPR), pp. 3141–3149 (2019)
6. He, K., Zhang, X., Ren, S., Sun, J.: Deep residual learning for image recognition. In: 2016 IEEE Conference on Computer Vision and Pattern Recognition (CVPR), pp. 770–778 (2016)
7. Hu, J., Shen, L., Sun, G.: Squeeze-and-excitation networks. In: 2018 IEEE/CVF Conference on Computer Vision and Pattern Recognition, pp. 7132–7141 (2018)
8. Liu, W., Wang, Z., Liu, X., Zeng, N., Liu, Y., Alsaadi, F.E.: A survey of deep neural network architectures and their applications. Neurocomputing **234**, 11–26 (2017). http://www.sciencedirect.com/science/article/pii/S0925231216315533
9. Loey, M., Manogaran, G., Taha, M.H.N., Khalifa, N.E.M.: A hybrid deep transfer learning model with machine learning methods for face mask detection in the era of the Covid-19 pandemic. Measurement **167**, 108288 (2021). http://www.sciencedirect.com/science/article/pii/S0263224120308289
10. Putro, M.D., Jo, K.: Fast face-CPU: a real-time fast face detector on CPU using deep learning. In: 2020 IEEE 29th International Symposium on Industrial Electronics (ISIE), pp. 55–60 (2020)

11. Putro, M.D., Jo, K.-H.: Real-time multiple faces tracking with moving camera for support service robot. In: Nguyen, N.T., Gaol, F.L., Hong, T.-P., Trawiński, B. (eds.) ACIIDS 2019. LNCS (LNAI), vol. 11432, pp. 639–647. Springer, Cham (2019). https://doi.org/10.1007/978-3-030-14802-7_55
12. Simonyan, K., Zisserman, A.: Very deep convolutional networks for large-scale image recognition. In: International Conference on Learning Representations (2015)
13. Zeiler, M.D., Fergus, R.: Visualizing and understanding convolutional networks. In: Fleet, D., Pajdla, T., Schiele, B., Tuytelaars, T. (eds.) ECCV 2014. LNCS, vol. 8689, pp. 818–833. Springer, Cham (2014). https://doi.org/10.1007/978-3-319-10590-1_53
14. Zhang, S., Wang, X., Lei, Z., Li, S.Z.: Faceboxes: a CPU real-time and accurate unconstrained face detector. Neurocomputing **364**, 297–309 (2019). http://www.sciencedirect.com/science/article/pii/S0925231219310719

Eye State Recognizer Using Light-Weight Architecture for Drowsiness Warning

Duy-Linh Nguyen(✉), Muhamad Dwisnanto Putro, and Kang-Hyun Jo

School of Electrical Engineering, University of Ulsan, Ulsan, Korea
{ndlinh301,dwisnantoputro}@mail.ulsan.ac.kr, acejo@ulsan.ac.kr

Abstract. The eye are a very important organ in the human body. The eye area and eyes contain lots of useful information about human interaction with the environment. Many studies have relied on eye region analyzes to build the medical care, surveillance, interaction, security, and warning systems. This paper focuses on extracting eye region features to detect eye state using the light-weight convolutional neural networks with two stages: eye detection and classification. This method can apply on simple drowsiness warning system and perform well on Intel Core I7-4770 CPU @ 3.40 GHz (Personal Computer - PC) and on quad-core ARM Cortex-A57 CPU (Jetson Nano device) with 19.04 FPS and 17.20 FPS (frames per second), respectively.

Keywords: Convolutional neural network (CNN) · Deep learning · Drowsiness warning · Eye detection · Eye classification · Eye state recognizer

1 Introduction

The traffic accident is a great threat to human beings all over the world. More than one million people die each year from road traffic crashes and 90% of the main cause is from drivers [3]. One of the main causes is driver drowsiness. This situation usually occurs when a driver lacks sleep, uses alcohol, uses drugs, or goes on a long trip. To detect driver's drowsiness, many methods have been specifically conducted such as analyzing human behavior, vehicle behavior, and driver physiology [21]. Human behavior can be surveillance through the extraction of facial features, eye features, yawning, or head gestures. Vehicle behavior can be monitored via vehicle movement in the lanes and relative to other vehicles operating nearby. Driver physiology can be estimated by sensors that measure heart rate, blood pressure, or sudden changes in body temperature. However, deploying applications to monitor vehicle behavior and examine human physiology requires huge complexity and costly techniques. In addition, it can cause uncomfortable and unfocus for the driver during road operation. From the above analysis, this paper proposes the light-weight architecture design supports the driver's drowsiness warning system. The system consists of two main stages based on extracting the eye area features: eye detection and classification.

N. T. Nguyen et al. (Eds.): ACIIDS 2021, LNAI 12672, pp. 518–530, 2021.
https://doi.org/10.1007/978-3-030-73280-6_41

Deploying the application is very simple on PC or on Jetson Nano device and a common camera.

The paper has two main contributions as follows:

- Proposed two light-weight Convolutional Neural Network architectures, includes eye detection and classification.
- Develop the eye state recognizer can run on small processor devices supporting for drowsiness warning system without ignoring the accuracy.

The remaining of this paper is organized: Sect. 2 present the previous related methodologies to eye state detection, drowsiness warning system, their strengths, and weaknesses. Section 3 shows detail about the proposed technique. Section 4 describes and analyzes the results. The paper finalizes by Sect. 5 with the conclusion and future work.

2 Related Work

In the related work section, the paper will show several methodologies applied to eye state detection and drowsiness warning system. These methodologies can be grouped into the untraining and training methodologies.

2.1 Untraining Methodologies

In the untrained method, sensors are often used to measure the signal obtained from parts of the human body or objects. In addition, several image processing algorithms are also used to extract characteristics on the image from which to make predictions. The techniques used in [4–6,9] rely on sensors arranged around the eyes to gather and analyze electrical signals. These techniques can collect signals very quickly but are uncomfortable for the user and may be subject to interference due to environmental influences. Therefore, it leads to low accuracy while expensive implementation. In the Computer Vision field, there are many methods to extract eye area and inside eye features without training. Specifically, the methods include iris detection based on calculating the variance of maximum iris positions [7], methods based on matching the template [13], and methods based on a fuzzy-system that using eye segmentation [11]. Scale Invariant Feature Transformation (SIFT) in [15] consider image information in continuous video, method based on the movement in facial and eyelid [19], method computes the variance in the values of black pixels in these areas [17]. These methods can provide powerful feature information but require complex computation and are very sensitive to illumination.

2.2 Training Methodologies

The training method is based on extracting the features and learning them. There are some traditional methods such as Support Vector Machine (SVM) [10],

Active Appearance Models (AAM) [27], Principal Component Analysis (PCA) subspace analysis [29]. These methods may achieve better eye classification accuracy than the untrained methods, but they need to be improved or combined with other techniques to adapt to variability in real-time.

With the explosive growth of machine learning, the widespread application of convolutional neural networks in image classification, object detection and recognition is increasing. Many typical CNNs can be used to classify eye state such as Lenet [16], Alexnet [12], VGG [24], Resnet [8] and so on. In these methods, the feature extracted automatically from the dataset through the training process and then classifies the images based on these features. Their performance is reasonable on accuracy and loss function. However, these models have heavy training time depend on the depth of models and size of input images. In addition, the complicated construction of the eye and eye area requires to improve to the CNN models to accommodate accuracy and loss function.

Recently years, several studies have used traditional face detection methods such as Viola-Jones [28], Haar-like feature [18], Adaboost [14] in combination with CNNs to detect eye status. However, these techniques often face illumination conditions, not frontal face, occluded or overlap, and oblique face position.

3 Methodology

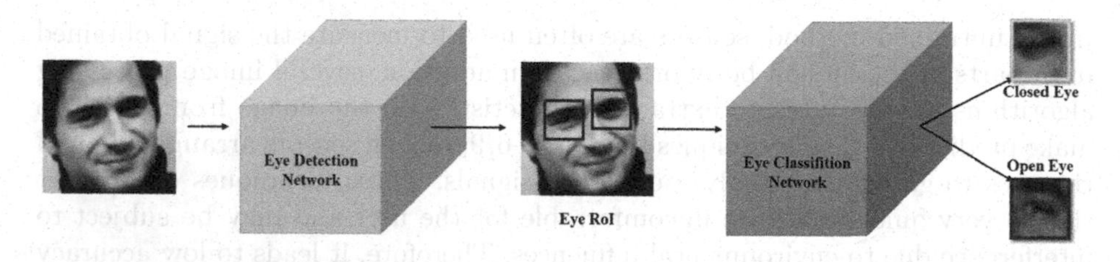

Fig. 1. The proposed pipeline of the eye state recognizer. It consists of two main networks: eye detection and eye classification network. The input image goes through the eye detection network and the RoI eye regions are generated in this network, then these regions will be classified by the eye classification network to predict eye state (open and closed for each eye).

The proposed pipeline of the detailed eye state recognizer is shown in Fig. 1. The pipeline consists of two networks which are eye detection and eye classification network. In the eye detection network, we proposed light-weight and efficient CNN to extract the Region of Interest (RoI) of the eye region and then crop these areas. The output of this network goes through the eye classification network, which is a simple CNN for classifying eyes. The output are eye states: closed eye and open eye in each eye region.

3.1 Eye Detection Network

This study proposes a convolutional neural network architecture that locates the eye areas in the images. This network extracts the feature maps by using the basic layers and components in CNN such as convolution and max-pooling layers, C.ReLU, and Inception modules. After that, two sibling convolution layers will be applied for classification and regression. The detail of the proposed network is described in Fig. 2.

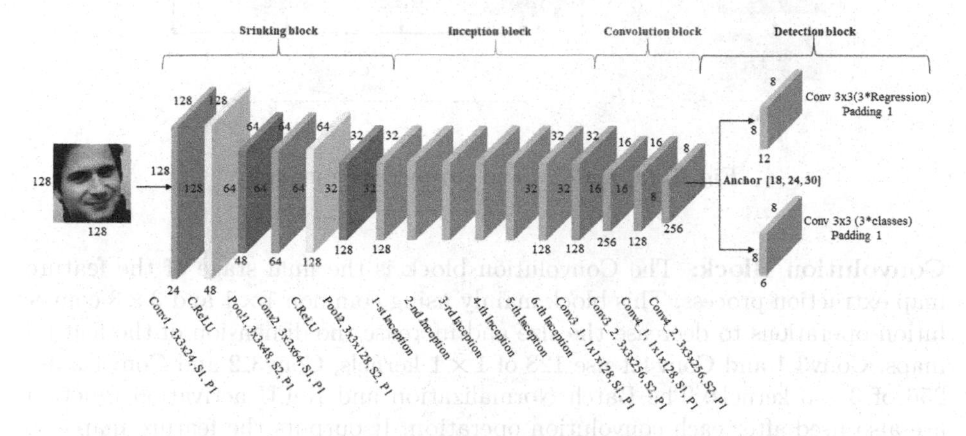

Fig. 2. The proposed eye detection network consists of four sequential blocks: Shrinking, Inception, Convolution, and Detection block.

Shrinking Block: This block quickly decreases the input image spatial by choosing the appropriate kernel size. The stride used in Conv1, Pool1, Conv2, and Pool2 is 1, 2, 1, and 2, respectively. The 3×3 kernel size is mainly used and the input image size is resized to 128×128. On the other hand, C. ReLu (Concatenated Rectified Linear Unit) [23] is also used to increase efficiency while ensuring accuracy. The C.ReLU module is described in Fig. 3(a). This block shrunk down the input image size to 32×32. In another word, the size is reduced by four times while maintaining the important information of the input image.

Inception Block: To build the Inception block, a combination of six Inception modules [26] is used. Each Inception module consists of four branches, using consequential convolution operations with kernel size 1×1, 3×3 and the number of kernels is 24, 32. Following each convolution operation, the Batch Normalization and ReLU activation function are used. In some branches, the max-pooling operation is also used and final by concatenation operation to combine the results of branches. With a multi-scale approach according to the width of the network, these branches can enrich receptive fields. Figure 3(b) shown the Inception module in detail. The feature map with a size is $32 \times 32 \times 128$ will be unchanged from $1 - st$ to $6 - th$ Inception and provided the various information of features when processed by this block.

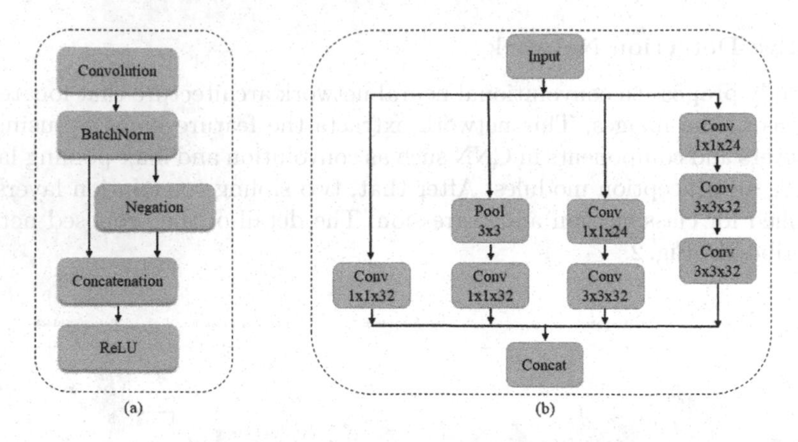

Fig. 3. (a) C.ReLU and (b) Inception module

Convolution Block: The Convolution block is the final stage of the feature map extraction process. This block mainly using common 1×1 and 3×3 convolution operations to decrease the size and increase the dimension of the feature maps. Conv3_1 and Conv4_1 use 128 of 1×1 kernels, Conv3_2 and Conv4_2 use 256 of 3×3 kernels. The Batch Normalization and ReLU activation function are also used after each convolution operation. It outputs the feature map size of $8 \times 8 \times 256$, which proves that the size of the feature map is further reduced by four times and the number of kernels is doubled from 128 to 256. The role of the Convolution block is the link between the feature map extraction and the detection stage.

Detection Block: The final of the eye detection network is the detection block. This block uses two 3×3 convolution operations for classification and bounding box regression. These layers apply on the 8×8 feature map which is an output feature map from the previous block. The various predefined square anchors use to predict the position of the corresponding eye in the original image. In this work, it uses three square anchors (18, 24, 30) for small eye sizes (18), medium eye sizes (24), and large eye sizes (30), respectively. In the end, the detector generates a four-dimensional vector (x, y, w, h) as the offset of location and a two-dimensional vector (eye or not eye) as label classification.

The loss function of the eye detection network is similar to RPN (Region Proposal Network) in Faster R-CNN [22], the softmax-loss to compute loss for classification and the smooth $L1$ loss for regression task. For an image, the loss function is defined as below:

$$L\left(\{p_i\}, \{t_i\}\right) = \frac{1}{N_{cls}} \sum_i L_{cls}\left(p_i, p_i^*\right) + \lambda \frac{1}{N_{reg}} \sum_i p_i^* L_{reg}\left(t_i, t_i^*\right), \quad (1)$$

where p_i is the predicted probability of anchor i being an object, p_i^* is ground-truth label. The anchor is positive when $p_i^* = 1$ and it is negative if $p_i^* = 0$. t_i is the center coordinates and dimension (Height and Width) of the prediction and t_i^* is the ground truth coordinates. $L_{cls}(p_i, p_i^*)$ is the classification loss using the softmax-loss shown as in Eq. (2), $L_{reg}(t_i, t_i^*) = R(t_i - t_i^*)$ with R is the Smooth loss $L1$ defined as in Eq. (3). The Eq. (1) is normalized by N_{cls} and N_{reg} and balancing by parameter λ. N_{cls} is normalized by the mini-batch size, N_{reg} is normalized by the number of anchor locations and λ is set by 10.

$$L_{cls}(p_i, p_i^*) = -\sum_{i \in Pos} x_i^p log(p_i) - \sum_{i \in Neg} log(p_i^0), \tag{2}$$

where $x_i^p = \{0, 1\}$ is indicator for matching the $i-th$ default box to ground-truth of category p, p_i^0 is the probability for non-object classification.

$$R(x) = \begin{cases} 0.5x^2 & if\,|x| < 1 \\ |x| - 0.5 & othcrwise \end{cases} \tag{3}$$

3.2 Eye Classification Network

Figure 4 shows a detailed description of the classification network architecture. Similar to the CNN of classification, this network is built based on sequential layers as convolution layers, average pooling layers, and uses Softmax function to classify the data.

This network architecture uses one group of two Convolution layers with 7×7 filter size, one group of two convolution layers with 5×5 filter size, two groups of two convolution layers with 3×3 filter size followed by each group of one average pool layer and one ReLU activation function. The feature extractor ends by one Convolution layer with a 3×3 filter size. The spatial dimensions of the feature map are reduced from 100×100 to 7×7. The global average pooling layer to further reduce the dimension of the feature map to 1×1. Finally, the network uses the Softmax activation function to generate the predicted probability of each class (open and closed eyes). Usage of Global average pooling can minimize the possibility of overfitting by reducing the total number of parameters in the network. On the other hand, to increase the ability to avoid network overfitting, the Batch Normalization method is also used after convolution operations. The classifier uses the Cross-Entropy loss function to calculate the loss during training.

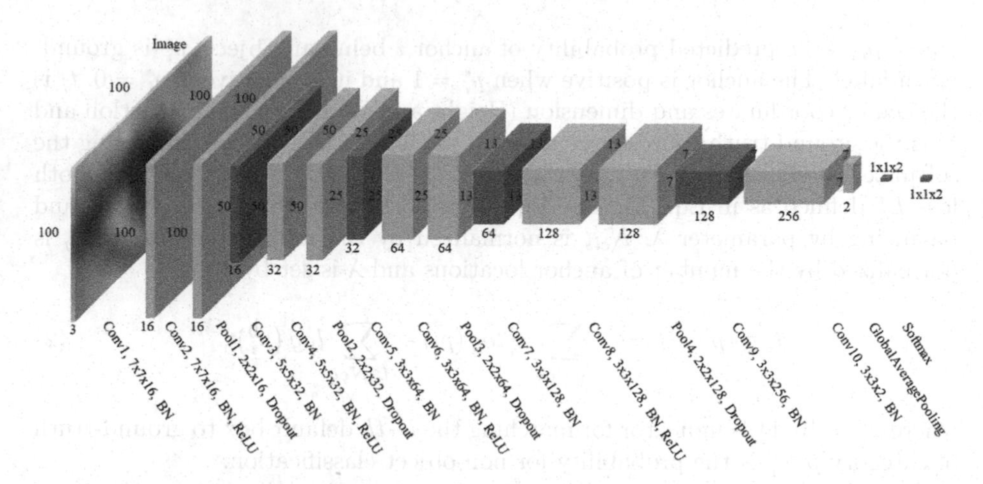

Fig. 4. The proposed eyes classification network. The network is based on nine sequential layers of convolution with filters of sizes 7×7, 5×5 and 3×3. Following the convolution layer groups are the ReLU activation functions and the average pooling layers. The global average pooling layer is used to quickly reduce the size of the feature map. Finally, it applies a Softmax activation function to compute the probability of open or closed eyes.

4 Experimental Result

4.1 Dataset Preparation

The eye detection network is trained on the CEW (Closed Eyes In The Wild) [25], BioID Face [1] and GI4E (Gaze Interaction for Everybody) dataset [2]. CEW dataset contains 2,423 subjects, among which 1,192 subjects with both eyes closed, and 1,231 subjects with eyes open. The image size is 100×100 (pixels) and it is extracted the eye patches of 24×24 from the central location of eye position. BioID Face dataset consists of 1,521 gray level images with a resolution of 384×286 pixel. Each one shows the frontal view of the face of 23 different persons. The eye position label is set manually and generated coordinate of ground-truth bounding box based on this position with 36×36 size. GI4E is a dataset for iris center and eye corner detection. The database consists of a set of 1,339 images acquired with a standard webcam, corresponding to 103 different subjects and 12 images each. The image resolution is 800×600 pixels in PNG format. It contains manually annotated 2D iris and corner points (in pixels). The coordinate of ground-truth of the bounding box also generated from the position of iris by size is 46×46. Each original data set is split into two subsets with 80% for training and 20% for the testing phase.

The eyes classification network is trained and evaluated on Closed Eyes In The Wild (CEW) dataset [22]. This dataset contains 2,384 eye images with closed eyes and 2,462 face images with open eye images. To improve the classification capacity, this dataset has been augmented by flipping vertically, changing the contrast and brightness. The dataset was divided into 80% images for the training set and 20% images for the evaluation set.

4.2 Experimental Setup

The training phase is implemented on GeForce GTX 1080Ti GPU, tested on Intel Core I7-4770 CPU @ 3.40 GHz, 8 GB of RAM (PC), and quad-core ARM Cortex-A57 CPU, 4 GB of RAM (Jetson Nano device). Many configurations have been used in the training phase. The network used the Stochastic Gradient Descent optimization method, the batch size of 16, the weight decay is 5.10^{-4}, the momentum is 0.9, the learning rates from 10^{-6} to 10^{-3}. In order to generate the best bounding box, the threshold of IoU (Intersection over Union) is set by 0.5. For the eye classification network which uses some basic configuration for image classification such as the Adam optimization method, batch size of 16, the learning rate is 10^{-4}.

4.3 Experimental Result

Fig. 5. The qualitative result of the eye detection network on CEW dataset (first row), on BioID Face dataset (second row), and on GI4E dataset (third row). This network can detect eye location in different situations: head poses, the glasses-wearing, laboratory environment. The number in each bounding box shows a confidence score of prediction.

Each network in the pipeline is individually trained and tested on the dataset of image and a comprehensive network was tested on a real-time system using a common camera connecting the PC based on CPU and Jetson Nano device with the quad-core ARM Cortex-A57 CPU. As a result, the eye detection network achieved results on CEW, BioID Face, and GI4E dataset with 96.48%, 99.58%, and 75.52% of AP, respectively. The testing result of the eye detection network on CEW, BioID Face, and GI4E test set shown in Table 1 and Fig. 5. The classification results of the eye classification network on the CEW dataset are shown in Table 2. The proposed classification network outperforms compared to popular classification networks with a very small number of parameters.

Table 1. The testing result of the eye detection network on CEW, BioID Face and GI4E test set.

Dataset	Average precision (%)
CEW	96.48
BioID Face	99.58
Gi4E	75.52

Table 2. The comparison of classification results of the eyes classification network with popular classification networks on the CEW dataset.

Network	Accuracy (%)	Number of paramenters
Proposed	97.53	632,978
VGG13	96.29	7,052,738
ResNet50	94.85	23,591,810
Alexnet	96.71	67,735,938
LeNet	96.70	15,653,072

Finally, the entire system was tested on a camera connected to a CPU-based PC and Jetson Nano device. In order to increase efficiency for eye state recognition, the pipeline adds the trained face detection network in previous work in [20]. The distance from the camera to the human face is equal to the distance in the car. Because the distance is set quite close (distance < 0.5 m), the images obtained via the camera are mainly images in the frontal face. This condition improves eye detection and open and closed eye classifier. Table 3 shown the speed testing result of the eye state recognizer on the camera.

Table 3. The speed testing results of eyes status recognizer on the camera.

Device	Face detection (fps)	Eye detection (fps)	Eye classification (fps)	Total (fps)
Jetson Nano	100.02	30.89	63.08	17.20
PC	198.22	42.52	41.73	19.04

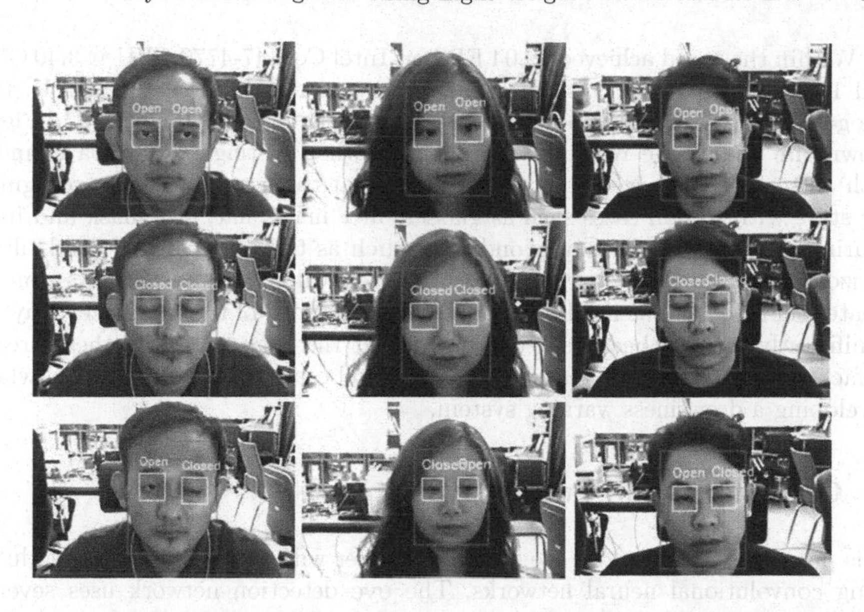

Fig. 6. The qualitative result when testing on camera connects with Jetson Nano device with three participants (two males and one female). The result shows in two open eyes (first row), two closed eyes (second row), one closed eye and one ope eye (third row).

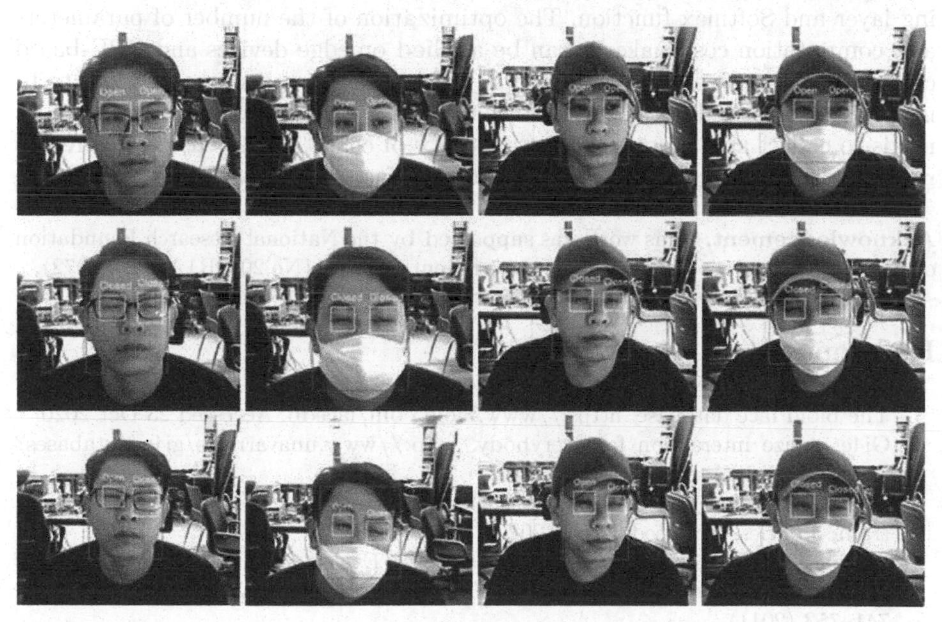

Fig. 7. The qualitative result when testing on camera connects with Jetson Nano device with four situations: glasses (first column), face mask (second column), hat (third column), face mask and hat-wearing (fourth column).

Within the speed achieved 19.04 FPS on Intel Core I7-4770 CPU @ 3.40 GHz and 17.20 FPS fps when tested on the quad-core ARM Cortex-A57 CPU, the recognizer can work well in normal conditions without delay. Figure 6 and Fig. 7 shown the qualitative result of the pipeline when testing on camera connect with Jetson Nano device. The result proves that the system also can recognize eye state with several cases such as glasses, face mask, hat, face mask and hat-wearing. However, under noise conditions such as the illumination, head tilted horizontally at an angle greater than 45°, vertical rotation head at an angle greater than 90°, or the head bows down, the efficiency of the recognizer may be significantly reduced because it can not detect the eyes to classify these areas. In fact, these cases can be referred to as unusual cases that can be alerted when developing a drowsiness warning system.

5 Conclusion and Future Work

This paper has proposed an eye state recognizer with two light-weight modules using convolutional neural networks. The eye detection network uses several basic layers in CNN, C.ReLu, Inception module. The eye classification network is a simple convolutional neural network that consists of convolution layers alternating with the average pooling layers, then ending by the global average pooling layer and Softmax function. The optimization of the number of parameters and computation cost makes it can be applied on edge devices and CPU-based computers. The eye state recognizer will be integrated with several modern techniques and advanced optimization in the future. On the other hand, the dataset needs to collect and annotate under a variety of conditions to ensure this recognizer works properly such as glasses, hat, and face mask-wearing.

Acknowledgement. This work was supported by the National Research Foundation of Korea (NRF) grant funded by the government (MSIT).(No.2020R1A2C2008972).

References

1. The bioid face database. https://www.bioid.com/facedb. Accessed 23 Oct 2020
2. Gi4e - gaze interaction for everybody. http://www.unavarra.es/gi4e/databases?languageId=1. Accessed 23 Oct 2020
3. Road traffic injuries. https://www.who.int/news-room/fact-sheets/detail/road-traffic-injuries. Accessed 22 Oct 2020
4. Bulling, A., Ward, J., Gellersen, H., Tröster, G.: Eye movement analysis for activity recognition using electrooculography. IEEE Trans. Pattern Anal. Mach. Intell. **33**, 741–753 (2011)
5. Champaty, B., Pal, K., Dash, A.: Functional electrical stimulation using voluntary eyeblink for foot drop correction. In: 2013 Annual International Conference on Emerging Research Areas and 2013 International Conference on Microelectronics, Communications and Renewable Energy, pp. 1–4 (2013). https://doi.org/10.1109/AICERA-ICMiCR.2013.6575966

6. Chang, W., Lim, J., Im, C.: An unsupervised eye blink artifact detection method for real-time electroencephalogram processing. Physiol. Meas. **37**(3), 401–17 (2016)
7. Colombo, C., Comanducci, D., Bimbo, A.D.: Robust tracking and remapping of eye appearance with passive computer vision. ACM Trans. Multimedia Comput. Commun. Appl. **3**, 2:1–2:20 (2007)
8. He, K., Zhang, X., Ren, S., Sun, J.: Deep residual learning for image recognition. CoRR abs/1512.03385 (2015). http://arxiv.org/abs/1512.03385
9. Hsieh, C.S., Tai, C.C.: An improved and portable eye-blink duration detection system to warn of driver fatigue. Instrum. Sci. Technol. **41**(5), 429–444 (2013). https://doi.org/10.1080/10739149.2013.796560
10. Jo, J., Lee, S., Jung, H., Park, K., Kim, J.: Vision-based method for detecting driver drowsiness and distraction in driver monitoring system. Opt. Eng. **50**, 7202 (2011). https://doi.org/10.1117/1.3657506
11. Kim, K.W., Lee, W.O., Kim, Y.G., Hong, H.G., Lee, E.C., Park, K.R.: Segmentation method of eye region based on fuzzy logic system for classifying open and closed eyes. Opt. Eng. **54**(3), 1–19 (2015). https://doi.org/10.1117/1.OE.54.3.033103
12. Krizhevsky, A., Sutskever, I., Hinton, G.E.: ImageNet classification with deep convolutional neural networks. Commun. ACM **60**(6), 84–90 (2017). https://doi.org/10.1145/3065386
13. Królak, A., Strumiłło, P.: Eye-blink detection system for human–computer interaction. Univers. Access Inf. Soc. **11**(4), 409–419 (2012). https://doi.org/10.1007/s10209-011-0256-6
14. Kégl, B.: The return of adaboost.mh: multi-class hamming trees (2013)
15. Lalonde, M., Byrns, D., Gagnon, L., Teasdale, N., Laurendeau, D.: Real-time eye blink detection with GPU-based sift tracking. In: Fourth Canadian Conference on Computer and Robot Vision (CRV 2007), pp. 481–487 (2007). https://doi.org/10.1109/CRV.2007.54
16. Lecun, Y., Bottou, L., Bengio, Y., Haffner, P.: Gradient-based learning applied to document recognition. Proc. IEEE **86**(11), 2278–2324 (1998). https://doi.org/10.1109/5.726791
17. Lee, W., Lee, E.C., Park, K.: Blink detection robust to various facial poses. J. Neurosci. Methods **193**, 356–72 (2010). https://doi.org/10.1016/j.jneumeth.2010.08.034
18. Mita, T., Kaneko, T., Hori, O.: Joint Haar-like features for face detection. In: Tenth IEEE International Conference on Computer Vision (ICCV 2005) Volume 1, vol. 2, pp. 1619–1626 (2005). https://doi.org/10.1109/ICCV.2005.129
19. Mohanakrishnan, J., Nakashima, S., Odagiri, J., Shanshan, Yu.: A novel blink detection system for user monitoring. In: 2013 1st IEEE Workshop on User-Centered Computer Vision (UCCV), pp. 37–42 (2013). https://doi.org/10.1109/UCCV.2013.6530806
20. Nguyen, D.L., Putro, M.D., Jo, K.H.: Eyes status detector based on light-weight convolutional neural networks supporting for drowsiness detection system. In: IECON 2020 The 46th Annual Conference of the IEEE Industrial Electronics Society, pp. 477–482 (2020). https://doi.org/10.1109/IECON43393.2020.9254858
21. Ramzan, M., Khan, H.U., Awan, S.M., Ismail, A., Ilyas, M., Mahmood, A.: A survey on state-of-the-art drowsiness detection techniques. IEEE Access **7**, 61904–61919 (2019). https://doi.org/10.1109/ACCESS.2019.2914373
22. Ren, S., He, K., Girshick, R.B., Sun, J.: Faster R-CNN: towards real-time object detection with region proposal networks. CoRR abs/1506.01497 (2015). http://arxiv.org/abs/1506.01497

23. Shang, W., Sohn, K., Almeida, D., Lee, H.: Understanding and improving convolutional neural networks via concatenated rectified linear units. CoRR abs/1603.05201 (2016). http://arxiv.org/abs/1603.05201
24. Simonyan, K., Zisserman, A.: Very deep convolutional networks for large-scale image recognition. CoRR abs/1409.1556 (2015)
25. Song, F., Tan, X., Liu, X., Chen, S.: Eyes closeness detection from still images with multi-scale histograms of principal oriented gradients. Pattern Recogn. **47**(9), 2825–2838 (2014). https://doi.org/10.1016/j.patcog.2014.03.024
26. Szegedy, C., et al.: Going deeper with convolutions. CoRR abs/1409.4842 (2014). http://arxiv.org/abs/1409.4842
27. Trutoiu, L.C., Carter, E.J., Matthews, I., Hodgins, J.K.: Modeling and animating eye blinks. ACM Trans. Appl. Percept. **8**(3) (2011). https://doi.org/10.1145/2010325.2010327
28. Viola, P., Jones, M.: Robust real-time face detection. Int. J. Comput. Vis. **57**, 137–154 (2004). https://doi.org/10.1023/B:VISI.0000013087.49260.fb
29. Wu, J., Trivedi, M.M.: An eye localization, tracking and blink pattern recognition system: algorithm and evaluation. TOMCCAP **6** (2010). https://doi.org/10.1145/1671962.1671964

NovemE - Color Space Net for Image Classification

Urvi Oza(✉), Sarangi Patel, and Pankaj Kumar

Dhirubhai Ambani Institute of Information and Communication Technology (DAIICT), Gandhinagar, Gujarat, India
{201921009,201601047,pankaj_k}@daiict.ac.in

Abstract. Image classification is one of the basic applications of computer vision where several deep convolutional neural networks (DCNN) have been able to achieve state of the art results. In this paper, we explore the efficacious effect of color space transformations on image classification. After empirically establishing that color space transforms indeed affects the accuracy of the classification in DCNN, we propose NovemE, an ensemble-based model made up of nine (novem) base learners. We use transfer learning with significantly reduced training time and improved accuracy of classification. This model integrates different color spaces and DCNNs in order to achieve a higher accuracy in classifying the given data. We experimented with CINIC10 and Street View House Number (SVHN) datasets and were successful in achieving significant improvement on the current state of the art results on these datasets.

Keywords: Color spaces · Image classification · Deep convolutional neural network

1 Introduction

Image classification refers to extracting information from images in order to segregate them into different labeled classes. These images are represented in a particular color space, usually RGB. Color is considered an important feature of image and color features of images have often been used in various applications such as category base image retrieval, classification, etc. In this paper we have analyzed whether performance of DCNNs varies in different color spaces or not.

There are multiple color spaces available having different color organization. As we cannot experiment with all of them, we have considered a subset that summarizes major available color spaces.

- **RGB** - It stores red(**R**), green(**G**) and blue(**B**) light intensity values separately and it is most commonly used in web-based application and to render images on computer screen.
- **rgb** - Normalized RGB color space preserves the color values and removes illumination dependence due to shadows and lighting information and thus used in robotics vision and object recognition tasks. It is denoted by **r**, **g** and **b** values, obtained by R/S, G/S and B/S respectively, where $S = R+G+B$.

N. T. Nguyen et al. (Eds.): ACIIDS 2021, LNAI 12672, pp. 531–543, 2021.
https://doi.org/10.1007/978-3-030-73280-6_42

- **YCbCr** - It stores color information and illumination information separately and it is most commonly used color space on luma chroma model in digital video and image systems. It is widely used in compression schemes such as MPEG, JPEG for digital encoding of color images. Here **Y** represents luma value whereas **Cb** and **Cr** stores chrominance values. YCbCr can be obtained by linear transformation from RGB [38].
- **HSV** - HSV is hue based color model derived by applying non-linear cylindrical transformation to RGB color model [20]. These color models are more natural to humans as it is based on how humans perceive color. Here **H** represents hue value, **S** represents saturation value or amount of gray and **V** represents brightness or intensity of color.
- **CIE-LAB** - The CIE color model is created by the International Commission on Illumination (CIE) to map all the colors which can be perceived by human eye. In CIE-Lab color space **L** is lighting component which closely matches with human perception of light, **a** and **b** components are (green - red) and (blue - yellow) color values respectively [4]. Conversion of RGB to CIE-LAB requires RGB to CIE-XYZ conversion followed by CIE-XYZ to CIE-LAB conversion.

Recently, DCNN are widely adopted in object recognition task due to their significant performance gain compared to traditional methods. Some of the classic DCNN architecture for object recognition application serve as general design guidelines for many computer vision tasks. So to investigate DCNNs performance with respect to color spaces, we have chosen to experiment with such classic networks. Deep CNN architectures used in our study are – VGG16 [36], InceptionV1 [37], ResNet18 [14] and DenseNet121 [15]

Our contributions in this paper are following:

1. We show that the performance of DCNNs vary with different color space transforms.
2. We propose a novel ensemble-based model that is a combination of nine base models with different color space transforms, NovemE. It proves that combining different models with different color space transforms yields a higher accuracy than just using one model in image classification.
3. We are able to achieve new state of the art results for SVHN dataset to the best of our knowledge.
4. By using transfer learning approach in NovemE we significantly reduce the training time and achieve performance as good as training from scratch for both datasets - CINIC10 and SVHN.

Hence forth the paper is organized as follows: In Sect. 2 we have briefly stated previous research works done on this topic. Section 3 reports the results of experimenting with different color spaces and analyzing their effect on different DCNN models for image classification. Our proposed model is described in Sect. 4. Results of our model are given in Sect. 5 and conclusion is given in Sect. 6

2 Previous Works

In recent time DCNNs have been highly used in computer vision tasks such as object recognition, object detection, segmentation, etc. [12,19] as they have achieved good score and less error rate compared to other traditional methods. There are many benchmark datasets available to train and test DCNN models. Street View House Number (SVHN) dataset is popular when it comes to image classification. Many deep neural networks have been successful in achieving state-of-the-art results on SVHN dataset [2,9,15,26]. CINIC10 dataset, though new, is being used increasingly as it provides a benchmark to test models between CIFAR-10 [18] and Imagenet dataset [11,30,33].

Though these results have achieved high accuracy, they did not take the effect of color space into consideration. In this paper, we propose a model that integrates different color space transforms in order to produce state-of-the-art results.

Similar research and analysis work have been done on how color spaces influences various algorithms: such as segmentation, object detection, recognition etc. An analysis of several color spaces for skin segmentation task was carried out [34] and it showed that skin pixel and hence skin segmentation, is not significantly affected by choice of color spaces. However, more recent studies shown that human skin pixel segmentation in YCbCr space is better than HSV space [35]. Analytical study of color spaces for plant pixel detection [21] suggested HSV as best color space to classify plant pixels. Study of foreground and shadow detection for traffic monitoring application [22] suggested that YCbCr is better color space for such applications. For traffic signal classification application deep neural network is being used [6] where input images are converted from RGB to LAB color space as preprocessing step to achieve better results compared to RGB color space.

An empirical examination of color spaces in DCNNs shows different performance results for image classification task [31]. ColorNet [13] is the only paper we found which comes very close to the research work we are presenting in this paper. ColorNet [13] proposes a network that takes into consideration several color spaces in order to improve accuracy of image classification but they have experimented with only one base model. Our proposed model NovemE, consists of different base models with color space transformations and thus we are able to achieve a higher accuracy. Moreover, we are using transfer learning which significantly reduces the time taken to train the base models. In recent time multiple object detection and classification based applications used transfer learning approach to save training time on different tasks such as monkey detection [23] and diabetic retinopathy classification [10].

3 Preliminary Analysis

In order to prove that the accuracy of image classification is affected by color space transforms, we did a preliminary analysis on datasets - CINIC10 and

SVHN using five different color spaces - RGB, HSV, YCbCr, normlized rgb, CIE-LAB. In this section detailed description of analysis on CINIC10 dataset is presented. As we are using transfer learning, all the DCNNs (Resnet18, Densenet121, InceptionV1) that we are experimenting with are pre trained on ImageNet [8] dataset. It is sufficient to use this method rather than training the model from scratch because the experimental data is similar to ImageNet dataset.

Following are details of preliminary analysis:-

- **Datasets** -
  - **CINIC10** - CINIC10 dataset has been released to public in 2018. CINIC10 is a benchmarking dataset made by combining images from CIFAR-10 and ImageNet. It is a balanced dataset as it consists of 270000 images (60000 images from CIFAR-10 and the rest from ImageNet). Split equally into train, test and validation sets. There are 10 classes and each class contains 3 sets (train, test, validation) of 9000 images each ($9000 \times 3 \times 10 = 270000$).
  - **SVHN** - Street View House Number is a real-world image dataset of image size 32×32 which is obtained from house numbers using Google Street View images [29]. It consists of 10 classes starting from digit 0 to digit 9. It consists of 73257 images for training, 26032 images for testing, and 531131 additional images which are significantly easier to classify. We merged and shuffled the training and extra images ($73257 + 531131 = 604388$) and divided into two to make training class and validation class, each consisting of 302194 images ($604388/2 = 302194$).
- **Model** - The pre-trained weights are used as initial weights and the fully connected layers are removed so that they can be fine-tuned again, according to given dataset. In all the models we have added fully connected layer of 256 neurons with 'RELU' as activation function. We have also added a dropout layer with dropout factor set to 0.4 in order to prevent overfitting of the network. Output layer of all these models are set to fully connected layer with 'softmax' activation function for output class predictions.
- **Pre-processing** - Following step-wise pre-processing has been applied on data before loading it to the network.
  - Color space transformation
  - Normalization
  - Shuffling of data
- **Training** - We have used Adam optimizer [16] and categorical cross-entropy as loss function for the model. As the networks we are using here are already initialized with Imagenet weights, we require small learning rate to fine-tune them. So, we have used 0.0001 as the starting learning rate and decreased the learning rate by 0.1 after every 7 epochs using learning rate scheduler. This ensured that the model converges properly. We have used a batch size of 128. To decide the number of epochs for training of all the three networks, we analyzed the training accuracy and validation accuracy at each epoch. Validation accuracy was considered as a parameter to prevent overfitting of the network. Based on these values, the network was trained for 15 epochs.

- **Validation and Verification** - Dataset is divided into three data subsets - train, validation, and test. We have used validation subset to monitor validation accuracy and loss while training models at each epoch.
- **Testing** - To measure performance of each trained network we have computed test accuracy (TA).
- **Analysis** - Test accuracy score of CINIC10 for different color space transforms on all three networks reported in Table 1. Table 1 shows that DCNNs are variant to color space transforms. In ResNet18, YCbCr give best results whereas in DenseNet121 and InceptionV1, RGB is the best performer followed closely by YCbCr. Normalized rgb gives the worst performance in all the networks. Similar results were observed on SVHN dataset. In SVHN, RGB was the best performer on all networks with YCbCr following closely. Another important observation is that each class performs differently in different color spaces. Table 2 shows test accuracy of 10 classes by DenseNet121 when color space transforms were applied on CINIC 10 dataset. It is quite evident from the table that though RGB gives best overall accuracy it does not necessarily give the best accuracy for each class. For example, classes like car, bird, cat and truck are better classified in HSV color space rather than any other color space. If the best results of each color space were to be combined then the overall resulting accuracy can be improved. On taking an average of the best accuracy scores for the 10 classes across all color space transforms in Table 2, we can show that the overall accuracy can be increased to 71.8 theoretically for DenseNet121. This lead us to the idea of combining different color spaces with different base learners to achieve even higher accuracy of classification. Our idea is one step ahead of the idea proposed in colornet [13]. They combined different color spaces with just one base learner. Here we are combining different base learners as well.

Table 1. Test accuracy of DCNNS in different color space on CINIC10 dataset.

Color space	ResNet18	DenseNet121	InceptionV1
RGB	63	**71**	**68**
HSV	62	66	62
YCbCr	**64**	68	67
n-rgb	51	58	53
CIE - Lab	55	57	57

Table 2. Test accuracy for 10 classes in various color spaces for DenseNet121 network.

Class	RGB	HSV	YCbCr	n-rgb	CIE-Lab
Plane	**85**	80	83	73	71
Car	**77**	**77**	76	69	64
Bird	67	**71**	65	51	51
Cat	50	**52**	44	41	33
Deer	**69**	52	51	65	46
Dog	47	43	**54**	45	38
Frog	72	70	**74**	72	69
Horse	**81**	75	77	65	65
Ship	76	76	**84**	56	63
Truck	65	**69**	65	63	58

4 Proposed Model

We propose a novel model, NovemE an ensemble-based model that uses different color space transforms to improve overall accuracy of image classification. Ensembles basically work on the theory that "the best performing model knows less about the dataset when compared to all the other weak models combined". So, the accuracy should improve on combining several models together. As proved earlier, different color space transforms work differently on each class of the dataset. Hence, it should result into an increase in accuracy if we combine the contributions of each color space transform in an ensemble model. Multiple DCNN models that contribute in making predictions for the ensemble model are known as base learners. Meta Learner is the final combined model that learns from output of these base learners.

The bias-variance tradeoff is important property of a model. A model should have enough degrees of freedom to understand the complexity of a dataset (low bias) but if the model has too much flexibility it may overfit the dataset (high variance). The fundamental rationale behind ensemble approach is that if base-learners do not perform well due to high bias or high variance, combination of these weak base-learners can reduce bias or variance in order to create an ensemble model (strong learner) which performs better.

4.1 Choice of Base Learners

The preliminary analysis (Sect. 3) was performed on all the available pre-trained models for each dataset. The top three models that gave the best accuracy for a dataset were used as base learners for the ensemble. Base learners used for SVHN dataset are **VGG16_bn, DenseNet121 and InceptionV1**. The base learners chosen for CINIC10 dataset are **ResNet18, DenseNet121 and InceptionV1**. Though InceptionV3 is the most advanced model, we chose InceptionV1 because the best results were noted by keeping the original size of images (32×32) whereas InceptionV3 requires the images to be at least 75×75. The reason for choosing ResNet18 for is that the same model was used to check accuracy of the CINIC10 dataset [7]. According to preliminary analysis results we have considered only three best color space transforms for each dataset. For CINIC10 dataset HSV, RGB and YCbCr color space transforms contributed to the best results 1. Similar analysis shows that RGB, HSV and YCbCr gave best results for SVHN dataset. Thus in total we have nine base learners (3 models with 3 different color space transforms) that are trained in the same way as described in Sect. 3. Experimenting with more or less than nine base learners did not result in an increase in the overall accuracy. Hence, an ensemble with nine base learners was an optimal choice.

4.2 Ensemble Approach

The choice of base learners should be coherent with the way these base learners are aggregated. There are different types of ensembles which combine base learners in different ways.

NovemE implements two types of ensembles - Weighted Average Ensemble and Stacking.

- **Weighted Average Ensemble:** Different weights are assigned to the nine base learners based on their performance on the validation set. These weights approximately sum up to 1. Higher weights are assigned to models that perform better and thus, these weights determine how much the model should contribute to the final output. Weights serve as the meta-learner. Figure 1 shows the model flow of Weighted Average Ensemble for CINIC10 dataset. Here the output vector of each base learner is multiplied with a weight and all the values are summed up to give a final prediction vector of the ensemble. The model flow is same for all datasets, but with different base learners. Over a number of iterations the ensemble weights are optimized and a weighted average is taken as the result. We have used Dirichlet distribution [28] to assign weights to the ensemble. Dirichlet distribution is a family of continuous multivariate probability distributions. The main reason to use this distribution is that it is a conjugate prior to a number of basic probability distributions like the categorical distribution and hence it works well when given as a prior distribution in Bayesian statistics. This process of randomly assigning weights using Dirichlet distribution is repeated for 10000 iterations. In each iteration, the performance of a hold-out validation set is evaluated by a metric - accuracy score [32]. The weights that give the best accuracy score are chosen to be the final weights of the ensemble.
- **Stacking:** Stacking is a technique which combines base learners by training a meta-learner that makes the final predictions. The meta-learner takes every single output of the base-learners as a training instance, learning how to best map the base-learner decisions into an improved output. The meta-learner can be any classic known machine learning model. We have used Random Forest Regression as meta-learner, as it can produce a highly accurate classifier and runs efficiently on large databases. We have experimented with two types of stacking - 1st Level Stacking and 3rd Level Stacking.
  - 1st Level Stacking - In this stacking method all the base learners run independently, then meta learner (Random Forest Regression in our case) aggregates all the outputs without any preferences. Some specifications of the random forest regressor used are listed below:
    1. n-estimators: The number of trees in the forest. We have considered 200 n estimators in this case.
    2. max-depth: The maximum depth that a tree can have. We have considered 15 in this case.
    3. n-jobs: It denotes the number of jobs that can be executed in parallel. We have considered 20 in this case.
    4. min-sample-splits: It denotes the minimum number of samples required for splitting a node. We have kept it 20 in this case.
  - 3rd Level Stacking - In this method, instead of one, three different meta learners are used to predict the final output. The model flow of 1st level stacking and 3rd level stacking ensemble is given in Fig. 2 and 3 respectively. The output of base learners is given to the 1st level meta learner

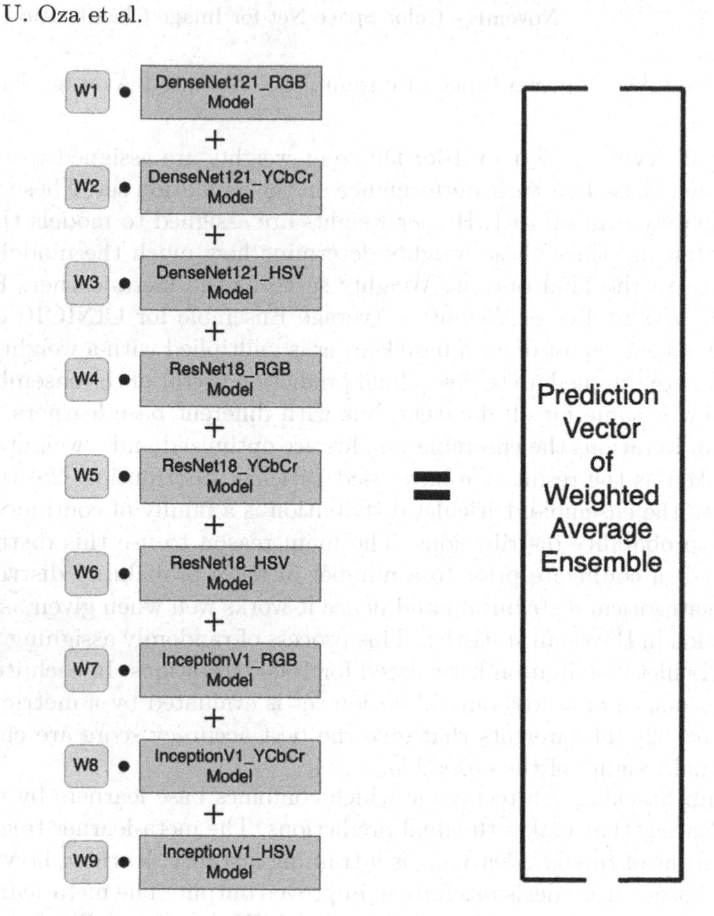

Fig. 1. Model flow of weighted average ensemble - output from each base learners are multiplied with different weights and final prediction vector is summation of these vectors

which is Random Forest Classifier in this case. It is followed by Extra Trees Classifier as the meta learner of the 2nd level. Finally the output of multiclass classification is predicted by the 3rd level meta learner which is Logistic Regression in our method. Extra Trees Classifier is different from Random Forest Regressor as it does not rely on best splits and apply random splits across all observations. So it prevents overfitting of the model. Extra Trees Classifier also have the same value of parameters as Random Forest Regressor but one important difference is Bootstrap parameter. Bootstrap is set to False in this classifier which means that samples are drawn without replacement, whereas in Random Forest Classifier samples are drawn with replacement.

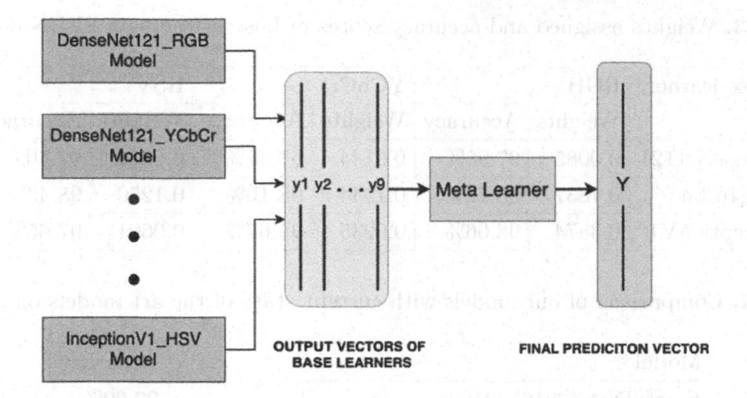

Fig. 2. Model flow of 1st level stacking ensemble: Meta learner predicts output vector based on output vectors of base learners

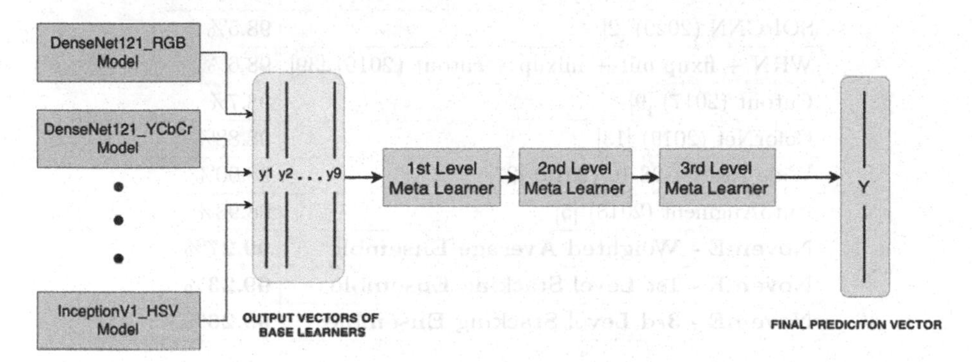

Fig. 3. Model flow of 3rd level stacking ensemble: Three different classifiers are used as meta learners working in sequential order to predict output vector

5 Results

Following are the results of NovemE on three datasets - SVHN and CINIC10.

- SVHN dataset - The results of base models and ensemble models tested on SVHN dataset are reported in Table 3 and Table 4 respectively. Table 4 shows the comparison of our ensemble models with other models which have given state of the art results on SVHN dataset. These values are taken from the top 35 papers that have done experiments on SVHN dataset. Weighted Average Ensemble performs the best and obtains a new state of the art result with an accuracy of 99.27%. 3rd level stacking ensemble also gives an accuracy of 99.26%. The base models that perform better are given more weight as seen in Table 3. The accuracy of image classification improves by 0.55% when compared to the best performing base learner (vgg16_bn in RGB) which is significant when we look at the recent current state of the art results (Table 4).

Table 3. Weights assigned and accuracy scores of base learners on SVHN dataset

Base learner	RGB		YCbCr		HSV	
	Weights	Accuracy	Weights	Accuracy	Weights	Accuracy
DenseNet121	0.0085	97.98%	0.0144	97.41%	0.0331	97.40%
vgg16_bn	0.1837	98.72%	0.1944	98.10%	0.1250	98.43%
InceptionV1	0.3574	98.06%	0.0235	95.67%	0.0601	97.65%

Table 4. Comparison of our models with current state of the art models on SVHN

Model	Accuracy
FractalNet (2016) [24]	97.99%
Densenet (2016) [15]	98.41%
CoPaNet-R-164 (2017) [3]	98.42%
SOPCNN (2020) [2]	98.5%
WRN + fixup init + mixup + cutout (2019) [39]	98.6%
Cutout (2017) [9]	98.7%
ColorNet (2019) [13]	98.88%
Wide-ResNet-28-10 (2019) [27]	98.90%
AutoAugment (2018) [5]	98.98%
NovemE - Weighted Average Ensemble	**99.27%**
NovemE - 1st Level Stacking Ensemble	**99.23%**
NovemE - 3rd Level Stacking Ensemble	**99.26%**

- CINIC10 dataset - The results of base models and ensemble models tested on CINIC10 dataset are reported in Table 5 and 6 respectively. For this dataset, Weighted Average Ensemble gives the best result with an accuracy of 76.66%. The accuracy of image classification improves by 5.35% when compared to the best performing base learner (DenseNet121-RGB). Our results are comparable to other research work done on CINIC10 dataset like the analysis of human uncertainty in classification (2019) [33]. Though experiments on CINIC10 dataset do not yield the highest accuracy but it is important to note that the highest increase in overall accuracy is seen in CINIC10 compared to other datasets (5.35%). Since CINIC10 is a comparatively new dataset, not much experimentation is done on it especially by transfer learning. Comparison of our model with the existing state of the art models is given in Table 6.

Table 5. Weights assigned and accuracy scores of base learners on CINIC10 dataset

Base Learner	RGB		YCbCr		HSV	
	Weights	Accuracy	Weights	Accuracy	Weights	Accuracy
DenseNet121	0.1694	71.31%	0.0708	68.52%	0.1070	66.48%
ResNet18	0.1094	63.12%	0.0711	64.47%	0.1009	62.83%
InceptionV1	0.1370	67.97%	0.1469	67.32%	0.0874	62.07%

Table 6. Comparison of our models with some current state of the art models on CINIC10

Model	Accuracy
Classification with human uncertainty (2019) [33]	60%
BatchBALD (2019) [17]	59%
Generalisation with knowledge distillation (2019) [1]	70.5%
DPDNet (2019) [25]	76.65%
NovemE - Weighted Average Ensemble	**76.66%**
NovemE - 1st Level Stacking Ensemble	**71.38%**
NovemE - 3rd Level Stacking Ensemble	**74.24%**

6 Conclusion

Empirical study of DCNN in different color spaces shows that their classification results are variant to color spaces and some color spaces performs equally good and sometimes better than RGB depending on dataset. One of the important observations we found during this study was that the test accuracy of each class differ from one color space to another. Some classes of images are better recognized in color spaces other than RGB. Based on this analysis we propose NovemE, a model that integrates different base models trained with different color space transforms to predict the final outputs. These ensemble-based models have two approaches - average weighted ensemble and stacking ensemble. Both approaches are successful in increasing the overall efficiency of the model. However we found that weighted average ensemble to perform better than two types of proposed stacking ensembles. This same idea of color space transforms can be applied to any model that has been developed earlier for image classification to increase the overall accuracy. Thus, this paper highlights the importance of color spaces for image classification. Future work comprises of testing NovemE on different larger datasets.

References

1. Arani, E., Sarfraz, F., Zonooz, B.: Improving generalization and robustness with noisy collaboration in knowledge distillation. arXiv preprint arXiv:1910.05057 (2019)

2. Assiri, Y.: Stochastic optimization of plain convolutional neural networks with simple methods. arXiv preprint arXiv:2001.08856 (2020)
3. Chang, J.R., Chen, Y.S.: Deep competitive pathway networks. arXiv preprint arXiv:1709.10282 (2017)
4. Connolly, C., Fleiss, T.: A study of efficiency and accuracy in the transformation from RGB to CIELAB color space. IEEE Trans. Image Process. **6**(7), 1046–1048 (1997)
5. Cubuk, E.D., Zoph, B., Mane, D., Vasudevan, V., Le, Q.V.: Autoaugment: Learning augmentation policies from data. arXiv preprint arXiv:1805.09501 (2018)
6. Dan, C., Ueli, M., Jonathan, M., Jürgen, S.: Multi-column deep neural network for traffic sign classification. Neural Netw. **32**, 333–338 (2012)
7. Darlow, L.N., Crowley, E.J., Antoniou, A., Storkey, A.J.: Cinic-10 is not imagenet or cifar-10. arXiv preprint arXiv:1810.03505 (2018)
8. Deng, J., Dong, W., Socher, R., Li, L.J., Li, K., Fei-Fei, L.: Imagenet: a large-scale hierarchical image database. In: 2009 IEEE Conference on Computer Vision and Pattern Recognition, pp. 248–255. IEEE (2009)
9. DeVries, T., Taylor, G.W.: Improved regularization of convolutional neural networks with cutout. arXiv preprint arXiv:1708.04552 (2017)
10. Doshi, N., Oza, U., Kumar, P.: Diabetic retinopathy classification using downscaling algorithms and deep learning. In: IEEE 7th International Conference on Signal Processing and Integrated Networks (SPIN), pp. 950–955 (2020)
11. Duan, M., et al.: Astraea: Self-balancing federated learning for improving classification accuracy of mobile deep learning applications. In: IEEE 37th International Conference on Computer Design (ICCD), pp. 246–254 (November 2019). https://doi.org/10.1109/ICCD46524.2019.00038
12. Gowda, S.N.: Human activity recognition using combinatorial deep belief networks. In: Proceedings of the IEEE Conference on Computer Vision and Pattern Recognition Workshops, pp. 1–6 (2017)
13. Gowda, S.N., Yuan, C.: ColorNet: investigating the importance of color spaces for image classification. In: Jawahar, C.V., Li, H., Mori, G., Schindler, K. (eds.) ACCV 2018. LNCS, vol. 11364, pp. 581–596. Springer, Cham (2019). https://doi.org/10.1007/978-3-030-20870-7_36
14. He, K., Zhang, X., Ren, S., Sun, J.: Deep residual learning for image recognition. In: IEEE Conference on Computer Vision and Pattern Recognition (CVPR), pp. 770–778 (2016)
15. Huang, G., Liu, Z., Van Der Maaten, L., Weinberger, K.Q.: Densely connected convolutional networks. In: Proceedings of the IEEE Conference on Computer Vision and Pattern Recognition, pp. 4700–4708 (2017)
16. Kingma, D.P., Ba, J.: Adam: a method for stochastic optimization. arXiv preprint :1412.6980 (2014)
17. Kirsch, A., Van Amersfoort, J., Gal, Y.: Batchbald: efficient and diverse batch acquisition for deep Bayesian active learning. In: Advances in Neural Information Processing Systems, pp. 7026–7037 (2019)
18. Krizhevsky, A.: Learning multiple layers of features from tiny images. University of Toronto (2009)
19. Krizhevsky, A., Sutskever, I., Hinton, G.E.: Imagenet classification with deep convolutional neural networks. In: Proceedings of the 25th International Conference on Neural Information Processing Systems, NIPS 2012, vol. 1, pp. 1097–1105. Curran Associates Inc., Red Hook, NY, USA (2012)
20. Kuehni, R.: Color Space and Its Divisions: Color Order from Antiquity to the Present. Wiley (2003), https://books.google.co.in/books?id=2kFVSRGC650C

21. Kumar, P., Miklavcic, S.J.: Analytical study of colour spaces for plant pixel detection. J. Imaging **4**(2), (2018). https://doi.org/10.3390/jimaging4020042, http://www.mdpi.com/2313-433X/4/2/42
22. Kumar, P., Sengupta, K., Lee, A.: A comparative study of different color spaces for foreground and shadow detection for traffic monitoring system. In: The IEEE 5th Proceedings of International Conference On Intelligent Transportation Systems, pp. 100–105. IEEE (2002)
23. Kumar, P., Shingala, M.: Native monkey detection using deep convolution neural network. In: Hassanien, A.E., Bhatnagar, R., Darwish, A. (eds.) AMLTA 2020. AISC, vol. 1141, pp. 373–383. Springer, Singapore (2021). https://doi.org/10.1007/978-981-15-3383-9_34
24. Larsson, G., Maire, M., Shakhnarovich, G.: Fractalnet: Ultra-deep neural networks without residuals. arXiv preprint arXiv:1605.07648 (2016)
25. Li, G., et al.: Psdnet and dpdnet: efficient channel expansion, depthwise-pointwise-depthwise inverted bottleneck block (2019)
26. Liang, S., Khoo, Y., Yang, H.: Drop-activation: Implicit parameter reduction and harmonic regularization. arXiv preprint arXiv:1811.05850 (2018)
27. Lim, S., Kim, I., Kim, T., Kim, C., Kim, S.: Fast auto augment. In: Advances in Neural Information Processing Systems, pp. 6665–6675 (2019)
28. Lin, J.: On the dirichlet distribution. Ph.D. thesis, Master's thesis, Department of Mathematics and Statistics, Queens University, Kingston, Ontario, Canada (2016)
29. Netzer, Y., Wang, T., Coates, A., Bissacco, A., Wu, B., Ng, A.Y.: Reading digits in natural images with unsupervised feature learning. In: NIPS Workshop on Deep Learning and Unsupervised Feature Learning 2011, p. 5 (2011)
30. Noy, A., et al.: Asap: architecture search, anneal and prune. In: International Conference on Artificial Intelligence and Statistics, pp. 493–503. PMLR (2020)
31. Oza, U., Kumar, P.: Empirical examination of color spaces in deep convolution networks. Int. J. Recent Technol. Eng. **9**(2), (2020)
32. Pedregosa, F., et al.: Scikit-learn: machine learning in python. J. Mach. Learn. Res. **12**, 2825–2830 (2011)
33. Peterson, J.C., Battleday, R.M., Griffiths, T.L., Russakovsky, O.: Human uncertainty makes classification more robust. In: The IEEE International Conference on Computer Vision (ICCV) (October 2019)
34. Phung, S.L., Bouzerdoum, A., Chai, D.: Skin segmentation using color pixel classification: analysis and comparison. IEEE Trans. Pattern Anal. Mach. Intell. **27**(1), 148–154 (2005). https://doi.org/10.1109/TPAMI.2005.17
35. Shaik, K., Packyanathan, G., Kalist, V., Sathish, B.S., Merlin Mary Jenitha, J.: Comparative study of skin color detection and segmentation in HSV and YCbCr color space. Procedia Comput. Sci. **57**, 41–48 (2015). https://doi.org/10.1016/j.procs.2015.07.362
36. Simonyan, K., Zisserman, A.: Very deep convolutional networks for large-scale image recognition (2014)
37. Szegedy, C., et al.: Going deeper with convolutions. arXiv preprint arXiv:1409.4842 (2014)
38. Tinku, A.: Integrated color interpolation and color space conversion algorithm from 8-bit bayer pattern RGB color space to 12-bit ycrcb color space (March 2006), http://www.freepatentsonline.com/7015962.html
39. Zhang, H., Dauphin, Y.N., Ma, T.: Fixup initialization: Residual learning without normalization. arXiv preprint arXiv:1901.09321 (2019)

Application of Improved YOLOv3 Algorithm in Mask Recognition

Fanxing Meng, Weimin Wei(✉), Zhi Cai, and Chang Liu

School of Computer Science and Technology, Shanghai University of Electric Power, Shanghai 200090, China
wwm@shiep.edu.cn

Abstract. With the influence of novel coronavirus, wearing masks is becoming more and more important. If computer vision system is used in public places to detect whether a pedestrian is wearing a mask, it will improve the efficiency of social operation. Therefore, a new mask recognition algorithm based on improved yolov3 is proposed. Firstly, the dataset is acquired through network video; secondly, the dataset is preprocessed; finally, a new network model is proposed and the activation function of YOLOv3 is changed. The average accuracy of the improved YOLOv3 algorithm is 83.79%. This method is 1.18% higher than the original YOLOv3.

Keywords: Mask detection · H-swish · FPN · Improved YOLOv3 · CNN

1 Introduction

Novel Coronavirus will require people to wear masks when traveling. In some public places, masks are tested mainly by staff supervision, which is time-consuming and laborious. At present, deep learning algorithm has been widely applied in the field of target detection. It is of great significance to use deep learning algorithm to detect whether pedestrians wear masks.

Early target detection was mainly based on manual feature design. This traditional target detection method is roughly divided into three parts: area selection (sliding window), feature extraction such as scale-invariant feature transform(SIFT), Histogram of Oriented Gradient(HOG [1]) and etc., classifier including support vector machine(SVM [2]) and Adaboost cascade classifier. Traditional target detection has two shortcomings: the region selection strategy based on sliding window is not targeted, has high time complexity and window redundancy, which seriously affects the speed and performance of subsequent feature extraction and classification. The characteristics of manual design have no good robustness to the changes of target diversity. In recent years, deep learning has achieved remarkable results in target detection. Methods of deep learning for target detection can be divided into two categories Proposal, among which the representative ones are Mainly R-CNN [3], Fast R-CNN [4], Faster R-CNN [5]

N. T. Nguyen et al. (Eds.): ACIIDS 2021, LNAI 12672, pp. 544–555, 2021.
https://doi.org/10.1007/978-3-030-73280-6_43

and MTCNN [6]. The second type is the deep learning target detection algorithm based on regression method. Typical algorithms include YOLO [7], YOLOv2 [8], YOLOv3 [9] and Single Shot MultiBox Detector(SSD [10]). The target detection algorithm based on regional Proposal cannot meet the requirements of real-time detection. For example, r-CNN training steps are tedious, which takes up a large amount of disk space, and it takes about 45S to process a picture under the condition of GPU acceleration. Even the mainstream Faster R-CNN cannot meet the requirement of real-time performance. The latter can realize end-to-end detection by using the regression target detection algorithm. Given an input image, the target position and category can be marked by regression directly in this image. However, the accuracy of this real-time detection algorithm is relatively low. Because the detection of wearing masks or not in this paper requires higher real-time performance of the algorithm, we chose the target detection algorithm based on regression.YOLOv3 processes about 45 images per second, but due to the absence of regional proposal mechanism, YOLOv3 has low detection accuracy, so we have made improvements on the basis of YOLOv3.

2 YOLOv3

2.1 The Principle of YOLOv3

YOLOv3 is a deep-learning target detection algorithm proposed by REDMON J et al. It attributes target detection to regression problem and proposes to complete target detection in one step and complete position detection and category prediction of all target objects in the picture in a CNN network model, which greatly improves the speed of target detection. YOLOv3 algorithm divides the input image into several feature graphs of different sizes of grids, and then predicts and locates the target object through the grid in each feature graph and its corresponding anchor box.

The backbone network of YOLOv3 algorithm uses Darknet53, which is a big improvement compared with YOLOv2 and YOLOv1. Darknet53 adopts the design concept of ResNet [11]. The advantage of ResNet is that it is easy to optimize and can solve the problem of training accuracy by increasing network depth. In solving the gradient disappearance problem, Darknet53 adopted the way of residual block jump connection. Figure 1 is Darknet53 network structure, which contains 53 convolutional layers. The network consists of a series of convolution layers of lx1 and 3 × 3(each layer is followed by a BN layer and a Leaky ReLU).

2.2 Improved Activation Function

Each convolution in the backbone network of YOLOv3 uses the unique DarknetConv2D structure, L2 regularization is performed during each convolution, batch normalization and LeakyReLU are performed after the convolution. ReLU sets all negative values to zero, and LeakyReLU assigns a non-zero slope to all

Type	Filters	Size	Output	
Convolutional	32	3x3	256x256	
Convolutional	64	3x3/2	128x128	
Convolutional	32	1x1		x1
Convolutional	64	3x3		
Residual			128x128	
Convolutional	128	3x3/2	64x64	
Convolutional	64	1x1		x2
Convolutional	128	3x3		
Residual			64x64	
Convolutional	256	3x3/2	32x32	
Convolutional	128	1x1		x8
Convolutional	256	3x3		
Residual			32x32	
Convolutional	512	3x3/2	16x16	
Convolutional	256	1x1		x8
Convolutional	512	3x3		
Residual			16x16	
Convolutional	1024	3x3/2	8x8	
Convolutional	512	1x1		x4
Convolutional	1024	3x3		
Residual			8x8	
Avgpool		Global		
Connected		1000		
Softmax				

Fig. 1. Darknet53 network structure drawing

negative values. The advantage of LeakyReLU is that it does not produce overfitting, and the calculation is simple and effective. In mathematical terms, we can express it as:

$$y_i = \begin{cases} x_i \; if(x_i \geq 0) \\ \frac{x_i}{a_i} \quad others \end{cases} \tag{1}$$

Where a_i is a fixed value. This activation function inherits the ReLU [12] activation function and has an efficient gradient descent while avoiding the phenomenon of neuronal death. Like ReLU, Swish has a lower bound but no upper bound, but its non-monotonicity is indeed different from other common activation functions. The Swish function is shown in the formula (2):

$$f(x) = x \cdot sigmoid(\beta x) \tag{2}$$

β is a constant or trainable parameter. The Swish function image is shown in Fig. 2. Google tests proved that Swish is suitable for local response normalization, and the effect of more than forty fully connected layers is far better than other activation functions, while the performance gap is not obvious within forty fully connected layers.

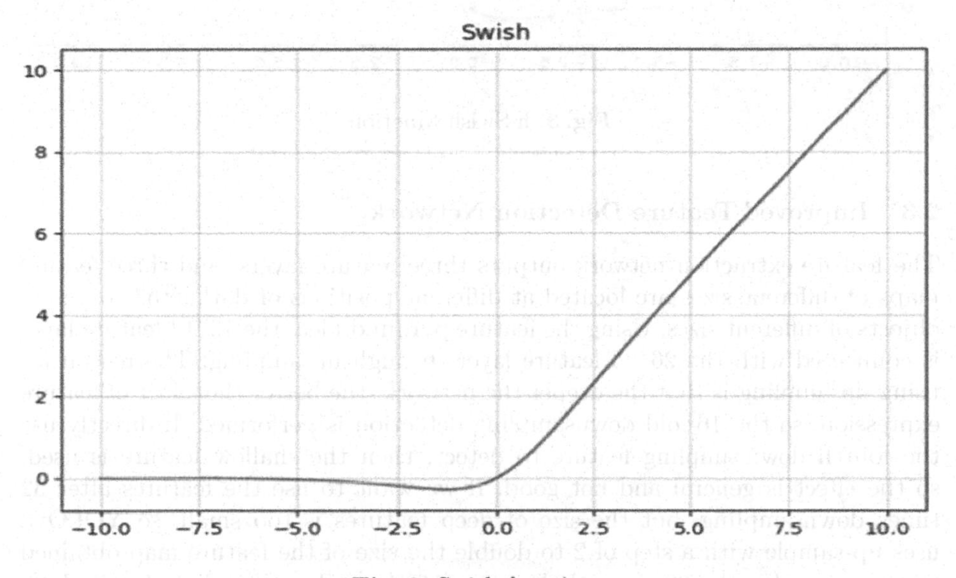

Fig. 2. Swish function

Although this Swish nonlinearity improves the accuracy, its cost is non-zero in an embedded environment, because it is much more expensive to calculate the sigmoid function on a mobile device. An approximate function can be used to approximate Swish. The paper [12] chooses to be based on ReLU6. Almost all software and hardware frameworks can implement ReLU6. Secondly, it can eliminate the potential loss of numerical accuracy caused by different implementations of approximate sigmoid in a specific mode. The Swish function is shown in the formula (3):

$$hSwish(x) = x\frac{ReLU6(x+3)}{6} \tag{3}$$

The h-Swish function image is shown in Fig. 3:

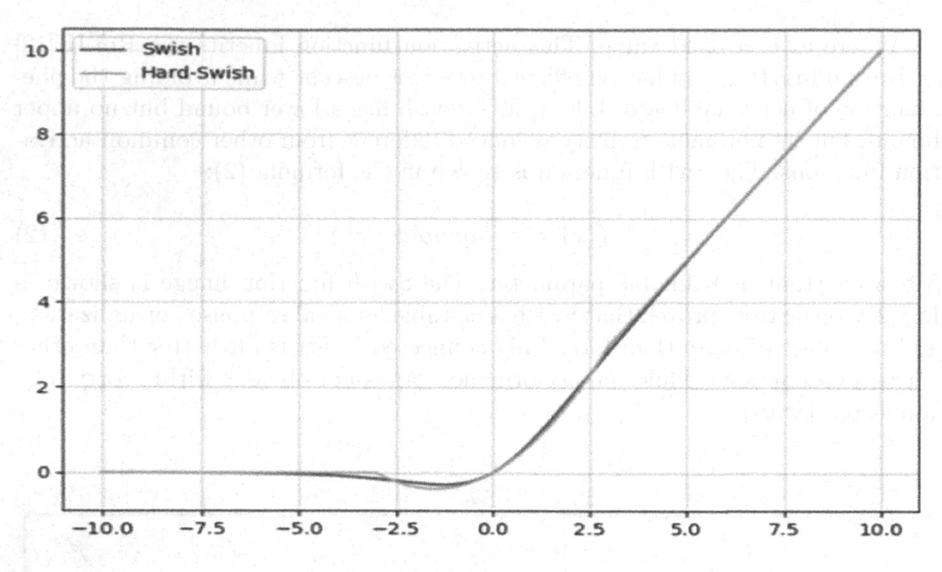

Fig. 3. h-Swish function

2.3 Improved Feature Detection Network

The feature extraction network outputs three feature layers, and three feature maps of different sizes are located at different positions of darknet53 to detect objects of different sizes. Using the feature pyramid idea, the 13*13 feature layer is connected with the 26*26 feature layer through upsampling. The reason for using upsampling is that the deeper the network, the better the effect of feature expression, so the 16-fold downsampling detection is performed. If directly use the fourth downsampling feature to detect, then the shallow feature is used, so the effect is general and not good. If we want to use the features after 32 times downsampling, but the size of deep features is too small, so YOLOv3 uses up-sample with a step of 2 to double the size of the feature map obtained by downsampling by 32 times, It becomes 16 times downsampling. Similarly, 8 times sampling is to perform upsampling with a step length of 2 for 16 times downsampling features, so that deep features can be used for detection.

Affected by the feature pyramid idea, we constructed a new type of feature pyramid network, as shown in Fig. 4, similar to the BiFPN [13]. First, the 13*13 feature layer is up-sampled and the 26*26 feature layer is connected. Then the 26*26 feature layer is up-sampled and then connected to the 56*56 feature layer; then down-sampling is performed in turn, and finally the feature map extracted by the backbone network is short-connected, and it is combined with the processed feature map connection. After the processing of the above steps, the feature information extracted by the backbone network can be reused.

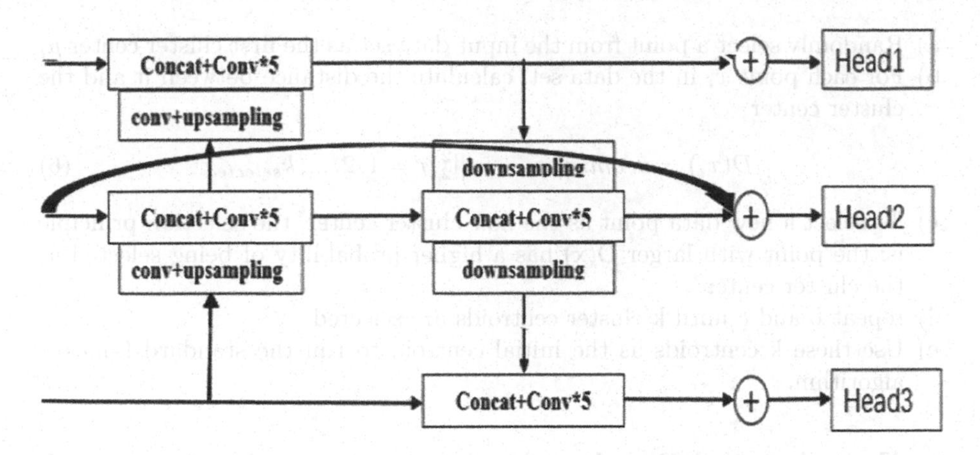

Fig. 4. Improved feature detection network

2.4 Improved Priors Anchor Algorithm

The anchor in YOLOv3 is obtained by clustering the COCO data set, and the mask data set in this experiment has a large gap between the target size of the COCO data set. If we use it to detect the target on our own data set, The part of the anchor size is unreasonable. Therefore, before the experiment, we need to re-cluster our own data set to find an anchor size suitable for the experimental data set to improve the detection rate of the bounding box. The clustering algorithm used by YOLOv3 is k-means. The K-means algorithm divides the samples into k clusters according to the distance between the samples for a given sample set. Make the points in the clusters as close together as possible, and make the distance between the clusters as large as possible. Assuming that the cluster is divided into $(C_1, C_2, ...C_k)$, our goal is to minimize the squared error E:

$$E = \sum_{i=1}^{k} \sum_{x \in C_i} || x - \mu_i ||_2^2 \tag{4}$$

Where ui is the mean vector of cluster Ci, and the expression is:

$$\mu_i = \frac{1}{|C_i|} \sum_{x \in C_i} x \tag{5}$$

However, the location of the initial centroid has a great influence on the final clustering result and running time, so it is necessary to choose a suitable centroid, and the k-means algorithm is a method of randomly initializing the centroid. In this experiment, an improved method of k-means algorithm—k-means++ algorithm is used. The steps of K-means algorithm are as follows:

a) Randomly select a point from the input data set as the first cluster center μ_i
b) For each point x_i in the data set, calculate the distance between it and the cluster center

$$D(x_i) = argmin||x_i - \mu_r||_2^2, r = 1, 2, ..., k_{selected} \tag{6}$$

c) To select a new data point as the new cluster center, the selection principle is: the point with larger D(x) has a higher probability of being selected as the cluster center.
d) repeat b and c until k cluster centroids are selected
e) Use these k centroids as the initial centroid to run the standard k-means algorithm.

3 Experimental Results

3.1 Data Collection and Processing

The data set mainly consists of two types of samples: samples with masks and samples without masks. Because there is no similar mask recognition algorithm before, the existing data set does not have enough samples to support the training of the model. The mask samples used in this paper were taken from various video websites, and the images were processed to meet the requirements of training and testing. The samples without masks were mainly randomly selected from the data-sets of Broadface [14] and VOC2007 [15]. Table 1 shows the distribution of sample categories.

Table 1. Category distribution of samples.

Sample classification	Number of training samples	Number of test samples
Samples wearing mask	2607	550
Samples not wearing mask	2558	604

3.2 Data Set Preprocessing

After obtaining samples with masks, supervised data enhancement should be carried out on the self-made data set. The main purpose is to make the model adapt to the detection under various scenes. The data enhancement methods used mainly include gray processing, changing the channel sequence of color images RGB, rotating, flipping, scaling and adding noise. The first two methods are to solve the problem of single color of sample. In order to adapt the model to more complex environmental detection, noise is added to part of the data set, and Gaussian noise is mainly used in this paper. Rotation and other operations in order to solve the real scene to be detected target location problems. Figure 5 is an example of channel sequence of color images, as shown in Fig. 5. Figure 5-a is the original image, Fig. 5-b, Fig. 5-c and Fig. 5-d are renderings of RGB channel sequence after being changed. Figure 6 shows an example of adding noise.

As shown in Fig. 6, Fig. 6-a is the original image and Fig. 6-b is the result of Gaussian blur. In order to prevent the model from overfitting, the uniform size Gaussian kernel was not used, but different Gaussian kernel was used to process the samples before shuffling them out of order. Table 2 shows the distribution of the categories of masks worn in the training set after treatment.

Table 2. Sample processing table.

Sample classification	Gray processing	Change channel order	Noise	Rotate and scale
Samples wearing mask	650	608	700	649

3.3 Network Model Training

The model training adopted a two-stage training. The learning rate of the first training was 0.0001. The EPOCH was set as 40 times, and the number of iterations per rotation was 110 times. During the training, the data sets were randomly divided into two parts, one for training and the other for testing, with a

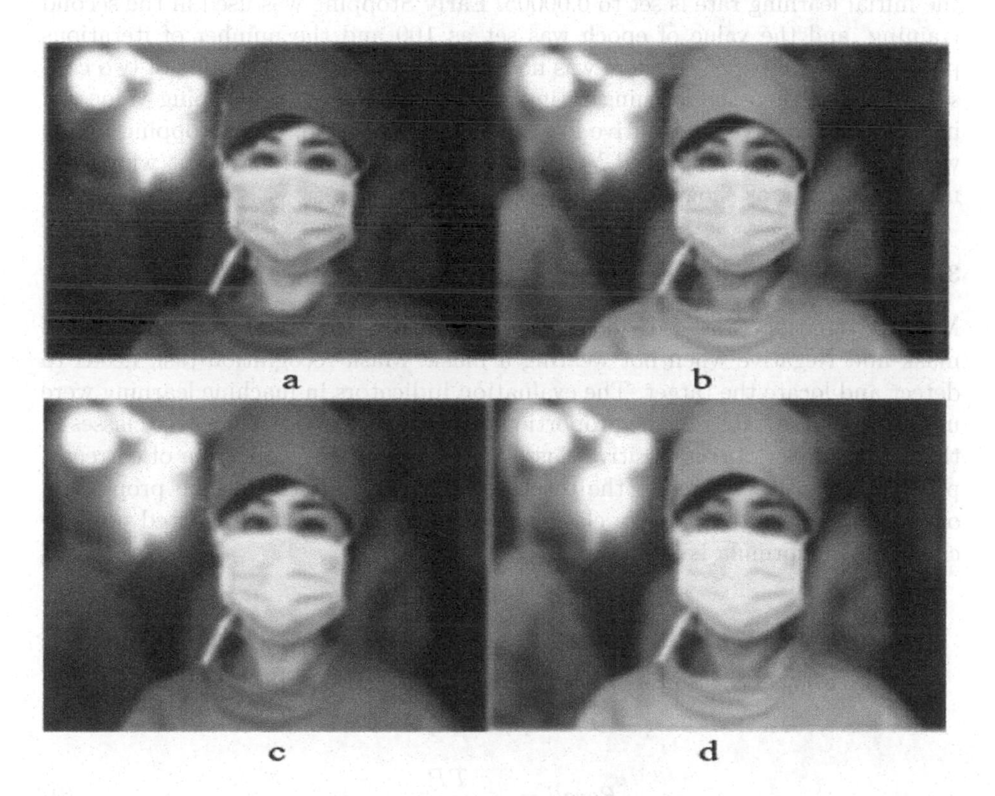

Fig. 5. Change the color image RGB order (Color figure online)

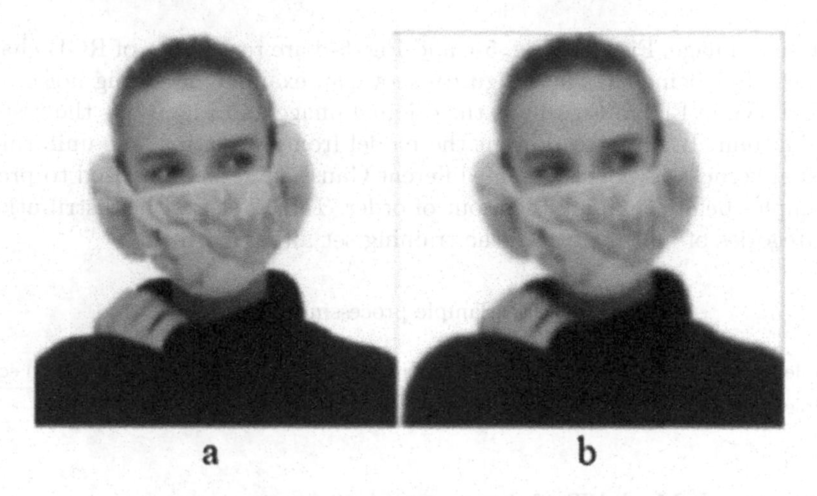

Fig. 6. Gaussian blur

ratio of 9:1. After the first training, load the weight file into the program, and then start the second training. The second training is a process of finishing, so the initial learning rate is set to 0.00005. Early Stopping was used in the second training, and the value of epoch was set as 100 and the number of iterations per revolution was 885. If there was no downward trend in the value of two consecutive loss, then the learning rating would be reduced. If stopping value did not decrease for five consecutive loss, then the training of early stopping model would be used. The improved YOLOv3 network was used to detect whether a mask was worn, as shown in Fig. 7.

3.4 Experimental Results

Mask recognition can be divided into two categories: Positive when wearing a mask and Negative when not wearing a mask. Mask recognition task needs to detect and locate the target. The evaluation indicators in machine learning were used to evaluate – Recall (the proportion of correctly predicted positive classes in the total number of true positive classes), Accuracy (the proportion of correctly predicted positive classes in the total number), and Precision (the proportion of correctly predicted positive classes in the total number of predicted positive classes). The formula is as follows:

$$Accuracy = \frac{TP + TN}{TP + TN + FP + FN} \tag{7}$$

$$P = \frac{TP}{TP + FP} \tag{8}$$

$$Recall = \frac{TP}{TP + FN} \tag{9}$$

Fig. 7. Test results

In this paper, the mean Average Precision (mAP) was used to evaluate the performance of the model. First, the confidence Score of all samples was obtained by using the trained model, and then the confidence Score was sorted, precision and recall were calculated, and AP was further calculated. Finally, the mean value was calculated according to AP to obtain the mAP. Table 3 shows the average accuracy and speed of using different network models to detect wearing masks.

Table 3. Performance comparison of different network models on the mask dataset.

Networks model	AP of negative sample	AP of positive sample	MAP
YOLOv3	83.64%	81.58%	82.61%
SSD	83.81%	81.75%	82.78%
Improved YOLOv3	84.78%	82.80%	83.79%

4 Conclusion

In this paper, an improved YOLOv3 network model is proposed based on YOLOv3 algorithm. Firstly, we change the activation function of yolov3 and use swish activation function instead of leakyrelu. Swish has the characteristics of no upper bound and lower bound, smooth and non monotone. Swish is better than relu in deep model, But the calculation of swish function is complex, so we use h-Swish instead of swish activation function; secondly, we modify the feature detection network, inspired by bifpn, we design a feature pyramid network with residual structure; finally, were cluster the data set to get a suitable anchor The experimental results show that compared with the original model, the improved model has higher accuracy and faster detection speed in the self-made dataset. In the future research, we will continue to improve the network structure of yolov3 to further improve the detection accuracy of the model.

References

1. Dalal, N., Triggs, B.: Histograms of oriented gradients for human detection. In. IEEE Computer Society Conference on Computer Vision and Pattern Recognition, pp. 886–893 (2005)
2. Burges, C.J.: A tutorial on support vector machines for pattern recognition. Data Min. Knowl. Disc. **2**, 121–167 (1998)
3. Girshick, R., Donahue, J., Darrell, T., et al.: Region-based convolutional networks for accurate object detection and segmentation. IEEE Trans. Pattern Anal. Mach. Intell. **38**(1), 142–158 (2015)
4. Girshick, R.: Fast r-cnn In: Proceedings of the IEEE International Conference on Computer Vision, pp. 1440–1448 (2015)
5. Ren, S., He, K., Girshick, R., et al.: Faster r-cnn: towards real-time object detection with region proposal networks. In: Advances in Neural Information Processing Systems, pp. 91–99 (2015)
6. Xiang, J., Zhu, G.: Joint face detection and facial expression recognition with MTCNN. In: 2017 4th International Conference on Information Science and Control Engineering (ICISCE), pp. 424–427. IEEE (2017)
7. Redmon, J., Divvala, S., Girshick, R., et al.: You only look once: unified, real-time object detection. In: Proceedings of the IEEE Conference on Computer Vision and Pattern Recognition, pp. 779–788 (2016)
8. Redmon, J., Farhadi, A.: YOLO9000: better, faster, stronger. In: Proceedings of the IEEE Conference on Computer Vision and Pattern Recognition, pp. 7263–7271 (2017)
9. Redmon, J., Farhadi, A.: Yolov3: An incremental improvement. arXiv preprint arXiv:1804.02767. (2018)
10. Liu, W., et al.: SSD: single shot MultiBox detector. In: Leibe, B., Matas, J., Sebe, N., Welling, M. (eds.) ECCV 2016. LNCS, vol. 9905, pp. 21–37. Springer, Cham (2016). https://doi.org/10.1007/978-3-319-46448-0_2
11. He, K., Zhang, X., Ren, S., et al.: Deep residual learning for image recognition. In: Proceedings of the IEEE Conference on Computer Vision and Pattern Recognition, pp. 770–778 (2016)

12. Ramachandran, P., Zoph, B., Le, Q.V.: Searching for activation functions. arXiv preprint arXiv:1710.05941. (2017)
13. Tan, M., Pang, R., Le, Q.V.: Efficientdet: scalable and efficient object detection In: Proceedings of the IEEE/CVF Conference on Computer Vision and Pattern Recognition, pp. 10781–10790 (2020)
14. Yang, S., Luo, P., Loy, C.C., et al.: Wider face: a face detection benchmark. In: Proceedings of the IEEE Conference on Computer Vision and Pattern Recognition, pp. 5525–5533 (2016)
15. Everingham, M., Eslami, S.M.A., Van Gool, L., et al.: The pascal visual object classes challenge: a retrospective. Int. J. Comput. Vis. **111**(1), 98–136 (2015)

Simulation and Experiment in Solving Inverse Kinematic for Human Upper Limb by Using Optimization Algorithm

Trung Nguyen[1](✉), Huy Nguyen[1], Khai Dang[1], Phuong Nguyen[1], Ha Pham[2], and Anh Bui[3]

[1] Hanoi University of Science and Technology, No. 1 Dai Co Viet Street, Hanoi, Vietnam
trung.nguyenthanh@hust.edu.vn
[2] Hanoi University of Industry, Hanoi, Vietnam
[3] Ministry of Public Security Vietnam, Hanoi, Vietnam

Abstract. The problem of inverse kinetics in the robotics field, especially for redundant driven robots, often requires the application of a lot of techniques. The redundancy in degree of freedom, the nonlinearity of the system lead to the inverse kinematics resolution more challenge. In this study, we proposed to apply the Differential Evolution optimization algorithm with the search space improvement to solve this problem for the human upper limb which is very typical redundancy model in nature. Firstly, the forward kinematic modeling and problem for the human arm was presented. Next, an Exoskeletal type Human Motion Capture (E-HMCS) device was shown. This device was used to measure the endpoint trajectories of Activities of Daily Living, together with desired joints' values corresponding. The measured endpoint trajectories then were put into the proposed optimization algorithm to solve the inverse kinematics to create a set of predicted joints value. The comparison results showed that the predicted joints' values are the same as the measured values. That demonstrates the ability to use the proposed algorithm to solve the problem of inverse kinematics for the natural movements of the human upper limbs.

Keywords: Proposed differential evolution · Inverse kinematics · Degree of Freedom · Exoskeleton type motion capture system · Activities of Daily Living

1 Introduction

The Inverse Kinematics (IK) problem for the robot involves determining the variable parameters that match the input information of the end effector's position and direction [1]. These matched values will ensure that subsequent robot control process will follow the desired trajectory. This is one of the important issues in the robotic field because it is related to other aspects such as motion planning, dynamic analysis and control [2]. Traditionally, there are some solutions to solve the IK problem for robots such as: geometry method is the method using geometric and trigonometric relationships to solve, the iteration method is the often required inversion of a Jacobian matrix to solve

N. T. Nguyen et al. (Eds.): ACIIDS 2021, LNAI 12672, pp. 556–568, 2021.
https://doi.org/10.1007/978-3-030-73280-6_44

the IK problem. However, when applying them to solve the problem of reverse kinetics for robots, especially redundant driving robots, it is often much more complicated and time-consuming. The reason is the nonlinearity of the formulas and the geometry transformation complexity between the working space (task space) and the joint space. In addition, the difficult point is in the singularity, the multiplicity of these formulas as well as the necessary variation of the formulas corresponding to the changes of different robot structures [3–5].

For upper exoskeleton rehabilitation robot type which has the same redundancy as the human arm, configuring the joint angles of the robot in accordance with those of humans' arm is one of the crucial control mechanisms to minimize the energy exchange between people and robot. In [6], the authors proposed the redundancy resolution algorithm uses a wrist position and orientation algorithm that allows the robot to form the natural human arm configuration as the operator changes the position and orientation of the end effector. In this way, the redundancy of the arm is mathematically expressed by the swivel angle. In addition to those existing methods of solving the IK problems, in recent years, the application of meta-heuristic optimization algorithm in solving IK problems has become increasingly common. The optimization-based approach has represented suitable ways to overcome the aforementioned problems. Indeed, these methods are generally based on the ways to choose objective function, constraints, and the chosen solution approach as well. There are many forms for objective functions based on some criteria such as singularity avoidance, obstacle avoidance, and physical constraints [7–9]. In [10], the author used a Differential Evolutionary algorithm to generate approximated solutions for a range of IK problems such as positions, orientation, physical factors like overall center-of-mass location. In [11], the redundant IK problem is formulated as the minimization of an infinity norm and translated to a dual optimization problem.

In this study, our improved DE algorithm [12] will be applied to solve the IK problem for the redundant driven model that is the human arm. The advantage of the proposed method is to significantly reduce the complexity and volume of calculations compared with the other methods. The results of the angles estimated from the application of the algorithm are compared with those measured by the Exoskeleton type Human Motion Capture System (E-HMCS) showing the similarity.

The remainder of the paper is divided into the following sections: Sect. 2 describes the model of the human arm. Next, the theory of DE algorithm with improvement search space (Pro DE) will be presented in Sect. 3. Section 4 covers the setup process to make experiment and run the proposed algorithm. The predicted results after applying the algorithm and comparing with the measured results are shown in Sect. 5. Finally, the conclusions are outlined in Sect. 6.

2 Human Arm with E-HMCS Model Description

2.1 Model Description

The human arm consists of hard links connected to each other by three anatomical joints (shoulder joint, elbow joint and wrist joint) while ignoring the movement of scapular and clavicle motions [12]. In this study, 7 Degree of Freedoms (DoFs) were analyzed including shoulder flexion/extension, shoulder abduction/adduction, shoulder lateral-medial rotation, elbow flexion/extension, elbow pronation/supination, wrist flexion/extension and wrist pronation/supination. In order to facilitate the research and manufacturing process of upper limb rehabilitation robot, we also proposed a 3D model that simulates the human arm to ensure the basic anthropometric parameters of Asians in general and especially pay attention to Vietnamese [14]. The human arm anthropometric parameters and Range of Motion (RoM) are shown in the Table 1 and Table 2, respectively. The segments' lengths of model are chosen based on the segments' length of the person who wore the E-HMCS.

For the purpose of evaluating the efficiency of the proposed algorithm, and at the same time determining the endpoint trajectory in ADL operations, we have also developed an Exoskeleton type Human Motion Capture System (E-HMCS) [17]. This is a measuring device with a structure similar to an exoskeleton fabricated with 3D printing technology and is cheap, simple calculation process. With these features, this device is also easily accessible to developing countries, limiting resources for expensive systems such as Video motion capture system. The human arm model wearing the E-HMCS device is shown in Fig. 1. The goal of this study is to use this measuring device to obtain the endpoint trajectory as well as the measured joint angle values. Endpoint trajectory is then put into Pro DE algorithm to predict the joints angle values and compare with the above measured values. With the aim of evaluating the coordinates of the end point

Table. 1. Some anthropometric parameters of human arm and human arm model

Segments	Average length of human (m) [15]	Length of model (m)
Upper arm	0,3	0.28
Lower arm	0,24	0.26
Hand	0,18	0.19
Arm	0,72	0.74

Table. 2. RoM of human arm and human arm model

Joints	Movements	Range of motion (ROM)	
		Human [16]	Model
Shoulder	Abduction–adduction	175° to −50°	175° to −50°
	Flexion–extension	165° to −50°	180° to −180°
	Internal–external rotation	70° to −45°	70° to −45°
Elbow	Flexion–extension	145° to 0°	145° to 0°
	Pronation–supination	75° to −110°	75° to −110°
Wrist	Flexion–extension	75° to −70°	73° to −73°
	Radial–ulnar deviation	20° to −40°	20° to −40°

position, the model will measure as well as predict for 4 measured angle values including q_1, q_2, q_3, q_4.

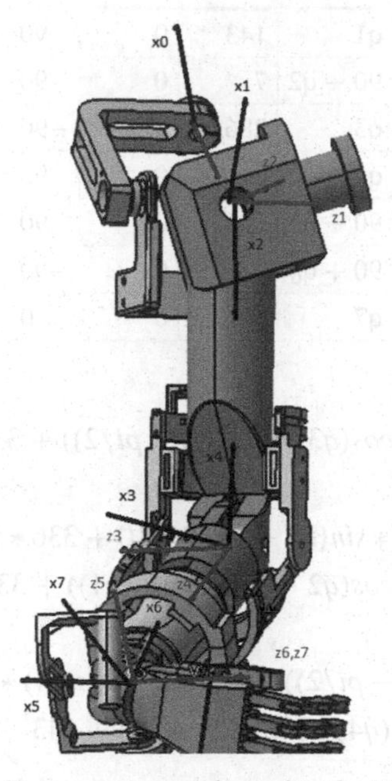

Fig. 1. Human arm model wearing the E-HMCS

2.2 Forward Kinematic Analysis

Forward kinematics analysis uses the kinematic equations of a robot to compute the position and orientation of the end-effector from specified values for the joint parameters. The homogeneous transformation matrix can be used to obtain the forward kinematics of the robot manipulator, using the DH parameters in Eq. 1:

$$T_{i-1}^{i} = \begin{bmatrix} C\theta_i & -S\theta_i & 0 & a_i \\ S\theta_i Ca_i & C\theta_i Ca_i & -Sa_i & -d_i Sa_i \\ S\theta_i Sa_i & S\theta_i Sa_i & -Ca_i & -d_i Ca_i \\ 0 & 0 & 0 & 1 \end{bmatrix} \tag{1}$$

where S and C denote the sine and cosine functions and their variables' value given as in Table 3.

The position and orientation of the end-effector can be determined by the following equation:

$$T_0^5 = T_0^1 * T_1^2 * T_2^3 * T_3^4 * T_4^5 \tag{2}$$

the position of end-effector coordinates in the working space are determined by:

$$xE = (2869 * cos(q1) * sin(q2 - pi/2))/10 - 336 * sin(q4) * (sin(q1)$$

Table. 3. D-H Parameters

θ	d(mm)	a(mm)	α(°)
q1	143	0	90
90 − q2	7	0	−90
q3	286.9	0	−90
q4	0	0	90
90 + q5	336	0	90
90 + q6	0	0	−90
q7	0	0	0

$$*sin(q3) - cos(q1) * cos(q3) * cos(q2 - pi/2)) + 336 * cos(q1)*cos(q4)$$
$$*sin(q2 - pi/2)$$
$$yE = (2869 * sin(q1) * sin(q2 - pi/2))/10 + 336 * sin(q4) * (cos(q1)$$
$$*sin(q3) + cos(q3) * cos(q2 - pi/2) * sin(q1)) + 336 * cos(q4) * sin(q1)$$
$$*sin(q2 - pi/2)$$
$$zE = (2869 * cos(q2 - pi/2))/10 + 336 * cos(q4) * cos(q2 - pi/2)$$
$$- 336 * cos(q3) * sin(q4) * sin(q2 - pi/2) + 143 \quad (3)$$

We can see that position of end-effector coordinates depends on joint variables q1, q2, q3, q4. Joint variables q5, q6, q7 determine the orientation orientation of the end-effector.

3 Proposed Optimization Algorithm

The mathematical process of calculating the variable joint parameters that achieve a specified position of the end-effector is known as inverse kinematics. Constraint of robot IK is non-linear equations and has a lot of local optimal solutions, the numerical solution is required to have a strong ability to obtain the global optimal value. There are some methodologies to solve the IK problem, for example the geometric method is the method using geometric and trigonometric relationships the iterative method often requires an inversion of a Jacobian matrix to solve the IK problem and Biomimicry Algorithm. In recent years, the evolutionary algorithms or bionic optimization algorithms have been mainly used, such as, Genetic Algorithm (GA), Differential Evolution (DE), Particle Swarm Optimization (PSO), Ant Colony Optimization (ACO), Artificial Bee Colony (ABC).

In this paper, Differential Evolution (DE) with searching space improvements is considered to be suitable method to resolve the proposed optimization problem. DE algorithm uses mechanisms inspired by biological evolution, such as reproduction, mutation, and selection. A mutant vector is generated according to:

$$V = X_{best} + F(X_{r1} - X_{r2}) \quad (DE/best/1) \quad (4)$$

In Eqs. (2), F is Scaling factor, r1, r2 is random solution, and $r1 \neq r2 \neq i$, X_{best} is population filled with the best member.

The crossover operation is applied to each pair of the generated mutant vector V_i, and its corresponding target vector X_i for the purpose of generating a trial vector:

$$U_i = \begin{cases} V_i, \text{ if } (\text{rand}[0, 1] \leq CR \text{ or } (j = j_{rand}) \\ X_i, \textit{otherwirse} \end{cases} \quad (5)$$

where CR is a user-specified crossover constant in the range [0, 1], j_{rand} is a randomly chosen integer in the range [1, n] to ensure that the trial vector U_i will differ from its corresponding target vector Xi by at least one parameter.

The method of selection used is:

$$X_{i,G+1} = \begin{cases} V_{i,G} \text{ if } f(U_{i,G}) \leq f(X_{i,G}) \\ X_{i,G}, \text{ otherwirse} \end{cases} \quad (6)$$

In this algorithm, firstly the robot moves from any position to the first point of the trajectory. With this first point, the initialization values for the particles are randomly selected in the full range of motion of joints. In addition, the target function in this case has the form:

$$Func.1 = a * \sqrt{\sum_{k=1}^{n} (q_i^k - q_0^k)} + b * \sqrt{(x_i - x_{ei})^2 + (y_i - y_{ei})^2 + (z_i - z_{ei})^2} \quad (7)$$

where the values q_0^k *and* q_i^k (i = 1) are the joint variable values at the original position and the 1st point on the trajectory respectively; (x_i, y_i, z_i) is the end-effector coordinates for the i-point (i = 1) found by the algorithm, (x_{ei}, y_{ei}, z_{ei}) is the desired end-effector coordinates; a, b are coefficient penalties. Cost function as the Eq. (7) ensures the energy spent on the joints to reach the 1st desired position is minimized. Besides, it also minimizes the distance error between the actual and desired end-effector position. The stopping condition of the trajectory points is that the Cost Func.1 value is less than the value of ε or the number of iterations reaches 400 and the number of times algorithm running <10.

After calculating the 1st point of the trajectory, the remaining points are calculated with a search limitation around the joints variables found by its trajectory point before. By this limitation, the program's searching space will be limited while ensuring the continuity of the joint variables. In this case, the target function will still be the same as the function of the 1st point, but it has coefficient a = 0.

4 Experiments and Proposed Algorithm Setup

4.1 Experiment Setup for ADL

As in the Fig. 2, in this study, E-HMCS measuring device was worn on a person with height 1.72 (m) and 22 years old. This person performed two complex ADL activities, drinking water and brushing teeth tasks. The drinking water task is divided into two stages of reaching and lifting. In order to perform these tasks, the experimental system should note a few points:

- Adjust the segments of the measuring system to match the length of the wearer's segments.
- The measuring device is firmly linked to the human hand by the strap band at upper arm.
- Testers need to wear and become familiar with the equipment. It is important to note that when performing ADL activities, the examiner needs to do it naturally and comfortably. If this person feels hindered by the measuring equipment, the adjustment should be made to get the most natural feeling in the joint position.

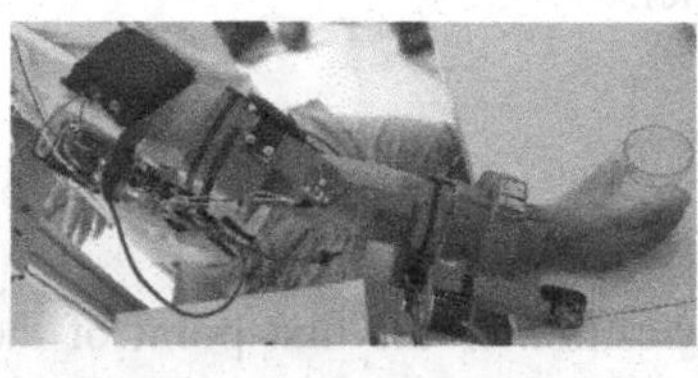
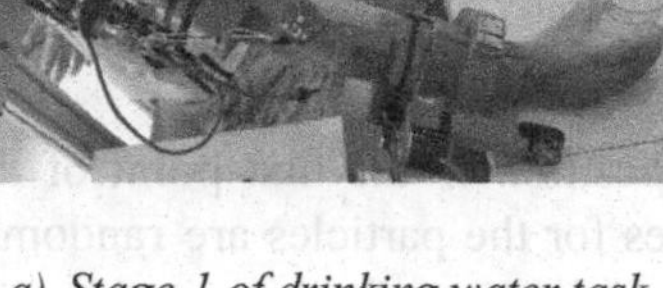
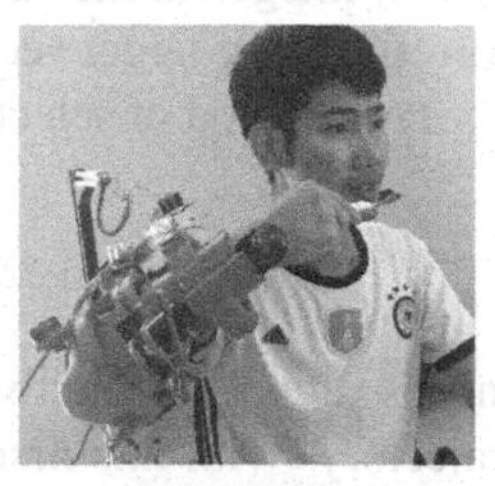
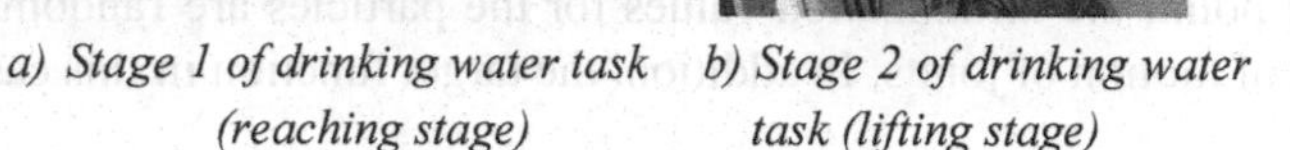

a) Stage 1 of drinking water task (reaching stage) *b) Stage 2 of drinking water task (lifting stage)* *c) Brushing teeth task*

Fig. 2. Experiment tasks

4.2 Proposed Algorithm Setup

When solving the IK problem for the 7-DOF serial robot manipulator of human arm, the study focused on three main aspects. The first of these is the sensitivity of the solution, or in the other word, the amount distance error of end effector. The second criterion was the execution time. In order to avoid the endless loop, the maximum numbers of generation ($iter_{max}$) was set as 600. And the final aspect is the searching space of joints' variables. Normally, almost studies have been using the Range of Motion (RoM) of joints for this searching space. Our algorithm 12 proposed to use the searching space of current generation is around previous optimal joints' values. In the Table 4 Optimization parameters used for proposed algorithm the ubs_{i+1} *and* lbs_{i+1} are the joints' upper and lower boundary of the current generation. When setting up the maximum distance error by the fitness value setting for the end effector position, the study set the value of $1e{-}8(m)$ for Pro DE algorithm. Running the IK problem in order to predict the 4 joints of q_1, q_2, q_3, q_4 were coded by Matlab version 2019a and run on the computer equipped with an Intel Core i5-4258U @2.4 GHz processor and 8 GB Ram memory.

Table 4. Optimization parameters used for proposed algorithm

$iter_{max}$	Max distance error (mm)	Searching space $[ubs_{i+1}\ lbs_{i+1}]$	Cross. rate	F_w
600	1e−8	$q_{oi} \pm \pi/100$	0.9	0.5

5 Results and Discussion

5.1 Drinking Water Task

As mentioned above, drinking water task is divided into two continues stages of reaching and lifting. Figures 3, 4 and 5 show the result for reaching stage. Meanwhile, the results for the lifting cup stage are shown in Figs. 6, 7 and 8. The endpoint movement trajectories, i.e., the wrist joint, of the two movements of reaching a cup and elevating the cup to mouth are shown in Fig. 3 and Fig. 6, respectively. These figures also partly show the quality of the Pro-DE algorithm when solving the IK problem for these orbits. This quality is demonstrated by the very close grip of the simulated trajectories compared to reference trajectories. The deviations of these two orbits over the two phases are shown more specifically in Fig. 4 and Fig. 6, respectively. Specifically, the average distance error between the end point orbits for stage 1 and stage 2 are $2.3e{-}6(mm)$ and $2.35e{-}6(mm)$, respectively. A very important result of this study is the quality of the predicted joints' angle values shown in Fig. 5 and Fig. 8. According to these results, the predicted values of joint variables are continuous and close to the measured values. Because the study only focused on the endpoint coordinates, the graph has four measurement angle values

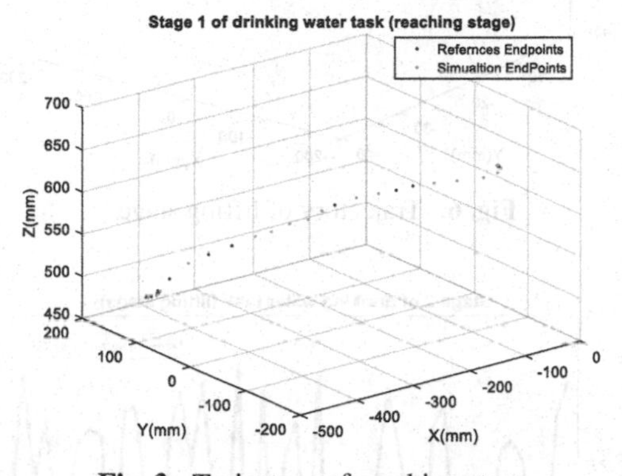

Fig. 3. Trajectory of reaching stage

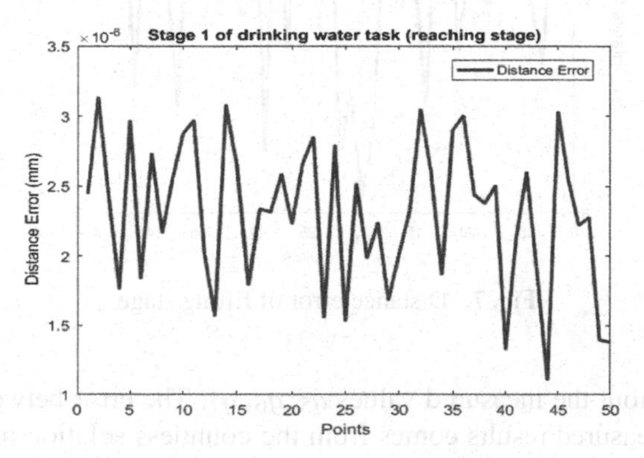

Fig. 4. Distance error of reaching stage

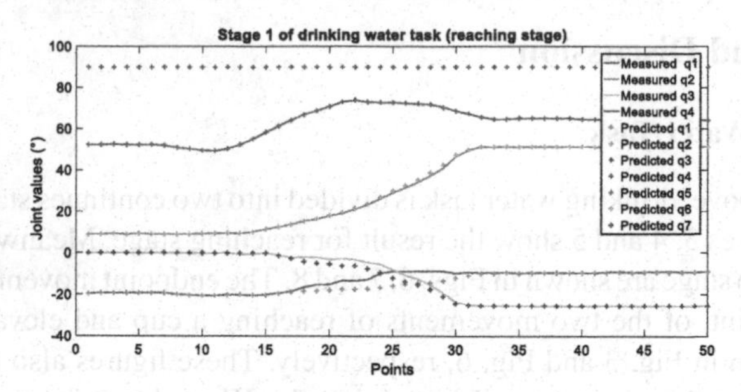

Fig. 5. Joint values of reaching stage

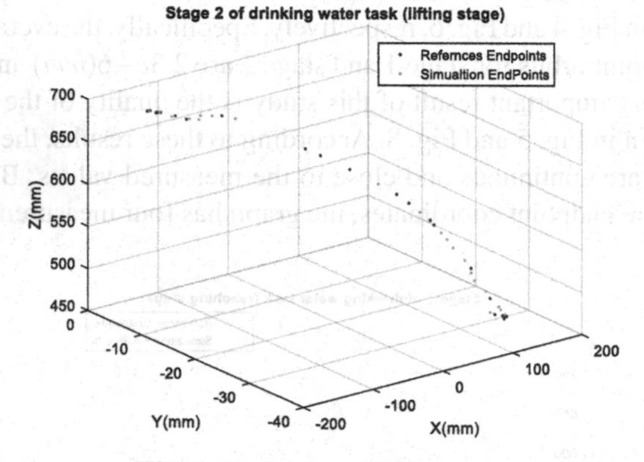

Fig. 6. Trajectory of lifting stage

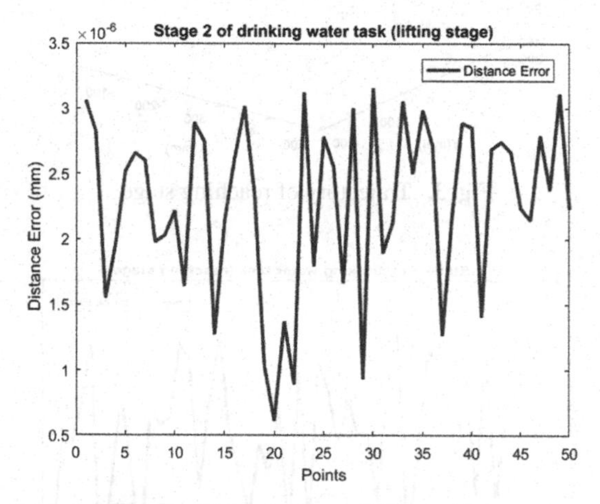

Fig. 7. Distance error of lifting stage

q_1, q_2, q_3, q_4 without the measured values q_5, q_6, q_7. The error between the predicted results and the measured results comes from the countless solution property of the IK

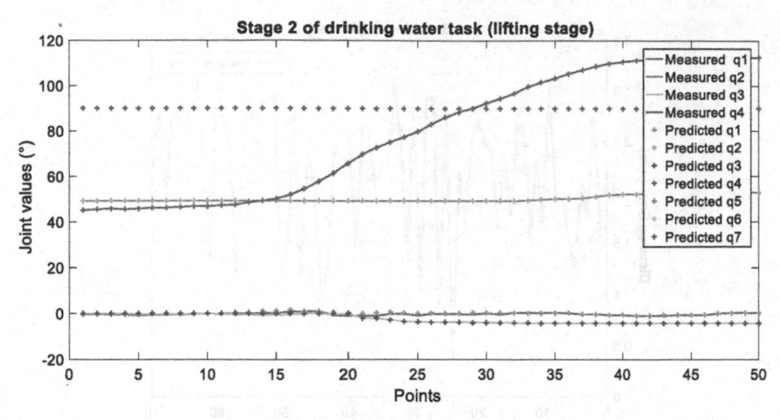

Fig. 8. Joint values of lifting stage

equation. The predicted values of q_5, q_6, q_7 are constant because they only affect the orientation of the end-effector but having no effect on the position.

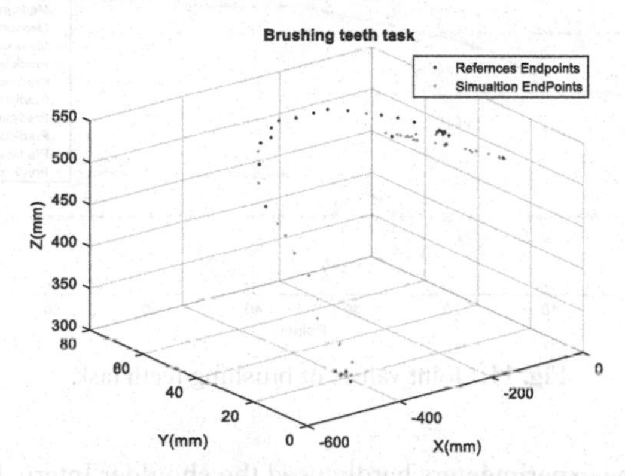

Fig. 9. Trajectory of brushing teeth task

5.2 Brushing Teeth Task

In this activity of ADL, the experimenter will lift the toothbrush from 1 point below the body to the mouth and perform the repetitive action of bringing the toothbrush to the left and right. Similar to the drinking water task, the endpoint trajectories results, distance error between simulation and reference trajectories, results of predicted and measured joints values are demonstrated in Fig. 9, Fig. 10, Fig. 11. Results after Appling the algorithm Pro-DE showed that the simulated end-effector trajectory closely followed the reference endpoint orbit. The average deviation between these 2 orbits is $2.2e{-}6(mm)$. In addition, the predicted joints variable values are also quite similar to the measured joints variable values. In these values the variable q_3 is constant while the last stage of angle q_2 is sinusoidal. The reason for this phenomenon is because in the brushing

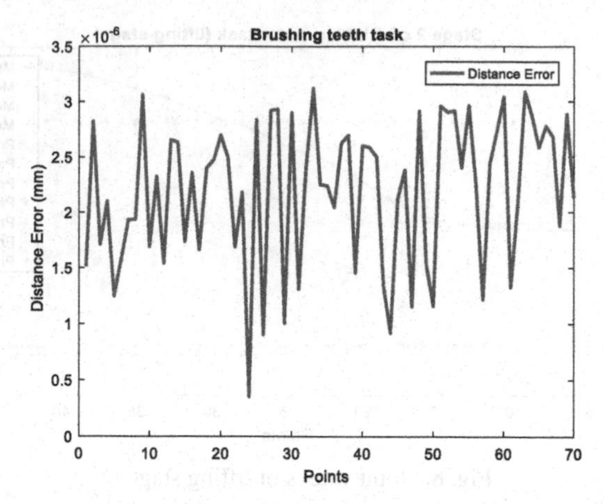

Fig. 10. Distance error of brushing teeth task.

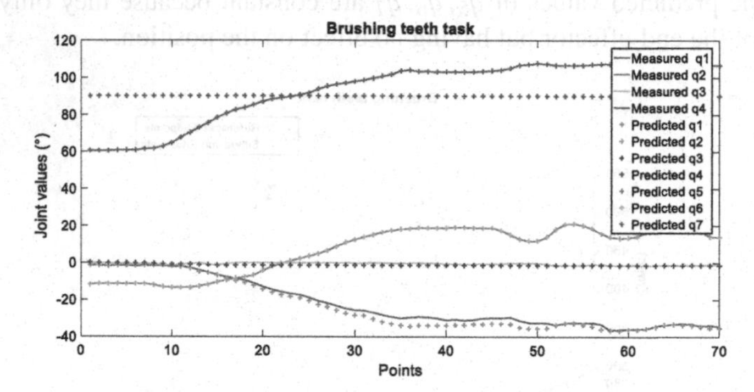

Fig. 11. Joint values of brushing teeth task

task as described, the experimenters hardly used the shoulder Internal-External rotation movement. That results in the angle q_3 being almost unchanged. Meanwhile, in order for the arm to perform the repetitive movement left to right, the q_2 joint will move forward and reverse again and again. The measured joints variable values are not available for the angles q_5, q_6, q_7 and they are also constants for the predicted values using the Pro-DE algorithm.

6 Conclusion

In this study, the DE algorithm with search space improvement (Pro-DE) was used to predict the values of joint variables of the human arm in two ADL activities of drinking water and brushing. To evaluate this algorithm, the study used exoskeleton type human motion capture system to measure the wrist joint trajectories (end effector) as well as the value of the joint variables in the above activities. In order to ensure the integrity of the operation, the experimenter should be familiar with measuring the equipment and perform the operations as naturally as possible. The measured endpoint orbits have been put into the suggested Pro-DE algorithm to predict the joints variable values. The experiment as well as the results after running the algorithm showed that the predicted endpoint and especially the predicted angle values of the IK problem closely follow the endpoint trajectory and the measured angle values. Deviations between the predicted angle and the measured angle in the results can be caused by unnaturalness when wearing a rigid measuring frame to obtain the results. The study demonstrates the ability to use the proposed algorithm to solve the inverse kinetic problem of redundant driven robots, and especially for the natural movements of the human arm. In the future, the study will investigate some other movements of the human arm to further confirm the effectiveness of the method.

Acknowledgement. Research reported in this paper was supported by Ministry of Science and Technology of Vietnam, under award number [**ĐTĐLCN.28/20**].

References

1. Köker, R., Çakar, T.: A neuro-genetic-simulated annealing approach to the inverse kinematics solution of robots: a simulation based study. Eng. Comput. **32**(4), 553–565 (2016). https://doi.org/10.1007/s00366-015-0432-z
2. Huang, H.C., Chen, C.P., Wang, P.R.: Particle swarm optimization for solving the inverse kinematics of 7-DOF robotic manipulators. In: IEEE International Conference on Systems, Man, and Cybernetics, Seoul, Korea, pp. 3105–3110 (2012)
3. Rubio, J.J., Bravo, A.G., Pacheco, J., Aguilar, C.: Passivity analysis and modeling of robotic arms. IEEE Lat. Am. Trans. **12**(8), 1381–1389 (2014)
4. Kou, Y.L., Lin, T.P., Wu, C.Y.: Experimental and numerical study on the semi-closed loop control of a planar robot manipulator. Math. Probl. Eng. **2014**, 1–9 (2014). https://doi.org/10.1155/2014/769038
5. Lopez-Franco, C., Hernandez-Barragan, J., Alanis, A.Y., Arana-Daniel, N.: A soft computing approach for inverse kinematics of robot manipulators. Eng. Appl. Artif. Intell. **74**, 104–120 (2018)
6. Hyunchul, K., Jacob, R.: Predicting redundancy of a 7DOF upper limb exoskeleton toward improved transparency between human and robot. J. Intell. Robot. Syst. **80**(1), 99–119 (2015)
7. Hwang, B., Jeon, D.: A method to accurately estimate the muscular torques of human wearing exoskeletons by torque sensors. Sensors (Basel, Switz.) **15**(4), 8337–8357 (2015). Ed. by V.M.N. Passaro
8. Whitney, D.: Resolved motion rate control of manipulators and human prostheses. IEEE Trans. Man-Mach. Syst. **10**, 4753 (2010)
9. Runarsson, T.P., Yao, X.: Search biases in constrained evolutionary optimization. IEEE Trans. Syst. Man Cybern. Part C **35**, 223–243 (2005)
10. Ben, K.: Inverse kinematic solutions for articulated characters using massively parallel 9 architectures and differential evolutionary algorithms. In: Workshop on Virtual Reality Interaction and Physical Simulation VRIPHYS (2017)
11. Karpinska, J., Tchon, K., Janiak, M.: Approximation of Jacobian inverse kinematics algorithms: differential geometric vs. variational approach. J. Intell. Robot. Syst. **68**, 211–224 (2012)
12. Thanh-Trung, N., Ngoc-Linh, T., Van-Tinh, N., Ngoc-Tam, B., Van-Huy, N., Dai, W.: Apply PSO algorithm with searching space improvements on a 5 degrees of freedom robot. In: 2020 3rd International Conference on Intelligent Robotic and Control Engineering (IRCE), Oxford, UK, 10–12 August 2020. IEEE (Indexed Scopus) (2020)

13. Korein, J.U.: A geometric investigation of reach. MIT Press, Cambridge (1986).ISBN 978-0-262-11104-1
14. Trung, N., Hiep, D., Thien, D., Tam, B., Dai, W.: Design a human arm model supporting the design process of upper limb rehabilitation robot. In: 14th South Asian Technical University Consortium Symposium 2020 (SEATUC 2020), pp. 2186–7631, Bangkok, Thailand, 27–28 February 2020 (2020)
15. David, A.W.: Biomechanics and Motor Control of Human Movement. Wiley, Hoboken (2009)
16. Atlas of Vietnamese anthropometric in working age, signs of joint activity range and visual field. Scientific and Technical Publisher (1997)
17. Thanh-Trung, N., Ngoc-Tam, B., Watanabe, D.: Design and manufacture a cheap equipment to measure human arm motion in developing countries. In: The 4th International Conference on Mechatronics Systems and Control Engineering (ICMSCE 2021) (2021)

Word Segmentation for Gujarati Handwritten Documents

Divya Bhatia[1(✉)], Mukesh M. Goswami[1,2], and Suman Mitra[1]

[1] Dhirubhai Ambani Institute of Information and Communication Technology, Gandhinagar, India
suman_mitra@daiict.ac.in

[2] Dharmsinh Desai Institute of Technology, Nadiad, India

Abstract. In this fast-evolving world, documents in numerous regional languages are finding a prominent place on the internet. This is evident from the increasing use of regional languages on hoardings of advertisements, boards of various stalls and shops, and even essential government instructions are found in regional languages. With the growing reach of the internet, the least privileged are also getting an opportunity to explore the world. Hence, more than a technological requirement, it has become a moral responsibility to put to the test research at the grass root levels in serving the ones who have remained aloof for a while. This work is a preliminary step in exploring the efficiency of various conventional techniques like morphology operations, connected components analysis, and finding contours in segmenting Gujarati handwritten words from scanned documents. The results obtained from the collected data are encouraging.

Keywords: Morphology operations · Word detection · Handwritten word segmentation · Automate annotations

1 Introduction

The rise of the internet and its ever-increasing reach has led to people trying to know the world more and consequently express themselves more. Work done in regional languages is very less, whether it is Translation, Word Recognition, Semantic Analysis, Keyword Spotting, or Ranking of Gujarati documents. The all-pervading problem of detection is of increased difficulty because of change in handwriting brings its own set of challenges. Noteworthy is that a decent amount of work has been done on most in use scripts like Hindi, Bangla or Tamil. However, for other regional scripts, Gujarati being in focus here, research is in a very elementary stage. This forms the sole motivation for embarking research in this regard for the authors. The primal task is to detect and segment handwritten Gujarati words. Focus on script independent lexicon free word segmentation system is made because of the unavailability of sufficient training data for the Gujarati language. Conventional methods like morphology operations,

N. T. Nguyen et al. (Eds.): ACIIDS 2021, LNAI 12672, pp. 569–580, 2021.
https://doi.org/10.1007/978-3-030-73280-6_45

finding contours, and connected component analysis are used because annotated and labeled Gujarati handwritten word dataset, which is essential for the deep learning-based approaches, is not available for Gujarati handwritten text. This method will be used as a preprocessing step to semi-automate the document annotation work for Gujarati handwritten documents so that more sophisticated deep learning-based techniques can be applied to further increase the accuracy of segmentation and recognition of handwritten Gujarati words, which potentially has many applications in word spotting, recognizing document, image indexing, or ranking because the approach being script independent can be applied to other Indian scripts.

1.1 Literature Review

A majority of the works on Indian scripts focuses only on isolated characters, and only a few consider offline handwriting recognition problems at the word level. Based on the information gathered, there is no available benchmark samples dataset of handwritten Gujarati words for comparison purposes. This type of work is categorized broadly based on methods employed: conventional methods or deep learning methods, such as Template Matching, HMM (Hidden Markov Model) on the conventional side, and Faster RCNN, CNN with Svm, or Deep Learning on the new evolving machine learning side. Deep Learning-based methods are popular because they find state-of-art architecture with unconstrained handwriting recognition and generic model recognition [12].

Some of the work includes feature extraction methods based on Wavelet, HOG (Histogram of Oriented Gradient), and PHOG (Pyramidal HOG) performed for handwritten word recognition in Bangla and Devanagari [8]. This technique extracts features from the entire word and uses lexicon dependent HMM decoding to recognize the whole word [13]. Lexicon dependent signifies that it uses the available vocabulary of thousands of Bangla and Devanagari words. A deep architecture of Convolutional Neural Network (CNN) for automatic feature generation and recognition of Recurrent Neural Network architecture (RNN) for handwritten Malayalam word recognition has used in some works [10]. The use of deep learning architecture was possible in the discussed case because of the attainable dictionary of Malayalam words, which in the present case for Gujarati words is unavailable.

Segmenting text from reports is a significant move prior to recognition. Some used semiotic based text segmentation on Marathi literature and pictures using ML or DL techniques. That work includes notable deep learning frameworks, i.e., U-Net and Residual U-Net (ResU-Net). U-Net was initially designed for medical image-related segmentation tasks. It is a basic encoder-decoder structure with skip connections to accurately include both spatial and labels information. This kind of design with skip connections dodges the vanishing gradient obstacle, and, also the feature addition in the model infers better on the segmentation task. However, this work performs character level segmentation, and our main focus is to develop word-level handwritten Gujarati segmentation [4].

Some work also used Region-based methods like R-CNN, Fast RCNN, and Faster RCNN. In RCNN regions are formed, which fed to Convolution Neural Network for classification and regression, creating bounding boxes on detected objects. The input layer provides methods such as selective search to propose regions forming ROI's, creating a bounding box, and sending them to CNN for future processing. For each image, separate convnet is there, and separate SVMs used to perform the task for classification and box regression [1]. In Fast RCNN, instead of running all the proposed image regions (ROI) through the network independently, the entire image is run through the network. A convolution feature map formed from a convolution layer used to generate the region of interest. In Faster RCNN, Region Proposal Network is used instead of a selective search method. It has a neural network that seems to propose regions. It slides a window upon feature maps and generates anchor boxes [3]. Faster RCNN considered to be the novel method in object detection and segmentation, but their disadvantage is that they use anchors as reference boxes, so they also need manually annotated ground truth data. However, there is one proposed solution of anchor free region proposal network(af-rpn) [2] it eliminates the need for this monotonous task of labeling the image and uses the efficient mechanism of regional proposal networks which works well with short-range text files, and handwritten documents are short-range files.

Labelimg is a graphical image annotation software tool, written in python and save annotations in XML. It used to create the bounding box and manual annotations on our dataset, which acts as a ground truth data for accuracy comparison.

The Apache PDFBox library is an open-source Java tool that works with PDF documents. It Saves PDF as image file formats such as PNG or JPEG and can extract content from documents. We have used it to convert all our scanned PDF documents into JPEG format.

2 Proposed Work

2.1 Methodology

As a first task, We collect 200 handwritten document samples. These were scanned using a scanner of 300 dpi in PDF format, then using a python script and the PDF Box tool kit of Apache, we have converted all the pdfs into jpeg format. Image segmentation includes distributing visual input into segments to analyze image analysis. Segments symbolize the target or parts of it (labels). Thus it considers the set of pixels as particular objects, not an individual unit; in other words, it is essentially the process of identifying the positions of one or more objects inside an image. The technique we used makes use of morphological operations. Morphology frames images based on shapes. In a morphological method, every pixel in the image is altered based on other pixels' value in its vicinity. It operates better for binary images, so it uses a binary image, and a structuring element as an input [6]. It frames objects in the input image based on characteristics of its shape, encoded in the structuring element, then convolves

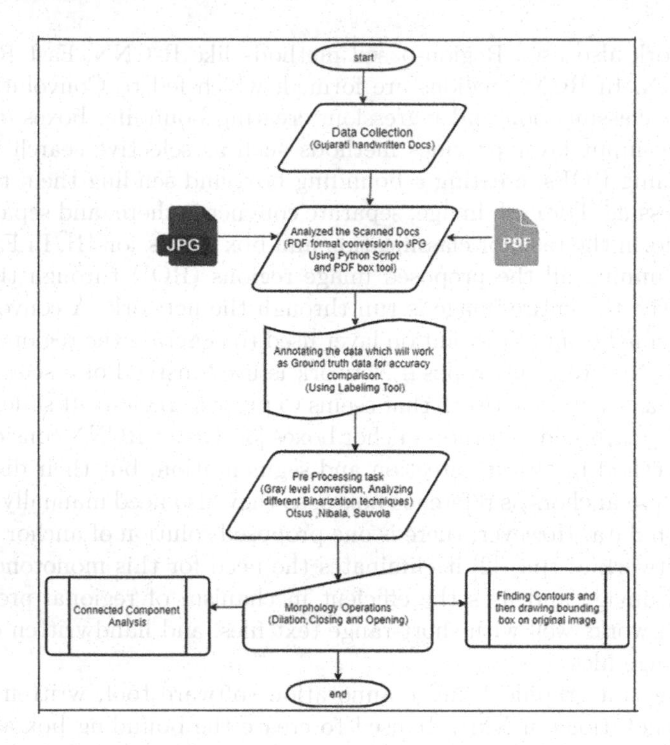

Fig. 1. Workflow of the proposed method

input image with the required structuring element based on set operation, hence generating the aspired result. Dilation, Erosion, Opening, Closing, Blackhat, Hit, and Miss Transform, Medial Axis Transform are most commonly used morphology operations [7].

First, the grayscale image is binarized, and then with a blackhat operator, the darkest image elements are enhanced, emphasizing less on noise. Image binarization is a process where a threshold value is chosen, and all pixels with values above this threshold are classified as white, and all other pixels as black. Further, for selecting threshold value, various techniques are used like the K-Means method to automatically identify the threshold for an image, which is popularly known as globally optimal Otsu's method. Interestingly, in some cases, the information is statistical and uses the mean, median, entropy, and at times, it is in the form of shape characteristics of the histogram. Binarization provides sharper and clearer contours of various objects present in the image; it discards the background noise and gives the contour of the image in the foreground. Experiments have performed with diverse binarization techniques (results shown below) like Ostus global thresholding, Nibala, and Sauvola local thresholding [5]. Among them, Ostus performs the best as per the final result shown below. When applying blackhat, a kernel used gives more importance to vertical image elements. The kernel has a fixed size, so the image is scaled, which also achieves performance improvement and favors normalization. Consequently, on the binarized image, multiple morphological operations are carried out, such as Dilation, Closing, and at last, followed by Opening. Opening morphology operation also filtered some noise in the input image.

After that, two things 1) Perform connected component analysis and 2) Find contours and use extracted rectangles to draw them, overlay on the original image. Contours can be described commonly as a trajectory uniting all the coherent points (along the boundary), having the identical color or intensity. The contours are a beneficial tool for shape study and object detection and recognition. The contour pixels are commonly a tiny subset of the absolute number of pixels depicting a pattern. Once the contour of a given pattern is obtained, its distinctive characters will be analyzed and used as features that will, later on, used in pattern classification. The volume of calculation is reduced when feature extraction algorithms are operated on the contour instead of on the entire pattern. Since the contour accords many features with the original pattern, the feature extraction process becomes much more effective when implemented on the contour than on the original pattern. Therefore, the accurate extraction of the contour will create more precise features, which will enhance the chances of correctly classifying a given pattern. The entire Workflow is shown in Fig. 1.

2.2 Disadvantages

Morphology operations depend on many parameters that are obliged to set manually by trial and error method; hence, they are very much prone to human errors. They sometimes will not work adequately with all the images because it is needed to tune the parameters and may require to change the dilation, erosion, and closing iterations and sequence for some images. While finding contours and fitting bounding boxes, the box's width and height need to set so the best fit can only be known by the trial and error method.

Handwritten documents have a distinctive style of writing and formatting, so perceived that for some of the documents parameter needs to be reset to obtain more reliable results.

Some words which have a single character are not recognizable because of the minimum width and height criteria. Those documents having less line space between two lines are seldom detected as one word. Also, a word sometimes detected half or double if there is slightly more space between the characters of one word (all such intricacies are shown in results below).

3 Results

3.1 Dataset

Dataset is collected manually; It has around 200 handwritten scanned documents in the Gujarati Script. All the documents have different handwriting written by the persons belonging to different sections of the population concerning age, sex, education, profession, and income. The dataset contains lots of different writing styles, variations, and possible complications. Initially, the documents scanned using the scanner of 300 dpi resolution in PDF format. Then, using the PDF-Apache tool kit and writing Python script, all the instances were converted into JPEG format. The instance of handwritten document is shown in Fig. 2.

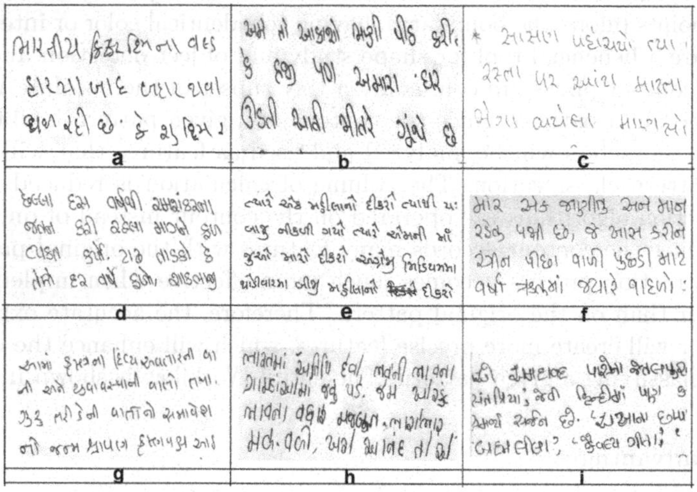

Fig. 2. The figure comprises diverse instances of datasets, In (a) (b) and (c) instances the separation distance between the words is more, (d), (e) and (f) are the instances of Good handwriting where words are properly aligned and clearly visible. In instances (g), (h), (i) the separation distance between words is less, the writing is poor also the words are not properly aligned in a line.

3.2 Binarization Results

There are various types of binarization techniques in image processing mainly classify as local binarization method and global binarization method. A window of NxN blocks thrust over the complete image, and a threshold value is calculated for every local area under the window for binarization in local methods. Otsu's thresholding method includes iterating through all the potential threshold values and computing a measure of spread for the pixel levels on each side of the threshold, i.e., the pixels that either fall in the foreground or the background, the threshold value estimated where the sum of foreground and background spreads is minimum. [5] We have used few Binarization techniques like Nibala, Savula, and Otsus. Among them, Nibala and Savula are local methods while Otsus is a global method. The results of all three techniques are shown in Fig. 3, 4 and 5 respectively. The majority of documents show that otsus binarization method performs better so we have applied it in our work.

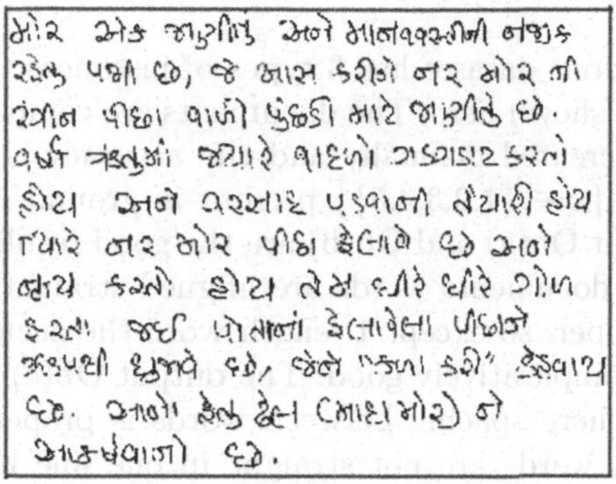

Fig. 3. Sauvola local Binarization technique [5]

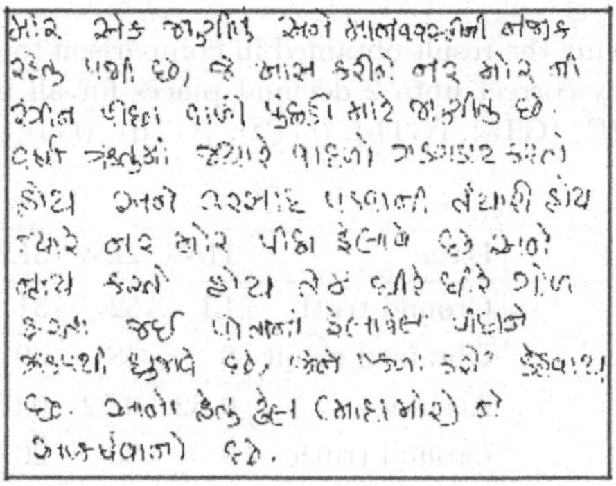

Fig. 4. Otsus global Binarization technique [5]

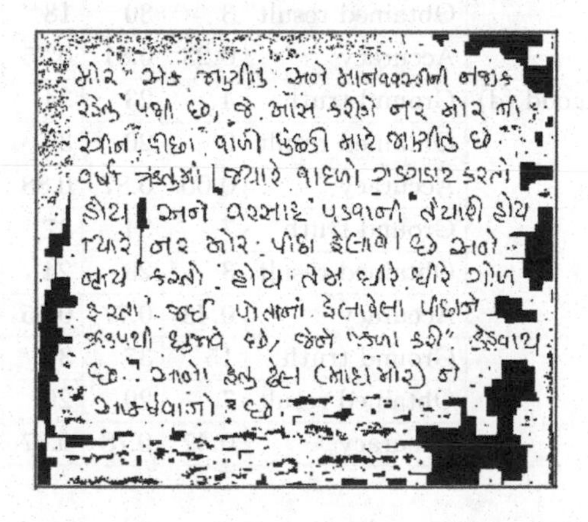

Fig. 5. Nibala local Binarization technique [5]

3.3 Final Output

As mentioned earlier our dataset has 3 types of instances the results of 2 documents of each type is shown here. The documents are compared with the ground truth data which is created manually and the accuracy is calculated based on no of x length word $\{x \in \{1,2,3,4,5\}\}$ present in ground data and detected in our result. The output Ob(a) and Ob(b) are the good results, the reason behind this is that in these documents words are aligned straight in one line, spacing between words is proper, so except 1 length word the accuracy for 2, 3, 4, and 5 length words are comparatively good. The output Ob(c) and Ob(d) are moderately good results here spacing between words is proper but space between the lines is less thus words are not straight in one line hence multiple words are detected as one. The output Ob(e) and Ob(f) are poor results the spacing between words is more even space between characters of a single word is also more so one word is detected as multiple words (Table 1).

Table 1. Table is showing the result obtained in comparison to ground truth data and corresponding Accuracy correct upto 2 decimal places for all images shown in figure (a), (b), (c), (d), (e), (f), (GTa), (GTb), (GTc), (GTd), (GTe), (GTf) Note:- xLW is x length of word

Type of Doc	Data	1LW	2LW	3LW	4LW	5LW
Good (a)	Ground truth	13	52	31	9	0
	Obtained result	8	48	30	9	0
	Accuracy	0.62	0.92	0.97	1.00	NA
Good (b)	Ground truth	9	30	20	5	3
	Obtained result	3	26	17	3	3
	Accuracy	0.33	0.87	0.85	0.6	1.00
Moderately good (c)	Ground truth	13	37	24	15	6
	Obtained result	3	30	18	12	5
	Accuracy	0.23	0.81	0.75	0.80	0.83
Moderately good (d)	Ground truth	4	23	34	26	17
	Obtained result	0	20	30	21	14
	Accuracy	0.00	0.87	0.88	0.80	0.82
Poor (e)	Ground truth	7	34	37	10	9
	Obtained result	3	26	24	0	4
	Accuracy	0.43	0.77	0.65	0.00	0.44
Poor (f)	Ground truth	15	37	30	7	3
	Obtained result	7	29	20	4	1
	Accuracy	0.47	0.78	0.67	0.57	0.33

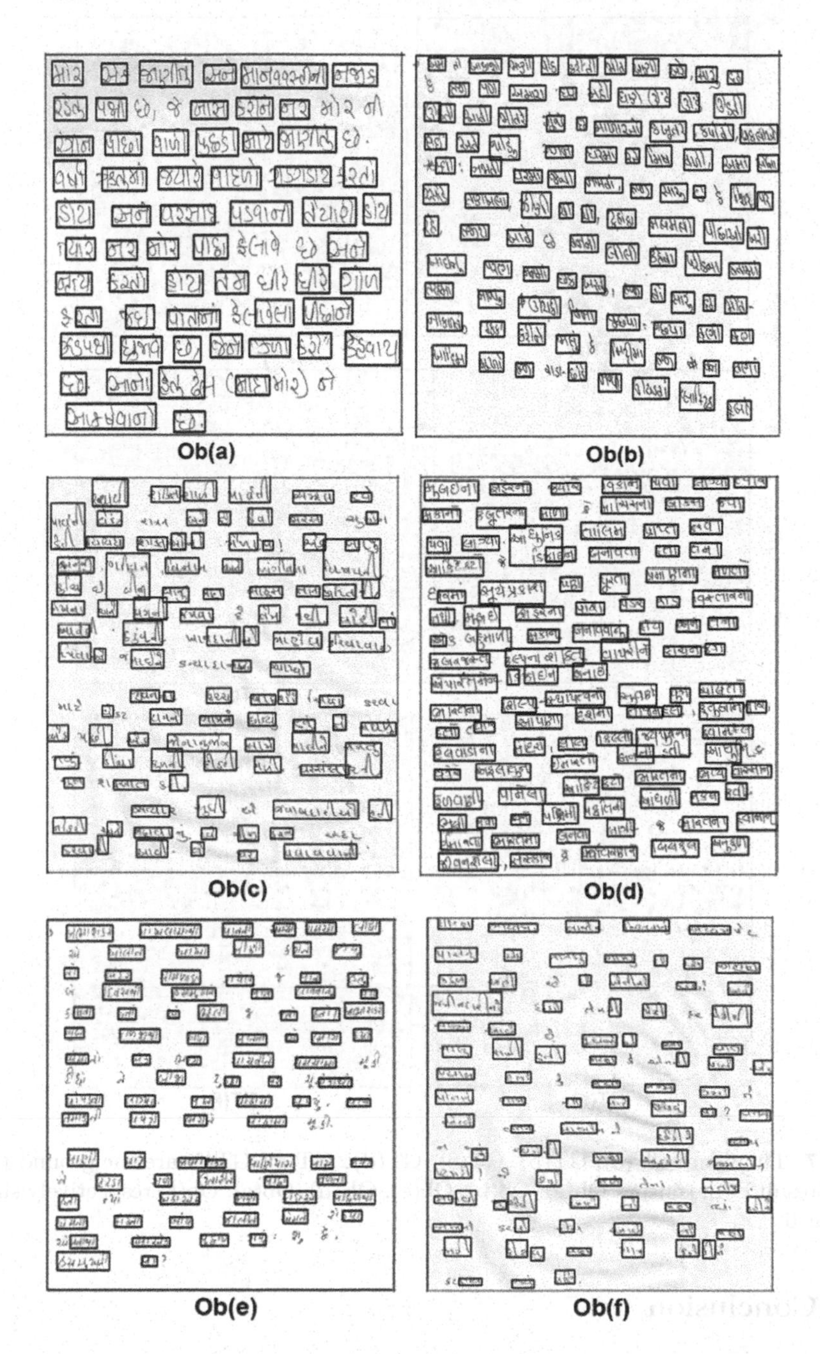

Fig. 6. Figure showing results of documents for which very good Ob(a) Ob(b), moderately good Ob(c) Ob(d) and poor Ob(e) Ob(f) results are obtained. The groung truth for these docs are shown in Fig. 7

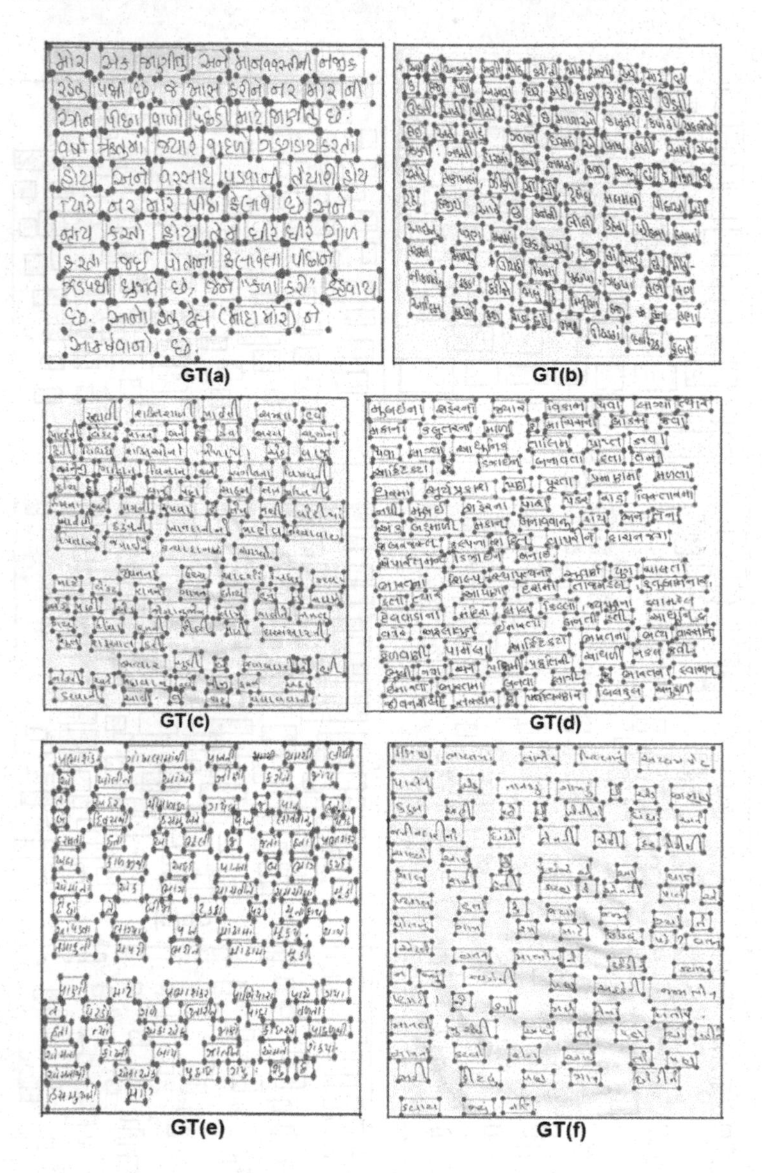

Fig. 7. The figure GT(a), GT(b), GT(c), GT(d), GT(e), GT(f) are the ground truth data result of the output Ob(a), Ob(b), Ob(c), Ob(d), Ob(e), Ob(f) respectively shown in Fig. 6

4 Conclusion

We demonstrate the conventional method for Gujarati handwritten word Segmentation from the scanned documents, collected from the native people. This method can be used as a pre-processing step in document annotation. As document annotation is very lengthy and time-consuming, this method can automate the process to some level, and the user is only required to correct the errors afterward, which could speed up the document annotation work. Hence this method is efficient in terms of time and resources as it will work on any machine with standard specifications GPUs are not necessarily required. It also does not demand any prior training on the dataset or availability of a dictionary of Gujarati words. It will work for all types of documents irrespective of length, or no of words. The accuracy depends on the type of writing and alignment of words as closely associated words sometimes detect one word and the words with more distance between the characters detected as a separate word, also it shows unsatisfactory performance to detect one length word in most documents. As a part of our future work, we would recognize the handwritten Gujarati words by forming an annotated dataset on all training documents and vocabulary of Gujarati words with the state-of-art methods like faster RCNN, YOLO, CNN-RNN hybrid architecture and CNN with SVM kind of deep learning techniques.

References

1. Ren, S., He, K., Girshick, R., Sun, J.: Faster R-CNN towards real-time object detection with region proposal networks, microsoft research. IEEE Trans. Pattern Anal. Mach. Intell. **39**(6), 1137–1149 (2016)
2. Zhong, Z., Sun, L., Huo, Q.: An anchor-free region proposal network for faster R-CNN-based text detection approaches. Int. J. Doc. Anal. Recogn. (IJDAR) **22**(3), 315–327 (2019). https://doi.org/10.1007/s10032-019-00335-y
3. Nagaoka, Y., Miyazaki, T., Sugaya, Y., Omachi, S.: In: 2017 14th IAPR International Text Detection by Faster R- CNN with Multiple Region Proposal Networks (2017)
4. Akhter, S.S.M.N., Rege, P.P.: Semantic segmentation of printed text from marathi document images using deep learning methods (2019)
5. Grdiet, P., Garg, N.: Binarization techniques used for grey scale images. Int. J. Comput. Appl. **71**(1), 8–11 (2013)
6. Chudasama, D., Patel, T., Joshi, S.: Image segmentation using morphological operations. Int. J. Comput. Appl. **117**(18), (2015)
7. Katona, M., Nyúl, L.G.: A novel method for accurate and efficient barcode detection with morphological operations (2012)
8. Retsinas, G., Stamatopoulos, N., Louloudis, G., Gatos, B.: Keyword spotting in handwritten documents using projections of oriented gradients (2016)
9. Dutta, K., Krishnan, P., Mathew, M., Jawahar, C.V.: Towards accurate handwritten word recognition for Hindi and Bangla. In: Rameshan, R., Arora, C., Dutta Roy, S. (eds.) NCVPRIPG 2017. CCIS, vol. 841, pp. 470–480. Springer, Singapore (2018). https://doi.org/10.1007/978-981-13-0020-2_41
10. Gan, J., Wang, W.: In-air handwritten English word recognition using attention recurrent translator. Neural Computing and Applications **31**(7), 3155–3172 (2017). https://doi.org/10.1007/s00521-017-3260-9
11. Bhattacharya, U., Chaudhuri, B.B.: Handwritten numeral databases of Indian scripts and multistage recognition of mixed numerals. IEEE Trans. Pattern Anal. Mach. Intell. (2008)
12. Shaw, B., Bhattacharya, U., Parui, S.K.: Offline handwritten devanagari word recognition: information fusion at feature and classier levels. In: 3rd IAPR Asian Conference on Pattern Recognition (ACPR) (2015)

13. Shaw, B., Parui, S.K., Shridhar, M.: A segmentation based approach. In: Pattern Recognition ICPR, Offline handwritten Devanagari word recognition (2008)
14. Garain, U., Mioulet, L., Chaudhuri, B.B., Chatelain, C., Paquet, T.: Unconstrained Bengali handwriting recognition with recurrent models. In: 13th International Conference on Document Analysis and Recognition (ICDAR) (2015)

Advanced Data Mining Techniques and Applications

A SPEA-Based Group Trading Strategy Portfolio Optimization Algorithm

Chun-Hao Chen[1], Chong-You Ye[2], Yeong-Chyi Lee[3], and Tzung-Pei Hong[4,5](✉)

[1] Department of Information and Finance Management, National Taipei University of Technology, Taipei, Taiwan
[2] Department of Computer Science and Information Engineering, Tamkang University, Taipei, Taiwan
[3] Department of Information Management, Cheng Shiu University, Kaohsiung, Taiwan
yeongchyi@gcloud.csu.edu.tw
[4] Department of Computer Science and Engineering, National Sun Yat-sen University, Kaohsiung, Taiwan
tphong@nuk.edu.tw
[5] Department of Computer Science and Information Engineering, National University of Kaohsiung, Kaohsiung, Taiwan

Abstract. Trading strategies are usually employed for finding trading signals for increasing returns as well as reducing risks. As a result, many approaches have been proposed for obtaining trading strategy portfolio. The group trading strategy portfolio (GTSP) optimization approaches that can be used to provide various trading strategy portfolios were also proposed. Because different criteria should be considered to derive GTSPs, a MOGA (multi-objective genetic algorithm) based approach has been presented for searching non-dominated solutions. In this paper, to extract a better set of non-dominated solutions, we propose a SPEA-based algorithm for deriving GTSPs with two objective functions. Since the goal of trading is to get profit, the first objective function is utilized to evaluate the return and risk of a candidate GTSP. The second objective function is used to evaluate whether the numbers of strategies between groups are similar and weights of groups as well. Experiments were conducted on a financial dataset to show the effectiveness of the proposed approach and comparison results of the proposed approach and the previous approach.

Keywords: Grouping genetic algorithm · Group trading strategy portfolio · Trading strategy · Multi-objective problem · Multi-objective optimization algorithms

1 Introduction

Trading strategies are commonly used to find suitable trading signals, including buying and selling signals, to achieve a more safety and profitable trading while various factors may affect the financial market. Many ways, such as technique, chip, and fundamental

N. T. Nguyen et al. (Eds.): ACIIDS 2021, LNAI 12672, pp. 583–592, 2021.
https://doi.org/10.1007/978-3-030-73280-6_46

indices can be utilized to form trading strategies. In the literature, many approaches have also been proposed to form trading strategies for different assets, e.g., stocks, futures [5, 7, 8, 14].

Since generating useful trading strategies is considered to be an optimization problem, some evolutionary-based approaches were proposed to handle it. For instance, Kim et al. derived a rule change trading system (RCTS) for the futures market taking advantage of the rough set and the genetic algorithm [6]. Ucar et al. proposed a two-level options trading strategy selection mechanism [13]. In their approach, the trend was found in the first level. Then, genetic algorithms (GA) and particle swarm optimization (PSO) were used to observe profitable option trading strategy. Drezewski et al. proposed a bio-inspired optimization approach for optimizing trading strategies for currency market [4].

To provide a more reliable way for investors to make profitable plans, trading strategy portfolio (TSP) is considered to reach the goal. A GA-based algorithm for discovering an appropriate investment strategy portfolio was introduced by Chen et al. [2]. In addition, Chen et al. first addressed problems of the group trading strategy portfolio optimization (GTSPO), and an approach with the grouping genetic algorithm (GGA) was then given to optimize a group trading strategy portfolio (GTSP) [9]. Since a GTSP was composed of trading strategy groups, various TSPs could be given for choosing satisfied TSPs to traders.

Many objectives should be considered to design more empirical algorithms and objectives usually have trade-off relationships in real application. Multi-objective evolutionary-based algorithms, therefore, have been proposed to handle the optimization problems of the multi-objective trading strategy [1, 10–12], such as Lohpetch et al. developed trading rules for stock market trading by utilizing multi-objective genetic programming algorithm [11].

With regard to the multi-objective GTSP optimization problem, an algorithm with the two objective functions for discovering a set of GTSPs using multi-objective genetic algorithm was presented by Chen et al. [3]. In that approach, a GTSP was represented by the three parts of grouping, weighting and trading strategy, and two objective functions were used to assess each of chromosomes in a population. Furthermore, the first objective function consisted of return and risk factors, and the second one contained group and weight balances. The non-dominated solutions were then copied to the Pareto solution set in every generation. Moreover, the final Pareto solutions were proffered to traders for making investment plans after evolution. The Pareto front derived by that approach, however, still remain space to be improved.

In this paper, we proposed an improved approach to derive a set of better Pareto solutions by using the strength Pareto evolutionary algorithm (SPEA) [15]. Based on the selected technical indices, our proposed approach first generates candidate trading strategies. A subset of the trading strategies, that are kept using the given ranking criterion, is then employed to form the initial population based on the encoding schema adopted in [3]. Since the goal of trading is to get profit, the first objective function is utilized to evaluate the return and risk of a candidate GTSP. The second objective function is used to evaluate whether the numbers of strategies between groups are similar and weights of groups as well. In every iteration, the non-dominated solutions are copied to Pareto set, and the reduction procedure will be performed if the size of Pareto set is larger than a threshold. After generations, the derived Pareto set will be outputted for users. To show the effectiveness of the proposed approach, experiments conducted on a real dataset were given.

2 Related Work

In the literature, to solve the multi-objective trading strategy optimization to discover a set of non-dominate solutions, several algorithms have been introduced [1, 3, 10–12]. Briza et al. presented a stock trading system by multi-objective particle swarm optimization to optimize the weights of used technical indicators with the objective functions, named percent profit and Sharpe ratio [1]. Chen et al. proposed a MOGA-based approach to handle the multi-objective GTSP optimization problem for finding a set of appropriate GTSPs (Pareto solutions) with two objective functions that were used for evaluating the profitable ability and the quality of trading strategy groups of possible solutions [3]. Pinto et al. designed a multi-objective evolutionary system for optimizing an investment strategy with the technical indictor, moving averages (MA), for stock trading. The objectives used in their approach were return and risk [10]. Lohpetch et al. adopted a multi-objective genetic programming (GP) algorithm to derive trading rules for operating on stock market [11]. It first defined a set of objectives. Then, combination of two or three objectives were used to test the performances. As a result, seven configurations of multi-objective GP algorithms were employed to find trading rules for analyzing the impacts on the stock market. Almeida et al. proposed an approach based on differential evolution to generate Pareto fronts the technical indicators with the objectives, maximizing return, minimizing risk and minimization of number of trades [12].

3 Components of the Proposed Algorithm

The encoding scheme, the two used objective functions, and genetic operations of the proposed approach are stated in this section.

3.1 Encoding Scheme

Encoding scheme is the important part of the genetic algorithm because it represents a solution to be searched during the optimization process. The main purpose of this study is to design an improved algorithm to find a set of GTSP which can meet the multi-criteria using the GGA. As mentioned above, a GTSP is a set of groups with TSs. And, each of trading strategy groups is assigned its own capital respectively. Thus, a GTSP is represented by the three parts of grouping, weighting and trading strategy, and is demonstrated in Fig. 1.

In Fig. 1, the first part of encoding scheme is the grouping part. It demonstrates the used m trading strategies can be divided into K groups. The second part of encoding scheme is the trading strategy part which expresses every TS should belong to which

Encoding scheme for a GTSP		
Grouping	Trading Strategy	Weight
Reserved cash		w_0
G_1	$\{[TS_1^1], [TS_2^1], \ldots, [TS_h^1]\}$	w_1
G_i	$\{[TS_1^i], [TS_2^i], \ldots, [TS_h^i]\}$	w_i
G_k	$\{[TS_1^k], [TS_2^k], \ldots, [TS_h^k]\}$	w_k

Fig. 1. CAST-based tennis motion clustering

group. The last part is the weight part. It is composed a weight sting, $w_0, w_1, \ldots, w_j, \ldots, w_k$, where w_i is a '1' string which reveals the invested capital of the *i*-th groups, and w_0 represents the reserved capital. A symbol '0' is inserted into every two weight strings, w_i and w_{j+1}, to separate the weights of two groups. Then, the initial population can be generated in accordance with the encoding scheme and updated during the evolution process. Hence, the length of the encoding scheme is summation of K (number of groups), m (number of trading strategies), and ($K + |w_0| + \ldots + |w_j| + |w_k|$).

3.2 Two Used Objective Functions

In this section, the two designed objective functions used in the proposed approach are defined below, and they includes the objective and subjective criteria. Frequently recognized objective factors, maximizing the returns and minimizing the risks, should be considered to be assessment of the quality of a trading strategy. The first objective function Obj_1 is shown in Formula (1).

$$Obj_1(C_q) = PR(C_q)^{\alpha} \times Risk(C_q), \tag{1}$$

where $PR(C_q)$ and $Risk(C_q)$ are the portfolio return and risk of the chromosome C_q, and α is a parameter used to show the influence of the portfolio return. The portfolio return of a chromosome $PR(C_q)$ is shown in Formula (2).

$$PR(C_q) = \frac{\sum_{j=1}^{numTSP} return(TSP_j)}{numTSP}, \tag{2}$$

where $return(TSP_j)$ is the return of a trading strategy portfolio TSP_j, and *numTSP* is the number of TSPs that can be generated from the chromosome. The return of a TSP is calculated by Formula (3).

$$return(TSP_j) = \sum_{i=1}^{K} avgReturnRate\left(TS_i^j\right) * weight_i * Capital, \tag{3}$$

where $avgReturnRate\left(TS_i^j\right)$ is the average return rate of a trading strategy in group G_i calculated from Formula (4), $weight_i$ is the weight of the *i*-th group, and the capital that a trader uses for trading is denoted as *Capital*.

$$avgReturnRate\left(TS_i^j\right) = \frac{\sum_{h=1}^{frequency_i} returnRate\left(TS_{ih}^j\right)}{frequency_i}, \tag{4}$$

where $frequency_i$ is the times of trading in the training period, and $returnRate\left(TS_{ih}^j\right)$ is shown in Formula (5).

$$returnRate\left(TS_{ih}^j\right) = \frac{sellPrice_h - buyPrice_h}{buyPrice_h} \tag{5}$$

where $sellPrice_h$ and $buyPrice_h$ are the selling and buying prices of the *h*-th trading using the *i*-th TS in TSP_j, respectively. The risk of a chromosome is determined by Formula (6).

$$risk\left(C_q\right) = \frac{\sum_{j=1}^{numTSP} risk\left(TSP_j\right)}{numTSP}, \tag{6}$$

where *numTSP* is the number of portfolios included in the chromosome. The $risk(TSP_j)$ is the risk of the *j*-th TSP and can be calculated using the Formula (7), and it illustrates the risk of a TSP with the minimum value of maximum draw down of TSs during the period of trading. Note that the maximum draw down of TSs will be normalized from 0 to 1.

$$risk\left(TSP_j\right) = min\left(MDD\left(TS_1^j\right) \ldots, MDD\left(TS_k^j\right)\right), \tag{7}$$

where $MDD\left(TS_i^j\right)$ indicates the maximum draw down of the chosen TS in TSP_j from the group G_i, and k is the number of groups. Thus, $MDD\left(TS_i^j\right)$ is calculated by the Formula (8).

$$MDD\left(TS_i^j\right) = min\left(returnRate\left(TS_{i1}^j\right) \ldots, returnRate\left(TS_{ifrequent_i}^j\right)\right), \tag{8}$$

where $returnRate\left(TS_{ih}^j\right)$ can be given in the same manner as in Formula (5). Both of group balance and weight balance are regarded as the subjective factors, since each of groups has its own trading strategies that have similar properties and is allocated weights in a GTPS. Hence, the second objective function Obj_2 is composed of both the group and weight balances, and is shown in Formula (9).

$$Obj_2\left(C_q\right) = GB\left(C_q\right)^{\beta} \times WB\left(C_q\right), \tag{9}$$

where β is the influence value of the group balance. The $GB(C_q)$ can be given from Formula (10) as the group balance of a chromosome.

$$GB\left(C_q\right) = \sum_{i=1}^{K} -\frac{|G_i|}{N} log \frac{|G_i|}{N}, \tag{10}$$

where $|G_i|$ is the number of the group G_i's TSs, and N is the number of the given TSs. For ensuring that the numbers of TSs among groups in a chromosome are similar, the group balance of a chromosome is calculated. The weight balance, $WB(C_q)$, is defined in Formula (11).

$$WB(C_q) = \sum_{i=1}^{K+1} -\frac{|w_i|}{T} log \frac{|w_i|}{T}, \tag{11}$$

where $|w_i|$ is the length of the string w_i, and T indicates the sum of the string length of w_i, $0 \leq i \leq K$. The utility of the weight balance is to keep from all capital if possible be situated in a particular group.

3.3 Genetic Operations

The genetic operators, including crossover, mutation, inversion and selection, are presented in this section, and they are performed to form new solutions and improve diversification. The crossover operation is applied on both parts of the grouping and weighting. Two individuals are first selected randomly from grouping part, and part of groups in the selected chromosome is chosen and implanted into another chromosome. The groups in the implanted chromosome are then dropped if redundant. Finally, merging or splitting on groups are repeated until the number of groups meets the chromosome length. To produce novel chromosomes, the two-point crossover operator is adopted on the weight part of a chromosome. Two points of a chromosome are first identified and their subsequences are then exchanged. Note that, to ensure that the numbers of '1's and '0's in a chromosome are correct, the appropriate arrangement should be made. For mutation operators, the one-point mutation operators are used in the grouping and weight parts. In inversion operator, groups are exchanged according to a random selected position. At last, the elitist strategy is employed to form next population.

4 Proposed SPEA-Based GTSP Optimization Algorithm

In this section, the pseudo code of the proposed SPEA-based GTSPO optimization algorithm with GGA for finding Pareto solutions is illustrated in Table 1.

At line 2, stock price series, technical indicators and desired number of trading strategies are given to the data-preprocessing procedure for generating the m trading strategies. Then, the generated m trading strategies are applied on the stock price series to identify trading signals at line 3. The initial population is then formed. The initial Pareto set is empty. Lines 6 to 29 start the evolution process. Lines 7 to 17 are used to calculate the two objective values of chromosomes. Then, based on the ranking result, the fitness values of chromosomes are derived from lines 18 to 19. The Pareto solutions are copied to *PSS* at line 21. After that, the genetic operators are executed to get offspring from lines 22 to 24. In the selection process, the chromosomes in *PSS* and *inverPopulation* are merged. The merged set is then used to generate next population using elitist strategy. Finally, after evolution, the derived Pareto set *PSS* will be outputted.

Table 1. Pseudo code of the proposed approach.

```
Input: A stock price series sp and technical indicators nTech.
Parameters:
    Number of groups K, number of trading strategies m, population size pSize, Pareto solution
    size ndSize, crossover rate p_c, mutation rate p_m, inversion rate p_i, number of iterations numIt-
    eration, and number of bits for weight part weightBits.
Output: A set of non-dominated solutions (Each solution is a GTSP).
----------------------------------------------------------------------------------------------
    Procedure: SPEA-GTSP Optimization
1.   SPEA_GTSP_Optimization (){
2.   processedTS = dataPreprocessing(sp, nTech, m);
3.   tradingSignals = generateTradingSignal(sp, processedTS);
4.   population ← initialPopulation(pSize, K, processedTS);
5.   PSS ← ϕ; //Pareto solution set
6.   FOR i = 0 to numIteration DO
7.     FOR q = 0 to pSize DO
8.     // First objective function (Objective Criteria)
9.         profit ← portfolioReturn (C_q, sp, tradingSignals);// Formula (2)
10.    risk ← riskofPortfolio(C_q, sp, tradingSignals); // Formula (6)
11.        objFunc_1 ← (profit)^α × risk; // Formula (1), Objective Criteria
12.    // Second objective function (Subjective Criteria)
13.        gb ← groupBalance(C_q); // Formula (10)
14.    wb ← weightBalance(C_q); // Formula (11)
15.    objFunc_2 ←(gb)^β × wb; // Formula (9), Subjective Criteria
16.        population ← updatePopulation(C_q, objFunc_1, objFunc_2);
17.    END q FOR LOOP
18.    rankingResult ← rankingChromosome(population);
19.    population ← fitnessEvaluation(rankingResult, population);
20.    // Copy chromosomes with Rank 1 to PSS
21.    PSS ← copyPSS(rankingResult, population);
22.    crossPopulation ← Crossover(p_c, population);
23.    mutaPopulation ← Mutation(p_m, crossPopulation);
24.    inverPopulation ← Inversion(p_i, mutaPopulation);
25.    mergedPopulation ← PSS ∪ inverPopulation;
26.    PSS ← reducingSizeofPSS(PSS, ndSize);
         //Average linkage method
27.    nextPopulation ← Selection(pSize, mergedPopulation);
28.    population ← nextPopulation;
29.  END i FOR LOOP
30.  Output the Pareto solution set PSS;
31. }
```

5 Experimental Evaluations

In this section, with a financial dataset collected in the period between 2013/01 and 2016/12, experimental results were given to demonstrate the effectiveness of the proposed approach, and the dataset is shown in Fig. 2.

From Fig. 2, we can know the trend of the collected dataset is basically sideways trend. In the experiments, the stock prices series from 2013 to 2015 and 2016 were used as training and testing datasets. The parameter setting is stated as follows: the number of groups was 3, the number of trading strategies was 15, the population size was 50,

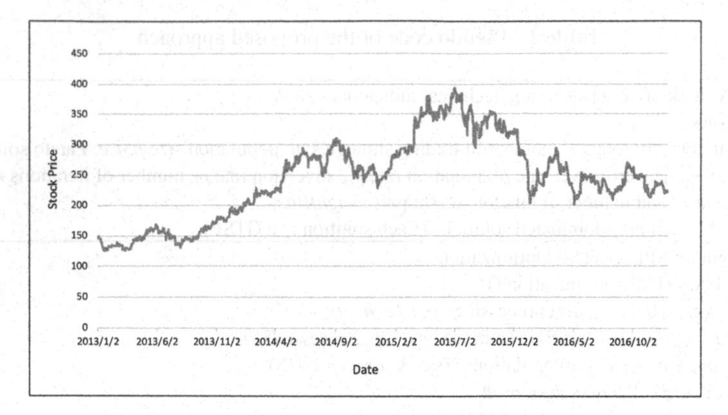

Fig. 2. The stock price series of the collected dataset.

the Pareto solution size was 200, the crossover rate was 0.8, the mutation rate was 0.3, the inversion rate was 0.8, the number of iterations was 200, and the number of bits for weight part was 100. Firstly, the Pareto fronts on the different generations are showed to present the effectiveness of the proposed approach in Fig. 3.

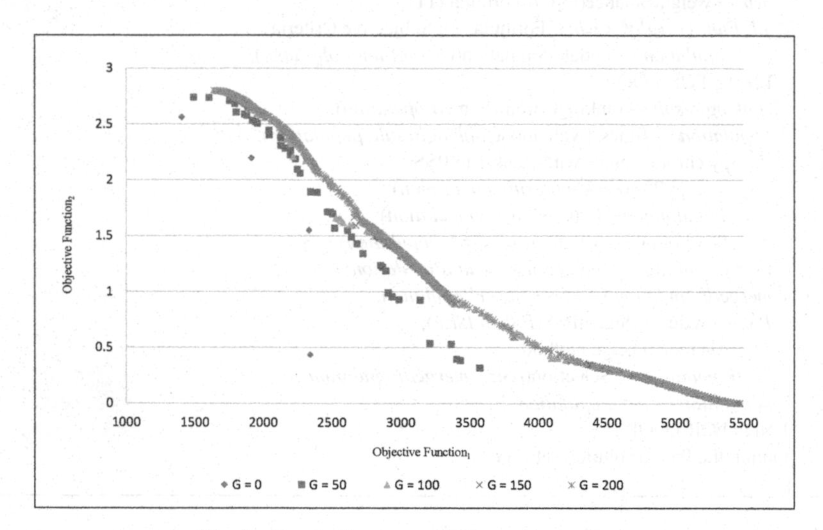

Fig. 3. Pareto fronts on different generations.

From Fig. 3, we can observe that the final Pareto fronts generated by the proposed approach is better than the initial one along with different generations. In other words, it indicated that the proposed approach can find set of appropriate non-dominated solutions.

At last, the experiments were made to compare the proposed approach to the previous approach in terms of the average and minimum returns of the three types of GTSPs (non-dominated solutions), $SolutionType_1$, $SolutionType_2$ and $SolutionType_3$, on the testing dataset. The $SolutionType_1$ and $SolutionType_3$ represented the solutions selected from both approaches with the highest values of the second and the first the objective functions, respectively. The $SolutionType_2$ meant the selected non-dominated solutions have both acceptable values of the two objective functions. The results are shown in Table 2.

Table 2. Comparison results of the proposed and previous approaches in terms of average and minimum returns.

	Previous approach		Proposed approach	
	Avg. return (%)	Min. return (%)	Avg. return (%)	Min. return (%)
$SolutionType_1$	0.13	0.06	0.14	0.06
$SolutionType_2$	0.15	0.06	0.17	0.08
$SolutionType_3$	0.2	0.08	0.22	0.1

From Table 2, according to $SolutionType_1$, we can observe that the average returns and minimum returns of the two GTSPs were smaller than other two types. The reason is that they were good at the second objective function. On the other hand, from GTSPs belong to $SolutionType_3$, we can see that their the average and minimum returns were the highest among three types. In addition, the GTSPs generated by the proposed approach were better than that derived by the previous approach no matter what solution types they belonged to. Based on the experimental results, it shows that the proposed approach is effective and gets better performances against the previous one.

6 Conclusions and Future Work

To provide various trading strategy portfolios to traders, the previous approach is presented to find non-dominated solutions using multi-objective genetic algorithm. Every non-dominated solution means a group trading stock portfolio (GTSP) which is made up of a set of trading strategy groups. To provide more useful GTSPs, in this paper, we have proposed an enhanced algorithm to solve GTSP optimization problem by combining the SPEA and the GGA. To show the effectiveness of the proposed approach, experiments made on a financial dataset showed that the proposed approach can get better Pareto front than the previous approach, and the profitable ability of the derived GTSPs also better than the previous approach. In the future, we will work on improving the proposed algorithm with different multi-objective optimization approaches and objective functions to discover more useful GTSPs.

Acknowledgments. This research was supported by the Ministry of Science and Technology of the Republic of China under grant MOSTs 109-2622-E-027-032 and 109-2221-E-390-013-MY2.

References

1. Briza, A.C., Naval, P.C.: Stock trading system based on the multi-objective particle swarm optimization of technical indicators on end-of-day market data. Appl. Soft Comput. **11**, 1191–1201 (2010)

2. Chen, J.S., Hou, J.L., Wu, S.M., Chang-Chien, Y.W.: Constructing investment strategy portfolios by combination genetic algorithms. Expert Syst. Appl. **36**(2), 3824–3828 (2009)
3. Chen, C.-H., Gankhuyag, M., Hong, T.-p., Wu, M.-E., Wu, J.M.-T.: A multiobjective-based group trading strategy portfolio optimization technique. In: Pan, J.-S., Lin, J.C.-W., Liang, Y., Chu, S.-C. (eds.) ICGEC 2019. AISC, vol. 1107, pp. 87–93. Springer, Singapore (2020). https://doi.org/10.1007/978-981-15-3308-2_10
4. Drezewski, R., Dziuban, G., Pająk, K.: The bio-inspired optimization of trading strategies and its impact on the efficient market hypothesis and sustainable development strategies. Sustainability **10**, 1–45 (2018)
5. Kim, Y., Enke, D.: Developing a rule change trading system for the futures market using rough set analysis. Expert Syst. Appl. **59**, 165–173 (2016)
6. Kim, Y., Ahn, W., Oh, K.J., Enke, D.: An intelligent hybrid trading system for discovering trading rules for the futures market using rough sets and genetic algorithms. Appl. Soft Comput. **55**, 127–140 (2017)
7. Prasetijo, A.B., Saputro, T.A., Windasari, I.P., Windarto, Y.E.: Buy/sell signal detection in stock trading with bollinger bands and parabolic SAR: with web application for proofing trading strategy. In: International Conference on Information Technology, Computer, and Electrical Engineering, pp. 41–44 (2017)
8. Wang, L.: Dynamical models of stock prices based on technical trading rules—Part III: application to Hong Kong stocks. IEEE Trans. Fuzzy Syst. **23**(5), 1680–1697 (2015)
9. Chen, C.H., Chen, Y.H., Lin, J.C.W., Wu, M.E.: An effective approach for obtaining a group trading strategy portfolio using grouping genetic algorithm. IEEE Access **7**, 7313–7325 (2019)
10. Pinto, J., Neves, R.F., Horta, N.: Multi-objective optimization of investment strategies based on evolutionary computation techniques, in volatile environments. In: The International Conference on Enterprise Information System, pp. 480–488 (2014)
11. Lohpetch, D., Corne, D.: Multi-objective algorithms for financial trading: multi-objective out-trades single-objective. In: IEEE Congress of Evolutionary Computation, pp. 192–199 (2011)
12. de Almeida, R., Reynoso-Meza, G., Steiner, M.T.A.: Multi-objective optimization approach to stock market technical indicators. In: The IEEE Congress on Evolutionary Computation, pp. 3670–3677 (2016)
13. Ucar, İ., Ozbayoglu, A.M., Ucar, M.: Developing a two level options trading strategy based on option pair optimization of spread strategies with evolutionary algorithms. In: IEEE Congress on Evolutionary Computation, pp. 25–28 (2015)
14. Syu, J.H., Wu, M.E., Lee, S.H., Ho, J.M.: Modified ORB strategies with threshold adjusting on Taiwan futures market. In: IEEE Conference on Computational Intelligence for Financial Engineering and Economics, pp. 1–7 (2019)
15. Zitzler, E., Thiele, L.: An evolutionary algorithm for multiobjective optimization: the strength Pareto approach. Technical report 43, Computer Engineering and Networks Laboratory (TIK), Swiss Federal Institute of Technology (ETH) Zurich (1998)

Develop a Hybrid Human Face Recognition System Based on a Dual Deep Neural Network by Interactive Correction Training

Pin-Xin Lee[1], Ding-Chau Wang[1(✉)], Zhi-Jing Tsai[2(✉)], and Chao-Chun Chen[2]

[1] MIS, Southern Taiwan University of Science and Technology, Tainan, Taiwan
{MA890101,dcwang}@stust.edu.tw
[2] IMIS/CSIE, NCKU, Tainan, Taiwan
{P96091115,chaochun}@mail.ncku.edu.tw

Abstract. In recent years, the rapid development of artificial intelligence, Internet of Things, and cloud technology has led to the widespread application of face recognition technology, especially in the face recognition system based on deep learning. At present, the face recognition systems only use a single face recognition model for the identity prediction. Furthermore, systems wouldn't perform data training automatically when facial images are captured under monitoring. In the other words, these systems are supposed to manually review the prediction result of face images in order to add training to improve the accuracy rate, because there is no other review method for data training to evaluate the prediction result in these systems. Therefore, this paper developed a method using mutual correction of classification models based on two human face neural network models, called the Dual Face Recognition Model Interactive Correction Training System. Inside the system, we use both FaceNet and OpenFace as the core, to build a module update algorithm. The method is to achieve each set of forecast identity label and confidence for a new face image by both Face Net training models individually. And then the method proceeds to compare with these two sets and mutual correct the results. The new face image with mutual correction identity is used for data training. Therefore, system can execute automatic and dynamic updates, and effectively improve classification accuracy. Moreover, thus system has the advantages of real-time training and limiting model size due to the use of its training process on the classification model. Experimental results show that our system has high performance with a small number of training samples. The system can also automatically improve the accuracy rate and AUC value, and lower error sample rate through the system's modular update algorithm.

Keywords: Face recognition · Deep learning · FaceNet · OpenFace

1 Introduction

The most widely known biometric technology on the market today is fingerprint recognition, which has become a regular feature in companies and governments, whether

N. T. Nguyen et al. (Eds.): ACIIDS 2021, LNAI 12672, pp. 593–605, 2021.
https://doi.org/10.1007/978-3-030-73280-6_47

it is a fingerprint scanner or an access control system, or even a smartphone. Despite the growing awareness of information security, fingerprint recognition has become an important entry point for secrecy, speed, popularity and technological maturity. However, since fingerprint recognition technology extracts fingerprint image features from human fingers to identify the identity of the fingerprint, for fingerprint recognition systems that require contact, it will lead to health problems and fingerprint theft, as well as human physiological changes (e.g. skin breakage, calluses, etc.). Face Recognition has the advantages of being non-contact and non-invasive, meaning that it does not have the same problems as fingerprint recognition in terms of usage. Moreover, in recent years, due to the rapid development of hardware, face recognition systems based on Machine Learning or Deep Learning have become one of the most popular technologies in the market, as the accuracy rate has greatly improved, and the accuracy rate of facial recognition systems has been greatly improved.

It is worth noting that in recent years, the emergence of Convolutional Neural Network (CNN) has promoted the development of deep learning, and many technology companies have developed various face recognition models due to it. For example, FaceNet developed by Google team is a neural network model using InceptionNet architecture for training. Compared with traditional recognition systems, FaceNet uses Triplet loss instead of Softmax classification method to map images into quantifiable 128-dimensional features, so that we can easily achieve face recognition function through European distance calculation [1]. In addition, OpenFace, which is developed using the FaceNet architecture, uses CASIA-WebFace and FaceScrub as two public training sets for dataset training. Therefore, it is suitable for mobile devices with limited memory [2]. These two recognition models obtained 99.63% and 92.92% accuracy in the LFW (Labeled Faces in the Wild) recognized test set respectively.

Although the current face recognition model has nearly 100% accuracy in the recognized test set, in real-world applications, the accuracy of face recognition may be significantly reduced due to various factors such as lighting, angle, expression and age, as well as the presence of a face recognition device that is capable of recognizing faces in real time. As such, some companies use machine learning classification methods based on KNN or SVM to train face classification models with quantitative features output from FaceNet, thus allowing us to add new data to the face model by collecting new data. However, in order to avoid feeding the wrong data for recognition errors, most of these methods use manual intervention to add data to the training process.

Therefore, in order to improve the performance of face recognition, we found that the classification models of different face neural networks may show different accuracy for each identity within the same training sample. If integrated, it allows the classification predictions between neural networks to be corrected for each other (The detailed information will be verified in Sect. 5.1). In addition, the prediction results of one classification model can be verified by a second classification model to reduce the chance of feeding the wrong samples or missing a large number of true samples with high values.

In summary, this paper develops a method of mutual correction of classification models trained based on two face neural network models, which is called Dual Face Recognition Model Interaction Calibration Training System. In this system, we use FaceNet and OpenFace as the core face neural networks and construct a module update algorithm to compare the results of the face classification models trained by the two face neural networks to obtain a set of predicted identity and confidence for each of the new face images. After obtaining the best prediction results, the new face images are fed with the correct identities to replace the manual screening and achieve the automatic dynamic update method, which effectively improves the classification accuracy.

2 Related Works

The earliest research on face recognition is Eigenface proposed by Kirby and Sirovich [3], and is also the first practical application of face recognition, which refers to how to obtain an optimal set of feature vectors in face recognition problems. Eigenface was later applied to the face classification problem by Turk and Pentland [4], whose method mainly uses principal components analysis (PCA) to extract the eigenvalues and eigenvectors of the face image, and retains the number p of eigenvectors (eigenface) to construct the eigenspace. After which the training sample set (x_i, $i = 1, 2, 3\ldots M$) is then projected into the feature space to form the training set. In the recognition process, the test image is normalized and projected to the feature spacc, and then the image x_i with the minimum distance is calculated by using the European distance for each sample projection of the training set, which is the identity of the image. The eigenface was later applied to Linear Discriminant Analysis (LDA) by Fisher's proposed Fisherface method [5].

The traditional machine learning method that is still widely used today is based on the method proposed by Ojala [6]. For example, Zhang [7] in his paper mainly divides the image into $N \times N$ areas with human faces, and then divides each area into a 3×3 nine-box grid, and uses the grayscale value of the pixel in the center of the grid to match the surrounding neighboring points. After matching, a binary grid is formed and the values are arranged clockwise from the top left corner to obtain a set of binary values, which are called LBP Mask and can be used to reflect the texture information of the region. Finally, these LBP masks are transformed into a LBP Histogram to obtain a set of feature vectors, which allows us to train a face classification model directly by classification method.

In the current popular open source deep learning models, Lei [8] uses FaceNet for face feature extraction and Dilb for pre-processing steps (face detection, alignment) and KNN for a classification model, and the system achieves high accuracy performance in predicting results for ASEAN children. Li [9] proposes a SVM incremental learning approach based on the Openface neural network model. The S-DDL (Self-Detection, Decision and Learning) algorithm is proposed in their paper, and the classification model trained by the current neural network model is used to output a set of classification labels for new images added during the actual use, and then the classification model is directly updated dynamically. In order to reduce a large number of new images to be added, they introduce incremental learning to reduce the learning time of each batch.

In the area of convolution neural network, Wang [10] has designed a real-time human contact method based on multiple convolution neural network, which proposes a self-built CNN neural network structure with input layers, three volume layers, three pool layers, one full interface and outputs. And use data enhancement technology to increase the revenue, diversify training and improve the internet's generalization. Sun [11] uses a convolution neural network and SVM to extract more hidden facial features by training CNN, and SVM also substituted the Sortmax method to categorize human facial features. In existing open source deep learning models, Openface has been improved by modifying neural network parameters to assess the right Triplet during the training process and using the Adam Optimizer to reduce the risk of early recovery [12].

Scholars have proposed ways to improve the accuracy and performance of the face recognition model, but they did not pay attention to the number of training samples required for the initial model. When the number of categories trained in a classification model increases and the sample size decreases, how to maintain a good accuracy rate despite this limitation. The models proposed by some scholars are based on a single neural network for classification [9]. Therefore, if the neural network recognizes the wrong situation, it is likely to mark the new image as the wrong identity and carry out the risk of training, which will lead to the recognition error. The probability increases greatly, and it is more likely to happen if it is based on the situation where the above categories are more and the samples are fewer, or the test samples belong to various environments. Therefore, this system is based on this relationship to establish a mutual correction method for the classification models trained by the two facial neural networks to assist the new face image to be fed with the correct identity, and moreover replace manual screening by the automatic dynamic update method to achieve the best improvement in accuracy.

3 Dual Face Recognition Model Interaction Calibration Training System

In this section, we introduce the system architecture and implementation proposed in this paper, as shown in Fig. 1. The difference between our system and traditional face recognition systems is that our system will continuously collect face data through the actual application process and generate the predicted data identified by the current face classification model, and then the system will use the self-constructed module update algorithm to determine the correctness of this predicted data in an automated manner, in order to select a suitable new sample to train and update the face classification model to improve the accuracy rate efficiently. The accuracy rate can be improved efficiently.

This system is mainly composed of four sub-modules, including (a) User Data Registration Module, (b) Face Recognition Module, (c) Module Update Module, and (d) Face Recognition Surveillance and depending on the use situation can be divided into three types:

Situation (I): Users register their identities with the system through application or web pages (APP, Web), and REST Services transmits the user data to User Data Registration Module via HTTP protocol, and writes to the User Registration Database when the registration is completed. ***Situation (II)***: The continuous local sensing data received by Rasberry Pi3 through the sensor WebCam, and after MQTT Services delivers images to the Face Recognition Module for analysis, then the Module Update to extract the predicted data generated by Face Recognition, executing the module update algorithm to automatically correct the system Within the face classification model and the predicted data is also written into the Face Recognition Database, providing situation (III)

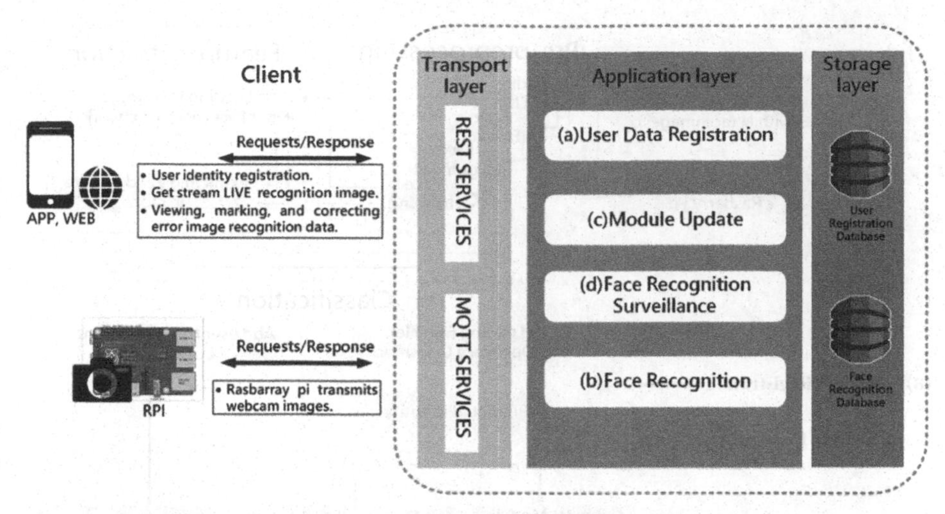

Fig. 1. Dual face recognition model interaction calibration training system

monitoring services. ***Situation (III)***: When users need monitor system Recognition via application or web page (APP, Web), REST Services transfers user requests to the Face Recognition Surveillance Module, which provides user identifiable data information and manual correction services when they receive the request.

3.1 Users Registration and Face Recognition

Situation (I) and Situation (II) are one of the core functions of the system, and according to the situation description, we can represent the detailed flow in Fig. 2. The two scenarios are mainly performed using User Data Registration Module and (b) Face Recognition Module respectively.

They include three main phases: pre-processing, feature extraction, and face classification. First, we will use MTCNN face detection method to find out the face coordinates in the image, export the Bounding Box and five key points (eyes, nose, mouth) of the face and then crop them, and then use the key points for Affine transformation to align the face, as shown in Fig. 3.

Then, in the feature extraction stage, the posed cropped faces are scaled to 182×182 and 96×96 respectively, and then input to FaceNet and OpenFace neural network to extract the features of the faces to obtain a 128-dimensional feature vector each. Finally, this 128-dimensional feature vector is used in the classification phase to train the classification model using SVM, and different tasks are performed depending on the function of the module. In the case of user data registration, the feature vectors and user data are imported into the current classification model and then trained to obtain a classification model that can predict this identity. For Face Recognition, the identity of this feature vector is forecast through the current classification model. It also performs a module update algorithm to perform a dynamic update service to identify data output from the classification model.

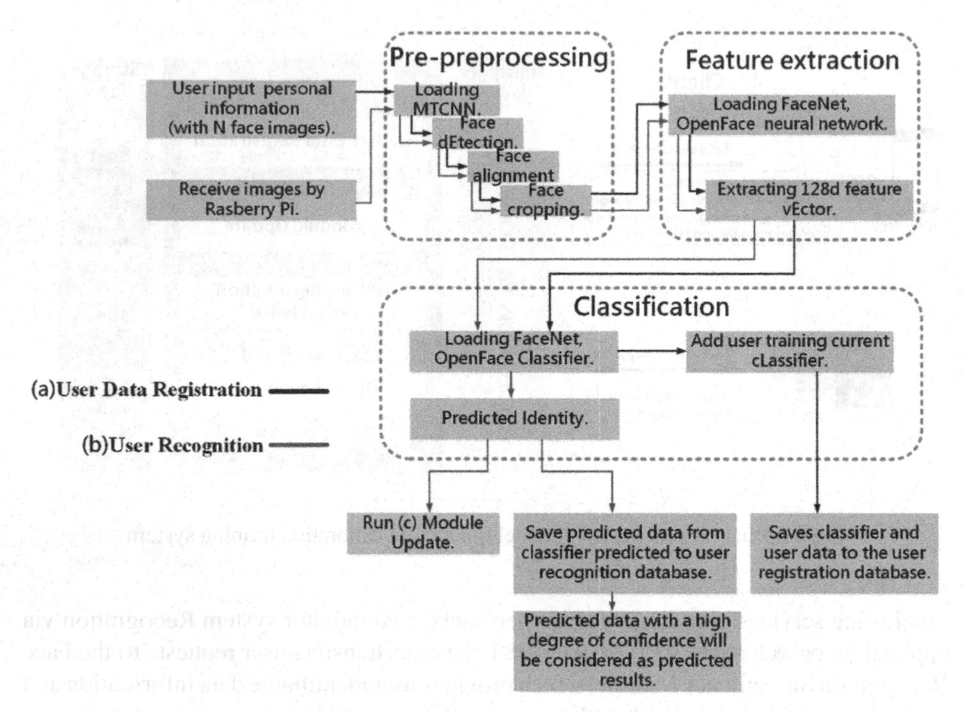

Fig. 2. (a) User data registration module and (b) Face recognition module flowchart

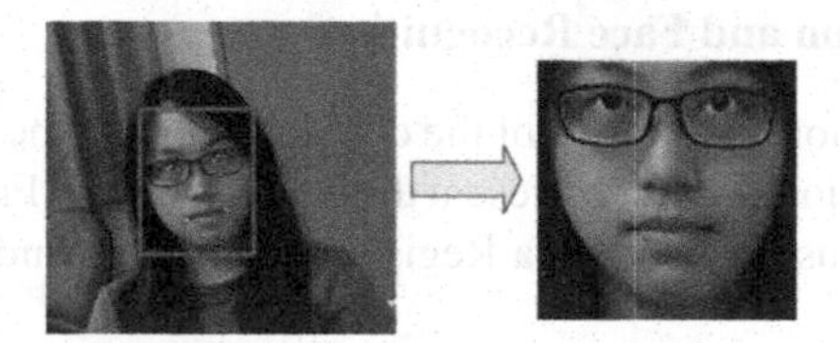

Fig. 3. Pre-processing process

3.2 Module Update Algorithm

In traditional face recognition systems, classification models are trained before they can be used. However, these methods tend to overlook valuable data collected or mistake worthless data for valuable data. Therefore, it would be helpful to identify the intrinsic value of these data to improve the system.

Figure 4 shows the workflow of a Module Update Algorithm. Firstly, the system retrieves the predicted data recorded by the Face Recognition Module in the Face Recognition Database and the corresponding set of face images, and these data are used to determine further processing through decision making. If the confidence level predicted by the FaceNet classification model is higher than the threshold and the confidence level predicted by the OpenFace classification model is lower than the threshold, the system will use the identity predicted by the FaceNet classification model as the identity of the face image; conversely, the identity predicted by the OpenFace classification model will be used.

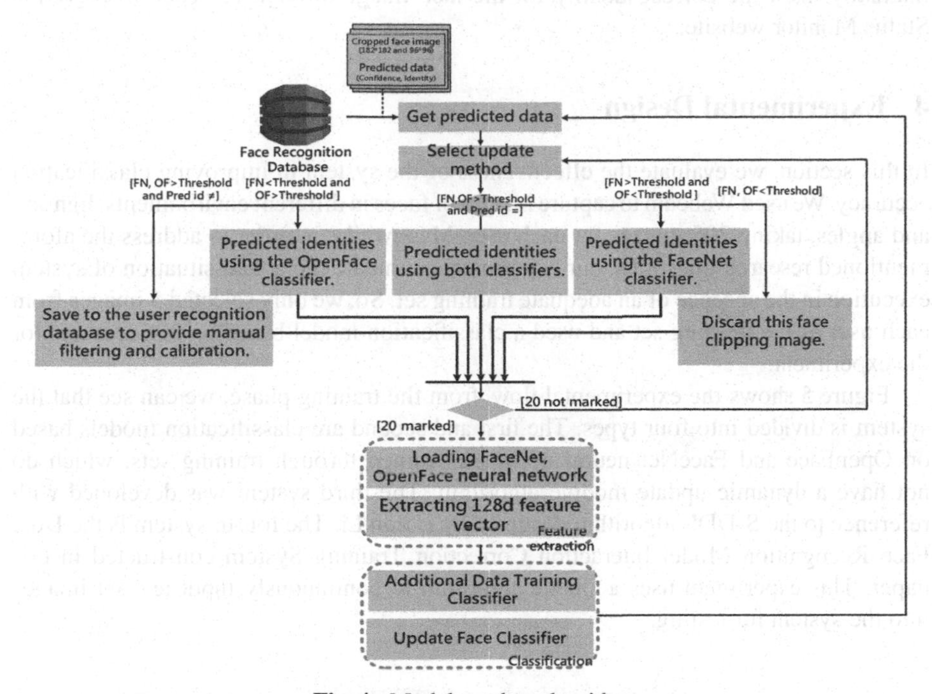

Fig. 4. Module update algorithm

If the confidence of both classification models is higher than the threshold, and the identity of the predicted is the same, then you can choose either. However, if confidence is higher than threshold, but the identity of the predicted is different, the system will write the data into the database, allowing us to determine and choose the right identity by hand. In addition, regardless of whether the identity of the forecast is the same, if the confidence of both classification models is lower than threshold, the system automatically excludes photos and does not join the training set because it does not believe that the image is valuable to the model (it may not be the identity of the model).

After the system has decided on the update method, if 20 predicted data and images have been marked, then the system will start the classification training of the image, where the face image is already done in the pre-processing stage because it is from the image recorded by the Face Recognition Module. Therefore, the system simply feeds the images into the nerve network, generates a 128-dimensional feature vector, and then trains the classification model, which replaces the current classification model.

3.3 Divergent Predicted Results

The processing of the predicted data by the update module algorithm may produce divergent predicted results, which represent predictions where both classification models of the system have above-threshold confidence in their predictions, but different identities. These disputed results will be stored in the Face Recognition database, allowing us to manually mark the correct identity for the face image through the Face Recognition Status Monitor website.

4 Experimental Design

In this section, we evaluate the effectiveness of the system in improving classification accuracy. We used Webcam to capture images of faces in different environments, lighting and angles, taking 205 images for each user. Meanwhile, in order to address the aforementioned research questions, our experiments simulate the actual situation of system execution in the absence of an adequate training set. So, we only selected 5 images from each user as the training set and used a classification model based on 20 identities for the experiment.

Figure 5 shows the experimental flow, from the training phase, we can see that the system is divided into four types. The first and second are classification models based on OpenFace and FaceNet neural networks, trained through training sets, which do not have a dynamic update module algorithm. The third system was developed with reference to the S-DDL algorithm designed by Lufan Li. The fourth system is the Dual Face Recognition Model Interaction Correction Training System constructed in this paper. This experiment uses a Queue algorithm to continuously input test set images into the system for testing.

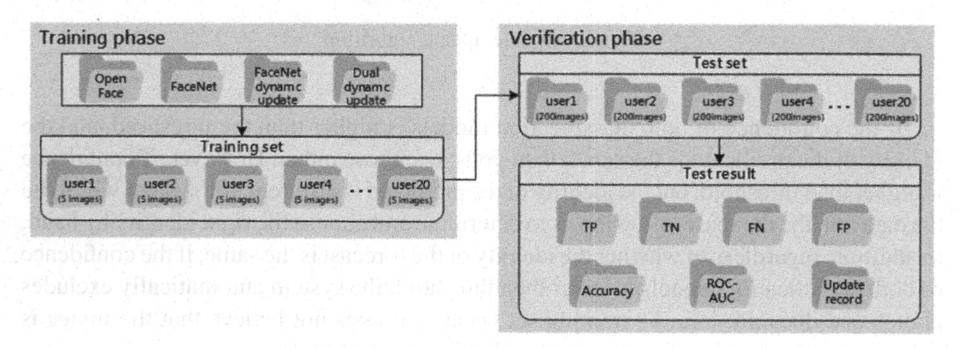

Fig. 5. Experimental process

From the validation phase, the test results to be observed in this experiment are shown. We classify the test results of the system into four values according to the Confusion Matrix: *TP (True Positive)*, predicted is Positive, actual is Positive; *TN (True Negative)*, predicted is Negative, actual is Negative; *FP (False Positive)*, predicted is Positive, actual is Negative; *FN (False Negative)*, predicted is Negative, actual is Positive. From these four values, we can obtain the classification accuracy by simply summing (TP + TN)/(TP + TN + FP + FN).

In addition, if we perform several experiments to set different thresholds, multiple test results (TP, FP, TN and FN) will be generated. For each test result, we can calculate the TPR (True Positive Rate) and FPR (False Positive Rate) to plot the ROC (Receiver operator characteristic) curve. The ROC curve is an indicator of the change between TPR and FPR, and is widely used in two or more categories of classification. The more the curve is toward the upper left corner of the y-axis, the better the classification performance of the model. We can also use the ROC curve to integrate into the AUC (Area Under the Curve) score evaluation index, which represents the meaning of a randomly selected Positive, the classification model will correctly determine the probability of Positive than the probability of Negative, the higher the AUC, the higher the classification model will be correct. However, ROC curves and AUC score were initially created under a binary classification question, so we had to integrate their metrics through micro-averaging.

Furthermore, the experiment concludes by recording the update process of the third system and our system using their respective algorithms to extract test sets as training data: *Input Error samples*, it is the test result of FP and FN; *Input Correct samples*, it is the test result of TP and TN; *Reserved samples*: It is a sample that is considered to be of value but has not been correctly predicted (This sample may be the identity of the model).

5 Experimental Results

According to the experimental design of the Sect. 4, we will compare the accuracy, ROC curve and AUC score of the system in this section.

5.1 Accuracy

Figure 6 shows a comparison of the classification accuracy of the four systems for 20 identities. It can be seen that the Dual Face Recognition Model Interaction Calibration Training System constructed in this paper has the best accuracy, achieving 93% accuracy when the number of test samples reached 4000 images, while the other three systems did not reach 90% accuracy. It is also interesting to note that during the training phase of this experiment, only 5 images of each identity were used to train the system, and if we add more training samples, it will advance to improve the accuracy of the system.

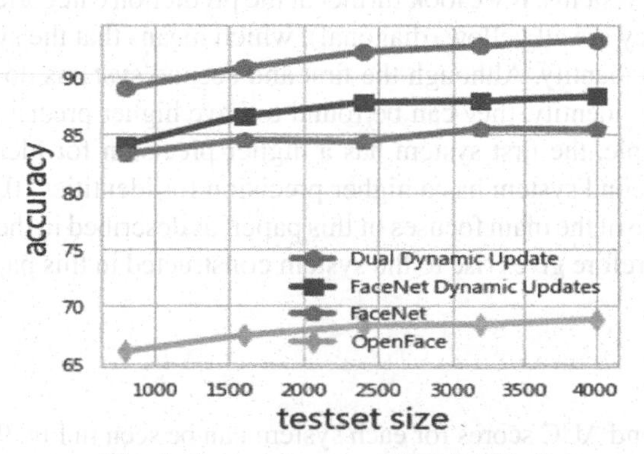

Fig. 6. System accuracy (20 identities)

The number of identities trained by the classification model affects the accuracy. As shown in Fig. 7, if we reduce the number of identities to 10, we observe that our system has an accuracy rate of 96.9%. The third system had an accuracy of 93.73%, but it would have a 94.5% accuracy rate when verifying 400 images, with a tendency for the accuracy rate to decrease. The reason for this decline can be explained that the system may have predicted the test sample incorrectly and fed the wrong sample to the system, resulting in subsequent misjudgements. This problem is significantly reduced compared to our system (more information in Sect. 5.3).

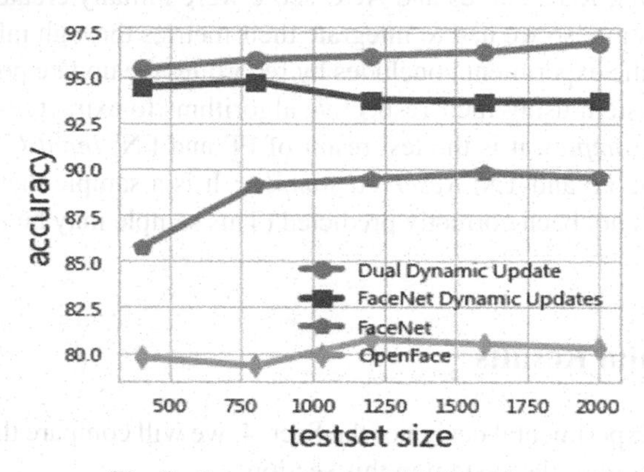

Fig. 7. System accuracy (10 identities)

In addition, we use a Confusion Matrix to show the system's detailed predicted profiles for each identity. Figure 8 shows, from top to bottom, the confusion matrix between the first, second and our system, with the horizontal coordinates being the predicted class and the vertical coordinates being the true class of the sample. We normalize the confusion matrix. Thus, the higher the accuracy, the more the diagonal lattice will be biased towards yellow (1.0) and the non-diagonal lattice will be biased towards dark blue (0.0); conversely, the lower the accuracy, the more the diagonal lattice will be biased towards dark blue (0.0) and the diagonal lattice will be biased towards yellow (1.0).

Therefore, it can be seen that the confusion matrix of our system has the best accuracy compared to other systems. If we look further at the prediction categories in each column, we can see that they are all yellow (diagonal), which means that the system has the best precision for each identity. Although the first and second systems do not have the best precision for every identity, they can be found to have higher precision for a particular identity, for example, the first system has a higher precision for identities (1–10, 13–17, 19) and the second system has a higher precision for identities (0, 11–12, 18). This phenomenon is one of the main focuses of this paper, as described in the fourth paragraph of Sect. 1, and therefore gives rise to the system constructed in this paper, which has the best accuracy rate.

5.2 ROC-AUC

The ROC curves and AUC scores for each system can be seen in Fig. 9. Our constructed system shows the best results in the experiment (AUC = 0.993). The second is an

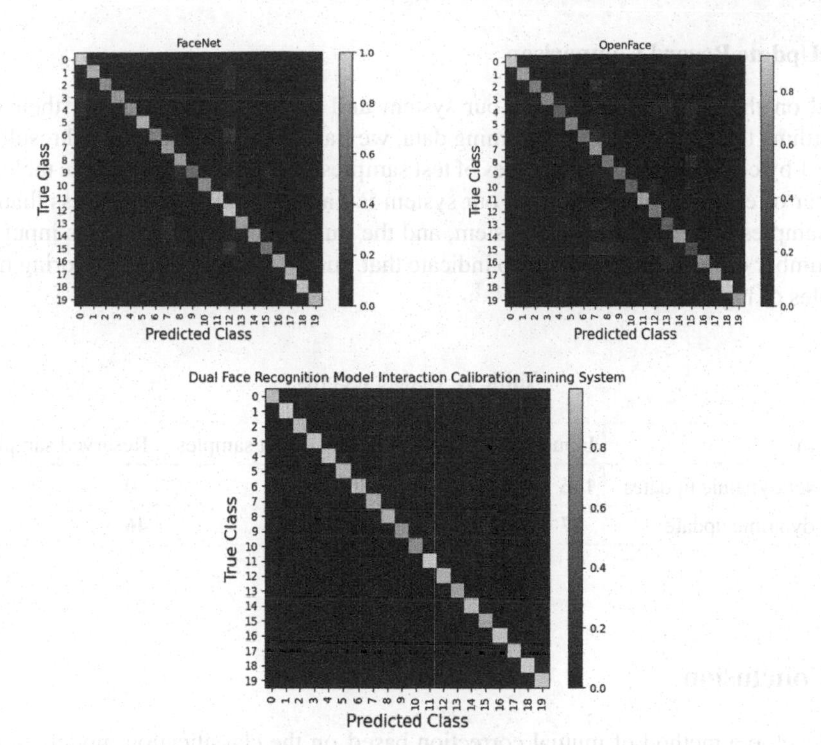

Fig. 8. Confusion matrix of the systems (Color figure online)

autonomously updatable face recognition system based on a single face neural network (Lufan Li's S-DDL algorithm) (AUC = 0.989). The lowest score was for the OpenFace system, which does not hold the dynamic update feature (AUC = 0.948).

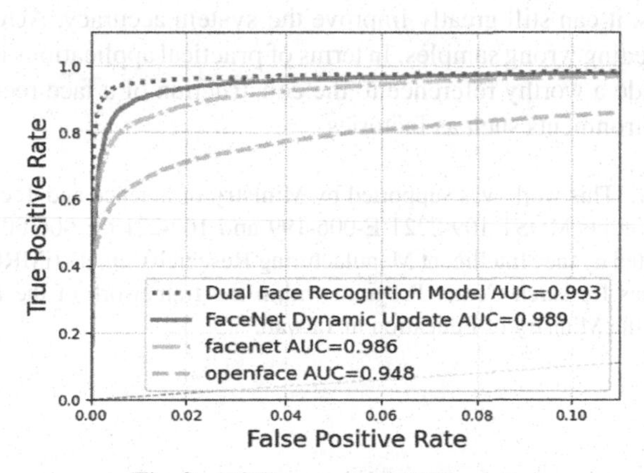

Fig. 9. ROC curves and AUC score

5.3 Update Record Comparison

Based on the updating process of our system and the third system, using their own algorithms to extract test sets as training data, we have recorded and collated results on Table 1 by comparing the actual types of test samples. As can be seen from the table, the number of error samples input into our system is only 27, which is much less than the 146 samples input into the third system, and the number of correct samples input and the number of retained samples also indicate that our system is capable of storing more samples of higher value.

Table 1. Update records

System	Input error samples	Input correct samples	Reserved samples
FaceNet dynamic updates	146	2304	0
Dual dynamic update	**27**	**2454**	**46**

6 Conclusion

We develop a method of mutual correction based on the classification models trained by two facial neural networks, which is called a Dual Face Recognition Interactive Correction Training System. The module update algorithm inside the system is used to compare the face classification models trained by the two facial neural networks and obtain a set of predicted identities and confidence levels for the newly input face images. After obtained the best prediction result, assist the new face image to be fed into the correct identity. Experimental results show that our system performs well. With a few training samples, it can still greatly improve the system accuracy, AUC score, and the probability of feeding wrong samples. In terms of practical applications in the future, this article can provide a worthy reference to the construction of a face recognition system for complex environments such as factories.

Acknowledgment. This work was supported by Ministry of Science and Technology (MOST) of Taiwan under Grants MOST 109-2221-E-006-199 and 109-2218-E-006-007. This work was financially supported by the "Intelligent Manufacturing Research Center" (iMRC) in NCKU from The Featured Areas Research Center Program within the framework of the Higher Education Sprout Project by the Ministry of Education in Taiwan.

References

1. Schroff, F., Kalenichenko, D., Philbin, J.: FaceNet: a unified embedding for face recognition and clustering. In: 2015 IEEE Computer Society Conference on Computer Vision and Pattern Recognition (CVPR), pp. 1–4 (2015)

2. Amos, B., Ludwiczuk, B., Satyanarayanan, M.: OpenFace: a general-purpose face recognition library with mobile applications (2016)
3. Sirovich, L., Kirby, M.: Low-dimensional procedure for the characterization of human faces. J. Opt. Soc. Am. A **4**(3), 519–524 (1987)
4. Turk, M., Pentland, A.: Eigenfaces for recognition. J. Cogn. Neurosci. **3**(1), 71–86 (1991)
5. Belhumeur, P.N., Hespanha, J.P., Kriegman, D.J.: Eigenfaces vs. Fisherfaces: recognition using class specific linear projection. IEEE Trans. Pattern Anal. Mach. Intell. **19**(7), 711–720 (1997)
6. Ojala, T., Pietikäinen, M., Harwood, D.: A comparative study of texture measures with classification based on featured distributions. Pergamon Pattern Recogn. **29**(1), 51–59 (1996)
7. Zhang, G., Huang, X., Li, S.Z., Wang, Y., Wu, X.: Boosting local binary pattern (LBP)-based face recognition. In: Li, S.Z., Lai, J., Tan, T., Feng, G., Wang, Y. (eds.) SINOBIOMETRICS 2004. LNCS, vol. 3338, pp. 179–186. Springer, Heidelberg (2004). https://doi.org/10.1007/978-3-540-30548-4_21
8. Oo, S.L.M., Oo, A.N.: ASEAN child face recognition system with FaceNet. In: Myanmar Universities' Research Conference (MURC), pp. 62–68 (2019)
9. Li, L., Jun, Z., Fei, J., Li, S.: An incremental face recognition system based on deep learning. In: 2017 Fifteenth IAPR International Conference on Machine Vision Applications (MVA), pp. 212–215 (2017)
10. Wang, Y.-K., Zheng, Y.-X., Lin, C.-S.: Classiface: real-time face recognition based on multi-task convolution neural network. Int. J. Sci. Eng. **8**(1), 15–28 (2018)
11. Sun, Y., Wang, X., Tang, X.: Hybrid deep learning for face verification. IEEE Trans. Pattern Anal. Mach. Intell. **38**(10), 1997–2009 (2016)
12. Santoso, K., Kusuma, G.P.: Face recognition using modified OpenFace. Proc. Comput. Sci. **135**, 510–517 (2018)

Forecasting Stock Trend Based on the Constructed Anomaly-Patterns Based Decision Tree

Chun-Hao Chen(✉), Yin-Ting Lin, Shih-Ting Hung, and Mu-En Wu

Department of Information and Finance Management, National Taipei University of Technology, Taipei, Taiwan
{chchen,mnwu}@ntut.edu.tw, {t108ab8003,t109749005}@ntut.org.tw

Abstract. Forecasting and modeling the trend of stock price is a crucial task in the field of financial market. Along with the advantage of more abundant and transparent data in recent years, how to find useful features to predict stock trend is important. Many works predicted stock price through technical indicators, and some of the works combined technical and chip analysis to forecast stock trend. In other words, there may have patterns that have influence on the stock trend. Because before the uptrend or downtrend happening, there may exist anomaly patterns. In this paper, we firstly define the anomaly patterns, and then propose an approach for constructing a classifier using decision tree based on the defined anomaly patterns for stock trend prediction. It first locates the stock trend periods from the given stock prices series. Then, for every stock trend period, it will try to identify the anomaly patterns from a given specific period before the stock trend period. The discovered anomaly patterns associated with the label of the stock trend period is formed an instance. Then, the generated instances are used to construct the anomaly-patterns based decision tree. Experiments one the real datasets were made to reveal the effectiveness of the proposed approach.

Keywords: Stock trend prediction · Anomaly pattern · Decision tree · Trading strategy

1 Introduction

Whether in the industry or academia, the research of stock market trend prediction has become a popular research topic [1]. Forecasting stock market is usually considered as one of the most challenging but important tasks among time series predictions [2, 3]. Zhang et al. illustrated the importance of stock market prediction [4]. Because stock market is a dynamic and nonlinear system, it is difficult to predict through few features. According to the principle of the efficient market hypothesis [5], the stock price will follow a random pattern. In fact, the stock market is full of information asymmetry. Therefore, to find anomaly patterns may be used to reveal a series of operations made by the specific societies who may try to affect the trend of stock markets.

N. T. Nguyen et al. (Eds.): ACIIDS 2021, LNAI 12672, pp. 606–615, 2021.
https://doi.org/10.1007/978-3-030-73280-6_48

Outlier detection has been studied in the context of a large of application fields [7]. In the financial market, anomaly detection usually applies in credit card fraud. For other aspects, outliers are often ignored or eliminated to avoid interference and ensure the accuracy of the forecast in time series data [8, 9]. Ma et al. used isolation forest which is an unsupervised learning algorithm to eliminated outlier data [10]. However, in the time series of stock market, anomaly data could be the indicators and should be magnified and viewed because the occurrence of these anomalies may affect the stock price.

The main reason for investing is to obtain positive returns, so many researchers focus on how to predict stock trends in order to obtain not only stable but also high returns. There are many ways to predict stock price trends, e.g., the technical analysis such as moving average (MA), moving average convergence divergence (MACD) and relative strength index (RSI), the fundamental analysis such as return on equity (ROE), return on assets (ROA) and earnings per share (EPS), or news analysis including news materials, community forums to make predictions [11, 13]. Maciel et al. through technical analysis, including simple moving average (SMA), exponential moving average (EMA), average true volatility (ATR), RSI and other technical indicators for price prediction [11]. Ren et al. calculated sentiment scores and combined price data through news and forum data, and used machine learning methods such as support vector machines to predict stock price trends [3]. However, there are few works using anomaly patterns for stock trend prediction. For example, Zhao et al. proposed a novel outlier mining algorithm to detect anomalies on the basis of volume of stock data [6].

Hence, this research devotes to define anomaly events to figure out the reason whether the stock price trend is rising or falling in the short-term period. As a result, in this paper, we firstly define the anomaly patterns, and then propose an approach for constructing a classifier using decision tree based on the defined anomaly patterns for stock trend prediction. It first locates the stock trend periods from the given stock prices series. Then, for every stock trend period, it will try to identify the anomaly patterns from a given specific period before the stock trend period. The discovered anomaly patterns associated with the label of the stock trend period is formed an instance. Then, the generated instances are used to construct the anomaly-patterns based decision tree. Experiments one the real datasets were made to reveal the effectiveness of the proposed approach.

In the following, the problem definitions are stated in Sect. 2. The proposed approach is introduced in Sect. 3. The experimental results are given as discussed in Sect. 4. Finally, the conclusion and future work is described in Sect. 5.

2 Definitions and Problem Statement

In this section, the related definitions are given. The two stock trends are first stated. Then, the definitions of anomaly patterns are given. At last, the problem to be solved is described. In this work, there are two types of the stock trends that are uptrend and downtrend, and they are shown in Definitions 1 and 2.

Definition 1. Uptrend (UT). UT is one of the labels which is defined as when stock prices go up to R percent in k days, it would be regarded as a label UT in this research.

It is shown in formula (1) as follows:

$$UT = \frac{P_c - P_{c-k}}{P_{c-k}} \geq R\%, \tag{1}$$

where P_c means today's close price, and P_{c-k} means close price of the day k before day c. For example, if k and R values are set at 5 and 15, and the close prices of P_c and P_{c-k} are 115 and 100, then this stock trend period is regarded as UT.

Definition 2. Downtrend (DT). DT is another label which is defined as when stock price fell down R percent in k days, it would be regarded as a label DT in this research. The formula is shown as follows:

$$DT = \frac{P_c - P_{c-k}}{P_{c-k}} \leq R\%, \tag{2}$$

where P_c means today's close price, and P_{c-k} means close price of the day k before day c. For example, if k and R values are set at 5 and 20, and the close prices of P_c and P_{c-k} are 80 and 100, then this stock trend period is regarded as DT.

Definition 3. The anomalies of Over Volume (OVA). OVA is one of the features which can be defined as if today's volume V_t is larger than the sum volume of last k days, it would be regarded as over volume anomaly in this research. The formula is given as follows:

$$OVA = V_t > \sum_{i=1}^{k} V_{t-i}, \tag{3}$$

where V_t is today's volume data, and V_{t-i} is the volume of the day $t - i$. For example, if the values of the V_t and $\sum_{i=1}^{k} V_{t-i}$ are 6000 and 2000, then it will be identified as the OVA.

Definition 4. The anomalies of Less Volume (LVA). LVA means if today's volume is less than the average volume of past k^2 days, it would be regarded as the OLV pattern in this research. The formula is shown as follows:

$$LVA : V_t < \frac{\sum_{i=1}^{k} V_{t-i}}{k^2}, \tag{4}$$

where V_t is today's volume data, and V_{t-i} is the volume of the day $t - i$. For example, if the values of the V_t and $\frac{\sum_{i=1}^{k} V_{t-i}}{k^2}$ are 1000 and 3000, then it will be identified as the LVA.

Definition 5. The anomalies of Higher Price (HPA). HPA means if today's close price higher than the average close price past n days plus the v times standard deviation of the average close price of past n days, it would be regarded as HPA in this research. The formula is shown as follows:

$$HPA : P_c > \mu(P_c, n) + \mu(P_c, n) \times \sigma \times v, \tag{5}$$

where P_c means close price of day c, $\mu(P_c, n)$ is the average prices of days c-n to c, σ is the standard deviation, and v is a parameter.

Definition 6. The anomalies of Lower Price (LPA). LPA means if today's close price lower than the average close price past *n* days minus the *v* times standard deviation of the average close price of past *n* days, it would be regarded as LPA in this research. The formula is given as follows:

$$LPA : P_c < \mu(P_c, n) - \mu(P_c, n) \times \sigma \times v \quad (6)$$

where P_c means close price of day *c*, $\mu(P_c, n)$ is the average prices of days *c-n* to *c*, σ is the standard deviation, and *v* is a parameter.

Definition 7. The anomalies of Buy Chip (BCA). When foreign investment and investment trust companies constantly bought a specific stock more than *n* days, it would be defined as a BCA pattern.

Definition 8. The anomalies of Sell Chip (SCA). When foreign investment and investment trust companies constantly sold a specific stock more than *n* days, it would be defined as a SCA pattern.

Definition 9. A stock trend anomaly (STA). A STA consists of a set of attribute values and a label extract from a given stock price period, where the attribute values could be OVA, LVA, HPA, LPA, BCA, SCA, etc., and the labels could be UT or DT. An example is given in Fig. 1.

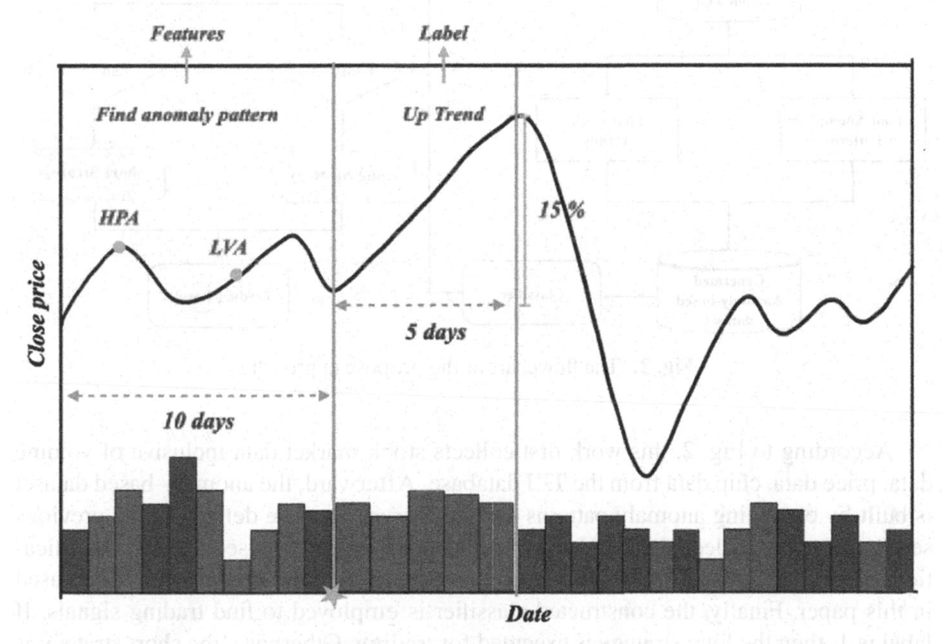

Fig. 1. A stock trend anomaly

From Fig. 1, it indicates that the anomaly patterns, including the HPA and LVA, are identified in the 10 days, and then the stock trend of next 5 days is the uptrend. Note that the stock trend anomaly will be used as a training instance, and the anomaly patterns and the uptrend are attributes and the label.

Hence, based on the abovementioned definitions, the purpose of this paper is to construct a stock trend prediction model based on those abnormal events, including OVA, LVA, HPA, LPA, BCA, and SCA.

3 Proposed Approach

In this section, the flowchart of the proposed approach is stated in Sect. 3.1. Then, the pseudo code of the proposed approach is described in Sect. 3.2.

3.1 Flowchart of the Proposed Approach

The flowchart of the proposed approach to construct the anomaly-pattern based classifier for stock trend prediction is shown in Fig. 2.

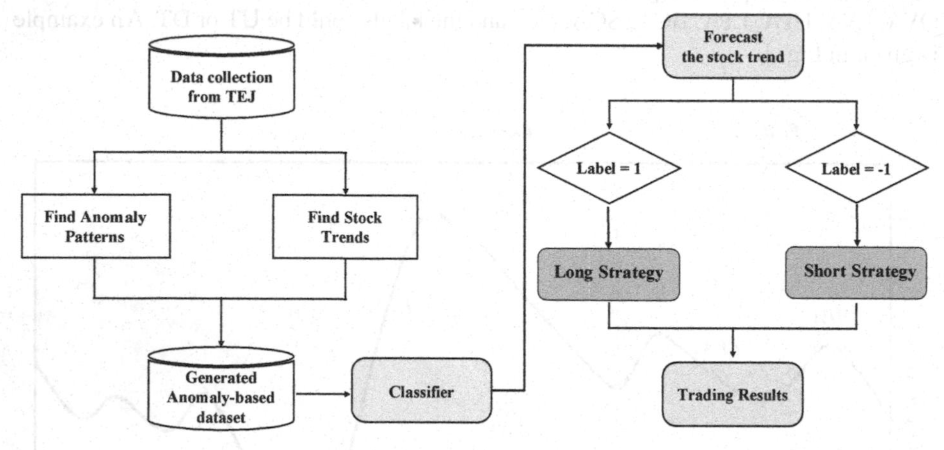

Fig. 2. The flowchart of the propose approach

According to Fig. 2, this work first collects stock market data inclusive of volume data, price data, chip data from the TEJ database. Afterward, the anomaly-based dataset is built by extracting anomaly patterns and stock trend that are defined in the previous section from the collected data. Using the generated anomaly- based dataset, classification models can be employed to find the classifier, and the decision tree has been used in this paper. Finally, the constructed classifier is employed to find trading signals. If label is 1, then the long strategy is executed for trading. Otherwise, the short strategy is executed.

Algorithm 1: Anomaly-Pattern-based Decision Tree Construction Algorithm

Input: A set of stock price series $SP = \{sp_i \mid 0 \le i \le n\}$, a set of chip series $CD = \{cd_i \mid 0 \le i \le n\}$ and a set of volume series $VD = \{sp_i \mid 0 \le i \le n\}$.
Parameters: Number of stock *numStock*, change rate *R*, days of a stock trend period *D*, days before a stock trend *aDate*
Output: The anomaly-pattern-based decision tree classifier (APDTC).

```
1.  ADE_DecisionTree (){
2.        stockTrendDates = findTrendDate(numStock, SP, R, D)
3.        anomalyDates = findAnomalyDate(numStock, SP, CD, VD);
4.        STAs = findAnomalyPattern(stockTrendDates, anomalyDates, aDate){
5.            FOR every series spi DO
6.              STAs = mergeAnomalyTrend(stockTrendDates, anomalyDates, aDate)
7.            END FOR LOOP
8.            return STAs
9.        }
10.       Output APDTC ← DecisionTreeClassifier(STAs);
11. }
```

Fig. 3. The pseudo code of the proposed approach.

3.2 Pseudo Code of Proposed Algorithm

Based on the flowchart, the pseudo code of the proposed algorithm that is utilized to construct an anomaly-based decision tree for stock trend prediction is shown in Fig. 3.

From Fig. 3, the proposed approach first generates the stock trend dates *stockTrendDates* and anomaly pattern dates *anomalyDates* (lines 2 and 3). The stock trend anomaly instances are then generated using the based on the stock trend dates, anomaly pattern dates, and a given specific period *aDate* (lines 4 to 9). Finally, the anomaly-pattern-based decision tree classifier is constructed using the generated stock trend anomaly instances (line 10). The procedure for finding anomaly pattern dates is shown in Fig. 4 as follows:

From Fig. 4, using the given stock price series, chip series and volume series, the procedure is executed to find the dates of six anomaly patterns according to the definitions of them (lines 2 to 9). Then, the discovered anomaly pattern dates are sorted based on the dates (line 10), and the result is returned (line 13).

Procedure: Find Anomaly Date

```
1.  findAnomalyDate(numStock, SP, CS, VD){
2.      FOR i = 0 to numStock DO
3.          overVolumeDate = generateOverVolumeAnomaly(vd_i);
4.          lessVolumeDate = generateLessVolumeAnomaly(vd_i);
5.          highBBandsDate = generateHighPriceAnomaly(sp_i);
6.          lowBBandsDate = generateLowPriceAnomaly(sp_i);
7.          buyChipDate = generateBuyChipAnomaly(cd_i);
8.          sellChipDate = generateSellChipAnomaly(cd_i);
9.      END i FOR LOOP
10.     anomalyDate = sortAllAnomaly(overVolumeDate, lessVolumeDate,
11.                                  highBBandsDate, lowBBandsDate,
12.                                  buyChipDate, sellChipDate)
13.     return anomalyDate
14. }
```

Fig. 4. The procedure to find anomaly pattern dates

4 Experimental Evaluations

4.1 Dataset Description

The stock price series, chip series and volume series of stocks in Top-fifty companies ETF in Taiwan was collected from 2008/1/2 to 2020/3/31 from TEJ database in Taiwan. The datasets from 2008/1/2 to 2019/3/31 are used for training, and others are used for testing.

4.2 The Generated Anomaly-Pattern-Based Decision Trees

Using the proposed algorithm, the generated anomaly-pattern-based decision trees is shown in Fig. 5.

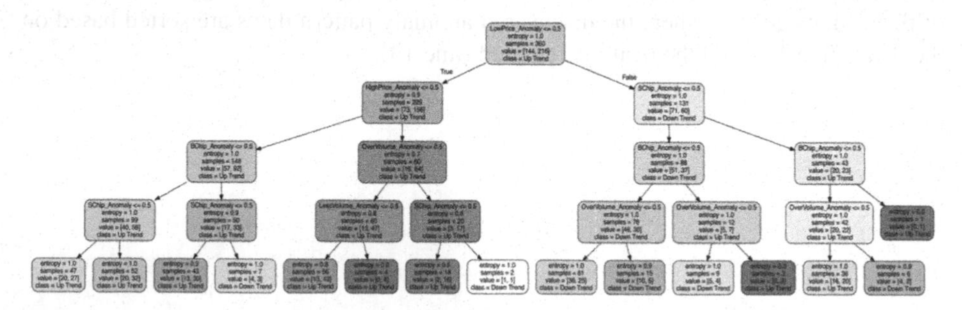

Fig. 5. The result of the decision trees

The best accuracy of the generated decision tree is *82.93%*. In addition, the five interesting rules shown in Fig. 5 are listed as follows:

- In case of the anomalies of higher price happens, it's about *80%* probability to become uptrend in the following next 5 days.
- In case of the anomalies of higher price and the anomalies of over volume happen, it's about *85%* probability to become uptrend in the following next 5 days.
- In case of the anomalies of higher price and the anomalies of less volume happen, it's about *100%* probability to become uptrend in the following next 5 days.
- In case of the anomalies of lower price and the anomalies of over volume happen, it's about *70%* probability to become downtrend in the following next 5 days
- In case of the anomalies of lower price, the anomalies of buy chip and over volume happen, it's about *100%* probability to become downtrend in the following next 5 days.

Experiments were then made to show the cumulative return of the classifier on five companies from 2019/4/1 to 2020/3/31, and the results are shown in Fig. 6.

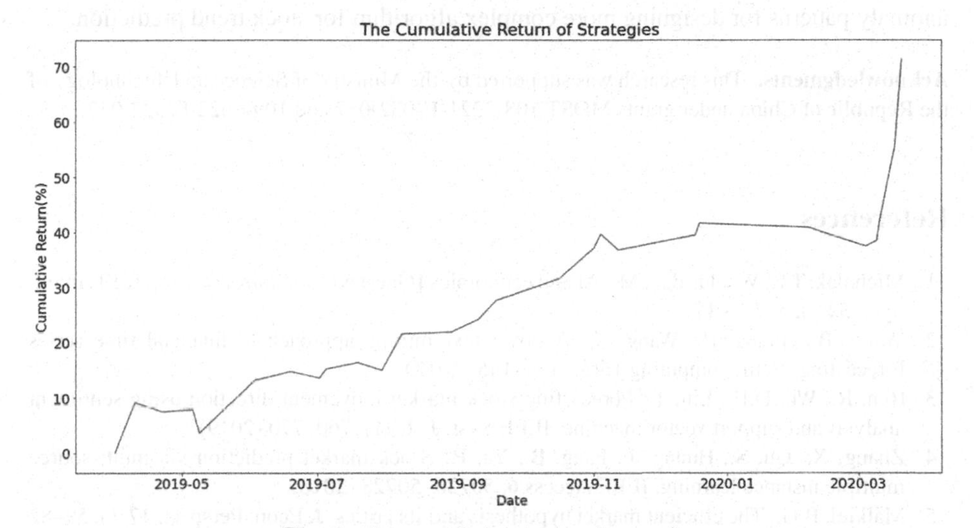

Fig. 6. The cumulative return of strategies

From Fig. 6, we can see that the cumulative return of the generated classifier is always positive and the range is from 0 to 70%, which means the generated anomaly-pattern-based decision tree is effective. Note that after deducting the transaction costs, the total cumulative return of strategies is *71.26%*. From the experimental results, we can conclude that the proposed approach is effective in terms of getting a positive cumulative return, and reveal that using anomaly pattern to predict stock trends is workable.

5 Conclusions and Future Work

Financial market is always an interesting research field for researchers because many investors' behaviors exist in different financial datasets, and one of the attractive research topics is stock trend prediction. Before the stock trend happening, including uptrend or downtrend, there may exist various anomaly patterns. Thus, to find the relationship between the stock trends and anomaly patterns is important. In this paper, an algorithm has thus been proposed for construct the anomaly-pattern based decision tree to reveal the relationship between stock trends and anomaly patterns and for stock trend prediction. Experiments on the real financial datasets were also made to show the merits of the proposed approach. The results indicate that the best accuracy of the proposed approach is *82.93%*. Furthermore, some interesting rules are also discovered, e.g., "When the Lower Price Anomaly, Buy Chip Anomaly, Over Volume Anomaly occurred simultaneously, the stock trend is downtrend." In addition, the results also show that the cumulative return of the proposed approach is above 70%. In the future, we will continue to consider more anomaly patterns for designing more complex algorithm for stock trend prediction.

Acknowledgments. This research was supported by the Ministry of Science and Technology of the Republic of China under grants MOST 108-2221-E-032-037 and 109-2622-E-027-032.

References

1. Michalak, T.P., Wooldridge, M.: AI and economics [Guest editors' introduction]. IEEE Intell. Syst. **32**(1), 5–7 (2017)
2. Wang, B., Huang, H., Wang, X.: A novel text mining approach to financial time series forecasting. Neurocomputing **15**(3), 136–145 (2012)
3. Ren, R., Wu, D.D., Liu, T.: Forecasting stock market movement direction using sentiment analysis and support vector machine. IEEE Syst. J. **13**(1), 760–770 (2019)
4. Zhang, X., Qu, S., Huang, J., Fang, B., Yu, P.: Stock market prediction via multi-source multiple instance learning. IEEE Access **6**, 50720–50728 (2018)
5. Malkiel, B.G.: The efficient market hypothesis and its critics. J. Econ. Perspect. **17**(1), 59–82 (2003)
6. Zhao, L., Wang, L.: Price trend prediction of stock market using outlier data mining algorithm. In: IEEE Fifth International Conference on Big Data and Cloud Computing, pp. 93–98 (2015)
7. Gupta, M., Gao, J., Aggarwal, C.C., Han, J.: Outlier detection for temporal data: a survey. IEEE Trans. Knowl. Data Eng. **26**(9), 2250–2267 (2014)
8. Ahmed, M., Choudhury, N., Uddin, S.: Anomaly detection on big data in financial markets. In: IEEE/ACM International Conference on Advances in Social Networks Analysis and Mining, pp. 998–1001 (2017)
9. Idrees, S.M., Alam, M.A., Agarwal, P.: A prediction approach for stock market volatility based on time series data. IEEE Access **7**, 17287–17298 (2019)
10. Ma, Y., Zhang, Q., Ding, J., Wang, Q., Ma, J.: Short term load forecasting based on iForest-LSTM. In: IEEE Conference on Industrial Electronics and Applications, pp. 2278–2282 (2019)
11. Maciel, L.: Technical analysis based on high and low stock prices forecasts: evidence for Brazil using a fractionally cointegrated VAR model. Empirical Econ. **58**, 1513–1540 (2020)

12. Shen, W., Guo, X., Wu, C., Wu, D.: Forecasting stock indices using radial basis function neural networks optimized by artificial fish swarm algorithm. Knowl.-Based Syst. **24**(3), 378–385 (2011)
13. Financial Stability Board: Artificial intelligence and machine learning in financial services. Financial Stability Board (2017)

A Transparently-Secure and Robust Stock Data Supply Framework for Financial-Technology Applications

Lin-Yi Jiang[1](✉), Cheng-Ju Kuo[1](✉), Yu-Hsin Wang[2], Mu-En Wu[3], Wei-Tsung Su[4], Ding-Chau Wang[5], O. Tang-Hsuan[1], Chi-Luen Fu[1], and Chao-Chun Chen[1]

[1] IMIS/CSIE, National Cheng Kung University, Tainan, Taiwan
{P96094189,P96084087,P96081128,chaochun}@mail.ncku.edu.tw
[2] CS, National Tsing Hua University, Hsinchu, Taiwan
s105062223@m105.nthu.edu.tw
[3] IFM, National Taipei University of Technology, Taipei, Taiwan
mnwu@ntut.edu.tw
[4] CSIE, Aletheia University, Taipei, Taiwan
suwt@mail.au.edu.tw
[5] MIS, Southern Taiwan University of Science and Technology, Tainan, Taiwan
dcwang@mail.stust.edu.tw

Abstract. Recently, program trading has become the mainstream of financial information technology. The current FinTech applications mainly encounter three technical issues, including scalability, long-time retrieval, and security. Many small-and-medium companies seek economical solutions, while the three above properties are still met. In this paper, we propose Send-and-Subscribe (SaS) framework, aiming at providing a secure and robust financial stock data retrieval repository. In our survey to the related industries, the proposed novel framework is the first work on addressing the above three issues for financial computing areas. Finally, we conduct a set of experiments to validate the proposed framework on the real-world Taiwan stock data. The test results show the proposed framework indeed satisfies the security, scalability, and long-query requirements, comparing to existing solutions.

Keywords: Financial technology (FinTech) · Security · Robustness · Financial data services · Granular computing

This work was supported by Ministry of Science and Technology (MOST) of Taiwan under Grants MOST 109-2221-E-006-199, 108-2221-E-034-015-MY2, and 109-2218-E-006-007. This work was financially supported by the "Intelligent Manufacturing Research Center" (iMRC) in NCKU from The Featured Areas Research Center Program within the framework of the Higher Education Sprout Project by the Ministry of Education in Taiwan.

N. T. Nguyen et al. (Eds.): ACIIDS 2021, LNAI 12672, pp. 616–629, 2021.
https://doi.org/10.1007/978-3-030-73280-6_49

1 Introduction

Creating FinTech applications is a critical issue of future financial industry. Big Data and Big Money: The Role of Data in the Financial Sector [1] point out that FinTech applications mainly consider three properties: security, scalability and long query. The security requirement requests the immediate results of a trading strategy shall not be leak out. Although the stock data is public to all, how a broker manipulate the stock data (called a trading strategy) is the key technique to earn profit. The scalability requirement requests the several thousands of applications can retrieve the financial data at the same time, which relies on the robustness of the financial data repository. The scalability crucially tests robustness of the request-hosting interface and the database server. The long execution time for data retrieval queries, which then occurs disconnection due to too long waiting. A qualified financial data repository shall support long queries. Summarizing these requirements, a secure and robust data access mechanism is the solution to support FinTech applications.

The FinTech data exchange, such as stock data retrieval, plays a central role in developing FinTech ecosystems. Industrial experts indicates three properties in the FinTech data exchange must be supported [2–6], including scalability, complex queries, and security, and they are briefly described as follows.

- **Scalability.** The FinTech data server needs to handle a huge amount of data access from hundreds of thousands of users or programs per day, which easily incurs service unavailable due to system crash or no response. One of frequently happened situations is that too many arrived requests occupy most computational resources and thus, no sufficient resources for handling the newly arrival requests, which incurs request failure due to time out. In this way, throughput of the FinTech data server decreases.
- **Complex Queries** over big FinTech data. Many data requests to the FinTech data server are complex queries, which means they need to look up the data table multiple times. In addition, the FinTech data server may store big data, such as stock trading data for past decades. Hence, such data requests need long response time to prepare results. With current web protocols (e.g., HTTP, HTTPS), they would incur service failures due to time out error. Same to the first case, throughput of the data server decreases.
- **Security.** The FinTech data server provides financial sensitive data to different users and programs in the associated FinTech ecosystems, and thus, the data security has to be sustained. Particularly, one most concerned data security in the internet-based FinTech applications is the data communication over internet, which has become the necessary function provided by the FinTech data server. Moreover, for promoting FinTech services, developers of FinTech ecosystems ask the data security transparent to the FinTech applications for reducing development load.

A qualified FinTech data server need to provide high availability to these three properties. Some venders develop such FinTech data server by the big data solutions, such as Hadoop or cloud computing servers, for fulfill the above three properties. However, such solutions spend enormous monetary cost on system development and maintenance. Many small-and-medium companies seek economical alternatives, while the three above properties are still met.

In this paper, we propose Send-and-Subscribe (SaS) framework, aiming at providing a secure and robust financial stock data retrieval repository. To achieve high scalability, our proposed framework reserves dedicated computational resources on dealing with arrival requests and the shared resources on processing data retrieval. In this way, the number of service rejection to arrival requests can be minimized. To support long-response requests from complex queries, our proposed framework adopts the message queue as the medium of replying queries, instead of waiting query results in traditional HTTP connections. The message queue-based request interaction needs few resources on handling arrival requests. FinTech applications will be informed to retrieve acquired data once the query results are ready. Such interaction protocol not only solves the long-response request issue, but also further increase scalability of the FinTech data server. A new protocol for such data retrieval needs to be designed in this work, and will be presented in Sect. 3.2. To achieve security, our proposed framework provides a transparently-secure communication mechanism, meaning that the financial data will be automatically encrypted/decrypted during transmission over internet. This mechanism greatly reduces the programming burdens for developers in the financial industry. In our survey to the related industries, the proposed novel framework is the first work on addressing the above three issues for financial computing areas. Finally, we conduct a set of experiments to validate the proposed framework on the real-world Taiwan stock data. The test results show the proposed framework indeed satisfies the security, scalability, and long-query requirements, comparing to existing solutions.

2 Proposed Send-and-Subscribe (SaS) Framework

2.1 System Architecture

In order to support the characteristics of scalability, security, and processing long queries in the financial data repository, the proposed system adopts multiprocessing and message queue as the underlying computation and communication infrastructure. In such architecture, components can be designed separately and communicates each other by message queue to achieve concurrent execution and resource sharing for improving efficiency of the overall system.

Figure 1 shows the structure of SaS Framework. The system we proposed is mainly composed of four modules: task-dispatching interfaces, message queue, and process unit. Each of which is described below.

Task-Dispatching Interfaces: is designed quickly respond to a large number of requests by using independent computing resources for achieving high scalability. This service is developed based on the Restful API forms, and APIs describe FinTech data retrieval requests. The REST design defines the unique URL locations for financial data, which reduces the interdependence of object resources and better handles kinds of business logic.

Message Queue: is used to solve HTTP time out problem in web environments for supporting long queries over the financial data repository. Clients and the FinTech data server can communicate through topics, which are switches of data exchanges in the message queue protocols. Through the subscribe/publish-style message delivery mechanism in the message queue protocol, the client (subscriber of the topic) receives data from the server (publisher of the topic). Our system adopts our previously developed SMQTT framework [7], by combined the automated encryption/decryption method, so that the FinTech data server provides secure data transmission channels.

Process Unit: is the main processing unit for arrival requests, which consists of certains data retrieval functions sharing the underlying hardware resources via multithreading techniques. While the requested data are completely retrieved from the financial data repository, the process unit then uses the message queue server to deliver data to the requested client as the role of the data publisher in the message queue protocol.

In summary, the financial data repository system is designed with a secure user authentication mechanism, and also has a dedicated thread to handle a large number of complex financial data retrieval requests for greatly improving the scalability and stability of the system. The proposed framework also avoids the situations that a large number of requests slow down the server as mechanisms efficiently handle batch data retrieval operations (add, delete, modify, search, etc.), so that chance of system crashes or painful delays is significantly reduced.

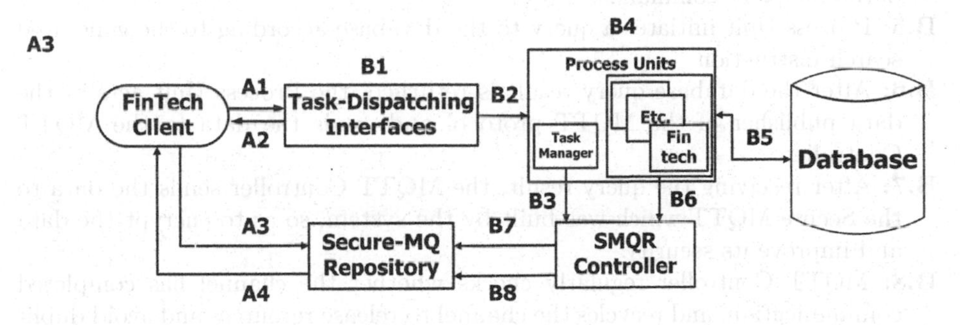

Fig. 1. Architecture of our proposed financial data retrieval framework with properties of security, scalability, and long queries.

The operational flow of retrieving financial data via our proposed framework is as follows, and these steps correspond to those in Fig. 1.

A.1: The client sends a financial data retrieval request to Task-Dispatching Interface.

A.2: The Task-Dispatching Interface returns the system-generated message-queue topic, which is used to deliver acquired data later.

A.3: The client subscribes the returned topic to the message queue server.

A.4: The client receives acquired financial data via the subscribed message queue topic, where the financial data is automatically encoded by the server and decoded by the message queue client embedded in the FinTech client. Note that the data secure is transparent to the FinTech client, as it is automatically done in the internet communication via message queue.

The retrieval protocol of FinTech applications is further explained in Sect. 2.2. Notice that considering secure data transmission and privacy among multiple clients in the data delivery issue with uncertain response time is not easily achieve directly by just adopting a message queue (MQ) repository. Some mechanisms on managing the MQ repository and integrating the retrieval processing units and the MQ repository are required to fulfill such retrieval paradigm and they will be presented in the next section.

The operational flow of processing financial data requests in our proposed framework is as follows, and these steps also correspond to those in Fig. 1.

B.1: The server-side Task-Dispatching Interfaces receives the financial data retrieval request from the client, distributes tasks according to the request, and returns the subscription topic generated randomly.
B.2: The server-side Task-Dispatching Interfaces transmits the request parameters and subscription topic to the Process Unit.
B.3: After the Task Manager of the Process Unit receives the subscription topic, it initiates a subscription to the MQTT Controller.
B.4: The Process Unit processes the parameters passed by the client to generate database query commands.
B.5: Process Unit initiates a query to the database according to the generated search instruction.
B.6: After the database query result is returned, the Process Unit acts as the data publisher of the MQTT protocol and sends the data to the MQTT Controller.
B.7: After receiving the query result, the MQTT Controller sends the data to the Secure MQTT which was built by the system, so as to encrypt the data and improve its security.
B.8: MQTT Controller regularly checks whether the channel has completed communication, and recycles the channel to release resources and avoid duplication of subscription topics.

The core functional mechanisms will be presented in details in Sect. 3.

2.2 Data Retrieval Example in FinTech Applications

Figure 2 shows an example of the data retrieval protocol for FinTech applications, where the protocol contains four steps corresponding to the operational flow of retrieving financial data mentioned in the last subsection. The example are coded in the Python language, and developers can use other preferred programming languages in their implementation of FinTech applications. The first step (Line 8–15) is to send a request of acquiring financial data from the remote financial data server. More request types are presented in the following section. The second step (Line 16–18) is to get the topics as the basis of MQTT subscription action initiating. The third step (Line 19–29) is to initiates the subscription to the MQTT server according to the topics got from the second step. Then in the last step (Line 30–32), we can receive the ciphertext form financial data.

```
1   import requests
2   import mqtt
3
4   url="http://sas_framework/api/v1/bargain/"
5   MQTTHOST ="http://sas_framework/"
6   MQTTPORT = 1883
7
8   # A.1: The client sends a financial data retrieval request
9   #      to Task-Dispatching Interface.
10  def get_bragain_data(stock_id,start_date,end_date):
11      params={'stock_id':stock_id,'start_date':start_date,
12              'end_date':end_date}
13      request=requests.get(url,params=params)
14      if request.status_code == requests.code.ok:
15          return request.topic
16  # A.2: The Task-Dispatching Interface returns the
17  #      system-generated message-queue topic.
18  topic=get_bragain_data('2330','20120704','20200831')
```

```
19  # A.3: The client subscribes the returned topic to
20  #      the message queue server.
21  def subscribe_bragain_topic():
22      client=mqtt.Client()
23      client.on_connect=on_connect
24      client.on_message=on_message
25      client.subscribe(topic)
26      return client
27  mqttClient=subscribe_bragain_topic()
28  mqttClient.connect(MQTTHOST,MQTTPORT,60)
29  mqttClient.loop_forever()
30  #A.4: The client receives financial data.
31  def on_message(client):
32      print(msg.topic+" "+"bragain_data:"+msg.payload)
```

A.1 (Lines 8–15), A.2 (Lines 16–18), A.3 (Lines 19–29), A.4 (Lines 30–32)

Fig. 2. An example of retrieving financial data from FinTech applications to our proposed framework.

On one hand, we observe the retrieval protocol is similar to ordinary ways in most data platforms, which does not increase programming burdens to FinTech application developers. On the other hand, it is worth emphasizing that the security and robustness have been guaranteed by our proposed financial data retrieval framework. The technical details and experimental evidences will be presented in the following sections.

3 Core Functional Mechanisms

In the proposed framework mentioned above, certain core functional mechanisms are needed for supporting scalability, security, and long query. We present each of them in details in the following subsections.

3.1 Scalability-Based Parallel Data Processing Mechanism

This proposed mechanism, called connection-biased hybrid resource allocation mechanism (CHRAM), is to preserve some *dedicated resources* for handling arrived requests to the short task-dispatching RESTful APIs, and on the other hand, use *shared resources* for finding results for these financial data requests, which may be long queries over big data. In this way, we maximize the number of arrived requests and still achieve high parallelism of request processing. With such design philosophy, the overall response time of processing a request might be prolonged due to preservation of the dedicated resources. This glitch can be alleviated by adopting the MQ-based repository in our framework, which can automatically push the results back to the request clients, for avoiding the data access failure due to disconnection of long waiting.

Figure 3 illustrates the design of the connection-biased hybrid resource allocation mechanism (CHRAM). The physical computational resources of the data server are divided into two classes: dedicated threads and shared threads. The dedicated threads always occupy the physical processors without releasing for other components or programs. These dedicated threads are allocated for processing the task-dispatching RESTful interfaces in this framework. On the other hand, the shared threads can be released for other components or programs with the kernel-level context switching mechanism inside the operating system, so that multiple processes or programs can be executed in parallel with virtual multi-processor. These dedicated threads are allocated for components of this framework other than the RESTful interfaces.

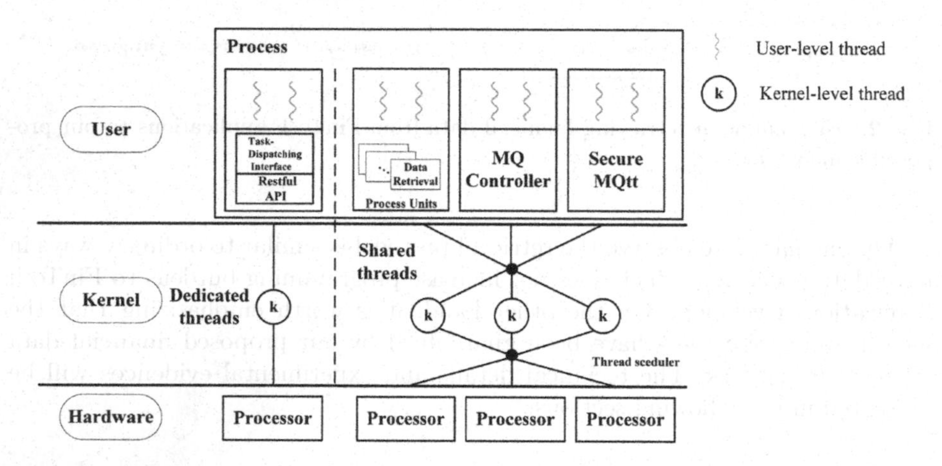

Fig. 3. Illustration of the connection-biased hybrid resource allocation mechanism (CHARM) for high connection-scalable and parallel financial data retrieval. Dedicated threads are allocated for providing the light-weight RESTful APIs; shared threads for other components shown in Fig. 1.

3.2 MQ-Based Data Retrieval Protocol

The message queue (MQ)-based data retrieval protocol is designed to cope with complex queries requiring long response time, and contains three components: task-dispatching RESTful interface, data retrieval-purpose topic management, and task routing mechanism. These components are presented as follows.

Task-Dispatching RESTful Interface. Figure 4 shows the RESTful interfaces for accessing financial data. Notice that these RESTful interfaces which provided in this framework are merely designed to dispatch tasks, instead of processing them. A task here stands for the internal requests of accessing financial data, which mainly consists of three steps. The first step is to create a corresponding topic of this request in system by sending a topic creation request to MQ controller. The sequential search is fast since it operates on the shared memory. An alternative is to use two slot sets "Used" and "Unused" (according to the OCPY field) for maintaining all slots of TOLUT. In this way, topic creation will randomly selects an unused slot from the "Unused" set, and then places this selected slot to the "Used" set. This alternative is faster, but it also consumes more storage space. Developers can choose the proper way according to their requirements. The second step is to dispatch the task and topic information to task manager. The third step is to deliver Send the IP of the SMQ repository with topic name RTN to the requesting application. In summary, a task-dispatching RESTful interface can be executed with very short response time and extremely low processing load since the design of each RESTful API contains only two message transmissions and only few searching operations are executed over the shared memory. In other words, this RESTful interfaces design reduces the chance that financial data retrieval is disconnected due to long waiting.

ID	API	Example	Explanation
1	/api/v1/stock_list	http://muen-fintech.imis.ncku.edu.tw/api/v1/stock_list	Get complet list of stocks.
2	/api/v1/bargain/{stock_id}	http://muen-fintech.imis.ncku.edu.tw/api/v1/bargain/2330	Get bargain data of a specifred stock.
3	/api/v1/bargain/{stock_id}/ {start_date}/{end_date}	http://muen-fintech.imis.ncku.edu.tw/api/v1 /bargain/2330/start_date/end_date	Get bargain data from start date to end date of a specified stock.

Fig. 4. List of provided RESTful APIs for accessing financial data.

Algorithm 1: Kernel Design for Task Dispatching APIs

Input: Query q, number of TOLUT slots n;
/* Step 1: find out the first unused topic number as the topic name for delivering query result. */
1 $RTN = 0$; // RTN: request topic number. 0 is the initial value. Used number is between 1 and N.
2 **for** $(s = 1$ *to* N *with step* $1)$ **do**
3 | **if** *(TOLUT[s].OCPY==0)* **then**
4 | | $RTN = s$; // found 1-st unused topic number.
5 | | Break; // Break the for-loop as RTN is found.
6 | **end**
7 **end**
8 **if** $(RTN \neq 0)$ **then**
9 | Encapsulate the query q as a task T_{RTN} for internal exchange;
10 **else**
11 | **return** "Exceeding the upper bound of SMQ repository.";
12 **end**
/* Step 2: dispatch the task and topic information to task manager. */
13 Send (T_{RTN}, RTN) to the task manager;
/* Step 3: return RTN to the requested application. */
14 Send the IP of the SMQ repository with topic name RTN to the requested application;
15 **return** "The query q has been dispatched.";

Data Retrieval-Purpose Topic Management. This topic management mechanism is designed for the MQ controller to automatically maintain the MQ repository to provide request result retrieval. More specifically, from the arrival of a data retrieval task to the end of result delivery, the MQ controller firstly creates a corresponding topic in the secure MQ repository, then publishes the data to the topic, and last revokes the topic after data retrieval. Algorithm 2 depicts the mechanism with the three above-mentioned key operations.

- **(OP1) Create topics:** When a data retrieval request is dispatched to MQ controller, the MQ controller will create a corresponding topic for the Fintech application which utters the request. This mechanism creates the UTN topic in the MQ repository, and then marks UTN.OCPY as 1.
- **(OP2) Data publishing** over topics: When the result of the data retrieval request has been acquired, the MQ controller will publish the result to the topic which is associated with the request (through looking up table). this operation give the current timestamp to the timestamp of the corresponding OTN slot in TOLUT.
- **(OP3) Revoke expired topics:** When the result of data retrieval request has been published in the associated topic within a predefined period (e.g., one hour), the MQ controller will revoke the topic in order to release system resources for future usage. The revocation of topics is executed periodically by MQ controller for keeping system's stability. For each OTN topic whose published-time is over θ, this mechanism will revoke the OTN topic in the MQ repository, except the server itself. Then the MQ controller will mark the OTN as UNN.

Algorithm 2: Data Retrieval-based Topic Control Algorithm.

```
   /* OP1: Topic Creation.                                                    */
 1 On receiving topic creation request with RTN from the task manager: //
 2 Create the RTN topic in the SMQ repository;
 3 Mark OCPY of the RTN-th slot in TOLUT as 1;
   /* OP2: Data Publishing.                                                   */
 4 On receiving the query request R_RTN associated to RTN from the task manager:
     //
 5 Publish R_RTN to the RTN topic in the SMQ repository;
 6 Fill the current timestamp to the TMESTMP of the RTN slot in TOLUT;
   /* OP3: Topic Revocation                                                   */
 7 On every checking time for revoking topics: //
 8 for each slot s in TOLUT do
 9 |   if (s.OCPY==1 and time(NOW) - TMESTMP > θ) then
10 |   |   Revoke the s-th topic in the SMQ repository;
11 |   |   Mark OCPY of the s-th slot in TOLUT as 0;
12 |   end
13 end
```

In summary, the data retrieval-purpose topic control mechanism provides data delivery with resource control to achieve the system robustness in the perspective of data exportation.

Task Routing Mechanism for Task Manager. The task routing mechanism is designed to process a dispatched task in the task manager. The task routing mechanism plays a key role in the proposed framework, as the task manager bridges the RESTful interfaces, the processing agents, and the MQ repository. The workflow of the task routing mechanism is sketched as follows.

1. send a "topic creation" message to the MQ maintainer. The topic is named as a random number for easy maintenance.
2. send "tasks" to corresponding processing units, e.g., stock-list processor, (api-name) processor, and receive task results from them. Note that each RESTful API should have a corresponding processing unit for providing complete financial data retrieval services.
3. send a "topic publish" message to the MQ maintainer for publishing the received task results to the created topic in the MQ repository.

Notice that the last published timestamp of each topic needs to be recorded, so that the MQ maintainer can revoke this topic later for better resource utilization.

Algorithm 3: Task Routing Algorithm.

```
  Input: Task T_RTN, request topic number RTN;
  /* Step 1: Topic Creation Request.                                  */
1 Send a RTN topic creation message to the SMQ controller;
  /* Step 2: Task Processing.                                         */
2 Send task T_RTN to the corresponding processing unit (say, u);
3 Receive the task result r from the processing unit u;
4 R_RTN = (r, RTN); // Compose task result with r and RTN.
  /* Step 3: Task Result Publishing Request.                          */
5 Send R_RTN to the SMQ controller;
```

3.3 Customized Coding Methods for Message Queue Systems

Figure 5 shows the operational flow of data exchange in the SaS Framework. First, when the financial data sender sends the financial data to the SMQTT module, the SMQTT module will encrypt the financial data using the method selected by clients, and create an MQTT message to encapsulate the encrypted financial data, encryption method, key and other related information. After that, the SMQTT module will send MQTT message to the financial data requester through SMQTT tunnel. The financial data in the entire communication process is protected, transmitted in cipher text. In the end, after the financial data requester receives the MQTT message, it then decrypts the message by identifying the encryption mechanism and key in the message.

We have implemented certain encryption methods for supporting the transparently-secure message routing mechanism presented in the last subsection. Among of these, two of them are customized and particularly fit for the high security requirement and the group sharing requirement. They are briefly presented below.

Codec 1: Disposable-Key-Based Coding Method (DKCM). The DKCM is based on the disposable-feature. For avoiding explicitly exploring the coded

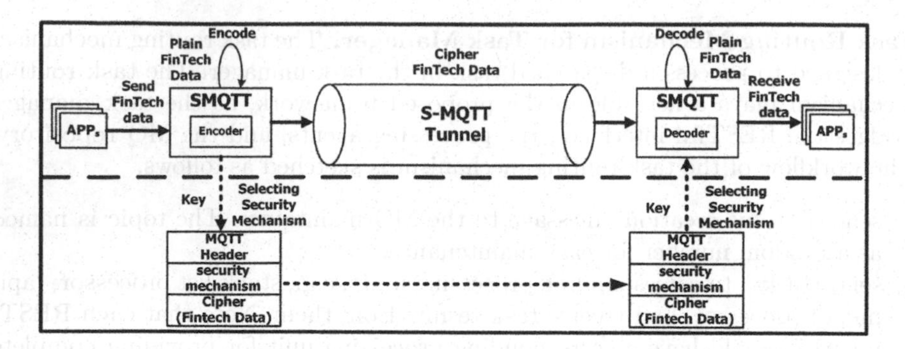

Fig. 5. Illustration of the transparently-secure message routing mechanism.

message, the DKCM would internally generate a disposable key (for one-time use only) for further protecting the coded message. The DKCM is particularly suitable for increasing security level to financial data as many stock records are similar. The technical details of the DKCM can be referred to [8].

Codec 2: CP-ABE Method. The CP-ABE [9] is a popular group sharing security mechanism recently. We consider a requirement that the financial client applications would share the refined stock information to a group of users, for promoting their stock business. In this situation, CP-ABE is a proper choice to achieve this purpose in our MQ-based data delivery repository. We have successfully implemented a solution for embedding CP-ABE [7] to the MQTT [10], which is the fundamental technique used in this framework.

4 Experimental Results

We develop a prototype of the proposed scheme on the personal computer with Intel E5-2660 v2 processor and 128 GB memory. The system prototype is developed with programming tools, including Python 3.5, GCC 7.5.0. The testing dataset in the database is the stock data from the Taiwan Capitalization Weighted Stock Index (TAIEX), which covers the listed stocks traded on the Taiwan Stock Exchange Corporation since 1967. The financial data contains 10,000,000 records over 2000 companies, whose data size is around 10 GB.

4.1 Scalability Test of the Proposed SaS Framework

This experiment is designed to test scalability of the proposed SaS framework, compared with existing HTTPS protocol, and Fig. 6 shows the experimental results. The horizontal axis is the request load represented by the number of requests per second, and the vertical axis is the success ratio, which is defined as the number of successful executed requests over the number of all requests. From the results, we can find that SaS framework has a higher success rate than the traditional HTTPS method. Even under the high-load conditions, SaS framework completely handles a large number of financial data retrieval requests, while the successful request ratio of the traditional HTTP-based method significantly drops after the number of requests per second reaches 7000 times.

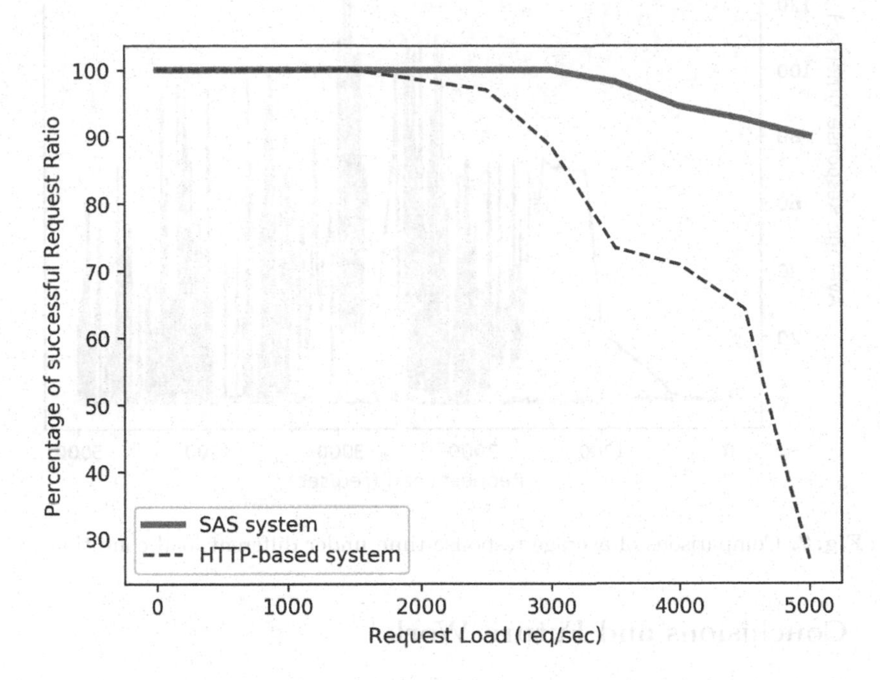

Fig. 6. Comparisons of the successful request ratio under different load conditions.

Figure 7 shows the average response time of different data retrieval schemes under various request loads. This experiment studies how much delay our Send-and-Subscribe framework would incur. Intuitively, one may think the SaS incurs long delay as the SaS framework adopts two communication protocols (HTTP and message queue) and certain mechanisms need to work collaboratively. From to the results, SaS framework has a lower request response time than traditional methods under high load conditions. This is because the request response time of traditional method fluctuates greatly under high load conditions. The unstable data transmission process takes more execution time and increases the failure rate. On the contrary, the SaS framework only provides clients an MQTT data access tunnel. Hence, clients can obtain the requested data stably via MQTT channels, and greatly reduces the failure rate of data access requests.

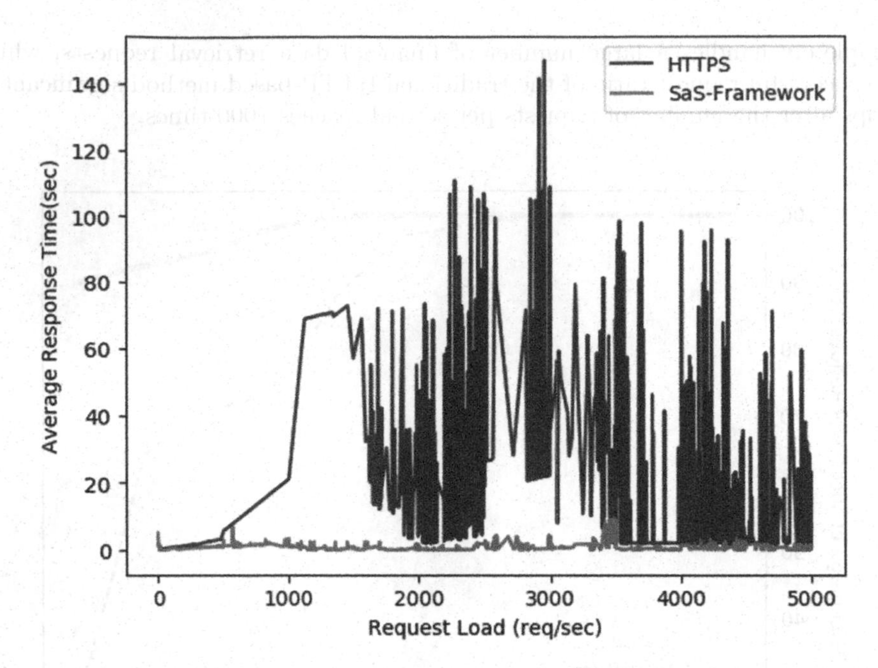

Fig. 7. Comparisons of average response time under different load conditions.

5 Conclusions and Future Work

In this paper, we proposed the Sas framework, which can provide transparently-security, scalability, and long queries to create FinTech applications. Our interfaces are provided in the popular RESTful style, and various mechanisms for security, scalability, and long queries, are embedded in the backend data repository server. We also provided an example of data retrieval protocol to access our proposed framework. We conduct experiments to have integrated performance tests. Comparing to existing solutions, the testing results show our SaS framework indeed satisfies security, scalability, and long-query requirements. An extension of this work is to apply the proposed framework to the large-scale distributed databases, e.g., HBase, for obtaining advantages of big data processing on handling more financial data from various markets.

References

1. Trelewicz, J.Q.: Big data and big money: the role of data in the financial sector. IT Prof. **19**(3), 8–10 (2017)
2. Gai, K., Qiu, M., Sun, X.: A survey on FinTech. J. Netw. Comput. Appl. **103**, 262–273 (2018)
3. Gomber, P., Kauffman, R.J., Parker, C., Weber, B.W.: On the Fintech revolution: interpreting the forces of innovation, disruption, and transformation in financial services. J. Manag. Inf. Syst. **35**, 220–265 (2018)

4. Gomber, P., Koch, J., Siering, M.: Digital finance and FinTech: current research and future research directions. J. Bus. Econ. **87**, 537–580 (2017)
5. He, D., et al.: Fintech and Financial Services: Initial Considerations (2017)
6. Gimpel, H., Rau, D., Röglinger, M.: Understanding fintech start-ups - a taxonomy of consumer-oriented service offerings. Electron. Mark. **28**, 245–264 (2018)
7. Su, W., Chen, W., Chen, C.: An extensible and transparent thing-to-thing security enhancement for MQTT protocol in IoT environment. In: Global IoT Summit (GIoTS), pp. 1–4 (2019)
8. Chen, C.-C., Hung, M.-H., Lin, Y.-C.: Secure data exchange mechanism in hybrid cloud-based distributed manufacturing systems. In: IEEE International Conference on Automation Science and Engineering (CASE 2016) (2016)
9. Bethencourt, J., Sahai, A., Waters, B.: Ciphertext-policy attribute-based encryption. In: 2007 IEEE Symposium on Security and Privacy. IEEE, May 2007
10. OASIS. MQTT version 3.1.1 OASIS standard. http://docs.oasis-open.org/mqtt/mqtt/v3.1.1/os/mqtt-v3.1.1-os.html

Comparison of Machine Learning Methods for Life Trajectory Analysis in Demography

Anna Muratova[1(✉)], Ekaterina Mitrofanova[1], and Robiul Islam[1,2]

[1] National Research University Higher School of Economics, Moscow, Russia
{amuratova,emitrofanova}@hse.ru, r_islam@ieee.org
[2] Innopolis University, Innopolis, Russia

Abstract. Nowadays there are representative volumes of demographic data which are the sources for extraction of demographic sequences that can be further analysed and interpreted by domain experts. Since traditional statistical methods cannot face the emerging needs of demography, we used modern methods of pattern mining and machine learning to achieve better results. In particular, our collaborators, the demographers, are interested in two main problems: prediction of the next event in a personal life trajectory and finding interesting patterns in terms of demographic events for the gender feature. The main goal of this paper is to compare different methods by accuracy for these tasks. We have considered interpretable methods such as decision trees and semi- and non-interpretable methods, such as the SVM method with custom kernels and neural networks. The best accuracy results are obtained with a two-channel convolutional neural network. All the acquired results and the found patterns are passed to the demographers for further investigation.

Keywords: Sequence mining · Demographics · Decision trees · SVM · Neural networks · Classification

1 Introduction

Nowadays researchers from different countries have access to a large amount of demographic data, which usually consists of important events, their sequences, and also different features of people, for example, gender, generation, and location, among others. Demographers investigate the relationship between events and identify frequently occurring sequences of events in the life trajectories of people [19]. This helps researchers to understand how the demographic behaviour of different generations in different countries has changed and also allows researchers to track changes in how people prioritise family and work, and compare the stages of growing up of men and women [1,2,10].

There are different works in this sphere. For example, in [3] authors used decision trees to find patterns that discern the demographic behaviour of people

N. T. Nguyen et al. (Eds.): ACIIDS 2021, LNAI 12672, pp. 630–642, 2021.
https://doi.org/10.1007/978-3-030-73280-6_50

in Italy and Austria, their pathways to adulthood. In [4], the authors discover association rules to detect frequent patterns that show significant variance over different birth cohorts.

Commonly, demographers rely on statistics, but it has some limitations and does not allow for sophisticated sequence analysis. The study of demographic sequences using data mining methods allows for extracting more information, as well as identifying and interpreting interesting dependencies in the data.

Demographers are interested in two main tasks. The first one is the prediction of the next event in personal life trajectories, based on the previous events in their life and different features (for example, gender, generation, location etc.) [20]. The second task is to find the dependence of events on the gender feature, that is, whether the behaviour of men differs from that of women in terms of events and other features [11]. Let us call this task gender prediction. So, the former problem resembles such fundamental problems in machine learning like the next symbol prediction in an input sequence [23], while the latter can be easily recast as a supervised pattern mining problem [27].

The main goal of this paper is to compare different methods, both interpretable and not-interpretable by accuracy. Interpretable methods are good for further interpreting and working with results. Not-interpretable methods allow us to find how accurate the prediction is and to find the best method for these types of data. Among our interpretable methods are decision trees with different event encoding schemes (binary, pairwise, time encoding, and different combinations of these encodings). Among our non-interpretable methods are special kernel variants in the SVM method (ACS, LCS, CP without discontinuities) and neural networks (SimpleRNN, LSTM [13], GRU [7], Convolutional). For all of the methods, we used our scripts and modern machine learning libraries in Python.

As a result of the work, we obtained patterns that are of interest to demographers for further study and interpretation. The best method by accuracy was also found, which is also important for event prediction. Among all of the considered methods, the best method in terms of accuracy is a two-channel Convolutional Neural Network (CNN). The previous results, mainly devoted to pattern mining and rule-based techniques, can be found in our works [11,12,14,16]. Since the considered dataset is unique and was extensively studied only by demographers in a rather descriptive manner, we compare our results on the level of machine techniques' performance and provide the demographers with interesting behavioral patterns and further suggestions on the methods' applicability.

The paper is organised as follows. In Sect. 2, we describe our demographic data. Section 3 contains results obtained with decision trees for the prediction of the next event in a personal life trajectory (or an individual life course) and events that distinguish men and women. Section 4 presents the results using special kernel variants in the SVM method, and Sect. 5 is devoted to the Neural Networks Results (SimpleRNN, LSTM, GRU, CNN). In Sect. 6, a comparison of different classification methods is made, and Sect. 7 concludes the paper.

2 Data Description

The data for the work was obtained from the Research and Educational Group for Fertility, Family Formation, and Dissolution. We used the three-wave panel of the Russian part of the Generations and Gender Survey (GGS), which took place in 2004, 2007, and 2011. After cleaning and balancing the data, it contained the results of 6,626 respondents (3,314 men and 3,312 women). In the database, the dates of birth and the dates of first significant events in respondents' live courses are indicated, such as completion of education, first work, separation from parents, partnership, marriage, the breakup of the first partnership, divorce after the first marriage and birth of a child. We also indicated different personal sociodemographic characteristics for each respondent, such as type of education (general, higher, professional), location (city, town, village), religion (religious person or not), frequency of church attendance (once a week, several times per week, minimum once a month, several times in a year, or never), generation (1930–1969 or 1970–1986), and gender (male or female).

A small excerpt of sequences of demographic events based on the life trajectories of five people is shown in Table 1. Events are arranged in the order of their occurrence. It can be seen that for the first respondent, the events work and separation from parents happened at the same time. These events are indicated in curly braces. Note that sequence 3 contains no events so far due to possibly young age (likewise, generation) of the respondent.

Table 1. An excerpt of life trajectories from a demographic sequence database.

Sequence ID	Demographic sequence
1	$\langle\{work, separation\}, partner, marriage, child\rangle$
2	$\langle work, separation\rangle$
3	$\langle\rangle$
4	$\langle work, partner\rangle$
5	$\langle separation, work, marriage, child\rangle$

3 Decision Trees

3.1 Typical Patterns that Distinguish Men and Women

Let us find different patterns for men and women (gender prediction task) based on previous events in their lives and other sociodemographic features.

We tested decision trees with different event encodings [3]:

1. Binary encoding (or BE), where value "1" means that event has happened in a personal life trajectory and "0" if the event has not happened yet.
2. Time encoding (or TE) with the age in months of when the event happened.

3. Pairwise encoding (or PE), which consists of event pairs coding to mark the type of mutual dependency. If the first event occurred before the second or the second event did not happen yet, then the pair of events is encoded with the symbol"<", if vice versa, then ">", if the events are simultaneous, then "=" and if none of the events has happened yet, then "n".

In addition to these encodings, we also used their different combinations.

Table 2 presents the results of accuracy for decision trees with different event encodings for the gender prediction task. It can be seen that time-based encoding is better than binary and pairwise. Also adding binary or pairwise encodings to time-based encoding slightly lower the accuracy than just using time-based encoding alone. The best accuracy result is obtained for all of the three encodings together with accuracy 0.692.

Table 2. Classification accuracy of different encoding schemes for gender prediction.

Encoding scheme	Classification accuracy
Binary	0.582
Time-based	0.676
Pairwise	0.590
Binary and time-based	0.665
Binary and pairwise	0.592
Time-based and pairwise	0.674
BE, TE, and PE	**0.692**

Let us consider this decision tree since it gives the best accuracy. Table 3 shows the difference between Russian men and women in demographic and socioeconomic spheres. The higher "speed" of reproductive events occurrence in women's life courses indicates the pressure of the "reproductive clocks" over women. During the Soviet era, women who got their first births after the age of 25 were stigmatised and called "older parturient". Men took more time to obtain all the events because they had such obligatory events as military service which made them delaying some other significant events of life.

Table 3. Patterns from the decision tree for men and women for all generations.

Gender	Rules for event sequences	Confidence
Men	First work after 21, marriage after 29.5	70.2%
	Child after 22.9, marriage in 23.9–29.5	69.3%
	Marriage in 23.3–23.9, child after 24.7	67%
Women	Marriage before 20.6 years	77.1%
	First work in 17.7–21, marriage after 29.5	62.8%
	Child before 22.9, marriage in 23.8–29.5	61.2%

Also, we obtained interesting patterns for men and women based only on different features. The main feature which shows the highest difference in patterns is religion. With the probability of 65.9 % if the person is not religious, it is more likely men. Among women the highest number of religious ones is in 1945–1949 with higher education, the probability is 67.6%. Among men the highest number of religious ones is in 1975–1979 with general education, the probability is 65.2%. Also in 1930–1954—the period which embraces Industrialization and the Second World War when the Soviet government declared atheism policy—there are more religious women, however, in 1980–1984—the period preceding ideologic liberalization in 1988—more religious people are among men.

3.2 Prediction of the Next Event in Personal Life Trajectories

Now let us look at the features and events to predict the next event in an individual life course. As in Sect. 3.1, we consider three types of encoding: binary, time-based, and pairwise, as well as all kinds of their combinations.

Table 4 presents results of accuracy for decision trees with different event encodings for the next event prediction task. It can be seen that the best classification accuracy of 0.878 is obtained with the binary encoding scheme. Also adding a pairwise encoding to time-based encoding as well as adding binary encoding to pairwise encoding slightly improves their own accuracy result. The time-based encoding scheme is the lowest by accuracy.

Table 4. Classification accuracy of different encoding schemes for the next life-course event prediction.

Encoding scheme	Classification accuracy
Binary	**0.878**
Time-based	0.359
Pairwise	0.701
Binary and time-based	0.746
Binary and pairwise	0.844
Time-based and pairwise	0.496
BE, TE, and PE	0.750

Let us consider decision tree with binary encoding, since it gives the best accuracy. In Table 5, several patterns, i.e. classification rules, from this decision tree for all generations, both men and women are presented. Note that events in the rules' premises are not indicated in their real order.

From the table, we can see that people tend to find first work after completion of education (event work after education with probability 98.2% vs. event education before work with probability 90.3%).

Table 5. Patterns from the decision tree for the next event's prediction task.

Premise: path in the tree	Conclusion: leaf	Confidence
Education, separation from parents, child birth	Work	98.9%
Work	Education	98.2%
Education, separation from parents, marriage	Child birth	95.7%
Education, child birth	Separation from parents	93.9%
Education	Work	90.3%
Child birth, education, marriage, partner, separation from parents, work	Breakup	78.1%
Work, education	Separation from parents	76.7%
Work, separation from parents, education, marriage, child birth	Divorce	72.9%

Also based on the features only, we obtained that the main feature which shows the highest difference for the last event in a person's life course is education type (general, higher, professional).

4 Using Customized Kernels in SVM

4.1 Classification by Sequences, Features, and the Weighted Sum of Their Probabilities

Using special kernel functions in the SVM method for sequence classification is discussed in [15]. In paper [16] the authors used the following sequence similarity measures: CP (common prefixes), ACS (all common subsequences), and LCS (longest common subsequence). Since demographers are interested in sequences of events without discontinuities (gaps) we derived new formulas, which are the modifications of the original ones [9]. Sequences of events without gaps preserve the right order in which events happened in a person's life.

Let s and t be given sequences and LCSP be the longest common sequence prefix, then similarity measure common prefixes without discontinuities can be calculated as:

$$sim_{CP}(s,t) = \frac{|LCSP(s,t)|}{\max(|s|,|t|)} \tag{1}$$

Let LCS be the longest common sequence, then similarity measure based on the longest common subsequence without discontinuities is calculated as:

$$sim_{LCS}(s,t) = \frac{|LCS(s,t)|}{\max(|s|,|t|)} \tag{2}$$

If k is the length of common subsequence, $\Phi(s,t,k)$ is the number of common subsequences of s and t without discontinuities of length k, then similarity measure all common subsequences without discontinuities is calculated as:

$$sim_{ACS}(s,t) = \frac{2\Sigma_{k\leq l}\Phi(s,t,k)}{l(l+1)}, \text{ where} \tag{3}$$

$$l = max(|s|,|t|) .$$

Let us consider special kernel variants in the SVM method (Support Vector Machines). We will combine two methods of classification: by sequences using special kernel functions based on sequence similarity measures without discontinuity CP, ACS, and LCS and by features using the SVM method with default parameters (the kernel function is RBF). This can be done using the probabilities of referring to a certain class (let us consider the case with two classes, men and women), calculated by the SVM method.

Having obtained the probability values for each method, we can perform classification based on the weighted sum of the probabilities of the two methods. Since the methods give different classification accuracies, the final probability of assigning an object $\mathbf{x} = (f, s)$ to a class is calculated by the formula:

$$P(class|\mathbf{x}) = \frac{A_s \cdot P_s(class|\mathbf{x}) + A_f \cdot P_f(class|\mathbf{x})}{A_s + A_f} \tag{4}$$

A_s is the accuracy by sequences, A_f is the accuracy by features, P_s is the conditional probability by sequences, and P_f is the conditional probability by features.

That formula takes into account the accuracy of the method for the final probability calculation. The probability calculated by each method will be included in the final result with a coefficient equal to the method accuracy. Note that probabilistic calibration of classifiers and sampling techniques for imbalanced classes are often used in machine learning applications in various domains, for example, in medicine [24].

The results for pattern prediction that distinguish men and women and for the next event prediction are presented in Table 6. We can see that the highest accuracy for the gender prediction is 0.678, which is obtained with kernel function CP using the weighted sum of probabilities. The weighted sum of probabilities

Table 6. Classification by sequences, features, and weighted sum of their probabilities (for gender and next event prediction).

Gender prediction	CP	ACS	LCS
Accuracy of sequence classification (SVM, custom kernel functions: CP, ACS, LCS)	0.648	0.659	0.490
Accuracy of classification by features (SVM default – RBF)	0.615	0.615	0.615
Accuracy of classification by weighted sum of probabilities	**0.678**	0.670	0.612
Next event prediction	CP	ACS	LCS
Accuracy of sequence classification (SVM, custom kernel functions: CP, ACS, LCS)	0.839	**0.911**	0.807
Accuracy of classification by features (SVM default – RBF)	0.367	0.367	0.367

for the case of next event prediction gives lower results due to the small accuracy of classification by features. The best result for this case is obtained with kernel function ACS with the accuracy 0.911.

4.2 Classification by Features and Categorical Encoding of Sequences

Another possible method of classification by sequences is by transforming each sequence to the feature. After that, existing methods of classification by features could be used.

There are 6,626 sequences in our dataset, where 1,228 sequences are unique. We consider the sequence as a feature taking 1,228 different values. Each unique sequence was encoded as an integer. Then scikit-learn SVM module with default parameters was used for classification.

We obtained the accuracy of 0.716 for the patterns that distinguish men and women and the accuracy of 0.775 for the next event prediction.

5 Neural Network Models

We performed classification using neural networks software Keras[1] with Tensorflow as backend. The simulation was performed on the GPU. Recurrent Neural Network (RNN) allows us to reveal regularities in sequences. Three types of recurrent layers were compared in Keras: SimpleRNN, GRU, and LSTM. All types of recurrent layers showed good performance with a little less accuracy for SimpleRNN.

Table 7. Neural networks performance for gender and next event prediction.

Network	Gender prediction			Next event prediction		
	Accuracy	Variance	Time, s	Accuracy	Variance	Time, s
GRU RNN	0.760	0.000090	14.11	0.930	0.000055	11.01
LSTM RNN	0.759	0.000123	14.54	0.930	0.000042	11.57
SimpleRNN	0.755	0.000135	19.81	0.927	0.000049	13.50
Convolutional	0.762	0.000193	11.31	0.931	0.000041	8.17

For the network with recurrent layer, accuracy 0.760 was obtained for the patterns that distinguish men and women, and 0.930 for the next event prediction.

Also, a two-channel model with a convolutional layer was implemented[2]. A 1D convolutional layer was used for sequences and dense layers for features. We

[1] https://keras.io.
[2] https://github.com/anya-m/2CNNSeqDem/.

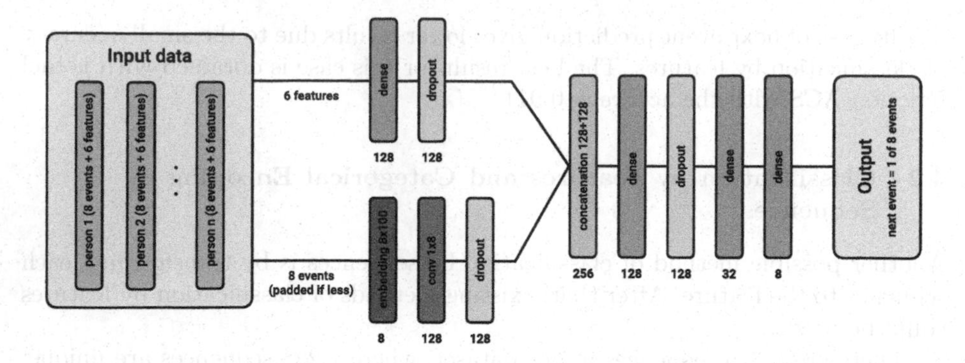

Fig. 1. The two-channel network structure for the next event prediction problem.

obtained the accuracy 0.762 for the patterns that distinguish men and women and the accuracy 0.931 for the next event prediction. We can see that all implemented layers give high accuracy. For the next event prediction, the accuracy is much higher than for the gender prediction in all cases.

Note that we employ 80-to-20 random cross-validation splits with 10 repetitions and report the averaged results.

The structure of the two-channel network layer for the next event prediction is shown in Fig. 1.

6 Comparison of Methods

The accuracies of all the methods for both tasks, gender prediction, and the next event prediction, are presented in Tables 7 and 8.

Table 8. Comparison of the methods.

Method	Gender prediction	Next event prediction
Decision trees	0.692 (BE, TE and PE)	0.878 (BE)
SVM with custom kernel function	0.678 (CP)	0.911 (ACS)
SVM with sequences transformed into features	0.716	0.775
Recurrent neural network	0.760	0.930
Convolutional neural network	**0.762**	**0.931**

From the table, it can be seen that the highest classification accuracy for both cases is obtained with convolutional neural networks, which means that it is an optimal method (among the considered) for these tasks. Also, the accuracy for the prediction of the next event in personal life trajectories is higher in all of the methods than the accuracy for gender prediction.

7 Conclusion

This work contains results of different machine learning methods for the two demographers' tasks in sequence mining, such as prediction of the next event in an individual life course and finding patterns that distinguish men and women. The interpretable machine learning models [21] such as decision trees are suitable for further interpretation by demographers. The best encodings of events for the cases of gender prediction and the next event prediction are different, so there is no universal best choice.

The SVM method with custom kernels and with sequences transformed into features has approximately the same accuracy for the gender prediction problem, however, for the next event prediction, the resulting accuracy is much higher for the custom kernel function ACS. Although, the prefix-based kernel (CP) in combination with feature-based prediction after their weighting by accuracies gave the best accuracy as well, which shows that starting events in the individual life-course may contain important predictive information. The best accuracy results are obtained with the two-channel convolutional neural network for both cases, especially, with the highest accuracy of 0.931 for the next event prediction. Recurrent neural networks also result in high accuracy, but slightly lower than that of CNN.

Among the future research directions we may outline the following ones:

1. What are the main demographic differences between modern and Soviet generations of the Russian population that machine learning and pattern mining algorithms can capture? Answering this question is very important for demographic theory because it either confirms or disproves a predictive potential of the current ideas about the stadiality of demographic modernisation.
2. Which of the proposed methods so far suits the best the demographer's needs? For example, prefix-based emerging sequential patterns without gaps[3] in terms of pattern structures [5] are good candidates for studies of the transition to adulthood [11]. Other methods and combinations of existing ones appear, which could be of interest for both data scientists and demographers [17,19].
3. Comparative studies of modern Russian and European generations are useful to prove or deny a hypothesis that Russia still follows a different demographic trajectory than European countries due to its Soviet past, for example, in contrast to Western vs. Eastern Germany [25]. One of the plausible hypotheses is that there exists a lag in about 20–25 years between Russia and European countries [18], in terms of demographic behavior patterns.

[3] For emerging patterns in classification setting cf. [8].

4. If like in our studies, neural networks result in high accuracy in different demographic classification problems, they need direct incorporation of interpretable techniques on the level of single events or their itemsets like Shapley value based approaches [6], which are mainly used for separate features on the level of single examples.
5. Further studies of similarity measures [9,22] is needed as well as that of the interplay between the complexity of sequences (cf. turbulence measure in [10]) and their interpetability [11].

Another promising direction, which is often implicitly present in real data science projects but remains unattended in sequence mining research, is outlier detection [26]. For example, in our previous pattern mining studies, we found the following emerging sequence peculiar for men,

$$\langle \{work\}, \{education\}, \{marriage, partner\}, \{divorce, break\text{-}up\} \rangle,$$

but the events divorce and break-up would rarely happen within one month (the used time granule) and they require different preceding events, namely marriage and partnership, which also cannot happen simultaneously. Thus, together with the involved demographers, we realised that there is a misconception of the survey's participants how they treat the terms marriage and partnership (they are not equal); further, we have eliminated the issue by employing an extra loop in the data processing via checking concrete dates and marital statuses.

Acknowledgment. The authors would like to thank Prof. G. Dong for his interest in our previous work on prefix-based emerging sequential patterns.

The study was implemented in the framework of the Basic Research Program at the National Research University Higher School of Economics and funded by the Russian Academic Excellence Project '5-100'. This research is also supported by the Faculty of Social Sciences, National Research University Higher School of Economics.

References

1. Aisenbrey, S., Fasang, A.E.: New life for old ideas: the "second wave" of sequence analysis bringing the "course" back into the life course. Soc. Meth. Res. **38**(3), 420–462 (2010). https://doi.org/10.1177/0049124109357532
2. Billari, F.C.: Sequence analysis in demographic research. Can. Stud. Popul. [Arch.] **28**, 439–458 (2001)
3. Billari, F.C., Fürnkranz, J., Prskawetz, A.: Timing, sequencing, and quantum of life course events: a machine learning approach. Eur. J. Popul. (Revue européenne de Démographie) **22**(1), 37–65 (2006). https://doi.org/10.1007/s10680-005-5549-0
4. Blockeel, H., Fürnkranz, J., Prskawetz, A., Billari, F.C.: Detecting temporal change in event sequences: an application to demographic data. In: De Raedt, L., Siebes, A. (eds.) PKDD 2001. LNCS (LNAI), vol. 2168, pp. 29–41. Springer, Heidelberg (2001). https://doi.org/10.1007/3-540-44794-6_3
5. Buzmakov, A., Egho, E., Jay, N., Kuznetsov, S.O., Napoli, A., Raïssi, C.: On mining complex sequential data by means of FCA and pattern structures. Int. J. Gen Syst **45**(2), 135–159 (2016). https://doi.org/10.1080/03081079.2015.1072925

6. Caruana, R., Lundberg, S., Ribeiro, M.T., Nori, H., Jenkins, S.: Intelligible and explainable machine learning: best practices and practical challenges. In: Gupta, R., Liu, Y., Tang, J., Prakash, B.A. (eds.) The 26th ACM SIGKDD Conference on Knowledge Discovery and Data Mining, KDD 2020, pp. 3511–3512. ACM (2020). https://dl.acm.org/doi/10.1145/3394486.3406707
7. Cho, K., et al.: Learning phrase representations using RNN encoder-decoder for statistical machine translation. In: Moschitti, A., Pang, B., Daelemans, W. (eds.) Proceedings of the 2014 Conference on Empirical Methods in Natural Language Processing, EMNLP 2014, pp. 1724–1734. ACL (2014). https://doi.org/10.3115/v1/d14-1179
8. Dong, G., Li, J.: Emerging pattern based classification. In: Liu, L., Özsu, M.T. (eds.) Encyclopedia of Database Systems, 2nd edn. Springer, Boston (2018). https://doi.org/10.1007/978-1-4614-8265-9_5002
9. Egho, E., Raïssi, C., Calders, T., Jay, N., Napoli, A.: On measuring similarity for sequences of itemsets. Data Min. Knowl. Discov. **29**(3), 732–764 (2015). https://doi.org/10.1007/s10618-014-0362-1
10. Elzinga, C.H., Liefbroer, A.C.: De-standardization of family-life trajectories of young adults: a cross-national comparison using sequence analysis. Eur. J. Popul. (Revue européenne de Démographie) **23**(3), 225–250 (2007). https://doi.org/10.1007/s10680-007-9133-7
11. Gizdatullin, D., Baixeries, J., Ignatov, D.I., Mitrofanova, E., Muratova, A., Espy, T.H.: Learning interpretable prefix-based patterns from demographic sequences. In: Strijov, V.V., Ignatov, D.I., Vorontsov, K.V. (eds.) IDP 2016. CCIS, vol. 794, pp. 74–91. Springer, Cham (2019). https://doi.org/10.1007/978-3-030-35400-8_6
12. Gizdatullin, D., Ignatov, D., Mitrofanova, E., Muratova, A.: Classification of demographic sequences based on pattern structures and emerging patterns. In: Supplementary Proceedings of 14th International Conference on Formal Concept Analysis, ICFCA, pp. 49–66 (2017)
13. Hochreiter, S., Schmidhuber, J.: LSTM can solve hard long time lag problems. In: Mozer, M., Jordan, M.I., Petsche, T. (eds.) Advances in Neural Information Processing Systems (NIPS), Denver, CO, USA, 2–5 December, vol. 9, pp. 473–479. MIT Press (1996)
14. Ignatov, D.I., Mitrofanova, E., Muratova, A., Gizdatullin, D.: Pattern mining and machine learning for demographic sequences. In: Klinov, P., Mouromtsev, D. (eds.) KESW 2015. CCIS, vol. 518, pp. 225–239. Springer, Cham (2015). https://doi.org/10.1007/978-3-319-24543-0_17
15. Lodhi, H., Saunders, C., Shawe-Taylor, J., Cristianini, N.: Watkins, C: Text classification using string kernels. J. Mach. Learn. Res. **2**, 419–444 (2002). http://jmlr.org/papers/v2/lodhi02a.html
16. Muratova, A., Sushko, P., Espy, T.H.: Black-box classification techniques for demographic sequences: from customised SVM to RNN. In: Tagiew, R., Ignatov, D.I., Hilbert, A., Heinrich, K., Delhibabu, R. (eds.) Proceedings of the 4th Workshop on Experimental Economics and Machine Learning, EEML 2017, Dresden, Germany, 17–18 September 2017, pp. 31–40. CEUR Workshop Proceedings, Aachen (2017). http://ceur-ws.org/Vol-1968/paper4.pdf
17. Piccarreta, R., Studer, M.: Holistic analysis of the life course: methodological challenges and new perspectives. Adv. Life Course Res. (2019). https://doi.org/10.1016/j.alcr.2018.10.004
18. Puur, A., Rahnu, L., Maslauskaite, A., Stankuniene, V., Zakharov, S.: Transformation of partnership formation in eastern Europe: the legacy of the past demographic

divide. J. Comp. Fam. Stud. **43**, 389–417 (2012). https://doi.org/10.3138/jcfs.43.3.389
19. Ritschard, G., Studer, M.: Sequence analysis: where are we, where are we going? In: Ritschard, G., Studer, M. (eds.) Sequence Analysis and Related Approaches. LCRSP, vol. 10, pp. 1–11. Springer, Cham (2018). https://doi.org/10.1007/978-3-319-95420-2_1
20. Rossignon, F., Studer, M., Gauthier, J.-A., Goff, J.-M.L.: Sequence history analysis (SHA): estimating the effect of past trajectories on an upcoming event. In: Ritschard, G., Studer, M. (eds.) Sequence Analysis and Related Approaches. LCRSP, vol. 10, pp. 83–100. Springer, Cham (2018). https://doi.org/10.1007/978-3-319-95420-2_6
21. Rudin, C.: Stop explaining black box machine learning models for high stakes decisions and use interpretable models instead. Nat. Mach. Intell. **1**(5), 206–215 (2019). https://doi.org/10.1038/s42256-019-0048-x
22. Rysavý, P., Zelezný, F.: Estimating sequence similarity from read sets for clustering next-generation sequencing data. Data Min. Knowl. Discov. **33**(1), 1–23 (2019). https://doi.org/10.1007/s10618-018-0584-8
23. Solomonoff, R.J.: The Kolmogorov lecture the universal distribution and machine learning. Comput. J. **46**(6), 598–601 (2003). https://doi.org/10.1093/comjnl/46.6.598
24. Tomczak, J.M., Zieba, M.: Probabilistic combination of classification rules and its application to medical diagnosis. Mach. Learn. **101**(1–3), 105–135 (2015). https://doi.org/10.1007/s10994-015-5508-x
25. Wahrendorf, M., et al.: Agreement of self-reported and administrative data on employment histories in a German cohort study: a sequence analysis. Eur. J. Popul. (2019). https://doi.org/10.1007/s10680-018-9476-2
26. Wang, T., Duan, L., Dong, G., Bao, Z.: Efficient mining of outlying sequence patterns for analyzing outlierness of sequence data. ACM Trans. Knowl. Discov. Data **14**(5), 62:1–62:26 (2020). https://doi.org/10.1145/3399671
27. Zimmermann, A., Nijssen, S.: Supervised pattern mining and applications to classification. In: Aggarwal, C.C., Han, J. (eds.) Frequent Pattern Mining, pp. 425–442. Springer, Cham (2014). https://doi.org/10.1007/978-3-319-07821-2_17

An Efficient Method for Multi-request Route Search

Eric Hsueh-Chan Lu(✉) and Sin-Sian Syu

Department of Geomatics, National Cheng Kung University, No. 1, University Road, Tainan 701, Taiwan (R.O.C.)
luhc@mail.ncku.edu.tw

Abstract. Location-based services are ubiquitous in our daily lives, which make our lives more convenient. One of the popular topics is the route planning for Point of Interest in the urban environments. In this article, we study on Multi-Request Route Planning. The previous research only provided an approximate solution to this problem. In this work, we investigate the possibility of finding the best route within a reasonable time and propose the One-way Search algorithm. This method expands the node in one direction in one search, and each search will reduce the candidate nodes for the next search. In the experiment, we compare our method with the Anytime Potential Search/Anytime Non-Parametric A* algorithm, which was proposed in recent years and is also an optimal solution. The evaluation is based on two real-world datasets, and the results show that our method outperforms the competition method in terms of execution time and memory usage.

Keywords: Multi-Request Route Planning (MRRP) · Shortest path problem · Anytime algorithm · One-way search algorithm · APTS/ANA*

1 Introduction

With the rapid development of wireless communication technology, Global Positioning System (GPS), and the popularization of smart mobile devices, research related to Location-Based Service (LBS) has received increasing attention. One of the popular topics is the route planning for Point of Interest (POI) in urban environments. State-of-the-art web map services such as Google Maps, Bing Maps, and OpenStreetMap (OSM) have provided route planning services, i.e., user specifies a starting point and destination, or several stops they want to pass, the route planning service will give a *good* route in terms of distance or time. In the past decade, new types of queries have been proposed, i.e., considering the type or category of POIs that users require, rather than specific locations [6, 14]. However, in real life, a POI may provide a variety of services that meet the needs of users, and route planning considering this feature has been proposed in recent years, namely *Multi-Request Route Planning* (*MRRP*) [9]. Nevertheless, regardless of the kind of route planning query, the user is mainly concerned with the accuracy and immediacy of the query, i.e., the travel cost of the route is as low as possible, and the response time of the service cannot be too long. In terms of service providers, in addition

N. T. Nguyen et al. (Eds.): ACIIDS 2021, LNAI 12672, pp. 643–652, 2021.
https://doi.org/10.1007/978-3-030-73280-6_51

to having complete POI data and road network information to give higher quality results, and because of the large number of queries, the computational cost cannot be too high. These are the main challenges in the study of route planning.

The following is a scenario for route planning. Suppose an office worker has worked hard for a day and is preparing to get off work. Before he goes home, he needs to post a business letter, withdraw some money, and buy some bread to reward himself. Therefore, this problem is to find a route that can satisfy all the needs of the user from the starting point to the destination, and the length of the route is as short as possible. For methods that consider POI as a type, the service may choose a post box, an ATM, and a bakery to form a route; however, considering that POI provides a variety of services, it will find that the post office can provide both mailing and withdrawal services, and therefore can simultaneously address the needs of the user, and this may be a shorter route but the previous methods are not aware of it.

This paper aims to solve *MRRP* query. *MRRP* is a generalization of the *Travel Salesman Problem (TSP)* [11], if all POIs provide only one service, and each POI provides different services. In addition, because *TSP* is NP-hard, *MRRP* is also NP-hard, which means that with the number of requests increases, the search time grows exponentially. However, as described above, the route planning service requires the immediacy of the query, because the user can't stand the response time of the service for too long, therefore only the approximate solutions are provided in the *MRRP* study. In this study, we propose the One-way search algorithm to answer *MRRP* query. This method is an anytime algorithm that combines branch and bound and greedy algorithms, so it can quickly return a route that satisfies all the needs of the user. Each search will update the bound information to reduce the candidate POI for the next search, and if there are no candidates in an iteration, it means that the optimal solution has been found. In the experiment, we evaluated our approach using two real-world datasets of different scales. The results show that our method can return the optimal solution of 90% ratio in the first second in both datasets, and if enough time is given, the shortest route can be obtained.

The remainder of this paper is organized as follows. In Sect. 2, we first briefly introduce the related studies on route planning. Next, we explain the detailed methodology in Sect. 3. In Sect. 4, we show the experimental evaluations based on two real-world datasets. Finally, we summarize this work and provide some possible future work in Sect. 5.

2 Related Work

The shortest path problem is a classic problem in graph theory, which aims to find the shortest path of two nodes in the graph. Dijkstra [2] and A* [4] are well-known search algorithms used to solve this problem, where A* is widely used due to its performance and accuracy. A* is a best-first search that expands the path with the lowest cost node by using the priority queue. The difference between A* and Dijkstra is that the former's cost function adds a heuristic function $h(n)$, which is the estimated cost from the current node to the goal node. A* search has several variants that improve memory usage or speed up search time, such as IDA* [5], SMA* [13], ARA* [7], and APTS/ANA* [16], where APTS/ANA* automatically adjusts the weight value of the heuristic function $h(n)$

for each search by the mechanism of potential [17]. Additionally, APTS/ANA* is our comparison method in this paper, because both are anytime algorithms that can find the optimal solution, namely the shortest path.

User-demand-oriented route planning has been discussed in the last decade, such queries will first determine the appropriate locations based on the user's needs or constraints, and then arrange the proper order of visits as the route planning result. The route planning which considers POI as a type can be traced back to the advanced research in 2005. *Trip Planning Query (TPQ)* [6] was proposed by Li *et al.*, in this problem, each POI is assigned a specific category, and the goal is to find a short route that passes through all the specified categories from the start point to the destination. *Optimal Sequenced Route (OSR)* query [14] is like *TPQ*, but the order of visits to categories is fixed. In *OSR*, they propose a method that progressively filtering out POIs that are unlikely to be the best route and picking unfiltered POIs from the destination to the starting point to construct the shortest route. There are several variants of route planning in [1, 3, 8, 10, 12, 15]. However, previous studies designed methods on consideration of POI as a type; on the other hand, Lu *et al.* proposed the concept of multiple requests, i.e., considering a POI that can simultaneously satisfy multiple requirements, and proposed *MRRP* [9]. In the paper, they designed the approximation algorithms to answer the *MRRP* query and proved that the proposed algorithm is better than the existing algorithms that do not consider POI to solve multiple requests.

3 Proposed Method

One-way search is an anytime algorithm that searches the route in a greedy way, i.e., in each search, it expands a POI that can solve at least one request which is unsolved in the current path, thus an available solution can be quickly generated; in addition, One-way search is also a branch-and-bound method. It restarts the search multiple times and updates the bound values of nodes, which reduces the candidate nodes for the next search. When there is no candidate node at the starting point, it means that the last found route is the optimal solution, and the algorithm terminates. The following is the description of the One-way search algorithm.

3.1 Data Pre-processing

This section introduces the fundamentals of the One-way search algorithm, including the structure of the graph used in our method, and the bound properties of the nodes of the graph.

1. Node Equivalence: This term describes the graph used for our algorithm and explains the node equivalence. A node records a POI reference and the requests that have been solved so far, and we say that the two nodes are equivalent if and only if the two attributes above are the same. At the beginning we are at the starting point s, then connects all POIs that can serve unsolved requests. The following nodes explore in this way. Finally, if the node has solved all requests, then directly connect to the destination e. It must be noted that all the nodes in the graph are not equivalent.

Equivalent nodes should have the same information, it is useless to store more than one equivalent node in the program, which will result in data redundancy and memory waste.
2. Forward & Backward: Each node records the bound information, i.e., the shortest length from the starting point s to the node as well as the shortest length from the node to the destination e, and is denoted as Forward and Backward, respectively. The formal calculations for Forward and Backward are Eqs. (1) and (2). In Eq. (1), the forward of node v_i is the forward value of an in-neighborhood v_j plus the distance between v_i and v_j, the smallest of all. Equation (2) is similar except that v_j is replaced by out-neighborhoods. It should be noted that this term only states that the minimum Forward and Backward are recorded on the node. In practice, the bound information of the node is updated as the search progresses.

$$forward(v_i) = min(v_j.forward + d(v_i, v_j)), \; where \; \forall v_j \in N^-(v_i) \tag{1}$$

$$backward(v_i) = min(v_j.backward + d(v_i, v_j)), \; where \; \forall v_j \in N^+(v_i) \tag{2}$$

3.2 Pruning and Wilting

In the previous section, we already know the structure of the graph used in our method, and the bound information for each node. This section will continue to introduce the *pruning* and *wilting* mechanisms based on these fundamentals. The purpose of pruning is to reduce the search space by eliminating candidates that can prove will not be an optimal solution. There are three pruning strategies for the One-way search algorithm, namely *filter*, *potential* and *petrifaction*. When a candidate is eliminated by any of the three pruning, a wilting value is generated, which is used to reduce the bound value of the node, more specifically, when the current node has no candidate nodes to expand the path, the *Forward* value of the current node will be assigned the maximum value of all wilting values. The following describes three kinds of pruning and their wilting calculations in detail.

1. Filter: The first one is filter, which is the most intuitive of all. As mentioned in the previous section, the Forward of a node records the shortest length from the starting point s to the node so far. If the length from the current node v_{curr} to its out-neighborhood v_{next} is greater than the Forward value of v_{next}, it means there is another path to v_{next} that is shorter, so we don't consider v_{next} to be a candidate for v_{curr}, because the shortest length of v_{next} to destination e is fixed. If we choose a longer path from starting point s to v_{next}, it will not be the shortest route. Equation (3) is the wilting rule of filter, and the wilting value is calculated as the minimum value among v_{next}'s Forward and minLength, minus the distance from v_{curr} to v_{next}, which means that if v_{curr} is considered as a candidate next time, the length to v_{curr} must be less than this bound value, where minLength is the shortest length of the valid route found so far, and the path in the search should not exceed this value.

$$wilt_{FT}(v_{curr}, v_{next}) = min(v_{next}.forward, minLength) - d(v_{curr}, v_{next}),$$
$$where\ \forall v_{next} \in N^{+}(v_{curr}), when\ v_{curr}.forward + d(v_{curr}, v_{next}) > min(v_{next}.forward, minLength) \quad (3)$$

2. Potential: The out-neighborhood v_{next} of the node v_{curr} does not necessarily have the potential to be the shortest route, even if v_{next} is not eliminated by the filter mechanism. If the length of the current path to v_{next} plus the distance of v_{next} directly to the destination e is greater than or equal to the route-length $L(v)$ recorded on v_{curr}, i.e., $v_{curr}.forward + d(v_{curr}, v_{next}) + d(v_{next}, e) \geq L(v_{curr})$, then v_{next} has no potential to be the shortest route. The wilting rule of potential as shown in Eq. (4).
$$wilt_{PT}(v_{curr}, v_{next}) = min(L(v_{curr}), minLength) - d(v_{curr}, v_{next}) - d(v_{next}, e),$$
$$where\ \forall v_{next} \in N^{+}(v_{curr}), when\ v_{curr}.forward + d(v_{curr}, v_{next}) + d(v_{next}, e) \geq min(L(v_{curr}), minLength) \quad (4)$$
3. Petrifaction: Petrifaction indicates that the Backward value of a node will not change afterward. In this section, we first explain the mechanism of petrifaction, and then explain how to use this characteristic to pruning. Petrifaction occurs in the reverse update of the search. For the use of pruning, a petrifaction node is equivalent to a destination. When a path arrives at a petrifaction node, we can directly take the Backward value of the node as the length of the remaining route and do not need to consider it as a candidate. Equation (5) is the wilting rule of petrifaction.
$$wilt_{PF}(v_{curr}, v_{next}) = min(v_{next}.forward, minLength) - d(v_{curr}, v_{next}),$$
$$where\ \forall v_{next} \in N^{+}(v_{curr}), when\ v_{next} \in Petrifaction \quad (5)$$

4 Experimental Evaluation

This section first describes the POI data sets used in the experiment, followed by comparison with other methods. The results are the average of 100 route queries, and for convenience, the destination of each query is the starting point and is randomly located in the dataset area. All experiments are implemented in Java SE 8 on a machine with CPU 3.40 GHz with 32 GB of memory running Linux Ubuntu 16.04.

4.1 Data Collection

We use the Google Places API [18] to collect two real-world POI datasets, namely Germany-Oldenburg and Taiwan-Tainan, as small and large datasets in the experiment. The two datasets contain 6,233 and 25,426 POIs (N_P), and 91 and 96 different services (N_S), respectively. Figure 1 shows the distribution of POIs for each service in the two regions. Most of the POIs are concentrated in the top 30% of all services. In the experiment, we randomly select the services as the user-specified requests and based on the number of POIs of the service as the probability of its selection, in line with the principle of fairness. Each POI has its own latitude and longitude, and the distance between two POIs is calculated as the Euclidean distance.

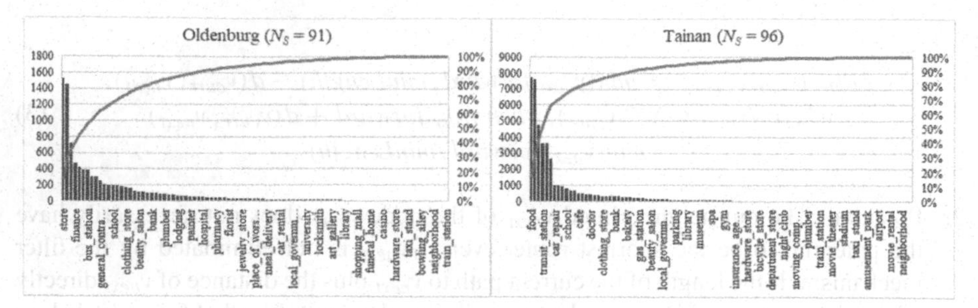

Fig. 1. The distribution of the number of POI for each service

4.2 Method Competition

The experiment is conducted with two scale datasets, namely Oldenburg and Tainan, and compared our method with APTS/ANA* [16]. There are three comparisons, including the impact of the number of requests N_R, the impact of the number of POIs N_P, and the ratio of optimal solution. Except for the experiment of the number of requests N_R, the default number of requests for other experiments is 7.

1. Impact of Number of Requests N_R: The purpose of this experiment is to examine the impact of the number of requests on the algorithm, while the number of requests ranges from 5 to 9. Figure 2 and Fig. 3 show the experimental results. The performance difference between the two methods is small before the number of requests 7. After 7, the calculation time and memory usage required by APTS/ANA* are increased more rapidly than the One-way search. The reason is that the APTS/ANA* search belongs to best-first. When its heuristic function is not effective, it may expand too many unnecessary nodes, which wastes not only time but also memory space. On the other hand, One-way search is a greedy method, which reduces the chance of expanding unnecessary nodes, but instead relies on the efficiency of the bound, which is the advantage of One-way search to best-first search.

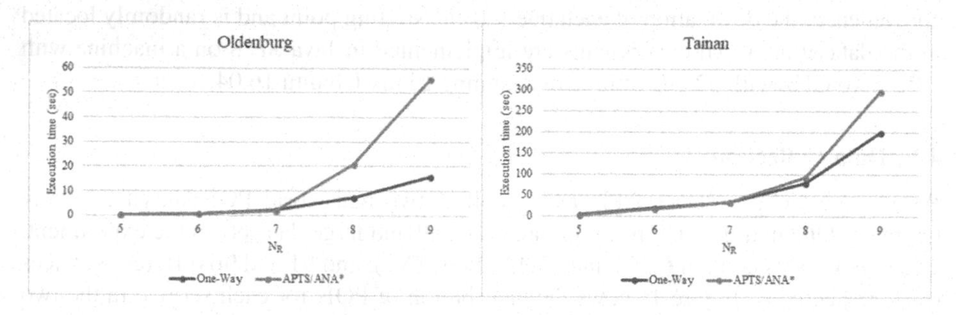

Fig. 2. Impact of number of requests N_R (execution time)

2. Impact of Number of POIs N_P: This experiment explores the impact of the POI scale (N_P) on the algorithm. To control the number of POIs, the POIs in the dataset are

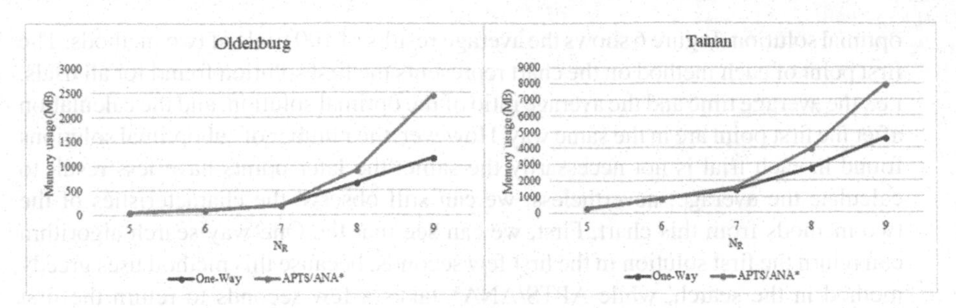

Fig. 3. Impact of number of requests N_R (memory usage)

randomly deleted to the specified number N_P, while the POIs with more services are more likely to be deleted. For the N_P range of the two datasets, the Oldenburg is from 2,000 to 6,000 and the Tainan is from 15,000 to 25,000. Figure 4 and Fig. 5 show the experimental results. Since Oldenburg is a small-scale dataset, the performance of the two methods is not much different. For the Tainan dataset, One-way search is always better than APTS/ANA* in terms of execution time and memory usage.

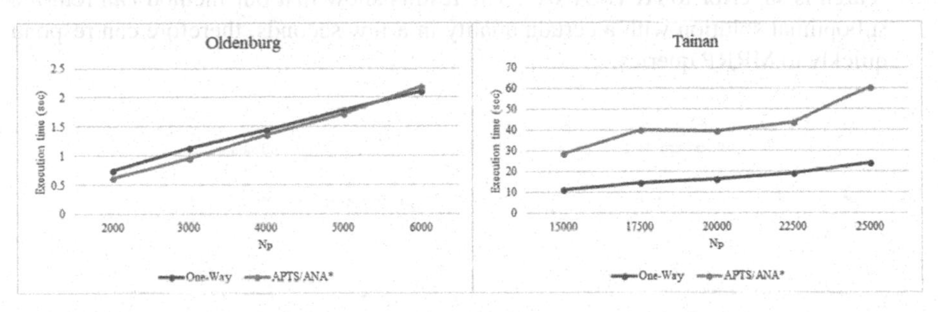

Fig. 4. Impact of number of POIs N_P (execution time)

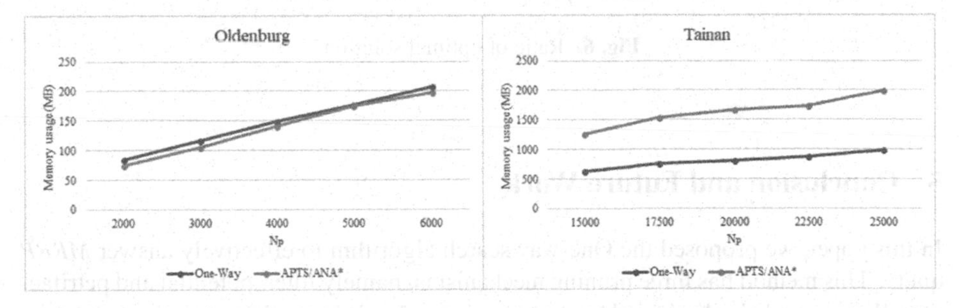

Fig. 5. Impact of number of POIs N_P (memory usage)

3. Ratio of Optimal Solution: Since both methods are anytime algorithms, it means that if there is a suboptimal solution before finding the optimal solution, it can be returned in advance, therefore the final experiment is to observe the ratio of the

optimal solution. Figure 6 shows the average results of 100 trials of two methods. The first point of each method on the chart represents the first solution found for all trials, i.e., the average time and the average ratio of the optimal solution, and the calculation after the first point are in the same way. However, the number of suboptimal solutions found in each trial is not necessarily the same, the later points have less result to calculate the average; nevertheless, we can still observe the characteristics of the two methods from this chart. First, we can see that the One-way search algorithm can return the first solution in the first few seconds, because this method uses greedy method in the search, while APTS/ANA* takes a few seconds to return the first solution. This is related to the characteristics of expanding node for the best-first search, causing APTS/ANA* to return the suboptimal solution slowly. For the One-way search algorithm, it can return 90% of the optimal solution for the two datasets in the first second, then the method will continue to search for the route and finally converge to the optimal solution. However, in the Oldenburg dataset, some trials require a convergence time of about 5 s, while in the Tainan dataset it takes 80 s, even within 10 s, the 99% ratio of the optimal solution has been reached. From this result, although the two methods require more time to find the optimal solution in some trials, they can quickly converge in the search process. Among them, One-way search is superior to APTS/ANA*. The results show that our method can return a suboptimal solution with a certain quality in a few seconds, therefore can respond quickly to MRRP queries.

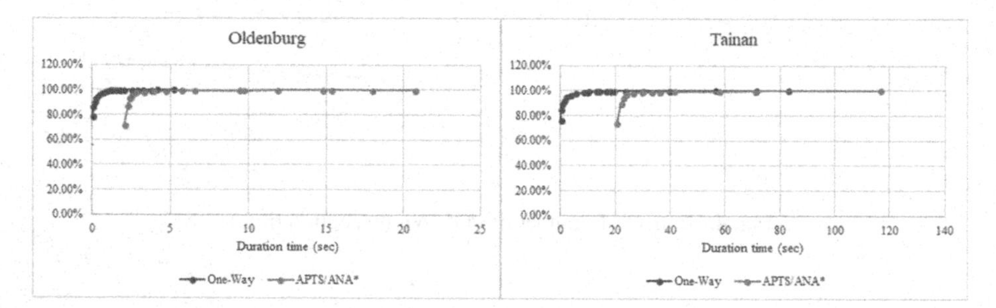

Fig. 6. Ratio of optimal solution

5 Conclusion and Future Work

In this paper, we proposed the One-way search algorithm to effectively answer *MRRP* query. This method has three pruning mechanisms, namely filter, potential and petrifaction. When a node is eliminated by the pruning mechanism, a wilting value is generated, which is used to narrow the bound value of the node to reduce the candidate for the next search. In the experimental part, we evaluated our method with two real-world datasets of different scales and compared it with the APTS/ANA* algorithm, where the comparison items include the impact of the number of requests and the number of POIs. The results show that when the query problem is more complex, our method is superior to the competition method in terms of execution time and memory usage. Since both methods are anytime algorithms, we also experimented with the ratio of optimal solution to observe the convergence efficiency of the two methods. The results show that our method can find the 90% ratio optimal solution in 1 s, while APTS/ANA* is delayed by a few seconds depending on the scale of the problem before the first solution is found. From this experiment, it is also shown that although our method requires tens of seconds to several minutes to find the optimal solution, it has a good performance in terms of convergence efficiency, and therefore can answer *MRRP* query. In the future, we will try to develop other heuristic pruning and experiment with larger real-world data sets to show the effectiveness and robustness of the One-way search algorithm.

Acknowledgement. This research was supported by Ministry of Science and Technology, Taiwan, R.O.C. under grant no. MOST 109-2121-M-006-013-MY2 and MOST 109-2121-M-006-005-.

References

1. Chen, H., Ku, W.S., Sun, M.T., Zimmermann, R.: The multi-rule partial sequenced route query. In: Proceedings of The 16th ACM SIGSPATIAL International Conference on Advances in Geographic Information Systems, p. 10 (2008)
2. Dijkstra, E.W.: A note on two problems in connexion with graphs. Numer. Math. **1**(1), 269–271 (1959). https://doi.org/10.1007/BF01386390
3. Eisner, J., Funke, S.: Sequenced route queries: getting things done on the way back home. In: Proceedings of The ACM 20th International Conference on Advances in Geographic Information Systems, pp. 502–505 (2012)
4. Hart, P.E., Nilsson, N.J., Raphael, B.: A formal basis for the heuristic determination of minimum cost paths. IEEE Trans. Syst. Sci. Cybern. **4**(2), 100–107 (1968)
5. Korf, R.E.: Depth-first iterative-deepening: an optimal admissible tree search. Artif. Intell. **27**(1), 97–109 (1985)
6. Li, F., Cheng, D., Hadjieleftheriou, M., Kollios, G., Teng, S.-H.: On trip planning queries in spatial databases. In: Bauzer Medeiros, C., Egenhofer, M.J., Bertino, E. (eds.) SSTD 2005. LNCS, vol. 3633, pp. 273–290. Springer, Heidelberg (2005). https://doi.org/10.1007/11535331_16
7. Likhachev, M., Gordon, G.J., Thrun, S.: ARA*: anytime A* with provable bounds on sub-optimality. In: Advances in Neural Information Processing Systems, pp. 767–774 (2004)
8. Lu, Q.: A best first search to process single-origin multiple-destination route query in a graph. In: The IEEE 2nd International Conference on Advanced Geographic Information Systems, Applications, and Services, pp. 137–142 (2010)
9. Lu, E.H.-C., Chen, H.S., Tseng, V.S.: An efficient framework for multirequest route planning in urban environments. IEEE Trans. Intell. Transp. Syst. **18**(4), 869–879 (2017)
10. Ma, X., Shekhar, S., Xiong, H., Zhang, P.: Exploiting a page-level upper bound for multi-type nearest neighbor queries. In: Proceedings of The 14th Annual ACM International Symposium on Advances in Geographic Information Systems, pp. 179–186 (2006)
11. Menger, K.: Das botenproblem. Ergebnisse Eines Mathematischen Kolloquiums **2**, 11–12 (1932)
12. Ohsawa, Y., Htoo, H., Sonehara, N., Sakauchi, M.: Sequenced route query in road network distance based on incremental Euclidean restriction. In: Liddle, S.W., Schewe, K.-D., Tjoa, A.M., Zhou, X. (eds.) DEXA 2012. LNCS, vol. 7446, pp. 484–491. Springer, Heidelberg (2012). https://doi.org/10.1007/978-3-642-32600-4_36

13. Russell, S.: Efficient memory-bounded search methods. In: Proceedings of The 10th European Conference on Artificial Intelligence, pp. 1–5 (1992)
14. Sharifzadeh, M., Kolahdouzan, M., Shahabi, C.: The optimal sequenced route query. VLDB J. Int. J. Very Large Data Bases **17**(4), 765–787 (2008). https://doi.org/10.1007/s00778-006-0038-6
15. Sharifzadeh, M., Shahabi, C.: Processing optimal sequenced route queries using Voronoi diagrams. GeoInformatica **12**(4), 411–433 (2008). https://doi.org/10.1007/s10707-007-0034-z
16. Stern, R., Felner, A., van den Berg, J., Puzis, R., Shah, R., Goldberg, K.: Potential-based bounded-cost search and anytime non-parametric A*. Artif. Intell. **214**, 1–25 (2014)
17. Stern, R.T., Puzis, R., Felner, A.: Potential search: a bounded-cost search algorithm. In: Proceedings of The 21 International Conference on Automated Planning and Scheduling, pp. 234–241 (2011)
18. Google Places API. https://cloud.google.com/maps-platform/places/

Machine Learning Model for the Classification of Musical Composition Genres

Javor Marosevic, Goran Dajmbic, and Leo Mrsic(✉)

Algebra University College, Ilica 242, 10000 Zagreb, Croatia
{javor.marosevic,goran.dambic}@racunarstvo.hr,
leo.mrsic@algebra.hr

Abstract. The aim of this paper is to validate the applicability of standard machine learning models for the classification of a music composition genre. Three different machine learning models were used and analyzed for best accuracy: logistic regression, neural network and SVM. For the purpose of validating each model, a prototype of a system for classification of a music composition genre is built. The end-user interface for the prototype is very simple and intuitive; the user uses mobile phone to record 30 s of a music composition; acquired raw bytes are delivered via REST API to each of the machine learning models in the backend; each machine learning model classifies the music data and returns its genre. Besides validating the proposed model, this prototyped system could also be applicable as a part of a music educational system in which musical pieces that students compose and play would be classified.

Keywords: Machine learning · Musical composition genre classification · Logistic regression · SVM · Neural networks

1 Introduction

Music is the generally accepted name for shaping the physical appearance of sound in a purposeful way. Music, although rooted in basic physical phenomena such as sounds, tones, and temporal relationships, is a very abstract phenomenon defined by subjective features and as such presents a challenge when attempting to analyze it using exact computer methods. Recognizing the musical characteristics of a piece of music is often a problem, both for people and especially for computers. Creating models and systems that can recognize these characteristics and classify a musical work in one of the predefined classes is not a simple undertaking and involves expertise in several scientific fields; from physics (to understand the mechanical properties of sound), through digital signal analysis (to overcome the problem of converting sound into its digital form) to knowledge of music theory (to analyze the relationships between the frequencies of sound (tones) so that useful characteristics specific to a particular genre of musical work can be identified). The genre of a musical work is a set of aspects that characterize that musical work so it can be paired with other, similar musical works, thus forming classes into which musical works can be classified. In order to find similarities between musical works,

N. T. Nguyen et al. (Eds.): ACIIDS 2021, LNAI 12672, pp. 653–664, 2021.
https://doi.org/10.1007/978-3-030-73280-6_52

common terminology is needed that describes the relationships between the physical phenomena of sound and time and allows for their objective comparison. The aim of this paper is to propose a model that can classify the genre of a musical work from a provided soundtrack, with high accuracy. Based on the proposed model, a system was developed that was used for recognition testing. The functionalities of the system are divided into two specific parts: functionalities related to the training of models and functionalities related to the use of trained models. When training models, the system must be able to receive a series of sound recordings, analyze them, generate features based on the analysis in a format suitable for training several types of different models, train several types of prediction models and evaluate trained prediction models. When using trained models, the system must be able to receive the soundtrack in a predefined format, analyze the received soundtrack and return information to the user about the genre of the musical work of the submitted soundtrack and how certain model is that the submitted soundtrack belongs to the recognized class. As the user interface of the system, a mobile application was created that offers the user the functionality of recording audio from the environment, submitting that audio to the backend application and then receiving and displaying feedback on the music class of the submitted recording and prediction reliability. There exist several works that deal with a similar topic as this paper does. In contrast, this paper attempts to identify the genre of a musical work using lower order features and introduces several higher order features that use elements of music theory to better describe the characteristics of individual genres. Bisharad and Laskar use the architecture of repeatable convolutional neural networks and the melspectrogram as input features (Bisharad and Laskar 2019). Nanni et al. have an interesting approach combining acoustic features and visual features of audio spectrograms in genre recognition (Nanni et al. 2016). Cataltepe et al. use MIDI as an input format for recognition which greatly facilitates the extraction of higher order features because it is not necessary to process the input signal to obtain tonal features (Cataltepe et al. 2007). Li and Ogihara apply a hierarchy of taxonomy to identify dependency relationships between different genres (Li and Ogihara 2005). Lidy and Rauber evaluate feature extractors and psycho-acoustic transformations for the purpose of classifying music by genre.

2 Background

Computer-assisted sound analysis is a subset of signal processing because the focus is on characteristics of signal with frequencies less than 20 kHz. Sound can be viewed as the value of air pressure that changes over time (Downey 2014). For computer-assisted sound analysis, it is practical to convert sound into an electrical signal. Such electrical signal can be represented as a series of values over time or graphically. The graphical representation of the signal over time is called the waveform (Fig. 1, top left). It is rare to find a natural phenomenon that creates a sound represented by a sinusoid. Musical instruments usually create sounds of more complex waveforms. The waveform of the sound produced by the violin is shown in Fig. 1, top right. It can be noticed that the waveform is still repeated in time, i.e. that it is periodic, but it is much more complex than the waveform of a sinusoid. The waveform of a sound determines the color or timbre of the sound, which is often described as sound quality, and allows the listener to recognize different types of musical instruments.

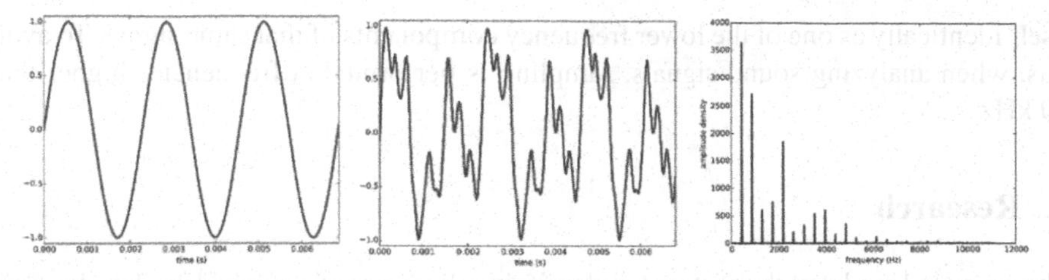

Fig. 1. Left: sound signal of 440 Hz; Middle right: violin sound waveform; Right: violin sound signal spectrum

Spectral Decomposition

One of the most important tools in signal analysis is spectral decomposition, the concept that each signal can be represented as the sum of sinusoids of different frequencies (Downey 2014). Each signal can be converted into its representation of frequency components and their strengths called the spectrum. Each frequency component of the spectrum represents a sinusoidal signal of that frequency and its strength. By applying the spectral decomposition method to the violin signal, the spectrum shown in Fig. 1 bottom will be obtained. From the spectrum it is possible to read that the fundamental frequency (lowest frequency component) of the signal is around 440 Hz or A4 tone. In the example of this signal, the fundamental frequency has the largest amplitude, so it is also the dominant frequency. As a rule, the tone of an individual sound signal is determined by the fundamental frequency, even if the fundamental frequency is not the dominant frequency. When a signal is decomposed into a spectrum, in addition to the fundamental frequency, the signal usually contains other frequency components called harmonics. In the example in Fig. 1, there are frequency components of about 880 Hz, 1320 Hz, 1760 Hz, 2200 Hz, etc. The values of the harmonic frequencies are multiples of the fundamental signal frequency. Consecutive harmonics, if transposed to one octave level from the fundamental tone, form intervals of increasing dissonance. The series of harmonics over the base tone is called the harmonic series or series of overtones and always follows the same pattern. Fourier transforms are a fundamental principle used in the spectral decomposition of signals because they are used to obtain the spectrum of the signal from the signal itself. In this paper, discrete Fourier transforms were used. Discrete Fourier transform takes as input a time series of N equally spaced samples and as a result produces a spectrum with N frequency components. The conversion of a continuous signal into its discrete form is performed by sampling. The level of the continuous signal is measured in equally spaced units of time, and the data on each measurement is recorded as a point in the time series. The appropriate continuous signal sampling frequency is determined by applying the Nyquist-Shannon theorem or sampling theorem: "If the function $x(t)$ does not contain frequencies higher than B hertz, the function is completely determined by setting the ordinate by a series of points spaced $1/(2B)$ seconds." (Shannon 1949). This means that the sampling frequency of the signal must be at least $2B$, where B is the highest frequency component of the signal. This frequency is called the Nyquist frequency. The problem of sampling a signal with a frequency lower than the Nyquist frequency occurs in the form of frequency aliasing, a phenomenon where the frequency component of a higher frequency signal manifests itself identically as one of the lower frequency components of that same signal. To avoid this, when analyzing sound signals, sampling is performed at frequencies higher than 40 kHz.

3 Research

The proposed model of the system for classifying the genre of a musical work is designed modularly. The system itself is designed in such a way that each module of the model acts as a separate unit (micro service) and is created as a separated functionality that allows scaling the system according to need and load. The system architecture is shown in Fig. 2. The process begins by analyzing a number of audio files and discovering their characteristic features selected for this paper. Feature detection is aided by the use of the open source LibROSA library, which enables basic and complex signal analysis used in sound analysis. The features used for the analysis are divided into lower order features and higher order features. Lower order features are characterized by general properties related to digital signal analysis and concern basic signal measurements without elements related to music and music theory. Higher order features characterize properties related to elements of human understanding of music and music theory such as tones, the relationship between tones, diatomicity, and chromatic properties. After the feature discovery process, the obtained features serve as an input to the prediction model training process. Model training also involves the process of normalizing and preparing data depending on which type of model is selected. Each model goes through a model evaluation process where the model is tested under controlled conditions and preliminary measures of the predictive accuracy of each model are taken. The trained model is used within an endpoint implemented as an HTTP API service. The service is in charge of receiving, processing, analyzing and returning responses to user queries. The service performs the same steps for analyzing the dedicated soundtrack that are used in the feature discovery process that precedes the prediction model training. The user accesses the service via the HTTP protocol and, respecting the defined contract, can develop her own implementation of the user application or use the mobile application developed for this paper and available on the Android operating system to access the service. The mobile application serves as a wrapper around the HTTP API and takes care of implementation details such as audio retrieval, data transfer optimization, feedback processing, session saving when analyzing longer audio tracks, and displaying service feedback. A GTZAN data set consisting of 1000 30-s audio tracks divided into 10 genres, each containing 100 sound tracks, was used to train the model. Audio recordings are in WAV format characterized as 22050 Hz, mono, 16-bit (Marsyas 2020). The genre classes of this data set are: Blues, Classical, Country, Disco, Hip-hop, Jazz, Metal, Pop, Reggae and Rock, and this also represents the classes used by trained models. Logistic regression, neural network and support vector methods (SVM) were used as machine learning methods. To train the model, it was necessary to select a set of features that describe the differences between the individual classes of elements to be classified, in our case genres. To support the process of classifying the genre of a musical work, several lower order features and several higher order features were selected. Lower order features used are: RMS of an audio signal, spectral centroid, spectral bandwidth, spectral rolloff, and zero crossing rate. The RMS of an audio signal is the root value of the mean value of the square of the signal amplitude calculated for each sampled frame of the input signal. The RMS of an audio signal is usually described as the perceived volume of sound derived from that same audio signal. The mean value of the RMS of all frames and the standard deviation of the RMS of all frames of the input audio signal were used as classification features. The reason for using these features is that certain genres typically contain high levels of differences in perceived volume within the same piece of music, and these features help to discover these characteristics. The spectral centroid is the weighted mean of the frequencies present in the signal calculated for each sampled input signal frame (Fig. 3). The spectral centroid of an audio signal is described as the "center of mass" of that audio signal and is characteristic of determining the "brightness" of sound. Sound brightness is considered to be one of the most perceptually powerful characteristics of timbre sounds.

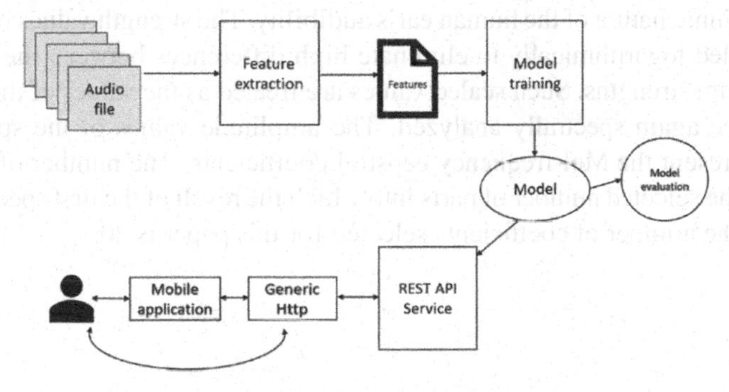

Fig. 2. The architecture of the proposed system

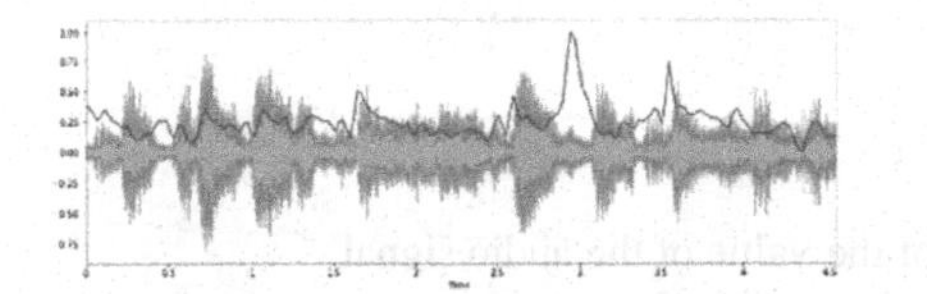

Fig. 3. Spectral centroid of individual audio signal frames

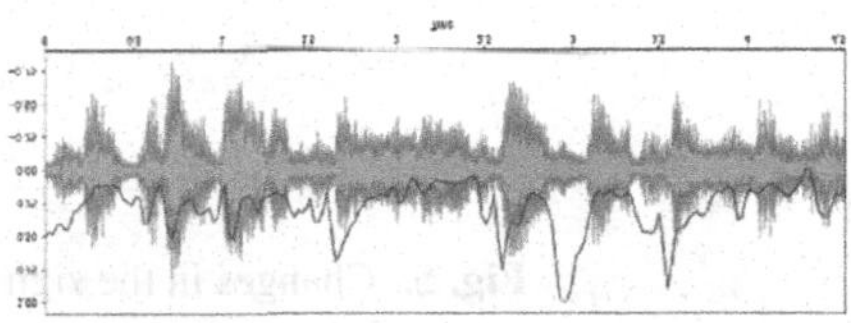

Fig. 4. Spectral rolloff of individual audio signal frames

The mean value of the spectral centroid of all frames and the standard deviation of the spectral centroid of all frames of the input audio signal were used as classification features. The reason for using these features is to reveal the timbre characteristics of individual instruments used in different genres of musical works. Spectral bandwidth is the bandwidth at half the peak of the signal strength (Weik 2000) and helps to detect audio signal frequency scatter. The mean value of the spectral bandwidth of all frames and the standard deviation of the spectral bandwidth of all frames of the input audio signal were used as classification features. Spectral rolloff is the value of the frequency limit below which a certain percentage of the total spectral energy of an audio signal is located (Fig. 4) and helps to detect the tendency of using high or low tones within a piece of music. The mean value of the spectral rolloff of all frames and the standard deviation of the spectral rolloff of all frames of the input audio signal were used as classification features. The zero crossing rate is the rate of change of the sign of the value of the audio signal (Fig. 5). This measure usually takes on more value in percussion instruments and is a good indicator of their use within musical works. The mean value of the zero crossing rate of all frames and the standard deviation of the zero crossing rate of all frames of the input audio signal were used as classification features. Of the higher order features, the following were used: Mel-frequency cepstrum coefficients, melodic change factor and chromatic change factor. Mel-frequency cepstrum coefficients are a set of peak values of logarithmically reduced strengths of the frequency spectrum of signals mapped to the mel frequency scale. These features are extracted by treating the frequency spectrum of the input signal as an input signal in another spectral analysis. The input signal is converted into a spectrum and such a spectrum is divided into several parts, and the frequency values are taken as the frequency values from the mel scale due to the logarithmic nature of the human ear's audibility. The strength values of each of the parts are scaled logarithmically to eliminate high differences between the peak values of the spectrum strengths. Such scaled values are treated as the values of the new signal which is once again spectrally analyzed. The amplitude values of the spectrum thus obtained represent the Mel-frequency cepstral coefficients. The number of coefficients depends on the selected number of parts into which the result of the first spectral analysis is divided. The number of coefficients selected for this paper is 20.

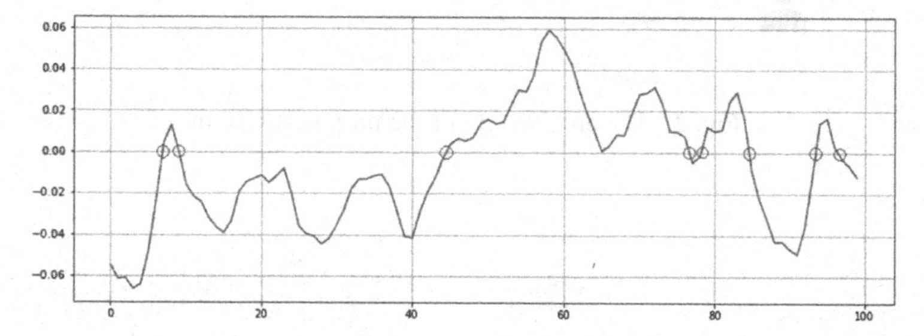

Fig. 5. Changes in the sign of the value of the audio signal

The melodic change factor represents the rate of melodic change of the predominant tones within the input audio signal. The melodic change factor is calculated by first generating a chromagram (Fig. 6) of the input signal. The chromagram represents the distribution of the tone classes of the input audio signal over time. Distribution is calculated for each input signal frame. The melodic change factor helps us to discover the rate of melodic dynamism of a musical work.

The chromatic change factor represents the rate of chromatic change of the predominant tones within the input audio signal. By chromatic change we mean the increase or decrease of two consecutive tones by one half-step. The chromatic change factor is calculated in a similar way as the melodic change factor. A chromagram is generated and the strongest predominant tone class of the input frames is extracted using a sliding window to reduce the error. After extracting the strongest tone class, the cases when two consecutive frames have a change of tone class in the value of one half-degree are

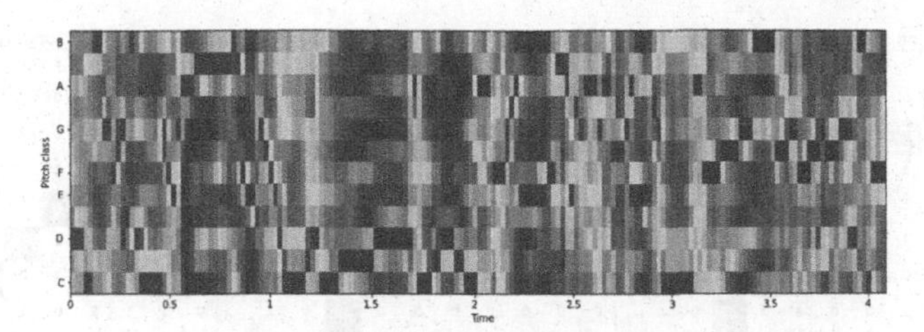

Fig. 6. Chromagram of audio signal

counted. This number is divided by the total number of input frames to obtain a change factor independent of the duration of the audio input signal. The chromatic shift is very characteristic for certain genres of musical works, such as works from the jazz genre or the classical genre, and at the same time rare in parts of the popular genre or rock genre.

4 Results and Discussion

In this paper, the term "controlled conditions" refers to the testing of a model on a test dataset separate from the original dataset on which the model was trained. These data are pseudo-randomly extracted from the original dataset and it is expected that this subset of data shares very similar characteristics as the set on which the models of the genre classification system of the musical work were trained. A rudimentary measure of model accuracy shows that all three models used in this paper have an average prediction accuracy between 70% and 80% (Table 1). The neural network model and the SVM method model show approximately equal accuracy while the logistic regression model is about 5% less accurate. The resulting confusion matrices of prediction models are shown in Fig. 7. It can be observed that all three models have lower accuracy of classification of musical works of the genres Rock, Disco, Country and Reggae. It is interesting to observe that Rock music is often misclassified as Disco music, and the reverse is not valid, i.e., Disco music is not classified as Rock music except in the case of models of SVM model. A similar situation can be noticed with Disco and Pop music. Disco music is more often classified as Pop music, while Pop music is not classified as Disco music.

Table 1. Accuracy of each model

Model	Model accuracy
Logistic regression	0.72
Neural network	0.77
SVM	0.76

A closer look at the confusion matrices of all three models leads to the conclusion that the neural network model has much less scattering of results than the other two

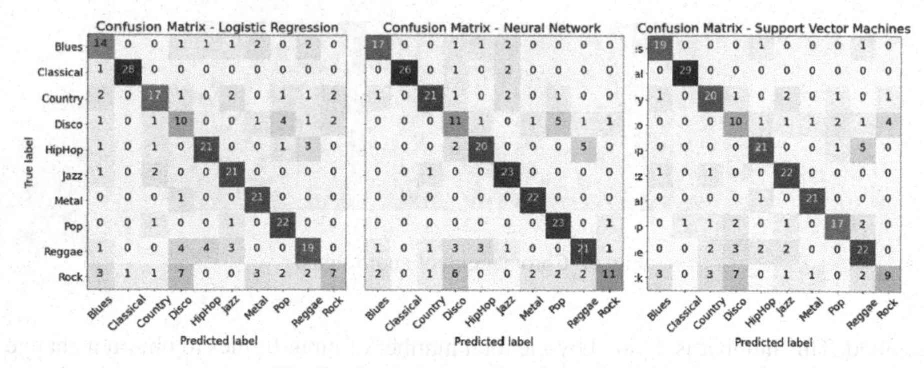

Fig. 7. Confusion matrix of used models

Table 2. Area under the ROC curve of the genres of individual models

Genre	Log. Regres. ROC AUC	Neur. Netw. ROC AUC	SVM ROC AUC
Blues	0.967	0.989	0.981
Classical	1	0.998	1
Country	0.946	0.969	0.971
Disco	0.875	0.917	0.939
Hip-hop	0.941	0.963	0.981
Jazz	0.975	0.988	0.990
Metal	0.999	1	0.997
Pop	0.989	0.992	0.983
Reggae	0.928	0.951	0.934
Rock	0.879	0.914	0.853
Average	**0.945**	**0.968**	**0.963**

models. In the case of most genres, the neural network model classifies the entities of a genre into fewer possible classes than the other two models do, i.e. when classifying works of a genre, there is a good chance that one of the two possible classes will be attributed to the work. This is not the case with the Rock and Reggae genres in which the neural network model also classifies entities into several different classes.

Figure 8 shows the ROC curves for all three models, separately for each prediction class. Areas below the ROC curves were also calculated for each of the prediction models and shown in Table 2 as well as the average area under the curves (AUC) of all genres.

Curves representing the genres Classical, Metal and Jazz have a very steep rise at the beginning which indicates a high certainty that, in case of recognition of these categories, musical works really belong to this category which can be attributed to the characteristics of these genres of musical works. The Blues and Pop genre curves also show good recognition reliability characteristics. In other genres, when classifying a

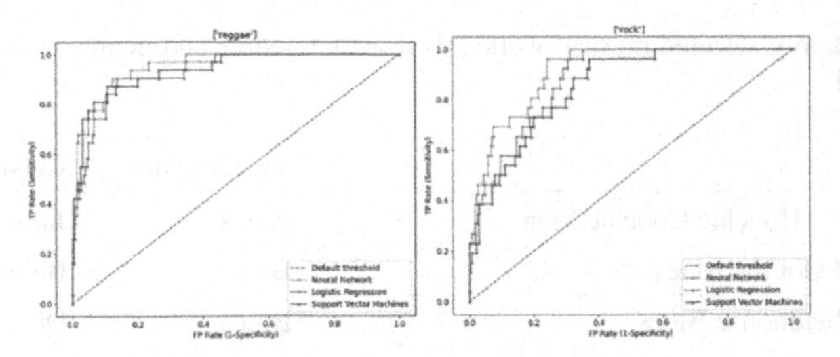

Fig. 8. ROC curves for selected genres

piece of music into one of these categories, we must be careful because the number of false identifications of these categories is slightly higher than the previously mentioned. Such results are expected because these musical genres often borrow elements that could be classified in some of the other genres of musical works, and it is not difficult to imagine a situation where one would misrecognize a genre based on a short clip of one of these musical genres.

In this paper, the term "uncontrolled conditions" is considered to be a set of user selected parts of musical works and their classification by the proposed system. Testing in uncontrolled conditions was performed by selecting three musical works from each genre. A 30-s sample was taken from each selected piece of music and this sample was passed through the system. The 30-s sample was taken so that the beginning of the sample corresponds to the time point of the musical work at 45 s, and the end of the sample corresponds to the point of the musical work at 75 s. This method of sampling was chosen in order to avoid the appearance of silence at the beginning of some musical works and to avoid intros that might not belong to the genre of the musical work. Audio recordings are submitted to processing to the HTTP API where the trained models are located. The result of classification was provided by the neural network model because this model showed the highest recognition accuracy. Selected musical works and their test results are shown in Table 3.

To make it easier to comprehend the test results in uncontrolled conditions, the results are visualized in Fig. 9.

It is interesting to note that certain genres have one hundred percent accuracy of classification, while some genres have not correctly recognized any of the cases. The category of classical music is characterized by the use of a very different spectrum of instruments from other categories so the high recognition accuracy in this category is not surprising. Likewise, the Hip-hop category is characterized by an extremely strong rhythmic and predominant vocal component and the result is expected. The unexpected result is the inability to recognize any randomly selected work in the Country, Metal and Pop genres. A possible explanation for these results is that these genres are relatively similar to other genres, in terms of the features selected and the instruments used. Also, within these genres there are several different subgenres that differ greatly from each other and it is possible that the dataset on which the models are trained is biased in favor of some of these subgenres and that randomly selected musical works that do not have the characteristics needed to classify that musical works into a trained model class. The low recognition accuracy in the Rock and Blues genres might be explained

Table 3. Randomly selected musical works, their actual genres and genres classified by neural network model

Musical work	Actual genre	Classified genre
Muddy Waters - Hoochie Coochie Man	Blues	Jazz
Sam Myers - I Got the Blues	Blues	Blues
The Doors - Roadhouse Blues	Blues	Disco
Bach - Brandenburg Concerto No. 1 in F major	Classical	Classical
Claude Debussy - Clair De Lune	Classical	Classical
Dmitri Shostakovich - Waltz No. 2	Classical	Classical
Hank Williams – Jambalaya	Country	Classical
Johnny Cash and June Carter – Jackson	Country	Jazz
N. Sinatra - These Boots Are Made For Walkin'	Country	Reggae
Bee Gees - Stayin' Alive	Disco	Disco
Boney M. – Rasputin	Disco	Rock
Earth, Wind & Fire - Boogie Wonderland	Disco	Disco
Dr Dre - What's the Difference	Hip-hop	Hip-hop
The Sugar Hill Gang - Rapper's Delight	Hip-hop	Hip-hop
Tupac Ft Elton John - Ghetto Gospel	Hip-hop	Hip-hop
Dave Brubeck - Golden Brown	Jazz	Jazz
John Coltrane - Giant Steps	Jazz	Jazz
Medeski, Martin & Wood – Kota	Jazz	Classical
Disturbed - Down With the Sickness	Metal	Hip-hop
Drowning Pool – Bodies	Metal	Hip-hop
Metallica - Fade to Black	Metal	Rock
Abba – Waterloo	Pop	Classical
Cyndi Lauper - Girls Just Want To Have Fun	Pop	Jazz
Michael Jackson - Billie Jean	Pop	Reggae
Bob Marley - Buffalo soldier	Reggae	Reggae
Peter Tosh - Wanted Dread & Alive	Reggae	Disco
UB40 -Kingston Town	Reggae	Reggae
Dire Straits - Romeo And Juliet	Rock	Reggae
Kansas - Carry on Wayward Son	Rock	Rock
ZZ Top - La Grange	Rock	Disco

in a similar way. Both genres are very general and their subgenres reflect some of the characteristics of other genres. These hypotheses are also supported by controlled test

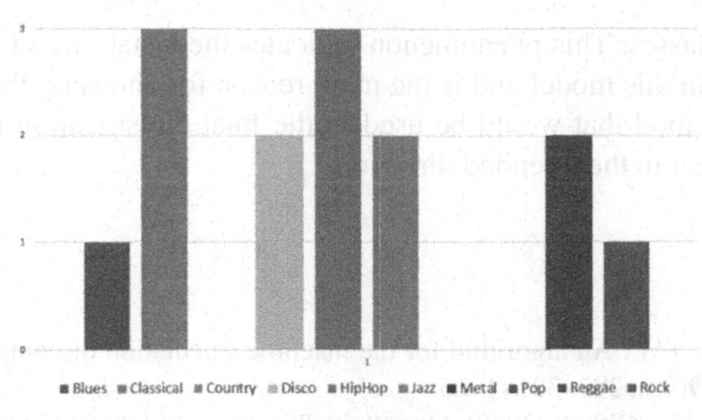

Fig. 9. Number of correct classifications by genre

results that show lower accuracy in recognizing these genres on test data extracted from the original training dataset.

5 Conclusion

The test results of the proposed system for classifying the genre of a musical work show that all considered models have an accuracy between 70% and 80%. The analysis of the results shows that the models show better performance in the classification of certain genres, and for some genres the accuracy is very close to one hundred percent (for example, in the case of classical music genre accuracy is greater than 90%) which shows predispositions to improve classification results in other genres also. Testing in controlled conditions and analysis of results with convolutional matrices and ROC curves showed slightly worse recognition results of the Rock, Disco and Reggae genres, while the results of the Classical, Metal and Jazz genres showed very good recognition results. Testing in uncontrolled conditions in most genres follows the results of controlled testing, except for the genres Metal, Pop and Country, which showed very poor results. Possible explanations for this discrepancy are the diversity of styles within the genres and the similarity of individual subgenres with other music genres and the bias of the input dataset to some of these subgenres, which leads to incorrect classification of individual musical works belonging to these genres. The obtained results suggest that by training the model on a larger dataset the accuracy might be higher and that by generating a larger number of higher orders features it would be possible to increase the accuracy of the classification using the current dataset. It should also be taken into account that the boundaries between genres of musical works are not clearly defined and many musical works belonging to a certain genre also contain elements characteristic for other genres. For example, it is sometimes difficult for a human being to decide whether a work is heavy metal or hard rock. This suggests that the accuracy of classification, from the point of view of classifying a musical work into only one of the available classes, will not be possible to achieve a very high accuracy (95% and more). Classifying works into one or more genres could potentially show better results. In the context of the proposed model and the developed system, the neural network model proved to be the most accurate classification model. The classifier based on the SVM shows the same average accuracy as the classifier based on neural networks, but with a greater scatter of results between several possible classes. This phenomenon indicates the instability of this classifier for the purpose used in this model and is the main reason for choosing the neural network as the primary method that would be used in the final classification model in case of further development in the intended direction.

References

Cooley, J.W., Tukey, J.W.: An algorithm for the machine calculation of complex Fourier series. Math. Comput. **19**(90), 297–301 (1965)

Dabhi, K.: Om Recorder Github (2020). https://github.com/kailash09dabhi/OmRecorder

Downey, A.B.: Think DSP. Green Tea Press, Needham (2014)

National Radio Astronomy Observatory: Fourier Transforms (2016). Preuzeto 2020 iz. https://www.cv.nrao.edu/course/astr534/FourierTransforms.html

Encyclopaedia Britannica: Encyclopaedia Britannica, 16 June 2020. https://www.britannica.com/art/consonance-music

Fawcett, T.: An introduction to ROC analysis. Science Direct (2005)

Hrvatska Enciklopedija: Hrvatska enciklopedija, mrežno izdanje, Leksikografski zavod Miroslav Krleža (2020). Preuzeto 2020 iz. https://www.enciklopedija.hr/natuknica.aspx?ID=67594

ITU: Universally Unique Identifiers (UUIDs) (2020). Preuzeto 3. 8 2020 iz ITU. https://www.itu.int/en/ITU-T/asn1/Pages/UUID/uuids.aspx

Marsyas: GTZAN dataset, 10 July 2020. http://marsyas.info/downloads/datasets.html

Serdar, Y.: What is TensorFlow? The machine learning library explained (2019). Preuzeto 25. 8. 2020 iz. https://www.infoworld.com/article/3278008/what-is-tensorflow-the-machine-learning-library-explained.html

Theano: Theano documentation (2017). http://deeplearning.net/software/theano/

Melo, F.: Area under the ROC curve. In: Dubitzky, W., Wolkenhauer, O., Cho, K.H., Yokota, H. (eds.) Encyclopedia of Systems Biology. Springer, New York (2013). https://doi.org/10.1007/978-1-4419-9863-7_209

Merriam-Webster.Com: Merriam Webster (2020). Dohvaćeno iz. https://www.merriam-webster.com/dictionary/music

Shannon, C.E.: Communication in the presence of noise. Proc. IRE **37**(1), 10–21 (1949)

Square, Inc.: Retrofit (2020). Preuzeto 05. 08 2020 iz. https://square.github.io/retrofit/

Ting, K.M.: Confusion matrix. In: Sammut, C., Webb, G.I. (eds.) Encyclopedia of Machine Learning. Springer, Boston (2011). https://doi.org/10.1007/978-0-387-30164-8_157

Weik, M.H.: Computer Science and Communications Dictionary. Springer, Boston (2000). https://doi.org/10.1007/1-4020-0613-6

Bisharad, D., Laskar, R.H.: Music genre recognition using convolutional recurrent neural network architecture. Expert Syst. **36**(4), e12429 (2019)

Nanni, L., Costa, Y.M.G., Lumini, A., Kim, M.Y., Baek, S.R.: Combining visual and acoustic features for music genre classification. Expert Syst. Appl. **45**, 108–117 (2016)

Cataltepe, Z., Yaslan, Y., Sonmez, A.: Music genre classification using MIDI and audio features. EURASIP J. Adv. Sig. Process. **2007** (2007). https://doi.org/10.1155/2007/36409. Article ID 36409

Li, T., Ogihara, M.: Music genre classification with taxonomy. In: Proceedings of the IEEE International Conference on Acoustics, Speech, and Signal Processing (ICASSP 2005), Philadelphia, PA, vol. 5, pp. v/197–v/200 (2005). https://doi.org/10.1109/icassp.2005.1416274

Lidy, T., Rauber, A.: Evaluation of feature extractors and psycho-acoustic transformations for music genre classification. In: ISMIR, pp. 34–41 (2005)

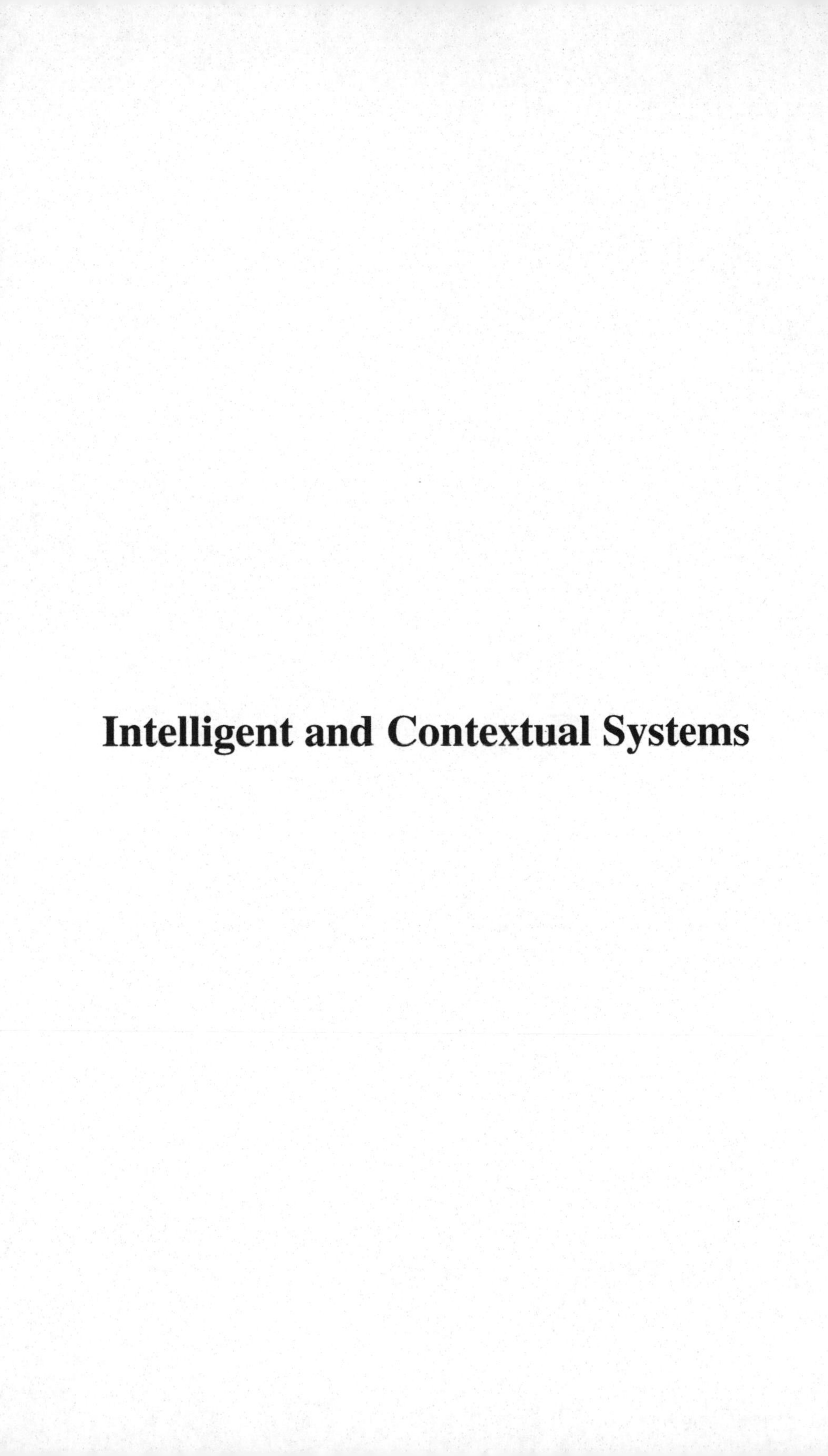

Intelligent and Contextual Systems

Emotional Piano Melodies Generation Using Long Short-Term Memory

Khongorzul Munkhbat[1], Bilguun Jargalsaikhan[1], Tsatsral Amarbayasgalan[1], Nipon Theera-Umpon[3,4], and Keun Ho Ryu[2,3](✉)

[1] Department of Computer Science, School of Electrical and Computer Engineering, Chungbuk National University, Cheongju 28644, Korea
{khongorzul,bilguun,tsatsral}@dblab.chungbuk.ac.kr

[2] Faculty of Information Technology, Ton Duc Thang University, Ho Chi Minh City 700000, Vietnam
khryu@tdtu.edu.vn, khryu@ieee.org

[3] Biomedical Engineering Institute, Chiang Mai University, Chiang Mai 50200, Thailand
nipon.t@cmu.ac.th

[4] Department of Electrical Engineering, Faculty of Engineering, Chiang Mai University, Chiang Mai 50200, Thailand

Abstract. One of the tremendous topics in the music industry is an automatic music composition. In this study, we aim to build an architecture that shows how LSTM models compose music using the four emotional piano datasets. The architecture consists of four steps: data collection, data preprocessing, training the models with one and two hundred epochs, and evaluation by loss analysis. From the result of this work, the model trained for 200 epochs give the lowest loss error rate for the composing of emotional piano music. Finally, we generate four emotional melodies based on the result.

Keywords: Music Information Retrieval · Automatic music generation · Deep Learning · Long Short-Term Memory

1 Introduction

The rapid development of Artificial Intelligence (AI) is advancing in many areas such as bioinformatics, natural language processing, speech and audio recognition, image processing, social network filtering, smart factory and so forth [1–4]. It is also bringing a new wave to the music industry. In Deep Learning (DL), the subfield of AI, music is one of the most demanding domains, moreover, it is called the Music Information Retrieval (MIR) [5]. There are some challenges for instrument recognition [6] and track separation [7], automatic music transcription [8], automatic categorization [9] and composition [10, 11], and music recommendation [12]. One of the famous topics in MIR is automatic music composition. It is a process of creating or writing new music pieces [13]. It is impossible to compose music without knowledge and theory of music, so there are only people with special feelings of art or professionals in the field of music. According to the

N. T. Nguyen et al. (Eds.): ACIIDS 2021, LNAI 12672, pp. 667–677, 2021.
https://doi.org/10.1007/978-3-030-73280-6_53

demands of society today, all types of entertainment such as movies, videos, computer games, marketing and advertisement need a new kind of hit music that leads the market and reaches the users. Music is based on human emotions. It can boost our mood, change the emotion, and influence a response, so creating a song for emotion and mood is very helpful. For instance, calming music can help relieve stress, while happy and energetic music can provide energy during activities such as exercise. Music therapists commonly use a variety of emotional music for the treatment [14].

In this work, we aim to generate new piano melodies for four different emotions using Long Short-Term Memory (LSTM) neural network. This work addresses the limitation of conventional music composition by automatically generating emotional melodies and lack of repeating melodic structure.

The paper is organized as follows: The literature reviews related to music generation task are provided in Sect. 2, while Sect. 3 presents our proposed architecture, dataset, and techniques we applied in this study. The experimental result is shown in Sect. 4, and we provide some conclusion and future work in Sect. 5.

2 Literature Review

A feed-forward network is incapable of storing any information about the past, which means it cannot perform the tasks to predict and generate the next step based on previous history. To address this issue, the Recurrent Neural Network (RNN) was created based on study of [15] and [16] explored the RNN in the music task for the first time in 1994. Although the RNN has hidden layers with memory cells, the results in music generation task were not enough for musical long-term dependency due to vanishing gradient problem [17]. Because long-term dependencies are a key expression of musical style, genre, and feeling [18] for music. [19], one of the old studies, used the LSTM that is a special kind of RNN and capable of learning long-term dependencies. They aimed to show that the LSTM can learn to reproduce a musical chord structure.

Therefore, we can mention many works used other techniques such as Variational Autoencoder (VA), and Generative Adversarial Network (GAN) [20–23] for music generation. There are interesting studies that combined the task of composing music with pictures, movies, and video games. In 2016, [24] used a multi-task deep neural network (MDNN) to jointly learn the relationship among music, video, and emotion from an emotion annotated MV corpus in order to solve the problem of the semantic gap between the low-level acoustic (or visual) features and the high-level human perception. It shows that creating songs by exploring emotions is the basis for art, films, and recordings. Another music generation study, based on facial expressions, is written by [25]. Two kinds of models, image classification and music generation, are proposed in that work and finally, the Mean Opinion Score (MOS) was used for evaluating.

3 Methods and Materials

3.1 Automatic Melody Generation Architecture

The architecture of emotion-based automatic melodies generation built in this work consists of four steps as shown in Fig. 1. Dataset is gathered and prepared in the first step and then the processes which convert audio data to symbolic data, extract notes and chords musical objects, and encode the features into the sequential list are performed as a preprocessing step. Specifically, we convert the text sequences to integers using the mapping function and normalize between zero and one to prepare network input displayed in Fig. 1. After preparing inputs of the model, we train the LSTM networks for one and two hundred epochs and evaluate using loss analysis in the third step. Finally, we generate four kinds of emotional melodies.

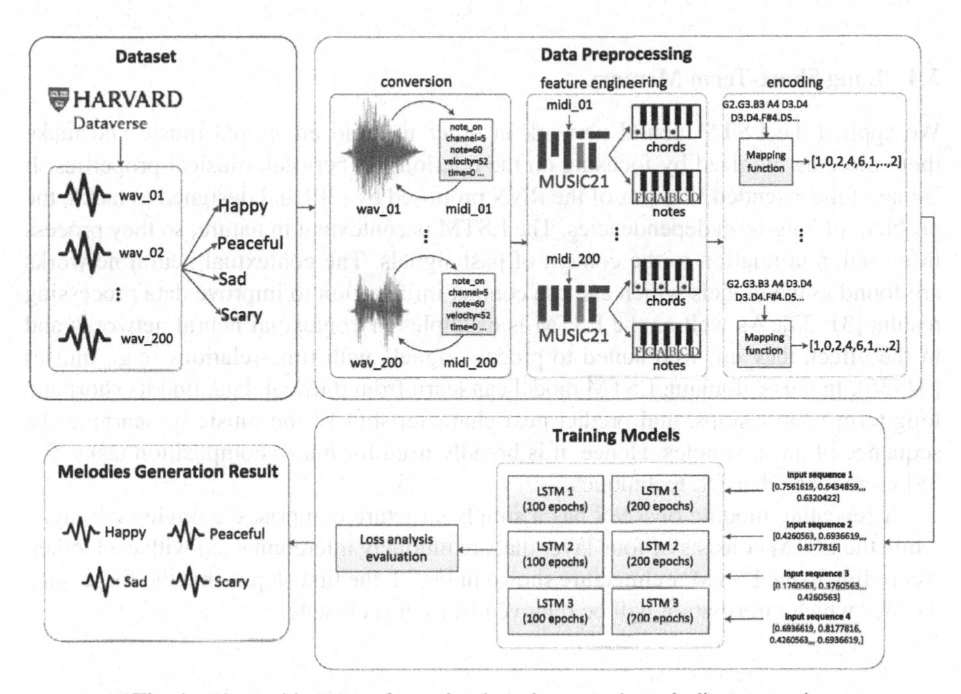

Fig. 1. The architecture of emotion-based automatic melodies generation

3.2 Data Collection and Preparation

We chose the "Music and emotion dataset (Primary Musical Cues)" proposed by [26]. It contains 200 piano songs with wav format and has four emotion semantics: happy, peaceful, sad, and scary. One of the most important issues in the psychology of music is how music affects emotional experience [27]. For instance, songs that express happy emotion have a fast rhythm, a high-pitch range, and a major mode, while songs with sad emotion have a slow rhythm, a low-pitch range, and a minor mode [28].

3.3 Data Preprocessing

Since the music dataset used in this work has an audio format, it is necessary to translate it into a midi symbolic format. Midi stands for Musical Instrument Digital Interface and is denoted by.mid or.midi. It is a standard protocol for exchanging musical information by connecting devices such as computers, synthesizers, electronics, and any other digital instruments and was developed to allow the keyboard of one synthesizer to play notes generated by another. It defines codes for musical notes as well as button, dial, and pedal adjustments, and midi control messages can orchestrate a series of synthesizers, each playing a part of the musical score. After converting all music data into midi format, chords and notes are separated from midi dataset using object-oriented Music21 toolkit [29].

3.4 Long Short-Term Memory

We applied the LSTM neural network in order to build emotional music and make them more harmonized by focusing on the relationship between musical properties. It is one of the extended versions of the RNN proposed by [30] and designed to tackle the problem of long-term dependencies. The LSTM is contextual in nature, so they process information in relation to the context of past signals. The contextual neural networks are found as the models which can use context information to improve data processing results [31–33]. As well as the LSTM is examples of contextual neural networks and in the effect, they are well suited to process signals with time-relations (e.g., music) [34–36]. In music domain, LSTM model can learn from musical data, find its short and long-term relationships, and predict next characteristics of the music by learning the sequence of musical notes. Hence, it is broadly used for music composition tasks [37, 38] more than other DL techniques.

A repeating module of RNN has a simple structure comprises a single tanh layer, while the LSTM consists of four-layer that are uniquely interconnected with each other. According to the LSTM architecture shown in Fig. 2, the first step is that the forget gate decides which information will be removed from the cell state.

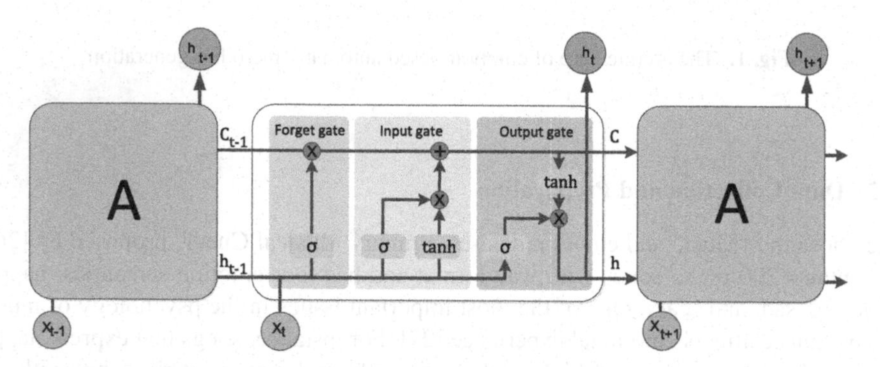

Fig. 2. The architecture of LSTM

This decision is made by a sigmoid layer and it outputs a number between 0 and 1. For instance, the "zero" value means information that will be forgotten, the "one" value indicates information that needs to go on. (1), (2), and (3) show the equations for forget

(f_t), input (i_t), and output (o_t) gates separately.

$$f_t = \sigma\left(w_f\left[h_{t-1}, x_t\right] + b_f\right) \quad (1)$$

$$i_t = \sigma\left(w_i\left[h_{t-1}, x_t\right] + b_i\right) \quad (2)$$

$$o_t = \sigma\left(w_o\left[h_{t-1}, x_t\right] + b_o\right) \quad (3)$$

Herein, σ is a sigmoid function, w is the weight of the three gates, h_{t-1} is the input of the previous timestep, x_t is the input of the current timestep, and b is the bias of the three gates.

The next step is that the input gate controls what information will be stored in the cell state. Here, the sigmoid layer decides which values will update, while a tanh layer builds a vector of new candidate values that could be added to the cell state. The latter step is that the output layer transfers data and information left in the state to the next layer. The output of the LSTM network is the input of the next step.

4 Experimental Result

We built three kinds of models which are LSTM with a single layer, LSTM with two layers, and LSTM with two layers has different configures and named the models as the LSTM 1, LSTM 2, and LSTM 3 in the following figures. The models trained for 100 epochs and 200 epochs separately, and the experimental results were compared later.

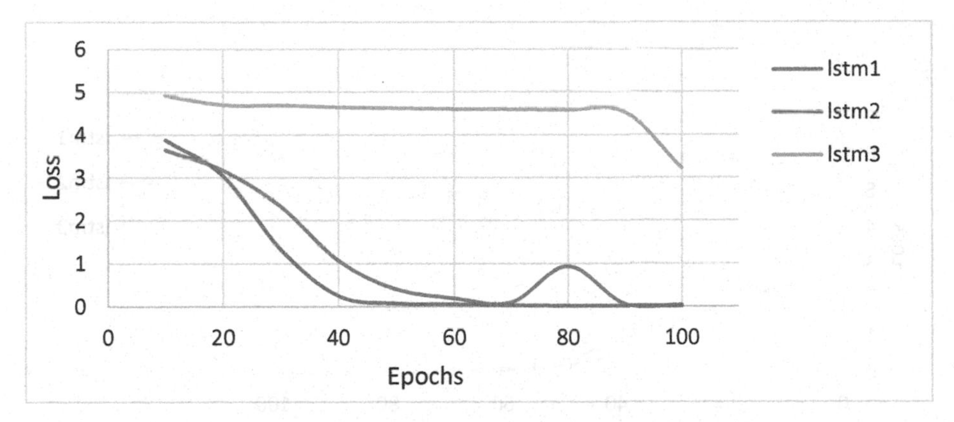

Fig. 3. Loss analysis of happy melody (100 epochs)

The LSTM models with a single layer and two layers are configured with 512 neurons followed by batch normalization, dropout 0.3, flatten, dense and softmax activation layers, while the third two-layered LSTM model is tuned with 512 neurons followed by batch normalization, dropout 0.3, dense 256, rectifier linear activation function, batch normalization, dropout 0.3, flatten, dense and softmax activation layers. Hyperparameters are important to the quality of DL models because they can manage the behaviors of training models [39]. The main idea of the dropout is to randomly drop units from the network during the training phase. It prevents overfitting problems and accelerates training neural networks [40]. We set up the softmax function as activation which can handle multiple classes. Therefore, it helps to normalize the output of each neuron to a range between 1 and 0 and returns the probability that the input belongs to a specific class. The categorical cross-entropy and RMSprop are chosen for the loss function and optimizer to the models' hyperparameter adjustment. In terms of model evaluation, the loss function is used to measure the difference between the predicted value and the true value of the model [41]. The lower the loss rate defines the better model.

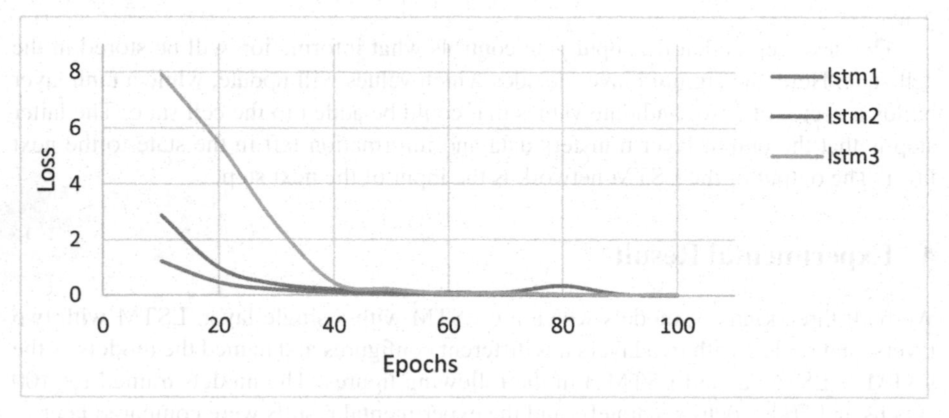

Fig. 4. Loss analysis of peaceful melody (100 epochs)

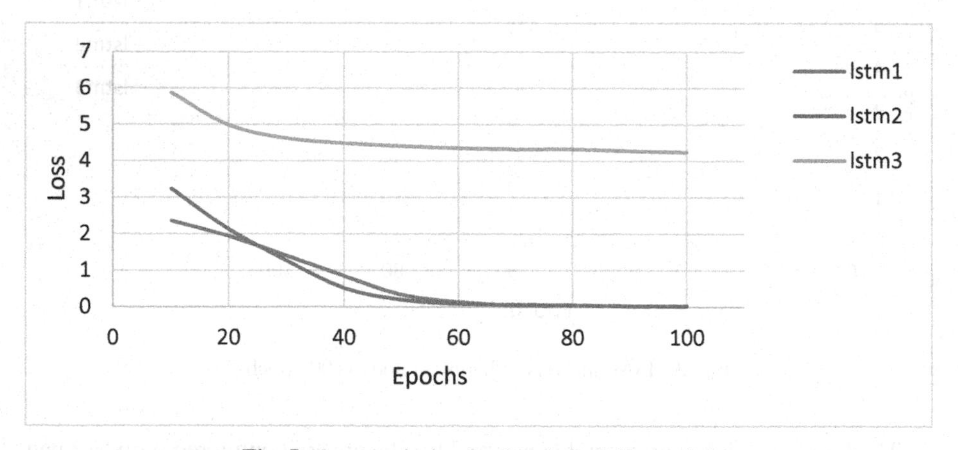

Fig. 5. Loss analysis of sad melody (100 epochs)

Firstly, we trained the three models during 100 epochs and the results of the experiment are shown by graph from Fig. 3, Fig. 4, Fig. 5 and Fig. 6. Herein, the LSTM model with a single layer gave us the better results for peaceful and sad melody generation as

0.005 and 0.0127, while the two-layered LSTM model showed the lowest loss rate for happy and scary melody generation as 0.0118 and 0.031.

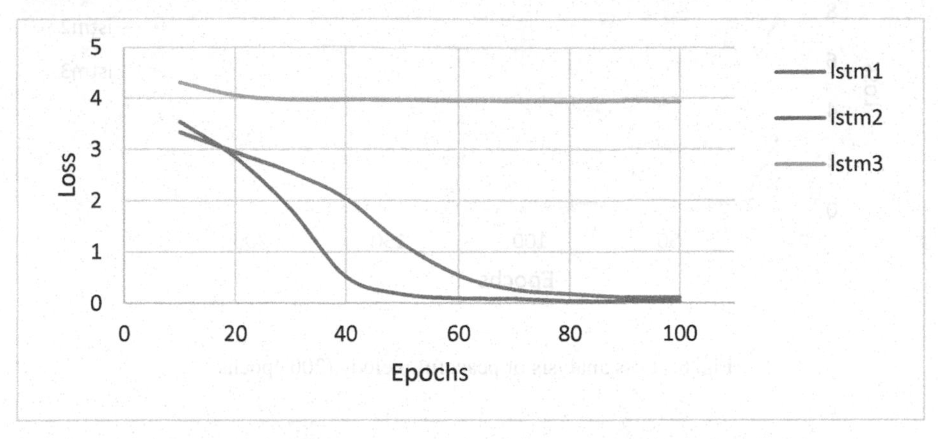

Fig. 6. Loss analysis of scary melody (100 epochs)

After training the models on 100 epochs, we conducted the experiments again during 200 epochs to show the results' comparison. The Fig. 7, Fig. 8, Fig. 9 and Fig. 10 displays the loss analysis results of emotional melody generation experiment during 200 epochs. In the experiment, the lowest error rates for happy, peaceful, sad, and scary are 0.0007, 0, 0.0002, and 0.0031, respectively.

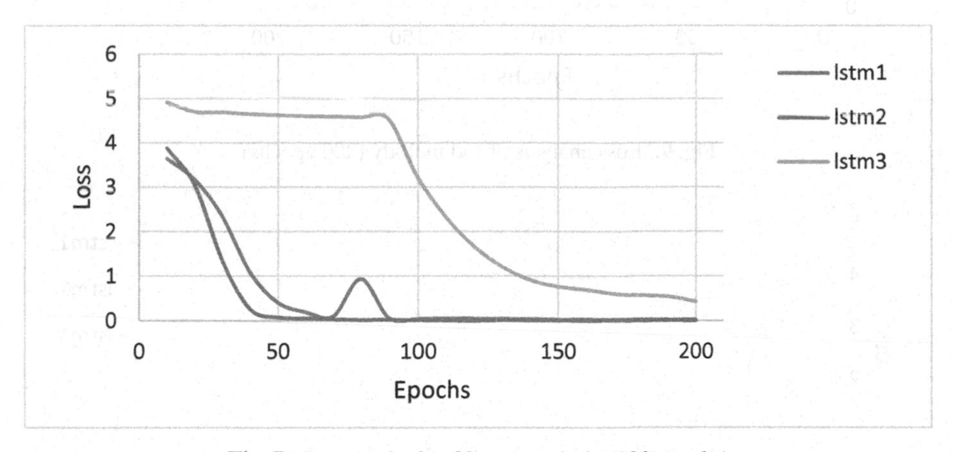

Fig. 7. Loss analysis of happy melody (200 epochs)

We show the comparison of experimental results for each emotion melody generation performed on different epochs in Table 1. Therefore, the experimental results on 200 epochs show that increasing the number of epochs can reduce the model loss error rate.

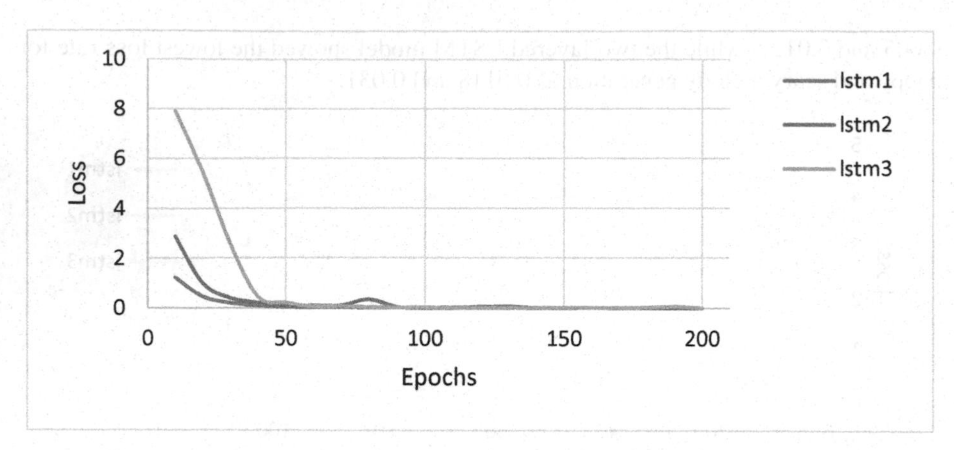

Fig. 8. Loss analysis of peaceful melody (200 epochs)

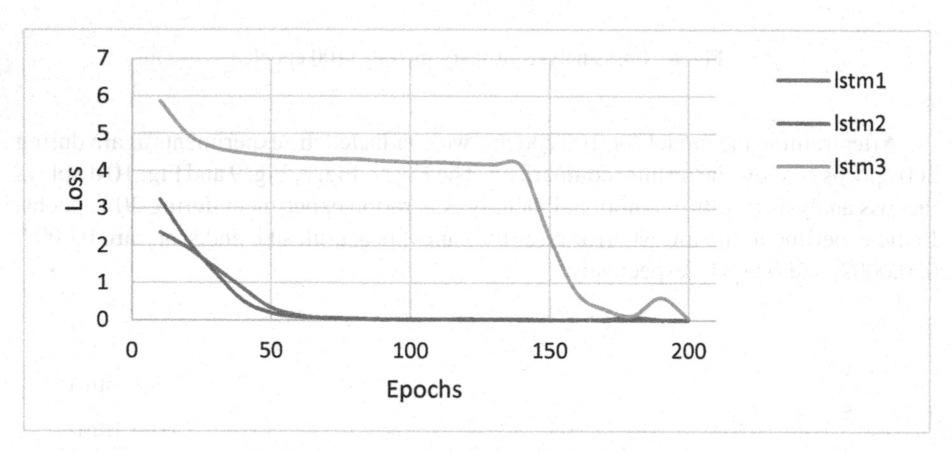

Fig. 9. Loss analysis of sad melody (200 epochs)

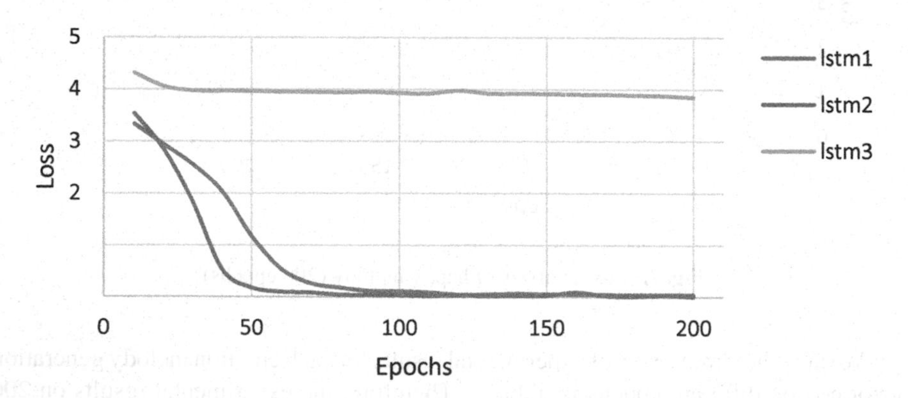

Fig. 10. Loss analysis of scary melody (200 epochs)

Table 1. Comparison of the lowest loss analysis results on each emotion

Emotion	100 epochs	200 epochs
Happy	0.0118	0.0007
Peaceful	0.005	0
Sad	0.0127	0.0002
Scary	0.031	0.0031

5 Conclusion and Future Work

Automatic music generation is a popular research area in the music industry. In this study, the architecture of emotion-based piano melodies generation using LSTM network was created. We built three kinds of models which are LSTM with a single layer, LSTM with two layers, and LSTM with two layers has different configure. The models were trained for 100 epochs and 200 epochs separately, and the experimental results were compared by loss error rate. The results of the model, trained with 200 epochs, show that increasing the number of epochs can reduce the model loss error rate.

In the future, this work can be improved by experimenting with other state-of-art techniques such as DL algorithms and Attention mechanism which captures a long-range structure within sequential data and more focuses on the relationship between musical properties. Even in the future, such kind of study will be a demanding and big leap in the world of the music industry.

Acknowledgment. This research was supported by Basic Science Research Program through the National Research Foundation of Korea (NRF) funded by the Ministry of Science, ICT & Future Planning (No. 2017R1A2B4010826), and (2019K2A9A2A06020672) and (No. 2020R1A2B5B02001717).

References

1. Young, T., Hazarika, D., Poria, S., Cambria, E.: Recent trends in deep learning based natural language processing. IEEE Comput. Intell. Mag. **13**(3), 55–75 (2018)
2. Deselaers, T., Hasan, S., Bender, O., Ney, H.: A deep learning approach to machine transliteration. In: Proceedings of the Fourth Workshop on Statistical Machine Translation, pp. 233–241 (2009)
3. Voulodimos, A., Doulamis, N., Doulamis, A., Protopapadakis, E.: Deep learning for computer vision: a brief review. Comput. Intell. Neurosci. **2018**, 1–13 (2018)
4. Shen, D., Wu, G., Suk, H.I.: Deep learning in medical image analysis. Annu. Rev. Biomed. Eng. **19**, 221–248 (2017)
5. Choi, K., Fazekas, G., Cho, K., Sandler, M.: A tutorial on deep learning for music information retrieval. arXiv preprint arXiv:1709.04396 (2017)
6. Han, Y., Kim, J., Lee, K.: Deep convolutional neural networks for predominant instrument recognition in polyphonic music. IEEE/ACM Trans. Audio Speech Lang. Process. **25**(1), 208–221 (2016)

7. Rosner, A., Kostek, B.: Automatic music genre classification based on musical instrument track separation. J. Intell. Inf. Syst. **50**(2), 363–384 (2017). https://doi.org/10.1007/s10844-017-0464-5
8. Sigtia, S., Benetos, E., Dixon, S.: An end-to-end neural network for polyphonic piano music transcription. IEEE/ACM Trans. Audio Speech Lang. Process. **24**(5), 927–939 (2016)
9. Pham, V., Munkhbat, K., Ryu, K.: A classification of music genre using support vector machine with backward selection method. In: 8th International Conference on Information, System and Convergence Applications, Ho Chi Minh (2020)
10. Sturm, B.L., Santos, J.F., Ben-Tal, O., Korshunova, I.: Music transcription modelling and composition using deep learning. arXiv preprint arXiv:1604.08723 (2016)
11. Munkhbat, K., Ryu, K.H.: Music generation using long short-term memory. In: International Conference on Information, System and Convergence Applications (ICISCA), pp. 43–44 (2019)
12. Cheng, Z., Shen, J.: On effective location-aware music recommendation. ACM Trans. Inf. Syst. (TOIS) **34**(2), 1–32 (2016)
13. Kratus, J.: Nurturing the songcatchers: philosophical issues in the teaching of music composition. In: Bowman, W., Frega, A. (eds.) The Oxford Handbook of Philosophy in Music Education. Oxford University Press, New York (2012)
14. Monteith, K., Martinez, T.R., Ventura, D.: Automatic generation of music for inducing emotive response. In: ICCC, pp. 140–149 (2010)
15. Rumelhart, D.H.: Learning representations by back-propagating errors. Nature **323**(6088), 533–536 (1986)
16. Mozer, M.C.: Neural network music composition by prediction: exploring the benefits of psychoacoustic constraints and multi-scale processing. Connect. Sci. **6**(2–3), 247–280 (1994)
17. Hochreiter, S., Bengio, Y., Frasconi, P., Schmidhuber, J.: Gradient flow in recurrent nets: the difficulty of learning long-term dependencies. In: Kremer, S.C., Kolen, J.F. (eds.) A Field Guide to Dynamical Recurrent Neural Networks. IEEE Press, New York (2001)
18. Cooper, G.W., Cooper, G., Meyer, L.B.: The Rhythmic Structure of Music. The University of Chicago Press, Chicago (1960)
19. Eck, D., Schmidhuber, J.: A first look at music composition using LSTM recurrent neural networks. Istituto Dalle Molle Di Studi Sull Intelligenza Artificiale **103**, 48 (2002)
20. Google Brain Magenta. https://magenta.tensorflow.org/. Accessed 06 May 2020
21. Clara: A neural net music generator. http://christinemcleavey.com/clara-a-neural-net-music-generator/. Accessed 06 May 2020
22. Mao, H.H.: DeepJ: style-specific music generation. In: 2018 IEEE 12th International Conference on Semantic Computing (ICSC), pp. 377–382 (2018)
23. Tikhonov, A., Yamshchikov, I.P.: Music generation with variational recurrent autoencoder supported by history. arXiv preprint arXiv:1705.05458 (2017)
24. Lin, J.C., Wei, W.L., Wang, H.M.: Automatic music video generation based on emotion-oriented pseudo song prediction and matching. In: Proceedings of the 24th ACM International Conference on Multimedia, pp. 372–376 (2016)
25. Madhok, R., Goel, S., Garg, S.: SentiMozart: music generation based on emotions. In: ICAART, vol. 2, pp. 501–506 (2018)
26. Eerola, T.: Music and emotion dataset (Primary Musical Cues) (2016)
27. Juslin, P.N.: Musical Emotions Explained: Unlocking the Secrets of Musical Affect. Oxford University Press, New York (2019)
28. Eerola, T., Friberg, A., Bresin, R.: Emotional expression in music: contribution, linearity, and additivity of primary musical cues. Front. Psychol. **4**, 487 (2013)
29. Cuthbert, M.S., Ariza, C.: music21: A toolkit for computer-aided musicology and symbolic music data. In: Proceedings of the 11th International Society for Music Information Retrieval Conference, pp. 637–642 (2010)

30. Hochreiter, S., Schmidhuber, J.: Long short-term memory. Neural Comput. **9**(8), 1735–1780 (1997)
31. Kamara, A.F., Chen, E., Liu, Q., Pan, Z.: Combining contextual neural networks for time series classification. Neurocomputing **384**, 57–66 (2020). https://doi.org/10.1016/j.neucom.2019.10.113
32. Huk, M.: Measuring the effectiveness of hidden context usage by machine learning methods under conditions of increased entropy of noise. In: 3rd IEEE International Conference on Cybernetics (CYBCONF 2017), Exeter, UK, pp. 1–6. IEEE Press (2017). https://doi.org/10.1109/CYBConf.2017.7985787
33. Huk, M.: Non-uniform initialization of inputs groupings in contextual neural networks. In: Nguyen, N.T., Gaol, F.L., Hong, T.-P., Trawiński, B. (eds.) ACIIDS 2019. LNCS (LNAI), vol. 11432, pp. 420–428. Springer, Cham (2019). https://doi.org/10.1007/978-3-030-14802-7_36
34. Maheswaranathan, N., Sussillo, D.: How recurrent networks implement contextual processing in sentiment analysis. arXiv preprint arXiv:2004.08013 (2020)
35. Mousa, A., Schuller, B.: Contextual bidirectional long short-term memory recurrent neural network language models: a generative approach to sentiment analysis. In: Proceedings of the 15th Conference of the European Chapter of the Association for Computational Linguistics: Volume 1, Long Papers, Spain, pp. 1023–1032 (2017)
36. Rahman, M.A., Ahmed, F., Ali, N.: Contextual deep search using long short term memory recurrent neural network. In: 2019 International Conference on Robotics, Electrical and Signal Processing Techniques (ICREST), pp. 39–42. IEEE (2019)
37. Svegliato, J., Witty, S.: Deep jammer: a music generation model. Small **6**, 67 (2016)
38. Huang, A., Wu, R.: Deep learning for music. arXiv preprint arXiv:1606.04930 (2016)
39. Wu, J., Chen, X.Y., Zhang, H., Xiong, L.D., Lei, H., Deng, S.H.: Hyperparameter optimization for machine learning models based on Bayesian optimization. J. Electron. Sci. Technol. **17**(1), 26–40 (2019)
40. Srivastava, N., Hinton, G., Krizhevsky, A., Sutskever, I., Salakhutdinov, R.: Dropout: a simple way to prevent neural networks from overfitting. J. Mach. Learn. Res. **15**(1), 1929–1958 (2014)
41. Book, M.S.: Generating retro video game music using deep learning techniques. Master's thesis, University of Stavanger, Norway (2019)

SE-U-Net: Contextual Segmentation by Loosely Coupled Deep Networks for Medical Imaging Industry

Lin-Yi Jiang[1(✉)], Cheng-Ju Kuo[1(✉)], O. Tang-Hsuan[1], Min-Hsiung Hung[2], and Chao-Chun Chen[1]

[1] IMIS/CSIE, NCKU, Tainan, Taiwan
{P96094189,P96084087,chaochun}@mail.ncku.edu.tw
[2] CSIE, PCCU, Taipei, Taiwan
hmx4@faculty.pccu.edu.tw

Abstract. We proposed a context segmentation method for medical images via two deep networks, aiming at providing segmentation contexts and achieving better segmentation quality. The context in this work means the object labels for segmentation. The key idea of our proposed scheme is to develop mechanisms to elegantly transform object detection labels into the segmentation network structure, so that two deep networks can collaboratively operate with loosely-coupled manner. For achieving this, the scalable data transformation mechanisms between two deep networks need to be invented, including representation of contexts obtained from the first deep network and context importion to the second one. The experimental results reveal that the proposed scheme indeed performs superior segmentation quality.

Keywords: Deep learning · Medical imaging segmentation · Computed tomography · Transpose convolution · Contextual computing

1 Introduction

Due to the advance of image processing and computer techniques, the medical imaging equipment become popular in hospitals for rapidly and precisely diagnosing various internal medicine symptoms. One frequently used medical image form is the computed tomography (CT), where the computer system controls the movement of the X-ray source to produce the image for further diagnosis.

This work was supported by Ministry of Science and Technology (MOST) of Taiwan under Grants MOST 109-2221-E-006-199, 108-2221-E-034-015-MY2, and 109-2218-E-006-007. This work was financially supported by the "Intelligent Manufacturing Research Center" (iMRC) in NCKU from The Featured Areas Research Center Program within the framework of the Higher Education Sprout Project by the Ministry of Education in Taiwan.

N. T. Nguyen et al. (Eds.): ACIIDS 2021, LNAI 12672, pp. 678–691, 2021.
https://doi.org/10.1007/978-3-030-73280-6_54

The medical imaging diagnosis technologies have been widely studied in past decades [9], and provide symptom identifying assistance. Thus, medical imaging industries are eager to add the intelligent diagnosis functions to medical systems to increase doctors' diagnosis performance [2].

Traditional medical imaging diagnosis applied image processing techniques to identify symptoms based on their appearant characteristics. These methods worked well for some symptoms, and are hard to be applied to other symptoms, unless the algorithms are modified according to the new symptoms. The advance of artificial intelligence technologies greatly improves medical image processing capacities. These works use convolutional neural network (CNN), fully-connected neural network (FCN), artificial neural network (ANN) to accomplish missions of object detection, classification, and segmentation. Recently, some works [4,12] provide the medical image diagnosis with the deep learning technologies. Notably, the popular works in deep learning techniques [4] performs detection or segmentation for images of general situations, while the medical images have properties of low color differences between foreground and background, making computer systems hard to distinguish segmentation targets. None of them provide solutions for improving quality of detection to medical images. These existing methods might give inaccurate results so that doctors may be with low confidence to the diagnosis support systems and still need complete effort to diagnose symptoms by themselves. On the other hand literature [1,2,5–8] shows that contextual information processing - especially with neural networks - is very effective in increasing quality of classification of data vectors.

In this work, we proposed a contextual segmentation method for medical images via two deep networks, aiming at providing segmentation semantics and achieving better segmentation quality. The semantics in this paper means the object labels for contextual blocks. Previous works achieve each of these separately. In this work, we solve the two issues in a uniform framework, where the object detection deep network (ODDN) is used to extract semantics, and the developed Semantic U-Net (SE-U-Net) is used to perform medical image segmentation. The kernel idea of our proposed scheme is to develop two mechanisms to elegantly transform object detection labels with associated bounding boxes into the deep segmentation network structure. The first is the scalable data exchange interface between ODDNs and SE-U-Net need to be clearly defined, where a context matrix is invented to connect two networks. The second is that the U-Net structure needs to be modified to fit the additional semantic information. We conduct experiments to validate our proposed scheme on the medical images for identifying symptoms of the coronavirus disease 2019 (COVID-19). The experimental results from the COVID-19 dataset reveal that the proposed scheme indeed performs superior segmentation quality.

2 Backgrounds

2.1 Object Detection Deep Networks

The object detection deep networks (ODDNs) have been studied in the last decade, and many famous methods, e.g., YOLOv3 [10], are proposed and widely recognized. Particularly, they are validated in crucial simulation tests and then open-sourced, so that the area of ODDNs rapidly grows up and become one of most important applications in deep network technologies. The main underlying techniques of ODDNs is the CNNs and the FCNs, as the formal ones extract object features automatically and the later ones map the generated features to specified object classes. Many ODDNs are further developed for fitting different application domains. Nowadays, the deep networks are well trained with data sets via graphic processing unit (GPU) for efficiency.

2.2 Segmentation Deep Networks for Medical Images

Segmentation for medical images has a requirement: the processed images need to preserve most properties of original ones. Thus, the deep networks used in medical imaging industries need to bring portions of the original images to outputs for ensuring high similarity. The popular segmentation deep network for medical images is U-Net [11]. The key design of the U-Net is that extracted features are directly copied and moved to the generation stage, so that many principle properties are incrementally added to the output to preserve the essential elements of original images. Many variants of the U-Net structure are also proposed in recent year, such as ResUNet [3], and they all reserved the copy-and-move components.

3 Proposed Contextual Segmentation Scheme

3.1 Overall Architecture

Figure 1 shows the architecture of our proposed contextual segmentation scheme, which includes two deep networks and they are loosely coupled in the architecture with merely data exchange. The first deep network is used to perform object detection for providing contexts, and the second one is used to perform segmentation over medical images. The advantage of the architecture is two-folded: one is easy to watch the effect of each deep network, the other is avoid complicated model creation procedure. The kernel idea of the scheme includes the loose-coupled of two deep networks and the mechanisms of integrating them. The technical details of the whole architecture, model creation, and network integration will be presented in the following subsections.

Assume a medical image, denoted $\mathbf{X}^{[i]}$, retrieved from the medical image database via the DICOM (Digital Imaging and Communications in Medicine) protocol has $w \times h$ pixels with ch channels (shortly, it denotes $w \times h \times ch$). For example, the medical images used in this study are of $416 \times 416 \times 3$. The

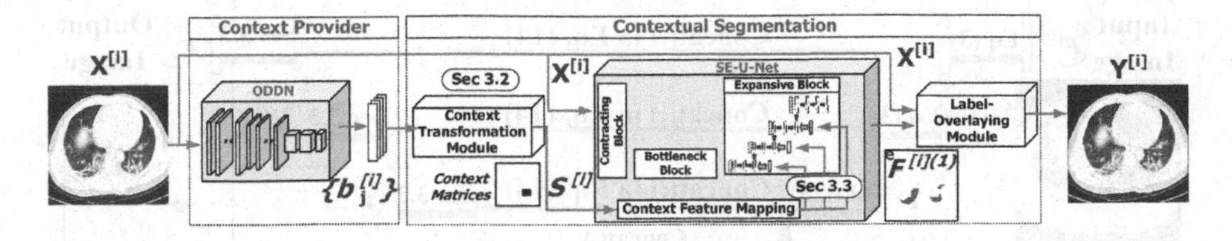

Fig. 1. The proposed contextual segmentation architecture for medical image processing.

output image of the SE-U-Net denotes $\mathbf{Y}^{[i]}$, whose dimension is the same to the input image, i.e., $dim(\mathbf{X}^{[i]}) = dim(\mathbf{Y}^{[i]})$, implying the image size remains after our developed deep network proceeds it. The first network shown in the left-hand side of the figure is an object-detection deep network (ODDN), which generates object detection results in a bounding-box set, denoted BBS, with the format $\{\hat{\mathbf{b}}_j^{[i]} = (\hat{x}_j^{[i]}, \hat{y}_j^{[i]}, \hat{w}_j^{[i]}, \hat{h}_j^{[i]}, \hat{c}_j^{[i]}) | j = 1, \cdots, \hat{k}^{[i]}\}$, where $(\hat{x}_j^{[i]}, \hat{y}_j^{[i]})$ indicates the center of the bounding box (shortly, BBox), $\hat{w}_j^{[i]}$ and $\hat{h}_j^{[i]}$ indicate the width and height, $\hat{c}_j^{[i]}$ is the classification of the BBox, and $\hat{k}^{[i]}$ is the number of BBoxes in $\mathbf{X}^{[i]}$. Since the output format is clearly defined in the architecture, developing deep-network components is flexible and many existing works, such as YOLO, Fast-RCNN, SSD, etc. are considerable candidates. For example, YOLO is used in the experimental study. Without loss of generality, we present the model operation of the ODDN that identifying $\hat{k}^{[i]}$ objects (represented in the form of BBS) for the input $\mathbf{X}^{[i]}$ as

$$M_{ODDN}(\mathbf{X}^{[i]} | \theta_{OD}) = \{\hat{\mathbf{b}}_j^{[i]} | j = 1, \ldots, \hat{k}^{[i]}\} \quad (1)$$

where θ_{OD} is the hyperparameters of the ODDN. For finding out θ_{OD}, the loss function used in the optimizer of the ODDN in this work can be expressed as

$$\begin{aligned} L_{ODDN}(\mathbf{b}^{[i]}, \hat{\mathbf{b}}^{[i]}) &= \sum_{j}^{k} \sum_{\hat{j}}^{\hat{k}} \mathbb{1}^{obj}(\mathbf{b}_j^{[i]}, \hat{\mathbf{b}}_{\hat{j}}^{[i]}) \times \left(\left(x^{[i]_j} - \hat{x}_{\hat{j}}^{[i]} \right)^2 + \left(y_j^{[i]} - \hat{y}_{\hat{j}}^{[i]} \right)^2 \right) \\ &+ \sum_{j}^{k} \sum_{\hat{j}}^{\hat{k}} \mathbb{1}^{obj}(\mathbf{b}_j^{[i]}, \hat{\mathbf{b}}_{\hat{j}}^{[i]}) \times \left(\left(\sqrt{w_j^{[i]}} - \sqrt{\hat{w}_{\hat{j}}^{[i]}} \right)^2 + \left(\sqrt{h_j^{[i]}} - \sqrt{\hat{h}_{\hat{j}}^{[i]}} \right)^2 \right) \\ &+ \sum_{j}^{k} \sum_{\hat{j}}^{\hat{k}} \mathbb{1}^{obj}(\mathbf{b}_j^{[i]}, \hat{\mathbf{b}}_{\hat{j}}^{[i]}) \times (c_j^{[i]} - \hat{c}_{\hat{j}}^{[i]})^2 + \sum_{j}^{k} \mathbb{1}^{noobj}(\hat{\mathbf{b}}^{[i]}, \mathbf{b}_j^{[i]}) + \sum_{\hat{j}}^{\hat{k}} \mathbb{1}^{noobj}(\mathbf{b}^{[i]}, \hat{\mathbf{b}}_{\hat{j}}^{[i]}) \end{aligned} \quad (2)$$

where $\mathbb{1}^{obj}(\mathbf{b}_j^{[i]}, \hat{\mathbf{b}}_{\hat{j}}^{[i]})$ returns 1 if BBoxes of two parameters predict the same object, 0 otherwise; $\mathbb{1}^{noobj}(\mathbf{b}_1, \mathbf{b}_2)$ returns 1 if the object in $\mathbf{b}_2$ is not inside $\mathbf{b}_1$, 0 otherwise.

The second deep segmentation network, shown in the right-hand side of the figure, is called Semantic U-Net (denoted SE-U-Net), which is a variant of the U-Net structure [11]. Figure 2 illustrates the core SE-U-Net structure, consisting of the constructing block for constructing features of medical images, the

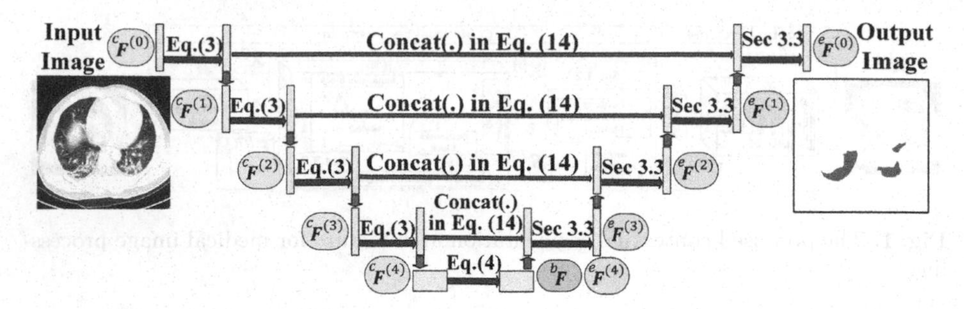

Fig. 2. Illustration of the Core SE-U-Net structure of four levels ($v = 4$) for extracting context from CT images (the feature ${}^{e}\mathbf{F}^{(0)}$ contains generated symptom regions.)

bottlenecking block for transforming features between different feature spaces, and the expensive block for expansive features to medical images. The expensive block will also accommodate the context obtain from the ODDN, which will be formally presented in Sect. 3.3. The key design of the SE-U-Net is that each layer of the constructing block and the expansive block are of the same size and the feature of the former block are directly copy and move to the later block, so that features in the construction block are incrementally added to the output to preserve the essential elements of original images.

The constructing block has v levels and each level performs two convolution operations for feature extraction with one maximum pooling operation for dimension reduction to extracted features. Thus, a feature ${}^{c}\mathbf{F}^{[i](i)}$ of $\mathbf{X}^{[i]}$ generated from level i-th of the constructing block can be expressed as

$$ {}^{c}\mathbf{F}^{[i](i)} = MaxPool(Conv(Conv({}^{c}\mathbf{F}^{[i](i-1)}))), \text{ where } i = 1, \ldots, v. \tag{3} $$

where $MaxPool(.)$ and $Conv(.)$ are the maximum pooling operation and the convolution operation, respectively, and ${}^{c}\mathbf{F}^{[i](0)}$ is the inputted image $\mathbf{X}_i$. The bottleneck block performs two convolution operations, and a feature ${}^{b}\mathbf{F}^{[i]}$ generated in the constructing block for $\mathbf{X}^{[i]}$ can be expressed as

$$ {}^{b}\mathbf{F}^{[i]} = Conv(Conv({}^{c}\mathbf{F}^{[i](v)})) \tag{4} $$

Note that the constructing block and the bottleneck block play as an encoder to extract key features by downsampling procedures via a couple of convolution and max pooling operations. The expansive block has the same number of levels to the constructing block (i.e., v levels) and each level performs two convolution operations and one transposed convolution operation for expanding features obtained from the previous level. A feature ${}^{e}\mathbf{F}^{(i)}$ generated from level i-th of the constructing block is determined by using the feature ${}^{e}\mathbf{F}^{(i+1)}$ and the context provided by the ODDN. Comparing to the previous two blocks, the expansion block plays as a decoder to generate an image from a feature by upsampling procedures. The technical details of ${}^{e}\mathbf{F}^{(i)}$ will be discussed in Sect. 3.3. The specification of the SE-U-Net used in this work are shown in Table 1. The loss function L_{SEU} is defined based on the idea of Dice coefficient for measuring similarity between pixel sets from a CT image and the labeling image, which can represented as the following equation:

$$L_{SEU} = \sum_{k=1}^{N} 1 - \frac{2 \times (\Sigma_{i=1}^{w} \Sigma_{j=1}^{h} \mathbf{Y}_{i,j}^{(k)} \times \hat{\mathbf{Y}}_{i,j}^{SEU(k)})}{(\Sigma_{i=1}^{w} \Sigma_{j=1}^{h} \mathbf{Y}_{i,j}^{(k)} \times \mathbf{Y}_{i,j}^{(k)}) + (\Sigma_{i=1}^{w} \Sigma_{j=1}^{h} \hat{\mathbf{Y}}_{i,j}^{SEU(k)} \times \hat{\mathbf{Y}}_{i,j}^{SEU(k)})} \quad (5)$$

After obtaining the output ${}^{e}\mathbf{F}^{[i](1)}$ of $\mathbf{X}^{[i]}$ from the SE-U-Net, the label-overlaying module would overlay the refined context on the original medical images, and it can be represented in the following equation:

$$\mathbf{Y}_{j,k,l}^{[i]} = \begin{cases} \mathbf{X}_{j,k,l}^{[i]} + {}^{e}\mathbf{F}_{j,k,l}^{[i](1)}, c \in \text{channel red} \\ \mathbf{X}_{j,k,l}^{[i]}, \text{otherwise} \end{cases} \quad (6)$$

For ease of study, all contexts are highlighted in the red channel, and the overlay function can be modified according to different application needs.

Table 1. The layer specification of SE-U-Net studied in this paper, where the input size is 416 × 416 × 3 and the output size is 512 × 512 × 1.

Contacting block	Kernel size	Depth	Strides	Padding
Convolution	3 × 3	64	[1, 1]	SAME
Convolution	3 × 3	64	[1, 1]	SAME
Max Pooling	2 × 2		[2, 2]	VALID
Convolution	3 × 3	128	[1, 1]	SAME
Convolution	3 × 3	128	[1, 1]	SAME
Max Pooling	2 × 2		[2, 2]	VALID
Convolution	3 × 3	256	[1, 1]	SAME
Convolution	3 × 3	256	[1, 1]	SAME
Max Pooling	2 × 2		[2, 2]	VALID
Convolution	3 × 3	512	[1, 1]	SAME
Convolution	3 × 3	512	[1, 1]	SAME
Max Pooling	2 × 2		[2, 2]	VALID
output size	26 × 26 × 512			
Convolution	3 × 3	1024	[1, 1]	SAME
Convolution	3 × 3	1024	[1, 1]	SAME
Convolution	3 × 3	1024	[1, 1]	SAME
Convolution	3 × 3	1024	[1, 1]	SAME
output size	26 × 26 × 1024			

Expansive block	Kernel size	Depth	Strides	Padding
Transpose convolution	2 × 2	512	[2, 2]	VALID
Convolution	3 × 3	512	[1, 1]	SAME
Convolution	3 × 3	512	[1, 1]	SAME
Transpose convolution	2 × 2	256	[2, 2]	VALID
Convolution	3 × 3	256	[1, 1]	SAME
Convolution	3 × 3	256	[1, 1]	SAME
Transpose convolution	2 × 2	128	[2, 2]	VALID
Convolution	3 × 3	128	[1, 1]	SAME
Convolution	3 × 3	128	[1, 1]	SAME
Transpose convolution	2 × 2	64	[2, 2]	VALID
Convolution	3 × 3	64	[1, 1]	SAME
Convolution	3 × 3	64	[1, 1]	SAME
Convolution	3 × 3	1	[1, 1]	SAME

Notice that two deep networks in our architecture are loosely coupled with merely data delivery via the context transformation module. Such design has two advantages. The first is that the two deep networks can be trained individually with less training data, and spend less model creation time compared to a single concated deep network. The second is that context extraction and segmentation are both performed with satisfied high quality in the separated deep networks. In case the single deep network solution is adopted, developers do not know quality of context extraction and segmentation separately as they are all mixtured encoded in the hyperparameters of the single deep network.

Operational Workflows of the Proposed Scheme:
Two operational workflows, model creation and medical image diagnosis, for the proposed context segmentation scheme are presented below.

Model Creation Workflow
Assume a set of medical images $\mathbf{X}^{[i]}$, $i = 1, \ldots, N$ are given and associated labels of $\mathbf{X}^{[i]}$ are collected from domain experts in the form: $\{\mathbf{Y}^{[i]}, \mathbf{c}^{[i]}\}$, where $\mathbf{Y}^{[i]}$ is the associated segmented image, $\mathbf{c}^{[i]}$ is the associated class label set.

Step 1. Create an optimal ODDN model (θ^*_{OD}).
Assume $BBox(\mathbf{Y}^{[i]}, \mathbf{c}^{[i]}) = \{\mathbf{b}^{[i](j)} = (x_b^{[i](j)}, y_b^{[i](j)}, w_b^{[i](j)}, h_b^{[i](j)}, c_b^{[i](j)}) | j = 1, \ldots, k^{[i]}\}$ is an extraction function that returns a set of minimal bounding boxes $\mathbf{b}^{[i](j)}$ covering irregular labeled segmentation shapes in $\mathbf{Y}^{[i]}$, including the center point, width, and height, where $k^{(i)}$ is the number of collected labels in $\mathbf{X}^{[i]}$. Let $M_{OD}(\mathbf{X}^{[i]}|\theta_{OD}) = \{\hat{\mathbf{b}}^{[i](j)} | j = 1, \ldots, \hat{k}^{[i]}\}$ be the output of the ODDN with hyperparameter θ_{OD} to the medical image $\mathbf{X}^{[i]}$. Given labeled dataset $D_{OD} = \{\mathbf{X}^{[i]}, BBox(\mathbf{Y}^{[i]}, \mathbf{c}^{[i]}) | i = 1, \ldots\}$, in this step, We use the training/testing processes in deep learning to find out the optimal hyperparameters for the ODDN network (i.e., θ^*_{OD}), so that the loss function L_{OD} is minimum. That is,

$$\theta^*_{OD} = \arg\min_{\theta; D_{OD}} L_{OD}(M_{OD}(\mathbf{X}^{[i]}|\theta), BBox(\mathbf{Y}^{[i]}, \mathbf{c}^{[i]}) = \{\mathbf{b}^{[i](j)}\}) \qquad (7)$$

Note that the definition of L_{OD} can refer to Eq. (2).

Step 2. Create an optimal SE-U-Net model (θ^*_{SEU}).
Let $M_{SEU}(\mathbf{X}^{[i]}, CM(\mathbf{Y}^{[i]})|\theta_{SEU}) = \hat{\mathbf{Y}}^{SEU[i]}$ be the output of the SE-U-Net with hyperparameter θ_{SEU} to the medical image $\mathbf{X}^{[i]}$, where $CM(\mathbf{Y}^{[i]})$ return the context matrices of $\mathbf{Y}^{[i]}$. Given $D_{SEU} = \{\mathbf{X}^{[i]}, BBox(\mathbf{Y}^{[i]}), \mathbf{Y}^{[i]} | i = 1, \ldots\}$, in this step, we use the training/testing processes to find out the optimal hyperparameters for the SE-U-Net (i.e., θ^*_{SEU}), so that the loss function L_{SEU} of the SE-U-Net, defined in Eq. (5) is minimum. That is,

$$\theta^*_{SEU} = \arg\min_{\theta} L_{SEU}(M_{SEU}(\mathbf{X}^{[i]}, BBox(\mathbf{Y}^{[i]})|\theta), \mathbf{Y}^{[i]}) \qquad (8)$$

The pair $(\theta^*_{OD}, \theta^*_{SEU})$ are the models used in our proposed scheme.

Medical Image Diagnosis Workflow
The dashed arrows in Fig. 1 indicate the medical image diagnosis workflow. Due to length limit, we ignore the detailed steps here.

3.2 Context Matrix Transformation

In this subsection, we present a key mechanism, Context Matrix Transformation (CMT), used in the context transformation module for the scalable data exchange interface between ODDNs and SE-U-Net. The context matrix is scalable as it is adaptive to the medical image size and also accommodate multiple labels in an image. Given a bounding box set $\{\hat{\mathbf{b}}_j^{[i]}\}$ obtained from an ODDN for a medical image, the CMT generates a context matrix, denoted $\mathbf{S}^{[i]}$, whose dimension is the same to the original image $\mathbf{X}^{[i]}$.

Figure 3 illustrates the context matrix transformation, which contains two steps: the matrix-rendering step and the label filling step. The matrix-rendering step is to map the bounding boxes to a matrix with size $w \times h$. Each bounding box $\hat{\mathbf{b}}_j^{[i]}$ of $\mathbf{X}^{[i]}$ has attributes $(\hat{x}_j^{[i]}, \hat{y}_j^{[i]}, \hat{w}_j^{[i]}, \hat{h}_j^{[i]})$, which indicates the corresponding sub-matrix, as shadowed ones in the figure. Each sub-matrix can be indicated by two coordinates: the top-left (tl) point and the bottom-right (br) point, and is represented as

$$tl_j^{[i]} = (tl_j^{[i]}.x, tl_j^{[i]}.y) = ((\hat{x}_j^{[i]} - \hat{w}_j^{[i]}/2) \times w, (\hat{y}_j^{[i]} - \hat{h}_j^{[i]}/2) \times h) \tag{9}$$

$$br_j^{[i]} = (br_j^{[i]}.x, br_j^{[i]}.y) = ((\hat{x}_j^{[i]} + \hat{w}_j^{[i]}/2) \times w, (\hat{y}_j^{[i]} + \hat{h}_j^{[i]}/2)) \times h) \tag{10}$$

Note that for medical imaging applications, each element only belong to one lable at most, that is, it is either marked or unmarked in the context matrix. The tl-br representation is easy to ensure such property, compared to the bounding-box representation. The label filling step is to filling elements in the sub-matrix $(\hat{x}_j^{[i]}, \hat{y}_j^{[i]}, \hat{w}_j^{[i]}, \hat{h}_j^{[i]})$ with the label value $\hat{c}_j^{[i]}$. Other matrix elements are filled with zero, meaning they are unlabeled. After the two steps, all labels of a medical image $\mathbf{X}^{[i]}$ are encoded into a context matrix $\mathbf{S}^{[i]}$ as follows

$$\mathbf{S}_{r,t}^{[i]} = \begin{cases} \hat{c}_j^{[i]}, & \text{if } r \in [tl_j^{[i]}.x, br_j^{[i]}.x] \text{ and } t \in [tl_j^{[i]}.y, br_j^{[i]}.y], \\ 0, & \text{otherwise.} \end{cases} \tag{11}$$

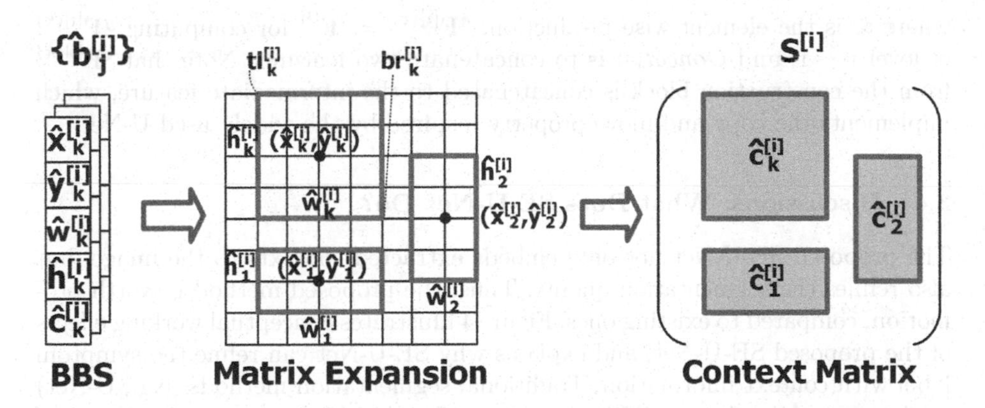

Fig. 3. Illustration of transforming a bounding-box set to a context matrix.

3.3 Computing Expansive Features with Context Matrice

We present another key mechanism in SE-U-Net, Context Feature Mapping (CFM), which is used to refine expansive features in this subsection. In other words, the CFM mechanism is to fuse a context matrix into the core SE-U-Net structure, The inputs of the CFM mechanism include the context matrix $\mathbf{S}^{[i]}$ and the generated features ${}^{c}\mathbf{F}^{[i](k)}$ of the construction block (referring to Fig. 2.)

Let $\mathbf{G}^{[i](k)}$ be the augmented context matrix of the expensive block in the k-th stage and ${}^{g}\mathbf{F}^{[i](k)}$ be the augmented context feature considering $\mathbf{G}^{[i](k)}$. The procedure include two steps. The first step is to produce the augmented context matrices $\mathbf{G}^{[i](k)}$ for $\mathbf{X}^{[i]}$. The purpose of producing $\mathbf{G}^{[i](k)}$ is to scale the context matrix $\mathbf{S}^{[i]}$ to fit the different feature sizes in the k levels of the expensive block. The matrix element $\mathbf{G}^{[i](k)}_{x,y}$ can be computed as the following equation.

$$\mathbf{G}^{[i](k)}_{x,y} = \begin{cases} \hat{c}^{[i]}_j, & \text{if } (x \times \frac{w}{w^{(k)}}) \in [tl^{[i]}_j.x, br^{[i]}_j.x] \text{ and } (y \times \frac{h}{h^{(k)}}) \in [tl^{[i]}_j.y, br^{[i]}_j.y], \\ 0, & \text{otherwise.} \end{cases} \tag{12}$$

The second step is to fuse the augmented context matrix to the expansive block of the core SE-U-Net structure. For level k, the feature ${}^{e}\mathbf{F}^{[i](k+1)}$ obtained in the $(k+1)$-th level is upsampling (via the transpose convolution operation), which is next fused with and the augmented context matrix $\mathbf{G}^{[i](k)}$. The fusion result is then concatenated with the feature ${}^{c}\mathbf{F}^{[i](k)}$, previously defined in Eq. 3, which are then performed convolutions for smoothing the extracted symptoms. The technical details can be represented in the following two formulae.

$${}^{g}\mathbf{F}^{[i](k)} = TransConv({}^{e}\mathbf{F}^{[i](k+1)}) \otimes \mathbf{G}^{[i](k)},\ k = v-1, \ldots, 0. \tag{13}$$

$${}^{e}\mathbf{F}^{[i](k)} = Conv(Conv(Concat({}^{g}\mathbf{F}^{[i](k)}, {}^{c}\mathbf{F}^{[i](k)}))),\ k = v-1, \ldots, 0. \tag{14}$$

where $\otimes$ is the element-wise production, ${}^{e}\mathbf{F}^{[i](v)} = {}^{b}\mathbf{F}^{[i]}$ for computing ${}^{g}\mathbf{F}^{[i](k)}$ of level $v-1$, and $Concat(.)$ is to concatenate two features. Note that ${}^{c}\mathbf{F}^{[i](k)}$ from the construction block is concatenated to the intermediate feature, which implements the copy-and-move property inspired by the widely used U-Net.

3.4 Discussions: What Does SE-U-Net Do?

The proposed SE-U-Net not only embeds extracted contexts to the image, but also refines the segmentation quality. Thus, the proposed method is worth promotion, compared to existing ones. Figure 4 illustrates conceptual working effects of the proposed SE-U-Net, and explains why SE-U-Net can refine the symptom label with context information. Traditional segmentation methods (e.g., U-Net) perform segment the possible symptom region with features extracted in construction levels, and due to the privacy issue concerned in most hospitals, the number of medical images used for creating models is limited. Thus, U-Net obtains less accurate symptom region, as shown in the middle of the figure.

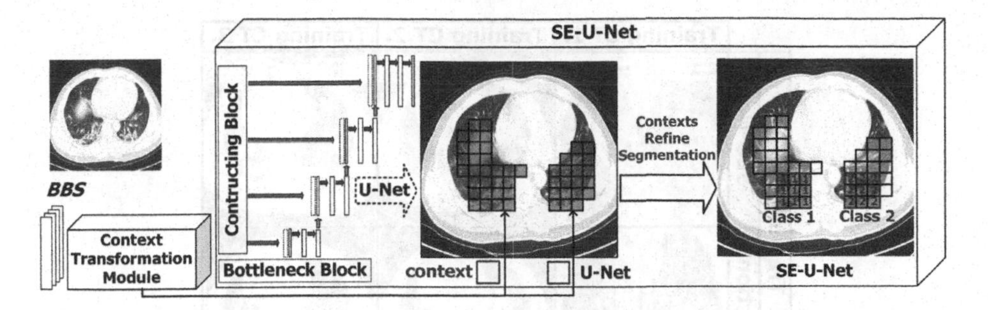

Fig. 4. Illustration of working effects of the proposed SE-U-Net.

For tackling the less quality of a segmentation model, SE-U-Net uses context obtained from another deep network (i.e., ODDN) to further refines the symptom region in the expansive block by means of feature fusion operations (i.e., Eqs. (13) and (14).) In this way, SE-U-Net improves quality of possible symptom regions, as shown in the right of the figure. The conceptual illustration also sustains design of the proposed loose-coupled architecture.

4 Case Study

4.1 System Deployment and Experimental Settings

We deploy the proposed SE-U-Net system on the personal computer with two graphic processing cards of Nvidia GTX 1080 Ti. The personal computer is with Intel i9-7920x processor and 128 GB memory. The system prototype is developed with programming tools mostly used in the deep-learning research, including Python 3.5, TensorFlow 1.6.0, CUDA v9.0.176, and cuDNN 70005. The COVID-CT-Dataset obtained from https://github.com/UCSD-AI4H/COVID-CT has 349 CT images of size 416 × 416 × 3 containing clinical findings of coronavirus disease 2019 (COVID-19) from 216 patients.

4.2 Expr. 1: Visualization Effects of Context Segmentation

Figure 5 shows the visualization effects of the SE-U-Net, where the red is the SE-U-Net, the blue is the human labels, and the pink is the intersection of both the SE-U-Net and the human experts. Three processed CT images are selected from both training and testing datasets for visual validation. We also show the associated object detection results, i.e., contexts provided by the ODDN, for understanding the where the SE-U-Net focuses and how it performs segmentation. From results of the training dataset, we can see most symptoms are marked (i.e., pink color), which indicate the created SE-U-Net model is well trained and performed sufficiently acceptable.

In results of the testing dataset with the created model, the first two cases are randomly selected and the last one is a relative worse instance. The first two

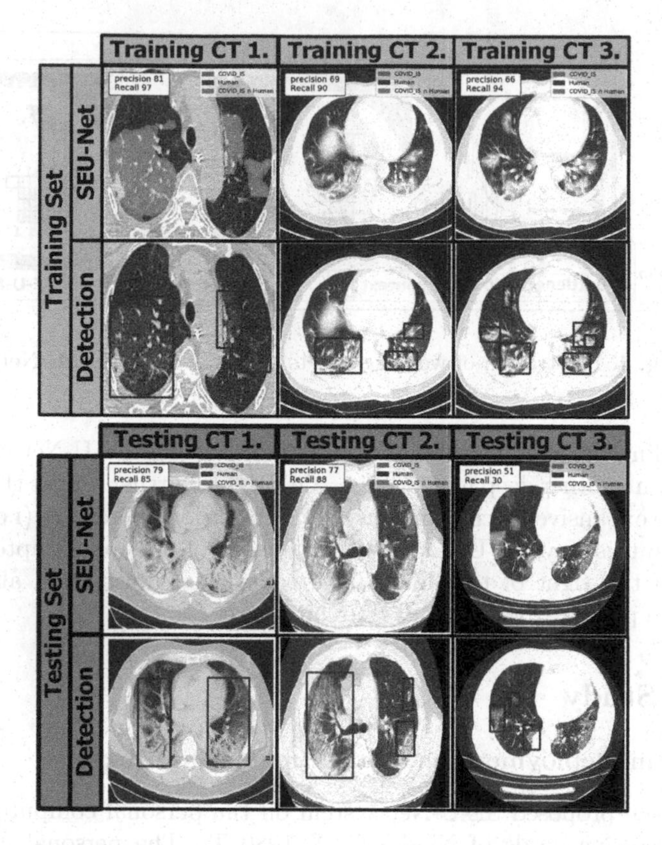

Fig. 5. Visualization effects of the SE-U-Net, where blue boxes are contexts, pink regions are true positives, and blue regions are false negatives. (Color figure online)

shows most marked parts are also recognized by the human experts (i.e., the pink shapes), indicating that our SE-U-Net quite successfully inspects the unseen CT images. In the third one, symptoms in the right lung are missed so that the inspection performance decreases. Note that the ODDN does not provide any context to the SE-U-Net in this instance. In this situation, the segmentation job is performed merely depending on the SE-U-Net model, and its performance in the case is similar to that of the traditional U-Net.

4.3 Expr. 2: Comparisons of Segmentation Methods

This experiment studies the effects of our SE-U-Net and existing U-Net and the results are shown in Fig. 6, where the same CT images adopted in the first experiment are used for fair comparisons. The two models are well trained as possible. From the results of the training dataset, we can see that the SE-U-Net and the U-Net mark most human labeling regions (i.e., high recall value), meaning that both models have sufficient capacity to inspect symptoms. By further analysis based on the precision value, we can see that the SE-U-Net has lower false positives than the U-Net, indicating that the SE-U-Net provides more accurate information to doctors. Results of the testing dataset have similar phenomena, which supports the above segmentation effects of the two methods.

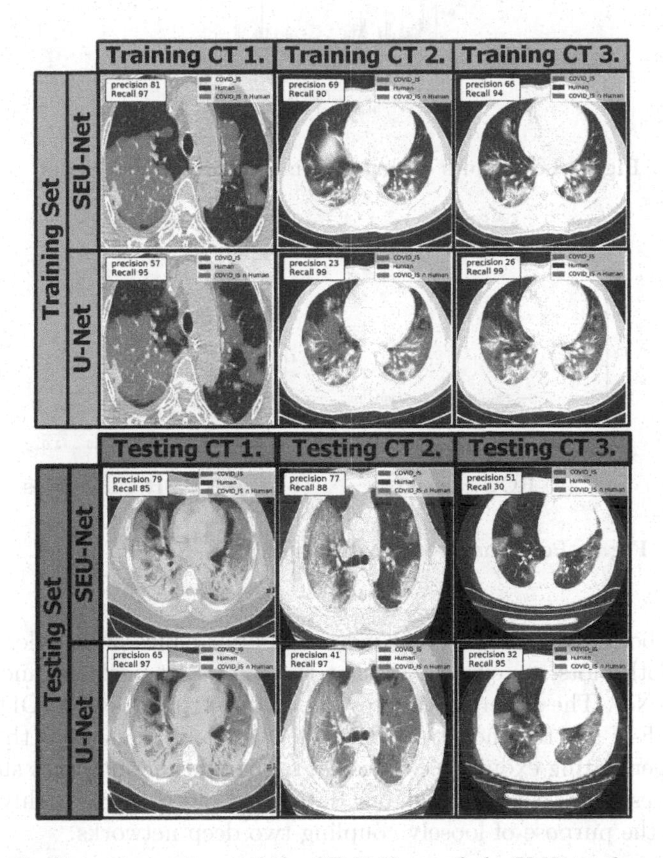

Fig. 6. Visualization comparisons of the SE-U-Net and the U-Net, where pink regions are true positives, red regions are false positives, and blue regions are false negatives. (Color figure online)

4.4 Expr. 3: Quantitative Comparisons

We present quality of the created model used in experiments and comparisons of deep networks in this experiment. Figure 7 shows the loss-function value in various epoches during creating the SE-U-Net model. We can see that the loss-function values of the training dataset are close to zero after 2000 epoches, meaning that the created model has capacity to identify symptoms. Values of the validation dataset are also quite low, verifying that the model is qualified to perform context segmentation tasks.

Figures 8 shows comparisons of our SE-U-Net and the existing U-Net via the receiver operating characteristic (ROC) curve and the precision-recall distribu-

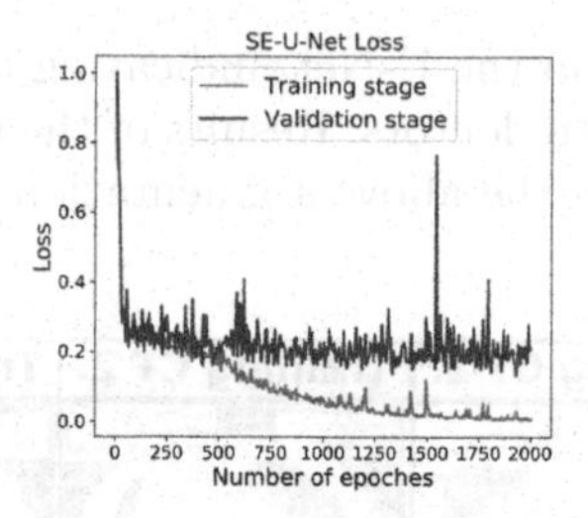

Fig. 7. Loss values during creating the SE-U-Net model.

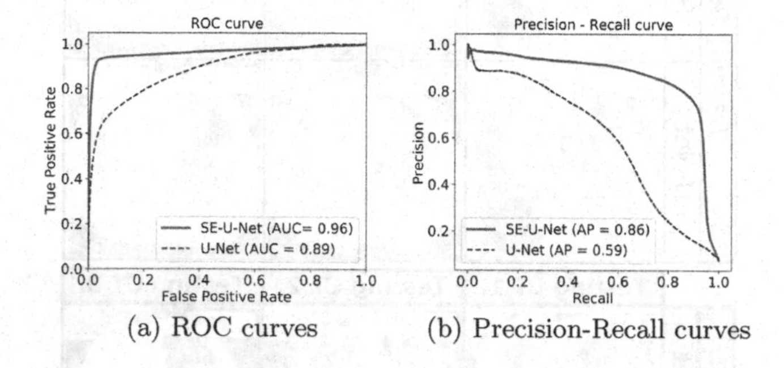

(a) ROC curves (b) Precision-Recall curves

Fig. 8. Performance comparisons of SE-U-Net and U-Net.

tion, which both are widely used to measure performance of deep learning models. From both plots, we clearly see that the SE-U-Net performs more superior than the U-Net. The experimental results confirm that the fore ODDN indeed assists the SE-U-Net to concentrate on the localized region, so that the SE-U-Net focuses on generating exquisite contextual segmentation outcomes shown previously. The results also verify that our developed mechanisms highly effectively accomplish the purpose of loosely coupling two deep networks.

5 Conclusions and Future Work

In this paper, we proposed SE-U-Net to provide contexts and high-quality segmentation to CT images for assisting doctors to diagnose symptoms. Traditional segmentation only consider symptom identification with a single deep networks, which could be less accurate. Our proposed SE-U-Net can employ object detection deep networks for acquiring contexts with bounding boxes, and then loosely combine those contexts to the attentioned U-Net for further refining the segmentation quality. We also give analysis to explain reasons that the SE-U-Net can refine the segmentation quality. We developed the SE-U-Net prototype and conducted experiments to test its performance. The experimental results revealed that the proposed SE-U-Net indeed performs superior than existing methods in metrics concerned by hospital experts. Our future work will extend the SE-U-Net to diagnose other organs, such as the pancreas, which may be less apparent in medical images. Certain mechanisms sensitive to such unapparent symptoms need to be additionally invented for the SE-U-Net.

References

1. Bovolo, F., Bruzzone, L.: A context-sensitive technique based on support vector machines for image classification. In: Pal, S.K., Bandyopadhyay, S., Biswas, S. (eds.) PReMI 2005. LNCS, vol. 3776, pp. 260–265. Springer, Heidelberg (2005). https://doi.org/10.1007/11590316_36
2. Cai, G., et al.: One stage lesion detection based on 3D context convolutional neural networks. Comput. Electr. Eng. **79**, 106449 (2019)
3. Diakogiannis, F.I., Waldner, F., Caccetta, P., Wu, C.: ResUNet-a: a deep learning framework for semantic segmentation of remotely sensed data (2019)
4. Ghamdi, M.A., Abdel-Mottaleb, M., Collado-Mesa, F.: DU-Net: convolutional network for the detection of arterial calcifications in mammograms. IEEE Trans. Med. Imaging **39**, 3240–3249 (2020)
5. Ghimire, B., Rogan, J., Miller, J.: Contextual land-cover classification: incorporating spatial dependence in land-cover classification models using random forests and the Getis statistic. Remote Sens. Lett. **1**, 45–54 (2010)
6. Huk, M.: Non-uniform initialization of inputs groupings in contextual neural networks. In: Nguyen, N.T., Gaol, F.L., Hong, T.-P., Trawiński, B. (eds.) ACIIDS 2019. LNCS (LNAI), vol. 11432, pp. 420–428. Springer, Cham (2019). https://doi.org/10.1007/978-3-030-14802-7_36
7. Huk, M., Mizera-Pietraszko, J.: Context-related data processing in artificial neural networks for higher reliability of telerehabilitation systems. In: 2015 17th International Conference on e-health Networking, Application & Services (HealthCom), pp. 217–221 (2015)
8. Kamara, A.F., Chen, E., Liu, Q., Pan, Z.: Combining contextual neural networks for time series classification. Neurocomputing **384**, 57–66 (2020)
9. Litjens, G., et al.: A survey on deep learning in medical image analysis. Med. Image Anal. **42**, 60–88 (2017)
10. Redmon, J., Divvala, S., Girshick, R.B., Farhadi, A.: You only look once: unified, real-time object detection. In: 2016 IEEE Conference on Computer Vision and Pattern Recognition (CVPR), pp. 779–788 (2016)
11. Ronneberger, O., Fischer, P., Brox, T.: U-net: convolutional networks for biomedical image segmentation. In: MICCAI (2015)
12. Wang, X., et al.: A weakly-supervised framework for COVID-19 classification and lesion localization from chest CT. IEEE Trans. Med. Imaging **38**, 2615–2625 (2020)

Contextual Soft Dropout Method in Training of Artificial Neural Networks

Tu Nga Ly[1], Rafał Kern[2](✉), Khanindra Pathak[3], Krzysztof Wołk[4], and Erik Dawid Burnell[5]

[1] Ho Chi Minh City International University, Ho Chi Minh City, Vietnam
ltnga@hcmiu.edu.vn

[2] Faculty of Computer Science and Management, Wroclaw University of Science and Technology, Wroclaw, Poland
rafal.kern@pwr.edu.pl

[3] Indian Institute of Technology, Kharagpur, India
khanindra@mining.iitkgp.ernet.in

[4] ALM Services Technology Group, Wroclaw, Poland

[5] Science Applications International Corporation, Cookeville, USA

Abstract. Research upon artificial neural networks has yielded regularization techniques in order to reduce overfitting, reduce weight increase rate, and in consequence improve model performance. One of the commonly used regularization technique is dropout. In order to make the technique more resistant to emerging interdependencies while learning, the dropout has been generalized to its improved version called Soft Dropout. In this paper we have proposed an extended version of Soft Dropout called Contextual Soft Dropout. This technique is dedicated to be used with Contextual Neural Networks and utilizing their multi-step aggregation functions. Experimental results have shown improvement in overfitting decrease and have achieved lower logloss values for used real-life and benchmark data sets.

Keywords: Dropout · Soft Dropout · Muting factor · Overfitting

1 Introduction

One of the most important computational models in contemporary computer science are artificial neural networks (ANN). They have been successfully incorporated into various fields of life, including biology [1], physics [2, 3], medicine [4, 5] and various other fields [6–11]. Multiple architectures of neural network-based models have been introduced in recent decades. These include e.g. Multi-Layer Perceptron (MLP) networks, Kohonen Self-Organizing Neural Networks [12], Recurrent Neural Networks [13–15], Convolutional Neural Networks [16]. New developments in this field include deep generative models, such as Generative Adversarial Networks (GAN) [17], Deep Belief Networks, Variational Autoencoders (VAE) [18].

The common feature of all aforementioned models is their static character in terms of connection activation between neurons. Typically, training and model construction

N. T. Nguyen et al. (Eds.): ACIIDS 2021, LNAI 12672, pp. 692–703, 2021.
https://doi.org/10.1007/978-3-030-73280-6_55

forces the network to activate each and every connection for all input vectors end eventually additional features based on their pre-processing [19, 20]. To overcome this issue contextual solutions are being considered [21, 22], including a new family of neural networks, namely Contextual Neural Networks (CxNN) [23–26].

The model of CxNN is based on MLP and has been verified both in terms of real-life and benchmark data. CxNN usage include such problems as: analysis of minutia groups in fingerprint detection systems [27, 28], predicting capacity of communication channels in cognitive radio [29] and analyses of cancer gene expression microarray data [30–33]. Other applications include rehabilitation and context-based text mining [34–36]. CxNNs' characteristics comprise the possibility to reduce average activity of neural links. The purpose for this is to restrict the time cost of calculation of neural networks' outputs. Moreover, such approach can increase accuracy, achieved by using context data to adaptively filter signals being processed [23, 31].

The aggregation of input data in contextual neural networks is conditional: the input signals need to fulfill specified criteria in order to be aggregated. This makes the aggregation function more complex than simple ones (e.g. weighted sum) – its evaluation is split into multiple steps [23, 24]. Each processing step results with aggregation of selected subset of inputs which is used not only to calculate output of a neuron but also to evaluate when aggregation should stop. Such approach results with contextual behavior of neurons: a neuron can set its output taking into consideration varying subsets of inputs for different input vectors. The definition and prioritization of the latter is performed during the training of a network. Resulting neuron groups, listed in a specific order have properties of scan-paths. The neurons of CxNN, taking their context into consideration, are practical usage of Stark's scan-path theory, in the same time mimicking basic elements of the sensoric system of humans [37].

In order to build a contextual neural network a modification of conventional training algorithm is required that adds possibility of construction of needed scan-paths. An example of such algorithm is generalized error backpropagation algorithm (GBP) [23, 24, 32, 33]. This method uses self-consistency paradigm in order to optimize weights (continuous parameters) and grouping (discrete parameters) of neuron inputs with gradient-based approach [38]. Authors of [24] have indicated that GBP can train CxNNs with different aggregation functions, including CFA, OCFA, RDFA, PRDFA and Sigma-if (SIF).

Compared to MLP models, CxNN with number of groups above than one can benefit with better accuracy of classification – as the literature shows for various standard test sets [23, 27–31]. This, possibly, happens due to the following facts. Firstly, multi-step conditional aggregation of inputs can constrain the impact of noise in the training data by not aggregating inputs considered as less relevant. Secondly, during the learning process scan-path modification occurs, and their activation change, when different training vectors are processed. As a result, for various data vectors, in different training epochs, the network decides, whether the input should be aggregated, based on current state of the neuron and portion of information being processed. This behavior of CxNNs can be characterized as partially similar to dropout regularization [24, 39].

Nevertheless, some of the common issues in neural-network based machine learning still occur in contextual models. These include vanishing or exploding gradients and overfitting to training data. To tackle them, methods of training regularization have been developed. Popular examples of such techniques are max-norm, L2 and dropout. The dropout is based on alternate disabling of random groups of neurons in subsequent steps of training [39–41]. Such approach transforms a learning process of a neural model into an ensemble learning, where the base classifiers are subnetworks, which share the information and partial outcomes for given data vectors. Finally, all the subnetworks are used together, as a model which has an initial architecture [42, 43]. Alternatively, during signal processing in Contextual Neural Networks there is higher degree of freedom, due to the fact that each connection can be enabled or disabled, opposed to disabling whole neurons by dropout. It is worth to notice that this operation in CxNNs is not random – it is determined by a context of signals being processed and knowledge represented by the whole neural network. Last, but not least, the adaptive disabling of neural links occurs not only during training, but also beyond it. This leads to considerable decrease of activity of links between neurons [23, 24].

It was also analyzed that regularization formed by contextual neural networks can be combined with improved version of dropout called Soft Dropout [44]. Its usage decreases overfitting and increases classification properties of neuronal models. This suggests the need to explore properties of similar but more advanced solutions. Thus in this work we propose improved version of the Soft Dropout regularization, called further Contextual Soft Dropout (CxSD). CxSD is dedicated to be used during training of Contextual Neural Networks and allows to exploit their multi-step aggregation functions for more precise and neuron-wise control of levels of the Soft Dropout. It is shown that such operation can improve properties of the training process and of resulting models.

The further parts of the paper are constructed as follows. After the introduction, in section two we recall basic principles of the Soft Dropout method. Next, in the third section the details of operation of proposed Contextual Soft Dropout method are discussed. Later, in section four results of experiments are presented showing effects of training Contextual Neural Networks with Generalized Backpropagation method and CxSD for selected benchmark [45] and real-life problems (ALL-AML leukemia gene expression microarray data) [46]. Finally a summary is given with conclusions and directions for further research on the usage of CxSD in training of Contextual Neural Networks.

2 Soft Dropout in Neural Networks Training

The classical dropout is applied to the whole neuron. However, partial dropout methods have been introduced. In such case, only some subsets of neurons are enabled or disabled. One can distinguish at least two approaches for this task. A first one incorporates the features of CxNN, where some connections do not have to change state. The second one – Soft Dropout – is based on a partial modification of neuron output values used for smooth control on the strength of the dropout [44]. In Soft Dropout neuron in a "dropped" state is not blocked. Instead its output value is multiplied by the muting factor $1 - \psi$, and as such is used in given training step. Thus the output function Y of the neuron applying Soft Dropout has been defined as follows:

$$Y\left(w_k^l, x_p, b_k^l\right) = F\left(\varphi\left(w_k^l, x_p, b_k^l\right)\right) * (1 - \psi), \tag{1}$$

where F is the selected activation function, ψ is the neuron' muting factor, φ is an aggregation function, w_k^l and b_k^l state for the sets of weights and biases of k-th neuron in layer l and x_p is a vector of neuron inputs values for training pattern p.

It was also shown that if the N-input neuron is using standard aggregation function given by the formula:

$$\varphi\left(w_k^l, x_p, \theta_k^l\right) = \sum_{j=1}^{N} (w_{kj}^l * x_{pj} + b_k^l), \tag{2}$$

then the weights update can be performed with the following equation:

$$\Delta w_{kj}^l = \eta * \frac{\partial E_p}{\partial w_{kj}^l} = -\frac{\partial Q}{\partial w_{kj}^l} * \left(\frac{\partial (F \circ \varphi)}{\partial w_{kj}^l} * (1 - \psi)\right), \tag{3}$$

where η is the length of the learning step, E_p is the output error for the pattern p and Q is loglos loss function. It can be observed that the derivative of the loss function Q multiplied by the derivative of the activation function F is regulated by a Soft Dropout fading factor $1 - \psi$. When the neuron is not dropped out the modification of weights is done as for the model without dropout. For "dropped" neurons weights are updated proportionally to the value of $1 - \psi$.

3 Contextual Soft Dropout Method

The important element of the Soft Dropout method analyzed earlier is the constant value of the activation fading factor [43]. Such solution is simple and can be applied also during training of not contextual neural networks. But the assumption that activation fading should be constant for all neurons and during the whole training process can be not optimal. Especially for contextual neural networks this would not take into account that the same neurons can aggregate signals from different subsets of theirs inputs for different data vectors. When contextual neurons are limiting their aggregation process to the most important groups of inputs they base their activation on the most important signals. In such situations their importance for the calculation of the outputs of the neural network can be higher than when they have problems with interpretation of the most important signals and need to analyze also the less important groups of inputs. Thus the influence of the Soft Dropout on the training process of the neural network can be different for various levels of activity of inputs of the given neuron.

In the result of the above we propose a Contextual Soft Dropout regularization method which assumes the relationship between activity of scan path represented by multi-step aggregation function of contextual neuron and the fading of its activation value. In this paper we are giving the solution for the Contextual Neural Networks using Sigma-if function. The Sigma-if aggregation is defined by the formula:

$$\varphi_{Sigma-if}(\mathbf{w}, \mathbf{x}, \boldsymbol{\theta}) = \sum_{g=1}^{g*} \sum_{i=1}^{N_{xn}} w_i x_i \delta(g, \theta_i) \tag{4}$$

where $\delta(g, \theta_i)$ is the Kroneckers' delta function for g-th step of inputs aggregation and i-th input of the neuron belonging to the group θ_i and $g*$ is the maximal index of the group analyzed by the neuron for the given input vector $\boldsymbol{x}$.

Following the notation used in (2), and omitting the bias term in the list of parameters, the output value of the contextual neuron when Contextual Soft Dropout is used can be proposed in the following form:

$$Y\left(w_k^l, x_p, \theta_k^l\right) = F\left(\varphi\left(w_k^l, x_p, \theta_k^l\right)\right) * (1 - \psi) * \left(1 + \beta / A\left(w_k^l, x_p, \theta_k^l\right)\right), \quad (5)$$

where $A(\cdot) \in [\frac{1}{G}, 1]$ is an activity of the neuron inputs for the given data vector $\boldsymbol{x}$ with G groups of inputs of equal size, and $\beta \in R$ is the constant factor controlling the strength of the contextual term of CxSD regularization. The activity A of the neuron inputs is defined with the formula:

$$A\left(w_k^l, x_p, \theta_k^l\right) = \frac{1}{G} \sum_{j=1}^{N} H\left(g_j^{*l} - \theta_j^l\right), \quad (6)$$

where G is the maximal number of the group of connections of the neuron and H is the Heaviside' step function. g_j^* and θ_j are the maximal index of the group analyzed by the neuron for the given input vector x and the index of the group to which belongs connection j-th, respectively. In the effect, for the given level of fading ψ and positive values of β near zero, neurons that analyze low number of groups of inputs are muted less in the following cycle of backpropagation than neurons that analyze high number of inputs.

Under above assumptions, to be able to update weight w_{kj}^l of the j-th connection of the k-th neuron in layer l during backpropagation training, there is a need to calculate modified partial derivative of the loss function $\frac{\partial E_p}{\partial w_{kj}^l}$. It was shown in [23] that the partial derivative of the Sigma-if aggregation function of the neuron k in the layer $l+1$ over the activation u of the neuron j in layer l is:

$$\frac{\partial \varphi_k^{(l+1),p}}{\partial Y_j^{l,p}} = w_{k,j}^{l+1} H(g_k^{*(l+1),p} - \theta_{k,j}^{l+1}) \quad (7)$$

The activity of connections A given by (6) is a multi-step function, which is constant in G consecutive intervals except the points where it is not defined. Thus its classical derivative over Y is zero in each interval where A is constant, and its' distributional derivative is zero in the points of discontinuity. This simplifies the final calulation of the derivatives needed to update the weights during the backward propagation phase, and allows to write that:

$$\Delta w_{k,j}^l = \eta \frac{\partial E_p}{\partial w_{kj}^l} \frac{\partial (Q \circ \delta)}{\partial w_{k,j}^l} == -\eta \frac{\partial Q}{\partial w_{k,j}^l} \left(\frac{\partial (F \circ \varphi)}{\partial w_{k,j}^l} * (1 - \psi) * \left(1 + \frac{\beta}{A}\right)\right). \quad (8)$$

Finally, when CxSD regularization is used, the detailed weight update rule for the j-th input of the k-th neuron in the l-th hidden layer of the Sigma-if contextual neural network for the data pattern p has the form:

$$\Delta w_{k,j}^{l} = \eta Y_j^{(l-1),p} H(g_k^{*l,p} - \theta_{k,j}^{l}) F'(\varphi_k^{l,p}) \cdot (1 - \Psi)(1 + \beta / A_j^{l,p}) \cdot \sum_{i=1}^{n_{l+1}} \delta_i^{(l+1),p} w_{i,k}^{l+1} H(g_i^{*(l+1),p} - \theta_{i,k}^{l+1})(1 - \Psi)(1 + \beta / A_i^{l+1,p}). \tag{9}$$

It can be noticed that formula (9) is generalization of the original weight update rule defined for genenralized error backpropagation algorithm [23]. Heaviside function H is element characteristic for contextual neural networks with multi-step aggregation functions. It causes that the backward pass of the GBP algorithm modifies weights of only those connections which were active during the forward phase. And the right-most terms Ψ, β and A appear due to the usage of Contextual Soft Dropout.

4 Results of Experiments

The proposed Contextual Soft Dropout method was verified during experiments in which MLP and contextual neural networks were trained with the usage of H2O framework [47]. Both types of models were constructed by the Generalized BP method to classify vectors of ALL-AML leukemia gene expression data collected by Golub [46] and of four selected benchmark sets from the UCI Machine Learning Repository: Breast Cancer, Crx, Heart Cancer and Sonar [45]. Table 1 includes the most important attributes of those data sets as well as related architectures of neuronal models used during experiments.

Considered settings of the GBP training algorithm were: weighted sum or Sigma-If aggregation, hard ReLU activation, training step $= 0.1$, maximal number of training epochs $= 1000$, loss function: Log-Loss. Settings for CxNNs: initial connections grouping: single group, number of groups $G = 7$, aggregation threshold $\varphi* = 0.6$, groups actualization interval $= 25$ epochs. Optimization: Stochastic Gradient Descent, mini-batch size $= 1$, weights initialization: Xavier. Probability of dropout of hidden neurons $= 0.5$. Analyzed values of muting factor $\psi = \{0.7, 0.9, 1.0, 1.3\}$. Strength of the contextual term of CxSD $\beta = 0.1$ was selected during initial experiments.

Table 1. Attributes of considered data sets and related architectures of neural networks.

Dataset	Hidden neurons	Inputs of neural network	Classes (network outputs)	Data vectors
Breast cancer	50	9	2	699
Crx	50	60	2	690
Heart cancer	10	28	2	303
Sonar	30	60	2	208
Golub	20	7130	2	72

For each set of parameters experiments were conducted 50 times – each measurement point is the result of repeated five times 10-fold cross-validation. Statistical analysis of results included calculation of average values and standard deviations of logloss measure for training and testing data.

Analysis of the Soft Dropout method given in [43] revealed that values of the muting factor ψ above 1.0 help to considerably decrease the overfitting of the trained neural networks. Most probably this is due to the fact, that in such situation neurons which are chosen to be "disabled" work in opposition to their functioning when they are "enabled". In the "disabled" state they create a distracting subnetwork that forces the "enabled" part of the model to deal with additional source of problematic signals. On the other hand outcomes presented in [44] can be also used as reference results for Contextual Soft Dropout method for strength of the contextual term of CxSD $\beta = 0$. Thus the following

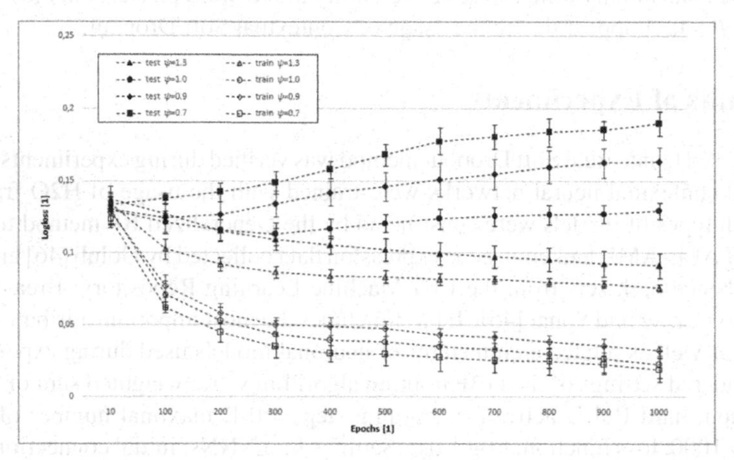

Fig. 1. The average levels of logloss measure observed with 5 × 10-fold CV used to train CxNN neural networks to solve "Breast Cancer" problem for selected values of muting factor ψ. Contextual Soft Dropout regularizer with $\beta = 0.1$ and GBP were used during training.

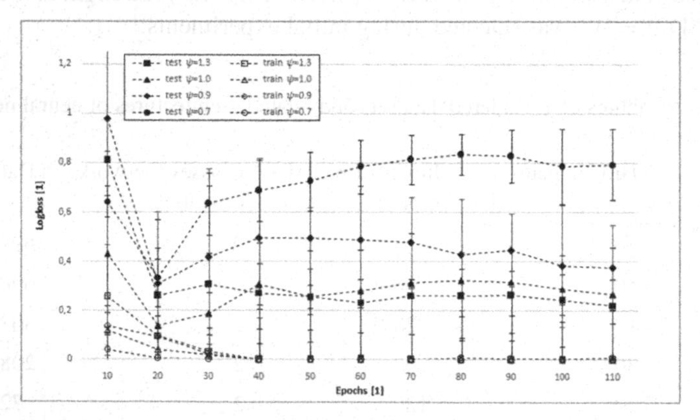

Fig. 2. The average levels of logloss measure observed with 5 × 10-fold CV used to train CxNN neural networks to solve "Golub" problem for selected values of muting factor ψ. Contextual Soft Dropout regularizer with $\beta = 0.1$ and GBP were used during training.

observations are extension of the earlier experiments, but in this case – with the example strength of the contextual term of CxSD $\beta = 0.1$. This enables the CxSD to show its' basic properties. The measurements for Sigma-if models are presented on Fig. 1 and Fig. 2.

It can be observed on Fig. 1 and Fig. 2 that with increasing value of muting factor ψ the overfitting decreases. And for $\psi > 1$ overfitting is lower than for standard dropout. This agrees with results given in [44]. But there are also differences. First is that at the beginning of the training the average logloss values are higher for $\beta = 0.1$ than when $\beta = 0$. The differences are high - between 120 and 400% for training data and 35% for test data. But this initial worsening of results vanishes during the draining. At the end of training the average values of logloss for test data are lower 14 to 19% for CxSD in comparison to plain Soft Dropout. This shows, as expected, that the Contextual

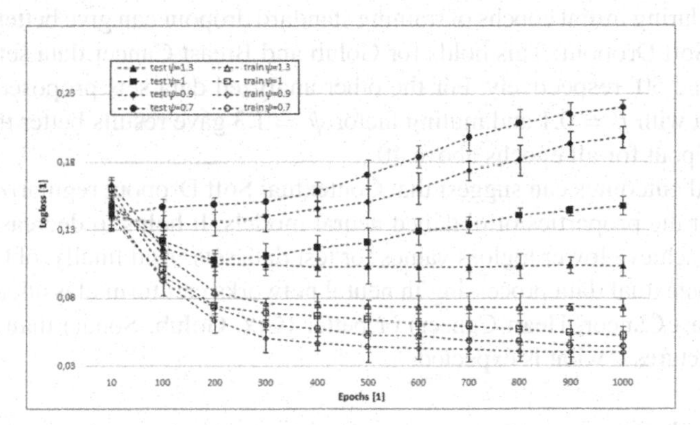

Fig. 3. The average levels of logloss measure observed with 5 × 10-fold CV used to train MLP neural networks to solve "Breast Cancer" problem for selected values of muting factor ψ. Contextual Soft Dropout regularizer with $\beta = 0.1$ and GBP were used during training.

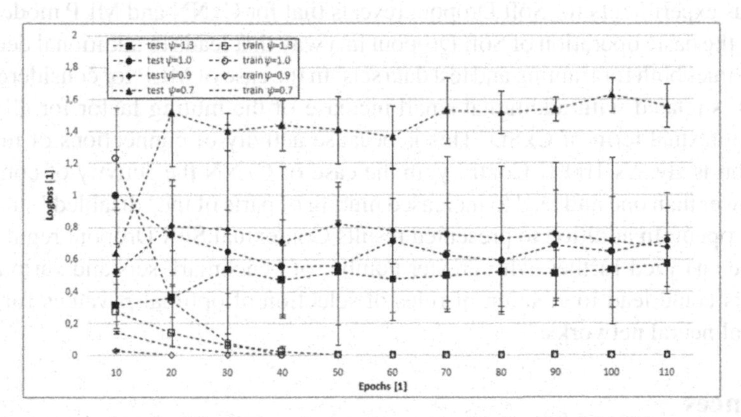

Fig. 4. The average levels of logloss measure observed with 5 × 10-fold CV used to train MLP neural networks to solve "Golub" problem for selected values of muting factor ψ. Contextual Soft Dropout regularizer with $\beta = 0.1$ and GBP were used during training.

Soft Dropout can lead to better results than Soft Dropout. Interestingly, for the "Breast Cancer" dataset, CxSD decreased classification errors both for test and training data.

In a second run it was verified how Contextual Soft Dropout changes properties of MLP neural networks. Outcomes for Breast Cancer and Golub data sets are given in Fig. 3 and Fig. 4, respectively.

The above figures indicate that also for MLP Networks, the proposed Contextual Soft Dropout method with $\beta = 0.1$ and muting factor $\psi = 1.3$ can decrease logloss values and the overfitting better than a standard dropout and Soft Dropout.

For the rest of analyzed data sets, i.e., Crx, Heart Cancer and Sonar, the outcomes were analogous as the measurements presented for Golub and Breast Cancer benchmarks. In each case, average values of loglos measure for test data are lower for Contextual Soft Dropout with $\beta = 0.1$ and muting factor $\psi = 1.3$ than when standard dropout was used, and the increasing value of ψ results with decreasing of overfitting. But it can be also noticed that during initial epochs of training standard dropout can give better results than Contextual Soft Dropout. This holds for Golub and Breast Cancer data sets till around epoch 200 and 50, respectively. For the other analyzed data sets proposed Contextual Soft Dropout with $\beta = 0.1$ and muting factor $\psi = 1.3$ gave results better than standard and Soft Dropout for all epochs above 40.

Presented outcomes can suggest that Contextual Soft Dropout regularization can be beneficial for the properties of obtained neural models. It helps to decrease overfitting as well as to achieve lower logloss values for test data sets. And finally, obtained results show that contextual data processing in neural networks can form classification models similar (Breast Cancer, Heart Cancer) or better (Crx, Golub, Sonar) than MLP of the same architectures – what is expected.

5 Conclusions

Experiments discussed in this paper suggest that Contextual Soft Dropout method can be used as efficient regularization technique. Additionally, comparison with results of analogous experiments for Soft Dropout reveals that for CxNN and MLP models CxSD modifies the basic operation of Soft Dropout in a way that leads to additional decrease of logloss values both for training and test data sets. In the case of MLP for considered values of β this is related with additional small increase of the muting factor for all neurons by the contextual term of CxSD. This is because activity of connections of multilayer perceptron is always 100%. Contrary, in the case of CxNN the activity of connections can be lower than one and lead to increased muting of parts of the "disabled" subnetwork during dropout. In addition to presented results Contextual Soft Dropout regularization should be analyzed further, with greater number of benchmark sets and various values of β. This could lead to creation of rules of selection of optimal β values for training contextual neural networks.

References

1. Mendez, K., Broadhurst, D., Reinke, S.: The application of artificial neural networks in metabolomics: a historical perspective. Metabolomics **15**(11), 1–14 (2019). https://doi.org/10.1007/s11306-019-1608-0

2. Zhang, Q., et al.: Artificial neural networks enabled by nanophotonics. Light Sci. Appl. **8**(1), 14 (2019)
3. Guest, D., Cranmer, K., Whiteson, D.: Deep learning and its application to LHC physics. Annu. Rev. Nucl. Part. Sci. **68**, 1–22 (2018)
4. Nasser, I.M., Abu-Naser, S.S.: Lung cancer detection using artificial neural network. Int. J. Eng. Inf. Syst. (IJEAIS) **3**(3), 17–23 (2019)
5. Suleymanova, I., et al.: A deep convolutional neural network approach for astrocyte detection. Sci. Rep. **8**(12878), 1–7 (2018)
6. Chen, S., Zhang, S., Shang, J., Chen, B., Zheng, N.: Brain-inspired cognitive model with attention for self-driving cars. IEEE Trans. Cogn. Dev. Syst. **11**(1), 13–25 (2019)
7. Liu, L., Zheng, Y., Tang, D., Yuan, Y., Fan, C., Zhou, K.: Automatic skin binding for production characters with deep graph networks. ACM Trans. Graph. (SIGGRAPH) **38**(4), 12 (2019). Article 114
8. Gao, D., Li, X., Dong, Y., Peers, P., Xu, K., Tong, X.: Deep inverse rendering for high-resolution SVBRDF estimation from an arbitrary number of images. ACM Trans. Graph. (SIGGRAPH) **38**(4), 15 (2019). Article 134
9. Tsai, Y.C., et al.: FineNet: a joint convolutional and recurrent neural network model to forecast and recommend anomalous financial items. In: Proceedings of the 13th ACM Conference on Recommender Systems RecSys 2019, pp. 536–537. ACM, New York (2019)
10. Batbaatar, E., Li, M., Ho Ryu, K.: Semantic-emotion neural network for emotion recognition from text. IEEE Access **7**, 111866–111878 (2019)
11. Wang, Z.H., Horng, G.J., Hsu, T.H., Chen, C.C., Jong, G.J.: A novel facial thermal feature extraction method for non-contact healthcare system. IEEE Access **8**, 86545–86553 (2020)
12. Dozono, H., Niina, G., Araki, S.: Convolutional self organizing map. In: 2016 IEEE International Conference on Computational Science and Computational Intelligence (CSCI), pp. 767–771. IEEE (2016)
13. Huang, X., Tan, H., Lin, G., Tian, Y.: A LSTM-based bidirectional translation model for optimizing rare words and terminologies. In: 2018 IEEE International Conference on Artificial Intelligence and Big Data (ICAIBD), China, pp. 5077–5086. IEEE (2018)
14. Athiwaratkun, B., Stokes, J.W.: Malware classification with LSTM and GRU language models and a character-level CNN. In: Proceedings of 2017 IEEE International Conference on Acoustics, Speech and Signal Processing (ICASSP), pp. 2482–2486. IEEE, USA (2017)
15. Munkhdalai, L., Park, K.-H., Batbaatar, E., Theera-Umpon, N., Ho, R.K.: Deep learning-based demand forecasting for Korean postal delivery service. IEEE Access **8**, 188135–188145 (2020)
16. Gong, K., et al.: Iterative PET image reconstruction using convolutional neural network representation. IEEE Trans. Med. Imaging **38**(3), 675–685 (2019)
17. Karras, T., Aila, T., Laine, S., Lehtinen, J.: Progressive growing of GANs for improved quality, stability, and variation. In: International Conference on Learning Representations, ICLR 2018, pp.1–26 (2018)
18. Higgins, I., et al.: β-VAE: learning basic visual concepts with a constrained variational framework. In: International Conference on Learning Representations, ICLR 2017, vol. 2, no. 5, pp. 1–22 (2017)
19. Gościewska, K., Frejlichowski, D.: A combination of moment descriptors, fourier transform and matching measures for action recognition based on shape. In: Krzhizhanovskaya, V.V., et al. (eds.) ICCS 2020. LNCS, vol. 12138, pp. 372–386. Springer, Cham (2020). https://doi.org/10.1007/978-3-030-50417-5_28
20. Frejlichowski, D.: Low-level greyscale image descriptors applied for intelligent and contextual approaches. In: Nguyen, N.T., Gaol, F.L., Hong, T.-P., Trawiński, B. (eds.) ACIIDS 2019. LNCS (LNAI), vol. 11432, pp. 441–451. Springer, Cham (2019). https://doi.org/10.1007/978-3-030-14802-7_38

21. Hrkút, P., Ďuračík, M., Mikušová, M., Callejas-Cuervo, M., Zukowska, J.: Increasing k-means clustering algorithm effectivity for using in source code plagiarism detection. In: Narváez, F.R., Vallejo, D.F., Morillo, P.A., Proaño, J.R. (eds.) SmartTech-IC 2019. CCIS, vol. 1154, pp. 120–131. Springer, Cham (2020). https://doi.org/10.1007/978-3-030-46785-2_10
22. Mikusova, M., Zukowska, J., Torok, A.: Community road safety strategies in the context of sustainable mobility. In: Mikulski, J. (ed.) TST 2018. CCIS, vol. 897, pp. 115–128. Springer, Cham (2018). https://doi.org/10.1007/978-3-319-97955-7_8
23. Huk, M.: Backpropagation generalized delta rule for the selective attention Sigma-if artificial neural network. Int. J. Appl. Math. Comp. Sci. **22**, 449–459 (2012)
24. Huk, M.: Notes on the generalized backpropagation algorithm for contextual neural networks with conditional aggregation functions. J. Intell. Fuzzy Syst. **32**, 1365–1376 (2017)
25. Huk, M.: Stochastic optimization of contextual neural networks with RMSprop. In: Nguyen, N.T., Jearanaitanakij, K., Selamat, A., Trawiński, B., Chittayasothorn, S. (eds.) ACIIDS 2020. LNCS (LNAI), vol. 12034, pp. 343–352. Springer, Cham (2020). https://doi.org/10.1007/978-3-030-42058-1_29
26. Szczepanik, M., et al.: Multiple classifier error probability for multi-class problems. Eksploatacja i Niezawodnosc - Maintenance Reliab. **51**(3), 12–16 (2011). https://doi.org/10.17531/ein
27. Szczepanik, M., Jóźwiak, I.: Data management for fingerprint recognition algorithm based on characteristic points' groups. In: New Trends in Databases and Information Systems, Foundations of Computer and Decision Sciences, vol. 38, no. 2, pp. 123–130. Springer, Heidelberg (2013). https://doi.org/10.2478/fcds-2013-0004
28. Szczepanik, M., Jóźwiak, I.: Fingerprint recognition based on minutes groups using directing attention algorithms. In: Rutkowski, L., Korytkowski, M., Scherer, R., Tadeusiewicz, R., Zadeh, L.A., Zurada, J.M. (eds.) ICAISC 2012. LNCS (LNAI), vol. 7268, pp. 347–354. Springer, Heidelberg (2012). https://doi.org/10.1007/978-3-642-29350-4_42
29. Huk, M., Pietraszko, J.: Contextual neural-network based spectrum prediction for cognitive radio. In: 4th International Conference on Future Generation Communication Tech-nology (FGCT 2015), pp. 1–5. IEEE Computer Society, London (2015)
30. Burnell, E., Wołk, K., Waliczek, K., Kern, R.: The impact of constant field of attention on properties of contextual neural networks. In: Nguyen, N.T., Jearanaitanakij, K., Selamat, A., Trawiński, B., Chittayasothorn, S. (eds.) Intelligent Information and Database Systems: 12th Asian Conference, ACIIDS 2020, Phuket, Thailand, March 23–26, 2020, Proceedings, Part II, pp. 364–375. Springer International Publishing, Cham (2020). https://doi.org/10.1007/978-3-030-42058-1_31
31. Huk, M.: Non-uniform initialization of inputs groupings in contextual neural networks. In: Nguyen, Ngoc Thanh, Gaol, Ford Lumban, Hong, Tzung-Pei., Trawiński, Bogdan (eds.) ACIIDS 2019. LNCS (LNAI), vol. 11432, pp. 420–428. Springer, Cham (2019). https://doi.org/10.1007/978-3-030-14802-7_36
32. Huk, M.: Training contextual neural networks with rectifier activation functions: role and adoption of sorting methods. J. Intell. Fuzzy Syst. **37**(6), 7493–7502 (2019)
33. Huk, M.: Weights ordering during training of contextual neural networks with generalized error backpropagation: importance and selection of sorting algorithms. In: Nguyen, N.T., Hoang, D.H., Hong, T.-P., Pham, H., Trawiński, B. (eds.) ACIIDS 2018. LNCS (LNAI), vol. 10752, pp. 200–211. Springer, Cham (2018). https://doi.org/10.1007/978-3-319-75420-8_19
34. Huk, M.: Context-related data processing with artificial neural networks for higher reliability of telerehabilitation systems. In: 17th International Conference on E-health Networking, Application & Services (HealthCom), pp. 217–221. IEEE Computer Society, Boston (2015)
35. Kwiatkowski J., et al.: Context-sensitive text mining with fitness leveling genetic algorithm. In: 2015 IEEE 2nd International Conference on Cybernetics (CYBCONF), Gdynia, Poland, 2015, electronic publication, pp. 1–6 (2015). ISBN: 978-1-4799-8321-6

36. Huk, M.: Measuring the effectiveness of hidden context usage by machine learning methods under conditions of increased entropy of noise. In: 2017 3rd IEEE International Conference on Cybernetics (CYBCONF), Exeter, pp. 1–6 (2017)
37. Privitera, C.M., Azzariti, M., Stark, L.W.: Locating regions-of-interest for the Mars Rover expedition. Int. J. Remote Sens. **21**, 3327–3347 (2000)
38. Glosser, C., Piermarocchi, C., Shanker, B.: Analysis of dense quantum dot systems using a self-consistent Maxwell-Bloch framework. In: Proceedings of 2016 IEEE International Symposium on Antennas and Propagation (USNC-URSI), Puerto Rico, pp. 1323–1324. IEEE (2016)
39. Srivastava, N., Hinton, G.E., Krizhevsky, A., Sutskever, I., Salakhutdinov, R.R.: Dropout: a simple way to prevent neural networks from overfitting. J. Mach. Learn. Res. **15**, 1929–1958 (2014)
40. Ko, B., Kim, H.G., Choi, H.J.: Controlled dropout: a different dropout for improving training speed on deep neural network. In: Proceedings of 2017 IEEE International Conference on Systems, Man, and Cybernetics (SMC), Canada. IEEE (2018)
41. ElAdel, A., Ejbali, R., Zaied, M., Ben Amar, C.: Fast deep neural network based on intelligent dropout and layer skipping, In: Proceedings of 2017 International Joint Conference on Neural Networks (IJCNN), Anchorage, USA (2017)
42. Salehinejad, H., Valaee, S.: Ising-dropout: a regularization method for training and compression of deep neural networks. In: Proceedings of 2019 IEEE International Conference on Acoustics, Speech and Signal Processing (ICASSP), Brighton, United Kingdom (2019)
43. Guo, J., Gould, S.: Depth dropout: efficient training of residual convolutional neural networks. In: 2016 International Conference on Digital Image Computing: Techniques and Applications (DICTA), Gold Coast, Australia. IEEE (2016)
44. Wołk, K., Palak, R., Burnell, E.D.: Soft dropout method in training of contextual neural networks. In: Nguyen, N.T., Jearanaitanakij, K., Selamat, A., Trawiński, B., Chittayasothorn, S. (eds.) ACIIDS 2020. LNCS (LNAI), vol. 12034, pp. 353–363. Springer, Cham (2020). https://doi.org/10.1007/978-3-030-42058-1_30
45. UCI Machine Learning Repository. https://archive.ics.uci.edu/ml
46. Golub, T.R., et al.: Molecular classification of cancer: class discovery and class prediction by gene expression monitoring. Science **286**, 531–537 (1999)
47. H2O.ai documentation. https://docs.h2o.ai/h2o/latest-stable/h2o-docs/index.html

Bone Age Assessment by Means of Intelligent Approaches

Maria Ginał and Dariusz Frejlichowski(✉)

Faculty of Computer Science and Information Technology,
West Pomeranian University of Technology, Szczecin, Żołnierska 52, 71-210 Szczecin, Poland
ginal-maria@zut.edu.pl, dfrejlichowski@wi.zut.edu.pl

Abstract. There are many practical applications of intelligent and contextual approaches. The development of systems designed for medical diagnosis is an example of a real world problem employing such approaches. The appropriate understanding of context of digital medical data is crucial in this case. In the paper the automatic bone age assessment by means of digital RTG images of pediatric patients is analyzed. The particular stages of the approach are presented and the algorithms for each of them are proposed. The approach consists of the localization of a hand area, localization of characteristic points, localization of regions of interest, feature extraction and classification. The discussion on each of them is provided. Particular algorithms are proposed and discussed.

Keywords: Image processing · Bone age assessment · Object identification · Feature extraction

1 Introduction and Motivation

Intelligent and contextual approaches have many important and practical applications nowadays (e.g. [1–3]). Automatic or semi-automatic medical diagnosis can be considered as one of them (e.g. [4]), supported usually by advanced imaging algorithms (e.g. [5]). The appropriate understanding of context of digital medical data (e.g. images) is crucial for efficient work of such systems. Therefore, for many years scientists have been trying to apply various algorithms in order to obtain more effective, and above all – more reliable systems that could support work of physicians or laboratory technicians.

In this paper an example of a problem connected with the analysis of medical images is investigated. The initial results are given on the problem of automatic bone age assessment by means of digital RTG images of pediatric patients. The main goal of the paper is to provide some discussion on the problem and propose several algorithm to be applied on the particular stages of the developed approaches. The problem is non-trivial, since even for practitioners the bone age estimation is not always obvious. Hence, the application of intelligent tools and algorithms in order to process the images automatically is even more challenging.

The first task that has to be performed in order to provide an efficient intelligent system is the design of particular steps and later the appropriate selection or development

N. T. Nguyen et al. (Eds.): ACIIDS 2021, LNAI 12672, pp. 704–716, 2021.
https://doi.org/10.1007/978-3-030-73280-6_56

of algorithms for each of them. This is the subject of the fourth section of this paper. And before that, the introduction to the problem of bone age assessment is analyzed in the second section, and brief description of related works is given in the third one. Finally, the last section concludes the paper and provides the future plans for works on the topic.

2 Bone Age Estimation

Bone Age Assessment Motivation and Methods
Bone age assessment is an examination which is performed on pediatric patients. It can be described as an indicator of biological maturity and it may differ from chronological age due to various reasons. It is possible to be delayed or accelerated in relation to the standards. 'Short stature', which can be defined as a height that is two or more standard deviations below the mean age for the specific age and gender [6], is a relevant reason for assessing bone age but not the only one. Short height can cause a parental anxiety, although in not every case it is caused by disease or disorders.

Bone age estimation might be necessary with the following conditions:

- Diagnosing diseases which result in short or tall stature in children [7, 8]. Note that bone age assessment comes in pair with clinical examinations, patient history gathering, including the prenatal and family history. In other words, bone age estimation is not the only factor that defines a diagnosis;
- Monitoring the effectiveness of a treatment [9];
- Predicting a child's adult height [10];
- In situations where accurate birth records are unknown and chronological age is unknown [11], or when actual age might be misrepresented, yet the authenticity of the information must be accurate and checked (e.g. frauds in competitive sports [12]). Note that bone age might not provide accurate information on calendar age. There are too many factors which may have impact on bone age aside from diseases (e.g. ethnic [13], genetic, and nutritional).

The methods of assessing bone age with the aid of medical imaging techniques include: ultrasonography (USG) [14–17] and magnetic resonance imaging (MRI) [18–20]. These two techniques eliminate necessity of additional radiation exposure, yet methods which involve usage of left hand and wrist radiographs are the most common, efficient and they will be the subject of discussion in this paper.

Greulich and Pyle Atlas Method
The Greulich and Pyle atlas method is still one of the most commonly applied atlases because of its simplicity of usage [6]. The book contains reference radiograph images of the left hand and wrist of children till 18 y.o. for females and 19 y.o. for males. Both groups are assessed separately as their skeletal development differs within the age (girls physically mature faster than boys). Standards were stated as an average from radiographs obtained serially in thousands of children as a part of The Brush Foundation Growth Study from 1931 to 1942. Hand and wrist radiographs belong to upper middle-class Caucasian children in Cleveland, Ohio, United States (healthy boys and girls)

[21]. Method itself consists in comparing an examined radiograph with the series of standards and the final bone age is equal to the standard that fits most closely. Besides illustrations, a proper explanation of gradual age related changes observed in the bone structure is attached in a text form [22]. Greulich and Pyle atlas includes not quite accurate assumption that skeletal maturation is uniform (as far as healthy children are concerned), i.e. all bones present in radiograph have the same skeletal age. In general, GP method can be called holistic as authors did not specify the process of assessment step by step with the distinction of various regions which sums the final bone age [23]. The last edition was published in 1959. Over 50 years passed by, hence many factors may result in changes in rate of human biological development in present-day population.

Tanner-Whitehouse – Scoring System Method

The Tanner-Whitehouse method is more accurate in description, but needs more time while assessment process is performed. The aim of this method is to obtain a sum of several regions assessed separately. Each bone under consideration starting from its time of appearance till its final shape has sequence of recognizable stages (from A to H or I) what can be defined and described as stages of development. Total score is transformed into the bone age [6]. TW scoring system was developed on a basis of radiographs of average socioeconomic class children in the United Kingdom. Data was collected in the 1950s and 1960s. The method was updated in 2001 in order to improve relations between bone maturity score and bone age (TW3) [21].

A total of 20 bones (regions of interest) are observed:

- TW RUS (radius-ulna-short bones): 13 bones including the radius, ulna and short bones of the thumb, middle and little fingers;
- TW carpus: 7 carpal bones;
- TW20: 20 bones, including combination of the two sets mentioned above.

Digital Atlases

Aside of publications relatively distant in time (GP and TW) there are also more 'up to date' sources – digital versions of bone age atlases [23, 24]. Apart from the illustrations, they contain adequate introduction to the subject and problems related to it, deviation tables and instructions for interpreting individual bone age indicators for a given period of life. The images in the Gilsanz and Ratib atlas [23] have been developed as Caucasian standards from 522 radiographs. The patterns were created on the basis of several radiographs of children considered healthy and being a good reference point. The atlas is divided into 29 age groups (from 8 months to 18 years) for both groups (boys and girls). The Gaskin and Kahn's atlas [24] contains patterns developed from many thousands of X-ray images by means of the University of Virginia PACS system. Each image was originally interpreted and assessed for bone age by pediatric radiologists. Many examinations of the images were supported by standards and descriptions stated by Greulich and Pyle. Selected patterns have been improved manually to enhance details. The book contains two sets of images – separately for girls and boys. For girls, 27 images of hand and wrist can be found (from newborn to 18 years old). For boys, the set is limited to 31 photos up to the age of 19. Most importantly, the atlas also includes illustrations with footnotes, i.e. the bone elements and structures which should be taken into account when assessing the age for each development stage separately are marked.

Bone Development

A level of skeletal maturity is measured taking under consideration the size, shape and degree of mineralization of bone which tends to achieve full maturity. Examination of skeletal stage development is a task that demands knowledge of processes by which bone forms and develops. The growth of the bones can be longwise and in width which occurs from two different processes [23, 25, 26]. Roughly speaking, these developments are visible on radiographs of hand and wrist in the form of shape, size and distances from particular bone structures relative to each other which change in time (Fig. 1).

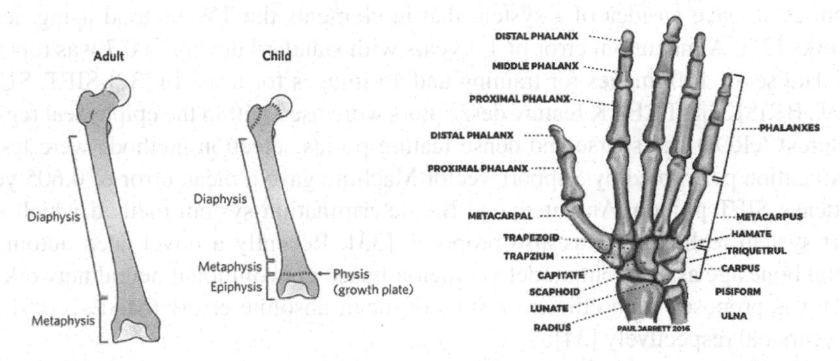

Fig. 1. From left to right: anatomical differences between adult and child long bone [27], illustration of bone anatomy of the hand and wrist [26].

The Purpose of Work

Human hand and wrist consist of 28 bones (not including the sesamoid bone): distal phalanges (5 bones), middle phalanges (4 bones), proximal phalanges (5 bones), metacarpus (5 bones), carpus (7 bones), radius and ulna. That is a lot of elements, which features have 'information encoded' about actual bone age. In the digitalization era it is still a challenging task to describe bone age estimation procedure in machine language due to many not obvious factors. Not only the characteristics of radiograms, but also appropriate process of evaluation upon elaborated universal set of rules based on experts knowledge is problematic. The main purpose is to create a system which in fully automatic way extracts crucial information about bone age hidden in radiograph of a hand and wrist and simulates the knowledge of the experts in the stage of classification. Moreover, such system ought to be independent of factors including ethnicity, what means that it could be efficiently used in different countries inhabited by various populations.

3 Related Works

The topic of the automatic bone age assessment system is not relatively new, however it is not fully investigated. In late 90s a system for automatic evaluation of the skeletal maturation of Japanese children was proposed [28]. The automatic measurements of epiphysis and metaphysis were performed on the distal, middle and proximal phalanges of the third finger. The obtained results indicated a significant correlation between chronological age and the ratio of epiphyseal width to metaphyseal width. Not much later a method providing information about four features (epiphyseal diameter, metaphyseal diameter, distance between lower diaphysis (or metacarpal) and epiphyseal diameter) was described in order to help radiologists to determine bone age [29]. This work was extended and an automated bone age determination system was made. It contains a collection of 1400 digitized left hand radiographs and a computer-assisted diagnosis (CAD) module. The features were extracted from seven regions of interest (ROIs): the carpal bone ROI and six phalangeal PROIs (the distal and middle regions of three middle fingers). Eleven category fuzzy classifiers were trained [30]. Leonardo Bocchi et al. gave an idea of a system that implements the TW method using neural networks [31]. A maximum error of 1.4 years with standard deviation 0.7 was reported for a data set of 120 images for training and 40 images for tests. In [32] SIFT, SURF, BRIEF, BRISK and FREAK feature descriptors were used within the epiphyseal regions of interest (eROI) and sparse and dense feature points selection methods were tested. Classification performed by Support Vector Machine gave a mean error of 0.605 years (for dense SIFT points). An automated BA determination system method which uses expert system techniques was also proposed [33]. Recently a novel deep automated skeletal bone age assessment model via region-based convolutional neural network (R-CNN) was proposed with efficient results of mean absolute errors (MAEs): 0.51 and 0.48 years old respectively [34].

The leading commercial solution for automated bone age estimation is BoneXpert which was introduced in 2009 and has been still developed. The software locates all bones in the hand (21 bones: radius, ulna, metacarpals and phalanges) and is based on the standard Greulich-Pyle and Tanner-Whitehouse methods. In addition, the seven carpal bones are considered [35]. The system uses the active appearance model for the bone reconstruction. BoneXpert consists of three layers: the layer A locates and reconstructs the borders of the bones using supervised and unsupervised machine learning, the layer B determines bone age for each bone and validates bone ages, and the layer C adjusts and calculates bone age according to GP or TW method [36]. In 2019 a new version was released and the root mean square error (RMSE) of BoneXpert relative to manual bone age rating is equal to 0.63 year [37].

4 Proposed Approach

Stages of the Proposed Solution

Simplified key stages of a proposed solution can be described using three steps: ROIs localization, selection of extracted features and rule-based inference. These steps are described generally as they could be solved in many ways. Each of the steps has an impact on the final result and experiments are required for each of them. Further part of the work will focus on the first step (Fig. 2).

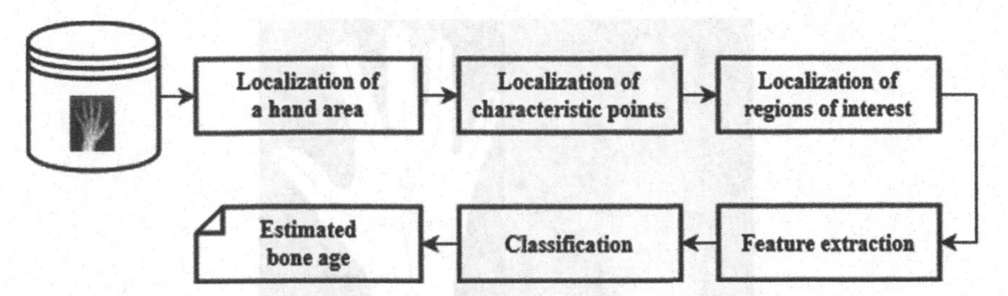

Fig. 2. Main schema of the proposed fully automated bone age assessment approach.

Selected Solutions

Certain steps of proposed processing scheme could be accomplished using the following approaches:

(a) **Hand area segmentation** - the main goal of this step is to separate the hand area from the background and undesirable objects present in the image (e.g. image noise, markers, tags) that provide no information about bone age. It is very important that input image represents a hand which is placed in a 'standard way', i.e. in accordance with X-ray image standard for bone age assessment (fingers laid relatively separately). It is possible that the background of the X-ray image is not homogeneous as far as structure and illumination go. As borders of the hand are supposed to be enhanced, noise reduction algorithm must take it under consideration. An anisotropic diffusion proposed by Perona P. and Malik J. [38] is characterized by such a property. The segmentation itself can be performed using Otsu method (locally) [39], edge detectors (e.g. Canny edge detector [40]) or a local entropy of grayscale image [41]. Regardless of the segmentation algorithm, morphological operations could be helpful in order to open or close some areas [42] (Fig. 3).

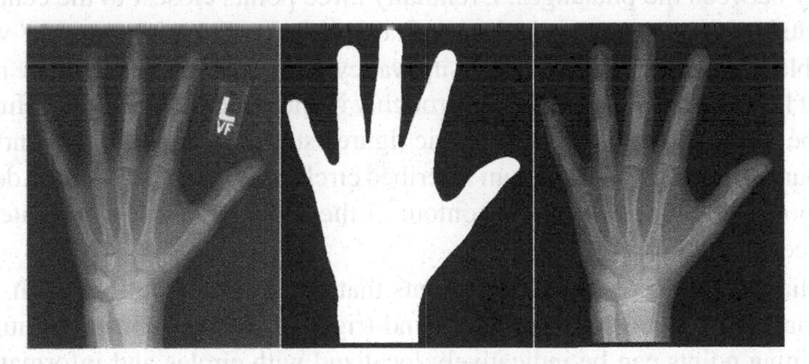

Fig. 3. From left to right: original input image; hand binary mask; masked input image (based on: [43]).

(b) **Characteristic points** – this step requires the definition of characteristic points. The points will be used to extract the regions of interest in the next stage (Fig. 4):

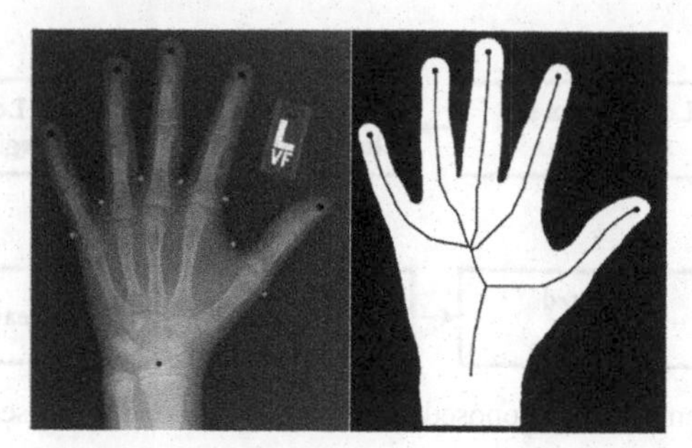

Fig. 4. From left to right: characteristic points (red – fingertips, green – finger valleys and auxiliary points, blue – wrist landmark); the binary mask of the hand area, its skeleton and the points of the fingertips (based on: [43]). (Color figure online)

I. Finger tips – five points located on the ends of each distal phalanges. Skeletonization is an operation which can easily handle this task [42]. As orientation of a hand is upward, fingertips are assigned from left to right. The operation must be used on smoothed binary hand mask to avoid redundant branches.
II. Finger valleys – four points located between fingers on a hand area. Two options are considered: skeletonization [44] and usage of a convex hull property [45]. This time skeleton is obtained from the background of an image (complementary image of binary hand mask). As far as convex hull goes, the first step is to subtract the binary image of the hand from the area limited by the convex hull. In result, an indefinite number of objects will be obtained for which spaces between the phalanges will be located. However, the number of designated objects cannot be estimated in advance due to the differences in the positioning of the hands on X-ray images. Objects with a significant (i.e. larger) surface area were selected for further operations (outliers). The distances from the points of objects contours to the centroid of the palm area were calculated. The point closest to the centroid indicates a potential point in the valley between the phalanges. Eventually three points closest to the centroid were selected, which are respectively 1st, 2nd and 3rd valley point as they were most reliable with this method. Last missing valley point will be located in the next steps.
III. Wrist landmark – one point located roughly in the middle of the wrist. Human hand can be simplified to simple geometric figures such as circles. Wrist landmark can be found in the area of maximum inscribed circle in a hand mask [46]. Additionally, one point of tangency with the contour of the hand and a circle indicates a valley between the thumb (Figs. 5 and 6).
IV. Auxiliary points – three missing points that could not be defined with the above criteria. One of them was already found (right restrictive point of a thumb). Two remaining points can be indicatively localized with circles and information about the center of a maximum inscribed circle in a hand contour. Radius would be equal to the distance between a center of this circle and 1st or 2nd valley point (Fig. 7).

(c) **ROI extraction** – regions of interest can be considered as wrist region of interest (WROI), phalangeal regions of interest (PROIs) and epiphyseal regions of interest

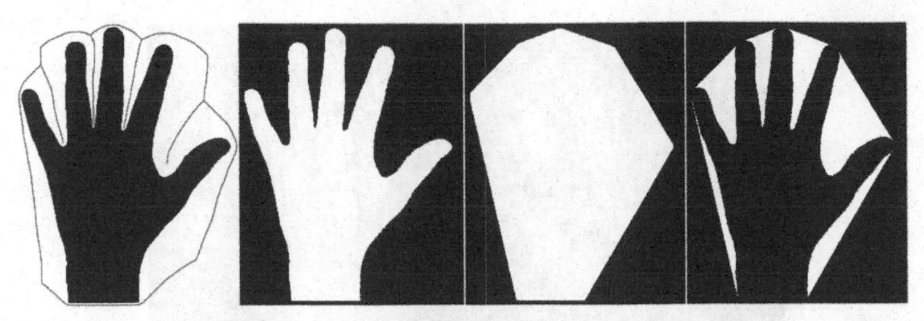

Fig. 5. From left to right: skeleton of image with valleys points, binary hand mask, convex hull area of a hand, binary hand mask subtracted from convex hull area (based on: [43]).

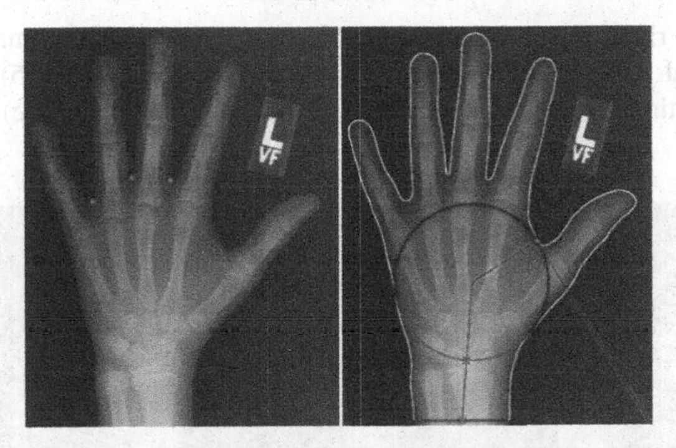

Fig. 6. From left to right: three selected valley points; maximum inscribed circle in a hand contour, wrist landmark (blue points) and two points limiting a thumb (green dots) (based on: [43]). (Color figure online)

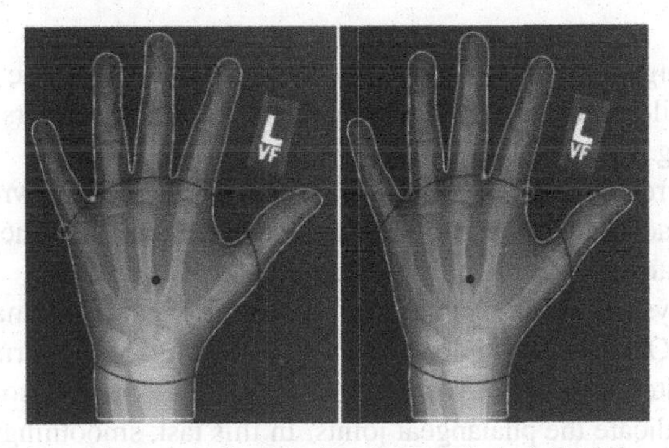

Fig. 7. Auxiliary circles and two auxiliary points (green circles) (based on: [43]). (Color figure online)

(EROIs). ROIs can be grouped based on anatomical similarity [33]. Eventually, it was assumed that regions of interest could be assigned to one of four groups: distal joints ROIs, proximal joints ROIs, metacarpal joints ROIs, carpal and wrist ROI (Figs. 8 and 9).

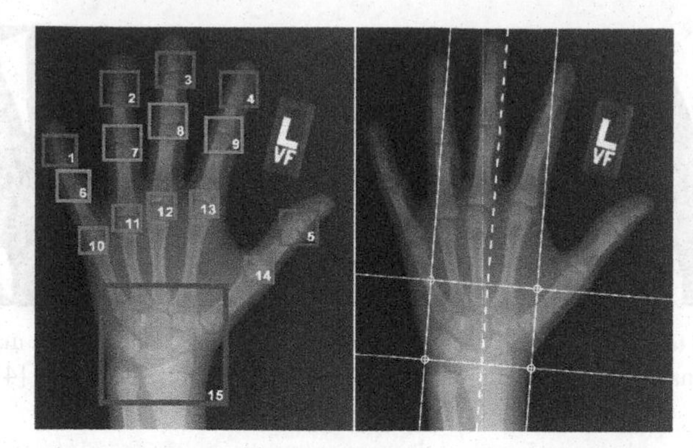

Fig. 8. From left to right: green – distal joints ROIs (1–5), yellow – proximal joints ROIs (6–9), orange – metacarpal joints ROIs (10–14), red – carpal and wrist ROI (15); perpendicular and parallel lines delimiting the wrist area (based on: [43]). (Color figure online)

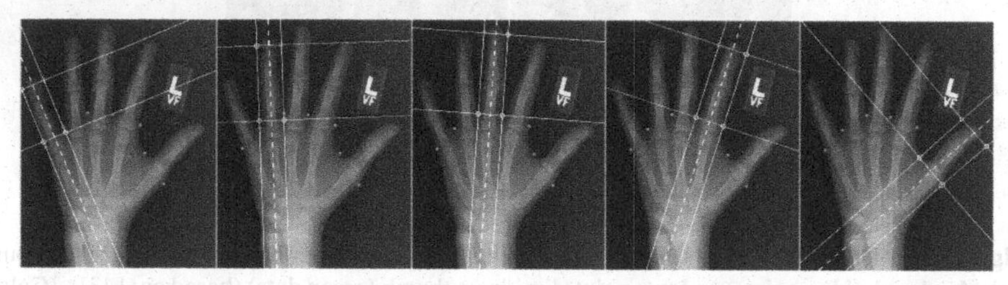

Fig. 9. Perpendicular and parallel lines delimiting the phalangeal areas (based on: [43]).

I. PROI (phalangeal region of interest) – area that contains three bone connections (EROIs). While having all characteristic points location of fingers it can be extracted by calculating perpendicular and parallel lines.
II. WROI (wrist region of interest) – area that contains carpal and wrist area. As above, it can be extracted by calculating perpendicular and parallel lines. An information about the center of maximum inscribed circle is also needed.
III. EROI (epiphyseal region-of-interest) – areas of distal, proximal and metacarpal joints (14 EROIs). Information about these joints might be determined by means of creating an intensity profile. Within properly processed signal some 'peaks' can be found that indicate the phalangeal joints. In this task smoothing filters that do not distort the signal tendency can be useful (e.g. Savitzky-Golay filter [47]). A second order derivative captures large variations of the intensity profile [48]. Cluster centers might indicate joints between the bones, as there may still be more peaks in the signal (three cluster for fingers 1–4 and two clusters for a thumb) (Figs. 10 and 11).

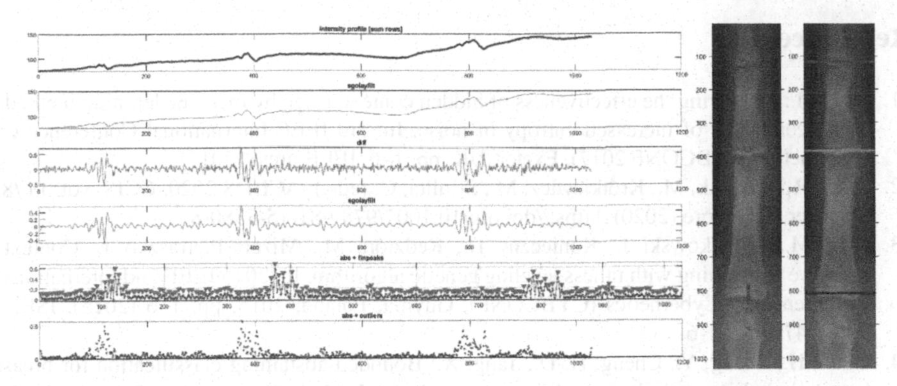

Fig. 10. From left to right: sample intensity profile of a finger and its stages of processing; potential finger joints points (red, green, blue) and its clusters centers (magenta) (based on: [43]). (Color figure online)

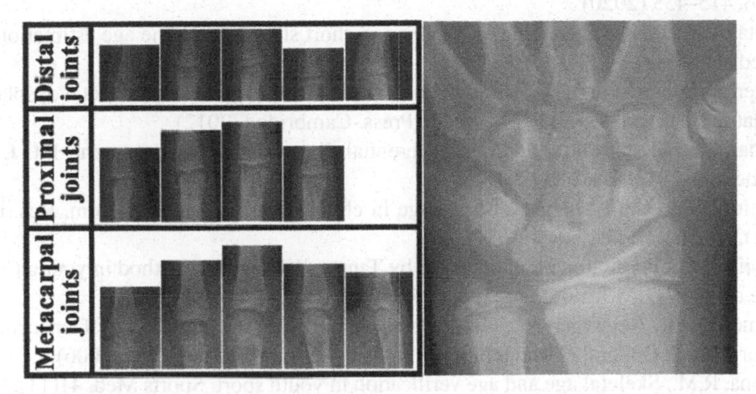

Fig. 11. Sample of final extracted regions of interest: 14 EROIs and WROI (based on: [43]).

5 Conclusions and Future Plans

In the paper the problem of eventual automatic bone age assessment by means of digital RTG images of pediatric patients was investigated. The application of some intelligent methods was proposed and discussed. The approach includes such stages as: localization of a hand area, localization of characteristic points, localization of regions of interest, feature extraction and classification. Particular algorithms for solving them were proposed and discussed.

Future plans include above all extensive experimental analysis of proposed in this paper approaches, what is assumed a step in the process of constructing fully automatic method for bone age assessment of pediatric patients. The approach is supposed to be based on the intelligent algorithms, mainly from the broad area of computer vision and image understanding methods.

References

1. Huk, M.: Measuring the effectiveness of hidden context usage by machine learning methods under conditions of increased entropy of noise. In: 3rd IEEE International Conference on Cybernetics (CYBCONF 2017), Exeter, UK, pp. 1–6. IEEE press (2017)
2. Sitek, P., Pietranik, M., Krótkiewicz, M., Srinilta, C. (eds.): ACIIDS 2020. CCIS, vol. 1178. Springer, Singapore (2020). https://doi.org/10.1007/978-981-15-3380-8
3. Huk, M., Kwiatkowski J., Konieczny D., Kędziora M., Mizera-Pietraszko J.: Context-sensitive text mining with fitness leveling genetic algorithm. In: 2015 IEEE 2nd International Conference on Cybernetics (CYBCONF), Gdynia, Poland, 2015, pp. 1–6 (2015). ISBN: 978-1-4799-8321-6
4. Guan, H., Zhang, Y., Cheng, H.-D., Tang, X.: Bounded-abstaining classification for breast tumors in imbalanced ultrasound images. Int. J. Appl. Math. Comput. Sci. **2**(30), 325–336 (2020)
5. Franchini, S., Gentile, A., Vassallo, G., Vitabile, S.: Implementation and evaluation of medical imaging techniques based on conformal geometric algebra. Int. J. Appl. Math. Comput. Sci. **3**(20), 415–433 (2020)
6. Gupta, A.K., Jana, M., Kumar, A.: Imaging in short stature and bone age estimation. Indian J. Pediatr. **86**, 1–13 (2019)
7. Unger, S., Superti-Furga, A., Rimoin, D.L.: A diagnostic approach to skeletal dysplasias. In: Pediatric Bone, pp. 403–437. Academic Press, Cambridge (2012)
8. Kumar, S.: Tall stature in children: differential diagnosis and management. Int. J. Pediatr. Endocrinol. **2013**(S1), P53 (2013)
9. Martin, D.D., et al.: The use of bone age in clinical practice–part 2. Horm. Res. Paediatr. **76**(1), 10–16 (2011)
10. Ostojic, S.M.: Prediction of adult height by Tanner-Whitehouse method in young Caucasian male athletes. QJM Int. J. Med. **106**(4), 341–345 (2013)
11. Schmeling, A., Reisinger, W., Geserick, G., Olze, A.: Age estimation of unaccompanied minors: part I. General considerations. Forensic Sci. Int. **159**, S61–S64 (2006)
12. Malina, R.M.: Skeletal age and age verification in youth sport. Sports Med. **41**(11), 925–947 (2011)
13. Ontell, F.K., Ivanovic, M., Ablin, D.S., Barlow, T.W.: Bone age in children of diverse ethnicity. AJR Am. J. Roentgenol. **167**(6), 1395–1398 (1996)
14. Mentzel, H.J., et al.: Assessment of skeletal age at the wrist in children with a new ultrasound device. Pediatr. Radiol. **35**(4), 429–433 (2005)
15. Khan, K.M., Miller, B.S., Hoggard, E., Somani, A., Sarafoglou, K.: Application of ultrasound for bone age estimation in clinical practice. J. Pediatr. **154**(2), 243–247 (2009)
16. Utczas, K., Muzsnai, A., Cameron, N., Zsakai, A., Bodzsar, E.B.: A comparison of skeletal maturity assessed by radiological and ultrasonic methods. Am. J. Hum. Biol. **29**(4), e22966 (2017)
17. Torenek Ağırman, K., Bilge, O.M., Miloğlu, Ö.: Ultrasonography in determining pubertal growth and bone age. Dentomaxillofacial Radiol. **47**(7), 20170398 (2018)
18. Urschler, M., et al.: Applicability of Greulich-Pyle and Tanner-Whitehouse grading methods to MRI when assessing hand bone age in forensic age estimation: A pilot study. Forensic Sci. Int. **266**, 281–288 (2016)
19. Terada, Y., et al.: Skeletal age assessment in children using an open compact MRI system. Magn. Reson. Med. **69**(6), 1697–1702 (2013)
20. Pennock, A.T., Bomar, J.D.: Bone age assessment utilizing knee MRI. Orthopaedic J. Sports Med. **5**(7_suppl6), 2325967117S0042 (2017)

21. Satoh, M.: Bone age: assessment methods and clinical applications. Clin. Pediatr. Endocrinol. **24**(4), 143–152 (2015)
22. Mughal, A.M., Hassan, N., Ahmed, A.: Bone age assessment methods: a critical review. Pak. J. Med. Sci. **30**(1), 211 (2014)
23. Gilsanz, V., Ratib, O.: Hand Bone Age: A Digital Atlas of Skeletal Maturity, 2nd ed. (2012)
24. Gaskin, C.M., Kahn, M.M.S.L., Bertozzi, J.C., Bunch, P.M.: Skeletal Development of the Hand and Wrist: A Radiographic Atlas and Digital Bone Age Companion. Oxford University Press, Oxford (2011)
25. 10.2020 r. Bone Formation and Development. https://open.oregonstate.education/aandp/chapter/6-4-bone-formation-and-development/
26. 10.2020 r. Hand and Wrist Anatomy. https://murdochorthopaedic.com.au/our-surgeons/paul-jarrett/patient-information-guides/hand-wrist-anatomy/
27. 10.2020 r. Anatomic differences: child vs. Adult. https://www.rch.org.au/fracture-education/anatomy/Anatomic_differences_child_vs_adult/
28. Sato, K., et al.: Setting up an automated system for evaluation of bone age. Endocr. J. **46**(Suppl), S97–S100 (1999)
29. Pietka, E., Gertych, A., Pospiech, S., Cao, F., Huang, H.K., Gilsanz, V.: Computer-assisted bone age assessment: Image preprocessing and epiphyseal/metaphyseal ROI extraction. IEEE Trans. Med. Imaging **20**(8), 715–729 (2001)
30. Gertych, A., Zhang, A., Sayre, J., Pospiech-Kurkowska, S., Huang, H.K.: Bone age assessment of children using a digital hand atlas. Comput. Med. Imaging Graph. **31**(4–5), 322–331 (2007)
31. Bocchi, L., Ferrara, F., Nicoletti, I., Valli, G.: An artificial neural network architecture for skeletal age assessment. In: Proceedings 2003 International Conference on Image Processing (Cat. No. 03CH37429), vol. 1, pp. I-1077. IEEE (2003
32. Kashif, M., Deserno, T.M., Haak, D., Jonas, S.: Feature description with SIFT, SURF, BRIEF, BRISK, or FREAK? a general question answered for bone age assessment. Comput. Biol. Med. **68**, 67–75 (2016)
33. Seok, J., Kasa-Vubu, J., DiPietro, M., Girard, A.: Expert system for automated bone age determination. Expert Syst. Appl. **50**, 75–88 (2016)
34. Liang, B., et al.: A deep automated skeletal bone age assessment model via region-based convolutional neural network. Futur. Gener. Comput. Syst. **98**, 54–59 (2019)
35. 10.2020. Automated bone age estimation. https://bonexpert.com/
36. Thodberg, H.H., Kreiborg, S., Juul, A., Pedersen, K.D.: The BoneXpert method for automated determination of skeletal maturity. IEEE Trans. Med. Imaging **28**(1), 52–66 (2008)
37. Thodberg, H.H., Martin, D.D.: Validation of a new version of BoneXpert bone age in children with congenital adrenal hyperplasia (CAH), precocious puberty (PP), growth hormone deficiency (GHD), turner syndrome (TS), and other short stature diagnoses. In: 58th Annual ESPE, vol. 92. European Society for Paediatric Endocrinology (2019)
38. Perona, P., Malik, J.: Scale-space and edge detection using anisotropic diffusion. IEEE Trans. Pattern Anal. Mach. Intell. **12**(7), 629–639 (1990)
39. Otsu, N.: A threshold selection method from gray-level histograms. IEEE Trans. Syst. Man Cybern. **9**(1), 62–66 (1979)
40. Canny, J.: A computational approach to edge detection. IEEE Trans. Pattern Anal. Mach. Intell. **6**, 679–698 (1986)
41. Yang, W., Cai, L., Wu, F.: Image segmentation based on gray level and local relative entropy two dimensional histogram. PLoS ONE **15**(3), e0229651 (2020)
42. Tadeusiewicz, R., Korohoda, P.: Computer image analysis and processing. Wydawnictwo Fundacji Postępu Telekomunikacji (1997). (in Polish)
43. 03.2020 r. Image Processing and Informatics Lab - Computer-aided Bone Age Assessment of Children Using a Digital Hand Atlas. https://ipilab.usc.edu/computer-aided-bone-age-assessment-of-children-using-a-digital-hand-atlas-2/

44. Mestetskiy, L., Bakina, I., Kurakin, A.: Hand geometry analysis by continuous skeletons. In: Kamel, M., Campilho, A. (eds.) Image Analysis and Recognition, pp. 130–139. Springer Berlin Heidelberg, Berlin, Heidelberg (2011). https://doi.org/10.1007/978-3-642-21596-4_14
45. Su, L., et al.: Delineation of carpal bones from hand X-ray images through prior model, and integration of region-based and boundary-based segmentations. IEEE Access **6**, 19993–20008 (2018)
46. Tolga Birdal. Maximum Inscribed Circle using Distance Transform (https://www.mathworks.com/matlabcentral/fileexchange/30805-maximum-inscribed-circle-using-distance-transform), MATLAB Central File Exchange (2020). Accessed 9 Oct 2020
47. Schafer, R.W.: What is a Savitzky-Golay filter?[Lecture notes]. IEEE Signal Process. Mag. **28**(4), 111–117 (2011)
48. Cunha, P., Moura, D.C., López, M.A.G., Guerra, C., Pinto, D., Ramos, I.: Impact of ensemble learning in the assessment of skeletal maturity. J. Med. Syst. **38**(9), 87 (2014)

Random Number Generators in Training of Contextual Neural Networks

Maciej Huk[1(✉)], Kilho Shin[2], Tetsuji Kuboyama[2], and Takako Hashimoto[3]

[1] Faculty of Computer Science and Management, Wroclaw University of Science and Technology, Wroclaw, Poland
maciej.huk@pwr.edu.pl
[2] Gakushuin University, Tokyo, Japan
kilhoshin314@gmail.com, ori-aciids2020@tk.cc.gakushuin.ac.jp
[3] Chiba University of Commerce, Chiba, Japan
takako@cuc.ac.jp

Abstract. Much care should be given to the cases when there is a need to compare results of machine learning (ML) experiments performed with the usage of different Pseudo Random Number Generators (PRNGs). This is because the selection of PRNG can be regarded as a source of measurement error, e.g. in repeated N-fold Cross Validation (CV). It can be also important to verify if the observed properties of a model or algorithm are not due to the effects of the use of a particular PRNG. In this paper we conduct experiments so that we can observe the possible level of differences in obtained values of various measures of classification quality of simple Contextual Neural Networks and Multilayer Perceptron (MLP) models for various PRNGs. It is presented that the results for some pairs of PRNGs can be significantly different even for large number of repeats of 5-fold CV. Observations suggest that when different ML models and algorithms are compared with the usage of 5-fold CV when different PRNGs were used, the confidence interval should be doubled or confidence level higher than 95% should be used. Additionally, it is shown that even under such conditions classification properties of Contextual Neural Networks are found statistically better than of not-contextual MLP models.

Keywords: Measures · Uncertainty · PRNG · Cross-validation · Training

1 Introduction

Among the various machine learning methods, artificial neural networks (ANNs) are currently very promising. They have been successfully incorporated into various fields of life, including biology [1], physics [2, 3], medicine [4, 5] and other fields [6–9]. A large number of architectures of ANN models is considered - from Convolutional Neural Networks (CNN) [9], Variational Autoencoders (VAE), Generative Adversarial Networks (GAN) [10, 11], recurrent architectures using Long Short-Term Memory (LSTM) and Gated Recurrent Units (GRU) [12, 13], as well as Multilayer Perceptrons (MLP) [14] and Self Organizing Maps [15]. There exist also solutions such as Contextual Neural

N. T. Nguyen et al. (Eds.): ACIIDS 2021, LNAI 12672, pp. 717–730, 2021.
https://doi.org/10.1007/978-3-030-73280-6_57

Networks (CxNNs), which are generalizations of MLPs and allow context-sensitive processing of the input data by usage of neurons with conditional aggregation functions [16]. This can lead to high reduction of signals processing cost and increase of the quality of classification [17] and was used e.g. in fingerprints detection in criminal investigations [18] and to classify cancer gene expression microarray data [19].

Important element of training of artificial neural networks (and in Machine Learning in general) is selection and usage of pseudo random number generator (PRNG). It is crucial for settings initial values of many parameters of NN models and related optimization algorithms [11, 20]. Pseudo random numbers are used during the training to guide stochastic search with e.g. stochastic gradient descent, and to perform dropout for Deep Learning [13, 14]. Randomized algorithms are often used to reduce the cost of computation or to obtain robust results by sampling, such as Markov Chain Monte Carlo (MCMC) or repeated N-fold Cross Validation [16, 21]. PRNGs are also applied to drive genetic operators of evolutionary algorithms – especially when NN models are the subject of evolution. Finally, it is worth underlining that PRNGs serve a vital role during training and evaluation of ANNs to:

- verify that the observed behavior of models and algorithms is independent of the selection of training/testing data and initial values of parameters,
- alleviate the vanishing/exploding gradient problems and increase the stability of ANN training (see e.g. Xavier, He or LeCun weights initialization), and
- get reproducible results for the same data.

PRNGs used for machine learning and training of ANNs are algorithms to produce pseudo random sequences expected to approximate the property of *true* random numbers. Actually, various PRNGs are developed so far [22, 23] and many of them are used in different machine learning and programming frameworks. For example Python language, CUDA and Matlab are using the Mersenne Twister algorithm as the core PRNG [24]. Object Pascal (Delphi) and many of the modern C++ variations by default are using simple Linear Congruential Generators as PRNGs [25]. Others, like Go and Rust, tend to favor PRNGs from XorShift and PCG families [23, 26, 27]. On top of that, various algorithms are in use to convert pseudo random sequences of integer values into pseudo random sequences of real values of given distributions (e.g. Box-Müller transform and Ziggurat algorithm for Gaussian distribution) [28, 29].

The above illustrates the situation in which Authors of Machine Learning studies can select many different PRNG solutions. They need also consider the importance of the long term independence of PRNGs in their simulation software from the evolution of third-party libraries. But what is even more important, much care should be given to the cases when there is a need to compare results of ML experiments performed with the usage of different PRNGs. This is because the selection of PRNG can beperceived as a source of measurement error, e.g. in typical procedures of calculation of measures of the quality of classifiers, such as e.g. R-times repeated N-fold Cross Validation. And finally, it seems to be crucial to verify that the observed properties of a model or algorithm are not due to the effects of the use of a particular PRNG.

Considering the above, it this paper we conduct experiments so that we can observe the possible level of differences in obtained values of various measures of classification quality of simple Contextual Neural Networks and MLP models for various PRNGs used in machine learning practice as well as few others – all listed in Table 1. Calculated measures include e.g. macro averaged classification accuracy, AUC, Gini Index [30] and macro averaged Modified Conditional Entropy (MCEN) [31]. In the result this paper contributes to answer the following questions:

- After how many repeats of CV the influence of PRNG on obtained results can be ignored?
- Are there measures of ANNs classification quality which are less affected by the selection of PRNG than the others?
- What level of PRNG-related error one should consider when comparing different classifiers with the usage of given number of repeats of N-fold CV?

The latter is finally applied to the example of Contextual Neural Networks as their selected classification properties are statistically compared with Multilayer Perceptron Networks of analogous architectures.

The further parts of the paper are constructed as follows. After the introduction, in section two we discuss the basic principles and problems of the usage of PRNGs in machine learning (training of ANNs). Next, in the third section we recall the details of operation of Contextual Neural Networks. Later, in section four results of experiments are presented showing effects of the usage of various PRNGs on the measures of classification properties of CxNNs and their comparison with MLP networks. Finally a summary is given with conclusions and directions for further research.

2 Principles of Contextual Neural Networks

The most important advantage of contextual neural networks (CxNN) when com-pared with MLP is caused from the property of CxNN that they include context neurons equipped with conditional multistep aggregation functions instead of the usual total aggregation function. The functions allow networks to aggregate values from specific inputs under predetermined conditions, and consequently, can improve the time costs and quality of the classification [16, 17].

Conditional multistep aggregation is performed in steps. In each step, the inputs from one selected group of inputs are aggregated. Aggregation of data from subsequent groups of inputs is performed on a basis of the input scan-path defined during CxNNs training. If the condition for stopping the aggregation is met, it is suspended, and consequently, some of the inputs in the calculation of the neuron's output will be completely omitted.

All contextual aggregation functions, that we encountered preparing the literature review, work on the same principle, including examples such as PRDFA, OCFA and SIF [17]. Figure 1 shows a diagram and formulas that were used to calculate the activation and output of a context neuron using SIF aggregation function.

During the aggregation, in each of the following $k \leq K$ steps, the values of the k-th group of inputs are aggregated and added to previously neuron accumulated that was defined in the previous step. If the accumulated activation is higher than the threshold value $\varphi *$ determined before the aggregation, the other groups (as well as all their inputs)

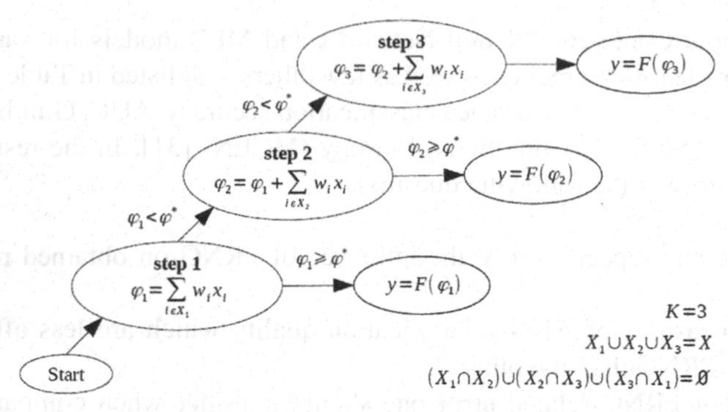

Fig. 1. Diagram of the example SIF aggregation function describing changes of attention field of contextual neuron with number of groups $K = 3$, aggregation threshold φ^*, activation function F and set of inputs X.

are skipped and do not appear in the next steps of the process. The number of steps required to reach the aggregation threshold depends on two basic factors: the scanning path defined by the neuron parameters and the signals from the actually processed input vector.

Scan paths learned during the training process are realized by aggregation functions of neurons. To incorporate constructing of scan paths into the processes of neural networks training, the traditional error back-propagation algorithm was extended to the Generalized error Back-Propagation algorithm (GBP) [16]. GBP joins Stochastic Gradient Descent method with the self-consistency paradigm, which is frequently used in physics to solve equation systems that include non-continuous and non-differentiable parameters [32].

It introduces only one additional parameter, namely the interval of groups update ω. It sets the number of epochs after which actual groupings of inputs of hidden neurons are calculated from weights of their connections. The value of this parameter plays an important role of controlling of the strength of relation between weights and scan paths and is crucial to make the self-consistency method stable. Extensive analysis of the GBP algorithm with examples of its usage can be found in [19].

Presented construction of the Contextual Neural Networks leads to properties which are different than in the case of MLPs or other non-contextual networks. The usage of conditional aggregation functions can significantly decrease the activity of the hidden connections of the neural network. The percentage of connections active during data processing change in relation to analyzed signals, but its average value drops easily below 10% and in some cases even lower [16, 17]. This can lead to significant reduction of the cost of data processing by the neural network both during the training and after it is finished.

Moreover, contextual behavior of CxNNs is not random and is not forced as in the case of dropout technique, but is also a kind of regularization. Instead of being directly randomized, it depends on what the NN model learned during the previous epochs of the training and on actually processed signals. This can lead to not processing signals which are not needed to calculate outputs of the network for the given input data vector. In the result the quality of data classification by the CxNN neural networks can be higher than for analogous MLP models. Independent observations of this effect report various levels of this improvement [16, 33]. But it can be noticed that in mentioned works different ML frameworks with different PRNGs were used. Thus it can be interesting to perform more precise measurements of this factor considering potential influences of different PRNGs.

3 Influence of PRNGs Using in Machine Learning

One of the most serious concerns in large scale Monte Carlo simulations and Neural Networks training in ML is proper approximation of real random numbers with pseudo random numbers generated by the software [21]. This is because performed simulations need to be not only repeatable, but the generation of pseudo random numbers should not influence the experiments with any artificial effects and should be time efficient.

To guarantee lack of artificial correlations within streams of pseudo random numbers one can use secure PRNG methods. Secure PRNG algorithms are defined rigidly in a mathematical manner in cryptography and are required to support the property called *output indistinguishability*: Polynomial-time probabilistic Turing machines (PPTM) cannot distinguish outputs of a secure PRNG algorithm from real random numbers with non-negligible advantage of probability [34]. Rigidly speaking, the *output indistinguishability* always assumes that the P-NP conjecture holds true. But in the reality, often the indistinguishability is also discussed assuming weaker conjectures, such as e.g. the RSA conjecture [35]. The RSA conjecture claims that the RSA encryption function $E(x) = x^e \bmod N$ is one-way. Till now it was not falsified thus it is assumed that the algorithm of the RSA generator is known to be secure.

But all the known secure PRNGs including the RSA generator are very slow when they are run with secure parameter settings. For example, the RSA generator generates a single bit by executing a single modular exponentiation. The time complexity of the modular exponentiation is evaluated by $O(N^3)$ in general, and even if one uses a constant exponent (usually, the fourth Fermat prime is used as the exponent), it is merely reduced to $O(N^2)$. With Karatsuba method it is possible to reduce the time complexity of the RSA by 35–42% for number of bits between 1000 and 2000 [36]. But even in such case generating random bits using the RSA generator is significantly inefficient.

In order to make the RSA conjecture true with the current practical computer architectures, it is recommended to set N, an RSA modulus, larger than 2,000 bits long [37]. It is reported that executing a modular exponentiation with this long modulus is a million times slower than executing AES-128, the most widely used block cipher [38]. Executing AES-128 in the counter mode can generate pseudo random numbers, and AES-128/CTR is actually used as a PRNG algorithm. This implies that the RSA generator is about 128×10^6 times slower than AES-128/CTR as PRNG algorithms. Although the AES-128/CTR generator can be efficient enough in some situations, its time complexity is also large enough to not use it in large scale simulations.

Therefore, in practice there is a need to use fast PRNG algorithms, which do not support the security of output indistinguishability, but are generating streams of pseudo random numbers with the amount of artificial imperfections as low as possible. Examples of such algorithms are e.g. ISAAC [39], PCG32 [23] and Mersenne Twister [24]. Typical considered PRNGs with their basic properties are listed in Table 1.

Table 1. Basic characteristic of considered PRNG algorithms. LCG – Linear Congruential Generator, LFSR – Linear Feedback Shift Register, PCG – Permuted Congruential Generator

Name	Context size [B]	Cycle length	Class	Crypt. Secure	Usage (with variations)
PasRand [25]	4	2^{32}	LCG	No	Delphi, Borland C/C++, MS Visual C/C++, WatcomC, C99, C11, C18, C++ 11, Java util.Random, POSIX
MT19937 [24]	2502	2^{19937}	LSFR	No	Python (2.2+), C++ 11, Free Pascal, Go, Julia, R, Matlab, Octave, CUDA, SPSS, SAS
PCG32 [23]	64	2^{64}	PCG	No	Rust, Implemented for C, C++, Java, Pascal, Haskell
Kiss123 [42]	20	2^{123}	Combination generator	No	STATA (deprecated)
XOR4096 [26]	522	2^{4096}	XorShift	No	Rust, Go (xoshiro), CUDA (xorwow)
SALSAR20 [40]	202	2^{70}	Stream cipher	Yes	ChaCha version in Linux /dev/urandom, Google TLS/SSL, Rust
ISAAC [39]	2066	2^{8295}	Stream cipher	Yes	GNU coreutils, Rust, translated to C++, Forth, Java, Delphi, Pascal, C#, Haskell, Lisp, PHP, JavaScript
AESR128 [38]	290	2^{128}	Block cipher in counter mode	Yes	NIST SP 800–90 standard, translated to C++, Pascal, used in Atmel and TI chips for RFID and immobilizers
WELL1024a [41]	134	2^{1024}	LSFR	No	See MT19937, OptaPlanner

The quality of PRNGs can be analyzed with sets of dedicated statistical procedures such as TestU01 [43] and related comparisons of fast PRNGs can be found e.g. in [23]. It is shown that most of considered PRNGs fail to pass some of the tests. But results of such tests do not show how particular PRNGs will influence experimental results of given machine learning experiments, for example training of ANNs with repeated N-fold cross-validation.

In practice it should be noticed that at least two major aspects of using different PRNGs can influence the variability of related results of ML experiments. First: each PRNG algorithm can generate only a finite set of pseudo random sequences and in the result, for a given machine learning model architecture, a finite set of possible combinations of initial values of model parameters. But what is even more important, for different PRNG algorithms the set of generated sequences is also different. Thus, it can happen that two distinct PRNGs applied to setting the initial parameters of ANNs will not be able to create the same initial and - in the result - trained models (Fig. 2).

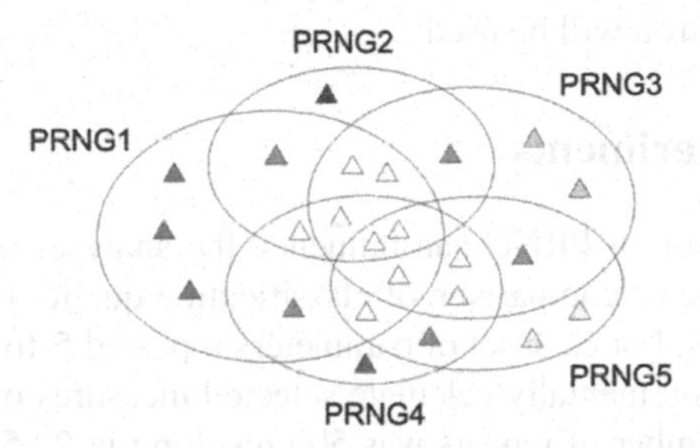

Fig. 2. Symbolic presentation of disjoint sets of possible initializations of parameters of the given ML model architecture by different pseudo random number generators. Many initial models can be created the same with various PRNGs (white triangles) but there are also models which will be specific for only two (grey triangles) or even one PRNG (other colors). Areas of ellipses are related to different lengths of the cycle of PRNGs. (Color figure online)

Thus even if large number of initial ML models would be considered to estimate average value of given measure, and those models would be ideally equidistributed over the space of models which can be created with given PRNG, some of the models can be specific only for this PRNG (Fig. 2). In the result, the value of selected measure of model quality can be slightly different for different PRNGs. By analogy this applies also to PRNG related training algorithms. And it is hard to estimate possible levels of differences in practical settings, especially when the differences between lengths of the cycle of considered PRNGs is huge (as e.g. between PasRand and MT19937).

The second aspect is strictly combinatorial and it will be shown on the example of the R-times repeated *N*-fold cross-validation. *N*-fold cross-validation is a procedure regarded as one of the best methods of estimation of measures of classifiers quality [44, 45]. In *N*-fold cross validation, the instances of a dataset are divided into *N* partitions. Then *N* iterations of training, prediction of classes and evaluation of measures are performed using one partition as a test dataset and the remaining *N*-1 partitions as a training dataset. When partitioning instances, for statistical correctness, it is important that the process of deliver instances into partitions is independently and uniformly random. Using real random numbers, fulfilling this statistical requirement is easy, but there is no guarantee for whether this requirement is fulfilled, when practical, fast PRNGs are used.

When 50 instances of data vectors are divided into five partitions with 10 instances, the total number of different ways of division is approximately 4.0×10^{29}. For 100 and 200 instances, it rises to 9.1×10^{63} and 1.8×10^{133}, respectively. In machine learning experiments, a dataset with a few hundred instances is rather small, and it is not rare to work with datasets with tens of thousands of instances. With such large datasets, the number of possible divisions is almost indefinitely large. In other words, the population of divisions can be viewed as infinite. Thus the small bias of outputs of PRNGs may cause a considerable bias in partitioning of instances, which possibly can significantly influence the results of experiments. Under such circumstances, the possibility that some divisions of data vectors with given PRNGs can be not possible plays second role.

In the following, using repeated N-fold cross validation we experiment with several well-known and widely-used practical PRNG algorithms to answer the aforementioned questions. As an example a comparison of key properties of Contextual Neural Networks and Multilayer Perceptron will be used.

4 Results of Experiments

To check if the selection of PRNG can influence the analyses of ML experiments we investigate the example of comparison of classification quality measures for MLP and CxNN neural networks. For each set of parameters repeated 5-fold stratified Cross Validation was used to incrementally calculate selected measures of classification quality of the models. The number of repeats was 500 resulting in 22.5 thousands of training processes for each combination of NN architecture and data set. Models were trained to solve selected benchmark problems from the UCI ML repository given in the Table 2. Architectures of NN models were set by the one-hot-encoding with two sizes of hidden layers. Winner-takes-all method was used for decoding of output classes. Numbers of neurons in the hidden layer were set to the values for which MLP networks achieved best results as suggested by earlier experiments [16].

Table 2. Attributes of considered data sets and related architectures of neural networks.

Dataset	Hidden neurons	Inputs of neural network	Classes (network outputs)	Data vectors
Iris	5, 10	4	3	150
Wine	5, 10	13	3	178
Lung cancer	10, 20	224	3	32
Sonar	5, 10	60	2	208
Soybean	10, 20	134	19	693

Both CxNNs and MLP networks were using neurons with unipolar sigmoid activation function. The aggregation function of neurons of Contextual Neural Networks was SIF with maximal number of groups $K = 7$ (as suggested by the Miller effect) [46]. GBP algorithm was used for training with training step $\alpha = 0.01$ and interval of groups actualization $\omega = 25$. Initial weights of connections were set with the Xavier algorithm [47]. Macro classification Accuracy and Macro Modified Conditional Entropy, PPV, TPR, AUC and Gini Index [30, 31] were calculated as averages after every 10 repeats of 5-fold CV for growing set of trained models. Experiments for given set of parameters were repeated with the same random sequence of seeds for all PRNGs. Considered PRNGs are given in Table 1. Margins of error for averages were calculated for 95% confidence level.

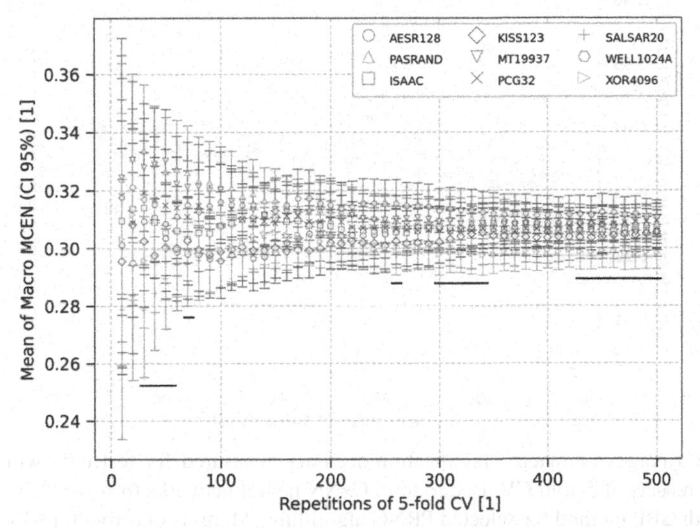

Fig. 3. The averages of macro Modified Conditional Entropy measured for test data with increasing counts of repeats of 5-fold CV used to train CxNN neural networks to solve "Lung Cancer" benchmark problem with GBP method for selected PRNG algorithms. Margins of errors for 95% confidence level. Cases of non-overlapping confidence intervals are underlined in black. (Color figure online)

As expected, among considered PRNGs there is no such PRNG which would cause generally better or generally worse results of generated NN models than any other PRNG. The means of considered measures appear to converge to the same narrow set of values regardless of the choice of PRNG (please see Fig. 3 and Fig. 4). This was expected as well as the fact that 500 repeats of 5-fold CV is not enough to observe statistical differences of results related to existence of models specific only for given PRNGs. Observation of such effect would be interesting from the theoretical point of view, but would require enormous computational resources. And in practice more important are effects that take place for low number or repeats of CV.

By closely analyzing the results presented on Fig. 3 and Fig. 4 one can notice that when NN models are generated with various types of PRNGs the averaged results of 10, 200 and even 500 times repeated 5-fold CV in some cases can be not enough to properly compare results when 95% confidence interval is used. In some cases the differences between PRNGs can lead to improper conclusions that given algorithms or types of models are statistically different with 95% confidence, even when they are identical. Problematic cases can be noticed for all analyzed measures and are marked with black horizontal line on Fig. 3 and Fig. 4. Presented observations suggest that in cases when results are being compared for models obtained with the use of different PRNGs the margin of error should be increased twice – or the confidence level higher than 95% should be used.

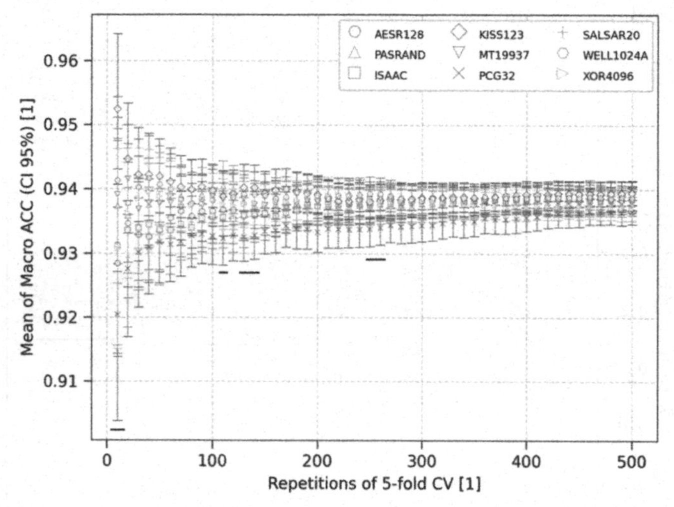

Fig. 4. The averages of macro classification accuracy measured for test data with increasing numbers of repeats of 5-fold CV used to train CxNN neural networks to solve "Iris" benchmark problem with GBP method for selected PRNG algorithms. Margins of errors for 95% confidence level. Cases of non-overlapping confidence intervals are underlined in black. (Color figure online)

The above finding can be especially important when the real differences between compared types of models or training algorithms seem to be small. In such cases simply using 10-fold CV and comparing averages with 95% of confidence level could give improper results.

Having the above knowledge it is possible to perform more precise comparison of basic classification properties of CxNN and MLP models. Example result for averaged Macro classification Accuracy is presented on the Fig. 5. For clarity results of only those three PRNGs are presented which for given type of neural network generated highest, lowest and close-to-average results. It can be noticed that fully randomized tests confirm earlier results that CxNNs allow to form models with higher Macro Accuracy than MLP models. Other analyzed measures such as Macro MCEN, PPV, TPR, AUC and Gini Index also show such relation (not depicted on figures). For example for the Lung Cancer data set Macro MCEN decreased from 0.32 ± 0.01 for MLP to 0.29 ± 0.01 for CxNNs. And for Sonar data analogous results were 0.71 ± 0.01 for MLP and 0.63 ± 0.01 for CxNNs, what is a considerable improvement.

The differences between classification quality of CxNN and MLP models are visible at 95% confidence level for 5-fold CV repetition count above 100 for mean of macro averaged ACC (Fig. 5).

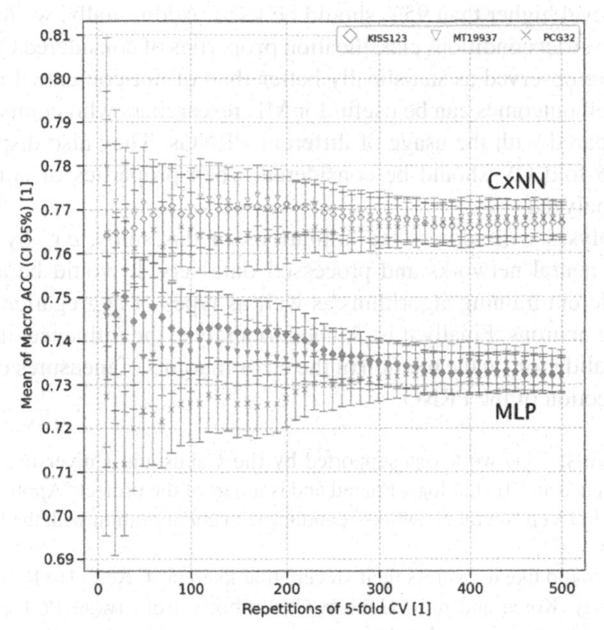

Fig. 5. The averages of macro classification accuracy measured for test data with increasing numbers of repeats of 5-fold CV used to train CxNN and MLP neural networks to solve "Lung Cancer" benchmark problem with GBP method for selected PRNG algorithms. Margins of errors calculated for 95% confidence level. Results for MLP are indicated with filled markers.

For Macro MCEN we noted that the difference is visible for at least 200 repeats of 5-fold CV. This suggests that further investigations related to CxNNs should consider at least 100-times repeated 5-fold CV and 200 – or, if possible, 300 – can be even better choice.

5 Conclusions

In this work nine widely used PRNG algorithms were tested with respect to whether execution of repeated 5-fold cross-validation with different PRNG can lead to different conclusions. The nine PRNGs were selected so that they are sufficiently efficient in time, which is a strong requirement to PRNGs nut typically used in machine learning research. During the tests five well known UCI ML benchmark datasets and two types of neural networks were used (MLP and CxNN) as classifiers. After performing multiple repeats of CV with the individual PRNGs, we investigated whether the obtained results are different in terms of such measures as means of macro modified conditional entropy and macro accuracy with 95% confidence level.

It is presented that there exist pairs of PRNGs for which mean results were statistically significantly different for selected numbers of repeats of 5-fold CV. Observations suggests that when different ML models and algorithms are compared with the usage of 5-fold CV when different PRNGs were used, the confidence interval should be doubled or confidence level higher than 95% should be used. Additionally, we have shown that even under such strict conditions classification properties of considered Contextual Neural Networks are observed as statistically better than of not-contextual models such as MLPs. Presented outcomes can be useful for ML researchers who compare models and algorithms prepared with the usage of different PRNGs. They also display how many repetitions of 5-fold CV should be considered when properties of contextual neural networks are analyzed.

Further analyses will address the limitations of this study, e.g. by increasing the size of trained neural networks and processed data sets. It would be also interesting to consider different training algorithms as well as types of aggregation and activation functions of the neurons. Finally it is planned to analyze the influence of the number of folds of cross validation on the relation of the error margins of measures of classification quality and selection of the PRNG.

Acknowledgements. This work was supported by the Gakushuin University, Japan, Wroclaw University of Science and Technology, Poland and is a part of the project "Application of Contextual Aggregation in Deep Neural Networks" conducted in collaboration with the Iwate Prefectural University, Japan.

The authors would like to express their sincere thanks to prof. Keun Ho Ryu from Chungbuk National University, Korea and prof. Goutam Chakraborty from Iwate Prefectural University, Japan for discussions on contextual and aware systems.

The authors also sincerely acknowledge Wolfgang Ehrhardt for preparation of his excellent open source library of cryptographic, hashing and PRNG algorithms.

Contribution. The idea of the presented experiments, related software, visualizations of results and simulations were developed and conducted by M. Huk. Analysis, interpretation and discussion of results as well as text preparation were done by all the Authors.

References

1. Mendez, K., Broadhurst, D., Reinke, S.: The application of artificial neural networks in metabolomics: a historical perspective. Metabolomics **15**(11), 1–14 (2019). https://doi.org/10.1007/s11306-019-1608-0
2. Zhang, Q., et al.: Artificial neural networks enabled by nanophotonics. Light Sci. Appl. **8**(1), 14 (2019)
3. Guest, D., Cranmer, K., Whiteson, D.: Deep learning and its application to LHC Physics. Annu. Rev. Nucl. Part. Sci. **68**, 1–22 (2018)
4. Nasser, I.M., Abu-Naser, S.S.: Lung cancer detection using artificial neural network. Int. J. Eng. Inf. Syst. (IJEAIS) **3**(3), 17–23 (2019)
5. Suleymanova, I., et al.: A deep convolutional neural network approach for astrocyte detection. Sci. Rep. **8**(12878), 1–7 (2018)

6. Chen, S., Zhang, S., Shang, J., Chen, B., Zheng, N.: Brain-inspired cognitive model with attention for self-driving cars. In: IEEE Transactions on Cognitive and Developmental Systems, vol. 11. no. 1, pp. 13–25. IEEE (2019)
7. Liu, L., et al.: Automatic skin binding for production characters with deep graph networks. ACM Trans. Graph. **38**(4), 12 (2019). Art. 114
8. Gao, D., Li, X., Dong, Y., Peers, P., Xu, K., Tong, X.: Deep inverse rendering for high-resolution SVBRDF estimation from an arbitrary number of images. ACM Trans. Graph. (SIGGRAPH) **38**(4), 15 (2019). Article 134
9. Gong, K., et al.: Iterative PET image reconstruction using convolutional neural network representation. IEEE Trans. Med. Imaging **38**(3), 675–685 (2019)
10. Higgins, I., et al.: Beta-VAE: learning basic visual concepts with a constrained variational framework. In: International Conference on Learning Representations, ICLR 2017, vol. 2, no. 5, pp. 1–22 (2017)
11. Karras, T., et al.: Progressive growing of GANs for improved quality, stability, and variation. In: International Conference on Learning Representations, ICLR 2018, pp. 1–26, (2018)
12. Huang X., et al.: A LSTM-based bidirectional translation model for optimizing rare words and terminologies. In: 2018 IEEE International Conference on Artificial Intelligence and Big Data (ICAIBD), China, IEEE, pp. 5077–5086 (2018)
13. Athiwaratkun, B., Stokes, J.W.: Malware classification with LSTM and GRU language models and a character-level CNN. In: Proceedings of 2017 IEEE International Conference on Acoustics, Speech and Signal Processing (ICASSP), pp. 2482–2486. IEEE, USA (2017)
14. Amato, F., et al.: Multilayer perceptron: an intelligent model for classification and intrusion detection. In: 31st International Conference on Ad-vanced Information Networking and Applications Workshops (WAINA), Taipei, Taiwan, pp. 686–691. IEEE (2017)
15. Dozono, H., et al.: Convolutional self organizing map. In: 2016 IEEE International Conference on Computational Science and Computational Intelligence (CSCI), pp. 767–771. IEEE (2016)
16. Huk, M.: Backpropagation generalized delta rule for the selective attention Sigma-if artificial neural network. Int. J. App. Math. Comp. Sci. **22**, 449–459 (2012)
17. Huk, M.: Notes on the generalized backpropagation algorithm for contextual neural networks with conditional aggregation functions. JIFS **32**, 1365–1376 (2017)
18. Szczepanik, M., Jóźwiak, I.: Fingerprint recognition based on minutes groups using directing attention algorithms. In: Rutkowski, L., Korytkowski, M., Scherer, R., Tadeusiewicz, R., Zadeh, L.A., Zurada, J.M. (eds.) ICAISC 2012. LNCS (LNAI), vol. 7268, pp. 347–354. Springer, Heidelberg (2012). https://doi.org/10.1007/978-3-642-29350-4_42
19. Huk, M.: Stochastic optimization of contextual neural networks with RMSprop. In: Nguyen, N.T., Jearanaitanakij, K., Selamat, A., Trawiński, B., Chittayasothorn, S. (eds.) ACIIDS 2020. LNCS (LNAI), vol. 12034, pp. 343–352. Springer, Cham (2020). https://doi.org/10.1007/978-3-030-42058-1_29
20. Knuth, D.E.: The Art of Computer Programming, 3rd edn, vol. 2. Seminum. Alg (1998)
21. Gentle, J.E.: Random Number Generation and Monte Carlo Methods, 2nd edn. Springer, New York (2003)
22. Klimasauskas, C.C.: Not knowing your random number generator could be costly: random generators - why are they important. PC AI Mag. **16**, 52–58 (2002)
23. O'Neill, M.E.: PCG: A Family of Simple Fast Space-Efficient Statistically Good Algorithms for Random Number Generation. Technical Report. Harvey Mudd College, pp. 1–58 (2014)
24. Matsumoto, M., Nishimura, T.: Mersenne twister: a 623-dimensionally equidistributed uniform pseudo-random number generator. ACM Trans. Model. Comput. Sim. **8**(1), 30 (1998)
25. Press, W.H., Teukolsky, S.A., Vetterling, W.T., Flannery, B.P.: Numerical Recipes: The Art of Scientific Computing, 3rd edn. Cambridge University Press, New York (2007)
26. Brent, R.P.: Note on Marsaglia's Xorshift random number generators. J. Stat. Softw. **11**(5), 1–5 (2004)

27. Vigna, S.: An experimental exploration of marsaglia's xorshift generators, scrambled. ACM Trans. Math. Softw. **42**(4), 1–23 (2016)
28. Balakrishnan, N., et al.: On box-muller transformation and simulation of normal record data. Communi. Stat. Simul. Comput. **45**(10), 3670–3682 (2016)
29. Marsaglia, G., Wan, T.W.: The ziggurat method for generating random variables. J. Stat. Softw. **5**, 1–7 (2000)
30. Sokolova, M., Lapalme, G.: A systematic analysis of performance measures for classification tasks. Inf. Process. Manag. **45**, 427–437 (2009)
31. Delgado, R., Núñez-González, D.: Enhancing Confusion Entropy (CEN) for binary and multiclass classification. PLoS ONE **14**(1), e0210264 (2019)
32. Glosser, C., Piermarocchi, C., Shanker, B.: Analysis of dense quantum dot systems using a self-consistent Maxwell-Bloch framework. In: Proceedings of 2016 IEEE Int. Symposium on Antennas and Propagation (USNC-URSI), Puerto Rico, pp. 1323–1324. IEEE (2016)
33. Wołk, K., Burnell, E.: Implementation and analysis of contextual neural networks in H2O framework. In: Nguyen, N.T., Gaol, F.L., Hong, T.-P., Trawiński, B. (eds.) ACIIDS 2019. LNCS (LNAI), vol. 11432, pp. 429–440. Springer, Cham (2019). https://doi.org/10.1007/978-3-030-14802-7_37
34. Katz, J., Yehuda, L.: Introduction to Modern Cryptography, 2nd edn., pp. 1–603. Chapman and Hall/CRC Press, Boca Raton (2015)
35. Steinfeld, R., Pieprzyk, J., Wang, H.: On the provable security of an efficient RSA-based pseudorandom generator. In: Lai, X., Chen, K. (eds.) ASIACRYPT 2006. LNCS, vol. 4284, pp. 194–209. Springer, Heidelberg (2006). https://doi.org/10.1007/11935230_13
36. Gopal, V., Grover, S., Kounavis, M.E.: Fast multiplication techniques for public key cryptography. In: IEEE Symposium on Computers and Communications, pp. 316–325. IEEE (2008). https://doi.org/10.1109/ISCC.2008.4625631
37. Barker, E., Dang, Q.: NIST Special Publication 800–57, Part 3, Rev. 1: Recommendation for Key Management: Application-Specific Key Management Guidance, National Institute of Standards and Technology, 12 (2015). https://doi.org/10.6028/NIST.SP.800-57pt3r1
38. Nechvatal, J., et al.: Report on the development of the advanced encryption standard (AES). J. Res. NIST **106**(3), 511–577 (2001)
39. Jenkins R.J.: Fast software encryption. In: ISAAC, pp. 41–49 (1996)
40. Tsunoo, Y., Saito, T., et al.: Differential cryptanalysis of Salsa20/8, SASC 2007: The State of the Art of Stream Ciphers, eSTREAM report 2007/010 (2007)
41. Panneton, F.O., L'eEcuyer, P., Matsumoto, P.: Improved long-period generators based on linear recurrences modulo 2. ACM Trans. Math. Soft. **32**(1), 16 (2006)
42. Rose, G.G.: KISS: A bit too simple. Cryptogr. Commun. **10**(1), 123–137 (2017). https://doi.org/10.1007/s12095-017-0225-x
43. L'ecuyer, P., Simard, R.: TestU01: AC library for empirical testing of random number generators. ACM Trans. Math. Softw. **33**(4), 1–40 (2007)
44. Rodriguez, J.D., et al.: Sensitivity analysis of k-fold cross validation in prediction error estimation. IEEE Trans. Pattern Anal. Mach. Int. **32**(3), 569–575 (2010)
45. Bouckaert, R.R.: Estimating replicability of classifier learning experiments. In: Proceedings of the 21st International Conference on Machine Learning, Banf, Canada (2004)
46. Miller, G.A.: The magical number seven, plus or minus two: some limits on our capacity for processing information. Psychol. Rev. **63**(2), 81–97 (1956)
47. Glorot, X., Bengio, Y.: Understanding the difficulty of training deep feedforward neural networks. J. Mach. Learn. Res. **9**, 249–256 (2010)

The Impact of Aggregation Window Width on Properties of Contextual Neural Networks with Constant Field of Attention

Miroslava Mikusova[1], Antonin Fuchs[2], Marcin Jodłowiec[3](✉), Erik Dawid Burnell[4], and Krzysztof Wołk[5]

[1] University of Zilina, Zilina, Slovakia
mikusova@fpedas.uniza.sk
[2] Rubi Consulting, Bratislava, Slovakia
[3] Faculty of Computer Science and Management, Wroclaw University of Science and Technology, Wroclaw, Poland
marcin.jodlowiec@pwr.edu.pl
[4] Science Applications International Corporation, Cookeville, USA
[5] ALM Services Technology Group, Wroclaw, Poland

Abstract. Artificial Neural Networks are quickly developing machine learning models with numerous applications. This includes also contextual neural networks (CxNNs). In this paper it is analyzed how the width of the aggregation window of the Constant Field of Attention (CFA) aggregation function changes the basic properties of the CxNN models, with the focus on classification accuracy and activity of hidden connections. Both aspects were analyzed in the custom H2O environment for real-life microarray data of ALL-AML leukemia gene expression and selected benchmarks hosted by the UCI ML repository. The results of presented study confirmed that increased length of the aggregation window of the CFA function can lead to similar or better classification accuracy than in the case of the Sigma-if aggregation. On the other hand, the activity of hidden connections in the majority of analyzed data sets resulted considerably higher. Presented analysis can be used as a guideline in further research on contextual neural networks.

Keywords: Contextual processing · Aggregation functions · Selective attention

1 Introduction

Technological progress related to development of artificial neural networks can be observed in many situations. Artificial neural networks are a vital part of many solutions for autonomous vehicles [1], recommendation systems [2], video rendering and images up scaling accelerators [3, 4]. The frequent use of neural networks and their importance is confirmed additionally by their adaptation in setups of scientific experiments [5] or medical diagnostic frameworks [6].

Widespread usage of neural networks was possible due to a great variety of architectures such as Variational Autoencoders (VAE), Generative Adversarial Networks (GAN)

N. T. Nguyen et al. (Eds.): ACIIDS 2021, LNAI 12672, pp. 731–742, 2021.
https://doi.org/10.1007/978-3-030-73280-6_58

[7–9] and general Convolutional Neural Networks (CNN) [10–14]. The list includes also recurrent neural networks, both using Long Short-Term Memory (LSTM) and its later modifications: Gated Recurrent Units (GRU) [15, 16]. Along those complicated solutions simpler ones are also in use – for example Self Organizing Maps (SOM) [17] and feedforward multilayered perceptron (MLP) [18]. It is noteworthy that despite the vast variation in training approaches and in developed architectures, most of analyzed models of artificial neural networks share the same property: all input signals are always processed. And such behavior is not typically observed in neural networks found in nature. Distinguishing importance of signals and limiting processing of not important information plays the key role in problems solving and energy preservation of humans and other living beings. It can be also used for the same purposes in the form of technical solutions of contextual information processing [19–21].

As a result of the above a different approach was developed in the form of Contextual Neural Networks (CxNN) [22–25]. This architecture uses contextual neurons [26–29] which can calculate neuron activation using inputs selected in relation to the data being processed. CxNNs can be considered as a generalization of multilayer perceptron. Contextual neural networks are used in applications such as predicting communicates in cognitive radio [30], classification of cancer gene expression microarray data [31–34] or finger-print detection [35, 36]. Some complex context-sensitive text mining solutions and telemedicine systems [37, 38] can be also based on CxNNs.

Stark scan-path theory [19] is related and can be used to interpret vital parts of CxNNs. In detail, CxNN model uses multi-step aggregation functions that represent trained scan-paths guiding both the calculation of neurons' output values, as well as the selection of inputs being processed. Therefore, a different subset of inputs can be used by CxNNs for different data vectors. This can lead to decrease of the activity of connections between neurons and to decrease of the data processing errors due to contextual filtering of input signals. Various contextual aggregation functions were defined for use within CxNNs, including Sigma-if (SIF) and CFA. It is also worth to notice that the weighted sum aggregation used in MLP networks is a special case of both the Sigma-if and CFA functions with the length of the scan-path equal to one.

The remaining part of this paper is organized as follows: the next section describes the state of the art in the field of contextual neural networks using multi-step aggregation functions. In Sect. 3, the CFA aggregation function is presented and the importance of the width of aggregation window is discussed. Section 4 includes the presentation of experiments and the analysis of results. The last section gives the conclusions and suggests possible usages of CFA aggregation and promising directions of related research.

2 Generalized BP and Contextual Neural Networks

Contextual neural network (CxNNs) is a type of artificial neural network utilizing the complex aggregation functions for neurons. The aggregation functions are characterized by multiple conditional steps [32, 39, 40]. The status of an input considering its participation in aggregation functions depends on certain conditions. This architecture differs from traditional MLP networks, which aggregate all the incoming inputs. Literature review shows that for some tasks like classification we can observe efficiency improvement [22, 33].

Multi-step aggregation functions operate in steps. Each of the steps incorporates the aggregation of a single, arbitrarily chosen group of inputs. Those groups are subsequently treated in accordance with the scan path, which is defined during the training. The aggregation ends when its condition is met. This results with rejection of some neuron inputs. This observation applies to all popular contextual aggregation functions (e.g. PRDFA, OCFA and SIF [26]), which are the objects of a separate research. In the Fig. 1 authors of [22] have shown formally how to calculate activation and output of a contextual neuron, utilizing the SIF aggregation function.

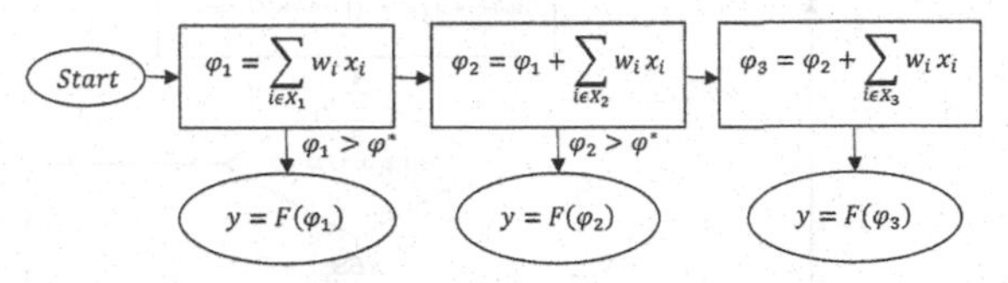

Fig. 1. Diagram presenting the SIF aggregation function which shows changes of attention field of contextual neuron with number of groups $K = 3$, aggregation threshold $\varphi*$, activation function F and set of inputs X.

The computation of SIF aggregation function is limited by a specific aggregation threshold $\varphi*$. This means, that when this threshold will be achieved or exceeded, the function will stop. The selection of groups is performed by splitting the neuron's inputs to K disjoint groups, of equal cardinality. If the equal division is impossible, the last group comprises less or more inputs. The aggregation proceeds in loop: for each step $k \leq K$ the values of the k-th group are aggregated. Then, the aggregation result is added to the accumulator, containing the previous activation value. If the exceeding of threshold will occur in the middle of the loop, the left groups will not be processed, thus will not be included in the aggregation. The number of steps which are needed to meet the threshold are specified by the scan path stemming from the neuron parameters. The signals from the input vector being processed affect this number as well.

Using GBP Algorithm to Train Contextual Neural Models

Different methods can be utilized to train contextual neural networks. However, the most popular is Generalized error backropagation algorithm (GBP). GBP extends the traditional error backpropagation algorithm. The extension incorporates the self-consistency method [22, 26], which is also used in the field of physics for analytically solving equations sets (with differentiable and non-differentiable parameters) [39, 40]. The GBP features the selection of values of discrete and not differentiable contextual neuron parameters which can be optimized by gradient-based methods. In the Fig. 2, the flowchart of GBP has been shown in the form of a block chart.

GBP starts with a random initialization of the connection weights. Then, it creates a single group of all inputs for each neuron. Before the groups are updated, the operation of aggregation function does not differ from standard weighted sum aggregation (SUM). Subsequently, through each training epoch, the outputs of network and batch output error E are calculated for each training data vector. In the end, the backpropagation occurs via all the network layers. The weights of connections, which are active in the forward

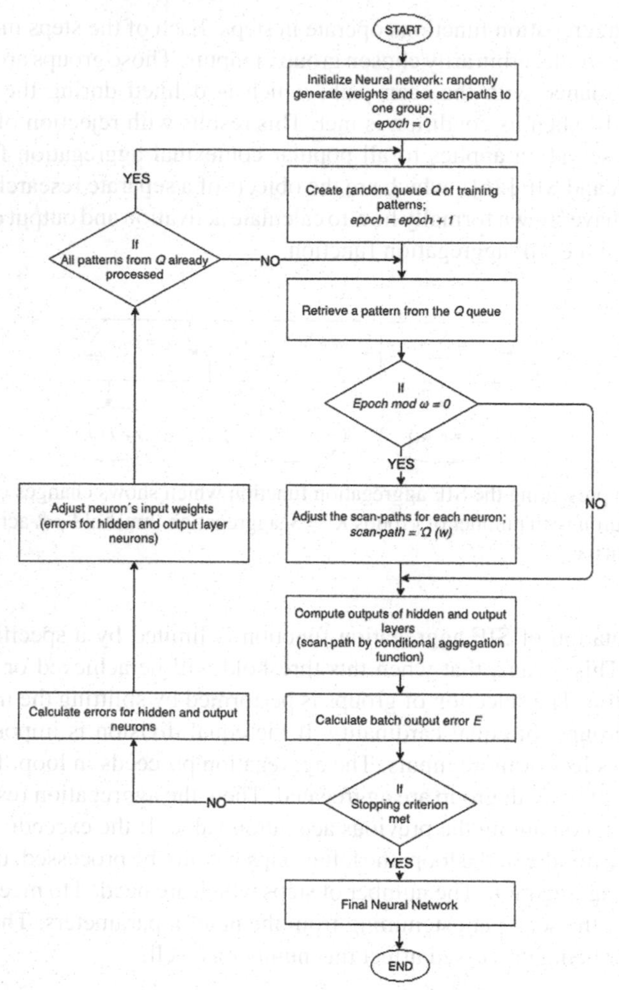

Fig. 2. Diagram of the Generalized error backpropagation algorithm (GBP). Ω stands for grouping function and ω for scan-paths update interval respectively.

phase, are modified in accordance with stochastic gradient descent method (SGD). Until the termination criterion is met, the process repeats.

The other aspect in GBP is to compute the scan-paths. In order to achieve that, after every ω epochs the grouping function $\Omega(w)$ is utilized to get the actual groupings of hidden neurons' inputs. The intensity of bound between the weights and neuron scan-paths is controlled by the group update interval ω. It is crucial for the self-consistency method stability. The low values of ω indicate too tight bound. In this case, the error space alters too frequently and SGD algorithm might not successively find the decent local optimum. In the opposite, high ω the pace of optimization is too fast, causing the selection of scan paths not being optimized. The infinite value of ω reduces the GBP method to a normal BP. In most cases, the ω value is selected from the range 10 to 25.

The grouping function $\Omega(w)$ for a given neuron sets the highest weighted connections to the most significant groups. As stated before, it also separates the inputs into K groups of similar size.

It can be observed, that the GBP characterizes each of the hidden neurons using two vectors: connection weights vector w and virtual scan-path vector. Both of them are interlinked: the scan-path is calculated with the $\Omega(w)$ function. The modifications of weights w rely on the construction of the scan-path. The tight connection between two vectors makes the self-consistency method to be well-suited for GBP. However, at the end of training, when the model approximates the acceptable local optimum, the modifications of both weights and scan-path are minimal. This results with the utilization of weight vectors to store information about virtual scan-path. Thus, they are not needed to be saved separately. This means, that in order to use contextual neural network, the only input that is needed to be known is the weights vector w, grouping function $\Omega$and an additional parameter K, which holds the number of groups in the scan-path.

3 Windowed Attention of Contextual Neurons

The attention of contextual neurons can be constrained in dynamic or static way. In dynamic approach the effective decision space is being increased in each iteration, which is called also the "evolving decision space". An example of this idea is a neuron which first attempts to solve problem in low-dimensional space, and if it fails, then it gradually expands the decision space (increases its dimensionality) until the problem is solved or all available inputs are used. The dimensionality might be increased by adding into consideration not only the signals from actual inputs, but also the inputs from all groups analyzed in previous steps. On the Fig. 1 there is the case of SIF function which is a good example of mentioned approach. SIF is one of the aggregation functions of a dynamic field of attention. It extends properties of neurons according to dynamic field of attention. It is known, that single neuron, which uses SIF aggregation function and as an activation function uses sigmoid function is able to solve linearly inseparable problems [22, 26]. The computational cost of neurons usage can be significantly decreased thanks to limitation of theirs connections activity. This can be gained by narrowing the field of attention of neurons. Furthermore, the accuracy of data processing can be increased by applying contextual filtering mechanisms.

In static approach signals from all inputs are used. Therefore, because the MLP networks with weighted SUM aggregation are always processing signals from all inputs, they have constant field of attention. The interesting fact is that in the case of SUM function the constant filed of attention can be considered both as negative and positive property. Neurons models with SUM aggregation, regardless processed data, have constant, 100% inputs activity. Even if some of the inputs are irrelevant for classification of given input vector, they will be used anyway. Irrelevant signals can be regarded as noise, and have negative impact on neurons output accuracy. An advantage of this feature is that it is not necessary to create any dedicated scan-paths, because the neurons already have access to all available data. Therefore, in this approach, training of such a network is much easier than in the previous method. Moreover, hardware implementations of neural networks are easier to prepare, because their construction is simpler that in the case of contextual models.

It is worth noting, that SIF and weighted sum functions behave the same way if we assume the groups number equal to one. In fact, the first one is a generalization of the second one. This means that is it possible to use a hybrid approach, which provides benefits from those mentioned previously: relatively low complexity level thanks to constant size of the field of attention, as well as the ability to contextual filtering of inputs signals. This makes mentioned hybrid approach extremely interesting therefore it is investigated further in the next parts of this paper.

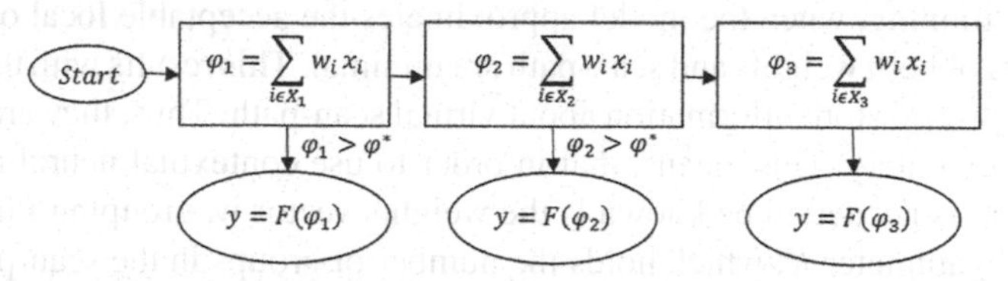

Fig. 3. Diagram presenting the CFA aggregation function which shows changes of attention field of contextual neuron with number of groups $K = 3$, aggregation window width = 1, aggregation threshold φ^*, activation function F and set of inputs X.

Figure 3 presents the idea of CFA aggregation in three steps for aggregation window width equal one. The result of aggregation function on each step is not passed to any of the next steps. If its value does not reach the given threshold it is dropped and the algorithm proceeds to next group of inputs. This is the general difference between CFA and SIF functions. The independency of each step has an important consequence: each neuron in each aggregation step has a field of attention not greater than the defined size of aggregation window. If the width of the aggregation window is greater than one, the fields of attention in the following steps of aggregation can overlap – what is a case of Overlapping CFA (OCFA) aggregation function [26]. In this work a basic CFA function is considered for which the fields of attention of following aggregation steps do not overlap. But even in such case there is still a contextual, indirect dependency between neighbor groups of inputs formed during the training by the selection of the importance and processing sequence of inputs.

The usage of the CFA aggregation function may bring following profits:

- reduction of the inter-neuronal connections activity,
- opportunity of contextual filtering applied to processed signals (more effective than for SIF aggregation function),
- easier parallelization of aggregation calculation thanks to greater independency of separate fields of attention of the neuron.

The last item on the lists looks especially promising because groups-activities dependency is the major drawback in SIF aggregation function approach. It results with problems to use modern processors optimized for parallel processing. Proposed approach changes the aggregation process therefore some additional constrains are need to be set in order to enable the possibility to create the scan-path of neurons using CFA function. Therefore following problems need to be discussed: does GBP method can be successfully applied in CxNNs training using CFA aggregation functions? What are the classification properties of this kind of contextual models? Both problems will be discussed and verified in the following parts of this paper.

Moreover, previous studies have shown that the limitation of the window width of the CFA aggregation function to one group of inputs has negative influence on properties of contextual models [41]. Resulting neural networks have increased average activity of hidden connections and decreased level of classification accuracy in comparison to Sigma-if and MLP models. Thus in this study it is especially examined how the length of the aggregation window changes the properties of contextual neural networks using the CFA aggregation.

4 Results of Experiments

In the presented study it is analyzed how the length of the window width of the CFA aggregation function changes the basic properties of the CxNN models. Thus the accuracy of classification and activity of hidden connections are measured for different sizes of the CFA window. Obtained results are also compared to the previously reported outcomes for the Sigma-if aggregation. Analyzed models were prepared for real-life microarray data of ALL-AML leukemia gene expression and for six selected benchmark data sets available from the UCI ML repository [42, 43]. The most important properties of those problems and related NN models are presented in the Table 1.

Reported outcomes were collected for the CxNN models created with the use of Generalized BP algorithm implemented within custom H2O machine learning framework based on its version Yates (3.24.0.3) [44–47]. The settings of the most important parameters were: aggregation = CFA or SIF, activation = tanh, threshold of aggregation = 0.6, number of epochs between updates of groupings = 10, number of hidden layers = 1. Number of connections groups and of neurons in the hidden layer used in presented study are provided in the Table 1. Connections weights initialization: Xavier. Input data standardization: unipolar Min-Max scaler. Inputs coding: one-hot. Outputs interpretation: winner-takes-all. Loss function: logloss. The width of the CFA window was in the range from 1 to 4 while for the Iris data set it was limited to 2 due to the selected number of groups. The statistical analysis was based on performed 10 times 10-fold stratified cross-validation.

The outcomes given in Table 2 display that increased length of the aggregation window of the CFA function can lead to similar or better classification accuracy than in the case of the Sigma-if aggregation. Moreover, it can be noticed that in the case of almost all considered problems the longer the CFA aggregation window is the lower are the classification errors. In the effect, among results of contextual neural networks with the CFA aggregation, in the case of six data sets the best classification accuracy was observed when aggregation window was equal four. This is expected because longer aggregation window allows CFA function to analyze in single aggregation step mutual relations of higher number of neuron inputs.

On the other hand, in the case of all analyzed data sets except Breast cancer and Crx, CFA aggregation leads to higher activity of hidden connections of contextual neural network than Sigma-if (Table 3). Most probably this is the effect of the limitation of the

Table 1. Properties of considered benchmark data sets and related neural networks considered in this study. Numbers of connections groups and of hidden neurons were preselected to maximize the classification accuracy of the CxNN models with the Sigma-if aggregation function.

Dataset	Samples	Attributes	Classes	Number of groups	Hidden neurons count
Breast cancer	699	10	2	7	35
Iris	150	4	3	2	4
Sonar	208	60	2	9	30
Crx	690	15	2	7	30
Golub	72	7130	2	1000	2
Soybean	307	35	19	10	20
Heart disease	303	75	5	7	10

Table 2. Average classification error and standard deviation for models with CFA and SIF functions and selected benchmark sets. The statistically best values are written in bold.

Dataset	CFA [1] width 1 $\times 10^{-3}$	CFA [1] width 2 $\times 10^{-3}$	CFA [1] width 3 $\times 10^{-3}$	CFA [1] width 4 $\times 10^{-3}$	SIF [1] x10^{-3}
Breast cancer	45 ± 38	41 ± 39	35 ± 29	**31 ± 32**	**32 ± 21**
Crx	142 ± 63	142 ± 59	140 ± 60	**136 ± 52**	**138 ± 13**
Golub	87 ± 0	79 ± 0	82 ± 0	76 ± 0	**51 ± 0**
Heart disease	174 ± 112	175 ± 108	169 ± 114	**163 ± 105**	191 ± 25
Iris	**27 ± 48**	32 ± 47	n.a	n.a	47 ± 48
Sonar	152 ± 104	148 ± 97	150 ± 96	141 ± 92	**137 ± 73**
Soybean	51 ± 26	50 ± 26	47 ± 21	**43 ± 28**	48 ± 43

CFA aggregation function which forces the neural network to analyze in each step of aggregation only small subsets of inputs. In the result neurons using CFA aggregation typically need to process more groups of inputs to find enough information to generate expected values of the outputs. This hypothesis can be additionally backed up by the fact, that in the case of Sonar, Soybean and Heart cancer problems the mean activity of connections is noticeably lower when the width of CFA aggregation is maximal. But such relation is not observed for Breast Cancer, Crx, Golub and Iris sets.

Additionally, it is interesting to notice in Table 2 that classification accuracy for Golub problem irrespectively of the length of the CFA aggregation window has standard deviation equal zero. But this holds also for the Sigma-if aggregation what suggests that such effect is related to the characteristics of the Golub set and not to properties of the CFA aggregation function.

Table 3. Average activity and standard deviation of hidden connections of neural models for CFA and SIF aggregation functions and selected benchmark sets. The statistically best values are written in bold.

Dataset	CFA [1] width 1	CFA [1] width 2	CFA [1] width 3	CFA [1] width 4	SIF [1]
Breast cancer	88.1 ± 8.2	**84.2 ± 9.5**	92.7 ± 7.4	89.5 ± 5.8	**84.9 ± 0.2**
Crx	**76.7 ± 21.1**	79.3 ± 21.3	81.6 ± 20.9	85.8 ± 19.7	83.3 ± 14.9
Golub	19.4 ± 16.3	19.7 ± 18.6	20.6 ± 16.1	23.1 ± 17.2	**7.2 ± 5.6**
Heart disease	67.6 ± 10.5	72.5 ± 9.1	64.7 ± 10.8	62.8 ± 11.0	**60.0 ± 6.0**
Iris	100 ± 0.0	100 ± 0	n.a	n.a	**82.7 ± 2.1**
Sonar	54.2 ± 14.4	52.6 ± 12.7	51.9 ± 14.8	48.7 ± 16.2	**46.4 ± 8.8**
Soybean	55.1 ± 1.42	56.0 ± 1.38	54.9 ± 1.52	53.3 ± 1.64	**52.5 ± 1.2**

5 Conclusions

This work includes discussion of the relation between the width of the aggregation window of the CFA function and basic properties of contextual neural networks. Measures of mean classification accuracy and mean activity of hidden connections were analyzed for real-life microarray data of ALL-AML leukemia gene expression and selected benchmarks hosted by the UCI ML repository. Experiments were conducted with the use of CxNN models trained with generalized error backpropagation algorithm implemented in the custom H2O environment.

The outcomes of experiments suggest that increasing the width of the window of the CFA aggregation function has positive influence on the classification accuracy of contextual neural networks. In case of some benchmark problems it leads also to decrease of the mean activity of the hidden connections. Additionally, for five of seven considered data sets observed classification accuracy of CxNNs with CFA aggregation was higher than results of analogous neural models using Sigma-if aggregation. But it is also clearly seen that except rare cases, usage of CFA aggregation creates contextual neural networks with considerably higher activity of hidden connections than neurons are based on SIF function. Most probably this is related to the fact that Sigma-if neurons can build their scan-paths being more adaptive than in the case of CFA models with static field of attention.

Presented analysis can be used as a guideline in further research on contextual neural networks. While CFA function can help to train contextual models with interesting classification properties, neural networks with dynamic field of attention achieve both high classification accuracies and low activity of hidden connections. Thus the latter can be analyzed with higher priority. One exception can be eventual hybrid contextual neural networks in which neurons of the same layer are using various aggregation mechanisms. During training of such models CFA function – due to its limitations – can potentially function as a regularizer, preventing too quick decrease of the activity of connections.

References

1. Chen, S., Zhang, S., Shang, J., Chen, B., Zheng, N.: Brain-inspired cognitive model with attention for self-driving cars. IEEE Trans. Cogn. Dev. Syst. **11**(1), 13–25 (2019)
2. Tsai, Y.-C., et al.: FineNet: a joint convolutional and recurrent neural network model to forecast and recommend anomalous financial items. In: Proceedings of the 13th ACM Conference on Recommender Systems RecSys 2019, pp. 536–537. ACM, New York (2019)
3. Liu, L., Zheng, Y., Tang, D., Yuan, Y., Fan, C., Zhou, K.: Neuro skinning: automatic skin binding for production characters with deep graph networks. ACM Trans. Graph. (SIGGRAPH) **38**(4), 12 (2019). Article 114
4. Gao, D., Li, X., Dong, Y., Peers, P., Xu, K., Tong, X.: Deep inverse rendering for high-resolution SVBRDF estimation from an arbitrary number of images. ACM Trans. Graph. (SIGGRAPH) **38**(4), 15 (2019). Article 134
5. Guest, D., Cranmer, K., Whiteson, D.: Deep learning and its application to LHC physics. Annu. Rev. Nucl. Part. Sci. **68**, 1–22 (2018)
6. Suleymanova, I., Balassa, T., et al.: A deep convolutional neural network approach for astrocyte detection. Sci. Rep. **8**(12878), 1–7 (2018)
7. Nankani, D., Baruah, R.D.: Investigating deep convolution conditional GANs for electrocardiogram generation. In: IEEE IJCNN 2020 under WCCI 2020, Glasgow, UK, pp. 1–8 (2019). https://doi.org/10.1109/IJCNN48605.2020.9207613
8. Higgins, I., et al.: Beta-VAE: learning basic visual concepts with a constrained variational framework. In: International Conference on Learning Representation, ICLR 2017, vol. 2, no. 5, pp. 1–22 (2017)
9. Karras, T., Aila, T., Laine, S., Lehtinen, J.: Progressive growing of GANs for improved quality, stability, and variation. In: International Conference on Learning Representation, ICLR, pp. 1–26 (2018)
10. Gong, K., Guan, J., Kim, K., Zhang, X., Yang, J., Seo, Y., et al.: Iterative PET image reconstruction using convolutional neural network representation. IEEE Trans. Med. Imaging **38**(3), 675–685 (2019)
11. Batbaatar, E., Li, M., Ho, R.K.: Semantic-emotion neural network for emotion recognition from text. IEEE Access **7**, 111866–211187 (2019)
12. Wang, Z.H., et al.: A novel facial thermal feature extraction method for non-contact healthcare system. IEEE Access **8**, 86545–86553 (2020)
13. Horng, G.J., Liu, M.X., Chen, C.C.: The smart image recognition mechanism for crop harvesting system in intelligent agriculture. IEEE Sensors J. **20**(5), 2766–2781 (2020)
14. Munkhdalai, L., et al.: Deep learning-based demand forecasting for Korean postal delivery service. IEEE Access **8**, 188135–188145 (2020)
15. Huang, X., Tan, H., Lin, G., Tian, Y.: A LSTM-based bidirectional translation model for optimizing rare words and terminologies. In: 2018 IEEE International Conference on Artificial Intelligence and Big Data (ICAIBD), China, pp. 5077–5086. IEEE (2018)
16. Athiwaratkun, B., Stokes, J.W.: Malware classification with LSTM and GRU language models and a character-level CNN. In: Proceedings of 2017 IEEE International Conference on Acoustics, Speech and Signal Processing (ICASSP), pp. 2482–2486. IEEE (2017)
17. Dozono, H., et al.: Convolutional self organizing map. In: 2016 IEEE International Conference on Computational Science and Computational Intelligence (CSCI), pp. 767–771. IEEE (2016)
18. Amato, F., et al.: Multilayer perceptron: an intelligent model for classification and intrusion detection. In: 31st International Conference on Advanced Information Networking and Applications Workshops (WAINA), Taipei, Taiwan, pp. 686–691. IEEE (2017)
19. Privitera, C.M., Azzariti, M., Stark, L.W.: Locating regions-of-interest for the Mars Rover expedition. Int. J. Remote Sens. **21**, 3327–3347 (2000)

20. Andreu, J., Baruah, R.D., Angelov, P.: Automatic scene recognition for low-resource devices using evolving classifiers IEEE International Conference on Fuzzy Systems (FUZZ-IEEE 2011), pp. 2779–2785. IEEE (2011)
21. Frejlichowski, D.: Low-level greyscale image descriptors applied for intelligent and contextual approaches. In: Nguyen, N.T., Gaol, F.L., Hong, T.-P., Trawiński, B. (eds.) ACIIDS 2019. LNCS (LNAI), vol. 11432, pp. 441–451. Springer, Cham (2019). https://doi.org/10.1007/978-3-030-14802-7_38
22. Huk, M.: Backpropagation generalized delta rule for the selective attention Sigma-if artificial neural network. Int. J. Appl. Math. Comput. Sci. **22**, 449–459 (2012)
23. Vanrullen, R., Koch, C.: Visual selective behavior can be triggered by a feed-forward process. J. Cogn. Neurosci. **15**, 209–217 (2003)
24. Huk, M.: Measuring the effectiveness of hidden context usage by machine learning methods under conditions of increased entropy of noise. In: 2017 3rd IEEE International Conference on Cybernetics (CYBCONF), Exeter, pp. 1–6 (2017). https://doi.org/10.1109/CYBConf.2017.7985787
25. Szczepanik, M., et al.: Multiple classifier error probability for multi-class problems. Eksploatacja i Niezawodnosc - Maintenance Reliab. **51**(3), 12–16 (2011). https://doi.org/10.17531/ein
26. Huk, M.: Notes on the generalized backpropagation algorithm for contextual neural networks with conditional aggregation functions. J. Intell. Fuzzy Syst. **32**, 1365–1376 (2017)
27. Mel, B.W.: The Clusteron: toward a simple abstraction for a complex neuron. In: Advances in Neural Information Processing Systems, vol. 4, pp. 35–42. Morgan Kaufmann (1992)
28. Spratling, M.W., Hayes, G.: Learning Synaptic clusters for nonlinear dendritic processing. Neural Process. Lett. **11**, 17–27 (2000)
29. Gupta, M.: Correlative type higher-order neural units with applications. In: IEEE International Conference on Automation and Logistics, ICAL2008, Springer Computer Science, pp. 715–718 (2008)
30. Huk, M., Pietraszko, J.: Contextual neural network based spectrum prediction for cognitive radio. In: 4th International Conference on Future Generation Communication Technology (FGCT 2015), pp. 1–5. IEEE Computer Society, London (2015)
31. Huk, M.: Non-uniform initialization of inputs groupings in contextual neural networks. In: Nguyen, N.T., Gaol, F.L., Hong, T.-P., Trawiński, B. (eds.) ACIIDS 2019. LNCS (LNAI), vol. 11432, pp. 420–428. Springer, Cham (2019). https://doi.org/10.1007/978-3-030-14802-7_36
32. Huk, M.: Training contextual neural networks with rectifier activation functions: Role and adoption of sorting methods. J. Intell. Fuzzy Syst. **38**, 1–10 (2019)
33. Huk, M.: Weights ordering during training of contextual neural networks with generalized error backpropagation: importance and selection of sorting algorithms. In: Nguyen, N.T., Hoang, D.H., Hong, T.-P., Pham, H., Trawiński, B. (eds.) ACIIDS 2018. LNCS (LNAI), vol. 10752, pp. 200–211. Springer, Cham (2018). https://doi.org/10.1007/978-3-319-75420-8_19
34. Huk, M.: Stochastic optimization of contextual neural networks with RMSprop. In: Nguyen, N.T., Jearanaitanakij, K., Selamat, A., Trawiński, B., Chittayasothorn, S. (eds.) ACIIDS 2020. LNCS (LNAI), vol. 12034, pp. 343–352. Springer, Cham (2020). https://doi.org/10.1007/978-3-030-42058-1_29
35. Szczepanik, M., Jóźwiak, I.: Fingerprint recognition based on minutes groups using directing attention algorithms. In: Rutkowski, L., Korytkowski, M., Scherer, R., Tadeusiewicz, R., Zadeh, L.A., Zurada, J.M. (eds.) ICAISC 2012. LNCS (LNAI), vol. 7268, pp. 347–354. Springer, Heidelberg (2012). https://doi.org/10.1007/978-3-642-29350-4_42
36. Szczepanik, M., Jóźwiak, I.: Data management for fingerprint recognition algorithm based on characteristic points' groups. In: New Trends in Databases and Information Systems, Foundations of Computing and Decision Sciences, vol. 38, no. 2, pp. 123–130. Springer, Heidelberg (2013)

37. Huk, M.: Context-related data processing with artificial neural networks for higher reliability of telerehabilitation systems. In: 17th International Conference on E-health Networking, Application & Services (HealthCom), pp. 217–221. IEEE Computer Society, Boston (2015)
38. Kwiatkowski J., et al.: Context-sensitive text mining with fitness leveling genetic algorithm. In: 2015 IEEE 2nd International Conference on Cybernetics (CYBCONF), Gdynia, Poland, pp. 1–6 (2015). ISBN: 978-1-4799-8321-6
39. Raczkowski, D., et al.: Thomas Fermi charge mixing for obtaining self-consistency in density functional calculations. Phys. Rev. B **64**(12), 121101–121105 (2001)
40. Glosser, C., Piermarocchi, C., Shanker, B.: Analysis of dense quantum dot systems using a self-consistent Maxwell-Bloch framework. In: Proceedings of 2016 IEEE International Symposium on Antennas and Propagation (USNC-URSI), Puerto Rico, pp. 1323–1324. IEEE (2016)
41. Burnell, E., Wołk, K., Waliczek, K., Kern, R.: The impact of constant field of attention on properties of contextual neural networks. In: Nguyen, N.T., Jearanaitanakij, K., Selamat, A., Trawiński, B., Chittayasothorn, S. (eds.) ACIIDS 2020. LNCS (LNAI), vol. 12034, pp. 364–375. Springer, Cham (2020). https://doi.org/10.1007/978-3-030-42058-1_31
42. Dua, D., Graff, C.: UCI Machine Learning Repository. University of California, School of Information and Computer Science, Irvine, CA (2019). https://archive.ics.uci.edu/ml
43. Golub, T.R., et al.: Molecular classification of cancer: class discovery and class prediction by gene expression monitoring. Science **286**, 531–537 (1999)
44. H2O.ai. H2O Version 3.24.0.4, Fast Scalable Machine Learning API For Smarter Applications (2019). https://h2o-release.s3.amazonaws.com/h2o/rel-yates/4/index.html
45. Janusz, B.J., Wołk, K.: Implementing contextual neural networks in distributed machine learning framework. In: Nguyen, N.T., Hoang, D.H., Hong, T.-P., Pham, H., Trawiński, B. (eds.) ACIIDS 2018. LNCS (LNAI), vol. 10752, pp. 212–223. Springer, Cham (2018). https://doi.org/10.1007/978-3-319-75420-8_20
46. Wołk, K., Burnell, E.: Implementation and analysis of contextual neural networks in H2O framework. In: Nguyen, N.T., Gaol, F.L., Hong, T.-P., Trawiński, B. (eds.) ACIIDS 2019. LNCS (LNAI), vol. 11432, pp. 429–440. Springer, Cham (2019). https://doi.org/10.1007/978-3-030-14802-7_37
47. Bouckaert, R.R., Frank, E.: Evaluating the replicability of significance tests for comparing learning algorithms. In: Dai, H., Srikant, R., Zhang, C. (eds.) PAKDD 2004. LNCS (LNAI), vol. 3056, pp. 3–12. Springer, Heidelberg (2004). https://doi.org/10.1007/978-3-540-24775-3_3

Towards Layer-Wise Optimization of Contextual Neural Networks with Constant Field of Aggregation

Miroslava Mikusova[1], Antonin Fuchs[2], Adrian Karasiński[3], Rashmi Dutta Baruah[4], Rafał Palak[5](✉), Erik Dawid Burnell[6], and Krzysztof Wołk[7]

[1] University of Zilina, Zilina, Slovakia
mikusova@fpedas.uniza.sk
[2] Rubi Consulting, Bratislava, Slovakia
[3] HSBC Digital Solutions, Hong Kong, China
[4] Indian Institute of Technology Guwahati, Guwahati, Assam, India
r.duttabaruah@iitg.ac.in
[5] Faculty of Computer Science and Management, Wroclaw University of Science and Technology, Wrocław, Poland
rafal.palak@pwr.edu.pl
[6] Science Applications International Corporation, Cookeville, USA
[7] ALM Services Technology Group, Wroclaw, Poland

Abstract. In this paper contextual neural networks with different numbers of connection groups in different layers of neurons are considered. It is verified if not-uniform patterns of numbers of groups can influence classification properties of contextual neural networks. Simulations are done in dedicated H2O machine learning environment enhanced with Generalized Backpropagation algorithm. Experiments are performed for selected UCI machine learning problems and cancer gene expression microarray data of bone marrow acute lymphatic and myeloid leukemia.

Keywords: Contextual processing · Selective attention · Numbers of groups

1 Introduction

Artificial neural networks form a set of highly evolved machine learning models and techniques that are actually used in many fields. Examples include biology [1], financial systems for forecasting and detecting anomalies [2], cognitive models for controlling the behavior of self-driving cars [3] and huge range of medical diagnostic systems [4–6]. Neural networks are used also for adjustment and regulation of values of parameters in scientific experiments [7, 8], as well as in the latest hardware accelerators for video rendering and image upscaling [9, 10]. All this has been achieved by the development of several types of artificial neural models. As the most important of them we can mention Convolutional Neural Networks (CNN) [11–13], Variational Autoencoders (VAE),

N. T. Nguyen et al. (Eds.): ACIIDS 2021, LNAI 12672, pp. 743–753, 2021.
https://doi.org/10.1007/978-3-030-73280-6_59

Generative Adversarial Networks (GAN) [14, 15], recurrent architectures using Long Short-Term Memory (LSTM) and Gated Recurrent Units (GRU) [16, 17]. In the majority of cases, these solutions require a significant amount of memory and time to be trained, so simpler models, such as multilayer perceptrons (MLP) [18] and Self Organizing Maps [19] are used whenever it is possible. But it can be noticed that all these types of neural models have a common feature – they are always processing signals from all inputs. In most cases, training and neural network architecture forces the model to activate all connections for all input vectors and features based on their pre-processing [20, 21]. On one hand, this simplifies their construction and training, but on the other hand it significantly limits their contextual data processing capabilities. In order to overcome this shortcoming, contextual neurons [22–24] and contextual neural networks (CxNN) have been developed, providing models with a set of unique properties [25–31].

Contextual neural networks can be considered as a generalization of MLP models [22]. In various scientific sources, we encounter their successful use in predicting communications in cognitive radio [32], classifying cancer gene expression microarray data [26–28], as well as detecting fingerprints through contextual analysis of groups of markers in criminal investigations [33, 34]. Contextual processing and CxNNs can also be considered as important elements of complex context-sensitive text mining solutions and an integral part of modern telemedicine systems [35–37].

Neurons in CxNNs are an implementation of the scan-path theory that was developed by Stark with the objective to capture the general properties of the human sensory system [38]. The conditional multistep neuronal aggregation functions accumulate progressively (step by step) the signals from the inputs until a given condition is met. The ordered list of groups forms a scanning path that which further controls which input signals are analyzed in relation to the content of the data actually being processed. This means that for different data vectors, a given contextual neuron can compute its output using different subsets of inputs.

The research presented in this paper is based on the usage of contextual neural networks with different numbers of connection groups in different layers of neurons. Specifically, constant (uniform), decreasing and increasing patterns of numbers of groups in hidden layers are considered. The main objective of our study was to find an answer to the question of whether or not the not-uniform patterns of numbers of groups can influence classification properties of contextual neural networks. The experiments were performed using selected reference problems from the UCI Machine Learning Repository [39], as well as real-life data (Golub cancer gene expression microarray data) [40].

The next parts of the paper are arranged as follows. Section 2 contains a description of contextual neural networks as well as the Generalized Error Backpropagation algorithm. Section 3 explains the idea and potential meaning of not-uniform patterns of numbers of groups in CxNNs. Section 4 presents the detailed plan as well as the results of experiments, along with a discussion of the measured outputs. Finally, in Sect. 5, the conclusions are summarized as well as suggestions for possible further directions of research in the field of the not-uniform patterns of numbers of groups in layers of neurons of contextual neural networks.

2 Contextual Neural Networks

Contextual neural networks (CxNNs) are machine learning models based on neurons witch multi-step aggregation functions [23]. Conditional character of those neurons allows them to aggregate signals from selected inputs under given condition. This is generalization of MLP based neural networks which always aggregate and process signals from all their inputs. It is known that conditional aggregation – in relation to MLP – can considerably decrease computational costs and improve classification accuracy [22, 26].

Multistep aggregation is realized as a sequence of conditional step. Inputs of neuron are divided to a given number of groups and in each step signals from inputs belonging to one group are aggregated. The sequence of groups defining the aggregation process is set during the training process and can be perceived as a scan-path considered earlier by Stark [38]. The process of input signals accumulation is performed until given condition is met. After that signals which were not read are ignored. This principle is common for the whole family of contextual aggregation functions such as e.g. Sigma-if (SIF), PRDFA and CFA (constant field of attention) [22, 23, 25, 27].

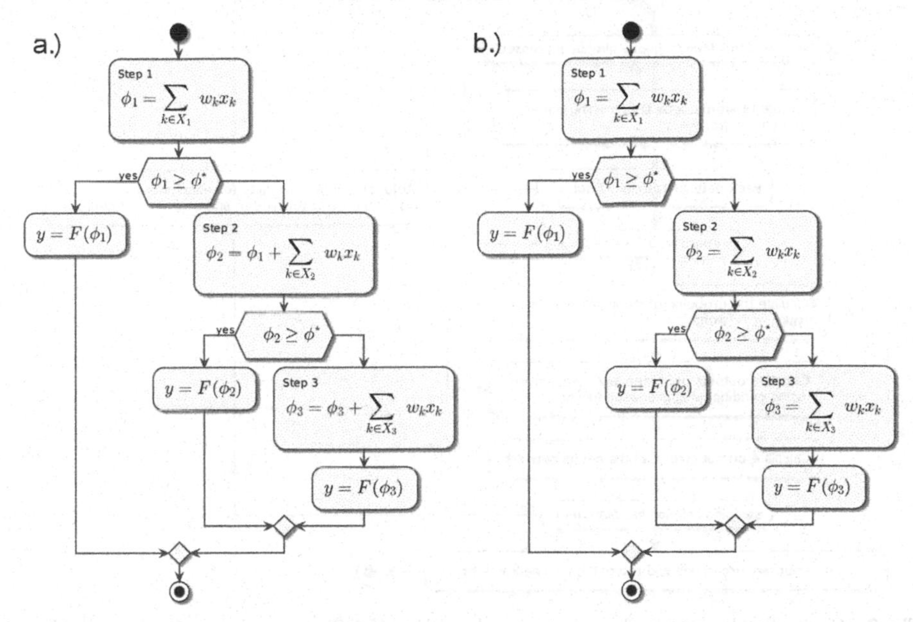

Fig. 1. Diagrams of the SIF (a) and CFA (b) aggregation functions. Examples given for contextual neuron with number of groups $K = 3$, aggregation threshold φ^*, activation function F and set of inputs X being sum of subsets X_1, X_2 and X_3.

Figure 1 presents diagrams of two example contextual aggregation functions: Sigma-if and CFA. In both cases the set of inputs X is divided into $K = 3$ subsets and the condition of aggregation is defined by the value of aggregation threshold φ^*. If in i-th step the aggregated value φ_i is greater than φ^* the aggregation is finished. The difference between presented functions is in the fact that SIF performs cumulative aggregation (value aggregated in the previous steps is added to the actual), and CFA ignores signals aggregated in previous steps.

One could conclude that the above solution will need for each neuron not only vector of connection weights and aggregation threshold φ^*, but also additional vector with assignment of all input connections to selected K groups. In practice this is not needed and assignment of connections to groups can be coded in the vector of connection weights. To do that an assumption is needed – for example such that connections with highest weights belong to the most important group which is analyzed in the first step, etc. This assumption is realized by grouping function $\acute{\Omega}$, generating grouping vector from weights vector and the number of groups K. Such solution is generalization of MLP network and behaves as it when the number of groups $K = 1$.

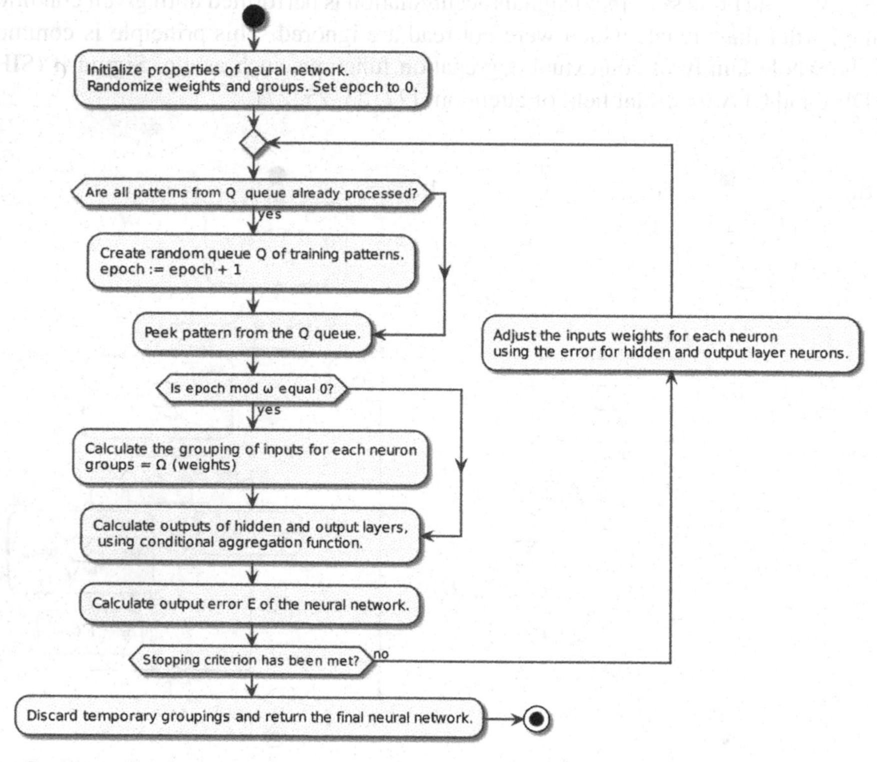

Fig. 2. Generalized error backpropagation algorithm (GBP) for training Contextual Neural Networks for error function E, grouping function $\acute{\Omega}$ and interval of groups update ω.

It was shown that training of neural networks constructed with the usage of contextual neurons can be done by the gradient algorithm [22, 24, 29]. In the case of CxNNs due to dualistic meaning of the weights vector the key to do that is the self-consistency paradigm known from physics [41]. It can be used to solve sets of equations with not-continuous and not-differentiable variables with the usage of gradient methods. This solution was merged with the error backpropagation algorithm in the form of generalized error backpropagation method (GBP) [22]. Block diagram of GBP algorithm is depicted on Fig. 2.

3 Not-Uniform Patterns of Connections Grouping in CxNNs

The analysis of literature related to contextual neural networks indicates that previous studies considered only such architectures of CxNNs where each hidden layer of neurons was assigned to the same number of groups K [22–34]. We have found this as a considerable limitation. In the effect in this paper we are experimenting with CxNNs having not-uniform patterns of numbers of groups in subsequent hidden layers.

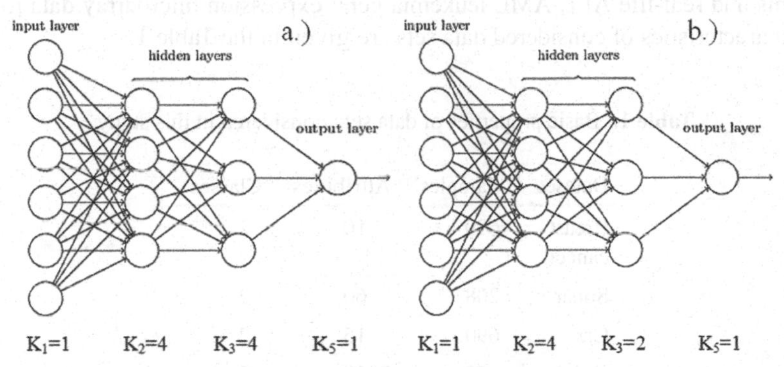

Fig. 3. Examples of CxNN neural networks with two hidden layers. Numbers of neurons in subsequent layers: N = [6,4,3,1]. Numbers of groups of connections in neurons of subsequent layers: network a.): K =[1,4,4,1] (uniform pattern in hidden layers), network b.): K = [1,4,2,1] (not-uniform pattern in hidden layers). Input and output layers have number of groups equal to one by default.

The basic example of the analyzed solution is presented on Fig. 3. Two contextual neural networks with the same architecture and numbers of neurons in hidden layers can differ with the number of groups K assigned to neurons in each hidden layer. Figure 3a shows the case where each neuron in each hidden layer has the same number of groups $K = 4$. Such uniform pattern of numbers of groups will be further designated as G1441. Contrary, Fig. 3b presents the same network but with $K = 4$ in first hidden layer and $K = 2$ in second hidden layer. Such not-uniform pattern of numbers of groups will be further named as G1421.

We are expecting that not-uniform patterns of numbers of groups in hidden layers can have considerable influence on properties of contextual neural networks. Moreover, even if each hidden layer would have the same number of neurons, by appropriately setting different numbers of groups in different layers one can construct structure analogous to encoder-decoder network [14, 15] but in the space of connection groupings. How this would influence the process of training and properties of the network is an open question. In this study we are limiting the set of considered patterns to the three basic kinds: uniform, increasing and decreasing.

4 Results of Experiments

During described experiment we have checked how layer-dependent setting of number of groups influences selected properties of four-layer CxNN neural networks with CFA aggregation functions. Each model included two hidden layers of 10 neurons each. Numbers of neurons in input and output layers were set with the usage of one-hot and one-neuron-per-class unipolar encodings, respectively. Hidden connections activity and overall classification accuracy were logged for constant (G1441, G1771), increasing (G1471) and decreasing (G1741, G1721) sequences of numbers of groups in subsequent hidden layers of CxNN models. Numbers of groups in the input and output layers, as well as the width of the CFA aggregation functions were set by default to one. Measurements were taken for NN models trained to solve selected benchmark UCI machine learning problems and real-life ALL-AML leukemia gene expression microarray data [39, 40]. Main characteristics of considered data sets are given in the Table 1.

Table 1. Basic properties of data sets considered in this study.

Dataset	Samples	Attributes	Classes
Breast cancer	699	10	2
Sonar	208	60	2
Crx	690	15	2
Golub	72	7130	2
Soybean	307	35	19
Heart disease	303	75	5

Obtained results were gathered for CxNN networks trained with dedicated version of the H2O 3.24.0.3 machine learning framework implementing GBP algorithm. [25, 42] Main settings used during creation of the models were as follows: neuron activation function = tanh, number of hidden layers = 2, number of neurons in each hidden layer = 10, groupings update interval = 5, aggregation function = CFA, aggregation threshold = 0.6. Xavier algorithm was used for connections weights initialization and unipolar Min-Max scaler was used for standardization of input data. The loss function was logloss. Stratified, ten times repeated 10-fold cross-validation was used to create statistical analysis of the models [43, 44].

Measured values for test data depicted in Table 2 show that non-uniform setting of the number of connections groups can have considerable influence on classification properties of CxNN models. It can be noticed that the average level of classification errors is lower when the number of connections groups decreases with increasing index of the layer of the network. This was observed both in G1741 and G1721 cases, but the latter seems to present stronger effect. Moreover, for those settings, five out of six data sets resulted with considerably decreased standard deviation of classification accuracy.

Table 2. Average classification error of CxNNs with CFA aggregation function for width = 1, test vectors of different datasets and considered patterns of numbers of groups in the hidden layers. The statistically best results are marked with bold.

Dataset	G1441 [%]	G1771 [%]	G1471 [%]	G1741 [%]	G1721 [%]
Breast cancer	4.5 ± 3.4	5.4 ± 3.6	4.8 ± 3.5	3.7 ± 3.1	**3.3 ± 2.3**
Crx	14.1 ± 5.4	14.1 ± 5.8	14.2 ± 5.6	13.8 ± 5.3	13.6 ± **1.4**
Golub	8.6 ± 0.3	8.7 ± 0.3	8.5 ± 0.4	**7.5 ± 0.3**	**5.7 ± 0.2**
Heart disease	17.1 ± 11.2	16.8 ± 10.9	17.3 ± 11.4	16.5 ± 10.2	16.3 ± **3.2**
Sonar	15.3 ± 10.5	16.4 ± 9.7	15.4 ± 10.7	14.2 ± 9.1	**13.8 ± 5.9**
Soybean	4.9 ± 2.9	5.1 ± 2.8	5.0 ± 2.6	4.4 ± 2.5	4.3 ± 2.8

It is interesting, that pattern of decreasing numbers of groups of connections in a characteristic way influences also activity of hidden connections of CxNNs (Table 3). For increasing (G1471) and constant groupings (G1441, G1771) there is no evident trend of changes of this quantity. But for considered decreasing patterns the average values of hidden connections activity is lower than for the others. And especially results for G1721 are statistically lower (better) in case of four out of six analyzed data sets. This can be perceived as an unexpected effect because lower number of groups of connections of contextual neuron often leads to their higher activity.

Table 3. Average activity of hidden connections of considered neural models, CFA aggregation function, test vectors of datasets and various patterns of numbers of groups in the hidden layers. The statistically best results are marked with bold.

Dataset	G1441 [%]	G1771 [%]	G1471 [%]	G1741 [%]	G1721 [%]
Breast cancer	89.5 ± 9.4	91.6 ± 9.2	89.4 ± 7.2	84.7 ± 5.9	**82.3 ± 3.3**
Crx	83.3 ± 21.5	85.3 ± 20.9	79.1 ± 20.8	72.7 ± 19.3	**67.4 ± 14.2**
Golub	20.9 ± 18.2	21.0 ± 16.8	20.5 ± 17.2	19.8 ± 15.7	18.7 ± 14.7
Heart disease	72.3 ± 9.4	68.7 ± 10.2	65.1 ± 8.3	63.2 ± 10.8	**62.5 ± 8.6**
Sonar	54.1 ± 12.8	51.8 ± 11.9	48.7 ± 12.6	47.0 ± 13.8	46.9 ± 8.3
Soybean	56.4 ± 1.4	55.9 ± 1.4	53.6 ± 1.6	54.4 ± 1.7	**53.2 ± 1.4**

The most evident case of such relation is contextual neuron with only one group, which by default has maximal, 100% activity of connections. And here we observe that decrease of the number of groups in the second hidden layer from four to two (change from G1741 to G1721) leads to the decrease of activity of hidden connections.

Finally it is worth to notice, that the presented measurements show that the GBP algorithm implemented in the H2O environment can train contextual neural networks with non-uniform setting of the number of groups of connections across hidden layers. Even if the design of GBP and the construction of CxNNs are general with the respect to the number of connections groups in hidden layers, to our best knowledge this is the first report in the related literature presenting this in practice.

5 Conclusions

In this paper we present outcomes of experiments with contextual neural networks using CFA aggregation function. The analyzed element was influence of the not-uniform settings of the numbers of groups across hidden layers of constructed CxNNs on their hidden connections activity and classification accuracy. All values were measured with the usage of 10 times repeated 10-fold cross validation for selected benchmark data sets known from UCI machine learning repository and real-life cancer gene expression microarray data of bone marrow acute lymphatic and myeloid leukemia (Golub). Training of contextual neural networks was done with the dedicated H2O machine learning framework.

The analysis of outcomes of performed experiments shows that not-uniform settings of the numbers of groups of connections in hidden layers of contextual neurons can considerably modify properties of CxNNs. Especially, when the numbers of groups decrease in consecutive hidden layers of neurons, observed classification accuracy can be statistically better for many problems. And such relation can be noticed also for activity of hidden connections. This result is interesting because lower numbers of groups of connections of contextual neurons typically are related with higher activity of their connections. But to find the reason of such effect, greater set of different patterns of not-uniform settings of the numbers of groups of connections should be analyzed. It would be also interesting to check if relations observed in this study are independent from the selection of aggregation functions of contextual neurons.

Finally, presented results confirm that H2O framework enriched with GBP algorithm can train CxNNs with not-uniform settings of numbers of connections groups in hidden layers of contextual neural models.

References

1. Mendez, K.M., Broadhurst, D., Reinke S.N., The application of artificial neural networks in metabolomics: a historical perspective. Metabolomics **15**(11), 142. Springer (2019). https://doi.org/10.1007/s11306-019-1608-0
2. Tsai, Y.C., et al.: FineNet: a joint convolutional and recurrent neural network model to forecast and recommend anomalous financial items. In: Proceedings of the 13th ACM Conference on Recom-mender Systems RecSys 2019, New York, USA, pp. 536–537. ACM (2019)
3. Chen, S., Zhang, S., Shang, J., Chen, B., Zheng, N.: Brain-inspired cognitive model with attention for self-driving cars. IEEE Trans. Cogn. Dev. Syst. **11**(1), 13–25. IEEE (2019)
4. Nasser, I.M., Abu-Naser, S.S.: Lung cancer detection using artificial neural network. Int. J. of Eng. Inf. Syst. (IJEAIS) **3**(3), 17–23 (2019)
5. Suleymanova, I., et al.: A deep convolutional neural network approach for astrocyte detection. Sci. Rep. **8**(12878), 1–7 (2018)
6. Wang, Z.H., Horng, G.J., Hsu, T.H., Chen, C.C., Jong, G.J.: A novel facial thermal feature extraction method for non-contact healthcare system. IEEE Access **8**, 86545–86553. IEEE (2020)

7. Qiming, Z., et al.: Artificial neural networks enabled by nanophotonics. Light: Sci. Appl. **8**(1), 14. Nature Publishing Group (2019)
8. Guest, D., Cranmer, K., Whiteson, D.: Deep learning and its application to LHC physics. Annu. Rev. Nucl. Part. Sci. **68**, 1–22 (2018)
9. Liu, L., Zheng, Y., Tang, D., Yuan, Y., Fan, C., Zhou, K.: Automatic skin binding for production characters with deep graph networks. ACM Trans. Graph. (SIGGRAPH) **38**(4), Article 114, 12 (2019)
10. Gao, D., Li, X., Dong, Y., Peers, P., Xu, K., Tong, X.: Deep inverse rendering for high-resolution SVBRDF estimation from an arbitrary number of images. ACM Trans. Graph. (SIGGRAPH) **38**(4), article 134, 15 (2019)
11. Gong, K., et al.: Iterative PET image reconstruction using convolutional neural network representation. IEEE Trans. Med. Imag. **38**(3), 675–685. IEEE (2019)
12. Munkhdalai, L., Park, K.-H., Batbaatar, E., Theera-Umpon, N., Ryu, K.H.: Deep learning-based demand forecasting for Korean postal delivery service. IEEE Access **8**, 188135–188145 (2020)
13. Batbaatar, E., Li, M., Ryu, K.H.: Semantic-emotion neural network for emotion recognition from text. IEEE Access **7**, 111866–111878. IEEE (2019)
14. Higgins, I., et al.: β-VAE: learning basic visual concepts with a constrained variational framework. In: International Conference Learning Representations. ICLR 2017, vol. 2, no. 5, pp. 1–22 (2017)
15. Karras, T., Aila, T., Laine, S., Lehtinen, J.: Progressive growing of GANs for improved quality, stability, and variation. In: International Conference on Learning Representations. ICLR 2018, pp. 1–26 (2018)
16. Huang, X., Tan, H., Lin, G., Tian, Y.: A LSTM-based bidirectional translation model for optimizing rare words and terminologies. In: 2018 IEEE International Conference on Artificial Intelligence and Big Data (ICAIBD), China, pp. 5077–5086. IEEE (2018)
17. Athiwaratkun, B., Stokes, J.W.: Malware classification with LSTM and GRU language models and a character-level CNN. In: Proceedings 2017 IEEE International Conference on Acoustics, Speech and Signal Processing (ICASSP), USA, 2017, pp. 2482–2486. IEEE (2017)
18. Amato, F., et al.: Multilayer perceptron: an intelligent model for classification and intrusion detection. In: 31st International Conference on Advanced Information Networking and Applications Workshops (WAINA), Taipei, Taiwan, pp. 686–691. IEEE (2017)
19. Dozono, H., Niina, G., Araki, S.: Convolutional self organizing map. In: 2016 IEEE International Conference on Computational Science and Computational Intelligence (CSCI), pp. 767–771. IEEE (2016)
20. Gościewska, K., Frejlichowski, D.: A combination of moment descriptors, fourier transform and matching measures for action recognition based on shape. In: Krzhizhanovskaya, V.V., et al. (eds.) ICCS 2020. LNCS, vol. 12138, pp. 372–386. Springer, Cham (2020). https://doi.org/10.1007/978-3-030-50417-5_28
21. Frejlichowski, D.: Low-level greyscale image descriptors applied for intelligent and contextual approaches. In: Nguyen, N.T., Gaol, F.L., Hong, T.-P., Trawinski, B. (eds.) ACIIDS 2019. LNCS (LNAI), vol. 11431. Springer, Cham (2019). https://doi.org/10.1007/978-3-030-14799-0
22. Huk, M.: Backpropagation generalized delta rule for the selective attention Sigma-if artificial neural network. Int. J. App. Math. Comp. Sci. **22**, 449–459 (2012)
23. Huk, M.: Notes on the generalized backpropagation algorithm for contextual neural networks with conditional aggregation functions. J. Intell. Fuzzy Syst. **32**, 1365–1376. IOS Press (2017)
24. Huk, M.: Stochastic optimization of contextual neural networks with RMSprop. In: Nguyen, N.T., Jearanaitanakij, K., Selamat, A., Trawiński, B., Chittayasothorn, S. (eds.) ACIIDS 2020. LNCS (LNAI), vol. 12034, pp. 343–352. Springer, Cham (2020). https://doi.org/10.1007/978-3-030-42058-1_29

25. Burnell, E.D., Wołk, K., Waliczek, K., Kern, R.: The impact of constant field of attention on properties of contextual neural networks. In: Nguyen, N.T., Trawinski, B., et al. (eds.) 12th Asian Conference on Intelligent Information and Database Systems, ACIIDS 2020. LNAI, vol. 12034, pp. 364–375, Springer (2020). https://doi.org/10.1007/978-3-030-42058-1_31
26. Huk, M., Non-uniform initialization of inputs groupings in contextual neural networks. In: Nguyen, N., Gaol F., Hong TP., Trawiński B. (eds) Intelligent Information and Database Systems. ACIIDS 2019. LNCS, vol. 11432, pp. 420–428. Springer, Cham (2019). https://doi.org/10.1007/978-3-030-14802-7_36
27. Huk, M.: Training contextual neural networks with rectifier activation functions: role and adoption of sorting methods. J. Intell. Fuzzy Syst. **37**(6), 7493–7502. IOS Press (2019)
28. Huk, M.: Weights ordering during training of contextual neural networks with generalized error backpropagation: importance and selection of sorting algorithms. In: Nguyen, N.T., Hoang, D.H., Hong, T.-P., Pham, H., Trawiński, B. (eds.) ACIIDS 2018. LNCS (LNAI), vol. 10752, pp. 200–211. Springer, Cham (2018). https://doi.org/10.1007/978-3-319-75420-8_19
29. Szczepanik, M., et al.: Multiple classifier error probability for multi-class problems. Eksploatacja i Niezawodnosc - Maintenance and Reliability **51**(3), 12–16 (2011). https://doi.org/10.17531/ein
30. Huk, M.: Measuring computational awareness in contextual neural networks. In: 2016 IEEE International Conference on Systems, Man, and Cybernetics (SMC), Budapest, pp. 002254–002259 (2016). https://doi.org/10.1109/SMC.2016.7844574
31. Huk, M., Measuring the effectiveness of hidden context usage by machine learning methods under conditions of increased entropy of noise. In: 2017 3rd IEEE International Conference on Cybernetics (CYBCONF), Exeter, pp. 1–6 (2017). https://doi.org/10.1109/CYBConf.2017.7985787
32. Huk, M., Pietraszko, J.: Contextual neural-network based spectrum prediction for cognitive radio. In: 4th International Conference on Future Generation Communication Technology (FGCT 2015). IEEE Computer Society, London, UK, pp. 1–5 (2015)
33. Szczepanik, M., Jóźwiak, I.: Data management for fingerprint recognition algorithm based on characteristic points' groups. In: New Trends in Databases and Information Systems. Foundations of Computing and Decision Sciences, vol. 38, no. 2, pp. 123–130, Springer (2013). https://doi.org/10.1007/978-3-642-32518-2_40
34. Szczepanik, M., Jóźwiak, I.: Fingerprint recognition based on minutes groups using directing attention algorithms. In: Rutkowski, L., Korytkowski, M., Scherer, R., Tadeusiewicz, R., Zadeh, L.A., Zurada, J.M. (eds.) ICAISC 2012. LNCS (LNAI), vol. 7268, pp. 347–354. Springer, Heidelberg (2012). https://doi.org/10.1007/978-3-642-29350-4_42
35. Kwiatkowski, J., et al.: Context-sensitive text mining with fitness leveling genetic algorithm. In: 2015 IEEE 2nd International Conference on Cybernetics (CYBCONF), Gdynia, Poland, 2015, pp. 1–6. Electronic Publication (2015). https://doi.org/10.1109/CYBConf.2015.7175957. ISBN: 978-1-4799-8321-6
36. Huk, M.: Using context-aware environment for elderly abuse prevention. In: Nguyen, N.T., Trawiński, B., Fujita, H., Hong, T.-P. (eds.) ACIIDS 2016. LNCS (LNAI), vol. 9622, pp. 567–574. Springer, Heidelberg (2016). https://doi.org/10.1007/978-3-662-49390-8_55
37. Huk, M.: Context-related data processing with artificial neural networks for higher reliability of telerehabilitation systems. In: 17th International Conference on E-health Networking, Application & Services (HealthCom). IEEE Computer Society, Boston, USA, pp. 217–221 (2015)
38. Privitera, C.M., Azzariti, M., Stark, L.W.: Locating regions-of-interest for the Mars Rover expedition. Int. J. Remote Sens. **21**, 3327–3347. Taylor and Francis (2000)
39. UCI Machine Learning Repository. https://archive.ics.uci.edu/ml
40. Golub, T.R., et al.: Molecular classification of cancer: class discovery and class prediction by gene expression monitoring. Science **286**, 531–537 (1999)

41. Glosser, C., Piermarocchi, C., Shanker, B.: Analysis of dense quantum dot systems using a self-consistent Maxwell-Bloch framework. In: Proceedings of 2016 IEEE International Symposium on Antennas and Propagation (USNC-URSI), Puerto Rico, pp. 1323–1324. IEEE (2016)
42. H2O.ai documentation. https://docs.h2o.ai/h2o/latest-stable/h2o-docs/index.html
43. Rodriguez, J.D., et al.: Sensitivity analysis of k-fold cross validation in prediction error estimation. IEEE Trans. Patt. Anal. Mach. Int. **32**(3), 569–575 (2010)
44. Bouckaert, R.R.: Estimating replicability of classifier learning experiments. In: Proceedings of the 21st International Conference on Machine Learning, Banf, Canada (2004)

Commonsense Knowledge, Reasoning and Programming in Artificial Intelligence

Community Detection in Complex Networks: A Survey on Local Approaches

Saharnaz Dilmaghani[1(✉)], Matthias R. Brust[1], Gregoire Danoy[1,2], and Pascal Bouvry[1,2]

[1] Interdisciplinary Centre for Security, Reliability, and Trust (SnT), University of Luxembourg, Esch-sur-Alzette, Luxembourg
{saharnaz.dilmaghani,matthias.brust,gregoire.danoy,pascal.bouvry}@uni.lu
[2] Faculty of Science, Technology and Medicine (FSTM), University of Luxembourg, Esch-sur-Alzette, Luxembourg

Abstract. Early approaches of community detection algorithms often depend on the network's global structure with a time complexity correlated to the network size. Local algorithms emerged as a more efficient solution to deal with large-scale networks with millions to billions of nodes. This methodology has shifted the attention from global structure towards the local level to deal with a network using only a portion of nodes. Investigating the state-of-the-art, we notice the absence of a standard definition of *locality* between community detection algorithms. Different goals have been explored under the *local* terminology of community detection approaches that can be misunderstood. This paper probes existing contributions to extract the scopes where an algorithm performs locally. Our purpose is to interpret the concept of locality in community detection algorithms. We propose a *locality exploration scheme* to investigate the concept of locality at each stage of an existing community detection workflow. We summarized terminologies concerning the locality in the state-of-the-art community detection approaches. In some cases, we observe how different terms are used for the same concept. We demonstrate the applicability of our algorithm by providing a review of some algorithms using our proposed scheme. Our review highlights a research gap in community detection algorithms and initiates new research topics in this domain.

Keywords: Complex networks · Local approaches · Local community detection · Local degree of community detection

1 Introduction

Densely connected components are inseparable from networks providing structural or functional roles of the applications represented by the network. Community detection algorithms aim to identify these densely connected components within a network. Each community consists of nodes that are similar or close

N. T. Nguyen et al. (Eds.): ACIIDS 2021, LNAI 12672, pp. 757–767, 2021.
https://doi.org/10.1007/978-3-030-73280-6_60

to each other more than other nodes outside the community. The existing community detection algorithms can be differentiated into categories of global and local approaches. Unlike global approaches, local methods are known to discover communities without the integral global structural information of the complex networks [17,30,31].

The primary goal of developing local community detection algorithms is to find a local community of a given node in the absence of global information of the network [3,12]. Utilizing the traditional global algorithms, that requires fetching a large-scale network, often produce structural hairballs and not meaningful communities [15] as studied in protein folding networks [28]. The initial solution of finding a local community structure for a given node has been further developed to detect all network communities. Therefore, locally detecting communities turn to answer today's large-scale networks that is one of the drawbacks of the global algorithms that tend to find all network communities using complete network information.

Motivation. The question is, then, what is defined as locality when it comes to community detection algorithms? Among various interpretations, one may define it as finding local community(s) [3,12] of a given node(s). In contrast, others infer it as a local approach by incorporating local information of a network to find all network communities [6,15,16]. The majority of the studies still lay down in a spectrum within these two classes. For instance, they exploit the entire network to extract information used in the core community detection operation, whereas the objectives that define a community are determined locally [4,10,19,24]. Considering all the above-mentioned points, we noted the absence of a comprehensive study that supports the need to define a standard terminology regarding the locality of community detection algorithms to analyze the existing approaches deeply in this field. Our goal is to address these gaps in this paper.

Contribution. In this paper, we raise a new research challenge on community detection approaches concerning the *locality* in different stages of the algorithm. We explore the corresponding concepts and terminologies in various references, yet often with different terms. We also investigate the working flow of community detection approaches that benefit from a locality level in their approach. We developed Locality Exploration Scheme (LES) to incorporate research questions on the *locality level* of an approach in each step of the algorithm. Our scheme surveys existing approaches and countermeasures from a broad perspective respecting the algorithm's input, the core workflow of community detection, and the resulting output. Employing our model, we analyze some of the references concentrating on the stages defined in our scheme and discuss the applied locality level.

To the best of our knowledge, no studies have previously addressed the mentioned challenges. Our scheme is the first model to assemble strategies and associated locality levels to develop a community detection algorithm with a predetermined level of locality.

Organization. Section 2 outlines the preliminaries and background required for the remainder of the paper. In Sect. 3, we describe our locality exploration scheme. Next, we provide a survey of some existing solutions and analyze them based on our scheme in Sect. 4. The paper is concluded with Sect. 5.

2 Preliminaries and Background

We assume G is a network denoted by $G = (V, E)$, where V is the set of nodes, and E is the set of edges representing links within pairs of nodes (v, u) such that $v, u \in V$. Each node $v \in V$ has a degree of k_v representing the number of its neighbours from $\Gamma(v)$, the neighbour list of v.

Definition 1. *Community Structure.* We define a community structure c as a sub-network of G, where the intra-connectivity is maximized compared to the inter-connections such that $c_i \cap c_j = \emptyset$ and $\bigcup c_i = V$.

Definition 2. *Local Community Structure.* As introduced by [12], a local community has no knowledge from outside of the community. It consists of core node(s) that are internal to the community such that they have no connection to the outside of the community, and border nodes that connect those core nodes to the unknown portion of the network (i.e., other communities).

Definition 3. *Community Detection Algorithm.* The algorithm that detect densely connected components of community structures in a network is known as community detection algorithms. *Local community detection* algorithms tend to discover local community structures of G. We define *Local Detection of Communities* as an algorithm that associate a level of locality in its process to detect all communities of a network.

Definition 4. *Source Node.* A set of nodes chosen according to a score (e.g., similarity and centrality) to represent a community structure are identified as source nodes. They are also sometimes referred as core, seed and central nodes in the literature. In most of the cases, identifying a source node initializes a community of a network.

Definition 5. *Locality Level.* Adopted from [18], we define a three-level spectrum of locality. Starting from the most relaxed level, *global-level* that has no constraints, then *community-level* that is limited to the information within the community, and finally, *node-level* locality as the most restricted level which incorporates only local information of a node (up to certain extension, e.g., second-neighborhood).

Definition 6. *Auxiliary Information.* Some approaches require extra information to operate, for instance a threshold value, or a node/link weight. In this paper, the extra information added to the process are considered as *auxiliary information.*

Definition 7. *Community Expansion.* A community detection algorithm often needs an expansion strategy to enlarge the initial source nodes or preliminary detected communities. It can be a fitness function to evaluate the membership of a node to a community or a modularity objective to measure the interconnectivity of the community.

3 Locality Exploration Scheme (LES)

The question of how much information algorithms need from a network for their operations is not a recent research question [2,5]. Stein et al. [33] have provided a classification of local algorithms in network research. They define a four-level model based on auxiliary information, non-constant run time, and functionality. We find this classification comprehensive in communication networks, however, it is limited when it comes to community detection criteria. Other comprehensive studies on community detection are confined to the two main categories of *local* and *global* algorithms [14,17,20,30,31,36,37] emphasizing global methods. Thus, we identify this absence of attention to local community detection in complex networks.

We provide Locality Exploration Scheme (LES) illustrated in Fig. 1 and combine the challenges raised in Sect. 1. The scheme considers a three-level model

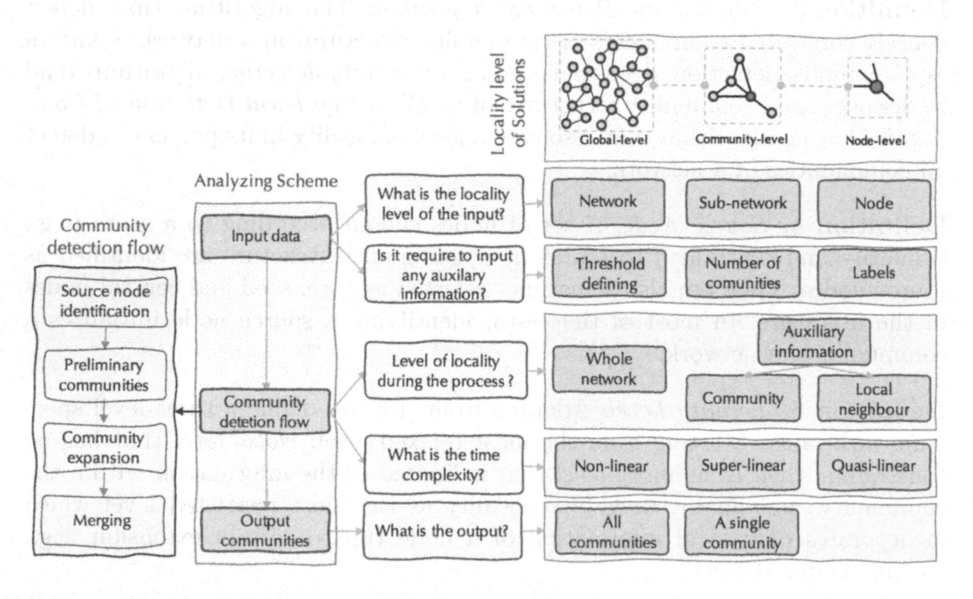

Fig. 1. An overview of proposed LES model. The *analysing scheme* represents a three-level structure of community detection approaches. In the middle rectangles, the main challenges that may raise at each stage is highlighted. The pink rectangles show the existing solutions in the literature sorted from left to right based on their locality level. (Color figure online)

for community detection algorithms: *Input data*, *Community detection flow*, and *Output communities*. In each step, we collect the possible solutions from the literature and sort them with the locality level described in Sect. 2. In the following subsections, we explain each stage provided in Fig. 1 for the possible solutions investigated from the literature.

3.1 Input Data

The initial step towards any community detection algorithm is the input provided to the algorithm. One of the main drawbacks of the global community detection is being dependent on the entire network to discover the communities. Although the community detection flow in local algorithms does not depend on the global structure, the input data includes the whole network for preliminary operations in several cases.

Considering large-scale networks, it is impossible to fetch the whole network for the next operations. Therefore, even if an algorithm offers an adequate locality level during some steps of the community detection flow (i.e., community expansion), it may fail to operate on a network if the input is entire network. Hence, this is an essential stage when investigating the locality of the algorithm. Besides the network's information, sometimes, the algorithm also expects to input certain auxiliary information, such as a threshold value. Depending on the auxiliary information, sometimes it may also impact the level of locality.

3.2 Community Detection Flow

We assembled the core operations towards a community detection procedure in this stage. Principally, the procedure is decomposed itself into four functions as described in Fig. 2. We noticed that when there is a discussion on community detection locality, it mainly refers to this stage. It is worth noting that not all algorithms follow the four-step model, in some instances some steps are combined (e.g., Source node identification and Preliminary communities). It is not a trivial task to decide about the locality of an approach based on this stage. A number of algorithms have included the entire network during the source node identification, however, they have increased the locality by incorporating local information while expanding the communities [9,10,19,35]. Thus, we analyze this stage given the workflow described in Fig. 2.

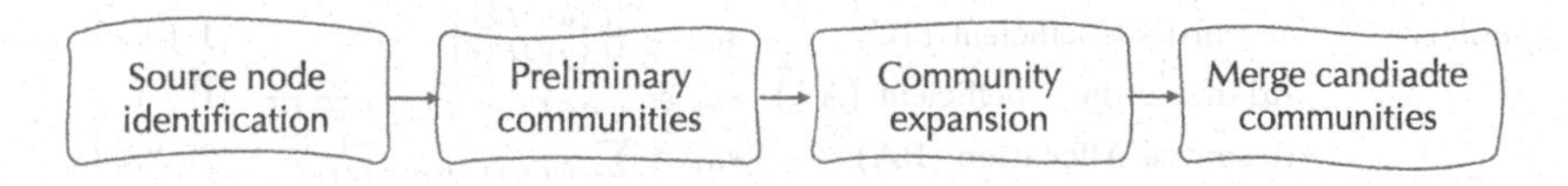

Fig. 2. Community detection flow.

Source Node Identification. Source node identification is one of the main steps that targets candidate nodes to be expanded later in order to shape the communities. The performance of the algorithm, however, depends highly on this step since the source nodes initiate output communities. Each contribution has introduced a slightly different approach to choose the source nodes. Besides algorithms that apply a random strategy (e.g., LPA) other tend to find the important nodes that is a good representation of its community to start their approach from. In this step, a dedicated score is first calculated for a particular set of nodes (or entire network) and then usually the list of scores is sorted to choose the best candidates as source nodes. We categorize the source node identification techniques into the following main classes:

- Network centrality metric [21,23]
- Node similarity score [34,38]
- Combination of topological measure (e.g., [35])

Table 1 summarized metrics used in reference for source node identification in some community detection approaches. It is noteworthy that the metrics, especially similarity scores, are not exploited only for identifying the source nodes but also as a similarity measure to quantify a node's belongingness to a community that we explore in the next subsections.

Besides the impact of the source node selection on the performance, it is also important to remind that not all of the metrics are local. Several metrics listed in Table 1 are required the knowledge from the entire network as the score that calculates both the degree and distance of a node from other high degree nodes. On the other, to choose source nodes, it is mostly observed that V is required to be sorted based on the chosen score (ref. Table 1).

Table 1. A summary of similarity scores of two nodes in the literature. ($\gamma(v)$ is the number of subgraphs with 3 edges and 3 vertices, one of which is v, $\tau(v)$ the number of triples on v, $\sigma_{st}(v)$ represents the shortest path from s to t through v s_{uv} represent a similarity score between node v and u.)

Categories	Metrics	Definition	References
Centrality	Degree	$k_v = \|\Gamma(v)\|$	[7,13,22,32]
	Clustering coeficient	$cc = \frac{\gamma(v)}{\tau(v)}$	[35]
	Betweenness centrality	$bc = \sum_{s,t \in V} \frac{\sigma_{st}(v)}{\sigma_{st}}$	[13,27]
	Edge density	$Den(G) = \frac{\|E\|}{\|V\|(\|V\|-1)/2}$	[25]
Similarity	Jaccard's Coefficient (JC)	$s_{uv} = \frac{\|\Gamma(u) \cap \Gamma(v)\|}{\|\Gamma(u) \cup \Gamma(v)\|}$	[4,10,19]
	Adamic-Adar Coefficient (AA)	$s_{uv} = \sum_{t \in \Gamma(u) \cap \Gamma(v)} \frac{1}{\log k(t)}$	[1]
	Resource Allocation (RA)	$s_{uv} = \sum_{t \in \Gamma(u) \cap \Gamma(v)} \frac{1}{k(t)}$	[29,32,38]
Combination	Degree - distance	$sc_i = k_v \times \sigma_v$	[9,35]

Preliminary Communities. After detecting source nodes, the initial communities are predefined and most of the time, they are considered preliminary communities. Thus, this step may not be taken as an independent stage in the community detection flow. In many references, yet, this step is developed to extend source nodes into preliminary communities. It can be operated as merely taking the first neighbourhood of a source node as its preliminary community [4], or choosing neighbours relying on a similarity score (ref. Table 1) [10,19]. The level of locality depends on the taken strategy. Per only the local neighbourhood provides a higher level of locality compared to other solutions.

Community Expansion. Several references have conducted an adequate locality level only during community expansion, regardless of previous input data. A list of local community expansion strategies developed in the literature mostly rely on community-level local information: internal connection and external connections of a community [36]. A list of important local fitness function are provided in Table 2. Moreover, some approaches do not depend on such objectives. Instead, they exploit techniques such as spreading of influence (i.e., LPA) [9,15,35] or random walk.

Moreover, many references measure these functions (ref Table 2) from a community perspective such that a community calculates a respective objective to decide about adopting a new node. In other words, it is the community that determines whether to accept the joining of a new node to its community (if that node maximized the objective function) or not. By slightly changing the perspective, one can operate any of these functions on a node to decide on surrounded communities to join [6,15,16]. By employing this strategy, the locality level of an approach increases to the node-level.

Table 2. Summary of existence fitness functions.

Fitness functions	Formula	Reference
Local modularity (Clauset) [11]	$R = \frac{\sum B_{ij}\sigma(i,j)}{\sum B_{ij}}$	[7]
Local modularity (Lou) [26]	$M = \frac{E_{in}}{E_{out}}$	[4,8]
Fitness function (Lancichinetti) [22]	$F_c = \frac{f_{in}^c}{(f_{in}^c + f_{out}^c)^{\alpha}}$	[19]

Merging Candidate Communities. In certain instances, communities are small or sparse. Hence, identifying The merging step is not always a requirement; however, it can increase the quality of resulted communities [4,10,15]. The approach is accomplished by considering each community as a node and pursuing a similar approach to finding the most similar communities to merge. In some cases, it also requires a given threshold to decide on the degree of similarities between two communities [10,15,19]. Overall, the function requires communities to operate and its locality level can be community-level if the threshold value is not relying on the global structure of a network.

3.3 Output Communities

Finally, the communities are identified that can either represent a set of communities from the entire network [10,15,16,35], or only a local community of a given node (subset of nodes) [3,12] depending on the purpose behind the community detection approach. The early work is primarily motivated by finding local communities of a given node [3,12]. Considering the underlying network, it is possible only to have meaningful local communities within a network [28] rather than global communities for the entire network.

4 Analyzing Existing Algorithms Based on LCE

In this section we provide a review analysis of some papers in regard to the scheme described in Sect. 3.

NSA. In [10], authors have proposed an algorithm (NSA) founded on Jaccard similarity. The algorithm requires a network G and a threshold value used during the merging process of community detection flow as input. The source identification relies on the high degree nodes. Afterwards, the preliminary communities shape, adding the most similar neighbours due to the Jaccard similarity scores. The produced small communities are then merged similarly based on a given threshold on the Jaccard score between two communities this time. NSA functions on time complexity of $O(n \log(n))$. The algorithm is global at the input level; however, it is operating locally given only nodes local neighborhood during the community detection flow. The outputs are all communities of a given network.

ECES. The algorithm [4] is motivated by the drawback of global community detection algorithms that require the network's global information to operate. However, the model itself needs a network to process the first step of community detection flow. That is to obtain the core nodes of the network using an extended Jaccard score. The score admits local information until the second-neighbour of a node. The highest score node is then extended, including its first neighbours forming preliminary communities. Next, each community is extended by the ratio of internal links to the external ones. Finally, candidate communities satisfying the condition relying on the sum of its nodes Jaccard score are merged. The output is a set of network communities. The algorithm operates in a super-linear time complexity of $O(n \log(n))$. Similar to NSA, this approach also depend on global information to detect the source nodes, however, it enjoys a level of locality (second-neighbourhood compared to the first-neighbourhood of NSA) while extending the communities in community detection flow.

InfoNode. In a recent approach [19], authors propose a model that concedes the increment of particular local community modularity as a condition of adopting a node. InfoNode requires both a network and a threshold value as input. Even though the source selection relies on a node degree, a local centrality metric, the approach needs to sort all nodes in regard to their degree. Therefore, this step is not local anymore. The high degree nodes are first enlarged to preliminary communities calculating the F fitness function [22], and then extended based on an internal force function defined by the authors. The growth of communities in the community detection flow is processed locally. The algorithm has a non-linear time complexity of $O(n^2)$.

DEMON. Slightly different than the previous papers, DEMON [15] defines locality as taking each node to be responsible for joining a community. However, the algorithm requires a global network and a threshold value to process. The source nodes are chosen randomly, and an ego network for each node is identified. The local expansion of the candidate nodes is operated similarly to the LPA technique. Finally, sparse communities are merged considering the threshold value. The time complexity of the algorithm is quasi-linear reported as $O(nk^{(3-\alpha)})$. Regardless of the input, all the steps during the community detection flow have been operated in a community-level locality.

LCDA-SSN. In another approach [16], authors developed a community detection algorithm that has increased the locality level compared to similar approaches. The proposed method is an iterative model taken only a node as input. They consider each node knowing its first neighborhood; hence, it discovers the network while operating on one node. The approach offers a self-defining source node selection giving a score to a visited node based on local structural information. The score is updated each time, as for the community cores. The community expansion is adopted from M local modularity [26]. However, it is applied to a node rather than a community. The output is all communities of a network in a quasi-linear time complexity of $O(nk)$.

5 Conclusion

In this paper, we bring up new research questions in the field of community detection algorithms in complex networks, highlighting two foremost challenges: first, the absence of a standard terminology when it comes to local community detection algorithms, and second, the gap between the interpretation of locality and community detection algorithms. We provide a Locality Exploration Scheme (LES) model based on the steps of the community detection approach and incorporate the research questions raised by the concept of locality in each step. By employing our LES, we could survey the existing techniques and strategies required for developing a community detection algorithm with an adequate locality level. Furthermore, we provide a thorough review of some of the references showing the applicability of our scheme. Our analysis can also be taken as a guideline to choose the most relevant functions while developing community detection. We show that ignoring the problem of defining locality can lead to misunderstandings, and if not addressed correctly. We plan to further extend our scheme by including evaluation metrics determined for these approaches.

Acknowledgment. This work has been partially funded by the joint research programme University of Luxembourg/SnT-ILNAS on Digital Trust for Smart-ICT.

References

1. Adamic, L.A., Adar, E.: Friends and neighbors on the web. Soc. Netw. **25**(3), 211–230 (2003)
2. Angluin, D.: Local and global properties in networks of processors. In: Proceedings of the Twelfth Annual ACM Symposium on Theory of Computing, pp. 82–93 (1980)
3. Bagrow, J.P., Bollt, E.M.: Local method for detecting communities. Phys. Rev. E **72**(4), 046108 (2005)
4. Berahmand, K., Bouyer, A., Vasighi, M.: Community detection in complex networks by detecting and expanding core nodes through extended local similarity of nodes. IEEE Tran. Comput. Soc. Syst. **5**(4), 1021–1033 (2018)
5. Brust, M.R., Rothkugel, S.: A taxonomic approach to topology control in ad hoc and wireless networks. In: International Conference on Networking (ICN 2007) (2007)
6. Brust, M.R., Frey, H., Rothkugel, S.: Adaptive multi-hop clustering in mobile networks. In: Proceeding of the 4th International Conference on Mobile Technology, Applications (2007)
7. Chen, Q., Wu, T.T.: A method for local community detection by finding maximal-degree nodes. In: 2010 International Conference on Machine Learning and Cybernetics, vol. 1, pp. 8–13. IEEE (2010)
8. Chen, Q., Wu, T.T., Fang, M.: Detecting local community structures in complex networks based on local degree central nodes. Phys. A Stat. Mech. Appl. **392**(3), 529–537 (2013)
9. Chen, Y., Zhao, P., Li, P., Zhang, K., Zhang, J.: Finding communities by their centers. Sci. Rep. **6**(1), 1–8 (2016)
10. Cheng, J., et al.: Neighbor similarity based agglomerative method for community detection in networks. Complexity **2019** (2019)
11. Clauset, A.: Finding local community structure in networks. Phys. Rev. E **72**(2), 026132 (2005)
12. Clauset, A., Newman, M.E., Moore, C.: Finding community structure in very large networks. Phys. Rev. E **70**(6), 066111 (2004)
13. Comin, C.H., da Fontoura Costa, L.: Identifying the starting point of a spreading process in complex networks. Phys. Rev. E **84**(5), 056105 (2011)
14. Coscia, M., Giannotti, F., Pedreschi, D.: A classification for community discovery methods in complex networks. Stat. Anal. Data Min. ASA Data Sci. J. **4**(5), 512–546 (2011)
15. Coscia, M., Rossetti, G., Giannotti, F., Pedreschi, D.: Demon: a local-first discovery method for overlapping communities. In: Proceeding of the 18th ACM SIGKDD International Conference on Knowledge Discovery and Data Mining (2012)
16. Dilmaghani, S., Brust, M.R., Danoy, G., Bouvry, P.: Local community detection algorithm with self-defining source nodes. In: Benito, R.M., Cherifi, C., Cherifi, H., Moro, E., Rocha, L.M., Sales-Pardo, M. (eds.) Complex Networks & Their Applications IX. COMPLEX NETWORKS 2020 2020. Studies in Computational Intelligence, vol. 943, pp. 200–210. Springer, Cham https://doi.org/10.1007/978-3-030-65347-7_17 (2020)
17. Fortunato, S.: Community detection in graphs. Phys. Rep. **486**(3–5), 75–174 (2010)

18. Gulbahce, N., Lehmann, S.: The art of community detection. BioEssays **30**(10), 934–938 (2008)
19. Guo, K., He, L., Chen, Y., Guo, W., Zheng, J.: A local community detection algorithm based on internal force between nodes. Appl. Intell. **50**(2), 328–340 (2019). https://doi.org/10.1007/s10489-019-01541-1
20. Harenberg, S., et al.: Community detection in large-scale networks: a survey and empirical evaluation. Wiley Rev. Comput. Stat. **6**(6), 426–439 (2014)
21. Hernández, J.M., Van Mieghem, P.: Classification of graph metrics. Delft University of Technology: Mekelweg, The Netherlands, pp. 1–20 (2011)
22. Lancichinetti, A., Fortunato, S., Kertész, J.: Detecting the overlapping and hierarchical community structure in complex networks. J. Phys. **11**(3), 033015 (2009)
23. Li, S., Huang, J., Zhang, Z., Liu, J., Huang, T., Chen, H.: Similarity-based future common neighbors model for link prediction in complex networks. Sci. Rep. **8**(1), 1–11 (2018)
24. Li, Y., He, K., Bindel, D., Hopcroft, J.E.: Uncovering the small community structure in large networks: a local spectral approach. In: Proceedings of the 24th International Conference on World Wide Web, pp. 658–668 (2015)
25. Lin, Z., Zheng, X., Xin, N., Chen, D.: CK-LPA: efficient community detection algorithm based on label propagation with community kernel. Phys. A: Stat. Mech. Appl. **416**, 386–399 (2014)
26. Luo, F., Wang, J.Z., Promislow, E.: Exploring local community structures in large networks. In: 2006 IEEE/WIC/ACM International Conference on Web Intelligence (WI 2006) (2006)
27. Mahyar, H., et al.: Identifying central nodes for information flow in social networks using compressive sensing. Soc. Netw. Anal. Min. **8**(1), 1–24 (2018). https://doi.org/10.1007/s13278-018-0506-1
28. Muff, S., Rao, F., Caflisch, A.: Local modularity measure for network clusterizations. Phys. Rev. E **72**(5), 056107 (2005)
29. Pan, Y., Li, D.H., Liu, J.G., Liang, J.Z.: Detecting community structure in complex networks via node similarity. Phys. A Stat. Mech. Appl. **389**(14), 2849–2857 (2010)
30. Porter, M.A., Onnela, J.P., Mucha, P.J.: Communities in networks. Not. AMS **56**(9), 1082–1097 (2009)
31. Schaeffer, S.E.: Graph clustering. Comput. Sci. Rev. **1**(1), 27–64 (2007)
32. Shang, R., Zhang, W., Jiao, L., Stolkin, R., Xue, Y.: A community integration strategy based on an improved modularity density increment for large-scale networks. Phys. A Stat. Mech. Appl. **469**, 471–485 (2017)
33. Stein, M., Fischer, M., Schweizer, I., Mühlhäuser, M.: A classification of locality in network research. ACM Comput. Surv. (CSUR) **50**(4), 1–37 (2017)
34. Wang, X., Sukthankar, G.: Link prediction in heterogeneous collaboration networks. In: Missaoui, R., Sarr, I. (eds.) Social Network Analysis - Community Detection and Evolution. LNSN, pp. 165–192. Springer, Cham (2014). https://doi.org/10.1007/978-3-319-12188-8_8
35. Wang, X., Liu, G., Li, J., Nees, J.P.: Locating structural centers: A density-based clustering method for community detection. PloS One **12**(1), e0169355 (2017)
36. Xie, J., Kelley, S., Szymanski, B.K.: Overlapping community detection in networks: the state-of-the-art and comparative study. Acm Comput. Surv. (csur) **45**(4), 1–35 (2013)
37. Yang, Z., Algesheimer, R., Tessone, C.J.: A comparative analysis of community detection algorithms on artificial networks. Sci. Rep. **6**, 1–18 (2016)
38. Zhou, T., Lü, L., Zhang, Y.C.: Predicting missing links via local information. Eur. Phys. J. B **71**(4), 623–630 (2009)

A Q-Learning Based Hyper-Heuristic for Generating Efficient UAV Swarming Behaviours

Gabriel Duflo[1(✉)], Grégoire Danoy[1,2], El-Ghazali Talbi[3], and Pascal Bouvry[1,2]

[1] SnT, University of Luxembourg, Esch-sur-Alzette, Luxembourg
{gabriel.duflo,gregoire.danoy,pascal.bouvry}@uni.lu
[2] FSTM/DCS, University of Luxembourg, Esch-sur-Alzette, Luxembourg
[3] University of Lille, CNRS/CRIStAL, Inria Lille, Lille, France
el-ghazali.talbi@univ-lille.fr

Abstract. The usage of Unmanned Aerial Vehicles (UAVs) is gradually gaining momentum for commercial applications. These however often rely on a single UAV, which comes with constraints such as its range of capacity or the number of sensors it can carry. Using several UAVs as a swarm makes it possible to overcome these limitations. Many metaheuristics have been designed to optimise the behaviour of UAV swarms. Manually designing such algorithms can however be very time-consuming and error prone since swarming relies on an emergent behaviour which can be hard to predict from local interactions. As a solution, this work proposes to automate the design of UAV swarming behaviours thanks to a Q-learning based hyper heuristic. Experimental results demonstrate that it is possible to obtain efficient swarming heuristics independently of the problem size, thus allowing a fast training on small instances.

Keywords: Hyper-heuristic · UAV swarming · Reinforcement learning

1 Introduction

In the past years the Unmanned Aerial Vehicles (UAVs) have found their way into an increasing number of civilian applications, such as delivery, infrastructure inspection or urban traffic monitoring. However, most of these rely on a single UAV, either remotely operated or autonomous, which comes with limitations due to the battery and payload capacity. A promising way to overcome these limitations while relying on current technology is to use multiple UAVs simultaneously as a swarm.

This work specifically considers the problem of area coverage by a swarm of UAVs. The latter finds applications in surveillance, search and rescue and smart agriculture to name a few. More precisely we tackle the *Coverage of a Connected-UAV Swarm* (CCUS) problem [8,9], which models the surveillance

N. T. Nguyen et al. (Eds.): ACIIDS 2021, LNAI 12672, pp. 768–781, 2021.
https://doi.org/10.1007/978-3-030-73280-6_61

problem as a bi-objective one, where both the area coverage and the swarm network connectivity [4] have to be maximised.

Many metaheuristics have been manually designed to address the problem of area coverage by a swarm of UAVs, for instance using ant colony methods [1]. This process can however be error-prone and time-consuming since it consists in predicting an emergent behaviour by designing local interaction.

Instead, this work proposes to automate the design of swarming behaviours, which remains an open research problem [3]. This principle of searching into a space of heuristics is also referred to as a hyper-heuristic. We here consider a Q-learning based hyper-heuristic (QLHH) to generate efficient distributed heuristic for the CCUS problem.

The remainder of this article is organised as follows. Section 2 first presents the related work on hyper-heuristics and their usage for swarming applications. Then, in Sect. 3 the proposed CCUS model is presented, followed by the designed QLHH in Sect. 4. The empirical performance of QLHH is then provided in Sect. 5 using coverage and connectivity metrics from the literature. Finally Sect. 6 provides our conclusions and perspectives.

2 Related Work

Hyper-heuristics were first mentioned as "heuristics to choose heuristics" [7]. More generally, they refer to high-level algorithms performing a searching process in a space of low-level heuristics. The purpose is thus to search a heuristic for a given problem, unlike metaheuristics which aim at searching a solution for a given instance.

Burke et al. made a classification of hyper-heuristics [5]. The latter is divided in 2 parts: the nature of the search space, and the nature of the feedback for the learning process. At the lowest level, the heuristics from the search space can be constructive or perturbative. A constructive heuristic modifies a non-feasible solution until it becomes feasible. At a higher level, two approaches can be used by the algorithm performing the searching process: selective and generative approaches. For the generative approach, the high-level algorithm is given a set of predefined "building-blocks" containing information relative to the problem. Its goal is thus to combine these blocks in order to make a new heuristic. The idea is to extract a dynamic part common to several known heuristics and to learn it. Finally, the last distinction is whether the feedback uses online or offline learning. For an online learning feedback, the learning process occurs while the model is executed on instances. This work proposes a hyper-heuristic for generating constructive heuristics using an online learning feedback.

Most generative hyper-heuristics use offline-learning, and more specifically genetic programming (GP). For instance Lin et al. evolve a task sequence for the multi-skill resource constrained project scheduling problem (MS-RCPSP) [13]. In [10], Duflo et al. evaluate the unvisited nodes at each iteration in the travelling salesman problem (TSP). Finally, Kieffer et al. sort the bundles in the bi-level cloud pricing optimisation problem (BCPOP) in [12].

When it comes to online learning, reinforcement learning (RL) is the most widely used technique for hyper-heuristics. It is however mainly used for selective approaches, while few generative ones have considered RL. For instance, Khalil et al. design in [11] a template of RL-based hyper-heuristic for several combinatorial optimisation problems.

Hyper-heuristics have also been used in the context of swarming. However, since the automation of swarming behaviours is a recent and still open research area, as mentioned in [3], the selective approach is prominently used in the literature. For instance Babic et al. design in [2] a hyper-heuristic where robots can choose at each iteration the best heuristic among a pool of predefined low-level heuristics to perform a certain number of predefined actions. To date, only Duflo et al. use a generative approach based on RL for defining the movement of UAVs in a swarm [8].

In the context of a swarm of UAV/robot, multiple agents are interacting. RL has then been applied on more general Multi-Agent Systems (MAS). Tuyls and Stone [14] provide a classification of paradigms, including "Online RL towards social welfare". In this paradigm, every agent of the MAS share the same policy, which is the context of this work.

Finally the CCUS problem is a bi-objective one, which implies that the generated heuristics must tackle bi-objective problems. However, for online approaches, Multi-Objective Reinforcement Learning (MORL) remains an open area [6]. One popular approach is to transform a multiple policy RL into a single one using a scalarisation function [15].

The work proposed in this article thus goes beyond the state-of-the-art as it proposes a generative hyper-heuristic, and more precisely a Q-learning hyper-heuristic (QLHH) to generate efficient UAV swarming behaviours. This QLHH is applied to tackle a problem introduced hereinafter for optimising the Coverage of a Connected UAV Swarm, so called CCUS. Since CCUS is a multi-objective optimisation problem (as described in the following section), one of the challenges with this work is to apply MORL techniques to the context of hyper-heuristics.

3 CCUS Model

This section provides a formal definition of the CCUS model. It considers the usage of a swarm of UAVs equipped with wireless communication interfaces, also referred to as flying ad hoc network (FANET), for covering an area. In that scenario, each UAV starts from a given point, executes its tour and returns to its starting point, which can be referred to as its base. CCUS aims at optimising both the coverage speed and the connectivity of the swarm.

3.1 Formal Expression

The CCUS model contains two main components: an environment graph $G_e = (V, E_e)$ and a communication graph $G_c = (U, E_c)$. Both are represented in Fig. 1. G_e is composed of a set V of vertices on which the UAVs can move. This graph

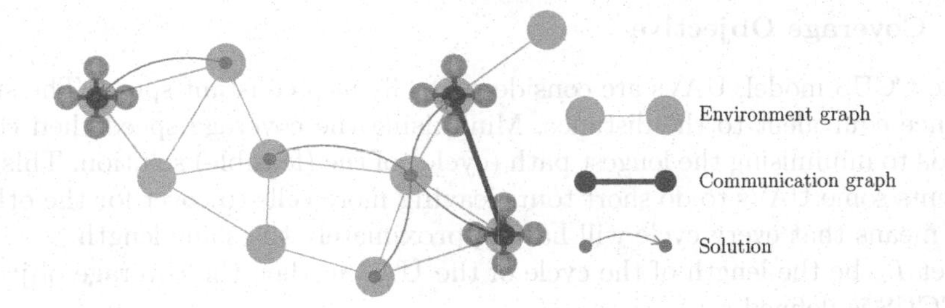

Fig. 1. CCUS solution at a certain time step

is weighted, non-directed and must be strongly connected so that every UAV can move to any other vertex. Otherwise, another constraint stating that there should be at least one UAV per connected component should be required. The weight corresponds here to the distance between two adjacent vertices. The function $dist : V^2 \to \mathbb{R}$ then returns the length of the shortest path between any two given nodes. G_c is composed of a set U of UAVs, which are linked if they are close enough for communicating (in a fixed communication range D_{com} which is specific to the FANET setup). This graph is not necessarily connected and especially dynamic since positions of UAVs evolve during the coverage. The function $pos : U \to V$ returns the current position of a given UAV.

An instance $I = (G_e, G_c)$ is then defined by an environment graph and a communication graph giving the initial positions of UAVs. Different initial positions would make a different instance. Finally, a solution for CCUS is a set of $|U|$ paths in G_e (one path per UAV). Each path must start from the initial vertex of the corresponding UAV (i.e., its base). Such a solution is feasible if and only if these paths are cycles and their union covers the G_e, so that the whole environment is covered and the UAVs have returned to their base.

3.2 Representation of Solutions

A solution S can be defined as a set of paths $\{S_u\}_{u \in U}$, where $S_u = (v_1, v_2, \cdots, v_{|S_u|})$ with $v_i \in V$. $\bar{S}$ then refers to the unvisited vertices from G_e.

$$\bar{S} = \{v \in V | \nexists u \in U, v \in S_u\}$$

If the coverage is paused at any moment, a solution will be obtained (not necessarily a feasible solution). Since CCUS is bi-objective, any solution S can be represented as a 2-dimensional vector $O(S) = (O^{cov}(S), O^{con}(S))^\top$, containing both objective values defined ealier. The two components $O^{cov}(S)$ and $O^{con}(S)$ respectively correspond to the biggest path at the current time step (see Sect. 3.3) and the average number of connected components in the communication graph until the current time step (see Sect. 3.4).

3.3 Coverage Objective

In the CCUS model, UAVs are considered to fly at a constant speed. The speed is hence equivalent to the distance. Minimising the coverage speed then corresponds to minimising the longest path (cycle) of the (feasible) solution. This also prevents some UAVs to do short tours leaving more cells to cover for the others. This means that every cycle will have approximately the same length.

Let L_u be the length of the cycle of the UAV u, then the coverage objective for CCUS is defined as:

$$\text{Minimise}\left\{\frac{|V|}{|V|-|\bar{S}|}\cdot \max_{u\in U} L_u\right\}$$

This objective value is slightly different from the one presented in [9]. The factor $|V|/(|V|-|\bar{S}|)$ has indeed been added. Since $(|V|-|\bar{S}|)/|V|$ corresponds to the rate of visited vertices, the length of the longest path is divided by this rate in order to penalise a non-feasible solution with vertices visited several times. This factor does not affect a feasible solution since it equals 1 in that case.

3.4 Connectivity Objective

Two UAVs can communicate if they are in a certain communication range D_{com}. They can exchange their local information. It is therefore possible for a UAV to access the local information of every UAV in its connected component in G_c. Maximising the connectivity then consists in minimising the global number of connected components in G_c over time.

Let C_t be the number of connected components in G_c at the time step t, then the connectivity objective for CCUS is defined as:

$$\text{Minimise}\left\{\sum_{t\in T} C_t\right\}$$

4 QLHH Algorithm

This section describes in detail QLHH, a Q-learning based algorithm for generating heuristics for the CCUS problem.

4.1 Structure

As a low-level heuristic, each UAV will asynchronously move to a new vertex until the whole environment graph is covered. Algorithm 1 gives an overview of the process. The choice of a new vertex is done thanks to a fitness function f which evaluates each possible destination. The UAVs then move to the vertex maximising f (line 4). The fitness function f is specific to the heuristic. Algorithm 1 then provides a template of low-level heuristics, where f is the dynamic part. The goal of the high-level algorithm, here Q-learning, is thus to find the best possible definition for f. Such a low-level heuristic runs in a quadratic time. Each UAV visits $\mathcal{O}(|V|)$ vertices during its tour. At each step, every unvisited vertice is evaluated, so there are $\mathcal{O}(|V|)$ operations. This gives a final time complexity of $\mathcal{O}\left(|V|^2\right)$ per UAV. Since the heuristic is distributed, the number of UAVs is not added to the complexity.

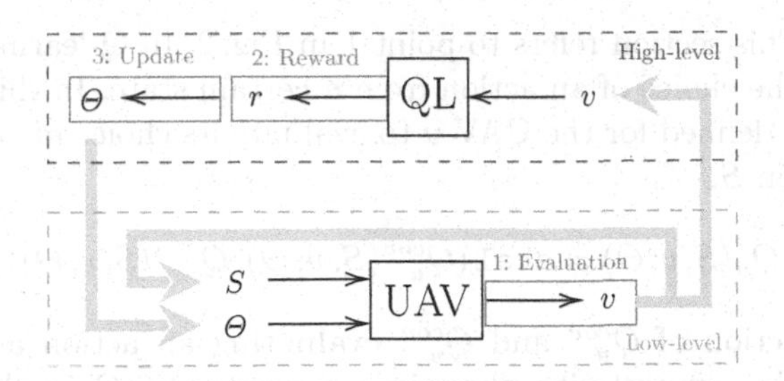

Fig. 2. Overview of the proposed QLHH algorithm

Algorithm 1: Low-level-heuristic template

```
   input  : Instance I = (G_e, G_c)
   output : Solution S = ⋃_{u∈U} S_u
 1 foreach UAV u ∈ U do                                    // asynchronously
 2 |  S_u ← (pos(u))
 3 |  while S̄ ≠ ∅ do
 4 |  |  next ← arg max_{v∈V} f(v)
 5 |  |  // u flies to next
 6 |  |  S_u ← S_u + shortest_path(pos(u), next)
 7 |  end
 8 |  // u flies back to its initial vertex
 9 |  S_u ← S_u + shortest_path(pos(u), S_u(0))
10 end
11 return S
```

For that purpose, it is needed to adapt Q-learning components, i.e. actions, states and policy, to the CCUS model. A state is a solution S at the time when the UAVs choose a new vertex. The latter corresponds to the action of UAVs. An action is thus represented by a vertex $v \in V$ on which the UAVs can move. The policy finally returns the node which maximises the fitness function f. Figure 2 shows an overview of the whole process. At each step, each UAV moves to a vertex v thanks to an evaluation function depending on the current solution S and a set of variables Θ. The latter defines a heuristic based on the template of low-level heuristics. It means that Θ is a parameter of the evaluation function. When every UAV has finished its tour, i.e. has returned to its initial node, a reward is given for each evaluation which has been made. This set of rewards is used to modify the value of Θ, and therefore to update the current low-level heuristic.

4.2 Detailed Steps

This section will detail the three steps shown in Fig. 2: firstly, the evaluation of vertices done by UAVs in order to choose their next destination; secondly, the reward given for such a choice made by a UAV; thirdly, the update of Θ according to the set of obtained rewards.

Evaluation. This section refers to point 1 in Fig. 2. In Q-learning, the Q function evaluates the choice of an action from a certain state. In this context, a Q_u function is thus defined for the UAV u to evaluate its choice of going to a vertex v from a solution S.

$$Q_u(S, v; \Theta) = \texttt{scal}\left(Q_u^{cov}(S, v; \Theta), Q_u^{con}(S, v; \Theta)\right)$$

Q_u is a function of Q_u^{cov} and Q_u^{con} evaluating an action according to the coverage and the connectivity objectives respectively. Θ is also divided into Θ^{cov} and Θ^{con}, which are both a collection of eight parameters $\Theta^o = \{\theta_j^o\}_{j=1}^8$, $\forall o \in \{cov, con\}$. Since both evaluations are not of the same order of magnitude, they are normalised in order to have a balance between both objectives. We here considered two possible scalarisation functions scal, i.e. linear and Chebyshev as introduced in [15].

With the linear function, Q_u is a linear combination of Q_u^{cov} and Q_u^{con}.

$$Q_u(S, v; \Theta) = \sum_{o \in \{cov, con\}} Q_u^o(S, v; \Theta^o)$$

A UAV u at the solution $S^{(t)}$ will thus move to the vertex $v^{(t+1)}$ which maximises Q_u.

$$v^{(t+1)} = \arg\max_{v \in V} \left\{Q_u\left(S^{(t)}, v; \Theta\right)\right\}$$

With the Chebyshev function, Q_u is the distance from an utopian point z according to the L_∞ metric (also called Chebyshev metric). z^{cov} and z^{con} are constantly adjusted during the learning process to represent the best evaluations for the coverage and the connectivity respectively.

$$Q_u(S, v; \Theta) = \max_{o \in \{cov, con\}} |Q_u^o(S, v; \Theta) - z^o|$$

A UAV u at the solution $S^{(t)}$ will thus move to the vertex $v^{(t+1)}$ which minimises Q_u, i.e. the distance from the best point z found so far.

$$v^{(t+1)} = \arg\min_{v \in V} \left\{Q_u\left(S^{(t)}, v; \Theta\right)\right\}$$

For each objective o, the evaluation $Q_u^o(S, v; \Theta)$ is a matrix computation. The state S and the action v are represented in the formula by p-dimensional vectors, respectively μ and μ_v. Their computation is detailed in the paragraph

Embedding structure below. The definition of $Q^o_u(S, v; \Theta)$ and the embedding structure differ from [9]. This new definition has shown that it is more stable.

$$Q^o_u\left(S, v; \Theta^o\right) = {\theta^o_2}^\top \texttt{relu}\left([\theta^o_3 \mu, \theta^o_4 \mu_v] + \theta^o_5 \delta^o_{uv}\right)$$

where $\theta^o_2, \theta^o_5 \in \mathbb{R}^{2p}$, $\theta^o_3, \theta^o_4 \in \mathbb{R}^{p \times p}$, $[\cdot, \cdot]$ is the concatenation operator and relu is the rectified linear unit, i.e. for any vector $X = (x_i)_i$, $\texttt{relu}(X) = (\max(0, x_i))_i$. The term δ^o_{uv} is useful for making each UAV u have a different evaluation of a vertex v. It is different according to the objective o. δ^{cov}_{uv} is the distance between the evaluating vertex and the UAV evaluating it, while δ^{con}_{uv} is the sum of distances between the evaluated vertex and the other UAVs.

$$\delta^{cov}_{uv} = dist(pos(u), v) \qquad \delta^{con}_{uv} = \sum_{u' \in U \setminus \{u\}} dist(pos(u'), v)$$

Embedding Structure. Each node $v \in V$ is represented by a p-dimensional feature μ_v. The latter is recursively computed according to the structure of the environment graph. $\forall v \in V$

$$\mu_v = \texttt{relu}\left(\theta^o_1 \cdot x^o_v\right)$$

where $\theta^o_1 \in \mathbb{R}^p$. The state variable x_v is different between the coverage and the connectivity. x^{cov}_v is a binary variable determining whether v has been visited or not, while x^{con}_v corresponds to the number of UAVs currently on v.

$$x^{cov}_v = \begin{cases} 1 \text{ if } \exists u \in U, v \in S_u \\ 0 \text{ otherwise} \end{cases} \qquad x^{con}_v = |\{u \in U \mid pos(u) = v\}|$$

Regarding the Q-learning aspect, any action is therefore represented as a p-dimensional vector μ_v. Similarly, any state is written as a p-dimensional vector $\mu = \sum_{v \in V} \mu_v$, by summing the embedding structure of every vertex in the environment graph. Thanks to this embedding structure, it is then possible to evaluate any action from any state of any instance.

Reward. This section details point 2 in Fig. 2. The asynchronous process performed by each UAV is shown in Algorithm 2. When a UAV u moves to a vertex $v^{(t+1)}$ from the solution $S^{(t)}$, the latter is added to the list S_u and a new solution $S^{(t+1)}$ is obtained. A reward r must then be given in order to value this choice according to the coverage and the connectivity objectives.

$$r\left(S^{(t)}, v^{(t+1)}\right) = O\left(S^{(t)}\right) - O\left(S^{(t+1)}\right)$$

The reward corresponds to the difference between the objective values of the solutions before and after the movement. Since these values must be minimised, the lower is the new one compared to the old one, the better is the action, and thus the greater must be the reward.

With such a reward, the objective value of a solution $S^{(t)}$ is equivalent to the cumulative reward $R_{0,t}$ since $O(S^{(0)}) = (0,0)^\top$. The cumulative reward $R_{i,j}$ defines the sum of every reward obtained by a UAV from the solution $S^{(i)}$ to reach the solution $S^{(j)}$.

$$R_{i,j} = \sum_{t=i}^{j-1} r\left(S^{(t)}, v^{(t+1)}\right) = O\left(S^{(i)}\right) - O\left(S^{(j)}\right)$$

When every UAV has finished its tour, a reward is given for each sliding window of τ movements made during the coverage. The action made by a UAV at time step t will be rewarded by the cumulative reward $R_{t,t+\tau}$.

Update. This section corresponds to point 3 in Fig. 2. The reward is therefore used for improving the future choices of action (which vertex to go). For that purpose, Θ^{cov} and Θ^{con} are updated when an instance has been processed, by performing a stochastic gradient descent (SGD) step to minimise the squared loss for every UAV u.

$$\left(y - Q_u^o\left(S^{(t)}, v^{(t+1)}; \Theta^o\right)\right)^2$$

with $y = \gamma \max_v \left\{Q_u^o\left(S^{(t+\tau)}, v; \Theta^o\right)\right\} + R_{t,t+\tau}$.

γ is the discount factor. Its value is between 0 and 1 and represents the importance of the future reward depicted by $\max_v \left\{Q_u^o\left(S^{(t+\tau)}, v; \Theta^o\right)\right\}$. $R_{t,t+\tau}$ is the cumulative reward obtained during the frame of τ movements. For each evaluation which has been made, a y is associated and can be assimilated to the rectified evaluation.

4.3 General Pseudo-code

Algorithm 2 describes the whole process of the proposed QLHH. The operation that each UAV executes until it comes back to its starting point appears between lines 7 and 25. If there are still unvisited vertices in G_e, every node is evaluated with Q_u according to the current solution S and Θ (line 13). The vertex maximising this evaluation is chosen. If every vertex has been visited, the UAVs will return to their initial position (line 15). The shortest path between the current position and destination of a UAV is then added to its path (line 18). The reward for such a movement is computed (line 19) and $S^{(t+1)}$ is the new current solution (line 20). Finally, the solution τ iterations ago, the current solution and the cumulative reward between both is registered in the memory $\mathcal{M}$ (line 23). This memory is then used to process the SGD step to modify Θ when every UAV has finished its tour (line 27).

5 Experiments

This section presents the experimental results of QLHH on the CCUS problem which have been conducted on the High Performance Computing (HPC) platform of the University of Luxembourg [17].

5.1 Comparison Heuristic

In order to evaluate the performance of the heuristic generated by QLHH, the results have been compared to those obtained with the manually-designed heuristic described in this section. This heuristic, so called *Weighted Objective* heuristic (WO), belongs to the space of low-level heuristic of QLHH. It means that it respects the template defined in Algorithm 1. At each step, every UAV evaluate every vertex in order to choose the next destination. In this case, this evaluation of the UAV u will be the sum of 2 values, $e^{cov}(u,v)$ and $e^{con}(u,v)$ evaluating the vertex v according to the coverage and the connectivity respectively.

$$e^{cov}(u,v) = \begin{cases} W - W \cdot dist(pos(u), v) & \text{if } x_v = 0 \\ 0 & \text{otherwise} \end{cases}$$

$$e^{con}(u,v) = \begin{cases} W & \text{if } D(u,v) \leq D_{com} \\ 2W - \frac{W}{D_{com}} D(u,v) & \text{otherwise} \end{cases}$$

where W is a given weight, representing the maximal value for both objectives, and $D(u,v) = \min_{u' \in U \setminus \{u\}} dist(pos(u'), v)$.

Algorithm 2: Algorithm of the proposed QLHH

```
   input  : Distribution 𝔻 of instances
   output : Updated Θ
 1 Randomly generate Θ
 2 I ← collection of instances i ↪ 𝔻
 3 M ← ∅                                          // initialisation of the memory
 4 foreach epoch e do
 5   foreach instance i ∈ I do
 6     S ← ∅                                      // initialisation of the solution
 7     foreach UAV u ∈ U do                       // asynchronously
 8       t ← 0
 9       S^(0) ← S
10       S_u ← (pos(u))
11       repeat
12         if S̄ ≠ ∅ then
13           v^(t+1) ← arg max_{v∈V} Q_u (S, v; Θ)
14         else                                   // every vertex has been covered
15           v^(t+1) ← S_u(0)
16         end
17         // u flies to v^(t+1)
18         S_u ← S_u + shortest_path(pos(u), v^(t+1))
19         Compute r(S^(t), v^(t+1))
20         S^(t+1) ← S
21         t ← t + 1
22         if t ≥ τ then
23           M ← M ∪ {(S^(t−τ), R_{t−τ,t}, S^(t))}
24         end
25       until v^(t+1) = S_u(0)
26     end
27     Update Θ with a SGD step for M
28   end
29 end
30 return Θ
```

5.2 Performance Metrics

QLHH Metrics. Two state-of-the-art metrics [4] have been used for assessing the performance of the heuristics generated with QLHH: one coverage metric (coverage speed) and one connectivity metric (number of connected components). Coverage speed expresses how fast the UAVs cover a certain rate r of vertices of the grid. It corresponds to the lowest time step t for which the rate of visited vertices exceeds r. For the experiments, $r = 0.95$, i.e. the coverage speed represents the time needed by the UAVs to cover 95% of the grid. The number of connected components gives the average number of connected components of the connectivity graph (represented in Fig. 1) at each time $t \in T$.

Table 1. Parameters used for the training executions

Parameter	Notation	Value
Problem-specific parameters		
Embedding dimension	p	8
Experience duration	τ	10
Maximum communication distance	D_{com}	4
Q-learning parameters		
Learning rate	α	0.01
Discount factor	γ	0.9

MO Metrics. Three metrics have been used to compare Pareto fronts in a bi-objective space, defined by both objective metrics explained earlier: HyperVolume (HV), Spread (Δ) and Inverted Generational Distance (IGD). HV represents the volume in the objective space covered by non-dominated solutions [18]. Δ defines how well the non-dominated solutions are spread in the front. It is the average of gaps between the distance between two adjacent solutions in the front and the mean of these distances. Finally IGD measures the average distance between the approximated front and the optimal one [16].

5.3 Experimental Setup

For the experiments, grid graphs have been used as environment graphs. Moreover, instances have been split into classes defined by their grid dimension and their number of UAVs. Within a class, instances thus differ in terms of the initial position of UAVs. A class is then written in the following format: `(dim_grid/nb_uavs)`. QLHH has been trained only on the smallest instance class (smallest number of vertices and UAVs). This not only permits to reduce the training time which can be considerably long on large instances, but it will also permit to demonstrate that the generated heuristics also perform well on larger problem instances. After training QLHH on the class ($5 \times 5/3$), the obtained heuristic will be executed 30 times on each instance class. Table 1 presents the parameterisation used for training QLHH. The first set of parameters depends on the problem, i.e. they are used in the context of CCUS. The second set of parameters is problem independent and is only used in the Q-learning process.

5.4 Experimental Results

After executing the generated heuristic (GH) and WO heuristic on 30 instances of each class, both Pareto fronts are compared according to the three MO metrics defined in Sect. 5.2. Results are presented in Table 2. For each class, bold values correspond to the best ones and relative differences are relative to highest values. The IGD metric shows that the Pareto front obtained with GH features a better convergence on all instances. On the contrary, the spread of solutions in the

Table 2. Comparison between WO heuristic and the heuristic generated with QLHH

Metrics	Instances		Heuristics		Relative
	# Vertices	# UAVs	QLHH	WO	Differences
HV	5 × 5	3	9.385e+1	**9.495e+1**	1.15%
	10 × 10	3	4.189e+2	**4.484e+2**	6.57%
		5	5.814e+2	**6.345e+2**	8.37%
		10	1.112e+3	**1.128e+3**	1.44%
		15	1.591e+3	**1.624e+3**	2.00%
	15 × 15	3	6.982e+2	**1.006e+3**	30.58%
		5	**1.493e+3**	1.342e+03	10.09%
		10	**2.581e+3**	2.386e+03	7.56%
Δ	5 × 5	3	**0.000e+0**	**0.000e+0**	/
	10 × 10	3	**4.203e–1**	5.616e–1	25.17%
		5	1.866e–1	**1.199e–1**	35.77%
		10	1.037e–1	**0.000e+0**	100.00%
		15	7.211e–1	**4.147e–3**	99.42%
	15 × 15	3	8.510e–1	**7.240e–2**	91.49%
		5	2.032e–1	**7.759e–2**	61.82%
		10	**1.281e–1**	2.421e–1	47.07%
IGD	5 × 5	3	**1.757e–1**	5.331e–1	67.04%
	10 × 10	3	**5.185e–1**	5.991e–1	13.45%
		5	**3.056e–1**	8.276e–1	63.07%
		10	**5.205e–1**	8.720e–1	40.31%
		15	**2.129e–1**	6.932e–1	69.29%
	15 × 15	3	**8.114e–1**	9.454e–1	14.18%
		5	**3.191e–1**	7.634e–1	58.20%
		10	**1.234e–1**	5.452e–1	77.37%

front (Δ) is worse for GH except for the two smallest and the largest instance classes ((5 × 5/3), (10 × 10/3) and (15 × 15/10)). Finally the HV metric, shows the good performance of GH with results comparable to WO almost every class, while GH performs better on the two largest instances.

Experimental results thus demonstrate that QLHH is able to generate competitive swarming heuristics for any of the instance classes. Moreover, they permit to outline that QLHH is able to generate scalable heuristics, since this good global performance is obtained while the model was trained on the smallest instances only (5 × 5/3). This is a strong asset since the training process would be very long for large instances. With that model, the training can be much faster while keeping good performances.

6 Conclusion

This work has presented a model for optimising the coverage of a connected swarm, so called *Coverage of a Connected-UAV Swarm* (CCUS). In order to generate efficient UAV swarming behaviours, a Q-Learning based Hyper-Heuristic (QLHH) has been designed for generating distributed CCUS heuristics. The experiments have shown a good stability of the model. It means that it is possible to fast train the model on small instances and maintain the good properties resulting on bigger instances. Future work will consist in conducting experiments on additional larger classes of instances. Another extension will consider extending QLHH with Pareto-based approaches, instead of the scalarisation currently used to balance both objectives.

References

1. Aznar, F., Pujol, M., Rizo, R., Rizo, C.: Modelling multi-rotor UAVs swarm deployment using virtual pheromones. PLoS ONE **13**(1), e0190692 (2018)
2. Babić, A., Mišković, N., Vukić, Z.: Heuristics pool for hyper-heuristic selection during task allocation in a heterogeneous swarm of marine robots. IFAC-PapersOnLine **51**(29), 412–417 (2018)
3. Birattari, M., et al.: Automatic off-line design of robot swarms: a manifesto. Frontiers Robot. AI **6**, 59 (2019)
4. Brust, M.R., Zurad, M., Hentges, L., Gomes, L., Danoy, G., Bouvry, P.: Target tracking optimization of UAV swarms based on dual-pheromone clustering. In: 2017 3rd IEEE International Conference on Cybernetics (CYBCONF), pp. 1–8. IEEE (2017)
5. Burke, E.K., et al.: Hyper-heuristics: a survey of the state of the art. J. Oper. Res. Soc. **64**(12), 1695–1724 (2013)
6. Liu, C., Xin, X., Dewen, H.: Multiobjective reinforcement learning: a comprehensive overview. IEEE Trans. Syst. Man Cybern. Syst. **45**(3), 385–398 (2015)
7. Cowling, P., Kendall, G., Soubeiga, E.: A hyperheuristic approach to scheduling a sales summit. In: Burke, E., Erben, W. (eds.) PATAT 2000. LNCS, vol. 2079, pp. 176–190. Springer, Heidelberg (2001). https://doi.org/10.1007/3-540-44629-X_11

8. Duflo, G., Danoy, G., Talbi, E.G., Bouvry, P.: Automated design of efficient swarming behaviours: a Q-learning hyper-heuristic approach. In: Genetic and Evolutionary Computation Conference Companion, pp. 227–228. ACM (2020)
9. Duflo, G., Danoy, G., Talbi, E.G., Bouvry, P.: Automating the design of efficient distributed behaviours for a swarm of UAVs. In: Symposium Series on Computational Intelligence - SSCI 2020. IEEE, Canberra, Australia (2020)
10. Duflo, G., Kieffer, E., Brust, M.R., Danoy, G., Bouvry, P.: A GP hyper-heuristic approach for generating TSP heuristics. In: 2019 IEEE International Parallel and Distributed Processing Symposium Workshops (IPDPSW), pp. 521–529. IEEE (2019)
11. Khalil, E., Dai, H., Zhang, Y., Dilkina, B., Song, L.: Learning combinatorial optimization algorithms over graphs. In: Guyon, I., (eds.) Advances in Neural Information Processing Systems. vol. 30, pp. 6348–6358. Curran Associates, Inc. (2017)
12. Kieffer, E., Danoy, G., Brust, M.R., Bouvry, P., Nagih, A.: Tackling large-scale and combinatorial bi-level problems with a genetic programming hyper-heuristic. IEEE Trans. Evol. Comput. **24**(1), 44-56 (2019)
13. Lin, J., Zhu, L., Gao, K.: A genetic programming hyper-heuristic approach for the multi-skill resource constrained project scheduling problem. Exp. Syst. Appl. **140**, 112915 (2020)
14. Tuyls, K., Stone, P.: Multiagent learning paradigms. In: Belardinelli, F., Argente, E. (eds.) EUMAS/AT -2017. LNCS (LNAI), vol. 10767, pp. 3–21. Springer, Cham (2018). https://doi.org/10.1007/978-3-030-01713-2_1
15. Van Moffaert, K., Drugan, M.M., Nowe, A.: Scalarized multi-objective reinforcement learning: novel design techniques. In: IEEE Symposium on Adaptive Dynamic Programming and Reinforcement Learning (ADPRL), pp. 191–199. IEEE (2013)
16. Van Veldhuizen, D.A.: Multiobjective Evolutionary Algorithms: Classifications, Analyses, and New Innovations. Ph.D. thesis, USA (1999), aAI9928483
17. Varrette, S., Bouvry, P., Cartiaux, H., Georgatos, F.: Management of an academic HPC cluster: the UL experience. In: Proceedings of the 2014 International Conference on High Performance Computing & Simulation (HPCS 2014), pp. 959–967. IEEE (July 2014)
18. Zitzler, E., Thiele, L.: Multiobjective evolutionary algorithms: a comparative case study and the strength pareto approach. IEEE Trans. Evol. Comput. **3**(4), 257–271 (1999)

Cross-Domain Co-Author Recommendation Based on Knowledge Graph Clustering

Tahsir Ahmed Munna[1,2](✉) and Radhakrishnan Delhibabu[3,4]

[1] Laboratory for Models and Methods of Computational Pragmatics, HSE University, Moscow, Russia

[2] CICANT – The Centre for Research in Applied Communication, Culture, and New Technologies, ULHT. Campo Grande, Lisbon, Portugal

[3] VIT University, Vellore, India

[4] Artificial Intelligence and Digitalization of Mathematical Knowledge Lab, Mathcenter, Kazan Federal University, Kazan, Russia

Abstract. Nowadays, due to the growing demand for interdisciplinary research and innovation, different scientific communities pay substantial attention to cross-domain collaboration. However, having only information retrieval technologies in hands might be not enough to find prospective collaborators due to the large volume of stored bibliographic records in scholarly databases and unawareness about emerging cross-disciplinary trends. To address this issue, the endorsement of the cross-disciplinary scientific alliances have been introduced as a new tool for scientific research and technological modernization. In this paper, we use a state-of-art knowledge representation technique named Knowledge Graphs (KGs) and demonstrate how clustering of learned KGs embeddings helps to build a cross-disciplinary co-author recommendation system.

Keywords: Knowledge graph · Embeddings · Recommender system · Cross-domain research · Clustering

1 Introduction

Research is turning into a progressively computerized and interdisciplinary process, furthermore, being information-driven it influences not only the academic community but also facilitates industry and government decisions. Presentation of research results, their dissemination, search and examination, as well as their perception, have reached a new level thanks to the extended use of web standards and computerised academic communication.

The number of scientific publications created by new researchers and the expanding digital accessibility of ancient academic rarities, and related datasets and metadata [12] are among current drivers of the generous development in academic correspondence.

N. T. Nguyen et al. (Eds.): ACIIDS 2021, LNAI 12672, pp. 782–795, 2021.
https://doi.org/10.1007/978-3-030-73280-6_62

Helping specialists with a more profound examination of insightful metadata and providing them with relevant suggestions can lead to new open doors in inquiring about the next research steps. Particularly, information availability and the recommendation about potential collaborations efforts between researchers can prompt better approaches for directing research.

To enhance recommendation quality, different information sources can be included in the recommendation process by the recommender system. To mention the most popular ones, recommendations can be derived from user preferences on items, user and item attributes (including textual and visual metadata), social ties between users and their interaction, ontology information, consumption history, and even expert knowledge [11].

As the creation of scientific papers extends, progressively more research papers are being distributed and partaken in specialised databases and on scholarly web services. Accordingly, this enormous academic information has prompted data over-burden both on human and technological level [2,27]. Thus, applying recommendation strategies to get a fast, exact, and adequate collaboration rundown can assist analysts with advancing their research, particularly when they move into another exploration field.

Social network analysis convenes to interpreting communication between people. Here, researchers have extensively studied different types of social network and their properties [5,10], link prediction techniques [16,17,23], and, finally, recommendations [6,15]. Even though recent advances in social networks demonstrate a rich variety of techniques, studying collaborations across two domains is still an emerging topic.

Meanwhile, interdisciplinary collaboration efforts have produced a tremendous effect on society. For instance, collaboration efforts among computer science and biology set up the field of bioinformatics. As a result of these cross-domain joint efforts, initially, incredibly costly errands such as DNA sequencing have gotten adaptable and available not only in laboratory conditions [1].

Presently data mining and medicine are cooperating in the field of medical informatics, which is a major development zone that is relied upon to have a tremendous effect on medication [7].

In our approach, we have applied knowledge graph embeddings for finding cross-domain fields by making clustering based on the researcher's working field then uncovering the most top-n similar researcher for a recommendation. Through this approach, a researcher can find a prospective co-author for collaborative work and pay attention to valuable works from a different part of expertise.

The paper is organised as follows. Section 2 describes the related work, while the methodology of our approach is described in Sect. 3. As the main goal of this paper is to recommend cross-disciplinary co-authorships, the proposed recommender model is described in Sect. 4. The obtained qualitative results are demonstrated in Sect. 5. Section 6 concludes the paper.

2 Background Study

Our co-author recommendation model conceptually inspired by Jie Tang et al. [22] where they dissect the cross-domain collaboration effort data by looking into its distributions and reveal the specific patterns compared to traditional collaborations in the same domain: 1) sparse connection: cross-domain collaborations are rare; 2) complementary expertise: cross-domain collaborators often have different expertise and interest; 3) topic skewness: cross-domain collaboration topics are focused on a subset of topics. They propose the Cross-domain Topic Learning (CTL) model to address these difficulties.

Tin Huynh et al. [13] examine a new methodology that utilizes extra information as new features to make a recommendation, i.e., the quality of the connection between organizations, the significance rating, and the activity scores of researchers. Thus, they propose another strategy for assessing the nature of collaborator recommendations and analyse a dataset obtained from the Microsoft Academic Search Web service.

Co-authorship recommendations are also performed via link prediction in [19]. The authors employed a tailored link embedding operator that learns edge embeddings via the inclusion of the neighbourhoods of incident nodes in the input co-authorship graph and include research interest information presented as an embedding of nodes in the keywords co-occurrence network connecting keywords related to a given research article [19]. They evaluate the proposed approach via related binary classification problems on the bibliographic dataset of a single university over the last 25 years period for research articles indexed in Scopus and other local research indices. The related recommendation problem is defined as the prediction of the connection expectation within a given co-authorship network.

Sie et al. [21] also proposed an interesting approach for finding co-author recommendations based on betweenness centrality and interest similarity. They present the COCOON CORE tool that matches an applicant and a co-author based on like-mindedness and power values. Like-mindedness guarantees that a prospective co-author shares a common ground, which is valuable for consistent collaboration. A more powerful co-author encourages the mediation of an article's idea with the respective community.

Yang et al. [29] explain how their method can help in scientific collaborator recommendations by taking care of the comparable problems for person-to-person matching. Their method produces personal profiles via incorporating personal research ability, co-authors' attributes, and academic institutional network (nearby and worldwide) through an SVM-Rank based consolidating system. The created exhaustive profiles mitigate data asymmetry and various similarity measures, which help to cope with data over-burdening. The proposed technique has been adopted in the ScholarMate research network[1].

[1] www.scholarmate.com.

Giseli et al. [18] presents a creative way to deal with recommending joint efforts in the setting of scholarly social networks employing two indices for global collaboration and global correlation of actors, i.e. researchers.

In the end, we can summarise that observed works on co-author recommender systems have been done via different approaches including graph embeddings, but, to the best of our knowledge, there are no works on cross-domain areas with knowledge graphs. Such a KGs-based approach is proposed in the next section.

3 Methodology

3.1 Our Approach

In our approach, we rely on knowledge graphs, where the term knowledge graph is understood as a knowledge base formally represented as a set of triples in the form object-predicate-subject.

To implement it, we use Ampligraph [8], an open-source Python library that predicts links between concepts in a knowledge graph. AmpliGraph[2] provides a user with a collection of neural-based models for relational learning, namely supervised learning on knowledge graphs. These models create knowledge graph embeddings, i.e. vector representations of entities in metric space, and then predict missing links with model-explicit scoring function (see Fig. 1). Figure 2 shows the schema of our approach that is described in detail below.

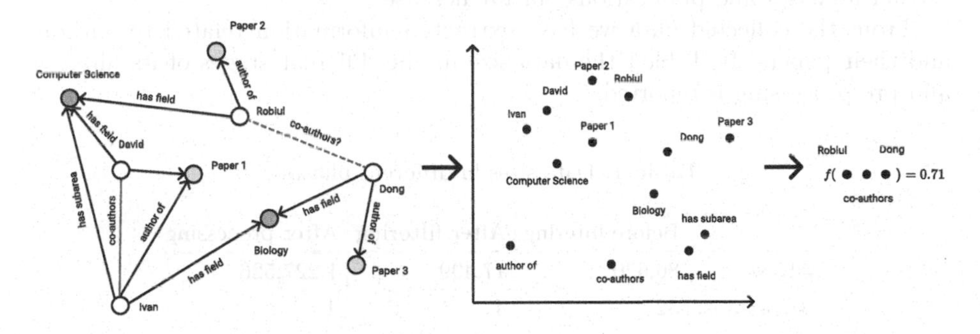

Fig. 1. A knowledge graph results in vector representations in a metric space and predicts unseen links by a scoring function

[2] https://github.com/Accenture/AmpliGraph.

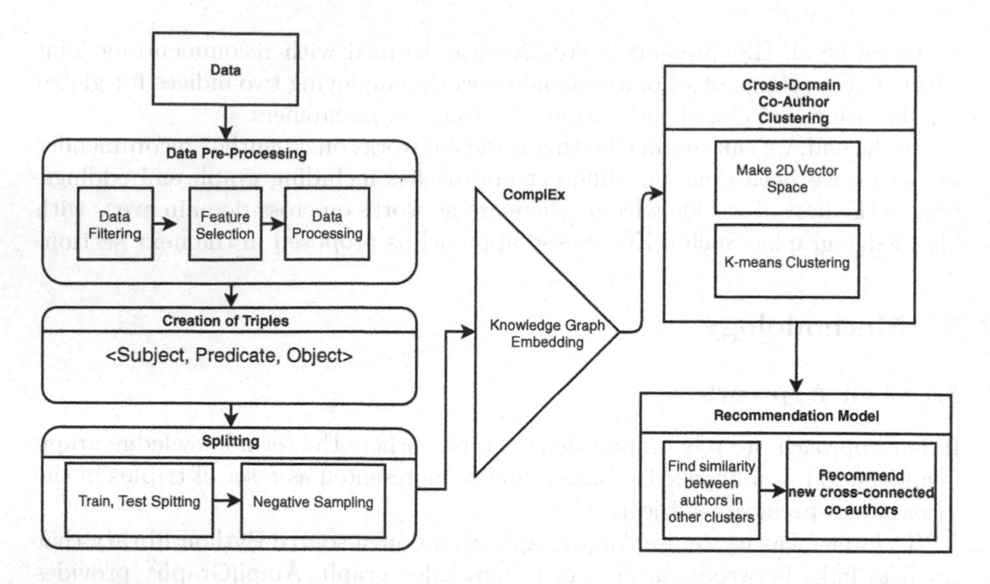

Fig. 2. The schema of our proposed approach

Data Description. There is a recent scholarly data collection[3] obtained from google Scholar[4], Scopus[5], and ACM Digital Library[6] in CSV (Comma Separated Values) format. We have been granted the permission for using this data for further analyses and publications. for further use.

From the collected data we have extracted information related to authors and their papers. In Table 1 the data size in the different stages of its filtering and pre-processing is reported.

Table 1. Data sizes in different phases

	Before filtering	After filtering	After processing
#Rows:	86,620	17,499	1,227,526
#Columns:	32	4	4

– Feature selection. We have manually selected several fields form our dataset, namely, "AuthordIDs", "Authors", "Fields", and "Subareas". "AuthorIDs" is the list of unique ids for several co-authors. "Authors" is the list of authors' names. The attribute "Fields" is the list of the main co-authors' expertise fields, while "Subareas" is the list of the co-author working sub-fields.

[3] https://www.mdhassan.com/about.
[4] https://scholar.google.com/.
[5] http://scopus.com/.
[6] https://dl.acm.org/.

- Before filtering data. Before filtering, we have a substantial amount of "NaN" values, duplicate values, and also many non-relevant features concerning the paper goal.
- After filtering data. In this phase we have dropped all "NaN" and all duplicate values. Moreover, we consider only four compound attributes named "AuthorIDs", "Authors", "Fields", "Subareas" from the cleaned dataset.
- After processing data. One can see that the data volume is seriously enlarged due to the presence of multiple values as lists in the original fields. So, we need to sort out them. For example, if there is a single row with 2 co-authors, 5 subareas and 2 fields then we create $2 \times 5 \times 2$ rows with attributes "AuthorID", "Author", "Field", "Subarea" for this record.

  By doing so we can further transform the data to a knowledge graph and thus enrich authors' fields and subarea labels by that of their co-authors for the particular paper, i.e. row.

Knowledge Graph Creation. We are going to build a knowledge graph based on the extracted researcher's information. The idea is that each author ID is an entity that will be associated with its name, subarea, field. The goal is to produce another representation of the dataset where data is stored in triples of the following structure: $\langle subject, predicate, object \rangle$.

We make our triples by means of connections between the entities, i.e. between the author's ID and areas, for example, $\langle 572017485, HasArea, Math \rangle$, and between authors and subareas, for example, $\langle 572017485, HasSubarea, Biology \rangle$.

Knowledge Graph Embedding Training. We split our graph into training, validation, and test sets such that the training set is utilised in learning knowledge graph embeddings, while the validation set is utilized in their evaluation. The test set is used to evaluate the classification algorithm based on the embeddings. In this split, the test set contains only entities and relations which also occur in the training set; it contains 644,451 triples. The training and validation sets contain 2,477,807 and 100,000 triples, respectively.

AmpliGraph encompasses several KGs embedding models, among those we use TransE [3], ComplEx [25], DistMult [28], HolE [20]. In our case, the ComplEx model has brought better performance in terms of average loss, though only the third best execution time. ComplEx [25] extends DistMult by presenting complex-valued embeddings in order to deal with a variety of binary relations, for example, symmetric and antisymmetric ones. In ComplEx, entity and relations embeddings $\mathbf{e}_s$, $\mathbf{e}_o$, $\mathbf{r}_p$ are vectors in complex space $\mathbb{C}^d$. The related scoring function of a triple (s, p, o) based on the trilinear Hermitian dot product in $\mathbb{C}^d$:

$$f_{ComplEx}(s, p, o) = Re(\langle \mathbf{r}_p, \mathbf{e}_s, \overline{\mathbf{e}_o} \rangle) = Re(\sum_{i=1}^{d} \mathbf{r}_{pi}, \mathbf{e}_{si}, \overline{\mathbf{e}_{oi}}),$$

where $Re(z)$ is the real part of a complex value z, and $\overline{\mathbf{e}_o}$ is the conjugate of $\mathbf{e}_o$. This scoring function is not symmetric and thus can handle non-symmetric relations.

Let us summarise the main model parameters given in Table 2: **k** is the dimensionality of the embedding space, **eta** (η) is the number of negative triples that must be created for every existing triple during the negative sampling phase, **batches_count** is the number of batches in which the training set must be split. The number of training epochs is set by **epochs**; **optimizer** sets the optimizer used to minimize the loss function from 'sgd', 'adagrad', 'adam', and 'momentum'; **loss** is the type of loss function to use during training; **regularizer** is the regularization strategy to use with the loss function, for example, L_2 regularization ($p = 2$). Note that optimizer and regularizer may have their own parameters.

Table 2. The best hyper-parameters of KG embedding model

batches_count=50, seed=0, epochs=300, k=100, eta=20, verbose=True
optimizer='adam', optimizer_params={'lr':1e-4},
loss='multiclass_nll', regularizer='LP', regularizer_params={'p':3, 'lambda':1e-5}

Table 2 contains the embedding model's best hyper-parameters found via grid-search scheme.

In Table 3, one can see the fitting times for the models on a commodity machine with the following hardware settings: 8 GB RAM, Core i5 CPU, 1.6 GHz × 8 cores.

Table 3. KG embedding model fitting: time measurements and average losses

Model	Execution time, h:m	Time per epoch, m:s	Average loss
ComplEx	3:19	2:30	**0.043**
TransE	**2:50**	**2:01**	0.047
DistMult	2:55	2:11	0.046
HolE	3:40	2:32	0.048

3.2 Research Area Clustering Based on Knowledge Graph Embedding

To qualitatively assess the subjective nature of the embeddings, we can project the obtained vector representations on a 2D space and further cluster them. In our case, we can cluster researcher working fields, which makes sense for the subsequent generation of recommendations.

We have 39 research fields (see Table 4) and utilize PCA [26] to project the input vectors (embeddings) from the 200D space into a 2D space.

We use K-means clustering [24], a centroid-based algorithm, where the decision on the assignment of a given point to a certain cluster is employed by the shortest distance rule.

Table 4. Researcher working fields

'Immunology', 'ComputerScience', 'Agricultural', 'Mathematics', 'BiologicalSciences', 'MolecularBiology', 'Multidisciplinary', 'Biochemistry', 'Medicine', 'Humanities', 'Astronomy', 'Chemistry', 'Physics', 'Engineering', 'Toxicology', 'Pharmaceutics', 'Nursing', 'Energy', 'MaterialsScience', 'Genetics', 'Microbiology', 'Neuroscience', 'EnvironmentalScience', 'Pharmacology', 'PlanetarySciences', 'ChemicalEngineering', 'Business', 'SocialSciences', 'Earth', 'Veterinary', 'Psychology', 'Dentistry', 'DecisionSciences', 'Arts', 'HealthProfessions', 'Accounting', 'Finance', 'Econometrics', 'Management', 'Economics'

The KElbowVisualizer[7] implements the "elbow" technique to select the optimal number of clusters. The junction point of the virtual elbow composed by the values of K-means cost function is usually a good candidate for choosing K, the number of clusters. To illustrate how KElbowVisualizer fits the K-Means model for a range of K from 1 to 10 on our two-dimensional dataset, we provide the reader with Fig. 3. In our case, three might be chosen as the optimal K.

The results of clustering are shown in Fig. 4 and Table 5.

Note that some of the clusters contain evident outliers like "Arts" in Cluster 0 or vague terms like "Multidisciplinary", so methods like DBScan may help to sort out them. However, we decided not to leave out any of the research interests and keep the trade-off between the number of clusters and experiment simplicity.

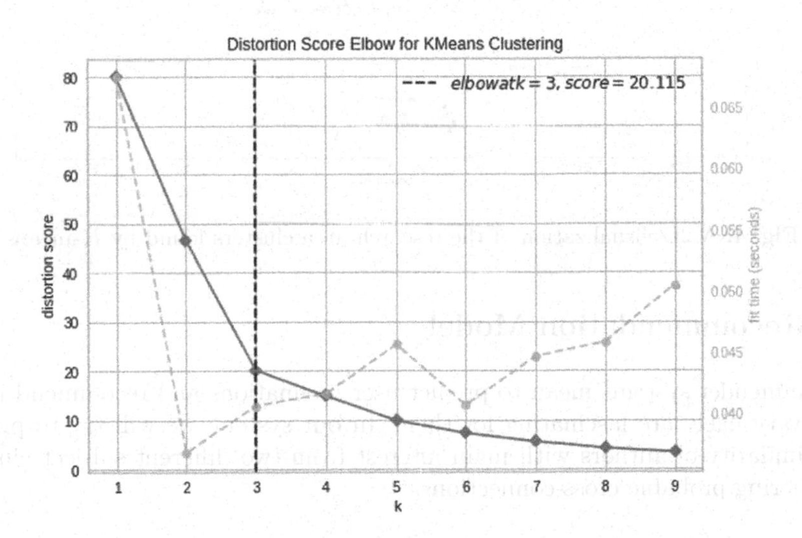

Fig. 3. Finding the optimal number of clusters by elbow method

[7] https://www.scikit-yb.org/en/latest/api/cluster/elbow.html.

Table 5. Cluster naming based on subject area

Name	Label	Cluster elements
Life sciences	0	'Immunology', 'Agricultural', 'Biological Sciences', 'Multidisciplinary', 'Medicine', 'Humanities', 'Toxicology', 'Pharmaceutics', 'Nursing', 'Microbiology', 'Neuroscience', 'Pharmacology', 'Social Sciences', 'Veterinary', 'Psychology', 'Dentistry', 'Arts', 'Health Professions'
Economics+	1	'Business', 'Accounting', 'Finance', 'Econometrics', 'Management', 'Economics', 'Earth', 'Planetary Sciences'
Computational Sciences	2	'ComputerScience', 'Mathematics', 'Molecular Biology', 'Biochemistry', 'Astronomy', 'Chemistry', 'Physics', 'Engineering', 'Energy', 'Materials Science', 'Genetics', 'Environmental Science', 'Chemical Engineering', 'Decision Sciences', 'Computer Science'

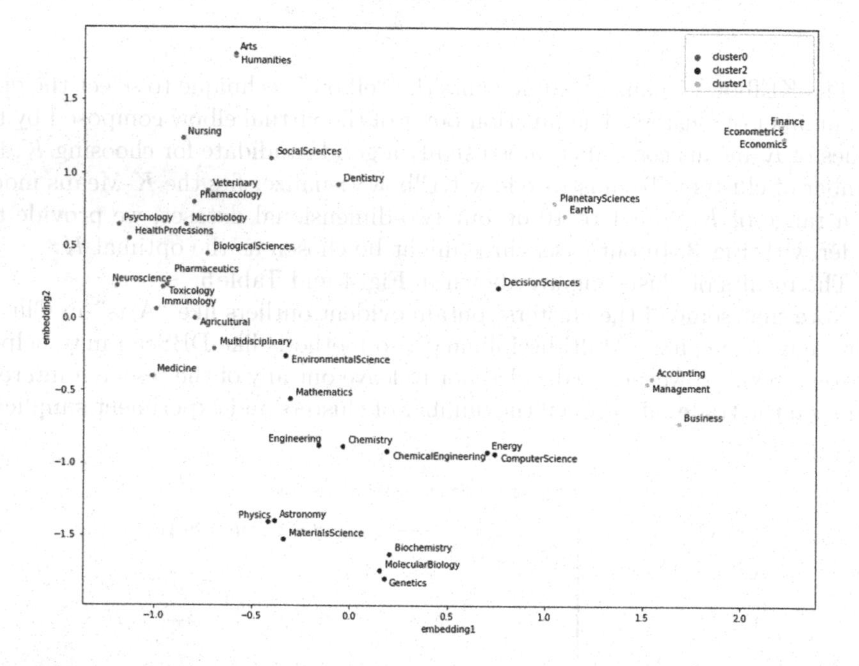

Fig. 4. A 2D visualization of the research area clusters found by K-means

4 Recommendation Model

Recommender systems mean to predict user inclinations and recommend items that very likely are fascinating for them. In our system, we will try to predict the similarity of authors with main interest from two different subject clusters for offering probable cross-connections.

4.1 Data Pre-processing

For our recommendation engine, we need to build a bag of words for finding the similarity of authors based on their research field and subareas. Thus, we concatenate "Field" and "Subarea" attributes for building a bag of words.

For building content-based recommendations we need to obtain a similarity matrix [30] using cosine similarity, which is computed as normalized dot product of two vectors **a** and **b**:

$$Cos(\widehat{\mathbf{a}, \mathbf{b}}) = \frac{\sum_1^n a_i b_i}{\sqrt{\sum_1^n a_i^2}\sqrt{\sum_1^n b_i^2}}.$$

Here, **a** and **b** are the vectors of authors' profiles composed from the counters of separate words extracted from the values of their 'Field" and "Subarea" attributes as strings.

Our aim is to find the cross-similarity of two authors with their main research interests from different clusters such that they can work together based on their common subareas.

4.2 Model Evaluation

Evaluation Metrics. To evaluate the quality of embedding models we use Mean Reciprocal Rank [9] and Hits@n.

For a set of triples T, the Mean Reciprocal Rank is the mean of $|T|$ reciprocal ranks, that is that of inverse positions in a vector of rankings:

$$MRR = \frac{1}{|T|} \sum_{(s,p,o) \in T} \frac{1}{rank_{(s,p,o)}}.$$

Hits@n shows how many elements of a vector of rankings are at the top-n positions:

$$Hits@n = \sum_{(s,p,o) \in T} [rank_{(s,p,o)} \leqslant n],$$

where $[z]$ is the Iverson predicate notation returning 1 if the predicate z is true and 0, otherwise, T is a set of triples. For example, Hits@3 demonstrates how often the existing triple (in the test set) was among three highly-ranked triples by a tested model.

Table 6. KG Embedding evaluation and comparison

Model	MRR	Hits@10	Hits@3	Hits@1
ComplEx	0.41	0.44	0.42	0.39
TransE	0.43	0.45	0.43	0.41
DistMult	**0.44**	**0.48**	**0.45**	**0.43**
HolE	0.43	0.44	0.43	0.40

From Table 6 we can see that model DistMult has the best performance.

Additional evaluation of our embedding models is done in the classification setting. In Table 7, we can see, the classification accuracy with embedding data is higher than that of the baselines methods on the original data. XGBoost results in 0.875 accuracy and it has managed to reach AUC=1 for class 0 (i.e. perfect prediction of the class). The second class AUC is 0.75, which means it might be the result of outliers' presence. Finally, class 3 has the highest accuracy compared to baseline algorithms, which is 0.91. Among the three baseline classifiers (Logistics Regression, Decision Tree, and Naive Bayes) performed in the original data, the highest accuracy comes for Naive Bayes, which is 0.772. On the other hand, we can observe that Decision Tree gives the second-best accuracy, 0.760, and its AUC score for the second class is 0.89, which is the highest among all the models. Logistic Regression is an outsider by both, AUC and accuracy.

From Tables 6 and 7 we can observe that learned embeddings are more appropriate for our data due to their better performance in terms of accuracy than that of the selected baselines.

Table 7. Result of the model and comparison with baseline

Model	Accuracy score	AUC accuracy score		
		0	1	2
XGBoost	0.875	1.0	0.75	0.91
Baseline with original data				
LogisticRegression	0.724	0.54	0.83	0.74
Decision tree	0.760	0.55	0.89	0.81
Naive bayes	0.772	0.64	0.85	0.78

5 Qualitative Results

In Table 8, we can observe the recommendation results for an exemplary target author, Scott Burleigh[8]. His research field is "Engineering" and subarea is "Computer Science", where the topmost recommended co-author's field is "Earth and Planetary Sciences" and the subarea is "Chemistry".

So, having only this meta-information in hands one cannot clearly see how these two authors could be cross-connected with each other. How they are similar? However, if we check the author's profile at Scopus, we can identify that Scott Burleigh has only one paper published entitled "Interplanetary overlay network an implementation of the DTN bundle protocol" [4] having 62 citations in there. Even though the authors did not have common keywords within their respective fields, for some reason "Planetary Science" is not in the target

[8] https://www.scopus.com/authid/detail.uri?authorId=7003872194.

Table 8. Recommendation result

Author ID	IDs of top-5 recommended authors
7003872194	55457352300, 15081871300, 8836762200, 57189952417, 35563455500
Author details	Details of the 1st recommended author
AuthorID: 7003872194 Authors: Burleigh, Scott Field: Engineering Subarea: Computer Science Cluster: 2	AuthorID: 55457352300 Authors: Rasmussen, Joseph T. Field: Earth and Planetary Sciences Subarea: Chemistry Cluster: 1

author's profile, the system has managed to identify the relevancy of recommendation in line with the field "Earth and Planetary Sciences" of the prospective co-author[9].

6 Conclusion

In this paper, we demonstrated the cross-domain co-author recommendations by means of knowledge graph embeddings with state-of-the-art techniques. For fitting the model we used approximate 2.5 million triples by incorporating information on the researchers' working fields and subareas. After that, we cluster the area interest embedding score by K-means clustering. Furthermore, we use classification techniques for evaluation of our models on the original relational data vs. KGs' embeddings, where the latter gave better accuracy ($\approx 87\%$) than the baselines. Finally, after getting the class of the target author's research area, we used cosine similarity between a prospective co-author from different clusters of research fields for making a recommendation on cross-domain co-authorship. As future work, we would like to further improve our approach by means of adding other relevant contextual information and compare it with the state-of-the-art recommendation techniques and icorporating such relevant techniques as OAC-triclustering and 3-FCA [14].

Acknowledgment. The study was implemented in the framework of the Basic Research Program at the HSE University and funded by the Russian Academic Excellence Project '5–100'. The work of the last author is supported by the Mathematical Center of the Volga Region Federal District (Project no. 075-02-2020-1478).

References

1. Bankevich, A., et al.: Spades: a new genome assembly algorithm and its applications to single-cell sequencing. J. Comput. Biol. **19**(5), 455–477 (2012). https://doi.org/10.1089/cmb.2012.0021

[9] https://www.scopus.com/authid/detail.uri?authorId=55457352300.

2. Beel, J., Gipp, B., Langer, S., Breitinger, C.: Paper recommender systems: a literature survey. Int. J. Digit. Libr. **17**(4), 305–338 (2016)
3. Bordes, A., Usunier, N., Garcia-Duran, A., Weston, J., Yakhnenko, O.: Translating embeddings for modeling multi-relational data. In: Advances in Neural Information Processing Systems, pp. 2787–2795 (2013)
4. Burleigh, S.: Interplanetary overlay network an implementation of the DTN bundle protocol. In: 2007 4th Annual IEEE Consumer Communications and Networking Conference, CCNC 2007, pp. 222–226 (2007). https://doi.org/10.1109/CCNC.2007.51
5. Chakrabarti, D., Faloutsos, C.: Graph mining: Laws, generators, and algorithms. ACM Comput. Surv. (CSUR) **38**(1), 2-es (2006)
6. Chen, H.H., Gou, L., Zhang, X., Giles, C.L.: Collabseer: a search engine for collaboration discovery. In: Proceedings of the 11th Annual International ACM/IEEE Joint Conference on Digital Libraries, pp. 231–240 (2011)
7. Chen, P.H.C., Liu, Y., Peng, L.: How to develop machine learning models for healthcare. Nature Mater. **18**(5), 410–414 (2019). https://doi.org/10.1038/s41563-019-0345-0
8. Costabello, L., Pai, S., Van, C.L., McGrath, R., McCarthy, N., Tabacof, P.: AmpliGraph: a Library for Representation Learning on Knowledge Graphs (March 2019). https://doi.org/10.5281/zenodo.2595043
9. Craswell, N.: Mean reciprocal rank. Encycl. Database Syst. **1703** (2009)
10. Faloutsos, M., Faloutsos, P., Faloutsos, C.: On power-law relationships of the internet topology. ACM SIGCOMM Comput. Commun. Rev. **29**(4), 251–262 (1999)
11. Grad-Gyenge, L., Filzmoser, P., Werthner, H.: Recommendations on a knowledge graph. In: 1st International Workshop on Machine Learning Methods for Recommender Systems, MLRec, pp. 13–20 (2015)
12. Henk, V., Vahdati, S., Nayyeri, M., Ali, M., Yazdi, H.S., Lehmann, J.: Metaresearch recommendations using knowledge graph embeddings. In: AAAI 2019 Workshop on Recommender Systems and Natural Language Processing (RECNLP) (2019)
13. Huynh, T., Takasu, A., Masada, T., Hoang, K.: Collaborator recommendation for isolated researchers. In: 2014 28th International Conference on Advanced Information Networking and Applications Workshops, pp. 639–644 (2014)
14. Ignatov, D.I., Gnatyshak, D.V., Kuznetsov, S.O., Mirkin, B.G.: Triadic formal concept analysis and triclustering: searching for optimal patterns. Mach. Learn. **101**(1–3), 271–302 (2015). https://doi.org/10.1007/s10994-015-5487-y
15. Konstas, I., Stathopoulos, V., Jose, J.M.: On social networks and collaborative recommendation. In: Proceedings of the 32nd International ACM SIGIR Conference on Research and Development in Information Retrieval, pp. 195–202 (2009)
16. Leskovec, J., Huttenlocher, D., Kleinberg, J.: Predicting positive and negative links in online social networks. In: Proceedings of the 19th International Conference on World Wide Web, pp. 641–650 (2010)
17. Lichtenwalter, R.N., Lussier, J.T., Chawla, N.V.: New perspectives and methods in link prediction. In: Proceedings of the 16th ACM SIGKDD International Conference on Knowledge Discovery and Data Mining, pp. 243–252 (2010)
18. Lopes, G.R., Moro, M.M., Wives, L.K., de Oliveira, J.P.M.: Collaboration recommendation on academic social networks. In: Trujillo, J. (ed.) ER 2010. LNCS, vol. 6413, pp. 190–199. Springer, Heidelberg (2010). https://doi.org/10.1007/978-3-642-16385-2_24
19. Makarov, I., Gerasimova, O., Sulimov, P., Zhukov, L.E.: Dual network embedding for representing research interests in the link prediction problem on co-authorship networks. PeerJ Comput. Sci. **5**, (2019). https://doi.org/10.7717/peerj-cs.172

20. Nickel, M., Rosasco, L., Poggio, T.: Holographic embeddings of knowledge graphs. In: Thirtieth Aaai conference on Artificial Intelligence (2016)
21. Sie, R.L.L., van Engelen, B.J., Bitter-Rijpkema, M., Sloep, P.B.: COCOON CORE: CO-author Recommendations based on betweenness centrality and interest similarity. In: Manouselis, N., Drachsler, H., Verbert, K., Santos, O.C. (eds.) Recommender Systems for Technology Enhanced Learning, pp. 267–282. Springer, New York (2014). https://doi.org/10.1007/978-1-4939-0530-0_13
22. Tang, J., Wu, S., Sun, J., Su, H.: Cross-domain collaboration recommendation. In: Proceedings of the 18th ACM SIGKDD International Conference on Knowledge Discovery and Data Mining, pp. 1285–1293 (2012)
23. Tang, L., Liu, H.: Relational learning via latent social dimensions. In: Proceedings of the 15th ACM SIGKDD International Conference on Knowledge Discovery and Data Mining, pp. 817–826 (2009)
24. Teknomo, K.: K-means clustering tutorial. Medicine **100**(4), 3 (2006)
25. Trouillon, T., Welbl, J., Riedel, S., Gaussier, É., Bouchard, G.: Complex embeddings for simple link prediction. In: International Conference on Machine Learning (ICML) (2016)
26. Wold, S., Esbensen, K., Geladi, P.: Principal component analysis. Chemometr. Intell. Lab. Syst. **2**(1–3), 37–52 (1987)
27. Xia, F., Wang, W., Bekele, T.M., Liu, H.: Big scholarly data: a survey. IEEE Trans. Big Data **3**(1), 18–35 (2017)
28. Yang, B., Yih, W.t., He, X., Gao, J., Deng, L.: Embedding entities and relations for learning and inference in knowledge bases. arXiv preprint arXiv:1412.6575 (2014)
29. Yang, C., Sun, J., Ma, J., Zhang, S., Wang, G., Hua, Z.: Scientific collaborator recommendation in heterogeneous bibliographic networks. In: 2015 48th Hawaii International Conference on System Sciences, pp. 552–561. IEEE (2015)
30. Zhou, Z., Cheng, Z., Zhang, L.J., Gaaloul, W., Ning, K.: Scientific workflow clustering and recommendation leveraging layer hierarchical analysis. IEEE Trans. Serv. Comput. **11**(1), 169–183 (2016)

Data Modelling and Processing for Industry 4.0

Trident: Change Point Detection for Multivariate Time Series via Dual-Level Attention Learning

Ziyi Duan, Haizhou Du(✉), and Yang Zheng

Shanghai University of Electric Power, Shanghai 200090, China
duhaizhou@shiep.edu.cn

Abstract. Change point detection is an important subset of anomaly detection problems. Due to the ever-increasing volume of time-series data, detecting change points has important significance, which can find anomalies early and reduce losses, yet very challenging as it is affected by periodicity, multi-input series, and long time series. The performance of traditional methods typically scales poorly.

In this paper, we propose Trident, a novel prediction-based change point detection approach via dual-level attention learning. As the name implies, our model consists of three key modules which are the prediction, detection, and selection module. The three modules are integrated in a principled way of detecting change points more accurately and efficiently. Simulations and experiments highlight the effectiveness and efficacy of the Trident for change point detection in time series. Our approach outperforms the state-of-the-art methods on two real-world datasets.

Keywords: Time series · Change point detection · Attention mechanism

1 Introduction

With the explosive development of big data analysis, anomaly detection in time-series is also increasingly important. Change point detection is an important subset of anomaly detection problems. Due to the ever-increasing volume of time-series data that must be efficiently analyzed, it is becoming a mainstream study in a wide variety of applications, including finance, energy, meteorology, medicine, aerospace, etc.

Change points are the moments when the state or property of the time series changes abruptly [2]. Increasing the detecting accuracy is beneficial to operational efficiency in many aspects of society [20,21], such as power load detection, online sales analysis, or weather forecasting. We could mine the potential mutations and take corresponding preventative measures early to reduce financial and time losses.

However, detecting change points in modern applications is particularly challenging, affected by the following complex factors: periodicity, multi-series, and

N. T. Nguyen et al. (Eds.): ACIIDS 2021, LNAI 12672, pp. 799–810, 2021.
https://doi.org/10.1007/978-3-030-73280-6_63

long series. Traditional methods cannot adaptively select relevant series and achieve feature extraction for multiple input series. And may lead to error accumulation and inefficient computation on the long-term series. Moreover, the real-world time-series data may contain a large number of change points and outliers. There are fundamental differences between them [22]. Among them, change points are the moments of the original series' property or state change abruptly. And outliers refer to the sudden single peak or decrease in the series [6]. Distinguishing them has always been one of the difficulties in change point detection.

The current methods of change point detection are mainly split up into probability and statistics-based, classification-based, and prediction-based. The traditional methods are ineffective in modeling complex non-linear time series data [16,23]. The prediction-based method is one of the most commonly used methods [19]. Recently, the deep learning-based approaches have demonstrated strong performance in time series modeling [4,10,17]. However, the existing methods only focus on improving the ability to learn nonlinear features, while they ignore the problem of feature and information loss.

To address these aforementioned issues, inspired by the multimodal features fusion [5,8] and the hierarchical attention networks [14,15], which are the latest progress of attention mechanisms [3,9], we proposed a prediction-based change point detection approach via dual-level attention learning, which we call Trident.

In this paper, Trident consists of three key modules: prediction, detection, selection. Accordingly, the key contributions can be summarized as follows.

- We propose Trident, a change point detection approach for time series employing dual-level attention learning. It could detect change points accurately and timely in long periodic series with multiple relevant input series.
- In the input attention stage, we integrate the novel multi-series fusion mechanism. To the best of our knowledge, this is the first time proposing the idea in the change point detection task, ensuring we can adaptively extract features from multivariate time series.
- In the temporal attention stage, in order to prevent error accumulation in long-series detection, we use the Bi-LSTM decoder to better capture the long-term temporal dependencies of the time series and improve accuracy.
- In the change point selection module, we propose a novel and simple algorithm. By setting the number of consecutive abnormal points in the series, we can identify change points and outliers, so that reduces the computation complexity and improves interpretability.

To demonstrate the effectiveness of Trident, we conducted experiments on two public datasets in different domains. Extensive experimental results show that our approach outperforms current state-of-the-art models on the two real-world datasets.

The remainder of the paper is organized as follows. We introduce the overview and the details of Trident in Sect. 2. Then we present the evaluation results and analyze the performance in Sect. 3. Lastly, we conclude our paper and sketch directions for the possible future work in Sect. 4.

2 Trident Design

The overall workflow of our approach is presented in Fig. 1. As we introduced in Sect. 1, Trident consists of three major modules.

We first predict the target series. Then, based on the deviation between the actual value and the predicted value, we determine the threshold and detect abnormal points. Lastly, we identify change points and outliers. The theories and details of the three core modules will be introduced below.

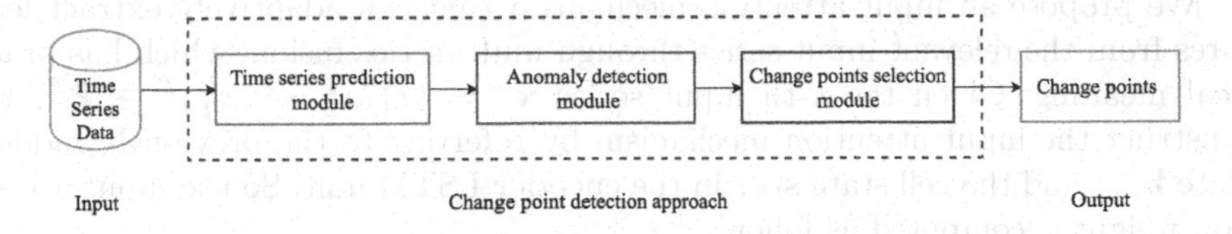

Fig. 1. The overall workflow of Trident

2.1 Time Series Prediction Module

In this module, we propose a novel dual-level attention-based approach for time series prediction. In the encoder, we employ a novel input attention mechanism and propose a multi-series fusion mechanism, which can adaptively learn the relevant input series and achieve feature extraction. In the decoder, a temporal attention mechanism is used to automatically capture the long-term temporal dependencies of the time series. The description of the proposed model is shown in Fig. 2.

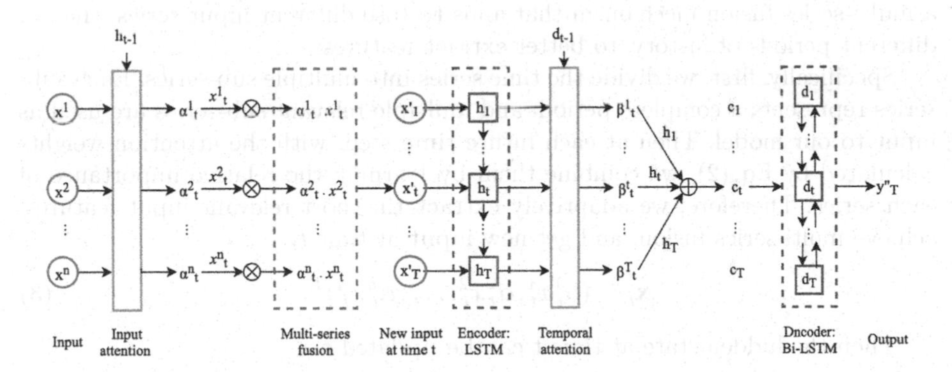

Fig. 2. The architecture of time series prediction module

Encoder with Input Attention. In our model, the encoder part is an LSTM, aiming to better capture the long-term dependencies of time series.

Given the input sequence $\mathbf{X} = (\mathbf{x}_1, \mathbf{x}_2, \ldots, \mathbf{x}_T)$ with $\mathbf{x}_t \in \mathbb{R}^n$, where n is the number of input series, T is the length of window size or time steps. The encoder can be applied to learn a mapping from $\mathbf{x}_t$ to $\mathbf{h}_t$ (at time step t) with: $\mathbf{h}_t = f_1(\mathbf{h}_{t-1}; \mathbf{x}_t)$, where $h_t \in \mathbb{R}^p$ is the hidden state of the encoder at time t, p is the size of the hidden state, x_t is the input of each time step, and f_1 is a non-linear activation function. Since we use an LSTM unit as f_1, so we can summarize the update as follows: $\mathbf{h}_t = LSTM(\mathbf{h}_{t-1}; \mathbf{x}_t)$.

We propose an input attention mechanism that can adaptively extract features from the relevant input series through multi-series fusion, which has practical meaning. Given the k-th input series $\mathbf{x}^k = (x_1^k, x_2^k, \ldots, x_T^k)^T \in \mathbb{R}^T$, we construct the input attention mechanism by referring to the previously hidden state $\mathbf{h}_{t-1}$ and the cell state $\mathbf{s}_{t-1}$ in the encoder LSTM unit. So the input attention weight is computed as follow:

$$e_t^k = \boldsymbol{\omega}_e^T tanh(\mathbf{W}_e[\mathbf{h}_{t-1}; \mathbf{s}_{t-1}] + \mathbf{U}_e\mathbf{x}^k + \mathbf{b}_e) \tag{1}$$

and

$$\alpha_t^k = \frac{\exp(e_t^k)}{\sum_{i=1}^n \exp(e_t^i)} \tag{2}$$

where $\mathbf{W}_e \in \mathbb{R}^{T \times 2p}$ and $\mathbf{U}_e \in \mathbb{R}^{T \times T}$ are matrices, $\boldsymbol{\omega}_e \in \mathbb{R}^T$ and $\mathbf{b}_e$ are vectors. They are parameters to learn.

The attention weight measures the importance of the k-th input feature (driving series) at time t. A softmax function is applied to α_t^k to ensure all the attention weights sum to 1.

Multi-series Fusion Mechanism. Based on the attention weights, we propose a multi-series fusion mechanism that aims to fuse different input series, that is, different periods of history, to better extract features.

Specifically, first, we divide the time series into multiple sub-series. Each sub-series represents a complete period, and multiple relevant sub-series are used as input to our model. Then at each future time step, with the attention weights calculated by Eq. (2), we combine them by learning the relative importance of each series. Therefore, we adaptively extract the most relevant input features, achieve multi-series fusion, and get new input at time t.

$$\mathbf{x}_t^{'} = (\alpha_t^1 x_t^1, \alpha_t^2 x_t^2, \ldots, \alpha_t^n x_t^n)^T \tag{3}$$

Then the hidden state at time t can be updated as:

$$\mathbf{h}_t = f_1(\mathbf{h}_{t-1}; \mathbf{x}_t^{'}) \tag{4}$$

where f_1 is an LSTM unit with $\mathbf{x}_t$ replaced by the newly computed $\mathbf{x}_t^{'}$.

With the proposed input attention mechanism and multi-series fusion mechanism, we can selectively focus on certain driving series rather than treating all the input driving series equally.

Decoder with Temporal Attention. Following the encoder with input attention, a temporal attention mechanism is used in the decoder to adaptively select relevant encoder hidden states across all time steps. Specifically, the attention weight of each encoder hidden state at time t is calculated based upon the previous decoder hidden state $\mathbf{d}_{t-1} \in \mathbb{R}^p$ and the cell state of the LSTM unit $\mathbf{s}_{t-1} \in \mathbb{R}^p$ with:

$$v_t^i = \boldsymbol{\omega}_v^T tanh(\mathbf{W}_v[\mathbf{d}_{t-1}; \mathbf{s}_{t-1}] + \mathbf{U}_v\mathbf{h}_i + \mathbf{b}_v) \tag{5}$$

and

$$\beta_t^i = \frac{\exp(v_t^i)}{\sum_{j=1}^{T} \exp(e_t^j)} \tag{6}$$

where $\mathbf{W}_v \in \mathbb{R}^{p\times 2q}$ and $\mathbf{U}_v \in \mathbb{R}^{p\times p}$ are matrices, $\boldsymbol{\omega}_v \in \mathbb{R}^p$ and $\mathbf{b}_v$ are vectors. They are parameters to learn.

Since each encoder hidden state $\mathbf{h}_i$ is mapped to a temporal component of the input, the attention mechanism computes the context vector $\mathbf{c}_t$ as a weighted sum of all the encoder hidden states $\mathbf{h}_1, \mathbf{h}_2, \ldots, \mathbf{h}_T$:

$$\mathbf{c}_t = \sum_{i=1}^{T} \beta_t^i \mathbf{h}_i \tag{7}$$

After getting the weighted summed context vectors, we can combine them with the target series:

$$\tilde{y}_t = \tilde{\boldsymbol{\omega}}^T[y_t; \mathbf{c}_t] + \tilde{\mathbf{b}} \tag{8}$$

The newly computed $\tilde{y}_t$ can be used for the update of the decoder hidden state at time t.

In our approach, a bi-directional LSTM (Bi-LSTM) is used as the decoder backbone. Because the traditional unidirectional LSTM will ignore the dynamic future information, which could have a strong influence on the time series forecast in practice. Therefore, the Bi-LSTM decoder aims to prevent error accumulation and improve accuracy for long-horizon forecasting.

It is composed of two LSTMs that allow both backward (past) and forward (future) dynamic inputs to be observed at each future time step. Then the hidden states of Bi-LSTM are fed into a fully-connected layer to produce final predictions. Formally, we define the formulations as follows:

$$\mathbf{d}_t = BiLSTM(\tilde{y}_t; \mathbf{d}_{t-1}, \mathbf{d}_{t+1}) \tag{9}$$

After updating the hidden state of the decoder, our model produces the final prediction result, denoted as $\hat{y}_t$.

Moreover, in the training procedure, we use the minibatch stochastic gradient descent (SGD) together with the adaptive moment estimation (Adam) optimizer to optimize parameters. We implemented our approach in the TensorFlow framework.

2.2 Anomaly Detection Module

After the final prediction result $\hat{y}_t$ is produced by the time series prediction module, the predicted values of the target series are delivered to the anomaly detection module. We calculate the deviation between the actual and the predicted value as $l_t = y_t - \hat{y}_t$, where y_t is the actual value, and $\hat{y}_t$ is the predicted value.

The absolute value of l_t is used as the anomaly score, denoted as e_t. The larger the anomaly score, the more significant the anomaly at the given time step. Therefore, we need to define a threshold based on the target series for anomaly classification.

In our approach, we adopt the Gaussian distribution-based method to determine the threshold. Extensive results in [13] show that anomaly scores fit the Gaussian distribution very well in a range of datasets. This is due to the fact that the most general distribution for fitting values derived from Gaussian or non-Gaussian variables is the Gaussian distribution according to the central limit theorem. Motivated by this, we define the following parameters based on Gaussian distribution:

$$(e_1, e_2, \ldots, e_n) \sim \boldsymbol{N}(\mu; \sigma^2) \tag{10}$$

$$\mu = \frac{1}{n}\sum_{t=1}^{n} e_t; \sigma = \sqrt{\frac{1}{n}\sum_{t=1}^{n}(e_t - \mu)^2} \tag{11}$$

Based on the Pauta criterion, we determine the range of the threshold according to the proportion of normal points in the dataset. Then we determine e as the threshold when the performance is optimal.

2.3 Change Points Selection Module

After the anomaly detection module classifies the data in the target series, this module will select the change points from the anomaly points. We propose a novel and simple approach to distinguish change points and outliers.

We have discussed the discrepancy between change points and outliers and the current research status in Sect. 1. In our approach, we use a simple strategy. We use the series containing the abnormal mark as input to this module. Once an abnormal point occurs, we declare it as an outlier. If a certain number of outliers are continuously found, the first point of this outlier series is declared as a change point. Thus, this method could reduce the computation complexity and improve interpretability. We define the parameter N (we set $N = 3$, which can achieve the best performance), which represents the minimum number of consecutive outliers.

3 Experiment and Evaluation

Based on the above approach, we designed the following experiment scheme. In this section, we conducted experiments on the three modules of the proposed method. Extensive results on two large real-world datasets show the effectiveness and superiority of Trident.

3.1 Experimental Environments

All the experiments were executed on a single computer with the following specifications: 3.1 GHz Intel Core i5 CPU, 16 GB 2133 MHz LPDDR3.

The software details are shown below: Model implementation: TensorFlow 1.13.2, Keras 2.1.0; Operating System: macOS Catalina 10.15.3.

3.2 Datasets Introduction

To demonstrate the effectiveness of Trident, we conducted experiments on two public datasets in different domains: The load Forecasting dataset and the Air-Quality dataset.

- **Load Forecasting Dataset:**[1]
  The dataset is hourly load data collected from utilities in 20 zones of the United States. The time period is from 2004 to 2008. It contains 33,000 instances with 29 attributes. This dataset was used in the Global Energy Forecasting Competition held in 2012 (GEFCom 2012) [7].
  We chose 4 areas of similar magnitude to conduct experiments. We took the power load in 2007 as the target series and the power load in 2004–2006 as the three relevant input series.
- **Air-Quality Dataset:**[2]
  The second dataset is the Air-Quality dataset, which can be found in the UCI Machine Learning Repository. It includes hourly data of 6 main air pollutants and 6 relevant meteorological variables at 12 air-quality monitoring sites in Beijing. This dataset covers the period from March 1st, 2013 to February 28th, 2017, and contains 420,768 instances with 18 attributes.
  We chose the ambient temperature as the prediction object. We used the 2016 data as the target series and the 2013–2015 data as the three relevant input series.
  The data in the dataset contains negative numbers and zeros, where zero will cause the metric MAPE to be unable to calculate. Therefore, we converted the representation of temperature in the dataset from degree Celsius to thermodynamic temperature: $T(K) = t(°C) + 273.15$, where T is thermodynamic temperature, and t is degree Celsius.

3.3 Results 1: Prediction Performance

In this section, we demonstrated the predicting performance of the time series prediction module of Trident, and proved that our approach can solve the three challenges we proposed well.

[1] https://www.kaggle.com/c/global-energy-forecasting-competition-2012-load-forecasting/data.

[2] http://archive.ics.uci.edu/ml/datasets/Beijing+Multi-Site+Air-Quality+Data.

Competing Methods and Evaluation Metrics. To demonstrate the effectiveness of Trident in predicting performance, we compared it against four baseline methods: Bi-LSTM [18], attention mechanism (Attn) [3], BiLSTM-Attn (the setting of Trident that does not employ the input attention mechanism), and DA-RNN (a dual-stage attention-based recurrent neural network) [15].

Moreover, to measure the effectiveness of the time series prediction module, we considered the three evaluation metrics: root mean squared error (RMSE), mean absolute error (MAE), and mean absolute percentage error (MAPE).

Results and Analysis. Firstly, we used the Load Forecasting dataset to verify the effectiveness of our approach for the proposed challenges.

For each experiment about the challenge, we set an experimental group and control groups. The experimental group uses 4 zones of data for 4 years (contains 3 input series and 1 target series). For the periodicity experiment, the control group C1 has no periodicity, predicting the data of 1 zone through the data of 3 zones. Similarly, for the long series experiment, the 3 control groups are denoted as C21–C23. Each sub-dataset use one-zone, two-zone, and three-zone data, respectively (with different length of the series in the sub-dataset). And for multi-series experiments, the control groups are denoted as C31 and C32, using two-year (1 input series) and three-year (2 input series) data, respectively. Other parameter settings are the same. The experimental results are shown in Fig. 3.

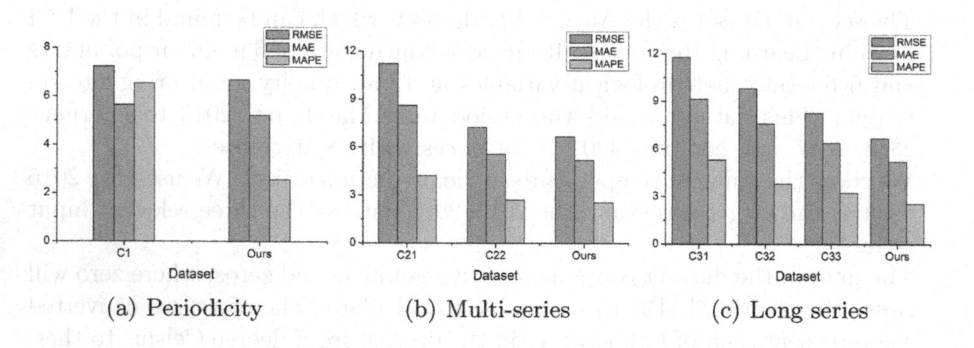

(a) Periodicity (b) Multi-series (c) Long series

Fig. 3. The comparative experiment results for the three proposed challenges.

Figure 3 shows that, with the increase of series length and the number of input series, or the improvement of data's periodicity, the three metrics have improved significantly. Trident shows good performance in solving each proposed challenge. Therefore, our approach has obvious advantages for the problems of long, periodic, and multiple input series.

Next, to measure the predicting performance of Trident, we conducted extensive experiments on Trident and all baseline methods on two real-world datasets. Table 1 summarizes the results.

Table 1. The predicting performance of Trident and all baseline methods.

Models	Load forecasting dataset			Air-quality dataset		
	RMSE	MAE	MAPE (%)	RMSE	MAE	MAPE (%)
Bi-LSTM	23.894	18.853	9.100	8.663	7.904	2.877
Attn	16.678	13.290	6.333	8.394	7.672	2.791
BiLSTM-Attn	14.432	10.806	5.336	8.286	7.561	2.746
DA-RNN	11.007	8.712	4.188	4.092	3.411	1.241
Ours	**6.665**	**5.196**	**2.547**	**3.239**	**2.615**	**0.950**

Table 1 shows that our approach outperforms the other baseline models markedly on the two datasets. The RMSE, MAE, and MAPE values improved by 40% and 22% on average on the two datasets, respectively. These experimental results demonstrated the superiority of Trident compared with the state-of-the-art methods.

For further comparison, we showed the prediction results of Trident on the two datasets in Fig. 4. The blue, orange, and green line represents actual value, training part, and test part respectively. We can observe that our approach performs well in prediction performance.

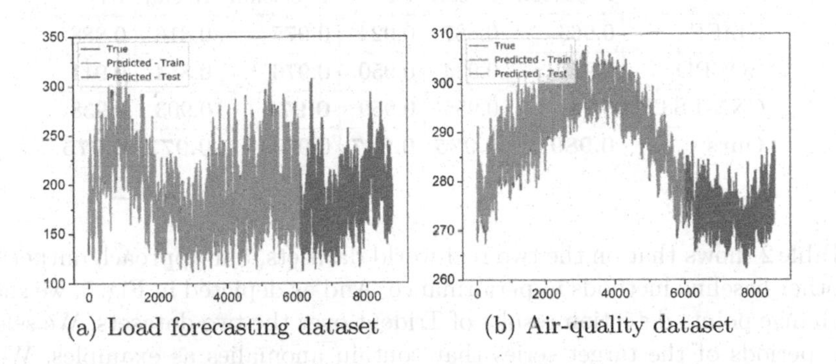

(a) Load forecasting dataset (b) Air-quality dataset

Fig. 4. The final prediction results of Trident over the two datasets.

This is because Trident integrated the dual-level attention mechanism and multi-series fusion mechanism, which can adaptively select relevant input series and extract features. Meanwhile, it employs a Bi-LSTM decoder to solve the problem of error accumulation in long-term series, thereby greatly improves the predicting accuracy.

Overall, extensive experiments show the effectiveness and superiority of Trident. It can comprehensively solve the three proposed challenges and shows good predictive performance.

3.4 Results 2: Change Point Detection Performance

In this section, we proved the effectiveness of the anomaly detection module and the change points selection module. We demonstrated the superiority of Trident by comparing it with the other three state-of-the-art approaches.

Competing Methods and Evaluation Metrics. We compared Trident against three baseline methods: CNN-LSTM (classification-based method) [12], Bayesian online change point detection (BOCPD) (probability method) [1], and KLIEP (an online density-ratio estimation algorithm) [11].

Moreover, we evaluated the efficiency of the change point detection module based on the following metrics: Precision, Recall, and F1 Score.

Results and Analysis. To measure the change point detecting performance of Trident, we conducted extensive experiments on Trident and other baseline methods on the two real-world datasets. Table 2 and Fig. 5 summarize the results.

Table 2. The change point detecting performance of Trident and all baseline methods.

Models	Load forecasting dataset			Air-quality dataset		
	Precision	Recall	F1	Precision	Recall	F1
KLIEP	0.963	0.887	0.924	0.975	0.816	0.888
BOCPD	0.967	0.934	0.950	0.976	0.854	0.911
CNN-LSTM	0.964	0.955	0.959	**0.977**	0.903	0.938
Ours	**0.989**	**0.985**	**0.987**	0.973	**0.977**	**0.975**

Table 2 shows that on the two real-world datasets, our approach outperforms the other baseline methods in performance. And as depicted in Fig. 5, we showed the change points detection results of Trident over the two datasets. We selected some periods of the target series that contain anomalies as examples. We can easily observe that performance of Trident in detecting change points is well. Trident can detect change points accurately and timely on the two datasets, and can identify change points and outliers.

In conclusion, Trident is a prediction-based change point detection approach. It employs the dual-level attention learning for time series prediction. And we used the Gaussian distribution-based method for the anomaly detection tasks, and combine the change point selection module. For the periodic long-term series and multiple relevant input series, our approach can detect change points accurately, and identify change points and outliers. Extensive experimental results prove the superiority of Trident compared with the state-of-the-art methods.

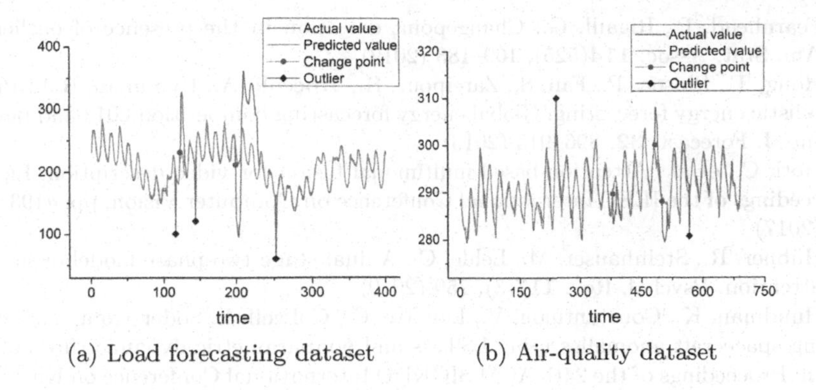

(a) Load forecasting dataset (b) Air-quality dataset

Fig. 5. The change point detection results of Trident on real-world datasets.

4 Conclusion

In this paper, we propose Trident, a novel change point detection approach in time series via dual-level attention learning. It consists of three key modules: time series prediction module, anomaly detection module, and change point selection module. In the time series prediction module, we use a dual-level attention learning model and integrate the multi-series fusion mechanism. It can adaptively extract features of input series. In the anomaly detection and change point selection module, we determine the threshold employing the Gaussian distribution-based method and identify change points and outliers. We verified the effectiveness of Trident on two public real-world datasets. Extensive experimental results show that our approach outperforms the state-of-the-art methods. In future work, we will further extend our approach to handle multivariate time series from different data sources. Due to the heterogeneity of data sources together with limited information about their interactions, exploring how to learn the complex dynamic correlations deserves our in-depth study.

References

1. Adams, R.P., MacKay, D.J.: Bayesian online changepoint detection. Stat. Optim. Inf. Comput. **1050**, 19 (2007)
2. Aminikhanghahi, S., Cook, D.J.: A survey of methods for time series change point detection. Knowl. Inf. Syst. **51**(2), 339–367 (2017)
3. Bahdanau, D., Cho, K., Bengio, Y.: Neural machine translation by jointly learning to align and translate. In: 3rd International Conference on Learning Representations, ICLR 2015 (2015)
4. Canizo, M., Triguero, I., Conde, A., Onieva, E.: Multi-head CNN-RNN for multi-time series anomaly detection: an industrial case study. Neurocomputing **363**, 246–260 (2019)
5. Fan, C., et al.: Multi-horizon time series forecasting with temporal attention learning. In: Proceedings of the 25th ACM SIGKDD International Conference on Knowledge Discovery & Data Mining, pp. 2527–2535 (2019)

6. Fearnhead, P., Rigaill, G.: Changepoint detection in the presence of outliers. J. Am. Stat. Assoc. **114**(525), 169–183 (2019)
7. Hong, T., Pinson, P., Fan, S., Zareipour, H., Troccoli, A., Hyndman, R.J.: Probabilistic energy forecasting: Global energy forecasting competition 2014 and beyond. Int. J. Forecast. **32**, 896–913 (2016)
8. Hori, C., et al.: Attention-based multimodal fusion for video description. In: Proceedings of the IEEE International Conference on Computer Vision, pp. 4193–4202 (2017)
9. Hübner, R., Steinhauser, M., Lehle, C.: A dual-stage two-phase model of selective attention. Psychol. Rev. **117**(3), 759 (2010)
10. Hundman, K., Constantinou, V., Laporte, C., Colwell, I., Soderstrom, T.: Detecting spacecraft anomalies using LSTMs and nonparametric dynamic thresholding. In: Proceedings of the 24th ACM SIGKDD International Conference on Knowledge Discovery & Data Mining, pp. 387–395 (2018)
11. Kawahara, Y., Sugiyama, M.: Sequential change-point detection based on direct density-ratio estimation. Stat. Anal. Data Min. **5**(2), 114–127 (2012)
12. Kim, T.Y., Cho, S.B.: Web traffic anomaly detection using C-LSTM neural networks. Exp. Syst. Appl. **106**, 66–76 (2018)
13. Kriegel, H.P., Kroger, P., Schubert, E., Zimek, A.: Interpreting and unifying outlier scores. In: Proceedings of the 2011 SIAM International Conference on Data Mining, pp. 13–24. SIAM (2011)
14. Liang, Y., Ke, S., Zhang, J., Yi, X., Zheng, Y.: GeoMAN: multi-level attention networks for geo-sensory time series prediction. In: IJCAI, pp. 3428–3434 (2018)
15. Qin, Y., Song, D., Cheng, H., Cheng, W., Jiang, G., Cottrell, G.W.: A dual-stage attention-based recurrent neural network for time series prediction. In: Proceedings of the 26th International Joint Conference on Artificial Intelligence, pp. 2627–2633 (2017)
16. Sadouk, L.: CNN approaches for time series classification. In: Time Series Analysis-Data, Methods, and Applications. IntechOpen (2018)
17. Salinas, D., Flunkert, V., Gasthaus, J., Januschowski, T.: DeepAR: probabilistic forecasting with autoregressive recurrent networks. Int. J. Forecast. **36**, 1181–1191 (2019)
18. Schuster, M., Paliwal, K.K.: Bidirectional recurrent neural networks. IEEE Trans. Sig. Process. **45**(11), 2673–2681 (1997)
19. Sezer, O.B., Gudelek, M.U., Ozbayoglu, A.M.: Financial time series forecasting with deep learning: A systematic literature review: 2005–2019. Appl. Soft Comput. **90**, (2020)
20. Su, Y., Zhao, Y., Niu, C., Liu, R., Sun, W., Pei, D.: Robust anomaly detection for multivariate time series through stochastic recurrent neural network. In: Proceedings of the 25th ACM SIGKDD International Conference on Knowledge Discovery & Data Mining, pp. 2828–2837 (2019)
21. Xu, H., et al.: Unsupervised anomaly detection via variational auto-encoder for seasonal KPIs in web applications. In: Proceedings of the 2018 World Wide Web Conference, pp. 187–196 (2018)
22. Zhang, A.Y., Lu, M., Kong, D., Yang, J.: Bayesian time series forecasting with change point and anomaly detection (2018)
23. Zhang, A., Paisley, J.: Deep Bayesian nonparametric tracking. In: International Conference on Machine Learning, pp. 5833–5841 (2018)

Facial Representation Extraction by Mutual Information Maximization and Correlation Minimization

Xiaobo Wang[1], Wenyun Sun[2], and Zhong Jin[1](✉)

[1] School of Computer Science and Engineering, Nanjing University of Science and Technology, Nanjing, China
{xiaobowang,zhongjin}@njust.edu.cn

[2] School of Artificial Intelligence, Nanjing University of Information Science and Technology, Nanjing, China
wenyunsun@nuist.edu.cn

Abstract. Facial expression recognition and face recognition are two amusing and practical research orientations in computer vision. Multi-task joint learning can improve each other's performance, which has rarely been studied in the past. This work proposes a joint learning framework to enhance emotional representation and identity representation extraction by incorporating a multi-loss training strategy. Specifically, we propose mutual information loss to ensure that the facial representation is unique and complete and offer correlation loss to extract identity representation using orthogonality constraints. Classification loss is used to learn emotional representation. As a result, we can obtain an unsupervised learning framework to reduce the identity annotation bottleneck using large-scale labelled emotional data for the face verification task. Our algorithm is verified on an artificially synthesized face database: Large-scale Synthesized Facial Expression Dataset (LSFED) and its variants. The identity representation obtained by the algorithm is used for face verification. The performance is comparable to some existing supervised face verification methods.

Keywords: Facial representation extraction · Mutual information · Correlation constraint

1 Introduction

In daily human communication, the information transmitted by face has reached 55% of the total information, plays an essential role in human-computer interaction (HCI), affective computing, and human behaviour analysis. Identity and emotion form the main components in the face domain. To extract discriminative representations, hand-crafted feature operators (i.e., histograms of oriented gradients (HOG), local binary pattern (LBP), and Gabor wavelet coefficients) are used in previous work.

N. T. Nguyen et al. (Eds.): ACIIDS 2021, LNAI 12672, pp. 811–823, 2021.
https://doi.org/10.1007/978-3-030-73280-6_64

However, in recent years, deep learning-based methods [3,15,18,21] are becoming more and more popular and have achieved high recognition accuracy beyond the traditional learning methods. Among them, few jobs consider both identity representation and emotional representation. Li et al. [10] proposed self-constrained multi-task learning combined with spatial fusion to learn expression representations and identity-related information jointly. The novel identity-enhanced network (IDEnNet) can maximally discriminate identity information from expressions. But the network is limited by the identity annotation bottleneck. Yang et al. [24] proposed a cGAN to generate the corresponding neural face image for any input face image. The neural face generated here can be regarded as an implicit identity representation. The emotional information is filtered out and stored in the intermediate layers of the generative model. They use this residual part for facial expression recognition. Sun et al. [17] proposed a pair of Convolutional-Deconvolutional neural networks to learn identity representation and emotional representation. The neutral face is used as the connection point of the two sub-networks, supervising the previous network to extract expression features and input to the latter network to extract identity features.

However, a significant drawback now is that these algorithms require identity supervision labels, among which neutral faces can be regarded as implicit identity labels. It is too strict for facial expression training data. To alleviate such shortcomings, we propose an unsupervised facial orthogonal representation extraction framework. On the premise that only emotional faces and emotion labels are provided, the emotional representation and the identity representation are evaluated using the linear irrelevance of facial attributes. The contributions of this paper are as follows:

- A lightweight convolutional neural network is proposed to extract the identity representation and the emotional representation simultaneously.
- A multi-loss training strategy is proposed, which is a weighted summation of the mutual information loss, the classification loss, and the correlation loss. The mutual information loss measures the relevance between input faces and the deep neural network's output representation. The classification loss is the cross-entropy function commonly used in facial expression recognition tasks. To make up for the lack of identity supervised information, correlation loss is utilized to constrain the linear uncorrelation between the identity representation and the emotional representation.
- The proposed algorithm for facial expression recognition and face verification has achieved outstanding performance on an artificially synthesized face database: Large-scale Synthesized Facial Expression Dataset (LSFED) [17] and its variants [16]. The performance is close to some supervised learning methods.

The rest of this article is organized as follows. Section 2 reviews related work. In Sect. 3, the main methods are proposed. The experiments and results are shown in Sect. 4. Section 5 gives the conclusion.

2 Related Work

2.1 Mutual Information Learning

Representation extraction is a vital and fundamental task in unsupervised learning. The methods based on the INFOMAX optimization principle [4,11] estimate and maximize the mutual information for unsupervised representation learning. They argue that the basic principle of a good representation should be complete and to be able to distinguish the sample from the entire database, that is, to extract the unique information of the sample, for which they introduce mutual information to measure for the first time.

Although mutual information is crucial in data science, mutual information has historically been difficult to calculate, especially for high-dimensional spaces. Mutual Information Neural Estimator (MINE) [2] presents that the estimation of mutual information between high dimensional continuous random variables can be achieved by gradient descent over neural networks. Mutual Information Gradient Estimator (MIGE) [22] argues that directly estimating MI gradient is more appealing for representation learning than estimating MI in itself. The experiments based on Deep INFOMAX (DIM) [4] and Information Bottleneck [1] achieve significant performance improvement in learning proper representation. Some recent works maximize the mutual information between images for zero-shot learning and image retrieval [6,19]. Generally speaking, the existing mutual information-based methods are mainly used to measure the correlation between two random variables, and they are mostly applied to the unsupervised learning of representations.

2.2 Orthogonal Facial Representation Learning

In the absence of identity labels, the algorithm proposed by Sun et al. [17] can obtain relatively clustered identity representation on the facial expression database. In the first half of the network, the neutral face and expression labels are taken as the learning objectives. Then, through the second pair of convolution deconvolution network, the neutral face and expression features are input to reconstruct the original emotional face.

However, in some tasks, the neutral face that belongs to the same person as the original emotional face is difficult to obtain. Excessive training data requirements have become the main disadvantage of the method [17]. Sun et al. [16] put forward an unsupervised orthogonal facial representation learning algorithm. Based on the assumption that there are only two variations in the face space. It should be noted that the emotional representation is invariant to identity change, and the identity representation is invariant to emotion change. To alleviate the dependence on the neutral face, they replace the supervision information with a correlation minimization loss to achieve a similar effect.

Although [16] solves too high database requirements to a certain extent, and the experimental performance on clean databases is also excellent. The reconstruction loss is too strict for facial representation extraction, and much task-independent information is compressed into the middle layer vector. Besides, the

Convolutional-Deconvolutional network will cause excessive expenses. We propose a similar unsupervised facial representation extraction framework to solve these problems, which only uses a lightweight convolutional neural network.

3 Proposed Method

3.1 Deep Neural Network Structure

First, we propose a learning framework consisting of a backbone network and a discriminant network. As shown in Fig. 1, an emotional face is fed into a self-designed VGG-like backbone network to extract identity representation and emotional representation. The network is stacked by 9 basic blocks, and each basic block contains a convolutional layer, a Batch Normalization layer, and an activation layer. The convolutional part consists of nine 3×3 convolutional layers and six pooling layers, and there is no fully connected layer. The convolutional part is formalized as f_θ, where θ represents the trainable parameters of the network. The forward propagation process of the network can be expressed as:

$$(d, l) = f_\theta(x) \tag{1}$$

where x represents an emotional face, d and l represent the corresponding identity representation and emotional representation, respectively. Compared with many complex and deep networks proposed in recent years, our network is simple and sufficient to meet facial expression recognition and face verification tasks. The network's input is 64×64, and the final output is a 519-dimensional global feature vector, where the 512-dimensional vector is identity representation, and the 7-dimensional vector is emotional representation. The specific configuration of the backbone network is shown in Table 1. The discriminant network is designed to estimate mutual information, which will be described in detail in the next section. The three grey squares in Fig. 1 represent three losses, namely mutual information loss L_{mi}, classification loss L_{cls}, and correlation loss L_{corr}.

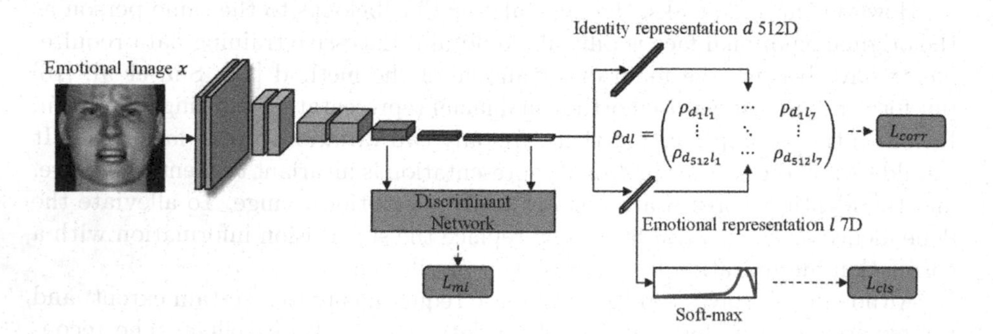

Fig. 1. The overall architecture of the proposed method

Table 1. Structure of the baseline network

Neural network layer	Feature map size	Number of parameters
Input layer	$64 \times 64 \times 1$	0
Convolutional layer	$64 \times 64 \times 16$	192
Convolutional layer	$64 \times 64 \times 16$	2352
Avg pooling	$32 \times 32 \times 16$	0
Convolutional layer	$32 \times 32 \times 32$	4704
Convolutional layer	$32 \times 32 \times 32$	9312
Avg pooling	$16 \times 16 \times 32$	0
Convolutional layer	$16 \times 16 \times 64$	18624
Convolutional layer	$16 \times 16 \times 64$	37056
Avg pooling	$8 \times 8 \times 64$	0
Convolutional layer	$8 \times 8 \times 128$	74112
Avg pooling	$4 \times 4 \times 128$	0
Convolutional layer	$4 \times 4 \times 256$	295680
Avg pooling	$2 \times 2 \times 256$	0
Convolutional layer	$2 \times 2 \times (512+7)$	1197333
Avg pooling	$1 \times 1 \times (512+7)$	0

A Batch Normalization layer (BN) and a Tanh activation function exist after each convolutional layer. The network uses BN technology to accelerate training and obtain centralized features, facilitating subsequent correlation loss calculations.

3.2 Mutual Information Loss

Previous work has shown that reconstruction is not a necessary condition for adequate representation. The basic principle of a good representation should be complete and to be able to distinguish the sample from the entire database, that is, to extract the unique information of the sample. We use mutual information to measure the correlation of two variables and maximize the correlation measure to restrict that the extracted information is unique to the sample. The overall idea is derived from Deep INFOMAX [4]. X represents the collection of emotional faces, Z represents the collection of encoding vectors and $p(z \mid x)$ represents the distribution of the encoding vectors generated by x, where $x \in X$and $z \in Z$. Then the correlation between X and Z is expressed by mutual information as:

$$I(X, Z) = \iint p(z \mid x)p(x) \log \frac{p(z \mid x)}{p(z)} \mathrm{d}x\mathrm{d}z \tag{2}$$

$$p(z) = \int p(z \mid x)p(x)\mathrm{d}x \tag{3}$$

A useful feature encoding should make mutual information as large as possible:

$$p(z \mid x) = \underset{p(z|x)}{\text{argmax}}\, I(X, Z) \tag{4}$$

The larger the mutual information means that the $\log \frac{p(z|x)}{p(z)}$ should be as large as possible, which means that $p(z \mid x)$ should be much larger than $p(z)$, that is, for each x, the encoder can find the z that is exclusive to x, so that $p(z \mid x)$ is much greater than the random probability $p(z)$. In this way, we can distinguish the original sample from the database only by z.

Mutual Information Estimation. Given the fundamental limitations of MI estimation, recent work has focused on deriving lower bounds on MI [20,23]. The main idea of them is to maximize this lower bound to estimate MI. The definition of mutual information is slightly changed:

$$\begin{aligned} I(X, Z) &= \iint p(z \mid x)p(x) \log \frac{p(z \mid x)p(x)}{p(z)p(x)} \mathrm{d}x\mathrm{d}z \\ &= KL(p(z \mid x)p(x) \| p(z)p(x)) \end{aligned} \tag{5}$$

To obtain complete and unique facial representation (i.e., identity representation and emotion representation), we maximize the distance between the joint distribution and the marginal distribution to maximize mutual information proposed in Eq. (5). We use JS divergence to measure the difference between the two distributions. According to the local variational inference of f divergence [13], the mutual information of the JS divergence version can be written as:

$$\begin{aligned} JS(p(z \mid x)p(x), p(z)p(x)) = \max_T \big(&E_{(x,z)\sim p(z|x)p(x)}[\log \sigma(T(x, z))] \\ &+E_{(x,z)\sim p(z)p(x)}[\log(1 - \sigma(T(x, z)))]\big) \end{aligned} \tag{6}$$

where T is a discriminant network, and σ is the sigmoid function. Refer to the negative sampling estimation in word2vec [9,12,14], x and its corresponding z are regarded as a positive sample pair (i.e., sampled from joint distribution), and x and randomly drawn z are regarded as negative samples (i.e., sampled from marginal distribution). As illustrated in Fig. 2. The discriminant network is trained to score sample pairs so that the score for positive samples is as high as possible, and the score for negative samples is as low as possible. Generally speaking, the right side of Eq. (6) can be regarded as the negative binary cross-entropy loss. For fixed backbone networks, mutual information is estimated (see Eq. (6)). Further, to train the discriminant network and the backbone network at the same time to evaluate and maximize the mutual information, respectively, Eq. (4) is replaced by the following objective:

$$\begin{aligned} p(z \mid x), T(x, z) = \underset{p(z|x),T(x,z)}{\text{argmax}} \big(&E_{(x,z)\sim p(z|x)p(x)}[\log \sigma(T(x, z))] \\ &+E_{(x,z)\sim p(z)p(x)}[\log(1 - \sigma(T(x, z)))]\big) \end{aligned} \tag{7}$$

where $p(z \mid x)$ is the backbone network proposed in Sect. 3.1, $T(x, z)$ is the discriminant network.

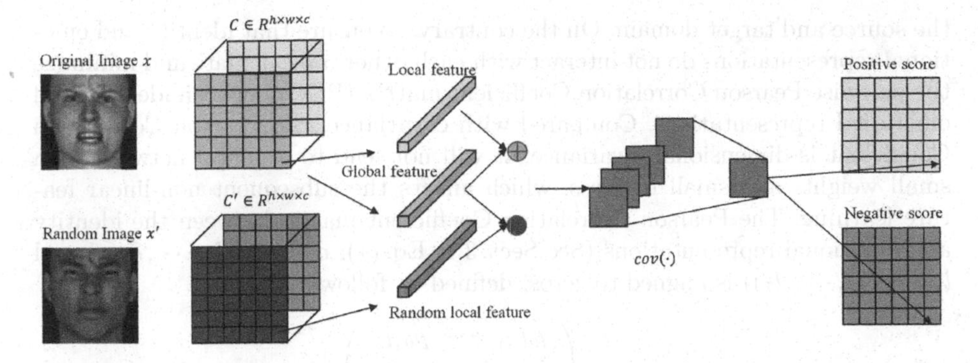

Fig. 2. The forward propagation process of the discriminant network. A random image is selected in a batch, $C \in R^{h \times w \times c}$ is the middle layer feature map. Cov $(\cdot)$ is a 2-layered 1×1 convolutional neural network and $\oplus$ indicates concatenate operation. The global feature is estimated from the original image. Local features and random local features are extracted from the same spatial position.

Mutual Information in a Neural Network. In a neural network, we can compute mutual information between arbitrary intermediate features. Therefore we figure another format of the mutual information in a neural network: $I(f_{\theta_1}(X), f_{\theta_2}(X))$, where f_{θ_1} and f_{θ_2} correspond to activations in different/same layers of the same convolutional network. When f_{θ_1} indicates the input layer and f_{θ_2} represents the top layer of the convolutional network, we call it global mutual information (GMI) because it considers the correlation between the entire faces X and its corresponding global representations Z. However, due to the original face's high dimensionality, it is challenging to directly calculate the mutual information between the network input and output features. And for face verification and facial expression recognition tasks, the correlation of face is more reflected in the local features. Therefore, it is necessary to consider local mutual information (LMI). Let $C \in R^{h \times w \times c}$ denotes the intermediate layer feature map, the mutual information loss is expressed by local mutual information as:

$$L_{mi} = I(C, Z) = \frac{1}{hw} \sum_{i,j} I(C_{i,j}, Z) \tag{8}$$

where $1 \ll i \ll h$ and $1 \ll j \ll w$. The mutual information between the vector of each spatial position of the feature map and the final global feature vector is calculated. Then the arithmetic mean of them, regarded as the local mutual information, is applied in the representation learning.

3.3 Correlation Loss

The second-order statistics of features has an excellent performance in face tasks and domain adaptation problems. The covariance alignment increases the correlation between the source and target domain by aligning the data distribution of the source and target domain. On the contrary, to ensure that identity and emotional representations do not interact with each other, we calculate and minimize the pairwise Pearson Correlation Coefficient matrix (PCC) between identity and emotional representations. Compared with covariance, the Pearson Correlation Coefficient is dimensional invariance. It will not lead to a neural network with small weights and small features, which affects the subsequent non-linear feature learning. The Pearson Correlation Coefficient matrix between the identity and emotional representations (See Sect. 3.1, Eq. (1), $d = (d_1, d_2, \cdots, d_{512})$ and $l = (l_1, l_2, \cdots, l_7)$) is aligned to zeros, defined as follows:

$$\rho_{dl} = \begin{pmatrix} \rho_{d_1 l_1} & \cdots & \rho_{d_1 l_7} \\ \vdots & \ddots & \vdots \\ \rho_{d_{512} l_1} & \cdots & \rho_{d_{512} l_7} \end{pmatrix} \tag{9}$$

The Pearson Correlation Coefficients of two random variables d_i and l_j are defined as follows:

$$\rho_{d_i l_j} = \frac{\mathrm{Cov}(d_i, l_j)}{\sqrt{\mathrm{Var}(d_\mathrm{i}) \mathrm{Var}(l_j)}} = \frac{E[(d_\mathrm{i} - E(d_\mathrm{i}))(l_j - E(l_j))]}{\sigma(d_\mathrm{i}) \sigma(l_j)} \tag{10}$$

where $\mathrm{Cov}(\cdot)$, $\mathrm{Var}(\cdot)$, $\mathrm{E}(\cdot)$, $\sigma(\cdot)$ are functions of covariance, variance, expectation, and standard deviation, respectively. The Eq. (10) shows that the PCC can also be regarded as a normalized covariance, and it varies from -1 to $+1$. -1 means a complete negative correlation, $+1$ means an absolute positive correlation, and zero indicates no correlation. Based on the PCC's properties, we define the correlation loss as follows:

$$L_{\mathrm{corr}} = \sum_{i,j} \left(\rho_{d_i l_j}\right)^2 \tag{11}$$

In a neural network, $\mathrm{E}(\cdot)$ is always estimated in a mini-batch. Here we propose a fairly simple method to obtain centralized features without additional computation. As shown in Fig. 1, we use the features after Batch Normalization and before the bias addition as the centralized identity representation, and the ground truth emotion label y in the form of C-dimensional one-hot code as the emotional representation. So $y - \frac{1}{C}$ is used as centralized emotional representations. $\mathrm{E}(\cdot)$ and $\sigma(\cdot)$ in Eq. (10) can be eliminated to achieve efficient and accurate forward/reverse calculation.

Finally, we use the cross-entropy function to define the expression classification loss L_{cls}:

$$L_{cls} = -\frac{1}{n} \sum_{i=1}^{n} \sum_{j=1}^{m} y_{i,j} \log \frac{e^{l_{i,j}}}{\sum_{j'=1}^{m} e^{l_{i,j'}}} \tag{12}$$

where n is the mini-batch size, m is the number of expression categories, y is the ground truth label, and l is the predicted probabilities in logarithmic space.

The total loss consists of mutual information loss, correlation loss, and classification loss:

$$L_{\mathrm{total}} = -\alpha L_{mi} + \beta L_{\mathrm{corr}} + L_{cls} \tag{13}$$

Among them, the non-negative α and β balance the importance of the three losses. We will discuss these hyperparameters in detail below.

4 Experiments

4.1 Databases and Preprocessing

To verify the superiority of our proposed algorithm, we used multiple facial expression databases to conduct experiments, including the LSFED, the LSFED-G, the LSFED-GS, and the LSFED-GSB. The generation of the LSFED are based on FaceGen modeller software [5], which strictly follows the definitions of FACS and EMFACS. The LSFED has 105000 aligned facial images. [16] proposed three variants. G represents Gaussian noise with Signal to Noise Ratio (SNR) = 20 dB. S represents random similarity transform. B represents random background patches from the CIFAR-10 database and CIFAR-100 database [7]. The samples in the LSFED, the LSFED-G, the LSFED-GS, and the LSFED-GSB are illustrated in Fig. 3. The four databases are roughly divided into training sets and testing sets with a ratio of 8:2.

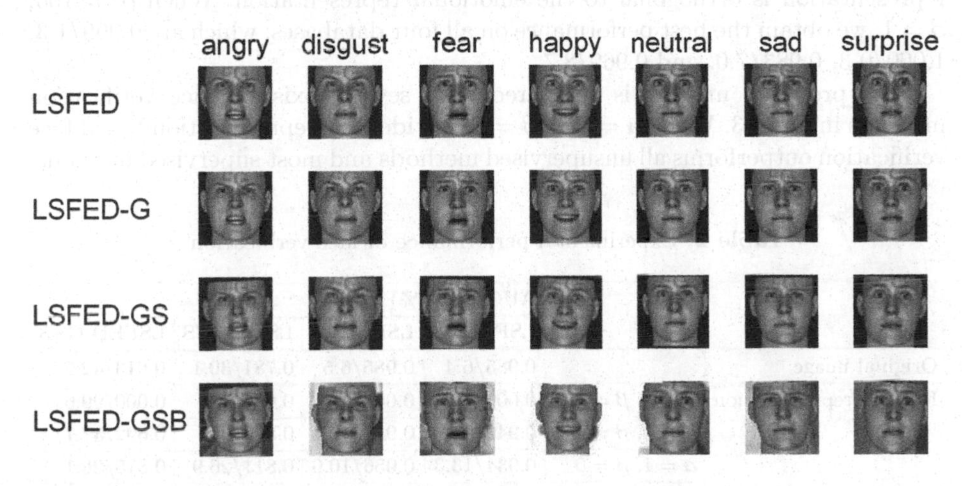

Fig. 3. The samples in the LSFED, the LSFED-G, the LSFED-GS, and the LSFED-GSB

In the training process, we use ADAM optimizer to minimize the total loss in a mini-batch. The experiments were carried out on GeForce GTX 1060. The learning rate is 0.001, and the momentum is 0.8. The whole training process stopped after 100 epochs.

4.2 Experiments Based on the Identity Representation

The learned identity representation is evaluated on a face verification task. We randomly choose 1000 positive pairs (same identity, different expressions) and 1000 negative pairs (different identities, same expression), compute the Euclidean distances between the learned identity representations of these pairs, use the median of the distances as the threshold for face verification. Areas under receiver operating characteristic curves (AUC) and Equal Error Rates (EER) are listed in Table 2 as indicators for evaluating the quality of face verification tasks.

When $\alpha = 0$, $\beta = 0$, there is no external loss to constrain the extracted facial representation except for expression classification loss. The learned identity representation cannot be used for face verification, and none of the 2000 pairs of faces selected randomly is correct. When $\alpha = 1$, $\beta = 0$, we use mutual information loss and facial expression classification loss to learn the identity representation, the face verification performance is better than the original image X on the LSFED-GS and the LSFED-GSB. It verifies the effectiveness of mutual information in extracting face unique information (i.e., emotional information and identity information), as described in Sect. 3.2. When $\alpha = 0$, $\beta = 1$, the identity representation is learned based on the assumption that the identity representation is orthogonal to the emotional representation. When $\alpha = 100$, $\beta = 1$, we obtain the best performance on all four databases, which are 0.999/1.3, 1.000/0.3, 0.983/7.0, and 0.969/8.7.

The proposed method is compared with several existing face verification methods in Table 3. When $\alpha = 100$, $\beta = 1$, the identity representation based face verification outperforms all unsupervised methods and most supervised methods.

Table 2. Experimental performance of face verification

		AUC/EER(%)			
		LSFED	LSFED-G	LSFED-GS	LSFED-GSB
Original image		0.985/6.4	0.985/6.5	0.781/30.1	0.613/42.2
Identity representation	$\alpha = 0$, $\beta = 0$	0.000/100	0.000/100	0.000/100	0.000/99.6
	$\alpha = 0$, $\beta = 1$	0.949/11.3	0.951/10.3	0.826/25	0.692/37.7
	$\alpha = 1$, $\beta = 0$	0.934/13.3	0.956/10.6	0.811/26.9	0.815/26.1
	$\alpha = 1$, $\beta = 1$	0.947/12.8	0.968/10.8	0.790/30.5	0;909/15.8
	$\alpha = 10$, $\beta = 1$	0.996/3.3	0.997/2.6	0.887/20.7	0.779/32.2
	$\alpha = 100$, $\beta = 1$	0.999/1.3	1.000/0.3	0.983/7.0	0.969/8.7

4.3 Experiments Based on the Emotional Representation

We directly use facial expression recognition (FER) accuracy to evaluate the learned emotional representation. When $\alpha = 0$, $\beta = 1$, mutual information loss is suppressed, we can guarantee that emotional representation and identity representation are orthogonal, and this constraint has side effects on facial

Table 3. Comparison with existing face verification algorithms

	Methods	AUC			
		LSFED	LSFED-G	LSFED-GS	LSFED-GSB
Unsupervised	Sun et al.'s unsupervised method [16]	1.000	1.000	0.920	0.768
	Proposed method	0.999	1.000	0.983	0.969
Supervised	Original image	0.985	0.985	0.781	0.613
	2-layered Neural Network	1.000	0.998	0.978	0.970
	AlexNet feature +2-layered NN	0.994	0.991	0.968	0.932
	Sun et al.'s supervised method [17]	1.000	1.000	0.999	0.998

expression recognition, especially the LSFED-GSB database, which decreased from 100% to 92.7%. When $\alpha \geq 1$ and $\beta = 1$, as the ratio of α to β increases (i.e., $\alpha/\beta = 1, 10, 100$), the facial expression accuracy of the LSFED and its variations also increases, which indicates that mutual information loss can assist in extracting more compact expression representation.

Compared with several existing methods, when $\alpha = 100$, $\beta = 1$, the accuracy outperforms the Nearest Neighbor Classifier, the PCA + LDA, and AlexNet [8]. It is comparable to Sun et al.'s methods [16,17] on four facial expression databases (Table 4).

Table 4. Experimental performance of facial expression recognition

Methods	Accuracy (%)			
	LSFED	LSFED-G	LSFED-GS	LSFED-GSB
Sun et al.'s unsupervised method [16]	100.0	100.0	99.9	99.1
Proposed method, $\alpha = 0$, $\beta = 0$	100.0	100.0	100.0	100.0
Proposed method, $\alpha = 0$, $\beta = 1$	99.4	99.3	97.9	92.7
Proposed method, $\alpha = 1$, $\beta = 0$	100.0	100.0	100.0	100.0
Proposed method, $\alpha = 1$, $\beta = 1$	99.7	99.7	98.4	96.5
Proposed method, $\alpha = 10$, $\beta = 1$	99.9	99.9	99.5	98.5
Proposed method, $\alpha = 100$, $\beta = 1$	99.9	99.9	99.5	99.4
Nearest Neighbor Classifier	95.6	94.2	70.8	61.9
PCA + LDA	97.8	99.4	93.1	89.8
Linear SVM	99.8	99.7	85.3	81.9
AlexNet [8]	91.9	97.4	96.3	98.1
Sun et al.'s supervised method [17]	100.0	100.0	99.8	98.7

5 Conclusion

In this paper, we present a novel approach for facial representation extraction (i.e., identity representation and emotional representation), which is based on a lightweight convolutional neural network and a multi-loss training strategy. First, based on the design idea of the VGG network, a lightweight convolutional neural network with only about 1.6 million parameters is proposed. Second, three losses are proposed to train the network. The mutual information loss is proposed to make sure that the facial representation is unique and complete, and the correlation loss is proposed to leverage orthogonality constraint for identity and emotional representation extraction. The classification loss is used to learn emotional representation. The learning procedure can capture the expressive component and identity component of facial images at the same time. Our proposed method is evaluated on four large scale artificially synthesized face databases. Without exploiting any identity labels, the identity representation extracted by our method is better than some existing unsupervised/supervised methods in the performance of face verification.

Acknowledgments. This work is partially supported by National Natural Science Foundation of China under Grant Nos 61872188, U1713208

References

1. Alemi, A.A., Fischer, I., Dillon, J.V., Murphy, K.: Deep variational information bottleneck. arXiv preprint arXiv:1612.00410 (2016)
2. Belghazi, M.I., et al.: MINE: mutual information neural estimation. arXiv preprint arXiv:1801.04062 (2018)
3. Cai, J., Meng, Z., Khan, A.S., Li, Z., O'Reilly, J., Tong, Y.: Island loss for learning discriminative features in facial expression recognition. In: 2018 13th IEEE International Conference on Automatic Face & Gesture Recognition, FG 2018, pp. 302–309. IEEE (2018)
4. Hjelm, R.D., et al.: Learning deep representations by mutual information estimation and maximization. arXiv preprint arXiv:1808.06670 (2018)
5. Singular Inversions: FaceGen Modeller (version 3.3) [computer software]. Singular Inversions, Toronto, ON (2008)
6. Kemertas, M., Pishdad, L., Derpanis, K.G., Fazly, A.: RankMI: a mutual information maximizing ranking loss. In: Proceedings of the IEEE/CVF Conference on Computer Vision and Pattern Recognition. pp. 14362–14371 (2020)
7. Krizhevsky, A., Hinton, G., et al.: Learning multiple layers of features from tiny images (2009)
8. Krizhevsky, A., Sutskever, I., Hinton, G.E.: ImageNet classification with deep convolutional neural networks. Commun. ACM **60**(6), 84–90 (2017)
9. Le, Q., Mikolov, T.: Distributed representations of sentences and documents. In: International Conference on Machine Learning, pp. 1188–1196 (2014)
10. Li, Y., et al.: Identity-enhanced network for facial expression recognition. In: Jawahar, C.V., Li, H., Mori, G., Schindler, K. (eds.) ACCV 2018. LNCS, vol. 11364, pp. 534–550. Springer, Cham (2019). https://doi.org/10.1007/978-3-030-20870-7_33
11. Linsker, R.: Self-organization in a perceptual network. Computer **21**(3), 105–117 (1988)
12. Mikolov, T., Chen, K., Corrado, G., Dean, J.: Efficient estimation of word representations in vector space. arXiv preprint arXiv:1301.3781 (2013)
13. Nowozin, S., Cseke, B., Tomioka, R.: f-GAN: training generative neural samplers using variational divergence minimization. In: Advances in Neural Information Processing Systems, pp. 271–279 (2016)
14. Rong, X.: word2vec parameter learning explained. arXiv preprint arXiv:1411.2738 (2014)

15. Shi, Y., Yu, X., Sohn, K., Chandraker, M., Jain, A.K.: Towards universal representation learning for deep face recognition. In: Proceedings of the IEEE/CVF Conference on Computer Vision and Pattern Recognition, pp. 6817–6826 (2020)
16. Sun, W., Song, Y., Jin, Z., Zhao, H., Chen, C.: Unsupervised orthogonal facial representation extraction via image reconstruction with correlation minimization. Neurocomputing **337**, 203–217 (2019)
17. Sun, W., Zhao, H., Jin, Z.: A complementary facial representation extracting method based on deep learning. Neurocomputing **306**, 246–259 (2018)
18. Sun, W., Zhao, H., Jin, Z.: A visual attention based ROI detection method for facial expression recognition. Neurocomputing **296**, 12–22 (2018)
19. Tang, C., Yang, X., Lv, J., He, Z.: Zero-shot learning by mutual information estimation and maximization. Knowl. Based Syst. **194**, 105490 (2020)
20. Tschannen, M., Djolonga, J., Rubenstein, P.K., Gelly, S., Lucic, M.: On mutual information maximization for representation learning. arXiv preprint arXiv:1907.13625 (2019)
21. Wang, K., Peng, X., Yang, J., Lu, S., Qiao, Y.: Suppressing uncertainties for large-scale facial expression recognition. In: Proceedings of the IEEE/CVF Conference on Computer Vision and Pattern Recognition, pp. 6897–6906 (2020)
22. Wen, L., Zhou, Y., He, L., Zhou, M., Xu, Z.: Mutual information gradient estimation for representation learning. arXiv preprint arXiv:2005.01123 (2020)
23. Xu, C., Dai, Y., Lin, R., Wang, S.: Deep clustering by maximizing mutual information in variational auto-encoder. Knowl. Based Syst. **205**, 106260 (2020)
24. Yang, H., Ciftci, U., Yin, L.: Facial expression recognition by de-expression residue learning. In: Proceedings of the IEEE Conference on Computer Vision and Pattern Recognition, pp. 2168–2177 (2018)

Localization System for Wheeled Vehicles Operating in Underground Mine Based on Inertial Data and Spatial Intersection Points of Mining Excavations

Artur Skoczylas(✉) and Paweł Stefaniak

KGHM Cuprum Research and Development Centre Ltd., gen. W. Sikorskiego 2-8, 53-659 Wroclaw, Poland
{askoczylas,pkstefaniak}@cuprum.wroc.pl

Abstract. An attitude and heading reference systems (AHRS) are widely used in tracking the fleet of wheeled vehicles as well as power tools supporting transport planning. Another functionality is post factum performance assessment for operations performed cyclically or in a certain sequence. In the literature, you can find a number of solutions of this type dedicated to wheeled transport, airplanes, agriculture, robotics, sports, film industry. Most of them require access to GPS and the question is how to localize in the underground condition where GPS access is unavailable? Of course, this is not a new problem. Many scientific works deal with inertial navigation dedicated to tunnels, underground mining excavations, sewers, and other similar applications. Most of them concern gyro drift and clearly underline the need to use the magnetometer signal to correct the estimated motion path with a Kalman filter or a complementary filter etc. The authors made many attempts to test the proposed solutions in the underground mine. Unfortunately, many factors translated into magnetic field disturbances, which had poor readings. The authors propose a low-cost solution based on the correction of the vehicle motion path estimated by integrating the gyro signal based on a digital map and the existing topology rules of the mine's road infrastructure. The article presents the methodology, along with the method of generalizing the digital map as well as an application of procedures on industrial data.

Keywords: Mining vehicles · Automation · Inertial sensors · Underground mining · Localization techniques

1 Introduction

In recent years an increasing interest in the monitoring of operational parameters of mining machines, particularly in the monitoring of load haul dump (LHD) machines [1, 2]. The data about operational parameters may get additional context by adding data about the positioning of the vehicles in the mine. The registration of machine positions allows relating the current state of the machine (for example overheating or operational overloads) with the particular mining region and infrastructure or environmental conditions

N. T. Nguyen et al. (Eds.): ACIIDS 2021, LNAI 12672, pp. 824–834, 2021.
https://doi.org/10.1007/978-3-030-73280-6_65

in this region. Also, accurate position monitoring of the vehicles significantly improves the safety of operations and can be used for example in anti-collision systems [3] or during the rescue operations in the mine in case of emergency [4]. Finally, position data can be used for the effectiveness assessment of the machines or operators. Unfortunately, the GPS signal, which is the most common method of tracking various objects is not available in the underground mine. It forces researchers to search for alternatives, one of which is inertial navigation. Inertial navigation systems for underground mining have been considered since the mid-1990s [5, 6]. In most cases, such systems were developed for autonomous navigation of LHD vehicles [7–9]. Also, some additional sensors, like laser sensors [10] or RFID anchoring [11] are usually used, which makes the system more expensive and harder to install. Advanced data analytics and the Internet of things (IoT) technologies nowadays penetrate different areas of industry [12]. Underground mining is not an exception. IoT technologies were used for example for ground monitoring [13], safety applications, or personnel tracking. Although the last one is also based on inertial sensors, the results cannot be transferred directly from personnel to mining equipment, because the algorithms based on step count estimation are often used, which means that dedicated algorithms for underground vehicles should be developed.

As it was shown in our previous article [13], the gyroscope drift can be a sufficient obstacle. The trajectories estimated for a shorter time period, for example for a single hauling cycle were in good agreement with the map and with themselves, but in a longer time perspective, significant deviations are observed. In general, the impact of gyroscope drift can be reduced by using the Kalman filter [14]. But usually for navigation problems in this case signal from the magnetometer is used, which is not reliable in the case of the underground mine.

In this article, we propose a method of correction of the trajectory, obtained using the inertial sensors and onboard machine monitoring system by integrating this data with the spatial map of the mine. We use the fact that the mining corridors in the majority of modern underground mines have so-called “Manhattan” structures – the paths are straight and mostly intersect at a 90° angle. Moreover, the machine can turn only on the intersections, so when a turn is observed, the probability that the machine is at a given intersection can be calculated and proper corrections into the machine trajectory can be done.

The structure of the paper is as follows: in Sect. 2 the input data and the methods of its collection are described, in Sect. 3 the preprocessing of data is discussed, then the procedure of path calculation is presented, the path is next integrated with the map of the mine, and finally main algorithm and the correcting one are presented.

2 Input Data Description and Pre-processing

The inertial navigation system uses signals from the IMU (Inertial Measurement Unit), which, depending on the version, measures: acceleration, angular velocity (6 DoF (Degrees of Freedom) IMU), and magnetic field (9 DoF IMU). These signals can be used for example to determine Euler angles and then, in conjunction with the information about machine speed or distance travelled, to determine the path that was travelled by the machine. In this article, we everywhere use a reference frame with a vertical Z-axis (so that the X-Y plane corresponds to an even terrain surface). In order to calculate the two-dimensional route, the yaw angle is used, which stands for rotation around Z-axis. In the vast majority of applications, this angle is determined in two different ways: by integrating the signal from the Z-axis of a gyroscope or using X and Y axes of the magnetometer. The yaw rotation angle determined from the gyroscope has greater accuracy in a short term perspective, but it drastically decreases over time due to the gyroscope drift phenomena. The yaw rotation angle determined by the means of a magnetometer and accelerometer is characterized by lower accuracy at the start, however, it stays at a relatively constant over time. Both of these methods are usually combined using a sensor fusion algorithm such as a Complementary filter or Kalman filter. Additionally, it is possible to attach to this fusion, measurements from other systems, such as GPS. Thanks to this application it is possible to create a system characterized by a fairly high level of accuracy (especially when using GPS).

However, the situation is different when we consider more difficult environments such as underground mines. In this situation, the GPS signal cannot be used. The use of a magnetometer is also very difficult due to the multiple sources of interference. However, by using the gyroscope itself, we put the system at risk of a significant error that after time will make subsequent measurements useless. To prevent this, readings should be combined with other information sources such as mine maps, for example.

Data available from the machine includes readings from the IMU 9 DoF sensor (NGIMU) and the SYNAPSA – the on-board measuring system that acquires the most important operational parameters of the vehicle (driving speed, engine speed, fuel consumption, etc.). In experiments, the IMU sensor was mounted on the articulated joint of the machine, and to protect against environmental threats it was additionally surrounded by a 1 cm thick stainless-steel housing. Although it prevents the sensor from the mechanical impact, it damps the magnetic field significantly. However, it should be highlighted that even without the housing, the magnetometer signal is useless in these conditions. That is why only signals from accelerometers and gyroscopes can be used, mainly we use a Z-axis gyroscope. Additionally, data about speed and currently selected gear from the SYNAPSA system was used. As a source of additional information not connected to the IMU or the machine itself, a mine map and additional key information such as the machine chamber location were used. In addition RFID tags technology can be used to support localization algorithms (some mines have already implemented such infrastructure in the most crucial regions).

Data from the gyroscope (Z-axis) are sampled with 50 Hz. The IMU itself calibrates the sensors and additionally, they are processed before saving using the AHRS fusion algorithms. Data from the SYNAPSA recorder include two variables: SPEED and SELGEAR. The SPEED variable contains information about the current machine speed stored as an integer. However, this variable is always positive, so it is impossible to get information about the direction of this speed. This information can be obtained using the SELGEAR variable describing how the machine's active gear and direction changed over time. If the SELGEAR variable is negative, then it means that the machine is going backward, thus the SPEED value is reversed. Despite the fact that the recorded SPEED variable is not burdened with larger errors, it is an integer and its decimal part is irretrievably lost in the registration process, which can be a source of some errors.

Data from the SYNAPSA recorder are sampled with a frequency of 1 Hz. Finally, the data from the machine recorder and IMU should be merged. The data from the IMU in described experiments is sampled with 50 Hz frequency, thus the data from SYNAPSA is interpolated using the forward filling method. The fragment of each used signal is shown on Fig. 1.

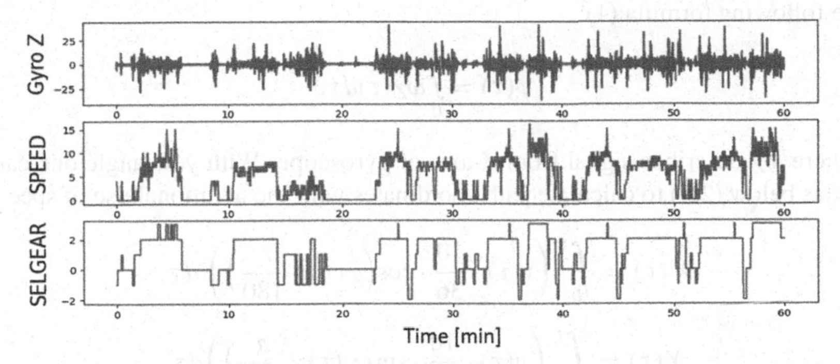

Fig. 1. Variables used for localization with the use of IMU sensor.

An additional source of information is the current map of the mining department where the experiments with the IMU sensor were carried out. The map has all possible paths marked in the mine, ventilation ducts (and therefore roads that are currently in use), and additional information, such as excavations or machine chambers. After being loaded, the map of the underground mine department must be converted into grayscale and then converted into a binary image using thresholding operation. Finally, some image compression is applied in order to reduce the image size (Fig. 2).

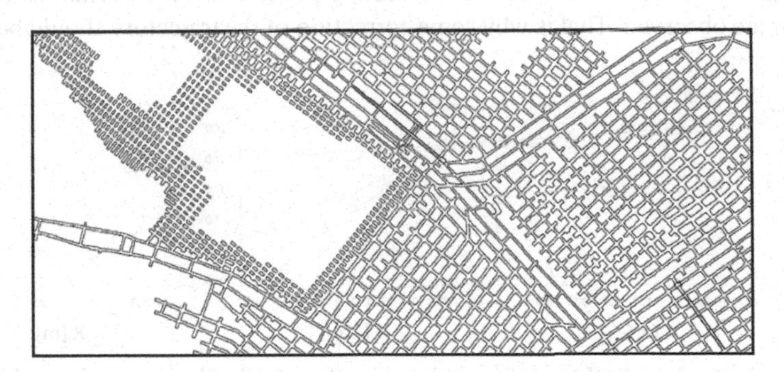

Fig. 2. The map of the mining department, where the experiments were performed.

3 Path Calculation and Its Integration with the Map

To calculate the trajectory distance the vehicle has travelled, first one needs to calculate the Yaw angle φ in any time τ given. This angle describes (looking at the vehicle from above) the turns of the machine (left, right). To calculate this value, the easiest way, it can be calculated is to integrate the gyroscope signal over time, what can be described by the following formula (1):

$$\varphi(\tau) = \int_{0}^{\tau} \omega_Z(\tau) d\tau, \tag{1}$$

where ω_Z describes signal from Z-axis of gyroscope. With yaw angle one can use formulas below (2,3) to calculate path coordinates with the additional use of speed v.

$$X(\tau) = \int_0^{\tau} \left(v(\tau) \cdot \frac{10}{36} \cdot \cos\left(\varphi(\tau) \cdot \frac{\pi}{180}\right)\right) d\tau, \tag{2}$$

$$Y(\tau) = \int_0^{\tau} \left(v(\tau) \cdot \frac{10}{36} \cdot \sin\left(\varphi(\tau) \cdot \frac{\pi}{180}\right)\right) d\tau. \tag{3}$$

Due to the discrete character of collected data, in this paper, the above integrals were calculated numerically using trapezoidal approximation.

Unfortunately, the gyroscope readings are affected by drift. Drift is a constantly changing error resulting from integrating two components: a low-frequency variable called bias (offset) and a high frequency one called random walk (integrated noise). This phenomenon can be seen on the road fragments shown in Fig. 3. In these sections, the machine traveled along a constant route from the excavation face to the grid, so the signal should be a straight line between two points with small exceptions for maneuvers and the machine should come back to the starting point, but some deviations from such behaviour are observed. That is why some correction of the trajectory should be applied.

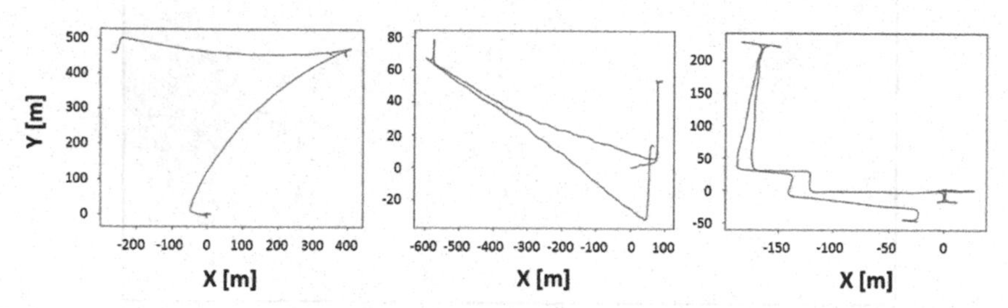

Fig. 3. The pass of the haul truck from the mining face to the damping point and back. The deviations of the trajectory caused by gyroscope drift are visible.

Such correction may be based on the fact, that underground mines have a structure of a network of connected tunnels. The machines can only move in the tunnels and cannot leave the road like a car. There is also the issue of closed or collapsed tunnels as well as a continuous change of excavations and thus also used roads. Finally, due to hard

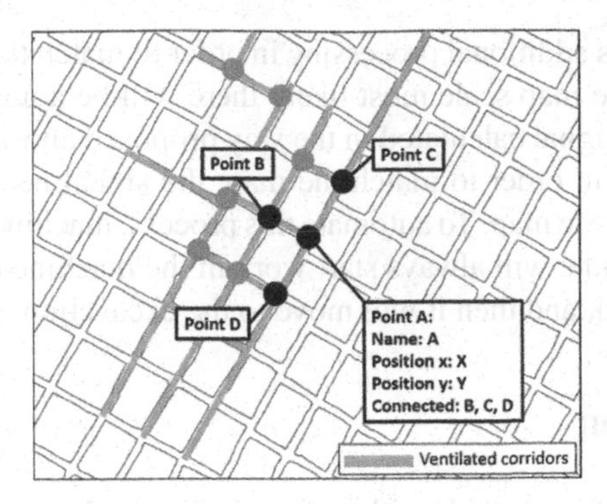

Fig. 4. A crossroad object example.

conditions in the underground mine, the machines use only ventilated tunnels, thus their possible routes are limited.

In order to connect the path signal with the map, we propose to look at the map as a series of connected points, where each point is one crossroad. This way one can define each of the intersections as an object from one class. This object proposed parameters are: name, position x, position y, and list with other objects connected to the current one. Such a situation is shown in Fig. 4. In this way knowing the initial starting point of the signal, all of the directions of motion will be reduced to only a few directions that are described by points connected to the analysed object. In most cases, modelling connection map will require manual creation of objects and can be hindered because many of the maps are stored on paper or in other forms unsuitable for that kind of needs. Nevertheless, such an effort is most likely one time. After the first creation, in case of any change in the mine road network, the change will most likely include only a few objects. New points can be added in a simple way by creating new instances of a class and adding that instance name to the connected object list parameter of connected intersections. Old points can be removed by doing a similar action.

Additionally around each crossroads, two decision boundaries were established. These boundaries define districts named *Proposed* and *Visited*. While the vehicle approaches specific crossroad, two situations shown in Fig. 5 may arise. Depending on the situation, that occurs, one of two algorithms of path estimation described in the next section will be used.

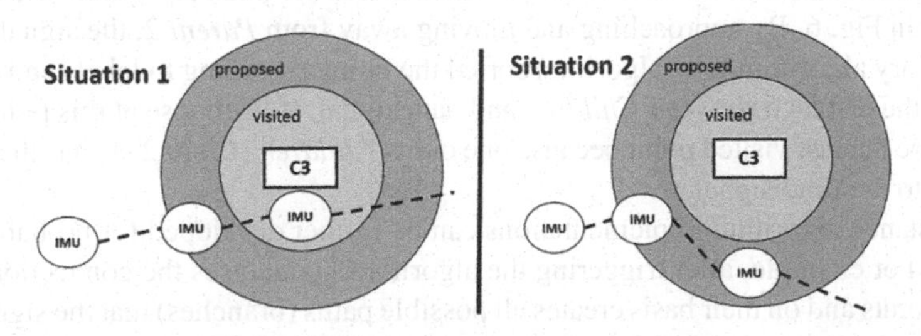

Fig. 5. Crossroad with decision boundaries.

Path signal requires additional processing in order to match the coordinates from the map. Depending on the map scale most likely there will be a need to scale path signal along both axes. Path signal calculated in the way proposed above will also always start around point (0,0), so in order to match the map, the signal needs to be moved to the starting coordinates on the map. To automate this process, machine chamber information can be used. The machine will always start work in the machine chamber, whose name and position are known, and then it will move to the excavation area.

4 Path Correction

As it was mentioned before, two algorithms for machine path correction were developed. First of all, switching between the algorithms should be considered. For computational reasons, the best method would be to operate a less complicated main algorithm, and if necessary, switch to a more advanced auxiliary algorithm. When the information about the current position of the vehicle is obtained, the auxiliary algorithm can be turned off and the whole process can again be based on the main algorithm.

When the vehicle leaves the starting point, the distances from the vehicle to all connected crossroads are being tracked. The distance is calculated from the coordinates of the machine, obtained by the use of Euclidean distance. Based on this distance we can tell if the machine crosses some of the decision borders, described in Fig. 5. If the machine first crosses the *proposed* border, and crosses the *visited* border, then the corresponding point is set as the current location of the vehicle, and the algorithm starts from the beginning.

But the second situation, described in Fig. 5, may occur: the machine can cross the *proposed* border, but then instead of crossing the *visited* border, it goes out of the *proposed* region. In this case, the auxiliary algorithm is used.

The operation of the auxiliary algorithm is based on a simple method of analysing the connection diagram and calculating the distance. The auxiliary algorithm analyses the connection grid and labels each of the points. The basic (starting) point is the *Last Visited* point. Each point connected to it (i.e. those points that were the objectives of the main algorithm) becomes a *Parent*. Then, from each *Parent*, a list of points, which is connected to it and have no assigned label yet, get the *Child* label. Then the algorithm from the position of *Children* tracks the signal, and the *Parent* whose *Child* approaches the signal the most is selected for the *Last Visited*. Then the algorithm returns to the point where it started and switches the process back to the main algorithm. This process is shown in Fig. 6. By approaching and moving away from *Parent* 2, the signal enabled the auxiliary algorithm. The algorithm sorted the point according to labels and began to measure the distance between *Children* and path signal. If to choose at this point *Parent* 2 would be the last visited point because one of his *Children* (*Child* 2–1) has the shortest distance to the road signal.

Assistance algorithm implementations can be further developed for probability calculation. For example, after triggering the algorithm, it analyses the connections of the nearest points and on their basis creates all possible paths (branches) that the signal could take. Then, based on the signal readings, the probability of taking is calculated for each of the possible branches, and points are selected on its basis.

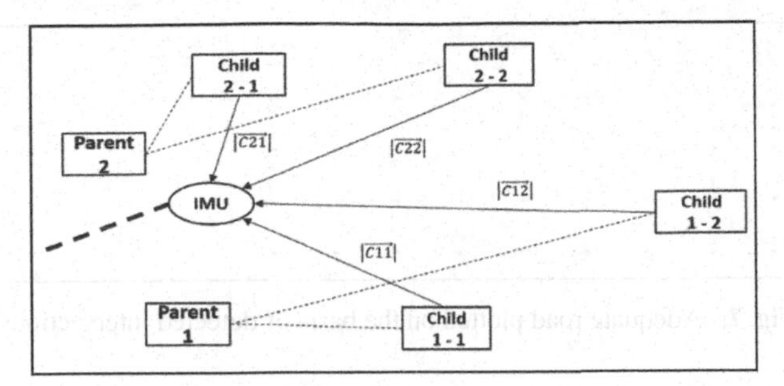

Fig. 6. Demonstration of the auxiliary algorithm.

5 Signal Correction

All the algorithms above describe a way to correct the trajectory of the machine, but not the gyroscope readings. Without it, even the best algorithms will have difficulty with detecting the path correctly when the gyroscope drift reaches a significant level. That is why based on this data, one can estimate the degree of gyroscope error at each visited point, which should reduce the drift enough for the correct path detection.

There are several ways to approach the problem of road correction. Simple curve snapping is one of them. When visiting a point, the road signal is analysed and its sample that is closest to the point just visited is selected. This sample serves as the signal intersection. The part that has already been analysed (from the beginning of the signal to the selected sample) remains unchanged. The non-analysed part, however, is snapped to the position of the point. More precisely, the entire signal is subjected to the same transformation that is created so that the first sample of the new signal moves to the position of the visited point.

Re-estimation of the Yaw angle is a completely different approach. Knowing the previously visited intersections as well as the initial location, it is possible to create a road from the beginning of the signal to the currently analysed point. From this, it is also possible to determine the angle of Yaw, which will not have bias and random walk errors. This angle can be used to repair the angle resulting from gyro integration. At each turn of the machine. Yaw angle expresses the sum of the angles by which the machine has rotated over time, this means that, for example, when the machine turns twice to the right, the Yaw angle will be 180 at the end. In order to repair the real signal, at the moment of turning, its angle Yaw should be reduced to that obtained by determining from the visited points. The path must then be recalculated using the integration method from the last visited point. The final result of the algorithm is presented in Fig. 7.

Correction of the road signal completes the whole algorithm. For clarity, it was also decided to present it in the form of a block diagram visible in Fig. 8. For the sake of simplicity, the presented diagram does not include the mapping of the road based on the passed intersections (which is shown in Fig. 7).

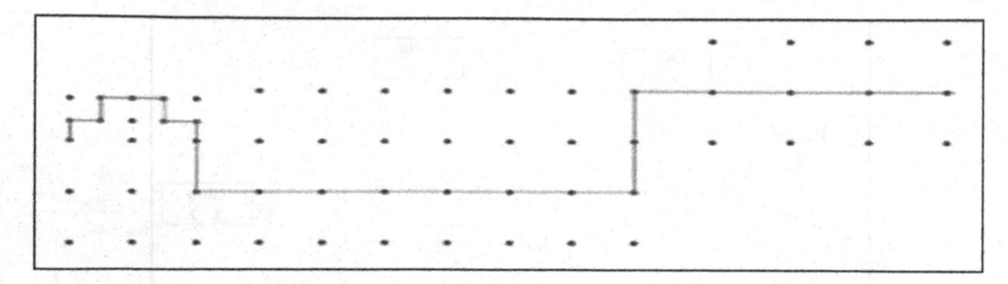

Fig. 7. Adequate road plotted on the basis of detected intersections.

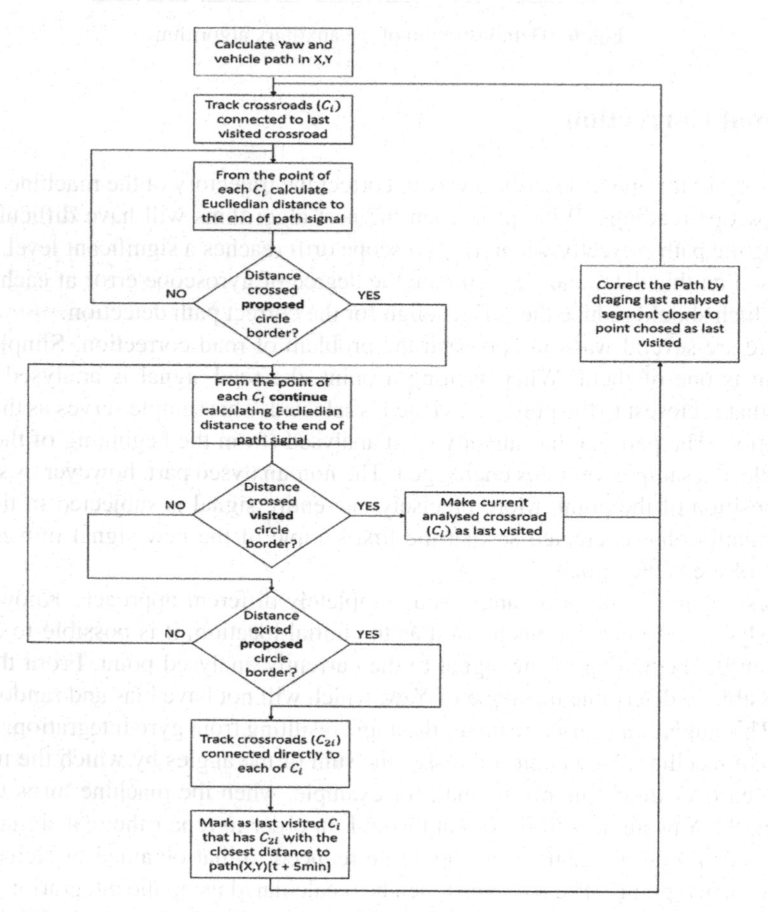

Fig. 8. Algorithm for localization of mining vehicles with the use of inertial sensors.

6 Summary

Tracking mining assets during the mining process is crucial for a number of reasons: machine location, material handling process assessment, and diagnosis, production optimization, anti-collision systems, as well as the development of inspection robots and autonomous vehicles. From the point of view of the underground mine, the available localization methods are difficult to implement, taking into account the local specific mine conditions, the problem of gyro drift, and the interference of the magnetometer signal. In this article, the authors proposed a simple method based mainly on the gyro signal and the topology of the road infrastructure in a mine with a room-and-pillar mining system. The article describes in detail the assumptions of the algorithm and all partial procedures. Finally, an example of the use of the algorithm on real data has been presented.

References

1. Koperska, W., Skoczylas, A., Stefaniak, P.: A simple method of the haulage cycles detection for LHD machine. In: Hernes, M., Wojtkiewicz, K., Szczerbicki, E. (eds.) Advances in Computational Collective Intelligence: 12th International Conference, ICCCI 2020, Da Nang, Vietnam, November 30 – December 3, 2020, Proceedings, pp. 326–337. Springer International Publishing, Cham (2020). https://doi.org/10.1007/978-3-030-63119-2_27
2. Wodecki, J., Stefaniak, P., Michalak, A., Wyłomańska, A., Zimroz, R.: Technical condition change detection using Anderson-Darling statistic approach for LHD machines–engine overheating problem. Int. J. Min. Reclam. Environ. **32**(6), 392–400 (2018)
3. Patnayak, S., Swain, A., Das, M.: Advance anti-collision device for vehicles using GPS and Zigbee. In: Chattopadhyay, J., Singh, R., Bhattacherjee, V. (eds.) Innovations in Soft Computing and Information Technology: Proceedings of ICEMIT 2017, Volume 3, pp. 117–123. Springer Singapore, Singapore (2019). https://doi.org/10.1007/978-981-13-3185-5_11
4. Zhang, K.F., Zhu, M., Wang, Y.J., Fu, E.J., Cartwright, W.: Underground mining intelligent response and rescue systems. Procedia Earth Planet. Sci. **1**(1), 1044–1053 (2009)
5. Scheding, S., Dissanayake, G., Nebot, E., Durrant-Whyte, H.: Slip modelling and aided inertial navigation of an LHD. In: Proceedings of International Conference on Robotics and Automation, vol. 3, pp. 1904–1909. IEEE (1997)
6. Scheding, S., Dissanayake, G., Nebot, E.M., Durrant-Whyte, H.: An experiment in autonomous navigation of an underground mining vehicle. IEEE Trans. Robot. Autom. **15**(1), 85–95 (1999)
7. Bakambu, J.N., Polotski, V.: Autonomous system for navigation and surveying in underground mines. J. Field Rob. **24**(10), 829–847 (2007)
8. Larsson, J., Broxvall, M., Saffiotti, A.: A navigation system for automated loaders in underground mines. In: Corke, P., Sukkariah, S. (eds.) Field and Service Robotics: Results of the 5th International Conference, pp. 129–140. Springer Berlin Heidelberg, Berlin, Heidelberg (2006). https://doi.org/10.1007/978-3-540-33453-8_12
9. Duff, E.S., Roberts, J.M., Corke, P.I.: Automation of an underground mining vehicle using reactive navigation and opportunistic localization. In: Proceedings 2003 IEEE/RSJ International Conference on Intelligent Robots and Systems (IROS 2003)(Cat. No. 03CH37453), vol. 4, pp. 3775–3780. IEEE (2003)
10. Larsson, J., Broxvall, M., Saffiotti, A.: Laser based intersection detection for reactive navigation in an underground mine. In: 2008 IEEE/RSJ International Conference on Intelligent Robots and Systems, pp. 2222–2227. IEEE (2008)
11. Xu, L.D., Xu, E.L., Li, L.: Industry 4.0: state of the art and future trends. Int. J. Prod. Res. **56**(8), 2941–2962 (2018)
12. Singh, A., Singh, U.K., Kumar, D.: IoT in mining for sensing, monitoring and prediction of underground mines roof support. In: 2018 4th International Conference on Recent Advances in Information Technology (RAIT), pp. 1–5. IEEE (2018)

13. Stefaniak, P., Gawelski, D., Anufriiev, S., Śliwiński, P.: Road-quality classification and motion tracking with inertial sensors in the deep underground mine. In: Sitek, P., Pietranik, M., Krótkiewicz, M., Srinilta, C. (eds.) Intelligent Information and Database Systems: 12th Asian Conference, ACIIDS 2020, Phuket, Thailand, March 23–26, 2020, Proceedings, pp. 168–178. Springer Singapore, Singapore (2020). https://doi.org/10.1007/978-981-15-3380-8_15
14. Lee, H.J., Jung, S.: Gyro sensor drift compensation by Kalman filter to control a mobile inverted pendulum robot system. In: 2009 IEEE International Conference on Industrial Technology, pp. 1–6. IEEE (2009

Innovations in Intelligent Systems

Evaluating Mutation Operator and Test Case Effectiveness by Means of Mutation Testing

Van-Nho Do[1], Quang-Vu Nguyen[2](✉), and Thanh-Binh Nguyen[2]

[1] Le Quy Don Gifted High School, Da Nang, Vietnam
[2] The University of Danang – Vietnam-Korea University of Information and Communication Technology, Da Nang, Vietnam
{nqvu,ntbinh}@vku.udn.vn

Abstract. Mutation analysis is one of the most popular testing techniques. However, this technique has some drawbacks in terms of time and resource consumption. Higher order mutation has been recently an object of interest by research community for overcoming such issues by reducing the number of equivalent mutants, generated mutants and simulating better realistic faults. In this paper, we focus on two problems for higher-order mutation. The first one is to evaluate the quality of mutation operators as well as generated mutants. The second one is to prioritize test cases by means of capability of killing mutants. We propose the evaluation of mutation operators basing on first-order mutants, and the quality operators are selected to create higher-order mutants. We assess also the quality of each test case and rank the test cases for testing. The results may help developers optimize their resource during planning testing activities.

Keywords: Mutation testing · Second-order mutant · Mutation operator evaluation · Test case prioritization

1 Introduction

Testing is one important activity for ensuring software quality. Amongst testing tasks, test case generation plays the crucial role because test cases should uncover faults in program under test. The objective of mutation testing is to evaluate the quality of test cases in terms of killing mutants, *i.e.* capability of revealing faults.

Mutation testing was first introduced by Lipton [1]. This field was later developed popularly by DeMillo *et al.* [2]. Mutation testing is considered as fault-based testing technique providing a test adequacy criterion. This criterion is utilized to evaluate the effectiveness of test set via its ability to uncover faults.

The principle of mutation testing is to micmic faults commited by programmers when writting programs. Each simple fault is injected into the original program to generate defective program, called *mutant*. For instance, a variable in program is replaced by another varibale of the same data type. This is named as *mutation operator*. Mutation operators can be designed basing on characteristics of programming language and programmers so that they represent programmers' mistakes in program. Each generated

N. T. Nguyen et al. (Eds.): ACIIDS 2021, LNAI 12672, pp. 837–850, 2021.
https://doi.org/10.1007/978-3-030-73280-6_66

mutant and the original program are run on a test set. The execution results are analysed. If the output of the mutant is different from the one of the original program, then the mutant is *killed* by that test set. If not, there may be two cases: the test set isn't good enough to kill the mutant; or the mutant and the original program always produce the same output on any test set. In the latter case, the mutant is said *equivalent*.

The main objective of mutation testing is to assess the quality of test set through a test adequacy criterion by defining a mutation score as follow: $MS = 100 * D/(N - E)$.

Where D is the number of killed mutants; N is the total of mutants; E is the number of equivalent mutants. The number of equivalent mutants is often difficult to be determined, a mutation score indicator (MSI) can be used: $MSI = 100 * D/N$, where D is the number of killed mutants; N is the total of mutants.

In 2009, Jia and Harman [3] introduced a new paradigm of mutation testing which is higher-order mutation. Tranditional mutation testing is considered as first-order mutation by inserting only one fault into original program, while higher-order mutants are generated by injecting multiple faults. According to [3, 4], higher-order mutation can address some fundamental drawbacks of first-order mutation, such as the reflection of real mistakes and the issue of equivalent mutant.

From the birth of higher-order mutation, many studies [5–17] have been introduced to improve its different aspects. In this paper, we concentrate on two problems regarding higher-order mutation testing: evaluating the quality of mutation operators as well as generated mutants and priotirizing test cases basing upon its capability of killing mutants. This may help developers allocate suitably their resources during testing phase.

The rest of the paper is organized as follows. Section 2 presents the related work. In Sect. 3, we introduce the research questions. The solutions are proposed in Sect. 4. Some experimental results are described and analysed in Sect. 5. Future work and conclusions are presented in Sect. 6.

2 Related Works

In this section, we focus on some major studies relating to higher-order mutation tesing. In [3, 4], traditional mutant, *i.e.* applying a single mutation operator, is called first-order mutant (FOM); in higher-order mutation testing, mutants are classified into two classes: second-order mutant (SOM) and higher-order mutant (HOM). SOMs are created by inserting two faults, while HOMs are generated by injecting more than two faults.

2.1 Second-Order Mutation Testing

In [5], Polo *et al.* proposed three algorithms to create SOMs: LastToFirst, DifferentOperators, and RandomMix. The result of this work showed that the number of SOMs was to 56% smaller compared with the number of FOMs, while the number of equivalent mutants was decreased to 74%.

Improving the algorithms in [5], Papadakis and Malevris [6] suggested five strategies for combining FOMs to generate SOMs: First2Last, SameNode, SameUnit, SU_F2Last and SU_DiffOp. Their results indicated that the number of generated SOMs reduced to half compared to the number of FOMs and the number of equivalent SOMs was reduced to 73%.

Kintis *et al.* [7] introduced two second order mutation testing strategies: the Relaxed Dominator strategy (RDomF) and the Strict Dominator strategy (SDomF). They proposed also the Hybrid Strategies (Hdom-20% and Hdom-50%), which generate both first and second order mutants. The experimental results showed that the number of generated equivalent SOMs of weak mutation, RDomF, SDomF, HDom (20%) and HDom (50%) strategies decreased about 73%, 85.4%, 86.8%, 81.4% and 65.5% respectively compared to the number of equivalent mutants of strong mutation.

Madeyski *et al.* [8] presented the NeighPair algorithm and JudyDiffOp algorithm. These algorithms generate SOMs by using diffrent strategies to combine FOMs. The experimentation results showed that these algorithms reduced the number of created SOMs to half comparing with the number of FOMs and the JudyDiffOp algorithm decreased the number of equivalent mutants to 66%.

2.2 Higher-Order Mutation Testing

Jia and Harman [3] defined a subsuming HOM if it is harder to be killed than the FOMs that produce it. They employed the three search-based algorithms including Greedy Algorithm, Genetic Algorithm, and Hill-Climbing Algorithm to generate subsuming HOMs. According to the approach, they have experimented on ten big programs and showed that 67% of generated HOMs were harder to be killed than FOMs producing them.

Langdon *et al.* [9, 10] proposed the use of the NSGAII algorithm (Non-Dominated Sorting Genetic Algorithm) with Pareto's multi-objective genetic programming and optimization to search for HOMs with two objectives, which are hard to be killed and make as small change as possible in program. Their approach targets to find HOMs with inserted faults that are semantically close to the original program instead of making syntax faults. The empirical results indicated that the approach was able to find harder to kill HOMs, and the number of equivalents mutants decreased considerably but the number of mutants grew exponentially by the mutation order.

Belli *et al.* [11] suggested the first-order mutation and higher-order mutation testing for graph-based models using only two mutation operators (insertion and omission). Their experimental results demonstrated that the number of HOMs was four times higher than the number of FOMs. The number of equivalent HOMs was reduced to 55% compared with the number of equivalent FOMs.

Akinde [12] was interested in the equivalent mutants and then utilized the MiLu tool to generate FOMs, SOMs and HOMs. They applied sampling technique to select FOMs, SOMs and random order mutants. The results concluded that the number of equivalent FOMs was reduced to 19% and the number of equivalent HOMs over the total HOMs was nearly 0%. However, higher order mutants generated by this method are easy to be killed.

Omar and Ghost [13] proposed four approaches to generate higher-order mutants for AspectJ programs in aspect-oriented programming: (1) inserting two faults in a single base class or two faults in a single aspect, (2) inserting two faults in two different base classes, (3) inserting two faults in two different aspects, and (4) inserting one fault in a base class and one in an aspect. They found that the first approach generated a bigger percentage of hard-to-kill HOMs than the constituent FOMs as compared to the last three approaches; and the last three approaches reduced the density of equivalent mutants.

In 2014, Ghiduk [14] suggested the application of genetic algorithms to generate test data for killing mutants. The experimental results showed that the generated test set killed 81.8% of FOMs, 90% of SOMs, and 93% of the third-order mutants over the total number of mutants.

Nguyen *et al.* [15–17] proposed a classification of 11 groups of HOMs, which cover all generated mutants. Among these groups, they focused especially on group High Quality and Reasonable HOMs – the HOMs of this group are more difficult to be killed than the FOMs creating them (Reasonable) and a test set that can kill these HOMs will also kill all FOMs creating them (High Quality). They concluded that this HOM group can replace all the FOMs without decreasing the effectiveness of mutation testing.

In our previous paper [18], we introduced different strategies combining FOMs to generate the number of HOMs (that less than number of FOMs) for a program under test, but keeping the quality of generated mutants. In this work, we conduct a further study to improve higher-order mutation testing by evaluating the quality of mutation operators as well as generated mutants and priotirizing test cases basing upon its capability of killing mutants.

3 Research Questions

In preceding section, we have seen that many recent studies are interested in HOMs. Some studies propose various strategies to produce HOMs from FOMs so that number of HOMs is reduced and HOMs still keep the quality in terms of uncovering defects. But no work bases on the quality of each mutation operator used to create mutants. This is our first objective in this paper.

Most of them evaluate the quality of set of test cases using mutation score as test adequacy criterion. However, no work evaluates individually the ability for reaving faults of each test case. It is our second objective in this work. To consider two above-mentioned objectives, in this paper, we pose the following research questions:

RQ1. Is it possible to assess the quality of mutation operator creating mutants through mutation score? Are quality mutation operators that generate quality HOMs?
RQ2. Is it posible to evaluate the quality of each test case through its capability of killing mutants? Is it possible to priotize test cases?
RQ1 aims at evaluating mutation operators basing on mutation score value to create quality HOMs from such mutation operators. Whilst, **RQ2** concentrates on analyzing the ability of realing fault of each test case, hence prioritzing test cases when running them.

4 The Proposed Approach

In this section, we first present the set of projects, which are employed for the experimentation. We then introduce the Judy tool for setting up the experimentation. The last part is reserved for the proposed process to conduct the evaluation of mutation operators as well as test cases.

4.1 Projects Under Test (PUTs)

In the experimentation, we have used 5 PUTs available on https://github.com. These Java open source projects, including the built-in set of test cases (TC), are described in Table 1.

Table 1. The projects under test

Project	Line of code	Number of TCs	URL
Antomology	1,073	22	https://github.com/codehaus/antomology
Commons-email	12,495	32	https://github.com/apache/commons-email
Commons-chain-1.2-src	17,702	73	https://github.com/apache/commons-chain
Commons-csv	9,972	80	https://github.com/apache/commons-csv
Commons-digester3-3.2-src	41,986	121	https://github.com/apache/commons-digester

4.2 Supporting Tool

The Java mutation testing tool, Judy [19], is chosen as a support tool to perform mutant generation as well as mutation testing execution and analysis in our experiment. Judy tool, written by Java programing language, is one of powerful tool in the field of mutation testing including first-order mutation testing (FOMT) and higher-order mutation testing (HOMT). With Judy, we can use a large set of Java mutation operators [19] to apply mutation testing to Java projects by using bytecodes translation technique (see Table 2).

4.3 The Approach

We design the process presented in Fig. 1 to answer the above-mentioned-research-questions. The process consists of 16 steps aiming at the both objectives: evaluating the quality of mutation operator and evaluating the quality of each test case. These steps are briefly described as follows:

(1) Getting the PUTs as the input (5 above-described PUTs in our experiment).
(2) Generating all possible first-order mutants (FOMs) by using the Judy tool with the full set of mutation operators available in Judy.
(3) Executing the built-in set of test cases of PUTs on original PUT and FOMs.

Table 2. Judy mutation operators

Group	Name of operator	Explain
AIR	AIR_Add	Replaces basic binary arithmetic instructions with ADD
	AIR_Div	Replaces basic binary arithmetic instructions with DIV
	AIR_LeftOperand	Replaces basic binary arithmetic instructions with their left operands
	AIR_Mul	Replaces basic binary arithmetic instructions with MUL
	AIR_Rem	Replaces basic binary arithmetic instructions with REM
	AIR_RightOperand	Replaces basic binary arithmetic instructions with their right operands
	AIR_Sub	Replaces basic binary arithmetic instructions with SUB
JIR	JIR_Ifeq	Replaces jump instructions with IFEQ (IF_ICMPEQ, IF_ACMPEQ)
	JIR_Ifge	Replaces jump instructions with IFGE (IFICMPGE)
	JIR_Ifgt	Replaces jump instructions with IFGT (IF_ICMPGT)
	JIR_Ifle	Replaces jump instructions with IFLE (IF_ICMPLE)
	JIR_Iflt	Replaces jump instructions with IFLT (IF_ICMPLT)
	JIR_Ifne	Replaces jump instructions with IFNE (IF_ICMPNE, IF_ACMPNE)
	JIR_Ifnull	Replaces jump instruction IFNULL with IFNONNULL and vice-versa
LIR	LIR_And	Replaces binary logical instructions with AND
	LIR_LeftOperand	Replaces binary logical instructions with their left operands
	LIR_OrReplaces	Replace binary logical instructions with OR
	LIR_RightOperand	Replaces binary logical instructions with their right operands
	LIR_Xor	Replaces binary logical instructions with XOR

(*continued*)

Table 2. *(continued)*

Group	Name of operator	Explain
SIR	SIR_LeftOperand	Replaces shift instructions with their left operands
	SIR_Shl	Replaces shift instructions with SHL
	SIR_Shr	Replaces shift instructions with SHR
	SIR_Ushr	Replaces shift instructions with USHR
Inheritance	IOD	Deletes overriding method
	IOP	Relocates calls to overridden method
	IOR	Renames overridden method
	IPC	Deletes super constructor call
	ISD	Deletes super keyword before fields and methods calls
	ISI	Inserts super keyword before fields and methods calls
Polymorphism	OAC	Changes order or number of arguments in method invocations
	OMD	Deletes overloading method declarations, one at a time
	OMR	Changes overloading method
	PLD	Changes local variable type to super class of original type
	PNC	Calls new with child class type
	PPD	Changes parameter type to super class of original type
	PRV	Changes operands of reference assignment
Java-specific features	EAM	Changes an access or method name to other compatible access or method names
	EMM	Changes a modifier method name to other compatible modifier method names
	EOA	Replaces reference assignment with content assignment (clone) and vice-versa
	EOC	Replaces reference comparison with content comparison (equals) and vice-versa
	JDC	Deletes the implemented default constructor
	JID	Deletes field initialization

(continued)

Table 2. (*continued*)

Group	Name of operator	Explain
	JTD	Deletes this keyword when field has the same name as parameter
	JTI	Inserts this keyword when field has the same name as parameter
Jumble-based	Arithmetics	Mutates arithmetic instructions
	Jumps	Mutates conditional instructions
	Returns	Mutates return values
	Increments	Mutates increments

(4) Filtering and grouping the generated mutants by each mutation operator.
(5) Calculating MSI (Mutation Score Indicator) [19] for each group filtered in step 4. Noting the groups of mutants (and also the mutation operator used to generate these mutants) that have MSI = 0.
(6) Generating All Possible (AP) second-order mutants (SOMs) by combining two generated FOMs guided by the LastToFirst algorithm [5].
(7) Executing the built-in set of test cases of PUTs on original PUT and constructed SOMs, named AP_SOMT (AP stands for All Possible).
(8) Calculating MSI of AP_SOMT, named AP_SOMT_MSI.
(9) Creating a new list of mutation operators, named Remaining Operators List, by removing the operators noted in step 5 (in this study, we call them "not interesting operators") from the full set of mutation operators available in Judy.
(10) Generating FOMs by using the Judy tool with the Remaining Operators List.
(11) Generating SOMs by combining two generated FOMs in step 10 guided by LastToFirst algorithm [5].
(12) Executing the built-in set of test cases of PUTs on original PUT and constructed SOMs in step 11, named EXP_SOMT (EXP stands for Experimental).
(13) Calculating MSI of EXP_SOMT in step 12, named EXP_SOMT_MSI.
(14) Evaluating the efficiency of two methods of generating SOMs (in steps 6 and 11) by comparing mutation score values: AP_SOMT_MSI and EXP_SOMT_MSI. **The results of this evaluation are used as the answer to research question RQ1**.
(15) Calculating mutant kill rate (number of killed mutants by each / total number of mutants generated), named TC_KILL_RATE, for each test case of EXP_SOMT. A mutant is called "killed" by a test case if its outputs are different from the ones of the original PUT when they were executed against this test case.
(16) Analyzing and evaluating the effectiveness of test cases based on the mutant kill rate of each test case (TC_KILL_RATE) calculated in step 15. **The results of this evaluation are used as the answer to research question RQ2**.

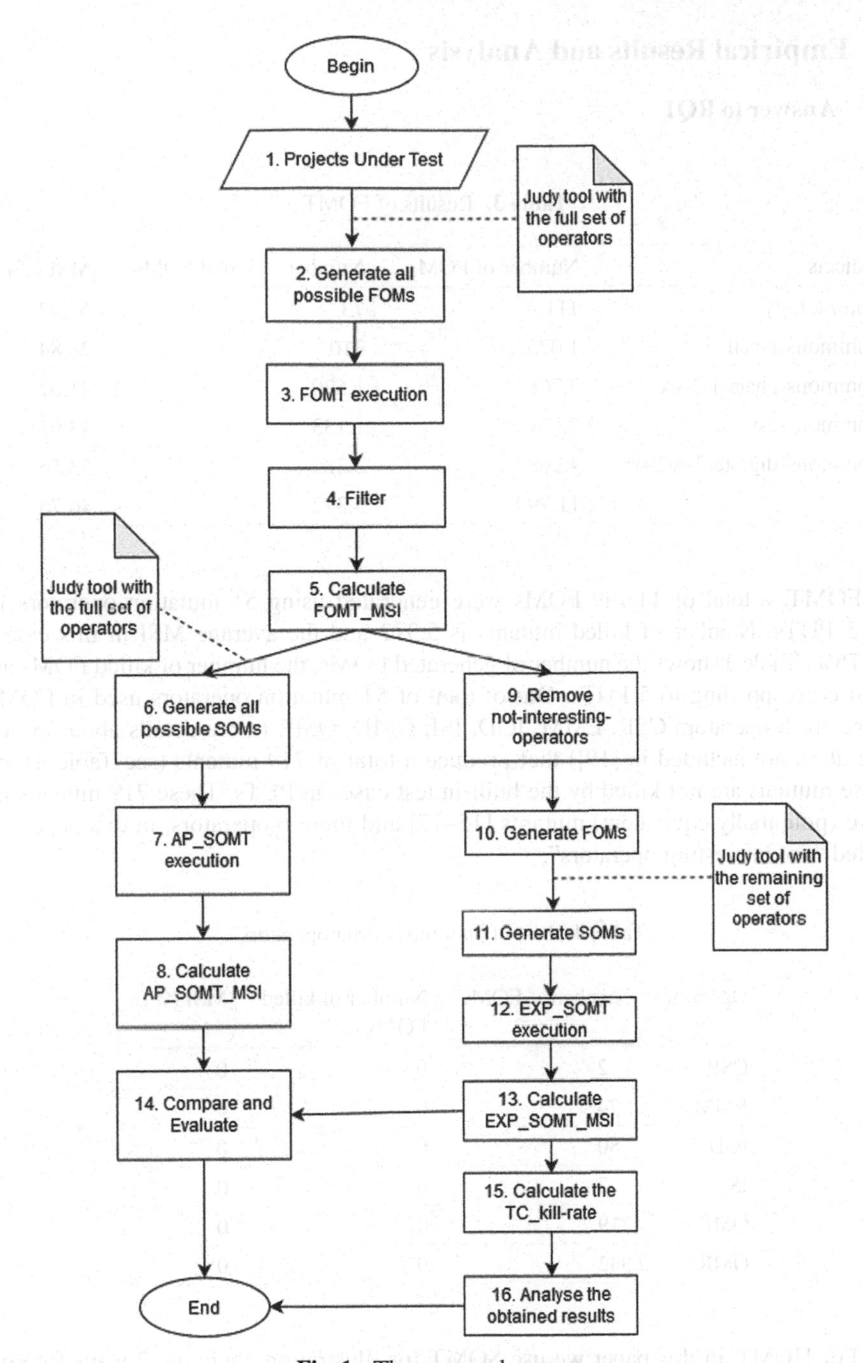

Fig. 1. The proposed process

5 Empirical Results and Analysis

5.1 Answer to RQ1

Table 3. Results of FOMT

Projects	Number of FOMs	Number of killed FOMs	MSI (%)
Antomology	111	73	65.77
Commons-email	1,075	310	28.84
Commons-chain-1.2-src	3,768	1,570	41.67
Commons-csv	2,577	1,133	43.97
Commons-digester3-3.2-src	4,268	2,286	53.56
	11,799	5,372	46.76

In FOMT, a total of 11,799 FOMs were generated using 51 mutation operators for all 5 PUTs. Number of killed mutants is 5,372 and the average MSI in this case is 46.76%. Table 3 shows the number of generated FOMs, the number of killed FOMs and MSI corresponding to 5 PUTs. Out of total of 51 mutation operators used in FOMT, there are 6 operators CSR, EMM, IOD, ISI, OMD, OMR (more details about these 6 operators are included in [19]) that produce a total of 719 mutants (see Table 4), but these mutants are not killed by the built-in test cases in PUTs. These 719 mutants are alive (potentially equivalent) mutants [15–17] and these 6 operators, in this paper, are called "not interesting operators".

Table 4. List of "not interesting operators"

Operator	Number of FOMs	Number of killed FOMs	MSI (%)
CSR	2	0	0
EMM	74	0	0
IOD	80	0	0
ISI	2	0	0
OMD	219	0	0
OMR	342	0	0

For HOMT, in this paper we use SOMT for illustration, there are 2 ways for constructing mutants. In the first way, named AP_SOMT, SOMs have been constructed by combining all possible FOMs guided by the LastToFirst algorithm [5] (see steps 2 and 6 of the process in Sect. 4.3). The second one, named EXP_SOMT, is described in detail by the steps 9, 10, 11 of the process in Sect. 4.3.

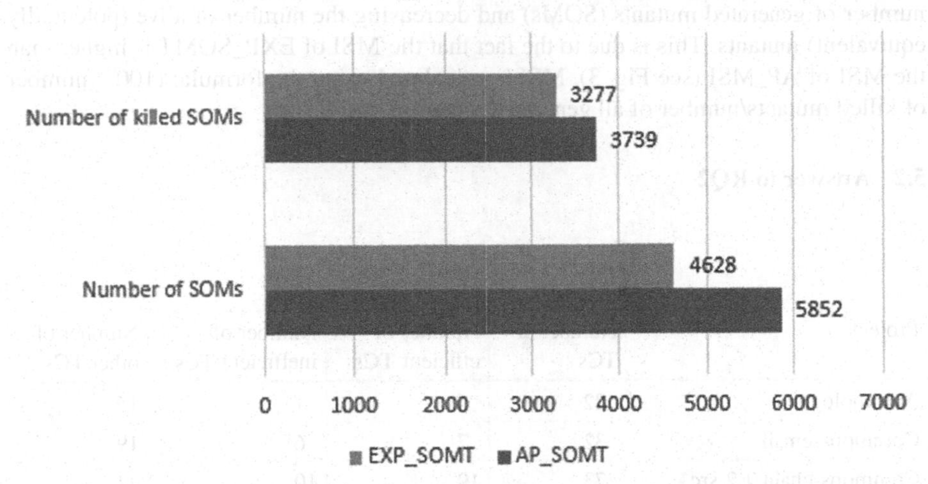

Fig. 2. Results of AP_SOMT and EXP_SOMT

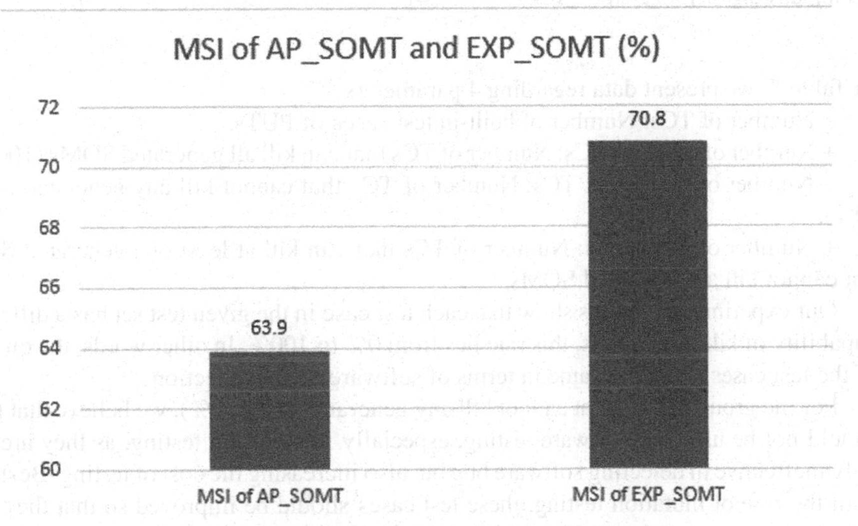

Fig. 3. MSI comparing of AP_SOMT and MSI_EXP

According to the results (see Fig. 2 and Fig. 3) of AP_SOMT, there are 5,852 constructed SOMs and the number of killed SOMs is 3,739 (MSI is 63.9%). While, with EXP_SOMT, the corresponding numbers are 4,628 and 3,277 (MSI is 70.8%).

In addition, the number of different pairs of operators, which are combined to construct the alive (potentially equivalent) mutants of AP_SOMT and EXP_SOMT are 56 and 28 respectively.

With the results presented above, we found that we can use the mutations score as a parameter to evaluate the quality of the mutation operators. And the quality of the operators has a direct influence on the quality of mutants in terms of reducing the number of generated mutants (SOMs) and decreasing the number of alive (potentially equivalent) mutants. This is due to the fact that the MSI of EXP_SOMT is higher than the MSI of AP_MSI (see Fig. 3). MSI is calculated using the formula: (100 * number of killed mutants/number of all generated mutants).

5.2 Answer to RQ2

Table 5. Classification of test cases

Project	Number of TCs	Number of efficient TCs	Number of inefficient TCs	Number of other TCs
Antomology	22	6	1	15
Commons-email	32	7	6	19
Commons-chain-1.2-src	73	19	10	44
Commons-csv	80	12	12	56
Commons-digester3-3.2-src	121	41	8	72

In Table 5, we present data regarding 4 parameters:

+ Number of TCs: Number of built-in test cases of PUTs.

+ Number of efficient TCs: Number of TCs that can kill all generated SOMs (100%).

+ Number of inefficient TCs: Number of TCs that cannot kill any generated SOM (0%).

+ Number of other TCs: Number of TCs that can kill at least one generated SOM but cannot kill all generated SOMs.

Our experimental results show that each test case in the given test set has a different capability of killing mutants, this rate lies from 0% to 100%. In other words, the quality of the test cases is not the same in terms of software defect detection.

For the group of TCs that cannot kill any generated SOM (0%), we believe that they should not be used for software testing, especially in mutation testing, as they are not only ineffective in detecting software bug but also increasing the cost of testing. Besides, with the role of mutation testing, these test cases should be improved so that they can help to detect faults.

On the contrary, TCs which can kill all generated SOMs (100%) are high quality TCs. These test cases are able to detect the difference of all the mutants from the original program. To put it optimistically, we can use them to replace all the remaining TCs in mutation testing as well as software testing without loss of testing effectiveness.

For the remaining group of TCs, it is entirely possible to prioritize them according to their capability of killing mutants from high to low. Then, testers can select the test cases for testing activities depending on their resources or their objectives.

The result of this ranking of test cases based on their quality, *i.e.* ability of revealing faults, are important to testers, because their resources are often limited or they can optimize their testing plan.

6 Conclusions and Future Work

Testing is not only one of the expensive activities in developing software, but also a key factor in software quality assurance. Improving this activity always attracts reserach community. In our study, we focus on enhancing mutation testing technique, especially higher-order mutation, which is one of the most popular techniques in software testing by its ease for automation.

We are interested in two main problems: evaluating the quality of mutation operator using to generate mutants, essentially higher-order mutants; and ranking test cases basing on their abililty of uncovering faults via mutation analysis. In this paper, we proposed an experimentation process to handle these problems. The process consisting of many steps allows selecting high quality mutation operators when creating higher-order mutants in mutation testing as well as prioritizing test cases to allocate better resources during testing activity.

The process was experimented on 5 real Java projects. The results indicate that our approach can reduce the cost but keep the quality of testing activity.

In the future work, we intend to expand the experimentation on industrial projects, primarily projects provided by software companies. In addition, we plan to improve higher-order mutation by applying machine learning techniques in classifying mutants without executing them.

References

1. Lipton, R.: Fault diagnosis of computer programs. Student report, Carnegie Mellon University (1971)
2. DeMillo, R.A., Lipton, R.J., Sayward, F.G.: Hints on test data selection: help for the practicing programmer. Computer **11**(4), 34–41 (1978)
3. Jia, Y., Harman, M.: Higher order mutation testing. Inf. Softw. Technol. **51**(10), 1379–1393 (2009)
4. Harman, M., Jia, Y., Langdon, W.B.: A manifesto for higher order mutation testing. In: Third International Conference on Software Testing, Verification, and Validation Workshops (2010)
5. Polo, M., Piattini, M., Garcia-Rodriguez, I.: Decreasing the cost of mutation testing with second-order mutants. Softw. Test. Verification Reliab. **19**(2), 111–131 (2008)
6. Papadakis, M., Malevris, N.: An empirical evaluation of the first and second order mutation testing strategies. In: Proceedings of the 2010 Third International Conference on Software Testing, Verification, and Validation, Workshops ICSTW 2010, pp. 90–99. IEEE Computer Society (2010)
7. Kintis, M., Papadakis, M., Malevris, N.: Evaluating mutation testing alternatives: a collateral experiment. In: 17th Asia Pacific Software Engineering Conference, APSEC (2010)
8. Madeyski, L., Orzeszyna, W., Torkar, R., Józala, M.: Overcoming the equivalent mutant problem: a systematic literature review and a comparative experiment of second order mutation. IEEE Trans. Softw. Eng. **40**(1), 23–44 (2014)
9. Langdon, W.B., Harman, M., Jia, Y.: Multi-objective higher order mutation testing with genetic programming. In: Proceedings Fourth Testing: Academic and Industrial Conference Practice and Research (2009)
10. Langdon, W.B., Harman, M., Jia, Y.: Efficient multi-objective higher order mutation testing with genetic programming. J. Syst. Softw. **83**, 2416–2430 (2010)

11. Belli, F., Güler, N., Hollmann, A., Suna, G., Yıldız, E.: Model-based higher-order mutation analysis. In: Kim, T.-h., Kim, H.-K., Khan, M.K., Kiumi, A., Fang, W.-c., Ślęzak, D. (eds.) ASEA 2010. CCIS, vol. 117, pp. 164–173. Springer, Heidelberg (2010). https://doi.org/10.1007/978-3-642-17578-7_17
12. Akinde, A.O.: Using higher order mutation for reducing equivalent mutants in mutation testing. Asian J. Comput. Sci. Inf. Technol. **2**(3), 13–18 (2012)
13. Omar, E., Ghosh, S.: An exploratory study of higher order mutation testing in aspect-oriented programming. In: IEEE 23rd International Symposium on Software Reliability Engineering (2012)
14. Ghiduk, A.S.: Using evolutionary algorithms for higher-order mutation testing. IJCSI Int. J. Comput. Sci. **11**(2), 93–104 (2014)
15. Nguyen, Q., Madeyski, L.: Searching for strongly subsuming higher order mutants by applying multi-objective optimization algorithm. In: Le Thi, H.A., Nguyen, N.T., Do, T.V. (eds.) Advanced Computational Methods for Knowledge Engineering. AISC, vol. 358, pp. 391–402. Springer, Cham (2015). https://doi.org/10.1007/978-3-319-17996-4_35
16. Nguyen, Q.V., Madeyski, L.: Higher order mutation testing to drive development of new test cases: an empirical comparison of three strategies. In: Nguyen, N.T., Trawiński, B., Fujita, H., Hong, T.-P. (eds.) ACIIDS 2016. LNCS (LNAI), vol. 9621, pp. 235–244. Springer, Heidelberg (2016). https://doi.org/10.1007/978-3-662-49381-6_23
17. Nguyen, Q.V., Madeyski, L.: Addressing mutation testing problems by applying multi-objective optimization algorithms and higher order mutation. J. Intell. Fuzzy Syst. **32**(2), 1173–1182 (2017)
18. Van Nho, D., Vu, N.Q., Binh, N.T.: A solution for improving the effectiveness of higher order mutation testing. In: 2019 IEEE-RIVF International Conference on Computing and Communication Technologies (RIVF), Danang, Vietnam (2019)
19. Madeyski, L., Radyk, N.: Judy – a mutation testing tool for Java. IET Softw. **4**(1), 32–42 (2010)

Usability Study of Mobile Applications with Cognitive Load Resulting from Environmental Factors

Beata Karczewska, Elżbieta Kukla, Patient Zihisire Muke, Zbigniew Telec, and Bogdan Trawiński(✉)

Faculty of Computer Science and Management, Wrocław University of Science and Technology, Wrocław, Poland
{elzbieta.kukla,patient.zihisire,zbigniew.telec, bogdan.trawinski}@pwr.edu.pl

Abstract. The main goal of this work was to design and conduct usability tests of mobile applications, with particular emphasis on cognitive load. Not only was the cognitive load due to user interface design errors considered, but also the cognitive load due to external factors such as the real world. The study involved 12 participants, who were young adults, i.e. people aged 20–30 who regularly use mobile devices. The participants were randomly divided into two groups of 6 people. These groups independently carried out the scenario of activities using two different applications, in two different environments: laboratory and field. The collected results were statistically analysed in terms of task completion time, number of actions, number of errors and satisfaction with regard to net promoters score and system usability scale. The analysis of these features allowed to determine the relationship between the quality of the graphical user interface, the impact of the environmental factors and user satisfaction.

Keywords: Usability testing · Cognitive load · Mobile applications · Environmental factors · Net promoter score · System usability scale

1 Introduction

Modern technological development puts more and more emphasis on mobility and adaptation to the constant movement of users. Programmers and application developers constantly forget about the need to adapt their products to people who use them in constant motion. As a result, when designing, you should not only remember about the standard aspects of usability, but also put more and more emphasis on the support of low resolutions, limitations in connectivity and high battery consumption. The critical thing here is to consider the context in which the applications are used.

Although technological and design advances can be seen in terms of the design of application graphical interfaces, their navigation and adaptation to mobility, many companies and application developers still do not take into account the aspect of the

N. T. Nguyen et al. (Eds.): ACIIDS 2021, LNAI 12672, pp. 851–864, 2021.
https://doi.org/10.1007/978-3-030-73280-6_67

inconvenience of the external environment. Overloading a sensory stimulus onto a person is called cognitive overload.

This term comes from psychology and assumes that the human mind, and more precisely memory, is able to process only a certain number of tasks [1]. After overloading memory with external factors, a person will have a difficult action, and his decision-making will also be disturbed. The theory dealing with cognitive load is still criticized in academia for the lack of conceptual clarity and the possibility of a methodological approach to the problem [2].

The main goal of this work is to test the usability of selected mobile applications in terms of cognitive load resulting from environmental factors. For this purpose, adequate user testing was carried out. The research was divided into 4 series, each of which assumed a different combination of tools and environments. So far, the authors of this paper have conducted a number of studies on the usability of mobile applications, including responsive web design [3–5] and data entry design patterns [6, 7].

2 Related Works

In the field of mobile devices, technological advances have led to the widespread use of smartphones to the detriment of desktop computers [8]. However, there are still many limitations and problems with the usability of the above-mentioned devices that affect the cognitive load of users depending on the environment in which they work.

Within the framework of software engineering, the usability of a system plays a key role while defining the perceived quality of the use of its users [9, 10]. In this context, usability is defined as a study of the intersection between users and systems, tasks to be performed as well as expectations in terms of use.

Harrison et al. [11] designed the People At the Center of Mobile Application Development (*PACMAD*) usability model developed to tackle existing usability models' challengers when it comes to mobile devices. The model comprises seven attributes demonstrating the application's usability: effectiveness, efficiency, satisfaction, learnability, memorability, errors and cognitive load. The innovation of the model concerns cognitive load as a new usability metric.

Mobile applications' usability model *PACMAD* includes cognitive load because it can directly affect and can be affected by the usability of the application. Indeed, it seems likely that mobile devices are particularly susceptible to the effects of cognitive overload due to their multiple configurations of tasks to be performed and size limitations [11].

Ejaz et al. [12] have argued that cognitive load is not only a supplementary attribute of usability, but one of main attributes when it comes to emergency scenarios. As a result, the authors made experimental comments and conclusions during the usability testing demonstrating the importance of cognitive load.

In context of mobile devices, cognitive load is related to the mental effort needed by a user to carry out tasks while utilizing a mobile device. Although it not a new concept, and does not rank first in usability research, but it is now gaining more popularity in the usability field due to the fact that the attention of users is usually divided between tasks which are being performed simultaneously [8].

Furthermore, Parente Da Costa et al. [10] provide another additional contribution by proposing a set of usability heuristics related to the software applications used in smartphones, taking into consideration the user, task and the context as usability factors and cognitive load as a crucial attribute of usability.

As mentioned before, the *PACMAD* usability model was developed to tackle the challengers of existing usability models such as Nielsen, Mobile usability heuristic and the ISO 9241-11 standard when it comes to mobile devices as shown in Table 1 [13].

Table 1. Models and usability metrics applied for mobile applications. Source [13]

Nielsen	Mobile usability heuristic	ISO 9241-11	PACMAD
Learnability	Findability and visibility of system status	Effectiveness	Effectiveness
Efficiency of use	Real word and match between system	Efficiency	Efficiency
Error frequency	Minimalist design and good ergonomics		Learnability
Memorability	Mapping and Consistency	Satisfaction	Satisfaction
Satisfaction	Screen readability, ease in input		Error
	Personalization and flexibility, efficient of use		Cognitive load
	Social conventions and aesthetic, privacy		
	Realistic error management		

In addition, the concept cognitive load is influenced by the characteristics of the subject, the characteristics of the task and the interactions between the two [14]. The characteristics of the task may include the time pressure, difficulty level of the task, the newness of the task, remuneration after performing the task, and the environment in which the task is being performed. In the other hand, the characteristics of the subject refer to stable factors that are unlikely to change with the task or environment, like the subjects' previous knowledge, preferences, cognitive capabilities and cognitive style and [15].

In that regard, it should be also highlighted that, the level of cognitive load of a participant who is taking part in a cognitive load experiment is primarily caused by the task complexity. But in reality, cognitive load is influenced by several factors like environmental disturbances, time pressure and task complexity [16]. In the same context, Ferreira et al. [17] argued that the actual magnitude of cognitive load that a person experiences is influenced by the environmental and social factors, individual differences and the tasks being performed.

3 Setup of Usability Experiments

3.1 Selection of Mobile Applications

In order to study the most important aspects of the usability of mobile applications with regard to cognitive load, the key is to select the appropriate applications. The aim of the study was to examine the influence of external factors on the efficiency of using the application and the positive reception of the application as well as the cognitive load caused by deficiencies in the design of the graphical user interface. It was important to study various types of actions ranging from simple and obvious activities and paths to follow to complicated activities such as entering data, preparing and making decisions, performing calculations.

For this reason, mobile home budget management applications seem to be particularly useful for the purposes of our research. In this type of apps, the key issue is the good design of human-computer interaction, due to the specificity of activities. There are many different actions available in the home budget applications, including categorizing, planning, entering and reporting expenses. This makes it necessary to simplify the action for the user. Moreover, there are many applications on the market that have taken up the topic of interface simplification too eagerly, resulting in unclear messages and the arrangement of action buttons.

Two home budget management applications available in the *Google Store* were selected for the research described in this article, differing from each other in terms of market maturity, popularity among users and the perceived usefulness of the graphical interface. These were the following mobile applications: "*Fast Budget - Expense & Money Manager*" [18] and "*Family Budget*" [19]. The former had over 1 million downloads, about 92 thousand ratings with an average rating of 4.6 on a 5-point scale. We will denote it as *App1* in the further part of the paper. In contrast, the latter had fewer than 1,000 downloads, few ratings and comments, and a smaller range of functionality and lower perceived ease of use. This application in the rest of the paper will be marked as *App2*. Sample screen shots illustrating the graphical user interfaces of *App1* and *App2* are shown in Figs. 1 and 2, respectively.

Fig. 1. Sample screenshots of the *Fast Budget* application (*App1*). Source [13].

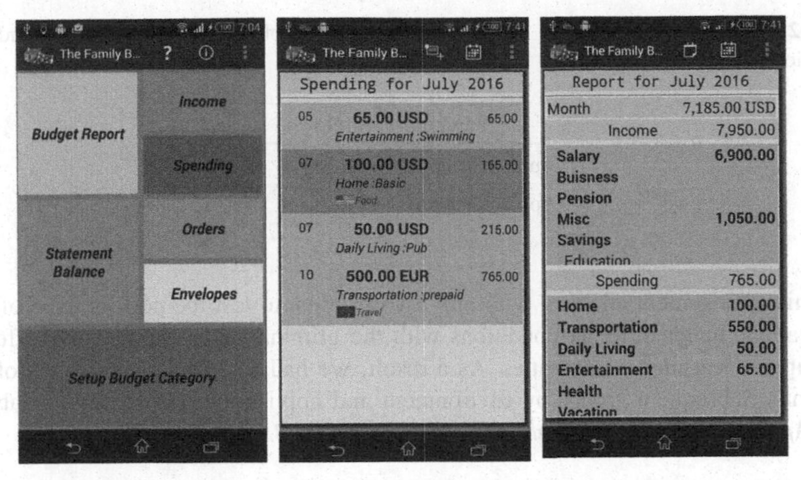

Fig. 2. Sample screenshots of the *Family Budget* application (*App2*). Source [14].

3.2 Environments and Groups of Participants

To adequately investigate the effect of cognitive load on the usability of selected mobile applications, the study was conducted in two different environments. The first was a laboratory environment, i.e. tests were conducted in a room where external stimuli were reduced [20]. The participants were able to use the apps without problems related to weather, noise, and other external stimuli.

The second environment was field-based, i.e. tests were carried out in the real, outdoor world. This maximized the cognitive load affecting the user and more accurately reflect the use of the app in natural settings [21]. This allowed us to maximize the cognitive load affecting the user and more accurately portray the use of the app in a real-world setting [21]. During this research, participants were not only exposed to naturally occurring distractions, but were also forced to participate in interactions with the moderator and answer additional questions that caused diversions.

A very important decision was the size of the study groups. Because of the need to perform the study free of charge, to prepare the subjects, and to systematize the external conditions in the field environment, the groups could not to be too large.

According to Nielsen [22], a sufficient number of testers is five. Following this logic, due to the small gain in the number of deficiencies found relative to the number of people taking the test, it is uneconomical to use more than five users [23]. However, as Faulkner [24] later demonstrated, the above statement is not accurate. There can be situations where a group of five people find 95% of the bugs in an application, but it may turn out that only 55% of errors will be found.

Twelve young people aged 22 to 31 were invited to the research, including 5 women and 7 men, as volunteers. All persons declared that they are advanced users of mobile devices. In order to minimize the influence of the learning factor on the research results, the participants were randomly divided into two groups A and B of 6 people. Table 2 shows the assignment of environments and applications to individual groups A and B.

Table 2. Arrangement of group of participants using applications in laboratory and field conditions

	Laboratory	Field
App1	Group A	Group B
App2	Group B	Group A

This arrangement allowed, with only two groups, to test the performance of both applications under different conditions with the elimination of the negative effect of participants learning the activities. As a result, we had 4 independent groups of participants. Subsequent groups by environment and application will be further labeled: *Lab_App1_A, Fld_App2_A, Lab_App2_B, Fld_App1_B.*

3.3 Task Scenario and Metrics Collected

In both applications, the number of tasks and their content were identical, the conditions under which the tests were performed were different. The research was conducted on the same new model of the Samsung Galaxy S10 smartphone with Android, in order to remove possible differences resulting from screen sizes of different brands and different operating systems. Tasks to be completed during the study are presented in Table 3.

Research for *Fld_App2_A* and *Fld_App1_B* groups, in order to reflect the standard use of the application as much as possible, was carried out in the outdoor world, i.e. in the middle of the day downtown. This allowed for a natural accumulation of environmental cognitive load. Additionally, participants were asked to walk continuously, and additional distracting questions were asked as they performed the tasks. Distracting questions to be answered during the study are shown in Table 4.

Table 3. Tasks to be completed during the study

No.	Task
1	Add an expense of PLN 20 to the Food category on January 10
2	Add an income of PLN 100 in the Salary category today
3	Add a spending limit of PLN 1000 for the month of January
4	Add a new expense category called "Birthday Party"
5	Read your January income/spending balance in the Report function

The study followed the ISO model of usability which is composed of three basic attributes: efficiency, effectiveness, and satisfaction. In total four metrics of efficiency and effectiveness were collected during completion of task scenarios. Moreover, two satisfaction questionnaires were administered after the whole scenario was accomplished. The metrics collected during the study are listed in Table 5.

Table 4. Distracting questions to be answered during the study

No.	Question	Comment
1	What did you eat for dinner yesterday?	Referring to the recent past requires a focus on the answer
2	What are your plans for the coming weekend?	The answer requires a longer elaboration, along with the use of the user's creativity
3	What color is my coat? (the moderator's coat)	The person will be forced to look away from the application, focus their attention on another object, and then return to the interrupted thread of activity

Table 5. Metrics collected during usability tests

Attribute	Metrics	Description	Unit
Efficiency	Time	Scenario completion time by a user excluding time for answering the distractive questions	[s]
	Actions	Number of clicks, scrolls, taps, swipes, etc. to complete the whole scenario	[n]
Effectiveness	Errors	Number of incorrect actions during completion of the whole task scenario	[n]
	Requests for help	Number of requests for assistance during the completion of the whole scenario	[n]
Satisfaction	NPS	Net Promoter Score - score of the single question survey administered after completion of the whole scenario	[−100,…,100]
	SUS	System Usability Scale - score of the 10 question survey administered after completion of the whole scenario	[0,…,100]

Time of Task Completion. Each of the respondents had time to perform their tasks measured. Time of scenario completion was measured for individual participant with an electronic stopwatch on a separate device, the examination was started with a voice announcement by the moderator. In the case of field tests, measuring was suspended while the respondent answered the distracting questions. In order to be able to make sure that the measured times were correct, the actions of each respondent were recorded using the *Screen Recorder* app provided by Google [25]. Due to the purpose of the research, which is to examine the impact of cognitive load on the usability of the application, the time of completing the entire task scenario was measured without considering the times of individual tasks.

Number of Actions Performed. This measure was obtained by analyzing the operation on the application using the *Screen Recorder* app. The action was touching the screen to tap on a given element, moving (scrolling), zooming in or out the screen. Later in the paper, the use of the term number of actions will refer directly to the term number of moves made and tapped elements.

Number of Errors. For each application, the optimal path with the fewest actions was determined. To obtain a measure of incorrect actions, the number of actions performed at a predetermined optimal path was subtracted from the number of actions performed by the user.

Number of Requests for Help. Despite the assumed failure to help the respondents, each of them had the opportunity to ask a question regarding the required activity. Each of these requests was recorded, including rhetorical questions without having to answer them. Due to the fact that the application should be designed so that the user can navigate in it without outside help, any, even rhetorical, requests for help were treated as negative reactions.

Net Promoter Score (NPS). The *Net Promoter Score* was adopted to measure participant satisfaction with the application. *NPS* is commonly used in business as a standard metrics of customer experience and approach to predict customer loyalty [26, 27]. We administered a question: "How likely is it that you would recommend this application to a friend or colleague?". The responses for this answer were scored on an 11 point scale from Very Unlikely (0) to Extremely Likely (10). According to the standard approach, the participants who responded with 9 or 10 were labelled Promoters; those who answered from 0 to 6 were termed Detractors. Responses of 7 and 8 were considered Neutrals and ignored. *NPS* was calculated by subtracting the percentage of Detractors from the percentage of Promoters according to Formula (1).

$$NPS = \left(\frac{No.\,of\ Promoters - No.\,of\ Detractors}{Total\ no.\,of\ Respondents} \right) * 100\% \tag{1}$$

NPS values vary between −100 and +100, with −100 occurring when each respondent critically evaluates the application and does not recommend it to other people, and +100 corresponds to a situation where each user recommends the application to friends. Positive *NPS* values are considered good results and values above 50 are considered excellent.

System Usability Scale (SUS). In our experiments, the user satisfaction was measured using the System Usability Scale [28–31]. *SUS* is a satisfaction questionnaire that has become the industry standard for over 30 years of use. *SUS* consists of 10 questions that users answer using a 5-point Likert scale from 1 (strongly disagree) to 5 (strongly agree). Questions marked with odd numbers refer to the positive aspects of using the application, whereas even-numbered questions refer to negative aspects. The *SUS* score ranges from 0 to 100. The mean *SUS* score of the 500 studies surveyed by Sauro [28] is 68. Thus, an SUS score above 68 can be considered above average, and anything below 68 is below average. Research to date shows that *SUS* is a reliable and important measure of perceived utility. The advantage of *SUS* is that it is relatively fast, easy and inexpensive, and at the same time provides a reliable means of measuring usability.

3.4 Statistical Analysis Approach

The values of metrics collected for individual participants of each group: *Lab_App1_A, Fld_App2_A, Lab_App2_B, Fld_App1_B* were used to examine differences in performance between individual groups. The parametric and non-parametric statistical tests for independent groups were employed as shown in Table 6. To determine what type of statistical significance tests should be used, the normality and equality of variance of the individual metrics are examined. In the event that if there is no evidence for rejecting null hypotheses in the Shapiro-Wilk test and Fisher-Snedecor, the parametric tests Student's t-test for independent groups should be employed. On the other hand, if the null hypotheses in the Shapiro-Wilk test is rejected, then the nonparametric Levene's test and Mann-Whitney U test should be applied. The level of significance was set to 0.05 in each test. All tests were performed using the *PQStat* software for statistical data analysis [29].

Table 6. Summary of statistical tests for independent groups in the *PQStat*

Feature examined	Test	
Normality of distribution	Shapiro–Wilk test	
Feature examined	**Normal distribution**	**Nonnormal distribution**
Equality of variances	Fisher-Snedecor test	Levene's test
Feature examined	**Parametric tests**	**Nonparametric tests**
Means/Medians	Student's t-test for independent groups	Mann-Whitney U test

4 Results of Usability Testing

The results of usability testing for individual groups of participants are presented in Figs. 3, 4, 5 and 6. It is clearly seen that average and median time of task completion, number of actions performed, number of errors committed, number of requests for help are lower for groups working in laboratory conditions as well as for groups utilizing *App1*. Statistical tests for independent groups revealed that the differences between mean values of following performance measures task completion time and error numbers were statistically significant in each case, but not for the number of actions. In all tests the distributions of collected data were normal and variances were equal. In consequence the parametric Student's t-tests for independent groups were carried out.

The results of the satisfaction questionnaires are presented in Figs. 7 and 8. In the case of both *NPS* and *SUS* Score measures, the subjective ratings of both applications were identical regardless of the environment in which they were used. Moreover, *App1* was positively assessed and *App2* was rejected by the research participants.

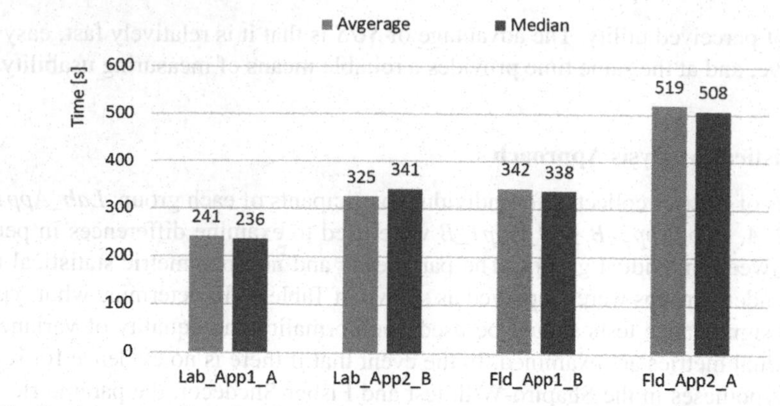

Fig. 3. Average and median time of task completion by individual groups of participants

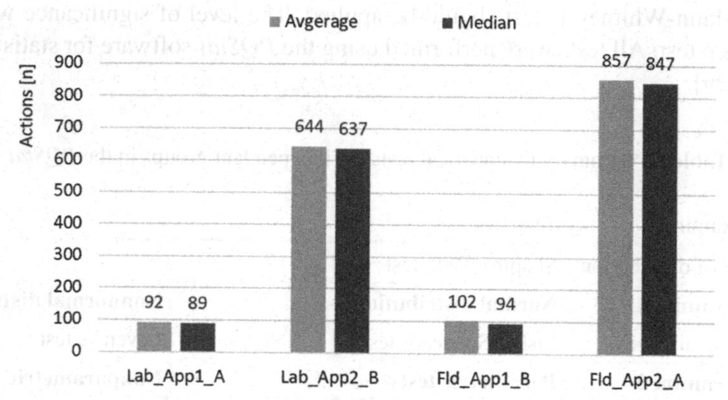

Fig. 4. Average and median number of actions performed by individual groups of participants

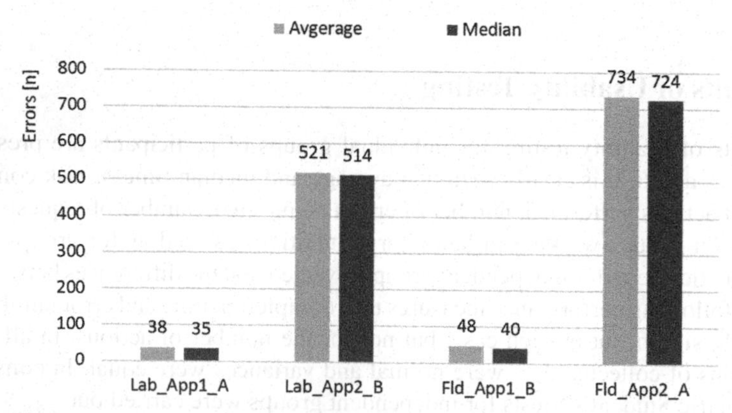

Fig. 5. Average and median number of errors committed by individual groups of participants

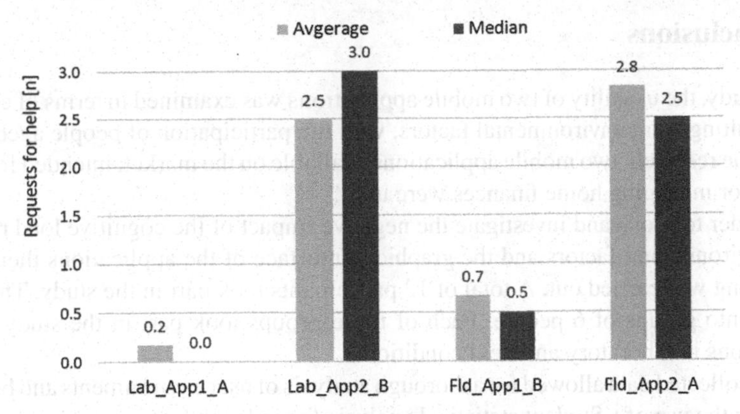

Fig. 6. Average and median number of requests for help by individual groups of participants

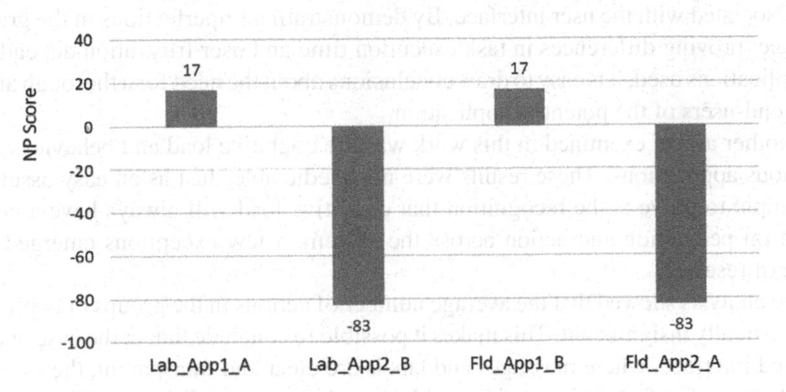

Fig. 7. *Net Promoter Score* for individual groups of participants

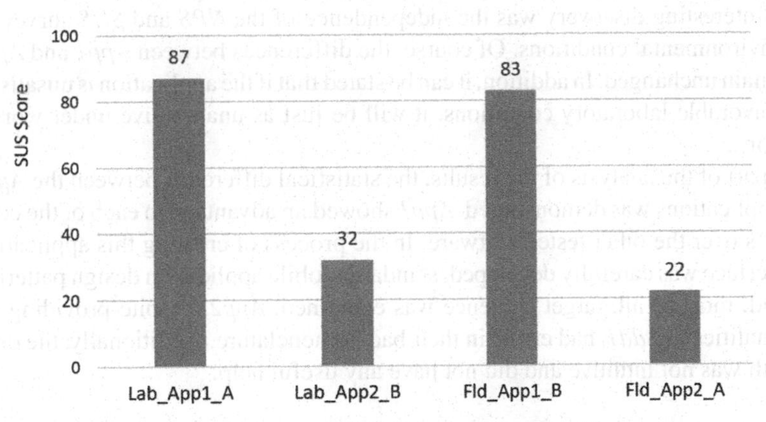

Fig. 8. *SUS* Score for individual groups of participants

5 Conclusions

In this study, the usability of two mobile applications was examined in terms of cognitive load resulting from environmental factors, with the participation of people aged 20–30. During the research, two mobile applications available on the market, intended for young people, for managing home finances were used.

In order to prove and investigate the negative impact of the cognitive load resulting from environmental factors and the graphical interface of the applications themselves, user testing was carried out. A total of 12 participants took part in the study. They were divided into groups of 6 people. Each of these groups took part in the study of both applications in laboratory and field conditions.

The collected data allowed for a thorough analysis of user environments and behavior, including the use of t-Student statistical analysis for independent groups.

The first obtained and easily measurable result was proving the problem of cognitive load associated with the user interface. By demonstrating imperfections in the graphical interface, proving differences in task execution time and user frustration depending on the applications used, it is easy to draw conclusions about the need for a thorough analysis of the end-users of the potential application.

Another aspect examined in this work was the cognitive load and behavior of users in various applications. These results were not predictable. Just as an easy assumption and simple to prove is the recognition that cognitive load will always have a negative impact on perception and action across the system, a few exceptions emerged in the course of research.

The analysis showed that the average number of actions in the groups of applications are statistically insignificant. This makes it possible to conclude that in the case of a well-designed interface, where messages and labels are clear and transparent, the user under the influence of unfavorable weather conditions, disturbances from real life, still is able to complete his/her task.

An interesting discovery was the independence of the *NPS* and *SUS* survey results from environmental conditions. Of course, the differences between *App1* and *App2* ratings remain unchanged. In addition, it can be stated that if the application is unsatisfactory under favorable laboratory conditions, it will be just as unattractive under worse field conditions.

As part of the analysis of the results, the statistical difference between the *App1* and *App2* applications was demonstrated. *App1* showed an advantage in each of the collected measures over the other tested software. In the process of creating this application, the user interface was carefully developed, standard mobile application design patterns were used and, most of all, target audience was examined. *App2*, despite providing similar functionalities to *Add1*, had errors in their bad nomenclature. Additionally, the designed user path was not intuitive and did not have any useful help.

References

1. Leppink, J.: Cognitive load theory: practical implications and an important. J. Taibah Univ. Med. Sci. **12**(5), 385–391 (2017). https://doi.org/10.1016/j.jtumed.2017.05.003
2. de Jong, T.: Cognitive load theory, educational research, and instructional design: some food for thought. Instr. Sci. **38**, 105–134 (2010). https://doi.org/10.1007/s11251-009-9110-0
3. Bernacki, J., Błażejczyk, I., Indyka-Piasecka, A., Kopel, M., Kukla, E., Trawiński, B.: Responsive web design: testing usability of mobile web applications. In: Nguyen, N.T., Trawiński, B., Fujita, H., Hong, T.-P. (eds.) ACIIDS 2016. LNCS (LNAI), vol. 9621, pp. 257–269. Springer, Heidelberg (2016). https://doi.org/10.1007/978-3-662-49381-6_25
4. Błażejczyk, I., Trawiński, B., Indyka-Piasecka, A., Kopel, M., Kukla, E., Bernacki, J.: Usability testing of a mobile friendly web conference service. In: Nguyen, N.-T., Manolopoulos, Y., Iliadis, L., Trawiński, B. (eds.) ICCCI 2016. LNCS (LNAI), vol. 9875, pp. 565–579. Springer, Cham (2016). https://doi.org/10.1007/978-3-319-45243-2_52
5. Krzewińska, J., Indyka-Piasecka, A., Kopel, M., Kukla, E., Telec, Z., Trawiński, B.: Usability testing of a responsive web system for a school for disabled children. In: Nguyen, N.T., Hoang, D.H., Hong, T.-P., Pham, H., Trawiński, B. (eds.) ACIIDS 2018. LNCS (LNAI), vol. 10751, pp. 705–716. Springer, Cham (2018). https://doi.org/10.1007/978-3-319-75417-8_66
6. Myka, J., Indyka-Piasecka, A., Telec, Z., Trawiński, B., Dac, H.C.: Comparative analysis of usability of data entry design patterns for mobile applications. In: Nguyen, N.T., Gaol, F.L., Hong, T.-P., Trawiński, B. (eds.) ACIIDS 2019. LNCS (LNAI), vol. 11431, pp. 737–750. Springer, Cham (2019). https://doi.org/10.1007/978-3-030-14799-0_63
7. Waloszek, S., Zihisire Muke, P., Piwowarczyk, M., Telec, Z., Trawiński, B., Nguyen, L.T.T.: Usability study of data entry design patterns for mobile applications. In: Nguyen, N.T., Hoang, B.H., Huynh, C.P., Hwang, D., Trawiński, B., Vossen, G. (eds.) ICCCI 2020. LNCS (LNAI), vol. 12496, pp. 888–901. Springer, Cham (2020). https://doi.org/10.1007/978-3-030-63007-2_70
8. Weichbroth, P.: Usability of mobile applications: a systematic literature study. IEEE Access **8**, 55563–55577 (2020). https://doi.org/10.1109/ACCESS.2020.2981892
9. Bashirl, M.S., Farooq, A.: EUHSA: extending usability heuristics for smartphone application. IEEE Access **7**, 100838–100859 (2019). https://doi.org/10.1109/ACCESS.2019.2923720
10. Da Costa, R.P., Canedo, E.D., De Sousa, R.T., De Oliveira Albuquerque, R., Garcia Villalba, L.J.: Set of usability heuristics for quality assessment of mobile applications on smartphones. IEEE Access **7**, 116145–116161 (2019). https://doi.org/10.1109/ACCESS.2019.2910778
11. Harrison, R., Flood, D., Duce, D.: Usability of mobile applications: literature review and rationale for a new usability model. J. Interact. Sci. **1**, 1 (2013). https://doi.org/10.1186/2194-0827-1-1
12. Ejaz, A., Rahim, M., Khoja, S.A.: The effect of cognitive load on gesture acceptability of older adults in mobile application. In: 2019 IEEE 10th Annual Ubiquitous Computing, Electronics and Mobile Communication Conference, UEMCON 2019, pp. 0979–0986 (2019). https://doi.org/10.1109/UEMCON47517.2019.8992970
13. Sunardi, Desak, G.F.P., Gintoro: List of most usability evaluation in mobile application: a systematic literature review. In: Proceedings of 2020 International Conference on Information Management and Technology, ICIMTech 2020, pp. 283–287 (2020). https://doi.org/10.1109/ICIMTech50083.2020.9211160
14. Paas, F., van Merriënboer, J.J.G.: Instructional control of cognitive load in the training of complex cognitive tasks. Educ. Psychol. Rev. **6**(4), 351–371 (1994). https://doi.org/10.1007/BF02213420
15. Sun, G., Yao, S., Carretero, J.A.: A pilot study for investigating factors that affect cognitive load in the conceptual design process. In: Proceedings of the Human Factors and Ergonomics Society, pp. 180–184 (2015). https://doi.org/10.1177/1541931215591037

16. Liang, Y., Liang, W., Qu, J., Yang, J.: Experimental study on EEG with different cognitive load. In: 2018 IEEE International Conference on Systems, Man, and Cybernetics, SMC 2018, pp. 4351–4356 (2019). https://doi.org/10.1109/SMC.2018.00735
17. Ferreira, E., et al.: Assessing real-time cognitive load based on psycho-physiological measures for younger and older adults. In: IEEE Symposium on Computational Intelligence, Cognitive Algorithms, Mind, and Brain, Proceedings, pp. 39–48 (2014). https://doi.org/10.1109/CCMB.2014.7020692
18. Fast Budget - Expense & Money Manager: https://play.google.com/store/apps/details?id=com.blodhgard.easybudget&hl=en&gl=US. Accessed 15 Jan 2021
19. Family Budget: https://play.google.com/store/apps/details?id=pl.com.cierniak.android.familybudget&hl=en&gl=US. Accessed 15 Jan 2021
20. Kaikkonen, A., Kekäläinen, A., Cankar, M., Kallio, T., Kankainen, A.: Usability testing of mobile applications: a comparison between laboratory and field testing. J. Usability Stud. **1**(1), 4–16 (2005)
21. Alshehri, F., Freeman, M.: Methods for usability evaluations of mobile devices. In: Lamp, J.W. (eds.) 23rd Australian Conference on Information Systems, pp. 1–10. Deakin University, Geelong (2012)
22. Nielsen, J.: Why You Only Need to Test with 5 Users: https://www.nngroup.com/articles/why-you-only-need-to-test-with-5-users/. Accessed 15 Jan 2021
23. Macefield, R.: How to specify the participant group size for usability studies: a practitioner's guide. J. Usability Stud. **5**(1), 34–45 (2009)
24. Faulkner, L.: Beyond the five-user assumption: benefits of increased sample sizes in usability testing. Behav. Res. Methods Instrum. Comput. **35**(3), 379–383 (2003)
25. Screen Recorder: https://play.google.com/store/apps/details?id=com.fragileheart.screenrecorder&hl=en_US&gl=US. Accessed 15 Jan 2021
26. Net Promoter Score. A Comprehensive Introduction. https://www.genroe.com/net-promoter-score. Accessed 15 Jan 2021
27. Pollack, B.L., Alexandrov, A.: Nomological validity of the net promoter index question. J. Serv. Mark. **27**(1), 118–129 (2013). https://doi.org/10.1108/08876041311309243
28. Sauro, J.: Measuring Usability with the System Usability Scale (SUS): https://measuringu.com/sus/. Accessed 15 Jan 2021
29. Thomas, N.: How to Use the System Usability Scale (SUS) to Evaluate the Usability of Your Website: https://usabilitygeek.com/how-to-use-the-system-usability-scale-sus-to-evaluate-the-usability-of-your-website/. Accessed 15 Jan 2021
30. Bangor, A., Kortum, P.T., Miller, J.T.: An empirical evaluation of the system usability scale. Int. J. Hum.-Comput. Interact. **24**(6), 574–594 (2008)
31. Brooke, J.: SUS: a retrospective. J. Usability Stud. **8**(2), 29–40 (2013)
32. Wieckowska, B.: User Guide - PQStat. Downloaded from https://download.pqstat.pl/UserGuideGeo.pdf. Accessed 15 Jan 2021

Author Index

Printed in the United States
by Baker & Taylor Publisher Services

Printed in the United States
by Baker & Taylor Publisher Services